Instrumental Methods of Analysis
Sixth Edition

HOBART H. WILLARD
University of Michigan

LYNNE L. MERRITT, JR.
Indiana University

JOHN A. DEAN
University of Tennessee at Knoxville

FRANK A. SETTLE, JR.
Virginia Military Institute

D. VAN NOSTRAND COMPANY
New York   Cincinnati   Toronto   London   Melbourne

D. Van Nostrand Company Regional Offices:
New York   Cincinnati

D. Van Nostrand Company International Offices:
London   Toronto   Melbourne

Copyright © 1981 by Litton Educational Publishing, Inc.

Library of Congress Catalog Card Number: 80-51096
ISBN: 0-442-24502-5

All rights reserved. No part of this work covered by the copyright hereon may be reproduced or used in any form or by any means—graphic, electronic, or mechanical, including photocopying, recording, taping, or information storage and retrieval systems—without written permission of the publisher. Manufactured in the United States of America.

Published by D. Van Nostrand Company
135 West 50th Street, New York, N.Y. 10020

10 9 8 7 6 5 4 3 2 1

Atomic Weights (Continued)

Element	Symbol	Atomic Number	Atomic Weight
Mercury	Hg	80	200.59
Molybdenum	Mo	42	95.94
Neodymium	Nd	60	144.24
Neon	Ne	10	20.179
Neptunium	Np	93	237.0482
Nickel	Ni	28	58.71
Niobium	Nb	41	92.9064
Nitrogen	N	7	14.0067
Nobelium	No	102	(255)
Osmium	Os	76	190.2
Oxygen	O	8	15.9994
Palladium	Pd	46	106.4
Phosphorus	P	15	30.97376
Platinum	Pt	78	195.09
Plutonium	Pu	94	(242)
Polonium	Po	84	(210)
Potassium	K	19	39.098
Praseodymium	Pr	59	140.9077
Promethium	Pm	61	(147)
Protactinium	Pa	91	231.0359
Radium	Ra	88	226.0254
Radon	Rn	86	(222)
Rhenium	Re	75	186.2
Rhodium	Rh	45	102.9055
Rubidium	Rb	37	85.4678
Ruthenium	Ru	44	101.07
Samarium	Sm	62	150.4
Scandium	Sc	21	44.9559
Selenium	Se	34	78.96
Silicon	Si	14	28.086
Silver	Ag	47	107.868
Sodium	Na	11	22.98977
Strontium	Sr	38	87.62
Sulfur	S	16	32.06
Tantalum	Ta	73	180.9479
Technetium	Tc	43	98.9062
Tellurium	Te	52	127.60
Terbium	Tb	65	158.9254
Thallium	Tl	81	204.37
Thorium	Th	90	232.0381
Thulium	Tm	69	168.9342
Tin	Sn	50	118.69
Titanium	Ti	22	47.90
Tungsten	W	74	183.85
Uranium	U	92	238.029
Vanadium	V	23	50.9414
Xenon	Xe	54	131.30
Ytterbium	Yb	70	173.04
Yttrium	Y	39	88.9059
Zinc	Zn	30	65.38
Zirconium	Zr	40	91.22

Numbers in parentheses are mass numbers of most stable or most common isotope.

Preface

Shortly after the publication of the Fifth Edition, our senior author, Dr. Hobart H. Willard, died. Dr. Willard was one of the pioneers in the establishment of courses in instrumental methods of analysis and was the instigator of this textbook. We sorely miss his wise counsel and guidance.

The Sixth Edition welcomes a new, younger co-author, Frank Settle. Dr. Settle is qualified by teaching and research experience in instrumental analysis, especially the electronic and instrumental design aspects, and he has completely rewritten these sections. The Sixth Edition contains five new or completely rewritten chapters entitled "Electronics: Fundamentals of Solid State Design," "Electronics: Commonly Used Signal Modifying Circuits," "Data Handling," "Computer-Aided Analysis," and "Process Instruments and Automatic Analysis."

The use of chromatographic methods of analysis has burgeoned since the last edition was published. Accordingly, the chapter on gas chromatography has been expanded from a single chapter to four chapters covering the general principles of chromatography, gas chromatography, liquid column chromatography instrumentation and methods, and high-performance liquid chromatography methods.

An entirely new chapter on the chemical analysis of surfaces is included in the Sixth Edition. The chapter departs somewhat from the approach of other chapters in that most other chapters describe specific methods and their application to many substances or their use in many situations. This new chapter describes the application of various methods, such as surface spectroscopy, sputter-etching, ion-scattering spectrometry, secondary ion mass spectrometry, ion microprobe mass analysis, Auger emission spectroscopy, and electron spectroscopy for chemical analysis to the specific task of characterization and analysis of surfaces.

An introduction to absorption and emission spectroscopy has been added. This chapter describes some relationships among representative optical phenomena that produce signals with chemical information. It gathers together many of the fundamental laws and principles that were previously dispersed in several chapters.

Some materials were consolidated or eliminated in order to accommodate the new content. Detectors for X rays and radioactivity are now gathered together in the chapter on X-ray methods. The discussion of refractometry and interferometry, methods that are not so widely used now, are shortened and combined into one chapter with polarimetry, circular dichroism, and optical rotatory dispersion. Separations by electrolysis and coulometric methods are combined into a single chapter since these methods have much in common. Voltammetry and amperometric titrations are also combined.

The sequence of the chapters in the Sixth Edition is somewhat changed to achieve a more logical order. Individual chapters are designed, in general, to stand alone, so that the order of presentation is not critical. Instructors may select materials for several levels of achievement and to suit their preferences for order of presentation. References to the literature and collateral readings are included in each chapter. The book should also be suitable as a reference.

Numerous examples are incorporated into the text, including those illustrating mathematical operations. These examples introduce the student to the units of measurement and reduce or eliminate dependence upon additional problem books. There are, in addition, a large number of problems at the end of most chapters. Selected answers are given separately at the end of the text. Many of these problems contain data that would be obtained in laboratory experiments and are thus of particular value for those unable to furnish equipment for specific areas of instrumentation, for supplementing experiments when laboratory periods are limited, or for self-study. An Instructor's Solutions Manual that provides solved problems is available from the publisher.

The experiments formerly included at the ends of the chapters have been collected together at the back of this edition. Some of the experiments are described in considerable detail for use by less experienced undergraduate students. Others are merely sketched outlines or suggestions for work to give instructors in advanced courses flexibility in eliciting from students a degree of independence and originality in the outline and execution of experimental work.

The Sixth Edition presents, as did the previous editions, a comprehensive overview of the field of instrumental analysis as commonly practiced today. We remain convinced that all chemistry students and, indeed, students in many of the other physical and biological sciences, will benefit from a comprehensive course reviewing the major methods available with a discussion of the basic principles, advantages and disadvantages, limitations, and applicability of each method. In their later work, the students should then be able to select the best method or a limited number of methods that will solve their immediate problem. The references contained herein will then lead them to more detailed and advanced discussions of the method or methods selected.

Separate listings of abbreviations and symbols are included in the front of the book. Whenever available, recommendations of concerned nomenclature commissions have been followed. In addition, the Appendixes provide a comprehensive tabulation of oxidation-reduction potentials in aqueous solution, polarographic half-wave potentials and diffusion-current constants, acid dissociation constants, formation constants of some metal complexes, flame emission and atomic absorption spectra, a conversion table involving values of absorbance for percent absorption, and a wavenumber-wavelength conversion table. A four-place table of common logarithms, a table of 1971 atomic weights, and a periodic chart of the elements facilitate computations and provide ready reference data.

The authors remain greatly indebted to the manufacturers who have generously furnished schematic diagrams, photographs, and technical information of their instruments. We would like to thank the following reviewers for their helpful comments: Arno Heyn, Boston University; George Morrison, Cornell University; Stanford Tackett, Indiana University of Pennsylvania; and Thomas Copeland, Northeastern University. Thanks are expressed also to many colleagues who have kindly helped with suggestions and improvements.

Lynne L. Merritt, Jr.
John A. Dean
Frank A. Settle, Jr.

Contents

Abbreviations		xiv
Symbols		xx
1	**An Introduction to Absorption and Emission Spectroscopy**	**1**
1.1	The Nature of Electromagnetic Radiation	1
1.2	The Electromagnetic Spectrum	3
1.3	Atomic Energy Levels	4
1.4	Molecular Electronic Energy Levels	6
1.5	Vibrational Energy Levels	10
1.6	Raman Effect	12
1.7	Nuclear Spin Behavior	13
1.8	Electron Spin Behavior	15
1.9	X-Ray Energy Levels	16
	Bibliography	18
2	**Ultraviolet and Visible Spectrophotometry—Instrumentation**	**19**
2.1	Radiation Sources	20
2.2	Detectors	22
2.3	Readout Module	30
2.4	Filters	30
2.5	Monochromators	34
2.6	Monochromator Performance	38
2.7	Grating Monochromator Systems	47
2.8	Instruments for Absorption Photometry	51
	Problems	63
	Bibliography	65
	Literature Cited	65

3 Ultraviolet and Visible Absorption Methods — 66

- 3.1 Fundamental Laws of Photometry — 66
- 3.2 Spectrophotometric Accuracy — 70
- 3.3 Photometric Precision — 73
- 3.4 Quantitative Methodology — 75
- 3.5 Differential or Expanded Scale Spectroscopy — 78
- 3.6 Difference Spectroscopy — 82
- 3.7 Derivative Spectroscopy — 82
- 3.8 Photometric Titrations — 83
- 3.9 Spectra of Solids — 86
- 3.10 Turbidity and Nephelometry — 90
- 3.11 Correlation of Electronic Absorption Spectra with Molecular Structure — 94
- Problems — 97
- Bibliography — 103
- Literature Cited — 104

4 Fluorescence and Phosphorescence Spectrophotometry — 105

- 4.1 Structural Factors — 107
- 4.2 Photoluminescence Intensity as Related to Concentration — 109
- 4.3 Instrumentation — 111
- 4.4 Instrumentation for Phosphorescence Measurements — 122
- Problems — 123
- Bibliography — 125
- Literature Cited — 126

5 Flame Emission and Atomic Absorption Spectrometry — 127

- 5.1 Nebulization — 127
- 5.2 Flames and Flame Temperatures — 129
- 5.3 Interferences — 134
- 5.4 Flame Spectrometric Techniques — 138
- Problems — 150
- Bibliography — 153
- Literature Cited — 153

6 Atomic Emission Spectroscopy — 154

- 6.1 Spectroscopic Sources — 155
- 6.2 Atomic Emission Spectrometers — 164
- 6.3 Photographic Detection — 170
- 6.4 Photoelectric Detection — 174
- Problems — 175
- Bibliography — 176
- Literature Cited — 176

7 Infrared Spectrophotometry — 177

7.1	Correlation of Infrared Spectra with Molecular Structure	178
7.2	Instrumentation	189
7.3	Sample Handling	201
7.4	Quantitative Analysis	209
	Problems	210
	Bibliography	216
	Literature Cited	216

8 Raman Spectroscopy — 217

8.1	Theory	217
8.2	Instrumentation	222
8.3	Sample Handling and Illumination	226
8.4	Diagnostic Structural Analysis	228
8.5	Polarization Measurements	231
8.6	Quantitative Analysis	232
	Problems	233
	Bibliography	238

9 X-Ray Methods — 239

9.1	Production of X Rays and X-Ray Spectra	240
9.2	Instrumental Units	245
9.3	Detectors for the Measurement of Radiation	249
9.4	Semiconductor Detectors	255
9.5	Direct X-Ray Methods	260
9.6	X-Ray Absorption Methods	262
9.7	X-Ray Fluorescence Methods	265
9.8	X-Ray Diffraction	270
	Problems	282
	Bibliography	287
	Literature Cited	287

10 Radiochemical Methods — 289

10.1	Nuclear Reactions and Radiations	289
10.2	Measurement of Radioactivity	294
10.3	Applications of Radionuclides	297
10.4	Activation Analysis	304
	Problems	310
	Bibliography	315

Contents

11 Nuclear Magnetic Resonance Spectroscopy   316

    11.1 Basic Principles   317
    11.2 Continuous-Wave NMR Spectrometers   324
    11.3 Pulsed Fourier Transform NMR Spectrometer   330
    11.4 Spectra and Molecular Structure   331
    11.5 Elucidation of Proton NMR Spectra   340
    11.6 Quantitative Analysis   347
           Problems   349
           Bibliography   356
           Literature Cited   356

12 Electron Spin Resonance Spectroscopy   357

    12.1 Electron Behavior   357
    12.2 ESR Spectrometer   358
    12.3 ESR Spectra   362
    12.4 Interpretation of ESR Spectra   367
    12.5 ENDOR   371
    12.6 ELDOR   371
    12.7 Quantitative Analysis   371
           Problems   372
           Bibliography   377
           Literature Cited   378

13 Chemical Analysis of Surfaces   379

    13.1 Ion Scattering Spectrometry (ISS)   383
    13.2 Secondary Ion Mass Spectrometry (SIMS)   386
    13.3 Auger Emission Spectroscopy (AES)   389
    13.4 Electron Spectroscopy for Chemical Analysis (ESCA)   394
           Bibliography   401

14 Refractometry and Interferometry; Polarimetry, Circular Dichroism, and Optical Rotatory Dispersion   403

    14.1 Theory   403
    14.2 Refractometers   406
    14.3 Polarimetry Theory   412
    14.4 Applications of Optical Rotatory Dispersion and Circular Dichroism   420
    14.5 The Polarimeter   421
    14.6 Instruments for Circular Dichroism Measurement   427
           Problems   427
           Bibliography   429
           Literature Cited   429

15	**Chromatography—General Principles**		**430**
	15.1	Classification of Chromatographic Methods	430
	15.2	Nature of Partition Forces	431
	15.3	Chromatographic Behavior of Solutes	432
	15.4	Column Efficiency and Resolution	436
	15.5	Column Processes and Band Broadening	440
	15.6	Reduced Variables	443
	15.7	Time of Analysis and Resolution	444
	15.8	Quantitative Analysis	446
		Problems	452
		Bibliography	453
		Literature Cited	453
16	**Gas Chromatography**		**454**
	16.1	Gas Chromatographs	454
	16.2	Detectors	464
	16.3	Optimization of Experimental Conditions	478
	16.4	Gas–Solid Chromatography	484
		Problems	486
		Bibliography	493
		Literature Cited	493
17	**Liquid Column Chromatography: Instrumentation and Optimization**		**495**
	17.1	Solvent Delivery System	495
	17.2	Sample Introduction	500
	17.3	Separation Column	502
	17.4	Detectors	504
	17.5	Optimization of Column Performance	515
		Problems	525
		Bibliography	527
		Literature Cited	527
18	**High-Performance Liquid Chromatography Methods**		**529**
	18.1	Adsorption Chromatography	530
	18.2	Liquid–Liquid Partition Chromatography	536
	18.3	Ion-Exchange HPLC	545
	18.4	Exclusion Chromatography	553
		Problems	561
		Bibliography	563
		Literature Cited	563

19 Mass Spectrometry — 565

19.1	Components of Mass Spectrometers	565
19.2	Resolution	577
19.3	Mass Spectrometers	578
19.4	Interfacing Chromatography and Mass Spectrometry	585
19.5	Quantitative Analysis of Mixtures	592
19.6	Use of Stable Isotopes	593
19.7	Leak Detection	594
19.8	Correlation of Mass Spectra with Molecular Structure	595
	Problems	600
	Bibliography	604
	Literature Cited	605

20 Thermal Analysis — 606

20.1	Differential Thermal Analysis and Differential Scanning Calorimetry	606
20.2	Thermogravimetry	609
20.3	Methodology of DSC (or DTA) and TG	611
20.4	Thermomechanical Analysis	616
20.5	Dynamic Mechanical Analysis	619
20.6	Thermometric Titrimetry	619
	Problems	622
	Bibliography	627
	Literature Cited	627

21 Introduction to Electrometric Methods of Analysis — 628

21.1	Types of Electrochemical Cells	630
21.2	Electrode Potentials	630
21.3	Electrochemical Cells	634
21.4	Reference Electrodes	634
	Bibliography	639
	Literature Cited	639

22 pH and Ion Selective Potentiometry — 640

22.1	Glass-Membrane Electrodes	641
22.2	Solid-State Sensors	643
22.3	Liquid-Membrane Electrodes	645
22.4	Gas-Sensing and Enzyme Electrodes	647
22.5	Interferences	648
22.6	Ion-Activity Evaluation Methods	650
22.7	The Measurement of pH	652
22.8	Glass Electrodes for pH Measurement	656

			CONTENTS	xi

	22.9	Electrometric Measurement of pH and pI	659
		Problems	661
		Bibliography	663
		Literature Cited	663

23 Potentiometric Titrations — 664

	23.1	Classification of Indicator Electrodes	665
	23.2	Location of the Equivalence Point	667
	23.3	Null-Point Potentiometry	679
	23.4	Classes of Potentiometric Titrations	679
		Problems	688
		Bibliography	689
		Literature Cited	689

24 Voltammetry, Polarography, and Related Techniques — 691

	24.1	Current–Voltage Relationships	692
	24.2	Characteristics of the Dropping Mercury Electrode	695
	24.3	The Half-Wave Potential	700
	24.4	Instrumentation	702
	24.5	Modern Voltammetric Techniques	707
	24.6	Applications	717
	24.7	Evaluation Methods	719
	24.8	Amperometric Titration Methods	720
	24.9	Two Indicator Electrodes	726
		Problems	729
		Bibliography	734
		Literature Cited	734

25 Electrogravimetry and Coulometry — 736

	25.1	Electroseparations	736
	25.2	Basic Principles	736
	25.3	Equipment for Electrolytic Separations	741
	25.4	Electrogravimetry	742
	25.5	Electrography	757
	25.6	Electrolytic Purification	759
	25.7	Coulometric Methods	759
		Problems	774
		Bibliography	779
		Literature Cited	779

26 Conductance Methods — 781

26.1	Electrolytic Conductivity	781
26.2	Measurement of Electrolytic Conductance	783
26.3	Direct Concentration Determinations	789
26.4	Conductometric Titrations	790
26.5	Measurement of Dielectric Constant	796
	Problems	798
	Bibliography	800
	Literature Cited	800

27 Electronics: Fundamentals of Solid-State Devices — 801

27.1	Basic Functions of Instrumentation	801
27.2	Semiconductor Components	803
27.3	Operational Amplifiers	810
27.4	Digital Integrated Circuits	813
	Bibliography	821

28 Electronics: Commonly Used Signal Modifying Circuits — 822

28.1	Development of Integrated Circuits	822
28.2	Digital MSI Circuits	823
28.3	Analog MSI Circuits	829
28.4	Composite Circuits	833
28.5	Large-Scale Integrated Circuits	843
	Bibliography	845

29 Data Handling — 846

29.1	Introduction	846
29.2	Signal-to-Noise Ratio	846
29.3	Sensitivity and Detection Limit	847
29.4	Sources of Noise	848
29.5	Hardware Components	851
29.6	Software Signal Enhancement	855
29.7	Evaluation of Results	861
29.8	Accuracy and Instrument Calibration	864
	Bibliography	867
	Literature Cited	867

30 Computer-Aided Analysis — 868

30.1	Introduction	868
30.2	Computer Organization—Hardware	871
30.3	Computer Organization—Software	875

	30.4	Implementation—Software Versus Hardware	879
	30.5	Data Representation	879
	30.6	Computerized Instrument Systems	881
	30.7	Microcomputer Interfacing	885
	30.8	Computer Controlled Laboratory Automation Systems	892
		Bibliography	895
		Literature Cited	896
31	**Process Instruments and Automatic Analysis**		**897**
	31.1	Introduction	897
	31.2	Industrial Process Analyzers	899
	31.3	Methods Based on Bulk Properties	901
	31.4	Infrared Process Analyzers	905
	31.5	Oxygen Analyzers	909
	31.6	On-Line Potentiometric Analyzers	913
	31.7	Process Gas Chromatography	915
	31.8	Continuous On-Line Process Control	919
	31.9	Automatic Chemical Analyzers	923
	31.10	Automatic Elemental Analyzers	931
		Problems	935
		Bibliography	936
		Literature Cited	936
		Experiments	937
		Answers to Problems	971
		Appendixes	995
		Index	1009

Abbreviations

absorption	Abs
alpha particle	α
alternating current	ac
American Society for Testing Materials	A.S.T.M.
American standard code for information interchange	ASCII
ampere	A
analog-to-digital converter	ADC, A/D
angstrom	Å
anodic	anod, a (subscript)
aqueous	*aq*
Association of Official Analytical Chemists	A.O.A.C.
atmosphere	atm
atomic absorption spectrometry	AAS
atomic emission spectroscopy	AES
atomic fluorescence spectrometry	AFS
atomic weight	at. wt.
attenuated total reflectance	ATR
Auger electron spectroscopy	AES
back scatter	BS
barn (10^{-24} cm^2)	b
beta particle	β
binary coded decimal	BCD
boiling point	bp
calorie	cal
capacitance	C
cathode ray tube	CRT
cathodic	cath, c (subscript)
centi- (prefix) (10^{-2})	c-
centimeter	cm
centipoise	cP
central processing unit	CPU
circa	*ca.*
citrate	Cit
complementary metal oxide semiconductor	CMOS

Compton edge	CE
conductance	1/R
coulomb	C
counts per minute (second)	cpm (cps)
cubic centimeter	cm^3
curie	Ci
cycles per second (hertz)	Hz
cylindrical mirror analyzer	CMA
decibel	dB
degree Celsius	°C
degree Kelvin	°K
deuteron	d
diameter	diam
differential scanning calorimeter	DSC
differential thermal analysis	DTA
digital-to-analog converter	DAC, D/A
digital voltmeter	DVM
dilution value of pH buffer	$\Delta pH_{1/2}$
diode transistor logic	DTL
direct current	dc
direct digital control	DDC
direct memory access	DMA
disintegrations per minute (second)	dpm (dps)
dropping mercury electrode	dme, de (subscript)
dual-in-line package	DIP
dyne	dyn
effective aperture ratio	f/number
electromotive force	emf
electron	e, e^-
electron capture detector	ECD
electron spectroscopy for chemical analysis	ESCA
electron spin resonance	ESR
electron volt	eV
equivalent weight	equiv wt
erasable programmable read-only memory	EPROM
et alii (and others)	*et al.*
ethyl	Et
ethylenediamine-N,N,N',N'-tetraacetate	EDTA, Y^{4-}
exclusion chromatography	EC
exempli gratia (for example)	e.g.
exponential	exp
external	ext
farad	f
fast Fourier transformation	FFT
field-effect transistor	FET

flame emission spectroscopy	FES
flame ionization detector	FID
flame photometric detector	FPD
formal (concentration)	F
Fourier transformation	FT
frequency	f
full width at half maximum	FWHM
gamma radiation	γ
gas (physical state)	g
gas chromatography	GC
gas chromatography/mass spectrometry	GC/MS
gas–liquid chromatography	GLC
gas–solid chromatography	GSC
gauss	G
Geiger–Müller	GM
geminal	*gem*
gram	g
hertz	Hz
hierarchial distributed control	HDC
high-performance liquid chromatography	HPLC
hour	hr
id est (that is)	i.e.
inch	in.
indicator	ind
inductance	L
induction coupled (argon) plasma	ICAP, ICP
infrared	ir
input/output	I/O
inside diameter	i.d.
integrated circuit	IC
integrated injection logic	IIL
internal	int
International Union of Pure and Applied Chemistry	IUPAC
ion-exchange chromatography	IEC
ion microprobe mass analyzer	IMMA
ion scattering spectroscopy	ISS
joule	J
kilo- (prefix) (10^3)	k-
kilocalorie	kcal
Kovats retention index	R.I.
large-scale integration	LSI
least significant bit	LSB
light emitting diode	LED
limiting	lim
liquid (physical state)	liq, l

liquid chromatography	LC
liquid chromatography/mass spectrometry	LC/MS
liquid–liquid (partition) chromatography	LLC
liquid–solid (adsorption) chromatography	LSC
liter	liter (alone), l (with prefixes)
logarithm (common or Briggsian or decadic)	log
logarithm (natural or Naperian)	ln
logical AND operation in Boolean algebra	· (center dot)
logical OR operation in Boolean algebra	+
lumen	lm
mass spectrometer	MS
maximum	max
medium-scale integration	MSI
mega- (prefix) (10^6)	M-
meta-	m-
metal oxide semiconductor	MOS
metastable (state)	m, m^*
meter	m
methyl	Me
micro- (prefix) (10^{-6})	μ-
micrometer (micron)	μm
microsecond	μsec
milli- (prefix) (10^{-3})	m-
milliampere	mA
milliequivalent	mequiv
milliliter	ml
millimole	mM
million electron volts	MeV
minimum	min
minute	min
molar (concentration)	M
mole	mol
molecular weight	mol wt
monolayer	ML
most significant bit	MSB
multiple internal reflectance	MIR
nano- (prefix) (10^{-9})	n-
nanometer (millimicron)	nm
Naperian base	e
negative	neg
nephelometric turbidity unit	NTU
neutron	n
normal (concentration)	N
normal (alkyl chain)	n-
not AND (results of AND operation negated)	NAND

not OR (results of OR operation negated)	NOR
nuclear magnetic resonance	NMR
numerical aperture	NA
ohm	Ω
operational amplifier	op amp
optical speed	f/number
optimum	opt
ortho-	o-
outside diameter	o.d.
oxidant	ox
oxide semiconductor field-effect transistor (MOSFET without metal gate)	OSFET
page(s)	p. (pp.)
para-	p-
parent ion	M
particle-induced X-ray emission	PIXE
parts per billion, volume	ppb, ng/ml
parts per billion, weight	ppb, ng/g
parts per million, volume	ppm, μg/ml
parts per million, weight	ppm, μg/g
pascal	Pa
percent	%
phenyl	ϕ
photoionization detector	PID
pico- (prefix) (10^{-12})	p-
positive	pos
positron	β^+
potential	E
programmable read-only memory	PROM
propyl	Pr
proton	p
proton magnetic resonance	PMR
quantum (energy)	$h\nu$
quantum efficiency	QE
radian	rad
radio frequency	rf
random access memory	RAM
read-only memory	ROM
reciprocal ohm	mho, Ω^{-1}
reductant	red
reference	ref
reset–set	R–S
resistance	R
reverse phase–ion pair partition	RP–IPP
revolutions per minute	rpm
sample and hold	S/H
saturated	satd

saturated calomel electrode	SCE
scanning Auger microprobe	SAM
scanning electron microscopy	SEM
second	sec
secondary ion mass spectrometry	SIMS
sigma	σ
small-scale integration	SSI
solid (physical state)	s
solvent (general)	S
specific gravity	sp gr
standard hydrogen electrode	SHE, NHE
standard temperature and pressure	STP
surface coated open tubular (column)	SCOT
Système International	SI
tesla	T
temperature	T, temp
tertiary	*tert-, t-*
tetramethylsilane	TMS
thermal conductivity detector	TCD
thermal gravimetry	TG
thermionic emission detector	TED
thermomechanical analysis	TMA
thousand electron volts	keV
torr (mm of mercury)	torr
transistor–resistor logic	TRL
transistor–transistor logic	TTL
tritium	t, ^{3}H
ultraviolet	uv
universal asynchronous receiver transmitter	UART
vacuum	vac
vacuum-tube voltmeter	VTVM
versus	vs.
very large-scale integration	VLSI
volt	V
volume	vol, V
volume per volume	v/v
volume per weight	v/w
wall coated open tubular (column)	WCOT
watt	W
wave number	cm^{-1}
X-ray absorption edge	K edge, L_I edge
X-ray absorption level	K, L_I
X-ray emission lines	$K\alpha, K\beta, L\alpha$
X-ray energy spectrometry	XES
year	yr

Symbols

A	absorbance; activity (radiochemistry); area; atomic weight
A_o	amplifier gain
a	specific absorptivity
a_i	hyperfine coupling constant (ESR)
a_x	activity of species x
AF	asymmetry factor
B	source brightness
b	distance; grating constant; optical path length; thickness
C	concentration
C_M	concentration of solute in mobile phase
C_S	concentration of solute in stationary phase
c	velocity of light in a vacuum
D	dielectric constant; diffusion coefficient
D_M	diffusion coefficient in mobile phase
D_S	diffusion coefficient in stationary phase
D^{-1}	linear reciprocal dispersion
D_c	concentration distribution ratio
d	diameter; distance; spacing
d_c	diameter of collimating mirror; cross section or column bore
d_f	effective thickness of stationary phase
d_p	particle diameter
E	electrode potential; energy of a photon; potential of half-reaction
$E°$	standard electrode potential
$E_{1/2}$	half-wave potential
E_b	core-electron binding energy
E_i	ionization energy
E_{ind}	indicator electrode potential
E_j	liquid-junction potential
E_k	kinetic energy
E_{ref}	reference electrode potential
e, e^-	electronic charge; Naperian base; base of natural logarithms (2.718...)
$e°$	solvent strength parameter
F	faraday; fluorescence
F_c	volume flowrate of mobile phase

f	focal length; fractional abundance; oscillator strength, frequency
f_x	activity coefficient of species x
$f(\theta)$	geometrical factor (fluorometers)
ΔG°	Gibbs free energy
g	spectroscopic splitting factor; statistical weights of particular species
$g(\lambda)$	detector efficiency
H	magnetic-field strength; plate height (chromatography)
ΔH	enthalpy change; peak-to-peak separation (ESR)
ΔH_s	molal heat of solution
ΔH_v	molal heat of vaporization
h	height; Planck's constant $[6.626\ 176(36) \times 10^{-34}\ \text{J} \cdot \text{sec}]$; reduced plate height
I	radiant intensity; spin quantum number of nuclei
I_d	diffusion-current constant
I_o	incident radiant energy; output intensity
I_v	emission line intensity
i	angle of incidence; current
i-	(prefix) iso-
i_d	diffusion current
$i_{\lim}$	limiting current
i_r	residual current
J	spin–spin coupling constant (nuclei)
j	compressibility factor (gas chromatography)
K_a	acid dissociation constant
K_{auto}	autoprotolysis constant
K_d, K	partition coefficient
K_f	formation constant
K_i	ionization constant (gaseous state)
K_{sp}	solubility product
K_w	ion product of water
k	Boltzmann constant $[1.380\ 662(44) \times 10^{-23}\ \text{J} \cdot \text{K}^{-1}]$; force constant (infrared); general constant
k'	partition ratio or capacity factor (chromatography)
$k_{\text{M/N}}$	selectivity coefficient for solutes M and N
k_o	column permeability
k_v	absorption coefficient (optical)
L	inductance; length or distance; lightness (color)
l	reduced column length (chromatography)
M	mass
M_I	spin quantum number (nucleus)
M_n	number–average molecular weight
M_s	angular momentum quantum number (electron)
M_w	weight–average molecular weight
m	mass; mass of mercury (polarography); order number (optical); metastable state (superscript)

Symbol	Description
m^*	metastable state
m^+	ionized mass fragment
m/e	mass-to-charge ratio
N	noise; plate number (chromatography); total number of something
N_A	Avogadro constant ($6.022\,045 \times 10^{23}$ mol^{-1})
N_{eff}	effective plate number
N_j, N_m, N^*	number of species in excited energy state
N_n, N_o	number of species in ground energy state
N_{req}	plates required
$N(E)$	energy distribution (Auger spectroscopy)
n	number of electrons transferred (electrochemistry); principal quantum number; unshared p-electrons
n-	semiconducting material containing a majority of negative charge carriers
n_{theor}	theoretical plate number
P	phosphorescence; pressure; radiant power
P_i	inlet gas pressure
P_M	parent mass peak
P_o	incident radiant power; outlet gas pressure
ΔP	pressure drop across a column
p	partial pressure of some gaseous material; depolarization ratio (Raman); type of electron
p-	semiconducting material containing a majority of positive charge carriers
p°	solute vapor pressure
Q	flowrate; heat capacity; number of coulombs
R	gas constant (molar) [$8.314\,41(26)$ J·mol^{-1}·K^{-1}; $1.987\,19(6)$ cal·mol^{-1}·K^{-1}]; resolution (chromatography); resolving power (optical)
R	retardation factor
R_L	load resistance
r	angle of diffraction; counting rate; radius; resolution (radiochemistry detectors)
r°	programmed rate of temperature increase (chromatography)
r_D	specific refraction
S	electron spin; saturation factor (radiochemistry)
S_1	first excited (singlet) electronic state
S_o	ground electronic state
ΔS	entropy change
S/N	signal-to-noise ratio
T	temperature; transmittance (optical)
T_1	first excited triplet (electronic) state; spin–lattice (or longitudinal) relaxation time (NMR)
T_2	spin–spin (or transverse) relaxation time (NMR)
T_b	boiling point
T_c	column temperature (chromatography)
t	time; prism base length
$t_{1/2}$	half-life
t_M	transit time of nonretained solute (chromatography)

t_p	time of solute passage through one plate
t_R	retention time
t'_R	adjusted retention time
u	reduced mass
$\bar{u}$	average linear velocity
V	volume
V_g°	specific retention volume (at 0°C)
V_g	volume of column occupied by gel matrix (exclusion chromatography)
V_i	internal volume within porous particles
V_M	volume of mobile phase
V_N	net retention volume
V_R	retention volume
V'_R	adjusted retention volume
V_S	cumulative internal volume within porous particles; volume stationary phase
V_t	total bed volume
v	velocity; volume
W	physical slitwidth (optical); weight; zone width at base line, 4σ (in chromatography)
$W_{1/2}$	zone width at 1/2 peak height
W_b	peak width at base line
w	effective aperture width
w_L	weight of stationary liquid phase
w_S	weight of adsorbent phase
X_C	capacitive reactance
X_L	inductive reactance
x	distance; general designation of species
Z	atomic number of an element; impedance
z	valence
z_+, z_-	ionic charge
α	degree of ionization; relative retention ratio
$[\alpha]$	specific rotation
α_i	degree of ionization
β	blaze angle; buffer value (pH); volumetric phase ratio (chromatography)
β_N	Bohr magneton
γ	activity coefficient; emulsion characteristic (photography); surface tension; obstructive (or tortuosity) factor (chromatography)
Δ	(prefix) symbol for finite change; spectral width (NMR)
δ	chemical shift (NMR); thickness of diffusion layer
ϵ	molar absorptivity
ϵ_{tot}	total porosity of column
η	index of refraction; viscosity
η_D	index of refraction (D line of sodium)
Θ	cell constant (conductance)
θ	angle; angle of diffraction

2θ	angular setting of diffraction angle (X ray)
$[\theta]$	molecular ellipticity
κ	specific conductance
Λ	equivalent conductance
Λ_∞	equivalent conductance at infinite dilution
λ	column packing uniformity (chromatography); decay constant (radiochemistry); wavelength
λ_+, λ_-	limiting equivalent ionic conductance
$\Delta\lambda$	base spectral width
λ_{max}	wavelength of an absorption maximum
μ	ionic strength; linear absorption coefficient; magnetic moment
μ_B	Bohr magneton [$9.274\,078(36) \times 10^{-24}$ J·T^{-1}]
μ_e	electron magnetic moment [$9.284\,832(36) \times 10^{-24}$ J·T^{-1}]
μ_m	mass absorption coefficient
μ_N	nuclear magneton [$5.050\,824(20) \times 10^{-27}$ J·T^{-1}]
μ/ρ	mass absorption coefficient
ν	frequency; reduced velocity (chromatography); designation of vibrational levels
$\bar{\nu}$	wave number
π	pi (3.1416...); type of electron or bond
ρ	density; resistivity
Σ	summation symbol
σ	reaction cross section; shielding constant (NMR, X ray); standard deviation
σ_{hkl}	reciprocal lattice vectors
τ	chemical shift (NMR); mean emission lifetime, resolving time; time constant
v	designation of vibrational level; velocity
Φ	number of bombarding particles or flux
ϕ	column flow resistance parameter; photoluminescence efficiency; work function
ω	angular frequency; chopping frequency; overpotential
ω_c	angular velocity
[ ]	molar concentration of species within brackets
*	(asterisk) metastable state

CHAPTER 1

An Introduction to Absorption and Emission Spectroscopy

The purpose of this chapter is to describe some relationships among representative optical phenomena which produce signals with chemical information. Chemical instrumentation does not create information; it refines the information already present in the signal from some transducer. Different optical signal sources give information that is so distinctive and so valuable that each has given rise to its own specialized literature and instrumentation.

Spectroscopy is the measurement and interpretation of electromagnetic radiation absorbed or emitted when the molecules, or atoms, or ions, of a sample move from one allowed energy state to another. Every atom, ion, or molecule has a unique and characteristic relationship with electromagnetic radiation. Our principal concern will be with the areas of spectroscopy that stem from changes in the rotational, vibrational, and electronic energies. In addition, energies resulting from energy differences that arise when a sample is placed in a magnetic or electric field are susceptible to spectroscopic studies. Nuclear magnetic resonance and electron spin resonance are two such studies.

1.1 THE NATURE OF ELECTROMAGNETIC RADIATION

A beam of radiation may be regarded as an electromagnetic waveform disturbance or photon of energy propagated at the speed of light. A photon has the properties of a microscopic particle of definite energy and at the same time has properties of a wave extending over a broad area of space. The Heisenberg uncertainty principle shows that it is not important or even possible to measure both the wave and particle properties of a photon simultaneously. However, it is useful to keep both properties in mind.

A photon originating at a point in space radiates from that point in a spherical wave characterized by electric field vectors which have periodic maxima perpendicular to the direction of propagation. The wavelength of the radiation, λ, can be visualized as the distance between these maxima; that is, from crest to crest in Fig. 1-1. In the figure the distance AB is one wavelength. Associated with wavelength is frequency, ν, the number of waves passing a fixed point, such as P, in a unit length of time. When a photon passes a particular region of space the electric field in that region oscillates with the frequency ν. Wavelength and frequency are related to the energy of a photon, E, by Planck's constant h,

$$\Delta E = h\nu = hc/\lambda \tag{1-1}$$

where c is the velocity of light in a vacuum, 3.00×10^8 m sec^{-1}.

Only frequency is truly characteristic of a particular radiation. Radiation is propagated through matter at velocities smaller than c owing to interactions between the electric vector and the bound electrons of the medium. Consequently, light can be refracted. The index of refraction of a medium, η, is the ratio of the speed of light in a vacuum to the speed of light in the medium. The index of refraction is also a function of wavelength; thus longer wavelengths have a smaller index of refraction in a transparent medium than do shorter wavelengths. When the light of a particular wavelength enters matter its velocity decreases but its frequency remains constant. In the ultraviolet, visible, and infrared regions of the spectrum, the velocity of radiation in air is within 0.1% of the velocity in a vacuum, making it satisfactory to use Eq. 1-1 to interrelate wavelength and frequency.

Wave numbers, $\bar{\nu}$, are sometimes used to express frequency. They are calculated as follows:

$$\bar{\nu} = 1/\lambda \tag{1-2}$$

Wave numbers, in units of cm^{-1}, express the number of waves that occur per centimeter; this number is directly proportional to the frequency:

$$\bar{\nu} = c\nu \tag{1-3}$$

A beam carrying radiation of only one discrete wavelength is said to be monochromatic. A polychromatic beam contains radiation of several wavelengths. Two light waves can

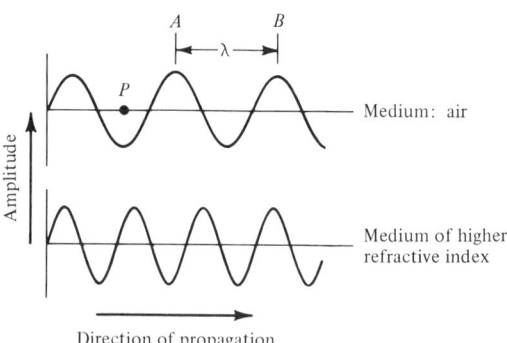

FIGURE 1-1 Some characteristics of electromagnetic radiation.

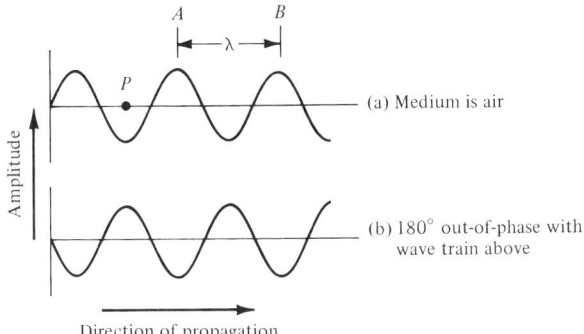

FIGURE 1-2 Two wave trains out-of-phase with each other.

combine to interfere with one another, either constructively or destructively. In Fig. 1-2 the two light waves have the same wavelength, frequency, and amplitude, but are out of phase by 180°. When combined they destructively interfere to cancel exactly one another. If they were in phase, they would constructively interfere to produce a wave having twice the amplitude of each single component.

A polarized beam of light is one in which the changes in amplitude that occur with time are all in the same plane. Polarization can be produced by selective absorption of light as it passes through certain substances.

1.2 THE ELECTROMAGNETIC SPECTRUM

The interaction of matter and radiation takes place throughout the entire electromagnetic spectrum, the name given to the broad range of radiations extending from cosmic rays with wavelengths as short as 10^{-9} nm all the way up to radio waves, which have a wavelength in excess of 1000 km. Encompassed within these extremes and moving from short to long wavelengths are gamma rays, X rays, far, middle, and near ultraviolet rays, the visible light portion of the spectrum, infrared rays, and microwaves. The nature of all these radiations is the same, and all move with the speed of light. They differ only in frequency and wavelength, and in the effects which they can produce.

The chemical and physical effects of various types of radiation are quite different, and these differences can be understood in terms of the differing energies of the photons. In the radio frequency range, the energy of one photon is very low, and the energy transitions are concerned with reorientation of nuclear spin states of substances in a magnetic field. In the slightly higher-energy microwave region, there are changes in electron spin states for substances with unpaired electrons when in a magnetic field. In the infrared region absorption causes changes in vibrational energy accompanied by changes in rotational energy. Changes in the visible and ultraviolet regions involve the electron energy of atoms or molecules; for molecules this is accompanied by changes in vibrational and rotational energy. These changes in electronic energy involve the most loosely held or so-called outer electrons. Finally, at the high-energy end, X rays will raise inner electrons to excited states. Gamma radiation originates within atomic nuclei. The various regions in the elec-

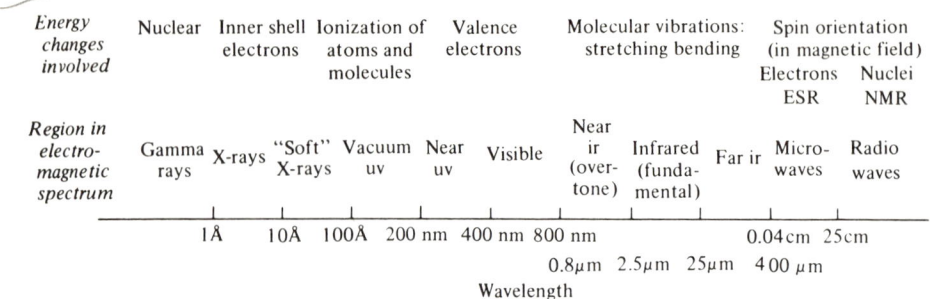

FIGURE 1-3 Schematic diagram of the electromagnetic spectrum. Note that the wavelength scale is nonlinear.

tromagnetic spectrum are displayed in Fig. 1-3 along with the nature of the changes brought about by the radiation.

1.3 ATOMIC ENERGY LEVELS

Following the quantum theory, atoms can exist only at discrete potential energy levels. The potential energy of an atom depends on the electron configuration and transition of outer electrons between fixed energy levels to emit or absorb radiation at discrete energies. The frequency of radiation absorbed or emitted is proportional to the change in potential energy involved, and is given by Eq. 1-1. Thus atomic spectra involve only transitions of electrons from one electronic energy level to another. Each transition accounts for the presence of a specific frequency of light and hence the presence of a spectral line either in absorption or in emission.

Grotrian developed a graphical method for presenting atomic energy levels and electronic transitions that is almost universally used. The diagram permits the representation of spectral terms and transitions, as shown in Fig. 1-4 for several elements. The vertical axis is an energy axis and the energy levels, or terms, are shown as horizontal lines. Absorption is represented by an upward line between adjacent energy levels. A spectral emission line results from a transition from a higher-energy level to a lower one. The vertical distance representing a transition is a measure of the energy of the transition. The energy of the transition from the ground electronic state to the first excited state is great enough that only a few atoms with labile valence electrons have absorption spectra in the visible region. The absorption by atoms is limited to a comparatively few resonance lines. When an atom is excited by absorbing light or by collision with excited electrons, ions, or molecules, it normally remains in an excited state for only a very short time, approximately 10^{-9} sec, before it loses all or part of its excitation energy by collisions or by emitting a photon.

The major contributions to the energy terms are associated with a principal quantum number $n = 1, 2, 3, \ldots$. For example, when the $3s$ valence electron of sodium is excited to $4s$, $5s$, $6s$, and larger orbitals, the series of energy levels labeled ^2S results. The separations of levels decreases as the value n increases. The continuum (ionization level) starts at $n = \infty$.

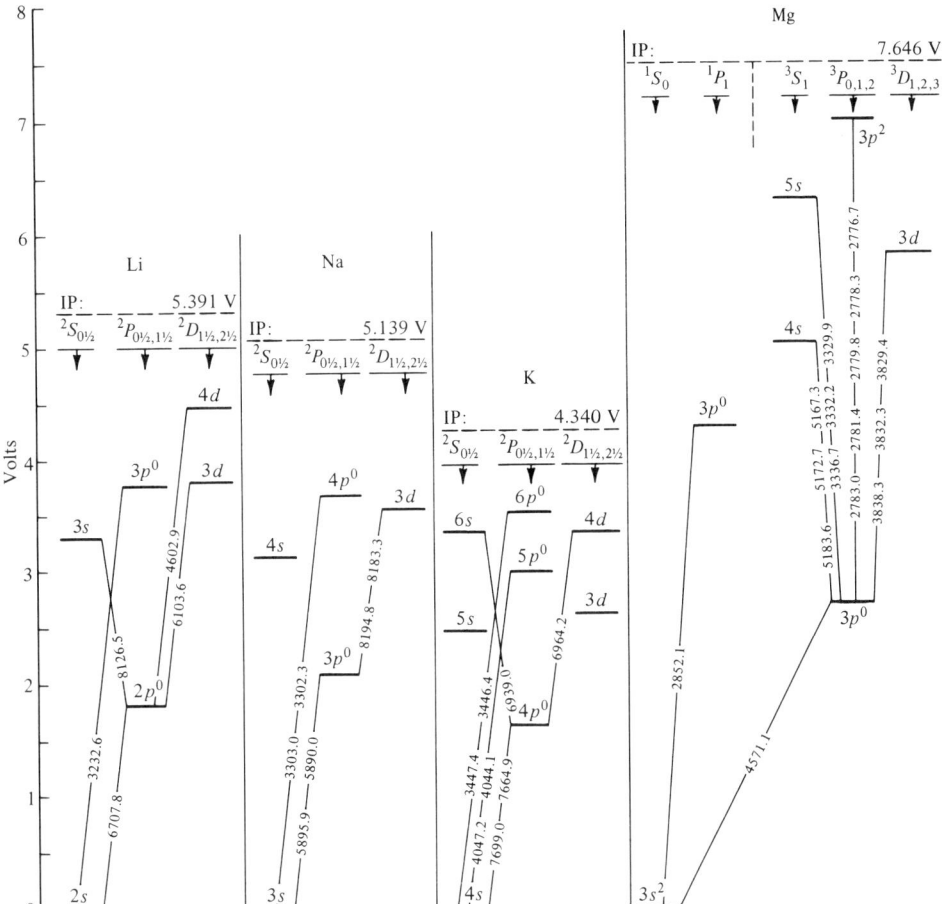

FIGURE 1-4 Atomic term diagram for Li, Na, K, and Mg. Wavelengths given in angstroms.

The resultant orbital angular momentum or resultant eccentricity of electronic orbitals having the same principal quantum number accounts for somewhat smaller energy differences which are classified as different series. The resultant orbital angular momentum is the vector sum of the orbital angular momenta of the individual electrons, $\ell = 0, 1, 2, 3, \ldots$ (symbol: $s, p, d, f, \ldots$) and is represented by the quantum number $L = 0, 1, 2, 3, \ldots$ (series symbol: $S, P, D, F, \ldots$). For example, a ^2P series results when the single valence electron of sodium is excited to p states of higher orbits.

The total angular momentum, the result of the different possible combinations of electron orbital angular momentum and electron spin angular momentum, accounts for still smaller energy differences which distinguish the slightly different terms of multiplets. For example, each of the energies in the ^2P series is actually a pair of levels of slightly different energy depending upon whether the total angular momentum is $\frac{1}{2}$ or $\frac{3}{2}$. The $^2S \leftrightarrow {}^2P$ transitions appear as doublets. The superscript to the left of the term symbol is the multiplicity, the number of different values of the total angular momentum for a particular resultant orbital angular momentum. The s-terms, though being singlets, always possess the same multiplicity index as the p-, d-, and f-terms belonging to them in a transition.

Selection rules are statements of transition probabilities that are based on experience or quantum-mechanical calculation. On the basis of selection rules, transitions are called allowed or forbidden. Selection rules are not absolute but serve to define the transitions which occur most strongly. For example, the selection rule $\Delta L = \pm 1$ means that the spectral transitions that are most likely to occur are those between neighboring term series. In the case of multiplet terms, the inner quantum number j takes one of the two values $\ell + \frac{1}{2}$ or $\ell - \frac{1}{2}$. The selection rule for j is $\Delta j = 0, \pm 1$ ($0 \leftrightarrow 0$ excluded).

The ionization energy or ionization limit of an atom is the energy to which the various series converge. The kinetic energy of the ejected electron is not quantized, and energy levels beyond this limit are continuous. The energy levels of the ion which remains correspond more closely to those of the preceding element in the periodic table than to those of the parent atom.

1.4 MOLECULAR ELECTRONIC ENERGY LEVELS

The electrons of a molecule can also be excited to higher-energy states, and the radiation that is absorbed in this process, or the energy emitted in the return to the ground state, can be studied. The energies involved are generally large, 50–150 kcal mol^{-1}; consequently, electronic spectra of molecules are usually found in the ultraviolet or visible region of the electromagnetic spectrum. In molecular spectra transitions between electronic states may be accompanied by transitions between rotational and vibrational energy levels which impart fine structure on the electronic absorption bands. As a result, the spectra of molecules are much more complicated than those of atoms. But this means that information may be obtained about molecular vibrations and rotations that reveal a great deal about molecular structure. Homonuclear diatomic molecules, which do not have vibration–rotation spectra in the infrared, do show vibrational and rotational structure in their electronic spectra.

Molecular electronic energies are represented by potential energy curves (or surfaces) in which the potential energy of each electronic state is plotted as a function of internuclear distance. Two-dimensional diagrams are inadequate, except in the case of diatomic molecules, but they are nevertheless useful in describing general phenomena that are observed. Examples of electronic energy levels are shown in Fig. 1-5. Excitation strong enough to excite the emission spectra of molecules usually dissociates the molecule. Most molecular electronic spectra are therefore observed by absorption methods.

An electronic energy level is a physically stable state of a molecule when the potential energy curve has a minimum. Most molecules have excited electronic states which are not stable. Excitation to these states leads to dissociation and the spectrum corresponding to these transitions is continuous. In some cases a stable excited state intersects an unstable excited state and dissociation can occur as a radiationless transition between the two.

Electronic states of simple molecules may be characterized by quantum numbers which are derived from the quantum numbers of the component atoms. For example, a state is

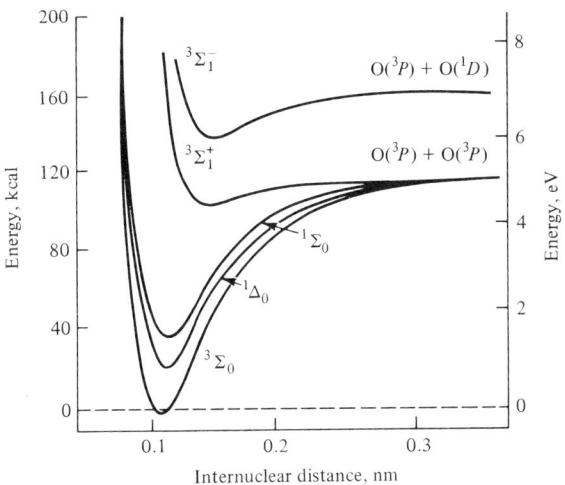

FIGURE 1-5 Potential-energy curves for some of the electronic states of the O_2 molecule.

characterized by the resultant orbital angular momentum quantum number $\Lambda = 0, 1, 2, \ldots$ (symbols: $\Sigma, \Pi, \Delta, \ldots$), and the multiplicity of a state is denoted by a superscript preceding this designation as in the case of atoms.

According to the Pauli exclusion principle, spins of two electrons in the same orbital are opposite to one another; that is, paired. A molecule with an even number of electrons has all electrons paired and is said to be in a singlet state. Whether the molecule is in the ground state or excited state, as long as the electrons are paired, the molecule is in a singlet state, $S_0, S_1, S_2, \ldots$.

Before the molecule can absorb radiation it must interact with it and this interaction must occur within the period of oscillation of the light wave, approximately 10^{-15} sec, the period during which the photon and the molecule are in contact. Consequently, exchange can occur only by interaction with the potential energy component of the molecule's total energy via movement of electrons. Franck and Condon pointed out that since electrons move much more rapidly than nuclei, it is a good approximation to assume that, in an electronic transition, the nuclei do not change their positions. Therefore, an electronic transition may be represented by a vertical line in an energy diagram. Furthermore, the electronic levels inside molecules are quantized so that the absorption bands can occur only at definite values corresponding to the energies required to promote electrons from one level to another.

Since a molecule vibrates, even when it is in the lowest vibrational energy level, a range of internuclear distances must be considered. In the lowest-energy state the most probable internuclear distance is that corresponding to the equilibrium position. For the higher-energy states the most probable configuration is at the ends of the vibration, where the atoms must stop and reverse their direction. For a solute molecule surrounded by solvent molecules, transitions are expected to have greater probability of starting near the midpoint of the lowest vibrational level of the ground electronic state and proceeding to the $v = 2$ vibrational level of the excited electronic state. Transitions to other vibrational levels of the excited state occur with lower probabilities. Thus, as Fig. 1-6 shows, an electronic

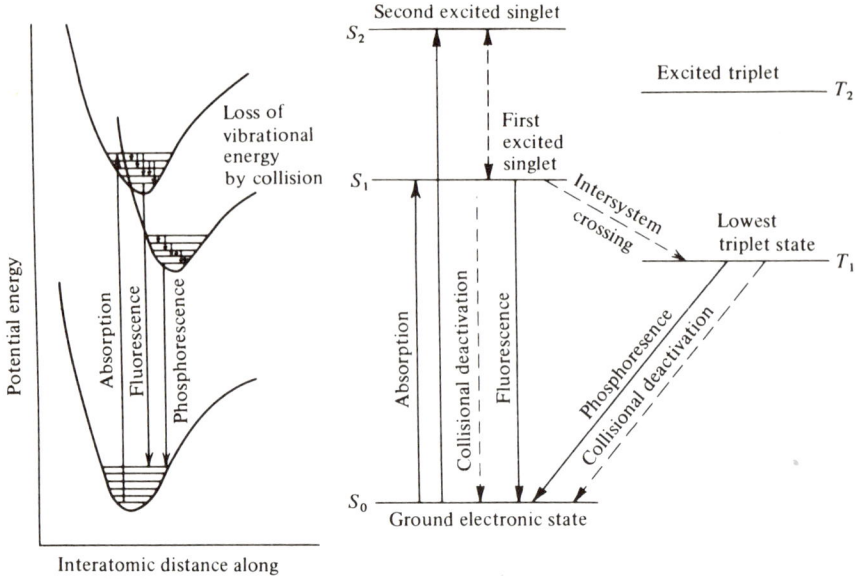

FIGURE 1-6 Schematic energy-level diagram for a diatomic molecule.

transition, in absorption, may show a series of closely spaced lines corresponding to different vibrational (and rotational) energies of the upper state.

Absorption terminates when the solute molecule arrives in any one of several possible vibrational levels in an excited electronic state that is still surrounded by the ground state equilibrium of the solvent molecules. An excited molecule can return to its ground state by any of several paths shown in Fig. 1-6. The favored route is the one that minimizes the lifetime of the excited state. Within 10^{-12} sec the electronically excited solute molecule drops to the lowest vibrational level of the lowest excited singlet state by means of radiationless processes. The excess energy is transferred to other molecules through collisions as well as by partitioning the excess energy to other possible modes of vibration or rotation within the excited molecule. The solvent molecules reorient themselves to a state of equilibrium compatible with the new molecular polarity. The vibrational and solvent relaxation processes are accompanied by a loss of thermal energy.

There is another route by which molecules in higher excited states can reach the lowest vibrational state of a lower electronic level. Where the vibrational levels from different excited electronic states overlap and have the same potential energy, internal conversion (a radiationless process) can occur. Here, the excited molecule proceeds from the higher electronic state to the lowest vibrational level of the lower excited state via a series of vibrational relaxations, an internal conversion, further relaxations, and so on. This process is not well understood; it occurs by direct vibrational coupling between electronic states and by quantum-mechanical tunneling.

When molecules reach the lowest vibrational level of the lowest excited singlet state, the radiation of fluorescence can occur when the electron returns to any of the vibrational levels of the ground electronic state. Each transition involves radiation of a specific wavelength. This radiative process ($S_1 \rightarrow S_0$) has a short natural lifetime (10^{-9} to 10^{-7} sec) so that in many molecules it can compete effectively with other processes capable of removing the excitation energy, such as internal conversion or intersystem crossing. The consequences of this mechanism are twofold. First, the fluorescence spectrum will approximately mirror the absorption (or excitation) spectrum. However, because the molecules relax to lower vibrational levels in the excited state and because of the solvent reorientation in the excited state and ground state, the electromagnetic radiation corresponding to fluorescence is of lower energy than the exciting radiation, and therefore appears at longer wavelengths. Second, although the intensity of the fluorescence spectrum depends upon excitation wavelength, its spectral pattern is independent of the excitation wavelength. Of course, if the absorption process leads to an electronic state in which the energy exceeds the bond strength of one of the solute's linkages, then excitation energy is lost by molecular dissociation before fluorescence can occur.

If the potential energy curve of the excited singlet state crosses that of the triplet state, some excited molecules may pass over to the lowest triplet state via an intersystem crossing which involves vibrational coupling between the excited singlet state, S_1, and the triplet state, T_1. A triplet state is one in which all the electrons in the molecule are paired except two. Although single-triplet transitions are forbidden processes, the internal conversion from the excited singlet to the triplet state may occur with some probability, since the energy of the lowest vibrational level of the triplet state is lower than that of the singlet state. The probability of intersystem crossing is greater when the potential energy curves cross at the lowest point on the excited singlet curve. Once indirect occupation of the triplet state has been achieved, the molecule will undergo a vibrational relaxation and solvent reorientation to arrive at the lowest vibrational level of the lowest excited triplet state. From this state, electromagnetic radiation can be emitted, or internal conversion can occur. If a radiative transition occurs, it is called phosphorescence. Phosphorescence will occur with low probability since spin reversal must once more occur. Consequently, the triplet state persists for a relatively long average lifetime. The decay time of phosphorescence is similar to the lifetime of the triplet state, approximately 10^{-4} to 10 sec. Spin-orbit coupling, which is a magnetic perturbation capable of flipping spins, is believed to be the main source of phosphorescence transitions back to the ground singlet state. The long lifetime of the triplet state greatly increases the probability of collisional transfer of energy with solvent molecules which is very efficient in solution at room temperature and is often the main pathway for loss of triplet state excitation energy. Because of this, phosphorescence is rarely observed at room temperature but can be easily observed by dissolving the solute in a solvent that freezes to form a rigid glass at the temperature of liquid nitrogen. In summary, phosphorescence is influenced by vibrational relaxation and solvent reorientations in the excited singlet state, the triplet state, and the ground state, as well as intersystem crossing. Accordingly, the electromagnetic energy corresponding to phosphorescence is of still lower energy than fluorescence and will appear at longer wavelengths.

In both fluorescence and phosphorescence the lower-energy photon is emitted in an arbitrary direction and at wavelengths longer than the excitation wavelength. Based on these phenomena we have the twin techniques of spectrofluorometry and spectrophosphorimetry which offer some unique advantages not possessed by absorption spectrophotometry.

1.5 VIBRATIONAL ENERGY LEVELS

The molecular motion that has the next lower energy after electronic transitions is the vibration of the atoms of the molecule with respect to one another. Vibrational energies are usually an order of magnitude smaller than the electronic energy. The simplest case involves the vibrational motion of the atoms of a diatomic molecule. In a diatomic molecule the only vibration is the stretching of the bond between the two atoms represented by the two-dimensional potential energy diagram in Fig. 1-6 (lowest electronic state). The allowed energies for a diatomic molecule, as given by the quantum-mechanical vibrational energy level, are

$$e_{vib} = (v + \tfrac{1}{2}) \frac{h}{2\pi} \sqrt{\frac{k}{u}} \qquad v = 0, 1, 2, \ldots \tag{1-4}$$

where k is the force constant and measures the force required to stretch a bond by a given distance (that is, the stiffness of the chemical bond) and u is the reduced mass for the two atoms, m_1 and m_2;

$$u = \frac{m_1 m_2}{m_1 + m_2} \tag{1-5}$$

Equation 1-4 indicates a pattern of energy levels with a constant spacing. As we have seen, at normal temperatures practically all molecules are in the ground electronic state. Except in cases where there are vibrational levels of very low energy, molecules are also usually in the vibrational ground state ($v = 0$).

Coupling with electromagnetic radiation occurs if the vibrating molecule produces an oscillating dipole moment that can interact with the electric field of the radiation. Homonuclear diatomic molecules like H_2, O_2, or N_2, which necessarily have a zero dipole moment for any bond length, will fail to interact. However, the dipole moment of molecules like HCl can be expected to be some function, usually unknown, of the internuclear distance. The vibration of such molecules leads to an oscillating dipole moment, and a vibrational spectrum can be expected which lies in the infrared region.

Even when interaction between a vibrating molecule and radiation occurs, a further selection rule applies which restricts transitions resulting from the absorption or emission of a quantum of radiation by the relation $\Delta v = \pm 1$. Only $\Delta v = 1$ pertains to absorption spectroscopy. The vibrational frequency is greater the smaller the mass of the vibrating atoms and the greater the force restoring the atoms to their equilibrium position (Eq. 1-4). Motions involving hydrogen atoms are found at much higher frequencies than are motions involving heavier atoms. For multiple bond linkages, the force constants of double and triple bonds are roughly two and three times those of the single bonds, and the absorption position becomes approximately two and three times higher in frequency. Interaction with neighboring atoms or groups may alter these values somewhat, as will resonating structures, hydrogen bonds, and ring strain.

The vibrational modes for a methylene group are illustrated in Fig. 1-7. In a symmetrical group such as methylene there are identical vibrational frequencies. For example, the asymmetric vibration denoted b occurs in the plane of the paper and also in the plane

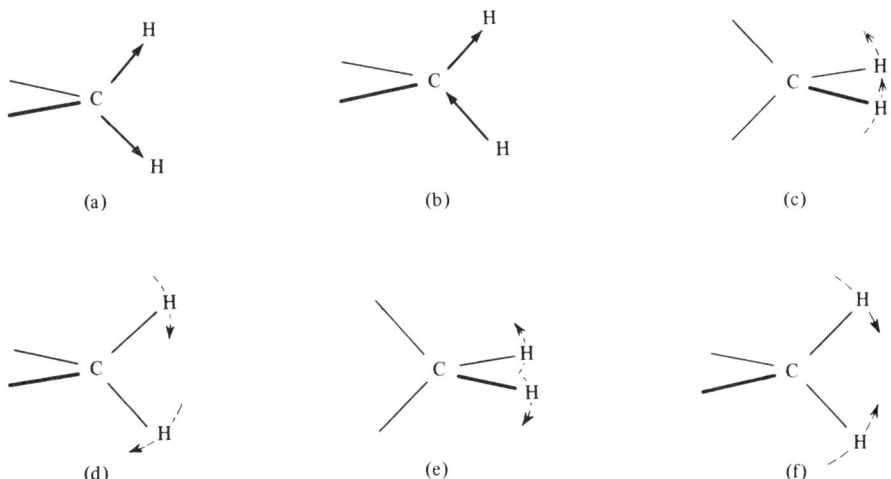

FIGURE 1-7 Vibrational modes of the H—C—H group. (a) Symmetrical stretching, (b) asymmetrical stretching, (c) wagging or out-of-plane bending, (d) rocking or asymmetrical in-plane bending, (e) twisting or out-of-plane bending, and (f) scissoring or symmetrical in-plane bending.

at right angles to the paper. In space these two are indistinguishable and said to be one "doubly degenerate" vibration. In the symmetric stretching mode denoted (a), there will be no change in the dipole moment as the two hydrogen atoms will move equal distances in opposite directions from the carbon atom, and the vibration will be infrared inactive. However, for the asymmetric vibrations, there will be a change in the dipole moment, since during these vibrations the centers of highest positive (hydrogen) and negative (carbon) charge will move in such a way that the electrical center of the group is displaced from the carbon atom. These vibrations will be observed in the infrared spectrum of the methylene group.

When a three-atom system is part of a larger molecule, it is possible to have bending or deformation vibrations. These are vibrations which imply movement of atoms out from the bonding axis. Four types can be distinguished:

1. Deformation (or scissoring) wherein the two atoms connected to a central atom move toward and away from each other with deformation of the valence angle.
2. Rocking or in-plane bending wherein the structure unit swings back and forth in the symmetry plane of the molecule.
3. Wagging or out-of-plane bending wherein the structure unit swings back and forth in the plane perpendicular to the molecule's symmetry plane.
4. Twisting wherein the structural unit rotates back and forth around the bond which joins it to the rest of the molecule.

Splitting of bending vibrations due to in-plane and out-of-plane vibrations is found with larger groups joined by a central atom. An example is the doublet produced by the *gem*-dimethyl group. Bending motions produce absorption at lower frequencies than fundamental stretching modes.

Molecules composed of several atoms vibrate not only according to the frequencies of the stretching modes and bending motions, but also at overtones of these frequencies. When one bond vibrates, the remainder of the molecule is also involved. The harmonic (overtone) vibrations possess a frequency which represents approximately integral multiples of the fundamental frequency. A combination band is the sum, or the difference, of the frequencies of two or more fundamental or harmonic vibrations. The uniqueness of an infrared absorption spectrum arises largely from these bands which are characteristic of the entire molecule. The intensities of overtone and combination bands are usually about one-hundredth those of fundamental bands.

The intensity of a fundamental vibrational absorption band is proportional to the square of the rate of change of dipole moment with respect to the displacement of the atoms. In some cases, the magnitude of the change in dipole moment may be quite small, producing only weak absorption bands, as in the relatively nonpolar $C \equiv N$ group. By contrast, the large permanent dipole moment of the $C=O$ group causes strong absorption bands, often the most distinctive feature of an infrared spectrum.

1.6 RAMAN EFFECT

In Raman spectroscopy the interaction stems from the electric field and an oscillating polarizability within the molecule. An electric field can act on a molecule to distort the electron distribution and thus to induce a dipole. If the ease of distortion, or polarizability, oscillates, so also will the induced dipole. For example, in the symmetric stretching mode of carbon dioxide there will be no change in the dipole moment as the two negative centers will move equal distances in opposite directions from the positive center. However, the electron cloud around the molecule alternately elongates and contracts, changing the polarizability accordingly. In such a case, the changing molecular polarizability will cause a modulation of the scattered light at the vibrational frequency. Hence, the induced classical oscillating dipole radiates not only at the frequency of the incident light but also at frequencies corresponding to the sum and the difference of this frequency and the molecular vibrational frequencies, as will be shown subsequently.

In the Raman effect this mechanism is utilized by irradiating the sample with an intense monochromatic beam of radiation. The wavelength or the samples are chosen so that this exciting radiation is not absorbed. It does, however, through the induced oscillating dipole(s) that it stimulates, lead to the transfer of energy with the rotation and vibration modes of the sample molecules.

Most collisions of the incident photons with the sample molecules are elastic; that is, Rayleigh scattering where radiation is scattered in all directions by interaction with atoms in its path. However, about one in every million collisions are inelastic and involve a quantized exchange of energy between the scatterer and the incident photon to give weak scattered lines which are separated from the exciting line by frequencies equal to vibrational frequencies of the scatterer. In the quantum-mechanical representation of the origin of Raman lines, the incident photon elevates the scattering molecule to a quasiexcited state whose height above the initial energy level equals the energy of the exciting radiation (Fig. 1-8). This quasiexcited state then radiates light in all directions except along the line of action of the dipole; that is, the direction of the incident radiation. On the

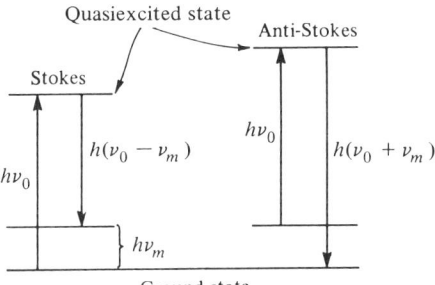

FIGURE 1-8 Quantum representation of energy interchange involved in the Raman effect.

return to the ground electronic level, a vibrational quantum of energy may remain with the scatterer; if so, there is a decrease in the frequency of the reemitted radiation. If the scattering molecule is already in an excited vibrational level of the ground state, a vibrational quantum of energy may be abstracted from the scatterer, leaving it in a lower vibrational level and thus increasing the frequency of the scattered radiation. For either case, the shift in frequency of the scattered Raman radiation is proportional to the vibrational energy involved in the transition. Thus, the Raman spectrum occurs as a series of discrete frequencies shifted symmetrically above and below the frequency of the exciting radiation, and in a pattern characteristic of the molecule. The shift is independent of the frequency of the incident radiation; however, the intensity of the scattered radiation varies with the fourth power of the frequency of the incident radiation. The Raman lines usually studied are those on the low frequency side of the incident radiation, the Stokes lines, which are more intense. By convention the positions of Raman lines are expressed as wave numbers, but more correctly they are wave number differences.

Vibrational Raman spectra have, for complex molecules, the general appearance of the corresponding infrared absorption spectra, and often the same vibrational energy-level separation shows up as a spectral line in both spectroscopic methods. However, when symmetric molecules are considered, a dramatic difference between the two spectral techniques becomes evident. A homonuclear diatomic molecule will exhibit a Raman vibrational spectrum because the molecule will be more, or less, polarizable when it is lengthened than when it is shortened. For molecules with a center of symmetry, Raman spectra provide information on the symmetric vibrations of molecules, and infrared absorption spectra on the antisymmetric.

1.7 NUCLEAR SPIN BEHAVIOR

In addition to charge and mass, about half of the known isotopes possess spin, or angular momentum. The spinning charge generates a magnetic field, and associated with the angular momentum is a magnetic moment. These nuclei resemble a tiny bar magnet, the axis of which is coincident with the axis of spin. When placed in a powerful, uniform magnetic field, such nuclei are acted upon by a torque and tend to assume an allowed orientation with respect to the external magnetic field. The field aligns the spinning nuclei against the disordering tendencies of thermal processes. However, the nuclei do not align perfectly parallel (or antiparallel) to the field. Instead, their spin axes will be inclined to the field and, like the top of a gyroscope, will precess about the field direction wherein each pole of the nuclear axis sweeps out a circular path in the *xy*-plane, as shown in Fig. 1-9. Increasing the strength of the field only makes the nuclei precess faster. By applying a second, much weaker radio frequency (rf) field at right angles to the uniform magnetic field, the nuclei can be made to undergo a transition to a higher-energy level. When the frequency of the rotating component of this second rf field reaches the precession frequency, the spinning nuclei will absorb energy and flip into a higher-energy level. For protons, and other nuclei with a spin of $\frac{1}{2}$, this means an energy level antiparallel to the uniform field.

The resonance frequency, ν, that will effect transitions between energy levels is derived by equating the Planck quantum of energy with the energy of reorientation of a magnetic dipole:

$$\Delta E = h\nu = \mu H_0/I \tag{1-6}$$

where H_0 is the uniform magnetic field, μ is the magnetic moment of the nuclei, and I is the spin quantum number in units of $h/2\pi$. There will be $2I + 1$ possible orientations, and

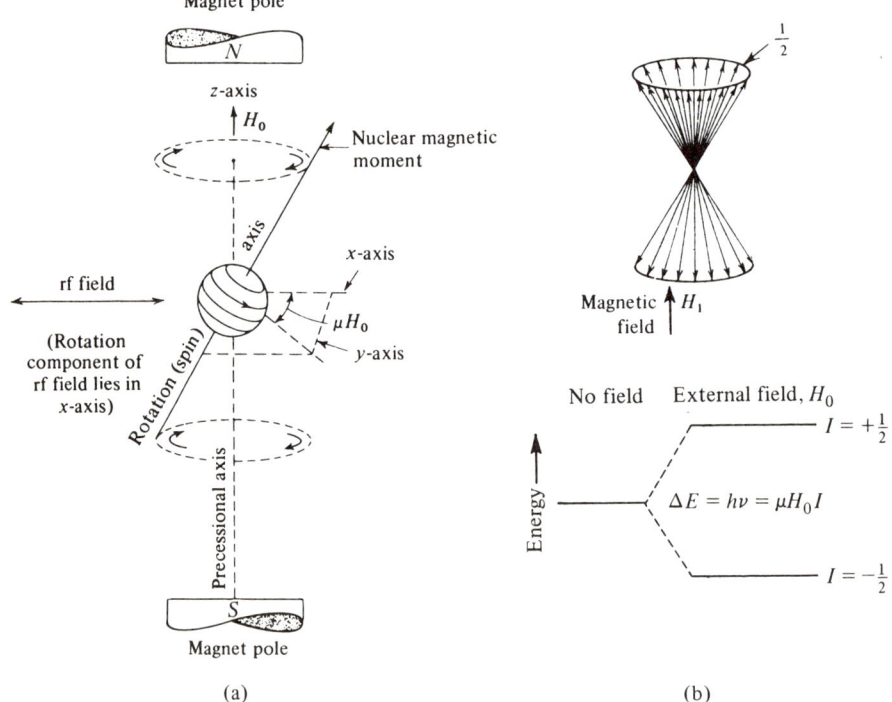

FIGURE 1-9 (a) Spinning nucleus in a magnetic field. (b) Energy-level diagram for a nucleus (lower) and nuclear orientation (upper).

corresponding energy levels. Nuclei with $I = \frac{1}{2}$ give the best resolved spectra because their electric quadrupole moment is zero. They act as though they were spherical bodies possessing a uniform charge distribution which circulates over their surfaces. These nuclei include ^{1}H, ^{13}C, ^{19}F, and ^{31}P. Nuclei with spins of 1 or greater also possess nuclear electric quadrupole moments, and so are readily disturbed by molecule electric field gradients. The result is a shortening of spin lifetime in a given state, and "smearing-out" of the spectroscopic signal. The electric quadrupole moment measures the electric charge distribution within a nucleus when it possesses nonspherical symmetry.

As indicated in Eq. 1-6, the frequency of the resonance absorption varies with the value of the applied field. For example, in a magnetic field of 14,092 G, the protons will precess at 60 MHz, the rf frequency required. In a field of 23,490 G, the precessional frequency rises to 100 MHz. Also since the strength of the absorption signal is roughly proportional to the square of the magnetic field strength, larger values of field strength lead to a stronger signal.

The energy difference between the two energy levels for a proton is not very large compared to thermal energies, only about 0.01 cal. Consequently, thermal agitation diminishes the slight excess of nuclei in the lower-energy state. At normal temperatures and with a magnetic field of 14 kG, only about 20 protons out of each 10 million serve as the effective participating population, because of the cancellation of opposing vectors for the remainder. If an absorption signal is to persist, some mechanism must be provided for replenishing the number of nuclei in the lower-energy state, otherwise in time the rf field would cause the populations of the energy levels to become equalized. The spin system would become saturated.

Energy absorbed and stored in the upper energy level can be dissipated and the nucleus returned to the lower-energy level by a process called spin–lattice relaxation. It is brought about by interaction of the spin with the fluctuating magnetic fields produced by the random motions of neighboring nuclei (called the "lattice," whether the material is crystalline, amorphous, or fluid). In solids and viscous liquids, the relaxation time is on the order of hours, but in typical organic liquids and dilute solutions the time is in the range of 1–20 sec.

1.8 ELECTRON SPIN BEHAVIOR

The electron, like the proton and other spinning nuclei, is a charged particle. It spins and hence has a magnetic field. But it spins much faster than nuclei and thus has a much stronger magnetic field than they have. Associated with the spin is a magnetic moment since the spinning electron behaves like a magnet with its poles along the axis of rotation. If the electron has not only an intrinsic magnetic moment along its own axis but also one associated with its circulation in an atomic orbit, the electron will possess a total magnetic moment equal to the vector sum of these magnetic moments. The ratio of the total magnetic moment to the spin value is a constant for a given atom in a given environment, and is called the gyromagnetic ratio or spectroscopic splitting factor for that particular electron. The fact that these ratios differ for various atoms and environments and the fact that local

16   CHAPTER 1

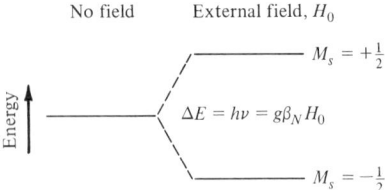

FIGURE 1-10   Energy-level diagram for an unpaired electron.

magnetic fields depend on the structure of matter permit the spectral separation and electron spin resonance spectroscopy.

In the absence of an external magnetic field, the free electron may exist in one of two states, $+\frac{1}{2}$ or $-\frac{1}{2}$, of equal energy, and thus is degenerate. Imposition of an external static magnetic field, H_0, removes the degeneracy and causes the electron to precess. Two energy levels are established, as shown in Fig. 1-10. The lower-energy state has the spin magnetic moment aligned in the direction of the magnetic field and corresponds to the quantum number, $M_s = -\frac{1}{2}$. The difference in energy between the two levels is given by

$$\Delta E = h\nu = g\beta_N H_0 \tag{1-7}$$

where g is the spectroscopic splitting factor and β_N is the Bohr magneton. Transitions from one state to the other can be induced by probing the electron with electromagnetic radiation in the microwave range, around 10,000 MHz. The interaction which causes the transitions is between the magnetic dipole of the electron and the oscillating magnetic field accompanying the electromagnetic radiation. Thus when microwaves travel down a rectangular waveguide, they produce a rotating magnetic field at any fixed point, which can serve to flip over electron magnets in matter. When the magnetic field is expressed in kilogauss, the resonance frequency in megahertz for a free electron is given by

$$\nu = 2800 H_0 \tag{1-8}$$

Typical energy involved is 1 cal mol^{-1}.

1.9   X-RAY ENERGY LEVELS

X-ray emission and absorption spectra are quite simple and all elements have a similar pattern. This relative simplicity of X-ray spectra is explained by the fact that the spectra result from transitions between energy levels of the innermost electrons in the atom. There are only a few electrons in these inner shells and the resulting energy levels are limited, thus giving rise to only a few permitted transitions. There is only one K shell. The L electrons are grouped according to their binding energy into three sublevels: $L_I, L_{II},$ and L_{III}. The complete M shell consists of five sublevels.

The lines of heavier elements fall at higher-energy positions in the spectrum. The relationship between the frequency of a given line and the atomic number of the element, Z, is

$$\nu = R(Z - \sigma)^2 \left(\frac{1}{n_2^2} - \frac{1}{n_1^2} \right) \tag{1-9}$$

where R is the Rydberg constant, n_2 and n_1 are the electron quantum numbers, and σ is a shielding constant (approximately equal to 1 for K electrons). The frequency of a given X-ray line therefore increases approximately as the square of the atomic number of the element involved.

A typical example of the X-ray energy levels that are involved in absorption and emission is shown in Fig. 1-11. Absorption by electrons in an inner shell requires an energy that is at least greater than that required to raise the electron to the ionization limit. As the energy of the incident radiation is increased, there is successive ionization first of electrons in the outermost shell, here the M shell, then of electrons in the L shells as the discrete L_{III}, L_{II}, and L_I absorption edge energies are progressively exceeded, and finally culminating in the ionization of the K shell electron. X-ray spectra show no absorption lines but only an absorption edge because the electron is ejected completely from the atom.

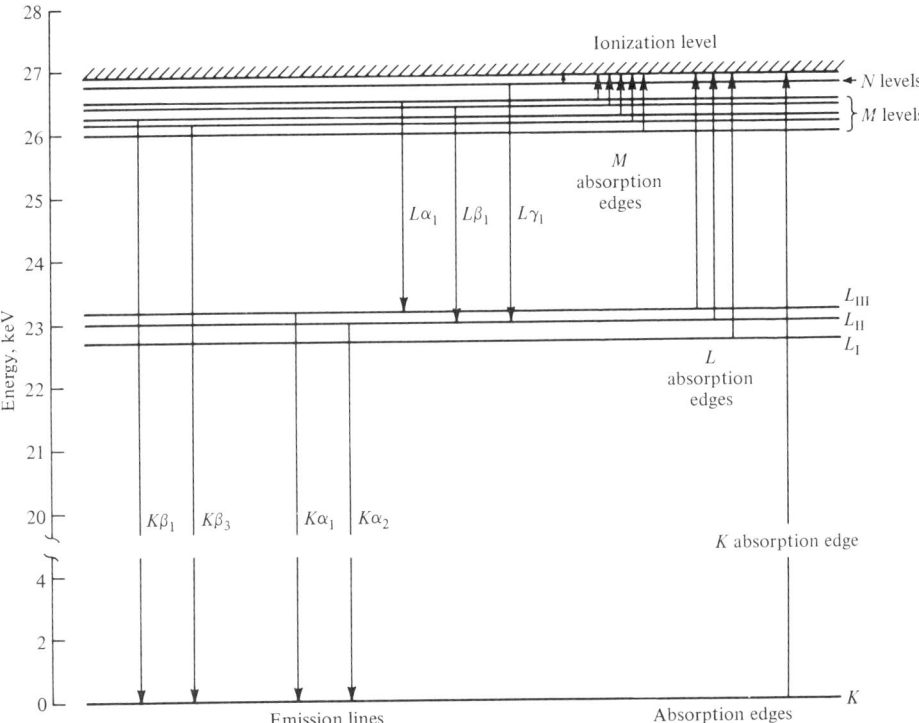

FIGURE 1-11 Energy-level diagram of cadmium for X-ray transitions.

X-ray emission lines of significant intensity are those for which the selection rules $\Delta \ell = \pm 1$ and $\Delta j = 0, \pm 1$ are both satisfied. That is, an X-ray line is emitted if the rules give the difference in the initial and final states for the electron transition that fills the hole created by the absorption step, and leads to the emission of the characteristic quantum of energy. To illustrate from Fig. 1-11, transitions from the L_{II} and L_{III} levels are permitted when the electron vacancy initially occurs in the K level, but the transition from the L_I level is not permitted because for it $\Delta \ell$ is zero. The permitted transitions give rise to the $K\alpha_2$ and $K\alpha_1$ emission lines, respectively. Permitted transitions from the M_{II} and M_{III} sublevels give rise to the $K\beta_3$ and $K\beta_1$ lines in emission, respectively. Of course, the electron vacancies created when an L electron falls back to the K shell lead to a series of L emission lines, with the electron originating from the M or other outer shells, and so on for successively existing outer electron shells.

BIBLIOGRAPHY

Bair, E. J., *Introduction to Chemical Instrumentation*, McGraw-Hill, New York, 1962.
Crooks, J. E., *The Spectrum in Chemistry*, Academic, New York, 1978.

CHAPTER 2

Ultraviolet and Visible Spectrophotometry — Instrumentation

In this chapter characteristic design and operational features of instruments suitable for use in absorption spectrophotometry of the visible and ultraviolet regions of the spectrum will be considered. The instrument modules are shown in schematic form in Fig. 2-1. A source of radiation must be provided with each spectral region having its own requirements. All spectrophotometers include some way to discriminate between different radiation frequencies either through use of filters, prisms, or gratings. The sample absorbs a portion of the incident radiation; the remainder is transmitted on to a detector where it is changed into an electrical signal and displayed, usually after amplification, on a meter, chart recorder, or some type of readout device.

The following definitions will be employed in the text:

Photometer An instrument that furnishes the ratio, or some function of the ratio, of radiant power of two electromagnetic beams.

Spectrometer, optical An instrument with an entrance slit, a dispersing device, and one or more exit slits, with which measurements are made at selected wavelengths within

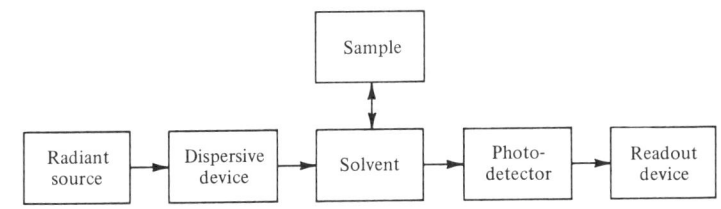

FIGURE 2-1 Instrument modules for measuring absorption of radiation.

the spectral range, or by scanning over the range. The quantity detected is a function of radiant power.

Spectrophotometer A spectrometer with associated equipment, so that it furnishes the ratio, or a function of the ratio, of the radiant power of the two beams as a function of spectral wavelength. These two beams may be separated in time, space, or both.

2.1 RADIATION SOURCES[1,2]

Radiation sources in absorption spectrophotometry have two basic functions. They must provide sufficient radiant energy over the wavelength region where absorption is to be measured. Second, they should maintain a constant light intensity over the time interval during which absorption measurements are made. If light intensity is low in the region where absorption is measured, monochromator slits must be relatively wide to obtain the necessary energy throughput. The resultant spectral bandwidth may cause errors in absorptivity measurements. Generally brightness is not a problem. In design considerations, however, it must be remembered that flux density of radiant energy varies inversely as the square of the distance from the source.

Hydrogen or Deuterium Discharge Lamps

Work in ultraviolet regions is done mainly with hydrogen or deuterium discharge lamps operated under low pressure (approximately 0.2-5 torr) and low voltage (approximately 40-V dc) conditions. Heated cathodes provide the essential function of maintaining the discharge. The discharge has negative temperature versus resistance characteristics so a current-regulated power supply is required. A vital feature of these lamps is a mechanical aperture between the cathode and the anode which constricts the discharge to a narrow path. Normally, the anode is placed close to the aperture which creates an intensely radiating ball of light about 0.6-1.5 mm in diameter on the cathode side of the opening. The use of deuterium in place of hydrogen slightly increases the size of the light ball and enhances brightness 3-5 times. Imaging the light source at the spectrometer entrance slit depends upon the aperture plate. Increased collection efficiency can be achieved by positioning the lamp arc at one of the foci of an elliptical reflector. This results in the collection of greater than 2π steradians, which translates into better than 60% collection efficiency as compared to typical housings that offer a maximum of 10% collected efficiency. Below 360 nm these discharge lamps provide a strong continuum which fulfills most needs in the ultraviolet region. With fused silica envelopes, work to about 160 nm is feasible. At wavelengths longer than about 380 nm, the discharge has emission lines superimposed on the continuum which present a nuisance.

Incandescent Filament Lamps

Measurements above 350 nm and into the near infrared to 2.5 μm are usually made with incandescent filament lamps which give continua over this range. In these lamps a wire filament, generally tungsten, is heated to incandescence by an electric current. The filament is enclosed in a hermetically sealed bulb of glass that is filled with an inert gas or a vacuum. Filaments are usually coiled to increase their emissivity, efficacy, and mean luminance. Incandescent lamps are rugged, low-cost units sufficiently bright for nearly all absorption work in the ultraviolet/visible region.

Tungsten–halogen lamps are a special class with iodine added to normal filling gases. The envelope is fabricated of quartz to tolerate a higher lamp operating temperature of 3500°K. The iodine combines chemically at the bulb wall with sublimed tungsten. The resulting WI_2 gas migrates back to the hot filament where it decomposes and tungsten is redeposited. The cycle is repeated, continuously cleaning the bulb. These lamps maintain over 90% of their initial light output throughout life.

The spectral distribution of an incandescent filament is basically that of a blackbody radiator. Therefore, measurements very far from the peak wavelength are susceptible to stray light effects. Unfortunately, the tungsten lamp emits the major portion of its energy in the near infrared, with a maximum at about 1000 nm and drops off very rapidly in the ultraviolet region to 1/100 of that value at about 300 nm. Only about 15% of the radiant energy falls within the visible region at the lamp's apparent color temperature of about 2850°K (Fig. 2-2). Often a heat-absorbing filter or cold dichroic mirror is inserted between the lamp and sample holder to remove the infrared radiation without seriously diminishing radiant energy at shorter wavelengths. The glass envelope absorbs strongly below 280 nm. Incandescent lamps are important sources in spectrometric applications because of their excellent stability, rather than because of their spectral radiance.

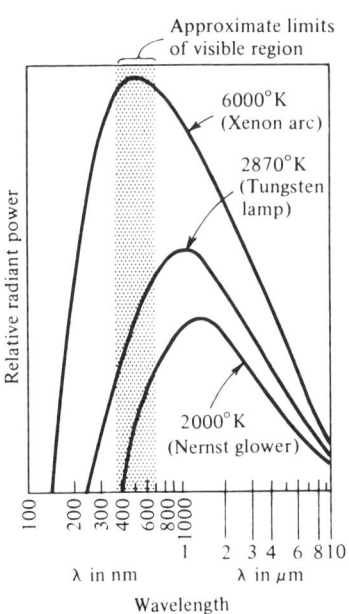

FIGURE 2-2 Spectral distribution curves of radiant energy sources.

Source Stability

High short-term stability of an incandescent or discharge source is required for single-beam spectrophotometers. The photocurrent generated within an illuminated detector is proportional to the lamp voltage raised to some power that is larger than unity (3–4 for incandescent lamps). To stabilize the photocurrent within 0.2%, which represents attainable spectrophotometric precision, the source voltage for incandescent lamps would have to be regulated within a few thousandths of a volt. Source stability is achieved by using storage batteries or constant-voltage transformers and electronic voltage regulators.

By placing a second detector in the optical path and sampling a portion of the radiant energy, the monitored signal may be used to correct the lamp output in a desired manner. This is achieved by feeding the signal back to a programmable power supply and either increasing or decreasing the output current. In this manner the optical ripple can be reduced to 0.1% peak-to-peak over the short term. This order of stability is impossible to obtain with gas discharge or arc lamps.

Modulation or Pulsing Modes of Operation

A feedback loop within the source power supply allows the power supply to be modulated or pulsed. Modulation with an external voltage source allows one to modulate or program the lamp system to follow a sine, square, or ramp function within the limits set by the various lamp operating parameters. By using optical feedback and external modulation, it is possible to obtain a very low distortion level in the optical signal. Since the lamp is now incorporated in the overall feedback path, any nonlinearities in the lamp characteristics will be accounted for. Such low level of distortion is almost impossible to achieve using mechanical choppers because of the very exact shape required for the chopper openings in relation to the beam geometry.

The pulsing mode of operation varies from modulation in that the lamp is raised to a level well above normal lamp operating conditions. The idle current is set at a low value and this current is briefly increased during the pulse. This pulsing to a level above normal lamp operating conditions results in an optical output many times that normally attainable. The greatest increase is experienced in the ultraviolet region; the least in the infrared. The minimum pulse duration is 300 μsec, the maximum can be several seconds or longer, limited by the type of lamp used. Both pulsing and modulation result in decreased lamp life.

2.2 DETECTORS[3,4]

A detector is a transducer, converting electromagnetic radiation into electrons and, subsequently, a current flow in the readout circuit. Many times the photocurrent will require amplification, particularly when measuring low levels of radiant energy. There are single-element detectors, such as photovoltaic cells, solid-state photodiodes, photoemissive tubes, and photomultiplier tubes, and multiple-element detectors such as solid-state array detectors. Important characteristics of any type of detector are spectral sensitivity, wavelength response, gain, and response time.

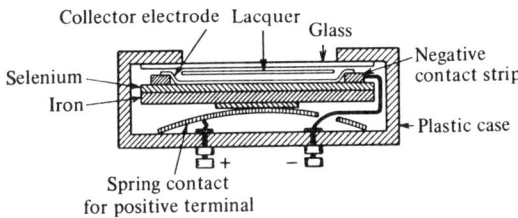

FIGURE 2-3 Construction of a photovoltaic cell.

Photovoltaic Cells

Photovoltaic, or self-generative, cells are simple and rugged in construction, require no auxiliary power supply, and can be connected directly to a microammeter or galvanometer to read their output. Their construction is shown in Fig. 2-3. A metal plate, such as iron, is used for one electrode, and a thin layer of a semiconductor, such as selenium, is deposited on this base electrode. Then a very thin, semitransparent layer of silver or gold is sputtered over the selenium to act as a second collector electrode.

Radiant energy falling upon the selenium semiconductor produces electron–hole pairs at the silver–selenium interface. The electrons pass to the silver collector electrode. A hypothetical barrier region appears to exist near the interface across which electrons pass easily from the semiconductor to the collector electrode, whereas a moderate resistance opposes the electron flow in the reverse direction. The migration and resulting separation of these holes and electrons cause a small difference in potential to develop between the base electrode and the collector electrode. If the external circuit has a resistance of 400 Ω or less, a short-circuit current will flow that is very nearly proportional to the radiant power of the incident light beam.

The spectral response of the selenium photovoltaic cell, with a glass protective cover, adequately covers the visible region, as shown in Fig. 2-4. Maximum response occurs for the green through yellow wavelengths. Because the cell impedance is low, the output photocurrent cannot be amplified unless a regenerative feedback type of amplifier is used. Consequently, the photovoltaic detector finds use mainly in inexpensive filter photometers that permit a fairly high level of illumination to strike the detector and obviate, therefore, the need to amplify the photocurrent.

Photovoltaic cells show fatigue effects. Upon illumination, the initial photocurrent may be appreciably higher than the steady-state value reached after a few minutes. Even

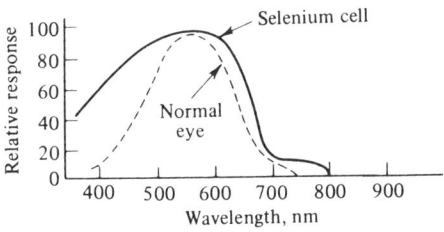

FIGURE 2-4 Spectral response of a selenium photovoltaic cell with glass cover.

the steady-state value will slowly decrease exponentially with time, particularly at high levels of illumination. The response time is slow. Thus the photovoltaic cell fails to respond immediately to change in levels of illumination such as would occur if the light beam were to be modulated or interrupted by a chopper. Also, the cell possesses a high temperature coefficient so that readings should not be taken before the cell reaches ambient temperature after the instrument is turned on. The photovoltaic cell is one of the least satisfactory photodetectors because of its limited sensitivity and relatively narrow linear dynamic range.

Photoemissive Tubes

Vacuum photoemissive tubes are simply photocathode–anode combinations contained in an evacuated envelope without the intervening multiplier chain of a photomultiplier tube. The typical single-stage vacuum phototube contains a light-sensitive cathode in the form of a half cylinder of metal, coated on its receiving surface with a light-sensitive layer, and an anode wire located along the axis of the cylinder or a rectangular wire that frames the cathode. The assembly is shown in Fig. 2-5, along with the simple phototube circuit.

When radiation strikes the photocathode, photoelectrons are ejected and are drawn to the positive anode, constituting a current. All the electrons are collected by maintaining the anode at about +90 V relative to the cathode. The photoelectric current flows through the load resistance, R_L, developing the signal voltage, $e_s = iR_L$. The load resistance in the external circuit is normally the input resistor in an amplifier circuit. The resulting current flowing in the external circuit is directly proportional to the rate of photoelectron emission, which is proportional in turn to the incident light flux. Photoemissive tubes are limited in sensitivity by spurious emission of electrons caused by thermal energy (dark current), and by the very low level of current produced by low light levels. Although photo-

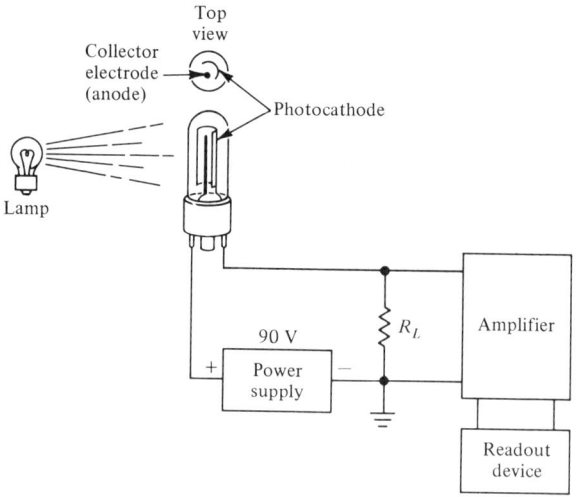

FIGURE 2-5 Photoemissive tube and its accessory circuit.

currents as small as 10 pA may be easily amplified, there is great difficulty in amplifying lower currents which may be smaller than the ohmic leakage across the tube envelope which shunts the load resistance. The time constant may become sufficiently large to degrade the effective response time of the detector.

For an accuracy of 1% or better, a calibration is required to overcome the slight non-linearity in dependence of photocurrent on illumination unless a suitable null method can be employed. Spectral response is the same as those to be discussed under photomultiplier tubes, while the gain is, of course, unity. The time response is about 150 psec. Photoemissive tubes are useful mainly for following low repetition rate, high-intensity sources.

The primary source of noise in photoemissive detectors at room temperature is generally shot noise. Shot noise is due to the fundamentally discontinuous (quantized) nature of light energy and electrical current. Photons arrive at the cathode randomly even though the overall intensity of the light beam is constant. Thus photoelectrons are emitted from the cathode and arrive at the anode also randomly, and yet the long-term rate of photo-electron pulses at the anode is constant and proportional to the light intensity. The same is true of thermionic electrons emitted from the cathode.

Photomultiplier Tubes

The electron multiplier phototube, or photomultiplier tube, as it is commonly called, is a combination of a photoemissive cathode and an internal electron multiplying chain of dynodes. The two popular designs, the circular cage and in-line configurations, are shown in Fig. 2-6. Incident radiation ejects photoelectrons (approximately 10^{-13} sec) from the cathode. The emitted photoelectrons are focused by an electrostatic field and accelerated toward a curved electrode, the first dynode, coated with a compound (BeO, GaP, or CsSb) that ejects several electrons when subjected to the impact of a high-energy electron. The overall rounded shape of the dynodes converge the electrons in one dimension, while the field-forming ridges at or near the dynode ends converge the electrons in the second dimension. Repeating this electron-multiplying process over successive dynodes maintained at higher voltages produces a current avalanche that finally impinges on the anode. Internal current amplification, or gain, is thereby achieved. The tube output may, of course, be further amplified. To prevent deterioration of dynode surfaces due to local heating effects and to prevent tube fatigue, the anode current must be kept below 1 mA. This in turn requires that voltage between the final dynode and the anode be restricted to 50 V or less. Response time is about 0.5 nsec with the GaP dynode surface; it is about 1–2 nsec with the other materials.

The resistor chain divides up the operating voltage so that a potential difference of about 75–100 V exists between adjacent dynodes. The most important interelectrode potential in a photomultiplier tube is the cathode-to-first-dynode voltage, which should always be maintained at the value recommended by the manufacturer of the tube. This can be done by using a constant-voltage (Zener) diode of the proper voltage in place of the cathode-to-first-dynode divider resistor.

Ideally, the total gain, G, of a photomultiplier tube having n stages and the secondary electron emission factor f per stage is $G = (f)^n$. The exact value of f depends upon both

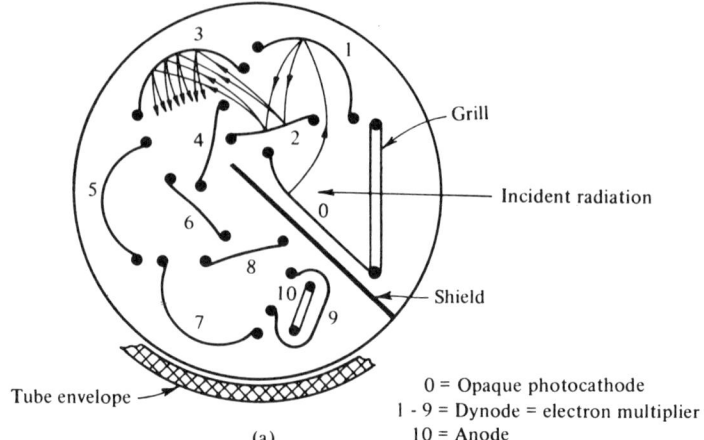

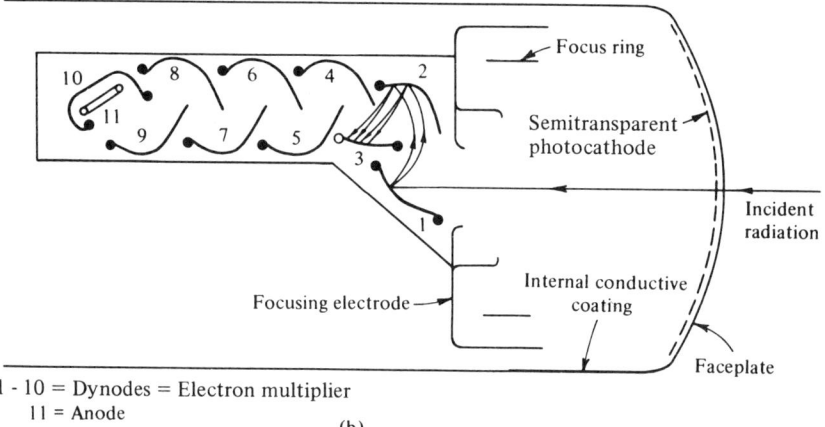

FIGURE 2-6 Photomultiplier design. (a) The circular-cage multiplier structure in a "side-on" tube and (b) the linear-multiplier structure in a "head-on" tube. (Courtesy of Radio Corporation of America.)

the nature of the dynode secondary-emitting material and the imposed electrical potential. For older dynode materials the value of f has ranged from 3 to 10. With GaP coatings it can easily reach 50 so that the total number of dynode stages can be reduced while still achieving high gain. As a result of their large internal amplification, photomultiplier tubes can be used only at low power levels of about 10^{-14}–10^{-4} lm. The ability to change the sensitivity over a wide range simply by changing the supply voltage is a unique advantage of photomultiplier tubes.

The photocathode operates on the principle that electrons are emitted from certain materials in direct proportion to the number of light quanta striking the surface containing these materials. For optimum efficiency a photocathode surface must have the highest possible absorption coefficient for the incident radiation and the lowest possible energy absorption coefficient for photoelectrons. The surface material must also have a low work function in order to extend its spectral coverage to longer wavelengths. A photoemissive detector cannot respond to light whose photons have an energy below the work function. Further demands concern the chemical and electrical properties of the cathode.

The transmission photocathode, which is superior from an electron-optical point of view, must be simultaneously thick enough to absorb most of the incident light, and thin enough so that the generated photoelectrons can traverse it while retaining enough energy to overcome the work function barrier at the vacuum interface. Similarly, in the reflection or opaque photocathode, those photons that are absorbed too deeply within the material will generate photoelectrons that can no longer get to the surface and escape.

The spectral sensitivity for several types of photocathode materials is shown in Fig. 2-7. Cathode quantum efficiency (QE), the parameter in the illustration, is the average number of photoelectrons emitted from the photocathode divided by the number of incident photons. There are essentially 11 different chemical compositions of photocathodes. These are offered in either semitransparent or opaque forms, which yield different responses. Either physical type can be combined with about 10 different window materials to generate the majority of the commercial offerings. The most sensitive cathode compositions are the bialkali types (K–Cs–Sb) which can yield tube response in the millions of amperes per watt range. Classical red response is produced by the multialkali types (Na–K–Cs–Sb) which can give usable sensitivities out to the vicinity of 850 nm, or the Ag–O–Cs type which is usable to 1.1 μm. A series of Ga–In–As compositions have excellent sensitivities out to 1.1 μm. To reach the ultraviolet end of the spectrum, excel-

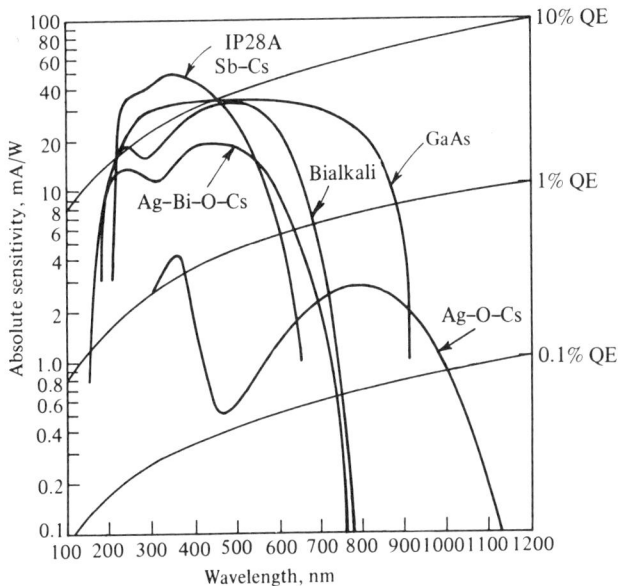

FIGURE 2-7 Spectral response curves of selected photoemissive surfaces. (Courtesy of Radio Corporation of America.)

lent responses can be obtained by coupling virtually any type of cathode with an appropriately transparent window. Cs–Te cathodes, the solar-blind-type, operate from 120 to 350 nm. Flat responses are available in the Ga–As series; such a composition generates a tube usable from 200 to 940 nm while varying in radiant response by only 10% from 440 to 880 nm. At short wavelengths the sensitivity of the detector may be impaired by absorption by the envelope material. The short wavelength limit is about 350 nm with a glass envelope or 200 nm with silica. Thus the shape of the spectral response is a function of the cathode composition and envelope material.

When a photoemissive tube is operated in darkness, a current still flows in the anode lead. This *dark current* is traceable to thermionic emission, field emission, ohmic leakage, and emission caused by natural radioactivity of ^{40}K in the glass envelope. Since the dark current also produces an amplified current at the output, it sets a low limit to the light intensity that can be directly detected. Thermal dark current can be minimized by cooling the photomultiplier tube. Because the dark current represents a steady component, it may be offset automatically by a potentiometer zeroing arrangement or manually subtracted.

All phototubes register a small residual current flow even when they are in total darkness. This dark current is related to tube construction details such as cathode and dynode materials, residual gas pressure, surface contamination, and background radiation. The purchase of selected tubes with guaranteed gain and dark current specifications is mandatory if the tubes are to be used in demanding applications.

Photodiodes

Photodiodes operate on a completely different principle from the previously discussed detectors. The construction of a planar-diffused silicon *p–n* junction diode is shown in Fig. 2-8. The process starts with a very high-resistivity intrinsic silicon material. Very shallow *p* and *n* diffusions are made in the top and bottom surfaces, respectively, and the

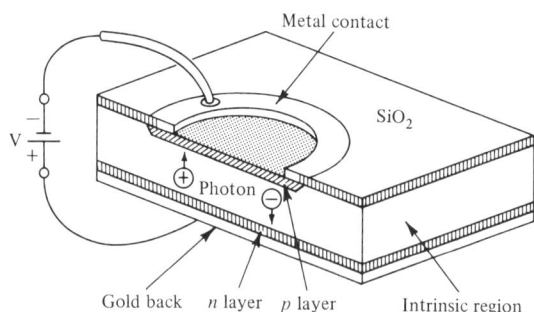

FIGURE 2-8 Construction of a planar-diffused *p–n* junction photodiode.

top surface is covered with a protective SiO_2 layer. Metal contacts formed on the top and bottom surfaces provide electrical connections. The diffused *p* regions determine the junction and optically active area. A photon must reach the active (or intrinsic) area to produce current flow in the external circuit.

A *p-n* semiconductor junction is reverse biased so that no current flows. When photons interact with the diode, electrons are promoted to the conduction band where they can act as charge carriers. Thus, the generated current is proportional to the incident light intensity. Most of the devices detect only visible and near infrared radiation. Diode responsivity is typically 250–500 mA W^{-1} across the visible spectrum. This is at least an order of magnitude better than vacuum photoemissive tubes. The output photocurrent is linear up to 10 decades versus illumination level. Typical spectral output is shown in Fig. 2-9.

In general, the speed of the photodiode is limited by the time constant formed between the amplifier input impedance and the intrinsic shunt capacitance of 2–5 pF. To keep the capacitance as low as possible, extremely small devices are used, and the optical signal is coupled to them via a lens. Rise times are about 5 nsec; however, fall times may be slower.

Linear diode arrays gather signals on many discrete detector elements simultaneously, with serial electronic scanning of each element. Solid-state arrays offer excellent stability coupled with low spreading of charge from channels in which a large signal is present into adjacent channels. Positional accuracy is maintained constant. Then wide-dynamic range, high-precision measurements are made possible.

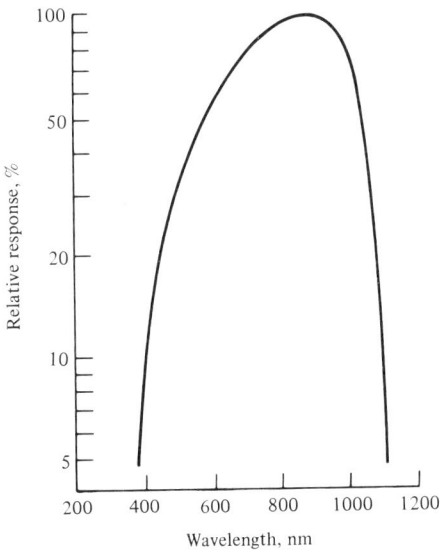

FIGURE 2-9 Spectral response of a typical *p–n* photodiode.

2.3 READOUT MODULE

In the simpler instruments, experiments produce direct-current (dc) signals which are amplified by dc amplifiers and displayed on meters, recorders, or digital voltmeters. However, high-gain dc amplifiers are subject to significant drift and offset errors. The presence of low-frequency ($1/f$) noise in the signal seriously restricts the extent to which the signal-to-noise ratio can be improved by simple low-pass filtering. For these reasons, it is often desirable to modulate the signal and thereby transform it into some alternating-current (ac) frequency high enough to avoid the drift and $1/f$ noise problems. After amplification by an ac amplifier the signal is converted back into dc by a demodulator or rectifier because all commonly used readout devices require dc signals.

Modulation is usually performed by interrupting (chopping) the light beam striking the detector by means of a rotating sectored disk. The position of the chopper in the optical path differs for different areas of application. Typical placements of the chopper are shown in Figs. 2-28 and 2-30 to 2-32.

2.4 FILTERS

Spectrophotometric methods call for the isolation of discrete bandwidths of radiation. Additionally, in an emission mode, the most favorable signal ratio between background and analytical line emission must be selected. To isolate a narrow band of wavelengths, filters or monochromators, or both, are used.

Filters provide high light throughput, approximately 50–80% efficiency. Generally a filter photometer is comparable to an $f/3$ grating system. Scattered light is relatively high, especially in an unfocused system. Assembly of filter instruments is relatively easy for perhaps as many as five wavelengths. Bandpass can equal that achieved with 0.25-m grating mounts when interference filters are involved.

Absorption Filters

Absorption filters derive their effects from bulk interactions of radiation within the material. Some types rely on selective scattering and in others true ionic absorption predominates. Transmission is a smoothly decreasing function of thickness described by the exponential law of absorption.

Absorption filters are produced in a variety of host materials: gelatin, glass, liquid, and plastic. Glass filters are extensively used in automated chemical analysis equipment and colorimetry. The scattering type depends on scattering crystals formed within the glass mass through a postreduction thermal treatment. Shorter wavelengths are scattered and absorbed while longer wavelengths are unaffected. The other absorption type attains its selectivity through actual absorbing ions in true solution. Cut-on and cutoff (or sharp-cut) filters enjoy wide use as blocking filters to suppress unwanted spectral orders from interference filters and diffraction gratings. One series consists of sharp cutoff filters that pass long wavelengths, the red and yellow series; the other comprises long-wavelength

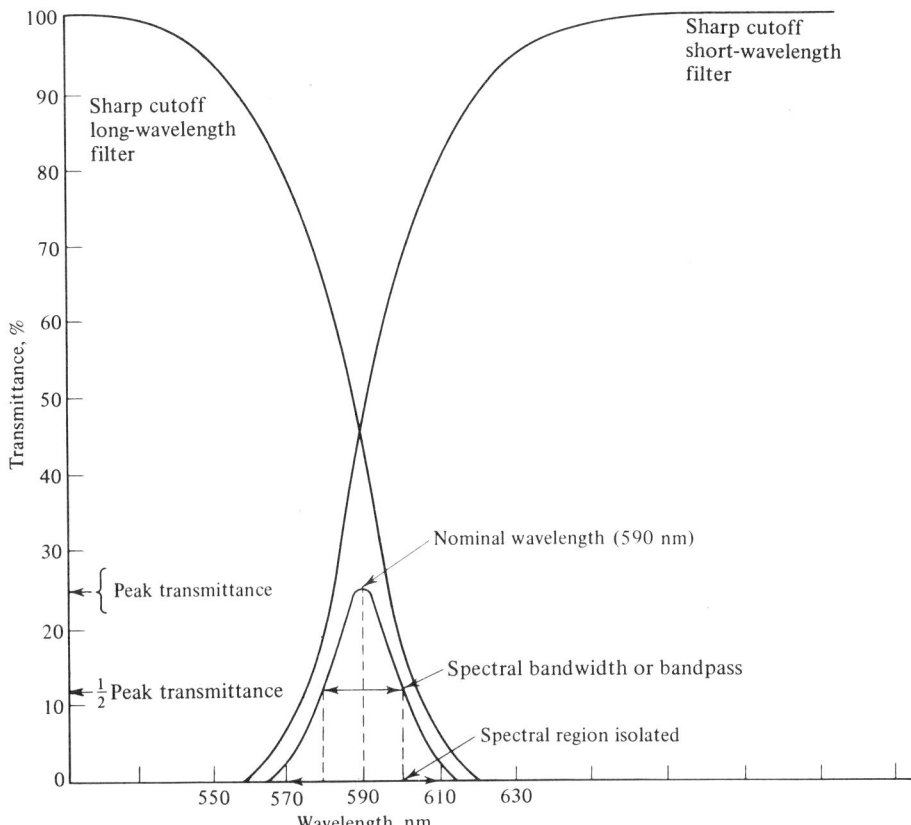

FIGURE 2-10 Spectral transmittance characteristics of a composite glass absorption filter and its components.

cutoff filters, the blue and green series. Composite glass absorption filters are constructed from unit sharp-cut filters, as shown in Fig. 2-10. The spectral separation between 50% cut-on and cutoff points of the bandpass curve are from 20 to 70 nm. Peak transmission is 5–20%, decreasing with improved spectral isolation. While such filters do not approach the extremely narrow bandpass and high peak transmission typical for the multilayer, all-dielectric interference filter, they have the advantage of relative insensitivity to input angle.

There is a wide variety of plastic filters, both sharp-cut and intermediate bandwidth types. Plastic filters may be produced either by bulk colorants introduced in the basic batch or through subsequent dye treatments of clear base stock. Cut-on types, unlike their glass counterparts, exhibit no fluorescence in the visible.

Interference Filters

A simple two interface (Fabry–Perot) filter consists of a dielectric spacer film (CaF_2, MgF_2, or SiO) sandwiched between two parallel, partially reflecting metal films, usually

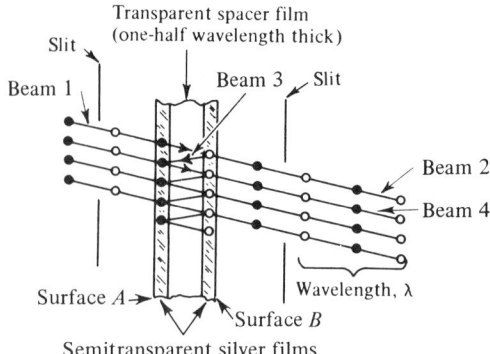

FIGURE 2-11 Schematic of an interference filter and path of light rays through the filter.

of silver (Fig. 2-11). Thickness of the dielectric film is controlled to be only one, two, or three half-waves in thickness. These are referred to as first-, second-, or third-order filters, respectively.

A portion of the incident light normal to the filter (beam 1) passes through (beam 2), while another portion (beam 3) is reflected from surface B back to surface A. A portion of this reflected light is again reflected from A through the dielectric layer and exits as beam 4 parallel (actually coincident) to beam 2. Thus the path traveled by beam 4 is longer than that of beam 2 by twice the product of dielectric spacer thickness and its refractive index. When the layer thickness b is one-half the wavelength of the light to be transmitted in the refractive index η of the dielectric, beams 2 and 4 will be in phase and will interfere constructively. The expression for central wavelengths at which full reinforcement will occur is

$$\lambda = 2\eta b/m \tag{2-1}$$

where m is the order number. Since partial reinforcement occurs for other path differences, the filter actually transmits a band of radiant energy. The bandpass is 10–15 nm, full width at half maximum (FWHM) transmission; the maximum transmission is usually 40% with this type of filter.

For example, a dielectric layer of $\eta = 1.35$ that is 185 nm in thickness will provide a first-order filter at a central wavelength of 500 nm. This filter also passes harmonic bands centered at 250 nm in the second order and 167 nm in the third order. Unwanted transmission bands can be eliminated by using an appropriate cut-on or cutoff absorption filter as one of the protecting glass covers. An excellent blocking filter is an interference filter whose first order matches the desired band when higher-order central wavelengths are used.

Second- and third-order bands are narrower, and most interference filters are arranged to transmit one of these. However, the free spectral range becomes shortened as the order is increased. Transmissions of Fabry–Perot filters are normally 10–100 times higher than for a monochromator with an equivalent bandpass.

Band center wavelength will vary with both temperature of the filter and the angle of incidence of radiation that is to be filtered. The transmission band will shift to longer wavelengths with increasing temperature and to shorter wavelengths with decreasing temperature for most filters. In the visible spectrum the shift is approximately 0.01 nm/°C. With increasing angle of incidence the band center wavelength shifts to shorter wavelengths.

A *wedge filter* consists of a wedge-shaped slab of dielectric deposited between the semi-reflecting metallic layers. A continuously variable transmission interference filter is obtained. At each point along the length of the filter a different wavelength is transmitted. Different wavelengths may then be isolated by moving either the wedge past a slit assembly or by passing a slit assembly along the filter. In Fig. 3-9 such a filter is used as the dispersing element in a monochromator. A circular wedge filter can also be constructed and employed to continuously sweep variable wavelengths past the slit assembly.

By replacing the metal films with a stack of all-dielectric films, as is done in *multilayer filters,* vastly improved performance may be obtained. Since the absorption of the dielectric layers is very nearly zero, a considerable variety of bandwidths may be produced with high transmission being maintained at the same time as low background. The reflecting stack of films is comprised of alternating high index of refraction, low index of refraction layers, each one-quarter-wave optical thickness. When a train of light waves strikes a multilayer optical coating, the beam will be divided at each film interface into a series of reflected and transmitted components. As shown in Fig. 2-12, as wavelengths are scanned away from the central wavelength in either direction, rather quickly the layers are no longer a quarter-wave thick or close to it. Hence general transmission occurs. These transmission "wings" must be eliminated over the spectral region covered by the

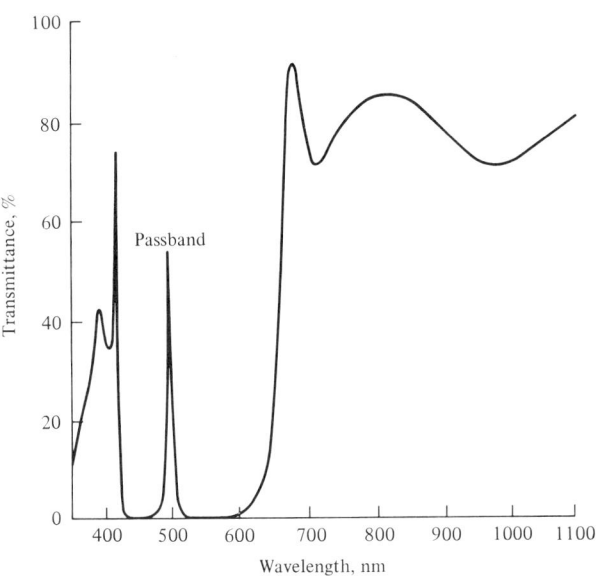

FIGURE 2-12 Transmittance of a multilayer interference filter with passband located at **500 nm.**

detector through the use of auxiliary blocking filters applied as a separate sandwich component or more directly as the multilayer substrate.

Extremely complex layer designs with 5–25 layers have been evolved, and now filters of this type can be made with bandpasses less than 0.1 nm in particular wavelength regions (1–5 nm is usual) and with background transmission less than 10^{-6}. Fully blocked filters with peak transmission of 80% are not uncommon and 55–60% can be considered standard. Multilayer interference filters are available over the wavelength region from 180 nm in the vacuum ultraviolet to 35 μm in the infrared.

A serious problem can arise if the incident light is either highly convergent or divergent. First, the wavelength being transmitted at any point on the filter is obviously a function of the angle of light incidence and one observes an integrating effect both broadening the bandwidth and lowering the peak transmission. Second, although only the effective index is usually considered, in actuality both the high and low index layers are shifting independently. Thus a mismatch in layer thickness ensues with consequent detrimental effects.

A *long-wave pass* type of interference filter (also called a *dichroic* filter) is designed to transmit wavelengths longer than some desired cutoff wavelength. Customarily the cutoff is referred to as that wavelength at which the filter first reaches 5% absolute transmission. Properties of prime interest to the user are the steepness of rise from the region of low transmission to that of higher transmission, the transmission in the passband, and the background transmission. Usually long-wave pass filters exhibit a slope of less than 5%, although steeper slopes can sometimes be achieved at the expense of smooth transmission in the passband. The passband transmission is generally not flat but exhibits an oscillation owing to the presence of secondary reflection bands. The long wavelength to which the filter will transmit is usually governed by the inherent absorption of the substrate material chosen since in most instances the film materials are nonabsorbing over a greater wavelength range. Background transmission is usually 10^{-3} or less.

2.5 MONOCHROMATORS

Much versatile optical instrumentation is designed around a monochromator, a module which consists in general of (1) an entrance slit that provides a narrow optical image of the radiation source, (2) a collimator that renders the light spreading from the entrance slit parallel, (3) a grating or prism for dispersing the incident radiation, (4) a collimator to reform images of the entrance slit, and (5) an exit slit to isolate the desired spectral band by blocking all of the dispersed radiation except that within a given resolution element. Typical optical arrangements will be considered shortly.

The primary function of a monochromator is to provide a beam of radiant energy of a given nominal wavelength and spectral bandwidth. The spectral output of any monochromator used with a continuous light source, regardless of focal length and slit width, consists of a wavelength range with an average wavelength of the value indicated on the monochromator wavelength dial. A secondary function of the monochromator is the adjustment of the energy throughput. The luminous flux emerging from the exit slit can be varied by adjusting the slit width. However, since slit width also controls spectral bandwidth, exces-

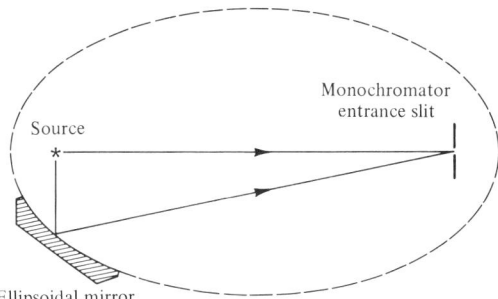

FIGURE 2-13 Off-axis ellipsoidal mirror for gathering light from radiation sources.

sively wide slit widths, with consequent large spectral bandwidths, will cause deviations from Beer's law (see Chapter 3). Excessively small slit width will result in low-energy throughput, and will affect analytical sensitivity as a result of signal-to-noise degradation from high photomultiplier dynode voltages that would be required.

Basic requirements imposed on a monochromator include design simplicity, resolution, spectral range, purity of exciting light, and dispersion. Each of these quantities will be discussed in subsequent sections in this chapter. The choice of a radiation source and a detector are closely interrelated. Large dispersion and high resolving power in monochromators will stress the importance of emission spectra with discrete lines or sharp absorption bands, whereas emission bands and usual absorption bands show up more clearly with instruments of medium dispersion.

Mirrors or lenses are used to collect light radiating from a source and direct it to the monochromator entrance slit. An effective collecting arrangement based on an off-axis ellipsoidal mirror is illustrated in Fig. 2-13. When the source is placed at focus B of the mirror, and the entrance slit at focus A, light gathered from a large solid angle is focused on the slit.

Slits

In practical spectrophotometry the monochromator module is not capable of isolating a single wavelength of light from the continuous spectrum emitted by the source. Rather, a definite band of radiation is passed by the monochromator.[5] This finite band arises from the slit distributions. The entrance or aperture of a monochromator is a long, narrow slit whose width is generally adjustable. Slits longer than 3 mm may have curved sides but we shall speak, for simplicity, of the bright rectangle of radiation formed by the entrance slit when illuminated by the source. Inside the monochromator the rays diverge from the entrance slit and illuminate the collimator mirror, which renders them parallel and focuses them upon the dispersing element. Leaving the collimator, the parallel set of rays is simply a broadened version of the entrance slit. This rectangle of light must be large enough to illuminate the entire side of the prism or the length of the grating. In turn, the dispersing device separates the incident polychromatic radiation into an array of monochromatic rectangles, each of which leaves the dispersing device at a slightly dif-

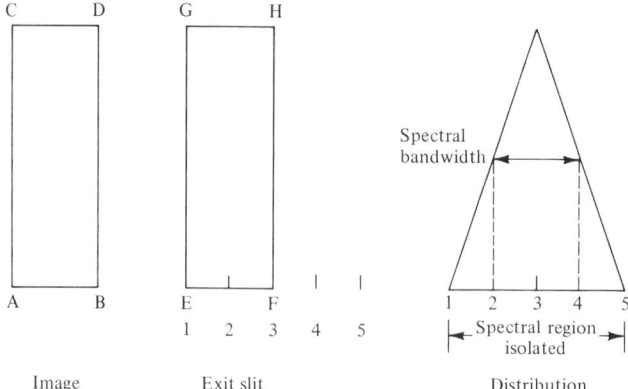

FIGURE 2-14 Slit distribution function when the image size and the exit slit are identical.

ferent angle. The monochromatic rectangles overlap badly. The dispersed beam is intercepted by a second collimator mirror identical to the first (or a different segment of the first collimator), which is used to focus and reduce each rectangle to an image of the entrance slit. These final images fall in a plane termed the *focal plane* in which a stationary exit slit is located. The exit slit fixes the dimensions through which the image must pass. The distance between the second collimator and the exit slit is termed the *focal length* of the monochromator.

Two small 45° mirrors enable the input and output to be in line. By removing one or both of these mirrors, parallel or right-angle configurations of input and output can easily be achieved.

Assuming that the radiation is of equal intensity (irradiance) throughout the image, we can use the analysis of Buc and Stearns[6] and Hogness et al.[7] for dependence of intensity at the exit slit on wavelength setting when employing symmetrical slit distributions. Referring to Fig. 2-14, let *ABCD* be the image of a monochromator band at the exit plane and let *EFGH* be the dimensions of the exit slit. The numbers 1, 2, 3, ... correspond to positions of wavelength setting. We now plot transmittance of the system as a function of the position of the image of the entrance slit as it moves along the focal plane and passes over the exit slit when the dispersing device is pivoted as in wavelength scanning. Up to the point when the leading edge *BD* of the image reaches position 1, no light is transmitted by the exit slit. When *BD* reaches position 2, half of the total light is passed. At the nominal wavelength, position 3, the exit slit aperture is filled and the transmitted intensity is 100% of that available. As the image reaches position 4 the transmittance of the system again falls to one-half, and at position 5 it falls to zero. This triangular plot of transmitted intensity versus the wavelength, which describes the position of the image, is the slit distribution observed for monochromatic light on scanning. For continuous radiation a monochromator also gives a triangular intensity pattern but the abscissa now represents the range of wavelengths passed at the setting of the central wavelength.

Generally, slits are characterized only by their width. The spectral bandwidth may be rigorously defined as the wavelength interval of the radiation leaving the exit slit of a monochromator between limits set at a radiant power level halfway between the continuous background and the peak of an absorption band of negligible intrinsic width. More simply defined, *spectral bandwidth* or *bandpass* is the difference in wavelength between the points where the transmittance is one-half the maximum or as the bandwidth containing 75% of the radiant energy leaving the monochromator. The spectral region isolated (along the baseline, or abscissa) is the sum of the image width and the exit slit width in wavelength units. In Fig. 2-14 the spectral region isolated corresponds to the distance from points 1 to 5.

The choice of slit width is basically a trade-off between intensity and resolution. For scanning molecular spectra, the slit width should be adjusted so that the spectral bandpass is about one-tenth the natural bandwidth of the spectral feature to be recorded. For atomic line spectra, the lines recorded are actually slit functions with half-width equal to one spectral bandwidth. Thus the choice of slit width depends on the separation of the spectral lines or isolation of the desired analytical line from adjacent spectral features.

Thin Film Coatings[8]

Within a monochromator the necessary collimating and focusing is performed by front-surface mirrors. By eliminating lenses, chromatic aberrations are minimized.

When light passes from one medium into a second of different index of refraction, part of the light is reflected at the boundary surface. This loss may be expressed as

$$\text{reflection} = \left(\frac{\eta_2 - \eta_1}{\eta_2 + \eta_1}\right)^2 \qquad (2\text{-}2)$$

Since the first medium is normally air ($\eta_1 = 1$), the loss becomes a direct function of the substrate index alone. The loss for high index materials, such as are customarily used in the infrared, is quite severe and can lead to extreme attenuation in multicomponent systems. Even for low index glasses the total loss can be serious if several optical components are involved in the optical path. Examination of a typical spectrophotometer reveals many glass-to-air interfaces in the mirrors and windows. A reflection loss of 4% occurs at each glass/air interface ($\eta_2 = 1.52$ for glass).

The reflection from a glass surface may be modified by the application of either dielectric or metallic thin film coatings. In many cases a single layer of optical thickness equal to one-quarter of the wavelength of the light concerned is employed. The normal reflectance at incidence of such a quarter wave film is given by

$$\text{reflection} = \left(\frac{\eta_f^2 - \eta_a \eta_g}{\eta_f^2 + \eta_a \eta_g}\right)^2 \qquad (2\text{-}3)$$

where η_f is the refractive index of the film, η_a that of the surrounding medium, and η_g that of the substrate. Application of antireflection coatings to each glass surface in the optical path reduces the reflectance to approximately 0.2% per surface.

Fiber Optics

Fiber optic bundles are composed of numerous strands of glass or plastic fused at the ends. A single fiber will transmit light; a bundle of fibers will transmit both light and images. Fiber optics transmit light by total internal reflection. Total reflection occurs whenever a ray traveling in a transparent medium strikes an interface with another medium of lower refractive index, provided that the angle of incidence is greater than a certain value known as the critical angle. This angle is defined by a ray which is refracted parallel to the interface. Thus any optical fiber must consist of a high refractive index core surrounded by a lower index sheath.

Light picked up within a limited cone or acceptance angle at one end of the fiber will emerge from the other end at the same angle. A measure of this cone or numerical aperture (NA) is given by

$$\mathrm{NA} = \eta_3 \sin \theta_w = (\eta_1^2 - \eta_2^2)^{1/2} \tag{2-4}$$

where η_1 is the refractive index of the core material, η_2 the index of the sheath, η_3 the index of the surrounding medium, and θ_w the maximum acceptance half-angle.

In glass fibers the two glasses are chosen not only to give the required numerical aperture, but also to have compatible melting points and expansion coefficients. Typical fibers have a core index of 1.64 and sheath index of 1.53, giving a numerical aperture of 0.54 which is equivalent to a full angle of 66°. Typical fiber diameters range from 10 to 150 μm. In plastic fibers a typical core is polymethylmethacrylate ($\eta = 1.49$) surrounded by a transparent polymer sheath with an index of 1.39. Fiber diameters range from 0.13 to 1.0 mm.

2.6 MONOCHROMATOR PERFORMANCE

The performance of a monochromator involves three interrelated factors: resolution, light-gathering power, and purity of light output. The resolution depends on the dispersion and on the perfection of the image formation, whereas the purity is determined mainly by the amount of stray or scattered light. Large dispersion and high resolving power in monochromators are necessary to measure accurately emission spectra with discrete lines or sharp absorption bands, whereas emission bands and usual broad absorption bands show up with instruments of medium dispersion.

Dispersion

Dispersion can be defined as the spread of wavelengths in space; that is, separation of a mixture of wavelengths into component wavelengths. This can be accomplished in a monochromator with a prism (refraction phenomenon) or with a grating (diffraction phenomenon). Linear reciprocal dispersion, D^{-1}, is defined as the range of wavelengths spread over a unit distance in the focal plane of a monochromator, or

$$D^{-1} = d\lambda/dx \tag{2-5}$$

The dimensions are nanometers per millimeter. The relation between angular dispersion, the angular range $d\theta$ over which a waveband $d\lambda$ is spread, and linear dispersion is

$$D = f(d\theta/d\lambda) \tag{2-6}$$

where f is the focal length of the monochromator. Bandpass now can be expressed in terms of the physical width of the slits, W, and reciprocal linear dispersion of the monochromator:

$$\text{bandpass} = WD^{-1} \qquad (2\text{-}7)$$

The base width $\Delta\lambda$ of the slit function is given by

$$\Delta\lambda = 2WD^{-1} \qquad (2\text{-}8)$$

For example, using slits 0.1 mm in width in a monochromator whose linear reciprocal dispersion is 1.6 nm/mm, the bandpass would be 0.16 nm. Under the same conditions, two spectral lines separated in wavelength by 0.6 nm would be 0.38 mm apart in the focal plane of the monochromator at the exit slit.

Resolution

The resolution, or resolving power, of a monochromator is the ability to distinguish as separate entities adjacent spectral features, perhaps absorption bands or emission lines. Resolution is determined by the size and dispersing characteristics of the grating or prism, the optical design of which the dispersing device is a part, the slit width of the monochromator, and, in recording spectrophotometers, resolution is also a function of the recording system at the scan speed.

The widely used definition for resolution, R, is

$$R = \lambda/d\lambda = w(d\theta/d\lambda) \qquad (2\text{-}9)$$

where λ is the wavelength under study, $d\lambda$ the wavelength difference measured between line (or peak) centers, and w is the effective aperture width. Definitions aside, just what is the criterion for calling resolved two adjacent spectral features? In practice any of several answers may be given. The most liberal statement is that peaks are resolved when the intensity falls at least 10% between them; a situation suitable perhaps for qualitative identification but hardly satisfactory for quantitation. Usually one would prefer a valley that extends to the background (baseline) between two discrete emission lines without any stipulation concerning line intensities; that is, the bases of the slit functions of the two lines may touch but not overlap. In this case, the resolution is the same as the base width of one slit function, as given by Eq. 2-7. The resolution of an actual monochromator is generally less than the theoretical value due to optical aberrations, imperfections, diffraction effects, and other deleterious effects. The effect of spectral bandwidth on observed band shapes is demonstrated in Fig. 2-15. Too little resolution depresses the peak heights, the observed bandwidth of the peaks increases, separation of the two bands is less well defined, and uncertainties arise due to increased peak height dependence on minor changes in slit width. On the other hand, with too much resolution, unnecessary noise is superimposed upon the signal with no noticeable improvement in peak height or separation of the two bands. Obviously, an optimum spectral bandwidth exists. When the spectral bandwidth is one-tenth of the true or natural bandwidth of the peak, deviation from the true peak height is less than 0.5%.

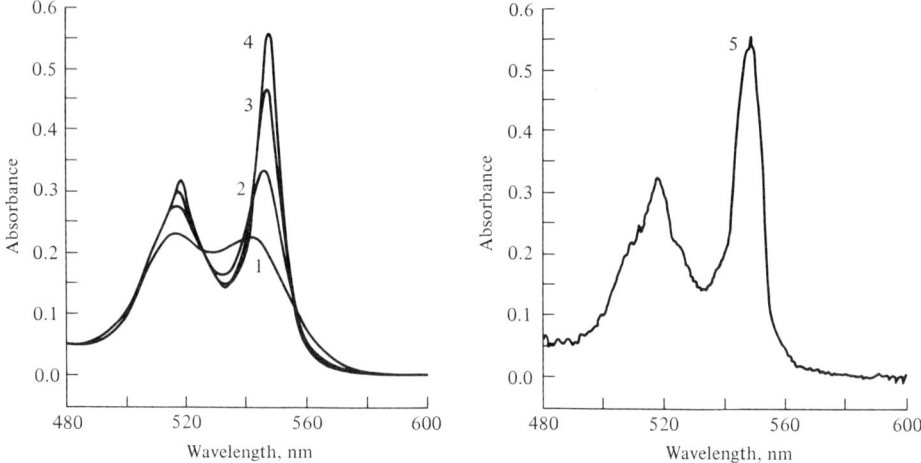

FIGURE 2-15 Effect of spectral bandwidth on observed absorption band shapes of cytochrome *c*. Spectral bandwidths are (1) 20 nm, (2) 10 nm, (3) 5 nm, (4) 1 nm, and (5) 0.08 nm.

The optical system of a monochromator produces an inherent curvature in slit image that becomes evident when slit height exceeds about 3 mm. In precision instruments the loss of resolution on lengthening slits is sometimes diminished by giving the entrance slit a radius of curvature that will oppose the one produced optically. This radius of curvature equals one-half the straight-line distance between the entrance and exit slits. At full slit height resolution is retained while light-gathering power is maximized.

Light-gathering Power

With narrow spectral slit widths spectral features quite close together can be resolved. However, the signal-to-noise ratio is important. Sufficient light must reach the detector to enable the signal to be distinguished above the background. Here enters the light-gathering power of the instrument. The *f*/number, or speed of a spectrometer, is an indication of the ability of the collimator mirror to collect light emerging from the entrance slit. It is expressed by

$$f/\text{number} = f_c/d_c \qquad (2\text{-}10)$$

where f_c and d_c are the focal length and diameter of the collimating mirror, respectively. The smaller the *f*/number, the greater the light-gathering ability.

Prisms as Dispersive Devices

The action of a prism depends on the refraction of light by the prism material. The dispersive power depends on the variation of the refractive index with wavelength. A light ray entering a prism at an angle of incidence *i* will be bent toward the normal (ver-

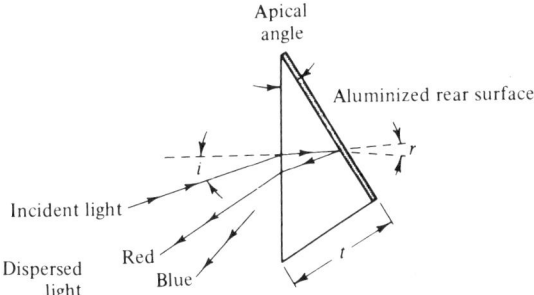

FIGURE 2-16 The prism as a dispersing medium. Littrow-type mounting: i is the angle of incidence, r is the angle of refraction, t is the base width of the prism, and the apical angle is $30°$.

tical to the prism face) and, at the prism–air interface, it is bent away from the vertical, as depicted in Fig. 2-16. To minimize astigmatism of a prism and achieve best definition, the prism should be illuminated by parallel light with the slit parallel to the prism edge and illuminated so that the light rays pass through a plane parallel to the prism base. The rays should pass through the prism symmetrically so that the incident and emergent beams form equal angles to the faces; the prism is then used at minimum deviation. The image of the entrance slit is projected onto the exit slit as a series of images ranged next to each other, caused by light of shorter wavelengths being more strongly bent than light of longer wavelengths. A nonlinear wavelength scale results. Linear dispersion for a prism is a function of wavelength. For a medium quartz prism monochromator of focal length 600 mm, typical values of reciprocal linear dispersion would be 0.6 nm/mm at 230.0 nm, 1.04 nm/mm at 270.0 nm, 1.56 nm/mm at 310.0 nm, 2.9 nm/mm at 370.0 nm, 5.4 nm/mm at 450.0 nm, and 12.0 nm/mm at 600.0 nm. Flint glass provides about threefold better dispersion than quartz or fused silica, and is the material of choice for the near infrared/visible region of the spectrum. Fused silica or quartz is required for work in the ultraviolet region.

The resolving power of a prism is given by

$$R = t(d\eta/d\lambda) = (d/f)(dx/d\lambda) \qquad (2\text{-}11)$$

where t is the base length of the prism. The resolving power is limited by the prism's base length and the dispersive power of the material, $d\eta/d\lambda$. The latter is not constant for a prism but increases from long wavelengths to shorter wavelengths. This requires a knowledge of the refractive index of the dispersing material and its rate of change as a function of wavelength, or the linear dispersion as a function of wavelength. A graph supplying this information should be provided each instrument by the vendor.

One of the most widely used prism monochromator designs is the Littrow mounting shown in Fig. 2-17. The particular advantages of the Littrow mount are a high degree of dispersion in a compact arrangement, a single collimator mirror serves to collimate the entrance beam and focus the dispersed beam, and avoidance of double refraction if an anisotropic material like quartz is used. Prisms are mostly found today in double monochromators, particularly in a prism-grating double monochromator where the prism serves as an "order sorter" as well as the dispersing device in the first monochromator.

The Littrow mount will accommodate several design variations: a 30° prism backed by a reflecting surface, a 60° prism and separate Littrow mirror, or a plane grating discussed in the next section. The source illuminates a condensing mirror which brings the reflected beam to a focus on the plane of the entrance slit of the monochromator. The image of the

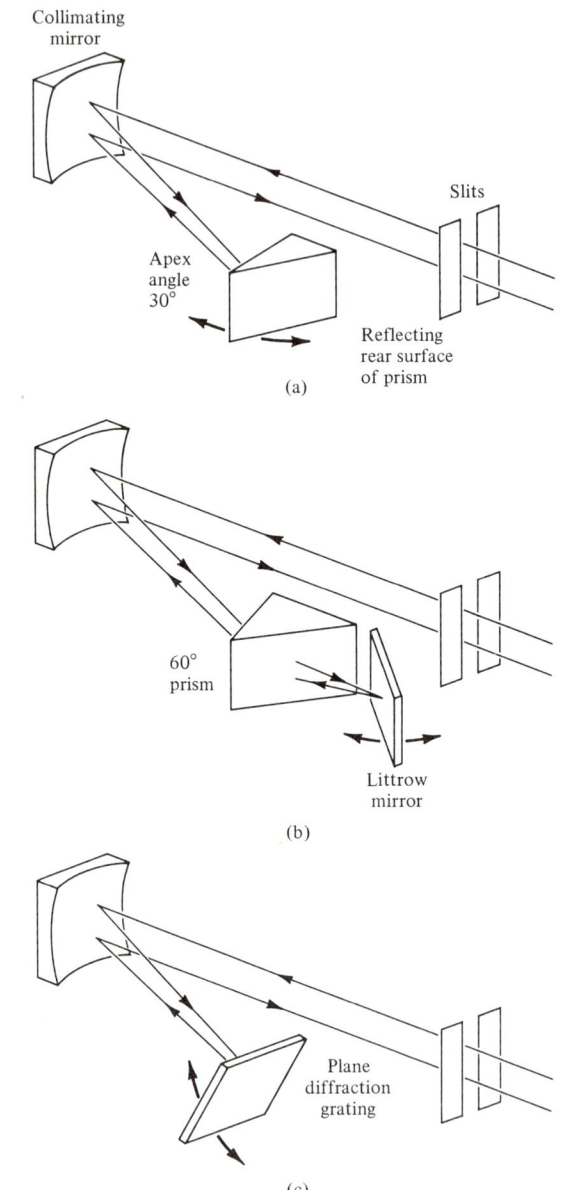

FIGURE 2-17 Littrow mounting in three configurations. For wavelength selection: (a) 30° prism rotates, (b) Littrow mirror rotates, and (c) diffraction grating rotates.

entrance slit is collimated by the parabolic mirror and directed onto the dispersing device. Then the refracted or diffracted beam is sent back to the same collimating mirror, but at a different height, and the beam is then projected and focused onto the exit slit which selects a portion of the dispersed spectrum for transmission through the sample and onto the detector. The upper and lower portions of the same slit assembly are used as entrance and exit slits, thus providing perfect correspondence of slit widths. The slit system can be fixed or continuously adjustable. In the 30° arrangement, the prism is rotated by means of a mount connected to the wavelength scroll. In the 60° arrangement, the Littrow (plane) mirror behind the prism reflects the light beam and returns it through the prism a second time, thus doubling the dispersion. The mirror is turned through a small angle to obtain the different wavelengths at the exit slit.

Dispersion by a Diffraction Grating

Virtually all gratings are of the reflection-type. A diffraction grating consists of a number of parallel, equally spaced grooves ruled by a properly shaped diamond tool directly into a highly polished surface. The quality of a grating is closely connected to the degree of precision with which the straightness, the parallelism, and the equidistance of the grooves are controlled. The profile of the grooves, shown in Fig. 2-18, is determined by the purpose for which the grating is intended. This must be maintained constant from the first to the last groove over the required ruling distances of from 10 to 25 cm. Ruling a master grating is a slow, arduous process that requires experience, skill, and unlimited patience. Only the possibility of producing several replicas from the master has allowed ruled diffraction gratings to become widely available.

The process of replicating a master grating involves (a) applying a film of parting agent to the master, (b) vacuum depositing a layer of aluminum, and (c) attaching a glass or quartz base to the aluminum layer with epoxy cement. After an appropriate time interval the replica is separated.

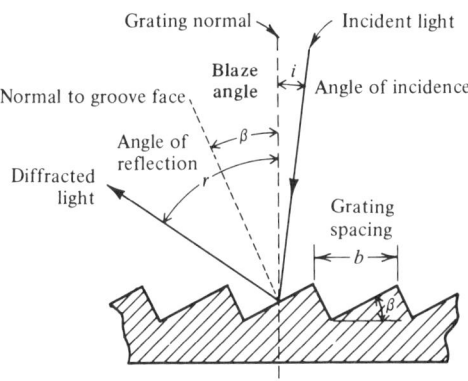

FIGURE 2-18 Cross-section diagram of a diffraction grating showing the "angles" of a single groove, which are microscopic in size on an actual grating.

Each groove of the grating has a broad face exposed to the incident light. Light incident on each groove is diffracted (that is, spread out) over a range of angles. At certain angles reinforcement, or constructive interference, occurs, as stated by the grating formula

$$b(\sin i \pm \sin r) = m\lambda \tag{2-12}$$

where b, the grating constant, is the distance between adjacent grooves, i is the angle of incidence, r is the angle of diffraction, and m is designated the order. A positive sign applies in the grating formula where incident and diffracted beams are on the same side of the grating normal. Rulings number from 20 grooves/mm in the far infrared to as many as 3600 or more grooves/mm for the visible and ultraviolet regions.

The grating formula shows that the incident energy is diffracted into several orders, shown in Fig. 2-19. The repartition of energy into these different orders depends on the groove profile. Modern ruled gratings have a rectangle–triangle groove profile, the echelette ruling, and, as a consequence, have the ability to concentrate most of the incident energy into a single order. Most gratings are used in the first order (except in the infrared) as this minimizes difficulties with overlapping wavelengths and gives high efficiency over a wide range. A portion of the incident light is simply reflected specularly by the grating (acting like a mirror) and forms the zeroth order of the direct image.

To illustrate the use of the grating equation, the primary angle at which light of 300 nm will be diffracted at normal incidence ($i = 0°$) by a grating ruled 1180 grooves/mm in the first order is given by

$$\begin{aligned}\sin r &= m\lambda/b - \sin i \\ &= [1 \times 3.00 \times 10^{-5} \text{ cm}] \, [(1/11{,}800) \text{ cm}] - 0 \\ &= 0.354\end{aligned}$$

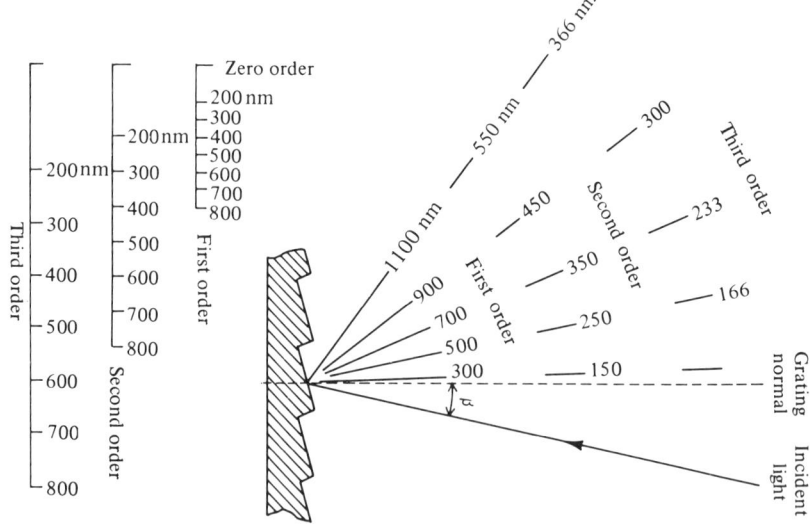

FIGURE 2-19 Overlapping orders of spectra from a reflection grating.

The angle having this sine is 20.8°. The second-order light of 150 nm will appear at the same angle. Some kind of filtering becomes necessary to prevent overlap of orders. It may take the form of wavelength cutoff by the detector, a bandpass filter, or cross-dispersion with another grating or prism.

To enhance the spectral efficiency of a diffraction grating, the groove angle is controlled so that a maximum amount of light is dispersed, or concentrated, into the wavelength (angular) region over which use is intended. As an approximation, efficiency will peak in the direction where the angle of diffraction is equal to the angle of specular reflection from the face of the grating groove. In a Littrow mount the angle of incidence will equal the angle of diffraction. This angle is known as the *blaze angle* whose value is usually specified as the first-order wavelength although, of course, the grating is also blazed in the second order for half the first-order wavelength, and so on. For unpolarized light this relation holds quite well for blaze angles up to 20°. In this case the grating equation becomes

$$m\lambda_\beta = 2b \sin \beta \qquad (2\text{-}13)$$

where λ_β is the blaze wavelength for a grating in a Littrow configuration. The actual blaze wavelength is nearly the nominal value with most other mountings.

The useful wavelength range of a grating monochromator is essentially determined by the fact that the efficiency, expressed as the ratio of reflected to incident radiation, drops on either side of the blaze wavelength. Maximum efficiency at the blaze wavelength is under 70%. Intensity drops to one-half the maximum intensity at the blaze wavelength when the diffraction angle is about one-half the blaze angle (on one side) or about three times the blaze angle on the long wavelength side. This is equivalent to a range in the first order extending from two-thirds the blaze wavelength to twice the blaze wavelength.

The bandpass is limited by groove density, focal length, and slit width. As an example, a 0.25-m mount with a 1200 grooves/mm grating and 1.0-mm slit offers about a 4-nm bandpass. Bandpass can be improved by limiting the entrance and exit slits but at a sacrifice in light throughput. In this respect, a 2400-groove/mm holographic plane grating (see below) can improve the bandpass to 2 nm without a change in throughput or in scattered light.

In grating monochromators the noise level (unwanted radiation) has two different names and origins: ghosts and stray light. Ghosts are related to periodic errors in the position of the grooves coming either from the ruling engine screw or from periodic vibrations originating outside the machine during the ruling process. Stray (scattered) light originates in nonperiodic errors in spacing of grooves and in nonperfect planeity of the grooves' reflective surface. However, the scattered light level of a grating monochromator is generally lower than that of a filter photometer. Precision ruled gratings or plane holographic gratings may be required for more stringent requirements.

Angular dispersion, $d\theta/d\lambda$, of a grating, used in the autocollimating (Littrow) mode, is given by the expressions

$$d\theta/d\lambda = m/(b \cos r) \qquad (2\text{-}14)$$
$$d\theta/d\lambda = (2/\lambda) \tan r \qquad (2\text{-}15)$$

Both are equivalent. In the plane of the exit slit, the linear dispersion is

$$\frac{dx}{d\lambda} = \frac{2f \tan r}{\lambda} = \frac{mf}{b \cos r} \tag{2-16}$$

Since $\cos r$ will be virtually constant for reflection angles up to 20°,

$$d\lambda/dx \simeq b/mf \tag{2-17}$$

Thus a grating monochromator has nearly constant dispersion throughout the spectrum and, consequently, a linear scale for wavelength. This feature constitutes one of the most important advantages of gratings over prisms.

When the order m is regarded as fixed, large dispersion can be obtained through use of gratings with a large number of grooves/mm. From Eq. 2-15 it is clear that for a given wavelength, dispersion will be a function only of $\tan r$. Changes in spacing and numbers of grooves have no effect on resolution and dispersion when a grating or echelle is used at a given angle. An echelle grating appears quite similar to a normal blazed plane grating except that the short side of the grooves is used. Echelle gratings are designed to be used at blaze angles greater than 45°. In this way, a spectrometer using an echelle grating gives high dispersion without a very long focal length, and high resolution without extremely fine groove spacings. An echelle spectrometer is shown in Fig. 6-12.

In a grating monochromator the effective aperture is simply the width of an individual groove, b, multiplied by the total number of rulings, N, and by $\cos r$; or $bN \cos r$. Assuming that the angle between the incident and diffracted rays is small, the theoretical resolution of any grating is described by two formulas:

$$R = \frac{\lambda}{d\lambda} = mN = \frac{2Nb \sin r}{\lambda} \tag{2-18}$$

$$R = \frac{2W}{\lambda} \sin r \tag{2-19}$$

where r (or θ) is the angle between the diffracted ray and the grating normal. Both equations are correct and equivalent to each other; the second is written for the autocollimating (Littrow) mode of operation. Equation 2-18 implies that for a given order m the resolution increases with the number of lines ruled on the grating. Equation 2-19 makes clear that for a grating of width W and at a given wavelength, resolution is purely a function of the sine of the diffraction angle. The latter is typically a 10° to 15° angle for ordinary gratings, but many echelles operate at 63°. Instead of using a large number of grooves to achieve high resolution, the echelle increases the blaze angle and the order to achieve very high resolution.

The ruled area of a grating should be large enough to intercept all the incident light even when the grating is turned to its extreme angular position. Any smaller area will decrease the useful light in the spectrum and increase that going into the zeroth order, or wasted altogether by missing the grating.

Holographic Gratings[9]

To fabricate regular holographic gratings, two collimated beams of monochromatic laser light are used to produce interference fringes in a photosensitive material which has been deposited on optically flat glass. The glass can be either plane or concave. The portion of the photoresist exposed to the laser light is then washed away, creating a groove structure in relief. The grating is then coated with an appropriate reflective layer and can be used in the same manner as a ruled grating. Because holographic gratings are recordings of perfect optical phenomena, they have absolutely no ghosts and a much lower stray light level. Holographic gratings can be produced with as many as 6000 grooves/mm in sizes up to 600 × 400 mm. By changing the wavelength of the laser beam, one obtains different groove spacings; the limiting factor is the availability of different wavelengths. When used with parallel beams of incident light, there are no aberrations, and aberrations can sometimes be eliminated when using nonparallel beam configurations.

A concave holographic grating, when illuminated through an entrance slit, will create, when rotated, a perfect or nearly perfect image onto a fixed exit slit. The concave diffraction grating is in itself a monochromator. No other optical components are required which eliminate all other elements and alignment of these components with respect to the grating. Compared with a plane ruled grating, the efficiency of a holographic grating generally is lower but is very uniform throughout the spectrum.

High-throughput systems incorporating holographic gratings have been built from $f/5$ to $f/1$. Accompanying bandpass of 0.4–4 nm is available in a physical configuration smaller than most 0.25-m plane grating systems. All dimensions and geometries are worked out by computer by the grating manufacturer and provided to the customer. Although the grating can be used in a mounting that rotates, the simplest mount for applications involving a finite number of wavelengths is a spectrographic mount wherein the grating would remain fixed and a photodetector would be positioned in the plane of the exit slit for each wavelength of interest. This provides an optical system with no moving parts, able to monitor one or several wavelengths simultaneously. This type of mount will be encountered in the discussion of spectrographic instruments and in instruments for monitoring liquid chromatographic effluents.

2.7 GRATING MONOCHROMATOR SYSTEMS

Popular grating systems usually involve some variation of the Ebert or Littrow monochromator configuration. Although continuously tunable, they are designed to pass only a single wavelength at a time. Therefore, they are less useful with multichannel detection systems since they generally vignette; that is, fade into the surrounding background, leaving no sharp edge. This effect reduces intensities of noncentral wavelengths because the optics are not physically large enough to prevent light loss at the edges. An f/number increase reduces vignetting. Light lost through vignetting is not completely out of the pic-

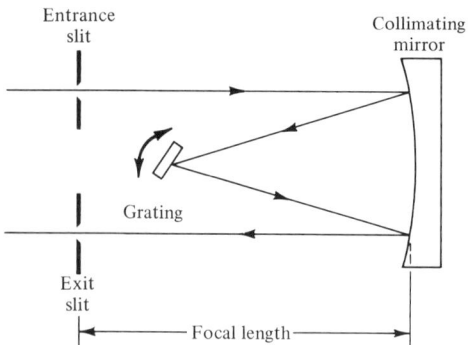

FIGURE 2-20 The Ebert mounting.

ture; instead, it reappears as stray light, limiting signal-to-noise performance of the spectrometer.

Ebert Mounting

The optical features of the Ebert mount are shown schematically in Fig. 2-20. Entrance and exit slits are on either side of the grating. A single concave spherical mirror is used as a collimating and focusing mirror. Light beams entering the monochromator strike the left side of the mirror, are collimated and reflected to the grating. The diffracted radiation goes to the right half of the same mirror and is focused on the exit slit. The slit image suffers no aberrations since the two reflections occur off axis. Since the entrance and diffracted beams use different portions of the mirror, no scattering into the optical path results from the mirror. The Ebert design offers high aperture in a relatively small package. The wavelength is selected by simple pivoting of the grating about the monochromator axis; the angle between the incident and diffracted rays remains constant. Focus is unaffected. A sine-bar drive produces a direct wavelength readout on a linear scale. A cosecant-bar drive provides a linear wave number scale.

Fastie suggested an "under-over" design in which the entrant beam passes below the grating and the diffracted emergent beam passes above. The Fastie–Ebert is one design that is commonly employed for small spectrophotometers. It offers high aperture, usually $f/3$ to $f/5$.

Czerny–Turner Mounting

In the side-by-side Czerny–Turner mount, shown in Fig. 2-21, two smaller concave mirrors replace the single large mirror of the Ebert mounting. All advantages of the Ebert mounting pertain to this one also. The Czerny–Turner design has a low aperture, usually $f/6$ to $f/10$, and may be more difficult to align. It is generally used in more expensive systems for better stray light and resolution characteristics.

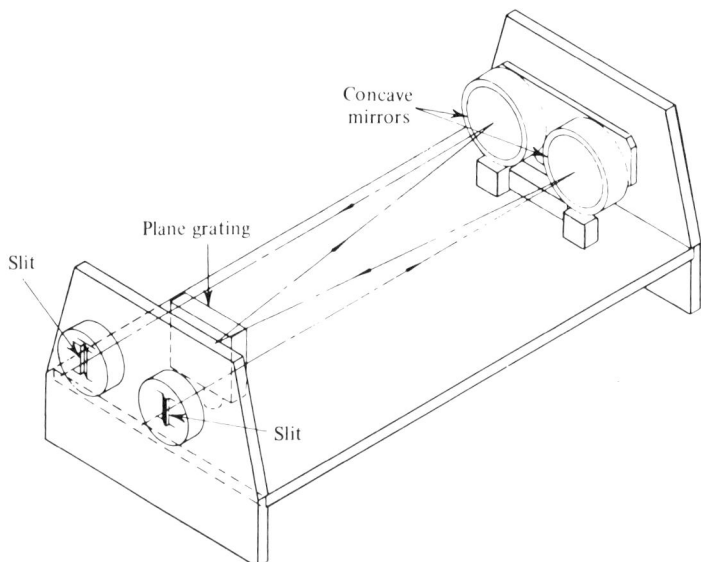

FIGURE 2-21 The Czerny–Turner mounting.

Littrow Mounting

The Littrow configuration requires a single mirror for focusing and collimating. It is commonly employed for small spectrophotometers. The Littrow mount is usually $f/3$ to $f/5$. In this autocollimating configuration (see Fig. 2-17), the grating's axis of rotation is parallel to and in the plane of the grooves. In addition the axis of rotation intersects the light at right angles. As a result, angles of incidence and diffraction are on the same side of the grating normal and are nearly equal, so that the grating equation becomes $m\lambda = 2b \times \sin i$. This mounting ensures high grating efficiency, good spectral purity, small aberrations, and compactness.

In a Littrow mount the source illuminates a single mirror on its upper portion. The beam is collimated, directed to the grating, where the beam is dispersed. The diffracted beam is sent back to the same mirror, but on its lower portion, where it is focused on the exit slit. The upper and lower portions of the same slit assembly are used as entrance and exit slits, thus providing perfect correspondence of slit widths. The slit width can be fixed or continuously adjustable.

Seya–Namioka Mounting

In the Seya–Namioka mount (Fig. 2-22) entrance and exit slits are placed on a circle whose diameter is the radius of the grating. For convenient use of the monochromator, two flat mirrors bend the entrance and exit beams. This mounting has the advantage, in particular, of being compact with a relatively high resolution. It is designed for use where a continuously tunable, pure monochromatic light is required. The use of a single active

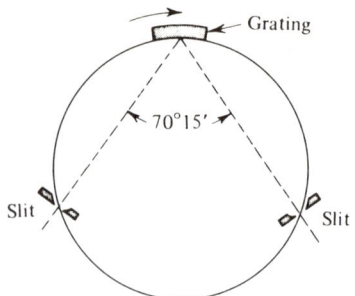

FIGURE 2-22 The Seya–Namioka mounting.

surface, such as a holographic grating, and just two reflective surfaces provides a very low (optical) noise level. A constant angle of about 70° is maintained between fixed entrance and exit slits. Scanning is accomplished by rotating the grating about its vertical axis.

Rowland Circle Mounting

The Rowland circle spectrometer (Fig. 6-9) is a better choice for use with multichannel detectors, since all first-order wavelengths are imaged at full grating efficiency with no vignetting. Problems with this spectrometer in the past were a nonflat focal plane coupled with high cost for concave diffraction gratings, with poor performance typically resulting from the difficulty of ruling a concave grating. However, holographic gratings provide a high-resolution, flat focal plane, low stray light, and nonvignetting spectrophotometer for multichannel applications.

The configurations of holographic systems are straightforward and similar for many applications. Figure 2-23 is a typical application for a small, general-purpose spectrophotometer. The optical portion of the spectrometer consists merely of the grating. It is a fairly large aperture ($f/3$) system for high throughput. Rotation is quite simple. Motion is transferred through a sine-bar linkage. Spectral coverage is 180–700 nm with a 4-nm bandpass, and less than 0.2% stray light at 210 nm.

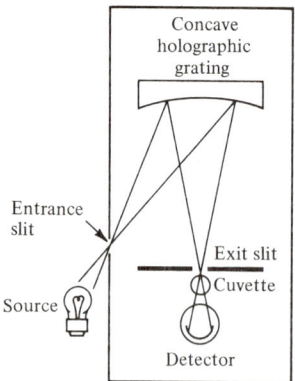

FIGURE 2-23 Small concave holographic grating configuration for a spectrophotometer.

2.8　INSTRUMENTS FOR ABSORPTION PHOTOMETRY

The state of ultraviolet/visible spectrometry recently has been described as "mature," and its 36-year history has been described.[10] Nevertheless modern advances in detector, spectrometer, and signal-processing technology have introduced new solutions to old problems. Historically, scanning monochromators have been used with single channel detectors on either single or mechanically complex double-beam systems for absorbance measurements over the 190–800-nm spectral range. These spectrometer systems often are used with analog calculation of percent transmittance and absorbance; they also may be interfaced to a microprocessor for simplified data reduction. Now a new generation of instruments, incorporating recent advances in multichannel detector technology, concave holographic diffraction gratings, and microcomputers, are appearing with heretofore unattainable capabilities.

The essentials of an analytical instrumental system were shown in Fig. 2-1. It consists simply of a source, focusing optics, unknown or standard sample cuvette, a wavelength isolation device, and a detector with amplifier and readout system. From an engineering standpoint, it is desirable that this type of system be detector limited; that is, the limiting factor should be the noise generated by the detector. Anything that can be done to increase signal levels at the detector is therefore desirable. The measure of performance is usually defined as precision, or photometric accuracy.

In terms of construction one recognizes the differences between single-beam and double-beam light paths, and whether the photometer module is direct reading or employs a balanced circuit. Special features include double monochromation and dual wavelength systems. In the final selection of an instrument, things to consider include initial cost, maintenance, flexibility of operation, resolution characteristics, wavelength range, accuracy, and perhaps auxiliary equipment available to expand into other areas of application. Some instrument designs have centered on performing sequential measurements rapidly and automatically through the use of rapid sampling accessories. These accessory units rapidly fill and evacuate the spectrophotometer cell and make it possible to analyze many samples per hour on a routine basis. Spectrophotometers are available to perform reaction-rate analysis, or to analyze solutions for several ingredients simultaneously. Many present advantages of spectrophotometers are direct results of the capabilities of the microprocessors used in automating the instruments. For example, one commercial instrument is a completely digital, interfaced spectrometer with a programmable statistical calculator that provides for unattended data acquisition, storage, and calculation, first and second derivative spectra, peak location, and peak area integration.

Instruments for absorption photometry may be classified as spectrophotometers (incorporating a monochromator) or filter photometers. Dispersive spectrometers are expensive but offer the advantage of continuous wavelength selectability. They are typically limited by light throughput, light scatter, and size. A monochromator is an inefficient device for transferring optical energy and energy loss is serious. For example, if an exit width of 0.1 nm with a height of 10 mm is used, the resultant aperture to the detector is on the order of 10^{-6} mm^2. To sense this small amount of energy a photomultiplier tube is generally required. The photomultiplier tube, inherently noisy, further degrades the signal-to-noise ratio and places severe demands on the detector if high precision is to be obtained.

Filter photometers offer an economic advantage over dispersive instruments, as well as increased luminosity, particularly when utilizing the interference-type filter. Of course, filters must be kept in stock for all wavelengths of measurement. If an analysis requires a measurement at a wavelength for which no filter is available, less than maximal sensitivity must be accepted. Because of the higher transmission efficiency, a more efficient optical system can be employed, markedly reducing effects of nearby emission or absorption features, and providing higher signal-to-noise ratios than can normally be obtained with a monochromator. The aperture to the detector can easily be opened to a centimeter in diameter, resulting in an effective detector aperture of 0.79 cm^2 (as compared to 10^{-6} mm^2 for monochromators). The detector receives a signal many magnitudes greater, allowing the use of such devices as vacuum photodiodes or semiconductor detectors rather than photomultiplier tubes. Not only do these detectors have better noise characteristics, but they are much less expensive. For routine or repetitive analyses, the stability of filters offers definite speed advantages. Pushbutton selection is possible. There is no danger of incorrect wavelength setting or of drift from the peak wavelength during operation, necessitating readjustment or recalibration.

Single-Beam Instruments[11]

The simplest type of absorption spectrometer is based on single-beam operation in which a sample is examined to determine the amount of light absorbed at a given wavelength. The results are then compared with a reference (usually the solvent alone) obtained in a separate measurement. Changes in source intensity and detector sensitivity with wavelength generally limit single-beam instruments to measurements at a single wavelength. The instruments are thus primarily employed for the quantitative determination of the concentration of a single component when a large number of similar samples are to be analyzed. In using a single-beam instrument, the absorption maximum of the analyte must be known in advance. The wavelength is set to this absorption peak, the reference material (solvent blank) is then positioned into the light path and the instrument is adjusted to read 0% transmittance when no light passes to the detector (light shutter closed) and to read 100% with the shutter open. After these adjustments have been made, the sample is placed in the light path and the absorbance (or transmittance) is read and related to the concentration either through the use of calibration curves or by the use of appropriate algebraic methods. Depending upon the instrument used, the readout may be viewed directly by observing the deflection of a meter, a "null-balance" scale setting, a recorder trace, or on a digital printout. An obvious requirement of single-beam instruments is a high degree of stability both of the light source and the detector system, as fluctuations would cause erroneous readings. However, in routine work where errors of ±1% of the measured quantity are insignificant, instrument stability is usually not a limiting factor.

Filter Photometers A relatively inexpensive, nonscanning absorption photometer can be designed around a set of filters. Such instruments are adequate for many methods,

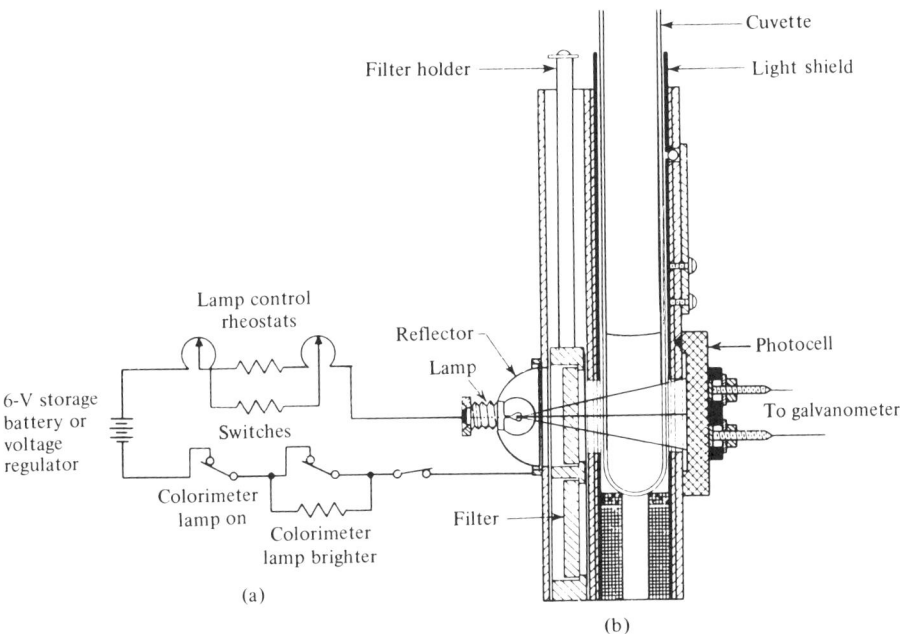

FIGURE 2-24 Schematic optical and electrical diagram of a single-beam photometer. Evelyn photoelectric colorimeter. (Courtesy of Rubicon Co.)

especially for absorbing systems with broad absorption bands. With a large energy throughput guaranteed by the filter and large aperture (often $f/1$), an amplifier is often unnecessary.

A single-beam, direct-reading, filter photometer is illustrated in Fig. 2-24. The optical path is simply from the light source, through the filter and sample holder, and to the detector. Light from the tungsten-filament lamp in the reflector is defined in area by fixed apertures in the sample holder and restricted to a desired band of wavelengths by an absorption or interference filter. After passing through the sample cuvette, the light strikes the surface of a photovoltaic cell, the output of which is measured by the deflection of a rugged light-spot galvanometer. The lamp is energized by a storage battery or by a constant-voltage transformer. Under optimum conditions an accuracy of about 2% in transmittance may be obtained.

To operate an instrument of this design, the solvent blank or reference solution is positioned in the light path and the instrument is adjusted to read 0% transmittance when no light passes to the detector (shutter closed or light off) and 100% with the shutter open and light on. The 100% adjustment can be accomplished in one of three ways: (1) by a diaphragm or guillotine somewhere in the light beam; (2) by a rheostat in the source circuit to alter the lamp brightness; or (3) by adjusting electrically the galvanometer pointer by a potentiometer "bucking" circuit. After these adjustments have been made, the sample is placed in the light path and the absorbance or transmittance read. Frequent resettings of the 100% transmittance value with the blank in the beam are usually necessary to minimize drift in the performance of the components.

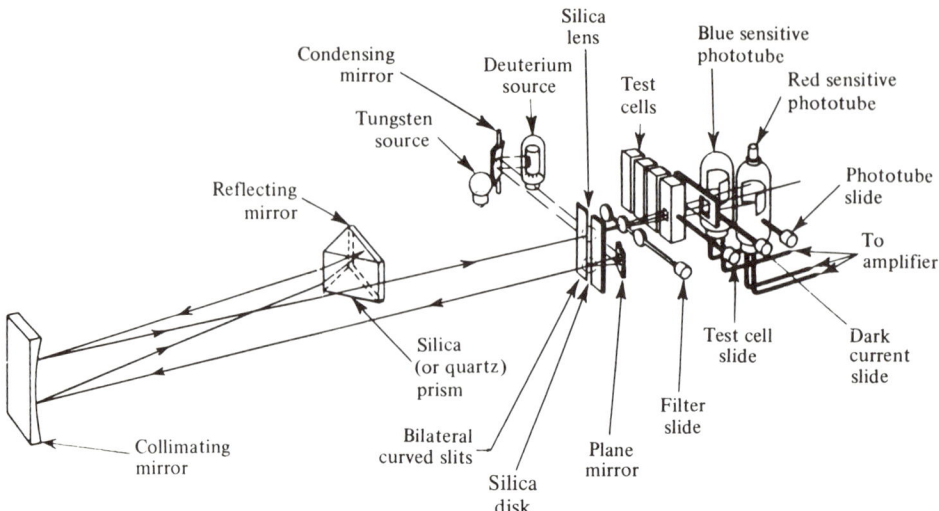

FIGURE 2-25 Schematic optical arrangement of a Littrow-mount, prism spectrophotometer. (Courtesy of Beckman Instruments, Inc.)

Spectrophotometer Although production is discontinued, the prism spectrophotometer, shown in Fig. 2-25, was a "classic" in the field for several decades. In some models a grating was substituted for the prism in the Littrow mount. Dual light sources permitted operation from 210 to about 950 nm where the red-sensitive phototube no longer provided adequate response. Phototubes were interchanged at about 625 nm by a simple slide mechanism. Auxiliary blocking filters placed in the filter slide reduced stray light. Four cuvettes positioned in the "test cell slide" could be successively brought into the light path reproducibly by means of notched index markers.

Double-Beam Instruments[12]

In double-beam instruments, a monochromatic light beam is split into two components, usually of equal intensity. One beam passes through the sample, and the other through a reference; the difference between the two components is determined simultaneously. Double-beam instruments are commonly employed for recording absorption spectra and for high sensitivity work. Since the intensity of light passing through the reference varies with source energy, monochromator transmission, reference material transmission, and detector response, all of which vary with wavelength, some means must be provided to maintain the output signal (corresponding to I_0) at a constant level. This may be accomplished by one of several alternative methods: (1) feedback loop to regulate photodetector sensitivity via dynode voltage; (2) control the monochromator slit width by means of servomotors and mechanical slit drives; and (3) position an optical wedge in the light path automatically to increase or attenuate radiation reaching the detector.

Automatic gain control is the least expensive of the three modes since it involves only electronic circuitry and no mechanical components. It provides constant slit width and thus constant resolving power during the scan when a grating monochromator is used. However, the noise level of the photodetector varies with gain and thus is not constant throughout the scan.

Automatic slit control is much more costly since complicated mechanical slit drives must be incorporated into the instrument. However, photodetector gain and therefore noise remain constant. As the slit width varies, resolution does likewise.

Optical wedge systems have intermediate utility. A less complex mechanical drive system is needed than for slit drives. Photodetector gain as well as resolution remain constant throughout the scan.

In double-beam operation, absorption spectra are automatically corrected for instrument response as a function of wavelength. Because the ratio I/I_0 is continuously compared, source instability and amplifier drift will affect both beams similarly and the effects should cancel. Solvent absorbance is automatically subtracted by placing solvent in the reference beam. Another benefit is the capability to place two samples into the instrument and have the absorbance of one subtracted from the other which reduces manual data manipulation. In difference measurements only spectral differences due to perturbations caused by temperature, solvent, pH, or ionic strength are recorded.

Double-beam filter photometers employing photovoltaic cells fall into two categories. In a bridge-potentiometer arrangement (Fig. 2-26) the null-balance galvanometer may be considered as receiving the photocurrent from each photocell through a universal shunt. Each shunt is a low-resistance, 400-Ω, linearly wound potentiometer. The beam of filtered light is divided, part passing through the solution in the sample cuvette before falling on the measuring photocell, and the other part deflected to a second photocell. The currents produced by the two photocells are brought into electrical balance by adjusting the contactor on the "slidewire scale" potentiometer, the latter usually calibrated both in 100 linear divisions and in logarithmic units.

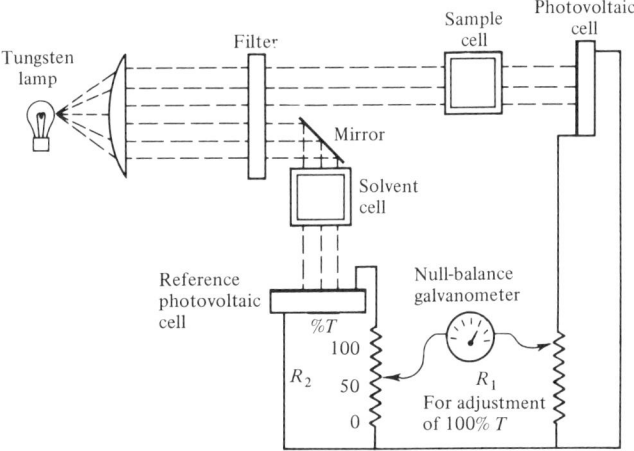

FIGURE 2-26 Bridge-potentiometer circuit for double-beam filter photometer equipped with photovoltaic cells.

To operate this type of double-beam photometer, the null-balance galvanometer is adjusted mechanically to position the pointer at midscale with the source off. With the source on, the reference solution in the light path, and the measuring slidewire set at 100, balance is restored either by adjusting the contactor on slidewire R_1 (Fig. 2-26), adjusting the intensity of the reference light beam by means of an iris diaphragm, or insertion of a neutral density wedge into the reference beam. Subsequently, these adjustments remain fixed while standards and unknowns are introduced into the sample cuvette, and the contactor on the slidewire R_2 is adjusted to obtain the scale reading. The observation precision may be as good as ±0.5%.

The optical diagram of a popular direct-reading double-beam spectrophotometer is shown in Fig. 2-27. A small replica grating provides the dispersion and, in conjunction with fixed slits, provides a bandpass of 20 nm for the wavelength range from 340 to 625 nm. The useful spectral range can be extended to 950 nm by substitution of a red-responsive phototube and blocking filter. A constant voltage for the incandescent lamp is ensured by a feedback system based on monitoring the lamp output by a second phototube. In the optical system the field lens focuses light from the entrance slit on the objective lens, thence to the plane diffraction grating. An image of the entrance slit appears at the exit slit. The 0% transmittance reading is established by balancing a differential amplifier while the occluder blocks the beam. The occluder slips into place whenever the sample cuvette is removed. To set the 100% transmittance reading the light control is adjusted with a blank in the beam. The light control is a vee-shaped slit that can be moved into or out of the beam. The reading scale is also scribed in absorbance units.

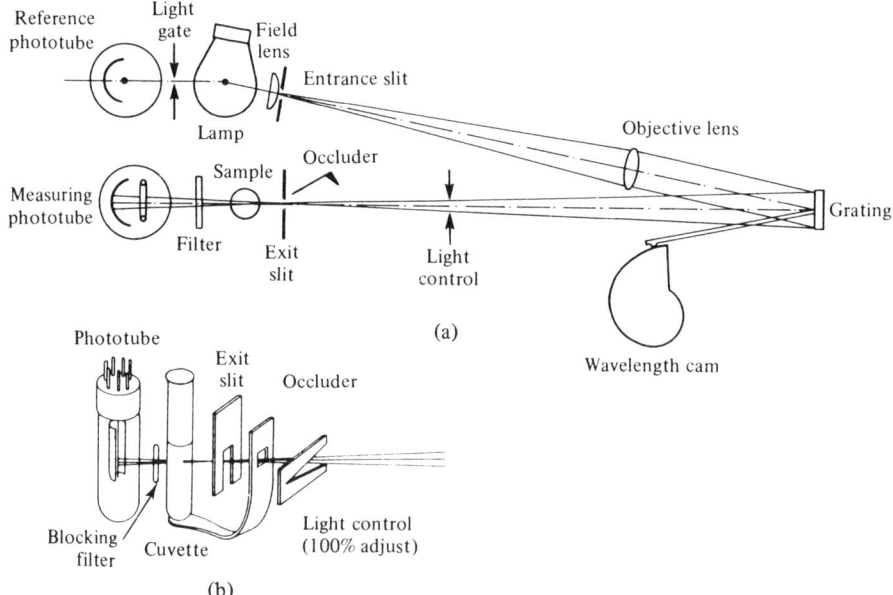

FIGURE 2-27 Bausch & Lomb Spectronic 20 colorimeter: (a) schematic optical diagram and (b) detail of photocell and cuvette. (Courtesy of Bausch & Lomb Optical Co.)

Direct-reading spectrophotometers are limited in reliability due to the lack of accuracy of the indicating meters, typically ±1–3%. By virtue of being direct reading, they feature fast operation. Simple in design, they are relatively inexpensive. Where moderate accuracy is acceptable, the direct-reading instruments are widely used.

Scanning Double-Beam Spectrophotometer

Spectrophotometers of this type feature a continuous change in wavelength. One beam permanently accommodates a reference or blank, and the other the sample. An automatic comparison of transmittance of sample and reference is made while scanning through the wavelength region. The ratio of sample to reference, after conversion to absorbance values, is plotted as a function of wavelength on a recorder. Automatic operation eliminates many time-consuming manual adjustments, especially for qualitative analysis where complex absorption curves are to be traced over a wide spectral range.

In an optical arrangement denoted double beam-in-time, shown in Fig. 2-28, light from either the visible or ultraviolet source enters the grating monochromator in the Czerny–Turner configuration. Broad-band filters contained in a filter wheel are automatically indexed into position at the required wavelengths to reduce the amount of stray light and unwanted orders from the diffraction grating. The optical beam is then directed alternately through the sample and reference cells by a system of rotating sector mirrors (choppers) and corner mirrors. The open sector passes the beam to one channel and reflects it (mirror sector) to the second channel. Each beam, a pulse of light separated in time by a dark interval (the matte black metal framing the mirror of the sector), is then directed onto a photomultiplier tube in a time-sharing procedure. After amplification, the reference signal is used to provide a signal to the dynode voltage regulator. The dynode voltage is varied to maintain a constant reference signal and keep all amplifiers within required operating limits. The zero correction (dark current) amplifier is in a feedback loop to ensure that the input amplifier's output is zero when there is no light falling on the photomultiplier during the dark interval. The sample signal is either processed through a loga-

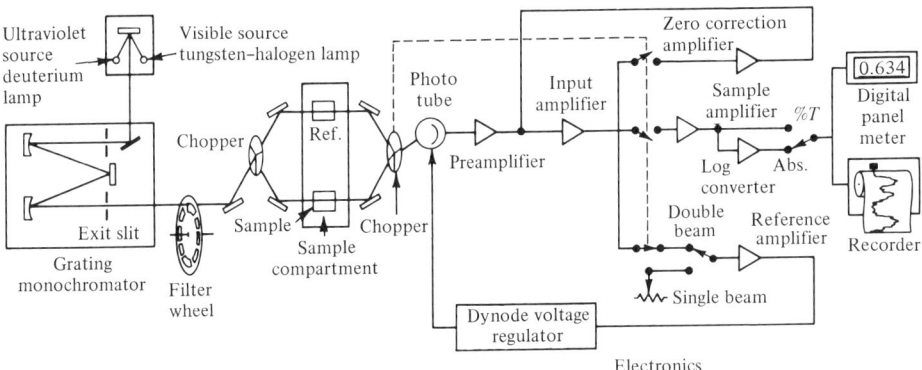

FIGURE 2-28 Schematic diagram of a double-beam spectrophotometer (Varian 634) with dual-source, single grating, Czerny–Turner monochromator. (Courtesy of Varian Associates, Inc.)

rithmic converter for absorbance measurements or is used directly for transmittance measurements. The result is fed to a recorder, meter, or digital indicator.

In an optical null procedure, the two chopped beams fall on a single detector as before. If intensities are identical, the amplifier has a dc output. Any difference in intensities will result in an ac signal of the chopping frequency. This unbalance signal is further amplified and used to drive an optical attenuator into or out of the reference beam. The fraction of open space the attenuator gives in a reference beam (compared to position with reference sample in both beams) corresponds to the percent transmittance of the sample. Linkage between the servomotor driving the attenuator and recorder pen provides the transmittance trace. The optical null procedure finds more use in infrared instruments where there exist energy-limited situations and detectors are slower in response time.

Diode-Array Rapid Scanning Spectrometer

In the diode-array scanning spectrometer a microcomputer-based signal analyzer functions as the system controller in addition to performing arithmetic functions. Using a modular approach, each individual data acquisition function is governed by a set of instructions recorded on read-only memories (ROM). Operating power and scanning signals are provided to either a 256- or 512-channel diode array. The diodes in the arrays are 50 μm wide by 0.4 mm long; coupled to the spectrometer they give a spectral coverage of 1.25 nm per diode. The overall wavelength coverage of the combination thus is 320 nm for the 256-element detector and 640 nm for the 512-element detector. Since the overall spectral range which each array can detect is 200–1100 nm, one can monitor a 320-nm window over the range with the 256-diode detector or a 640-nm window with the 512-diode detector. Since the holographic-grating Rowland circle spectrometer is corrected to focus on a flat field from 200 to 800 nm over 2.54 cm, the spectral range from 200 to 840 nm is easily handled in a single pass with the 512-diode array. Detectors are scanned at the rate of 10 μsec per diode. Therefore, the total time between scans of the spectral region from 200 to 840 nm is a minimum of 5.12 msec. An additional 40 μsec is used between scans to reset the array. As each diode is scanned in sequence by an electronically driven shift register, the field-effect transitor switches are closed, completing the circuit from the +5-V bias supply through the diode to the input of the charge-sensitive preamplifier. The diode and its associated capacitance are recharged by closing this circuit, and the recharging signal amplified, peak-detected, and presented to a 12-bit analog-to-digital converter. Dynamic range of 12 bits (4096:1) is accomplished with low noise. Output of the analog-to-digital converter is recorded for signal averaging in the data memory. Once exposure time (in spectroscopy) and number of scans have been preset, the spectrometer, data memory, and CRT display serve as a self-supporting signal-acquisition–signal-averaging display system without operator intervention.

It is possible to acquire ultraviolet/visible spectra in one segment of the data memory while processing or outputting data from another data memory segment. Processing includes wavelength calibration, spectral smoothing, peak integration, background stripping, transmittance and absorbance calculation, differentiation, integration, and other arithmetic functions.

Reversed-Optics Technique (Parallel Wavelength Acquisition)[13]

In the reversed-optics technique, all wavelengths pass through the sample. The beam then returns, by means of a unique cube corner, to a holographic grating, then to the detector, as illustrated in Fig. 2-29. This arrangement is also called parallel wavelength acquisition. During the analysis, the beam director directs the analytical beam to any one of several sample positions, as instructed by the microprocessor. The device sweeps from the reference position through the other sample positions, making calculations to lock in within a 46-msec time span. After the light has passed through the sample, the beam returns to the detector. The beam director aligns the beam with the spectrometer entrance slit, 60 μm in width, and the beam is optically locked in position by means of a servomechanism. A holographic grating disperses the light beam and a diode-array detector allows 1-sec scans over the entire spectral range from 200 to 800 nm.

Cell path lengths can extend up to 10 cm. The minimum unapertured sample volume, for a 1-cm path, is 50 μl. The instrument can perform multicomponent analysis for up to seven solutes across the full spectral range. Over a single wavelength span, simultaneous analysis for up to 12 components is possible. During analysis, the system takes for each sensor three readings per second of both standard and sample, plus one reading of the dark current. For the entire instrument, this amounts to 2800 readings per second. Thus, the instrument is able to display statistical data, such as standard deviation, independence of standards, and relative fit error. Sample rates as high as 360 per hr can be attained for operation at a single wavelength or in multicomponent analysis.

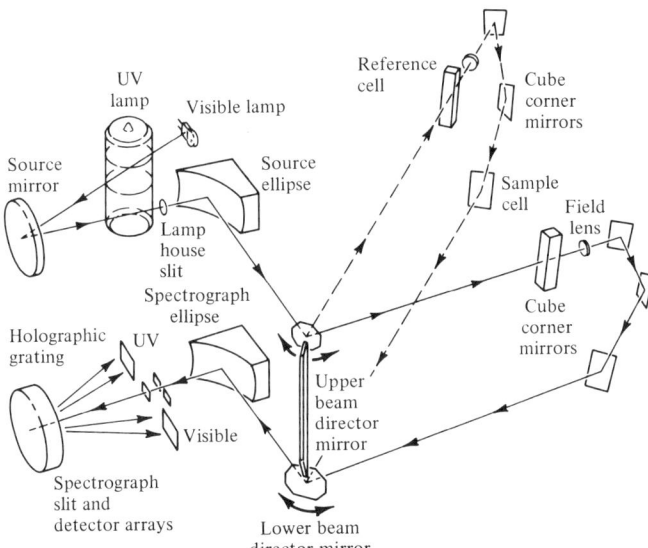

FIGURE 2-29 Schematic diagram of a rapid scanning spectrophotometer (Hewlett-Packard HP 8450A) employing the reversed-optics configuration. (Courtesy of Hewlett-Packard.)

In the configuration shown in Fig. 2-29, the sample is ahead of the monochromator. Therefore the spectrophotometer detects fluorescence where it occurs, not where the analyst is trying to measure absorbance. This distorts base lines whereas the conventional arrangement of sample and spectrometer distorts peaks but not base lines. The prevention of peak distortion gives different, and more accurate, molar absorbances. The instrument matches, but does not optically remove scattered radiation; but the system can calculate the scattering and remove it mathematically.

Double Monochromation

A double monochromator consists of two dispersing systems used in series. The intervening slit is simply the exit slit of the first monochromator and entrance slit of the second. Both wavelength drives must track together perfectly. In some double monochromators, different dispersing elements are used in each half, usually the first is a prism and the second a grating, as shown in Fig. 2-30. The quartz prism acts as an order sorter to present a single order to the second grating monochromator. Sampling geometry is double beam.

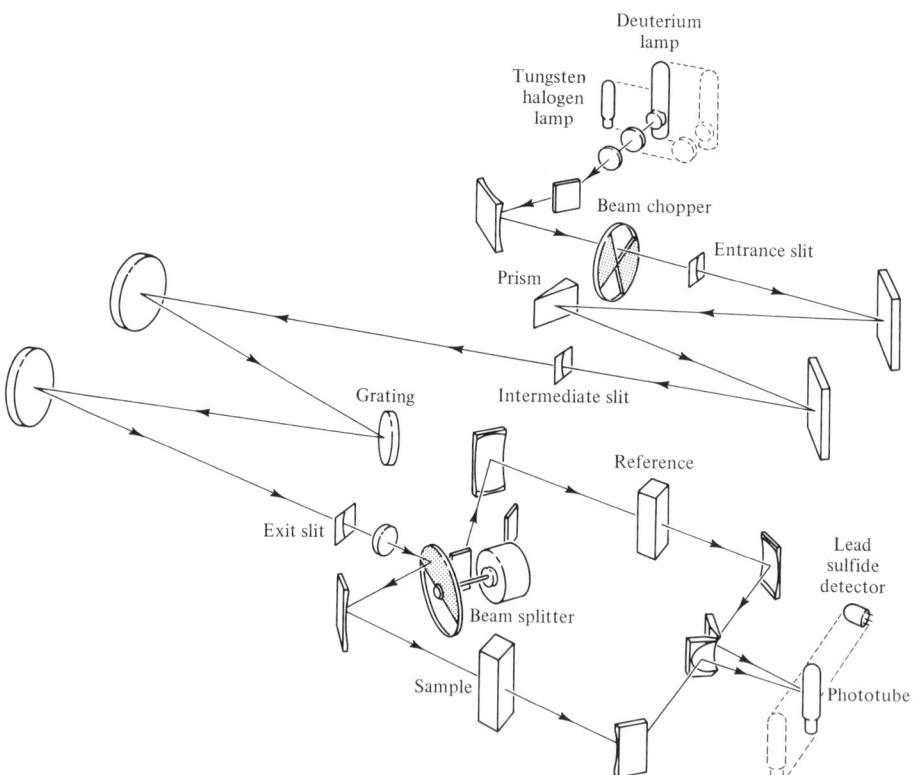

FIGURE 2-30 Schematic optical diagram of a double beam-in-time spectrophotometer with double monochromation (Cary Model 17D). (Courtesy of Varian Associates, Inc.)

With two identical monochromators, the dispersion and resolution are approximately doubled, and stray radiation is greatly reduced. For example, if the intensity of the stray radiation in each monochromator is 0.1% of that of the primary beam, the usual situation in the ultraviolet and visible, then the double monochromator would reduce this to 0.0001% (0.1% of 0.1%). In the near infrared the stray light reduction is less, about 0.1% overall. Of course, the transmission factor of the double monochromator is about half that of either monochromator by itself. If desired, the increased resolution can be traded off for a gain in energy throughput. For example, by opening the slits to twice the width needed if only one monochromator were used, a resolution comparable to that of a single monochromator can be secured and, at the same time, a quadrupling of energy at the exit slit is secured in the double unit.

It is also possible to get some of the effect of two dispersing elements by passing the light through a single dispersing element twice. Such a double-pass monochromator is shown in Fig. 2-31. This arrangement is considerably less expensive than a true double monochromator, but is also less effective in reducing scattered light.

Double Wavelength Spectrophotometer[14,15]

Double wavelength spectrophotometry refers to the photometric measurement of a material by passing radiation of two different wavelengths through the same sample before

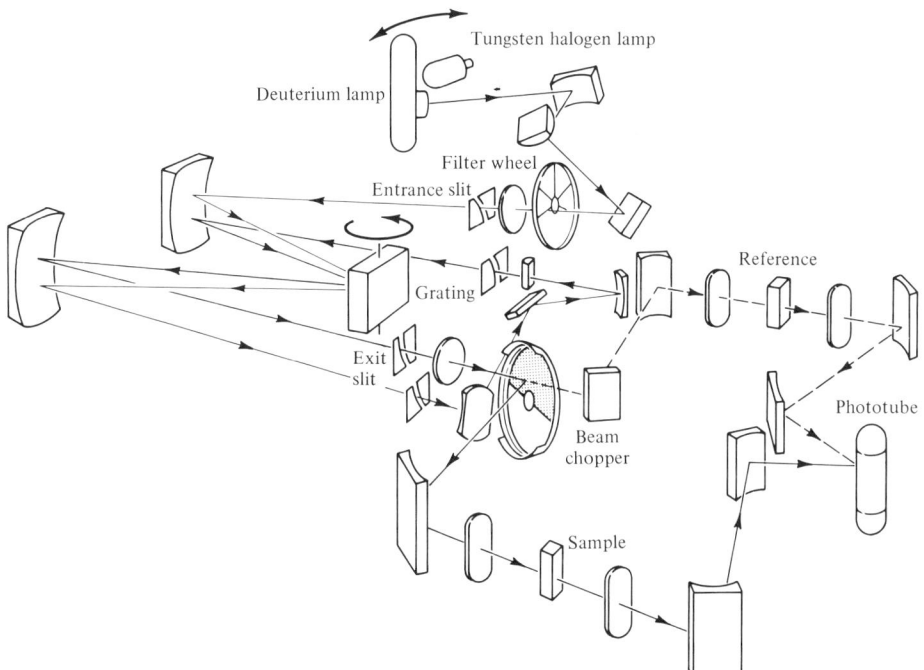

FIGURE 2-31 Schematic optical diagram of a double-pass monochromator in a double beam-in-time spectrophotometer (Varian Model 219). (Courtesy of Varian Associates, Inc.)

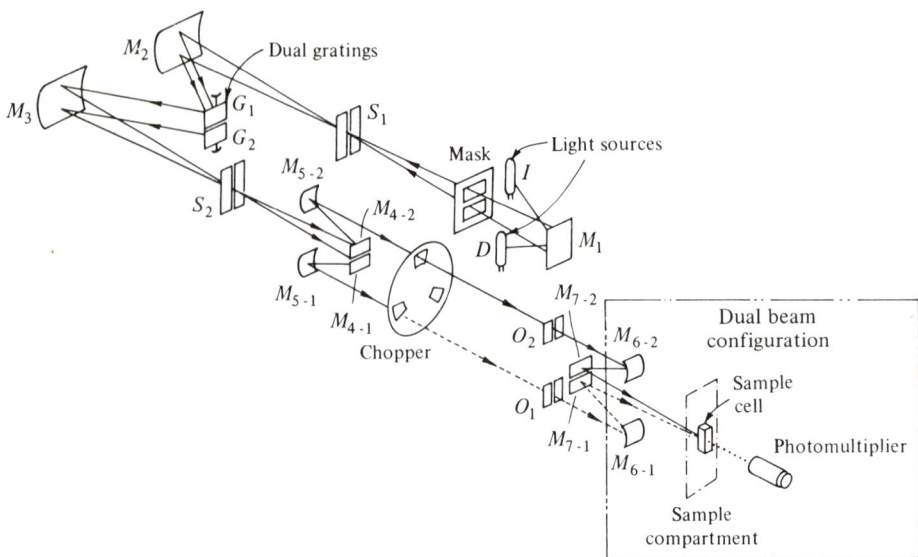

FIGURE 2-32 Optical schematic of a double wavelength spectrophotometer (Courtesy of Perkin-Elmer Corp.)

reaching the detector, as shown in Fig. 2-32. The sample is positioned close to the detector to compensate better for any turbidity or scattering. Light from the source passes into a Czerny–Turner monochromator through slit S_1 to form an image of the mask on gratings G_1 and G_2. The dispersed radiation of both monochromators is focused by the collimator mirror, M_3, split, and chopped. Bilateral optical attenuators, O_1 and O_2, situated at the pupil image of each beam, are used to vary the intensity of radiation of each beam continuously to compensate for intensity differences. Mirrors M_{7-1} and M_{7-2} converge the two beams through a single cuvette in the sample compartment. The time-separated sample, reference, and zero signals are compared.

Double wavelength spectrophotometry provides information from two wavelengths per unit time. All other factors being equal, the resultant data should be more useful than data from a double-beam absorbance measurement. This is the fundamental principle underlying the application of double wavelength spectrophotometry. The measurement compensates for the presence of one parameter, be it an interfering impurity, scattering sample, or an indistinct shoulder on the side of a band.

The sample and reference beams may be set at different wavelengths (fixed) chosen such that the absorbance of an interfering component at both wavelengths either is identical or their absorbance difference is exactly zero. This mode is used where the interfering substance highly overlaps the analyte or when the effect of turbidity must be minimized.

The output of two fixed, but different, wavelengths may be measured independently. This mode is particularly suitable in reaction kinetic studies where absorbance changes of two species can be monitored simultaneously. When compounds of protein or nucleic acid origin are studied, the relative absorbances at 254 and 280 nm are often used. Proteins absorb more at 280 nm, and nucleic acids absorb more strongly at 254 nm.

In a third mode, the reference beam is fixed and the sample beam scans. This mode is used to determine the spectral characteristics of a highly turbid sample. Light is scattered in a random (zigzag) manner; both wavelengths are scattered to the same extent and therefore subject to the same path length.

Derivative absorption measurements can be made by scanning with the two monochromators operating with a fixed, small wavelength difference between them. This mode is useful where the analyte in a two-component system occurs on the side of the absorption band of the interferent.

PROBLEMS

1. Show that the violet of the third-order spectrum overlaps the red of the second-order spectrum.

2. Assuming that the limits of the visible spectrum are approximately 380 and 700 nm, find the angular breadth of the first-order visible spectrum produced by a plane grating having 780 grooves/mm with the light incident normally on the grating.

3. For each individual plane reflection grating, supply the missing information.

GRATING	GROOVES/mm	BLAZE WAVELENGTH, nm	BLAZE ANGLE
A		500	26.4°
B	1180		8.1°
C	1180	300	
D		400	13.7°
E	1180		17.2°
F	1180	600	
G		500	8.5°
H	590		17.2°
I		1.0 μm	8.5°
J	197	2.6 μm	
K		5.0 μm	21.6°
L	74	10.0 μm	

4. With a single-stage vacuum photoemissive detector, values of R_L up to 10^{10} Ω are commonly used. (a) What problem develops if the detector leads are long and the stray capacitance approaches 100 pF or more? (b) Suggest remedies for this problem.

5. Calculate the thickness of dielectric spacer required for individual interference filters whose nominal wavelength is to peak in the first order at these values: (a) 422.7 nm, (b) 455 nm, (c) 590 nm, and (d) 767 nm. Assume that the dielectric material is magnesium fluoride ($\eta = 1.38$).

6. An interference filter is peaked at 656.0 nm in the second order. What are the third-order passband and the first-order passband?

7. If a third-order, 500-nm interference filter is desired, from what bordering transmission bands must it be isolated? What type of absorption filter would be suitable? How thick a dielectric layer is required, if calcium fluoride ($\eta = 1.35$) is the dielectric material?

8. If magnesium fluoride ($\eta = 1.38$) is applied as a film to a glass mirror ($\eta = 1.52$ for the glass), to what value is the reflectance in air reduced at the wavelength of green light? More sophisticated multilayer coatings form a beamsplitter. A film of zinc sulfide ($\eta = 2.37$) applied to glass raises the reflectance and transmittance to what values?

9. Assume that the image of the entrance slit is one-half the size of the exit slit in case A. What is the slit distribution function? In case B, the image size is twice the exit slit size. What slit distribution is now obtained? Finally, what is the slit distribution if either the image width approaches zero for a finite exit slit size, or the exit slit width approaches zero for a finite image size? For the last model, ignore the increase in diffraction which occurs as the slits are narrowed.

10. The yellow sodium line at 589.3 nm is actually a doublet of 0.59-nm peak separation. (a) What is the minimum number of lines that a grating must have to resolve this doublet in the first three orders? (b) What must be the spectral bandwidth to achieve base line resolution? (c) What slit width is necessary if the monochromator has a reciprocal linear dispersion of 1.6 nm/mm in the first order?

11. Assuming ideal bandpass characteristics, would it be feasible to employ an interference filter with a bandpass of 10.0 nm to isolate the calcium emission line at 422.7 nm from the potassium emission line at 404.4 nm?

12. A description of a commercial spectrophotometer follows: Diffraction grating with 600 grooves/mm. Focal length 330 mm. Blaze angle, first order, 500 nm. (a) What is the reciprocal linear dispersion? (b) Could the two emission lines of hydrogen, 656.28 and 656.10 nm, be resolved using a slit width of 0.010 mm?

13. For a bridge-potentiometer circuit arrangement, derive an expression for the opposing currents through the galvanometer and show that the potentiometer setting on the "calibrated slidewire" is directly proportional to the transmittance.

14. Calculate the effect of stray light on absorbance, assuming no stray light is absorbed, when the true absorbance is 0.5, 1.0, 1.5, and 2.0. Assume three levels of stray light: 0.1, 1.0, and 5.0%. Calculate the observed absorbance at each level of stray light (or the percent change in absorbance).

15. Show mathematically why double-beam methods do not eliminate errors caused by improper dark current compensation. How would an extraneous factor have to affect the I/I_0 ratio if it can be eliminated by double-beam methods?

BIBLIOGRAPHY

Bair, E. J., *Introduction to Chemical Instrumentation*, McGraw-Hill, New York, 1962.
Kolthoff, I. M. and P. J. Elving, Eds., *Treatise on Analytical Chemistry*, Part I, Vol. 5, Wiley-Interscience, New York, 1964.
Olsen, E. D., *Modern Optical Methods of Analysis*, McGraw-Hill, New York, 1975.
Strobel, H. A., *Chemical Instrumentation*, 2nd ed., Addison-Wesley, Reading, Mass., 1973.

LITERATURE CITED

1. Hell, A., *Anal. Chem.*, **43**, 79A (1971).
2. Lewin, S. Z., "Luminous Gas Light Sources," *J. Chem. Educ.*, **42**, A165 (1965).
3. Lytle, F. E., *Anal. Chem.*, **46**, 545A (1974).
4. Talmi, Y., "TV-Type Multichannel Detectors," *Anal. Chem.*, **47**, 658A, 697A (1975).
5. Alman, D. H. and F. W. Billmeyer, Jr., "A Review of Wavelength Calibration Methods for Visible-Range Photoelectric Spectrophotometers," *J. Chem. Educ.*, **52**, A281, A315 (1975).
6. Buc, G. L. and E. I. Stearns, *J. Opt. Soc. Am.*, **35**, 458 (1945).
7. Hogness, T. R., F. P. Zscheile, Jr., and E. A. Sidwell, Jr., *J. Phys. Chem.*, **41**, 379 (1937).
8. Baumeister, P. and G. Pincus, "Optical Interference Coatings," *Sci. Am.*, pp. 59–75 (December 1970).
9. Flamand, J., A. Grillo, and G. Hayat, *Am. Lab.*, p. 47 (May 1975).
10. Beckman, A. O., W. S. Gallaway, W. Kaye, and W. F. Ulrich, "History of Spectrophotometry at Beckman Instruments, Inc.," *Anal. Chem.*, **49**, 280A (1977).
11. Lott, P. F., *J. Chem. Educ.*, **45**, A185, A273 (1968).
12. Lott, P. F., *J. Chem. Educ.*, **45**, A89, A169 (1968).
13. Thomas, H. L., *Industrial Research/Development* (July 1979), p. 86.
14. Porro, T. J., *Anal. Chem.*, **44**, 93A (1972).
15. Sellers, R. L., G. W. Lowry, and R. W. Kane, *Am. Lab.*, p. 61 (March 1973).

CHAPTER 3

Ultraviolet and Visible Absorption Methods

When an electromagnetic wave of a specific wavelength impinges upon a substance, the fraction of the radiation absorbed, ignoring losses due to reflections and scattering, will be a function of the concentration of the substance in the light path and the thickness of the sample. The complication of reflection and window absorption can be avoided by defining I_0 as the radiant power after passing through a blank contained in the same sample cuvette. The transmittance T is defined as the ratio of the intensity (or radiant power) of unabsorbed radiation (relative to the blank), I, to the intensity of the incident radiation; thus $T = I/I_0$. Absorbance, A, is the base-ten logarithm of the reciprocal of the transmittance:

$$A = \log \frac{1}{T} = -\log \frac{I}{I_0} \tag{3-1}$$

Percent transmittance is $100T$; percent absorption is $100(1 - T)$.

Analytical applications of the absorptive behavior of substances can be either qualitative or quantitative. The qualitative applications of absorption spectrometry depend on the fact that a given molecular species absorbs light only in specific regions of the spectrum, and in varying degrees characteristic of that particular species. Such a display is called an absorption spectrum of that molecular species, and serves as a fingerprint for identification purposes. The quantitative aspect will be considered in the subsequent sections of this chapter.

3.1 FUNDAMENTAL LAWS OF PHOTOMETRY

As a beam of photons passes through a system of absorbing species, the rate of photon absorption with distance traversed is directly proportional to the power of the photon beam, I (sometimes symbolized by P, since the intensity has units of energy per unit time, or power). The reduction in intensity, $-dI$, can be stated mathematically as

$$\frac{-dI}{dx} = kI \tag{3-2}$$

where k is a proportionality constant characteristic of the nature of the absorbing species and of the energy of the photons, and I represents the radiant power at any distance x in the absorbing medium. Rearranging and separating variables in Eq. 3-2 gives

$$-\frac{dI}{I} = -d(\ln I) = k\,dx \tag{3-3}$$

which is a mathematical statement of the fact that the fraction of radiant power absorbed is proportional to the thickness traversed. Now if it is stipulated that I_0 is the radiant power at $b = 0$, and that I represents the radiant power of the transmitted radiation emerging from the absorbing medium at $x = b$, Eq. 3-3 can be integrated along the entire radiation path:

$$-\int_{I_0}^{I} d\ln I = k \int_{0}^{b} dx \tag{3-4}$$

obtaining

$$\ln I_0 - \ln I = \ln \frac{I_0}{I} = kb \tag{3-5}$$

In simple terms, Lambert's law (Eq. 3-5) states that, for a given concentration of absorber, the intensity of transmitted light, previously rendered plane parallel and entered the absorbing medium at right angles to the plane, decreases logarithmically as the path length increases arithmetically.

Of much greater interest is the dependence of intensity on the concentration of absorbing species in solution. Beer found that increasing the concentration of absorber had the same effect as a proportional increase in the radiation-absorbing path length. Thus, the proportionality constant k in Eq. 3-5 is in turn proportional to the concentration of absorbing solute, or

$$k = aC \tag{3-6}$$

Use of base-ten logarithms instead of natural logarithms requires only that the value of k (or a) be changed. Thus the combined law becomes

$$\log \frac{I_0}{I} = abC \tag{3-7}$$

where a incorporates the conversion factor to base-ten, namely, 2.303. This is the most familiar expression of the combined Lambert–Beer law, which is usually simply called Beer's law. If the sample path length is expressed in centimeters and the concentration in grams of absorber per liter of solution, the constant a, designated as specific absorptivity or specific absorption coefficient, has the units of liter g^{-1} cm^{-1}.

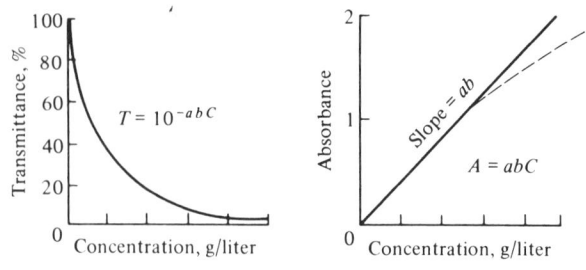

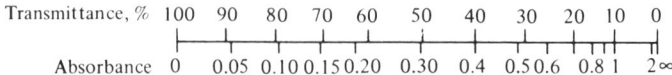

FIGURE 3-1 Representation of Beer's law and comparison between scales in absorbance and transmittance.

Frequently, it is desirable to specify C in terms of molar concentration, with b remaining in units of centimeters. Then Eq. 3-7 is rewritten as

$$\log \frac{I_0}{I} = \epsilon b C \tag{3-8}$$

where ϵ, in units of liter mol^{-1} cm^{-1}, is called the molar absorptivity or molar absorption coefficient. In the older literature, molar absorptivity may have been called molar extinction coefficient or molar absorbancy index.

A plot of absorbance versus concentration will be a straight line passing through the origin, as shown in Fig. 3-1. Readout scales and meter scales on spectrophotometers are usually calibrated to read absorbance as well as transmittance.

Absorption of radiation by molecules at specific wavelengths is frequently used for quantitative analysis owing to the direct relationship between absorbance and concentration. Sensitivity of spectrometric analysis is dictated by the magnitude of the absorptivity and the minimum absorbance which can be measured with the required degree of certainty. For example, if the molar absorptivity for iron(II)-1,10-phenanthroline complex is 12,000 liter mol^{-1} cm^{-1} and the minimum detectable absorbance is 0.01, then, for a 1.00-cm path length, the minimum molar concentration which can be detected is

$$C = \frac{A}{\epsilon b} = \frac{0.01}{(12{,}000 \text{ liter mol}^{-1} \text{ cm}^{-1})(1.00 \text{ cm})} = 1.20 \times 10^{-6} M$$

Deviations from Beer's Law

Deviations from Beer's law fall into three categories: real, instrumental, and chemical. Real deviations arise from changes in the refractive index of the analytical system. Kortum and Seiler[1] pointed out that Beer's law should be expected to apply only at low concentrations. It is not absorptivity which is constant and independent of concentration, but the expression

$$a = a_{\text{true}} \frac{\eta}{(\eta^2 + 2)^2} \qquad (3\text{-}9)$$

where η is the refractive index of the solution. At concentrations $10^{-3}M$ or less, the refractive index is essentially constant, but at high concentrations the refractive index may vary considerably and so will absorptivity. This does not rule out quantitative analyses at high concentrations, since bracketing standard solutions and a calibration curve can provide sufficient accuracy. The refractive index effect may be encountered in high-absorbance differential spectrophotometry.

The derivation of Beer's law assumed monochromatic light, but truly monochromatic light is approached only in specialized line emission sources. All monochromators, regardless of quality and size, have a finite resolving power and therefore minimum instrumental bandwidth. However, if absorptivity is essentially constant over the instrumental bandwidth, then Beer's law will be followed within close limits. The relative rate of change of absorptivity at the wavelength interval at which a measurement is taken determines how close to monochromaticity the light must be; that is, the instrumental bandwidth required. Thus, if the absorptivity is not constant over the range of wavelengths used, Beer's law will fail. Fundamentally, this departure is due to the fact that with all photometers it is the radiant power of the component wavelengths which are additive (or nearly so), whereas Beer's law requires the logarithms be additive. The range of wavelengths passed through the sample at any particular wavelength setting is generally sufficiently large so that the absorptivity will be different for the different frequencies contained in the light beam. Departure from Beer's law is most serious for wide bandpasses and narrow absorption bands, and is less significant for the broad bands and narrow slits. Often the deviation from Beer's law becomes evident at higher concentration. On a plot of absorbance versus concentration, the curve bends toward the concentration axis. This lack of adherence to Beer's law in the negative direction, dotted line in Fig. 3-1, is always undesirable because of the rather large increase in the relative concentration error.

Chemical deviations from Beer's law are caused by shifts in the position of a chemical or physical equilibrium involving the absorbing species. Consider, for example, the following equilibria:

$$2CrO_4^{2-} + 2H^+ \rightleftarrows 2HCrO_4^- \rightleftarrows Cr_2O_7^{2-} + H_2O \qquad (3\text{-}10)$$

The dichromate(VI) ion absorbs in the visible region at 450 nm. Upon diluting a dichromate solution the equilibrium shifts to the left. The equilibrium can be controlled by converting all the chromium(VI) species to $Cr_2O_7^{2-}$ by making the solution $0.1M$ in sulfuric acid, or converting all the chromium(VI) species to CrO_4^{2-} by making the solution $0.05M$ in potassium hydroxide. Beer's law will then be followed. If an absorbing species is involved in an acid–base equilibrium, Beer's law will fail unless the pH and ionic strength are kept constant. Here the pH should be adjusted at least three more, or three units less than, the pK_a value of the monoprotic acid. Alternatively, the wavelength corresponding

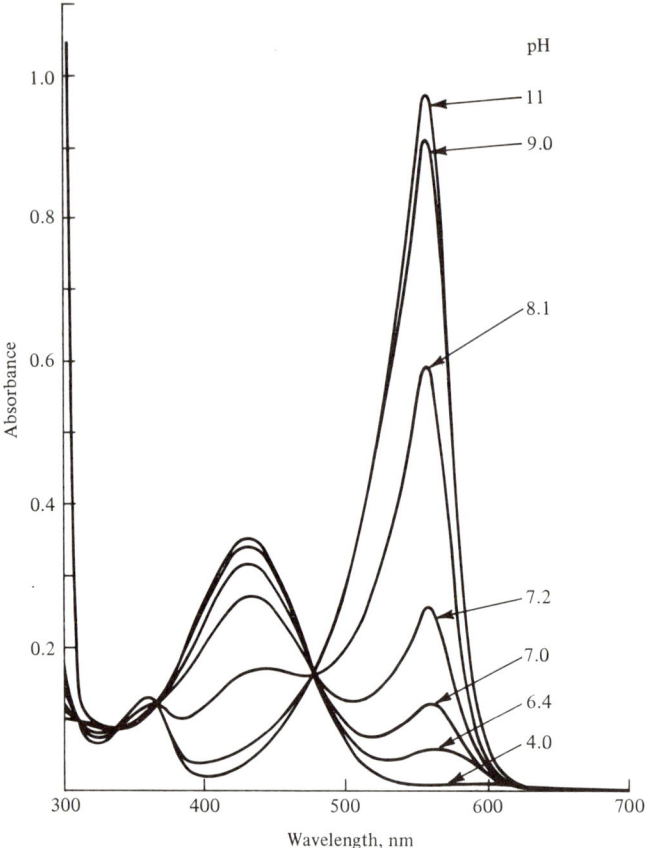

FIGURE 3-2 Chemical equilibrium between two solution components, the conversion of phenol red (pK_a = 7.9) from the yellow (acidic) to the red (basic) form. Absorption maxima are at 433 and 558 nm, respectively, for the acidic and basic forms. Isosbestic points are recorded at 338, 367, and 480 nm.

to an isosbestic point can be used (Fig. 3-2). An isosbestic point is a wavelength where the molar absorptivity is the same for two materials that are interconvertible without regard to the equilibrium position of the reaction between them. When the absorbing species is a complex ion, the concentration of free ligand must be constant and large in comparison to the amount of the analyte being sought, usually 100-fold excess.

3.2 SPECTROPHOTOMETRIC ACCURACY

Accuracy denotes the nearness of a measurement to its accepted value. Precision, discussed in the following section, describes the reproducibility of results. A third term, photometric linearity, is defined as the ability of a photometric system to yield a linear relationship between radiant power incident upon its detector and some measurable quantity provided by the system.

Accuracy and linearity often are confused since an instrument with high photometric accuracy must also have good photometric linearity; however, accuracy and linearity are not synonymous. Linearity, in its simplest concept, is based upon the additivity of empirical (not absolute) absorbance values. This will not necessarily be equivalent to accuracy because of the possibility of exponential nonlinearity. Absolute accuracy is based upon the relationship between the measured transmittance and some absolute change in light intensity at the detector. Accuracy generally is determined by comparing the radiant energy flux through apertures mounted in the sample beam, which are opened and closed at various points in the calibration process.

Instrumental Parameters Affecting Photometric Accuracy[2]

The spectral bandwidth selected and the observed bandwidth of the recorded absorption band are directly related. If the slits are too wide, the absorbance peak height will be depressed and the observed bandwidth will be greater than the natural bandwidth, thus giving erroneous absorbance values. When the spectral bandwidth is 10% or less of the natural bandwidth, the ratio of the observed peak height to the true peak height will be at least 99.5%. Thus, if possible, the bandwidth should be decreased until there is no further change in recorded spectra, as was shown in Fig. 2-15. However, in order to maintain the best signal-to-noise ratio, the slits should not be narrowed more than necessary. If one is working with wide band instruments, it is possible that a spectral bandwidth cannot be selected which fully resolves the absorption bands.

Too rapid a scan speed is another source of error when using scanning spectrophotometers. When the scan rate is too fast for a given spectral bandwidth and pen period, the absorbance maxima are shifted to lower wavelengths and the band is broadened and depressed. In order to obtain a "true" spectrum, the scan speed must be operated at a rate no faster than one-tenth of the natural bandwidth per pen period.

A third source of error is wavelength inaccuracy. The degree of error will vary with the magnitude of wavelength inaccuracy, and with the natural bandwidth. Wavelength calibration standards are available.

Stray light can be a major problem.[3,4] In general, stray light becomes a serious problem at the extreme ends of an instrument's spectral range where the photodetector sensitivity is low or the monitoring light beam is weak. Principal sources of stray light include light scattered by dust or smudges on optical surfaces; optical surface imperfections of mirrors, lenses, and cuvette walls; light scattered by diffraction at the slit–jaw edges, a natural phenomenon which occurs no matter how sharp the slit edges are made; off axis reflections from interior surfaces of the monochromator which can be attenuated by multiple velvet-black baffle surfaces; and light scattered to the extent to which the grating acts as a simple mirror instead of a perfect diffracting surface. As a result, a small fraction of undispersed radiant energy may pass through the exit slit of the monochromator and reach the detector. The absolute intensity of this stray light will be constant at any given time, but can increase as the instrument ages and the optical materials in the monochromator deteriorate.

Stray light affects photometric accuracy by causing deviations from Beer's law. Positive deviations from Beer's law occur if the stray light is absorbed, and negative deviations occur if it is not. The latter is usually the case, and observed absorbances are reduced markedly as the stray light increases and the absorbance of the solution is increased. The apparent absorbance, assuming that none of the stray light, I_s, is absorbed by the sample or reference material, is given by

$$A_{\text{apparent}} = \log \frac{I_0 + I_s}{I + I_s} \tag{3-11}$$

For example, if an instrument has stray light of 1% at a particular wavelength, the instrument can never read more than two absorbance units irrespective of sample concentration. With a double monochromator it is possible to achieve stray radiation levels of less than 10^{-6} relative to the true signal. In the past low cost spectrophotometers were usually limited to a dynamic range from 0.001 to 2.000 absorbance units. However, recent improvements in stray light filters and circuit design extend this range by at least one order of magnitude to 3.000 absorbance units. The ability to measure high absorbance with good linearity helps to increase sample throughput because it avoids the use of calibration curves, and reduces the number of reruns necessitated by samples exceeding the upper limit of the instrument's range.

The attainment of low stray light does not in itself guarantee good linearity. In addition, it is necessary to adjust the spectrophotometer reading very accurately to 0% transmittance when the light beam is blocked at the detector. The benefits of stray light levels as low as 0.01% cannot be realized unless the error of the 0% T adjustment is 0.005% T, or less.

The lower limit of the dynamic range of a spectrophotometer is determined by instrument stability and noise, which are primarily dependent upon the electromechanical stability of the entire system, including the optics as well as the electronics. Very rigid mounts for optical components and minimization of circuit susceptibility to ambient temperature and line voltage changes permit measurements of low absorbance solutions with scale expansion of the absorbance signal.

Sample Handling

Since all cuvettes have slight imperfections, reflection and/or scattering losses will change as different parts of the cuvette face are exposed to the light beam. Therefore, it is important to reposition a cuvette as precisely as possible when duplicating an analysis.

To clean cuvettes (and other optical surfaces), lens paper soaked in spectrograde methanol, which is held by hemostats or a similar device, should be employed. By cleaning cuvette faces in this manner a film of methanol is left which evaporates, leaving the faces free of contaminants. If maximum precision is required when duplicating photometric measurements, the best method is to leave the cuvette in place and change the solution by means of a syringe. Reproducibility is almost twice as good as that obtained when removing the cuvette. With instrumentation that is capable of precisely reading a 0.0001 absorbance unit change, very small differences are measurable.

3.3 PHOTOMETRIC PRECISION

The ultimate precision of a photometric measurement is determined by instrumental noise; that is, the statistical fluctuations of the signal reaching the detector, which in turn is a function of the type of detector being used. A quantitative analysis should be conducted within the range of transmittance for which a given uncertainty in transmittance, ΔT, or the equivalent quantity, ΔI (the noise associated with the intensity of radiant energy reaching the detector), will cause the least uncertainty in concentration, ΔC. Instruments with detectors subject only to Johnson noise, which is noise independent of light intensity, will display optimum precision in one range, and spectrophotometers with photoemissive detectors that are limited by shot noise will perform best in a somewhat different concentration range.

Uncertainty of the transmittance setting for most instruments will be on the order of 0.01–0.002 of the total scale; the latter value is considered a practical limit in ordinary work. The actual value of ΔT may be ascertained by measuring the transmittance of perhaps 30 portions of a solution and calculating the standard deviation of the measurements. Each measurement should include the operations of emptying, refilling, and repositioning the cuvette in its holder. Usually ΔT is taken as twice the average deviation of the replicate readings in order to include the uncertainty involved in setting the scale to 0% T with the light beam occluded and to 100% T with the pure solvent. With modern spectrophotometric instruments, capable of scale expansion or containing digital readout devices, the influence of reading error on the precision can be considerably lessened or made completely negligible with respect to noise on the readout signal.

Relative Concentration Error

In absorption measurements with a constant source, the light quanta arrive at such a high rate that the response of the detector does not depend upon their discrete nature. The minimum relative concentration error for detectors where the noise is independent of the intensity of radiant energy reaching the detector can be derived as follows. Assuming Beer's law is obeyed, rearranging the expression for the law gives

$$C = \frac{A}{ab} = \frac{1}{ab} \log \frac{I_0}{I} = -\frac{\log T}{ab} \qquad (3\text{-}12)$$

Differentiation of Eq. 3-12 gives

$$dC = -\frac{0.434}{ab}\left(\frac{dI}{I}\right) \quad \text{or} \quad \frac{dC}{dT} = -\frac{0.434}{Tab} \qquad (3\text{-}13)$$

Replacing the quantity ab by its equivalent from Beer's law, and rearranging, gives

$$\frac{dC}{C} = \frac{-0.434}{A}\left(\frac{dI}{I}\right) \quad \text{or} \quad \frac{dC}{C} = \frac{0.434}{\log T}\left(\frac{dT}{T}\right) \tag{3-14}$$

Thus the relative concentration error, dC/C, depends inversely on the product of absorbance and transmitted radiant intensity.

The transmittance at which the propagation of error is smallest is found by differentiating Eq. 3-14 and setting the derivative equal to zero:

$$\frac{d}{dI}\left(\frac{dC}{C}\right) = \frac{-dT}{2.3}\frac{d}{dI}\left(I \log \frac{I_0}{I}\right)^{-1} = \frac{dT}{2.3}\left(I \log \frac{I_0}{I}\right)^{-2}\left(\log \frac{I_0}{I} - 0.434\right) = 0 \tag{3-15}$$

This yields the nontrivial solution

$$\log \frac{I_0}{I} = 0.434 = A \quad \text{or} \quad T = 0.368$$

The mimimum error then becomes

$$\frac{dC}{C} = \frac{-0.434}{(0.368)(0.434)} dT = -2.72\, dT \tag{3-16}$$

Equation 3-16 is strictly true only when differentials are involved. Thus, it is reasonable to say that a 0.1% error in transmittance produces a 0.27% error in sample concentration. A plot of relative concentration error (and also error due to path length) as a function of absorbance (Fig. 3-3) is rather flat between 25% T and 50% T so that careful adjustment of solution concentration to read 36.8% T is of little value.

The minimum relative concentration error for photoemissive detectors where shot noise predominates can be determined in a similar manner. In this situation, the noise $(dI$ or $dT)$ is proportional to the square root of the light intensity; that is, $dT = k\sqrt{I}$. Replacement of dT by $k\sqrt{I}$ in Eq. 3-15 yields

$$\frac{dC}{C} = -0.434\, k \left(\sqrt{I} \log \frac{I_0}{I}\right)^{-1} \tag{3-17}$$

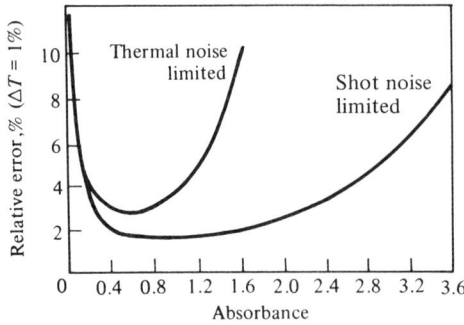

FIGURE 3-3 Relative concentration error, $\Delta C/C$, in percent, for a constant transmittance error.

Differentiation of Eq. 3-17 gives

$$\frac{d}{dI}\left(\frac{dC}{C}\right) = 0.434\, k \left(\sqrt{I} \log \frac{I_0}{I}\right)^{-2} \left[\frac{1}{2}\left(\frac{1}{\sqrt{I}} \log \frac{I_0}{I}\right) - \frac{\sqrt{I}}{I}0.434\right] \quad (3\text{-}18)$$

The minimum for this function is when

$$\log \frac{I_0}{I} = 2(0.434) = 0.868 = A \quad \text{or} \quad T = 0.135$$

For detectors where shot noise predominates, the region of minimum error extends over a very wide range of transmittance values from 0.01 to 0.50, as shown in Fig. 3-3. Thus, more concentrated solutions may be used which, in turn, reduces solution preparation errors, and also the effect of errors from cuvette matching. Window cleanliness and freedom from scratches on cuvette faces are less significant.

The foregoing derivations apply to single-beam spectrophotometers. The noise level for double-beam instruments is the quadratic sum of the noise levels of both the sample beam and the reference beam. The calculations for a double-beam instrument therefore are more complex. However, the final results are identical.

3.4 QUANTITATIVE METHODOLOGY

In developing a quantitative method for determining an unknown concentration of a given species by absorption spectrometry, the first step is the choice of the absorption band at which absorbance measurements are made. An ultraviolet/visible absorption spectrum of the species to be determined is obtained either from the literature or experimentally by means of a scanning double-beam spectrophotometer. From an inspection of the absorption spectrum, a suitable absorption band is selected. Absorptivity at any given wavelength is constant and an inherent characteristic of the absorbing species. The path length is made constant by using carefully matched cuvettes.

The numerical value of the absorptivity will determine the slope of the analytical curve and will influence the concentration range over which determinations can be made. When several absorption bands of suitable absorptivity are present, the band selected should favor wavelength regions that correspond to relatively high output of the light source and high spectral sensitivity of the photodetector. The absorption band should not overlap absorption bands of the solvent or contaminants, including excess reagents, that might be present in the sample.

Although many organic compounds absorb quite strongly, only a limited number of inorganic ions do, and it is the normal procedure of inorganic absorption spectrophotometry to add a molecule or reagent species to the solution of the inorganic ion which will react with it and, in the process, bring about a marked change in the spectral absorption characteristics of the reagent. Formation of metal–organic complexes is well known. For example, since the iron(II) ion is very weakly colored, a complexing agent, 1,10-phenanthroline, is added to form an ion-association species that is suitable for the determination of very small amounts of iron. A few moments' reflection will also bring to mind a number of possibilities among organic compounds. For example, although alcohols possess no absorption spectra between 200 and 1000 nm, treatment of an alcohol with phenyl isocyanate yields the corresponding phenyl alkyl carbamate which absorbs at about 280 nm. Semicarbazones display maxima that are shifted to longer wavelengths by 30–40 nm, with an average increase in molar absorptivity of 10,000 compared with the original carbonyl compound. Conversely, the strong absorption of anthracene can be eliminated by a Diels–Alder reaction with 1,2-dicyanoethylene.

Although very few reactions are specific for a particular substance, many reactions are quite selective, or can be rendered selective through the introduction of masking agents, control of pH, use of solvent extraction, adjustment of oxidation state, or by prior removal of interferents. Both the color-developing reagent and the absorbing product must be stable for a reasonable period of time. It is often necessary to specify that the color comparisons be made within a definite period of time, and it is always advisable to prepare standards and unknowns on a definite time schedule.

It is necessary that, along the wavelength axis, the absorption spectrum of the analyte or reagent–ion adduct should be well separated in at least one place from the absorption spectrum of the reagent itself, although it generally matters little if it is separated from that of the ion because the intensity of the absorption band due to the latter is low. It is also desirable, though not strictly essential, that the separation between the two spectra should be good with respect to the absorbance axis so that the maximal sensitivity may be utilized.

Solvents

Solvents used in spectrophotometry must meet certain requirements to assure successful and accurate results. First, the solvent chosen must dissolve the sample, yet be compatible with cuvette materials. Solubility data for common substances are available in reference works such as the *Merck Index* and Lange's *Handbook of Chemistry*. The solvent must also be relatively transparent in the spectral region of interest. To avoid poor resolution and difficulties in spectrum interpretation, a solvent should not be employed for measurements near or below its ultraviolet cutoff; that is, the wavelength at which absorbance for the solvent alone approaches one absorbance unit. Ultraviolet cutoffs for solvents commonly used are given in Table 3-1.

Once a solvent is selected, based on physical and spectral characteristics, its purity must be considered. The absorbance curve of a solvent, as supplied, should be "smooth," that is, not exhibit extraneous impurity peaks in the spectral region of interest. Solvents especially purified and certified for spectrophotometric use are available from suppliers.

Selection of Analytical Wavelength

When filter photometers are employed, the proper filter can be selected during the course of preparing the calibration curve. A series of standard solutions is prepared, including a blank. Using one filter at a time, a series of calibration curves is plotted. The filter that permits closest adherence to linearity over the widest absorbance interval and

TABLE 3-1 Ultraviolet Cutoffs of Spectro-Grade Solvents
(10-mm path vs. distilled water)

Solvent	Wavelength, nm	Solvent	Wavelength, nm
Acetic acid	260	Glycerol	207
Acetone	330	Hexadecane	200
Acetonitrile	190	Hexane	210
Benzene	280	Methanol	210
1-Butanol	210	2-Methoxyethanol	210
2-Butanol	260	Methylcyclohexane	210
n-Butyl acetate	254	Methyl ethyl ketone	330
Carbon disulfide	380	Methyl isobutyl ketone	335
Carbon tetrachloride	265	2-Methyl-1-propanol	230
1-Chlorobutane	220	N-Methylpyrrolidone	285
Chloroform (stabilized		Pentane	210
with ethanol)	245	Pentyl acetate	212
Cyclohexane	210	1-Propanol	210
1,2-Dichloroethane	226	2-Propanol	210
1,2-Dimethoxyethane	240	Pyridine	330
N,N-Dimethylacetamide	268	Tetrachloroethylene	
N,N-Dimethylformamide	270	(stabilized with thymol)	290
Dimethysulfoxide	265	Tetrahydrofuran	220
1,4-Dioxane	215	Toluene	286
Diethyl ether	218	1,1,2-Trichloro-1,2,2-tri-	
Ethanol	210	fluoroethane	231
2-Ethoxyethanol	210	2,2,4-Trimethylpentane	215
Ethyl acetate	255	o-Xylene	290
Ethylene chloride	228	Water	191

yields the largest slope, but with a small or zero intercept, will constitute the best choice. Naturally, if a spectrophotometer is available, the wavelength of maximum absorbance is quickly ascertained from a wavelength scan.

Computer software programs can handle multicomponent mixtures and deviations from Beer's law. Options involving concentration calculations include linear least squares with forced zero intercept, a method that uses several stored standards and the origin to construct the calibration curve; ordinary least squares in which the constructed calibration curve does not necessarily pass through the origin; second order with or without zero intercept, a method that handles deviations from Beer's law because of nonlinearity at high absorption; multicomponent analysis of up to 12 components by the single wavelength method or seven components over a wavelength range in which the computer uses stored standards to construct and fit a synthetic spectrum of the unknown mixtures.

Simultaneous Spectrophotometric Determinations

When no region can be found free from overlapping spectra of two chromophores, it is still possible to devise a method based on measurements at two wavelengths. Two dis-

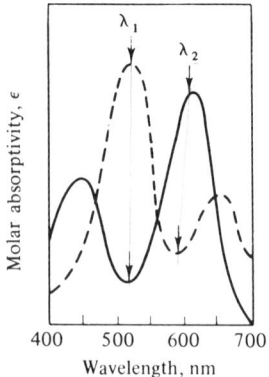

FIGURE 3-4 Simultaneous spectrophotometric analysis of a two-component system. Selection of analytical wavelengths indicated by arrows.

similar chromophores must necessarily have different powers of light absorption at some point or points in their absorption spectra. If, therefore, measurements are made on each solution at two such points, a pair of simultaneous equations may be obtained from which the two unknown concentrations may be obtained. First, it is necessary to select two points on the wavelength scale where the ratio of the molar absorptivities are maximal. For the system illustrated in Fig. 3-4, $(\epsilon_1/\epsilon_2)_{\lambda_1}$ and $(\epsilon_2/\epsilon_1)_{\lambda_2}$ are maximal. Neither of these wavelengths need necessarily coincide with an absorption maximum for either component. Next, it is necessary to calculate the molar absorptivity for each component at each wavelength selected. Now two simultaneous equations may be written:

$$C_1(\epsilon_1)_{\lambda_1} + C_2(\epsilon_2)_{\lambda_1} = A_{\lambda_1} \tag{3-19}$$

$$C_1(\epsilon_1)_{\lambda_2} + C_2(\epsilon_2)_{\lambda_2} = A_{\lambda_2} \tag{3-20}$$

The equations are solved for the concentration of each component. Simultaneous determinations rest on the assumption that the substances concerned contribute additively to the total absorbance at an analytical wavelength.

3.5 DIFFERENTIAL OR EXPANDED SCALE SPECTROSCOPY

In the *ordinary* spectrophotometric method, two adjustments are required before the actual measurement of standards and unknowns is made. First the zero point of the transmittance scale must be adjusted to read zero with no light reaching the detector. This is done by placing an occluder in the light beam; the occluder, which may simply be a shutter, represents a completely opaque species. The second manipulation brings the adjustment of the 100% transmittance on to the scale by placing pure solvent in the light beam and balancing the instrument to read 100 (or full scale). After these operations the instrument is capable of the measurement of any light intensity falling between total darkness and one equal to the light intensity passing the pure solvent. To complete the analysis the transmittance of at least one solution of known concentration is measured to establish the proportionality between absorbance and concentration from Beer's law; then the transmittance of the unknown solution is measured.

In the derivation of the relative concentration error, the incident intensity, I_0, was considered a constant and was disregarded in finding the optimum conditions. Setting $(d/dI_0) \times (dC/C) = 0$ shows that I_0 should be infinite. However, under ordinary spectrophotometric methods, the maximum value of I_0 is limited by the length of the potentiometer slidewire, since one end corresponds to zero intensity and the other end to the full intensity at zero concentration, or I_0. This is an artificial requirement. Calibration of the transmittance scale can also be made, for example, by using two reference solutions containing the absorbing species in different concentrations. Two neutral density filters would be equally suitable. The only condition is that one reference absorber shall transmit more radiant energy than the sample to be measured, and the other reference absorber transmit less radiant energy than the sample. From these possible alterations, three differential or scale-expansion techniques arise which can be used to increase precision: the high-absorbance method, the trace-analysis method, and the maximum-precision method.

In the *high-absorbance* method the dark current is still measured using a shutter to occlude the light beam while adjusting the scale to read 0% T. To make the 100% T adjustment, however, a finite reference solution replaces the pure solvent. The reference solution is more dilute than the unknown. For example, if the sample shows a 20% transmittance by the ordinary method and a standard reference solution reads 36%, the latter solution is now used to set the instrument to read 100% T. A threefold scale expansion is accomplished, as illustrated in Fig. 3-5. The unknown will now read 56% T relative to the reference solution; that is, the apparent absorbance is 0.26 and to this must be added

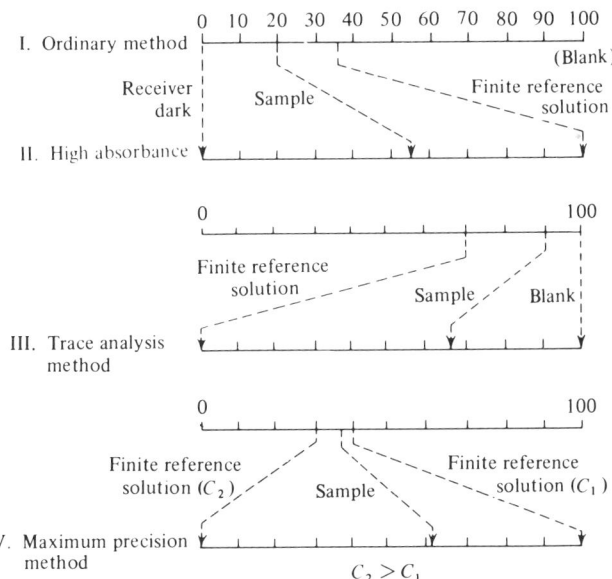

FIGURE 3-5 Differential (or expanded scale) spectrophotometry.

the absorbance (0.44) of the reference solution for a total of 0.70 absorbance units for the sample (equivalent to 20% T).

To compensate for the lesser amount of transmitted light reaching the detector in the scale-expansion method, the instrument must have one or more provisions for reserve sensitivity. One approach is to increase the amplification of the detector output without, however, increasing the noise component of the signal. A second method involves increasing the slit width if such an increase is compatible with required spectral purity and the natural bandwidth of the absorption band being measured. The presence of stray radiation in the instrument will limit the usefulness of the high-absorbance method.

The artificial I_0 does increase the accuracy in the measurement and thus justifies the expanded scale that automatically results. The relative concentration error becomes finite at the high end of the transmittance scale. In fact, the error becomes dependent in a pronounced manner upon the actual absorbance of the reference standard used to make the 100% T setting, as illustrated in Fig. 3-6. The error function is given by

$$\frac{dC}{C} = \frac{0.434 \Delta T}{(I/I_0)(\log I_0/I)_{sple} + (\log I_0/I)_{ref}} = \frac{0.434 \Delta T}{(TA)_{sple} + A_{ref}} \quad (3\text{-}21)$$

The position of minimum error gradually shifts to 100% T on the transmittance scale as the reference concentration increases; this amounts to comparing the reference and unknown at the same scale setting.

The optical path of the cuvettes must be known with a precision equal to the best precision expected for the differential method, or else one cuvette must be used for all measure-

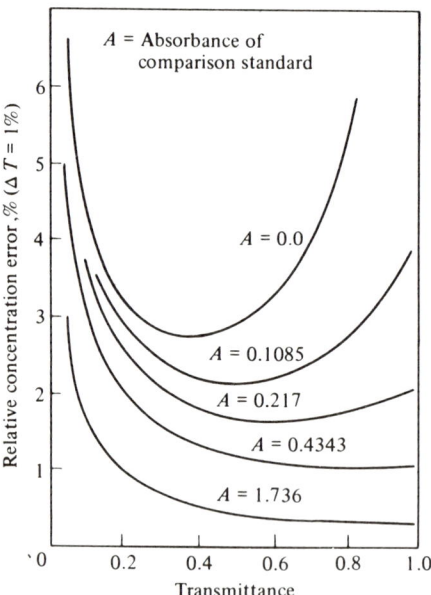

FIGURE 3-6 Plot of the relative concentration error. (From C. F. Hiskey, *Anal. Chem.*, 21, 1440 (1949). Courtesy of *Analytical Chemistry*.)

ments. To minimize volumetric errors, aliquots should be taken by weight. The precision attainable by the high-absorbance method may approach 0.01%. A calibration curve is often desirable since increasing the sensitivity may cause some deviations from Beer's law. For interesting discussions of the differential method and a rigorous treatment of the error function, the reader is referred to papers by Bastian,[5-7] Hiskey,[8,9] and Reilley and Crawford.[10,11]

In the *trace-analysis* method a large increase in sensitivity is achieved by "arranging" a positive deviation from Beer's law. The opaque reference standard is abandoned. The transmittance scale is set to 100% T with the solvent blank as in the ordinary method, but the zero energy is "faked" by placing in the light beam either a reference solution, a screen, or a neutral density filter. Whichever is used should transmit a finite amount of radiant energy, but an amount less than that of the most concentrated sample solution. For example, and following Fig. 3-5, suppose in the ordinary method a standard reference shows 70% transmittance and the sample reads 90%. Using the 70% T reference in the light beam to make the 0% T setting results in a several fold scale expansion, and the sample now reads 67% of the full scale.

In order to use this method, the instrument must have a zero-suppression (dark-current, or "bucking") control, with which the response obtained through a finite reference solution can be made to read 0% T. A calibration curve must be constructed since the sensitivity increase is achieved at the expense of a positive deviation from Beer's law. Interaction between the controls used for the 0% T and 100% T scale settings necessitates making several trials before the transmittance scale will be adjusted. The error at the low transmittance end of the transmittance scale now becomes finite. The expression for the relative concentration error becomes

$$\frac{dC}{C} = \frac{0.434\,(1 - T_{\text{ref}})}{TA}\,\Delta T \tag{3-22}$$

The *maximum-precision* method is simply a combination of the preceding two methods. Both ends of the transmittance scale are calibrated with standard reference solutions spaced around transmittance values which lie within the relatively flat portion of the curve of relative concentration error. In this method a standard solution having a concentration somewhat higher than the unknown is used to set the scale at 0% T. A second standard solution that is more dilute than the unknown is placed in the light beam for the 100% T adjustment. Figure 3-5 illustrates the procedure for reference solutions that read 30% transmittance (used to set the 0% T) and 40% (used to make the 100% T setting). The maximum-precision method improves an already favorable transmittance reading (at 36.3% T) by making the reading 10 times more precise at 63.0% T. A calibration curve will be necessary, as positive deviations from Beer's law occur. The instrument requirements for this method are the combined requirements of the previous two methods, and the limitations likewise tend to be a combination. The relative concentration error is given by

$$\frac{dC}{C} = \frac{0.434\,(T_{100\%\,\text{ref}} - T_{0\%\,\text{ref}})}{TA}\,\Delta T \tag{3-23}$$

In theory the precision of this method can be increased by making the difference between the two reference solutions small. But this assumes the availability of an instrument with very high sensitivity and stability.

3.6 DIFFERENCE SPECTROSCOPY

Difference spectroscopy provides a sensitive method for detecting small changes in the environment of a chromophore, or it can be used to demonstrate ionization of a chromophore leading to identification and quantitation of various components in a mixture. In difference spectroscopy, absorption spectra of two samples of slightly different composition or physical state are compared. Common features in the spectra cancel, and bands which are recorded can be interpreted in terms of known differences between the samples. Recording accurate difference spectra requires a spectrophotometer with low stray light, good resolution, wavelength accuracy and repeatability, and an overall system designed to minimize noise when working with samples that significantly attenuate the signal in both reference and sample channels.

Difference spectroscopy is utilized in toxicology laboratories for analysis of dangerous drugs. Barbiturates, for example, display characteristic spectral changes when the structure is converted from keto to enol forms by pH changes. Biochemists routinely use difference spectroscopy to study the conformation of globular proteins in solution. Informative difference spectra have been obtained using solvent perturbation, pH difference, temperature difference, and concentration difference.

3.7 DERIVATIVE SPECTROSCOPY[12]

In derivative spectroscopy the first or higher derivative of absorbance (or spectral intensity) with respect to wavelength is recorded versus wavelength. In a derivative spectrum the ability to detect and to measure minor spectral features is considerably enhanced. This enhancement of characteristic spectral detail can distinguish very similar spectra and follow subtle changes in a spectrum. Moreover, it can be of use in quantitative analysis to measure the concentration of an analyte whose peak is obscured by a larger overlapping peak due to something else in the sample (and thus avoid prior separations), as shown in Fig. 3-7. In this particular example one cannot draw a unique tangential baseline, but if a reasonable guess is made, the reading of 0.4 is far too low. Referring to the derivative spectra in the lower right of Fig. 3-7, one can take as the measure of the analyte intensity the vertical distance between the adjacent maximum and minimum of the first derivative. Now our estimate of the analyte intensity is low by only 12%. Whenever the interfering band is at least a factor of 2 broader than the analyte band, it will usually be advantageous to base the measurement on the derivative spectra.

A variety of different experimental techniques have been used to obtain derivative spectra. If the spectrum has been recorded digitally or is otherwise available in computer-readable form, then the differentiation can be done numerically. Alternatively, the deriva-

ULTRAVIOLET AND VISIBLE ABSORPTION METHODS   83

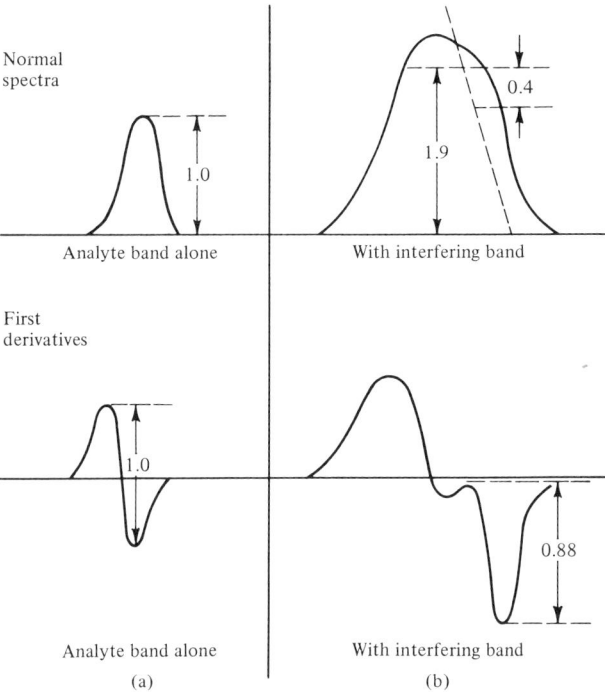

FIGURE 3-7 First derivative spectrometry for the quantitative measurement of intensity of a small band (a) alone and (b) obscured by a broader overlapping band (Reprinted with permission from T. C. O'Haver, *Anal. Chem.*, 51, 91A (1979). (Copyright 1979 of American Chemical Society.)

tive spectra may be recorded directly in real time, either by wavelength modulation or by obtaining the time derivative of the spectrum scanned at a constant rate. In the latter case, a quite simple electronic differentiator can be used. Several commercial spectrometers capable of recording derivative spectra are now available. Dual-wavelength spectrophotometers can obtain first derivative spectra by wavelength modulation.

3.8 PHOTOMETRIC TITRATIONS[13,14]

The change in absorbance of a solution may be used to follow the change in concentration of a light-absorbing constituent during a titration. The absorbance is linearly proportional to the concentration of absorbing constituent rather than logarithmically as in potentiometric methods. This means that in a titration in which the titrant, the reactant, or a reaction product absorbs, the plot of absorbance versus titrant will consist, if the reaction is complete, of two straight lines intersecting at the end point—similar to amperometric and conductometric titrations. For reactions that are appreciably incomplete, extrapolation of the two linear segments of the titration curve establishes the intersection and

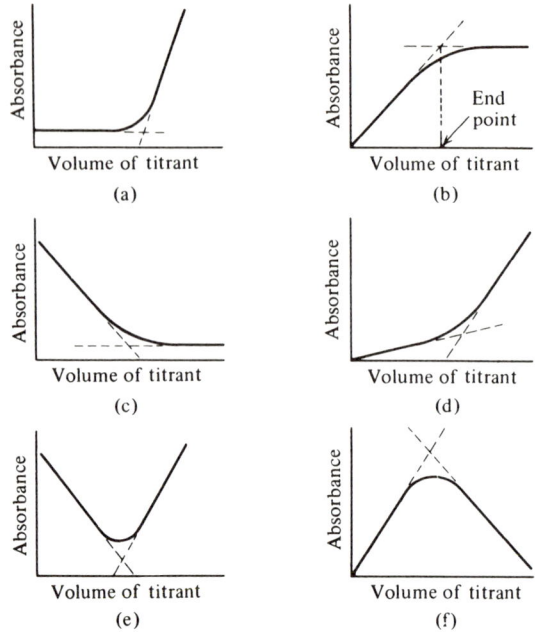

FIGURE 3-8 Possible shapes of photometric titration curves.

end-point volume. Possible shapes of photometric titration curves are shown in Fig. 3-8. Curve (a), for example, is typical of the titration where the titrant alone absorbs, as in the titration of arsenic(III) with bromate-bromide, where the absorbance readings are taken at the wavelength where the bromine absorbs. Curve (b) is characteristic of systems where the product of the reaction absorbs, as in the titration of copper(II) with EDTA. When the analyte is converted to a nonabsorbing product—for example, titration of p-toluidine in butanol with perchloric acid at 290 nm—curve (c) results. When a colored analyte is converted to a colorless product by a colored titrant—for example, bromination of a red dyestuff—curves similar to (e) are obtained. Curves (d) and (f) might represent the successive addition of ligands to form two successive complexes of different absorptivity.

Photometric titrations have several distinct advantages over a direct photometric determination. The presence of other absorbing species at the analytical wavelength, as in curve (a) of Fig. 3-8, does not necessarily cause interference, since only the change in absorbance is significant. However, the absorbance of nontitratable components (color or turbidity) must not be intense because, if so, the absorbance readings will be limited to the undesirable upper end of the scale unless the slit width or amplifier gain can be increased. Only a single absorber needs to be present from among the reactant, the titrant, or the reaction products. This extends photometric methods to a large number of nonabsorbing constituents. Precision of 0.5% or better is attainable because a number of pieces of information are pooled in constructing the segments of the titration curve.

The analytical wavelength is selected on the basis of two considerations: (1) avoidance of interference by other absorbing substances and (2) need for a molar absorptivity which

ULTRAVIOLET AND VISIBLE ABSORPTION METHODS   85

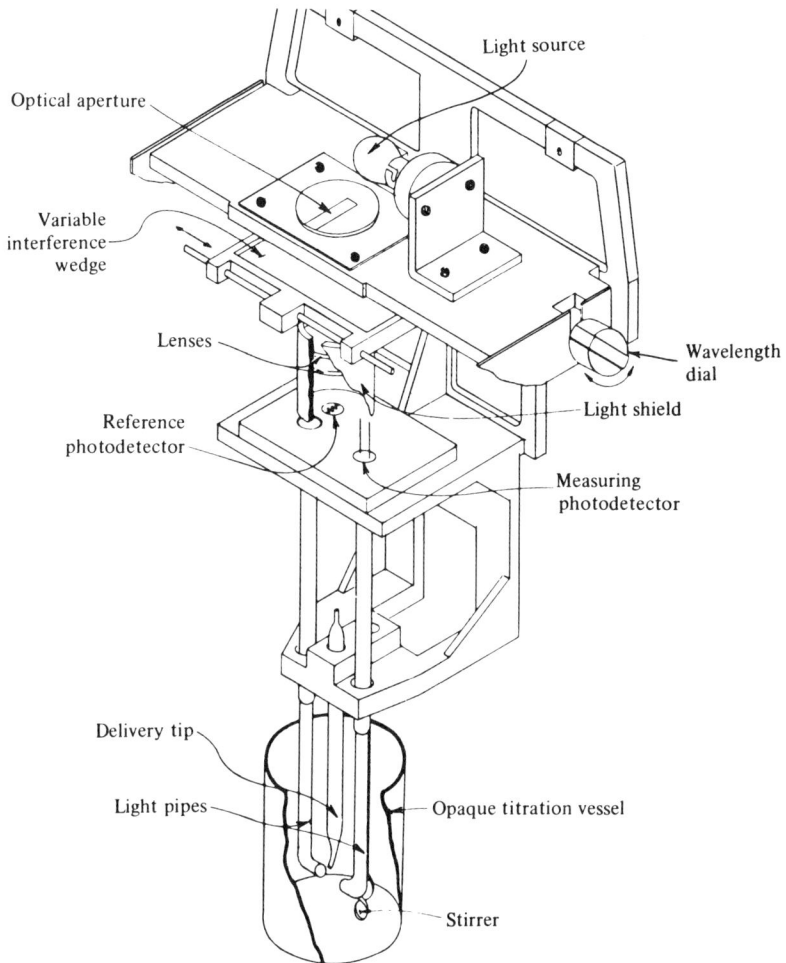

FIGURE 3-9 The Fisher photometric accessory to the Titralyzer for automatic photometric titrations. (Courtesy of Fisher Scientific Co.)

will cause the change in absorbance during the titration to fall within a convenient range. Often the chosen wavelength lies well apart from an absorption maximum.

Volume change is seldom negligible, and straight lines are obtained only if correction is made. This is done simply by multiplying the measured absorbance by the factor $(V + v)/V$ where V is the volume initially and v is the volume of titrant added up to any point. If the correction is not made, the lines are curved down toward the volume axis and erroneous intersections are obtained. Use of a microsyringe and relatively concentrated titrant is desirable. Stray-light error also affects the linearity of the titration curve. The upper limit of concentration permissible is found by delivering from the buret into a beaker of transparent liquid measured portions of a colored substance known to obey Beer's law.

After correcting for dilution, the plot of absorbance versus concentration will be a straight line up to the absorbance value where the stray-light error becomes detectable.

Areas of particular applicability are for solutions so dilute that the indicator blank is excessive by other methods, or when the color change is not sharp due perhaps to titration reactions which are incomplete in the vicinity of the equivalence point, or when extraneous colored materials are present in the sample. Ordinarily there is no difficulty in working in solutions of either high or low ionic strength or in nonaqueous solvents. One of the attractive features of photometric titrations is the ease with which the sensitivity of measurements can be changed, simply by changing the wavelength or the length of the light path. When self-indicating systems are lacking, an indicator can be deliberately added, but in a relatively large amount to provide a sufficient linear segment on the titration curve beyond the equivalence point.

All one needs to carry out photometric titrations in the visible region is a light source, a series of narrow bandpass filters, a titration vessel (which can be an ordinary beaker), a receptor, and a buret or other titrant delivery unit. The entire assembly is housed in a light-tight compartment. Photometers or spectrophotometers with provision for inclusion of a suitable titration vessel from 5- to 100-ml capacity are suitable. It is imperative that the titration vessel remain stationary throughout the titration. By the use of Vycor beakers and an appropriate spectrophotometer, titrations may be conducted in the ultraviolet region. Provision for magnetic stirring from underneath or some type of overhead stirrer is desirable; otherwise, manual agitation after the addition of each increment of titrant is necessary. For the transmission of photometric end points, the ends of two fiber optic light pipes can be located facing each other across an area below the buret tip in the titrating vessel. One pipe conducts dispersed light to the sample, and the other pipe carries the transmitted light to the photodetector (Fig. 3-9).

3.9 SPECTRA OF SOLIDS

Absorption measurements on nonclear solutions and solids are precluded with standard absorption spectrophotometers. Yet these materials are found in the real world. With the proper instruments or accessories it is possible to obtain useful data from turbid liquids, powders, and opaque and translucent solids.

Reflection occurs whenever a light ray encounters a boundary between two media. The light reflected from the first surface of contact is called the specular (gloss, sheen) component. These encounters are repeated over and over again in granular or fibrous structures where a light beam will encounter a new interface every few millionths of a centimeter. These repeated encounters result in thorough diffusion such that the surface tends to appear uniformly bright in all directions; this is the diffuse component that is responsible for color where color exists. Particle size plays an important role. Whiteness and reflectance gain as the diameter is reduced to about one-half of the wavelength of the incident light. At very small particle diameters, less than one-fourth the light wavelength, scattering takes over and diffuse reflection falls off.

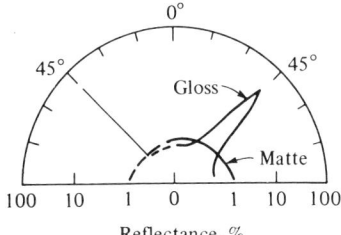

FIGURE 3-10 Goniophotometric curves. (Courtesy of Hunter Associates Laboratory, Inc.)

Specular Reflection (Gloss)

Specular reflection, or gloss, may be defined as the degree to which a surface possesses the light-reflecting property of a perfect mirror. This is mirrorlike reflection for which the angle of reflection equals the angle of incidence. Surfaces with intermediate properties between zero gloss (matte) and a perfect mirror can have their characteristics measured with a goniophotometer. The gloss of any surface increases as the angle of incidence increases; that is, departs from a line perpendicular to the surface.

In a gloss meter, light from a prefocused incandescent source falls on the sample at a specific angle. Light specularly reflected at an equal but opposite angle falls on a photocell and the intensity is indicated on a meter. Gloss classification over a wide range from high to low is best made with the incidence angle of 60°. A multiangle glossmeter consists of five prefocused systems at angles of 20°, 45°, 60°, 75°, and 85°. Small angles serve best to differentiate among surfaces of high gloss, and large angles work best among surfaces of low gloss. Instruments with 45° geometry meet special requirements of the plastics and ceramics industries; the 75° geometry meets the needs of the book and paper industry.

With a goniophotometer the reflected light is measured as a function of a variable angle of incidence. Goniophotometric curves for gloss (Fig. 3-10), analogous to spectrophotometric curves for color, have a maximum in the direction of mirror reflection. The narrower and higher the peak, the higher the gloss.

Diffuse Reflection

Diffuse reflection is seen with dull surfaces. The reflected light is dispersed through a wide range of angles. A perfectly diffusing surface (matte), even under unidirectional illumination, has a constant luminance regardless of the angle from which it is viewed. The radiation reflected from the sample surface is detected and recorded as a function of wavelength. Instruments fall into three categories: colorimeter, reflectometer, and spectroreflectometer. The colorimeter employs three filter–photocell combinations, each with its own readout dial calibrated for the color coordinate values of the system selected. A reflectometer employs a series of filters to obtain approximately monochromatic radiation. The spectroreflectometer embodies a monochromator.

Optical arrangements for reflectance measurements are of three types: integrating sphere; annular, ellipsoidal mirror; and reflection configuration. The reflection-type in-

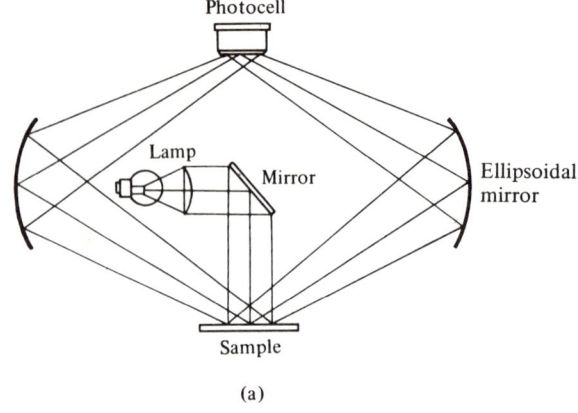

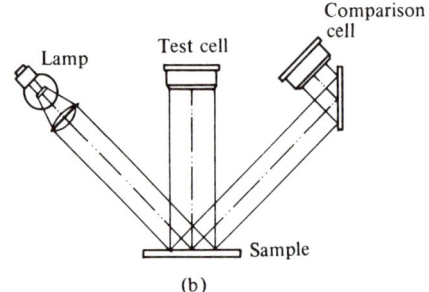

FIGURE 3-11 Types of reflectance instruments. (a) Annular, ellipsoidal, mirror-type and (b) reflection-type.

strument is a simple arrangement (Fig. 3-11b) where the reflected radiation is detected by a photodetector without any type of collector device being used. Radiation from the slit and field lens strikes the sample surface at an angle of 45° to the perpendicular. The reflected radiation, at 90° to the surface of the sample, is measured. Reflected light is distributed to three phototubes, each having a different filter inserted in front of the detector.

The annular, ellipsoidal mirror configuration is shown in Fig. 3-11a. The sample is illuminated at a zero angle of incidence. The reflected radiation from the sample is collected by an annular, ellipsoidal mirror, and then detected by the photodetector. A screen attenuator is sometimes located in the reference beam to permit the reference beam to be standardized against a reference sample.

In the integrating sphere (Fig. 3-12), the light from the source lamp passes through a monochromator. A chopper splits the initial beam, sending it alternately through the sample and reference ports. Diffusely reflected light from either the reference plate (barium sulfate or magnesium oxide) or the sample is trapped in the sphere and is detected by the photomultiplier tube. The sphere is coated on the interior with magnesium oxide or

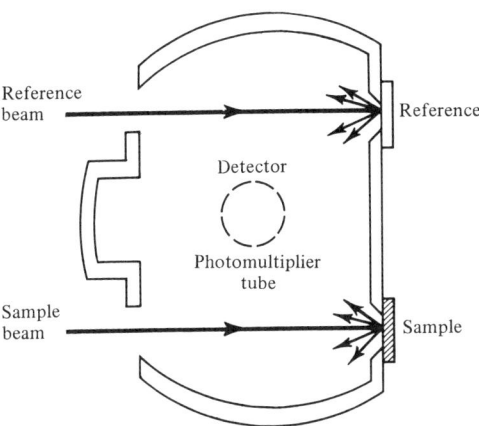

FIGURE 3-12 Integrating sphere attachment (placed in the sample compartment of a spectrophotometer) for measurement of diffusely reflected light. (Courtesy of Perkin-Elmer Corp.)

barium sulfate, an efficient diffuse reflector. The specular components from both the sample and the reference are reflected out of the sphere through their respective entrance ports in the configuration illustrated. The specular components may also be rejected by using light traps of black velveteen located at an angle of 90° to the sample and reference materials. When the instrument operates in the transmittance mode, its output at any wavelength gives the percentage of light diffusely reflected from the sample relative to that from the reference.

The same sphere also can make scattered transmittance measurements. This measurement is achieved by placing the sample against the outside of the sample beam entrance port and replacing the diffusely reflecting sample by a second barium sulfate plate. In this configuration, the sphere collects transmitted light over a large solid angle, thus reducing the adverse effects of the scatter. The ability to reduce light-scattering effects becomes more pronounced as the sample thickness becomes less. The technique is especially useful in the study of thin film substrates. An important application of scattered transmittance is in the quality control of containers used for pharmaceutical products wherein it is necessary to restrict the amount of light entering the containers.

Color Measurements[15]

Probably the best established application of diffuse reflectance is color measurement. This application depends on the relationship of color differences to differences in the reflectance spectra of objects. For quantitative purposes, color may be represented by a set of three numbers (tristimulus values) for red, blue, and green spectral colors (CIE system of XYZ coordinates); of green, amber, and blue reflectances (GAB system); of R_d, a, and b or its closely related system: L, a_L, and b_L. With these scales it is possible

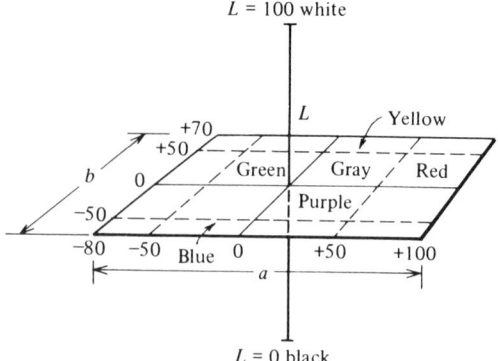

FIGURE 3-13 The GAB system in which the colors of specimens may be specified and visually interpreted. (Courtesy of Hunter Associates Laboratory, Inc.)

to represent colors by position in a three-coordinate system (Fig. 3-13). The diffuse reflectance term, R_d, is the percentage of light reflected by the sample relative to that reflected by a magnesium oxide standard; it is equal to the value Y (CIE system) and G (GAB system). The term L measures lightness and corresponds closely to visual estimates of this quantity: $L = 10\sqrt{Y} = 10\sqrt{R_d}$. In the rectangular diagram, a and b are the chromaticity coordinates; the vertical axis is the L value. If a sample has zero value for a and b, it is black, gray, or white, depending on the value of L. A plus value for a indicates redness; a minus value, greenness. A plus value for b indicates yellowness; a minus value, blueness. When the information being sought is color difference rather than the colors themselves, the color difference is computed as follows:

$$\text{color difference} = \sqrt{(\Delta L)^2 + (\Delta a)^2 + (\Delta b)^2} \tag{3-24}$$

Reflectance curves of color materials may look somewhat like a transmittance curve (Fig. 3-14). A perfectly white sample would reflect all of the light, and its curve would be a horizontal line at 100%. A theoretically perfect black sample would absorb all of the light and its curve would be a horizontal line at zero. A neutral gray would give a horizontal line, its position on the plot depending upon the depth of shade. The brighter a color, the more nearly vertical the reflectance band, while duller shades will produce flatter curves (approaching a gray).

3.10 TURBIDITY AND NEPHELOMETRY

Turbidity is an expression of the optical property of a sample which causes light to be scattered and absorbed rather than transmitted in straight lines through the sample. Scattering is elastic so that both incident and scattered light are of the same wavelength. Turbidity is caused by the presence of suspended matter in a liquid. A scattering center is actually an optical inhomogeneity in an otherwise homogeneous medium. An atom, a molecule, a thermal density fluctuation, a colloidal particle, or a suspended solid all can

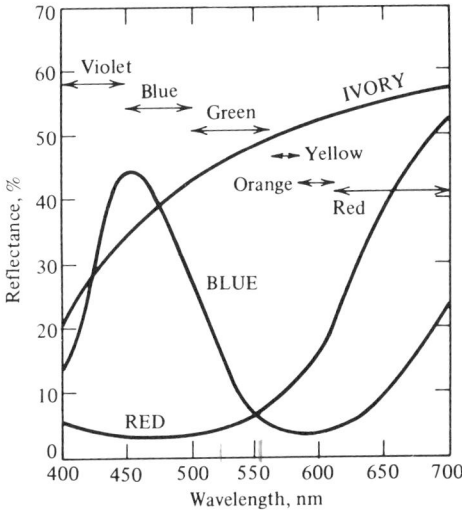

FIGURE 3-14 Spectral reflectance curves for selected colors.

produce an optical inhomogeneity resulting in light scatter. The intensity of the perpendicularly polarized component of scattered light and the parallel polarized component of scattered light are functions of the relative refractive index, the size parameter, and the angle of observation (relative to the incident wavelength), as well as to the concentration of scatterer. When no molecular absorption is present in the sample, the refractive index becomes the conventional value. The size parameter ($\alpha = 2\pi r/\lambda$) involves the radius of the scattering center and the wavelength of the incident radiation. It is this ratio that determines the phase distribution of the scattered radiation over the scattering center. This phase distribution shapes the scattering envelope and determines the resulting angular dependence. When the size parameter is smaller than one-tenth the wavelength of the incident light, the scattering envelope is symmetrical and is termed Rayleigh scattering. As the size parameter becomes approximately one-fourth the wavelength of the incident light, scattering is concentrated in the forward direction. Ultimately angular oscillations begin to appear for particles larger than the wavelength of incident light accompanied with extreme concentration of scattering in the forward direction. For very large values, there is no wavelength dependence. Between these two limiting cases, the scattering efficiency is a complex oscillatory function of the size parameter and refractive index, where any wavelength can be preferentially scattered. Light-scattering theory is further complicated by other sample parameters such as particle shape, molecular absorption, sample concentration, size distributions of scatterers, and other physical and optical anisotropies. Consequently, the relationship between any measurable indication of scattered light intensity and concentration of scatterer is not simple. Analytical determinations must therefore be empirical. In fact, differences in the physical design of an instrument will cause differences in measured values for turbidity, even though the same calibration material was used for each instrument.

Standards

A suspension resulting from accurately weighing and dissolving 5 g of hydrazinium(2+) sulfate ($N_2H_6SO_4$) and 50 g of hexamethylenetetramine in 1 liter of distilled water is defined as 4000 nephelometric turbidity units (NTU). After standing 48 hr the insoluble polymer (formazin) formed by the condensation reaction develops a white turbidity. This turbidity can be repeatedly prepared within an accuracy of ±1%. The mixture can be diluted to prepare standards of any desired value.

Instrumentation[16,17]

For the measurement of very small amounts of turbidity (sample transmittance greater than 90%), as in water and waste-water analysis, the nephelometric method is the one of choice. A typical instrument with a 90° detection angle is shown in Fig. 3-15. While this angle is not the most sensitive to concentration, it is probably least sensitive to variations in particle size. It also affords a simple optical system that is relatively free from stray light. If the sample is entirely free of scatterers, no scattered light will reach the photodetector, and the indicating meter will read zero. Increasing turbidity gives an increase in meter reading. A linear response is obtained around zero reading and extending to a certain turbidity, after which the response begins to level off. Further increases in turbidity cause a decrease in response and finally the instrument goes blind at a high turbidity value. Both sensitivity and linearity are functions of the path length traversed by the scattered

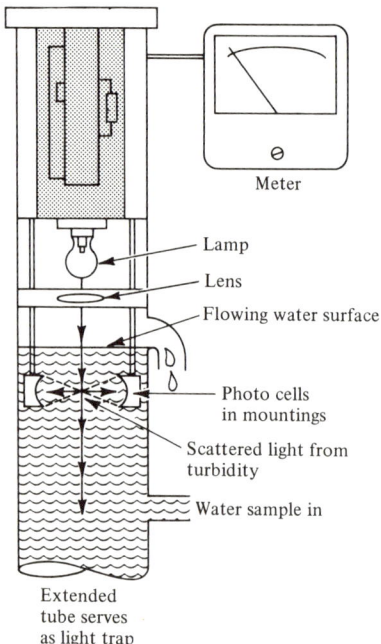

FIGURE 3-15 Low range turbidimeter. (Courtesy of Hach Chemical Co.)

light. While sensitivity increases as path length increases, linearity is sacrificed at high concentrations as the sample becomes increasingly opaque and the light cannot penetrate. The shorter the light path in the nephelometer, the higher the turbidity that can be measured, but sensitivity is lost at low concentrations. This trade-off can be eliminated with an adjustable path length. Stray light becomes an added complication with the use of a short path length. Any scratches, imperfections in the cell windows, dirt, film, or condensation on the walls will scatter light, some of which usually reaches the detector and gives a positive error to the turbidity measurement.

To overcome the problem of stray light caused by cell windows in a nephelometer when attempting to measure very low turbidities, the surface scatter instrument was designed (Fig. 3-16). Light is admitted through the upper surface of the sample in a nephelometer. These instruments are capable of accurately measuring trace turbidities in the hundredths of NTU. The surface scatter approach can also be used for high turbidities. When a very narrow beam of light strikes the surface of the sample at a very low angle, part of the beam is reflected by the water surface and escapes to a light trap. The remaining portion enters the sample at approximately a 45° angle. If particles of turbidity are present, light scattering will occur and some of the scattered light will reach the detector located slightly above the surface of the sample. These instruments provide a practical method for continuous monitoring of turbidity in water control and industrial operations.

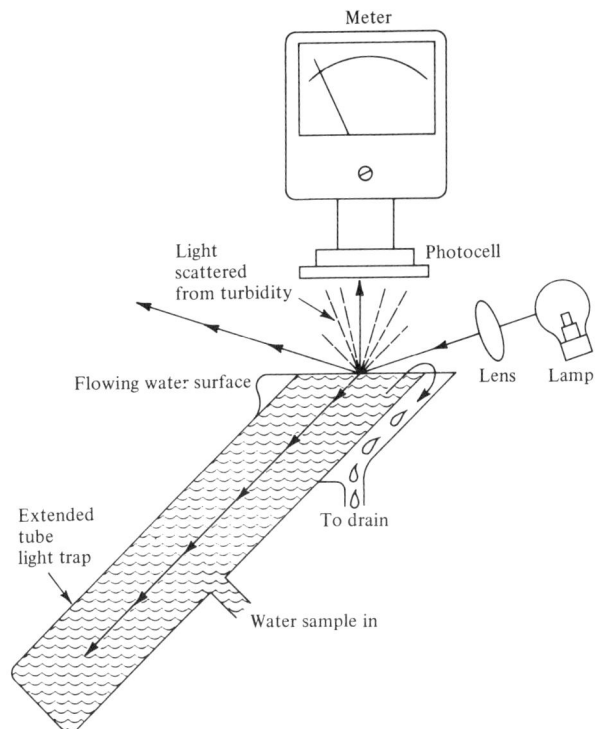

FIGURE 3-16 Surface scatter turbidimeter. (Courtesy of Hach Chemical Co.)

Applications of nephelometry and turbidimetry are widely varied. Some determinations involve systems that are turbid prior to entering the analytical laboratory, such as in the determination of suspended material in waters. Measurements of the clarity of beverages and pharmaceuticals is typical of the simple kind of nephelometric determination. This is essentially an appearance measurement designed to evaluate the amount of haze, or cloudiness, present in a sample. Clarity and sparkle are important characteristics of product quality; the presence of suspended materials, even in amounts so small as to be invisible at the bottling point, will, after bottling and storage, ultimately result in an unsightly and unpalatable sediment. The suitability of industrial process waters and clarity of boiler feed waters and condensates are typical examples of determinations in everyday use.

3.11 CORRELATION OF ELECTRONIC ABSORPTION SPECTRA WITH MOLECULAR STRUCTURE

When molecules interact with radiant energy in the visible and ultraviolet region, the absorption of energy consists in displacing an outer electron in the molecule. Rotational and vibrational modes will be found combined with electronic transitions. Broadly, the spectrum is a function of the whole structure of a substance rather than of specific bonds. No unique electronic spectrum will be found; this is a poor region for the product identification by the "fingerprint" method. Information obtained from this region should be used in conjunction with other evidence to confirm the identity of a compound; for example, previous history of a compound, its synthesis, auxiliary chemical tests, and other spectroscopic methods. On the other hand, electronic absorption often has a very large magnitude. Molar absorptivity values frequently exceed 10,000, whereas in the infrared they rarely exceed 1000. Thus, dilute solutions are adequate in visible-ultraviolet spectrophotometry.

Structural Features

We will consider only those molecules capable of absorption within the wavelength region from 185 to 800 nm. Compounds with only single bonds involving σ-valency electron exhibit absorption spectra only below 150 nm and will be discussed only in interaction with other kinds. In covalently saturated compounds containing heteroatoms—for example, nitrogen, oxygen, sulfur, and halogen—unshared p-electrons are present in addition to σ-electrons. Excitation promotes a p-orbital electron into an antibonding σ orbit, that is, an $n \rightarrow \sigma^*$ transition, such as occurs in ethers, amines, sulfides, and alkyl halides. In unsaturated compounds absorption results in the displacement of π-electrons. Molecules containing single absorbing groups, called *chromophores,* undergo transitions at approximately the wavelengths indicated in Table 3-2.

Molecules with two or more isolated chromophores will absorb light of nearly the same wavelength as a molecule containing only a single chromophore of a particular type, but the intensity of the absorption will be proportional to the number of that type of chromophore present in the molecule. Appreciable interaction between chromophores does not occur unless they are linked to each other directly; interposition of a single methylene group, or *meta*-orientation about an aromatic ring, is sufficient to insulate chromophores almost completely from each other. However, certain combinations of functional groups afford chromophoric systems which give rise to characteristic absorption bands.

TABLE 3-2 Electronic Absorption Bands for Representative Chromophores

Chromophore	System	λ_{Max}	ϵ_{Max}	λ_{Max}	ϵ_{Max}	λ_{Max}	ϵ_{Max}
Ether	$-O-$	185	1000				
Thioether	$-S-$	194	4600	215	1600		
Amine	$-NH_2$	195	2800				
Thiol	$-SH$	195	1400				
Disulfide	$-S-S-$	194	5500	255	400		
Bromide	$-Br$	208	300				
Iodide	$-I$	260	400				
Nitrile	$-C\equiv N$	160					
Acetylide	$-C\equiv C-$	175–180	6000				
Sulfone	$-SO_2-$	180					
Oxime	$-NOH$	190	5000				
Azido	$>C=N-$	190	5000				
Ethylene	$-C=C-$	190	8000				
Ketone	$>C=O$	195	1000	270–285	18–30		
Thioketone	$>C=S$	205	strong				
Esters	$-COOR$	205	50				
Aldehyde	$-CHO$	210	strong	280–300	11–18		
Carboxyl	$-COOH$	200–210	50–70				
Sulfoxide	$>S\rightarrow O$	210	1500				
Nitro	$-NO_2$	210	strong				
Nitrite	$-ONO$	220–230	1000–2000	300–4000	10		
Azo	$-N=N-$	285–400	3–25				
Nitroso	$-N=O$	302	100				
Nitrate	$-ONO_2$	270 (shoulder)	12				
	$-(C=C)_2-$ (acyclic)	210–230	21,000				
	$-(C=C)_3-$	260	35,000				
	$-(C=C)_4-$	300	52,000				
	$-(C=C)_5-$	330	118,000				
	$-(C=C)_2-$ (alicyclic)	230–260	3000–8000				
	$C=C-C\equiv C$	219	6500				
	$C=C-C=N$	220	23,000				
	$C=C-C=O$	210–250	10,000–20,000			300–350	weak
	$C=C-NO_2$	229	9500				
Benzene		184	46,700	202	6900	255	170
Diphenyl				246	20,000		
Naphthalene		220	112,000	275	5600	312	175
Anthracene		252	199,000	375	7900		
Pyridine		174	80,000	195	6000	251	1700
Quinoline		227	37,000	270	3600	314	2750
Isoquinoline		218	80,000	266	4000	317	3500

The unsubstituted conjugated *diene,* heteroannular or open chain, absorbs nominally near 214 nm unless both double bonds are contained within the same ring. A homoannular diene absorbs at 253 nm for 1,3-cyclohexadiene, at 228 nm for cyclopentadiene, and at 241 nm for 1,3-cycloheptadiene. An alkyl or O-alkyl group displaces the absorption band to the red about 5 nm, as does an exocyclic bond, a Cl- or a Br-substituent. A S-alkyl or another double bond in conjugation displaces the band 30 nm to the red. A red shift of 60 nm is associated with a N-alkyl group. Change of solvent has little effect. Heteroannular and acyclic dienes display molar absorptivities in the 8000–20,000 range, whereas homoannular dienes display values in the 5000–8000 range.

α, β-Unsaturated carbonyl compounds absorb in a similar spectral range to conjugated dienes. From the parent system, absorbing at 215 nm,

$$\underset{}{\overset{O}{\|}}\;\;\overset{\alpha}{C}-\overset{\beta}{C}=\overset{\gamma}{C}-\overset{\delta}{C}=C$$

the wavelength shifts to the red by 10 nm for an α-alkyl substituent, 12 nm for a β-alkyl substituent, 18 nm for γ- or δ-alkyl substituent, 39 nm if a homoannular enone system is present, and 30 nm for an exocyclic double bond or for a double bond extending the conjugation. Subtract 10 nm for a five-membered ring ketone, 5 nm for aldehydes, and 20 nm for carboxylic acids and esters. Although the influence of other substituents is not so predictable, typical red shifts in ethanol medium (in nm) are: —OH, α(35), β(30), δ(50); —OCOCH$_3$(6); —OCH$_3$, α(35), β(30), γ(17), δ(31); —S-alkyl, β(85); —Cl, α(15), β(12); —Br, α(25), β(30); and —N(alkyl)$_2$, β(95). Enones can be distinguished from the dienes since only the enones display significant solvent effects which take the form of a blue shift of the carbonyl band in water of 8 nm, and a red shift of 11 nm for aliphatic hydrocarbons and 5 nm for dioxane (with ethanol as reference). The molar absorptivity of cisoid enones is usually less than 10,000, whereas it is greater than 10,000 for transoid enones.

In an inert solvent, benzene shows absorption bands at 180–185 nm (ϵ = 40,000), 193–204 nm (ϵ = 5000), and 230–270 nm (B-band, ϵ = 250). In polynuclear aromatic systems all the bands are displaced to longer wavelengths and become more intense. Ring substitution affords red shifts and intensification of the spectrum. The absorption of the B-band moves to longer wavelengths in cases where the new substituent is electron donating or capable of conjugation. With electron-withdrawing substituents, practically no change in the maximum position is observed. The vibrational fine structure of the benzenoid spectrum tends to disappear on substitution, and if the spectrum is determined in solvents more polar than aliphatic hydrocarbons. Extremely useful and identifying solvent shifts are observed in the spectra of phenol between neutral and alkaline media, and in the aniline spectra between neutral and acidic media.

When electronically complementary groups are situated *para* to each other in disubstituted benzenes, there is a pronounced red shift in the main absorption band due to the extension of the chromophore from the electron-donating group through the ring to the electron-withdrawing group. When the *para* groups are not complementary, or when the groups are situated *ortho* or *meta* to each other, the observed spectrum is usually similar to that of the separate, noninteracting chromophores.

Multiple bonded oximes, nitriles, nitro, and azo compounds show no or only weak absorption. If conjugated, the transitions and absorption wavelengths resemble those of the dienes and enones.

Stereochemical Effects

In predicting the wavelength of maximum absorption, allowance must be made for the likely shape of the molecule and for any unusual strain. Distortion of the chromophore may lead to red or blue shifts, depending on the nature of the distortion. In extreme cases conjugation is almost completely inhibited through rotation of the system out of planarity, and the molecule absorbs as two distinct entities. Steric interactions can also cause a change in the absorption intensity of the main conjugation band. In *cis–trans* isomerism, generally the *cis*-configuration of a conjugated compound absorbs with lower intensity due to shorter distance between the ends of the chromophore.

A great deal of "negative" information may be deduced regarding molecular structures. If a compound is highly transparent throughout the region from 220 to 800 nm, it contains no conjugated unsaturated or benzenoid system, no aldehyde or keto group, no nitro group, and no bromine or iodine. If the screening indicates the presence of chromophores, the wavelength(s) of maximum absorbance are ascertained and tables are searched for known chromophores. Further information may be deduced from the shape, intensity, and detailed location of the bands. Finally, the absorption spectrum is compared with the spectra in standard compilations of ultraviolet–visible spectra.[18,19] Oftentimes, structural details can be inferred from the close resemblance of a compound's spectrum with that of a compound of known and related structure—for example, in petroleum ether, the spectra of toluene and chlorobenzene are similar.

PROBLEMS

1. In the transmission curve of an NaCl plate, opaque below the cutoff at 450 cm^{-1}, the presence of stray light is noted at 379 cm^{-1}. A similar tracing obtained using a CaF$_2$ plate which is opaque below 900 cm^{-1} shows the absence of the stray light. From the difference in the sample cutoff wavelengths, where did the stray light originate?

2. With a certain filter photometer, using a 510-nm filter and 2.00-cm cuvettes, the reading on a linear scale for I_0 was 85.4. With a $1.00 \times 10^{-4} M$ solution of a chromophore in the cuvette, the value of I was 20.3. Calculate the molar absorptivity.

3. The simultaneous determination of titanium and vanadium, each as their peroxide complex, can be done in steel. When 1.000-g samples of steel were dissolved, colors developed, and diluted to 50 ml exactly, the presence of 1.00 mg of Ti gave an absorbance of 0.269 at 400 nm and 0.134 at 460 nm. Under similar conditions 1.00 mg of V gave an absorbance of 0.057 at 400 nm and 0.091 at 460 nm. For each of the following samples, 1.000 g in weight and ultimately diluted to 50 ml, calculate the percent titanium and vanadium from these absorbance readings:

SAMPLE	A_{400}	A_{460}	SAMPLE	A_{400}	A_{460}
1	0.172	0.116	5	0.902	0.570
2	0.366	0.430	6	0.600	0.660
3	0.370	0.298	7	0.393	0.215
4	0.640	0.436	8	0.206	0.130
			9	0.323	0.177

4. The absorption spectra for tyrosine and tryptophan in $0.1F$ NaOH are shown in the illustration below. Select appropriate wavelengths for the simultaneous determination of each component in mixtures.

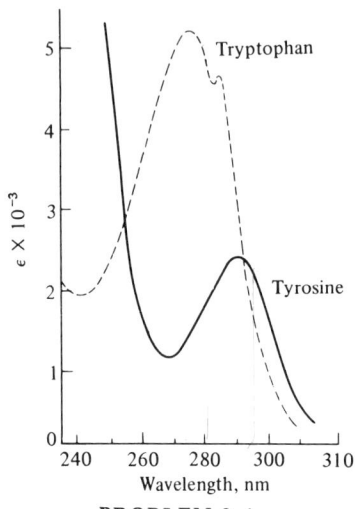

PROBLEM 3-4

5. A mixture of sodium acetate and o-chloroaniline solution, 10 ml each, were titrated in glacial acetic acid at 312 nm with a $0.1010N$ $HClO_4$ solution. Sodium acetate does not absorb in the ultraviolet portion of the spectrum, but it is a stronger base than o-chloroaniline. These results were obtained (corrected for dilution):

VOLUME OF TITRANT, ml	ABSORBANCE	VOLUME OF TITRANT, ml	ABSORBANCE
0.00	0.68	8.25	0.37
1.00	0.68	8.50	0.32
2.00	0.68	8.75	0.26
3.00	0.68	9.00	0.20
4.00	0.67	9.25	0.14
5.00	0.66	9.50	0.09
6.00	0.63	10.50	0.02
7.00	0.56	11.00	0.02
8.00	0.42	11.50	0.02

Plot the results and calculate the concentration of sodium acetate and of o-chloroaniline in the original aliquots.

6. In a photometric titration of magnesium with $0.00130M$ EDTA at 222 nm, the following procedure was employed. All reagents except the magnesium-containing solution were placed in the titration cell, and the slit width was adjusted to give zero absorbance. The following readings were observed after additions of the standard EDTA:

ABSORBANCE	EDTA ADDED, ml	ABSORBANCE	EDTA ADDED, ml
0.000	0.00	0.429	0.60
0.014	0.10	0.657	0.80
0.200	0.40	0.906	1.00

At this point, the magnesium solution was added and the absorbance fell to zero. The titration was continued with the following results:

ABSORBANCE	EDTA ADDED, ml	ABSORBANCE	EDTA ADDED, ml
0.000	1.00	0.360	3.00
0.020	1.50	0.580	3.20
0.065	2.00	0.803	3.40
0.160	2.50	1.000	3.60
0.240	2.80	1.220	3.80

Plot the results, explain the curves obtained, and calculate the number of micrograms of magnesium in the sample.

7. Sketch the photometric titration curve of phosphoric acid with sodium hydroxide, using methyl orange and phenolphthalein as indicators and monitoring the red color at 522 nm for the former, and at 553 nm for the latter.

8. Graph the data and determine the acid dissociation constant for these materials.

	p-NITROPHENOL $\epsilon \times 10^{-3}$			PAPAVERINE (CATION FORM) $\epsilon \times 10^{-4}$	
pH	3170 Å	4070 Å	pH	2390 Å	2510 Å
3.0		0.33	2.0	3.36	5.90
4.0	9.72	0.33	3.0	3.36	5.90
5.0	9.72	0.50	4.0	3.39	5.83
6.0	9.03	1.66	5.0	3.48	5.63
6.2	8.61	2.28	5.6	3.86	5.19
6.4	8.19	3.99	5.8	3.93	4.91
6.6	7.36	5.14	6.0	4.30	4.61
6.8	6.39	7.22	6.2	4.61	4.15
7.0	5.55	9.16	6.4	4.86	3.71
7.2	4.45	11.65	6.6	5.22	3.30
7.4	3.61	13.40	6.8	5.46	2.77
7.6	2.92	15.00	7.0	5.66	2.51
7.8	2.08	16.90	7.4	6.03	2.00
8.0	1.81	17.50	8.0	6.27	1.63
9.0	1.39	18.33	11.0	6.43	1.56
10.0	1.39	18.33	12.0	6.44	1.56

9. Determine the acid dissociation constant for each indicator from the absorbance, at the wavelength of maximum absorbance, measured as a function of pH. Ionic strength was 0.05.

BROMOPHENOL BLUE λ_{max} = 592 nm		METHYL RED λ_{max} = 530 nm		BROMOCRESOL PURPLE λ_{max} = 591 nm	
ABSORBANCE	pH	ABSORBANCE	pH	ABSORBANCE	pH
0.00	2.00	2.00	3.20	0.00	4.00
0.18	3.00	1.78	4.00	0.24	5.40
0.58	3.60	1.40	4.60	0.66	6.00
0.98	4.00	0.92	5.00	0.87	6.20
1.43	4.40	0.48	5.40	1.13	6.40
1.75	5.00	0.16	6.00	1.37	6.60
2.10	7.00	0.00	7.00	1.72	7.00
				2.00	8.00

10. A series of chromium(III) nitrate solutions were measured according to the ordinary method with 0% T set with the phototube darkened and 100% T set with the pure solvent. From the results obtained at 550 nm, (a) calculate the relative concentration error for each measurement, assuming $\Delta T = 0.004$; and (b) plot the results as relative concentration error versus absorbance. Calculations should be done for the two limiting cases: shot-noise limited and Johnson-noise limited.

CONCENTRATION, M	ABSORBANCE	CONCENTRATION, M	ABSORBANCE
Blank	0	0.0300	0.357
0.0050	0.060	0.0400	0.476
0.0100	0.119	0.0500	0.595
0.0150	0.179	0.0600	0.714
0.0200	0.238	0.0800	0.952
0.0250	0.298	0.1000	1.190
		0.1100	1.309

11. From the series of solutions used in Problem 10, the 0.0500M solution was used to set the 100% T reading. These results were obtained:

CONCENTRATION, M	TRANSMITTANCE	CONCENTRATION, M	TRANSMITTANCE
0.0500	1.000	0.110	0.230
0.0600	0.775	0.120	0.162
0.0700	0.584	0.130	0.120
0.0800	0.453	0.140	0.097
0.0900	0.380	0.150	0.074
0.1000	0.295		

Assuming $\Delta T = 0.004$, plot the relative concentration error versus transmittance.

12. In using the low absorbance method a 0.100M solution of chromium(III) nitrate was used as the standard with which the scale was set at 0% T by means of the zero-

suppressor (dark-current) control. Pure solvent was used for setting the 100% T point. These results were obtained:

CONCENTRATION, M	TRANSMITTANCE	CONCENTRATION, M	TRANSMITTANCE
Blank	1.000	0.0500	0.197
0.0100	0.745	0.0600	0.130
0.0200	0.546	0.0700	0.077
0.0300	0.420	0.0800	0.037
0.0400	0.286	0.0900	0.012

Assuming $\Delta T = 0.004$, plot the relative concentration error versus transmittance.

13. With the maximum precision method, a 0.0500M chromium(III) nitrate solution was used to set the 100% T point, and a 0.100M solution was used to set the 0% T point. These results were obtained:

CONCENTRATION, M	TRANSMITTANCE	CONCENTRATION, M	TRANSMITTANCE
0.0500	1.000	0.0800	0.247
0.0600	0.695	0.0900	0.128
0.0700	0.437	0.100	0.000

Assuming $\Delta T = 0.004$, plot the relative concentration error versus transmittance.

14. The absorbance of a series of phosphate solutions, using the phosphovanadomolybdate complex at 420 nm, are as follows when referred to a 5.0-mg phosphate solution as reference standard (actual $A = 1.075$):

mg P_2O_5/100 ml	ABSORBANCE
5.0	0.000
5.2	0.046
5.4	0.092
5.6	0.138
5.8	0.184
6.0	0.230
6.2	0.276

(a) Determine the increase in precision for the measurement of the 5.2-mg phosphate solution as compared with the normal method. (b) Estimate the relative concentration error for the 5.6-mg phosphate solution when the 6.0-mg phosphate solution is used to set the zero transmittance reading in addition to the 5.0-mg phosphate solution being used for the 100% transmittance reading.

15. A series of solutions is prepared in which the amount of iron(II) is held constant at 2.00 ml of $7.12 \times 10^{-4} M$, while the volume of $7.12 \times 10^{-4} M$, 1,10-phenanthroline is varied. After dilution to 25 ml, absorbance data for these solutions in 1.00-cm cuvettes at 510 nm are as follows:

1,10-PHENANTHROLINE, ml	ABSORBANCE
2.00	0.240
3.00	0.360
4.00	0.480
5.00	0.593
6.00	0.700
8.00	0.720
10.00	0.720
12.00	0.720

(a) Evaluate the composition of the complex. (b) Estimate the value of the formation constant of the complex.

16. The method of continuous variation was used to investigate the species responsible for the absorption at 510 nm when the indicated volumes of $6.72 \times 10^{-4} M$ iron(II) solution were mixed with sufficient $6.72 \times 10^{-4} M$, 1,10-phenanthroline to equal a total volume of 10.00 ml, after which the entire system was diluted to 25 ml. Cuvettes were 1.00 cm.

IRON(II), ml	ABSORBANCE	IRON(II), ml	ABSORBANCE
0.00	0.000	5.00	0.565
1.00	0.340	6.00	0.450
1.50	0.510	7.00	0.335
2.00	0.680	8.00	0.223
3.00	0.794	9.00	0.108
4.00	0.680	10.00	0.000

(a) Elucidate the composition of the ion-association complex. (b) Calculate the molar absorptivity of the complex.

17. Evaluate the composition of the iron(II)-1,10-phenanthroline complex with an absorption peak at 510 nm on the basis of the following absorbance data obtained, after dilution to 25 ml, in 1.00-cm cuvettes.

IRON(II) CONSTANT AT 5.00 ml of $7.00 \times 10^{-4} M$		LIGAND CONSTANT AT 10.00 ml of $2.10 \times 10^{-3} M$	
$7.00 \times 10^{-4} M$ LIGAND, ml	A	$7.00 \times 10^{-4} M$ IRON(II), ml	A
1.00	0.177	0.50	0.177
2.00	0.235	1.00	0.352
3.00	0.352	1.50	0.530
4.00	0.470	2.00	0.706
5.00	0.585	2.50	0.883

18. The dissociation of the complex between thorium and quercetin can be expressed as $ThQ_2 \rightleftharpoons Th + 2Q$ (omitting formal charges). For a solution that was $2.30 \times 10^{-5} M$ in thorium and contained a large excess of quercetin, sufficient to ensure that all of the

thorium is present as the complex, the absorbance was 0.780. When the same amount of thorium is mixed with a stoichiometric amount of quercetin, the absorbance was 0.520. Calculate (a) the degree of dissociation and (b) the value of the formation constant of the complex.

19. Mesityl oxide exists in two isomeric forms: $CH_3-C(CH_3)=CH-CO-CH_3$ and $CH_2=C(CH_3)-CH_2-CO-CH_3$. One exhibits an absorption maximum at 235 nm with a molar absorptivity of 12,000; the other shows no high-intensity absorption beyond 220 nm. Identify the isomers.

20. Assign the structures shown to the respective isomer on the basis of this information: the α-isomer shows a peak at 228 nm (ϵ = 14,000) while the β-isomer has a band at 296 nm (ϵ = 11,000).

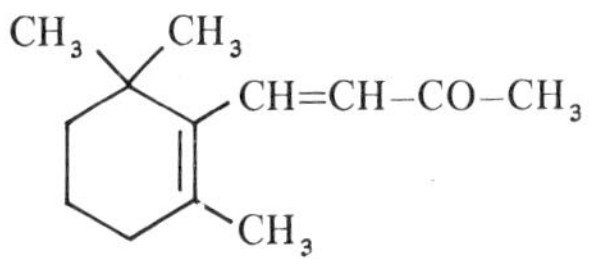

structure I 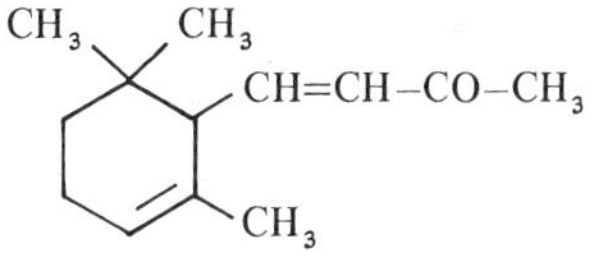   structure II

21. Explain how the ultraviolet spectrum can be used to decide between the following isomeric systems:

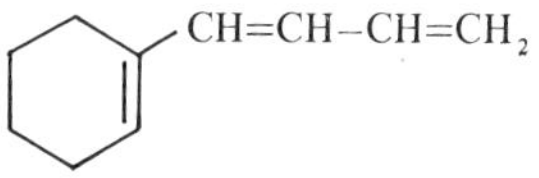

(a) 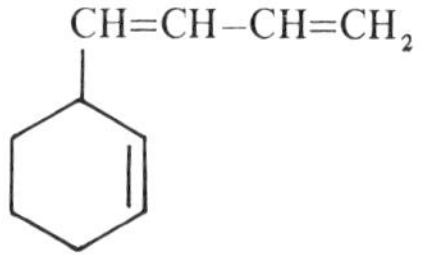   (b)

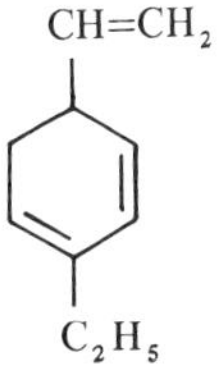

(c) 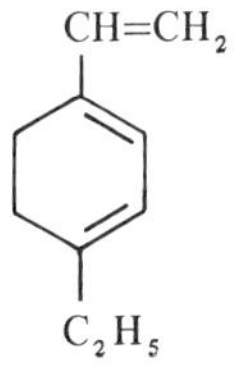   (d)

BIBLIOGRAPHY

Bauman, R. P., *Absorption Spectroscopy*, Wiley, New York, 1962.
Forbes, W. R., in S. K. Freeman, Ed., *Interpretive Spectroscopy*, Chap. 1, Reinhold, New York, 1965.
Jaffe, H. H. and M. Orchin, *Theory and Applications of Ultraviolet Spectroscopy*, Wiley, New York, 1962.
Olsen, E. D., *Modern Optical Methods of Analysis*, McGraw-Hill, New York, 1975.
West, W., Ed., *Chemical Applications of Spectroscopy*, Vol. IX, Part 1, 2nd rev. ed., Wiley, New York, 1968.

LITERATURE CITED

1. Kortum, G. and M. Seiler, *Angew. Chem,* **52**, 687 (1939).
2. Erickson, J. O. and T. Surles, *Am. Lab.,* p. 41 (June 1976).
3. Cook, R. B. and R. Jankow, *J. Chem. Educ.,* **49**, 405 (1972).
4. Slavin, W., *Anal. Chem.,* **35**, 561 (1963).
5. Bastian, R. *Anal. Chem.,* **21**, 972 (1949).
6. Bastian, R., R. Weberling, and F. Palilla, *Anal. Chem.,* **22**, 160 (1950).
7. Bastian, R., *Anal. Chem.,* **23**, 580 (1951).
8. Hiskey, C. F., *Anal. Chem.,* **21**, 1440 (1949).
9. Hiskey, C. F., J. Rabinowitz, and J. G. Young, *Anal. Chem.,* **22**, 1464 (1950).
10. Crawford, M., *Anal. Chem.,* **31**, 343 (1959).
11. Reilley, C. N. and C. M. Crawford, *Anal. Chem.,* **27**, 716 (1955).
12. O'Haver, T. C., *Anal. Chem.,* **51**, 91A (1979).
13. Goddu, R. F. and D. N. Hume, *Anal. Chem.,* **26**, 1679, 1740 (1954).
14. Headridge, J. B., *Photometric Titrations,* Pergamon, New York, 1961.
15. Hunter, R. S., *Off. Dig. Feder. Soc. Paint Technol.,* **35**, 350 (1963).
16. Surles, T., J. O. Erickson, and D. Priesner, *Am. Lab.,* p. 55 (March 1975).
17. Wendlandt, W. W., *J. Chem. Educ.,* **45**, A861, A947 (1968).
18. Lang, L., *Absorption Spectra in the Ultraviolet and Visible Region,* Academic, New York, 1961.
19. Sadtler Research Laboratories, *Ultraviolet Reference Spectra,* Philadelphia.

CHAPTER 4

Fluorescence and Phosphorescence Spectrophotometry

Luminescence is the term applied to the reemission of previously absorbed light. This chapter is concerned with molecular photoluminescence wherein photons of electromagnetic radiation excite molecules. Since molecules that have absorbed light are in a higher electronic state, they must lose their excess energy to return back to the ground state. If the excited molecule returns to the ground state by emitting light, it exhibits photoluminescence. The decay time of fluorescence is of the same order of magnitude as the lifetime of an excited singlet state (10^{-9} to 10^{-7} sec). Phosphorescence lifetimes usually fall in the range from 10^{-4} to 10 sec. The most striking difference is the conditions under which each type of photoluminescence is observed. Fluorescence usually is seen at moderate temperature in liquid solution. Phosphorescence is seen in rigid media, usually at very low temperatures.

As discussed in Chapter 1, a rough "mirror-image" relationship exists between the excitation spectrum and the photoluminescence spectrum (see Fig. 4-1). The intensity of the photoluminescence spectrum depends on the excitation wavelength, although its spectral position does not. The photoluminescence spectrum appears at longer wavelengths than the excitation spectrum. This phenomenon arises because the excitation process requires an amount of energy equal to the electronic energy change plus a vibrational energy increase; conversely, each deexcitation yields the electronic excitation energy minus a vibrational energy increase. Fluorescence and phosphorescence differ in that the wavelength of the fluorescence emission is always shorter than that of phosphorescence. Emphasis in this chapter will be on fluorescence spectroscopy, which is the more widely used of the two photoluminescence methods.

Fluorescence spectroscopy has assumed a major role in analysis, particularly in the determination of trace contaminants in our environment, industries, and bodies, because

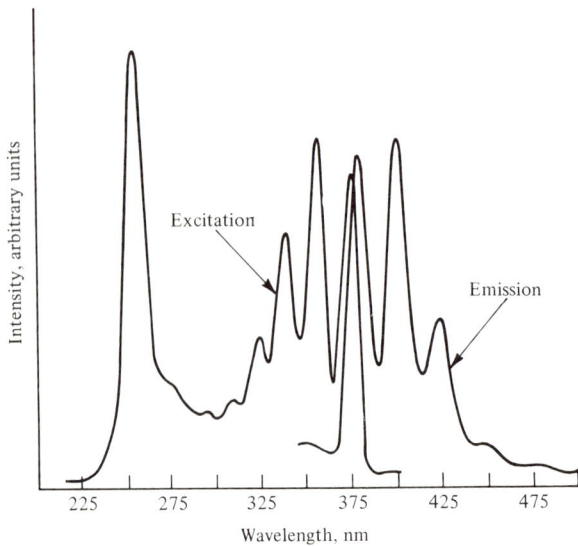

FIGURE 4-1 Excitation and fluorescence emission spectra of 0.3 µg/ml anthracene in methanol.

for applicable compounds fluorescence gives high sensitivity (in the low parts per trillion) and high specificity. High sensitivity results from the difference in wavelength between the exciting and fluorescence radiation. This results in a signal contrasted with essentially zero background; it is always easier to measure a small signal directly than as a small difference between two large signals as is done in absorption spectrophotometry. High specificity results from dependence on two spectra: the excitation and the emission spectrum. Two compounds that are excited at the same wavelength but emit at different wavelengths are readily differentiated without the use of chemical separation techniques. Also, a fluorescent compound in the presence of one or more nonfluorescent compounds is readily analyzed fluorometrically even when the compounds have overlapping absorption spectra. Even nonfluorescent or weakly fluorescent compounds can often be reacted with strong fluorophores enabling them to be determined quantitatively (see derivatization in Chapter 17). The phenomenon of fluorescence itself is subject to more rigorous constraints on molecular structure than is absorption. Many drugs possess rather high quantum efficiencies for fluorescence, such as quinine and lysergic acid diethylamide (LSD). As little as 1 ng/ml of the latter can be detected in a 5-ml sample of blood plasma or urine. Carcinogens, such as benzopyrene, are easily determined fluorometrically in air-pollution analysis.

To ascertain the excitation and emission spectra (Fig. 4-1) the following procedure is commonly followed. The excitation monochromator is varied until fluorescence occurs; often this can simply be observed visually. The excitation monochromator is then set at this wavelength (or at any point within the excitation wavelength band) and the emission monochromator is allowed to scan, recording the emission spectrum. The emission monochromator is then set at the wavelength at which maximum fluorescence occurred, and the excitation monochromator is now allowed to scan and the excitation spectrum is recorded. In turn, the final emission spectrum is obtained by setting the excitation monochromator at the maximum excitation wavelength and again scanning with the emission monochromator. Also one can obtain the conventional absorption spectrum of the compound by scanning the excitation monochromator through the absorption wavelengths with the emission monochromator set at the maximum wavelength of the emission spectrum. For analytical applications the emission spectrum is used. Often an excitation spectrum is first made in order to confirm the identity of the substance and to select the optimum excitation wavelength.

4.1 STRUCTURAL FACTORS[1-3]

Fluorescence may be expected generally in molecules that are aromatic or contain multiple-conjugated double bonds with a high degree of resonance stability. Both classes of substances have delocalized π-electrons that can be placed in low-lying excited singlet states. In polycyclic aromatic systems where the number of π-electrons available is greater than in benzene, these compounds and their derivatives are usually much more fluorescent than benzene and its derivatives. Substituents strongly affect fluorescence. A substituent that delocalizes the π-electrons, such as $-NH_2, -OH, -F, -OCH_3, -NHCH_3$, and $-N(CH_3)_2$ groups, often enhances fluorescence because they tend to increase the transition probability between the lowest excited singlet state and the ground state. Electron-withdrawing groups containing $-Cl, -Br, -I, -NHCOCH_3, -NO_2$, or $-COOH$ decrease or quench completely the fluorescence. Thus, aniline fluoresces but nitrobenzene does not.

Molecules with a nonbonding pair of valence electrons, for example, an amine with a lone electron pair on its nitrogen atom, often fluoresce. Such electrons can be promoted without disruption of bonding. In general, a delocalized π-system must also be part of this type of a molecule to ensure easy fluorescence.

Molecular rigidity lessens the possibility of competing nonradiative transitions by decreasing vibrations; this minimizes intersystem crossing to the triplet state and collisional heat degradation. For example, fluorescein and eosin are strongly fluorescent, but a similar compound phenolphthalein, which is nonrigid and where the conjugate system is disrupted, is not fluorescent. Given a series of aromatic compounds, those that are the most planar, rigid, and sterically uncrowded are the most fluorescent. The formation of chelates with metal ions, in general, also promotes fluorescence by promoting rigidity and minimizing internal vibrations. In this same sense, substances fluoresce more brightly in a glassy state or in viscous solution. The probability of intermolecular energy transfer between the fluorescer and other molecules tends to be reduced when working at low temperature and in a medium of high viscosity in which the rotational relaxation time of the fluorescer is much longer than the lifetime of the excited state.

Fluorescence intensity and wavelength often vary with solvent. Solvents capable of exhibiting strong van der Waal's binding forces with the excited-state species will prolong the lifetime of a collisional encounter and favor deactivation. Solvents which possess molecular substituents such as Br, I, NO_2, or —N=N— groups are undesirable because the strong magnetic fields which surround their bulky atomic cores promote spin decoupling of electrons and triplet state formation, giving rise to marked fluorescence quenching although these same solvents may promote phosphorescence. Indole illustrates the wavelength shifts that may occur in different solvents. Although the excitation wavelength remains at 285.0 nm in each solvent, the wavelength of maximum fluorescence is 297.0 nm in cyclohexane, 305.0 nm in benzene, 310.0 nm in 1,4-dioxane, 330.0 nm in ethanol, and 350.0 nm in water.

Changes in the system pH, if it affects the charge status of the chromophore, may influence fluorescence. Both phenol and anisole fluoresce at pH 7, but at pH 12 phenol is converted into the nonfluorescent anion, whereas anisole remains unchanged. Similarly aniline fluoresces in the visible at pH 7 and 12, but the protonated cation is nonfluorescent at pH 2. These observations can be explained by comparing resonance forms of anions and cations. For example, aniline in acid solution has the positive charge fixed at the nitrogen atom, and the anilinium ion has only the same resonance forms as benzene which fluoresces only in the ultraviolet region. However, in neutral or basic solution aniline has three additional resonance structures, resulting in a more stable excited singlet state and a longer wavelength of fluorescent radiation. In fact, some substances are so sensitive to pH that they can be used as indicators in acid–base titrations. The merit of such indicators is that they can be employed in turbid or intensely colored systems.

Frequently, weakly fluorescent or nonfluorescent aromatic compounds are strongly phosphorescent. Usually this indicates the involvement of $n-\pi$ absorption transitions. Such $n-\pi$ excited states are observed to have smaller energy gaps between the excited singlet and triplet levels and have longer excited-state lifetimes. These conditions favor population of the triplet state and lead to phosphorescence. Carbonyl-substituted aromatic compounds frequently exhibit this behavior.

Introduction of paramagnetic metal ions, such as copper(II) and nickel(II), gives rise to phosphorescence but not fluorescence in metal complexes. By contrast, magnesium and zinc compounds show only strong fluorescence. Generally only those cations which are diamagnetic when coordinated and are nonreducible will form fluorescent complexes. The transition metals with unfilled outer d orbitals will quench fluorescence completely. On the other hand, whereas paramagnetic species quench fluorescence they strongly promote intersystem crossing so that at low temperature those cations will be observed to promote phosphorescence.

Phosphorescence lifetimes are also affected by molecular structure. Unsubstituted cyclic and polycyclic hydrocarbons and their derivatives containing CH_3, NH_2, OH, COOH, and OCH_3 substituents have lifetimes in the range of 5–10 sec for most benzene derivatives, and 1–4 sec for many naphthalene derivatives. The nitro group diminishes the intensity of phosphorescence and the lifetime of the triplet state to about 0.2 sec. Aldehydic and ketonic carbonyl groups diminish the lifetime to about 0.001 sec. Introduction of bulky substituents which force a planar configuration to become nonplanar markedly shortens lifetimes.

4.2 PHOTOLUMINESCENCE INTENSITY AS RELATED TO CONCENTRATION

The quantitative relationship between fluorescence (or phosphorescence) intensity and concentration may be derived from Beer's law. The fraction of light transmitted through the solution is

$$I/I_0 = e^{-\epsilon b C} \tag{4-1}$$

and the corresponding fraction of light absorbed is

$$1 - (I/I_0) = 1 - e^{-\epsilon b C} \tag{4-2}$$

Rearranging, the fraction of light absorbed by the sample is

$$I_0 - I = I_0(1 - e^{-\epsilon b C}) \tag{4-3}$$

The total photoluminescence intensity is proportional to the quanta of light absorbed and to the photoluminescence efficiency, ϕ, which is the ratio of quanta absorbed to the quanta emitted; that is, the fraction of excited species that fluoresces rather than undergoes intersystem crossing. Thus, for fluorescence

$$F = (I_0 - I)\phi_F f(\theta) g(\lambda) = I_0 \phi_F (1 - e^{-\epsilon b C}) f(\theta) g(\lambda) \tag{4-4}$$

where $f(\theta)$ represents the geometrical factor and is determined by the solid angle of the fluorescing radiation subtended by the detector (photoluminescence is emitted in all directions but is viewed only through a limited aperture), and $g(\lambda)$ defines the efficiency of the detector as a function of the fluorescent wavelength incident on it. Equation 4-4 can be written in the following exponential power series as

$$F = I_0 \phi_F f(\theta) g(\lambda) \left[2.3\epsilon b C - \frac{(2.3\epsilon b C)^2}{2!} + \frac{(2.3\epsilon b C)^3}{3!} - \ldots + \frac{(2.3\epsilon b C)^n}{n!} \right] \tag{4-5}$$

For very dilute solutions in which not over 2% of the total excitation energy is absorbed, and the term $\epsilon b C$ is not greater than 0.05 in fluorescence (and 0.01 in phosphorescence), Eq. 4-5 simplifies to

$$F = 2.3 I_0 \phi_F f(\theta) g(\lambda) \epsilon b C \tag{4-6}$$

A similar expression can be written for phosphorescence.

Of particular interest in Eqs. 4-5 and 4-6 is the linear dependence of fluorescence on the excitation intensity. This means sensitivity can be increased by working at high excitation intensities to give large signal-to-noise ratios. Since the source intensity can change from time to time, fluorescence signals are not measured as absolute parameters; rather they are expressed in terms of relative fluorescence. All measurements are made relative to reference standards of known concentration. Analytical curves are constructed to relate relative fluorescence to concentration. All readings must be corrected for background fluorescence. A major advantage of photoluminescence instrumentation is the ability to adjust the sensitivity of instruments, so that after a standard of known concentration is

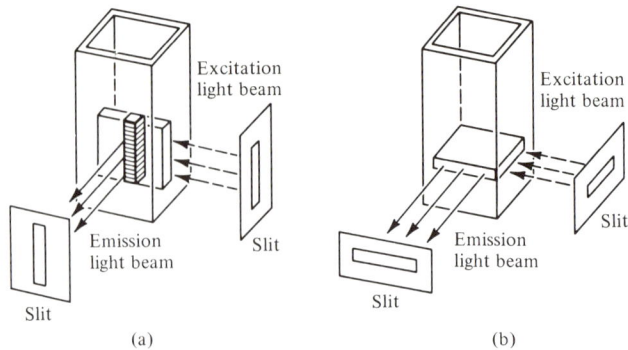

FIGURE 4-2 Illumination of a sample solution for a given spectral slit width: (a) conventional configuration and (b) horizontal beam focused on the sample cell after image rotation, an arrangement requiring only 0.6 ml of sample. (Courtesy of Perkin-Elmer Corp.)

measured, the instrument can be readily made to read directly in concentration. In photoluminescence photometry, the b-term in Eq. 4-5 is not the path length of the cell, but is the solid volume defined by the excitation and emission slit widths together. Therefore, slit widths are the critical factor and not the cell dimension (Fig. 4-2).

A plot of fluorescence (or phosphorescence) versus concentration, shown in Fig. 4-3, is often found to be linear over two or more decades, but there are limiting factors; namely, the blank fluorescence and quenching. The minimum detectable quantity of an analyte is generally limited by the magnitude of the blank. Solvent fluorescence and light scattering produce signals that at some point obscure the fluorescence of the analyte. According to the general expression for fluorescence, a sharp negative deviation will be exhibited at high absorbance values and concentrations (Eq. 4-4). Although a sufficiently dilute concentration has been stated as a concentration having an absorbance less than 0.05, a stricter limitation would limit the value to 0.02 or less. There is slight curvilinearity between 0.02 and 0.05 absorbance units. At higher concentrations self-quenching and self-absorption may also be responsible for negative deviation. Self-quenching results when

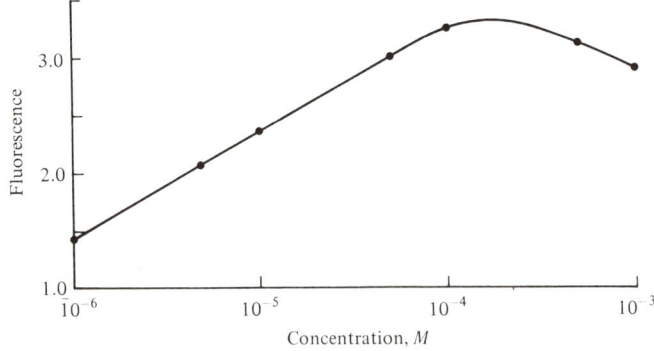

FIGURE 4-3 Fluorescence/concentration graph for the coenzyme NADH in distilled water solution. The linear portion of the curve extends from about 10^{-4} to $10^{-8} M$.

fluorescing molecules collide with each other and lose their excitation energy by radiationless transfer. Collisional impurity quenching leads to loss of fluorescence because of the formation of an excited complex between the excited analytical species and a ground-state impurity molecule and subsequent nonradiative energy losses. Dissolved oxygen, being paramagnetic, is a particularly serious offender. Heavy atoms or paramagnetic species strongly affect the rate of intersystem crossing which, in turn, alters the quantum efficiency for fluorescence or phosphorescence. To avoid quenching in most fluorometric procedures, paramagnetic species must be excluded from the sample solution.

Energy transfer quenching occurs when an impurity is present whose first excited singlet state is at an energy below that of the excited singlet state of the analytical species. A nonradiative transfer can occur followed by a further radiationless loss by the impurity. Aromatic substances are prime offenders in this category. "Spec-pure" solvents are essential for careful work. Self-absorption occurs when the emission and absorption wavelengths have some overlap, so that the emitted fluorescence radiation undergoes absorption in the solution, the so-called inner filter effect. When the sample is sufficiently concentrated to observe nonlinearity, the fluorescence will occur only in the earlier portion of the excitation path. If the excitation energy is absorbed before reaching the region where the fluorescing sample is observed by the emission photometer, the observed fluorescence is decreased proportionately in the region of the cell viewed by the emission photometer. A sample concentration this high would require front surface measurement, if dilution is not feasible.

4.3 INSTRUMENTATION

A generalized luminescence instrument is illustrated in Fig. 4-4. It consists of (1) a source of light, (2) a primary filter or excitation monochromator, (3) a sample cell, (4) a secondary filter or emission monochromator, (5) a photodetector, and (6) a data readout device. In contrast to ultraviolet visible instrumentation, two optical systems are necessary.

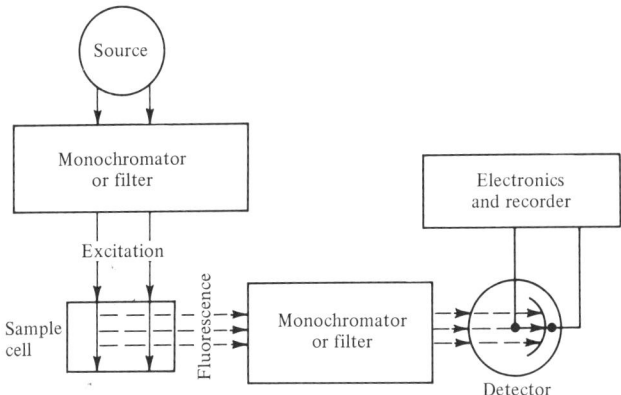

FIGURE 4-4 Basic components of fluorescence instrumentation.

The primary filter or excitation monochromator selects specific bands or wavelengths of radiation from the light source and directs them through the sample in the sample cell. The resultant luminescence is isolated by the secondary filter or emission monochromator and directed to the photodetector which measures the intensity of the emitted radiation. For the observation of phosphorescence a repetitive shutter mechanism will be required.

In the following discussion, fluorescence instruments will be categorized as filter fluorometers (also spelled fluorimeters but not the official IUPAC recommendation), spectrofluorometers, and compensating spectrofluorometers. The prefix "spectro-" implies that at least one dispersive monochromator is used in the instrument, usually as the excitation monochromator. A spectrofluorometer has the advantage over a filter instrument of being able to measure the spectral distribution of fluorescence emission (emission spectrum) and the variation in emission spectral radiance with excitation wavelength (excitation or fluorescence spectrum).

Sample Cell Geometry

There are three arrangements for illuminating and viewing the sample: the right-angle (90°) method, the frontal (37°) method, and the straight-through (transmission) method. Figure 4-5 illustrates the first two methods. The right-angle geometry is used almost exclusively in commercial instruments. The 90° geometry is efficient, because none of the sample cuvette surfaces directly illuminated by the excitation beam are viewed by the emission monochromator, hence no cuvette fluorescence (from trace uranium content of some glasses and from quartz) or light reflected from these surfaces enters the emission monochromator. Scattered light can originate only from the bulk of the solution itself. In the 90° viewing mode the excitation radiation does pass through a fairly long solution path so that there is an upper concentration limit that may be observed before the inner filter effect disrupts the linear relationship between luminescence intensity and solute

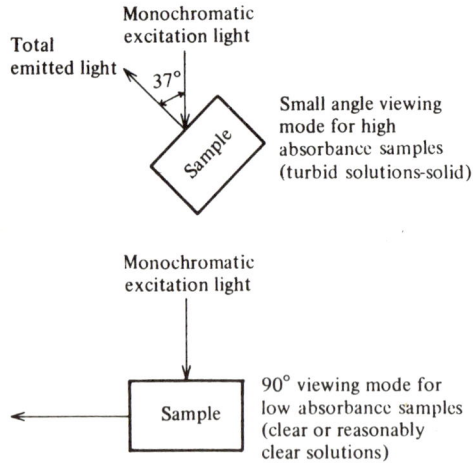

FIGURE 4-5 Viewing modes in fluorescence.

concentration. The arrangement of entrance and exit slits, as shown in Fig. 4-2, will influence the effective path length (or critical volume viewed).

The frontal method of cell illumination is used primarily for semiopaque materials or solids, or for solutions that are highly absorbing. The disadvantage of this configuration is that the emission monochromator views directly illuminated cell surfaces with their own residual fluorescence and unavoidable reflected light. Reflected light is minimized by the choice of 37° as the take-off angle.

The straight-through method is seldom used. It formerly enjoyed some popularity when working with fluorescing pellets after cooling the fusion fluxes in the determination of uranium with $LiF-Na_2CO_3$.

Sources

The primary factors to consider when selecting a light source for luminescence instrumentation are lamp intensity, wavelength distribution of emitted light, and stability. Of particular interest from Eq. 4-4 is the linear dependence of luminescent signal on the intensity of the exciting radiation. Consequently, it is advantageous to use a source as powerful as possible. Scanning spectrofluorometers require light sources that emit light continuously over a wide spectral range. Filter fluorometers or spectrofluorometers, used specifically for analytical measurements at fixed wavelengths, may utilize atomic spectral line sources.

High-pressure dc xenon arc lamps are used in nearly all commercial spectrofluorometers. The xenon lamp emits an intense and relatively stable continuum of radiation which extends in a continuous fashion from 300 to 1300 nm. Several strong emission lines lie between 800 and 1100 nm. The spectral output approximates that of a blackbody radiator with the equivalent color temperature of about 6000°K (see Fig. 2-1). During lamp operation the arc discharge is compressed within the narrow gap between the electrodes. Arc flicker determines the short-term stability, about 0.3%. Long-term stability is 1% drift per hour and is limited by electrode wear and arc wander. In contrast with mercury arcs, the spectral distribution is not as strongly dependent on operating gas pressure and voltage.

A xenon flash lamp provides a compact, low cost source. The sample is excited by a high-energy flash produced by discharge of a charged capacitor through a lamp filled with xenon. By making the flash repetitive, ac methods of amplification can be used and full advantage taken of the high peak flash intensity. Use of a 0.8-mm diameter capillary flash lamp has obvious advantages for microcell and continuous flow arrangements since the image produced at the sample position is about 2 mm wide and 18 mm high.

Low-pressure mercury vapor lamps are most frequently used in the filter fluorometers. The fill gas pressure is quite low (approximately 10 torr). The resulting arc discharge is spatially very diffuse and much less intense than that of a high-pressure arc. The stability is generally better. These lamps may be coated with a phosphor to emit a more nearly continuous spectrum, or may have a clear bulb of ultraviolet-transmitting material to permit use of the individual mercury lines which appear at (intensity) 253.7 (very strong), 313 (m), 365 (m), 404.7 (m), 407.8 (w), 435.8 (s), 546.1 (s), 577.0 (m), and 579.1 nm (m). Interference filters can be used to select individual mercury lines for excitation, or bandpass filters may be used to select for several lines. The discrete line emission output of the mercury lamp may show considerable sensitivity for a material whose excitation spectrum coincides with a mercury emission line but less sensitivity for a more strongly fluorescent compound which is excited more effectively with another excitation wavelength but one not appearing in the mercury lamp output.

Clearly, a laser would also be an acceptable source, provided monochromatic light were needed for excitation. Ultratrace inorganic ion determination is an area where laser excited fluorescence is applicable.[4]

Filter Fluorometers

Filter fluorometers offer the advantage of lower cost and convenience for repetitive, quantitative determinations. In this class of instruments, excitation and emission wavelengths are chosen by absorption or interference filters. For repetitive routine analyses the lack of versatility is not a major drawback and the high sensitivity, if the excitation wavelength is at an emission line of the light source, may be a distinct advantage. The small physical size of filters, as compared with monochromators, results in a compact instrument.

A filter fluorometer consists usually of a mercury lamp as an excitation source, a primary filter to transmit the desired excitation wavelength, and a sample cuvette. A photomultiplier tube measures the fluorescence emission. The secondary filter between the sample and photodetector is selected to transmit the fluorescence and to absorb scattered excitation radiation. A typical arrangement is shown in Fig. 4-6. If flow cells are incor-

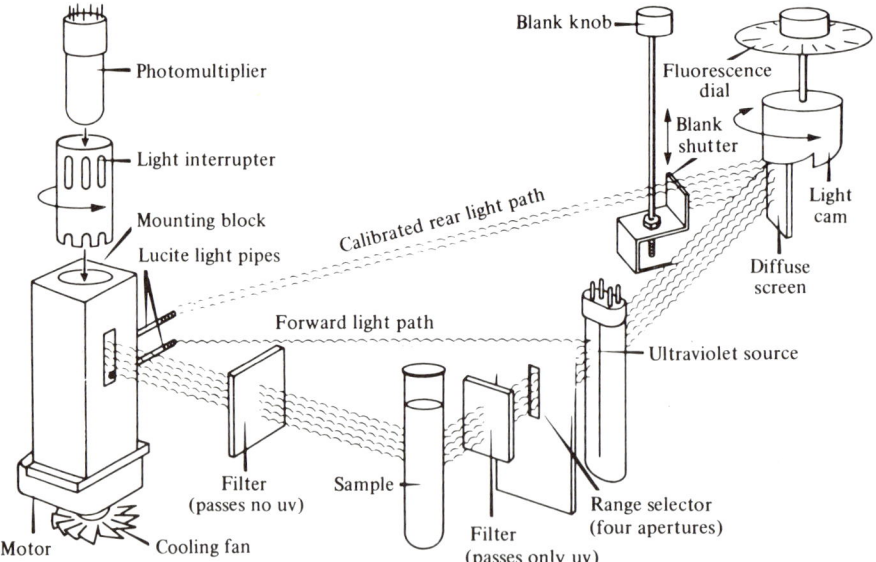

FIGURE 4-6 Optical diagram of a filter fluorometer, Turner model 110. (Courtesy of Turner Instrument Co.)

porated into the sample compartment, they can be part of continuous flow instrumentation such as high-performance liquid chromatographic (HPLC) detection systems (Chapter 17).

The use of optical filters results in very high excitation light levels and efficient light detection, but sacrifices the selectivity that can be obtained by more precise selection of excitation and emission wavelengths accomplished with monochromators. Excitation filters are generally bandpass-types which transmit a rather broad band of wavelengths although interference filters find use to isolate single mercury lines. Off-band transmission, hence stray light level, is usually quite high for interference filters when compared to absorption filters. This is attributable to unavoidable pin holes and defects in the film coatings. Emission filters are usually of the sharp cutoff-type which pass long wavelengths and attenuate shorter wavelengths. A point to remember is that bandpass filters have more than one transmission band or window. Photodetectors often have a low level response over a wide wavelength range. This may result in a high stray light level. Sharp cutoff filters, particularly glass filters, are frequently fluorescent themselves.

Filter fluorometers almost always use a partial double-beam arrangement to lessen effects of fluctuations and drift in source intensity and detector response. A variety of ingenious ways for obtaining the monitoring channel have been devised. One circuit arrangement which cancels out variations in line voltage, excitation radiation, and photomultiplier sensitivity is shown in Fig. 4-6. The optical bridge circuit measures the light differential between the fluorescent emission and a standard calibrated (rear light path) beam. Rotation of the fluorescence readout dial and connected diffuse screen adjusts the calibration beam to equal the sample emission intensity. When balanced, the photomultiplier detects no difference in signal intensity and thus provides an optical null balance. Another ratio system is shown in Fig. 4-7. The optical path of the reference beam is shown on the right side of the diagram. A portion of the lamp radiation passes through the primary optical system and is attenuated by means of the reference aperture disk before

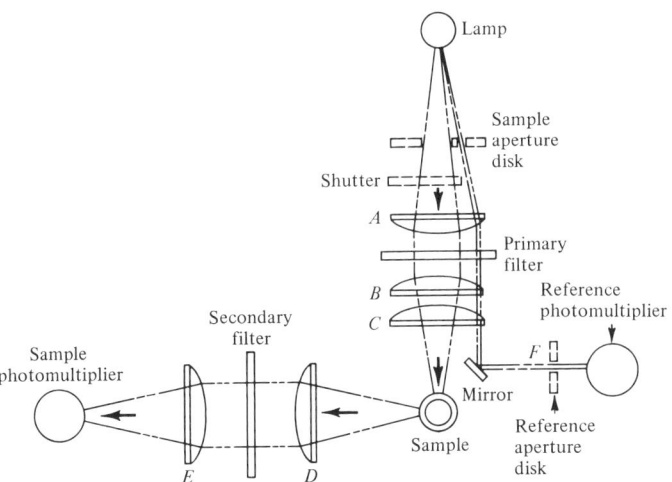

FIGURE 4-7 Optical diagram for Farrand Ratio Fluorometer. (Courtesy of Farrand Optical Co., Inc.)

reaching the reference photodetector. The electrical signal from the reference photodetector and the sample signal from the sample photodetector are fed into a solid-state electronic divider which performs the computation of ratioing the sample-to-reference signals.

For a filter fluorometer (or for any spectrofluorometer operated at constant excitation and emission wavelengths), Eq. 4-6 is reduced to $F = kC$. The constant k may be established by calibration with standards. The dial or meter reading is proportional to concentration. Measurements may be extended to extremely low concentrations by increasing the sensitivity of the photodetector or the intensity of the light source.

Fluorescence measurements are usually made by reference to some arbitrarily chosen standard. The standard is placed in the instrument and the circuit balanced with the reading scale at some chosen setting. Without readjusting any circuit components, the standard is replaced by standard reference solutions of the analyte and the fluorescence of each recorded. Finally, the fluorescence of the solvent and cuvette alone is measured to establish the true zero concentration. Some fluorometers are equipped with a zero-adjust circuit (see Fig. 4-6). A plot of fluorescence readings against concentration of the reference solutions furnishes the calibration curve. Some of the commonly used fluorescence standards are rhodamine B in ethylene glycol, quinine bisulfate in 0.1 N H_2SO_4, tryptophan in water, and anthracene in cyclohexane or ethanol. Glass reference filters are also suitable.

Spectrofluorometers

The greatest analytical scope of fluorescent analysis is achieved by replacing filters with grating monochromators to give scannable wavelength selection throughout the region from 200 to 800 nm, the most useful region for the fluorescence technique. Another advantage of spectrofluorometry is that the analyst can see the presence of scattered light by comparing the sample spectrum to a blank spectrum and examining for distortion of the fluorescence peak or the presence of additional peaks. A filter photometer cannot optically resolve the various Rayleigh and Raman scatter peaks, and the analyst does not observe their presence.[5]

Spectrofluorometers usually incorporate grating monochromators of the Czerny–Turner-type with a 0.25-m focal length and $f/4$ or $f/5$ aperture. Both monochromators use gratings with 600 grooves/mm, blazed (in the first order) for 300 nm in the excitation unit and for 500 nm in the emission unit. Filters are used to block out higher-order diffracted light. The basic concepts of a spectrofluorometer are illustrated in Fig. 4-8. The excitation monochromator is located between the light source and the sample and the emission monochromator between the sample and the photomultiplier tube. For quantitative analysis, one selects the desired excitation and emission wavelengths and compares the relative fluorescence intensities of standard and unknown samples. For good spectral selectivity and ability to resolve spectral fine structure, the emission monochromator should be able to resolve two lines 1.0 nm apart.

Instruments (Fig. 4-9a) that measure the characteristics of the sample without a continuous comparison to a reference standard possess the same major drawbacks which appear in attempting to record absorption spectra with a single-beam instrument. If the light source is unstable and varies in intensity, it can cause the appearance of false peaks

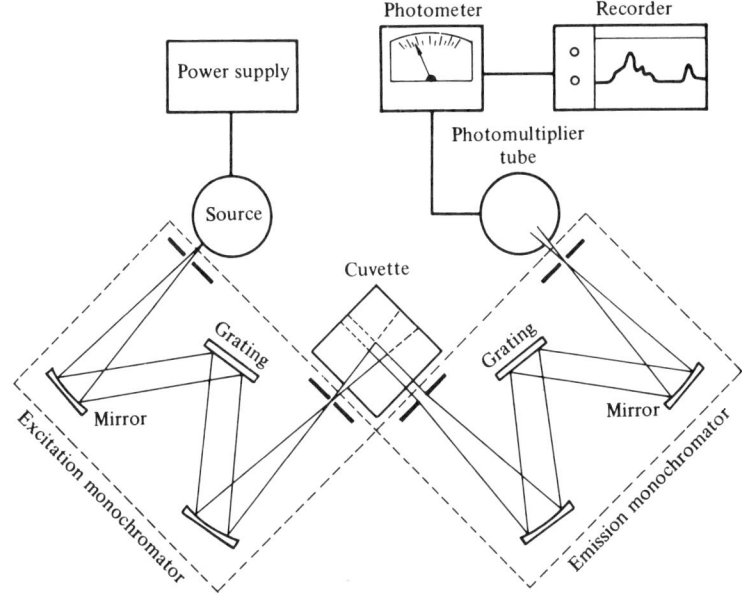

FIGURE 4-8 Schematic diagram of a fluorescence spectrophotometer.

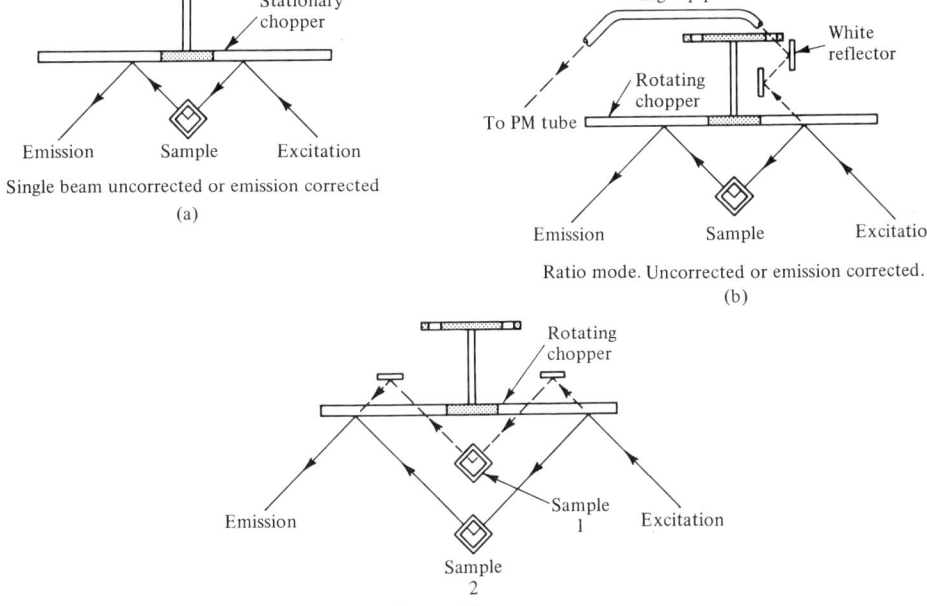

FIGURE 4-9 Different operational modes in fluorescence instrumentation shown schematically.

in the spectrum. Equally serious is the variation of the sensitivity of the photodetector with regard to wavelength. The term "uncorrected" is applied to this type of spectrofluorometer because the excitation and emission spectra presented are a combination of the true spectra of a compound and various instrumental artifacts. When excitation spectra are plotted, no attempt is made to hold I_0 constant. Excitation spectra are hence a composite of the true excitation spectra, the spectral distribution of lamp output with wavelength, and excitation monochromator efficiency with wavelength. Likewise, emission spectra are a composite of the true emission spectra, the spectral distribution of detector response, and emission monochromator efficiency with wavelength. In certain regions of the spectrum these instrumental factors become dominant. Despite these drawbacks, the uncorrected spectrofluorometer may be used for quantitative analyses with the convenience that wavelengths can be dialed and may be used for rough comparisons of spectra with other laboratories and direct comparison within one laboratory.

Instrumental limitations caused by the instability of the xenon source have been overcome by the *ratio mode* operation (Fig. 4-9b). A small fraction of the exciting light is directed to a reference photodetector which is chosen primarily for wide wavelength response. The emitted light is isolated by the emission photodetector specially selected for low dark current and high sensitivity. The output signal of the reference phototube is amplified and fed into the photomultiplier dynode voltage control circuit where it is used as a monitor to control the dynode voltage of both photomultiplier tubes in an inverse relationship. As the excitation radiation increases or decreases, due to arc fluctuations in the xenon lamp, there is a concomitant increase or decrease in relative fluorescence. The burden of correction is placed on the reference detector because it is the only component in the system with dynamic range and fast response time necessary for monitoring and adjusting the signal generated by the monochromatic excitation radiation. Since the signal is monitored after the excitation monochromator, excitation spectra are compensated for wavelength dependent energy fluctuations. It does not provide a true corrected excitation spectra however.

As the chopper rotates, all radiation is blocked from one or the other photomultiplier tube. Any signal now leaving the blocked-off detector is dark current inherent in the detector. The two signals, now separated, are fed into a differential amplifier which effectively subtracts dark current and presents only the difference signal to the electronics of the photometer. Thus a difference signal is quite simply obtained both from the photomultiplier tube monitoring the excitation source and the photomultiplier tube measuring the fluorescence intensity.

Double Monochromation

The optical section of the instrument comprises four grating monochromators. Two are used in tandem for dispersing excitation energy, and the other two are used for examination of fluorescent spectra. An optical filter inserted into the light path between the xenon source and the excitation monochromator rejects second-order excitation when excitation wavelengths above 400 nm are being used. The double monochromation drastically

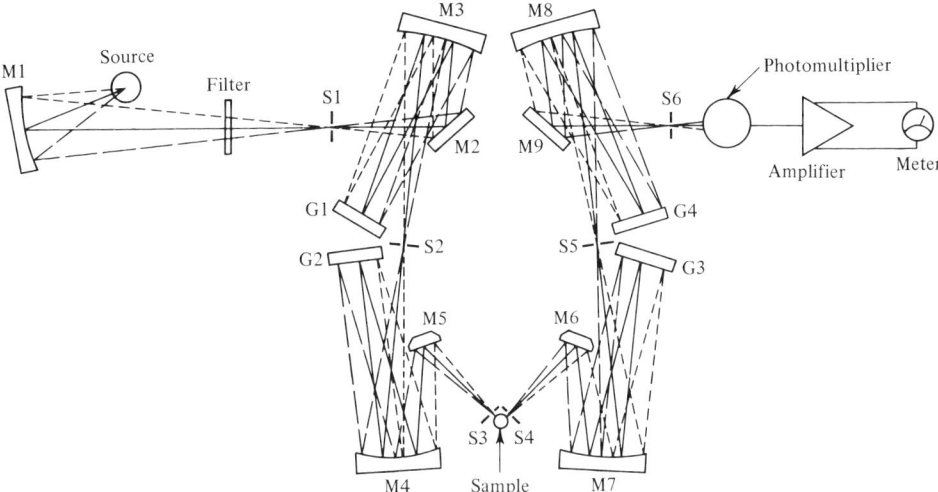

FIGURE 4-10 Optical diagram of a fluorescence spectrophotometer employing double monochromation. (Courtesy of Baird-Atomic.)

reduces scattered light and also permits the use of larger slits to achieve the same resolution as achieved in conventional monochromators using smaller slit widths. The use of larger slits allows more light to pass through the excitation monochromator to excite the sample. This is very advantageous when working with samples at low concentrations because a maximum amount of energy is made available without any sacrifice in resolution. The optical diagram of one such instrument is shown in Fig. 4-10.

Double-Beam Fluorescence Spectrophotometer[6]

In the use described, double beam refers to an optical-electrical system that permits the use of a second optical beam for essentially simultaneous compensation of a reference sample (Fig. 4-9c). Double-beam spectrophotometry has the most important advantage of providing increased convenience over single-beam measurements in that reference sample compensation can be performed by the instrument automatically and accurately. Figure 4-11 shows the optical schematic of one instrument. The design concept utilizes a classic time-sharing electro-optical system with equivalent optical paths for sample and reference beams and an optical chopper to create an ac signal as well as to split the beam. The arc of a 150-W xenon lamp is focused on the entrance slit of the excitation monochromator after the light is optically chopped with a bow tie-shaped disk. A sector mirror located beyond the exit slits of the excitation monochromator is driven by a shaft common to the chopper.

In the fluorescence double-beam mode the emitted light is viewed at right angles to the exciting light for both the sample and reference beams. The light from each beam is focused alternately on the entrance slit of the emission monochromator using common or equiva-

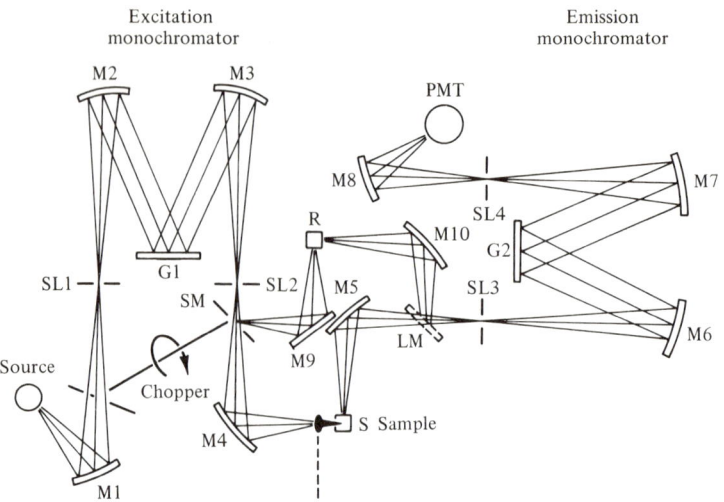

FIGURE 4-11 Optical diagram of a double-beam fluorescence spectrophotometer. SM, sector mirror; G, grating; LM, lattice mirror; M, mirror; PMT, photomultiplier tube; R, reference cuvette; SL, entrance and exit slits. (Courtesy of Perkin-Elmer Corp.)

lent front surface mirrors. The lattice mirror is a half-aluminized half-transmitting flat mirror which reflects the sample beam and transmits the reference beam to the emission monochromator. Both beams emerging from the exit slits of the emission monochromator are imaged on the photomultiplier detector 180° out-of-phase.

When the chopper disk is open the first time, the sector mirror reflects the beam to the reference sample position. When the chopper disk is open the second time, the sector mirror is also open and therefore allows light to pass to the sample position. The remaining time of each revolution of the chopper measures the optical zero during the opaque portion of the disk.

The double-beam technique provides greater analytical precision, convenience, and speed over comparable single-beam techniques in canceling out interfering scatter and fluorescence of reagents. It also enables small differences to be measured between two very similar fluorescent samples.

Complete correction for the excitation spectrum is made by recording the ratio of the output signal from the sample detector to that of a reference detector which monitors the emission of a quantum counter. The quantum counter is a special cell containing a high concentration of a fluorescent material, often rhodamine B (3 g/liter in ethylene glycol). Over a considerable range of exciting wavelengths, all incident energy is absorbed and a fixed quantum fraction reemitted at essentially a fixed wavelength (640 nm). A portion of this emitted beam is then allowed to fall alternately on the same photomultiplier that is used to measure the fluorescence emission from the sample. Signals due to the light from the quantum counter and to the sample are electrically separated and used to operate a ratio recorder. Since the light emitted by the quantum counter is proportional to the quanta falling on the sample, corrected excitation spectra are obtained automatically.

The *corrected spectra computer* is a microcomputer which automatically stores correction factors when rhodamine B or known reference samples are used.[7] During the calibration mode the operator only has to scan the spectrophotometer in a prescribed manner. For subsequent wavelength scans, in either corrected excitation or corrected emission modes, the information stored in memory during the calibration mode scan is recalled and used to correct automatically spectral data. In other words, an error curve is generated and stored for both the excitation and emission spectra.

Corrected fluorescence measurements can yield useful information concerning energy transfer studies (inter- and intramolecular), energy of the Stoke's shifts, and studies of macromolecules. Because a corrected fluorescence excitation spectrum is identical to an absorption spectrum, but at 1000 times the sensitivity, it can be used advantageously for trace analysis applications normally employing ultraviolet/visible absorption spectrophotometry.

Total Luminescence Spectroscopy[8]

Total luminescence spectroscopy is a method for measuring and displaying spectral intensity as a function of *all* useful excitation and emission wavelengths. A computer-controlled spectrofluorometer is used to scan the emission spectrum of a sample repeatedly as the excitation wavelength is increased in small increments. Intensity values are recorded and plotted on a digital plotter. The display is a three-dimensional representation comprising a series of closed, continuous, equal-intensity contours. It provides spectral data much as a topographic map displays height above sea level on a plane surface. The result is a graphic pattern in which spectral peaks, anomalies, and symmetries can be readily observed. In many cases known signature patterns of specific compounds can be immediately discerned or deduced. In the case of complex mixtures in which overlapping patterns prevent the identification of individual components, considerable information is still available. Classes of constituent compounds can often be identified; differences between closely similar compounds can be noted; changes in a particular mixture, sampled at different times or under different conditions, become readily apparent.

Once the data are stored in the computer, many different types of mathematical manipulations are possible. One involves the subtraction of the solvent background. In the case where contour maps of individual components are available and a set of calibrated matrices exist, it is straightforward to solve for the concentration of the individual components using a set of simultaneous equations. Several different pairs of excitation/emission wavelengths are required to solve for the concentration of the individual components in the mixture. The maximum emission intensity for a component with the least interference from the others would be selected. The total luminescence contour of the mixture aids in the determination of excitation/emission wavelength pairs that maximize the response to the desired materials while minimizing the response to other materials. Working from a bank of standard spectra, once quantitation has been made, one component at a time can be subtracted from the mixture, eventually resulting in no contour to confirm an adequate description of the mixture.

4.4 INSTRUMENTATION FOR PHOSPHORESCENCE MEASUREMENTS

Instrumentation for phosphorescence investigations is identical to that described for fluorescence measurements with the addition of a light interrupter and provision for immersion of the sample in a Dewar for liquid nitrogen temperatures. Excitation light from a xenon source, after dispersion by the excitation monochromator, is admitted to the sample via a fixed slit system and a rotating can chopper (Fig. 4-12) (or set of slotted disks with equally spaced ports) that permits periodic excitation of the sample and periodic out-of-phase measurement of phosphorescence. This allows measurement of the phosphorescent signal without interference from scattered light and short-lived fluorescence. The sample cuvette is a small Dewar made of fused silica and silvered, except in the region where the optical path traverses the Dewar. The solvent frequently used is a mixture of diethyl ether, isopentane, and ethanol in a volume ratio of 5:5:2. When cooled to liquid nitrogen temperature it gives a clear transparent glass.

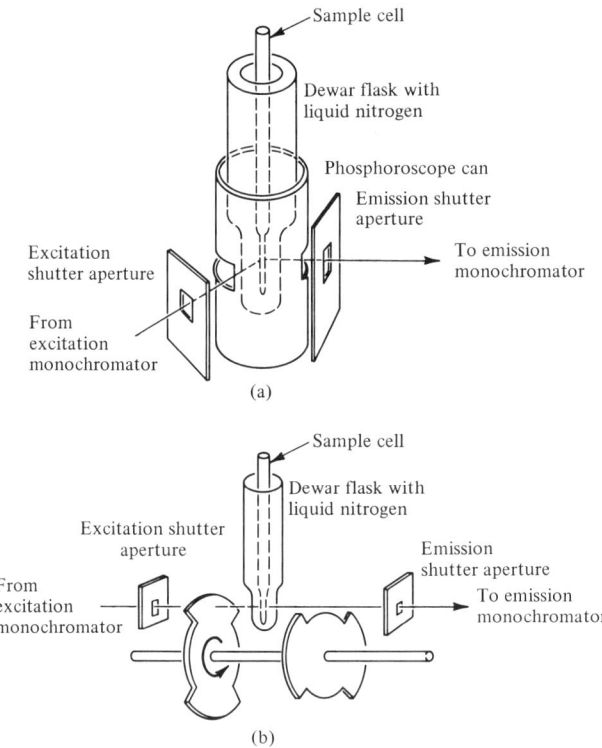

FIGURE 4-12 Schematic diagram of the interruption of excitation radiation and phosphorescence emission: (a) rotating can device and (b) rotating shutter. (Reprinted with permission from T. C. O'Haver and J. D. Winefordner, *Anal. Chem.*, 38, 602 (1966). Copyright 1966 American Chemical Society.)

The resolution time of the instrument is the length of time between the cutoff of each pulse of excitation light admitted to the sample and the clearing of the optical path by the second opening to allow the phosphorescence emission to enter the emission monochromator. The time is a function of the motor speed, the size and spacing of the openings, and the relative radial positions of the ports to each other. Decay curves can be recorded if the detector circuit is equipped with an oscillograph. Considerable improvement in time resolution in phosphorimetry is achieved using a microsecond-duration pulsed source in place of a rotating cam. In this case, the photomultiplier detector is gated after the flash in order to avoid overloading the detector from large scattered light levels. A single pulse phosphorescence shutter provides fast cutoff of excitation light and allows phosphorescence decay times down to milliseconds to be observed and recorded.

Despite the obvious advantages of time resolution, phosphorimetry is not as widely utilized as fluorometry. This fact is due to the added complexity of phosphorimetric instrumentation, the smaller number of species that phosphoresce, and the inconvenience of having to cool the sample to obtain adequate phosphorescence quantum efficiencies.

PROBLEMS

1. A 1.00-g sample of a cereal product was extracted with acid and treated so as to isolate the riboflavin plus a small amount of extraneous material. The riboflavin was oxidized by the addition of a small amount of $KMnO_4$, the excess of which was removed by H_2O_2. The solution was transferred to a 50-ml volumetric flask and diluted to the mark. A 25-ml portion was transferred to the sample holder and the fluorescence measured. Initially the fluorometer had been adjusted to read 100 scale divisions with a solution of quinine bisulfate. The solution read 6.0 scale divisions. A small amount of solid sodium dithionite was added to the cuvette to convert the oxidized riboflavin back to riboflavin. The solution now reads 55 scale divisions. The sample was discarded and replaced in the same cuvette by 24 ml of the oxidized sample plus 1 ml of a standard solution of riboflavin which contains 0.500 μg/ml of riboflavin. A small amount of solid sodium dithionite was added. The solution read 92 scale divisions. Calculate the micrograms of riboflavin per gram of cereal.

2. Solutions of varying amounts of aluminum were prepared, 8-quinolinol was added, and the complex was extracted with chloroform. The chloroform extracts were all diluted to 50 ml and compared in a fluorometer. These readings were obtained:

ALUMINUM μg/50 ml	FLUOROMETER READING	ALUMINUM μg/50 ml	FLUOROMETER READING
2	10	12	53
4	19	14	60
6	28	16	66
8	37	18	71
10	45		

Plot the fluorometer reading versus the aluminum concentration. Over what concentration range is Eq. 4-6 valid?

3. From the spectrum for phenanthrene, select the optimum wavelength for excitation and for fluorescence emission. Do this (a) assuming only a filter fluorometer is available and (b) when a grating monochromator spectrofluorometer is available.

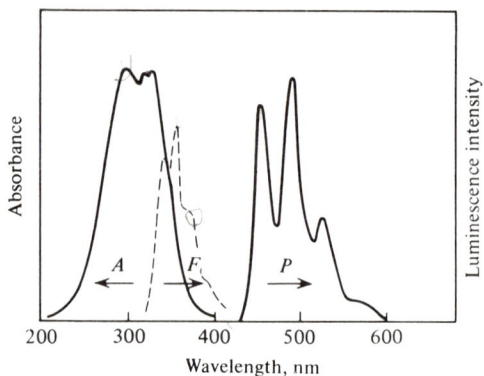

PROBLEM 4-3 Excitation and emission spectra of phenanthrene. *A*, excitation; *F*, fluorescence; *P*, phosphorescence.

4. From the appropriate spectra for phenanthrene and naphthalene, devise a method of analysis for each component assuming (a) only a spectrofluorometer is available and (b) assuming both a spectrofluorometer and a phosphorescence spectrometer is available.

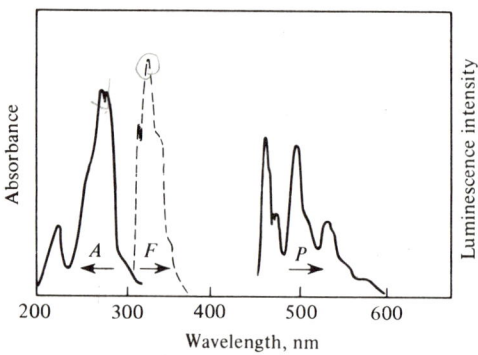

PROBLEM 4-4 Excitation and emission spectra of naphthalene. *A*, excitation; *F*, fluorescence; *P*, phosphorescence.

5. (a) Although the early optical designs involved front face fluorescence with solution cells, why is this arrangement impractical? (b) The right-angle arrangement for solutions introduces what two phenomena?

6. (a) How may Raman and Rayleigh scatter be identified when included within a fluorescence emission spectrum? (b) What is the limiting effect of Raman scatter when added to the sample measurement? (c) When Raman scatter does interfere, what four alternatives does the analyst possess for eliminating it? (d) Raman shifts of various solvents

are as follows: cyclohexane (2880 cm^{-1}), water (3380 cm^{-1}), ethanol (2920 cm^{-1}), and chloroform (3020 cm^{-1}). What would be the wavelength at which the Raman effect would appear for each solvent when the excitation wavelength was one of these mercury emission lines: 254, 313, 365, and 436 nm?

7. Compare the sensitivity increase with concentration for fluorescence and absorption measurements at constant signal-to-noise (theoretical). Do this for a 1-cm path length and molar absorptivity of 5000 liter cm mol^{-1}. $\phi_F C$ is assumed to be 0.20.

8. From the phosphorescence plus fluorescence spectrum and the phosphorescence spectrum alone, the following peaks were obtained for trivalent metal chelates of dibenzoylmethane. Construct approximate energy-level diagrams for the fluorescence and phosphorescence transitions to the vibrational levels in the ground electronic state.

	METAL ION		
	Al	Sc	Y
λ_{ex} for F, nm	417.5	425.0	429.0
λ_{ex} for P, nm	478.5	485.0	491.0
Phosphorescence, nm	478.3	484.0	491.0
	495.0	500.0	508.0
	512.0	515.0	525.0
	530.8	537.0	548.0
	548.0	556.0	580.0
	575.5	573.5	
Phosphorescence plus fluorescence, nm	417.1	434.0	428.3
	440.0	450.0	452.0
	467.6	484.0	489.0
	481.0	515.0	508.0
	495.0	520.0	525.0
	512.0	537.0	548.0
	530.3	approx. 556.0	approx. 569.0
	approx. 550.0		

BIBLIOGRAPHY

Bowman, R. L., "History and Development of the Spectrophotofluorometer," *Fluorescence News*, 3 (1), 1 (February 1974).

Guilbault, G. G., *Practical Fluorescence: Theory, Methods and Technique*, Marcel Dekker, New York, 1973.

Hercules, D. M., "Some Aspects of Fluorescence and Phosphorescence Analysis," *Anal. Chem.*, 38, 29A (1966).

Lott, P. F. and R. J. Hurtubise, "Instrumentation for Fluorescence and Phosphorescence," *J. Chem. Educ.*, 51, A315, A358 (1974).

Passwater, R. A., *Guide to Fluorescence Literature*, Vols. I–III, Plenum, New York, 1967, 1970, 1974.

Wehry, E. L., *Modern Fluorescence Spectroscopy,* Vols. I–II, Plenum, New York, 1976.
Winefordner, J. D., S. G. Schulman, and T. C. O'Haver, *Luminescence Spectrometry in Analytical Chemistry,* Wiley-Interscience, New York, 1972.

LITERATURE CITED

1. Schulman, S. G., *Fluorescence News,* **7** (4), 25 (1973); **7** (5) 33 (1973).
2. Wehry, E. L., *Fluorescence News,* **6** (1), 1 (1971).
3. Williams, R. T. and J. W. Bridges, *J. Clin. Pathol.,* **17,** 371 (1964).
4. Wright, J. C. and F. J. Gustafson, *Anal. Chem.,* **50,** 1147A (1978).
5. Passwater, R. A., *Fluorescence News,* **7** (3), 17 (1973).
6. Porro, T. J. and D. A. Terhaar, *Anal. Chem.,* **48,** 1103A (1976).
7. DiCesare, J. L. and T. J. Porro, *Trends in Fluorescence,* **1,** 16 (1978).
8. Giering, L. P., *Industrial Res/Dev.,* p. 134 (September 1978).

CHAPTER 5

Flame Emission and Atomic Absorption Spectrometry

Combustion flames provide a remarkably simple means for converting inorganic analytes in solution into free atoms. It is only necessary to introduce an aerosol of the sample solution into an appropriate flame, and a fraction or all of the metallic ions in the aerosol droplets are eventually converted into free atoms. Once the free atoms are formed, they may be detected and determined quantitatively at the trace level by atomic flame emission (FES), atomic absorption (AAS), or atomic fluorescence spectrometry (AFS). Flame methods, properly applied, now provide results for a number of elements by which all other methods must be judged, especially at low levels. This is clearly the case for zinc, cadmium, the alkali metals, and the alkaline earths. In many situations, FES or AAS will be preferred over all others for elements such as Al, Cr, In, Mn, Pb, and the heavier rare earths.

5.1 NEBULIZATION[1]

Although solids have been volatilized directly using induction heating or electron bombardment, virtually all commercial flame spectrometers rely on pneumatic nebulization of a liquid sample to deliver a steady flow of aerosol to a flame. The sample orifice is positioned either concentric with (Fig. 5-1), or at right angles to, the annulus from which the aspirating gas exits. Liquid, drawn through the sample capillary by the pressure differential generated by the high-velocity gas stream as it passes over the sample orifice, is set into oscillation. Filaments of liquid are drawn out from the bulk of the solution. The filaments collapse to form droplets. The cloud of droplets strikes an obstruction in the spray cham-

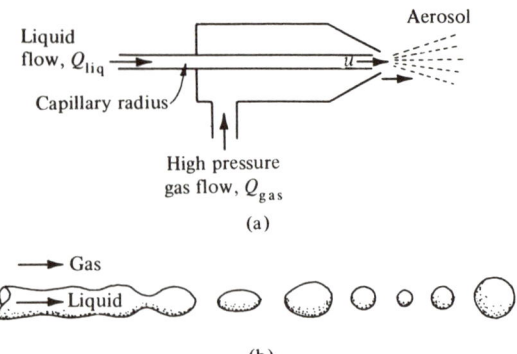

FIGURE 5-1 (a) Construction of pneumatic nebulizer and (b) breakdown of a liquid filament into droplets.

ber termed a spoiler or impact bead (or even a counter-jet of gas), which breaks the larger droplets into smaller ones. The final aerosol, really a very fine fog, is mixed with the oxidizer/fuel mixture and carried into the burner. A typical distribution range of droplet diameters is shown in Fig. 5-2. Droplets larger than about 20 μm are trapped in the spray chamber and flow to waste, or fail to desolvate completely in the flame before reaching the observation light path (see Fig. 5-5).

Nebulization is governed by such parameters as the viscosity, η, density, ρ, and surface tension, γ, of the sample solution, the flow rate of the nebulizer gas, Q_{gas}, and aspi-

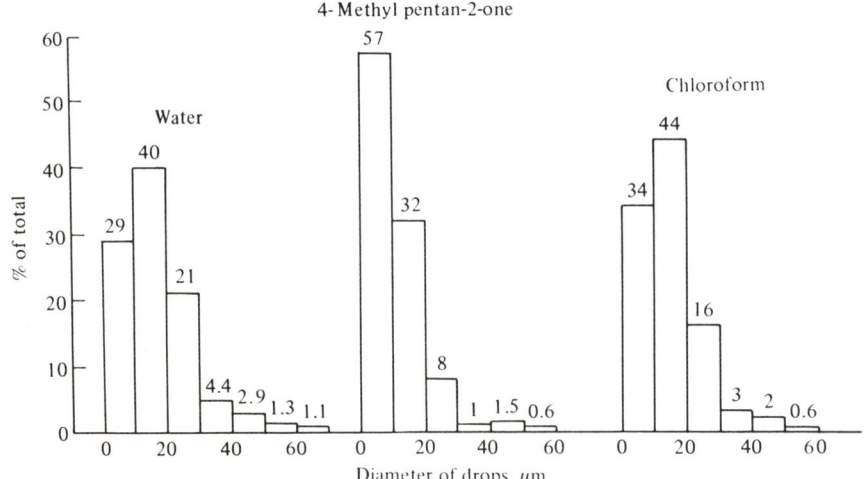

FIGURE 5-2 Representation of a drop-size distribution from a pneumatic nebulizer. (After J. A. Dean and W. J. Carnes, *Anal Chem.*, 34, 192 (1962). Courtesy of American Chemical Society.)

rated solution, Q_{liq}, and the velocity of the nebulizing gas, v. The volume–surface, or Sauter mean, diameter of the droplets, d_0, in micrometers, is given by an empirical expression[2]:

$$d_0 = \frac{585}{v}\left(\frac{\gamma}{\rho}\right)^{0.5} + 597\left[\frac{\eta}{(\gamma\rho)^{0.5}}\right]^{0.45} 1000\left(\frac{Q_{liq}}{Q_{gas}}\right)^{1.5} \quad (5\text{-}1)$$

Unanticipated changes in viscosity and surface tension can be avoided by preparing samples and standards similarly, and avoiding total acid or salt concentrations greater than about 0.5%.

5.2 FLAMES AND FLAME TEMPERATURES

As soon as the aerosol produced by nebulization in the spray chamber is transported into the flame, the following sequence of events occurs in rapid succession:

1. The solvent is evaporated, leaving minute particles of dry salt(s).
2. The dry solids are converted into the gaseous state.
3. A part of all of the gaseous molecules are progressively dissociated to give neutral atoms or radicals—the atomization step. These neutral atoms are the species which absorb in AAS and AFS, and are the potentially emitting species in FES. The efficiency with which the flame produces neutral atoms of the analyte is of equal importance in each of the flame techniques.
4. A portion of the neutral atoms may be thermally excited by collisions with partially burnt components in the flame gases, or even ionized. The fraction excited is important in FES, but a nuisance in AAS.
5. Some of the neutral atoms may combine with radicals in the flame gases to form new gaseous compounds such as metal monoxides. This gives rise to chemical interferences in all three flame techniques.

Taking as an example an aerosol containing calcium chloride, the sequence is

Aerosol: $CaCl_2$ (aq)
↓
Desolvation: $CaCl_2$ (s)
↓
Vaporization: $CaCl_2$ (g)   Ionization → Ca^+ (emission lines: 393.3, 396.8 nm)
↓
Atomization: $Cl(g) + Ca(g)$ Excitation → Ca^* (emission line: 422.7 nm)
                      + O ⇌   CaO (band heads: 606, 622 nm)
                      (from flame
                       gases)
                      + OH ⇌ CaOH   (band head: 554 nm)

The wavelengths at which the several entities emit their characteristic radiation is given in parentheses after each species.

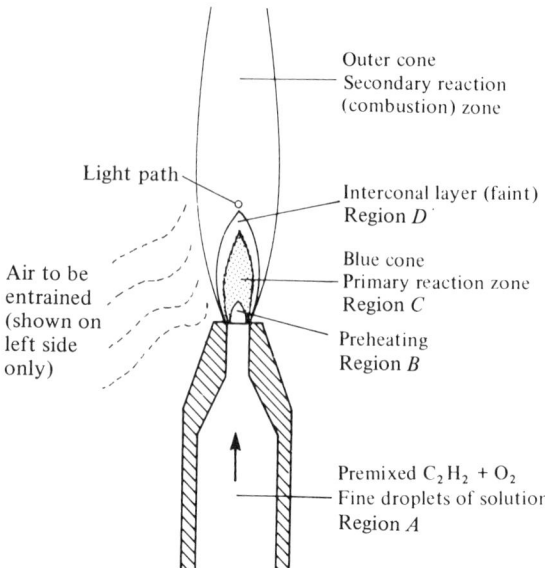

FIGURE 5-3 Schematic structure of a laminar flow flame.

The main requirements of a satisfactory flame are that it has the proper temperature and fuel/oxidant ratio to carry out the enumerated functions of the flame, and that the spectrum of the flame itself does not interfere with observation of the emission or absorption features being measured. Components of the flame gases limit the usable range to wavelengths longer than 210 nm.

The structure of a premixed flame, supported on a laminar flow burner, is shown in Fig. 5-3. Emerging from region A, the unburned hydrocarbon gas mixture passes into a region of free heating about 1 mm in thickness (region B) in which it is heated by conduction and radiation from reaction region C and by diffusion of radicals into it, which initiate the combustion. Flame gases travel upward from the reaction zone with velocities in the order of 1–10 m sec^{-1}. Gases emerging from region C consist mainly of CO, CO_2, and H_2O (and N_2, if air is one of the original gases), with lesser amounts of H_2, H, O, OH, and NO. The actual composition of the gases varies with the initial mixture composition. In the inner cone, the gases are not in thermal equilibrium, the amounts of the radicals (C_2, CH, H_3O^+, HCO^+) being too high. Some of these radicals are precursors of chemi-excitation reactions. Thermal equilibrium will be achieved almost completely in region D. Combustion is completed in the outer mantle, assisted by entrainment of surrounding air.

The temperature of the flame (Table 5-1) is the important characteristic that determines its possibilities in FES. The exact value depends on the fuel/oxidant ratio and is generally highest for a stoichiometric mixture. When selecting the flame, the signal-to-noise ratio (which is due to emission of interfering elements), the stability of the source (which determines the precision of measurements), and the economy of the analysis are also taken into account. The optimum temperature depends on the excitation and ionization potentials of the analyte. The theoretical calculations that are in accordance with experiments

TABLE 5-1 Characteristics of Common Premixed Flames

Fuel	Oxidant	Temperature,[a] °C	Burning velocity,[b] cm sec^{-1}
Acetylene	Air	2400	160–266 (160)
Acetylene	Nitrous oxide	2800	260
Acetylene	Oxygen	3140	800–2480 (1100)
Hydrogen	Air	2045	324–440
Hydrogen	Nitrous oxide	2690	390
Hydrogen	Oxygen	2660	900–3680 (2000)
Propane	Air	1925	43

[a] Stoichiometric mixture.
[b] Values in parentheses are probably the ones most applicable to laboratory burners.

prove that the acetylene/air flame can be used in practically all cases when detecting alkali elements. For the determination of alkaline-earth elements, as well as Ga, In, Tl, Cu, Co, Cr, Ni, and Mn, the acetylene/nitrous oxide flame is used to produce a higher flame temperature. The hotter flame allows the sensitive analysis of additional elements whose refractory oxides are not reduced to the atomic state in the acetylene/air flame. The acetylene/nitrous oxide flame is somewhat exceptional in combining high temperature with a propagation rate and residence time not much above that of the cooler acetylene/air flame. Special burner heads (5-cm slot in length, 0.5 mm wide) and control units for safely igniting and extinguishing this flame are needed to eliminate any risk of flashback to the spray chamber and consequent explosion.

A fuel-rich acetylene flame (ratio of fuel to oxidant exceeds that needed for stoichiometric combustion) provides a reducing atmosphere necessary for the production of a large free-atom population of those elements that have a tendency to form refractory oxides. The region of optimum emission is the interconal zone because of the presence of many carbon-containing radical species.

Shielding the premixed acetylene/nitrous oxide flame with either nitrogen or argon produces a large vertical temperature gradient in this flame. This gradient provides the required wide range of excitation conditions over a small flame volume, particularly useful when doing multielement determinations. Of course, means for adjusting the burner height (relative to the observation light path) are required.

Disadvantages of Flame Atomization

The premixed or laminar flow burner is limited to the use of solutions or very fine suspensions. Rarely can solid samples be atomized directly. Sensitivity is limited because the nebulizer-mixing chamber unit is inherently wasteful of sample. Only about 10% of the sample effectively reaches the flame as a fine aerosol that ultimately becomes atomized.

The residence time of an atom within the optical beam of the spectrometer, which is the effective absorption or emission volume of the flame gases, is extremely short (approximately 10^{-3} sec), depending on the velocity of the flame gases. If one assumes that the time required for attaining a steady-state atomization and for the measurement of the absorption or emission signal is 10 sec, then to achieve an equilibrium content of the element in the optical beam during this period of time, the required amount of substance will be $10/10^{-3} = 10^4$ times that present in the beam at any given moment. Furthermore, the attainable atom concentration in flames is limited by the dilution effect of the relatively high flowrate of unburnt gas used to support the flame and to transport the aerosol to the flame. Atom concentration is also limited by the flame gas expansion that occurs on combustion.

The flicker and general instability of a flame leads to flame noise, which limits the detectability of elements. The detection limit is usually quoted as the signal which equals the noise of the flame (peak to peak). The background spectrum of the flame, along with the noise, will limit or prevent scale expansion of the output signal due to analyte.

Distribution Patterns of Atomic Concentration in Flames

The concentration of unexcited and excited atoms varies in different parts of a flame and with the fuel/oxidant ratio. Studies of atomic or molecular distributions within the

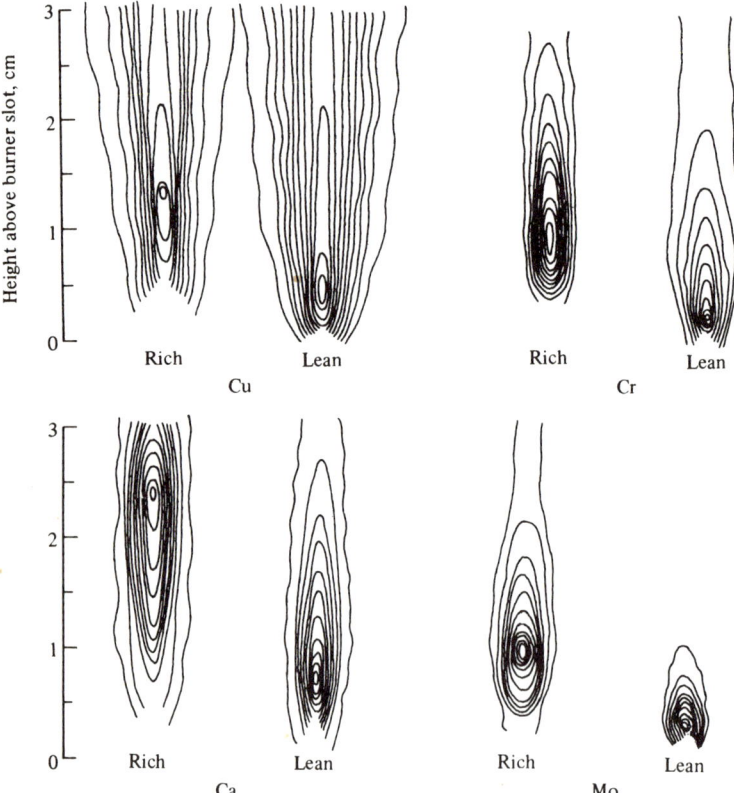

FIGURE 5-4 Distribution of atoms in a 10-cm air/acetylene flame. Fuel-rich and fuel-lean results are shown. Contours are drawn at intervals of 0.1 absorbance unit with maximum absorbance in center. (After C. S. Rann and A. N. Hambly, *Anal. Chem.*, 37, 879 (1965). Courtesy of American Chemical Society.)

flame envelope may be made by measuring absorption, emission, or fluorescence as the flame is moved vertically (or horizontally) relative to the light path of the optical system. Figure 5-4 shows the distribution of atoms in a 10-cm-long acetylene/air flame by absorption measurements.[3] Contours are drawn at intervals of 0.1 absorbance unit with maximum absorbance in the center. Neither the area of observation nor the fuel/oxidant ratio is critical for silver, whereas for molybdenum the region of maximum absorption is sharply localized. The height of maximum absorption marks the level where the increased atomization with height is just balanced by the rate of decrease in the concentration of free atoms through dilution by the flame gases and formation of oxides and hydroxides.

The distribution pattern obtained in emission very often differs from that obtained from absorption measurements.[4] For example, emission lines of boron (249.7 nm) and antimony (259.8 nm) are absent or very weak in the outer mantle of a stoichiometric flame but appear in unusual strength in the reaction zone of a fuel-rich acetylene flame.

Pressure Regulators and Flow Meters

To maintain a constant thermal environment in the flame, it is imperative that the gas pressures and gas flows be held constant while the flame spectrometer is operating. With two-channel instruments this requirement can be relaxed somewhat. Usual flows range from 1-4 liter min^{-1}. A knowledge of the individual flows of fuel and oxidant enables an operator to choose various mixtures ranging from lean to fuel-rich types of flames.

Burners

Burner configurations need not be any different in FES or AAS (Figs. 5-6 and 5-7). In AAS the obvious analogy with solution spectrophotometry early suggested a laminar flow slot burner with a slot 5 cm long and 0.5 mm wide, providing a rather long absorbing path. Line emission intensities are influenced in the same way by flame length. Increasing FES intensities in this way is of little use in the presence of strong flame background intensity. However, the greater line intensities obtained with slot burners permit the use of narrower slit widths, with worthwhile gain in the line-to-background ratios in certain practical situations. Self-absorption in the long flame is not a serious problem, at least not with a 5-cm slot burner. To analyze solutions at higher concentrations, which would show strong self-absorption and pronounced flattening of the calibration curves, it is a simple matter to turn the burner head crosswise to the optical path. This reduces sensitivity in both AAS and FES.

Since the optimum flame region varies with different elements, it is essential that the burner height be adjustable to secure the maximum emission or absorption signal. It is advisable to remove the products of combustion and heat by means of an exhaust hood over the burner.

In the laminar flow burner, shown in Fig. 5-5, the aerosol is produced within a mixing chamber where the coarse and fine droplets are separated. The fine droplets, virtually a fog, are mixed with the flame gases and pass on into the flame. The burner head can consist either of a rectangular array of holes or a slot at which the flame burns. No memory effect should exist; that is, the content of one sample should not affect the result from the next. The laminar flow burner serves well for flames with low burning velocities, such

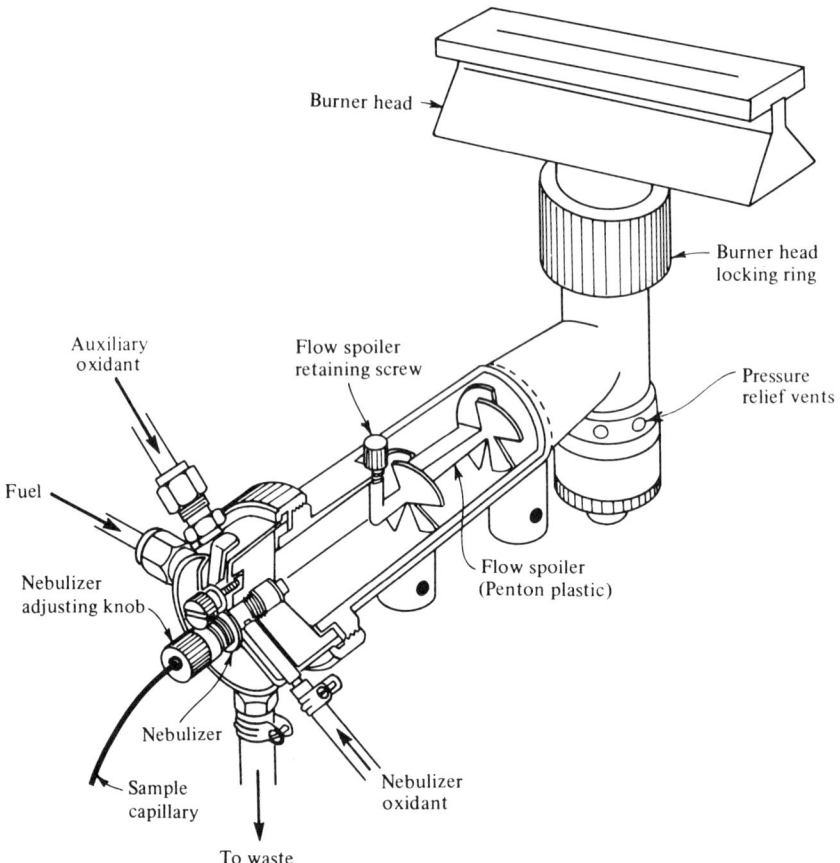

FIGURE 5-5 Slot burner and expansion chamber. (Courtesy of Perkin-Elmer Corp.)

as acetylene with air or nitrous oxide. To prevent flashback down the burner port and an explosion in the mixing chamber, the streaming velocity of the fuel/oxidant mixture through the burner port must be at least equal to the burning velocity, and to ensure a margin of safety and achieve a reasonably stiff flame, several times the burning velocity.

A needle burner may be used with the acetylene/oxygen flame and with an acetylene/nitrous oxide flame safely. The flame appears at the needle end. The needles of 0.2-2 mm i.d. are pressed onto the cap placed on the burner body, the distance between the needles being within the limit of 0.3-3 mm. A needle length of 5-10 mm allows for efficient cooling by outside diffusing air.

5.3 INTERFERENCES

Essentially the same interferences occur in FES and AAS for the same reasons, but to somewhat different extents. The literature, unfortunately, is replete with a welter of contradictory statements and hasty generalizations, based on inadequate measurements and misunderstandings. Interferences can be separated into four general classes: (1) background absorption, (2) spectral interference, (3) vaporization interference, and (4) ionization effects.

Background Absorption

The brightness of an acetylene/nitrous oxide flame and the complexity of its spectrum should not discourage anyone from considering this flame for trace metal analysis. The success of this flame depends on the high resolution of modern grating spectrometers. Corrections for background radiation from the flame itself and the additional background contributed by the sample matrix must be made. A common correction method is to slowly scan for a few nanometers in the vicinity of the line while recording the spectrum. A baseline is drawn beneath the emission line from the background (extrapolated) on both sides. Unfortunately, this method is a relatively slow process, consuming both time and sample, and the peak line intensity is observed only very briefly, thereby reducing the precision of the measurement. A more precise method, less wasteful of sample, can be employed after a preliminary wavelength scan has been made. The total intensity (line plus background) is first measured for a group of samples at the wavelength of the line. The wavelength setting is then changed by about two times the bandpass, and the adjacent background contributed by the matrix and flame is measured on both sides of the line and the average reading subtracted.

Background absorption occurs in AAS as molecular absorption and as light scattering by particles in the flame. Molecular absorption occurs when matrix species are vaporized along with the analyte atomic species and absorb a portion of the analyte atomic resonance line emitted from the light source. Many molecules, and in particular the alkali halides, possess comparatively strong absorption bands in the ultraviolet and visible spectral regions. Atomic fluorescence enjoys a considerable advantage over AAS as far as the error associated with matrix molecular absorption is concerned. Particulate matter in the flame may arise from unevaporated droplets, but more likely from unevaporated (refractory) salt particles left following desolation of the aerosol. Scattering need not be considered in emission measurements although the presence of hot particles in the flame would increase the background continuum. In AFS light scattered from particles would appear as a signal that could not be distinguished from that due to the analyte when a monochromatic source is used.

In some ways background absorption in AAS is a more serious problem than in FES because it is likely to go unnoticed. The monochromatic nature of the source prevents scanning. Correction for both background absorption and scattering can be accomplished by measuring the total absorption signal and subtracting from it the nonatomic portion. This is possible because the molecular absorption bands are broad. Atoms absorb only at specific wavelengths by contrast; at other wavelengths the atomic absorption will be zero even when the molecular absorption is still significant.

Automatic background correction in AAS is accomplished by making the correction at the same wavelength as the element of interest while still maintaining the one- or two-channel optical configuration (see Figs. 5-7 and 5-8). The hydrogen (or deuterium) hollow cathode lamp is preferred as the continuum source since the geometry and spectral output more closely match those of the metal lamp. The use of the modulation frequency to drive and discriminate the hydrogen lamp signal provides for fast correction, up to three corrections each 10 msec. Speed such as this is required because both sample and background are two ever-changing signals with nonflame atomizers. At wavelengths in the visible region problems arise because the hydrogen continuum source intensity is weaker than the output of many hollow cathode sources operating at optimum current levels. Substitution of a 150-W xenon–mercury arc lamp provides adequate radiant intensity from 220 to 600 nm.

Spectral Line Interferences

Spectral line interferences occur when a line of interest cannot be readily resolved from a line of another element or from a molecular band. Interference of this type is closely associated with the resolving power of the monochromator. Atomic line interferences will be more serious in FES, even with the best modern spectrometers. Their bandpass is still an order of magnitude broader than the line profiles of lamp and absorbing atoms in AAS. Molecular spectral interference is also more severe in FES. Part of a molecular spectral interference in any type of flame emission measurement will be chemiluminescence of molecular fragments formed in the flame gases. In AFS this can be discriminated against by amplitude modulation of the light source. The same is true for stray light which would be produced at the wavelength of interest by flame chemiluminescence in AAS. No such possibility exists in FES.

Instances of serious spectral interference in FES involve the manganese triplet (403.1, 403.3, 403.5 nm), the gallium line (403.3 nm), the potassium doublet (404.4, 404.7 nm), and the lead line at 405.8 nm. The orange band system of CaOH extending from 543 to 622 nm interferes with the sodium doublet (589.0, 589.6 nm) and the barium line at 553.6 nm.

Vaporization Interference

Vaporization interference occurs when some sample component influences the rate of vaporization of salt particles containing the desired analyte. It can arise from a chemical reaction that alters the vaporization behavior of the solid, or it can be a physical process in which the vaporization of the matrix controls the release of analyte atoms trapped within. This type of interference can be dealt with by informed choices of flames, burners, nebulizers, and chemical additives.

Hotter flames provide less vaporization interference. Use of the acetylene/nitrous oxide flame may often be justified for this reason alone, as it is better at decomposing the thermally stable phosphates, sulfates, silicates, and aluminates than the cooler acetylene/air flame. The shifting in equilibrium of reactions involving these thermally stable molecules, due to small uncontrolled changes in temperature, influences AAS and FES to similar degrees, and is perhaps the chief reason that AAS results apparently are about as dependent on flame temperature as those of FES.

Metal compounds in the flame are usually simple diatomic molecules, such as CaO, or triatomic molecules, such as CaOH. Elements such as Na, Cu, Tl, Ag, and Zn are practically completely atomized in the flame; they do not form molecular compounds with flame partners in noticeable proportions. A major fraction of the alkaline-earth elements is present as monoxides unless very fuel-rich flames are used. Metals such as La, Al, and Ti form refractory oxides which are extremely stable. As a consequence, the free-atom concentrations of these elements are virtually negligible in flames of stoichiometric composition and moderate temperature. However, in fuel-rich, acetylene/nitrous oxide flames, these oxides may sufficiently dissociate to enable these elements to be analyzed by either emission or absorption.

Releasing agents should be used whenever necessary. A few hundred parts per million of lanthanum or strontium are added to samples and standards to overcome such interferences as phosphates. These levels of lanthanum and strontium will often allow them to serve also as ionization suppressants, as they have ionization potentials lower than most of the elements listed above. A releasing agent combines with the interferent, or simply by mass action denies the test element to the interferent, thereby leaving the test element free to vaporize in the flame. In protective chelation or masking, the element sought is masked to prevent it from combining with the interferent in the solution phase although, of course, the masked species must be promptly decomposed subsequently in the flame. EDTA has been used to mask calcium in the presence of phosphate.

Ionization Effects

At the temperature of the acetylene/nitrous oxide flame many elements, especially the alkali metals, are appreciably ionized (Table 5-2). Ionization depopulates the neutral atom levels, both ground and excited, and lowers sensitivity. This problem is readily overcome by adding an excess of an easily ionized element, such as potassium, cesium, or strontium,

TABLE 5-2 Percent Ionization of Selected Elements in Flames[a]

Element	Ionization potential, eV	Acetylene/ air, 2400°C[b]	Acetylene/ oxygen, 3140°C	Acetylene/ nitrous oxide, 2800°C
Lithium	5.391	0.01	16.1	
Sodium	5.139	1.1	26.4	
Potassium	4.340	9.3	82.1	
Rubidium	4.177	13.8	88.8	
Cesium	3.894	28.6	96.4	
Magnesium	7.646		0.01	6
Calcium	6.113	0.01	7.3	43
Strontium	5.694	0.01	17.2	84
Barium	5.211	1.9	42.3	88
Manganese	7.43			5

[a]Partial pressure of metal atoms in the flame assumed to be 1×10^{-6} atm for acetylene/air and acetylene/oxygen flames, and approximately 10^{-8} atm for the acetylene/nitrous oxide flame.

as an ionization suppressant, to sample and standard solutions. This should be done in any event for any of the flame techniques with samples containing variable amounts of the alkali metals, even with acetylene/air flame. It is much more important to do so in the hotter acetylene/nitrous oxide flame.

The ionization suppressant should always be added in the range of 100–1000 μg/ml in using the hotter flames to determine elements that have ionization potentials below 7.5 eV. As the ionization constant, K_i, of the ionization suppressant decreases, a smaller quantity is effective in repressing the analyte ionization. To ensure that ionization interference is suppressed, the product $K_i C_i$ of the suppressant should be 100 times the product for the analyte.

The degree of ionization, α_i, is defined as

$$\alpha_i = \frac{[M^+]}{[M^+] + [M]} \tag{5-2}$$

At equilibrium, when the ionization and recombination rates are balanced, K_i (in atm) is given by

$$K_i = \frac{[M^+][e^-]}{[M]} = \left(\frac{\alpha_i^2}{1-\alpha_i^2}\right) p_{\Sigma M} \tag{5-3}$$

where $p_{\Sigma M}$ is the total atom concentration of metal in all forms in the burned gases.* The ionization constant can be calculated from the Saha equation:

$$\log K_i = \frac{-5040 E_i}{T} + \frac{5}{2}\log T - 6.49 + \log\left(\frac{g_{M^+} g_{e^-}}{g_M}\right) \tag{5-4}$$

where E_i is the ionization energy of the metal (in electron volts), T is the absolute temperature of the flame gases, and the "g terms" are the statistical weights of the ionized atom, the electron, and the neutral atom. For the alkali metals the final term is zero; for the alkaline-earth metals it is 0.6.

5.4 FLAME SPECTROMETRIC TECHNIQUES

Common to all three flame spectrometric methods are the characteristics of atom formation in the flame and the existence of several possible types of interferences. Yet each of the methods has its own particular instrumentation, and possesses its unique capabilities and limitations. In the following discussion, the methods will be considered independently.

Flame Emission Spectrometry[5]

In flame emission spectrometry a fine aerosol of the sample solution, following nebulization, is introduced into a flame where it is desolvated, vaporized, and atomized. Subse-

---

*For a derivation, see M. N. Saha, *Philos. Mag.*, **40**, 472 (1920), or M. W. Zemansky, *Heat and Thermodynamics*, Addison-Wesley, Reading, Mass., 1957.

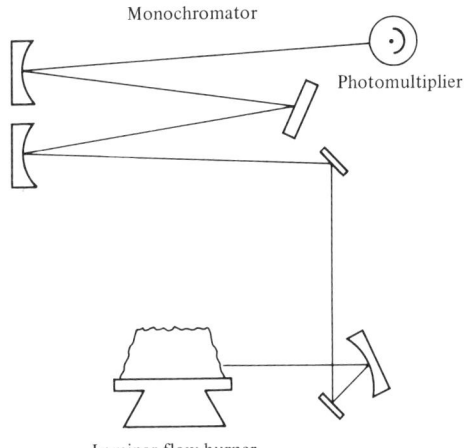

FIGURE 5-6 Schematic arrangement of a flame emission spectrophotometer.

quently, atoms and molecules are raised to an excited electronic state via thermal collisions with the constituents of the partially burned flame gases. Upon their return to a lower or ground electronic state, the excited atoms and molecules emit radiations that are characteristic for each. The emitted radiation passes through a monochromator (or suitable filters) that isolates the desired spectral feature which is then registered by a photodetector whose output is amplified and read on a meter or recorder. An instrument arrangement for flame emission is shown in Fig. 5-6.

The intensity of a spectral emission line, I_v, is determined by the number of atoms whose identical transitions occur simultaneously. It is given by the expression

$$I_v = \frac{VA_T h v_0 N_0 g_u}{B(T)} e^{-E/kT} \tag{5-5}$$

where V is the flame volume (aperture ratio) viewed by the detector, A_T is the number of transitions each excited atom undergoes per second, N_0 is the number of free metal atoms present in the ground electronic state per unit volume (and proportional to the analyte concentration in the sample solution nebulized), g_u is the statistical weight of the excited atomic state, $B(T)$ is the partition function of the atom over all states, E_u is the energy of the excited state, k is the Boltzmann constant, and T is the absolute temperature. From Eq. 5-5 it can be seen that a flame of higher temperature will produce a greater number of atoms in the excited state. Under conditions of thermal equilibrium, the ratio of the occupational number of atoms in an excited level, N^*, to N_0 is given in Table 5-3 for selected emission lines.

The modern grating spectrometer, equipped with a laminar flow burner, and a good recorder, serves equally well for FES and AAS. For FES one needs a grating spectrometer capable of giving a bandpass of about 0.05 nm or less in the first order. Slits should be adjustable to give greater intensity in situations not requiring high resolution. Spectrom-

TABLE 5-3 Values of N^*/N_0 for Various Resonance Lines

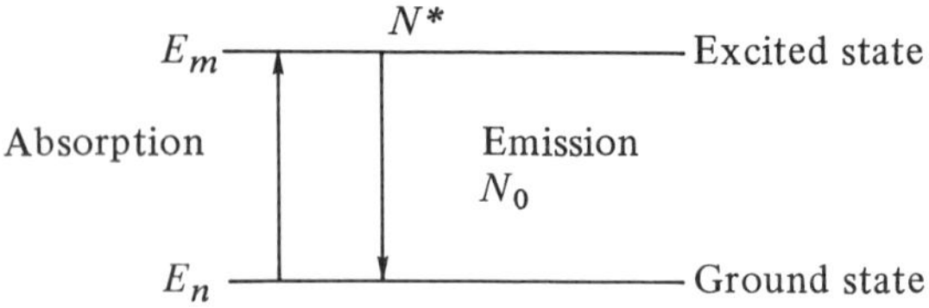

$$N^*/N_0 = (g_m/g_n)\, e^{-\Delta E_u/kT}$$

Resonance Line		g_m/g_n	ΔE (in eV)	N^*/N_0 2000°K	N^*/N_0 3000°K
Cs	8521	2	1.45	4.44×10^{-4}	7.24×10^{-3}
Na	5890	2	2.10	9.86×10^{-6}	5.88×10^{-4}
Ca	4227	3	2.93	1.21×10^{-7}	3.69×10^{-5}
Fe	3720		3.33	2.29×10^{-9}	1.31×10^{-6}
Cu	3248	2	3.82	4.82×10^{-10}	6.65×10^{-7}
Mg	2852	3	4.35	3.35×10^{-11}	1.50×10^{-7}
Zn	2139	3	5.80	7.45×10^{-15}	5.50×10^{-10}

eters of 0.33- to 0.5-m focal length with adjustable or exchangeable slits meet the requirements quite well. These instruments should be able to minimize the flame background emission and resolve atomic emission lines from nearby lines and molecular fine structure maxima. Emission band spectra from molecular species will show up more clearly with instruments of low dispersion. For scanning the spectrum, the instrument should be equipped with a scanning wavelength drive and a strip chart recorder having a pen speed of 1 sec, or less.

Background correction in emission can be done on a two-channel instrument shown in Fig. 5-8. One channel is tuned to the emission wavelength for the analyte, while the second channel is tuned to a nearby wavelength where analyte emission is not seen, but the background emission from the flame or interferent is seen. The need for proper positioning of the flame to ensure sampling of the optimum flame zone is important. The best entrance optics design is that which just fills the monochromator optics with a solid angle of radiation. At a high aperture ratio, the limit of detection is restricted not by shot noise of the photodetector but by both the instability of the flame noise and the flicker noise of matrix emission.

Atomic Absorption Spectrometry

Atomic absorption in flames is carried out by use of the principle originally discussed by Walsh.[6] This entails the determination of the absorption at the line center by using a narrow-line source emitting the given resonance line of the element, whose emission line profile is less than the absorption line profile of the analyte in the flame. The flame gases

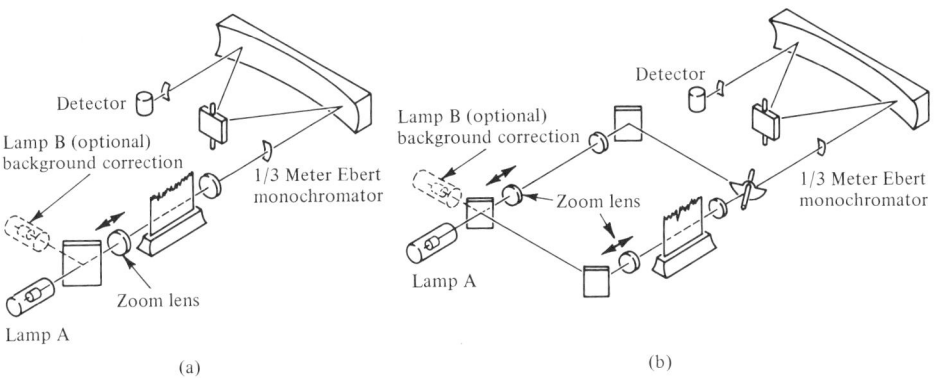

FIGURE 5-7 Optical diagrams of (a) a single-beam and (b) a double-beam atomic absorption spectrometer. (Courtesy of Instrumentation Laboratories, Inc.)

are treated as a medium containing free, unexcited atoms capable of absorbing radiation from an external source when the radiation corresponds exactly to the energy required for a transition of the test element from the ground electronic state to an upper excited electronic state. Unabsorbed radiation passes through a monochromator that isolates the exciting spectral line and into a photodetector. Absorption is measured by the difference in transmitted signal in the presence and absence of the test element. Instrumentation for AAS is shown in Fig. 5-7. Within the last few years significant developments in atomic absorption have been concerned with improvements in atomization techniques. Much effort has been centered around the development of electrothermal atomizers as substitutes for the conventional flame-based systems.

Transitions from the ground electronic state to the first excited state take place when radiation of frequency exactly equal to the resonance frequency passes through the flame gases into which the analyte has been transported as an aerosol. A part of the radiant energy of the incident light beam, I_0, will be absorbed. The transmitted intensity, I, may be written

$$I = I_0 \exp(-k_\nu d) \tag{5-6}$$

where k_ν is the absorption coefficient and d is the average thickness of the absorbing medium, that is, the path length of the flame horizontally. Around the central frequency there will exist a finite bandwidth due to absorption line broadening within the flame gases and a broadening of the emission source. The principal causes of absorption line broadening are Doppler (discussed under Sources) and Lorentz, or pressure, broadening. Lorentz broadening is due to collisions of the absorbing atoms with other molecules or atoms present in the flame gases.

For Eq. 5-6 to be valid, the bandwidth of the radiation to be absorbed by the atoms must be narrower than the absorption line for the absorbing species. This means that the linewidth of the primary radiation source must be less than 0.001 nm, the usual width of the narrow lines found in the absorption spectra of atomic species. This source requirement arises because all but the most expensive monochromators have bandpasses of 0.01 nm or more.

Specifications for a typical atomic absorption spectrophotometer might include a 0.33- to 0.5-m focal length Czerny–Turner monochromator with a 64 × 64-mm grating ruled with 2880 grooves/mm and blazed at 210 nm to cover the wavelength range of 190–440 nm. A second grating, ruled with 1440 grooves/mm and blazed at 580 nm, covers the wavelength range 400–900 nm. The two gratings are often mounted back-to-back on a turntable. Spectral bandwidths should cover 0.03–7 nm in the ultraviolet, or 0.06–14 nm in the visible; they should be switch selectable. The geometry of the light beam must be designed to provide optimum performance with both flame and electrothermal devices. Entrance optics are needed to focus an image of the source lamp upon the flame and to collect the smallest solid angle of radiation from the flame, and yet provide sufficient radiant flux at the photodetector. Such a system minimizes thermal emission pickup. One design employs a zoom lens system adjusted by the operator to produce optimum light transmission at any analytical wavelength. An $f/8$ or $f/10$ optical system gives narrow-beam geometry with realistic light-gathering ability.

Shown in Fig. 5-8 is a two-channel, double-beam, microcomputer-equipped spectrophotometer, capable of background correction in either or both channels. The instrument can determine two elements simultaneously, thereby doubling analytical speed. Alternatively, it can improve analytical accuracy by using the element in one channel as the internal standard. Instruments of this type can extend the analytical range for a given element; the same element is determined in both channels but with resonance lines of different sensitivity. The direct determination of ratios between the concentration of two elements is possible. This operating mode corrects for errors resulting from changes in flame conditions, aspiration rate, sample viscosity, and temperature of the sample solution.

A microcomputer system enables an atomic absorption instrument to correct the analytical curves in one or both channels with up to five standards, compute ratios, present

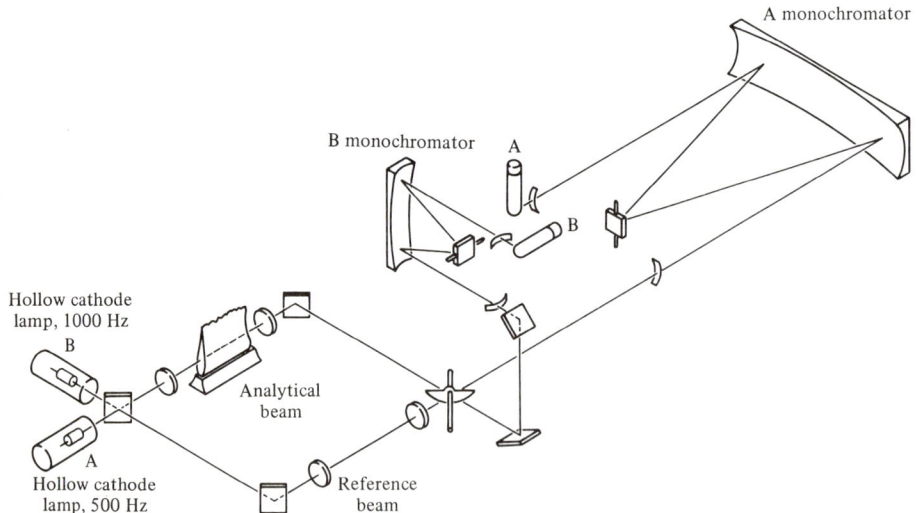

FIGURE 5-8 Optical schematic of a two-channel, double-beam atomic absorption spectrometer. (Courtesy of Instrumentation Laboratories, Inc.)

statistics, and do various housekeeping chores (selection of wavelength, slit width, fuel and oxidant flows, and ignition of flame) to make the instrument easy to operate. Deuterium arc background correction can be applied independently to either or both channels.

Sources for Atomic Absorption As external light sources, both hollow cathode lamps and electrodeless discharge tubes are used. A hollow cathode lamp has a Pyrex body and an end window of quartz. Within there is an anode wire along the outside of a cylindrical cathode, as shown in Fig. 5-9. The lamp is evacuated and filled with an ultrapure monoatomic gas (to avoid molecular continuum spectra), usually neon, occasionally argon, to a few torr. Lamps operate at currents below 30 mA and at voltages up to 300 V. Discharge occurs between the two electrodes. The cathode (4.0 mm i.d.) is bombarded by the energetic positively charged inert gas ions that are accelerated toward its surface by the potential existing in the discharge. The energy of these filler gas ions causes cathode material to be ejected or sputtered even though the element may be extremely involatile. Sputtered atoms are ejected into the plasma and they may be excited to emit their atomic spectrum by further collisions with excited filler gas atoms. When the cathode is formed into a cylinder, the discharge will tend to concentrate in the hollow, and more efficient sputtering and excitation occur. A protective shield (nonconductive) of mica around the outside of the cathode just below the lip causes the lamp to radiate more intensely because the shield prevents a spurious discharge to the outside of the cathode.

Cathode construction differs for various metals. If the metal is easily worked, the whole cathode may be made from the metal. If expensive, a thin liner is inserted into a copper cathode. When the metal's melting point is low, a cuplike carrier cathode is used. For hard

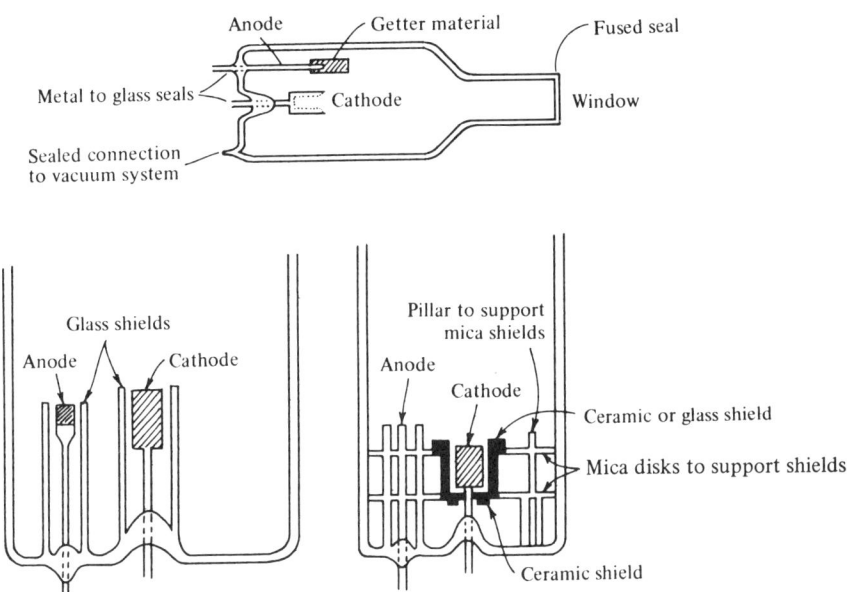

FIGURE 5-9 Schematic diagram of shielded-type hollow cathode lamp.

or brittle metals, an alloy or sinter of pressed metal powder is used, as is done often in multielement lamps.

Neon is the preferred filler gas since it gives up to three times better intensity and because it tends to suppress the ionic spectrum of some elements. Argon is substituted only when a neon spectral line occurs in close proximity to a resonance line of the metal liner.

The shape of the spectral line emitted by the source is important. Increasing the lamp current will increase its output intensity but sensitivity is reduced through line broadening and/or self-reversal for some elements. *Doppler broadening* is due to the motions of the radiating atoms as a result of thermal activity. For a given atomic line the broadening is proportional to the square root of the temperature. For narrow spectral lines from the source the temperature of the radiating plasma should be kept as low as possible. This is done by keeping the lamp current low. *Self-absorption broadening* is due to absorption of radiation by nonabsorbing atoms in the source and depends on the length of the non-emitting cloud through which the radiation must pass. It can be reduced by shortening the path length and the concentration of vapor through which the emitted light must pass. Good lamp design is necessary.

The fact that a hollow cathode lamp is required for each element is sometimes seen as a disadvantage of AAS. However, these lamps with their narrow band emissions provide virtually complete specificity for each element. Multielement lamps are available for certain combinations of elements. The cathode comprises sections or rings of the different metals, or from an alloy or pressed powder containing the elements, blended in the proportions best suited to obtaining emission lines of approximately equal intensity from each element. Some sensitivity is thereby sacrificed.

Several lamps may be located in a lamp turret accessory. Each lamp is kept at its correct operating current, thus eliminating warm-up delays. Rotation of the turret brings the chosen lamp into position and no further adjustment is required. This accessory is invaluable when several elements are to be determined in the same solution.

The electrodeless discharge lamp is well suited for AAS, and particularly for AFS where the higher output intensity is welcome. Lamps are made from silica tubing 1 cm in diameter and 7 cm in length. The sealed tube contains a low pressure of argon and a few milligrams of metal plus a small amount of iodine (or the more volatile iodide salt). Excitation is achieved by inserting the lamp in the cavity of a microwave generator (2450 MHz, 200 W). A discharge will occur in the gas due to acceleration of ions and electrons by the alternating electromagnetic field. No electrodes are required to convey the power through the bulb to the gas, hence the name. Operating tube temperature is very important. Lamp output often increases 1000-fold for a change of $130°C$; optimum temperature varies with different elements.

Modulated light sources are used to minimize the response to thermal emission in the flame. By modulating the light source either mechanically with a chopper or electronically, and using an ac amplifier or a phase-sensitive amplifier locked into a modulating frequency greater than about 100 Hz, drift and low-frequency noise are minimized. Of course, the signal is generally reduced by at least 50%.

Detector-Readout Unit Signals from nonflame devices are transient and of short duration. Consequently, the readout unit must possess high-speed electronics that can track the signals and signal processing that allows measurement of rapid events. Both peak height and peak area measurements should be possible since the latter compensates for atomization rate changes.

Care must be taken never to exceed the saturation limit of the photodetector by flooding it with light emanating from within the flame and arising from sample components present in high concentration even though their signals are eliminated by modulation. When the electronic noise in the detector–amplifier system can be reduced to a negligible level, scale expansion can profitably be employed. The zero point is displaced off scale by applying a potential counter to the incoming signal. In this way the readout device can be used as the upper end of a much longer scale and the reading for small decreases in transmitted light may be increased manyfold.

Atomic Fluorescence Spectrometry

In atomic fluorescence spectrometry the exciting source is placed at right angles to the flame and optical axis of the spectrometer. Some of the incident radiation is absorbed by atoms of the test element. Immediately afterwards this energy is released as atomic fluorescence of characteristic wavelength upon return of the excited atom to the ground electronic state. So far atomic fluorescence has been applied largely to research applications with no commercial exploitation. The schematic diagram of instrumental components is shown in Fig. 5-10.

The best burner system in AFS is probably a combination of acetylene/nitrous oxide and hydrogen/oxygen/argon using a rectangular flame with a premix laminar flow burner. The flame should possess a low background and a low quenching cross section, in addition to being efficient in producing a large free-atom population. Whereas the influence of flame background on the detection limit in AFS is most severe with an unmodulated source and dc detection, the presence of intense flame background is detrimental even for systems employing modulation. Although the unmodulated flame background is not amplified directly with ac detection, its presence results in noise at the output of the amplifier.

The intensity of fluorescence is linearly proportional to the exciting radiant flux. When no metal is present, only background radiation is seen from the flame. This difference

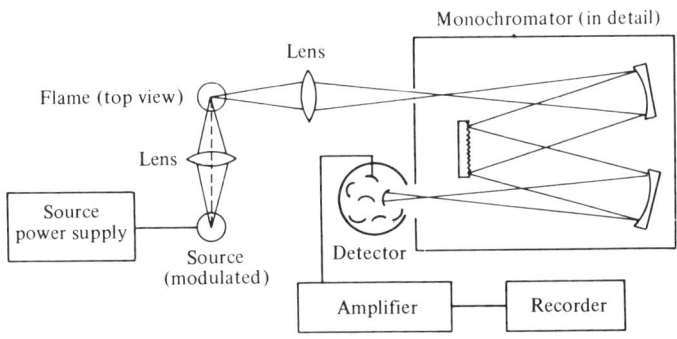

FIGURE 5-10 Schematic diagram of equipment for atomic fluorescence spectrometry.

between AFS and AAS is significant near the detection limit and thus for trace analysis. Atomic fluorescence spectrometry exhibits its greatest sensitivity for elements having high excitation energies.

Comparison of Flame Emission and Atomic Absorption Techniques

Comparing the analytical performances of both AAS and FES, it can be seen that these methods supplement each other in many respects. The analytical performance of FES is better for alkali, alkaline-earth, and rare earth elements, as well as for Ga, In, and Tl. Absorption flame spectrometry permits Ag, Al, Au, Bi, Cd, Cu, Hg, Pb, Te, Sb, Se, and Sn to be detected with high sensitivity. The performance of both methods for other elements is very similar. The choice of the method depends upon the matrix to be analyzed and the analysis-selectivity desired. In routine analysis it is reasonable to combine both methods. However, FES possesses one very important advantage in that it allows simultaneous quantitative multielement analyses to be performed. Atomic fluorescence spectrometry shows its greatest sensitivity for elements having their best lines at short wavelengths and thus competes more strongly with AAS.

Alkemade[7] has compared the theoretical basis of FES and AAS. He shows that AAS can be more sensitive than FES only if the brightness of the lamp exceeds that of a blackbody at the temperature of the flame, both measured at the wavelength of the analytical line. For nonthermal sources, as in various electrical discharges such as hollow cathode lamps, the spectral radiance of the lamp may be much larger than the flame as a blackbody radiator, permitting the great superiority of AAS at short wavelengths. Much hotter flames are required to improve FES in the deep ultraviolet. For lines at longer wavelengths, FES can be more sensitive than AAS. Populations of energy levels do not enter the argument. The reader should consult Alkemade's paper for more detailed discussion of this and other commonly accepted fallacies concerning AAS. With respect to hotter flames for FES, argon plasmas and induction-coupled plasmas, discussed in Chapter 6, do provide the answer.

For most elements, the flame probably is the chief source of the noise content of the signal by either AAS or FES. Since background intensity is considerably greater in FES than in AAS, the flame noise may influence detection limits more strongly in the former. In AAS, the primary light source is an additional source of noise. No consistent difference in reproducibility of results exists between the two methods. Relative standard deviations of 0.5-1.0% are attainable in favorable cases. Except for work near the detection limit, the instability of the nebulizer and flame makes the major contribution to the scatter in both methods.

Electrothermal Atomization[1,8]

Heated graphite furnaces and carbon rod analyzers are major types of flameless atom cells that fill an important ancillary role in AAS. As we have seen, the conventional nebulizing system is wasteful of sample and the residence time of metal atoms in the light path in a conventional flame is very short. The nonflame cell's most attractive features are high sensitivity (10^{-8} to 10^{-11} g absolute), the capability of handling very small sample volumes (5–100 μl) or solid samples directly without (sometimes) pretreatment, and low noise. Matrix effects are often more severe than in flame systems, and the precision, typically 5–10%, compares unfavorably with that of flames.

The auxiliary apparatus for electrothermal atomization consists of three parts: the workhead, the power unit, and the controls for the inert gas supply. The workhead is inserted in the spectrometer in place of the burner–nebulizer assembly. The power unit supplies operating current at the proper voltage to the workhead and provides automatic control of the entire heating program. The gas control unit provides metering and control over the flow of inert gas through the workhead, and for hydrogen gas in units which employ a hydrogen diffusion flame during a reductive ash cycle.

The heated graphite atomizer, shown in Fig. 5-11, consists of a hollow graphite cylinder, 28 mm in length by 8 mm in diameter, placed horizontally with the light path passing through longitudinally. The interior of the cylinder is coated with pyrolytic graphite. Electrodes at the ends of the cylinder are connected to a low-voltage, high-current supply capable of delivering up to 3.6 kW to the cylinder walls. Liquid samples are inserted through the orifice in the top center of the cylinder with a microsyringe. Solid samples can be introduced through one end with a special sampling spoon or inserted, after drying liquid samples, as a microboat constructed from tungsten. A metal housing around the furnace assembly is water-cooled to restore quickly the atomizer to ambient temperature after each atomization. Inert gas (argon) flow enters the graphite cylinder at the ends and exits at the sample introduction port. This gas flow ensures that the matrix components vaporized during the ashing step are quickly expelled and nothing deposits inside the cylinder where subsequent vaporization during the atomization step would produce a large background absorption signal. Removable quartz windows at each end of the graphite cylinder prevent ambient air from entering. A separate inert gas flowing around the outside of the cylinder, but inside the metal housing, prevents oxidation of the graphite.

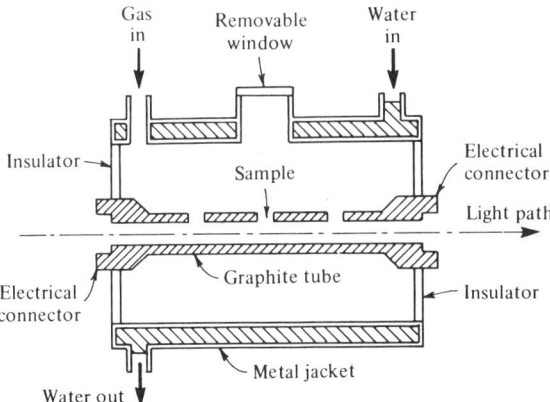

FIGURE 5-11 Cross section of a heated graphite atomizer. (Courtesy of Perkin-Elmer Corp.)

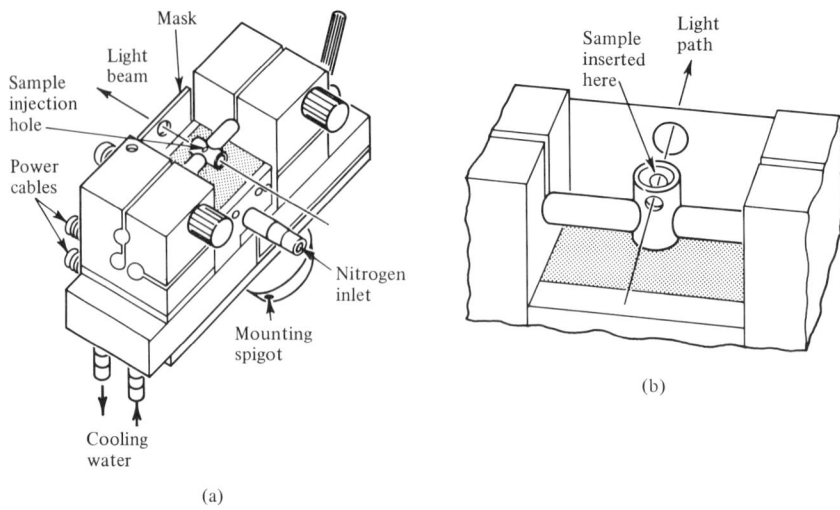

FIGURE 5-12 Carbon rod atomizer: (a) horizontal rod version and (b) vertical cup version. (Courtesy of Varian Associates, Inc.)

A miniature version, the carbon rod atomizer, consists of a three-piece tube or cup unit (Fig. 5-12). The workhead containing the miniature furnace is supported between two graphite electrodes in water-cooled terminal blocks. The furnace itself is 9 mm in length by 3 mm in diameter. Sample capacity is up to 10 μl for standard smooth tubes, and to 25 μl for tubes with grooves (or threads). The central unit can be replaced with a vertical cup held between the electrodes; this version is useful for solid samples or samples requiring preliminary chemical treatment that can be done directly in the cup. All units are coated with pyrolytic graphite. In normal use the carbon rod atomizer is protected from oxidation by a sheath of inert gas directed on the rod from a chimney beneath. When hydrogen gas is added to generate a reducing environment, the gas ignites spontaneously when the tube or cup reaches incandescence.

Following insertion or injection of the sample into the electrothermal atomizer, the heating sequence (independently programmable steps with temperature ramp and hold times in each step) is initiated to carry the sample through three stages: dry, ash or char, and atomize. In the dry cycle, the system is heated for 20-30 sec at 110-125°C to evaporate any solvent or extremely volatile matrix components. A dry residue of the sample remains, appearing as a slight stain or crust on the interior of the graphite tube or rod. The ash or char cycle is conducted at a predetermined intermediate temperature sufficient to accomplish several different things. One is to volatilize the higher boiling matrix components. Another is to pyrolyze matrix materials such as fats and oils, which will crack and carbonize. This cycle often converts the analyte to a different chemical state. The obvious problem at this stage is loss of analyte if the ashing temperature is too high or is maintained for too long a period. Finally, in the third stage, the optimum maximum power is applied to raise the furnace unit to the desired atomization temperature. The sudden power surge of the atomize cycle dissociates the analyte residue and volatilizes the atoms, thus creating the atom cloud responsible for the atomic absorption. The transient signal is recorded on a rapid-response recorder or encoded in a microprocessor for ultimate readout of peak height or peak area. The total atomization time is usually 2–3 sec with the carbon rod atomizer and 4–8 sec with the heated graphite atomizer.

A wide variety of samples can be handled directly. These include organic solvents, viscous liquids, liquids with a high dissolved solids content, and pulverized solids. Sample pretreatment is often not necessary, especially for biological and other organic materials, since organic material is destroyed during the ashing step. In contrast to the open flame of the burner system, the liberated atoms frequently remain in the optical beam longer than 1 sec. Hence the detection limits for the flameless technique is generally 2–3 orders of magnitude better.

Each step in the cycle of dry, ash, and atomize is important and the proper time and temperature parameters must be carefully selected. In the dry cycle the evaporation of solvent must be smooth and gentle to avoid mechanical losses by foaming or splattering. The course of the drying cycle should be followed by recording the absorption signal as it is attenuated by the escaping solvent vapors, and without background correction; there should be a smooth decline in the signal to baseline without any appearance of humps or spikes which indicate too rapid a heating rate. The ashing cycle is followed in the same manner; usually no loss of analyte will be observed up to some finite temperature, after which the atomic absorption signal begins to increase rapidly. Most organic materials pyrolyze at 350°C to leave a residue of amorphous carbon. Then, if a stream of air or oxygen is allowed to enter the furnace at this temperature, the carbon residue reacts rapidly to form carbon dioxide.

During the final atomization step, proper background correction with a continuum source is necessary to eliminate any molecular absorption contribution to the atom peak. Spectrometers with a fast response, not the conventional slow atomic absorption units, should be used. The latter units were designed for the measurement of stable signals generated from the chemical flame environment. When the temperature for a given element is too low, the sensitivity is drastically reduced. However, once a certain threshold atomization temperature has been reached, a higher temperature neither adds to nor detracts from the results. To determine two elements at once with a two-channel spectrometer, it is only necessary to set the atomization temperature to that required for the more refractory of the two elements.

Automatic pipetting eliminates the nuisance of manual operation where pipetting microsamples with exquisite care every 3 min creates tedium and chances for error. Autosamplers are commercially available as auxiliary options.

Chemical Vaporization[9]

A vapor generation accessory is a system which chemically treats the sample to generate a volatile product to be subjected to analysis for AAS. The technique has been applied to a series of elements (As, Bi, Ge, Sb, Se, Sn, and Te) that form volatile hydrides, and to the analysis of mercury. The system consists of a vapor generation unit in which mercury vapor or metallic hydride is generated, and a quartz absorption cell, attached to a standard atomic absorption burner, in which the atomic absorption is measured. Gaseous hydrides may be generated with sodium borohydride, dispensed in pellet form, as the reducing agent added to an acid solution. A flow of inert gas, either nitrogen or argon, transports the metallic hydride from the generation unit to the absorption cell where the hydride is decomposed using a normal acetylene/air flame. Chemical evolution of mercury vapor does not require the use of a flame for analysis.

PROBLEMS

1. For the analysis of cement samples, a series of standards were prepared and the emission intensity for sodium and potassium was measured at 590 and 768 nm, respectively. Each standard solution contained 6300 μg/ml of calcium as CaO to compensate for the influence of calcium upon the alkali readings. The results are shown below:

CONCENTRATION, μg/ml	EMISSION READING Na_2O	K_2O
100	100	100
75	87	80
50	69	58
25	46	33
10	22	15
0	3	0
Cement A	28	69
Cement B	58	51
Cement C	42	63

For each cement sample 1.000 g was dissolved in acid and diluted to exactly 100 ml. Calculate the percent of Na_2O and K_2O.

2. In Problem 1, what contributed to the emission reading of the blank at the analytical wavelength for sodium, but did not for potassium? A small quartz spectrometer was employed to obtain the results. Would the blank reading be larger, smaller, or the same if a filter photometer equipped with glass absorption filters had been employed? [See *Anal. Chem.*, **21**, 1296 (1949).]

3. Boron gives a series of fluctuation bands due to the radical BO_2 that lie in the green portion of the spectrum. Although the overlapping band systems present a problem in the measurement of the flame background, the minimum between adjacent band heads can be used. These results were obtained:

BORON PRESENT, μg/ml	EMISSION READING 518-nm PEAK	505-nm MINIMUM
0	36	33
50	44	36
100	52	39
150	60.5	42.5
200	68.5	45.5

What are the concentrations of boron in these unknowns?

A	45	36.5
B	85	65
C	66	50

[See *Anal. Chem.*, 27, 42 (1955).]

4. A calibration curve for strontium, taken at 460.7 nm, was obtained in the presence of 1000 μg/ml of calcium as CaO and also in the absence of added calcium. These results were

STRONTIUM PRESENT, μg/ml	EMISSION READING NO CALCIUM	CALCIUM ADDED
0	0	13
0.25	2	18.5
0.5	6	24
1.0	16	36
2.5	44	70
5.0	94	125
7.5	150	181
10.0	200	238

(a) Graph the calibration curve on rectilinear graph paper and also on log–log paper. (b) What might be the cause of the upward curvature in the region of low concentrations on the rectilinear graph when calcium is absent? (c) Why does the addition of calcium straighten the calibration curve and increase the net emission reading for strontium?

5. Calculate the iron content in a diethyldithiocarbamate extract using the following data:

ABSORBANCE UNITS BLANK	SAMPLE	IRON ADDED, μg/200 ml
0.0020	0.0090	None
0.0214	0.0284	2.00
0.0414	0.0484	4.00
0.0607	0.0677	6.00

6. A sample of mineral ash gave a meter reading of 37. Solutions B and C, containing the same quantity of unknown solution plus 40 and 80 μg/ml of added potassium, respectively, gave net meter readings of 65 and 93. Calculate the quantity of unknown potassium in the original solution.

7. A metal naphthenate sample, ashed and diluted to a fixed volume, gave a reading of 29. Solutions B and C, containing the same quantity of unknown solution plus 25 and 50 μg/ml of barium, gave readings of 53 and 78, respectively. Calculate the quantity of barium in the original solution.

8. To illustrate the effect of aqueous–organic solvents upon droplet size, calculate the mean droplet diameter for (a) water, (b) 50% methanol–water, and (c) 40% glycerol–water. Pertinent data follow:

SYSTEM	SURFACE TENSION, dyn/cm	VISCOSITY, dyn/cm^2	DENSITY, g/cm^3	VELOCITY OF ASPIRATING GAS, m/sec	Q_{air}/Q_{liq}
Ethanol, 50%	28	0.029	0.934		
Glycerol, 40%	68.6	0.039	1.102	279	2540
Methanol, 50%	30.6	0.027	0.946	198	9540
Methyl isobutyl ketone	24.6	0.0051	0.801		
Water	73	0.010	1.00	198	6400

9. For (a) water, (b) 50% (v/v) ethanol–water, and (c) methyl isobutyl ketone as solvents, plot the droplet diameters for solution flowrates ranging from 0.1 to 5 ml/min. As values for a typical nebulizer, assume the velocity of the aspirating gas to be 333 m/sec and Q_{gas} to be 8.5 liter/min. Other data are given in Problem 8.

10. Calculate the fraction of cesium atoms ionized in a flame at 2000°K when the total cesium concentration in the flame gases is (a) 10^{-4} atm, (b) 10^{-6} atm, (c) 10^{-7} atm.

11. Calculate the fraction of lithium atoms ionized in a premixed laminar flame at 3000°K when the concentration sprayed into the flame is (a) $10^{-2} M$, (b) $10^{-3} M$, and (c) $10^{-4} M$.

12. What individual amounts of (a) cesium, (b) rubidium, (c) potassium, or (d) lithium should be added individually to a flame at 2500°K and at 2800°K to suppress the ionization of a solution containing 0.23 μg/ml of sodium?

	K_i, atm	
ELEMENT	2500°K	2800°K
Li	1.48 × 10^{-9}	2.63 × 10^{-8}
Na	4.8 × 10^{-9}	7.40 × 10^{-8}
K	1.8 × 10^{-7}	2.08 × 10^{-6}
Rb	3.9 × 10^{-7}	3.98 × 10^{-6}
Cs	1.45 × 10^{-6}	1.32 × 10^{-5}

13. If lithium (670.8 mm) is used as an internal standard for the determination of sodium (589.0 nm) and potassium (766.5 nm), what is the maximum permissible temperature variation of the flame if one desires to maintain deviations in intensity ratios less than 1%? Assume a flame temperature of 2000°K.

14. From the absorbance traces shown for arsenic by atomic absorption, estimate (a) the signal-to-noise ratio, (b) the sensitivity of the method, and (c) the detection limit.

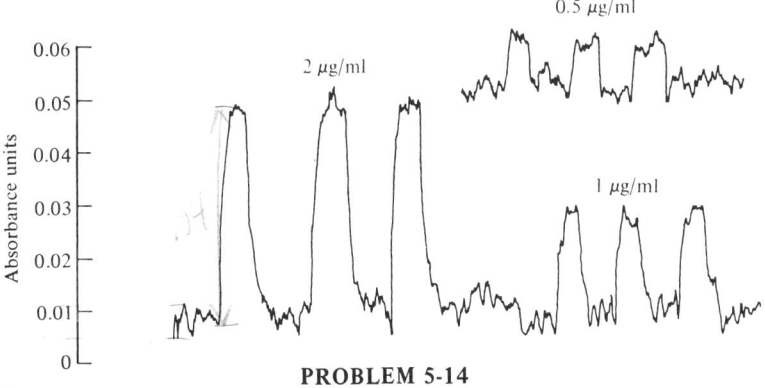

PROBLEM 5-14

BIBLIOGRAPHY

Alkemade, C. Th. J. and R. Herrmann, *Fundamentals of Analytical Flame Spectroscopy,* Hilger, Bristol, 1979.

Christian, G. D. and F. J. Feldman, *Atomic Absorption Spectroscopy,* Wiley-Interscience, New York, 1970.

Dean, J. A., *Flame Photometry,* McGraw-Hill, New York, 1960.

Dean, J. A. and T. C. Rains, Eds., *Flame Emission and Atomic Absorption Spectrometry: Theory,* Vol. 1, 1969; *Components and Techniques,* Vol. 2, 1971; *Elements and Matrices,* Vol. 3, 1975, Marcel Dekker, New York.

Fuller, C. W., *Electrothermal Atomizer for AAS,* The Chemical Society, London, 1977.

Kirkbright, G. F. and M. Sargent, *Atomic Absorption and Fluorescence Spectroscopy,* Academic, London, 1974.

Mavrodineanu, R., Ed., *Analytical Flame Spectroscopy: Selected Topics,* Macmillan, New York, 1970.

Mavrodineanu, R. and H. Boiteux, *Flame Spectroscopy,* Wiley, New York, 1965.

Pinta, M., *Atomic Absorption Spectrometry,* Hilger, London, 1975.

Sychra, V., V. Svoboda, and I. Rubeska, *Atomic Fluorescence Spectroscopy,* Van Nostrand Reinhold, London, 1975.

LITERATURE CITED

1. Syty, A., *Crit. Rev. Anal. Chem.,* **4**, 155 (1974).
2. Nukiyama, S. and Y. Tanasawa, *Trans. Soc. Mech. Eng. Japan,* Repts. 4, 5, 6 (1938-1940).
3. Rann, C. S. and A. N. Hambly, *Anal. Chem.,* **37**, 879 (1965).
4. Dean, J. A. and J. E. Adkins, *Analyst,* **91**, 709 (1966).
5. Pickett, E. E. and S. R. Koirtyohann, *Anal. Chem.,* **41**, 28A (1969).
6. Walsh, A., *Spectrochim. Acta,* **7**, 108 (1955).
7. Alkemade, C. Th. J., *Appl. Opt.,* **7**, 1261 (1968).
8. Sturgeon, R. E., *Anal. Chem.,* **49**, 1255A (1977).
9. Robbins, W. B. and J. A. Caruso, *Anal. Chem.,* **51**, 889A (1979).

CHAPTER 6

Atomic Emission Spectroscopy

Atomic emission spectroscopy has long been the standard for routine metal analysis in many analytical applications. The basic principle underlying the atomic emission spectrographic method may be described as follows: A minute part of the sample is vaporized and excited to the point of light emission. This may be done by means of an electric arc or spark, a dc argon plasma, an induction-coupled argon plasma, or a laser. Having selected the source, the emission spectrum is focused onto the spectrometer's entrance slit. The light derived from the vaporized, excited material is dispersed into its component parts in the spectrometer. Off axis mirror optics permits site selection within the source discharge without the need to move the excitation assembly; it also eliminates the chromatic and spherical aberrations encountered with lens optics. At the exit aperture this light is either photographed on a plate or film, or recorded by a photodetector. Since each element produces a series of spectral lines of specific wavelengths, which is characteristic of itself, the identification of an element is possible by studying the lines according to their respective locations. Determination of quantity is made according to the intensity of the lines in almost any matrix ranging from alloys to ores, or from ashes of organic materials to atmospheric dusts.

Atomic emission spectroscopy provides simultaneous multielement capability of emission while retaining the detection limits of the graphite furnace (see Chapter 5), effective analysis of refractory and rare earth elements and nonmetals, and ease of operation by trained technicians on a variety of real samples. Where atomic emission spectroscopy had lost ground, mostly to atomic absorption, was in the analysis of liquids. Techniques involving the traditional arc and spark sources proved unwieldy for the excitation of liquids. However, the development of dc argon plasma and induction-coupled argon plasma sources has caused a virtual renaissance of atomic emission spectroscopy in handling solutions.

6.1 SPECTROSCOPIC SOURCES

No single excitation source is best for all applications. The analyst should have available a wide variety of sources which can be selected in accordance with the analytical requirements. Factors which influence the type of excitation required include: the concentration of the elements being determined, the vapor pressures or volatilities of these elements and the matrix, the excitation potentials of the spectrum lines that must be used, and the physical condition of the sample.

For solid samples generally arc excitation is more sensitive and spark excitation more stable. The high-temperature plasma sources will be the choice for solutions and for gaseous samples; their sensitivity enable trace analyses to be carried out in the part-per-billion level.

Direct-Current Arc

The dc arc source consists of a high-current (5–30 A), low-voltage (10–25 V) electrical discharge supported between two electrodes and operating in air. Vaporization occurs from the heating caused by the passage of current. Arc temperatures range from 4000 to 6000°K. Excitation of a sample by a dc arc discharge is partly thermal and partly electrical in origin as a result of the concentration of high-velocity ions, electrons, and atoms. The energy available for excitation varies along the length of the arc discharge. Near the cathode (often the sample electrode) the plasma energy is highest, and here the sample is quickly vaporized into the high-temperature region.

It is difficult to control the arc current and arc resistance, two variables that govern the temperature of the discharge and the excitation of the sample. Generating intense heat, an arc causes the tiny sample to burst into a vapor so rapidly and erratically that portions can escape outside the arc column. Even light from the elements in the arc may not be directed to the slit reproducibly. Another difficulty with an arc is its tendency to flicker and wander on the electrode surfaces during the discharge. In addition there is a problem involving volatility wherein the more volatile components may be selectively vaporized during the early portion of the arcing period. To obviate this latter difficulty the sample may be burned to completion. All these problems make the dc arc source better suited to qualitative or semiquantitative analyses, rather than to quantitative work. The dc arc usually has higher sensitivity than other arc or spark sources, and consequently it is frequently used for the analysis of rocks, minerals, soils, and plant tissues, particularly for qualitative survey analysis.

The necessary components for a dc arc are a direct-current power supply, a variable resistor, and a discharge gap (Fig. 6-1). 230 V from the supply main through an isolation transformer is rectified by two mercury rectifier tubes, back-to-back. Ripple is reduced to suitable analytical tolerances by means of a smoothing choke coil. The current is varied on the primary side of the isolation transformer by means of a tapped reactor. Ignition is provided by means of a high-voltage ac spark. After starting the arc, the spark can be turned off to eliminate this type of excitation. If the spark is left on, it adds stability to the dc arc by reducing the arc wandering. In addition, sparklike or arclike lines can be reduced or enhanced as a function of the exposure requirements by means of the relative mixture of arc and spark excitation.

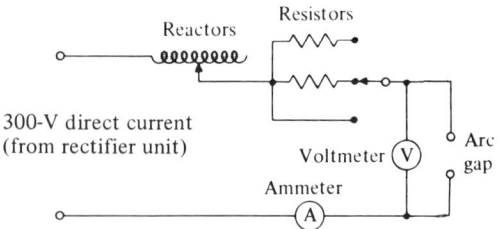

FIGURE 6-1 Circuit for a dc arc.

One popular electrode configuration, shown in Fig. 6-2, consists of a lower cup electrode which contains the sample and a counter electrode. Electrodes are carbon or graphite. Solid samples, usually in the form of powders, are placed in the cup electrode, alone or mixed with powdered graphite to enhance conduction. Conductive metals are usually cast or machined into appropriate electrodes which are used directly in the dc arc discharge.

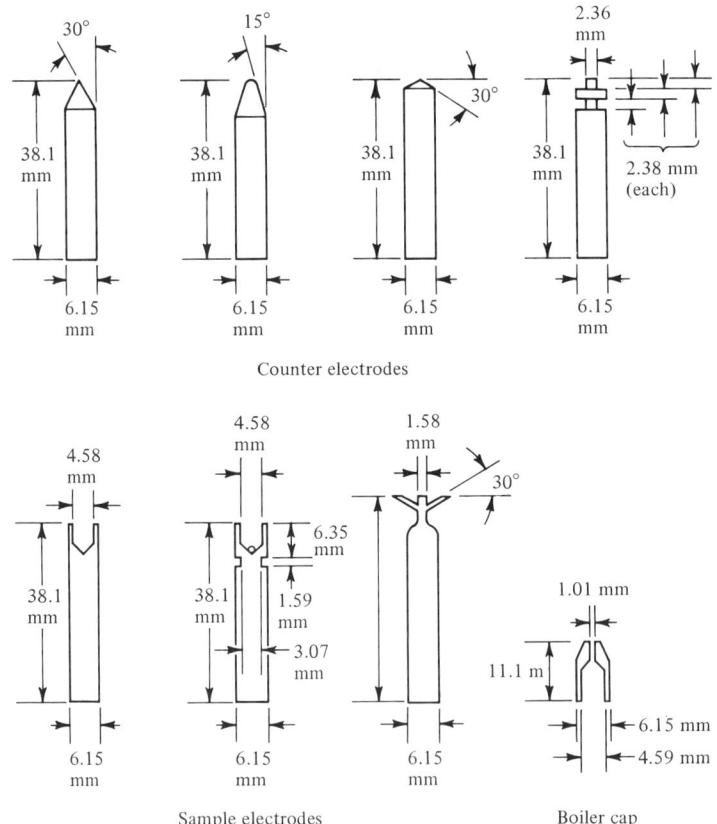

FIGURE 6-2 Examples of counter electrodes and sample electrodes for arc or spark sources.

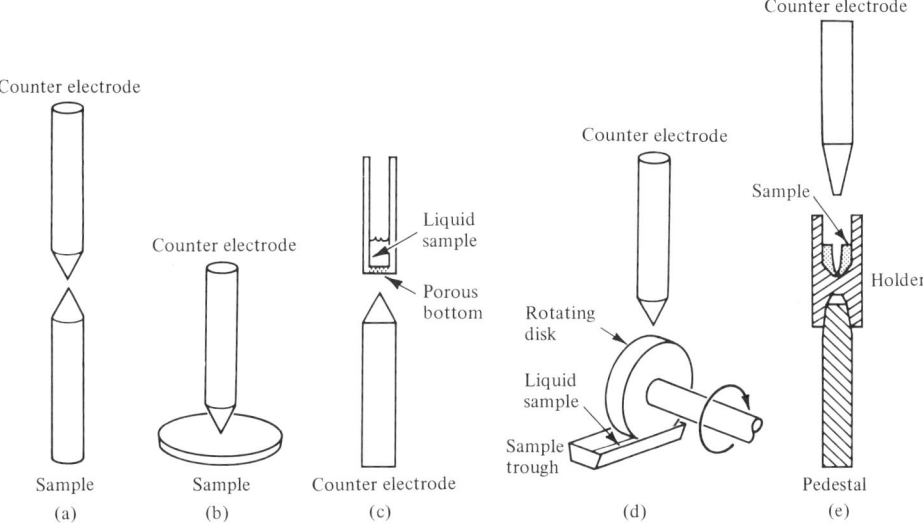

FIGURE 6-3 Electrode configurations: (a) point-to-point, (b) point-to-plane, (c) porous cup, (d) rotating disk, and (e) carrier distillation.

In the arrangement termed point-to-plane excitation (Fig. 6-3b), the sample electrode is cast into the form of a disk, and the arc is struck between the planar disk and a pointed counter electrode. In the point-to-point configuration (Fig. 6-3a), the sample electrode is cast in the form of a rod, the tip of which is sharpened to increase the electric field. Liquid samples may be transferred to an electrode and evaporated to dryness, before striking the arc.

Fractional volatilization of samples can be utilized for separation of more volatile species, such as the alkali metals and the alkaline earths, from the heavier or more refractory elements. By successive exposure of several spectra, above and below each other on the same photographic plate, the emission spectra of the more volatile elements can be detected before the complex spectra of the heavier elements can interfere. An extension of selective volatilization is found in the "carrier distillation" method sometimes utilized in the analysis of refractories. A low-boiling material is added to the sample to sweep (carry) the more volatile trace constituents up into the arc column while the base material remains largely behind unexcited. In order to heat the entire sample, it is placed in a special electrode perched atop a pedestal (Fig. 6-3e) to reduce the heat loss by convection. A few of the more popular carriers are copper hydroxy fluoride, silver fluoride, silver chloride, lithium fluoride, gallium oxide, or combinations thereof.

Radiation emitted by a dc arc discharge will be rich in emission lines and molecular spectra. Because the electrodes are heated to incandescence, they emit intense and spectrally continuous background radiation, resembling that from a blackbody. Atomic line spectra will be complex as the thermal and electrical energy of the dc arc is sufficient to populate many higher electronic energy levels, in contrast to the lower-temperature flames. Further complications stem from the many ionic emission lines, notably from elements having relatively low ionization energies (see Table 5-3). When an arc is operated between carbon electrodes in air, some cyanogen molecules are formed and, being excited by the arc, emit typical molecular band spectra in the region from 360 to 420 nm.

The light from the center portion of the dc arc is the portion usually focused on the entrance slit of the spectrometer because there may be a local concentration of certain ions near the electrodes. However, since the cathode region gives higher excitation energy, especially for lines arising from nonionized atoms, it is more sensitive for these atoms and is sometimes employed for illuminating the slit of the spectrometer.

The effects of arc wandering can be mitigated by using a narrow (3.18 mm) diameter electrode and setting the current to a value where the anode spot just covers the electrode's top surface. The anode spot is the area of intense light and heat where the arc strikes the electrode. Under such conditions the radiation reaching the slit is relatively constant and the sample volatilizes smoothly.

An attachment that can be used with the dc arc to enhance sensitivity is the Stallwood jet, shown in Fig. 6-4. Gas is forced into a swirl chamber and then upward through an orifice surrounding the sample electrode to keep the arc from wandering. As the electrode burns away, it may be advanced with respect to the curtain of gas. The assembly is inside a quartz enclosure to exclude ambient air. Typically the gas is a combination of oxygen: argon (30:70). Here the exclusion of nitrogen virtually eliminates the cyanogen bands which blanket such large portions of the ultraviolet spectrum. The Stallwood jet also reduces selective volatilization by cooling the bottom of a deep-cratered electrode with the forced flow of the gas.

The intensity of atomic emission in a dc arc discharge is dependent on the sample matrix because the matrix strongly affects the resistance and mode of excitation, whether thermal or electrical. Not only must samples and standards be virtually identical chemically, but physically as well. One popular scheme is to flux the sample with a material such as lithium tetraborate. To promote uniformity, lithium compounds are widely added as fluxing agents, buffers, and internal standards. Lithium is an element not ordinarily sought, and its excitation potential is approximately median for many materials. Lithium chloride is

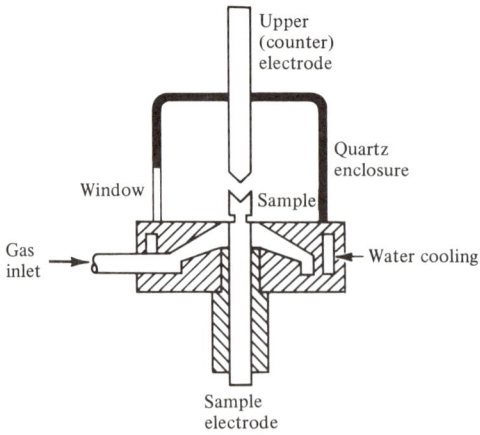

FIGURE 6-4 Schematic diagram of the Stallwood jet.

a low-boiling substance and, when added in large quantity to a sample, elements tend to behave uniformly. The intensity of their lines is independent of other minor elements present and the nature of the sample matrix.

Another way to minimize the effects of arc instability and the sample matrix is through use of an internal standard, an element whose vaporization and excitation characteristics very closely match those of the element to be determined. Use of an internal standard provides improvements in precision up to an order of magnitude.

Alternating-Current Arc

Alternating currents can be used to sustain an arc, although higher voltages are required than for a dc arc. The discharge is more uniform than that of the dc arc since its polarity reverses 120 times/sec if a 60-Hz source is used. The ac arc provides better reproducibility than the dc arc but the sensitivity is decreased. Thus, the sustaining ac arc represents a good compromise between stability and sensitivity.

The ac arc finds use for the analysis of residuals in steels and other metal samples as well as in handling of liquid samples which can be evaporated on the flat ends of the (copper) electrodes. The selective volitilization that is a problem with the dc arc does not occur with the ac arc.

High-Voltage, Alternating-Current Spark

The high-voltage, ac spark is normally the source providing the greatest precision and stability among arc and spark sources in spectrochemical analysis. The oscillatory, condensed spark is produced by connecting a high-voltage transformer (10–50 kV) across two electrodes. The frequency of oscillation, duration of a spark, and the amount of current will depend on the values of capacitance and inductance in the circuit, and on electrode separation. A condenser in parallel with the spark gap increases the current. An inductor in the circuit decreases the excitation of lines and bands of the air molecules. Large values of inductance decrease the excitation energy and make the spark more arclike in its characteristics. Rather elaborate source units are available for allowing variations in capacitance and inductance values.

The circuit shown in Fig. 6-5 employs an auxiliary rotating spark gap which is driven by a synchronous motor. A jet of air has also been used to trigger the breakdown of the

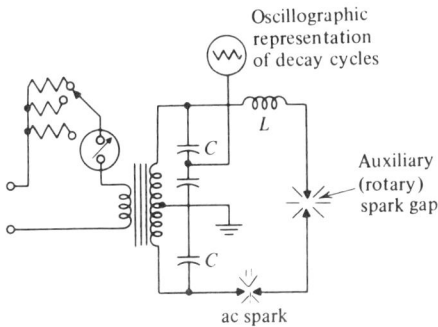

FIGURE 6-5 Circuit diagram for an ac spark source.

gap. The gap is closed for only a brief instant at the peak of each half-cycle. Thus the number of decay cycles in the spark is controlled. Oscillations are of a damped type and cause the polarity of the gap electrodes to reverse rapidly during the short (10–100 μsec) pulses, which may occur 1000–2000 times/sec. During oscillations of the spark, the electrodes change polarity so that the sample is recycled between sample and counter electrode to provide vaporization by positive-ion bombardment and efficient excitation.

Although the duration of a single spark is quite short, the period between discharges is relatively long. This delay gives the electrodes a reasonably long time to cool off between spark discharges; thus heating effects during excitation are largely eliminated, resulting in less fractional distillation and sample consumption as compared with an arc.

The high-voltage, ac spark source populates the very high-energy electronic levels of atoms. Thus the average available energy is distributed over a large number of excitation possibilities, including many ionic emission lines. The emission spectra, while more intense, are also more complex than those produced by the dc arc. This renders qualitative analysis more tedious. Thus, although excellent for precision, the ac spark is limited in sensitivity. In metal analysis the spark is commonly limited to concentrations down to 0.01%, depending on the sensitivity of the element being analyzed. With solution techniques or the copper spark method, spark excitation can be used successfully down into the part-per-million range.

Conducting samples are usually ground flat and used as one electrode with a pointed graphite counterelectrode (point-to-plane technique). Powdered samples are mixed with graphite powder and pressed into a pellet which is used as the plane electrode. Solutions are usually determined using a porous cup graphite electrode (Fig. 6-3c) or a rotating disk electrode (Fig. 6-3d). The former consists of a porous-bottom graphite cup containing the sample solution. The counterelectrode beneath the cup discharges to the wet bottom of the porous cup. The rotating disk electrode consists of a rotating graphite disk, the lower edge of which dips into the sample solution rendered conducting by the addition of acid, and carries it to the spark discharge region at the top of the disk.

Microprobes

The laser microprobe[1] is well suited to the analysis of very small samples or very small areas of a sample. No sample preparation is required. An optical pulsed ruby laser is focused via a microscope onto a very small area of the sample; visual focusing is used to select the sample area. The microprobe is shown in Fig. 6-6. The intense laser pulse vaporizes a small amount of sample, leaving a hemispherical crater about 50 μm in diameter. The vapor plume, on passing between two closely spaced electrodes that are charged to a high voltage, causes the electrode gap to break down. This results in sparklike excitation of the sample vapor.

The laser microprobe furnishes a tool for examining the interior of individual cells, even in living organisms, or inclusion areas in alloys. Resolution less than 50 μm is presently limited by the lenses used to focus the laser beam. The immense energy in such a small area may melt the cement between lenses or shatter the lens itself.

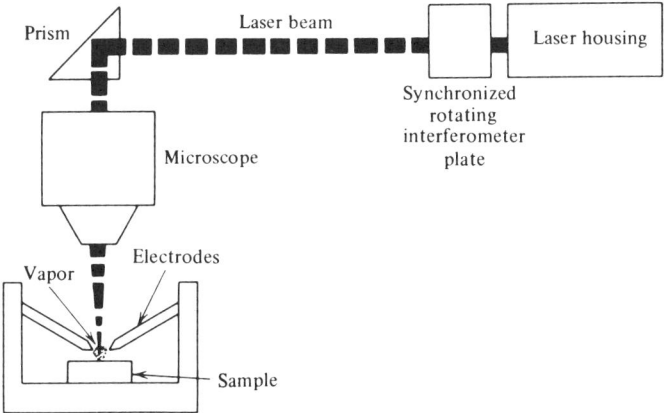

FIGURE 6-6 Schematic diagram of a laser microprobe. (Courtesy of Jarrell-Ash Co.)

A spark microprobe uses an electrode whose tip is polished to a 1-μm point.[2,3] It is positioned about 25 μm from the sample, which must conduct electrically or be deposited as a thin film on a conductor, and an ac spark is used to volatilize and excite to emission an area beneath the electrode point.

Plasma Emission Sources[4]

The inherent physical properties of a plasma system offer performance and operational advantages over traditional arc and spark emission sources. Sensitivity and accuracy are exceptional. Linear dynamic range is four or more orders of magnitude. Gases are simple to handle; among the substances that have been introduced in this manner are phosphine, metal hydrides, and ammonia. Liquids are easily introduced as aerosols generated by a pneumatic nebulizer. When running samples of high salt content, it is necessary to employ an argon saturator/nebulizer tip washer system to prevent clogging. There is no need to dilute samples just to prevent nebulizer clogging.

A plasma source provides the analyst with an extremely rich spectrum. One is not limited to ground-state transitions but can choose from first or even second ionization state lines. A problem that arises with such a catalog of spectral lines is an increased instance of spectral interference with a spectrometer of anything less than superior resolution. The property that makes a plasma most suitable as an excitation source is that large quantities of electrical energy can be transferred to it once it is sufficiently ionized, raising its temperature as high as 9000°K.

Inductively Coupled Argon Plasma[5-8]

The inductively coupled argon plasma (ICAP) torch is a special type of plasma that derives its sustaining power by induction from a high-frequency magnetic field. The plasma, actually a partially ionized gas, is formed electromagnetically by radio frequency induction-

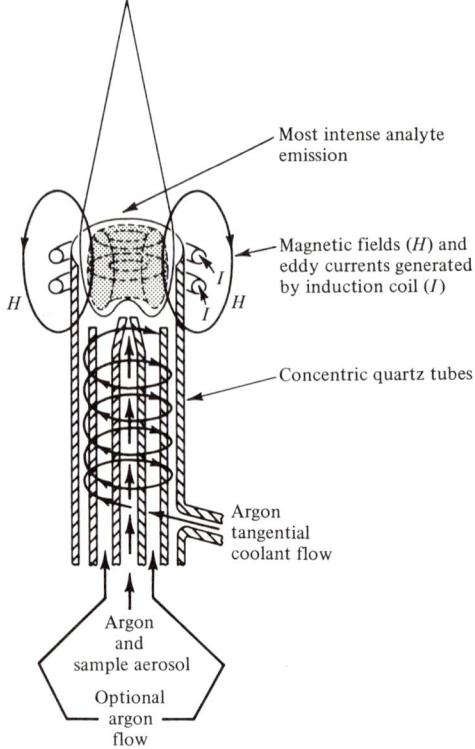

FIGURE 6-7 Schematic configuration of an induction-coupled argon plasma (ICAP) torch.

coupling of argon gas. The torch configuration is shown in Fig. 6-7. Two concentric silica tubes, the inner tube stopping below the induction coil, contain the argon plasma, while a third tube is inserted into the center to inject an aerosol through the plasma after it has formed. The work coil (water cooled) of a high-frequency generator is positioned around the top of the outer silica tube (20 mm in diameter).

To operate the torch, argon gas is fed tangentially into the inner silica tube and is ionized by the magnetic field produced by the induction coil (up to 2 kW of energy at 27.14 MHz). Since the argon plasma has a high density of free electrons, it is a good electrical conductor and will interact readily with the magnetic field that is created once power is supplied to the induction coil. The argon gas stream is then seeded with free electrons provided directly by a tesla discharge into the stream. Seed electrons in the coil space interact with the magnetic field and quickly gain sufficient energy to ionize the gas stream by collisional excitation. Once the plasma gains a sufficient free-electron concentration, an eddy current, induced by the magnetic field, flows in azimuthal circular closed paths around the discharge periphery. Power transfer between the induction coil and the discharge is similar to the transformer principle, in which the induction coil is equivalent to a two-turn primary winding and the plasma is equivalent to a one-turn secondary winding. After the ionization occurs, a flame-shaped plasma forms near the top of the torch.

If a second stream of gas (called the coolant) is made to flow tangentially around the plasma gas stream, the shape of the spheroidal discharge is modified by a flattening of its base. This makes it easier for a third stream, containing the sample aerosol, to be injected up the axis of the torch and into the core of the plasma, further modifying the shape of the fireball from a prolate spheroid to an annulus. A long, narrow, well-defined tail flame now emerges from the fireball. This tail flame becomes the spectroscopic source; it contains all the analyte atoms which are heated as they pass through the tunnel in the center of the annulus. The optical window used for analysis falls just above the apex of the primary plasma cone and just under the base of the flamelike afterglow. In this optical window the high background or current-carrying region of the plasma is excluded from the spectrometer. Typical argon flow rates are 1 liter/min in the carrier, 0–1 liter/min in the auxiliary plasma, and 15 liter/min in the coolant plasma.

There is no electrode contact in the ICAP source. Therefore, excitation and emission zones are resolved spatially, producing a relatively clean background spectrum which consists of argon lines and some weak band emission from OH, NO, NH, and CN molecules. This low background, combined with a high signal-to-noise ratio of analyte emission, produces low detection limits, typically in the parts-per-billion range. It is the geometry of the ICAP source that renders it particularly useful. The analyte is carried directly through a narrow channel in the center of the plasma discharge where it experiences high excitation temperature and long residence times (approximately 2 msec). The definitive boundary that exists between the analyte column and the current-carrying part of the inert plasma is one of the main reasons interelement and matrix effects are minimal. The high temperatures of the radiation zone ensure the complete breakdown of chemical compounds and impedes the formation of other interfering compounds.

For many analyte species, ion line emission is considerably more intense than neutral atom line emission in the ICAP source. For calcium, the neutral atom line at 422.7 nm cannot even be observed relative to the two ion lines. This is true for many other elements such as Ba, Be, Fe, Mg, Mn, Sr, Ti, and V. For these elements, ion lines provide the best detection limits.

Direct-Current Argon Plasma[9]

In the three-electrode direct-current argon plasma source, the plasma jet is formed between two spectrographic carbon anodes and a tungsten cathode in an inverted wye configuration (Fig. 6-8). The sample excitation region and observation area is centered in the crook of the wye where spectral contribution from the plasma continuum is minimal. The plasma requires about 1 kW of power and consumes about 8 liter/min of welder's grade argon. Of particular significance is the plasma's stability in the presence of varying solvent types such as those containing large amounts of dissolved solids (as high as 25%), organics, and high acid/alkaline concentrations. The degree to which aqueous calibration standard matrices must be matched to sample matrices is governed only by viscosity and surface effects on sample introduction rate (similar to ICAP).

To operate, the three electrodes are moved into contact by argon-actuated pistons, and plasma ignition is initiated automatically without a high-voltage spark as the electrodes are withdrawn. The wye configuration improves stability, as compared to the inverted

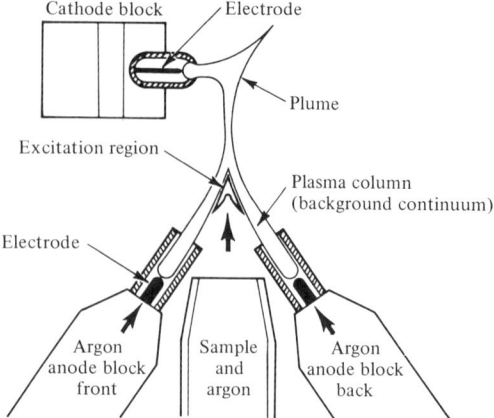

FIGURE 6-8 Schematic of the dc argon plasma source in the wye configuration. (Courtesy of Spectrametrics, Inc.)

vee of earlier versions, in that it stabilizes the position of the plasma and sample excitation area. The electrodes are cooled by a flowing water system. Once ignited, the plasma is sustained by a low voltage (40 V at 7.5 A) and produces temperatures as high as 9000 to 10,000°K. The excitation region formed at the juncture of the plasma is approximately 6000°K. Samples are nebulized and introduced into the excitation area in aerosol form. The entire cycle of desolvation, molecular dissociation, and excitation takes place in this extremely high-temperature excitation region during the residence time of the sample. Operations are conducted in an inert atmosphere. Because the excited sample is observed in a region separated from the main plasma core, an optimum signal-to-background noise is attained.

The only major problem-solving limitation of the dc argon plasma source is its unsuitability for totally automated operation. The plasma-supporting electrodes are consumed (approximately 6 mm) to the point of requiring reshaping after about 2 hr of continuous operation.

6.2 ATOMIC EMISSION SPECTROMETERS

In emission spectrometry the two major sources of noise that limit the sensitivity are from the source (flicker) and from the photomultiplier detector (dark current). Since dark-current noise will remain constant, it is important to decrease the source noise from the background by decreasing the signal bandpass until detector noise predominates or until the spectral linewidth is reached. Therefore, it is an advantage to have a high-resolution, high-luminosity monochromator so that one can easily isolate a spectral line profile from its background without loss of light throughput and consequent decrease of the source noise relative to the detector noise.

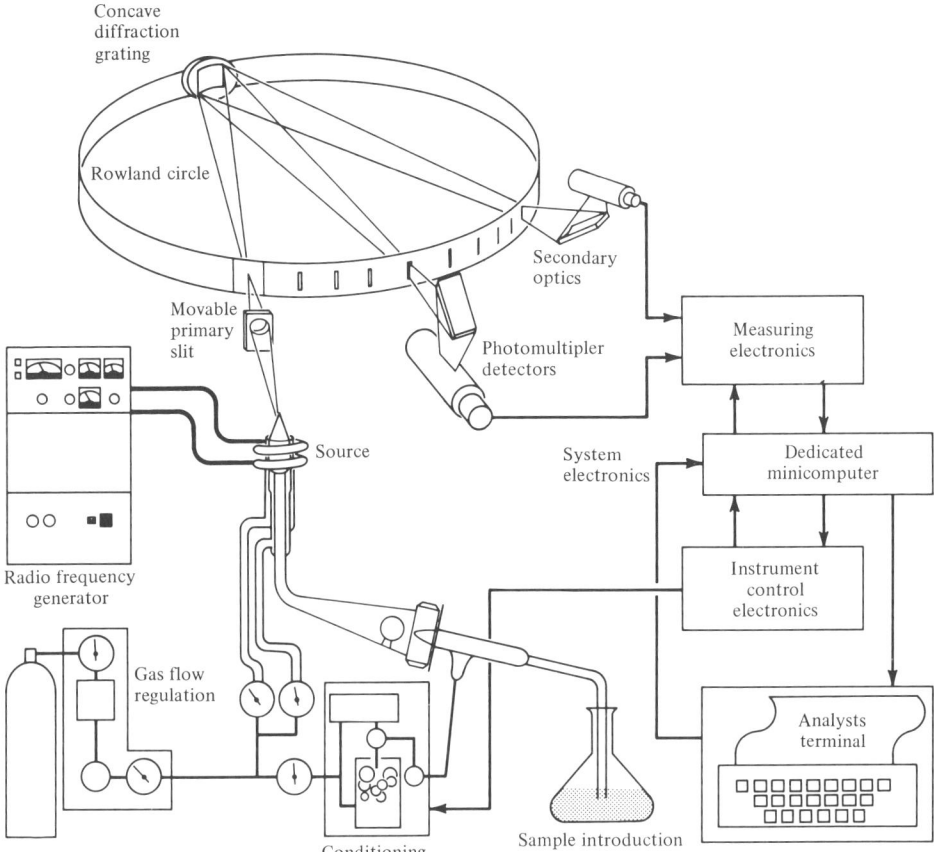

FIGURE 6-9 Schematic diagram of a nonscanning (direct-reader) spectrometer employing a holographic concave grating in the Rowland circle configuration. (Courtesy of Applied Research Laboratories.)

Concave Grating Instruments

The spectrometer shown in Fig. 6-9 is used for multielement (nonscanning) analyses and employs a concave holographic grating in a Rowland circle configuration. In this configuration slit, grating, and focal plane are arranged so as to lie on the circumference of the Rowland circle, which is just the focal curve of the concave grating. The circle has a diameter equal to the radius of curvature of the grating. The grating is the only optical surface between the spectrometer's entrance and exit slits. A 1-m focal length assures good spectral dispersion and resolution. A vacuum unit allows determinations of sulfur, phosphorus, boron, iodine, and other elements with spectral emissions down to 170 nm. The photomultiplier tubes are external to the spectrometer housing which reduces secondary array clutter and avoids secondary array scattered light. Secondary mirrors project

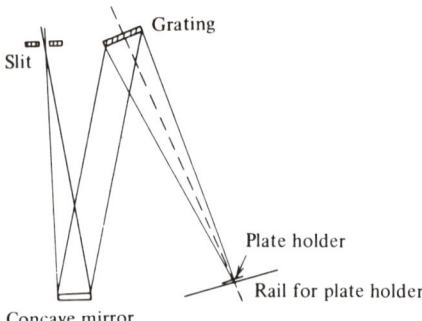

FIGURE 6-10 Wadsworth mounting for a concave grating.

and focus the light onto photomultiplier tubes either above or below the secondary optics. For most analytical tasks only the upper array is used. However, for extended programs for up to 60 elements, the lower array is available. Exit slits are positioned to pass only specific wavelengths for the elements of interest. Mirrors positioned behind the exit slits gather diverging light and focus it on the cathodes of the photomultiplier tubes.

The Wadsworth mounting (Fig. 6-10) requires a collimating concave mirror so that the concave grating may be illuminated by parallel light. The speed and light-gathering power of the arrangement is high since the grating is used at about half the image distance. Furthermore, the arrangement is stigmatic; that is, light arising from horizontal lines and from vertical lines is brought to a focus at the same distance from the grating. Most other grating mountings are astigmatic and it is then necessary to find some position beyond the slit in which to place any device that would limit the length of the lines produced by the slit. There are two positions in an astigmatic mounting in which lines will be in focus at the camera, one position for vertical lines (the slit edges) and one position for the horizontal lines (slit limiting devices). Bulk is a disadvantage of the Wadsworth mounting. An example is the Jarrell-Ash 1.5-m spectrograph. The grating has 600 grooves/mm and covers the range from 220 to 780 nm in the first order with a dispersion of 1.09 nm/mm. The camera photographs 50 cm of spectrum per exposure.

The Eagle mounting (Fig. 6-11) or modifications thereof is popular in spite of the fact that rather complicated adjustments are needed for the film and grating. All components are on one side of a Rowland circle, giving a layout that resembles a Littrow mount. Astigmatism is slight. An example is the Bausch & Lomb 1.5-m spectrograph with fixed slits of 10, 20, or 50 μm that provides a dispersion of 1.6 nm/mm in the first order. It covers the range from 225 to 625 nm.

The Paschen–Runge mounting is popular for large concave gratings. The Rowland circle is the basis of the design, with the grating and entrance slit at fixed positions in the circle to give a fixed angle of incident radiation striking the grating. The light from the entrance slit focuses along the Rowland circle and photographic plates or photodetectors are mounted along the circle. This mounting does not provide linear dispersion except near the grating normal, but it does provide wide spectral range. The Jarrell-Ash ICAP 0.75-m spectrometer employs this mounting.

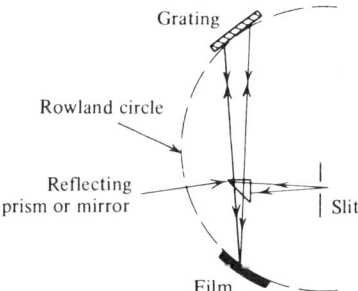

FIGURE 6-11 Eagle mounting for a concave grating.

Plane Grating Instruments

Almost all large spectrographs (3-m instruments) have plane gratings with modified Ebert mountings (see Fig. 2-20). The Ebert mounting is stigmatic and also achromatic, so that the rays of all wavelengths are brought to focus at the camera without changing the camera-to-mirror distance. This makes it easy to change wavelengths merely by rotating the grating. Higher grating orders are readily accessible. Standard gratings will have 600 or 1200 grooves/mm with resolution ranging from 0.51 nm/mm in the first order to 0.07 nm/mm in the third order and covering the spectral range of 180–3000 nm. Some instruments have an order sorter, a fore-prism arrangement that stands between the source and the main slit of the instrument and serves to place the various grating orders one above another on the photographic plate.

The ARL scanning spectrometer uses a Czerny–Turner monochromator with 1-m focal length. With slits 20 μm and a grating ruled with 1500 grooves/mm, the resolution is 0.022 nm/mm over the wavelength region from 190 to 670 nm. The rotation of the grating, driven by a computer-controlled stepping motor, permits selection of the desired wavelength and sequential examination of the spectral lines. It is possible to record the spectrum or to measure the intensities of the lines. The computer not only monitors the measuring process, but will find the desired spectral line on its own. The computer will check the analytical parameters for each line analyzed and regularly recalibrate the instrument. It finally performs all calculations and prints out results.

Echelle Grating/Prism Spectrometer[10,11]

Echelles are coarse but precisely ruled diffraction gratings having broad flat grooves, often 79 grooves/mm, which are used at high angles of incidence. Typically echelles are blazed at 63°26′. The ratio of groove width to groove height is 2:1 usually. Other groove spacings available are 316 and 31.6 grooves/mm for the same blaze angle. In echelles the spacing from one step to the next is many times greater than the wavelength of the light being dispersed. Consequently, many different wavelengths will coincide.

Echelles are used in high grating orders, often 40–120. In an echelle grating spectrometer the many higher orders must be separated from each other in some way, which is not

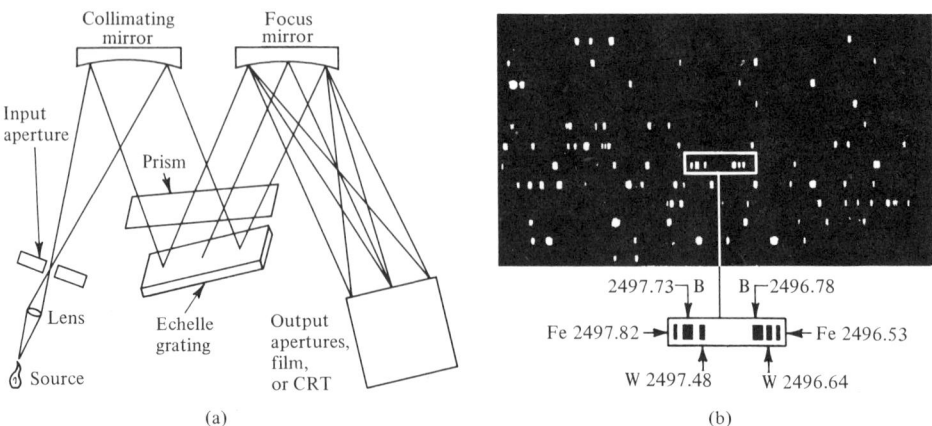

FIGURE 6-12 (a) Typical configuration for an echelle grating/prism spectrometer. (b) Echelle spectral pattern; each horizontal spectral pattern is a grating order. (Courtesy of Spectrametrics Inc.)

generally necessary in conventional spectrometers. This is accomplished by placing an auxiliary dispersing element, usually a prism (or low dispersion grating), in the spectrometer so that the prism disperses wavelengths at right angles to the echelle and thus effectively separates the orders. The optical arrangement is shown in Fig. 6-12 for a modified Czerny–Turner mount. The different orders appear as horizontal lines, with the lowest order (longest wavelength) appearing at the bottom and the highest order (shortest wavelength) appearing at the top. Here the prism-cross dispersing element is separating the wavelengths in the vertical direction, while the echelle grating separates them horizontally. Each raster line is equivalent to a segment (one order) of a conventional high-resolution spectrum, and successive raster lines are adjacent segments (and successive orders). The free spectral range, or wavelength range best covered by one order, varies from 1.8 nm at 200 nm to 11.1 nm at 500 nm for an echelle with 79 grooves/mm, blaze angle 63°26′, 128 mm in width.

The two-dimensional display pattern permits high dispersion of all wavelengths from 190 to 800 nm in a compact array, typically 10 × 13 mm, whereas a linear distance of about 2 m would be required for the same coverage. Since the spectrometer operates in multiple orders, the angular change in the dispersed rays in any one order is relatively small; therefore all wavelengths are measured at or near their optimum blaze angle. Operating at the blaze angle is very important because this is where maximum energy throughput is achieved; thus the efficiency and sensitivity of the total system is greatly increased. Another feature that substantially increases energy transfer is the short focal length required, only 0.75 m. Thus very low light levels are detectable that are normally lost to spectrometers having longer focal lengths.

There is a limitation on slit height, generally 1 mm or less, so that interference between orders is avoided. Since the echelle grating is physically somewhat larger than a conventional grating, this will make up for any loss of luminosity due to the higher angle of diffraction. This leaves the slit height as the principal limitation on luminosity in an echelle spectrometer when the unit is used in the photoelectric mode. However, balanced against this is the favorable grating efficiency of the echelle. For a given spectral resolution, a high-resolution echelle spectrometer can have much wider slits than a medium-resolution conventional spectrometer, thus increasing the relative luminosity of the echelle instrument.

Significant improvements in detection limits are achievable, particularly when high background "atom reservoirs" such as the nitrous oxide/acetylene flame or various plasma-type systems are used. Also spectral interferences, such as CaOH band emission on Ba 553-nm atomic emission, or Fe 249.782 and W 249.748 nm on B 249.773-nm atomic emission, are eliminated when a high-resolution echelle monochromator is used. This important practical property of an echelle spectrometer is of great importance in emission spectroscopy when high temperature and a dc argon plasma or an ICAP source is used.

When performing qualitative analysis, the echelle spectrometer operates in the photographic mode as a high-resolution spectrograph. A special camera attachment interfaces directly with the instrument. When used as a monochromator in the photoelectric mode, two wavelength dials are required, one to change orders and the other to select the wavelength within an order. Sometimes a spectral line will appear in more than one order, but there will be a most desired order with respect to relative signal intensity. For example, the Cd 228.8-nm line exhibits these (in parentheses) intensities for the order number: 110 (11), 111 (28), 112 (100), 113 (39), and 114 (5). The selected element's wavelength is detected by a single photomultiplier tube through a single slit in a cassette (Fig. 6-13). Changeover from element to element typically takes less than 2 min, a time that includes complete recalibration of the instrument under the control of a microprocessor. Multi-element measurement (up to 20 individual elements simultaneously) requires the addition of a multichannel module and interchanging the program cassettes. The position of the grating is fixed at a midpoint and the cassette passes the selected wavelengths through to 20 separate photomultiplier detectors arranged in a two-dimensional format to take advantage of the two-dimensional nature of the echelle spectral pattern. The detectors are in fixed positions and a series of lines is focused through a multichannel cassette (Fig. 6-13).

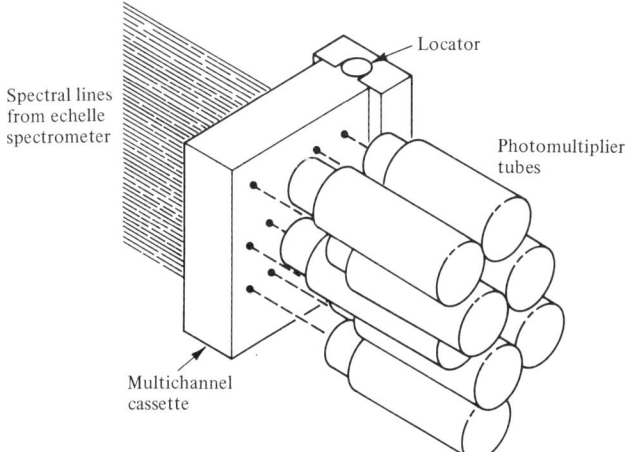

FIGURE 6-13 Components in direct-reader approach to multielement echelle spectrometry. (Courtesy of Spectrametrics Inc.)

Some of the lines are passed directly through the cassette to the appropriate detector, while others are deflected via mirrors within the cassette unit to a particular photomultiplier tube. A large number of prealigned cassette systems exist, and thus a wide variety of analytical programs may be selected. Each detector's output is under computer control and all sampling computation and reporting is totally automatic. Manual operations only require entering standard values for each element, injecting the blank, standard, and unknown samples for plasma sources, and then the computer takes over all operations and displays the results at the end of the analysis. One channel can be reserved for the background correction and another channel for the internal standard value.

6.3 PHOTOGRAPHIC DETECTION

Photographic materials consist of a light-sensitive emulsion coated on a glass plate or plastic film. The emulsion contains light-sensitive crystals of silver halides suspended in gelatin. On exposure, the silver halide crystals receiving radiation form a latent image. Subsequent chemical treatment converts the exposed silver halide crystals into a black deposit of silver at the site of the latent image. After development, the emulsion must be fixed in a solution which dissolves (by complexation) the unexposed silver halides. Finally, the photographic material must be washed thoroughly to remove the chemicals used in developing and in fixing. The entire series of operations must follow rigidly controlled conditions with respect to time, temperature, and chemicals. These operations are best carried out in automatic processing machines.

Emulsions are characterized by their useful wavelength range, speed, contrast, and granularity. The radiation, which is absorbed by silver halides, is restricted to blue, violet, and shorter-wavelength radiation. However, the emulsion can be sensitized optically by the addition of suitably chosen dyes that absorb radiation of longer wavelengths. The size of the silver halide crystals, or granularity, relates to the sensitivity of the emulsion and the resolution of the image. The smaller the granularity, the better the signal-to-noise ratio. For a faint line, a longer exposure with a low-granularity, high-contrast emulsion (which will be slow) is preferable to a short exposure with a fast film of higher granularity.

As many as 16 exposures may be placed on a 35-mm film and as many as 40 on a 10 × 25 cm plate. Hence the time-consuming step of processing the photographic emulsion is divided among the number of sample exposures on it. As many as 66 determinations in less than 30 min is possible.

The response of an emulsion is graphed in Fig. 6-14; the plot serves to calibrate the emulsion. The extent of blackening of the exposed film is related to the logarithm of exposure, which is the product of radiation intensity and time. There is a region (B-C) over which the response is linear with respect to the logarithm of exposure; this is the useful range. From the graph, the contrast of an emulsion is proportional to the slope of the curve. Speed is related to the range over which the curve is linear. For example, a slow, high-contrast emulsion will yield a curve of high slope whose linear range occurs at moderate to high exposures. In general, speed and contrast cannot both be maximized. Thus an emulsion appropriate to the radiation level to be measured must be ascertained by pre-

ATOMIC EMISSION SPECTROSCOPY   171

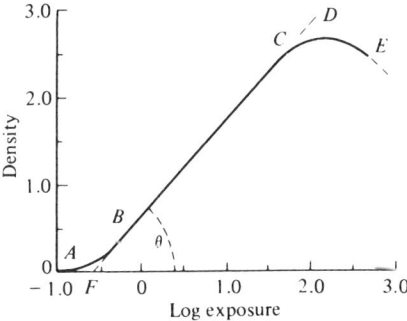

FIGURE 6-14   Characteristic curve of a photographic emulsion: *A*, threshold exposure; *B-C*, linear portion of curve; *D-E*, reversal region; *F*, inertia of emulsion.

liminary testing. Furthermore, each batch of film will differ slightly from previous batches of the same type; this requires recalibration each time a new batch is used.

For a very low intensity to be recorded, and not lie outside the linear range of an emulsion, either a long exposure will be necessary or the emulsion should be subjected to prefogging before exposure to the signal in order to bring the overall response into the linear range. If the intensity of the incident radiation is too great, attenuation can be achieved with neutral density filters or a stepped rotating sector. An excellent way of evaluating exposure is to measure an emission line as a logarithmic step sector rotates. The sector provides several graduated levels of illumination along the slit image. At least one level striking the photographic emulsion should fall in the range of proper exposure for every incident line. A rotating sector may be placed immediately before the slit of a stigmatic spectrograph, or at the horizontal focus position of an astigmatic spectrograph. Shown in Fig. 6-15 is a logarithmic step sector and its relationship to the entrance slit. This particular sector disk has four circular steps cut away. The length of the arc of each step is twice the previous steps; step 1 will get maximum exposure, step 2 will get half the exposure, step 3 one-fourth the exposure, and so on.

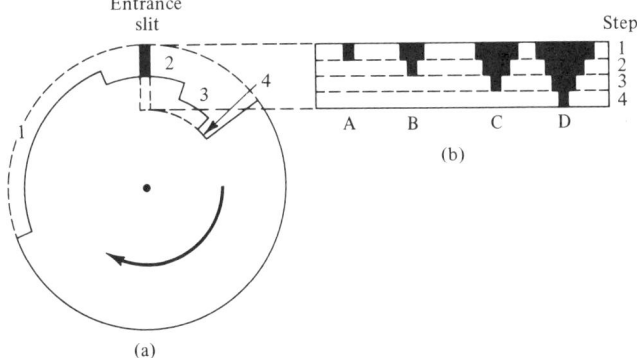

FIGURE 6-15   (a) Logarithmic step-sector disk rotating in front of entrance slit of spectrograph. (b) Schematic illustration of a four-line spectrum obtained with the sector. Line intensity (blackness on emulsion) is indicated by the width of the line.

Microphotometer-Comparator

The comparison of line intensities on a photographic emulsion requires the use of an instrument that will measure the relative transmittance of the line images. Figure 6-16 shows a commercial instrument; its optical schematic is shown in Fig. 6-17. Essential components include a carefully controlled light source (tungsten lamp), a slit to define the desired portion of the emulsion, a holder for the photographic emulsion, a photodetector, and a readout system. A racking mechanism permits moving the emulsion horizontally (scanning) and vertically from one spectrum to another.

As the microphotometer scans a spectrum, the amount of light impinging on the detector will change, and different photocurrents will result. These may be presented as a digital readout (%T). The amount of light passing through the slit and a clear section of the plate or film is arbitrarily considered 100% transmittance. When an opaque object is inserted between the light source and the slit, the transmittance falls to 0%. More intense light emission by the excited sample results in denser (blacker) lines than less intense light emission. Therefore, the relative amount of light falling from the light source in the microphotometer through the line on the emulsion and to the photodetector indicates a relative blackening of the emulsion on the scale between 0 and 100% T. The intensity of the light striking the plate and creating the line is then calculated by means of the characteristic curve for the emulsion under the chosen operating conditions. The microphotometer

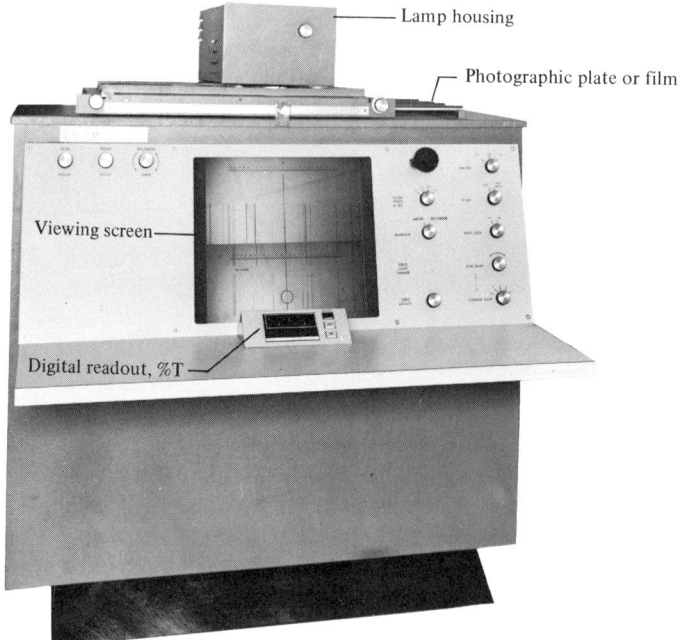

FIGURE 6-16 Microdensitometer comparator. (Courtesy of Baird Corp.)

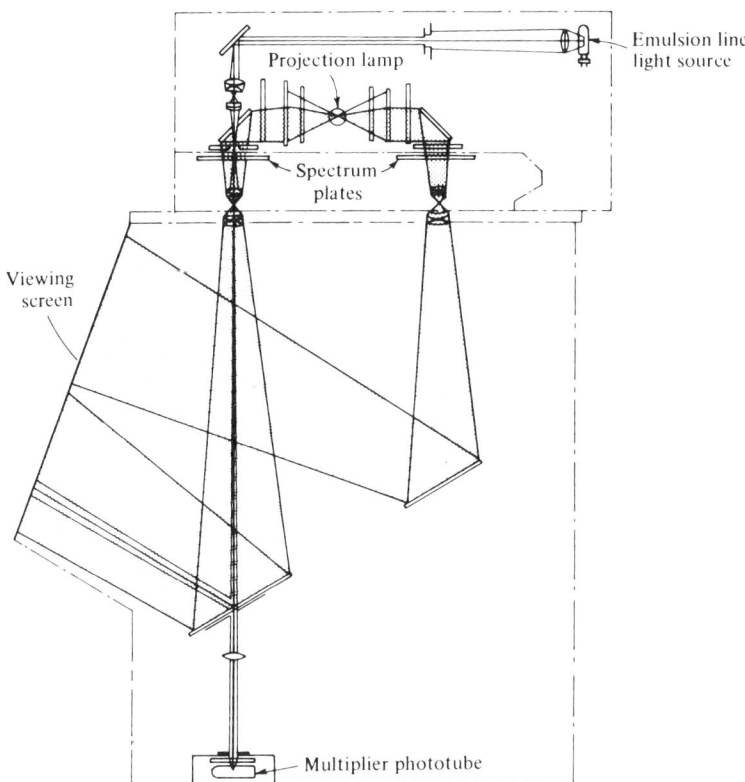

FIGURE 6-17 Optical schematic of the microdensitometer comparator. (Courtesy of Baird Corp.)

measurements are more precise when the blackening of two lines are nearly equal. Since the ratios of the intensities of the various steps in a logarithmic rotating sector disk are accurately known, one can easily calculate the original intensities of the lines from the steps where the two lines are of nearly equal blackening.

In quantitative work, either the ratio of the intensities of the analyte line to internal standard line is plotted against concentration, or the logarithm of the ratio is plotted against the logarithm of the concentration to establish the working (calibration) curve. Once this working curve has been established, similar unknown samples can be quickly analyzed. A new working curve must be established for each type of material, and whenever a new component has been introduced into the general type of material. In Fig. 6-15, the line density ($\log 1/T$) is indicated schematically by relative linewidths. A line that is not very intense, such as A, would have time to blacken the emulsion only in the small area that is exposed almost constantly, that is, step 1. By contrast, the line at D would be sufficiently intense to prepare a four-point graph like Fig. 6-14, plotting the density of steps 1–4 versus the arbitrary units of exposure of the respective steps.

6.4 PHOTOELECTRIC DETECTION

The analytical speed attainable with photographic plates is normally quite satisfactory. However, in a laboratory where large numbers of samples must be processed this speed is no longer sufficient. For samples which must be analyzed in a total elapsed time of a few minutes, direct-reading spectrographs were developed.

The optical system and excitation equipment of the direct readers are very much the same as for the photographic units. The difference occurs at the exit opening of the spectrograph. Here the light from individual element lines is isolated either by slits of the precise widths and in the precise locations where the desired spectral lines appear or by means of a complicated mirror system. Both methods focus the light of one spectrum line onto the cathode of a photomultiplier tube.

The intensity of spectral lines is converted by the photomultiplier tube into an electrical current used to charge a capacitor–resistor circuit. Two integration modes are employed; typical integration time is 25–40 sec. In the older method, capacitor voltage is the quantity read out. The voltage across the capacitor at the end of the exposure is a function of the accrued charge (detector current × time); thus the capacitor voltage is proportional to the time integral of the line intensity. Intensities can be compared by forming voltage ratios electrically. In integration with constant time, an electronic circuit converts the accrued charge into regular electric time pulses, the number of which is proportional to the value of the charge obtained during measurement. By measuring the capacitor charge, a large dynamic measuring range extending from 0 to 10^6 can be covered and takes advantage of large dynamic concentration ranges of certain atomic emission sources. The number of pulses is recorded by computer together with the wavelength involved with this acquisition.

Standardization of direct-reading spectrometers is carried out using a high and a low standard. In real situations, the relationship between the analytical signal, R, and the solution concentration, C, is described by

$$R = C \tan A + R_0 \qquad (6\text{-}1)$$

where R_0 is the background intensity and A is the slope of the calibration curve. Rearrangement of this equation, in which C_0 is the concentration equivalent to background (the concentration required to give a line-to-background ratio of unity), gives

$$C = R \cot A - C_0 \qquad (6\text{-}2)$$

After standardization is completed, the instrument is ready to analyze samples. Typically, sample analysis, data printout, and any associated data management functions are completed within 2 min after an analysis is initiated. Since the value of C_0 is dependent upon the spectral line and the element, standardization can be used effectively to diagnose spectrometer performance and standard preparation. The precision of results is in most cases better than 0.5% relative above concentrations where the element intensity is equal to the background signal, and the reproducibility for 8 hr yields a coefficient of variation lower than 2% relative.

PROBLEMS

1. A sample of an unknown light-metal alloy was placed on a spark stand and a spectrum was recorded. Observation of the spectogram revealed lines at the following wavelengths: 643.8, 518.4, 517.3, 481.0, 472.2, 468.0, 383.8, 383.2, 382.9, 361.1, 346.6, and 340.3 nm, plus many lines of aluminum. What elements, besides aluminum, are present?

2. In the spectrographic determination of lead in an alloy, using a magnesium line as internal standard, these results were obtained:

SOLUTION	DENSITOMETER READING Mg	DENSITOMETER READING Pb	CONCENTRATION OF LEAD, mg/ml
1	7.3	17.5	0.151
2	8.7	18.5	0.201
3	7.3	11.0	0.301
4	10.3	12.0	0.402
5	11.6	10.4	0.502
A	8.8	15.5	
B	9.2	12.5	
C	10.7	12.2	

(a) Prepare a calibration curve on a log–log paper. (b) Evaluate the concentrations for solutions A, B, and C.

3. A step sector with arc lengths (or angle subtended by each step at the center) in the ratio of 1:2:4:8:16:32 was rotated in front of the slit of a spectrograph while a sample of a tin alloy containing lead was being arced in the source unit. After the plate was developed, fixed, and dried, the density of a suitable tin line was measured at each step with a microphotometer.

The values of I_0/I obtained for each step were 1.05, 1.66, 4.68, 13.18, 37.15, and 52.5. Plot the characteristic curve for the film and determine the γ and the inertia of the emulsion.

4. Several standard samples of a tin alloy were prepared by chemical analysis for the lead content. These alloys were then employed as electrodes. The ratio of the density of the tin line at 276.1 nm and the density of the lead line at 283.3 nm were measured on the microphotometer. The results are listed below:

SAMPLE NO.	% LEAD	$D_{\text{tin line}}$	$D_{\text{lead line}}$
1	0.126	1.567	0.259
2	0.316	1.571	1.013
3	0.708	1.443	1.546
4	1.334	0.825	1.427
5	2.512	0.447	1.580

5. Using the results of Problem 3, plot a "working curve" of log percent lead as abscissa and log $(I_{\text{Pb}}/I_{\text{Sn}})$ as ordinate. An unknown tin alloy sample was treated in the

same way as the standards. The 276.1-nm tin line had a density of 0.920 on the photographic plate, while the 283.3-nm lead line had a density of 0.669. What was the percentage of lead in the alloy?

BIBLIOGRAPHY

Barnes, R. M., Ed., *Emission Spectroscopy,* Halsted, New York, 1976.
Boumans, P. W. J. M., *Theory of Spectrochemical Excitation,* Plenum, New York, 1966.
Fassel, V. A., *Anal. Chem.,* **51**, 1290A (1979).
Grove, E. L., Ed., *Applied Atomic Spectroscopy,* Vols. 1 and 2, Plenum, New York, 1978.
Mitteldorf, A. J., "Emission Spectrochemical Methods," Chap. 6, in *Trace Analysis,* G. H. Morrison, Ed., Wiley-Interscience, New York, 1965.
Schrenk, W. G., *Analytical Atomic Spectroscopy,* Plenum, New York, 1975.
Scribner, B. F. and M. Margoshes, "Emission Spectroscopy," in *Treatise on Analytical Chemistry,* Vol. 6, Part I, Chap. 64, I. M. Kolthoff and P. J. Elving, Eds., Wiley-Interscience, New York, 1965.
Slavin, M., *Emission Spectrochemical Analysis,* Wiley-Interscience, New York, 1971.

LITERATURE CITED

1. Brech, F., in *Analysis Instrumentation,* Vol. 6, p. 215, Plenum, New York, 1969.
2. Chaplenko, G., *The Spex Speaker,* **13** (3), 1 (October 1968).
3. Chaplenko, G., D. O. Landon, and A. J. Mitteldorf, *The Spex Speaker,* **11** (3), 1 (October 1966).
4. Skogerboe, R. K. and G. N. Coleman, *Anal. Chem.,* **48**, 611A (1976).
5. Fassel, V. A. and R. N. Kniseley, *Anal. Chem.,* **46**, 1110A, 1155A (1974).
6. Greenfield, S., *The Spex Speaker,* **22** (3), 1 (September 1977).
7. Greenfield, S., I. L. Jones, and C. T. Berry, *Analyst (London),* **89**, 713 (1964).
8. Ward, A. F., *Am. Lab.,* p. 79 (November 1978).
9. Reednick, J., *Am. Lab.,* p. 53 (March 1979).
10. Keliher, P. N., *Research/Development,* **27** (6), 26 (June 1976).
11. Keliher, P. N. and C. C. Wohlers, *Anal. Chem.,* **48**, 333A (1976).

CHAPTER 7

Infrared Spectrophotometry

The infrared region of the electromagnetic spectrum extends from the red end of the visible spectrum out to the microwave region. The region includes radiation at wavelengths between 0.7 and 500 μm or, in wave numbers, between 14,000 and 20 cm^{-1}. The spectral range of greatest use is the mid-infrared region, which covers the frequency range from 200 to 4000 cm^{-1} (50 to 2.5 μm). Infrared spectrophotometry involves the twisting, bending, rotating, and vibrational motions of atoms in a molecule (see Fig. 1-7). Upon interaction with infrared radiation, portions of the incident radiation are absorbed at particular wavelengths. The multiplicity of vibrations occurring simultaneously produces a highly complex absorption spectrum, which is uniquely characteristic of the functional groups comprising the molecule and of the overall configuration of the atoms as well.

Atoms or atomic groups in molecules are in continuous motion with respect to each other. The possible vibrational modes in a polyatomic molecule can be visualized from a mechanical model of the system, shown schematically in Fig. 7-1. Atomic masses are represented by balls, their weight being proportional to the corresponding atomic weight, and arranged in accordance with the actual space geometry of the molecule. Mechanical springs, with forces that are proportional to the bonding forces of the chemical links, connect and keep the balls in positions of balance. If the model is suspended in space and struck, the balls will appear to undergo random chaotic motions. However, if the vibrating model is observed with a stroboscopic light of variable frequency, certain light frequencies will be found at which the balls appear to remain stationary. These represent the specific vibrational frequencies for these motions.

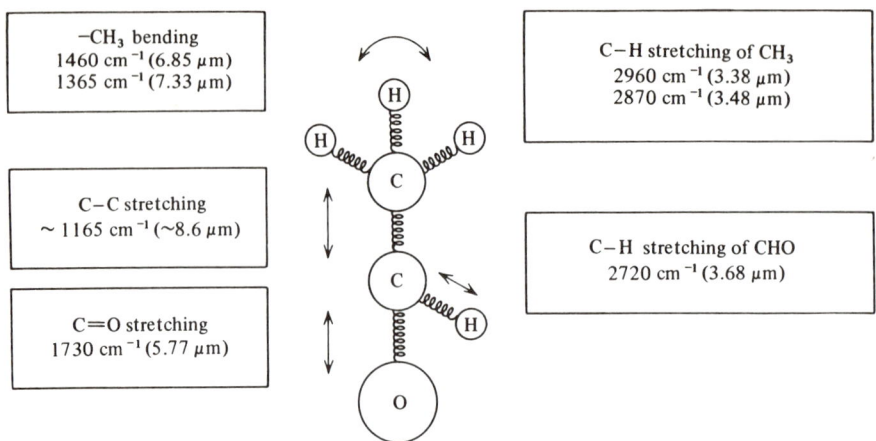

FIGURE 7-1 Vibrations and characteristic frequencies of acetaldehyde.

7.1 CORRELATION OF INFRARED SPECTRA WITH MOLECULAR STRUCTURE

The infrared spectrum of a compound is essentially the superposition of absorption bands of specific functional groups, yet subtle interactions with the surrounding atoms of the molecule impose the stamp of individuality on the spectrum of each compound. For qualitative analysis, one of the best features of an infrared spectrum is that the absorption or the *lack of absorption* in specific frequency regions can be correlated with specific stretching and bending motions and, in some cases, with the relationship of these groups to the remainder of the molecule. Thus, by interpretation of the spectrum, it is possible to state that certain functional groups are present in the material and that certain others are absent. With this one datum, the possibilities for the unknown can be sometimes narrowed so sharply that comparison with a library of pure spectra permits identification.

Near-Infrared Region

In the near-infrared region, which meets the visible region at about 12,500 cm^{-1} (0.8 μm) and extends to about 4000 cm^{-1} (2.5 μm), are found many absorption bands resulting from harmonic overtones of fundamental bands and combination bands often associated with hydrogen atoms. Among these are the first overtones of the O–H and N–H stretching vibrations near 7140 cm^{-1} (1.4 μm) and 6667 cm^{-1} (1.5 μm), respectively, combination bands resulting from C–H stretching, and deformation vibrations of alkyl groups at 4548 cm^{-1} (2.2 μm) and 3850 cm^{-1} (2.6 μm). Thicker sample layers (0.5–10 mm) compensate for lessened molar absorptivities. The region is accessible with quartz optics, and this is coupled with greater sensitivity of near-infrared detectors and more intense light sources. The near-infrared region is often used for quantitative work.

Water has been analyzed in glycerol, hydrazine, Freon, organic films, acetone, and fuming nitric acid. Absorption bands at 2.76, 1.90, and 1.40 μm are used depending on the concentration of the test substance. Where interferences from other absorption bands are severe or where very low concentrations of water are being studied, the water can be extracted with glycerol or ethylene glycol.

Near-infrared spectrometry is a valuable tool for analyzing mixtures of aromatic amines. Primary aromatic amines are characterized by two relatively intense absorption bands near 1.97 and 1.49 μm. The band at 1.97 μm is a combination of N—H bending and stretching modes and the one at 1.49 μm is the first overtone of the symmetric N—H stretching vibration. Secondary amines exhibit an overtone band but do not absorb appreciably in the combination region. These differences in absorption provide the basis for rapid, quantitative analytical methods. The analyses are normally carried out on 1% solutions in CCl_4, using 10-cm cells. Background corrections can be obtained at 1.575 and 1.915 μm. Tertiary amines do not exhibit appreciable absorption at either wavelength. The overtone and combination bands of aliphatic amines are shifted to about 1.525 and 2.000 μm, respectively. Interference from the first overtone of the O—H stretching vibration at 1.40 μm is easily avoided with the high resolution available with near-infrared instruments.

Mid-Infrared Region[1]

Many useful correlations have been found in the mid-infrared region (Fig. 7-2). This region is divided into the "group frequency" region, 4000–1300 cm^{-1} (2.5–8 μm), and the "fingerprint" region, 1300–650 cm^{-1} (8.0–15.4 μm). In the group frequency region the principal absorption bands may be assigned to vibration units consisting of only two atoms of a molecule; that is, units which are more or less dependent only on the functional group giving the absorption and not on the complete molecular structure. Structural influences do reveal themselves, however, as significant shifts from one compound to another. In the derivation of information from an infrared spectrum, prominent bands in this region are noted and assigned first. In the interval from 4000 to 2500 cm^{-1} (2.5–4.0 μm), the absorption is characteristic of hydrogen stretching vibrations with elements of mass 19 or less. When coupled with heavier masses, the frequencies overlap the triple-bond region. The intermediate frequency range, 2500–1540 cm^{-1} (4.0–6.5 μm), is often termed the *unsaturated* region. Triple bonds, and very little else, appear from 2500 to 2000 cm^{-1} (4.0 to 5.0 μm). Double-bond frequencies fall in the region from 2000 to 1540 cm^{-1} (5.0 to 6.5 μm). By judicious application of accumulated empirical data, it is possible to distinguish among C=O, C=C, C=N, N=O, and S=O bands. The major factors in the spectra between 1300 and 650 cm^{-1} (7.7 and 15.4 μm) are single-bond stretching frequencies and bending vibrations (skeletal frequencies) of polyatomic systems which involve motions of bonds linking a substituent group to the remainder of the molecule. This is the fingerprint region. Multiplicity is too great for assured individual identification, but collectively the absorption bands aid in identification.

Far-Infrared Region

The region between 667 and 10 cm^{-1} (15 and 1000 μm) contains the bending vibrations of carbon, nitrogen, oxygen, and fluorine with atoms heavier than mass 19, and

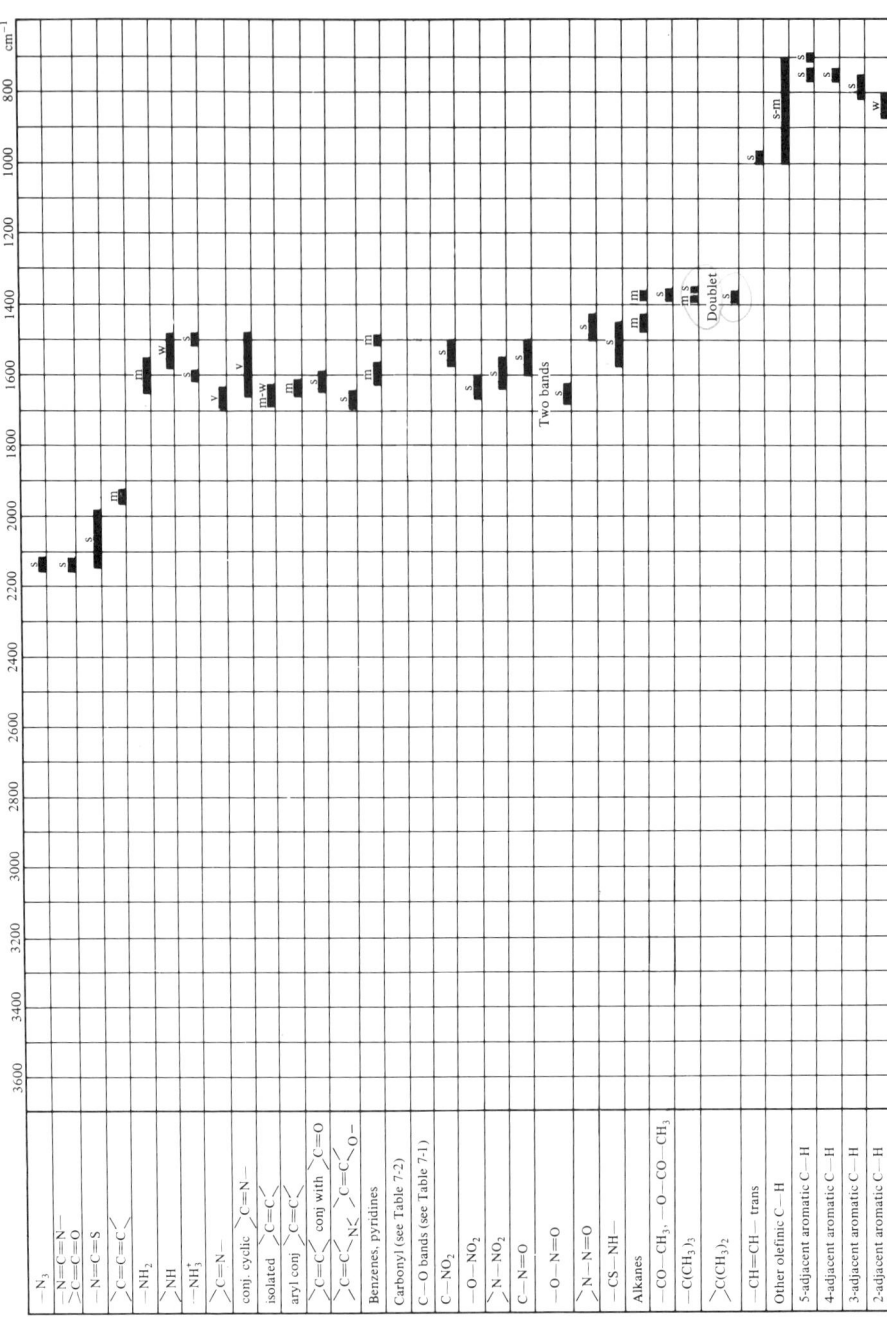

FIGURE 7-2 Some characteristic infrared absorption bands. Band positions are given for dilute solution in nonpolar solvents. Intensities are expressed as strong (s), medium (m), weak (w), and variable (v).

INFRARED SPECTROPHOTOMETRY 181

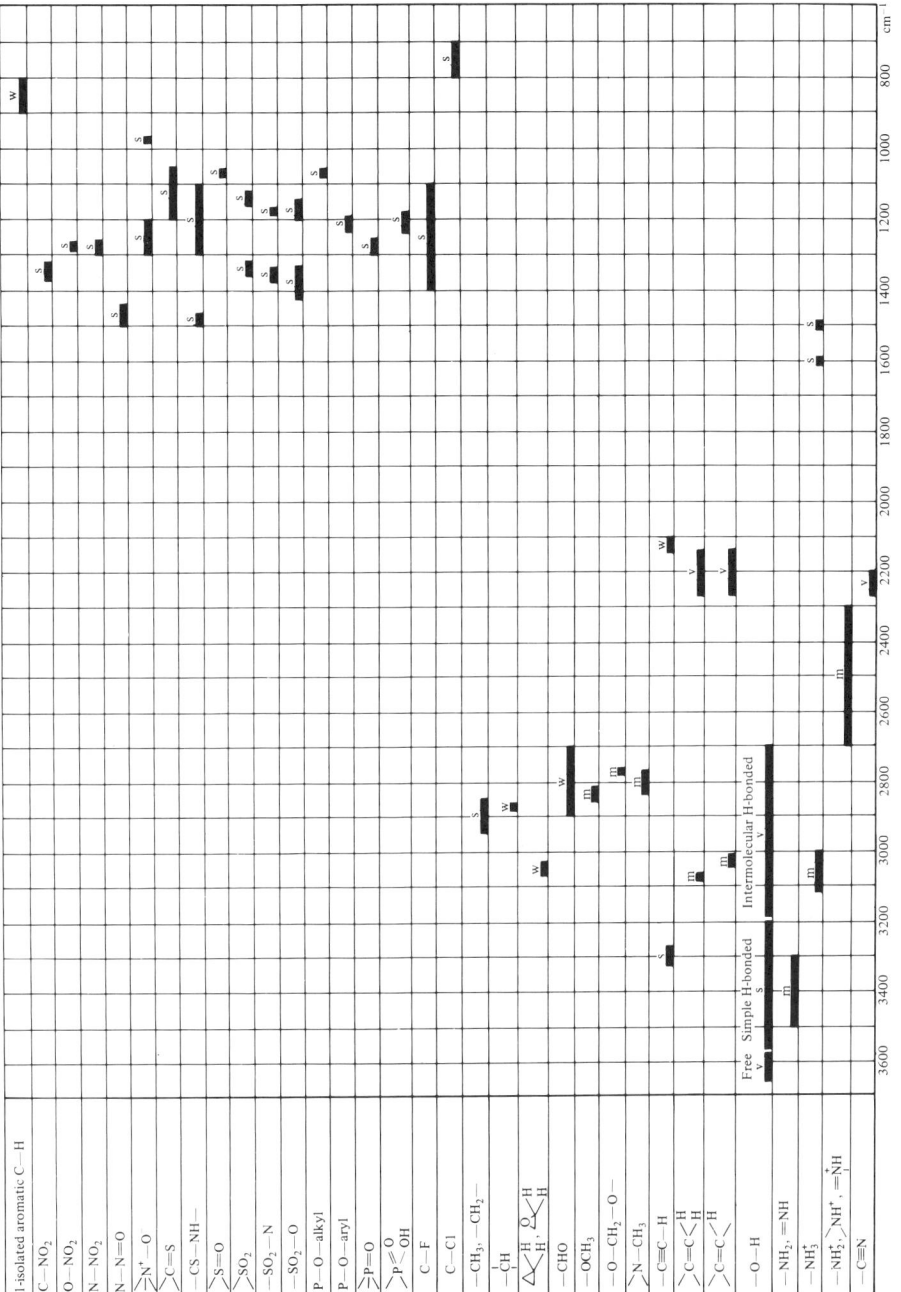

FIGURE 7-2 Continued

additional bending motions in cyclic or unsaturated systems. The low-frequency molecular vibrations found in the far-infrared are particularly sensitive to changes in the overall structure of the molecule. When studying the conformation of the molecule as a whole, the far-infrared bands differ often in a predictable manner for different isomeric forms of the same basic compound. The far-infrared frequencies of organometallic compounds are often sensitive to the metal ion or atom, and this, too, can be used advantageously in the study of coordination bonds. Moreover, this region is particularly well suited to the study of organometallic or inorganic compounds whose atoms are heavy and whose bonds are inclined to be weak.[2]

Structural Analysis

After the presence of a particular fundamental stretching frequency has been established, closer examination of the shape and exact position of an absorption band often yields additional information. The shape of an absorption band around 3000 cm^{-1} (3.3 μm) gives a rough idea of the CH group present. Alkyl groups have their C–H stretching frequencies lower than 3000 cm^{-1}, whereas alkenes and aromatics have them slightly higher than 3000 cm^{-1}. The CH$_3$ group gives rise to an asymmetric stretching mode at 2960 cm^{-1} (3.38 μm) and a symmetric mode at 2870 cm^{-1} (3.48 μm). For –CH$_2$– these bands occur at 2930 cm^{-1} (3.42 μm) and 2850 cm^{-1} (3.51 μm).

Next, one should examine regions where characteristic vibrations from bending motions occur. For alkanes, bands at 1460 cm^{-1} (6.85 μm) and 1380 cm^{-1} (7.25 μm) are indicative of a terminal methyl group attached to carbon exhibiting in-plane bending motions; if the latter band is split into a doublet at about 1397 and 1370 cm^{-1} (7.16 and 7.30 μm), geminal methyls are indicated. The symmetrical in-plane bending is shifted to lower frequencies when the methyl group is adjacent to >C=O (1360–1350 cm^{-1}), –S– (1325 cm^{-1}), and silicon (1250 cm^{-1}). The in-plane scissor motion of –CH$_2$– at 1470 cm^{-1} (6.80 μm) indicates the presence of that group. Four or more methylene groups in a linear arrangement gives rise to a weak rocking motion at about 720 cm^{-1} (13.9 μm). Figure 7-3 illustrates the typical spectrum of an alkane.

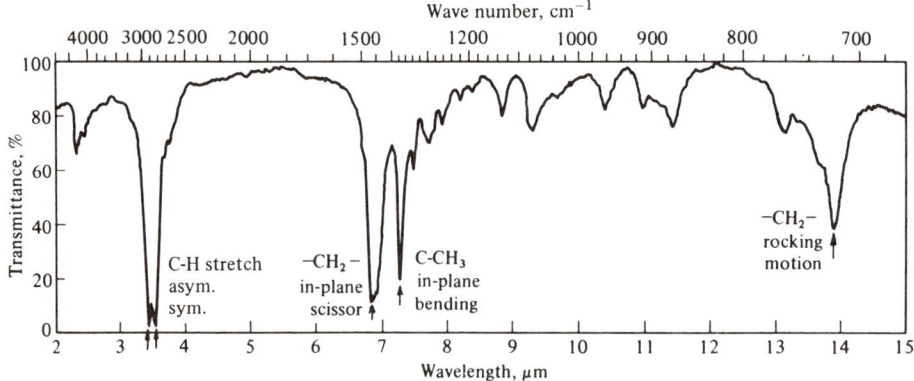

FIGURE 7-3 Typical infrared spectrum of a saturated n-alkane.

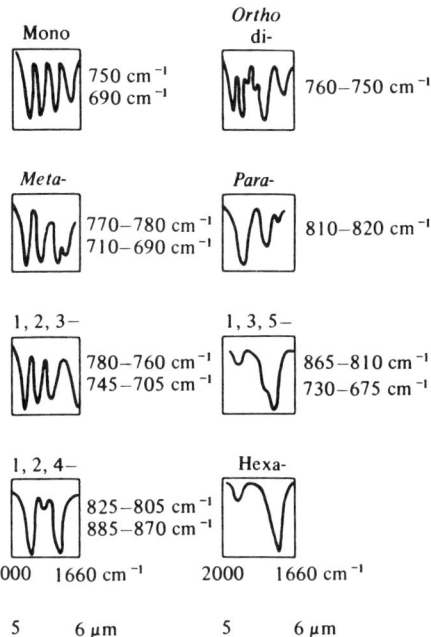

FIGURE 7-4 Benzene ring substitution—pattern of combination bands between 2000 and 1670 cm^{-1} (5 to 6 μm). To the right of each curve are the approximate positions of the C—H out-of-plane bending bands between 900 and 650 cm^{-1} (11.1–15.4 μm). (After C. W. Young, R. B. Duvall, and N. Wright, *Anal. Chem.*, 23, 709 (1951). Courtesy of American Chemical Society.)

The substitution pattern of an aromatic ring can be deduced from a series of weak but very useful bands in the region 2000–1670 cm^{-1} (5–6 μm) coupled with the position of the strong bands between 900 and 650 cm^{-1} (11.1 and 15.4 μm) which are due to the out-of-plane bending vibrations (Fig. 7-4). Absence of the symmetrical breathing mode at 690–710 cm^{-1} in the spectra of *para-* and *ortho-*substituted rings is helpful. The spectrum of *o*-xylene, shown in Fig. 7-5, is characteristic of aromatic systems. Ring stretching modes are observed near 1600, 1570, and 1500 cm^{-1} (6.25, 6.37, and 6.67 μm). These characteristic absorption patterns are also observed with substituted pyridines and polycyclic benzenoid aromatics.

The presence of an unsaturated C=C linkage introduces the stretching frequency at 1650 cm^{-1} (6.07 μm), shown in Fig. 7-6, and which may be weak or nonexistent if symmetrically located in the molecule. Mono- and trisubstituted olefins give rise to more intense bands than *cis-* or *trans-*disubstituted olefins. Substitution by a nitrogen or oxygen functional group greatly increases the intensity of the C=C absorption band. Conjugation with an aromatic nucleus causes a slight shift to lower frequency, but with a second C=C or C=O, the shift to lower frequency is from 40 to 60 cm^{-1} with a substantial increase in intensity. The out-of-plane bending vibrations of the hydrogens on a C=C linkage are very valuable. A vinyl group gives rise to two bands at about 990 cm^{-1} (10.1 μm) and 910 cm^{-1}

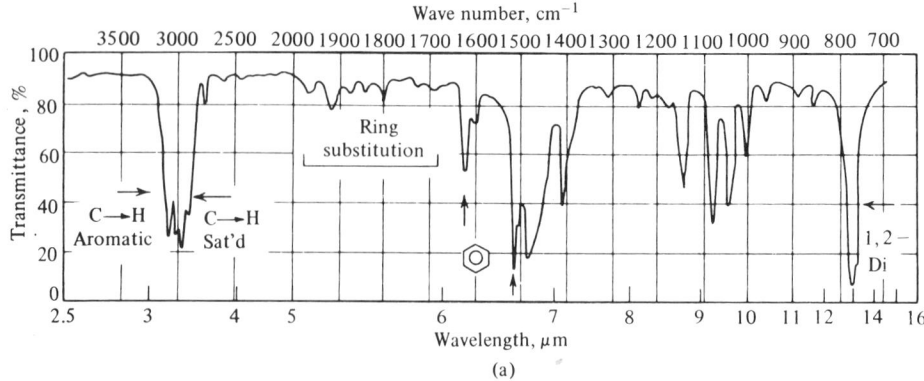

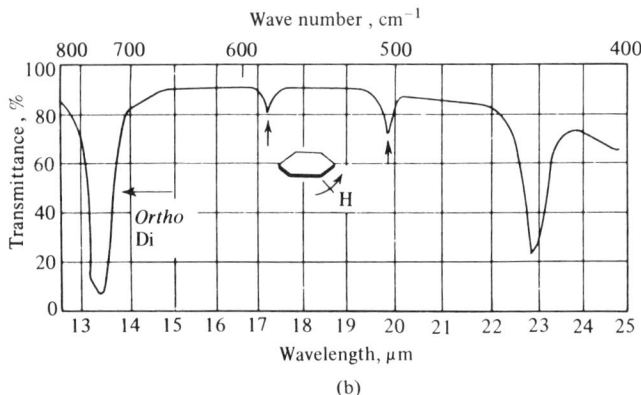

FIGURE 7-5 Infrared spectrum of *o*-xylene. (a) Sodium chloride region and (b) potassium bromide region.

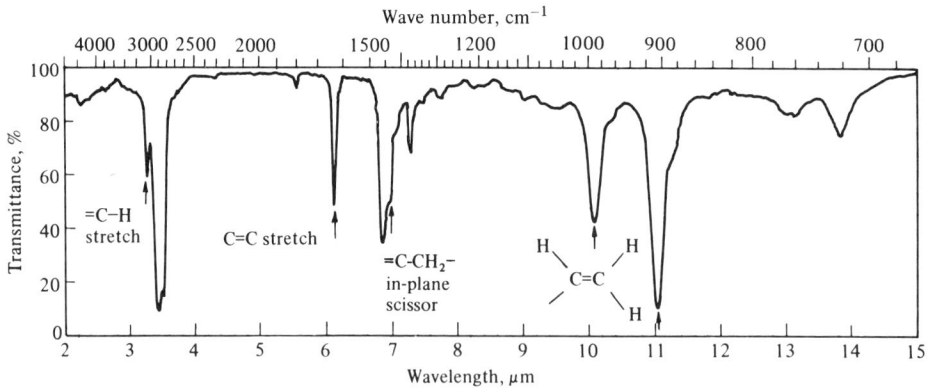

FIGURE 7-6 Infrared spectrum of octene-1.

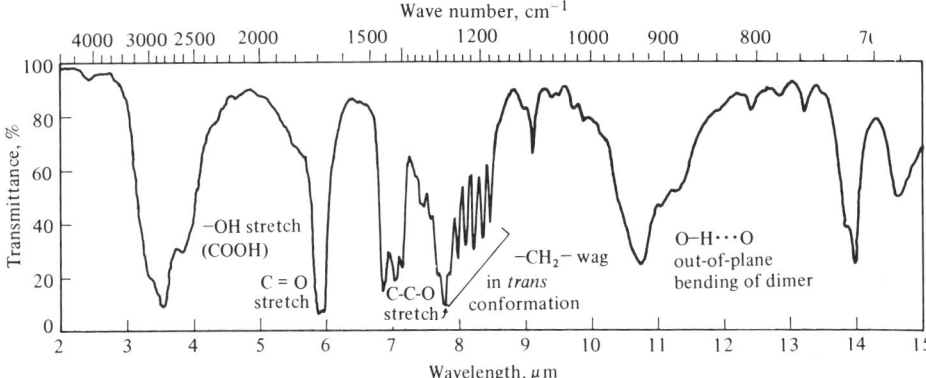

FIGURE 7-7 Infrared spectrum of stearic acid. Solid spectra of long-chain *n*-alkyl compounds exhibit a series of evenly spaced bands in the region 1350–1180 cm^{-1} that are characteristic of the chain length: 2 × number of bands = number of —CH_2— groups.

(11.0 μm). The =CH_2 (vinylidene) band appears near 895 cm^{-1} (11.2 μm) and is a very prominent feature of the spectrum. *Cis*- and *trans*-disubstituted olefins absorb near 685–730 cm^{-1} (13.7–14.6 μm) and 965 cm^{-1} (10.4 μm), respectively. The single hydrogen in a trisubstituted olefin appears near 820 cm^{-1} (12.2 μm).

In alkynes the ethynyl hydrogen appears as a needle-sharp and intense band at 3300 cm^{-1} (3.0 μm). The absorption band for —C≡C— is located in about the range from 2100 to 2140 cm^{-1} (4.67–4.76 μm) when terminal, but in the region from 2260 to 2190 cm^{-1} (4.42–4.56 μm) if nonterminal. The intensity of the latter type band decreases as the symmetry of the molecule increases; it is best identified by Raman spectroscopy. When the acetylene linkage is conjugated with a carbonyl group, however, the absorption becomes very intense.

For ethers the one important band appears near 1100 cm^{-1} (9.09 μm) and is due to the antisymmetric stretching mode of the —C—O—C— links. It is quite strong and may dominate the spectrum of a simple ether.

For alcohols the most useful absorption is that due to the stretching of the O—H bond. In the free or unassociated state, it appears as a weak but sharp band at about 3600 cm^{-1} (2.78 μm). Hydrogen bonding will greatly increase the intensity of the band and move it to lower frequencies and, if the hydrogen bonding is especially strong, the band becomes quite broad. Intermolecular hydrogen bonding is concentration dependent, whereas intramolecular hydrogen bonding is not concentration dependent. Measurements in solution under different concentrations are invaluable. The spectrum of an acid is quite distinctive in shape and breadth (Fig. 7-7) in the high-frequency region. The distinction between the several types of alcohols is often possible on the basis of the C—O stretching absorption bands, as indicated in Table 7-1. A spectrum of an alcohol is shown in Fig. 7-8.

The carbonyl group is not difficult to recognize; it is often the strongest band in the spectrum. Its exact position in the region, extending from about 1825 to 1575 cm^{-1} (5.48 to 6.35 μm), is dependent upon the double-bond character of the carbonyl group

TABLE 7-1 Classification of Various Types of Alcohols

Type of Alcohol	Position of C—O Bands	
	cm^{-1}	μm
Saturated tertiary Highly symmetrical secondary	1200–1125	8.30–8.90
Saturated secondary α-Unsaturated or cyclic tertiary	1125–1085	8.90–9.22
α-Unsaturated secondary Alicyclic secondary (5- or 6-membered ring) Saturated primary	1085–1050	9.22–9.52
Highly α-unsaturated tertiary Di-α-unsaturated secondary α-Unsaturated and α-branched secondary Alicyclic secondary (7- or 8-membered ring) α-Branched and/or α-unsaturated primary	<1050	>9.52

SOURCE: M. Gianturco, in S. K. Freeman, Ed., *Interpretive Spectroscopy,* Van Nostrand Reinhold, New York, 1965, p. 56, by permission.

(Table 7-2). Anhydrides usually show a double absorption band. Aldehydes are distinguished from ketones by the additional C—H stretching frequency of the CHO group at about 2720 cm^{-1} (3.68 μm). In esters (Fig. 7-9) two bands related to C—O stretching and bending are recognizable between 1300 and 1040 cm^{-1} (7.7 and 9.6 μm) in addition to the carbonyl band. The carboxyl group, in a sense, shows bands arising from the superposition of C=O, C—O, C—OH, and O—H vibrations (Fig. 7-7). Of five characteristic bands, three of these (2700, 1300, and 943 cm^{-1}; 3.7, 7.7, and 10.6 μm) are associated with vibrations of the carboxyl OH. They disappear when the carboxylate ion is formed. When the acid exists in the dimeric form, the O—H stretching band at 2700 cm^{-1} disappears, but the absorption band at 943 cm^{-1} due to OH out-of-plane bending of the dimer remains.

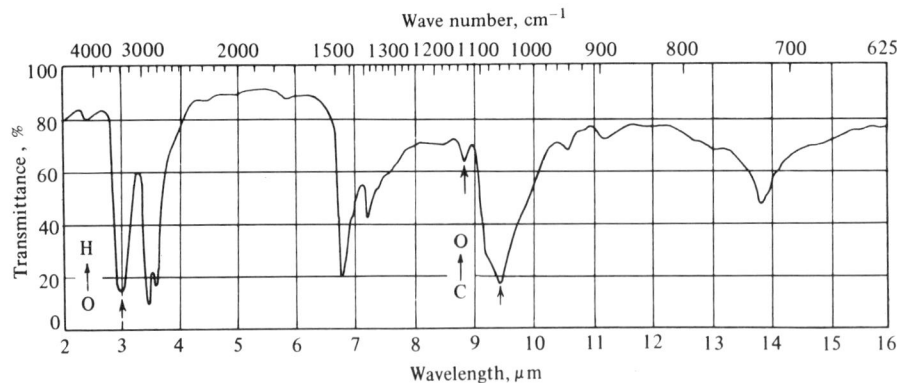

FIGURE 7-8 Infrared spectrum of lauryl alcohol, $CH_3(CH_2)_{10}CH_2OH$.

TABLE 7-2 Carbonyl Absorptions[a]

Typical of	Wave Number, cm^{-1}	Wavelength, μm
Anhydrides of carboxylic acids:		
aliphatic	1825[b]	5.48[b]
	1754	5.70
aromatic	1802[b]	5.55[b]
	1754	5.70
Chloride of carboxylic acids	1812	5.52
Carboxylic acids (monomers)	1776	5.63
Phenyl esters	1770	5.65
Vinyl esters of carboxylic acids	1770	5.65
Vinylidene esters of carboxylic acids	1764	5.67
Vinyl-type carbonates	1761	5.68
Normal carbonates	1751	5.71
Methyl esters of carboxylic acids	1748	5.72
Esters of carboxylic acids	1736	5.76
Esters of formic acid	1733	5.77
Aldehydes	1736	5.76
Ketones:		
aliphatic	1724	5.80
aromatic	1680–1645	5.95–6.08
Carboxylic acids (dimers)	1720–1700	5.81–5.88
Carbamates	1689	5.92
Amides (1°) of carboxylic acids	1718 sh	5.82 sh
	1684	5.94
Amides (2°) of carboxylic acids	1669	5.99
Amides (3°) of carboxylic acids	1667	6.00
Salts of carboxylic acids	1575	6.35

SOURCE: M. Gianturco, in S. K. Freeman, Ed., *Interpretive Spectroscopy,* Van Nostrand Reinhold, New York, 1965, p. 86, by permission.
[a]All values in CCl$_4$.
[b]Weakens as colinearity is approached.

The spectrum of an amine is shown in Fig. 7-10. Of particular interest in a primary amine (or amide) are the N–H stretching vibrations at about 3500 and 3400 cm^{-1} (2.86 and 2.94 μm), the in-plane bending of N–H at 1610 cm^{-1} (6.2 μm), and the out-of-plane bending of –NH$_2$ at about 830 cm^{-1} (12.0 μm), which is broad for primary amines. By contrast, a secondary amine exhibits a single band in the high-frequency region at about 3350 cm^{-1} (2.98 μm). The high-frequency bands broaden and shift about 100 cm^{-1} to lower frequency when involved in hydrogen bonding. When the amine salt is formed, these bands are markedly broadened and lie between 3030 and 2500 cm^{-1} (3.3 and 4.0 μm), resembling the –COOH bands in this region.

The nitro group is characterized by two equally strong absorption bands at about 1560 and 1350 cm^{-1} (6.41 and 7.40 μm), the asymmetric and symmetric stretching frequencies. In an N-oxide, only a single very intense band is present in the region from 1300 to 1200 cm^{-1} (7.70 to 8.33 μm). In addition there are C–N stretching and various bending vibra-

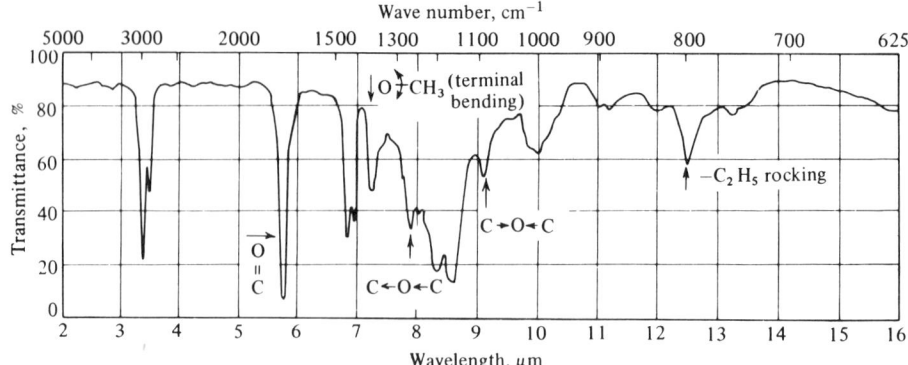

FIGURE 7-9 Infrared spectrum of dimethyl-2,5-diethyladipate.

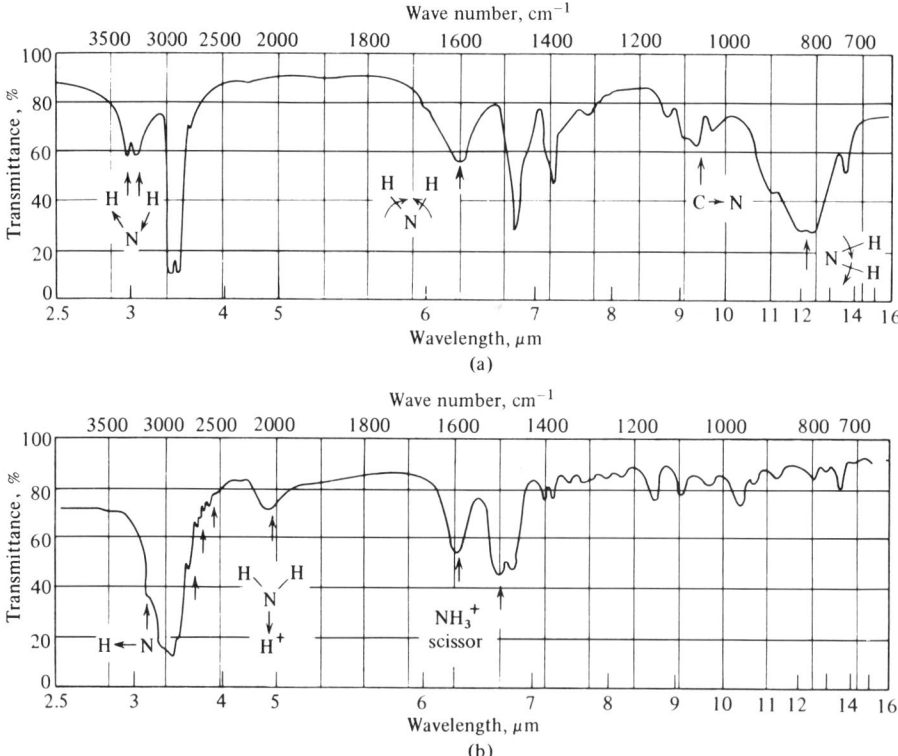

FIGURE 7-10 Infrared spectrum of n-hexylamine. (a) Free amine and (b) hydrochloride.

tions whose positions should be checked in Fig. 7-2. Quite analogous bands are observed for bonds between S and O; all are intense. Stretching frequencies for SO_2 appear around 1400–1310 and 1230–1120 cm^{-1} (7.14–7.63 and 8.13–8.93 μm); for S=O at 1200–1040 cm^{-1} (8.33–9.62 μm); and for S—O around 900–700 cm^{-1} (11.11–14.28 μm).

Compound Identification

In many cases the interpretation of the infrared spectrum on the basis of characteristic frequencies will not be sufficient to permit positive identification of a total unknown, but perhaps the type or class of compound can be deduced. One must resist the tendency to overinterpret a spectrum; that is, to attempt to interpret and assign all of the observed absorption bands, particularly those of moderate and weak intensity in the fingerprint region. Once the category is established, the spectrum of the unknown is compared with spectra of appropriate known compounds for an exact spectral match. If the exact compound happens not to be in the file, particular structure variations within the category may assist in suggesting possible answers and eliminating others. Several collections of spectra are available commercially.[3,4]

7.2 INSTRUMENTATION

It is convenient to divide the infrared region into three segments with the dividing points based on instrumental capabilities (Table 7-3). Different radiation sources, optical systems, and detectors are needed for the different regions. The standard infrared spectrophotometer is a filter-grating or prism-grating instrument covering the range from 4000 to 650 cm^{-1} (2.5 to 15.4 μm). Grating instruments offer high resolution that permits separation of closely spaced absorption bands, accurate measurements of band positions and intensities, and high scanning speeds for a given resolution and noise level. Modern spectrophotometers generally have attachments that permit speed suppression, scale expansion, repetitive scanning, and automatic control of slit, period, and gain. Often these are under the control of a microprocessor. Accessories such as beam condensers, reflectance units, polarizers, and microcells can usually be added to extend versatility or accuracy.

Spectrophotometers for the infrared region are composed of the same basic components as instruments in the ultraviolet–visible region, although the sources, the detectors, and the materials used in the fabrication of the optical components are different, except in the near-infrared. Radiation from a source emitting in the infrared region is interrupted (chopped, pulsed, or modulated) at a low frequency, often 10–26 Hz, and is passed alternately through the sample and the reference placed before the monochromator. This minimizes the effect of stray radiation emanating from the sample and cell before it reaches the detector, a serious problem in most of the infrared region. Temperature and relative humidity in the room housing the instrument must be controlled.

Radiation Sources

In the region beyond 5000 cm^{-1} (2.0 μm), blackbody sources without envelopes commonly are used. The same spectral characteristics cited for the tungsten incandescent lamp

TABLE 7-3   Components of Infrared Spectrophotometers

	REGION OF ELECTROMAGNETIC SPECTRUM		
	Near-Infrared	Mid-Infrared	Far-Infrared
Wave number, cm^{-1}	12,500 4000	200	10
Wavelength, μm	0.8 2.5	50	1000
Source of Radiation	Tungsten filament lamp	Nernst glower, Globar, or coil of Nichrome wire	High-pressure mercury-arc lamp
Optical System	One or two quartz prisms or prism-grating double monochromator	Two to four plane diffraction gratings with either a fore-prism monochromator or infrared filters	Double-beam grating instruments for use to 700 μm; interferometric spectrometers for use to 1000 μm
Detector	Lead sulfide photoconductive	Thermopile, thermistor, or pyroelectric	Golay, pyroelectric

apply to these as well (see Fig. 2-2). Unfortunately, the emission maximum lies in the near-infrared. A fraction of the shorter-wavelength radiation will be present as stray light, and this will be particularly serious for long-wavelength measurements.

A close-wound *Nichrome coil* can be raised to incandescence by resistive heating. A black oxide film forms on the coil which gives acceptable emissivity. Temperatures up to 1100°C can be reached. The Nichrome coil requires no water-cooling and little or no maintenance and gives long service. This source is recommended where reliability is essential, such as in nondispersive process analyzers and inexpensive spectrophotometers or filter photometers. Although simple and rugged, this source is less intense than other infrared sources.

A hotter, and therefore brighter, source is the *Nernst glower,* which has an operating temperature as high as 1500°C. Nernst glowers are constructed from a fused mixture of oxides of zirconium, yttrium, and thorium, molded in the form of hollow rods 1–3 mm in diameter and 2–5 cm in length. The ends of the rods are cemented to short ceramic tubes to facilitate mounting; short platinum leads provide power connections. Nernst glowers are fragile. They have a negative coefficient of resistance and must be preheated to be conductive. Therefore, auxiliary heaters must be provided as well as a ballast system to prevent overheating. A glower must be protected from drafts, but at the same time adequate ventilation is needed to remove surplus heat and evaporated oxides and binder.

The energy output is predominantly concentrated between 1 and 10 μm, with relatively low energy beyond 10 μm. Radiation intensity is approximately twice that of Nichrome and Globar sources except in the near-infrared.

The *Globar,* a rod of silicon carbide 6–8 mm in diameter and 50 mm in length, possesses characteristics intermediate between heated wire coils and the Nernst glower. It is self-starting and has an operating temperature near 1300°C. The temperature coefficient of resistance is positive and may be conveniently controlled with a variable transformer. Its resistance increases with the length of time used so that provision must be made for increasing the voltage across the unit. It is often encased in a water-cooled brass tube, with a slot provided for emission of radiation. The spectral output of the Globar is about 80% that of a blackbody radiator. In comparison with the Nernst glower, the Globar is a less intense source below 10 μm, the two sources are comparable out to about 15 μm, and the Globar is superior beyond about 15 μm. It finds some use out to about 50 μm.

In the very far-infrared, beyond 50 μm (200 cm^{-1}), blackbody-type sources lose effectiveness since their radiation decreases with the fourth power of wavelength. High-pressure mercury arcs, with an extra quartz jacket to reduce thermal loss, give intense radiation in this region. Output is similar to that from blackbody sources, but additional radiation is emitted from a plasma which enhances the long-wavelength output.

Detectors[5]

At the short-wavelength end, below about 1.2 μm, the preferred detection methods are the same as those used for visible and ultraviolet radiation. The detectors used at longer wavelengths can be classified into two groups: (1) thermal detectors, in which the infrared radiation produces a heating effect that alters some physical property of the detector, and (2) photon detectors, which use the quantum effects of the infrared radiation to change the electrical properties of a semiconductor.

Thermal Detectors

The active element in any thermal detector is as small as possible to maximize its temperature change for any level of infrared radiant energy. For the same reason, the element is blackened and thermally insulated from its substrate. When radiation ceases, the element returns to the temperature of the substrate with a decay time determined by finite thermal conductance of the insulation. Material properties affected in thermal detectors include an expansion of a solid or fluid (Golay cell), electrical resistance (thermistor), voltage induced at the junction of two dissimilar materials (thermocouple and thermopile), and electric polarization (pyroelectric).

Thermal detectors are usable over a wide range of wavelengths (Fig. 7-11), which includes both visible and infrared radiation, and they operate at room temperature. Their main disadvantages are slow response time (milliseconds) and lower sensitivity relative to other types of detectors. Their response time sets an upper limit to the frequency at which the radiation can be usefully modulated, chopped, or pulsed. The total mass represented by

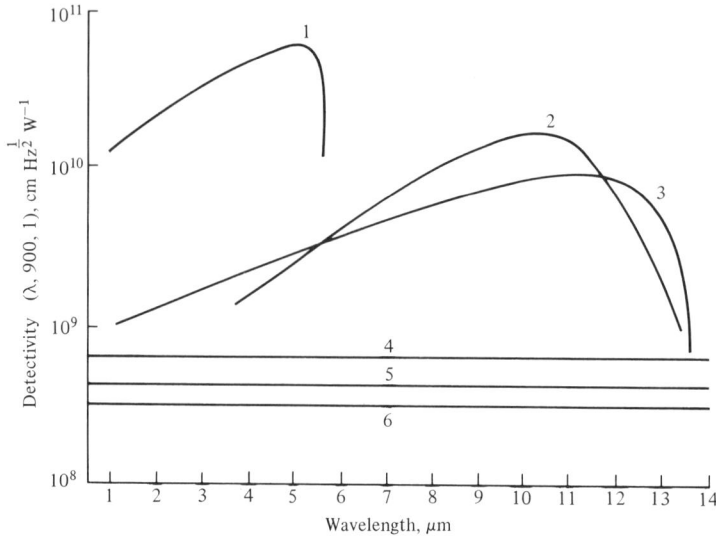

FIGURE 7-11 Wavelength response of some infrared detectors: (1) InSb at 77°K, (2) PbSnTe at 77°K, (3) PbSnTe at 4.2°K, (4) pyroelectric at 300°K, (5) thermistor at 300°K, and (6) thermopile at 300°K. Detectivity obtained by irradiating the detector with monochromatic power at the same wavelength as that where the detector produces its peak output, chopping frequency is 900 Hz, and noise is measured in a 1-Hz bandwidth. (Courtesy of Barnes Engineering Co.)

receiver, absorbing material, and temperature-sensing element must heat during each half-cycle of radiation striking the detector and cool when the detector is occluded.

A *thermocouple,* fabricated from two dissimilar metals like bismuth and antimony, will produce a small voltage proportional to the temperature of the junction. The surface receiving the incident radiation is coated with a metal oxide, such as gold or bismuth black which has little thermal mass. A *thermopile* consists of several (often six) thermocouples connected in series so their outputs add. Half the junctions are called "hot" and make up the active element. Alternate junctions, the "cold" ones, are thermally bonded to the substrate and remain at a relatively stable temperature. Thin-film techniques have miniaturized thermopiles. The entire assembly (Fig. 7-12) is mounted in an evacuated enclosure with an infrared-transmitting window so that conductive heat losses are minimized. Thermopiles offer the simplest and most direct means for converting radiant energy into an electrical signal. Frequency response is flat below 0.35 Hz. Response time is about 80 msec. To prevent the faint signals from being lost in the stray (noise) signals picked up by the lead wires, a preamplifier is located as close to the detector element as possible.

A *thermistor* functions by changing resistance when illuminated by infrared radiation. To minimize noise and drift, an infrared thermistor detector contains two closely spaced thermistor flakes; the *therm*ally sensitive res*istors* are sintered oxides of manganese, cobalt, and nickel which have a high temperature coefficient of resistance (approximately 4%/°C). One of these 10-μm-thick flakes is an active detector while the other acts as a compensating (or reference) detector. The active flake is coated with black material to increase its

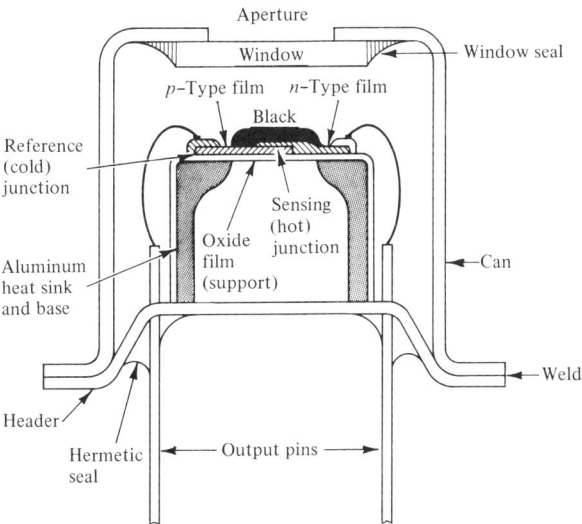

FIGURE 7-12 Schematic construction of a thermocouple (or thermopile) in cross section.

infrared absorption, whereas the compensating flake is optically shielded to prevent its exposure to the incident infrared radiation. The flakes are separately mounted on an insulating substrate that is placed on a heat sink. Two configurations may be used. In the solid backed configuration, the active element of the detector is centered under a window (Fig. 7-13a). By optically coupling the active thermistor wafer to the plano surface of an infrared transmitting hemisphere or hyperhemisphere (Fig. 7-13b), incident radiation is concentrated on the detector. Sensitivity is increased by a factor of 3.5 to 20 in this manner, depending upon the characteristics of the optical material. In either configuration the shielded wafer, when connected in a bridge circuit like Fig. 7-13c, compensates for ambient temperature changes. A steady bias voltage is applied across the detector elements; the drift due to ambient temperature change tends to cancel because both the active and compensating flakes are connected in series. By judicious choices of substrate material and the thickness of the flakes, detectors can be constructed with time constants on the order of a few milliseconds. In general, however, the response time and responsivity must be traded. Greater thermal contact assures faster response time, but since good thermal contact also prevents the flake from reaching higher temperatures, the detector responsivity decreases.

A *pyroelectric* detector contains a noncentrosymmetrical crystal which, below its Curie temperature, exhibits an internal electric field (or polarization) along the polar axis. The electric field results from the alignment of electric dipole moments. Heat resulting from radiation absorption produces thermal alteration of the crystal lattice spacing, which in turn changes the value of the electric polarization. The surfaces of the crystal normal to the polarization axis then develop a polarization charge. If electrodes are applied to these surfaces (one of which is infrared transparent) and connected through an external circuit, free charge will be brought to the electrodes to balance the polarization charge,

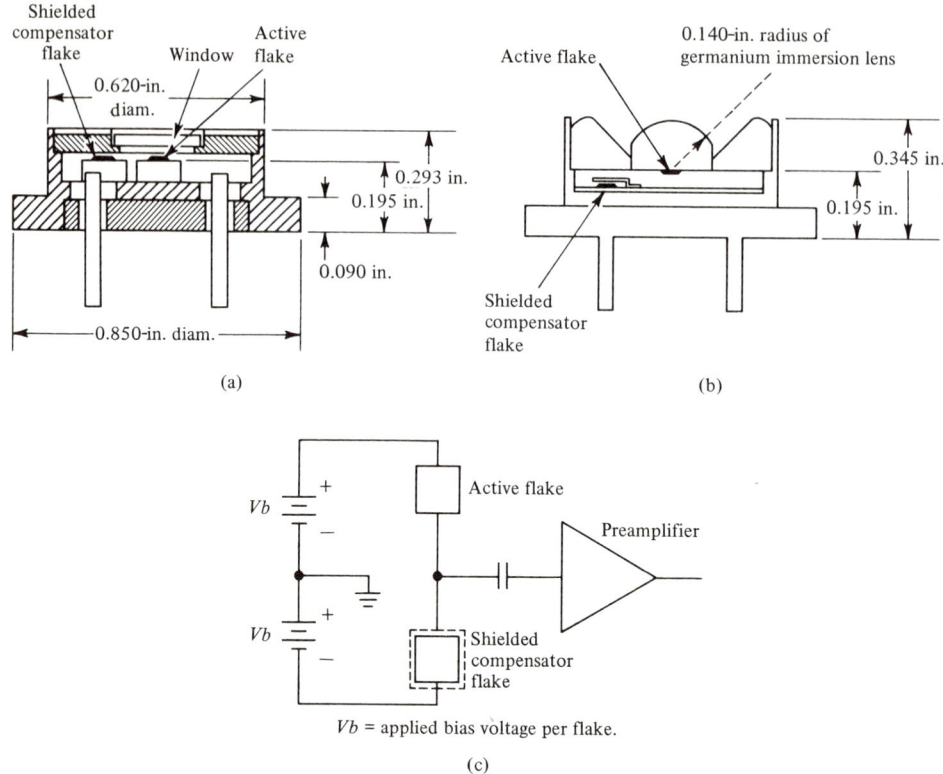

FIGURE 7-13 Thermistor–bolometer infrared detector: (a) solid backed, (b) optical coupled, and (c) bridge circuit. (Courtesy of Barnes Engineering Co.)

thus generating a current in the external circuit. Unlike other thermal detectors, the pyroelectric effect depends on the rate of change of the detector temperature rather than on the value of the temperature itself. This allows the pyroelectric detector to operate with a much faster response time. It also means that this type of detector will respond only to changing radiation that is chopped, pulsed, or otherwise modulated, but will ignore steady background radiation. The detector is a thin plate of pyroelectric material between two electrodes, forming a capacitor. The high impedance is reduced by mounting the subassembly in an enclosure containing a field-effect transistor connected as source follower and a matched load resistor. Pyroelectric materials include triglycine sulfate (TGS), deuterated triglycine sulfate (DTGS), $LiTaO_3$, $LiNbO_3$, and some polymer materials. Generally TGS and DTGS show superior features, but are limited in use because of their hygroscopicity and low Curie points (approximately 49°C). For ease in handling and higher Curie points, $LiTaO_3$ or $LiNbO_3$ is frequently used. With a 100-MΩ load resistor, the response time is 1 msec and responsivity (ratio of detector output to incident radiation) is about 100; whereas with a 1-MΩ load resistor, the response time is 10 μsec and responsivity is 1. The absence of a bias voltage means there is minimal low-frequency noise that would interfere with low-frequency scanning signals.

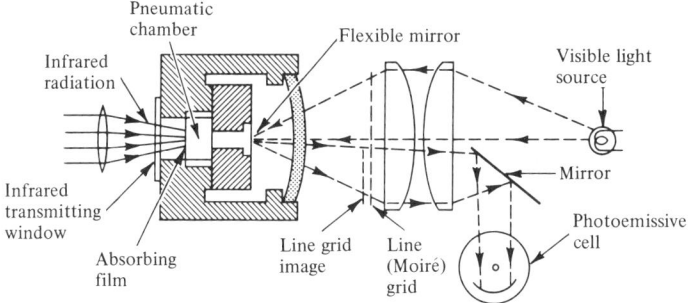

FIGURE 7-14 Golay pneumatic infrared detector.

The *Golay* pneumatic detector, shown in Fig. 7-14, utilizes the expansion of a gas as the measuring device. The unit consists of a small metal cylinder closed by a rigid blackened metal plate (2-mm square) at one end and by a flexible silvered diaphragm at the other end. The chamber is filled with xenon. Radiation passes through a small infrared-transmitting window and is absorbed by the blackened plate. Heat, conducted to the gas, causes it to expand and deform the flexible diaphragm (mirror). To amplify distortions of the mirror surface, light from a lamp inside the detector housing is focused upon the mirror which reflects the light beam onto a phototube. With the flexible mirror in its rest position, an image of half the Moiré grid falls on the other half so that no light passes through. Flexing of the mirror moves the image of the grid laterally so that varying amounts of light can reach the phototube. In an alternate arrangement, the rigid diaphragm is used as one plate of a dynamic condenser; a perforated diaphragm a slight distance away serves as the second plate. The distortion of the solid diaphragm relative to the fixed plate alters the plate separation and hence the capacity. Response time is approximately 20 msec. The Golay detector has a sensitivity similar to that of a thermocouple; it is significantly superior as a detector for the far-infrared. Since the angular aperture is 60°, the detector must be used with a system of condensing mirrors to concentrate the incident radiation.

Photon Detectors

The more sensitive infrared detectors rely on a quantum interaction between the incident photons and a semiconductor—the result producing electrons and holes. This is the internal photoeffect. A sufficiently energetic photon that strikes an electron in the detector can raise that electron from a nonconducting state into a conducting state. The excitation of electrons requires a definite minimum energy in the photon; the detectors thus exhibit a sharp cutoff toward the far-infrared. As conductors, electrons contribute to the current flow in one of two ways, depending on the configuration of the semiconductor. These are referred to as photovoltaic or photoconductive.

In a *photoconductive* detector, consisting of a homogeneous semiconductor chip, the presence of electrons in the conduction band will lower the chip's resistance. Intrinsic hole–electron pairs are created by raising an electron from the valence band to the conduc-

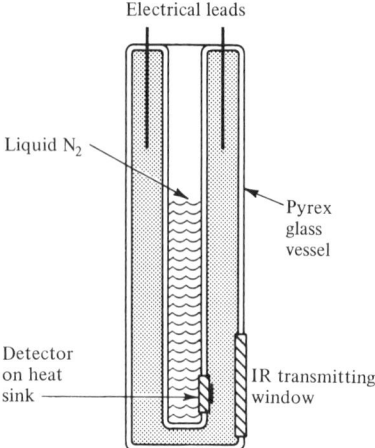

FIGURE 7-15 Simplified construction diagram of an indium antimonide detector cooled by liquid nitrogen. Electrical leads not shown connected to detector.

tion band of the semiconductor. Extrinsic excitation refers to electrons raised from or to impurity doping levels within the forbidden band of the semiconductor. In either case a bias current or voltage registers this change at the output.

Photovoltaic detectors generate a small voltage when exposed to radiation. InSb detectors use a diffused *p–n* junction in single-crystal indium antimonide. The *p*-type InSb is in a thin layer over the *n*-type material; radiation is incident upon the *p*-type surface. Photons having sufficient energy generate hole–electron pairs which are then separated by the internal field existing at the *p–n* junction. The result is a voltage that can be electronically processed as required by the application. The valence-to-conduction band gap of InSb is 0.23 eV at liquid nitrogen temperature. This accounts for the detector's sensitivity cutoff at a wavelength of 5.5 μm. The detector forms an integral part of its Dewar-type vacuum bottle cooling unit (Fig. 7-15). The cooling wall has sufficient volume to permit the use of either liquid nitrogen (4 hr/charge) or Joule–Thomson coolers using compressed nitrogen gas. Single-element detectors, as well as mosaic arrays, are available. Signal output is linear with irradiance over nine orders of magnitude. The time constant is less than 1 μsec.

Lead tin telluride detectors extend spectral sensitivity to considerably longer wavelengths than InSb. Two types are available. The first is cooled by liquid nitrogen and has optimum sensitivity throughout the 5–13 μm region. The second type is liquid helium-cooled and has optimum performance in the 6.6–18 μm region. When used with a current mode amplifier, the speed of response is not compromised for detector sensitivity and response times as fast as 20 nsec can be obtained. Bias currents are not required, resulting in very low low-frequency noise.

Spectrophotometers

Most infrared spectrophotometers are double-beam instruments in which two equivalent beams of radiant energy are taken from the source. By means of a combined rotating mirror and light interrupter, the source is flicked alternately between the reference and sample paths. In the optical-null system, the detector responds only when the intensity of the two beams is unequal. Any imbalance is corrected for by a light attenuator (an optical wedge or shutter comb) moving in or out of the reference beam to restore balance. The recording pen is coupled to the light attenuator. Although very popular, the optical-null system has serious faults. Near zero transmittance of the sample, the reference-beam attenuator will move in to stop practically all light in the reference beam. Both beams are then blocked, no energy is passed, and the spectrometer has no way of determining how close it is to the correct transmittance value. The instrument will go dead. However, in the mid-infrared region the electrical beam-ratioing method is not an easy means of avoiding the deficiencies of the optical-null system. To a large extent it is trading optical and mechanical problems for electronic problems.

Monochromators employing prisms for dispersion utilize a Littrow 60° prism–plane mirror mount (Fig. 2-17). Mid-infrared instruments employ a sodium chloride prism for the region from 4000 to 650 cm^{-1} (2.5 to 15.4 μm), with a potassium bromide or cesium iodide prism and optics for the extension of the useful spectrum to 400 cm^{-1} (25 μm) or 270 cm^{-1} (37 μm), respectively. Quartz monochromators, designed for the ultraviolet–visible region, extend their coverage into the near-infrared (to 2500 cm^{-1} or 4 μm).

Plane-reflectance grating monochromators dominate today's instruments. To cover the wide wavelength range, several gratings with different ruling densities and associated higher-order filters are necessary. This requires some complex sensing and switching mechanisms for automating the scan with acceptable accuracy. Because of the nature of the blackbody emission curve, a slit programming mechanism must be employed to give near constant energy and resolution as a function of wavelength. The principal limitation is energy. Resolution and signal-to-noise ratio are limited primarily by the emission of the blackbody source and the noise-equivalent power of the detector. Two gratings are often mounted back-to-back so that each need be used only in the first order; the gratings are changed at 2000 cm^{-1} (5.0 μm) in mid-infrared spectrometers. Grating instruments incorporate a sine-bar mechanism to drive the grating mount when a wavelength readout is desired, and a cosecant-bar drive when wave numbers are desired. Undesired overlapping orders can be eliminated with a fore-prism or by suitable filters.

The optical arrangement for a filter-grating spectrometer is shown in Fig. 7-16. The filters are inserted near a slit or slit image when the required size of the filter is not excessive. The circular variable filter is simple in construction. It is frequently necessary to use gratings as reflectance filters when working in the far-infrared in order to remove unwanted second and higher orders from the light incident on the far-infrared grating. For this purpose small plane gratings are used which are blazed for the wavelength of the unwanted radiation. The grating acts as a mirror reflecting the wanted light into the instrument and diffracting the shorter wavelengths out of the beam; a grating "looks" like a good mirror to wavelengths longer than the groove spacing.

Probably the most elegant filter is a prism because it provides a narrow band of wavelengths with high efficiency over a relatively broad spectral range. The prism and grating must track together over consecutive grating orders. Light from the parabolic mirror enters the fore-prism where it is dispersed so that only a relatively narrow band of wavelengths

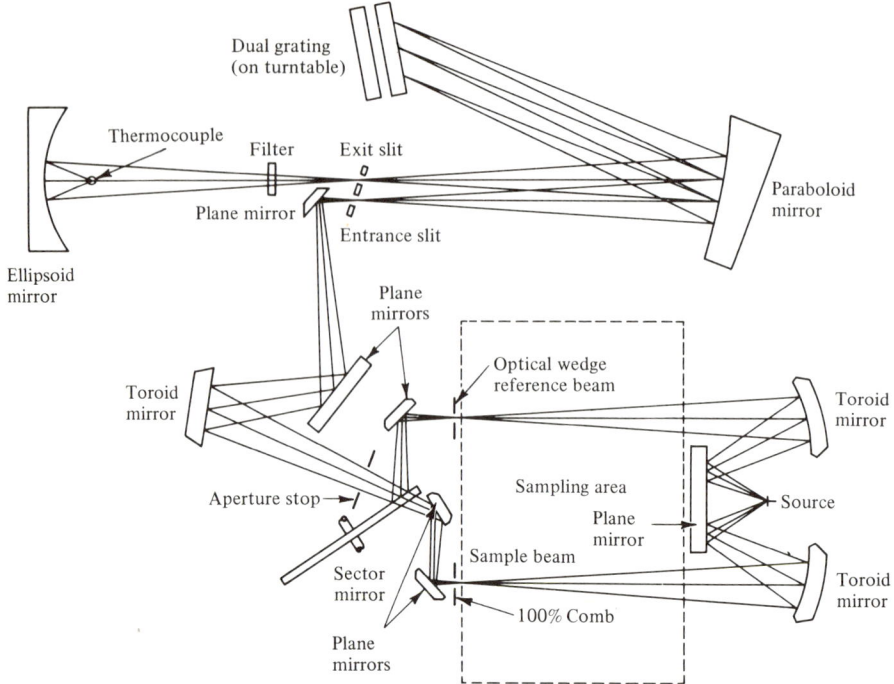

FIGURE 7-16 Optical schematic of a filter-grating, double-beam infrared spectrophotometer. (Courtesy of Perkin-Elmer Corp.)

is allowed to fall on the grating. The resolution of the prism can be quite low, because it need only exclude the adjacent orders, but for the higher orders, the interval gets successively narrower. Thus, it is preferable to use two gratings and confine their application to lower orders.

The use of microprocessors has alleviated many of the tedious requirements necessary to obtain usable data. Integrated scan controls allow the operator to select a single recording parameter, such as scan time, slit setting, pen response, and the microprocessor automatically optimizes the other conditions. Even under high-resolution conditions where the noise might obscure spectral detail, the operator can improve the S/N ratio by selecting the appropriate multiplier. When this occurs, the instrument automatically changes the scanning parameters to optimize recording conditions.

Single-beam photometers possess the capability for accurate measurement in quantitative analysis. Until recently they were not extensively utilized, primarily as a result of the need for extensive reduction of data produced. However, a built-in microcomputer results in the combined system's programmability as well as the virtually instantaneous reduction of data and availability of results. The optical schematic of a very simple infrared analyzer is shown in Fig. 7-17. Different wavelengths are discriminated by three circular-variable interference filters covering the spectral range from 2.5 to 14.5 μm. These filters are mounted on the shaft of a high-resolution potentiometer which senses its position. This

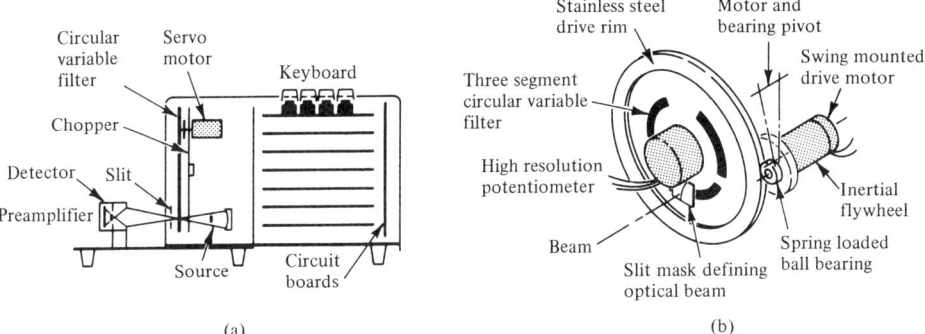

FIGURE 7-17 Infrared analyzer, single-beam: (a) optical schematic and (b) circular interference filter drive system. (Courtesy of Foxboro/Wilks, Inc.)

position information is compared to a fixed value from the microcomputer in the feedback loop to a dc servomotor that drives the filter wheel to the desired position. The instrument has a Nichrome wire source and a $LiTaO_3$ pyroelectric detector. Using an $f/1.5$ optical system with the low-resolution filter yields a high S/N ratio, allowing the detection of absorption as small as 0.0001 absorbance units.

Simple filter infrared analyzers have been designed around the same optical schematic shown in Fig. 7-17. Wavelength selection is made through the use of interchangeable, slide-mounted fixed-wavelength interference filters. Several filters may even be mounted on a filter wheel for convenient selection. Equipped with a micro-flow-through cell designed for continuous monitoring, these infrared analyzers can be coupled to liquid chromatographs for monitoring effluents, or can be employed for the quantitative determination of fiber finishes and lubricating oils, or dissolved hydrocarbons in water after extraction into CCl_4 or Freon solvent. Strongly absorbing liquids can be handled with a fixed flow-through internal reflection cell.

Fourier Transform Interferometer[2,6]

Instead of using a monochromator, the infrared radiation, after passage through a sample, can be analyzed by means of a scanning Michelson interferometer. Referring to Fig. 7-18, this consists of a moving mirror, 4, a fixed mirror, 3, and a beamsplitter, C. Radiation from the infrared source, B, is collimated by mirror 2 and the resultant beam is divided at the beamsplitter, half of the beam passing to mirror 3 and half reflected to the moving mirror. After reflection the two beams recombine at the beamsplitter and, for any particular wavelength, constructively or destructively interfere depending on the difference in optical paths between the two arms of the interferometer. With a constant mirror velocity the intensity of the emerging radiation at any one particular wavelength modulates in a regular sinusoidal manner. In the case of a broadband infrared source the emerging beam is a complex mixture of modulation frequencies, which, after passing through the sample compartment, is focused onto the detector, G.

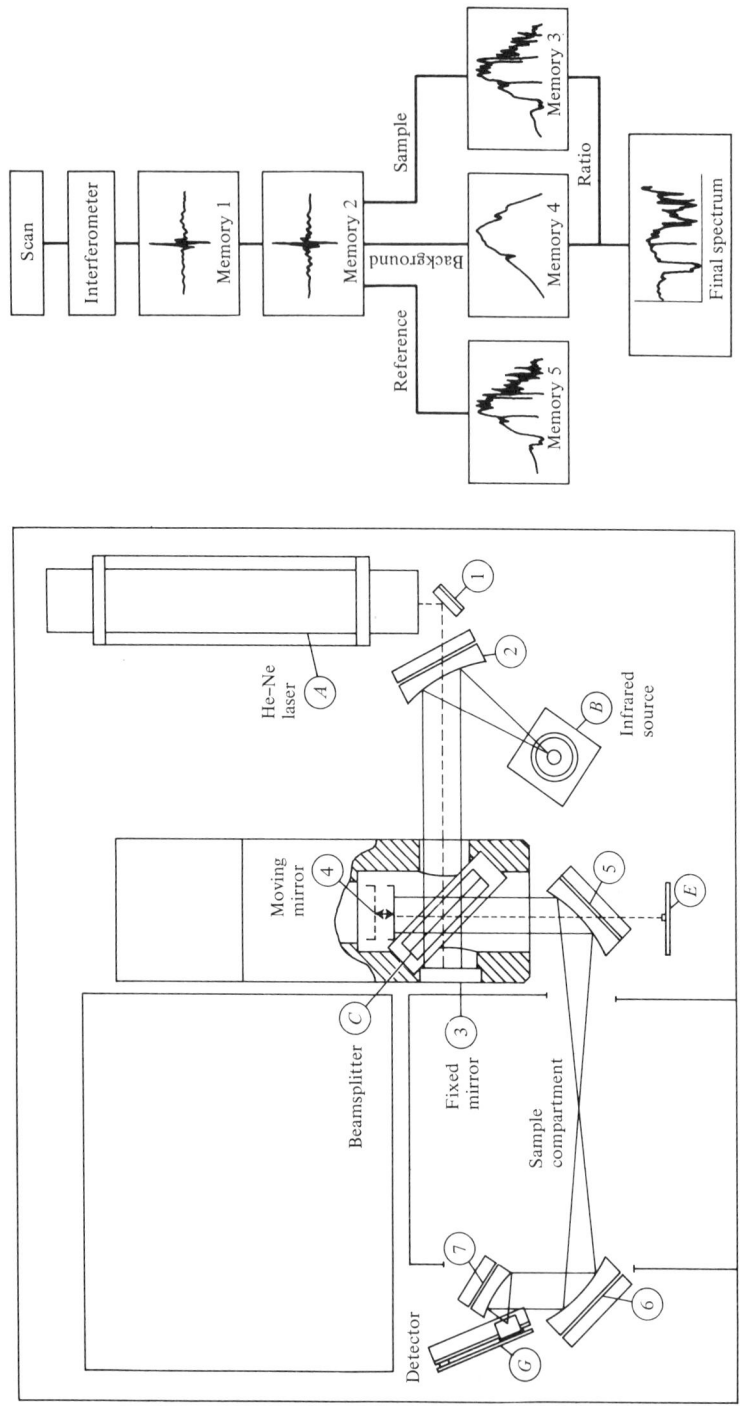

FIGURE 7-18 Infrared Fourier transform interferometric spectrometer: (a) optical path diagram and (b) block diagram of instrument's functions. (Courtesy of Nicolet Instrument Corp.)

This detector signal is sampled at very precise intervals during the mirror scan. Both the sampling rate and mirror velocity are controlled by a reference signal from detector E produced by modulation of the beam from the helium–neon laser A. The resulting signal is known as an interferogram (memory 1) and contains all the information required to reconstruct the spectrum via a mathematical process known as Fourier transformation. This technique has several distinct advantages over conventional dispersive techniques. There is only one moving part involved, mirror 4; this is mounted on a frictionless air bearing. Dispersion or filtering is not required, so that energy-wasting slits are not needed—a major advantage, particularly with energy a premium in the far-infrared. The use of a helium–neon laser as a reference results in near absolute frequency accuracy, better than 0.01 cm^{-1} over the range of 4800–400 cm^{-1}. Because all wavelengths are detected throughout the scan, the scanning interferometer achieves the same spectral signal-to-noise ratio as a dispersive spectrometer in a fraction of the time (Fellgett's advantage).

The automatic process between the initiation of the scan and the final plot is outlined in Fig. 7-18b. The interferogram recorded with each scan is stored in memory 1. This interferogram is then automatically aligned with, and added to, the averaged interferograms in memory 2. At the same time annotation of the plot is begun in preparation for the final spectrum. After 32 scans (approximately 60 sec), this averaged interferogram is Fourier transformed to produce a single-beam spectrum which, in the standard sample mode, is stored in memory 3. This single-beam spectrum is then ratioed against the stored background (run once a day) in memory 4 and the resulting "double-beam" spectrum plotted on the high-speed digital plotter. Memory 5 is additional space which is available for storage of a reference spectrum which would be used in the spectral subtraction technique. This memory is also used to store a newly measured spectrum while maintaining the sample and background spectra for further manipulation. The time from insertion of the sample to completed plot is about 2 min.

7.3 SAMPLE HANDLING

Infrared instrumentation has reached a remarkable degree of standardization as far as the sample compartment of various spectrometers is concerned. Sample handling itself, however, presents a number of problems in the infrared region. No rugged window material for cuvettes exists that is transparent and also inert over this region. The alkali halides are widely used, particularly sodium chloride, which is transparent at wavelengths as long as 16 μm (625 cm^{-1}). Cell windows are easily fogged by exposure to moisture and require frequent repolishing. Silver chloride is often used for moist samples, or aqueous solutions, but it is soft, easily deformed, and darkens on exposure to visible light. Teflon has only C–C and C–F absorption bands. For frequencies under 600 cm^{-1}, a polyethylene cell is useful. Infrared transmission materials are compiled in Table 7-4. Materials of high refractive index produce strong, persistent interference fringes.

Gases

In the analysis of gases the usual path length is 10 cm. When this is too short to measure the spectra of minor components or substances encountered in trace analysis, a variable-

TABLE 7-4 Infrared Transmitting Materials

Material	Wavelength Range, μm	Wave number Range, cm^{-1}	Refractive Index at 2 μm
NaCl, rock salt	0.25–16	40,000–625	1.52
KBr, potassium bromide	0.25–25	40,000–400	1.53
AgCl, silver chloride	0.40–23	25,000–435	2.0
AgBr, silver bromide	0.50–35	20,000–286	2.2
CaF$_2$, calcium fluoride	0.15–9	6670–1110	1.40
BaF$_2$, barium fluoride	0.20–11.5	50,000–870	1.46
CsBr, cesium bromide	1–37	10,000–270	1.67
CsI, cesium iodide	1–50	10,000–200	1.74
TlBr-TlI, KRS-5	0.50–35	20,000–286	2.37
ZnSe, zinc selenide (vacuum deposited)	1–18	10,000–55	2.4
Ge, germanium	0.50–11.5	20,000–870	4.0
Si, silicon	0.20–6.2	50,000–1613	3.5
Al$_2$O$_3$, sapphire	0.20–6.5	50,000–1538	1.76
Polyethylene	16–300	625–33	1.54

path cell provides path lengths in steps of 1.5 m for 20-, 40-, and 120-m cells. The light path is folded using internal gold-surfaced mirrors and gold-plated or stainless steel metal components. Further gains in sensitivity can be realized by increasing the pressure of the gas sample in the cell to 10 atm. Pressure broadening of absorption bands can be troublesome in quantitative work. The long-path gas cells are intended for measurements in the range of a few parts per million and lower, concentration ranges encountered with problems in air pollution, air monitoring, process instrumentation, and purity determinations. When working in a spectral region where water vapor or carbon dioxide absorption occurs, a dual cell system for compensation is desirable.

Liquids and Solutions

Samples that are liquid at room temperature are usually scanned in their neat form, or in solution. The sample concentration and path length should be chosen so that the transmittance lies between 15 and 70%. For neat liquids this will represent a very thin layer, about 0.001–0.05 mm in thickness. For solutions, concentrations of 10% and cell lengths of 0.1 mm are most practical. Unfortunately, not all substances can be dissolved in a reasonable concentration in a solvent that is nonabsorbing in regions of interest. When possible, the spectrum is obtained in a 10% solution of CCl$_4$ in a 0.1-mm cell in the region 4000–1333 cm^{-1} (2.5–7.5 μm), and in a 10% solution of CS$_2$ in the region 1333–650 cm^{-1} (7.5–15.4 μm). Transparent regions of selected solvents are given in Fig. 7-19. To obtain solution spectra of polar materials which are insoluble in CCl$_4$ or CS$_2$, chloroform, methylene chloride, acetonitrile, and acetone are useful solvents. Sensitivity can be gained by going to longer path lengths if a suitably transparent solvent can be found. In a double-beam spectrophotometer a reference cell of the same path length as the sample cell is

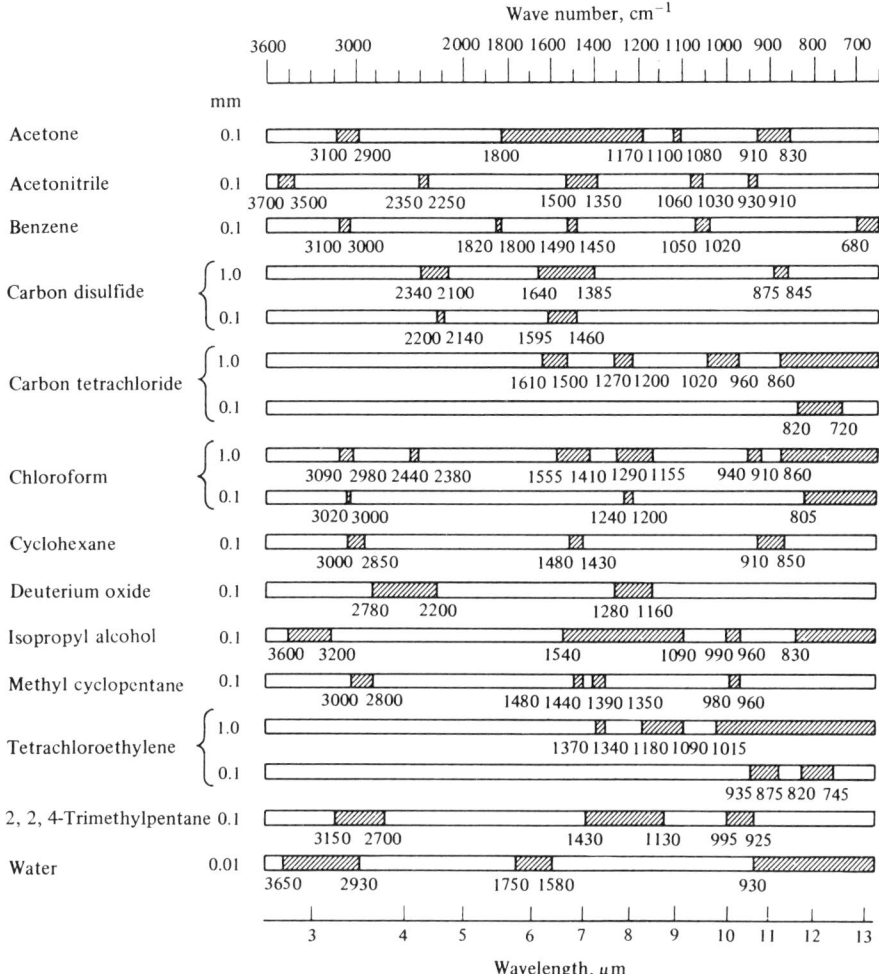

FIGURE 7-19 Transmission characteristics of selected solvents. The material is considered transparent if the transmittance is 75% or greater. Solvent thickness is given in millimeters.

filled with pure solvent and placed in the reference beam. Moderate solvent absorption, now common to both beams, will not be observed in the recorded spectrum. However, solvent transmittance should never fall under 10%.

The possible influence of a solvent on the spectrum of a solute must not be overlooked. Particular care should be exercised in the selection of a solvent for compounds which are susceptible to hydrogen-bonding effects. Hydrogen bonding through an —OH or —NH— group alters the characteristic vibrational frequency of that group; the stronger the hydrogen bonding, the greater is the lowering of the fundamental frequency. To differentiate between inter- and intramolecular hydrogen bonding, a series of spectra at different dilutions, yet having the same number of absorbing molecules in the beam, must be obtained. If, as the dilution increases, the hydrogen-bonded absorption band decreases while the unbonded absorption band increases, the bonding is intermolecular. Intramolecular bonding shows no comparable dilution effect.

Infrared solution cells are constructed with windows sealed and separated by thin gaskets of copper and lead which have been wetted with mercury. The whole assembly is securely clamped together and permanently mounted in a stainless steel holder. As the mercury penetrates the metal, the gasket expands, producing a tight seal. The cell is provided with tapered fittings to accept the needles of hypodermic syringes for filling. Each cell is labeled with its precise path length as measured by interference fringes. An unassembled (demountable) cell involves two window pieces and a Teflon fitting (which also accommodates a syringe for filling) which forms the leak-proof seal when the cell is slipped into a mount and knurled nuts are turned down until finger-tight.

Flow-through cells are useful for the continuous analysis of liquids. In repetitive sampling applications, quantitative accuracy is increased and sample handling facilitated. Coupling the infrared instrument to a chromatograph with a micro-flow-through cell makes it possible to monitor a column effluent on a functional group basis.

The variable path length cell consists of a cylindrical, Teflon-lined, stainless steel chamber with parallel windows. The path length may be continuously adjusted from 0.005 to 5 mm and reproduced to within 0.001 mm. A vernier and scale provided on the cylinder permits the cell thickness to be read within ±0.0005 mm. Accuracy of the thickness settings are ±0.001 mm or 1%, whichever is larger.

The variable path length liquid cell serves two very useful purposes in the infrared laboratory. The first is solvent compensating and differential analysis. When filled with solvent and mounted in the reference beam of the spectrophotometer, its path length may be adjusted to compensate for unwanted solvent absorption in the sample beam. In differential analysis, two liquids are examined which may differ only slightly in the concentration of the minor component. Careful adjustment of the variable cell will tend to enhance the spectrum of the minor component contained in the fixed path cell in the sample beam.

The second application is to use the variable path length cell in the sample beam. In order for weaker absorption bands of many liquids to appear with the proper intensity, the stronger bands will be totally absorbing and their exact location thus obscured. In such situations, the variable path length cell may be precisely adjusted to provide just the right intensity for exact location of both weak and strong bands. The variable path length cell can eliminate the requirement for a large collection of different thickness fixed path length cells.

The minicell is an economical approach for obtaining qualitative infrared spectra of liquids. It consists of a threaded, two-piece plastic body, and two silver-chloride cell windows, disks of special design. The windows fit into one portion of the cell. The second portion of the cell is then screwed in to form the seal. The AgCl cell windows each contain a 0.025-mm circular depression (also available as 0.100-mm depression). The rim of the window is flat, and the circumference is beveled to ensure proper sealing. Because AgCl flows slightly under pressure, a tight seal is formed. The circular depression in each window enables the cell path length to be varied: (1) the windows can be placed back to back for a conventional smear; (2) arranged with one back to the circular depression for 0.025-mm path length; or (3) positioned with facing circular depressions for 0.050-mm path length.

Films

Spectra of liquids not soluble in a suitable solvent are best obtained from capillary films. A large drop of the neat liquid is placed between two infrared transmitting windows which are then squeezed together and mounted in the spectrometer in a suitable holder. Plates need not have high polish, but must be flat to avoid distortion of the spectrum.

For polymers, resins, and amorphous solids, the sample is dissolved in any reasonably volatile solvent, the solution poured onto a rock-salt plate, and the solvent evaporated by gentle heating. If the solid is noncrystalline, a thin homogeneous film is deposited on the plate which then can be mounted and scanned directly. Sometimes polymers can be "hot pressed" onto plates.

Mulls

Powders, or solids reduced to particles, can be examined as a thin paste or mull. A small amount of the sample is thoroughly ground in a clean mortar until the powder is almost polished into the mortar. When grinding is completed, the mulling agent is introduced in a small quantity just sufficient to take up the powder. The resulting mixture approximates the consistency of toothpaste. The mixture is then transferred to the mull plates and the plates squeezed together to adjust the thickness of the sample. Sample thickness should be adjusted so that the strongest bands display between 60 and 80% absorption.

Multiple reflections and refractions off the particles are lessened by grinding the particles to a size in order of magnitude less than the analytical wavelength and surrounding the particles by a medium whose refractive index more closely matches theirs than does air. Liquid media include mineral oil or Nujol, hexachlorobutadiene, perfluorokerosene, and chlorofluorocarbon greases (fluorolubes). The latter are used when the absorption by the mineral oil masks the presence of C–H bands. For qualitative analysis the mull technique is rapid and convenient, but quantitative data are difficult to obtain, even when an internal standard is incorporated into the mull. Polymorphic changes, degradation, and other changes may occur during grinding.

Pellet Technique

The pellet technique involves mixing the finely ground sample (1–100 μg) and potassium bromide powder, and pressing the mixture in an evacuable die at sufficient pressure (60,000 –100,000 psi) to produce a transparent disk. Potassium bromide becomes quite plastic at high pressure, and will flow to form a clear disk. Grinding–mixing is conveniently done in a vibrating ball-mill (Wig-L-Bug®). Other alkali halides may also be used, particularly CsI or CsBr for measurements at longer wavelengths. Good dispersion of the sample in the matrix is critical; moisture must be absent. Freeze-drying the sample is often a necessary preliminary step.

KBr wafers can be formed, without evacuation, in a Mini-Press. Two highly polished bolts, turned against each other in a rugged stainless steel cylinder, produce a clear wafer in the cylinder which is also the holder. Pressure is applied with wrenches for about 1 min to 75–100 mg of powder, the bolts are removed, and the cylinder is installed in its slide holder.

By the pellet technique quantitative analyses are readily performed since an accurate measurement can be made of the weight ratio of sample to internal standard in each disk or wafer.

Cell Thickness

One of two methods may be used to measure the path length of infrared absorption cells: the *interference fringe* method and the *standard absorber* method. The interference fringe method is ideally suited to cells whose windows have a high polish. With the empty cell in the spectrophotometer on the sample side and no cell in the reference beam, the spectrophotometer is operated as near as possible to the 100% line. Enough spectrum is run to produce 20–50 fringes. The cell thickness, b, in centimeters, is calculated from the expression

$$b = \frac{1}{2\eta_D} \left(\frac{n}{\bar{\nu}_1 - \bar{\nu}_2} \right) \tag{7-1}$$

where n is the number of fringes (peaks or troughs) between two wave numbers $\bar{\nu}_1$ and $\bar{\nu}_2$, and η_D is the refractive index of the sample material. If measurements are made in wavelength (micrometers), the equation is

$$b = \frac{1}{2\eta_D} \left(\frac{n \lambda_1 \lambda_2}{\lambda_2 - \lambda_1} \right) \tag{7-2}$$

where λ_1 is the starting wavelength and λ_2 the finishing wavelength. The fringe method also works well for measurement of film thickness.

The standard absorber method may be used with a cell in any condition and with cavity cells whose inner faces do not have a finished polish. The 1960 cm^{-1} (5.10 μm) band of benzene may be used for calibrating cells which are less than 0.1 mm in path length, and the 845 cm^{-1} (11.8 μm) band for cells 0.1 mm or longer in path length. At the former frequency, benzene has an absorbance of 0.10 for every 0.01 mm of thickness; at 845 cm^{-1}, benzene has an absorbance of 0.24 for every 0.1 mm of thickness.

Multiple Internal Reflectance

The scope and versatility of infrared spectrophotometry as a qualitative analytical tool have been increased substantially by the technique of multiple internal reflections (also known as attenuated total reflectance or ATR). When a beam of radiation enters a plate (or prism), it will be reflected internally if the angle of incidence at the interface between sample and plate is greater than the critical angle (which is a function of refractive index). On internal reflection, all the energy is reflected. However, the beam appears to penetrate slightly (from a fraction of a wavelength up to several wavelengths) beyond the reflecting surface, and then return. When a material is placed in contact with the reflecting surface, the beam will lose energy at those wavelengths where the material absorbs due to an interaction with the penetrating beam. This attenuated radiation, when measured and plotted as a function of wavelength, will give rise to an absorption spectrum characteristic of the material which resembles an infrared spectrum obtained in the normal manner. Multiple internal reflectance both simplifies infrared sampling and opens new areas of practical investigation. Using the technique, qualitative infrared absorption spectra are easily obtained from the vast majority of solid materials without need for grinding or dissolving, or making a mull.

Most multiple internal reflectance work is done by means of an accessory readily inserted in, and removed from, the sampling space of a conventional infrared spectrophotometer (Fig. 7-20a). The accessory consists of a mirror system that sends the source radiation through the attachment and a second mirror system which directs the radiation into the monochromator. The mirrors are front surfaced aluminum and accommodate the full width of the instrument beams. Sample holders may be mounted in any of three positions, in order to change the external angle. The three standard positions are 30°, 45°, and 60°. The length-to-thickness ratio of the plate determines the number of reflections once the angle of incidence is selected; plate dimensions vary from 0.25 to 5 mm in thickness, and 1 to 10 cm in length. Twenty-five internal reflections are standard for a 2-mm plate. Par-

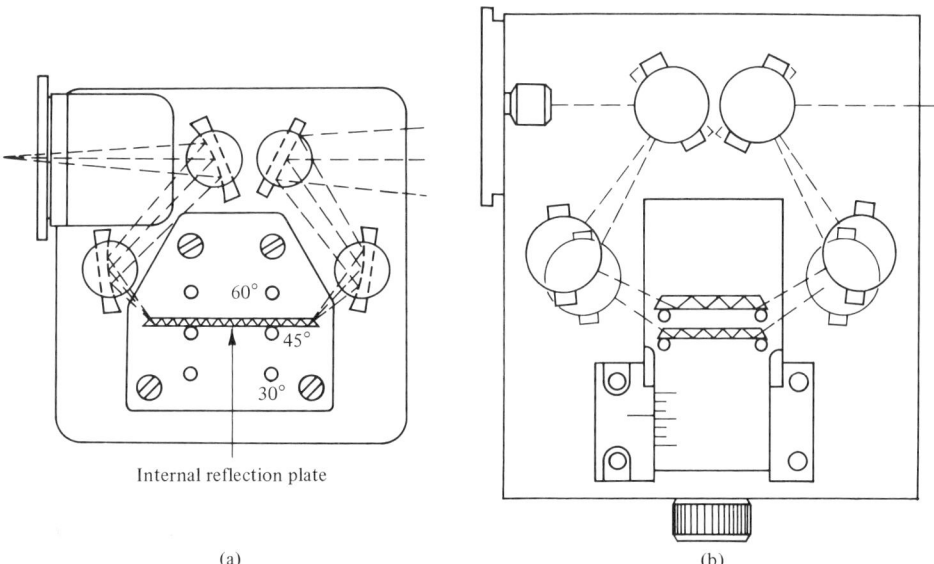

FIGURE 7-20 Multiple internal reflection attachment: (a) three-position pin plate for 30°, 45°, and 60° plates and (b) variable angle model. (Courtesy of Foxboro/Wilks, Inc.)

allelism and flatness of sampling surfaces and surface polish are critical. Where the best performance is required, it is desirable to have matching optical units in both beams of the monochromator. In this way a much flatter I_0 curve can be obtained since absorptions such as those in the reflector plates and from the Teflon "O" rings in liquid cells can be compensated out and there will be a minimum of interference from atomspheric absorption.

In the single-pass plate, light is introduced through an entrance aperture consisting of a simple bevel at one end of the plate and, after propagation via multiple internal reflections down the length of the plate, leaves by means of an exit aperture either parallel or perpendicular to the entrance aperture. The angle of the bevel determines the interior angle of incidence. This type of plate is useful for bulk materials, thin films, and surface studies. In the double-pass plate, light enters as before, propagates down the length of the plate, is totally reflected at the opposite end from a surface perpendicular to the sample surfaces, and returns to leave the plate via the exit aperture. The free end of the plate can be dipped into liquids or powders or placed in closed systems requiring only one optical window.

The apparent depth to which the radiation penetrates the sample is only a few micrometers and is independent of sample thickness. Consequently, ATR spectra can be obtained for many samples that cannot be studied by normal transmission methods. These include samples which show very strong absorptions, resist preparation in thin films, are characteristic only as thick layers, and are available on a nontransparent support. Aqueous solutions can be handled without compensating for very strong solvent absorptions. Samples containing suspended matter, such as dispersed solids or emulsions, that produce high backgrounds in transmission spectra due to scatter, give better results by multiple internal reflectance. Various sample holders are available. Some samples will self-adhere to the reflector plate. In other cases, some pressure is required to bring the sample in contact with the plate. Analyses of liquids are handled using a germanium reflector plate which has both the required insolubility and the short effective path length. The heated sample holder is a modified solid sample holder equipped with two wafer heaters and operates to 250°C in order to study thermal effects *in situ* or as a means of heat-softening various plastic samples to improve optical contact.

The appearance and intensity of a reflectance spectrum will depend upon the difference of the indices of refraction between the reflector plate (Table 7-4) and the rarer medium containing the absorber, and upon the internal angle of incidence. Thus a reflection plate of relatively high index of refraction should be used. The material found to perform most satisfactorily for the majority of liquid and solid samples is KRS-5. Its refractive index is high enough to permit well-defined spectra of nearly all organic materials, although it is soluble in basic solutions. AgCl is recommended for aqueous samples because of its insolubility and lower refractive index; germanium could also be used but is brittle. By varying the angle of incident radiation from 30° to 60°, the depth of penetration of energy into the sample may be changed. A commercial unit is shown in Fig. 7-20b; it utilizes a scissor-jack assembly linking the four mirrors and sample platforms in a pantograph system. At steep angles (near 30°) the depth of penetration of energy into the sample is considerably greater (perhaps an order of magnitude greater than at grazing angles of 60°). Being able to vary the angle of incidence, and thus the depth of radiant penetration, is of considerable significance to the study of surfaces. Much can be learned about the surfaces of

7.4 QUANTITATIVE ANALYSIS

The application of infrared spectroscopy as a quantitative analytical tool varies widely from one laboratory to another. However, the use of high-resolution grating instruments materially increases the scope and reliability of quantitative infrared work. Quantitative infrared analysis is based on Beer's law; apparent deviations arise from either chemical or instrumental effects. In many cases the presence of scattered radiation makes the direct application of Beer's law inaccurate, especially at high values of absorbance. Since the energy available in the useful portion of the infrared is usually quite small, it is necessary to use rather wide slit widths in the monochromator. This causes a considerable change in the apparent value of the molar absorptivity; therefore, molar absorptivity should be determined empirically.

The base line method (Fig. 7-21) involves selection of an absorption band of the substance under analysis which does not fall too close to the bands of other matrix components. The value of the incident radiant energy P_0 is obtained by drawing a straight line tangent to the spectral absorption curve at the position of the sample's absorption band. The transmittance P is measured at the point of maximum absorption. The value of log (P_0/P) is then plotted against concentration.

Many possible errors are eliminated by the base line method. The same cell is used for all determinations. All measurements are made at points on the spectrum which are sharply defined by the spectrum itself, thus there is no dependence on wavelength settings. Use of such ratios eliminates changes in instrument sensitivity, source intensity, or changes in adjustment of the optical system.

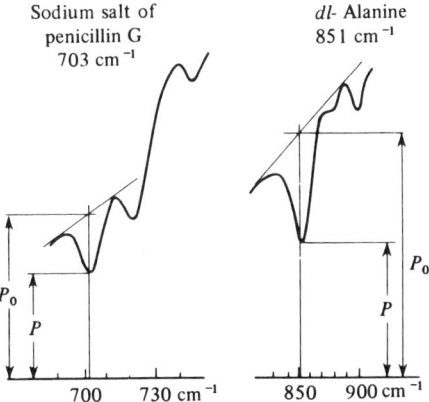

FIGURE 7-21 Base line method for calculation of the transmittance ratio in quantitative analysis.

Pellets from the disk technique can be employed in quantitative measurements. Uniform pellets of similar weight are essential, however, for quantitative analysis. Known weights of KBr are taken, plus a known quantity of the test substance from which absorbance data a calibration curve can be constructed. The disks are weighed and their thickness measured at several points on the surface with a dial micrometer. The disadvantage of measuring pellet thickness can be overcome by using an internal standard. Potassium thiocyanate makes an excellent internal standard. It should be preground, dried, and then reground, at a concentration of 0.2% by weight with dry KBr. The final mix is stored over phosphorus pentoxide. A standard calibration curve is made by mixing about 10% by weight of the test substance with the KBr–KSCN mixture and then grinding. The ratio of the thiocyanate absorption at 2125 cm^{-1} (4.70 μm) to a chosen band absorption of the test substance is plotted against percent concentration of the sample.

For quantitative measurements, the single-beam system has some fundamental characteristics that can result in greater sensitivity and better accuracy than the double-beam systems. All other things being equal, a single-beam instrument will automatically have a greater signal-to-noise ratio. There is a factor of 2 advantage in looking at one beam all the time rather than two beams half the time. Electronic switching gives another factor of 2 advantage. Thus, in any analytical situation where background noise is appreciable, the single-beam spectrometer should be superior.

PROBLEMS

1. What would be the frequency of the fundamental absorption if its first overtone was observed at 1820 cm^{-1}?

2. The molecular heterotope ^{35}Cl^{37}Cl has a fundamental band at 554 cm^{-1} in the gaseous state. Where would one expect the first and second overtones? What window material would be suitable?

3. Assuming a simple diatomic molecule, obtain the frequencies of the absorption band from the force constants given. Compare your answers with the tabulated positions in Fig. 7-2.

(a) $k = 5.1 \times 10^5$ dyn cm^{-1} for C–H bond in ethane.
(b) $k = 5.9 \times 10^5$ dyn cm^{-1} for C–H bond in acetylene.
(c) $k = 4.5 \times 10^5$ dyn cm^{-1} for C–C bond in ethane.
(d) $k = 7.6 \times 10^5$ dyn cm^{-1} for C–C bond in benzene.
(e) $k = 17.5 \times 10^5$ dyn cm^{-1} for C≡N bond in CH$_3$CN.
(f) $k = 12.3 \times 10^5$ dyn cm^{-1} for C=O bond in formaldehyde.

4. The apparent specific absorptivities are given for various infrared absorbers. Calculate the minimum liquid concentrations determinable (mg/ml) in 0.025-mm cells (for an absorbance reading of 0.005).

(a) $\alpha = 900$ for CHCl$_3$ at 1216 cm^{-1}.
(b) $\alpha = 1320$ for CH$_2$Cl$_2$ at 1259 cm^{-1}.
(c) $\alpha = 4900$ for C$_6$H$_6$ at 1348 cm^{-1}.

(d) $\alpha = 6080$ for $COCl_2$ at 1810 cm^{-1}.
(e) $\alpha = 4400$ for $CH_2ClCOCl$ at 1821 cm^{-1}.
(f) $\alpha = 1010$ for water at 1640 cm^{-1}.

5. The presence of ethylene in samples of ethane is determined by using the absorption band of ethylene at 2080 cm^{-1} (5.2 μm). A series of standards gave the following data:

% Ethylene:	0.50	1.00	2.00	3.00
Absorbance:	0.120	0.240	0.480	0.719

Calculate the percentage of ethylene in an unknown sample that had an absorbance of 0.412 when using the same cell and the same instrument.

6. The incorporation of an allyl group into one or both of the side chains of a barbiturate is always associated with the appearance of strong absorption bands at 10.1 and 10.8 μm. What alteration in these absorption bands would be expected by replacement of the hydrogen atom attached to the central carbon atom of the unsaturated allyl group by bromine?

7. Estimate the minimum concentration detectable ($A = 0.005$) in 0.05-mm cells for each of the following compounds, given their molar absorptivities:

(a) Phenol at 3600 cm^{-1}, $\epsilon = 5000$.
(b) Aniline at 3480 cm^{-1}, $\epsilon = 2000$.
(c) Acrylonitrile at 2250 cm^{-1}, $\epsilon = 590$.
(d) Acetone at 1720 cm^{-1}, $\epsilon = 8100$.
(e) Isocyanate (in polyurethane foam) monomer at 2100 cm^{-1}, $\epsilon = 17,000$.

8a. Calculate the thickness of the four cells from their interference fringes.

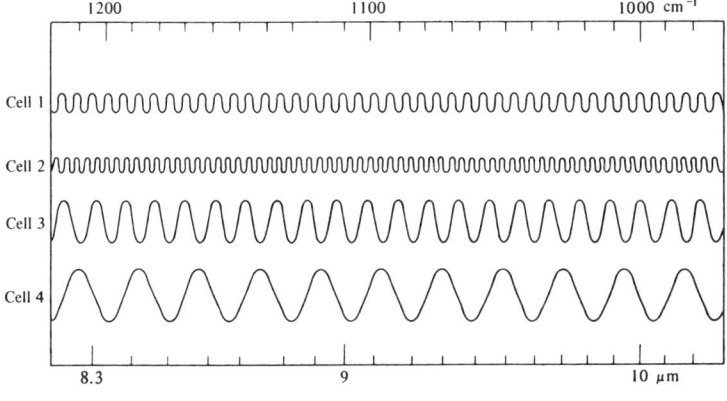

PROBLEM 7-8a

b. Calculate the path length of the infrared absorption cell from the transmittance curve for benzene shown.

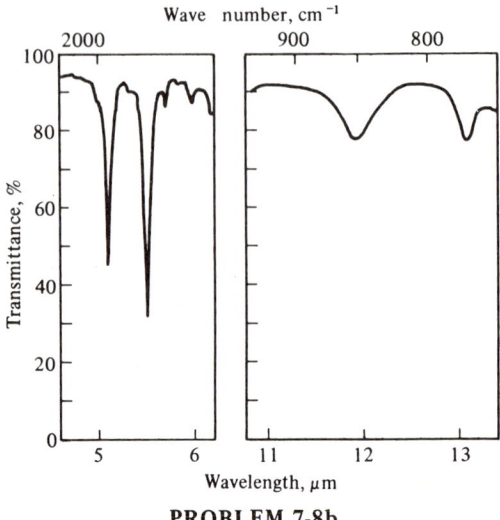

PROBLEM 7-8b

9. Identify the particular xylene from the infrared data:
   Compound A: Absorption bands at 767 and 692 cm^{-1} (13.0 and 14.4 μm).
   Compound B: Absorption band at 792 cm^{-1} (12.6 μm).
   Compound C: Absorption band at 742 cm^{-1} (13.5 μm).

10. A bromotoluene, C_7H_7Br, has a single band at 801 cm^{-1} (12.50 μm). What is the correct structure?

11. A chlorobenzene exhibits no absorption bands between 900 and 690 cm^{-1} (11.1 and 14.5 μm). What is the probable structure?

12. An aromatic compound, C_7H_8O, has these features in its infrared spectrum: Absorption bands present at 3040 cm^{-1} (3.30 μm), 1010 cm^{-1} (9.90 μm), 3380 cm^{-1} (2.96 μm), 2940 cm^{-1} (3.4 μm), 1460 cm^{-1} (6.85 μm), and 690 and 740 cm^{-1} (14.5 and 13.5 μm), whereas bands were absent at 1735 cm^{-1} (5.77 μm), 2720 cm^{-1} (3.68 μm), 1380 cm^{-1} (7.25 μm), and 1182 cm^{-1} (8.45 μm). Identify the mode of each absorption band present (and absent), and write the structure of the compound.

13. The only significant absorption bands observed in the infrared spectrum were stretching of saturated C–H at 2960 and 2870 cm^{-1} (3.38 and 3.49 μm), methylene bending at 1461 cm^{-1} (6.85 μm), terminal methyl at 1380 cm^{-1} (7.25 μm), and the rocking of ethyl groups at 775 cm^{-1} (12.9 μm). Deduce the structure of this compound, C_6H_{14}.

14. A crystalline material is believed to be either a substituted hydroxylethyl cyanamide(I) or an imino oxazolidine(II):

N≡C−NH₂⁺−CH−CH₂OH   HN=CH−NH−C(=O)−CH₂−
        I   II

Sharp bands are located at 3330 cm⁻¹ (3.0 μm) and 1600 cm⁻¹ (6.25 μm), but there are no bands at 2300 cm⁻¹ (4.35 μm) or 3600 cm⁻¹ (2.78 μm). Which structure fits the infrared data?

15. Deduce the structure of the compound whose empirical formula is C_4H_5N. Sharp, distinctive absorption bands, and virtually nothing else, occur at 3080 cm⁻¹ (3.25 μm), 2960 cm⁻¹ (3.38 μm), 2260 cm⁻¹ (4.43 μm), 1865 cm⁻¹ (5.36 μm), 1647 cm⁻¹ (6.08 μm), 1418 cm⁻¹ (7.05 μm), 990 cm⁻¹ (10.1 μm), and 935 cm⁻¹ (10.7 μm). The band at 1865 cm⁻¹ is weak.

16. As far as possible, deduce the structural formula from the following information contained in the infrared spectrum. Cell thickness: 0.05 mm. Absorption bands and their transmittance) at: 3080 cm⁻¹ (0.28); 2950, 2910, 2835 cm⁻¹ (0.28–0.37); 1848 cm⁻¹ (0.71); 1650 cm⁻¹ (0.65); 1455, 1440, 1410 cm⁻¹ (0.32–0.45); 990 cm⁻¹ (0.30); and 910 cm⁻¹ (0.29).

17. Deduce the structure of the compound whose empirical formula is C_7H_5OCl using the following data. The infrared spectrum was obtained using a cell of thickness 0.1 mm. Absorbances are given in parentheses after each frequency (in cm⁻¹): 3080 (0.07), 2810 (0.19), 2720 (0.17), 1705 (1.0), 1593 (0.27), 1573 (0.30), 1470 (0.20), 1438 (0.22), 1383 (0.22), 1279 (0.14), 1196 (0.82), 1070 (0.14), 900 (0.20), and 871 (0.20). The NMR spectrum showed a singlet at δ 9.95 and a 4-line pattern, symmetrical in appearance, with centers at δ 7.45 and δ 7.75 ($J = 7$ Hz in both cases); the number of protons was in the ratio of 1:2:2.

18. Deduce the product that gave the infrared spectrum shown, taken as a liquid film. The compound has a boiling point of about 101°C, $n_D^{20} = 1.3890$, and a molecular formula $C_6H_{12}O_2$.

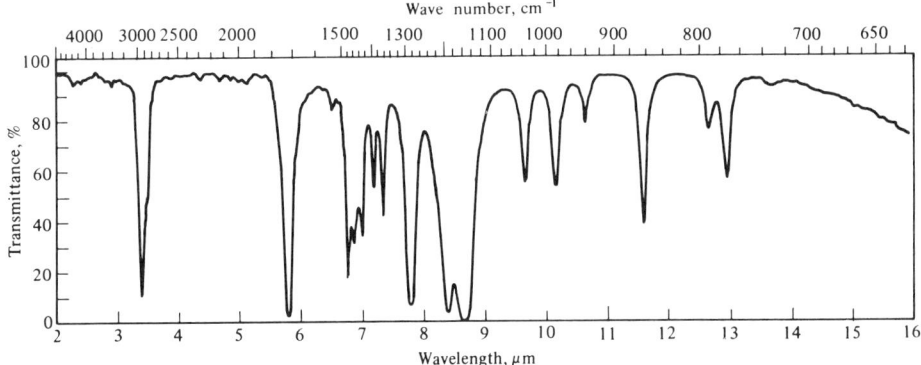

PROBLEM 7-18

19. Deduce the structure of the compound with molecular formula $C_6H_{12}O$ whose infrared spectrum is shown.

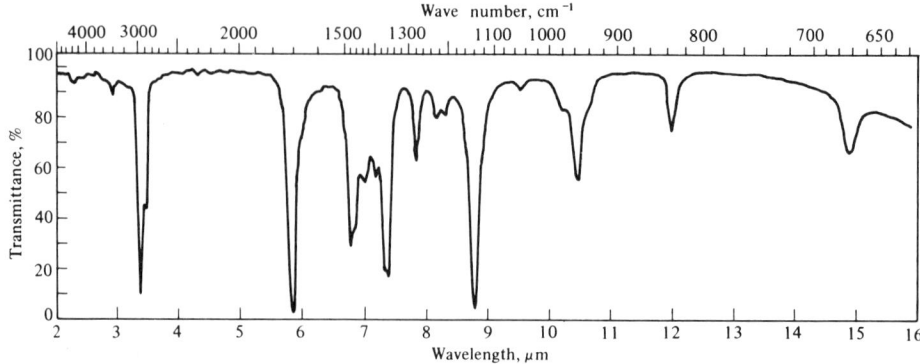

PROBLEM 7-19

20. From the infrared spectrum shown, deduce the molecular structure. The compound has a boiling point of 69°C.

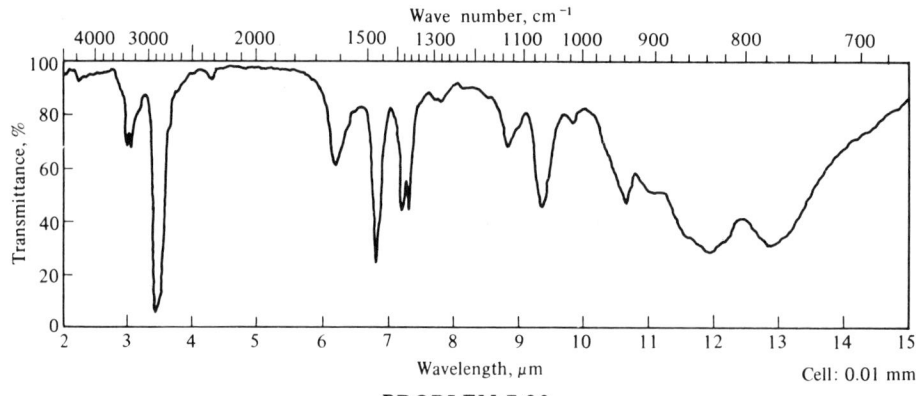

PROBLEM 7-20

21. From the infrared spectrum shown, and the molecular weight of 131, deduce the compound.

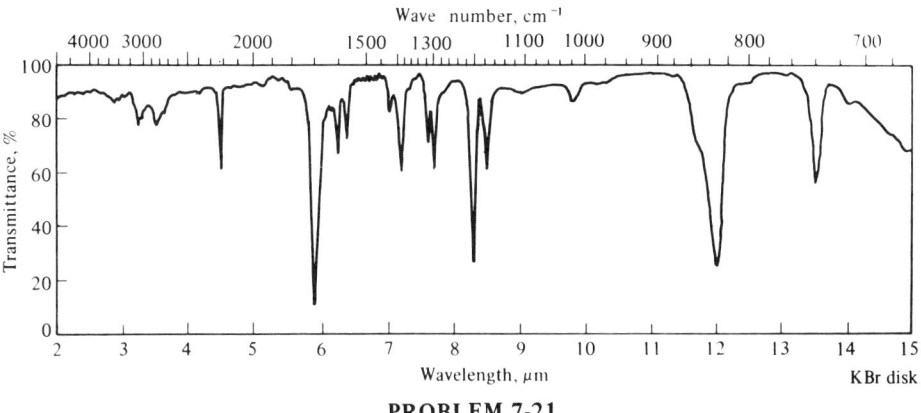

PROBLEM 7-21

22. Deduce the molecular structure of the compound, molecular weight 176, whose infrared spectrum shown in the figure was taken as a liquid film.

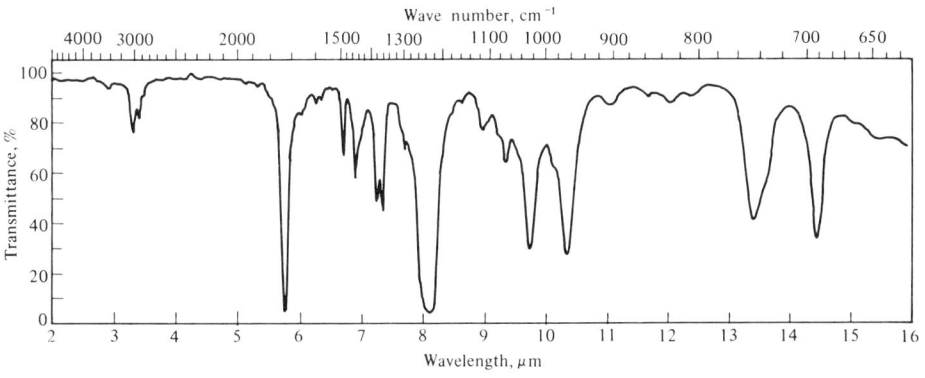

PROBLEM 7-22

23. From the infrared spectrum of the liquid film which is shown in the figure, and the molecular weight of 118, write the structure of the compound.

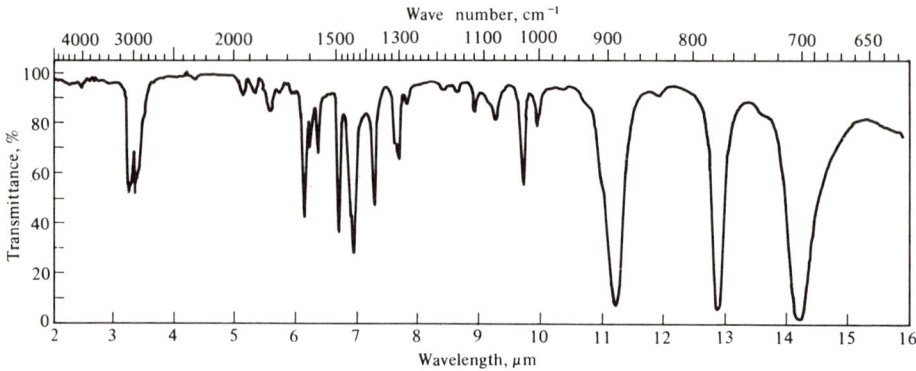

PROBLEM 7-23

BIBLIOGRAPHY

Bauman, R. P., *Absorption Spectroscopy,* Wiley, New York, 1962.
Brame, E. G., Jr., and J. G. Grasselli, Eds., *Infrared and Raman Spectroscopy,* Vol. 1, Parts A, B, and C, in *Practical Spectroscopy Series,* Marcel Dekker, New York, 1977.
Colthup, N. B., L. H. Daly, and S. E. Wiberley, *Introduction to Infrared and Raman Spectroscopy,* 2nd ed., Academic, New York, 1975.
Griffiths, P. R., *Chemical Infrared Fourier Transform Spectroscopy,* Wiley, New York, 1975.
Harrick, N. J., *Internal Reflection Spectroscopy,* Wiley, New York, 1967.
Silverstein, R. M., G. C. Bassler, and T. C. Morrill, *Spectrometric Identification of Organic Compounds,* 3rd ed., Wiley, New York, 1974.
Williams, D. H. and I. Fleming, *Spectroscopic Methods in Organic Chemistry,* 2nd ed., McGraw-Hill, New York, 1973.

LITERATURE CITED

1. Stewart, J. E., in *Interpretive Spectroscopy,* S. K. Freeman, Ed., pp. 131–169, Van Nostrand Reinhold, New York, 1965.
2. Low, M. J. D., *Anal. Chem.,* **41**, 97A (1969); *J. Chem. Educ.,* **47**, A163, A255, A349, A415 (1970).
3. Pouchert, C. J., Ed., *The Aldrich Library of Infrared Spectra,* 2nd ed., Aldrich Chemical Co., Milwaukee, 1975.
4. Sadtler Research Laboratories, *Catalog of Infrared Spectrograms,* Philadelphia. A continually updated subscription service.
5. Ewing, G. W., *J. Chem. Educ.,* **48**, A521 (1971).
6. Koenig, J. L., *Appl. Spectrosc.,* **29**, 293 (1975).

CHAPTER 8

Raman Spectroscopy

When monochromatic light is scattered by molecules, a small fraction of the scattered light is observed to have a different frequency from that of the irradiating light; this is known as the *Raman effect*. Since its discovery in 1928, the Raman effect has been important as a method for the elucidation of molecular structure, for locating various functional groups or chemical bonds in molecules, and for the quantitative analysis of complex mixtures, particularly for major components. Although Raman spectra are related to infrared absorption spectra, a Raman spectrum arises in a quite different manner and thus provides complementary information. Vibrations that are active in Raman may be inactive in the infrared, and vice versa. A unique feature of Raman scattering is that each line has a characteristic *polarization,* and polarization data provide additional information related to molecular structure.

8.1 THEORY

The Raman effect arises when a beam of intense monochromatic light passes through a sample that contains molecules that can undergo a change in molecular polarizability as they vibrate. It is strictly a quantum effect. Most collisions of the incident photons with the sample molecules are elastic (Rayleigh scattering). The electric field produced by the polarized molecule oscillates at the same frequency as the passing electromagnetic wave, so that the molecule acts as a source sending out radiation of that frequency in all directions. As shown in Fig. 8-1, the incident radiation does not raise the molecule to any particular quantized level, rather the molecule can be considered as in a virtual excited state.

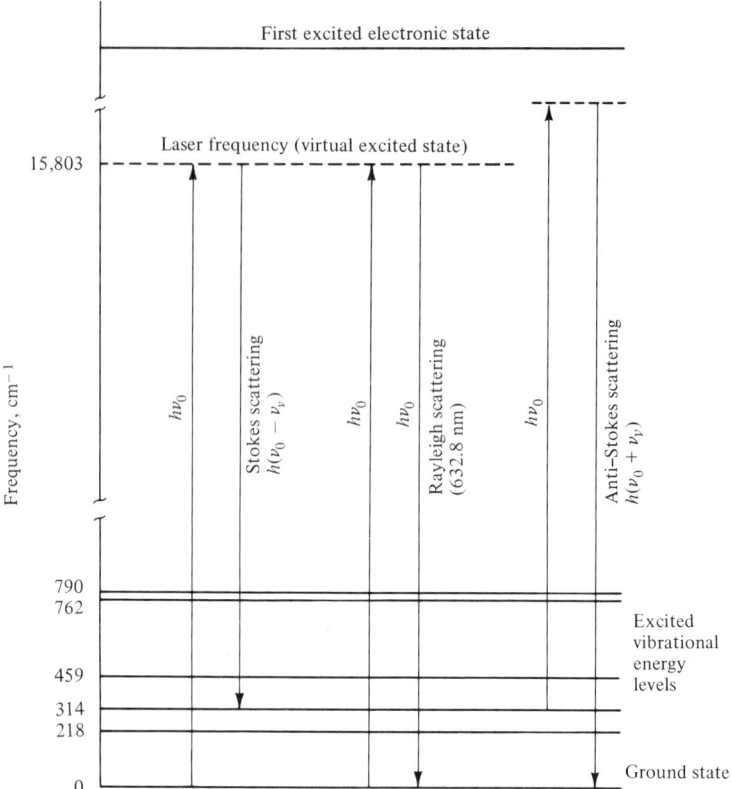

FIGURE 8-1 Energy interchange involved in Rayleigh and Raman scattering; the molecule involved is CCl_4 and the source is a He–Ne laser.

As the electromagnetic wave passes, the polarized molecule ceases to oscillate and returns to its original ground level in a very short time (approximately 10^{-12} sec).

A small proportion of the excited molecules (10^{-6} or less) may undergo a change in polarizability during one of the normal vibrational modes. This provides the basis for the Raman effect. Usually incident radiation, ν_0, is absorbed by a molecule in the lowest vibrational state. If the molecule reemits by returning not to the original vibrational state, but to an excited vibrational level, ν_ν, of the ground electronic state, the emitted radiation is of lower energy, or lower frequency ($\nu_0 - \nu_\nu$) than the incident radiation. The difference in frequency is equal to a natural vibration frequency of the molecule's ground electronic state. Several such shifted lines (the *Stokes* lines) normally will be observed in the Raman spectrum, corresponding to different vibrations in the molecule. This provides a richly detailed vibrational spectrum of a molecule (Fig. 8-2).

A few of the molecules initially will absorb radiation while they are in an excited vibrational state and will decay to a lower energy level, so that their Raman scattered light will have a higher frequency than the incident radiation. These are called *anti-Stokes* lines.

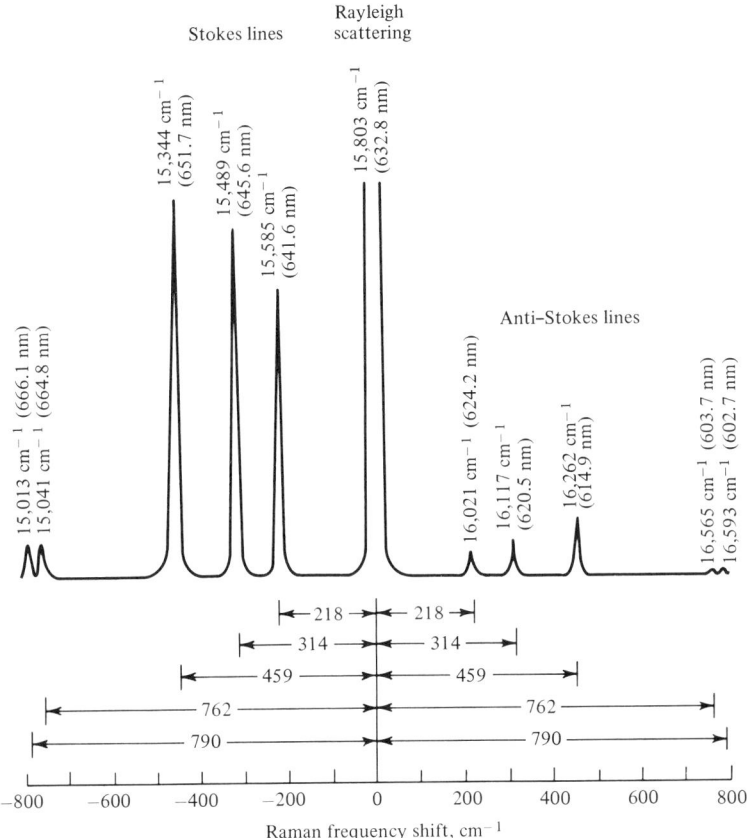

FIGURE 8-2 Raman spectrum of CCl_4 obtained with a He–Ne laser.

Thus the spectrum of the scattered light consists of a relatively strong component with frequency unshifted (Rayleigh scattering), corresponding to photons scattered without energy exchange, and the two components of the Raman spectrum: the Stokes lines and the anti-Stokes lines. Normally, for chemical analysis, only the Stokes lines are considered. These are more intense because under usual circumstances most molecules are initially in the lowest vibrational level.

In the usual Raman method the excitation frequency of Raman sources are selected to lie below most S–S^* electronic transitions and above most fundamental vibrational frequencies, but this need not be the case as the next section will show.

Resonance Raman Spectroscopy

Thus far we have explored what might be termed the ordinary Raman effect; that is, Raman scattering induced by excitation far removed from any electronic transitions. If, however, the laser frequency is permitted to fall near or become coincident with an electronic absorption, ν_{abs}, a very significant intensity enhancement may occur such that

$$\text{Intensity} \propto (\nu_{laser} - \nu_v)^4 \left(\frac{\nu_{abs}^2 + \nu_{laser}^2}{\nu_{abs}^2 - \nu_{laser}^2} \right)^2 \tag{8-1}$$

The advantage of the *resonance* Raman effect lies in its great sensitivity and selectivity as a probe of chromophore structure. The enhanced Raman lines may have intensities 10^2–10^6 times greater than normal Raman intensities. Consequently, resonance Raman spectra have low detection limits (10^{-6}–10^{-8} M) and are much simpler than normal Raman spectra since only vibrational modes associated with the "chromophore" are enhanced.

The resonance Raman effect results from the promotion of an electron into an excited vibronic state, accompanied by immediate relaxation into a vibrational level of the ground state. The process is not preceded by prior relaxation to the lowest vibrational level of the excited state as in ordinary fluorescence. The distinction is shown in Fig. 8-3. Consequently, the resonance Raman emission process is essentially instantaneous, and the resulting spectra consist of narrow bands. Resolution is good, 10–20 cm^{-1} (0.3–0.6 nm at 500 nm). For molecules in solution, electronic states are broadened by many closely spaced vibrational states. Excitation with radiation anywhere within this continuum will give rise to the same resonance Raman spectrum, with an intensity proportional to the absorption intensity.

Spectra may also be obtained using an excitation frequency just below the absorption band. Such spectra display smaller resonance enhancement, typically less than tenfold; they are called preresonance Raman spectra.

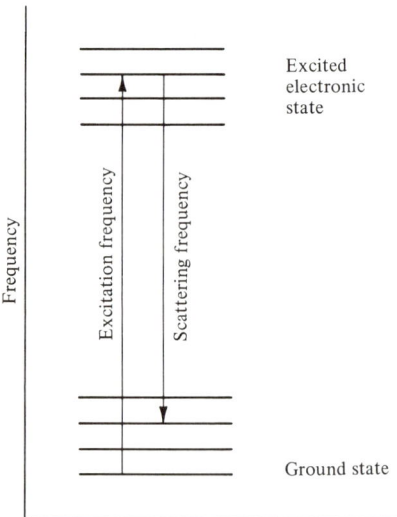

FIGURE 8-3   Resonance Raman effect shown schematically.

Not all of the normal Raman bands are equally enhanced. Only those vibrations that exhibit a large change in equilibrium geometry upon electronic excitation will produce strongly resonance-enhanced Raman bands. In practical terms this means that two classes of vibrational modes will produce intense resonance-enhanced spectra: total symmetric vibrations and those nontotally symmetric vibrations that vibronically couple two electronic states. Since an electronic transition is often more or less localized in one part of a complex molecule, the resonance Raman effect provides highly detailed information about the vibrational modes of chromophores that have an absorption band near the wavelength of the incident radiation. This selectivity, for example, is quite apparent in the heme proteins where resonance Raman bands are due solely to vibrational modes of the tetrapyrrole chromophore.

Comparison of Raman with Infrared Spectroscopy

Raman spectroscopy offers distinct advantages over the more direct infrared absorption measurement. First, Raman spectroscopy can be used to detect and analyze molecules with infrared inactive spectra, such as homonuclear diatomic molecules. For complicated molecules whose low symmetry does not forbid both Raman and infrared activity, certain vibrational modes are inherently stronger in the Raman effect and weaker in, or apparently absent from, the infrared spectrum. Raman activity tends to be a function of the covalent character of bonds. Hence a Raman spectrum reveals information regarding the backbone structure of the molecule, whereas the strong infrared features are indicative of polar segments.

Raman spectra can be used to study materials in aqueous solutions, a medium that transmits infrared very poorly. A third advantage is the ability to examine the entire vibrational spectrum with one instrument, unlike infrared spectroscopy in which the far-infrared is usually scanned separately from the mid-infrared. Finally, sample preparation for Raman is generally considerably simpler than for the infrared.

The sensitivity of resonance Raman spectroscopy to only chromophore vibrational modes may be considered either a strength or a weakness. Spectra are greatly simplified and a series of molecules containing slightly different chromophores will give spectra that are easily distinguished. On the other hand, if a series of molecules contains the same chromophore with, for example, different aliphatic side chains, the resonance Raman spectra will be nearly identical.

Some shortcomings of the Raman technique exist. Both liquid and solid samples must be free from dust particles or the Raman spectrum may be masked by Tyndall scattering. The primary disadvantage of Raman spectroscopy is the fluorescent background that accompanies intense laser irradiation of many biological materials. Relative to the Raman signal, the background can be enormous, completely obliterating the spectrum. Even if one could observe the Raman spectrum superimposed on the fluorescence background, the noise contribution of the fluorescence emission degrades the signal-to-noise ratio of the Raman spectrum. While the problem can be attacked by careful sample preparation, time-resolved spectroscopy, or coherent anti-Stokes Raman spectroscopy (CARS), there will always be experiments which remain highly intractable.

8.2 INSTRUMENTATION

The primary function of a Raman spectrometer is clean rejection of the intense Rayleigh scattering and detection of the weak Raman-shifted components. A computer adds significantly to the power and versatility of the spectrometer. An intense monochromatic light source, sensitive detection, and high light-gathering power, coupled with freedom from extraneous stray light, must be built into a Raman spectrometer. A double grating monochromator keeps stray light from the unshifted laser wavelength to a minimum. Visual-type optics are used throughout, and the entire spectral region is covered by a single-type grating.

The laser Raman spectrometer, shown in Fig. 8-4, consists of two basic units: the laser excitation unit and the $f/6.7$ spectrometer unit. The laser beam enters from the rear of the spectrometer into the depolarization autorecording unit and, after passing through this unit, it illuminates the sample. The Raman scattering, collected at 90° to the exciting laser beam, is focused on the entrance slit of a 0.5-m focal length, Czerny–Turner grating double monochromator. Immediately ahead of the spectrometer is a polarization scrambler which overcomes grating bias caused by polarized light. A polarization analyzer is placed between the condenser lens and the polarization scrambler when measuring the polarized Raman spectrum of single crystals. The spectrometer is arranged in a back-to-back configuration. The Raman scattered light is dispersed using gratings with 1200 grooves/mm

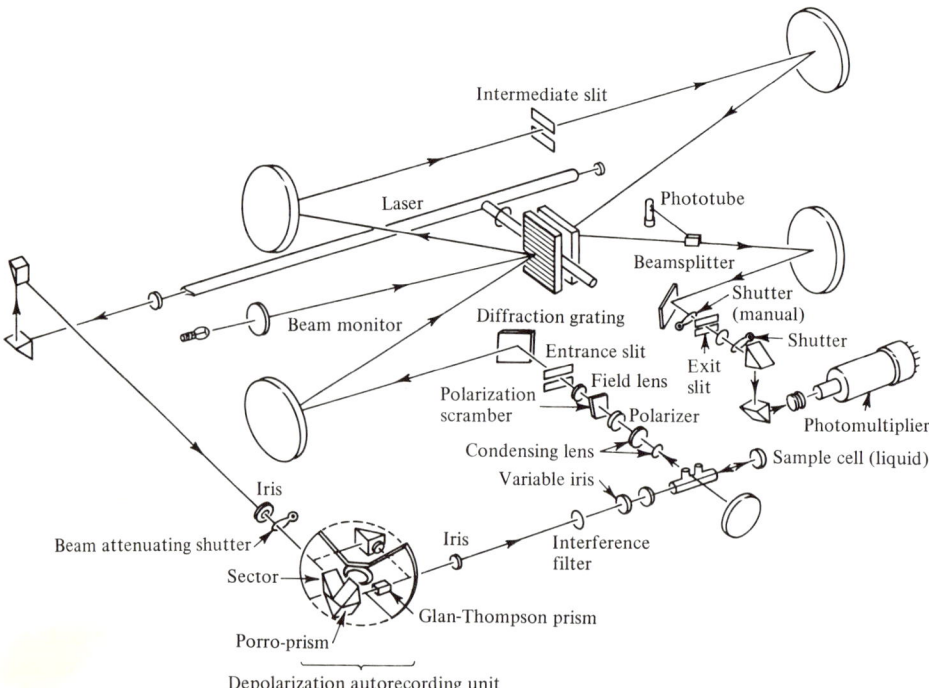

FIGURE 8-4 Optical schematic of a laser Raman spectrophotometer. (Courtesy of Jeol, Ltd.)

and finally passes through the monochromator exit slit and onto the photocathode of the photomultiplier tube. The photomultiplier tube is placed in a thermoelectric cooler (–30°C), markedly lowering the dark current and reducing noise, thus providing a high sensitivity and signal-to-noise ratio. In this particular instrument, the signals from the detector are amplified and counted with two photon counting systems, one of which is for the detection of ordinary Raman spectrum, while the other is for the calculation of the depolarization ratio. These signals are displayed on the two-pen digital X–Y recorder which is linked to the monochromator driven by the master pulse clock.

Photon counting has long been recognized as the most effective means of recovering low level Raman signals. However, it suffers from limited range and high cost. On the other hand, inexpensive dc amplifiers are excellent for strong signals, and provide the required range, but suffer from low signal-to-noise ratios for weak signals. One can combine the best of both these techniques. Operate the amplifier as a photon counter with a built-in discriminator for low level signals. When the signal level is strong enough, switch in automatically the dc system for the remaining levels of amplification.

The intermediate slit has a vertical mask which allows the height to be shortened when one is running a solid sample or utilizing the transverse illumination technique for microsamples. Masking of the intermediate slit reduces the spectral background and enhances the signal-to-noise ratio.

The operation of the Raman spectrometer, except for selection of slit widths and heights, is controlled automatically. The slit width selector wheel is linked directly to the electronics, so after determining scanning speed and slit width, the time constant is automatically selected to exactly match the criterion: time constant = (spectral bandwidth) /(scanning speed × 4). A precise electronic master clock provides pulses to the stepping motor operating the scanning system and to the stepping motor operating the X–Y recorder. Through its solid-state divider circuits, the master clock maintains the correct ratio between these pulses so that scanning speeds can be varied independently from the range of the recorder.

The Jobin–Yvon instrument (Fig. 8-5) combines the throughput of a double system with the rejection levels of a triple system. The $f/8$ monochromator is designed for stray

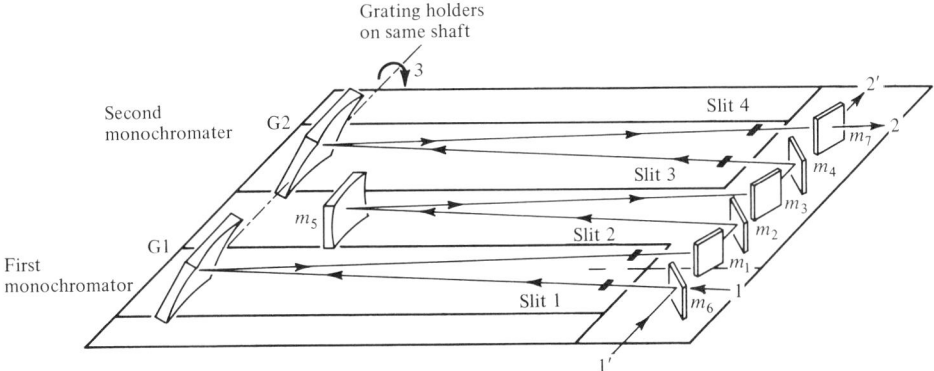

FIGURE 8-5 Laser Raman spectrophotometer. (Courtesy of Instruments SA, Inc.)

light reduction by using two concave, aberration-corrected, holographic gratings with 2000 grooves/mm. Both gratings are on the same shaft; thus tracking error is zero. The scattered radiation from the sample is gathered by a high aperture objective lens which focuses the light onto the straight horizontal entrance slit F1 of the first monochromator. The concave grating G1 diffracts the beam and focuses the selected wavelength onto the exit slit F2. The monochromatic light exiting from the first monochromator through F2 is now imaged onto the entrance slit F3 of the second monochromator by mirrors M1 through M5. These mirrors are not in the monochromator cavities and work in monochromatic light; therefore, they cannot contribute to scattered stray light. Light entering F3 is then diffracted and focused onto the exit slit F4 of the second monochromator by the concave grating G2. All four slits are individually controlled by the stepping motor; this permits complete versatility when using a computer in order to obtain either constant slit width spectra or constant bandpass spectra.

Laser Sources

A laser provides an almost ideal monochromatic source of narrow linewidth. It emits radiant energy that is coherent, parallel, and polarized. A laser beam can be kept as a very slim cylinder only a few micrometers in cross section. Laser operation involves three principles of physics: stimulated emission, population inversion, and optical resonance. Stimulated emission occurs when a photon strikes an excited atom or molecule and thereby causes that atom or molecule to emit its photon prematurely. This can occur only when the impinging photon has exactly the energy of the "stored" photon that would ultimately have been emitted spontaneously. The resulting emission falls precisely in phase with the electromagnetic wave that triggered its release and is identical in wavelength. Thus, the incoming photon is now joined by a second photon from the excited atom or molecule, resulting in a gain or amplification of photons and giving a perfectly coherent (in-phase) beam of radiation.

The precedence of stimulated emission over spontaneous emission is the basis for achieving laser action. For this to occur, a population inversion must take place; that is, there must be more molecules (or atoms) in the excited state than the ground state of the lasing material. This is possible only for a multilevel system, as illustrated in Fig. 8-6 for the He–Ne laser. A helium atom, excited by an electrical discharge to a long-lived triplet level, can only lose its energy by collision with another atom which has a comparable energy level available. One excited state of neon lies only 313 cm^{-1} below the excited helium state. As a result, a radiationless excitation of neon can occur on collision of the excited helium and ground-state neon atoms. The excited level of neon has a relatively long lifetime before spontaneously decaying to the ground state with the emission of photons at 1153 and 632.8 nm. Thus, it is possible to build up a larger population of neon atoms in the excited state than in the ground state (population inversion). The spontaneous emission of a single photon at either of these wavelengths can trigger a whole cascade of similar photons by the process of stimulated emission. Optical pumping is employed to build up a population of atoms in an excited state. Population inversion is relatively easy for many organic molecules since their energy levels constitute a multilevel system.

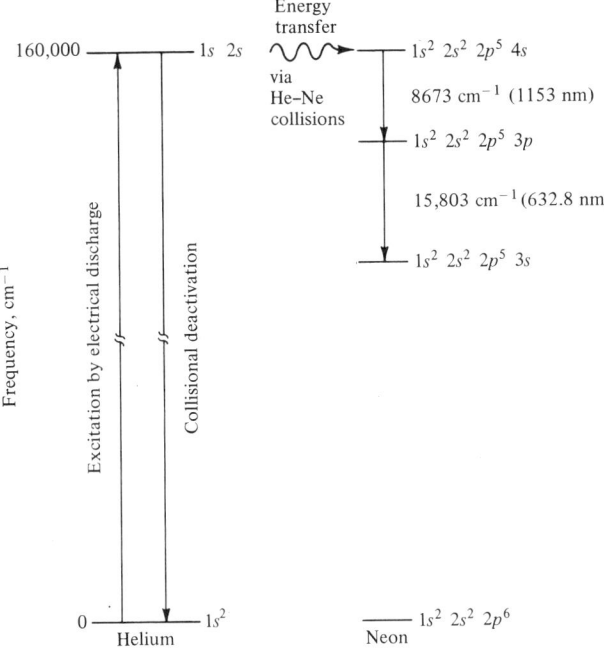

FIGURE 8-6 Energy levels involved in the He–Ne laser.

Optical resonance is the third principle essential to laser operation. Resonance is achieved by placing the lasing medium in a cavity situated between a pair of parallel, plane mirrors, as shown in Fig. 8-7. Making the spacing between the mirrors an integral multiple of the desired wavelength means that there will be a buildup of energy at the desired wavelength. If the resulting light is collinear with the optic axis and of a frequency falling within the bandwidth of one of the discrete optical frequencies, it will be reflected back and forth through the cavity. Growth of the wave will continue and, if the gain on repeated passages through the lasing medium is sufficient to compensate for losses within the cavity,

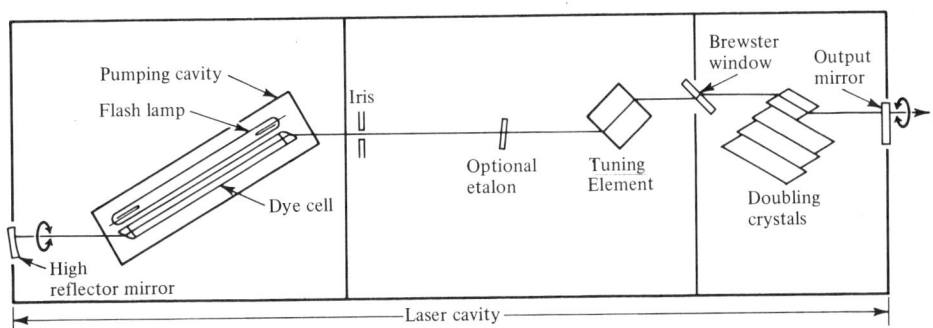

FIGURE 8-7 Optical schematic of a tunable dye laser. (Courtesy of Chromatix Inc.)

a steady wave will be built up. Any wave that is inclined at an angle to the long axis of the cavity will be lost after only a few reflections, or perhaps without ever striking one of the mirrors. If one of the mirrors is semitransparent, a portion of the wave can escape through it, constituting the output of the laser. The output is radiation of low divergence, all of the same frequency, and all in phase. A crystal of $LiIO_3$, internal to the laser cavity, may be used to frequency double the available wavelengths.

To pump the laser, flash lamps, inert gas arc lamps, or another laser have been used. Flash lamps are frequently used and are of two configurations: coaxial and linear. In either configuration, high voltage from a low inductance capacitor is rapidly pulsed through the lamp. Coaxial arrangement offers very high pulse power but low repetition rates. Linear flash lamps do not provide peak powers comparable to coaxial flash lamps but can operate at higher repetition rates to yield the same average power. All pulsed systems generate copious amounts of radio frequency interference. Proper design and construction must be used to reduce radio frequency interference to acceptable levels, and equipment must be shielded to prevent upsetting nearby electronics.

The He–Ne laser line at 632.8 nm is favorably located in the spectrum where the least amount of fluorescent problems appears in routine analyses. An argon laser possesses intense lines at 488.0 and 514.5 nm; coupling with krypton adds two other major lines at 568.2 and 647.1 nm. The Ar–Kr laser is ideal for many experiments. Chances are that with at least one of the exciting lines, problems of photodecomposition, fluorescence, or absorption will be successfully circumvented. Tunable dye lasers, whose output frequency may be varied over a short range, are almost indispensable for studies involving resonance Raman. Several dyes are necessary to cover the entire wavelength range accessible to dye lasers; the appropriate dye solution is continuously pumped through the laser cavity. Prisms and gratings have been used to tune dye lasers.

Detectors

Photographic detection has given way almost entirely to photoelectric detection. Choice of phototube response depends on which laser line is used. The trialkali photocathode has about 7% quantum efficiency at 632.8 nm and falls in efficiency fourfold for every 1000 cm^{-1}. Raman shifts of 3700 cm^{-1} require response to approximately 826.6 nm for the He–Ne laser. The extended red-sensitive multialkali cathode and gallium arsenide photocathode designs have much higher quantum efficiency in the red portion of the spectrum out to 900 nm. By contrast, photomultiplier tubes are near their peak sensitivity at 488.0 nm, one of the emission lines of the Ar–Kr laser. Obviously, laser selection and detector choice are interwoven, as shown in Fig. 8-8.

8.3 SAMPLE HANDLING AND ILLUMINATION

The use of laser excitation allows Raman spectroscopy to be performed on specimens in almost any state: liquid, solution, transparent solid, translucent solid, powder, pellet, or gas. Examination of liquids is usually performed using a single pass of the laser beam

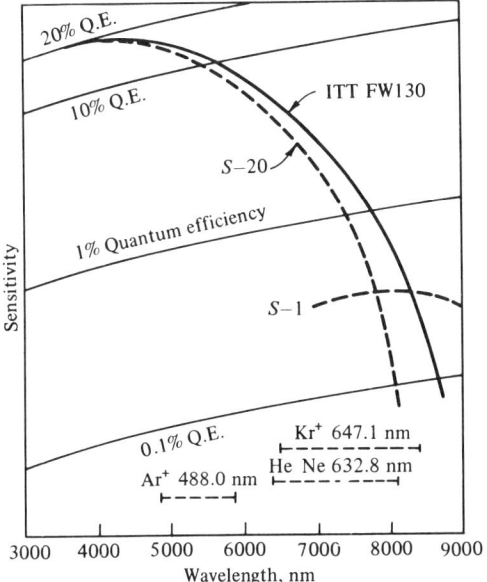

FIGURE 8-8 Sensitivity of several photomultiplier tubes. The dashed horizontal lines represent the range of 3500 cm^{-1}, which is the Raman shift from the designated laser exciting lines.

either axially or transverse to the neat liquid sample contained in a glass capillary tube. If the liquid is clear, the beam focused to a diffraction-limited point in a small sample will pass through the liquid and can be reflected back again for another pass. Considerable gain in Raman intensity can be achieved beyond a single pass. It permits work with extremely small samples, in the microliter or even nanoliter range. The volume of the focused He–Ne laser beam is about 8 nl at the diffraction-limited point. When greater volumes of sample are available, the exciting radiation may be multipassed through larger cells. Photo- or heat-labile materials may be studied in spinning cells which reduce exposure time in the laser beam. Water is a weak scatterer and therefore an excellent solvent for Raman work. This has important consequences in studies of biochemical interest and in the pharmaceutical industry. Other widely used solvents are carbon disulfide, carbon tetrachloride, chloroform, and acetonitrile. Their obscuration ranges are shown in Fig. 8-9.

Gas samples can be handled with powerful laser sources and efficient multipassing or intra-laser-cavity techniques, but are difficult to study because of their low scattering. Nevertheless, pollutant studies of emissions are being made from considerable distances away from the point of emanation.

Powders are tamped into an open-ended cavity for front surface illumination, or into a transparent glass capillary tube for transverse excitation. Forward (180°) sample illumination provides higher collection efficiency (better S/N ratio); however, Raman lines in the low-frequency region are more easily observed with right-angle viewing because the ratio of Raman to Tyndall and Rayleigh scattering is improved. For a translucent solid, the laser beam is focused into a conical cavity on the face of the sample, cut either into a cast

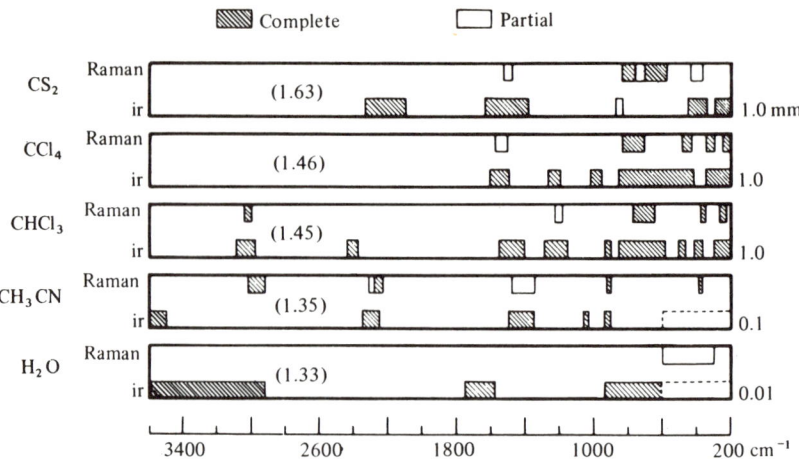

FIGURE 8-9 Obscuration ranges of the most useful solvents for Raman spectrometry in solution. The infrared obscuration at the indicated path lengths is given for comparison; the refractive index of the solvent is given in parentheses.

piece or a pellet formed by compression of powder. The cavity functions as a light trap, producing multiple scattering of the incident photons via the intrinsic reflectivity and transmittance of the specimen to the laser frequency. These arrangements are shown in Fig. 8-10. A transparent solid sample preserves the directionality of the laser beam as it passes through the specimen, producing a scattering image which is collinear with the monochromator slit aperture for scattering observed at 90° to the direction of incidence. Bulk polymer samples and also single, very fine fibers may be run intact. Polymerization studies on such individual fibers often yield information about their orientation and crystalline properties.

If fluorescence arises from impurities in the sample, one can clean up the sample, often without a great deal of effort, and generally observe a rather marked improvement in the quality of the spectrum. Techniques such as gas chromatography fractionations, recrystallization, distillation, and filtration are suitable for this purpose. The so-called "drench quenching" technique also works quite well, but can be rather time-consuming in some cases. This technique simply involves soaking the sample in the laser beam until the transient luminescence background decays to some reasonable level. If the fluorescence arises from the sample itself, one has little choice but to select a different excitation line.

8.4 DIAGNOSTIC STRUCTURAL ANALYSIS

In Raman spectroscopy those vibrations originating in relatively nonpolar bonds with symmetrical charge distributions and which are symmetrical in nature produce the greatest polarizability changes and are the most intense. Vibrations from $-C=C-$, $-C\equiv C-$, $-C\equiv N$, $-C=S$, $-C-S-$, $-S-S-$, $-N=N-$, and $-S-H$ bonds are readily observed. Raman lines are more characteristic than infrared absorption bands of the skeletal vibrations of finite chains and rings of saturated and unsaturated hydrocarbons.

The position of the symmetric ring stretching vibration of cyclic compounds is characteristic of the type and size of ring present in the compound. Aromatic compounds have particularly strong spectra; all have a strong ring deformation mode at 1600 ± 30 cm^{-1}. Monosubstituted compounds have an intense symmetric ring stretching vibration at about 1000 cm^{-1}, a strong in-plane hydrogen bending vibration at about 1025 cm^{-1}, and a weak

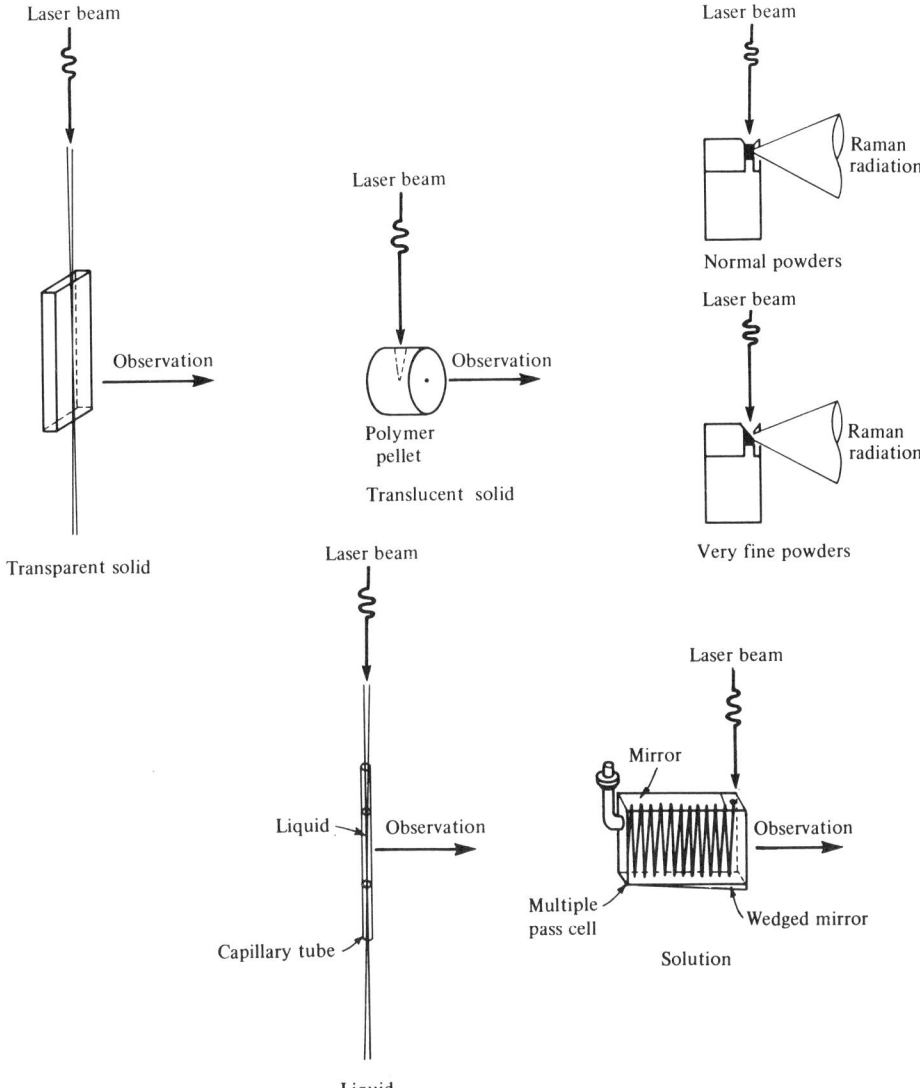

FIGURE 8-10 Experimental arrangements for laser excitation of specimens in various physical forms.

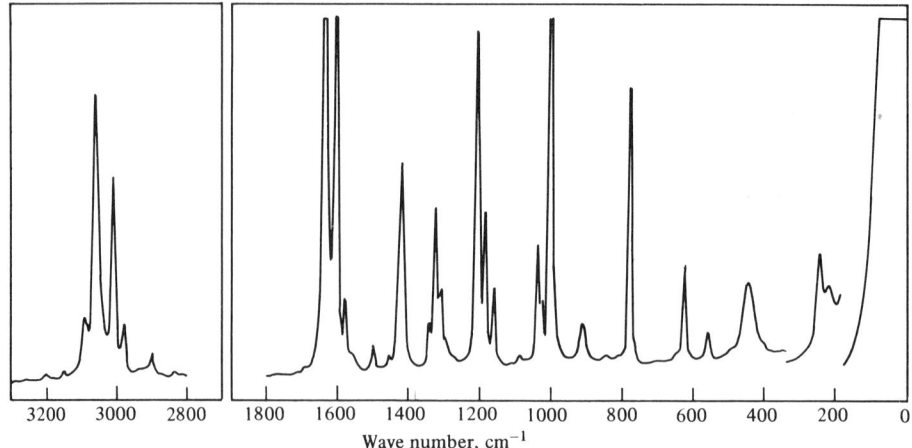

FIGURE 8-11 Raman spectrum of styrene monomer.

depolarized in-plane bending vibration at about 615 cm^{-1} (Fig. 8-11). *Meta-* and 1,3,5-trisubstituted compounds have only the line at 1000 cm^{-1}. *Ortho*-substituted compounds have a line at 1037 cm^{-1}, and *para*-substituted compounds have a weak line at 640 cm^{-1}.

The intense Raman band near 500 cm^{-1} is characteristic of the —S—S— linkage, and the band near 650 cm^{-1} derives from —C—S— stretching (Fig. 8-12). Also, the —S—H stretching band near 2500 cm^{-1}, normally weak in the infrared, shows high intensity in the Raman.

Although the —C≡N stretch appears in both the infrared and Raman spectra, it is markedly diminished in the infrared when an electronegative group such as chloride is α-substituted. The intensity is retained in the Raman spectrum.

Whenever the 3300 cm^{-1} region in the infrared is badly obscured by intense OH absorption, Raman spectroscopy is helpful because the OH band is weak whereas the NH and CH stretching frequencies still exhibit moderate intensity.

In the Raman spectrum the symmetrical methyl deformation frequency near 1380 cm^{-1} is sensitive to the environment of the methyl group. It is quite weak in alkyl com-

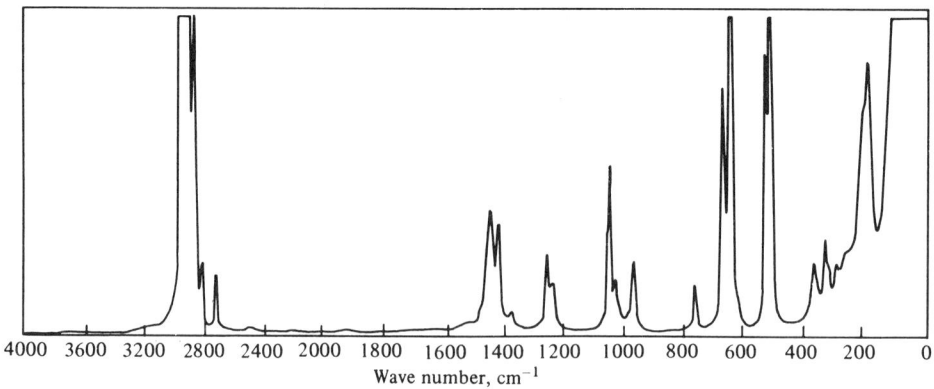

FIGURE 8-12 Raman spectrum of diethyldisulfide, $C_2H_5SSC_2H_5$.

pounds, but the band intensity is considerably enhanced when the methyl is attached to an aromatic ring and some types of double bonds.

Skeletal motions are very characteristic and highly useful for cyclic and aromatic rings, steroids, and long chains of methylenes. The spectrum in the region 800–1500 cm^{-1} is quite characteristic. In solid samples, sharpening and intensifying of certain bands appear to be a function of crystallinity.

For all molecules that have a center of symmetry, a band allowed in the infrared is forbidden in the Raman and vice versa. In molecules with symmetry elements other than a center of symmetry, certain bands may be active in the Raman, infrared, both, or neither. For a complex molecule that has no symmetry save the identity element, all of the normal vibrational modes are allowed in both the infrared and Raman spectra. The symmetry selectivity in Raman spectra results in a generally simpler spectrum than the corresponding infrared, a characteristic that is frequently useful. One other obvious difference is the tendency for the peaks in a Raman spectrum to have a greater range of intensities, from very weak to very strong. These factors, plus the contrasting relative intensities for a given group, are the main basis for making more confident assignment of chemical structure through the combination of infrared and Raman.

Raman spectroscopy has a distinct advantage in the detection of low-frequency vibrations; the lower limit is dictated by the nature of the sample. With gases, information can be taken to within 2 cm^{-1} of the exciting line. In less favorable cases, within 20–50 cm^{-1} is more typical. This corresponds to the far-infrared region where measurements are difficult. Most of the important vibrations involved in metal bonding of inorganic and organometallic compounds fall in the low-frequency region as a result of the large masses of the metal atoms. Raman spectroscopy has been applied to the analysis of strong acids and other aqueous solutions, and to the determination of the degree of dissociation of strong electrolytes and their corresponding activity coefficients.

The Raman technique has proved to be particularly valuable in the study of single crystals where the infrared technique has greater limitations on sample size and geometry. In addition, polarization data obtained from Raman spectra allow unambiguous classification of fundamentals and lattice modes into the various symmetry classes. Although Raman spectroscopy will never challenge X-ray diffraction as a tool for quantitative structural analysis, it is the preferred technique when qualitative information is sufficient, because it is faster and less expensive.

8.5 POLARIZATION MEASUREMENTS

The output of the laser is linearly polarized by passage through the Brewster's angle windows at each end of the plasma tube. This property can be put to use to determine the *depolarization ratio*, defined by

$$p = I_a/I_b \tag{8-2}$$

where I_a is the Raman intensity when the direction of polarization of the incident beam is parallel to the direction of observation, and I_b is the Raman intensity when the direction of polarization of the incident beam is perpendicular to the direction of observation (with the analyzer and polarizer crossed). The observation is made in a direction perpendicular to that of the incident beam. The I_b component is always preponderant. The depolarization ratio approaches a value of zero for highly symmetrical types of vibrations. For all nontotally symmetrical vibrations, the depolarization ratio will have a value of 0.86, the theoretical maximum, when the Raman line is said to be depolarized. Results are valuable in assigning frequencies to particular modes of vibration. A vibrational frequency that is antisymmetric or degenerate to one or more symmetry elements will be depolarized if observable, or else will be inactive.

The depolarization unit consists of an analyzer prism and a depolarization compensator in the path of the Raman scattered light. With this accessory unit the analyzer prism is set at 0°, then at 90°, and the ratio of the band intensities measured. Polarization characteristics of the monochromator are eliminated by the polarization compensator, perhaps a quartz wedge, which completely scrambles the polarized Raman scattering entering the spectrometer. Rotation of the plane of polarization is achieved by a half-wave retardation plate between the laser source and sample.

The instrument shown in Fig. 8-4 automatically records both the depolarization ratios and the Raman spectrum at the same time. This is very useful for the detection of a weak Raman band overlapped by a strong band, as, for example, the study of intermolecular interaction in solution, a quick analysis of substances, and the study of polarized Raman spectrum of a single crystal. The laser beam is chopped at 21.4 Hz by the rotating sector mirror, then alternately transmitted to and reflected by two different illumination systems. The reflected beam has the same polarization plane as the laser, and the transmitted beam has a polarization plane perpendicular to the laser beam through a half-wave plate and Glan Thompson prism. The polarization plane is rotated 90° with respect to the light axis. Raman scattering caused by the transmitted beam and that caused by the reflected beam are converted into electrical signals and recorded on the chart.

Depolarization is illustrated by the partial Raman spectrum of $CHCl_3$, shown in Fig. 8-13. Nonspherical symmetry exists for the molecule. The antisymmetric $C–Cl_3$ bending vibration at 261 cm^{-1} and the C–Cl stretching mode at 760 cm^{-1} are depolarized. At 366 cm^{-1} the polarized $C–Cl_3$ symmetric bending mode appears, as does the polarized symmetric C–Cl stretching mode at 667 cm^{-1}. The C–H stretching mode is not shown.

8.6 QUANTITATIVE ANALYSIS

The determination of the absolute intensities of Raman bands is even more difficult than the determination of the absolute intensity of infrared absorption bands. For this reason the intensity of a Raman line is usually measured in terms of an arbitrarily chosen reference line, usually the line of CCl_4 at 459 cm^{-1}, which is scanned before and after the spectral trace of the sample. Scattering intensities, or peak heights on the spectrum, are then converted to *scattering coefficients* by dividing the recorded height of the sample peak by the average of the heights of the dual traces of the CCl_4 peak. Both standard and sample must be recorded in cells of the same dimension.

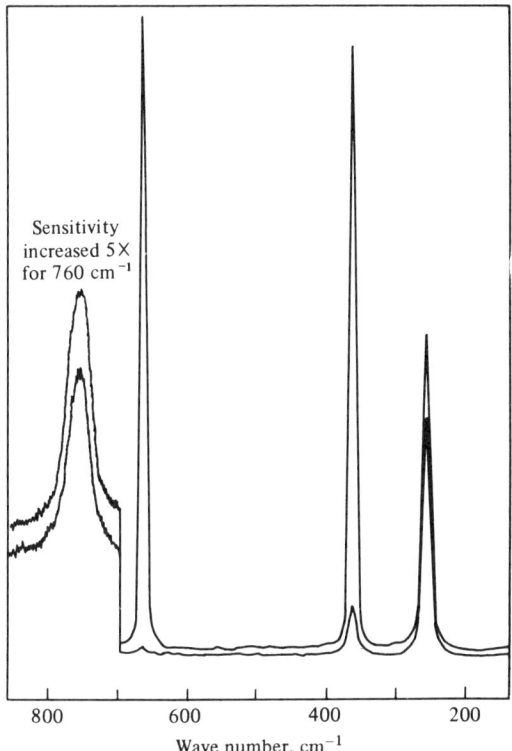

FIGURE 8-13 Partial Raman spectrum of $CHCl_3$ illustrating depolarization. Lower trace is when the direction of polarization of the incident beam is perpendicular to the direction of observation; the upper trace is when the incident beam is parallel to the direction of observation.

For quantitative analysis the intensity of Raman lines is directly proportional to the number of scatterer molecules and thus to the scattering coefficient. For mixtures in which the components are all of the same molecular type, there is a direct proportionality between the scattering coefficient and the volume fraction of the compound present. For mixtures of dissimilar type, Raman shifts will vary among the various compounds, and a broad band is recorded at the position characteristic of these bond types. The area under the recorded peak can be used as a measurement of scattering intensity.

PROBLEMS

1. What instrumental factors have lead to a false but widely held belief that C–H stretching vibrations are weak in Raman spectroscopy?

2. Suppose we wish to scan a Raman spectrum at the fastest possible speed. The minimum time constant of the electronics is 0.1 sec, and the described resolution (bandpass)

is 8 cm^{-1}. (a) What should the scan speed be? (b) If the dispersion of a double spectrometer with 1200-grooves/mm grating, working in the first order, is 0.55 nm/mm at the exit slit, what can the maximum slit opening be?

3. By what factor is the scattered intensity of a given band reduced in changing the excitation frequency from an argon laser (488.0 nm) to a neodymium-doped laser (1065 nm)? Ignore changes in detector response, as well as grating and reflector efficiencies.

4. What are the relative intensities of a Raman line excited by a He–Ne laser at 632.8 nm and one excited by an argon laser at 488.0 nm?

5. When excited by the mercury line at 435.8 nm, the spectral trace of benzene contains Raman lines at 606, 850, 991, 1176, 1584, 1605, 3047, and 3063 cm^{-1}. At what wavelengths will these Raman lines appear if benzene is irradiated with (a) a He–Ne laser (632.8 nm), (b) an argon ion laser (488.0 and 514.5 nm), and (c) a krypton laser (568.2 and 647.1 nm)?

6. If unfiltered laser excitation from either the krypton or argon laser were used to excite the Raman spectrum of benzene (Problem 5), to what extent would each suite of Raman lines overlap?

7. For carbon disulfide, all vibrations that are Raman active are infrared inactive, and vice versa, whereas for nitrous oxide (N_2O) the vibrations are simultaneously Raman and infrared active. What can one conclude concerning the structures of N_2O and CS_2?

8. For CCl_4, four principal Raman lines appear at 218, 314, 458, and 791 cm^{-1}. None of these infrared frequencies is absorbed. The depolarization ratios are 0.86, 0.86, 0.046, and 0.83, respectively. What can be concluded about the symmetry of the molecule? Is its spatial configuration planar or tetrahedral?

9. For each unknown mixture of the trimethylbenzenes, compute the volume percent of each. The scattering coefficient of the pure compound at each analytical wave number is tabulated:

	1,2,3-		1,2,4-		1,3,5-	
	625 cm^{-1}	0.627	716 cm^{-1}	0.208	570 cm^{-1}	0.555
Mixture A		0.209		0.069		0.185
B		0.251		0.054		0.189
C		0.157		0.077		0.211

10. Deduce the structure of the compound whose spectra is shown in the accompanying figure. The molecular weight is 140.

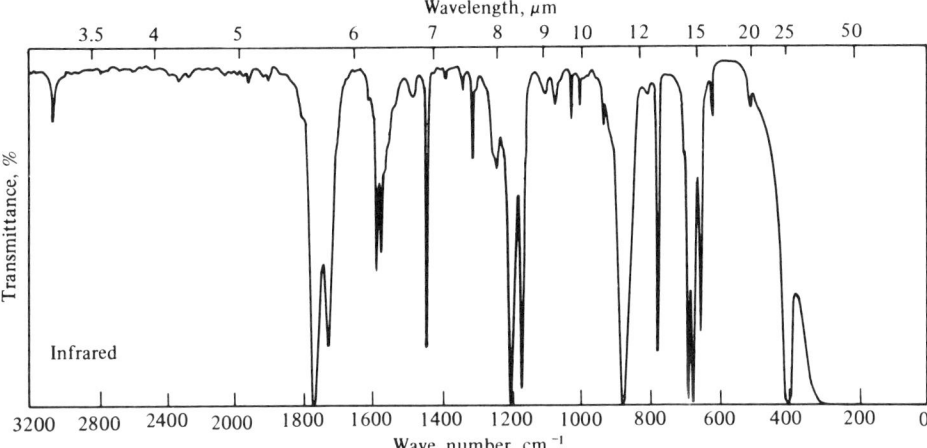

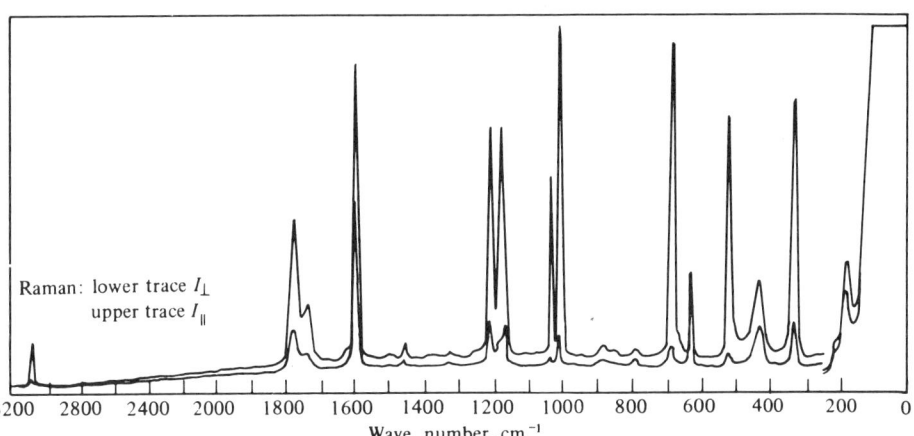

PROBLEM 8-10

11. The molecular weight of the compound is 54. Using the spectra in the accompanying figure, determine the structure of the compound.

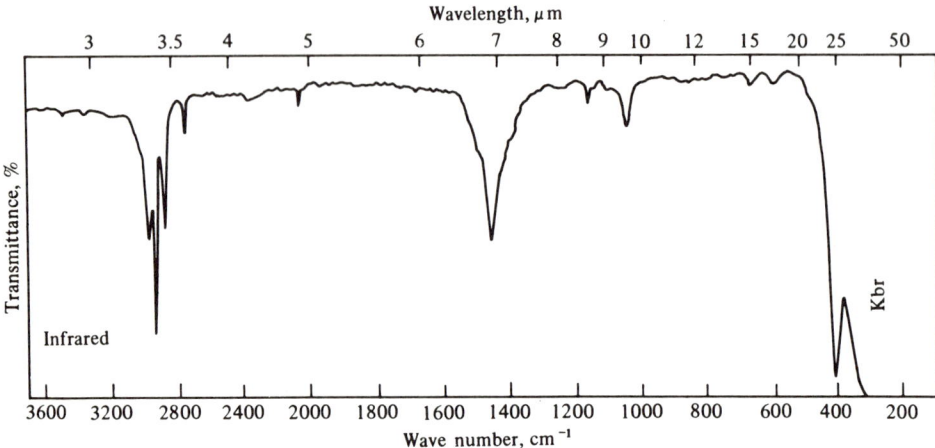

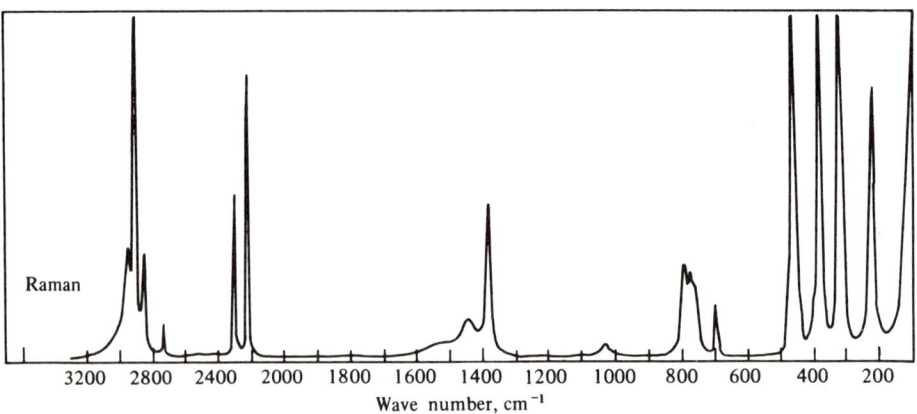

PROBLEM 8-11

12. From the infrared and Raman spectra shown, and the molecular weight of 123, deduce the structure of the compound.

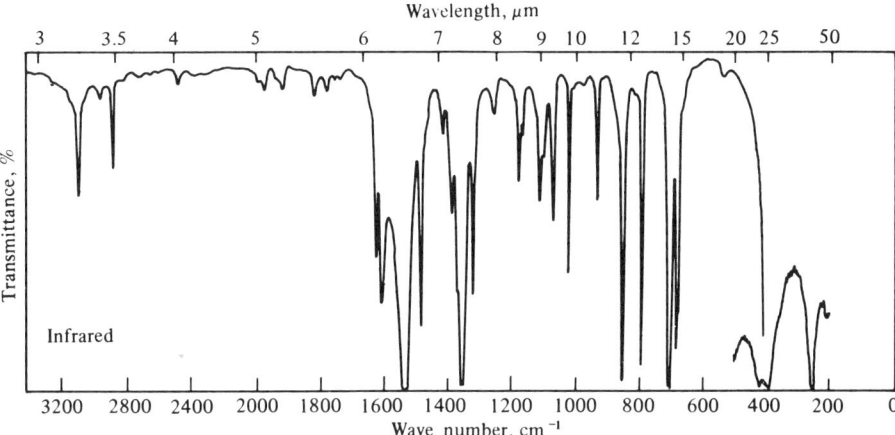

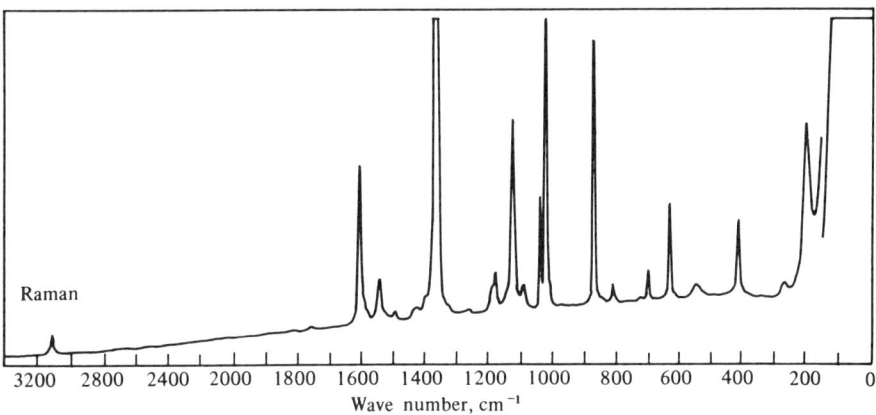

PROBLEM 8-12

BIBLIOGRAPHY

Allkins, J. R., "Tunable Lasers in Analytical Chemistry," *Anal. Chem.*, **47**, 752A (1975).
Bulkin, B. J., "Raman Spectroscopy," *J. Chem. Educ.*, **46**, A781, A859 (1969).
Green, R. B., "Dye Laser Instrumentation," *J. Chem. Educ.*, **54**, A365 (1977).
Haber, H. S., "Fluorescence Problems in Raman Spectroscopy," *Am. Lab.*, p. 67 (November 1973).
Harvey, A. B., "Coherent Anti-Stokes Raman Spectroscopy (CARS)," *Anal. Chem.*, **50**, 905A (1978).
Morris, M. D. and D. J. Wallan, "Resonance Raman Spectroscopy," *Anal. Chem.*, **51**, 182A (1979).
Sloane, H. J., "The Technique of Raman Spectroscopy: A State-of-the-Art Comparison to Infrared," *Appl. Spectrosc.*, **25**, 430 (1971).
Tobin, M. C., *Laser Raman Spectroscopy*, Wiley-Interscience, New York, 1971.
Washburn, W. H., "Synergistic Use of Infrared and Raman Spectroscopy," *Am. Lab.*, p. 47 (November 1978).

CHAPTER 9

X-Ray Methods

When an atom is excited by removal of an electron from an inner shell, it usually returns to its normal state by transferring an electron from some outer shell to the inner shell with consequent emission of energy as X rays; that is, photons of high energy and short wavelengths in the order of tenths of angstroms to several angstroms. Eventually a free electron will be captured by the ion.

X Rays can be used in chemical analysis in several ways. One method uses the fact that the X rays emitted by an excited element have a wavelength characteristic of that element and an intensity proportional to the number of excited atoms. Thus emission methods can be used for both qualitative and quantitative work. The excitation can be carried out in several ways: by direct bombardment of the material with electrons (direct emission analysis and electron probe microanalysis) or by irradiation of the material with X rays of shorter wavelength (fluorescent analysis).

A second method of X-ray analysis utilizes the differing absorption of X rays by different materials (absorption analysis). Major discontinuities in the absorption of X rays by an element occur when the energy of the X rays becomes sufficient to knock an electron out of the inner levels of an atom.

A third method of using X rays in analytical work is the diffraction of X rays from the planes of a crystal (diffraction analysis). This method depends upon the wave character of X rays and the regular spacing of planes in a crystal. Although diffraction methods can be used for quantitative analysis, they are most widely used for qualitative identification of crystalline phases.

In 1913 Moseley first showed the extremely simple relation between atomic number, Z, and the reciprocal of the wavelength, $1/\lambda$, for each spectral line belonging to a particular series of emission lines for each element in the periodic table. This relationship is expressed as

$$\frac{c}{\lambda} = a(Z - \sigma)^2 \tag{9-1}$$

where a is a proportionality constant and σ is a constant whose value depends on the particular series.

X-Ray emission and absorption spectra are quite simple because they consist of very few lines compared to the emission (Chapter 6) or absorption spectra observed in the visible or ultraviolet regions (Chapters 2 and 3). This relative simplicity arises because the X-ray spectra result from transitions between energy levels of the innermost electrons in the atom. There are only a few electrons in these inner shells and the resulting energy levels are limited, thus giving rise to only a few permitted transitions. There is only one K shell. The L electrons are grouped according to their binding energy into three sublevels: L_I, L_{II}, and L_{III}; the complete M shell consists of five sublevels. X-Ray emission or absorption spectra are dependent only on atomic number and not on the physical state of the sample nor on its chemical composition, except for the lightest atoms, because the innermost electrons are not involved in chemical binding and are not significantly affected by the behavior of the valence electrons.

9.1 PRODUCTION OF X RAYS AND X-RAY SPECTRA

An X-ray tube is basically a large vacuum tube containing a heated cathode (electron emitter) and an anode, or target (Fig. 9-1). Electrons emitted by the cathode are accelerated through a high-voltage field between the target and cathode. Upon impact with the target, the stream of electrons is quickly brought to rest. The electrons transfer their kinetic energy to the atoms of the material making up the target. Part of the kinetic en-

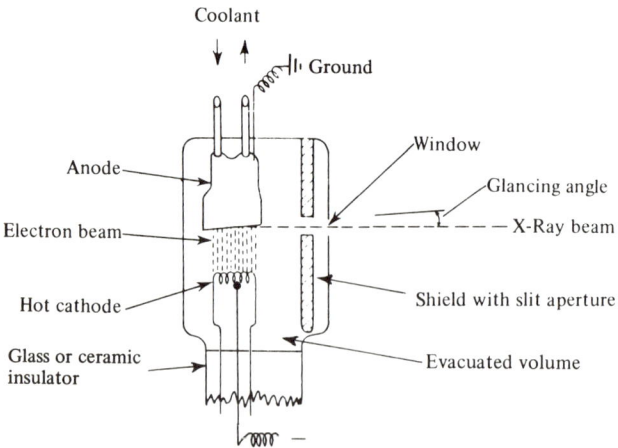

FIGURE 9-1 Schematic of an X-ray tube.

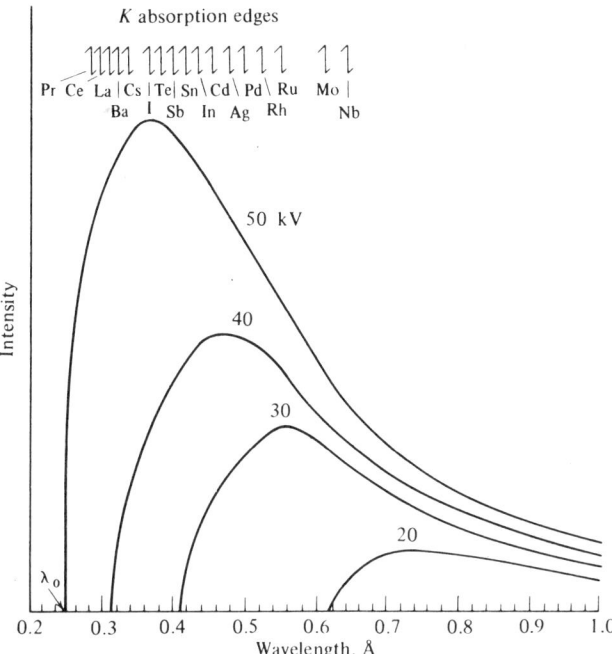

FIGURE 9-2 X-Ray continuum from a target operated at voltages specified. Along the top edge are indicated the wavelengths of K absorption edges for elements Z 41–59.

ergy is emitted in a continuous spectrum of X rays covering a broad wavelength range (Fig. 9-2) with a broad maximum in intensity and falling off to a definite short-wavelength limit. The broad range of X-ray photon energies is due to deceleration of the impinging electrons by successive collisions with the atoms of the target material. Consequently, the emitted quanta are of longer wavelength than the short-wavelength cutoff, λ_0, which is independent of the target element and depends only upon the voltage across the X-ray tube. At the cutoff wavelength all of the energy of the electron is converted, at one impact, to a photon. The relationship between the voltage and λ_0 is given by the Duane–Hunt equation[1]

$$\lambda_0 \text{ (in Å)} = \frac{hc}{eV} = \frac{12,400}{V} \tag{9-2}$$

where V is the X-ray tube voltage in volts, e is the charge on the electron, h is Planck's constant, and c is the velocity of light. An increase in tube voltage results in an increase in the total energy emitted and a movement of the spectral distribution toward shorter wavelengths. The wavelength of maximum intensity is about 1.5 times the short-wavelength limit. The intensity of the spectrum increases with atomic number of the target element.

If sufficient energy is available, the transfer of energy from the impinging electron beam may eject an electron from one of the inner shells of the atoms constituting the target material. The place of the ejected electron will then promptly be filled by an electron from an outer shell whose place, in turn, will be taken by an electron coming from still farther out. Thus the ionized atom returns to its normal state in a series of steps in each of which an X-ray photon of definite energy is emitted. These transitions give rise to the characteristic line spectrum of the material in the anode or of a specimen pasted on the target. When originating in an X-ray tube, these lines will be superimposed on the continuum. The K series of lines is observed when an electron in the innermost K shell is dislodged; it arises from electrons dropping down from L or M orbitals into the vacancy in the K shell. Corresponding vacancies in the L shells are filled by electron transitions from outer shells and give rise to the L series.

The characteristic X-ray spectrum of an element can also be excited by irradiation of a sample with a beam of X rays provided that the primary X radiation is sufficiently energetic to remove an electron from an inner shell of the element. Because the inner electron must be completely removed from the element, the energy required is greater than that of any emission lines in the element's spectra, for emission lines result when that series of electrons falls into a vacant inner level from higher energy levels within the atom. When the energy of the exciting radiation just equals the energy required to remove an electron from the element, the exciting radiation is strongly absorbed; that is, there is a sharp rise in the absorption of the exciting radiation (Fig. 9-3). This is known as an *absorption edge*. If an X-ray tube is used to produce the exciting radiation, λ_0, the short-wavelength cutoff must equal or be shorter than the wavelength of the absorption edge and thus there is a critical potential which must be applied to the tube.

Besides bombardment of a target with electrons, as in an X-ray tube, and irradiation of a target with energetic photons, as in the production of fluorescent radiation from a sample, X rays are also produced during the decay of certain radioactive isotopes. Many isotopes emit gamma rays which are the same as short-wavelength X rays. Other isotopes decay by K capture. In this process the nucleus captures a K electron, thus becoming an

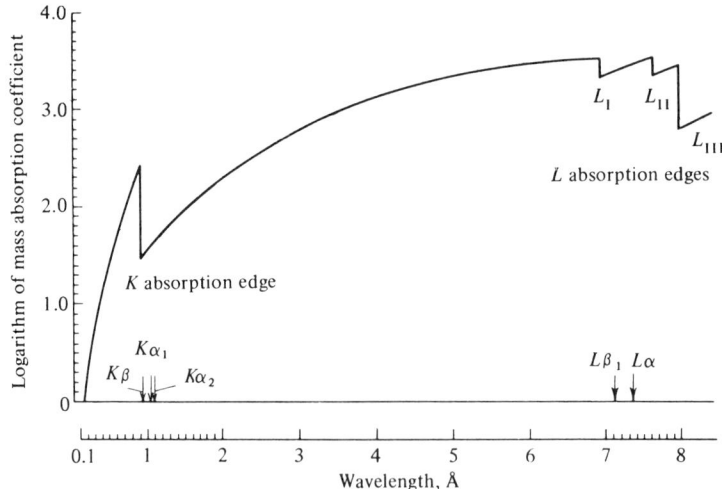

FIGURE 9-3 X-Ray absorption spectrum of bromine. The characteristic emission lines of the K and L series are shown with arrows.

element with one less atomic number. The vacant K shell is filled by electrons falling in from outer shells, thus emitting characteristic X rays. An example of a radioactive isotope that decays by K capture and may be useful as a source of essentially monoenergetic X rays is ^{55}Fe. The K lines of manganese are emitted. (See Chapter 10 for a more complete description of radioactive isotopes.)

Example 9-1

To calculate the short-wavelength limit for an X-ray tube operated at 50 kV,

$$\lambda_0 = \frac{12,400}{50,000} = 0.248 \text{ Å}$$

Upon irradiation with this energy europium (Z = 63), whose K absorption edge lies at 0.255 Å, would emit its characteristic K series of lines, although with low intensity. However, the energy is insufficient for excitation of the K lines of gadolinium (Z = 64), whose K absorption edge lies at 0.247 Å.

As the wavelength of the incident radiation is decreased or as the potential across an X-ray tube is increased, there is successive ionization: first of electrons in the M shells of the sample or target, then of electrons in the L shells as the L_{III}, L_{II}, and L_I absorption edges are progressively exceeded, and finally culminating in the K shell's absorption edge. The K spectra are generally used for the detection and analysis of elements up to about neodymium (Z = 60); the L spectra are used from lanthanum to *trans*-uranium elements when an X-ray tube that has a maximum rating of 50 kV produces the exciting radiation. The wavelengths of selected spectral lines and absorption edges of a number of elements are shown in Table 9-1.

TABLE 9-1 Characteristic Wavelengths of Absorption Edges and Emission Lines for Selected Elements

Element	Minimum Potential for Excitation of K Lines, kV	K Absorption Edge, Å	$K\beta$, Å	$K\alpha_1$, Å	L_{III} Absorption Edge, Å	$L\alpha_1$, Å
Magnesium	1.30	9.54	9.558	9.889	247.9	251.0
Titanium	4.966	2.50	2.514	2.748	27.37	27.39
Chromium	5.988	2.070	2.085	2.290	20.7	21.67
Manganese	6.542	1.895	1.910	2.102	19.40	19.45
Cobalt	7.713	1.607	1.621	1.789	15.93	15.97
Nickel	8.337	1.487	1.500	1.658	14.58	14.57
Copper	8.982	1.380	1.392	1.541	13.29	13.33
Zinc	9.662	1.283	1.295	1.435	12.13	12.26
Molybdenum	20.003	0.620	0.632	0.709	4.912	5.406
Silver	25.535	0.484	0.497	0.559	3.698	4.154
Tungsten	69.51	0.178	0.184	0.209	1.215	1.476
Platinum	78.35	0.158	0.164	0.186	1.072	1.313

SOURCE: J. A. Dean, Ed., *Lange's Handbook of Chemistry*, 12th ed., McGraw-Hill, New York, 1979.

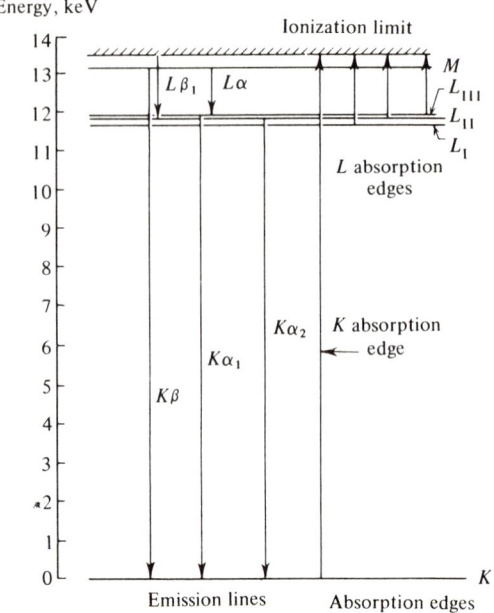

FIGURE 9-4 Energy-level diagram of bromine ($Z = 35$) showing the transitions that give rise to the absorption discontinuities and the emission lines.

Example 9-2

Consider a vacant orbital in a bromine atom produced by the ejection of an electron from the innermost K shell of electrons. The energy required just to lift a K electron out of the environment of the atom must exceed the energy of the K absorption edge at 0.918 Å, or

$$V = \frac{12{,}400}{0.918} = 13{,}475 \text{ V } (13.475 \text{ kV})$$

The wavelength of the K absorption edge is always shorter than that of the K emission lines. The $K\beta_1$ line at 0.934 Å arises when an electron drops from the M shell; the $K\alpha_1$ and $K\alpha_2$ lines, a closely spaced doublet at 1.048 and 1.053 Å, arise from sublevels of slightly different energies within the L shell. In energy units, the $K\alpha_1$ line represents the difference: K edge minus L_{III} edge. Thus, for bromine

$$K\alpha_1 \text{ (in keV)} = 13.475 - 1.522 = 11.953$$

The absorption and emission spectra for bromine are shown in Fig. 9-3, and the energy-level diagram is shown in Fig. 9-4.

The bond character in molecules and solids affects the X-ray spectra of the light elements whose emission lines originate from the valence electron shell, and even those lines

TABLE 9-2 Mean Wavelengths and Shifts of K Lines of Sulfur for the Different Oxidation States of Sulfur

Oxidation State	λ, XU[a]		Mean Shift	
	$K\alpha_1$	$K\alpha_2$	$\Delta\lambda$, XU	ΔE, eV
S^{6+}	5358.08	5360.89	−2.76	+1.19
S^{4+}	5358.63	5361.47	−2.20	+0.95
S^{2+}	5360.13	5362.93	−0.72	+0.31
S^0	5360.83	5363.66	0	0
S^{2-}	5361.15	5363.99	+0.33	−0.14

[a]One angstrom ≡ 1002.02 XU, where 1 XU = 1/3029.45 the spacing of the cleavage planes of a calcite crystal, a former standard wavelength unit.
SOURCE: A. Faessler, "X-Ray Emission Spectra and the Chemical Bond," in *Proceedings of the Xth Colloquium Spectroscopicum Internationale,* Spartan Books, Washington, 1963, pp. 307–319.

and absorption edges from the next innermost shell. In general, relative to the lines of the free element, the lines of the atom in a compound are shifted toward shorter wavelengths if the atom has a positive charge, and toward longer wavelengths if it has a negative charge in the compound. Similar fine structure may be observed at an absorption edge if a high-resolution spectrometer is employed. Mean wavelengths and shifts in the emission lines for the different oxidation states of sulfur are given in Table 9-2.

9.2 INSTRUMENTAL UNITS

Instrumentation associated with X-ray methods in general is outlined schematically in Fig. 9-5. Many of the components will be discussed more fully in subsequent sections.

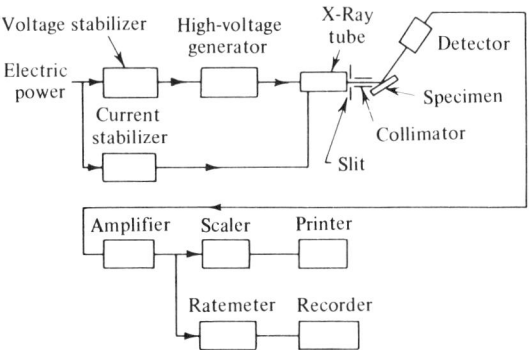

FIGURE 9-5 Instrumentation for X-ray spectroscopy.

X-Ray Generating Equipment

The modern X-ray tube is a high-vacuum, sealed-off unit, shown schematically in Fig. 9-1, usually with a copper or molybdenum target, although targets of chromium, iron, nickel, silver, and tungsten are used for special purposes. The target is viewed from a very small angle above the surface. If the focal spot is a narrow ribbon, the source appears to be very small when viewed from the end, which leads to the sharper definition demanded in diffraction studies. For fluorescence work the focus is much larger, about 5×10 mm and is viewed at a larger angle (about $20°$). Because it becomes very hot, the target is cooled by water and is sometimes rotated when a very intense X-ray beam is generated. The X-ray beam passes out of the tube through a thin window of beryllium or a special glass. For wavelengths from 6 to 70 Å, ultrathin films (1-μm aluminum or cast Parlodion films), separate the X-ray tube from the remainder of the equipment, which must be evacuated or flushed with helium.

Associated equipment includes high-voltage generators and stabilizers. Voltage regulation is accomplished by regulating the main ac supply. Current regulation is achieved by monitoring the dc X-ray tube current and controlling the filament voltage. Either full-wave rectification or constant high potential may be used to operate the X-ray tube. In full-wave rectification the voltage reaches its peak value 120 times a second but only persists at that value for a small fraction of the time. Constant high potential, obtained through electronic filtering, increases the output of characteristic X rays from a specimen, particularly with elements emitting at short wavelengths. With a tube operated at 50 kV the gain is twofold for elements up to about atomic number 35 (Br), increasing to fourfold for atomic number 56 (Ba). Commonly, X-ray tubes are operated at 50 or 60 kV. Tubes of 100 kV rating are available and extend the range of elements whose K series can be excited and the sensitivity, because on increasing the voltage the intensities of all lines increase.

Collimators

Radiation from an X-ray tube is collimated either by a series of closely spaced, parallel metal plates or by a bundle of tubes, 0.5 mm or less in diameter. In a fluorescence spectrometer, one collimator is placed between the specimen and the analyzer crystal to limit the divergence of the rays that reach the crystal. The second collimator, usually coarser, is placed between the analyzer crystal and the detector, where it is particularly useful at very low goniometer angles for preventing radiation that has not been reflected by the crystal from reaching the detector. Increased resolution can be obtained by decreasing the separation between the metal plates of the collimator or by increasing the length of the unit (usually a few centimeters), but this is achieved at the expense of intensity.

Filters

When the wavelengths of two spectral lines are nearly the same and there is an element with an absorption edge at a wavelength between the lines, that element may be used as a filter to reduce the intensity of the line of shorter wavelength. In X-ray diffractometry it

TABLE 9-3 Filters for Common Targets of X-Ray Tubes

Target Element	$K\alpha_1$, Å	$K\beta$, Å	Filter	K Absorption Edge Filter, Å	Thickness,[a] mm	Percent Loss of $K\alpha_1$
Mo	0.709	0.632	Zr	0.689	0.081	57
Cu	1.541	1.392	Ni	1.487	0.013	45
Cr	2.290	2.085	V	2.269	0.0153	51
	$L\alpha_1$, Å	$L\beta_1$, Å				$L\alpha_1$
Pt	1.313	1.120	Zn	1.283	—	—
W	1.476	1.282	Cu	1.380	0.035	77

[a] To reduce the intensity of the $K\beta$ line to 0.01 that of the $K\alpha_1$ line.

is common practice to insert a thin foil in the primary X-ray beam to remove the $K\beta$ line from the spectrum while transmitting the $K\alpha$ lines with a relatively small loss of intensity. Filters for the common targets of X-ray tubes are listed in Table 9-3. Background radiation (the continuum) is reduced by the same method. Usually it makes no difference whether the filter is placed before or after the specimen unless the specimen fluoresces; if so, the filter is placed at the entrance slit of the goniometer.

Analyzing Crystal

Virtually monochromatic radiation is obtained by reflecting X rays from crystal planes. The relationship between the wavelength of the X-ray beam, the angle of diffraction θ, and the distance between each set of atomic planes of the crystal lattice, d, is given by the Bragg condition[2]:

$$m\lambda = 2d \sin \theta \qquad (9\text{-}3)$$

where m represents the order of the diffraction. The geometric relations are shown in Fig. 9-6. For the ray diffracted by the second plane of the crystal, the distance $\overline{CBD}$

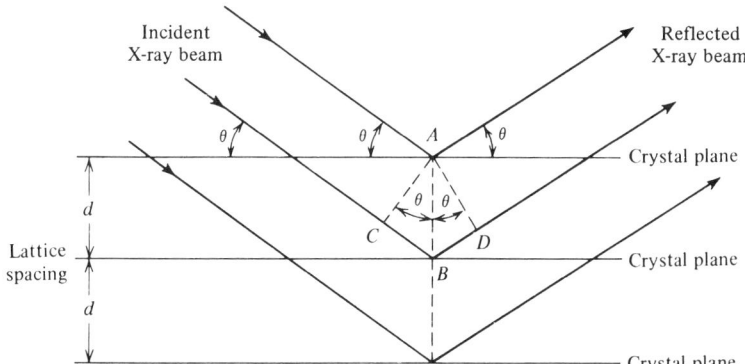

FIGURE 9-6 Diffraction of X rays from a set of crystal planes.

represents the additional distance of travel in comparison to a ray reflected from the surface. Angles CAB and BAD are both equal to θ. Therefore,

$$\overline{CB} = \overline{BD} = \overline{AB} \sin \theta \tag{9-4}$$

and

$$\overline{CBD} = 2\overline{AB} \sin \theta \tag{9-5}$$

where AB is the interplanar spacing, d. In order to observe a beam in the direction of the diffracted rays, $\overline{CBD}$ must be some multiple of the wavelength of the X rays so that the diffracted waves will be in phase. Note that the angle between the direction of the incident beam and that of the diffracted beam is 2θ. In order to scan the emission spectrum of a specimen, the analyzing crystal is mounted on a goniometer, an instrument for measuring angles, and rotated through the desired angular region, as shown in the schematic diagram of a fluorescent spectrometer (Fig. 9-21).

The range of wavelengths usable with various analyzing crystals is governed by the d-spacings of the crystal planes and by the geometric limits to which the goniometer can be rotated. The d-value should be small enough to make the angle 2θ greater than approximately 8° even at the shortest wavelength used, otherwise excessively long analyzing crystals would be needed in order to prevent the incident beam from entering the detector. A small d-spacing is also favorable for producing a larger dispersion, $\partial\theta/\partial\lambda$, of the spectrum, as can be seen by differentiating the Bragg equation:

$$\frac{\partial \theta}{\partial \lambda} = \frac{m}{2d \cos \theta} \tag{9-6}$$

On the other hand, a small d-value imposes an upper limit to the range of wavelengths that can be analyzed, because at $\lambda = 2d$ the angle 2θ becomes 180°. Actually, the upper limit to which goniometers can be rotated is mechanically limited to a 2θ value of around 150°. For longer wavelengths a crystal with a larger d-spacing must be selected. Crystals commonly used are listed in Table 9-4. These crystals are all composed of light atoms; only sodium chloride, quartz, and the heavy metal fatty acids* have elements heavier than $Z = 9$, so that their own fluorescent X rays will not interfere with measurements. Higher-order reflections, m greater than 1, from the analyzing crystal may result in the overlap of lines originating from different elements.

Analyzing crystals have presented problems in the extension of X-ray analysis beyond a few angstroms. Potassium acid phthalate has made the determination of magnesium and sodium more practical. To extend the analytical capabilities beyond 26 Å, multiple

---

*The lead palmitate and strontium behenate analyzers are prepared by repeatedly dipping an optical flat into the film of the metal fatty acid, that is, the Langmuir–Blodgett technique.

TABLE 9-4 Typical Analyzer Crystals

Crystal	Reflecting Plane	Lattice Spacing d in Å	USEFUL RANGE IN Å	
			Maximum[a]	Minimum[b]
Topaz	303	1.356	2.62	0.189
Lithium fluoride	200	2.014	3.89	0.281
Aluminum	111	2.338	4.52	0.326
Sodium chloride	200	2.821	5.45	0.393
Calcium fluoride	111	3.16	6.11	0.440
Quartz	10$\bar{1}$1	3.343	6.46	0.466
Ethylenediamine d-tartrate (EDDT)	020	4.404	8.51	0.614
Ammonium dihydrogen orthophosphate (ADP)	200	5.325	10.29	0.742
Pyrolytic graphite	002	6.71	12.96	0.936
Gypsum	020	7.60	14.70	1.06
Mica	002	9.963	19.25	1.39
Lead palmitate		45.6	78.3	6.39
Strontium behenate		61.3	121.7	8.59

[a] Maximum $2\theta = 150°$, $m\lambda = 2d \sin 75°$.
[b] Minimum $2\theta = 8°$. $m\lambda = 2d \sin 4°$.

monolayer soap film "crystals" are used. So far, a lead stearate decanoate "crystal" has proved to be superior for elements $Z = 9$ to $Z = 5$.

9.3 DETECTORS FOR THE MEASUREMENT OF RADIATION

Photographic Emulsion

Photographic film can be used to measure the intensity of radiation and is often used in diffraction studies. It is also used to measure the distribution of radioactive material in a thin section of a substance; that is, autoradiography or the distribution of different X-ray absorbers in a thin section of material, that is, microradiography. Film badges are used to measure the total exposure of workers to ionizing radiation.

The nature of the photographic process is discussed in Chapter 6. Suffice it to say here that with high-energy radiation such as X rays and radiation from radioactive nuclides each particle of silver halide which absorbs radiation becomes developable and, therefore, there exists a direct, linear relationship between blackening of the developed film and the intensity of radiation.

The Ionization Chamber

In the *ionization chamber* (Fig. 9-7) an electric field is applied between two electrodes across a volume of gas–air for alpha particles, krypton, or xenon under pressure for X or

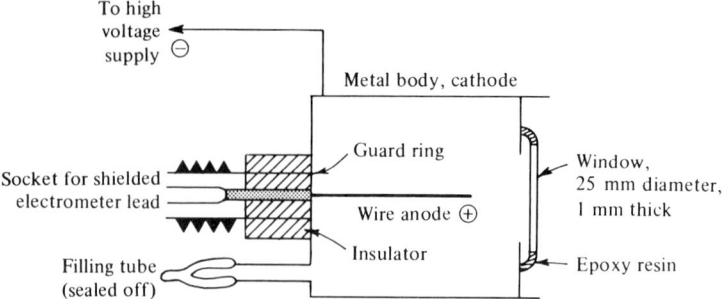

FIGURE 9-7 Schematic diagram of an ionization chamber.

gamma radiation. The potential across the electrodes is adjusted to minimize recombination of the ion pairs without causing gas amplification (Fig. 9-8). An ionization chamber is an accurate, quick-acting detector even for very weak radiation. The sample is placed outside the window or inserted in a well extending into the chamber volume. For each ionizing event *pulse ion chambers* produce an electronic pulse proportional to the number of electrons released by the ionizing radiation. *Current ion chambers* integrate the events and provide a dc current.

The Geiger Counter

The Geiger counter, also called a Geiger–Müller or G–M tube, is shown schematically in Fig. 9-9. A potential of 800–2500 V is applied to a central wire anode surrounded by a cylindrical cathode—a glass wall which has been silvered or a brass cylinder. The two electrodes are enclosed in a gas-tight envelope typically filled to a pressure of 80 mm of argon gas plus 20 mm of methane or ethanol or 0.1% of chlorine. A thin end-window of mica, about 2.5 cm in diameter and 2–3 mg/cm^2 in thickness, or a glass wall in dipping counters, is the point of entry of the radiation.

When an ionizing particle enters the active volume of the Geiger counter, collision with the filling gas produces an ion pair. This is followed by migration of these particles toward the appropriate electrodes under the voltage gradient. The mobility of the electron is quite high, and under the influence of the potential gradient it soon acquires sufficient velocity to produce a new pair of ions upon collision with another atom of argon. Under these conditions, which are repeated many times, each original ionizing particle entering the active volume of the counter gives rise to an avalanche of electrons traveling toward the central anode. Photons, emitted when the electrons strike the anode, spread the ionization throughout the tube. These processes produce a continuous discharge which fills the whole active volume of the counter in less than a microsecond. Each discharge builds up to a constant pulse of maximum amplitude (10 V) and 50–100 μsec duration. These pulses can be counted precisely with the aid of scaling circuits or a ratemeter with no intermediate amplification; this is the principal advantage of the Geiger counter.

During the time the electron avalanche is collected on the anode, the positive ions, being much heavier, have progressed only a short distance on their way to the cathode. Their travel time is about 200 μsec and during most of this time their presence as a virtual sheath

X-RAY METHODS   251

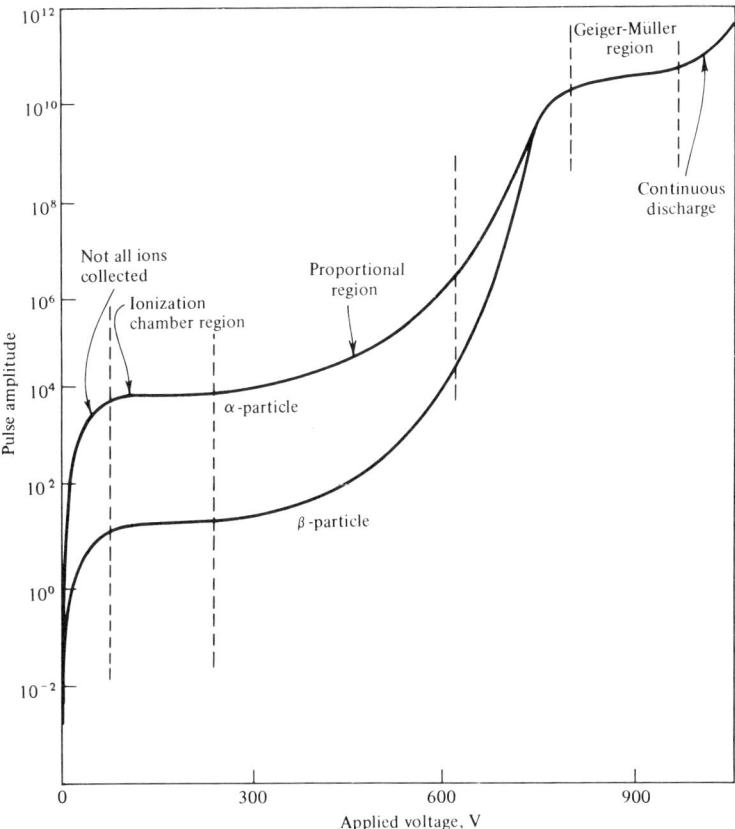

FIGURE 9-8   Pulse amplitude as a function of applied voltage for the ionization type of detectors.

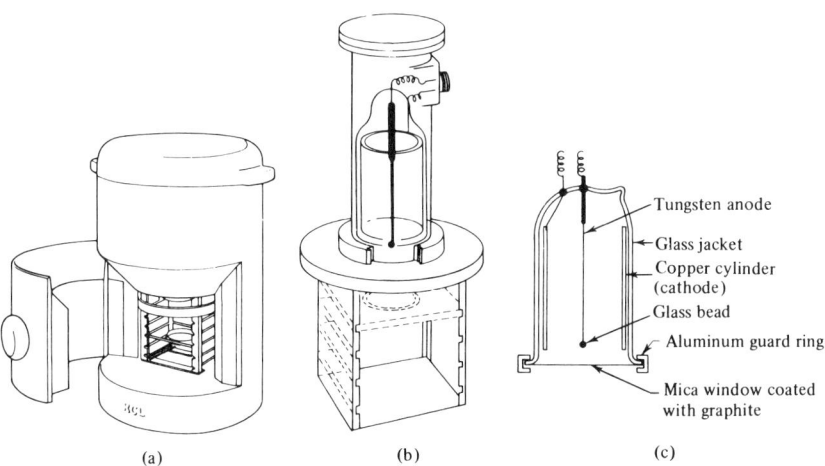

FIGURE 9-9   (a) End-window type of Geiger counter, (b) counter and sample holder with shielding removed, and (c) schematic of counter.

around the anode effectively lowers the potential gradient to a point where the counter is insensitive to the entry of further ionizing particles—the *dead time* of the counter. Because the halogen or organic gas molecules have a lower ionization potential than argon, after a few collisions the ions moving toward the cathode consist only of these entities. In contrast to argon ions, these positive ions do not produce photons when neutralized at the cathode. Consequently, photons which could initiate a fresh discharge are prevented from forming and the counter is self-quenching. Upon being neutralized, the organic filling gas dissociates to various molecular fragments and, eventually, the quenching agent is exhausted. Counter life is limited to about 10^{10} counts. Because chlorine atoms merely recombine, the quencher is available for further use. A halogen-quenched counter has a life in excess of 10^{13} counts.

Counting rates are limited to about 15,000 counts/min because of the long dead time of 200–270 μsec and the large reduction in true count rate as the maximum count rate is approached. Sensitivity for beta particles is excellent, but for X and gamma radiation the sensitivity is less than that of the scintillation counter. There is no possibility of using pulse height discrimination with a Geiger counter because the pulses are all the same amplitude, nor is there any practical correction for coincidence losses when scanning across a spectrum.

The argon-filled Geiger counter, using halogen as a quenching gas, has a sensitive volume wide enough to detect nearly the entire large-area beam used in some X-ray optics. The tube is relatively insensitive to scattered hard radiation and thus its background intensity is low. Its quantum efficiency is about 60–65% in the range from 1.5 to 2.1 Å, and decreases to 40% below 1.4 Å and above 2.9 Å (Fig. 9-10).

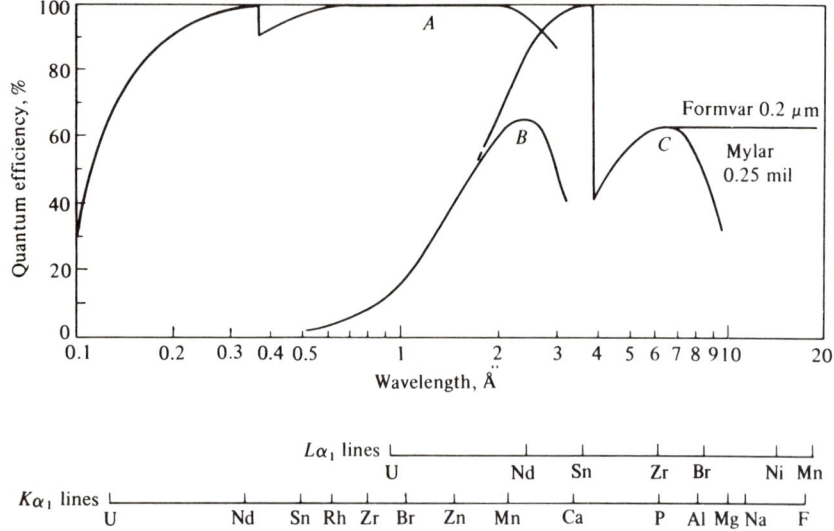

FIGURE 9-10 Quantum efficiencies of detectors commonly used in X-ray spectrometry. *A*, Scintillation counter with Tl activated, NaI scintillator. *B*, Argon-filled Geiger counter with Be window. *C*, Gas-flow proportional counter; 90% Ar, 10% CH_4; 0.25-mil Mylar window. Lower scales indicate the wavelengths of representative emission lines.

Proportional Counters

When the electric field strength at the center electrode of an ionization chamber is increased above the saturation level, but under that of the Geiger region (Fig. 9-8), the size of the output pulse from the chamber starts to increase but is still proportional to the initial ionization. A device operated in such a fashion is called a *proportional counter*.

Pulse formation is identical with that described for Geiger counters, but gas amplification is approximately 1000 times less. Consequently, a preamplifier ($\times$ 10) is needed and is mounted together with the detector to avoid reduction in pulse size through capacitance in connecting cables. In the proportional region few, if any, photons are released. Consequently, the total number of secondary electrons is proportional to the number of primary ion pairs produced by the original ionizing particle. Furthermore, the discharge is limited to the immediate environment of the entering ionizing particle and the path traversed by the ion pair plus their secondary electrons and positive ions. The dead time is thus very short, about 0.25 μsec. Multiplication factors from 10 to 10^5 are possible; they are dependent on applied voltage, gas pressures, and counter dimensions.

Proportional counters are useful for counting at extremely high counting rates—50,000 –200,000 counts/sec; the upper limit is imposed by the associated electronic circuitry. The signal produced is extremely small and requires both a preamplifier and a second stage of amplification before the signal can be fed to a scaler. Excellent plateaus of about 100 V can be obtained whose slopes are as low as 0.1% counting-rate change per 100 V (whereas values of less than 1% variation are uncommon with Geiger counters).

The proportional counter has about the same spectral sensitivity characteristics as the Geiger counter (Fig. 9-10). The window material and thickness have a great influence on the spectral characteristics. Detector windows present challenging problems in work at very long wavelengths (to 70 Å), because these windows must be transparent to very low energy photons and, in addition, must be capable of supporting atmospheric pressure. Typical windows include 1-μm (sign painter's) aluminum dipped in Formvar (usable for sodium and magnesium X rays), 1-μm hydrocarbon (cast Formvar, Parlodion, or collodion) films, and 0.1-μm hydrocarbon films. The 0.1-μm films must be supported on a 70% optical transmission grid or on the 0.5-mm spacing blade on a flow detector collimator. Their use is required for X radiation from oxygen, nitrogen, and boron. The lifetimes of unsupported, 1-μm films never exceed 8 hr.

A sample of radioactive material can actually be placed inside the active volume of a flow proportional counter (Fig. 9-11), thus avoiding losses due to window absorption. With this type of counter, the chamber is purged with a rapid flow of counter gas, often 10% methane in argon, and a steady flow of gas is maintained during counting. Counter life is virtually unlimited since the filling gas is constantly replenished. Such a counter is particularly suited for distinguishing and counting low-energy alpha and beta particles. For X-ray detection the flow proportional counter can be equipped with an extremely thin window, usually 0.25-mil Mylar film, which naturally decreases the losses due to window absorption of very soft X rays. Its range extends to 12 Å and is the counter of choice for long-wavelength X radiation. Windows of 0.1-μm Formvar or "thin" nitrocellulose film (often supported by screens) extend the transmission to approximately 120 and 160 Å, respectively.

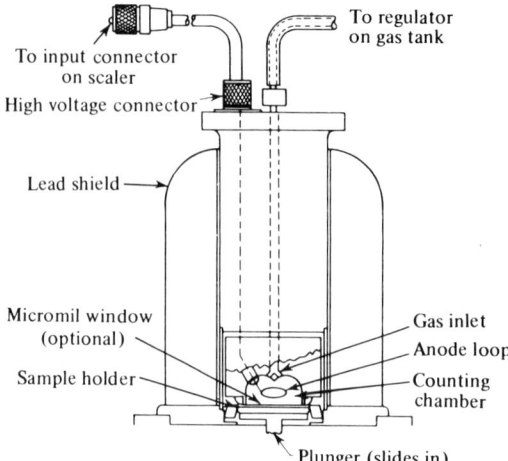

FIGURE 9-11 Schematic diagram of a flow proportional counter mounted in a shield; the very thin window is optional. The sample is inserted into the active volume by means of the lateral slide holder. (Courtesy of Nuclear-Chicago Corp.)

Scintillation Counters

Scintillators are chemicals used to convert radiation energy into light. When an ionizing particle is absorbed in any one of several transparent scintillators, some of the energy acquired by the scintillator is emitted as a pulse of visible or near-ultraviolet light. The light is observed by a photomultiplier tube, either directly or through an internally reflecting light pipe. The combination of a scintillator and photomultiplier tube is called a *scintillation counter* (Fig. 9-12). A good match should exist between the emission spectrum of

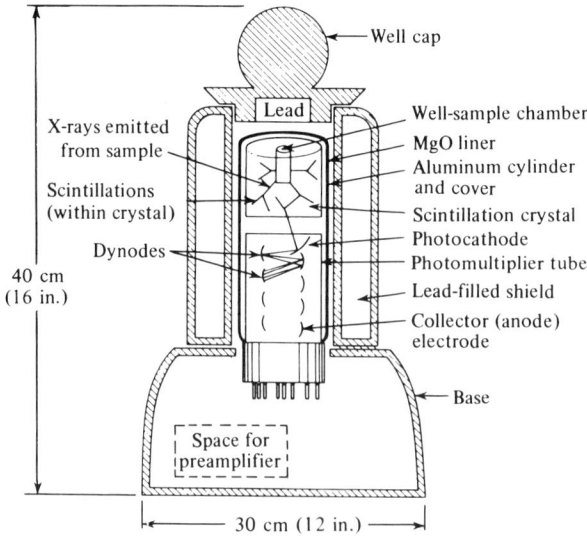

FIGURE 9-12 Well-type crystal scintillation counter and shield.

the scintillator and the response curve of the photocathode. The decay time for scintillators is very short: 250 nsec for a sodium iodide crystal, 20 nsec for anthracene, and 10 nsec for liquid organic systems. The signal from a scintillation counter is proportional to the energy dissipated by the radiation in the scintillator so that this counter may be used with pulse height discrimination.

For counting alpha particles the best scintillator is a thin layer of silver-activated zinc sulfide, which may be coated on the envelope of the photomultiplier tube.

Scintillation crystals of anthracene or stilbene (wavelength of emission: 445.0 and 410.0 nm, respectively) affixed by a good optical liquid to an end-window photomultiplier tube are suitable for beta particles of moderate and high energy. However, organic liquid scintillators are often preferred because of their shorter decay times. Low-energy beta emitters, such as ^{3}H, ^{14}C, and ^{35}S, are commonly counted by dissolving the compound containing the radionuclide in the liquid scintillator.

To measure X rays and gamma radiation and bremsstrahlung from high-energy beta emitters, an inorganic scintillator, such as sodium iodide crystal doped with 1% thallium (I) iodide, is best. This scintillator has a large photoelectric cross section, a high density which provides a high probability of absorption, and a high transparency to its own radiation (the optical emission lines of thallium) which enables large thicknesses to be used for absorption of X and gamma radiation. When such radiation interacts with a NaI (Tl) crystal, the transmitted energy excites the iodine atom and raises it to a higher-energy state. When the iodine atom returns to its ground electronic state, this energy is reemitted in the form of a light pulse in the ultraviolet which is promptly absorbed by the thallium atom and reemitted as fluorescent light at 410.0 nm. The crystal is sealed from atmospheric moisture and protected from extraneous light by an enclosure of aluminum foil, which also serves as an internal reflector. Such a scintillation counter has a nearly uniform and high quantum efficiency throughout the important X-ray region, 0.3–2.5 Å, and is usable to possibly 4 Å (Fig. 9-10). Longer wavelengths are absorbed in the coating covering the crystal.

The "well-type" scintillation crystal increases the counting efficiency by surrounding the sample with the detector crystal. The sample is placed in a well drilled into a crystal 5–10 cm in diameter; the size is chosen so that it contains the entire path of the ionizing particle or radiation and so measures the total energy. The resolution of a NaI (Tl) counter spectrometer is relatively poor (peak width of 6% at 1 MeV and 18% at 100 keV), but the efficiency approaches 100%, and the instrument is a multichannel device because the entire spectrum can be recorded at one time.

9.4 SEMICONDUCTOR DETECTORS

Semiconductor detectors have revolutionized X- and gamma-ray spectroscopy by providing energy resolution unattainable with previous methods. In these detectors the charge carriers produced by ionizing radiation are electron–hole pairs rather than ion pairs. The ionizing radiation lifts electrons into the conduction band and these electrons travel toward the positive electrode with high mobilities. The positive charge travels in the opposite direction by successive exchanges of electrons between neighboring lattice sites. Two general types of semiconductor detectors will be discussed: the surface barrier silicon detector and the lithium-drifted silicon and germanium detectors.

A surface barrier detector consists of a *p–n* junction formed at the surface of a slice of silicon. At the junction there is a planar region, in which no charge carriers are present, where an electric field exists. This region is called the *depletion region*. A thin depletion depth is present when no bias voltage is applied across the *p–n* junction. If a reverse bias is applied, the depletion depth is increased, and is given by $d \simeq 0.5 \sqrt{\rho V}$ (in micrometers) where ρ is the silicon resistivity in ohm-centimeters and V is the bias in volts. Thus, with higher bias and higher resistivity, deeper depletion depths are formed. If charges are injected into the depletion region, they will be swept out of it by the electric field, and a voltage pulse will appear across the *p–n* junction. If energy measurements are to be made, selection of a detector for a particular purpose requires selecting a depletion depth greater or equal to the range of the particle of interest.

The lithium-drifted germanium detector consists of a virtually windowless Ge(Li) crystal, a vacuum cryostat maintained by cryosorption pumping, a liquid nitrogen Dewar, and a preamplifier. The Ge(Li) crystal is fabricated by drifting lithium ions (a donor) into and through *p*-type germanium. This is performed under the influence of a high electric field at 400°C. This process results in compensation of all acceptors within the bulk material, yielding a very high-resistivity (or intrinsic) region which acts like ultrapure germanium within the bulk material. The drifting process is discontinued while a layer of *p*-type germanium still remains (Fig. 9-13). This intrinsic or compensated volume becomes the radiation-sensitive region. When ionizing radiation enters the intrinsic layer, electron–hole pairs are created, and the charge produced is rapidly collected under the influence of the bias voltage. The completed detector must be maintained at 77°K at all times to prevent precipitation of the lithium, since the lithium drift process is not stable at normal room temperature. At this low temperature thermal noise is greatly reduced and the resolution capabilities are vastly increased.

Lithium-drifted silicon detectors are prepared in a similar manner by drifting lithium ions into *p*-type silicon. They have become quite popular in recent years. These detectors come in a multitude of sizes and can be used at room temperature or down to liquid nitrogen

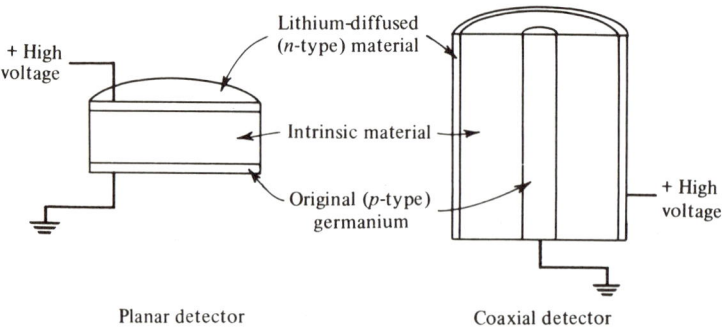

FIGURE 9-13 Schematic diagrams of two common types of lithium-drifted germanium detectors.

temperature. At the lower temperatures the detectors, of course, show lower background noise and higher resolution than when operated at room temperature.

The energy resolution of semiconductor detectors is intrinsically good because of the large number of electron–hole pairs formed in comparison with the number of ion pairs formed in a gas ionization chamber. The average energy for electron–hole pair production is 2.95 eV for germanium and 3.65 eV for silicon; this is far less than the 500 eV required per photoelectron in a NaI(Tl) scintillation detector. Thus, for a given amount of energy absorbed about 170 times as many electron–hole pairs are formed as ion pairs. Since the relative resolution is proportional to the square root of the signal, the resolution of the Ge(Li) detector is about a factor of 13 better than the NaI(Tl) detector (see illustrations accompanying Problems 22 and 23, Chapter 10). Typical quoted figures for energy resolution of a semiconductor detector 3 mm thick are 3.8 keV for electrons, 0.6 keV for X rays, and 20 keV for photons. This is the full width at half-maximum (FWHM) of a peak in the energy spectrum. The rise time is about 10 nsec. However, the semiconductor detector efficiency falls short of that of NaI(Tl) detectors, and is approximately 1% per millimeter of thickness.

Especially the lithium-drifted silicon detectors have increased the popularity of what has become known as energy-dispersive analysis. In energy-dispersive methods of X-ray analysis, the sample is irradiated with X rays or gamma radiation from a radionuclide source or ions to produce secondary X radiation characteristic of the elements present in the sample. These characteristic X rays pass through a hollow shield onto a Si(Li) detector. The purpose of the shield is only to prevent any X rays from hitting the edges of the detector where they may not be completely absorbed. The output of the Si(Li) detector is amplified by a preamplifier and then passes into a pulse height analyzer (Fig. 9-14) and associated electronic circuits which eventually can present counts versus X-ray energy and thus give qualitative and even quantitative evidence of the various elements present in the sample.

This method, sometimes known as X-ray energy spectrometry (XES), has the advantage of being able to give a complete qualitative analysis in one operation. Depending upon the primary radiation source used, the method can be applied to bulk analyses or just to analyses of surfaces and thin films down to a thickness of about 2 nm. Elements down to about carbon can be detected. The apparatus is also simple although the detector must be cooled to liquid nitrogen temperatures for best resolution and the whole must be contained in a vacuum when elements of low atomic number ($Z < 12$) are being sought.

Auxiliary Instrumentation

Detectors require auxiliary electronic equipment including a high-voltage supply, an amplifier (often plus a preamplifier), a scaler, and a count-registering unit. The required stability of the high-voltage supply and the required sensitivity and the linearity of the amplifier are dictated by both the detector and the application. The signals produced when an X-ray quantum is absorbed by proportional, scintillation, and semiconductor detectors are extremely small and require both a preamplifier and a second stage of amplification before the signal can be fed to a scaler or recorder. To diminish noise pickup, the preamplifier is located immediately after the detector in the latter's housing.

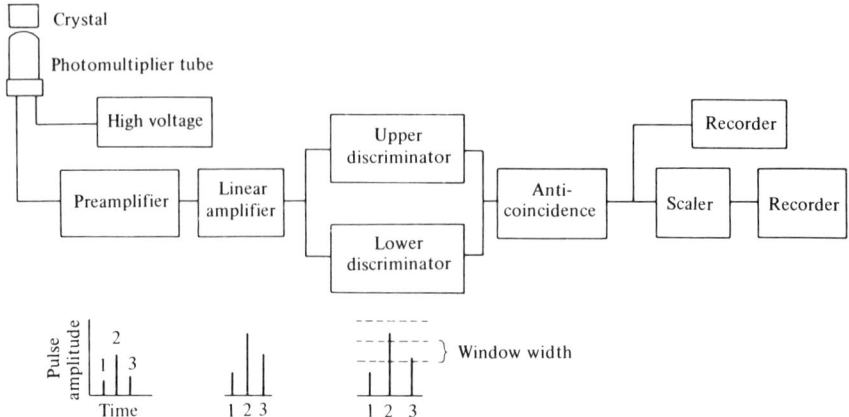

FIGURE 9-14 Block diagram of a single-channel pulse height analyzer.

From the detector the output pulse, after amplification, is fed into a scaling circuit which, in reality, is an electronic divider. The circuit is arranged so that the output is a single pulse for each $2, 4, 8, \ldots, 2^n$ incident particles. By a system of glow lamps the events withheld can be numbered. In the scale of two (binary) type, the overall scaling factor is 2^n, where n is the number of binary stages incorporated. More convenient and rapid reading is achieved on decade scalers. Each stage passes on every tenth pulse so that the instrument reads decimally. With either type the output from the scaler operates a mechanical register. Timing is done with built-in electric clocks which start and stop the count for a preset time interval, or, after a predetermined number of counts have been accumulated, the elapsed time is noted.

A significant amount of radiation from natural radioactive elements and cosmic rays is always present in the vicinity of a detector. Insertion of the counter into a shield of lead 2–3 in. in thickness reduces the background counting rate appreciably (Fig. 9-9). Further improvement can be achieved with anticoincidence circuits.

Pulse Height Discrimination

Whenever the amplitude of the pulse is proportional to the energy dissipation in the detector, the measurement of pulse height is a useful tool for energy discrimination. Current pulses are fed into a linear amplifier of sufficient gain to produce voltage output pulses in the amplitude ranges of 0–100 V.

One method of analysis of the pulse spectra is by use of a single-channel analyzer. The base line discriminator passes only those pulses above a certain amplitude and eliminates pulses below this amplitude. It is useful for excluding scattered radiation and amplifier noise. Pulses associated with a particular energy must be amplified sufficiently so that their amplitudes exceed the discriminator setting. In practice this is accomplished by a combination of adjustment of the gain of the amplifier and the dc voltage applied to the detector.

A pulse height analyzer also contains a second discriminator called variously the window width, the channel width, or the acceptance slit. Now all pulses above the sum of the base line and window setting are also rejected. Only pulses with an amplitude within the confines of these settings will be passed on to the counting stages. These operations are outlined schematically in Fig. 9-14. With circuits for pulse height discrimination, it is possible to discriminate electronically against unwanted wavelengths of different elements. Discrimination between elements 8–10 atomic numbers apart is possible with a scintillation detector. A proportional detector, because of its narrower pulse amplitude distribution, can discriminate between elements 4–6 atomic numbers apart. With semiconductor detectors even better resolution is possible, perhaps 1 or 2 atomic numbers apart. Unlike a filter, a pulse height analyzer can be used to pass either line of superposed spectral lines, serving, in effect, as a secondary monochromator. It is particularly useful for rejecting higher-order scattered radiation from elements of higher atomic number when determining the elements of lower atomic number. Modern pulse height analyzers with ever-improving resolution make finer and finer nondispersive analyzers.

Example 9-3

The use of a pulse height analyzer for Si $K\alpha_1$ radiation will illustrate the step-by-step operations. A relatively pure sample of silicon is inserted into the sample holder (see Fig. 9-15). Scanning with the goniometer from 106° to 110° provides the graph shown in Fig. 9-15. From an ethylenediamine d-tartrate crystal, Si $K\alpha_1$

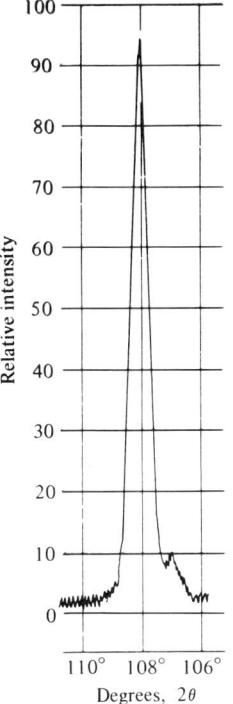

FIGURE 9-15 Relative intensity of the Si $K\alpha_1$ line as a function of goniometer setting (2θ). Analyzing crystal: ethylenediamine d-tartrate.

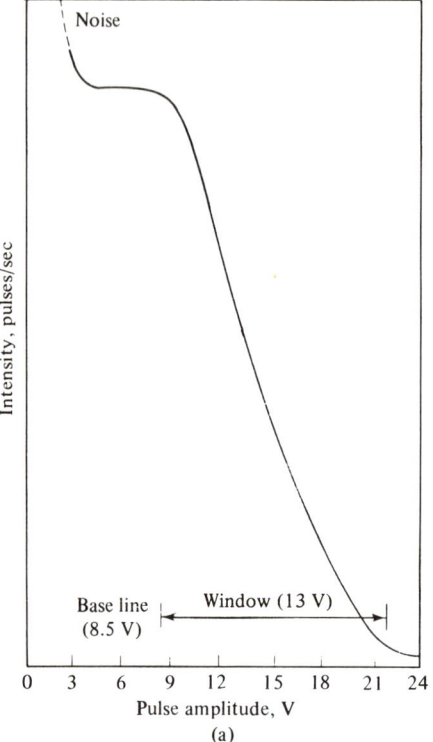
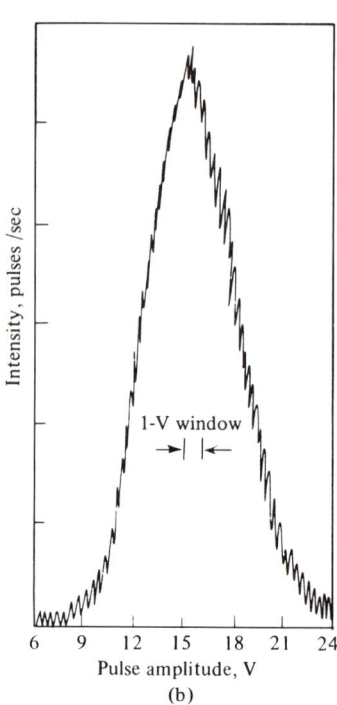

FIGURE 9-16 Pulse amplitude distribution of Si $K\alpha_1$ radiation. (a) Integral curve and (b) differential curve. Goniometer set at 108°.

radiation is reflected at $2\theta = 108°$. Next, with the goniometer set manually at the peak of the silicon radiation, the distribution of pulses due to the silicon X-ray quanta is obtained by scanning the pulse amplitude base line and using a 1-V window. The integral curve of intensity versus pulse amplitude is shown in Fig. 9-16. From this information, the base of the pulse height discriminator would be set at 8.5 V and the window width at 13.0 V, since it is noticed that no pulses are detected until the upper-line setting approaches 21 V. In this example, the silicon radiation was peaked at 15 V. Naturally, if the silicon pulses were peaked at a lower or higher voltage, then the window and base line settings would be different. The peak distribution (in volts) is a function of the dc voltage on the counter and the amplifier gain.

9.5 DIRECT X-RAY METHODS

The process of exciting characteristic spectra by electron bombardment was applied many years ago in the investigation of characteristic spectra of the elements by Siegbahn and others. In this manner the element hafnium was discovered by Von Hevesy and Coster in 1923. The specimen must be plated or smeared on the target of the X-ray tube. This has disadvantages: the X-ray tube must be reevacuated each time the specimen is changed; a demountable target is required; and the heating effect of the electron beam may cause chemical reaction, selective volatilization, or melting. These difficulties virtually prohibit the large-scale application of the direct method to routine analysis, except for electron probe microanalysis.

Electron Beam Probe

Electron probe microanalysis, developed by Castaing[3] (1951), is a method for the nondestructive elemental analysis from an area only 1 µm in diameter at the surface of a solid specimen. A beam of electrons is collimated into a fine pencil of 1-µm cross section and directed at the specimen surface exactly on the spot to be analyzed. This electron bombardment excites characteristic X rays essentially from a point source and at intensities considerably higher than with fluorescent excitation. The limit of detectability (in a 1-µm size region) is about 10^{-14} g. The relative accuracy is 1–2% if the concentration is greater than a few percent and if adequate standards are available.

Three types of optics are employed in the microprobe spectrometer: electron optics, light optics, and X-ray optics (Fig. 9-17). Of these, the most complex is the electron optical system, a modified electron microscope, which consists of an electron gun followed by two electromagnetic focusing lenses to form the electron beam probe. The specimen is mounted as the target inside the vacuum column of the instrument and under the beam. A focus-

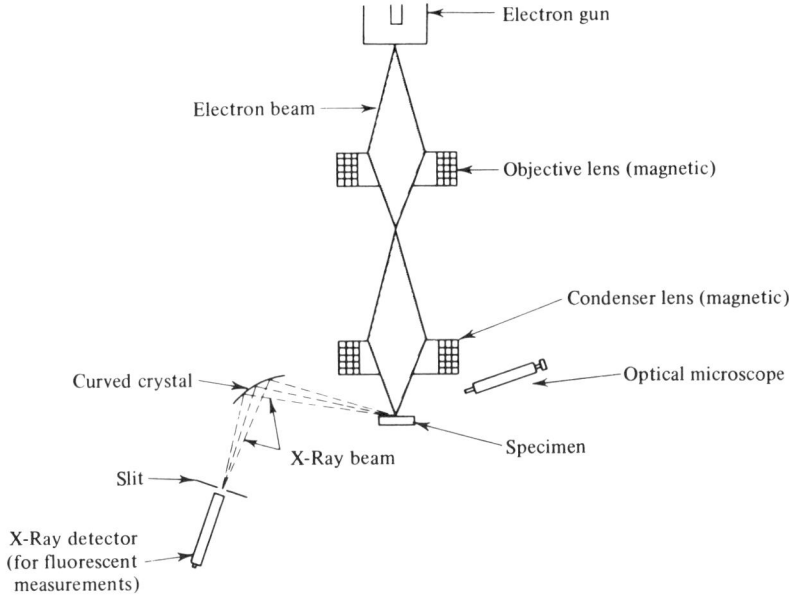

FIGURE 9-17 Schematic of an electron probe microanalyzer. The X-ray beam can be passed directly into the detector or reflected from the analyzer crystal.

ing, curved-crystal X-ray spectrometer is attached to the evacuated system with the focal spot of the electron beam serving as the source of X radiation. A viewing microscope and mirror system allow continuous visual observation of the exact area of the specimen where the electron beam is striking. Point-by-point microanalysis is accomplished by translating the specimen across the beam.

The method is used in the study of variations in concentration occurring near grain boundaries, the analysis of small inclusions in alloys or precipitates in a multitude of products, and corrosion studies where excitation is restricted to thin surface layers, because the beam penetrates to a depth of only 1 or 2 μm into the specimen.

9.6 X-RAY ABSORPTION METHODS

Because each element has its own characteristic set of K, L, M, and other absorption edges, the wavelength at which a sudden change in absorption occurs can be used to identify an element present in a sample, and the magnitude of the change can be used to determine the amount of the particular element present. The fundamental equation for the transmittance of a monochromatic, collimated X-ray beam is

$$P = P_0 e^{-(\mu/\rho)\rho x} \tag{9-7}$$

where P is the radiant power of P_0 after passage through x cm of homogeneous matter of density ρ and whose linear absorption coefficient is μ. The parenthetical term μ/ρ is the mass absorption coefficient, often expressed simply as μ_m. It depends upon the wavelength of the X rays and the absorbing atom; that is,

$$\mu_m = CZ^4 \lambda^3 \frac{N_A}{A} \tag{9-8}$$

where N_A is Avogadro's number, A is the atomic weight, and C is a constant over a range between characteristic absorption edges. It is significant that the mass absorption coefficient is independent of the physical or chemical state of the specimen. In a compound or mixture it is an additive function of the mass absorption coefficients of the constituent elements, namely,

$$\mu_{m_T} = \mu_{m_1} W_1 + \mu_{m_2} W_2 + \ldots \tag{9-9}$$

where μ_{m_1} is the mass absorption coefficient of element 1 and W_1 is its weight fraction, and so on for all the elements present. Because only one element has a change in mass absorption coefficient at the edge, the following relationship is obtained for the ith element:

$$2.3 \log \frac{P}{P_0} = (\mu''_{m_1} - \mu'_{m_1}) W_i \rho x \tag{9-10}$$

where the term in the parentheses represents the difference in mass absorption coefficient at the edge discontinuity. Thus, the logarithm of the ratio of beam intensities on the two sides of an absorption edge depends only upon the change in mass absorption coefficients of the element characterized by this edge and on the amount of the particular element in the beam; ρx is the mass thickness of the sample in grams per square centimeter. There is no matrix effect, which gives the absorption method an advantage over X-ray fluorescence analysis in some cases.

In analogy with absorption measurements in other portions of the electromagnetic spectrum, one would expect to obtain a representative set of transmittance measurements on each side of an absorption edge with an X-ray spectrometer and extrapolate to the edge. However, X-ray absorption spectrophotometers that provide a continuously variable wavelength of X radiation are not commercially available. Instead, only a single attenuation measurement is made on each side of the edge. A multichannel instrument is required.

The general procedure will be illustrated by the determination of lead tetraethyl and ethylene dibromide in gasoline. Four channels are required. One channel is used as a reference standard; the other three channels provide the analyses for lead, bromine, and a correction for variations of the C/H ratio and the presence of any sulfur and chlorine. Primary excitation is provided by an X-ray tube operated at 21 kV. The secondary targets for each channel are as follows, with the fluorescent X-ray lines employed:

Channel 1: RbCl, Rb $K\alpha_1$
Channel 2: RbCl, Rb $K\alpha_1$
Channel 3: SrCO$_3$, Sr $K\alpha_1$
Channel 4: NaBr, Br $K\alpha_1$

The relationship between the pertinent absorption edges and the target fluorescent emission lines is shown in Fig. 9-18. In operation, a nominal sample is sealed in the sample

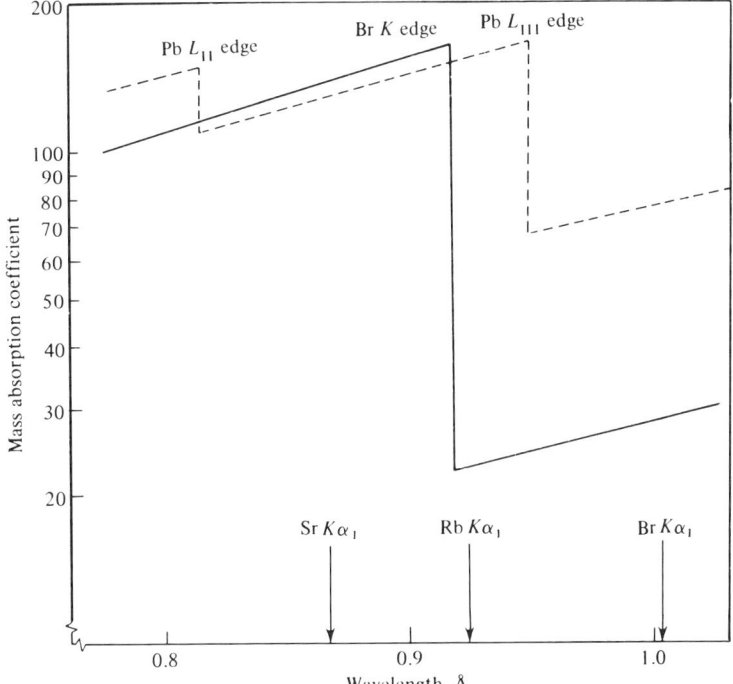

FIGURE 9-18 Absorption edges and emission lines pertinent to the X-ray absorption analysis of lead tetraethyl and ethylene dibromide in gasoline.

cell in Channel 1; the sample to be analyzed is placed in the cells in the remaining channels. The exposure is started and automatically terminated when the integrated intensity in Channel 1 reaches a predetermined value (perhaps 100,000 counts in a time interval of 100 sec). The integrated intensities accumulated in the other channels are then recorded. Initially the four channels are adjusted to reach 100% transmittance with nominally pure gasoline. Results for bromine are computed from the difference in counts between Channels 3 and 2; for lead from Channels 2 and 4.

Microradiography

Another application employing the different absorbing powers of different elements toward an X-ray beam permits the gross structure of various types of small specimens to be examined under high magnification. Positions where there are elements that strongly absorb the X rays will appear light, and positions where there are elements which do not absorb the X rays will appear dark on a film placed behind the sample.

Clark and Gross[4] developed a method employing ordinary X-ray diffraction equipment. Any of the commonly employed targets operated at 30–50 kV can be used. No vacuum camera is necessary. The microradiographic camera, shown in Fig. 9-19, is designed to fit as an inset in the collimating system of any commercial X-ray equipment. Special photographic film, which possesses an extremely fine grain, makes magnifications up to 200 times possible without loss of detail from graininess. Sample thicknesses vary from 0.075 mm for steels up to 0.25 mm for magnesium alloys. Only a few seconds of exposure is necessary.

Various techniques are possible depending upon the specimen. Biological specimens may be impregnated with a material of high molecular weight to characterize particular structures. Occasionally the necessary density variations are initially present within the sample. More often, the use of various selective monochromatic wavelengths from different target elements must be employed.

Another technique uses a tube with a very small focal spot as an X-ray source. If the focal spot approaches a point source, magnification in the microradiograph is obtained by simple geometry with considerable sharpness. Actually, focal spots as small as 5 μm in diameter can be obtained and magnifications up to 50 times or so are possible. The magnification is the ratio of the distance of film to target to the distance of object from the target.

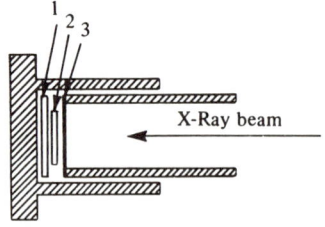

FIGURE 9-19 Schematic of microradiographic camera: (1) film, (2) sample, and (3) black paper.

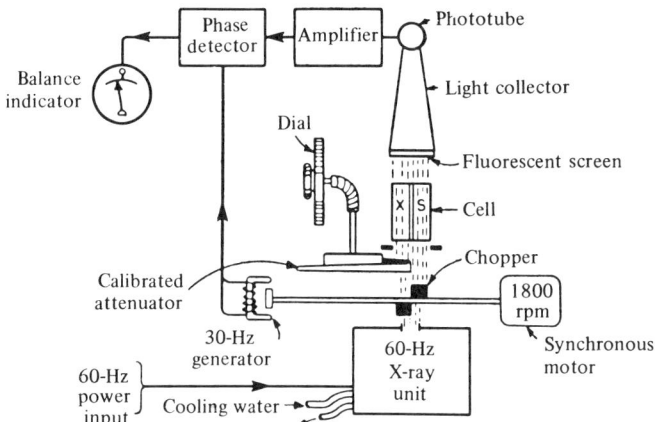

FIGURE 9-20 Nondispersive X-ray absorptiometer. (Courtesy of General Electric Co.)

Nondispersive X-Ray Absorptiometer

The general arrangement of a nondispersive X-ray absorptiometer is shown in Fig. 9-20. A tungsten target X-ray tube is operated at 15–45 kV. In the X-ray beam is a synchronous motor-driven chopper which alternately interrupts one-half of the X-ray beam. A variable-thickness aluminum attenuator (in the shape of a wedge) is placed between the chopper and reference sample compartment. Duplicate reference and sample cells up to 65 cm in length can be accommodated; those for liquids and gases can be arranged for continuous flow of process streams. Both halves of the X-ray beam fall on a common phosphor-coated photomultiplier tube which is protected from visible light by a thin metallic filter.

In operation, a reference sample is placed in the appropriate cell and the specimen to be analyzed in the sample tube. The attenuator is adjusted until the absorption in the two X-ray beams is brought into balance. The change in thickness of aluminum required for different samples is a function of the difference in composition. Prior calibration enables a determination in terms of the solute in an unknown. Liquids are simplest to handle. The thickness of solid specimens, and the density of powder samples, must be uniform to a precision greater than that expected in the result.

Polychromatic absorptiometry can be used to determine chlorine in hydrogen. Sulfur in crude oil can be distinguished from the carbon-hydrogen residuum. Other examples include barium fluoride in carbon brushes, barium or lead in special glass, and chlorine in plastics and hydrocarbons. In fact, the method is applicable to any sample that contains one element markedly heavier than the others and when the matrix is essentially invariant in concentration.

9.7 X-RAY FLUORESCENCE METHOD

Characteristic X-ray spectra are excited when a specimen is irradiated with a beam of sufficiently short-wavelength X radiation. Intensities of the resulting fluorescent X rays are smaller by a factor of roughly 1000 than an X-ray beam obtained by direct excitation with a beam of electrons. Only availability of high-intensity X-ray tubes, very sensitive detectors, and suitable X-ray optics renders the fluorescent method feasible. The intensity is important because it influences the time that will be necessary to measure a spectrum. A certain number of quanta has to be accumulated at the detector in order to reduce sufficiently the statistical error of the measurement. The sensitivity of the analysis, that is, the lowest detectable concentration of a particular element in a specimen, will depend on the peak-to-background ratio of the spectral lines. Relatively few cases of spectral interference occur because of the relative simplicity of X-ray spectra.

X-Ray Fluorescence Spectrometer

The general arrangement for exciting, dispersing, and detecting fluorescent radiation with a plane-crystal spectrometer is shown diagrammatically in Fig. 9-21. The specimen in the sample holder (often rotated to improve uniformity of exposure) is irradiated with an unfiltered beam of primary X rays, which causes the elements present to emit their characteristic fluorescence lines. A portion of the scattered fluorescence is collimated by the entrance slit of the goniometer and directed onto the plane surface of the analyzing crystal. The line radiations, reflected according to the Bragg condition, pass through an auxiliary collimator (exit slit) to the detector, where the energy of the X-ray quanta is converted into electrical impulses, or counts.

The primary slit, the analyzer crystal, and secondary slit are placed on the focal circle so that Bragg's law will always be satisfied as the goniometer is rotated, the detector being rotated at twice the angular rate of the crystal. The analyzer crystal is a flat single-crystal

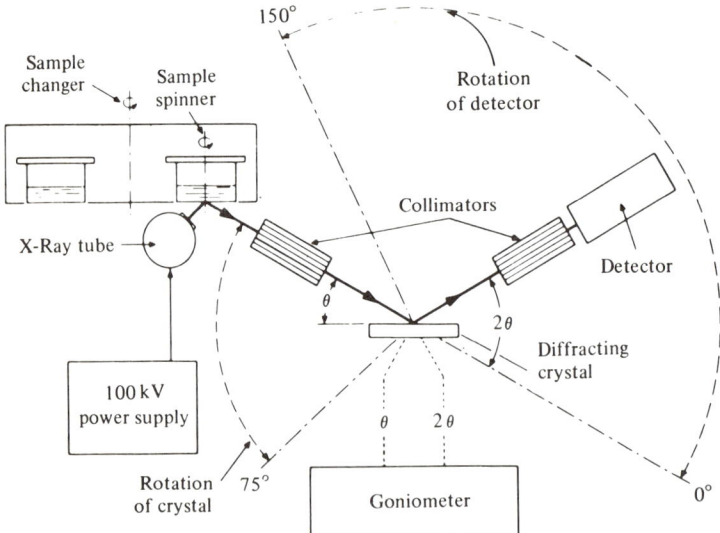

FIGURE 9-21 Geometry of a plane-crystal X-ray fluorescence spectrometer. (Courtesy of Philips Electronic Instruments.)

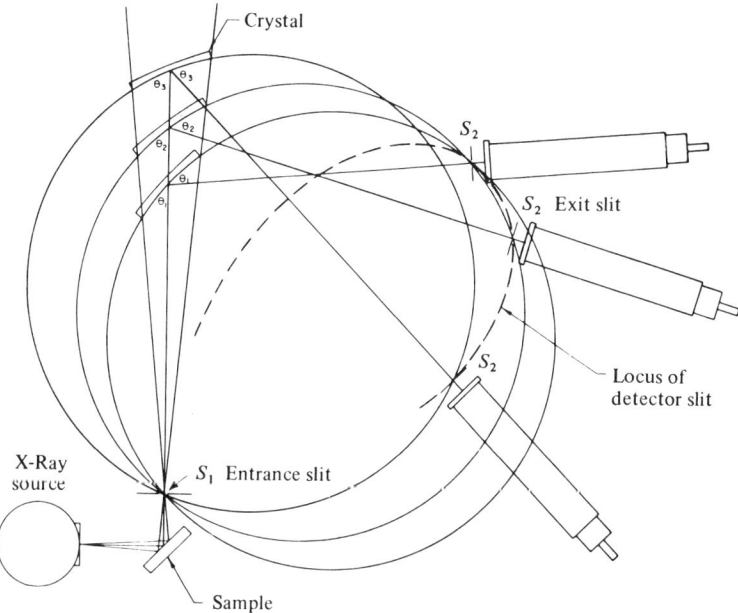

FIGURE 9-22 Focusing X-ray optics, employing a curved analyzing crystal. (Courtesy of Applied Research Laboratories, Inc.)

plate, 2.5 cm in width and 7.5 cm in length. The specimen holder is often an aluminum cylinder, although plastic material is used to examine acid or alkaline solutions. A thin film of Mylar supports the specimen, and an aluminum mask restricts the area irradiated (often a rectangle 18 mm by 27 mm). Intensity losses caused by the absorption of long-wavelength X rays by air and window materials can be reduced by evacuating the goniometer chamber. Another method for reducing losses is to enclose the radiation path in a special boot which extends from the sample surface to the detector window and then displace the air by helium, which has a low absorption coefficient. Vacuum spectrometers are used where helium is scarce and for elements boron ($Z = 5$) to sodium ($Z = 11$).

Focusing spectrometers that involve reflection from or transmission through a 10-cm or 28-cm curved crystal have been described.[5] Collimators are not required, and the increase in intensity obtained by focusing the fluorescence lines makes the technique suitable for the analysis of small specimens. In the curved-crystal arrangement (Fig. 9-22) the analyzing crystal is bent to a radius of curvature twice that of the focal circle, and then the inner surface is ground to the radius of curvature of the focal circle. A slit on the focusing circle acts as a divergent source of polychromatic radiation from the specimen. All of the radiation of one wavelength diverging from the slit will be diffracted at a particular setting of the crystal, and the diffracted radiation will converge to a line image at a symmetric point on the focusing circle. The angular velocity of the detector is twice that of the crystal and, as the two of them move along the periphery of the circle, the X-ray spectral lines are dispersed and detected just as in the flat-crystal arrangement.

In order to excite fluorescence the primary radiation must obviously have a wavelength shorter than the absorption edge of the spectral lines desired. Continuous as well as characteristic radiation of the primary target can serve the purpose. To get a continuous spectrum of short enough wavelength and of sufficient intensity, one may calculate the required voltage of the X-ray tube from Eq. 9-2, remembering that the wavelength of maximum intensity is approximately $1.5\lambda_0$. In qualitative analyses it is usually desirable to operate the X-ray tube at the highest permissible voltage in order to ensure that the largest possible number of elements in the specimen will be excited to fluoresce. It will also ensure the greatest possible intensity of fluorescence for each element in quantitative analyses. In two cases, however, the X-ray tube voltage should be made lower than the available maximum: (1) when it is desirable not to excite fluorescence of all elements in the specimen, but rather employ selective excitation conditions, and (2) when very long-wavelength spectral lines are to be excited—in order to minimize scattering of primary radiation through the system by holding down the intensity of the short-wavelength continuum. Besides X rays, electron bombardment such as used in the scanning electron microscope (SEM), the electron probe (EP), and the transmission electron microscope can be used to excite the characteristic fluorescent X rays of the elements present in a sample. For this reason many SEM, EP, and transmission electron microscopes are now available with optional crystal analyzers and detectors or Si(Li) detectors and pulse height analyzers to permit better X-ray identification of elements.

Certain radioisotopes are X-ray emitters and thus can be used as excitation sources for fluorescence. Since no high-voltage supply nor high-vacuum equipment is necessary, radioisotope sources are often used in portable equipment used for such applications as the monitoring of mine waters, stream pollution, and other field testing, especially for pollutants. X-ray emitting isotopes also give monochromatic X rays without the continuous background involved in ordinary X-ray tubes. On the other hand, the radiation cannot be turned off and constant, bulky shielding is required and the intensity is usually low. Some sources may have a rather short half-life and need frequent replacement and care must always be exercised in disposing of the old sources.

Bombardment of materials by ions such as protons can also lead to the emission of fluorescent X rays. Protons of several MeV energy from ion accelerators can have a flux density about two orders of magnitude greater than that for a standard X-ray tube. Protons do not penetrate deeply into matter and thus the use of protons or other ionized particles is particularly useful for very thin samples such as air particulate samples collected on a thin substrate. This method of analysis is sometimes known by the acronym PIXE (particle-induced X-ray emission).

Analytical Applications

For qualitative analysis, the angle θ between the surface of the crystal and the incident fluorescence beam is gradually increased; at certain well-defined angles the appropriate fluorescence lines are reflected. In automatic operation the intensity is recorded on a moving chart as a series of peaks, corresponding to fluorescence lines, above a background that arises principally from general scattering. The angular position of the detector, in degrees of 2θ, is also recorded on the chart. Additional evidence for identification may be obtained from relative peak heights, the critical excitation potential, and pulse height analysis.

For quantitative analysis, the intensity of a characteristic line of the element to be analyzed is measured. The goniometer is set at the 2θ angle of the peak, and counts are collected for a fixed period of time, or the time is measured for the period required to collect a specified number of counts. The goniometer is then set at a nearby portion of the spectrum where a scan has shown that only the background contributes. For major elements, 200,000 counts can be accumulated in 1 or 2 min. Background counts will require much longer time—a very low background may require 10 min to acquire 10,000 counts. The net line intensity, that is, peak minus background, in counts per second is then related to the concentration of the element via a calibration curve.

Particle size and shape are important and determine the degree to which the incident beam is absorbed or scattered. Standards and samples should be ground to the same mesh size, preferably finer than 200 mesh. Errors from differences in packing density can be handled by addition of an internal standard to the sample. Powders are pressed into a wafer in a metallurgical specimen press or converted into a solid solution by fusion with borax. Samples are best handled as liquids. If they can be conveniently dissolved, their analysis is greatly simplified and precision is greatly improved. Liquid samples should exceed a depth that will appear infinitely thick to the primary X-ray beam—about 5 mm for aqueous samples. The solvent should not contain heavy atoms; in this respect HNO_3 and water are superior to H_2SO_4 or HCl.

Before relating the intensity of fluorescent emission to concentration of emitting element, it is usually necessary to correct for matrix effects. Matrix dilution will avoid serious absorption effects. The samples are heavily diluted with a material having a low absorption, such as powdered starch, lithium carbonate, lampblack, gum arabic, or borax (used in fusions). The concentration, and therefore the effect, of the disturbing matrix elements is reduced, along with a reduction of the measured fluorescence. However, the most practical way to apply a systematic correction is by an internal standard. Even so, the internal standard technique is valid only if the matrix elements affect the reference line and analytical line in exactly the same way. The choice of a reference element depends on the relative positions of the characteristic lines and the absorption edges of the element to be determined, the reference element, and the disturbing elements responsible for the matrix effects. If either the reference line or the analytical line is selectively absorbed or enhanced by a matrix element, the internal standard line to analytical line ratio is not a true measure of the concentration of the element being determined. Preferential absorption of a line would occur if a disturbing element had an absorption edge between the comparison lines. The intensity of a line can be enhanced if a matrix element absorbs primary radiation and then, by fluorescence, emits radiation which, in turn, is absorbed by a sample element and causes the sample to fluoresce more strongly. Thus, if the matrix fluorescence lies between the absorption edges of the analytical and internal standard elements, selective enhancement might result.

The X-ray fluorescence method, inherently very precise, rivals the accuracy of wet chemical techniques in the analysis of major constituents. On the other hand, it is difficult to detect an element present in less than one part in 10,000. The method is attractive for elements that lack reliable wet chemical methods; for example, elements such as niobium, tantalum, sodium, and the rare earths. It often serves as a complementary procedure to optical emission spectrography, particularly for major constituents, and also for the analysis of nonmetallic specimens, because the sample need not be an electrical conductor. To overcome air absorption for elements of atomic number below 21, operating pressure must be 0.1 torr. Even so, below magnesium the transmission becomes seriously attenuated although the method has been extended to boron. The ultimate limit of X-ray fluorescence (XRF) in absolute terms is about 10^{-8} g while that of the PIXE method is around 10^{-12} g.

Simultaneous analysis of several elements is possible with automatic equipment, such as the Applied Research Laboratory Quantometer. Instruments of this type have semifixed monochromators with optics mounted around a centrally located X-ray tube and sample position. Each crystal is adjusted to reflect one fluorescence line to its associated detector. A compatible recording unit permits both optical and X-ray units to be recorded with the same console.

Energy Dispersion Spectrometers

For some samples where only very few elements are present and their X-ray lines are widely separated in wavelength, the crystal analyzer may be eliminated and pulse height discrimination employed in its place. For compositions greater than about 1%, and elements separated by a few atomic numbers, energy dispersion analysis is very useful because the intensities are increased about 1000-fold. With such an increase in radiation intensity emitted by the sample due to the larger acceptance angle of the detector, weaker primary sources can be used, for example, radioisotopes. The resolution, however, of an energy dispersion instrument is as much as 50 times less than the wavelength dispersion spectrometer using a crystal; thus overlapping of lines from nearby elements may occur.

9.8 X-RAY DIFFRACTION

Every atom in a crystal scatters an X-ray beam incident upon it in all directions. Because even the smallest crystal contains a very large number of atoms, the chance that these scattered waves would constructively interfere would be almost zero except for the fact that the atoms in crystals are arranged in a regular, repetitive manner. The condition for diffraction of a beam of X rays from a crystal is given by the Bragg equation, Eq. 9-3. Atoms located exactly on the crystal planes contribute maximally to the intensity of the diffracted beam; atoms exactly halfway between the planes exert maximum destructive interference and those at some intermediate location interfere constructively or destructively depending on their exact location but with less than their maximum effect. Furthermore, the scattering power of an atom for X rays depends upon the number of electrons it possesses. Thus the position of the diffraction beams from a crystal depends only upon the size and shape of the repetitive unit of a crystal and the wavelength of the incident

X-ray beam, whereas the intensities of the diffracted beams depend also upon the type of atoms in the crystal and the location of the atoms in the fundamental repetitive unit, the unit cell. No two substances, therefore, have absolutely identical diffraction patterns when one considers both the direction and intensity of all diffracted beams; however, some similar, complex organic compounds may have almost identical patterns. The diffraction pattern is thus a "fingerprint" of a crystalline compound and the crystalline components of a mixture can be identified individually.

Reciprocal Lattice Concept

Diffraction phenomena can be interpreted most conveniently with the aid of the reciprocal lattice concept. A plane can be represented by a line drawn normal to the plane; the spatial orientation of this line describes the orientation of the plane. Furthermore, the length of the line can be fixed in an inverse proportion to the interplanar spacing of the plane that it represents.

When a normal is drawn to each plane in a crystal and the normals are drawn from a common origin, the terminal points of these normals constitute a lattice array. This is called the *reciprocal lattice* because the distance of each point from the origin is reciprocal to the interplanar spacing of the planes that it represents. Figure 9-23 shows, near the origin, the traces of several planes in a unit cell of a crystal, namely, the (100), (001), (101),

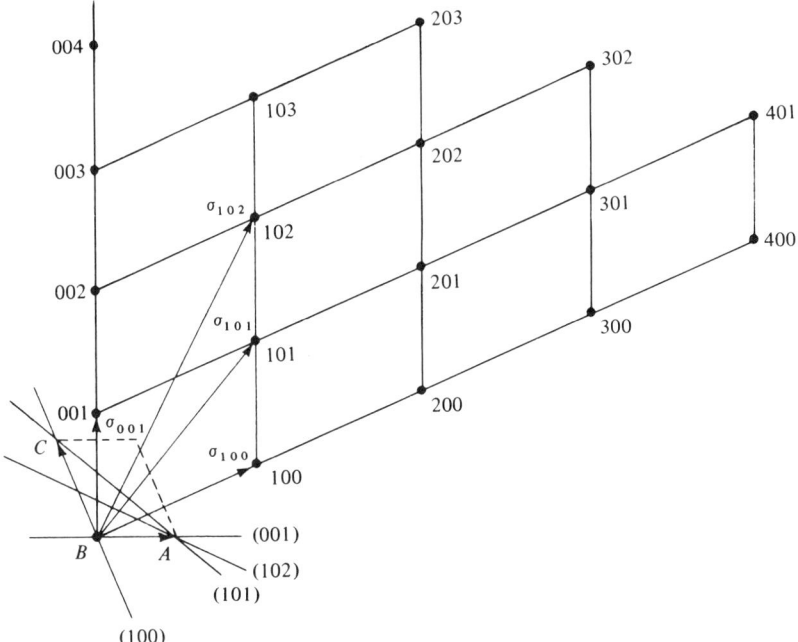

FIGURE 9-23 Side view of several planes in the unit cell of a crystal with the normals to the planes indicated.

and (102) planes. The normals to these planes, also indicated, are called the reciprocal lattice vectors σ_{hkl} and are defined by

$$\sigma_{hkl} = \frac{\lambda}{d_{hkl}}$$

In three dimensions, the lattice array is described by three reciprocal lattice vectors whose magnitudes are given by

$$a^* = \sigma_{100} = \frac{\lambda}{d_{100}}$$

$$b^* = \sigma_{010} = \frac{\lambda}{d_{010}}$$

$$c^* = \sigma_{001} = \frac{\lambda}{d_{001}}$$

and whose directions are defined by three interaxial angles $\alpha^*, \beta^*, \gamma^*$.

Writing the Bragg equation in a form that relates the glancing angle θ most clearly to the other parameters, we have

$$\sin \theta_{hkl} = \frac{\lambda/d_{hkl}}{2} \tag{9-11}$$

The numerator can be taken as one side of a right triangle with θ as another angle and the denominator as its hypotenuse (Fig. 9-24a). Because of the physical meaning of the quantities in Eq. 9-11, the construction can be interpreted as shown in Fig. 9-24b. The diameter of the circle $(\overline{ASO})$ represents the direction of the incident X-ray beam. A line through the origin of the circle, parallel to $\overline{AP}$ and forming the angle θ with the incident beam, represents a crystallographic plane that satisfies the Bragg diffraction condition. The line $\overline{SP}$, also forming the angle θ with the crystal plane and 2θ with the incident beam, represents the diffracted beam's direction. Then the line $\overline{OP}$ is the reciprocal lattice vector to the reciprocal lattice point P_{hkl} lying on the circumference of the circle. The vector σ_{hkl} originates at the point on the circle where the direct beam leaves the circle. The Bragg

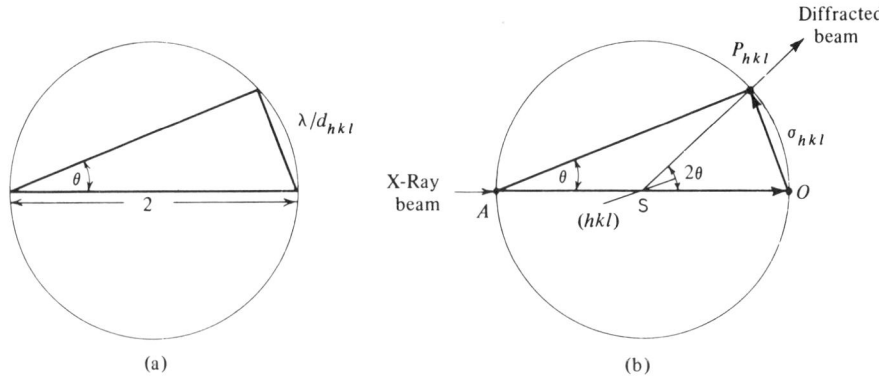

FIGURE 9-24 Representation of the diffraction condition.

X-RAY METHODS   273

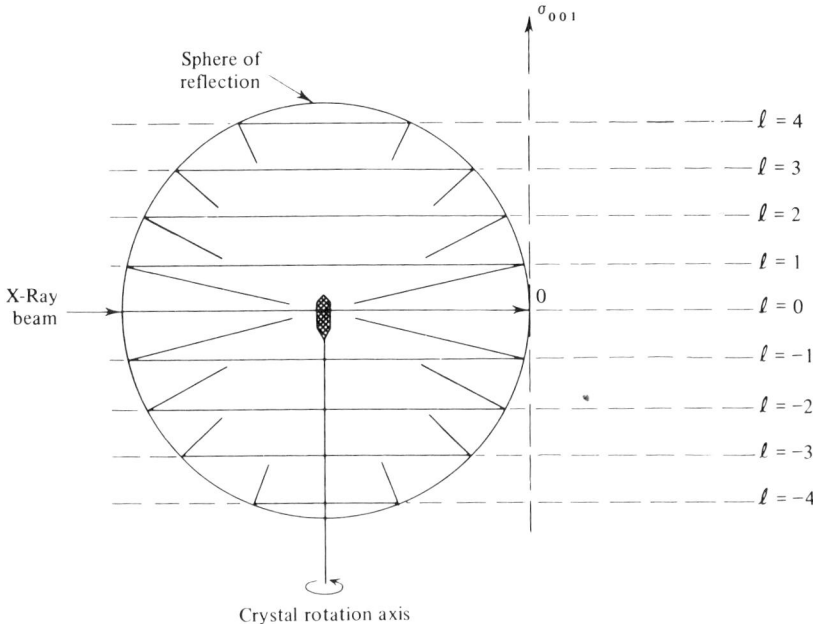

FIGURE 9-25 Reciprocal lattice construction for a rotating crystal.

equation is satisfied when and only when a reciprocal lattice point lies on the "sphere of reflection," a sphere formed by rotating the circle upon its diameter $\overline{ASO}$.

Thus, the crystal in a diffraction experiment can be pictured at the center of a sphere of unit radius, and the reciprocal lattice of this crystal is centered at the point where the direct beam leaves the sphere, as shown in Fig. 9-25. Because the orientation of the reciprocal lattice bears a fixed relation to that of the crystal, if the crystal is rotated, the reciprocal lattice can be pictured as rotating also. Whenever a reciprocal lattice point intersects the sphere, a reflection emanates from the crystal at the sphere's center and passes through the intersecting reciprocal lattice point.

Diffraction Patterns

If the X-ray beam is monochromatic, there will be only a limited number of angles at which diffraction of the beam can occur. The actual angles are determined by the wavelength of the X rays and the spacing between the various planes of the crystal. In the *rotating crystal method*, monochromatic X radiation is incident on a single crystal which is rotated about one of its axes. The reflected beams lie as spots on the surface of cones which are coaxial with the rotation axis. If, for example, a single cubic crystal is rotated about the (001) axis, which is the equivalent to rotation about the c^* axis, the sphere of reflection and the reciprocal lattice are as shown in Fig. 9-25. The diffracted beam directions are determined by intersection of the reciprocal lattice points with the sphere of reflection. All the reciprocal lattice points lying in any one layer of the reciprocal lattice layer perpendicular to the axis of rotation will intersect the sphere of reflection in a circle.

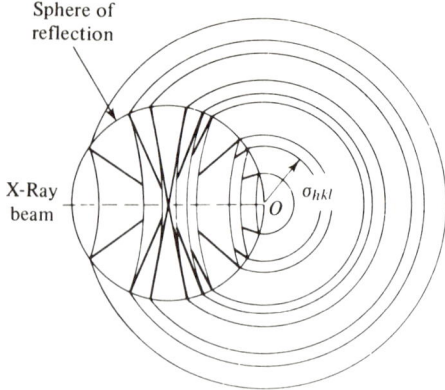

FIGURE 9-26 Origin of powder diffraction diagrams in terms of the series of concentric spheres generated from the reciprocal lattice points about the origin, O, of the reciprocal lattice and their intersection with the sphere of reflection.

The height of the circle above the equatorial plane is proportional to the vertical reciprocal lattice spacing c^*. By remounting the crystal successively about different axes, one can determine the complete distribution of reciprocal lattice points. Of course, one mounting is sufficient if the crystal is cubic, but two or more may be needed if the crystal has lower symmetry.

In a modification of the single-crystal method, known as the *Weissenberg method*, the photographic film is moved continuously during the exposure parallel to the axis of rotation of the crystal. All reflections are blocked out except those which occur in a single layer line. This results in a film that is somewhat easier to decipher than a simple rotation photograph. Still other techniques are used; one, the *precession method*, results in a photograph which gives an undistorted view of a plane in the reciprocal lattice of the crystal.

In the *powder method*, the crystal is replaced by a large collection of very small crystals, randomly oriented, and a continuous cone of diffracted rays is produced. There are some important differences, however, with respect to the rotating-crystal method. The cones obtained with a single crystal are not continuous because the diffracted beams occur only at certain points along the cone, whereas the cones with the powder method are continuous. Furthermore, although the cones obtained with rotating single crystals are uniformly spaced about the zero level, the cones produced in the powder method are determined by the spacings of prominent planes and are not uniformly spaced. The origin of a powder diagram is shown in Fig. 9-26. Because of the random orientation of the crystallites, the reciprocal lattice points generate a sphere of radius σ_{hkl} about the origin of the reciprocal lattice. A number of these spheres intersect the sphere of reflection.

Camera Design

Typical cameras for X-ray powder diffraction work are shown in Fig. 9-27. Cameras are usually constructed so that the film diameter has one of the three values 57.3, 114.6,

X-RAY METHODS   275

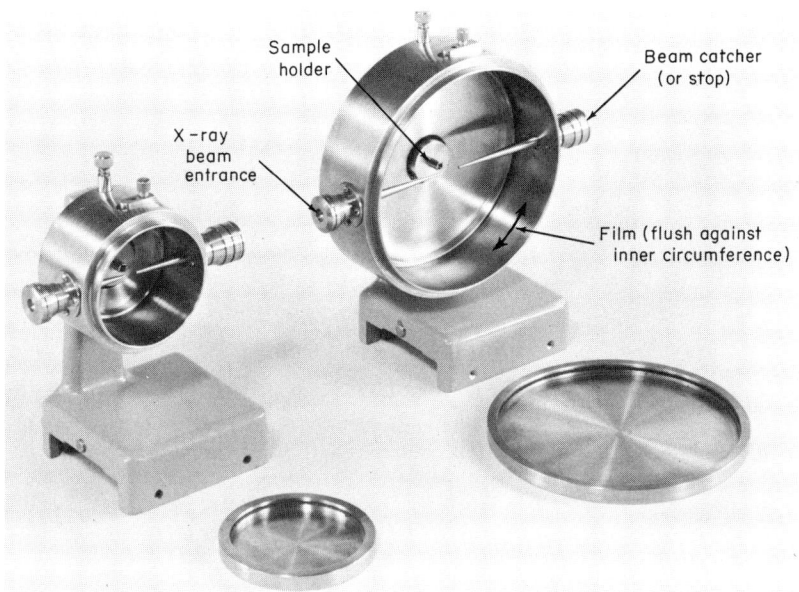

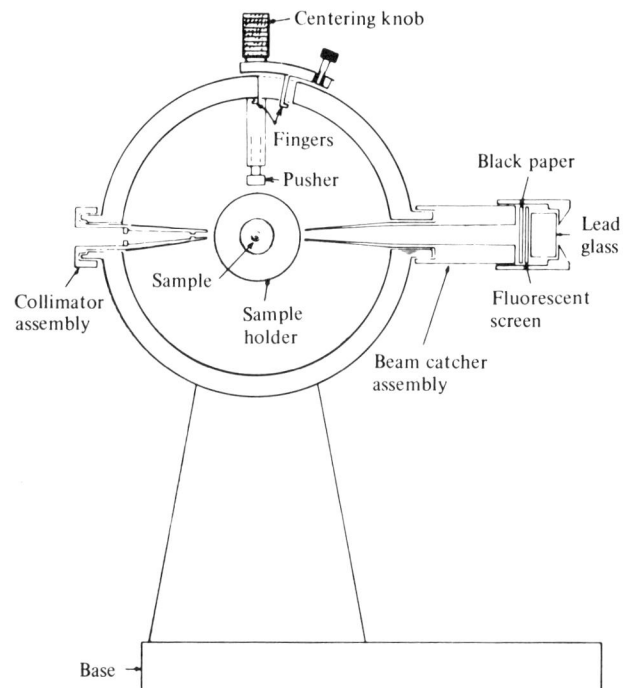

FIGURE 9-27 X-Ray powder diffraction cameras, 57.3-mm and 114.6-mm diameter, and the schematic of interior. (Courtesy of Philips Electronic Instruments.)

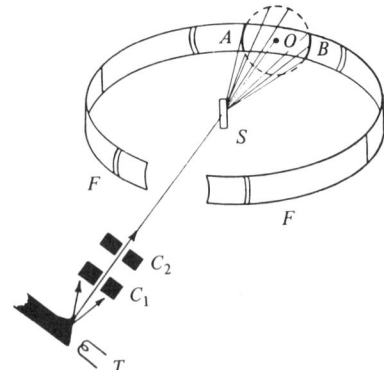

FIGURE 9-28 Schematic of powder diffraction patterns. T, X-ray tube; C_1 and C_2 collimating slits; S, powdered crystalline sample; A, a line on film, left portion, and B, right portion; O, intersection of undiffracted beam with film. Angle $ASO = 2\theta$.

or 143.2 mm. The reason for these values can be understood by considering the calculations involved. If the distance between corresponding arcs of the same cone of diffracted rays, for example, the distance between points A and B of Fig. 9-28, is measured and called S, then

$$4\theta_{\text{rad}} = \frac{S}{R} \qquad (9\text{-}12)$$

where θ_{rad} is the Bragg angle measured in radians, and R is the radius of the film in the camera. The angle, θ_{deg}, measured in degrees, is then

$$\theta_{\text{deg}} = \frac{57.295 S}{4R} \qquad (9\text{-}13)$$

where 57.295 equals the value of a radian in degrees. Therefore, when the camera diameter ($2R$) is equal to 57.3 mm, $2\theta_{\text{deg}}$ may be found by measuring S in millimeters. When the diameter is 114.59 mm, $2\theta_{\text{deg}} = S/2$, and when the diameter is 143.2 mm, $\theta_{\text{deg}} = 2(S/10)$.

Once angle θ has been calculated, Eq. 9-3 can be used to find the interplanar spacing, using values of wavelength λ from Table 9-1. Sets of tables[6] are available that give the interplanar spacing for the angle 2θ for the types of radiation most commonly used.

Cameras of larger diameter make it easier to measure the separation of lines provided that the lines are sharp. The sharpness of the lines depends to a large extent upon the quality of the collimating slits and the size of the sample. The slits should produce a fine beam of X rays with as small a divergence as possible. The sample size should be small so that it will act as a small source of the diffracted beam. On the other hand, smaller samples, finer pencils of incident X rays, and the cameras of larger diameter all tend to require longer exposure times, so that in practice a compromise must be made.

For very precise measurements of interplanar spacings, the diameter of the film and the separation of the lines must be very accurately known. Several methods of measurement have been proposed. The effective camera diameter can be determined by calibration with a

FIGURE 9-29   X-Ray powder diffraction photograph of sodium chloride. Film mounted in Straumanis method.

material such as sodium chloride whose interplanar spacings are accurately known. In another method, the Straumanis method,[7] the film is inserted in the camera so that the ends of the film are at about 90° from the point of emergence of the beam from the camera. The developed film appears as in Fig. 9-29. If a_1, a_2 and b_1, b_2 represent the two sides of arcs on the left and right sides of the film, then the two averages determine the positions of the entering and emerging beams.

Therefore,

$$\frac{b_1 + b_2}{2} - \frac{a_1 + a_2}{2} = \text{distance corresponding to } 180° \qquad (9\text{-}14)$$

or

$$360° = b_1 + b_2 - a_1 - a_2 \qquad (9\text{-}15)$$

and the angle 4θ associated with any pair of lines can then be calculated:

$$\frac{a_2 - a_1}{b_1 + b_2 - a_1 - a_2} = \frac{4\theta}{360} \qquad (9\text{-}16)$$

Choice of X Radiation

Two factors control the choice of X radiation, as can be seen by rearranging the terms of the Bragg equation

$$\theta = \sin^{-1}\left(\frac{\lambda}{2d}\right) \qquad (9\text{-}17)$$

Because the ratio in the parentheses cannot exceed unity, the use of long-wavelength radiation limits the number of reflections that can be observed. Conversely, when the unit cell is very large, short-wavelength radiation tends to crowd individual reflections very closely together.

The choice of radiation is also affected by the absorption characteristics of a sample. Radiation having a wavelength just shorter than the absorption edge of an element contained in the sample should be avoided, because then the element absorbs the radiation strongly. The absorbed energy is emitted as fluorescent radiation in all directions and increases the background (which would result in darkening on a film, making it more difficult to see the diffraction maxima sought). It is obvious, then, why one commercial source provides radiation sources from a multiwindow tube with anodes of silver, molybdenum, tungsten, or copper.

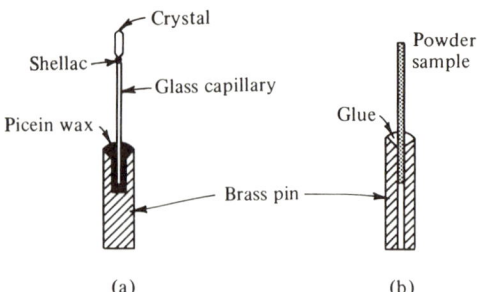

FIGURE 9-30 Specimen mounts for X-ray diffraction. (a) Single crystal and (b) powdered sample.

Specimen Preparation

Single crystals are used for structure determinations whenever possible because of the relatively large number of reflections obtained from single crystals and the greater ease of their interpretation. A crystal should be of such size that it is completely bathed by the incident beam. Generally, a crystal is affixed to a thin glass capillary which, in turn, is fastened to a brass pin, as shown in Fig. 9-30a.

When single crystals of sufficient size are not available or when the problem is merely the identification of a material, a polycrystalline aggregate is formed into a cylinder whose diameter is smaller than the diameter of the incident X-ray beam. Metal samples are machined to a desirable shape, plastic materials can often be extruded through suitable dies, and all other samples are best ground to a fine powder (200–300 mesh) and shaped into thin rods after mixture with a binder (usually collodion). The mount is shown in Fig. 9-30b.

Although liquids cannot be identified directly, it is frequently possible to convert them to crystalline derivatives that have characteristic patterns. Many of the classical derivatives can be used: identification of aldehydes and ketones as 2,4-dinitrophenylhydrazones, fatty acids as p-bromoanilides, and amines as picrate derivatives.

In order to obtain diffraction patterns from large, dense samples, a back-reflection camera can be used. The geometry of the camera is shown in Fig. 9-31. The X rays pass through an opening in the center of the film and impinge on the sample. Beams diffracted over a range of Bragg angles extending from 59° to 88° are registered on the circular film.

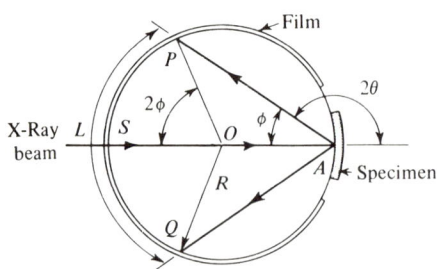

FIGURE 9-31 Geometry of the back-reflection symmetrical focusing camera.

Automatic Diffractometers

Results are achieved rapidly and with much better precision when automatic diffractometers are used to record diffraction data. A diffractometer to record data from powdered samples is built much like the one shown in Fig. 9-21. The X-ray tube furnishes the radiation directly (filters are generally used to get more nearly monochromatic radiation). The diffracting crystal is replaced by the powdered or metallic sample. To increase the randomness of orientation of the crystallites, the sample may be rotated in its own plane; that is, the plane perpendicular to the bisector of the angle between the source and detector beams. Note also that as the sample is rotated in the other plane to sweep through various θ angles, the detector must be rotated twice as rapidly to maintain the angle 2θ with the irradiating beam.

Proportional, scintillation, or semiconductor detectors, with their associated circuitry, are far superior to photographic film in regard to the number of reflections per day that can be recorded. They can achieve a precision of 1% or better. Even with the best darkroom and photometric procedures the relative degree of blackness of each spot or cone on the film cannot be estimated with an accuracy of much more than 10%, and often the error in estimation is greater than this. The principal advantage of photographic film over counters is that it provides a means of recording many reflections at one time.

Automatic single-crystal diffractometers are quite complex. A cradle assembly provides a wide angular range for orienting and aligning the crystal under study. A precision-diffractometer assembly allows the detector to traverse a spherical surface from longitude −5° to 150° and from latitude −6° to 60° on the Norelco instrument. The complete unit provides four rotational degrees of freedom for the crystal and two for the detector. Various crystal and counter angles are set on the basis of programmed information and the resulting diffraction intensity is measured as a function of angle. Single-crystal diffractometers controlled by computers which set the angles for the crystal and detector and record the data are widely used for gathering the information for crystal structure determinations.

X-Ray Powder Data File

If only the identification of a powder sample is desired, its diffraction pattern is compared with diagrams of known substances until a match is obtained. This method requires that a library of standard films be available. Alternatively, d values calculated from the diffraction diagram of the unknown substance are compared with the d values of over 25,000 entries, which are listed on plain cards, Keysort cards, and IBM cards in the X-ray powder data file.[8] An index volume is available with the file. The cataloging scheme[9] used to classify different cards lists the three most intense reflections in the upper left corner of each card. The cards are then arranged in sequence of decreasing d values of the most intense reflections, based on 100 for the most intense reflection observed. A typical card is shown in Fig. 9-32.

To use the file to identify a sample containing one component, the d value for the darkest line of the unknown is looked up first in the index. Since more than one listing containing the first d value probably exists, the d values of the next two darkest lines are then matched against the values listed. Finally, the various cards involved are compared.

5-0628

d	2.82	1.99	1.63	3.258	NaCl					★
I/I_1	100	55	15	13	SODIUM CHLORIDE		HALITE			

Rad. Cu λ 1.5405 Filter Dia.	dÅ	I/I_1	hkl	dÅ	I/I_1	hkl
Cut off I/I_1	3.258	13	111			
Ref. Swanson and Fuyat, NBS Circular 539, Vol. II, 41 (1953)	2.821	100	200			
	1.994	55	220			
Sys. Cubic S.G. O_H^5 – Fm3m	1.701	2	311			
a_0 5.6402 b_0 c_0 A C	1.628	15	222			
α β γ Z 4 Dx 2.164	1.410	6	400			
Ref. Ibid.	1.294	1	331			
	1.261	11	420			
	1.1515	7	422			
εa n $\omega\beta$ 1.542 $\varepsilon\gamma$ Sign	1.0855	1	511			
2V D mp Color	0.9969	2	440			
Ref. Ibid.	.9533	1	531			
	.9401	3	600			
	.8917	4	620			
An ACS reagent grade sample recrystallized twice from	.8601	1	533			
hydrochloric acid.	.8503	3	622			
X-ray pattern at 26°C.	.8141	2	444			
Replaces 1-0993, 1-0994, 2-0818						

FIGURE 9-32 X-ray data card for sodium chloride. (Courtesy of American Society for Testing Materials.)

A correct match requires that all the lines on the card and film agree. It is also good practice to derive the unit cell from the observed interplanar spacings and to compare it with that listed in the card.

If the unknown contains a mixture, each component must be identified individually. This is done by treating the list of d values as if they belonged to a single component. After a suitable match for one component is obtained, all the lines of the identified component are omitted from further consideration. The intensities of the remaining lines are rescaled by setting the strongest intensity equal to 100 and repeating the entire procedure.

Reexamination of the cards in the file is a continuing process in order to eliminate errors and remove deficiencies. Replacement cards for substances bear a star in the upper right corner.

X-Ray diffraction furnishes a rapid, accurate method for the identification of the crystalline phases present in a material. Sometimes it is the only method available for determining which of the possible polymorphic forms of a substance are present—for example, carbon in graphite or in diamond. Differentiation among various oxides—such as FeO, Fe_2O_3, and Fe_3O_4, or between materials present in such mixtures as KBr + NaCl, KCl + NaBr, or all four—is easily accomplished with X-ray diffraction, whereas chemical analysis would show only the ions present and not the actual state of combination. The presence of various hydrates is another possibility.

Quantitative Analysis

X-Ray diffraction is adaptable to quantitative applications because the intensities of the diffraction peaks of a given compound in a mixture are proportional to the fraction of the material in the mixture. However, direct comparison of the intensity of a diffraction peak in the pattern obtained from a mixture is fraught with difficulties. Corrections are frequently necessary for the differences in absorption coefficients between the compound being determined and the matrix. Preferred orientations must be avoided. Internal standards help but do not overcome the difficulties entirely.

Structural Applications

A discussion of the complete structural determination for a crystalline substance is beyond the scope of this text. It will suffice to point out that, with careful work, atoms can be located to a precision of hundredths of an angstrom or better.

In polymer chemistry a great deal of information can be obtained from an X-ray diffraction diagram. Fibers and partially oriented samples will show spotty diffraction patterns rather than uniform cones; the more oriented the specimen, the spottier the pattern. Figure 9-33 shows the fiber diagram of polyethylene. The center row of spots in the pattern is called the equator, and the horizontal rows parallel to the equator are called the layer lines. The equatorial spots arise by diffraction from lattice planes that are parallel to the fiber axis. The layer line spots arise by diffraction from planes that intersect the fiber axis. The repeat distance along the polymer chain can be calculated from the distances of the layer lines from the equator and their separation from one another. In the simplest cases the repeat distance will correspond to that of a fully extended chain of the known chemical composition.

Crystal Topography

There are a number of experimental diffraction techniques, developed in recent years, by which the microscopical defects in a crystal can be shown. Most crystals are far from perfect crystals and exhibit regions (grains) with somewhat differing orientations, or they may contain individual defects such as dislocations or faults distributed throughout the

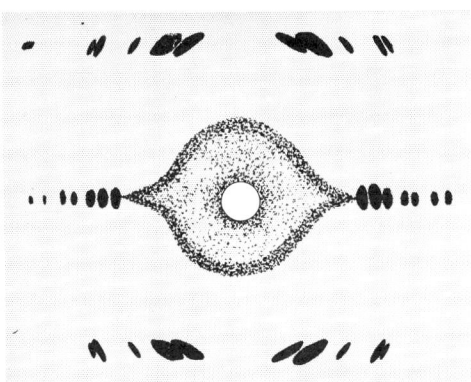

FIGURE 9-33   Fiber diagram of polyethylene. (From A. Ryland, *J. Chem. Educ.*, 35, 76 (1958). Reproduced by permission.)

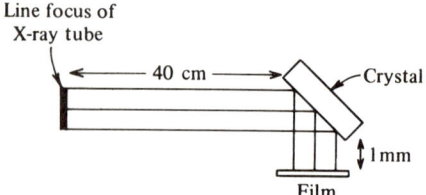

FIGURE 9-34 Experimental arrangement for the Berg–Barrett method of X-ray diffraction topography.

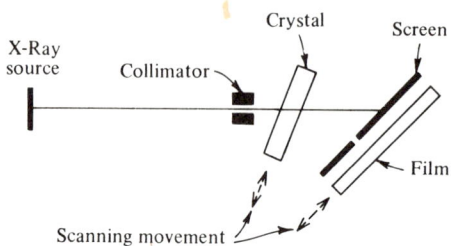

FIGURE 9-35 Experimental arrangement for the Lang method of X-ray diffraction topography.

crystal. Studies of these defects are important in understanding the nature of stress in metals, the nature and behavior of "doped" crystals used in transistors, the production of "perfect" crystals, and other phenonena.

Microradiographic methods are based on absorption and the contrast in the images is due to differences in absorption coefficients from point to point. X-ray diffraction topography depends for image contrast upon point-to-point changes in the direction or the intensity of beams diffracted by planes in the crystal.

One much used method of X-ray diffraction topography is known as the Berg–Barrett method. The experimental arrangement is shown in Fig. 9-34. The crystal is set so as to reflect the X rays at the Bragg angle for some plane. Geometric resolutions of about 1 μm can be achieved and single dislocations can be resolved. The contrast on the film is due to variations of the reflecting power due to imperfections in the crystal.

Another method for X-ray diffraction topography is known as the Lang method. The experimental setup is shown in Fig. 9-35. A ribbon X-ray beam is collimated to such a small angular divergence that only one characteristic wavelength is diffracted by the crystal. Simultaneous movement of the crystal and film allow a large area of the crystal to be investigated.

PROBLEMS

1. What is the short-wavelength limit for a 100-kV X-ray tube? What is the atomic number of the element for which just insufficient energy is available for excitation?

2. Write the transition relations for each of these lines: (a) $K\alpha_2$, (b) $K\beta_1$, and (c) $L\alpha_1$.

3. Calculate the critical excitation potentials for the K and L series of these elements:

ELEMENT	K ABSORPTION EDGE, Å	L_{III} EDGE, Å
Al	7.951	170
Cr	2.070	20.7
Zr	0.688	5.58
Nd	0.285	1.995
W	0.178	1.215
U	0.107	0.722

4. Calculate the wavelengths of the $K\alpha_1$ lines for the elements in Problem 3.

5. For what elements will Mo $K\alpha_1$ prove sufficiently energetic to excite their L_{III} spectra? What is the limit for K spectra?

6. The following measurements were obtained with a proportional flow counter:

APPLIED VOLTAGE, V	OBSERVED COUNT RATE, COUNTS/MIN	APPLIED VOLTAGE, V	OBSERVED COUNT RATE, COUNTS/MIN
1200	225,000	1600	231,000
1300	231,700	1700	231,500
1400	232,400	1800	233,000
1500	231,400	1900	271,000

Plot the results on graph paper and select an operating voltage.

7. Using a gamma emitter, these measurements were obtained with an ionization chamber:

APPLIED VOLTAGE, V	OBSERVED COUNT RATE, COUNTS/MIN	APPLIED VOLTAGE, V	OBSERVED COUNT RATE, COUNTS/MIN
1200	19,400	1470	41,000
1245	29,100	1500	40,900
1290	35,700	1545	41,100
1335	38,900	1590	41,500
1380	40,500	1635	42,200
1425	40,600	1680	45,100
		1725	48,700

Plot the results on graph paper and select an operating voltage.

8. What causes the discontinuity in the efficiency curve of the NaI scintillation counter and of the argon-filled flow proportional counter (Fig. 9-10)?

9. Compute the goniometer setting (2θ) for the $K\alpha_1$ lines of Al, S, Ca, Cr, Mn, Co, Br, Sr, Ag, Mo, and W, when the analyzing crystal is (a) LiF, (b) CaF_2, or (c) EDDT.

10. Discuss four ways that might be employed for the separation of interfering spectral lines.

11. For the determination of uranium in aluminum by measurement of $U L\alpha_1$ fluorescence, these counting relationships were obtained:

U (wt%)	COUNT RATE, cps
2	436
5	835
10	1262
15	1533
20	1720

Determine the slope of the calibration curve (cps/% U) over each interval of uranium concentration and the counting time (in minutes) required to achieve a 1% precision at the 95% confidence level.

12. Sulfur (0.4–6.0%) has been determined in carbon materials by X-ray fluorescence using the S $K\alpha$ line (5.36 Å) which under a particular set of operating conditions gave 1 cps equivalent to 0.014% S. Background radiation is equivalent to 0.05% S. Select a proper analyzing crystal, the goniometer setting, the excitation conditions (X-ray tube voltage), and counting times to achieve results whose deviation does not exceed 5% at the 95% confidence level.

13. What is the reflection angle (2θ) for Cu $K\alpha_1$ radiation from each analyzing crystal in Table 9-4?

14. The pulse amplitude distributions for these elements: Mg, Al, Si, P, S, and Ca, show a peak of pulse distribution at 11.0, 13.0, 15.0, 17.4, 19.0, and 31.8 V, respectively. For each the width at one-half peak height is 2.5 V, and the base width is 9.5 V. What base line and window settings (in volts) would be employed in these situations?

(a) The determination of Mg in the presence of P.
(b) The determination of S in the presence of Mg.
(c) The determination of P in the presence of Al.
(d) The separation of calcium from all the others.
(e) The total Al plus Si in a sample, all other elements absent.

15. Suggest methods for handling each pair of overlapping X-ray spectral lines whose wavelength in angstroms is enclosed in brackets:

(a) Mn $K\alpha_1$ [2.103] – Cr $K\beta$ [2.085]
(b) Zn $K\alpha_1$ [1.435] – Re $L\alpha_1$ [1.433]
(c) Nb $K\alpha_1$ [0.746] – W $L\alpha_1$ [1.476]

16. Graphically represent the following disturbing effects in the use of an internal standard element *(S)* for the determination of element *(E)*, the disturbing element being *(D)*. Plot all absorption edges and emission lines. (a) Selective absorption of "S," (b) selective absorption of "E," (c) enhancement of "S," and (d) enhancement of "E."

17. Strontium has been determined in sediments of oil-bearing formations using yttrium as an internal standard. The spectral characteristics of these elements are:

$$\text{Sr } K\alpha_1, 0.877; \quad K\beta, 0.783; \quad K \text{ edge}, 0.770$$
$$\text{Y } K\alpha_1, 0.831; \quad K\beta, 0.740; \quad K \text{ edge}, 0.727$$

(a) Compute the critical voltage for each element. (b) Using a LiF crystal, compute the Bragg angle (2θ) for the emission lines. (c) Calibration data obtained are tabulated:

Sr, WT. %	MEASUREMENT OF Y $K\alpha_1$; TIME (SEC) FOR 6400 COUNTS	MEASUREMENT OF Sr $K\alpha_1$; TIME (SEC) FOR 6400 COUNTS
0.0000	41.1	80.1
0.1000	40.1	60.4
0.2000	40.2	49.5
0.3000	40.0	41.6
0.4000	42.4	38.3

Plot intensity ratio versus concentration of strontium. Unknown samples gave these intensity ratios—Sr:Y—*A*, 0.8860; *B*, 0.7802; *C*, 0.6011.

18. What is the difference in wavelength (and Bragg angle) between Hf $L\alpha_1$ and the second order of Zr $K\alpha_1$, 1.566 Å and 0.784 Å (in first order, respectively), with a LiF analyzing crystal? Suggest a method for eliminating the second-order Zr line.

19. Oil paintings have been tested for authenticity by examining the pigment composition with the electron probe and a hypodermic needle core. With an ammonium dihydrogen orthophosphate crystal (101 plane), $d = 10.62$, the following characteristic X-ray spectra (and relative intensity) were obtained:

TOP WHITE LAYER, 2θ	BOTTOM WHITE LAYER, 2θ
27.6° (8)	34.0° (12)
30.0° (60)	36.0° (100)
58.0° (4)	71.0° (2)
63.0° (18)	

Was the painting produced before or after the time (~1900 A.D.) when titanium-white pigments were available?

20. From the data given in Table 9-3, estimate the mass absorption coefficients (a) of Zr for the Mo $K\alpha_1$ radiation, (b) of Ni for the Cu $K\alpha_1$ radiation, and (c) of Cu for the W $L\alpha_1$ radiation.

21. Calculate the reduction in intensity of an X-ray beam from the Mo $K\alpha_1$ line (0.709 Å) resulting from 1 ml of 1% TEL liquid in *n*-octane. The aviation mixture consists of 61.5%

$Pb(C_2H_5)_4$ and 38.5% $C_2H_4Br_2$ per milliliter of TEL. The mass absorption coefficients are C, 0.64 cm^2/g; H, 0.38 cm^2/g; Pb, 140 cm^2/g; and Br, 79 cm^2/g. Density = 0.72 g cm^{-3}. Do the calculations for (a) an *n*-octane blank and (b) versus air in the reference path.

22. Repeat Problem 21 using the Cu $K\alpha_1$ line (1.542 Å). The mass absorption coefficients are C, 4.6 cm^2/g; H, 0.43 cm^2/g; Pb, 241 cm^2/g; and Br, 88 cm^2/g. Note the increase in absorption by lead, but also by the carbon atoms.

23. Identify the emission lines and, from these, the base wire and plate metal in each sample. Spectrometer employed a LiF crystal and tungsten target operated at 50 kV.

SAMPLE	PLATE METAL, 2θ	BASE WIRE, 2θ
1	48.64°	41.30° (2nd)
2	99.87° (2nd)	63.88° (3rd)
3	31.19°	110.92° (2nd)
4	16.75°	110.92° (2nd)

24. A sample ground and pelleted with lithium carbonate and starch gave these emission lines (2θ) using a LiF crystal. Identify each line.

$$111.0°, 100.2°, 57.6°, 48.4°, 45.1°, 44.0°, 40.4°$$

25. Suggest an X-ray method for each of these determinations: (a) the thickness of electroplated metal films, such as successive layers of Cu, Ni, and Cr on steel (chrome plate), (b) the thickness of SrO and BaO on evaporated electrode coatings, and (c) the concentration of fillers and impregnants, such as BaF_2 in carbon brushes.

26. Iron-55 decays by *K*-electron capture to stable ^{55}Mn with the attendant emission of *K*-line X rays of Mn. The half-life of ^{55}Fe is 2.93 years. For which elements would this isotope be a convenient source of X radiation for absorption analysis?

27. An unknown material was placed in the sample holder of an X-ray fluorescence unit which used a tungsten target tube operated at 60 kV to furnish the exciting radiation. A mica crystal was used in the analyzer. The lattice spacing of mica is 9.984 Å. Reflections were observed at angles (2θ) of 9°34′, 12°8′, 19°12′, 24°24′, and 38°58′. Calculate the wavelength of the fluorescent lines and identify the elements present.

28. An unknown powder was placed in a sample tube, and the X-ray diffraction pattern was observed in a camera of radius 57.3 mm. The X-ray unit was fitted with a nickel target tube with a cobalt filter. The distances between corresponding arcs of the three strongest lines observed on the developed film were measured as 77.5, 89.9, and 130.4 mm. These lines seemed to the eye to be of about equal intensity. Calculate the spacings *d* of the crystal in angstrom units and identify the substance by reference to the A.S.T.M. card file.

29. In the case of polyethylene, the repeat distance obtained from the X-ray diffraction pattern of a fiber diagram was 2.54 Å. What type of chemical structure is implied?

BIBLIOGRAPHY

Bertin, E. P., *Principles and Practice of X-Ray Spectrometric Analysis,* 2nd ed., Plenum, New York, 1975.
Birks, L. S., *X-Ray Spectrochemical Analysis,* 2nd ed., Wiley-Interscience, New York, 1969.
Birks, L. S., *Electron Probe Microanalysis,* 2nd ed., Wiley-Interscience, New York, 1971.
Buerger, M. J., *X-Ray Crystallography,* Wiley, New York, 1942.
Bunn, C. W., *Chemical Crystallography,* 2nd ed., Oxford University, New York, 1961.
Clark, G. L., *Applied X-Rays,* 4th ed., McGraw-Hill, New York, 1955.
Liebhafsky, H. A., H. G. Pfeiffer, and E. H. Winslow, "X-Ray Methods: Absorption, Diffraction and Emission," in *Treatise on Analytical Chemistry,* Vol. 5, Part I, I. M. Kolthoff and P. J. Elving, Eds., Chap. 60, Wiley-Interscience, New York, 1964.
Liebhafsky, H. A., H. G. Pfeiffer, E. H. Winslow, and P. K. Zemany, *X-Ray Absorption and Emission in Analytical Chemistry,* Wiley, New York, 1960.
Liebhafsky, H. A., H. G. Pfeiffer, E. H. Winslow, and P. K. Zemany, *X-Rays, Electrons, and Analytical Chemistry,* Wiley-Interscience, New York, 1972.
Macdonald, G. L., "X-Ray Spectrometry," *Anal. Chem.,* **50** *(Fundamental Reviews)* 135R (1978) and earlier reviews.
McMurdie, H. F., C. S. Barrett, J. B. Newkirk, and C. O. Ruud, *Advances in X-Ray Analysis,* Vol. 21, Plenum, New York, 1978, and other volumes in this series.
Müller, R. O., *Spectrochemical Analysis by X-Ray Fluorescence,* Plenum, New York, 1972.
Pfluger, C. E., "X-Ray Diffraction," *Anal. Chem.,* **50** *(Fundamental Reviews)* 161R (1978) and earlier reviews.
Robertson, J. M., *Organic Crystals and Molecules,* Cornell University, Ithaca, New York, 1953.
Sproull, W. T., *X-Rays in Practice,* McGraw-Hill, New York, 1946.
Wittry, D. B., "X-Ray Microanalysis by Means of Electron Probes," in *Treatise on Analytical Chemistry,* Vol. 5, Part I, I. M. Kolthoff and P. J. Elving, Eds., Chap. 61, Wiley-Interscience, New York, 1964.

LITERATURE CITED

1. Duane, W. and F. L. Hunt, *Phys. Rev.,* **66**, 166 (1915).
2. Bragg, W. L., *The Crystalline State,* Macmillan, New York, 1933.
3. Castaing, R., Thesis, University of Paris, 1951.
4. Clark, G. L. and S. T. Gross, *Ind. Eng. Chem., Anal. Ed.,* **14**, 676 (1942).
5. Birks, L. S., E. J. Brooks, and H. Friedman, *Anal. Chem.,* **25**, 692 (1953).
6. Switzer, G., J. M. Axelrod, M. L. Lindberg, and E. S. Larsen, "Tables of Spacings for Angle 2θ, Cu $K\alpha_1$, Cu $K\alpha_2$, Fe $K\alpha$, Fe $K\alpha_1$, Fe $K\alpha_2$," Circular 29, Geological Survey, U. S. Dept. of Interior, Washington, D. C., 1948; "Tables for Conversion of X-Ray Diffraction Angles to Interplanar Spacings," Publication AMS 10, Government Printing Office, Washington, D. C.
7. Ievins, A. and M. E. Straumanis, *Z. Krist.,* **94**A, 40, 48 (1936); *see also* M. E. Strau-

manis, *Anal. Chem.*, **25**, 700 (1953).
8. "Index to the Powder Data File," *Am. Soc. Testing Materials, Spec. Tech. Publ.* **48L** (1962). *See also,* Morris, M. C. *et al., Natl. Bur. Stand. (U. S.), Monogr.* **25**, Sect. 13 (1976); and Morris, M. C. *et al.,* "Powder Diffraction Data from the Joint Committee on Powder Diffraction Standards Associateship at the National Bureau of Standards," Joint Committee on Powder Diffraction Standards, Swarthmore, Pa., 1976.
9. Hanawalt, J. D., H. W. Rinn, and L. K. Frevel, *Ind. Eng. Chem., Anal. Ed.,* **10**, 457 (1938).

CHAPTER 10

Radiochemical Methods

The advent of linear accelerators, the chain-reacting pile, and portable activation sources has made it possible to obtain artificially produced radionuclides of most of the elements. Over 1000 radioisotopes are now known. Many of these are available in large quantities and with extremely high activity for use in tracer studies. Naturally the availability of radionuclides has provided great impetus to their use. Under favorable conditions, radiochemical techniques provide the possibility of analyzing concentrations as low as 10^{-14} mol/ml.

10.1 NUCLEAR REACTIONS AND RADIATIONS

A radionuclide is characterized by its half-life, the type of transition involved when it decays, and the type and energy of the radiation emitted. Such information is essential for the recognition and understanding of the problems associated with the measurement of radionuclides.

Particles Emitted in Radioactive Decay

The heavy, naturally occurring radioactive elements, such as thorium, uranium, and the like, emit, among other products, doubly ionized helium particles known as *alpha particles*. Alpha particles have only a slight penetrating power, being stopped by thin sheets of solid materials and penetrating only 5–7 cm of air. Their energies, however, are generally very high, and may exceed 10 MeV (million electron volts). As a result, the ionizing power of an alpha particle is high; that is, on passing through material, a large number of ion pairs are produced along the linear path traversed by an alpha particle. For example, a 5-MeV alpha particle, which is stopped in 3.5 cm of air, produces about 25,000 ion pairs per centimeter of travel. Due to the greater ionizing power of alpha particles, they can generally be distinguished from beta or gamma radiation on the basis of pulse amplitude. The ionization chamber is the preferred detector. Alpha particle activity can be measured in the presence of considerable beta activity by first measuring the total activity, then interposing a filter to absorb the alpha particles, and measuring the beta activity.

A radioactive element formed by neutron capture usually has a higher neutron/proton ratio than its stable isobars, and therefore it frequently decays by beta particle emission. A *beta particle* is a very energetic electron or positron. Beta particle spectra are continuous. Few have the energy, E_{max}, which is the upper limit of the spectrum corresponding to the transition energy, often expressed as penetrating power or range. A 0.5-MeV beta particle has a range of 1 m in air and produces about 60 ion pairs per centimeter of its path. Above 0.4 MeV beta particles have sufficient energy to penetrate the windows of most counting devices, and measurement is not difficult. Below this energy value, however, special techniques are required. Very thin-window counters may be employed; the sample may be introduced directly into the active volume of the counter; or it may be dissolved in a liquid scintillator. Because corrections for self-absorption, self-scattering, and back-scattering are necessary but difficult to obtain, beta particle counting should be avoided when possible.

Gamma rays, actually high-energy photons, are monoenergetic, and gamma-ray spectra consist of discrete lines. The penetrating power of gamma radiation is much greater than that of either alpha or beta particles, but the ionizing power is less. The sensitivity of the detector must be increased by using longer chambers, by gas fillings under pressure that possess higher atomic number, or by thicker scintillator material. A filter of sufficient thickness to absorb all beta particles, and inserted between the sample and detector, permits the measurement of gamma radiation exclusively from a mixture of activities.

A long-lived positron-emitting nucleus may decay by capturing one of its own orbital K electrons; this is called *internal conversion.* The excess energy is emitted as a gamma ray, and the resulting ion with a vacant K orbital then emits X radiation characteristic of the new element.

Interaction of Nuclear Radiation with Matter

Since the radiations from radionuclides are detected by means of their interactions with matter, a brief summary of these modes of interaction is presented here.

Beta particles interact primarily with the electrons in the material traversed by the particle. The molecules may be dissociated, excited, or ionized. It is the ionization, however, which is of primary interest in the detection of beta particles. As a beta particle slows down while moving through matter, the specific ionization, that is, the number of ion pairs produced per unit track length, increases and reaches a maximum near the end of the track. On the average, each ion pair produced represents a loss of 35 eV. Actually, the beta particle may lose a large part of its energy in a single interaction, but if it does, the ions produced have so much excess energy that they in turn produce additional ion pairs. The

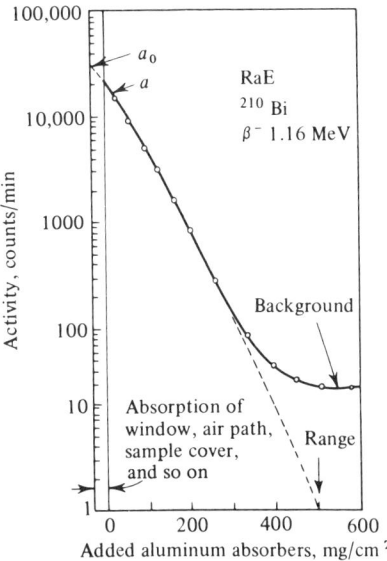

FIGURE 10-1 Absorption of beta particles.

absorption of beta particles in matter follows approximately the exponential relation given by Eq. 10-2 for gamma radiation, up to a certain thickness where the absorption finally exceeds that predicted by the exponential law and soon becomes infinite. This maximum thickness is known as the range of the beta particle (Fig. 10-1). For energies 0.5 to 3 MeV, the following range–energy relationship gives the maximum energy of the beta particle:

$$\text{range (in mg cm}^{-2} \text{ of absorber)} = 520 \, E_{max} - 90 \qquad (10\text{-}1)$$

or approximately $\frac{1}{2} E_{max}$ in MeV. In lower energy regions it is best to use a range–energy curve.

Whenever a beta particle comes close to a nucleus, it may have its direction of travel greatly changed, so that the path of a single beta particle in traversing matter may be a very devious one. Scattering due to the material supporting the radionuclide, called *backscattering*, increases the lower the energy of the beta particle and the higher the atomic number of the scatterer. Increasing the thickness of the support up to 0.2 range of a beta particle increases the back-scattering factor to a maximum value. Lead walls and doors of counting and shielding equipment should be lined with Lucite. Self-absorption and self-scattering also occur with beta particles. For very thin sources, 0.1–0.2 mg cm^{-2}, both factors are negligible—this situation prevails with carrier-free sources mounted on thin plastic film. For thicker samples, the same amount of inert carrier must be used or a correction curve prepared for different quantities of carrier.

The continuous X radiations produced when electrons are decelerated in the coulomb fields of atomic nuclei are called *bremsstrahlung*. This type of radiation is produced whenever fast electrons pass through matter; the efficiency of the conversion of kinetic energy into bremsstrahlung goes up with increasing electron energy and with increasing atomic number of the material.

Alpha particles lose energy by the same mechanisms as beta particles. However, owing to their large relative mass and higher charge, the specific ionization is much larger than for beta particles. Because the alpha particles are emitted from the nucleus with discrete energies and because the alpha particles lose only a small fraction of energy in collision with electrons, nearly all the alpha particles from a given nucleus will traverse the same distance in an absorber. Thus a determination of the range of alpha particles is a simple matter, and the energy of the alpha particles can be found by reference to plots. Although the range–energy curve is not linear, a useful rule of thumb is that for alpha emitters between 5 and 8 MeV in energy, the range in centimeters of air is approximately equal numerically to the energy in MeV.

Gamma rays lose energy on passage through matter in three ways: by the *photoelectric effect,* by the *Compton effect,* and by *pair production.* The photoelectric effect is important for heavy absorbing elements and for low gamma-ray energies. In this process gamma rays are absorbed by inner electrons bound in an atom, and the energy carried by the gamma ray is transferred completely to the electron with the resultant ejection of the electron from the atom (compare X-ray absorption). The Compton effect consists of a gamma ray interacting with an electron and transferring part of its energy to the electron. The electron is ejected from the atom and a new photon of lower energy proceeds from the collision in an altered direction. The Compton effect is important with light target elements and with gamma rays possessing energies less than 3 MeV. Pair production of a positron and an electron results when a high-energy gamma ray is annihilated following interaction with the nucleus of a heavy atom. Such a process is important with heavy elements and gamma rays of energies greater than 1.02 MeV, the energy corresponding to twice the rest mass of an electron. Conversely, when a positron and electron meet, the two are annihilated, and two gamma rays with energies of 0.51 MeV each are produced. Some of these spectral features are illustrated in Fig. 10-5.

Each ion pair which results from the passage of radiation through air represents an average energy loss of about 35 eV. The number of ion pairs per centimeter of travel is known as the *specific ionization.* In the case of gamma rays, the ion pairs formed are almost entirely produced by secondary processes—the photoelectrons, the Compton electrons, and the positrons and electrons. The attenuation of gamma radiation is given by

$$P = P_0 e^{-\mu x} \qquad (10\text{-}2)$$

where P is the radiant power of P_0 which is transmitted through an absorber of thickness x, and μ is the linear absorption coefficient. Often the energy of a gamma ray is expressed as the thickness of absorber required to diminish P to $\frac{1}{2} P_0$; this half-thickness is given by

$$x_{1/2} = \frac{0.693}{\mu} \qquad (10\text{-}3)$$

Absorber thicknesses are frequently given in units of surface density, g/cm²; that is, ρx where ρ is the density of the absorber. Equation 10-2 becomes

$$P = P_0 e^{-(\mu/\rho)\rho x} \tag{10-4}$$

where μ/ρ is known as the mass absorption coefficient.

Neutrons do not carry a charge and therefore they do not interact with electrons. Because of their lack of charge, neutrons can more readily enter the nucleus of an absorber element, and it is by secondary reactions that the neutron is detected and measured; for example,

$$^{10}B + n \rightarrow {^7}Li + \alpha \tag{10-5}$$

It is the alpha particle which is actually detected.

Radioactive Decay

The decay of a radionuclide follows the well-known first-order rate law, which may be written in differential form as follows:

$$\frac{dN}{dt} = -\lambda N \tag{10-6}$$

where N is the number of radionuclides remaining at time t, and λ is the characteristic decay constant (in time^{-1}). The activity A is related to N by the equation

$$A = \lambda N \tag{10-7}$$

and is usually the quantity observed or computed. The rate equation may be integrated to yield

$$A = A_0 e^{-\lambda t} \tag{10-8}$$

or

$$\ln A = \ln A_0 - \lambda t \tag{10-9}$$

where A_0 is the activity at some initial time, and A is the activity after elapsed time t.

The time required for one-half of the radioactive material to decay—the *half-life* of the radionuclide—is generally used in describing radioactive emitters, namely,

$$t_{1/2} = \frac{1}{\lambda} \ln \frac{A}{A/2} = \frac{0.693}{\lambda} \tag{10-10}$$

The half-life of the radioactive element is that time required for half of the radioactive nuclei of the element to disintegrate and release its energy. For example, strontium-90 has a half-life of approximately 30 years. In 30 years half of the nuclei will disintegrate and release their energy, in another 30 years half of the remainder of the nuclei will decay. This process continues such that every 30 years half of the remainder of the element decays and the other half remains stable for future decay. After 10 half-lives only 0.1% of

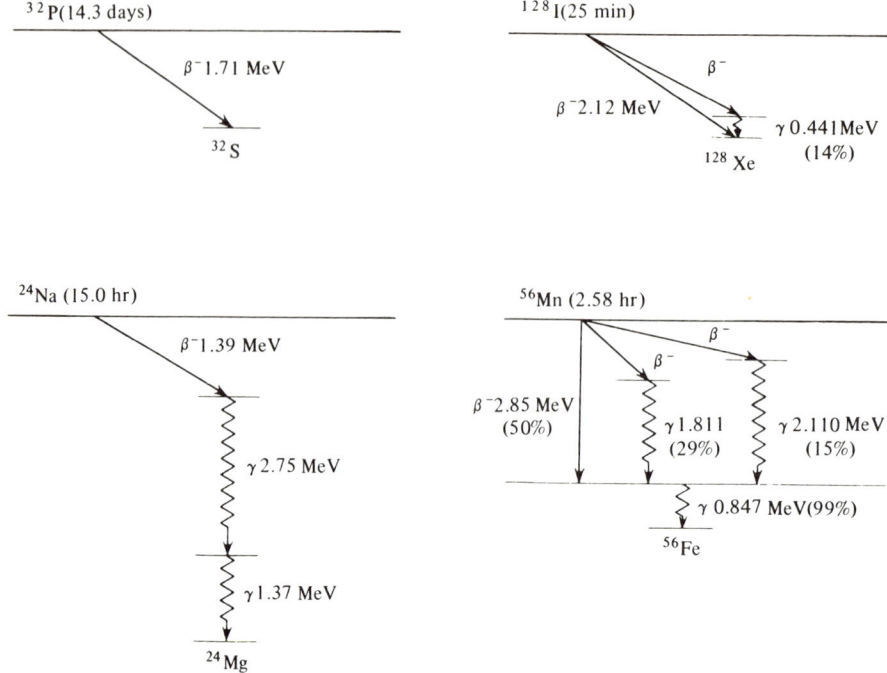

FIGURE 10-2 Some decay schemes of radioisotopes.

the radioactive nuclei remain. An accurate knowledge of the characteristic decay constant (λ) is essential when working with short-lived radionuclides in order to correct for the decay while the experiment is in progress. Selected decay schemes are shown in Fig. 10-2.

Units of Radioactivity

Activity is expressed in terms of the curie, where 1 Ci is 3.700×10^{10} disintegrations per second (dps). Specific activity, the activity per unit quantity of radioactive sample, is expressed in a variety of ways by dps per unit weight or volume, or in units such as microcurie or millicurie per milliliter, per gram, or per millimole. The last is preferable for labeled compounds.

10.2 MEASUREMENT OF RADIOACTIVITY

Radiation from radionuclides can be detected and measured in many ways. The best method to employ in any particular situation depends upon the nature of the radiation and and the energy of the radiation or particles involved. Specific detectors have been discussed in Chapter 9.

Statistics in Measurement of Radioactivity

The random nature of nuclear events requires that a large number of individual events be observed to obtain a precise value of the count rate. Several factors must be considered when attempting to measure any activity. The ionizing particle may never reach the active volume of the detector; instead it may be absorbed in the walls or air path. The detector may not be perfectly efficient because of several conditions—the detector may not have recovered from a previous event (the dead time), the particle from the source may not produce an ion in the sensitive volume of the detector, and some regions of the detector are more sensitive than others.

Statistical laws will predict the magnitude of the deviations about a mean value to be expected, as well as the probability of occurrence of deviations of a given magnitude. One standard deviation, or 1σ (sigma), is the maximum deviation from the "true" value that may be expected in 68% of a large series of measurements. It serves as a measure of the precision of a single observation or a series of observations. In activity measurements the standard deviation in the total count is equal to the square root of the number of counts N taken:

$$\sigma = \sqrt{N} \qquad (10\text{-}11)$$

The standard deviation in the counting rate r is

$$\sigma_r = \frac{1}{t}\sqrt{N} = \sqrt{\frac{r}{t}} \qquad (10\text{-}12)$$

Expressed in this manner, the standard deviation is a measure of the scatter of a set of observations around their mean value. The relative standard deviation (fractional 0.68 error), often given in percent, is

$$100\sigma = \frac{100}{\sqrt{N}} \qquad (10\text{-}13)$$

It is a measure of the precision of an observation. It can be expected that one standard deviation (or smaller) will arise in about 7 out of 10 determinations. The following confidence limits apply to representative multiples of the standard deviation:

Deviation	$\pm 0.68\sigma$	$\pm\sigma$	$\pm 2\sigma$	$\pm 3\sigma$
Population mean (limits)	$\pm 0.68\sigma$	$\pm\sigma$	$\pm 1.96\sigma$	$\pm 2.58\sigma$
Probability that observation (or confidence level) lies within this deviation, %	50	68	95	99

Example 10-1

If 8100 counts are timed, the standard deviation is

$$\sigma = \sqrt{8100} = 90 \text{ counts}$$

Expressed as relative standard deviation,

$$100\sigma = \frac{90}{8100}(100) = \frac{100}{\sqrt{8100}} = \pm 1.11\%$$

The probable error will be 0.68σ or 0.74%.

Example 10-2

To ascertain the number of counts which must be taken so that the deviation in 95% of the determinations (2σ) will not exceed 2.0%, proceed in this manner.

$$\frac{2\sqrt{N}}{N} = 0.02$$

$$\sqrt{N} = 100$$

$$N = 10,000 \text{ counts}$$

If a source has an activity of 200 counts/sec, it will take 50 sec to accumulate 10,000 counts.

A counter always shows some background activity. The statistical fluctuation of the background B must be included in any estimate of the standard deviation of the source whose activity A will include any background activity recorded simultaneously. The standard deviation of the net source activity, that is, $A - B$, is given by

$$\sigma = \sqrt{\sigma_A^2 + \sigma_B^2} \qquad (10\text{-}14)$$

Background radiation becomes significant when the peak-to-background ratio, $A:B$, is less than 20, and it becomes very difficult to measure a source accurately when the counting rate is just a little greater than the background rate. The optimum division of available time between background and source counting is given by

$$\frac{t_B}{t_A} = \sqrt{\frac{r_B}{r_A}} \qquad (10\text{-}15)$$

Here counting rates have to be obtained from a preliminary run.

Example 10-3

A sample gave a counting rate of 200 counts/min in a 10-min count. The background gave a counting rate of 40 counts/min in a 5-min count. What is the fractional 0.95 error of the sample corrected for background? How much time should be devoted to counting the sample and the background if the standard deviation of each measurement, corrected for background, is to be 5%?

The error of the sample corrected for background is $\sigma = \sqrt{\sigma_s^2 + \sigma_B^2}$ and $\sigma = \sqrt{N}$ for 68% probability. To express the error in terms of counts per minute, rt is substituted for N and the equation is divided by time, to give $\sigma = \sqrt{r/t}$. To express the error as a fractional error, the equation is divided by r: $F_y = 1/\sqrt{rt}$. Upon substituting these values of σ_s and σ_B into the first expression, and dividing both sides of the equation by the counting rate of the sample corrected for background, the fractional error of the sample is

$$F_y = \frac{1}{r_s - r_B}\sqrt{(r_s/t_s) + (r_B/t_B)} = \frac{1.96}{200 - 40}\sqrt{(200/10) + (40/5)} = 0.066$$

There are 95 chances out of 100 that the error of the sample corrected for background is less than 6.6%.

To derive the relation which gives the amount of time required to count the sample, t_B is made equal to $t_s \sqrt{r_B/r_s}$, and substituted into the fractional error equation. Rearranged,

$$t_s = \frac{1}{F_y^2} \left[\frac{r_s + (r_{B/C})}{(r_s - r_B)^2} \right] \quad \text{where } c = \sqrt{r_B/r_s}$$

$$= \frac{1}{(0.05)^2} \left[\frac{200 + (40/0.45)}{(200 - 40)^2} \right] = 4.5 \text{ min} \quad \text{and } t_B = 0.45(4.5) = 2.0 \text{ min}$$

Coincidence Correction

In order to correct for counting losses at high counting rates caused by the finite resolution time of counters, two courses are open. One method is to construct a calibration curve from a series of dilutions of a strong sample or from measurements of a series of standards of known strengths. A second procedure is to measure two samples separately and then measure the two samples together. The resolving time of the counter τ is given by

$$\tau = \frac{r_1 + r_2 - r_{1,2} - r_B}{2r_1 r_2} \tag{10-16}$$

where r_1 and r_2 represent the counting rate of source 1 plus background and source 2 plus background, respectively; $r_{1,2}$ is the measured counting rate of source 1 plus source 2 and the background r_B.

If the resolving time of the counter is greater than that of any other part of the measuring circuit, then the true counting rate r_0 is related to the observed counting rate r and the resolving time τ, namely,

$$r_0 = \frac{r}{1 - r\tau} \tag{10-17}$$

Geometry

The arrangement of counter and source should always be reproducible so that the solid angle subtended by the counter with respect to the source remains unchanged. The "geometry" of any arrangement may be obtained by measurement of standard sources. Insertion of the source into the volume counter shown in Fig. 9-11 approximates 2π-geometry; 4π-geometry is approximated in well-type scintillation detectors.

10.3 APPLICATIONS OF RADIONUCLIDES

The introduction of a radioactive-labeled material into a sample system or a measurement of the natural or induced radioactivity of a system become very useful techniques for rapid and economical methods of analysis for elements or materials. Isotope dilution with radioactive tracers, labeled reagents, activation analysis, or the use of radioactive tracers for procedure development have much use in analytical chemistry.

The chain-reacting pile is without a peer as a tool for the general quantitative production of radioactivity because of the magnitude and spatial extent of the thermal neutron flux it is able to sustain. The pile excels the cyclotron in respect to both intensity and the magnitude of the effective flux produced.

Preformed generator sources, such as a radionuclide cobalt-60 source or a beryllium neutron source, can be useful in meeting many analytical requirements when a pile is not conveniently near at hand.

A selection of radionuclides and their characteristics is given in Table 10-1. Nuclides with very short half-lives will decay too rapidly to be generally useful; on the other hand, a long-lived nuclide will be difficult to measure because disintegrations are too infrequent.

TABLE 10-1 Nuclear Properties of Selected Radioisotopes

		Target Isotope			Major Radiations, Energies in MeV (γ Intensities, %)
Radio-isotope	Half-life		Natural Abundance, %	Thermal Neutron Cross Section, barns	
^{3}H	12.26 yr				β^- 0.0186; no γ
^{14}C	5730 yr	^{13}C	1.108	0.0009	β^- 0.156; no γ
^{15}O	123 sec				β^+ 1.74; γ 0.511
^{22}Na	2.62 yr				β^+ 1.820, 0.545; γ 0.511, 1.275(100)
^{24}Na	14.96 hr	^{23}Na	100	0.53	β^- 1.389; γ 1.369(100), 2.754(100)
^{28}Al	2.31 min	^{27}Al	100	0.235	β^- 2.85; γ 1.780(100)
^{32}P	14.28 day	^{31}P	100	0.19	β^- 1.710; no γ
^{35}S	87.9 day	^{34}S	4.22	0.27	β^- 0.167; no γ
^{36}Cl	3.08 × 10^5 yr	^{35}Cl	75.53	44	β^- 0.714; γ 0.511
^{38}Cl	37.29 min	^{37}Cl	24.47	0.4	β^- 4.91; γ 1.60(38), 2.17(47)
^{40}K	1.26 × 10^9 yr		0.118	70	β^- 1.314; β^+ 0.483; γ 1.460(11)
^{42}K	12.36 hr	^{41}K	6.77	1.2	β^- 3.52; γ 0.31, 1.524(18)
^{45}Ca	165 day	^{44}Ca	2.06	0.7	β^- 0.252
^{51}Cr	27.8 day	^{50}Cr	4.31	17	γ 0.320(9); e^- 0.315
^{56}Mn	2.576 hr	^{55}Mn	100	13.3	β^- 2.85; γ 0.847(99), 1.811(29), 2.110(15)
^{55}Fe	2.60 yr	^{54}Fe	5.84	2.9	Mn X-rays
^{59}Fe	45.6 day	^{58}Fe	0.31	1.1	β^- 1.57, 0.475; γ 0.143(1), 0.192(3), 1.095(56), 1.292(44)
^{60}Co	5.263 yr	^{59}Co	100	19	β^- 1.48, 0.314; γ 1.173(100), 1.332(100)
^{63}Ni	92 yr	^{62}Ni	3.66	15	β^- 0.067; no γ
^{65}Ni	2.564 hr	^{64}Ni	1.16	1.5	β^- 2.13; γ 0.368(5), 1.115(16), 1.481(25)

TABLE 10-1 Continued

Radio-isotope	Half-life	Target Isotope			Major Radiations, Energies in MeV (γ Intensities, %)
		Isotope	Natural Abundance, %	Thermal Neutron Cross Section, barns	
^{64}Cu	12.80 hr	^{63}Cu	69.1	4.5	β^- 0.573; β^+ 0.656; e^- 1.33; γ 0.511, 13.4(1)
^{65}Zn	245 day	^{64}Zn	48.89	0.46	β^+ 0.327; e^- 1.106; γ 0.511, 1.115(49)
^{69m}Zn	13.8 hr	^{68}Zn	18.6	0.10	γ 0.439(95); e^- 0.429
^{76}As	26.4 hr	^{75}As	100	4.5	β^- 2.97; γ 0.559(43), 0.657(6), 1.22(5), 1.44(1), 1.789, 2.10(1)
^{80}Br	17.6 min	^{79}Br	50.52	8.5	β^- 2.00; β^+ 0.87; γ 0.511, 0.61(7), 0.666(1)
^{80m}Br	4.38 hr				γ 0.037(36); e^- 0.024, 0.036, 0.047
^{82}Br	35.34 hr	^{81}Br	49.48	3	β^- 0.444; γ 0.554(66), 0.619(41), 0.698(27), 0.777(83), 0.828(25), 1.044(29), 1.317(26), 1.475(17)
^{90}Sr	27.7 yr				β^- 0.546; no γ
^{90}Y	64.0 hr				β^- 2.27; no γ
^{110m}Ag	255 day	^{109}Ag	48.65	89	β^- 1.5; γ 0.658(96), 0.68(16), 0.706(19), 0.764(23), 0.818(8), 0.885(71), 0.937(32), 1.384(21), 1.505(11)
^{122}Sb	2.80 day	^{121}Sb	57.25	6	β^- 1.97; β^+ 0.56; γ 0.564(66), 1.14(1), 1.26(1)
^{124}Sb	60.4 day	^{123}Sb	42.75	3.3	β^- 2.31; γ 0.603(97), 0.644(7), 0.72(14), 0.967(2), 1.048(2), 1.31(3), 1.37(5), 1.45(2), 1.692(50), 2.088(7)
^{128}I	24.99 min	^{127}I	100	6.4	β^- 2.12; γ 0.441(14), 0.528(1), 0.743, 0.969
^{137}Cs	30.0 yr				β^- 0.511, 1.176; γ 0.662 (85)
^{137m}Ba	2.554 min				γ 0.662(89); e^- 0.624, 0.656
^{198}Au	2.697 day	^{197}Au	100	98.8	β^- 0.962; γ 0.412(95), 0.676(1), 1.088
^{204}Tl	3.81 yr	^{203}Tl	29.5	11	β^- 0.766

SOURCE: J. A. Dean, Ed., *Lange's Handbook of Chemistry*, 12th ed., McGraw-Hill, New York, 1979.

Preparation and Mounting of Samples

The active source must be free of interfering substances, in a suitable chemical and physical form, and disposed in a definite and fixed position relative to the detector. In virtually all applied radiochemistry, relative intensities of two or more samples are all that need be determined, thus making absolute measurements unnecessary. The principal requirement is reproducibility and, if this is not possible, the effects of the variable factors must be determined and a correction applied.

A chemical separation from inactive, and occasionally active, contaminants generally precedes the activity measurement. Ordinary analytical techniques and reactions form the basis for most radiochemical separations. Carrying a microconstituent by coprecipitation with a macroconstituent (a carrier), solvent extraction, volatilization, adsorption, ion exchange, electrodeposition, and chromatography are useful for handling very low concentrations of material. If the radiations are intense and penetrating, shielding and remote-control equipment are required. All work with appreciable quantities of radioactivity should be done with rubber gloves, protective clothing, and adequate ventilation to remove active vapors, dusts, and sprays. Of scarcely less importance is the danger of contamination by active materials of laboratories, equipment, and detecting instruments. Establishment and enforcement of suitable regulations and careful attention to cleanliness of operation are integral aspects of the proper technique of working with radioactivity.

The active material must be spread in a uniform layer over a definite area unless 4π-geometry is employed. Thin uniform layers of solids may be spread as slurries in water or other solvent which is later evaporated. If the deposit is not coherent but tends to fall apart and flake off, it may be stabilized by the use of a suitable binder (collodion), or it may be held in place with a covering layer of Scotch tape, aluminum foil, or similar material. The sample is frequently obtained in a solid form by precipitation from solution and separation by centrifugation. Crystalline precipitates are best, whereas flocculent precipitates, which are highly hydrated, tend to give unsatisfactory deposits because on drying they contract into a number of isolated, dense particles.

Evaporation of solutions in cup-shaped containers, or for small amounts, on flat surfaces, is convenient. The method is limited to nonvolatile activities. Electrodeposition onto a flat surface gives excellent deposits for many metals; a thin film of plastic sprayed with gold can serve as the active electrode.

Samples can be mounted on flat foils or disks; several tenths of a milliliter of a liquid can be held in place by surface tension or within a ring of silicone grease. When backscattering by the mount is objectionable, a thin film of low atomic number may be used—Mylar, polystyrene, and similar plastics. An arrangement is needed for holding samples in a definite position relative to the detector during measurement. Counters and ion chambers are generally equipped with an arrangement having one or more shelves or sets of slots in which sample holders of a standard size can be placed. Radionuclides emitting sufficiently penetrating beta and gamma radiations are conveniently assayed in solution with dipping counter tubes or counters surrounded by hollow jackets, or by insertion in the well of an ionization chamber or scintillation counter. Active gases can be introduced into proportional counters or ionization chambers equipped with stopcocks and pressure manometers.

Tagging Compounds

Since the radioactive isotopes are chemically identical with their stable isotope counterpart, they may be used to "tag" a compound. The tagged compound may then be followed through any analytical scheme, industrial system, or biological process. It is essential that a compound be tagged with an atom, however, which is not readily exchangeable with similar atoms in other compounds under normal conditions. For example, tritium could not be used to trace an acid if it were inserted on the carboxyl group where it is readily exchanged by ionization with the solvent.

It is not always true that the radionuclide will resemble exactly the normal isotope. Differences in weight do cause slight changes in the reactivity of molecules. For ordinary purposes, however, these isotope effects are slight except for the lightest elements (such as tritium). Many multiple decays are found in radionuclides produced in chain-reacting piles. If the parent (^{140}Ba, $t_{1/2}$ = 12.8 days) is longer-lived than the daughter (^{140}La, $t_{1/2}$ = 40 hr), a state of radioactive equilibrium is reached after about eight half-lives (13 days). Thus, if ^{140}Ba is to be used as a tracer for barium, the isolated samples either should be freed chemically of the daughter ^{140}La and counted without delay, or the samples should be kept for two weeks before counting.

In biological investigations the question of purity of the isolated material is important. One must be certain that the radioactivity of the compound isolated is due to the compound itself and not to some minor contaminant which may have a high specific activity.

In ordinary analytical work, radionuclides have been used to study errors resulting from adsorption and occlusion in gravimetric methods, and to devise methods of preventing coprecipitation, adsorption, and occlusion.

Analyses with Labeled Reagents

Radiometric methods employing reagent solutions or solids tagged with a radionuclide have been used to determine the solubility of numerous organic and inorganic precipitates, or as a radioreagent for titrations involving the formation of a precipitate. In this type of application it is necessary to establish the ratio between radioactivity and weight of radionuclide plus carrier present. This may be established by evaporating an aliquot to dryness, weighing the residue, and measuring the radioactivity.

In solubility studies, the compound of interest is synthesized, using the radionuclide, and a saturated solution of the compound is prepared. A measured volume of the saturated solution is evaporated to dryness, and the radioactivity of the residue determined. From the previously established relationship between weight and radioactivity, the amount of the compound present can be calculated.

Procedures have been described for the use of radioreagents as titrants. For example, phosphorus-32 was converted into a soluble phosphate and incorporated into a standard solution of disodium hydrogen ortho-phosphate. This solution was used to titrate several inorganic ions. After each addition of phosphate a sample of the clear, filtered solution was withdrawn (by means of a filter-stick) and the activity was determined. After the equivalence point was passed, the activity rose rapidly with additions of radioreagent. From the intersection of the activity curves, the end point was determined.

The efficiency of an analytical procedure can be determined by adding a known amount of a radioisotope before analysis is begun. After the final determination of the element in question, the activity of the precipitate is determined and compared with the activity at the start.

Chemical yields need not be quantitative in an analytical procedure when the results are corrected by the recovery of the radioisotope. To the mixture of ions, a known amount of radioisotope is added, or to a mixture of activities a known amount of carrier element is added, then separated in the necessary state of chemical and radiochemical purity but without attention to yield. The isolated sample is determined by any suitable method, and the activity is measured.

Isotope Dilution Analyses

This technique measures the yield of a nonquantitative process, or it enables an analysis to be performed where no quantitative isolation procedure is known. To the unknown mixture, containing a compound with inactive element P, is added a known weight W_1 of the same compound tagged with the radioactive element P^*. The specific activity A_1 of the tagged compound of weight W_1 is known. A small amount of the pure compound is isolated from the mixture and the specific activity measured (A). The amount isolated need be only a very small fraction of the total amount present, merely a sufficient quantity for weighing or determining accurately. The extent of dilution of the radiotracer shows the amount W of inactive element (or compound) present, as given by the expression

$$W = W_1 \left(\frac{A_1}{A} - 1\right) \tag{10-18}$$

The method has proved valuable in the analysis of complex biochemical mixtures and in the radiocarbon dating of archaeological and anthropological specimens.

Liquid Scintillation Counting

Low-energy beta emitters, such as ^{3}H, ^{14}C, and ^{32}P, are commonly counted by dissolving the compound containing a small amount of the radionuclide (called the *tracer*) in a liquid scintillator. The tracer labels the sample and allows the molecules to be traced or followed by the liquid scintillation spectrometer. The basic objective of the technique is to arrange for emitted beta particles to collide with the solvent molecules. The energy resulting from the collisions excites the solvent molecules and is transferred to other molecules until it is finally transferred to scintillator compounds. The scintillator molecules absorb the energy, a portion of which is then emitted in the form of photons of visible light, termed fluorescence. The photons are detected by photomultiplier tubes and converted into electrical energy for counting by the spectrometer.

The bulk solvent must efficiently transfer energy to a scintillator molecule, and be capable of dissolving the scintillators and the sample material. Aromatic solvents, such as toluene or xylene, are favored because of their efficiency in energy transfer. 1,4-Dioxane is employed when large amounts of water are involved; naphthalene is often added to improve the energy-transfer process and reduce quenching. Sometimes incorporation of aqueous sample solutions into a toluene-based system is possible by adding a nonionic surfactant such as Triton X-100. Glycol ethers and alcohols are also used as secondary solvents to improve water miscibility, and to allow counting at low temperatures.

The scintillator must be capable of absorbing light at one wavelength and reemitting it at a longer wavelength. Most scintillation spectrometers are sensitive to the fluorescent emission of the primary scintillator. However, if an older model is being used, it may be necessary to add a secondary scintillator. The latter absorbs the light emitted by the primary scintillator and emits it at a yet higher wavelength. When used, it is added to the extent of one-tenth or less of the primary scintillator. The most popular primary scintillator is 2,5-diphenyloxazole (PPO). The most widely used secondary scintillators are 2,2'-p-phenylene-bis (5-phenyloxazole), or POPOP, and 2,2'-p-phenylene-bis (4-methyl-5-phenyl-oxazole). For the primary scintillator the fluorescence emission maximum lies in the range from 360 to 365 nm, whereas that for POPOP lies around 410–420 nm. As shown in Fig. 10-3, the POPOP absorption spectrum overlaps well with the PPO emission spectrum,

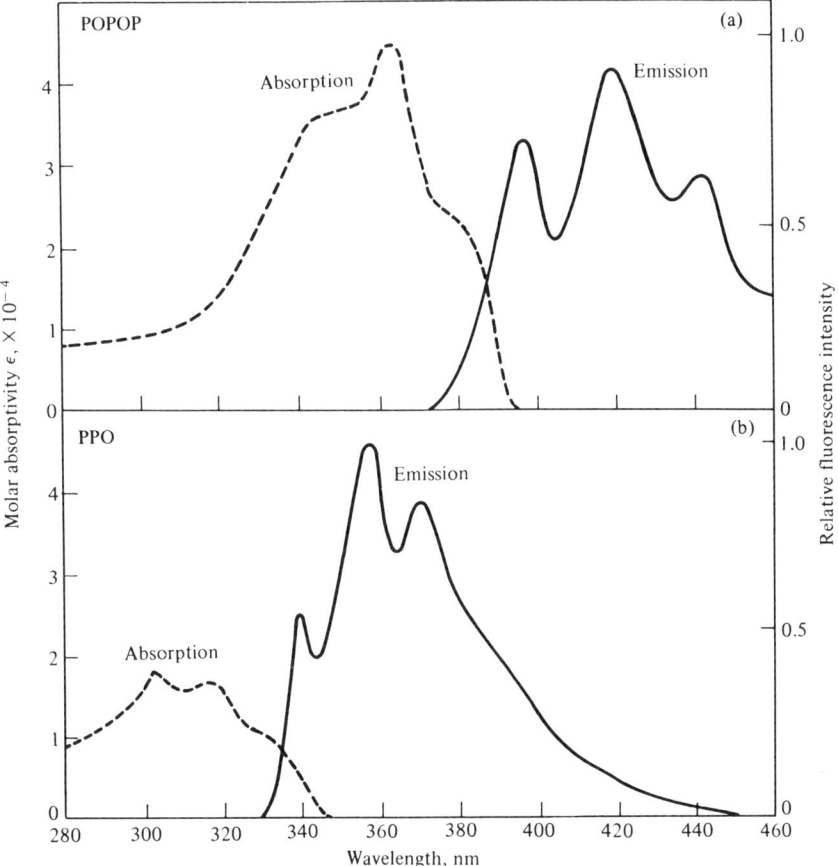

FIGURE 10-3 (a) Absorption and fluorescence emission spectra of POPOP. (b) Absorption and fluorescence emission spectra of PPO.

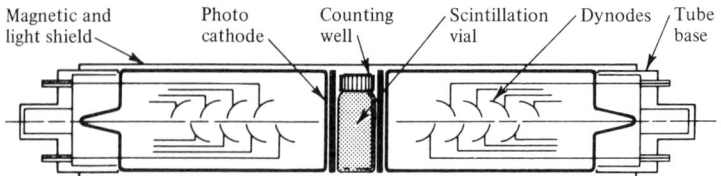

FIGURE 10-4 Diagram of a typical beta scintillation counter showing only the counting well and photomultiplier tube detectors.

and since POPOP has a very large molar absorptivity in the overlap region, only a very small amount of POPOP is necessary for nearly complete absorption of the PPO fluorescence. Also the POPOP fluorescence closely matches the response of the blue-sensitive photomultiplier tubes.

Liquid scintillation procedures can be extended to various samples. If radioactive carbon dioxide is being measured, the scintillator solution should contain a trapping agent, perhaps 1-amino-2-phenylethane (phenethylamine). Quaternary ammonium hydroxides in methanolic solution find use as tissue solubilizers. For materials that cannot be solubilized, a suitable gelling agent is used to prevent settling; this technique is termed suspension counting. Finely divided amorphous silica is widely used as a gelling agent.

Chemical impurities may interfere with the transfer of energy from solvent to solute to produce chemical quenching, or they may absorb the light emitted from the solute molecules to produce color quenching. Chemiluminescence will give yet a third change in the spectrum. Color quenching can be reduced or eliminated by digestion with hydrogen peroxide and perchloric acid. For ^{3}H and ^{14}C, complete combustion of the sample to water and a soluble carbonate produces a simple counting method.

A diagram of a typical beta scintillation counter is shown in Fig. 10-4. Glass and plastic liquid scintillation vials contain 20 ml in the regular size and 7 ml in the smaller size. High density polyethylene vials are economical; they should have counts of less than 10 cpm (for tritium). Borosilicate glass vials are impervious to aromatic hydrocarbons and best for sample storage; the vials should possess a low background count from potassium (less than 20 cpm when determining tritium).

10.4 ACTIVATION ANALYSIS

The basic principle of activation analysis consists of placing the sample to be analyzed in a flux of energetic charged particles or neutrons for a sufficient time to produce enough radionuclide product to measure with the desired statistical precision. The rate of production of radioactive atoms, N^*, is given by

$$\frac{dN^*}{dt} = \Phi \sigma N - \lambda N^* \qquad (10\text{-}19)$$

where Φ is the number of bombarding particles, or flux (in cm^{-2} sec^{-1}), σ is the reaction cross section expressed in units of cm^2/target atom (10^{-24} cm^2/nucleus, also denoted barns), N is the number of target nuclei available, and λ is the characteristic decay constant. The number of target nuclei is given by

$$N = wN_A f/M \qquad (10\text{-}20)$$

where w is the weight of the parent nuclide, N_A is Avogadro's number, M is the atomic weight of the nuclide, and f is the fractional abundance of the target nuclide. Integration of Eq. 10-19 over the time of irradiation yields the number of radioactive nuclei at the end of the irradiation:

$$N^* = \frac{\Phi \sigma N}{\lambda}(1 - e^{-\lambda t}) \qquad (10\text{-}21)$$

The initial activity just at the termination of the irradiation, A_0, is the product λN^*. Substituting $0.693/t_{1/2}$ for the decay constant in the exponential term in Eq. 10-21 yields

$$A_0 = \Phi \sigma N \left[1 - \exp\left(-\frac{0.693 t}{t_{1/2}}\right)\right] \qquad (10\text{-}22)$$

where t is the duration of the irradiation period. During the irradiation period some of the radionuclide produced will decay (Eq. 10-8). The term within the brackets of Eq. 10-22 is the saturation factor, S, and represents the ratio of the amount of activity produced during the irradiation period to that produced in infinite time. At $t/t_{1/2}$ values of 1, 2, 3, 4, 5, 6, ..., ∞, S has corresponding values of 0.5, 0.75, 0.87, 0.94, 0.97, 0.98, ..., 1.0. The induced activity will reach 98% of the saturation value for irradiation periods equal to six half-lives.

Example 10-4

Calculate the activity for a 10.0-mg sample of an aluminum alloy containing 0.041% manganese after a 0.50-hr irradiation in a flux of 5×10^{13} neutrons cm^{-2} sec^{-1}. Other necessary information is obtained from Table 10-1 for insertion in Eq. 10-22:

$$A_0 = \frac{(0.00041)(0.0100 \text{ g})(1.00)(6.02 \times 10^{23} \text{ nuclei mol}^{-1})}{54.94 \text{ g mol}^{-1}}$$

$$\times (5 \times 10^{13} \text{ } n \text{ cm}^{-2} \text{ sec}^{-1})(13.3 \times 10^{-24} \text{ cm}^2 \text{ nuclei}^{-1})$$

$$\times \left[1 - e^{-(0.693)(0.50 \text{ hr})/2.58 \text{ hr}}\right]$$

$$A_0 = 3.68 \times 10^6 \text{ disintegrations/sec}$$

Neutron Sources

A reactor is an unexcelled source of slow or thermal neutrons. Thermal neutrons are neutrons in kinetic-energy equilibrium with the surrounding temperature. Fluxes available range from 10^{11} to 10^{14} neutrons cm^{-2} sec^{-1}. Portable sources, for laboratories not having access to atomic reactors, depend on the reaction

$$^9\text{Be}(\alpha, n)^{12}\text{C}$$

and consist of an intimate mixture of an alpha emitter, such as ^{226}Ra, ^{210}Po, or ^{239}Pu, with beryllium. These sources provide only 10^7 neutrons cm^{-2} sec^{-1}. If nanogram sensitivity is not required, a californium-252 source supplies 10^{12} neutrons cm^{-2} sec^{-1}.

Some elements can be detected with more sensitivity or more conveniently by activation with fast neutrons than with thermal neutrons. The list of applicable elements includes N, O, F, Al, Si, P, Cr, Mn, Fe, Cu, Y, Mo, Nb, and Pb. In particular, the nondestructive determination of oxygen is now frequently done by activation with 14-MeV neutrons to form 7.35-sec ^{16}N via the ^{16}O$(n, p)^{16}$N reaction. The gamma radiation emitted by ^{16}N is mostly 6.13 MeV, with a little 7.13 MeV, which is exceptionally high so that there are essentially no interferences (except for F, which also undergoes the ^{19}F$(n, \alpha)^{16}$N reaction). Total analysis time for oxygen is less than 1 min; the limit of detection is 1 μg/g in a 10-g sample. For six elements, O, Si, P, Fe, Y, and Pb, the sensitivity is greater using fast rather than thermal neutrons.

As a neutron source, the fast-neutron (fission-spectrum) flux that is present in the core of a nuclear reactor can be used. Samples are wrapped with cadmium foil to prevent thermal neutrons from reaching the sample. A small accelerator source is a 200-keV Cockcroft–Walton deuteron accelerator with a metallic tritium target and a D$^+$ beam which forms neutrons by the ^{3}H$(d, n)^4$He reaction with energies of about 14 MeV. Pulsed generators provide very high, fast-neutron fluxes, well over 10^{16} neutrons cm^{-2} sec^{-1}, of about 30-msec duration. Pulsed operation favors the formation of short-lived (less than 50 sec) radioisotopes. The amount of a particular activity A_p produced in a pulse, relative to the amount of activity produced in the usual steady-state operation of the same reactor to saturation activity A_s, is approximately $A_p/A_s = 70/t_{1/2}$.

Irradiation Time

Because of the rapid asymptotic approach of the saturation factor to unity, in practice a sample is seldom irradiated for a period of time longer than one or a few half-lives of the induced activity of interest. Short-lived activities are enhanced, relative to longer-lived activities, by the use of a short irradiation period, followed quickly by counting. Similarly, longer-lived activities are enhanced by the use of a longer irradiation period, followed by an appreciable delay for the decay of interfering short-lived activities, before counting.

Example 10-5

In the determination of iron in aluminum alloys, the 1.29-MeV gamma of iron-59 was measured. A decay period of one week before chemical processing allowed for the decay of the 15.0 hr sodium-24 formed from the reaction ^{27}Al $(n, \alpha)^{24}$Na.

$$A = A_0 e^{-(0.693)(168)/(15.0)} = 0.00235 A_0$$

After an interval of seven days, the sodium-24 activity has decayed to 0.00235 of its original activity. The gamma radiation from sodium-24 at 1.369 and 2.754 MeV would no longer constitute an interference.

Since there is usually a significant interval between the cessation of the irradiation and the measurement of the induced activity, the activity at any time after irradiation must be corrected for the intervening decay of the radionuclide being counted.

Example 10-6

At 120 min after discharge from the reactor, all the aluminum-28 (2.3 min) and most of the other short-lived isotopes will have decayed to negligible activity. The activity of manganese-56, from Example 10-4, has dropped to 2.153×10^6 counts sec^{-1}. The activity at the moment of removal from the reactor is

$$2.153 \times 10^6 \text{ counts sec}^{-1} = A_0 e^{-(0.693)(2.0 \text{ hr})/2.58 \text{ hr}}$$
$$A_0 = 3.68 \times 10^6 \text{ counts sec}^{-1}$$

Analytical Procedures

In quantitative determinations the comparative method is used. The comparators (standard samples) are weighed and packed in duplicate, together with samples of appropriate size. Both sets of samples are irradiated in the same physical location and under the same flux conditions. The relative activities are directly proportional to the respective concentrations of parent nuclide:

$$\frac{A_{\text{unknown}}}{A_{\text{standard}}} = \frac{N_{\text{unknown}}}{N_{\text{standard}}} \quad (10\text{-}23)$$

There are two forms of the activation analysis method: (1) the purely instrumental (nondestructive) form and (2) the postirradiation, radiochemical-separation form. The purely instrumental technique involves only activation of the sample, followed by gamma-ray spectrometry of the activated sample. It requires that the induced activities of interest emit gamma rays, characteristic X rays, or positrons in an appreciable fraction of the disintegrations. A gamma-ray spectrometer is used to count a series of activated samples and standards under exactly the same conditions, but at different decay times and perhaps for different lengths of time. To take full advantage of the energy resolution of modern detectors, it is necessary to use hundreds of energy channels. Instead of using hundreds of single channel analyzers and scalers, modern systems achieve multichannel analysis by digitizing the detected pulse height (see Fig. 9-15). This is a measure of gamma-ray energy; the analyzer stores the data in a digital memory module.

Sodium iodide scintillation detectors are still used when the sample composition is known qualitatively and the readouts from the elements do not interfere with each other excessively. Where interference is a problem, it is often helpful to use a sharper energy-resolution solid-state detector.

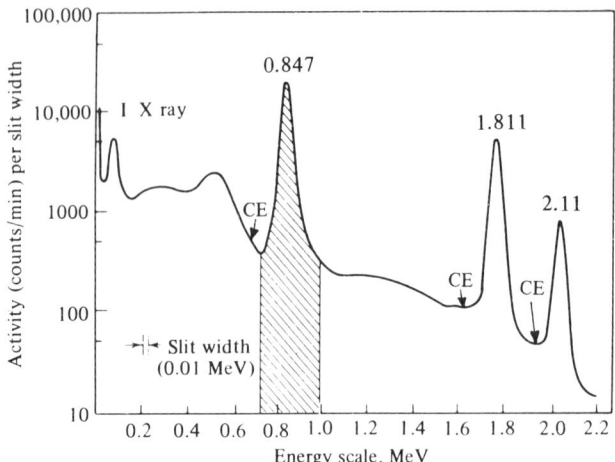

FIGURE 10-5 Gamma spectrum of manganese-56. The area indicated under the photopeak at 0.847 MeV would be used in quantitative work. CE indicates the Compton edges for each photopeak.

An illustrative pulse height spectrum, that of ^{56}Mn, is shown in Fig. 10-5. Manganese-56 follows the disintegration scheme illustrated in Fig. 10-2. The major feature in its spectrum is the sharp symmetrical peak at 0.847 MeV, the result of total absorption of the gamma-ray energy by the detector, and is normally referred to as the full-energy peak or photopeak. The location of the full-energy peak on the energy axis is the basis for the qualitative application of gamma-ray spectrometry. The area under the full-energy peak is related to the number of photons interacting with the detector, and is the basis for the quantitative applications. In addition to the photopeak at 0.847 MeV, two less intense photopeaks appear at 1.811 and 2.11 MeV. The continuous curve below the full-energy peaks is the Compton-continuum region. The maximum energy, in MeV, that a gamma-ray photon can lose by Compton scattering (the elastic scattering of photons by electrons) is given by

$$E_{CE} = \frac{E}{1 + 0.511/2E} \qquad (10\text{-}24)$$

where E_{CE} is the energy of the Compton edge. For the 0.847-MeV photopeak, E_{CE} is 0.627 MeV. Analogous Compton edges lie at 1.59 and 1.88 MeV for the less intense photopeaks. Less prominent features are a small peak at 0.511 MeV from an annihilation photon resulting from pair production in the shielding material by the gamma-ray peaks exceeding 1.02 MeV, peaks at full energy minus 0.511 MeV and full energy minus 1.02 MeV, and several back-scatter peaks of various sizes and at various energies (0.196, 0.223, and 0.227 MeV). The back-scattered energy, E_{BS}, in MeV is given by

$$E_{BS} = \frac{E}{1 + 2E/0.511} \qquad (10\text{-}25)$$

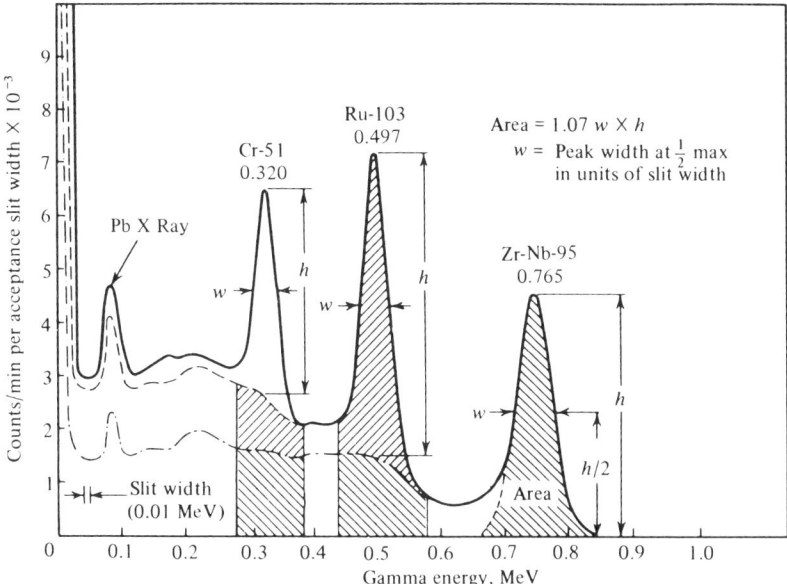

FIGURE 10-6 Typical three-component gamma spectrum illustrating two methods of quantitative analysis.

The net area under each photopeak is directly proportional to the absolute gamma emission rate of the corresponding isotope. One usually computes the net photopeak area above the Compton continuum, using a linear-base approximation, then corrects all of these counting rates to the desired reference time. For the analysis of mixed gamma-emitting isotopes, the characteristic lower-energy curve of the most energetic full-energy peak must be drawn in from previously recorded standard curves detailing the Compton continuum region, as shown in Fig. 10-6. These operations, termed spectrum stripping, may be performed automatically on multichannel analyzers with standard curves stored on magnetic tapes. Multichannel analyzers also have the capability of automatically adding together the counts in any groups of channels selected, such as those included within a photopeak.

If the induced activity of interest emits only beta particles or the interferences from other gamma-emitting induced activities prevent the purely instrumental detection of the activity of interest, the analyst may profitably resort to a postirradiation radiochemical separation procedure. After irradiation, the sample is dissolved and chemically equilibrated with a relatively large amount (but accurately known, and typically about 10.0 mg) of the elements of interest, as a carrier. If high levels of high specific activity nuclides of interferents are present, one usually dilutes them with similar amounts of hold-back carriers of these elements. Then the element of interest is separated and purified by any suitable separation procedure. Finally it can be counted on a beta-sensitive counter, or on a gamma-ray spectrometer. The amount of carrier element recovered is measured quantitatively so that the results can be normalized to 100% recovery. This makes it possible to employ fairly rapid, nonequilibrium separations. Recoveries of 50% or better are desirable since the higher the recovery the better the counting statistics.

One of the worst types of interference is typified by a mixture of nitrogen and zinc, each of which is most easily measured at 0.511 MeV, the positron annihilation peak; no particularly good photopeaks are available other than 0.511 MeV. The solution in such a case is to count the sample twice, using different decay times following irradiation. Then, calculation of the contribution of the two elements is facilitated by knowledge of the different decay properties of the radioisotopes.

PROBLEMS

1. A sample of ^{35}S contains 10 mCi. After 174 days, how many disintegrations per minute occur in the sample?

2. How much activity of the ^{32}P, the ^{131}I, and the ^{198}Au remains (a) after 14 days, (b) after 30 days, and (c) after 60 days?

3. Calculate the probable error, the 1σ, and the 2σ, variations (in percent) for each of these total number of counts: (a) 3200, (b) 6400, (c) 8000, (d) 25,600, and (e) 102,400.

4. Compute the dead time of the Geiger counter and the corresponding counting losses from this information: sample A gave a count rate of 9728 counts/min, sample B gave a rate of 11,008 counts/min, and together samples A plus B gave a rate of 20,032 counts/min.

5. Assuming that the dead time of a Geiger counter is 200 μsec, and that there are no other counting losses, what is the efficiency of the counter for (a) 2500 ionizing particles per second, (b) 1000, (c) 200, and (d) 5?

6. What is the useful range of counting rates if the dead time of the detector is (a) 0.25 μsec, (b) 1.0 μsec, (c) 5 μsec, and (d) 270 μsec?

7. The decay of a particular halogen, subjected to several hours of irradiation, provided the following data:

TIME, MIN	ACTIVITY, COUNTS/MIN	TIME, MIN	ACTIVITY, COUNTS/MIN
10	1800	50	650
18	1400	60	550
24	1215	80	430
32	970	120	330
36	880	180	270
40	800	240	230

Plot the decay curve on semilog paper and analyze it into its components. What are the half-lives and the initial activities of the component activities? Can you identify the particular halogen?

8. In a certain measuring arrangement, the beta particles of ^{136}Cs are absorbed as follows (correction made for gamma radiation):

Thickness of Aluminum, mg/cm^2	Activity, counts/min	Thickness of Aluminum, mg/cm^2	Activity, counts/min
0	10,000	53	270
12	4700	72	45
27	1700	85	10
41	730	100	10

Find the maximum energy of the beta radiation. What is the aluminum half-thickness?

9. The absorption of RaE (^{210}Bi) beta radiations produced the data below, uncorrected for counting losses or background. Dead time of counter is 200 μsec. Assume the absorption of the sample, air path, and counter window to be 34 mg/cm^2. Background counting rate is 30 counts/min.

Thickness of Aluminum, mg/cm^2	Activity, counts/min	Thickness of Aluminum, mg/cm^2	Activity, counts/min
0	19,100	163	1620
25	13,680	200	850
57	8720	265	270
90	4820	335	82
123	3080	1000	30

Determine the maximum energy of the beta radiation.

10. The visual range of absorber thickness for ^{32}P is 780 mg Al/cm^2. Determine the maximum energy of the beta particle.

11. Absorption data taken for ^{36}Cl indicated an aluminum half-thickness of 28 mg cm^{-2}. What is its maximum beta energy?

12. To a crude mixture of organic compounds containing some benzoic acid and benzoate was added 40.0 mg of benzoic acid-7-^{14}C (activity = 2000 counts/min). After equilibration, the mixture was acidified and extracted with an immiscible solvent. The extracted solid, following removal of solvent, was purified by recrystallization of the benzoic acid to a constant melting point. The purified material weighed 60.0 mg and gave a count rate of 500 counts/min. Compute the weight of benzoic acid in the crude mixture.

13. A fermentation broth was known to contain some Aureomycin. To a 1000-g portion of the broth was added 1.00 mg of Aureomycin containing carbon-14 (specific activity = 150 counts/min/mg). From the mixture, 0.20 mg of crystalline Aureomycin was isolated which had a net activity of 400 counts in 100 min. Calculate the weight of Aureomycin per 1000 g of broth.

14. Argon ionization detectors, employed in gas–liquid chromatography, utilize as radioisotope source ^{90}Sr and its daughter ^{90}Y. Estimate the range of the beta particles emitted in air and in iron (the material of construction of cell walls).

15. A 10.0-ml volume of a chloride-ion solution was added to a 50-ml volumetric flask and precipitated with 10.0 ml of 0.0440 N silver nitrate solution which contained silver-110. After the precipitate coagulated, the flask was filled to the mark and mixed thoroughly. A 20-ml aliquot of the clear supernatant liquid, after filtration or centrifugation, was counted and it gave a count rate of 924 counts/min. A 5.0-ml volume of the standard silver solution, diluted to 20 ml and counted, gave a count rate of 7555 counts/min. The background amounted to 100 counts/min. What is the chloride-ion concentration in the unknown solution?

16. In the analysis of mixtures of sodium and potassium carbonates, the half-lives are too nearly the same to permit a resolution of the composite gross decay curve if the total beta radiations were counted. Suggest a method for the analysis of this binary mixture. (*Hint:* the beta particle emitted in the decay of the sodium activity possesses a maximum energy of 1.39 MeV compared with the potassium activity, where beta particles of a maximum energy of 3.52 MeV are radiated.)

17. If a 10.0-mg sample of aluminum foil were irradiated for 30 min in a neutron flux of 5×10^{11} neutrons cm^{-2} sec^{-1}, how long should the sample be allowed to "cool" before chemical processing or counting in order that the strong aluminum activity will have decayed to less than 1 count/min?

18. For the irradiation time and flux stated in Problem 17, what is the limit of detection (40 counts/sec) for traces of sodium as sodium-24 in "pure" aluminum foil after the aluminum activity has decayed to less than 1 count/min. Counting geometry is 100%. Assume no other activities are present and ignore corrections for absorption of sodium beta particles by the aluminum foil.

19. What weight of sample should be taken for the activation analysis of an aluminum alloy which contains 0.019% zinc if the irradiation time is 62 hr with a flux of 5×10^{11} neutrons cm^{-2} sec^{-1}, followed by a cooling period of 24 hr? A counting rate of 1000 counts/min is desirable.

20. In a particular aluminum alloy, these elements are present in the following percentages: Cu, 0.30; Mn. 0.30; Ni, 0.59; Co, 0.0053. If all samples were 10.0 mg in weight, how long should the irradiations be continued for the determination of each element? Assume a counting rate of 10,000 counts/min in a 5-min counting period is desirable after a cooling period of 0.7 day. Flux is 5×10^{11} neutrons/cm^2/sec.

21. Following the irradiation conditions for copper in Problem 20, how many hours should elapse before a direct determination of copper (without intermediate chemical processing) is attempted? What thickness of aluminum absorber will attenuate completely all beta radiation from other elements when measuring the gamma radiation from radiocopper?

22. From the gamma-ray spectrum of neutron-activated sea water taken with a Ge(Li) detector, identify the elements present. Photopeak energies are expressed in keV.

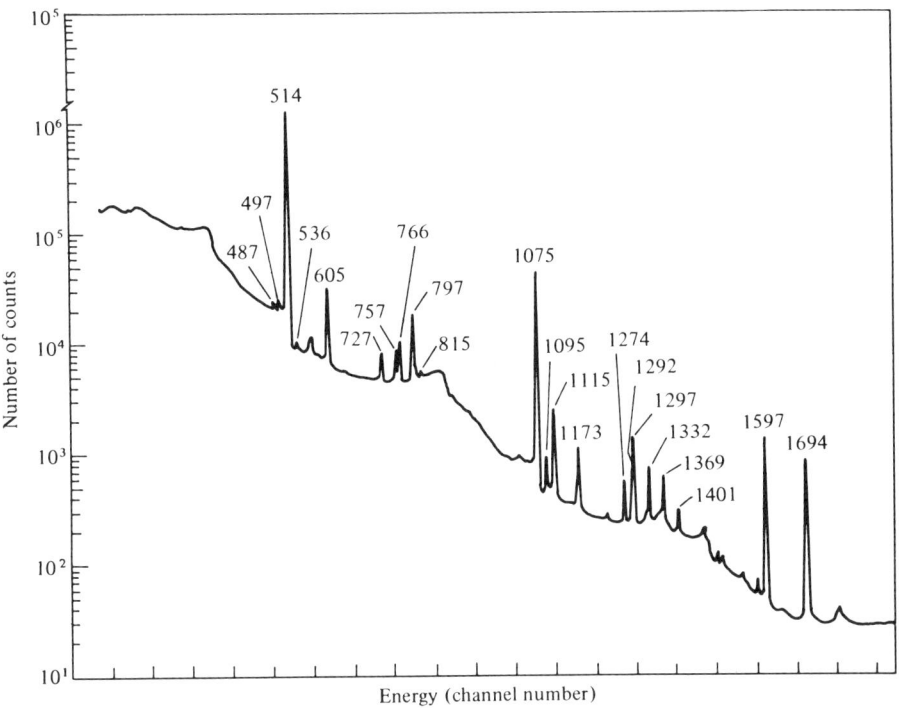

PROBLEM 10-22

23. From the gamma-ray spectrum taken with a Ge(Li) detector, decide whether a bullet produced the circular opening from which the sample was taken.

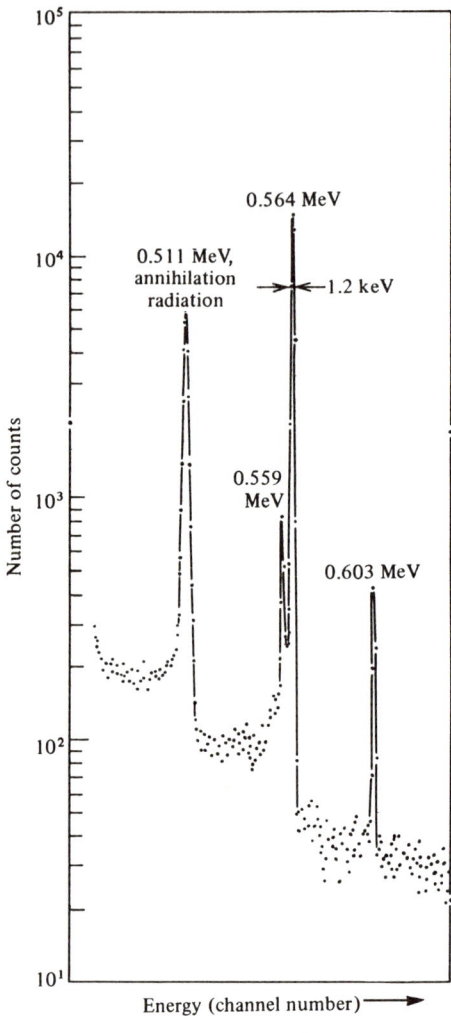

PROBLEM 10-23

BIBLIOGRAPHY

Barkouskie, M. A., "Liquid Scintillation Counting," *Am. Lab.*, p. 101 (May 1976).

Bernstein, K., "Activation Analysis," *Ind. Research,* p. 87 (September 1976).

Crouthamel, C. E. and R. R. Heinrich, "Radiochemical Separations," in *Treatise on Analytical Chemistry,* Vol. 9, Part 1, Chap. 96, I. M. Kolthoff and P. J. Elving, Eds., Wiley-Interscience, New York, 1971.

DeSoete, D., R. Gybels, and J. Hoste, *Neutron Activation Analysis,* Wiley-Interscience, New York, 1972.

Finston, H. L., "Radioactive and Isotopic Methods of Analysis: Nature, Scope, Limitations and Interrelations," in *Treatise on Analytical Chemistry,* Vol. 9, Part 1, Chap. 94, I. M. Kolthoff and P. J. Elving, Eds., Wiley-Interscience, New York, 1971.

Finston, H. L., "Nuclear Radiations: Characteristics and Detection," in *Treatise on Analytical Chemistry,* Vol. 9, Part 1, Chap. 95, I. M. Kolthoff and P. J. Elving, Eds., Wiley-Interscience, New York, 1971.

Friedlander, G., J. W. Kennedy, and J. M. Miller, *Nuclear and Radiochemistry,* 2nd ed., Wiley, New York, 1964.

Guinn, V. P., "Activation Analysis," in *Treatise on Analytical Chemistry,* Vol. 9, Part 1, Chap. 98, I. M. Kolthoff and P. J. Elving, Eds., Wiley-Interscience, New York, 1971.

Lukens, H. R., "Neutron Activation Analysis," *J. Chem. Educ.,* **44**, 668 (1967).

Lyon, W. S., Ed., *Guide to Activation Analysis,* Van Nostrand Reinhold, New York, 1964.

Ouseph, P. J. and M. Schwartz, "Nuclear Radiation Detectors," *J. Chem. Educ.,* **51**, A139, A209 (1974).

Seaman, W., "Tracer Techniques" in *Treatise on Analytical Chemistry,* Vol. 9, Part 1, Chap. 97, I. M. Kolthoff and P. J. Elving, Eds., Wiley-Interscience, New York, 1971.

CHAPTER 11

Nuclear Magnetic Resonance Spectroscopy

In nuclear magnetic resonance (NMR) spectroscopy, the characteristic absorption of energy by certain spinning nuclei in a strong magnetic field, when irradiated by a second and weaker field perpendicular to it, permits identification of atomic configurations in molecules. Absorption occurs when these nuclei undergo transitions from one alignment in the applied field to an opposite one. The amount of energy required to cause a particular nucleus to realign depends upon such factors as field strength, electronic configuration around the particular nucleus, anisotropy, type of molecule, and intermolecular interaction. The spectra obtained answer many questions such as (referring to specific nuclei): Who are you? Where are you located in the molecule? How many of you are there? Who and where are your neighbors? How are you related to your neighbors? The result is often the delineation of complete sequences of groups or arrangement of atoms in the molecule. Consequently, organic chemists have enthusiastically embraced NMR spectroscopy to identify and characterize molecules.

Analytical chemists, on the other hand, were more reluctant to accept NMR instrumentation. The sensitivity of NMR, compared with optical techniques, gas and liquid chromatography, and mass spectrometry, was down by several orders of magnitude, and usually precluded the use of NMR as a method for trace analysis. Also the cost and complexity of maintaining standard operating conditions frequently turned the tide in favor of other methods. Now, however, a new family of NMR instruments, called Fourier transform NMR and built around a small, high-speed digital computer, has revolutionized the practice of NMR in organic chemistry and firmly entrenched NMR techniques in the analytical chemist's arsenal of weapons.

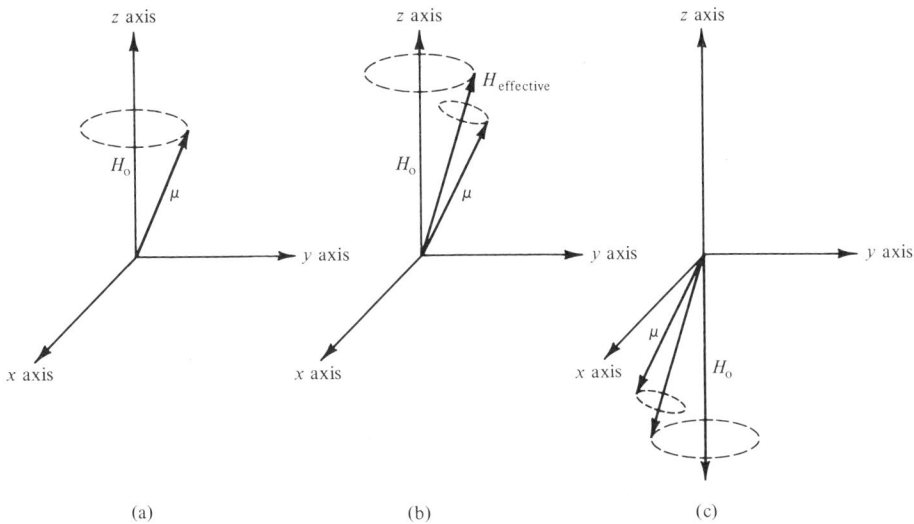

FIGURE 11-1 (a) Precession of a magnetic moment μ on application of a steady magnetic field H_0 along the z-axis. (b) Precession of μ on application of both H_0 and an rf magnetic field H_1 along the y-axis. (The fields add to give an effective field $H_{\text{effective}}$.) (c) At resonance the precessing nucleus flips to an antiparallel orientation relative to H_0.

11.1 BASIC PRINCIPLES

The nuclei of certain isotopes possess an intrinsic spinning motion around their axes. The spinning of these charged particles, or their circulation, generates a magnetic moment along the axis of spin (Fig. 11-1). If the nuclei are placed in an external magnetic field, their magnetic moment can align with or against the field. The individual nucleus spins around its axis and precesses about the force line of the applied magnetic field. These precessions are actually circular movements with respect to the force line and are restricted to a distinct number of angles between the field line and axis, as shown in Fig. 11-2. The field aligns the spinning nuclei against the disordering tendencies of thermal processes. However, the nuclei do not align perfectly parallel (or antiparallel) to the imposed magnetic field. Instead, their spin axes are inclined to the field and precess about the field direction, behaving like a gyroscope in a gravitational field. Each pole of the nuclear axis sweeps out a circular path in the xy-plane. Increasing the strength of the field only makes the nuclei precess faster. The frequency of precession, ν_0, is known as the Larmor frequency of the observed nucleus.

Nuclear Magnetic Energy Levels

For a nucleus to be magnetic, it must possess spin angular momemtum whose magnitude is $(h/2\pi)\sqrt{I(I+1)}$, where I is the spin quantum number of the particular nucleus and h is Planck's constant. Nuclei with $I = 0$ are nonmagnetic and will not concern us. Those

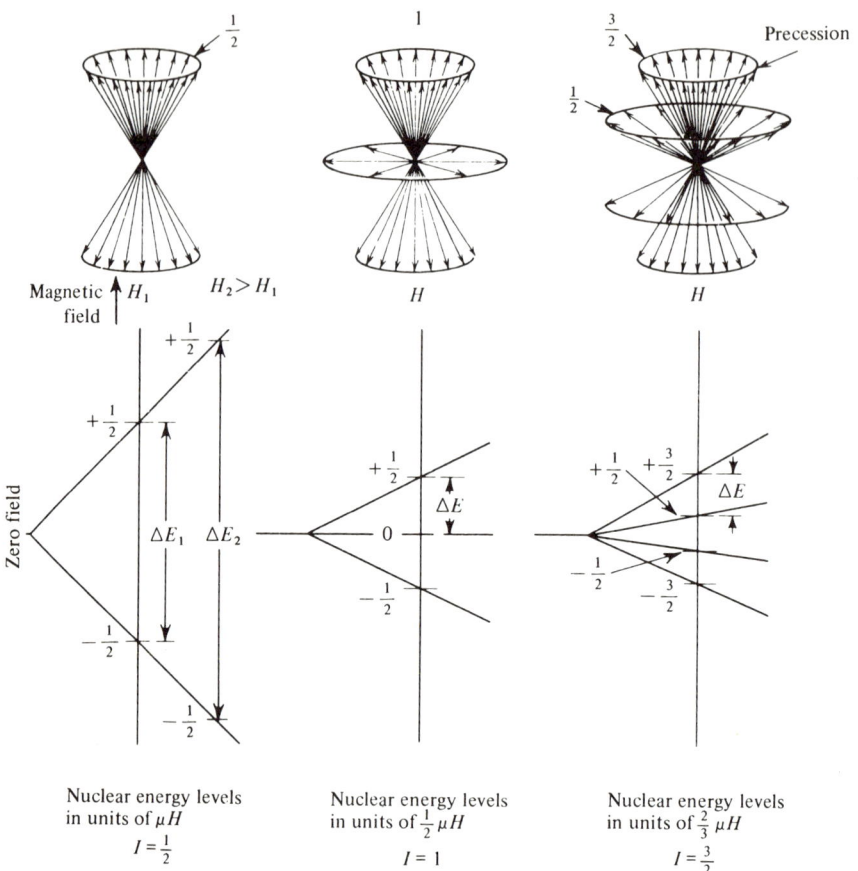

FIGURE 11-2 Nuclear orientation and energy levels of nuclei in a magnetic field for different spin numbers.

with $I = \frac{1}{2}$ give the best resolved spectra; important examples are ^{1}H, ^{13}C, ^{19}F, and ^{31}P nuclei. Nuclei of interest with $I > \frac{1}{2}$ include ^{2}H ($I = 1$), ^{14}N ($I = 1$), and ^{11}B ($I = \frac{3}{2}$). The magnetic axis of the nucleus can assume $2I + 1$ orientations with respect to the external magnetic field, and each orientation corresponds to a discrete energy level, given by

$$E = \frac{m\mu}{I} \beta H_0 \qquad (11\text{-}1)$$

where I is the spin number, m is the magnetic quantum number, E is the energy of transition, μ is the magnetic moment of the nucleus expressed in nuclear magnetons, β is a constant called the nuclear magneton (5.049 × 10^{-24} erg G^{-1}) and H_0 is the external magnetic field strength in gauss. The spectrum of allowed values, in terms of spin quantum number, is: $I, I - 1, \ldots, -(I - 1), -I$. Each value corresponds to a discrete orientation (and energy level). Hence, a nucleus with spin $\frac{1}{2}$ has two orientations; with spin 1, three orientations, and so on (Fig. 11-2).

At equilibrium the population of the various nuclear energy levels is predictable by use of a Boltzmann distribution

$$\frac{n_{upper}}{n_{lower}} = e^{-\mu H_0/IkT} \tag{11-2}$$

where k is the Boltzmann constant and T is the absolute temperature. For a magnetic nucleus of spin $\frac{1}{2}$ in a field of 14.09 kG, the distribution predicts a population ratio of 0.9999904 at room temperature. The lower energy level (orientation parallel to the applied magnetic field) is favored to the extent of approximately 9.5 excess nuclei out of every million. Thus, for a sample containing approximately 10^{19} nuclei, the effective participating population will be about 10^{14} nuclei.

Magnetic Resonance

The resonance frequency, ν_0, that will effect transitions between energy levels is derived by equating the Planck quantum of energy with the energy of reorientation of a magnetic dipole (Eq. 11-1):

$$\Delta E = h\nu_0 = \mu \beta H_0/I \tag{11-3}$$

If only H_0 is applied, the nuclear moments precess without any phase coherence. A radio frequency (rf) magnetic field, H_1, applied along the y-axis (Fig. 11-1), and represented as an oscillation along the x-axis, forces the nuclei to precess in phase. The rf field is the resultant of two superposed circularly varying fields of equal amplitude, one rotating clockwise and one counterclockwise, that add vectorially to give H_1. Only the component rotating in the same sense as the precessing nuclear magnetic moment can interact with the nucleus. As the frequency approaches that of the nuclear resonance frequency (Eq. 11-3), there is increasing interaction of the rf field H_1 and the precessing magnetic moment. When the frequencies are identical, resonance absorption occurs and the nuclei flip from the lower energy level to an upper energy level, that is, the spins originally precessing with H_0 flip over, and now precess against H_0. If the frequency of the rf field is swept through the region of the resonance frequency, peak absorption of energy from the rf oscillating field will be observed at the resonance frequency. Since there is a linear relation between resonance frequency and magnetic field, H_0, spectra may be expressed as intensity of absorption versus resonance frequency at fixed H_0, or against H_0 at fixed resonance frequency if the rf frequency is fixed and the magnetic field swept.

For a proton, $\mu = 1.41 \times 10^{-30}$ J G^{-1} or 2.7927 nuclear magnetons. From Eq. 11-3,

$$\nu = \frac{\Delta E}{h} = \frac{(1.41 \times 10^{-30} \text{ J G}^{-1})(14{,}092 \text{ G})}{(6.626 \times 10^{-34} \text{ J sec})(\frac{1}{2})} = 60 \times 10^6 \text{ sec}^{-1}$$

or

$$\nu = \frac{2.7927 (5.05 \times 10^{-24} \text{ erg G}^{-1})(14{,}092 \text{ G})}{(6.626 \times 10^{-27} \text{ erg sec})(\frac{1}{2})} = 60 \times 10^6 \text{ sec}^{-1}$$

TABLE 11-1 Magnetic Resonance Properties of Selected Nuclei

Isotope	Magnetic Moment, μ/μ_N*	Relative Sensitivity at Constant H_0†	NMR FREQUENCY, MHz		
			at 14.09 kG	at 21.14 kG	at 23.49 kG
^{1}H	2.7927	100	60.000	90.000	100.000
^{2}H	0.8574	0.96	9.210	13.815	15.352
^{13}C	0.7024	1.59	15.086	22.629	25.147
^{19}F	2.6288	83.4	56.444	84.666	94.087
^{31}P	1.1317	6.64	24.288	36.432	40.485

*In multiples of the nuclear magneton, $eh/4\pi Mc$.
†Sensitivity relative to the proton, assuming equal numbers of nuclei and the same relaxation time ratio, T_2/T_1.

Thus, in a magnetic field of 14,092 G, the protons will precess 60 million times per second, or 60 MHz. Consequently, 60 MHz is the resonance frequency required to flip the excess population in the lower-energy state to the higher-energy state.

Properties of nuclei frequently encountered in NMR spectroscopy are listed in Table 11-1. Because the strength of the absorption signal is roughly proportional to the square of the magnetic field, larger values of field strength lead to a stronger signal. Nuclei with $I = \frac{1}{2}$ act as though they are spherical bodies possessing a uniform charge distribution, which circulates over their surfaces. Their electric quadrupole moment is zero. Nuclei with spins of 1 or more possess nuclear quadrupole moments; the latter measures the electric charge distribution within a nucleus when it possesses nonspherical symmetry. Nuclei possessing nuclear quadrupole moments are readily disturbed by molecular electric field gradients. The result is a shortening of spin lifetime in a given state, and smearing out of the NMR signal.

Relaxation Processes

Upon irradiation of a particular nucleus, the rate of absorption of energy is initially greater than the rate of emission because of the slight excess of nuclei in the lower-energy state. However, the absorption signal rapidly attains some finite value. Only if relaxation back to the lower-energy state can occur at least as rapidly as absorption, will the intensity of nuclear absorption at a given frequency remain constant. Otherwise, in time, the rf field would equalize the populations of the energy levels and the spin system would become saturated. Since the difference in populations among the various energy levels is small, saturation is easy to bring about experimentally.

Two types of relaxation processes are involved. One, *spin–lattice* (or longitudinal) *relaxation,* is brought about by interaction of the spin with fluctuating magnetic fields produced by random motions of neighboring nuclei. The term spin–lattice relaxation has originated from considering an experimental set of nuclei as a spin system embedded in a lattice of other nuclei and electrons. The energy transferred from the nucleus in an upper-energy state is given to the lattice as extra translational or rotational energy. Relaxation occurs, in part at least, from thermal motions of other magnetic nuclei. Their Brownian motion gives rise to magnetic fields that occasionally have a fluctuation whose frequency is equal to the precession frequency of the nucleus to be relaxed. These oscillating components can therefore induce transitions and provide a mechanism by which nuclei lose their excess magnetic energy as thermal energy to the lattice. Basically, the relaxation process is first order and has a rate constant, called the spin–lattice relaxation time, T_1, that decreases exponentially. It is defined by the expression

$$(n - n_{eq})_t = (n - n_{eq})_0 \, e^{-t/T_1} \tag{11-4}$$

where n is the initial excess population in the lower-energy level and n_{eq} its equilibrium value in the presence of the rf field, H_1. Operationally T_1 represents the time, t, required for the Boltzmann distribution to be reestablished in the presence of H_0. When a time equal to T_1 has elapsed, the difference between the excess population and the equilibrium value has been reduced to 37% ($1/e$) of its original value. In solids and viscous liquids, the relaxation time is in the order of hours, but in typical organic liquids and dilute solutions the time is in the range of 0.01–100 sec.

A second time constant T_2 is assigned to *spin–spin* (or transverse) *relaxation* processes. Involved here are the transverse magnetization components in the *x*- and *y*-directions. When the rf field is off from the precession frequency, the spins tend to fall out of phase through dipole–dipole interaction and the net magnetization in the *x*- or *y*-direction falls toward zero. A nucleus in the upper-energy state can transfer its energy to a neighboring nucleus by a mutual exchange of spin. Recording the dispersion mode signal (see Fig. 11-5) as a function of time provides a curve from which T_2 can be calculated. Actually in liquids and gases where molecular diffusion and rotation are rapid processes, the local field around the magnetic nuclei changes sufficiently rapidly so that the mechanisms responsible for relaxation processes are essentially identical, and $T_1 = T_2$. If T_2 is long, the spectral lines are narrow; if it is short, they are broad.

Pulsed (Fourier Transform) NMR

The conventional NMR spectrometer scans the spectrum at a slow rate in order to avoid passing over a spectral line too rapidly since the lines are usually narrow. The spectrometer spends most of its time recording background, only occasionally does it record the desired information. Efficiency and consequently the sensitivity of such a system is far from optimum. The time required to observe a NMR spectrum by the continuous-wave method is Δ/r (in seconds), where Δ is the spectral width and r is the resolution desired. For ^{13}C at 25 MHz, where Δ is typically about 5 kHz and the linewidths are about 1 Hz, one must scan the 5-kHz region at a rate of 1 Hz sec^{-1}, or slower. This requires a minimum time of 5000 sec (or 83 min).

If a spectrum is thought of as a large number of small increments in frequency, each increment being just large enough to contain a typical spectral line, these increments can be examined simultaneously. This removes the constraint on scanning rate, since there is no scan. If the spectrum contains N increments, where N is just the spectral width (in hertz), the signal-to-noise ratio attainable is improved by approximately the factor $(N)^{1/2}$

since the signals add linearly while the noise adds as the square root of the number of pulses. For the proton NMR spectrum, this factor can be typically around 30, while for ^{13}C NMR it can be as large as 100.

Naturally the instrumentation must be modified appropriately to accomplish the simultaneous excitation of all the spectral lines and to sort the resulting information into the conventional representation of spectral lines. This is accomplished by applying a strong pulse of rf energy (H_1) to the sample for a very short time (1–1000 μsec). Under the influence of the pulse of rf energy, the magnetic moment spirals away from the z-axis in the direction of the static field. If the pulse is of the proper strength and duration, the magnetic moment is tipped by 90° and comes to rest in the xy-plane (Fig. 11-3). After the pulse has terminated, the restoring torque of the static field, H_0, causes a precession around the z-axis at the resonance frequency. The free precession of the nucleus under the influence of only the static field induces a decaying sinusoidal voltage in a coil of wire surrounding the sample. The voltage decays partly because the nuclear magnetic moment vector slowly spirals back up to its original position with an exponential time constant, T_1, the spin–lattice relaxation time. After several time constants ($3T_1$ to $5T_1$), the nuclei will have regained equilibrium and a second pulse can be applied to repeat the process. A second reason for the decay of the free induction signal is magnet inhomogeneity over the entire sample volume. Thus, some otherwise identically precessing nuclei begin to precess at slightly different rates and slowly lose phase coherence. Here we are speaking of the spin–spin relaxation time constant, T_2.

The free induction decay (impulse response) and the conventional continuous-wave display (steady state equilibrium) of the NMR spectrum form a Fourier transform pair. That is, the time response of the spins can be calculated from their frequency domain spectrum, and vice versa. The response of the entire spin system is picked up in the normal manner, amplified, and detected in the spectrometer. The free induction decay signal following each repetitive pulse is digitized by a fast analog-to-digital (ADC) converter, and the successive digitized transient signals are coherently added in the computer until an

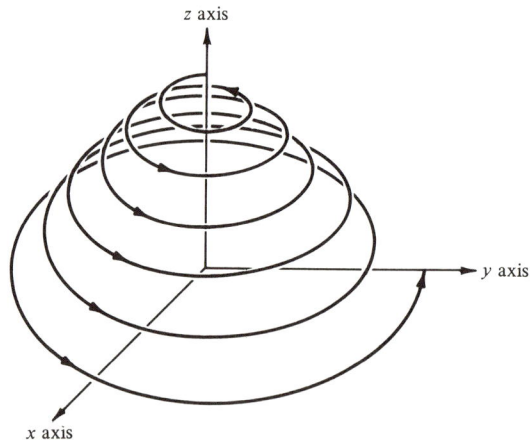

FIGURE 11-3 Spiral motion of nuclear magnetism in an oscillating magnetic field.

adequate signal-to-noise ratio is obtained. Using the Cooley–Tukey algorithm, the computer then performs a fast Fourier transformation to the frequency domain to plot a normal spectral presentation of the NMR absorption versus frequency in a matter of 10–20 sec. In a practical spectrometer a 16K computer can devote 8192 channels (8K) to data storage, the rest being used for the program to control the spectrometer and process the data.

The sampling time during which data points must be collected to obtain the true NMR spectrum after Fourier transformation depends on the spectral width Δ. The sweep time per data point (or dwell time) must be $(2\Delta)^{-1}$ sec/point. Thus, for a spectral width of 5 kHz, a dwell time of 100 μsec is required. Now multiplying the dwell time by the number of data points N to be collected during the free induction decay yields the time required for recording the interferogram digitally, which is $N/2\Delta$. For an 8K interferogram (8192 points), the time will be 0.82 sec. This is also the minimum repetition time between two pulses when several pulse interferograms must be accumulated in order to improve the signal-to-noise ratio. The resolution for this 8K interferogram (in Hz) is $2\Delta/N$, or 1.23 Hz. Hence, for a full data table, 8192 = (ADC rate) × (acquisition time) = 2Δ × (acquisition time).

The pulsed-Fourier transform technique makes possible the study of less sensitive nuclides, such as ^{14}N, ^{15}N, ^{17}O, ^{31}P, and unstable species, in addition to ^{13}C. Due to chemical shielding, each nucleus may resonate within a range of Larmor frequencies, depending on the chemical environment. In order to rotate all nuclear spins within that range by the same angle, the strength of the rf pulse must meet the requirement: $\mu H_1/I \gg 2\pi\Delta$. Furthermore, the pulse width, t_p, must be much shorter than the relaxation times, that is, $t_p \ll T_1, T_2$, so that relaxation is negligible during the pulse.

Proper selection of pulse sequences under computer control allows measurement of the various T_1 values. In one method, a 90° pulse is used to orient the spins. During a delay time, τ, a portion of these spins relax along the z-axis and some portion of the net nuclear magnetization along the z-axis ($\vec{M}_z$) is reestablished. A second 90° pulse will flip the remaining vectors out of the xy-plane in a $-z$ direction; while those which have reconstituted $\vec{M}_z$ are flipped back along y where they produce a second signal. The amplitudes of the first (A_0) and the second (A_τ) signals are related to the delay time by

$$\log(A_0 - A_\tau) = \log A_0 - (0.434/T_1) \quad (11\text{-}5)$$

A plot of $\log(A_0 - A_\tau)$ versus τ enables T_1 to be calculated from the slope. These measurements provide additional information of value to chemists. T_1 depends upon the average distance of the nucleus from magnetic neighboring nuclei, and upon the types of molecular motion which the functional group is undergoing. Internal rotations and segmental motions can be detected in this way.

Wide Line NMR

Wide line spectra are those in which the observed width of the resonance line is as large or larger than the major resonance shifts caused by differences in the chemical environment of the observed nucleus. Thus, wide line NMR supplies information regarding the concentration and physical environment of an observed isotope, but not its chemical environment. It is applicable to solids as well as liquids; sample sizes can range from 0.1 to 50 ml.

An early and continuing application has been the quantitative analysis of materials for particular isotope content from the integrated area under the NMR absorption band. It is a rapid, nondestructive method of analyzing for proton content of fats and oils, and of moisture in many types of materials. The determination of fluorine content in plastics and chemical compounds is another area of application. Calibration with a standard is required, and accuracy and precision are limited by environmental factors.

The width and shape of the resonance lines are indicative of the physical environment of the isotope. In particular, the width reveals the degree of motional freedom of the isotope in its physical environment, valuable information with respect to high polymer chemistry and solid-state physics. If a magnetic nucleus exists in an isolated magnetic field, its spectrum appears as a single line. However, if two hydrogen nuclei are contiguous to each other in a rigid state, the respective hydrogen nucleus is affected by the local magnetic field, which is caused by the magnetic dipole–dipole interaction, in addition to the external magnetic field. Stretched Teflon, for example, exhibits a doublet spectrum. If the temperature dependence of the linewidth is measured, the transition point of molecular motion can be obtained.

11.2 CONTINUOUS-WAVE NMR SPECTROMETERS

Continuous-wave NMR instrumentation involves six basic units: (1) a magnet to separate the nuclear spin energy states; (2) at least two rf channels, one for field/frequency stabilization and one to furnish rf irradiating energy; a third may be employed for each nucleus to be decoupled; (3) a sample probe containing coils for coupling the sample with the oscillating rf field(s); (4) a detector to process the NMR signals; (5) a sweep generator for sweeping either the static or oscillating field through the resonance frequency of the sample; and (6) a recorder to display the spectrum. These are schematically shown in Fig. 11-4. The spectrum can be scanned by the field-sweep method or the frequency-sweep method. If the static magnetic field, H_0, is held constant, which keeps the nuclear spin energy levels

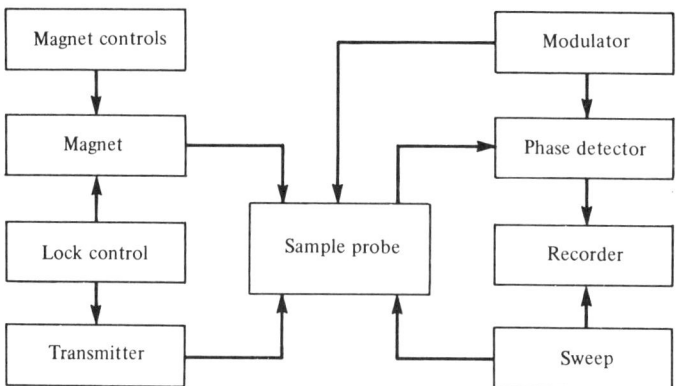

FIGURE 11-4 Block diagram of a high-resolution, continuous-wave, NMR spectrometer.

constant, then the rf signal can be swept (varied continuously over a spectral range) to determine the frequencies at which energy is absorbed; this is the *frequency-sweep* method. If the oscillatory rf signal, H_1, is held constant, then the magnetic field can be swept, which varies the energy levels, to determine the magnetic field strengths which produce resonance; this is the *field-sweep* method.

The Magnet

The strength of the magnetic field, H_0, determines the Larmor frequency of any nucleus. Because chemical shifts and spectrometer sensitivity are field dependent, it is often desirable to operate at the highest field strength commensurate with homogeneity and stability. The stronger the magnetic field, the better the line separation of chemically shifted nuclei in the frequency scale. Since coupling constants remain unaffected by the magnetic field strength, multiplet overlapping decreases with increasing field strength, and homonuclear couplings become small compared to chemical shift differences. The homonuclear multiplets approach first-order systems assignable by following the multiplicity rule. Moreover, the population of the lower spin level increases with increasing field, leading to a corresponding increase in the sensitivity of the NMR experiment.

For high-resolution work the magnetic field over the entire sample volume must be maintained uniform in space and time. Effective homogeneity of the field is promoted by (1) the use of large pole pieces composed of a very homogeneous alloy, (2) the polishing of pole faces to optical tolerances, and (3) the use of a narrow pole gap, that is, a smaller sample cross section and consequently a compromise with decreased sensitivity. Permanent magnets are simple and inexpensive to operate but require extensive shielding and must be thermostatted to ±0.001°C. Commercial units using electromagnets (or permanent magnets) operate at 14.09, 21.14, or 23.49 kG. An electromagnet requires elaborate power supplies and cooling systems, but these disadvantages are offset by the opportunity to employ different field strengths to disentangle chemical shifts from multiplet structures and to study different nuclei.

Cryogenic superconducting solenoids produce homogeneous fields at 50 and 70.5 kG and, as a result, make possible high-resolution experiments with 220- and 300-MHz rf fields, respectively. In the cryogenic solenoids many turns of copper-clad niobium–tantalum superconducting wire are immersed in a Dewar holding liquid helium. The Dewar is surrounded by another holding liquid nitrogen. Aided by shim coils positioned in the probe, the 220-MHz spectrometer achieves a resolution of 1.1 Hz (full line width at half maximum amplitude) and a 55:1 signal-to-noise ratio for the ethyl benzene quartet in a 1% solution. The 300-MHz instrument achieves resolution of 1.5 Hz and a 65:1 *S/N* ratio under the same conditions.

The Probe Unit

The probe unit is the sensing element of the spectrometer system. It is inserted between the pole faces of the magnet, in the *xy*-plane of the magnet–air gap by an adjustable probe holder. The probe unit houses the sample, the rf transmitter(s), output attenuator, receiver, and phase-sensitive detector. The sample is contained in a cylindrical, thin-walled, precision-bore, glass tube having an outer diameter of 5 mm. To average small magnetic field inhomogeneities in the xz-plane, an air-bearing turbine rotates the sample tube at a rate of several hundred revolutions per minute. This spinning produces sidebands in the spectrum because the NMR peaks are modulated at the spinning frequency. Since most NMR samples, except for ^{13}C solids and for wide line technique, are liquid solutions, the sample tube is filled until the length/diameter ratio is about five, which approximates that of an infinite cylinder. For minute samples, a tube with a capillary bore that widens to a spherical cavity at the position of the rf coil is used. If the ^{1}H spectrum is being studied, an amount between a few micrograms and a few milligrams of sample is dissolved in a solvent which has had all protons replaced by deuterium atoms. Chloroform-d, acetone-d_6, and benzene-d_6 are commonly used. The deuterium serves two purposes. It replaces hydrogen nuclei which would otherwise generate a background solvent signal and would overwhelm the signals from the sample. In addition, the magnetic resonance response of the deuterium nuclei can be used to lock the ratio of the magnetic field and frequency of the instrument over long periods of time. If the ^{13}C spectrum is desired, an amount of sample between a few milligrams and a few hundred milligrams is dissolved in one of the same deuterated solvents used for proton NMR. The deuterium in this case serves only the purpose of locking the spectrometer.

Two probe designs are used. A *single-coil* probe has one coil which not only supplies the rf radiation to the sample but also serves as a part of the detector circuit for the NMR absorption signal. To detect the resonance absorption and to separate the NMR signal from the imposed rf field, a rf bridge is used. The exciting signal is balanced against an equal amplitude reference, with the modulation appearing as bridge unbalance and extractable in that form. Alternatively, the signal can be amplified and then subjected to diode detection to extract the resonance spectrum. *Crossed-coil* (nuclear induction) probes have two coils, one coil for irradiating the sample and a second coil mounted orthogonally for signal detection. The irradiating coil is split in two halves with the sample inserted between. This coil is oriented with its axis perpendicular to the magnetic field (that is, along the y-axis). The detector coil is wound around the sample tube with its axis (the x-axis) perpendicular to both the field H_0 (y- axis) and the rf field (H_1) axis. Since magnetic resonance produces a net magnetization in the xy-plane, a current is generated at resonance in the receiver coils from an indirect coupling between the rf field and receiver coils, with the coupling produced by the sample itself. This design permits selective pickup of the resonance signal while virtually excluding the applied rf field.

Instrument Stabilization

The NMR spectrum is recorded directly on precalibrated chart paper. This demands that the ratio of rf frequency to field strength be very stable. To obtain an NMR peak having a linewidth of 0.1 Hz at 60 MHz requires an overall stability of $0.1/(60 \times 10^6)$, or about 2 parts in 10^9. There is no problem with the oscillating rf frequency, but magnet stability is only about 10^{-7}/hr. Independent stabilization of the two units is difficult. The

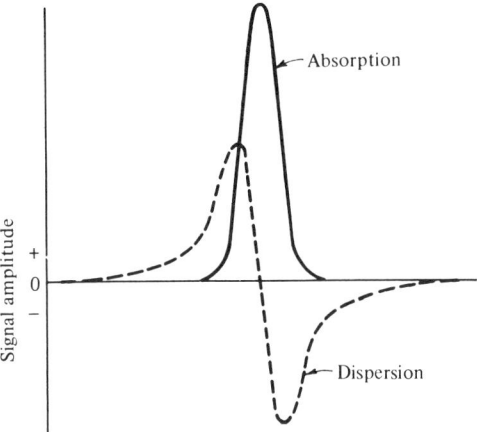

FIGURE 11-5 Line shapes of the two observable NMR signals.

field-frequency (H/ν) relation of Eq. 11-3 is more reliably maintained by means of servo loops that lock them together. Since a NMR signal is an ac signal with two components 90° out of phase, the detectable signals are an absorption component and a dispersion component, as shown in Fig. 11-5. By using a phase-sensitive detector which can be finely tuned to sense only one component, either the absorption or the dispersion mode can be observed. Since NMR spectra are usually observed in the absorption mode, the dispersion mode is available for field-frequency control and for measuring T_1, the spin–lattice relaxation time constant. The field/frequency control loop can be produced in two ways. An external lock uses a signal from a separate, adjacent sample, whereas the internal lock uses the sample under analysis for the locking signal. An external reference nucleus is continuously irradiated at its resonance frequency, and the resultant NMR dispersion signal is continuously monitored while the spectrum is being swept. If the frequency of irradiation exactly equals the frequency at the center of the dispersion signal, the NMR error signal will be zero. But if the magnetic field should change, the resonance condition is no longer fulfilled and an output signal will be produced. The output error signal is amplified and fed back into an array of very thin, multiple-turn coils of precise geometric shapes. These "shim" coils are manufactured of special material without soldering and encased in epoxy resin. They are mounted permanently in the pole cap covers of the magnet. Since inhomogeneities of a magnetic field take the form of gradients, corrections are made by creating small corrective fields that oppose these gradients. Corrections are available along the x-, y-, and z-axes, in the xy- and yz-planes, and for curvature.

In an internal lock system, a suitable reference such as tetramethylsilane (TMS) is added to the sample, so that the analytical sample nuclei and reference nuclei experience exactly the same magnetic field. An automatic shim control continuously compensates for the several gradients. The important y-axis shim control provides y-axis stability equivalent to that given to the xz-plane by spinning the sample, and permits signal accumulations of weak samples.

Sweeping Modes

To flip the rotating nuclear axes with respect to the magnetic field in the field-sweep method, a linearly oscillating rf field is imposed at right angles (x-axis) to the magnetic field (z-axis). Auxiliary coils, wound around the pole pieces, allow a sweep to be made through the applied magnetic field.

In the frequency-sweep mode, the magnetic field and the locking frequency are held constant while the portion of the spectrum of interest is scanned by sweeping the observing rf. It is then possible to introduce a third rf for the purposes of spin decoupling, or spin tickling experiments. This third rf field is held at a fixed separation from the reference line in the spectrum, and can thus be held at exact resonance for a chosen signal while the rest of the spectrum is examined by the rf field used for examining the spectrum (H_1, ν_1). The resultant decoupled spectra may then be observed by sweeping ν_1 through any desired part of the spectrum. This is the most immediate advantage of frequency sweep; namely, the effect on all other resonances of irradiating a chosen resonance may be observed in a single experiment. The frequency sweep is better than field sweeping because of the inherent advantage of a steady state experiment over a transient one. The sweep range for protons is approximately from 2000 Hz downfield to 500 Hz upfield; for other nuclei, such as ^{13}C and ^{19}F, the sweep range must be increased to more than 10 kHz. At the usual sweep rate of about 1 Hz, a ringing, envelopelike pattern appears over the trailing edge of an absorption band. It arises from rapid sweeping through the resonance condition. The frequency of the rotating component of magnetization varies with the changing sweep field and, as a result, the induced signals are alternately in and out of phase with the applied rf field, the frequency of which is constant. The observation of ringing is a good indication of a homogeneous field.

Minimal-Type NMR Spectrometer

Among the families of continuous-wave NMR spectrometers, the minimal type has stressed reliability, ease of operation, and a cost/performance trade-off. This basic instrument often utilizes a permanent magnet of 14, 21, or 23 kG field strength, and rf fields of 60, 90, or 100 MHz, respectively. Emphasis is on the measurement of proton NMR spectra. This family might be designated as a high-resolution, minimal-type NMR spectrometer lacking most of the auxiliary accessories found on the more sophisticated instruments.

A schematic diagram of a typical instrument is shown in Fig. 11-6. Each frequency needed for the selected magnetic nuclei is synthesized from a suitable harmonic of a 5-MHz crystal oscillator and mixed with the output of an appropriate low-frequency incremental oscillator. A control (dual) channel "locks" on the proton resonance of a water sample at 60.005 MHz. The frequency synthesizer uses the twelfth harmonic of the base oscillator (59.9 MHz), a 0.1-MHz incremental oscillator, and a single sideband modulator that mixes the two outputs and selects the upper sideband: 59.9 + 0.1 = 60.0 MHz. The final increment of 5 kHz is supplied by a 5-kHz audio-frequency oscillator that also modulates the H_0 field of the permanent magnet. For other nuclei, plug-in units provide a different multiplier circuit for selecting the desired harmonic, incremental oscillator, sweep oscillator parameters.

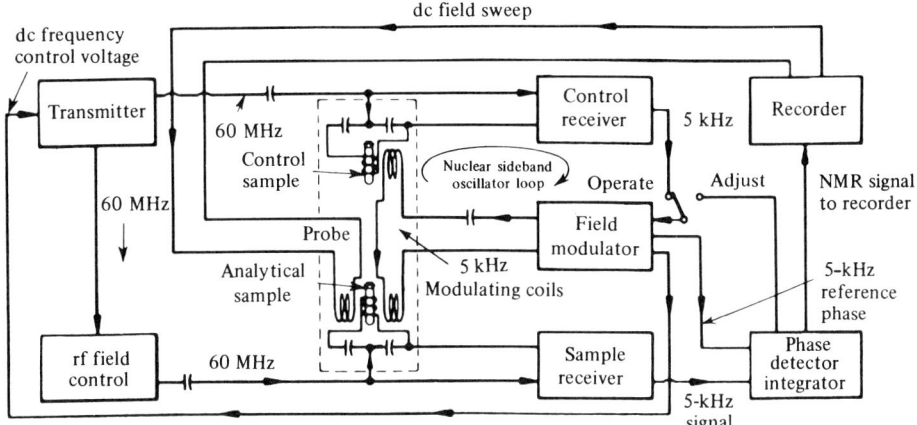

FIGURE 11-6 Schematic circuit diagram of Varian A-60 NMR spectrometer. (Courtesy of Varian Associates.)

Multipurpose NMR Spectrometers

The second family of NMR spectrometers is more diverse. Designed primarily for research, emphasis is on high performance and versatility; cost is a secondary consideration. These instruments are capable of high precision through the use of homonuclear and heteronuclear lock systems and frequency synthesizers. They are also characterized by high intrinsic sensitivity and ability to study a variety of nuclei.

The strength of the magnetic field is quite important since sensitivity, resolution, and the separation of chemically shifted peaks increase as the field strength increases. In addition, the complexities of spin–spin coupling are reduced at higher fields. These instruments may employ cryogenic solenoids and rf fields of 220 or 300 MHz.

Wide Line NMR Spectrometer

The wide line NMR spectrometer may use a frequency synthesizer to generate the rf field and a permanent magnet or a compact and light-weight electromagnet. Slowly varying scan voltages are directly injected in the regulator for the magnet power supply for the electromagnet. These signals cause a corresponding change in magnet energizing current, thereby creating a scan of the static field. Sample probe temperatures may be varied over the range −170° to 200°C. Sample tubes will be 15 or 18 mm in outer diameter. The standard magnetic field is 9.4 kG for protons and 10.0 kG for ^{19}F; the rf field is 40 MHz.

Instruments are also available wherein the rf applied field is continuously adjustable over a basic frequency range of 300 Hz–31 MHz, usually in steps of 10 Hz. The continuously adjustable H_1 level allows a quantitative determination of T_1 relaxation values.

For signal detection a sweep unit generates sinusoidal audio-modulation voltages having selectable frequencies of 20, 40, 80, 200, and 400 Hz. The output is amplified for simultaneous application to the probe modulation coils and to the x-axis of the oscilloscope. The sweep unit also provides a reference voltage having the same frequency as the selected modulation frequency. This is applied to a phase-sensitive detector to guarantee clear differentiation between absorption and dispersion modes. The detected signal contains a superposed audio frequency. Following either selective or broad band amplification, the signal can be recorded in first derivative mode, or according to the sideband procedures, in the undifferentiated form. An integrating circuit measures the area under the absorption band.

11.3 PULSED FOURIER TRANSFORM NMR SPECTROMETER

An NMR spectrometer capable of pulsed Fourier transform measurements is a combination of a continuous-wave circuit, as found in conventional NMR spectrometers, a computer controllable pulse generator, and a small digital computer of core memory size 8–20K. A simplified block diagram of a pulsed Fourier transform unit is shown in Fig. 11-7. The computer controllable pulse programmer generates dc pulses. For ^{13}C NMR resonance, the output of a 25-MHz rf oscillator in the Fourier transform unit is fed to an rf gate and sent to the power amplifier units only when the gate is open. The digital pulse programmer controls gate on/off (that is, the rf input signal of the H_1 transmitter) and determines the pulse width, interval, and repetition rate. The timing of each pulse train is controlled by the clock pulse from the built-in, highly stable, crystal-controlled oscillator to assure high resettability. The widths of the resulting rf pulses are adjustable in 1-μsec steps for 90° or 180° phase. In addition various pulse sequences can be programmed. After the rf pulse is amplified, the intense rf pulse excites the sample to be investigated. The free induction decay signal is then filtered, amplified, detected by the phase-sensitive detector, and digitized. This signal may be accumulated while simultaneously decoupling all protons. For ^{13}C experiments, the ^{2}H signal of the deuterated solvent is used for the internal lock of the magnetic field. The lock signal is observed on the oscilloscope to adjust the magnetic homogeneity during signal accumulation. The oscilloscope can also be used for monitoring the free induction decay signal and quick display of accumulated signals and the Fourier transformed spectra to determine if the signal-to-noise ratio is satisfactory. The rest of the operation is under computer control. The operator can set the values of the spectral width, acquisition time, number of transients, and pulse width. For the first look at a sample,

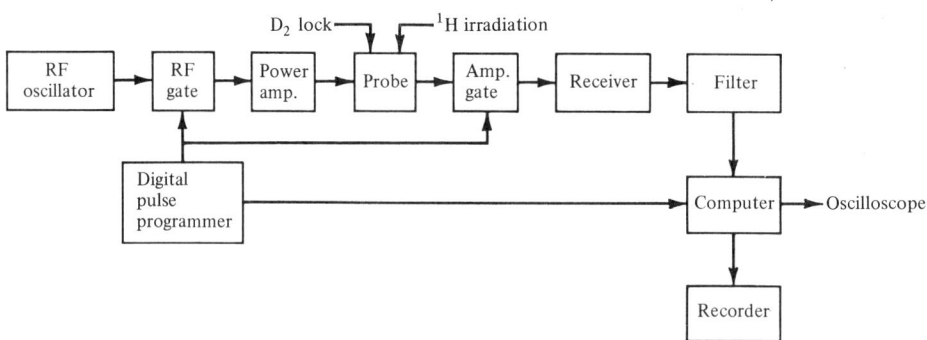

FIGURE 11-7 Block diagram of pulsed Fourier transform NMR spectrometer.

these and other parameters are optimized. For repetitive or routine analyses, however, optimum parameters can be stored in a tape cassette and the experiment can be set up simply by loading the parameter set into the computer. Acquisition of data is automatic and terminates when the number of transients requested is reached. A printout of the parameter set and the number of completed transients can be obtained for a permanent record. The fact that data are stored in digital form in the computer makes possible the numerical integration of the areas under each of the lines. This results in particularly accurate, drift-free integral values, and greatly decreases the requirements of operator skill. Spectra and integrals are plotted simply by entering the commands.

Magic Angle Spinning and Cross Polarization

No longer are chemists restricted to examining NMR spectra from liquid solutions. For nuclei, such as ^{13}C, line broadening for a polycrystalline material arises from the fact that the molecules are oriented in all possible directions. The shielding any particular nucleus receives from its electronic environment is thus a function of the orientation of the molecules containing it. This type of line broadening due to chemical shift anisotropy can largely be removed by rotating the sample very rapidly (>2 kHz) about an axis oriented at an angle of approximately 54.7° (the magic angle) with respect to the external magnetic field. The averaging that occurs is similar to tumbling in liquids; the quality of the NMR spectrum can approach that of the liquid state spectrum. The angle is a property of the local fields that electrons exert on nuclei and of the tensors that describe such behavior.

The accessory to accomplish NMR experiments with crystalline materials on 220-MHz superconducting NMR spectrometers includes a probe and a high-power, 100-W, 220-MHz amplifier capable of performing dipolar decoupling and cross polarization. Solids may be machined into rotors or packed as powders into hollow rotors. Samples are introduced into the probe by dropping the rotor into a tube at the top of the magnet. The rotor automatically positions itself at the magic angle.

Because ^{13}C has a low natural abundance, NMR measurements on this nucleus generally require signal averaging. Cross-polarization techniques overcome the problems of long T_1 times and consequent limited sensitivity. The technique relies on the presence of a system of abundant nuclear spins (^1H) in order to observe the NMR signal from the dilute nuclear spin (^{13}C). The procedure consists of four basic timed sequences of rf pulses. The four-part procedure consists of (1) polarization of the ^1H spin system by applying a 90° rf pulse at the ^1H resonance frequency; (2) spin locking in the rotating frame by applying a 90° phase shift to the foregoing field; (3) establishing $^{13}C-^1H$ contact by applying a rf field at the ^{13}C resonance frequency; and (4) observation of the ^{13}C free induction decay, while the ^1H field is maintained for decoupling. The entire sequence is repeated many times until a suitable signal-to-noise ratio for ^{13}C is achieved. Details are given by Miknis et al.[1]

11.4 SPECTRA AND MOLECULAR STRUCTURE

For most purposes, high-resolution NMR spectra can be described in terms of *chemical shifts* and *coupling constants.* Two other parameters sometimes involved are the *spin–*

lattice (T_1) and the *spin–spin* (T_2) *relaxation times* of the nuclei. Internal rotation, chemical exchange, and other rate processes can affect relaxation times so as to produce pronounced temperature dependent effects on the spectra. In solids, direct magnetic dipole–dipole interactions dominate, relaxation times are long, and the NMR spectra consist of very broad lines. In liquids and gases, the direct dipole–dipole interactions usually are averaged to zero by rapid intra- and intermolecular motions, the relaxation times are much shorter, and narrow line NMR spectra are observed.

Chemical Shifts

An important feature of high-resolution NMR spectra is *chemical shift*. In different chemical environments the same type of nucleus will be shielded slightly from the applied field in a manner depending on the distribution of the surrounding electrons. For a fixed external field, H_0, different screening factors cause slightly different resonant frequencies.

The magnitude of the effective field felt by each group of nuclei can be expressed as follows:

$$H_{\text{eff}} = H_0(1 - \sigma) \tag{11-6}$$

where σ is a nondimensional shielding constant, and may be either a positive or negative number. Thus, the protons at various sites in a molecule are spread out into a spectrum according to the values of their shielding parameters. A field or frequency sweep will bring protons at each particular site into resonance one after another. The more the field induced by the circulating electrons shields the nucleus and opposes the applied field, the higher must be the applied field to achieve resonance if the field is varied, or the lower must be the resonance frequency if the frequency is varied. The specific locations of the shifted resonant frequencies can be used to characterize the neighbors of a given nucleus. The value of the shielding constant depends on several factors, among which are the hybridization and electronegativity of the groups attached to the atom containing the nucleus being studied. Shielding effects seldom extend beyond one bond length except with very strong electronegative groups.

Because NMR spectrometers employing different field strengths are in use, it is desirable to express the position of resonance in field independent units and with respect to the resonance of a reference compound. For proton spectra in nonaqueous media, the reference material is tetramethyl silane, $(CH_3)_4Si$, abbreviated TMS, whose position is assigned as exactly 0.0 on the δ scale. TMS contains 12 protons but these are all chemically equivalent and therefore give rise to a single sharp signal. The magnitude of the *chemical shift* is expressed in parts per million:

$$\delta = \frac{H_{\text{sample}} - H_{\text{TMS}}}{\nu_1} \times 10^6 \tag{11-7}$$

where H_{sample} and H_{TMS} are the positions of the absorption lines for the sample and reference, respectively, expressed in frequency units (hertz); ν_1 is the operating frequency of the spectrometer. A positive δ value represents a greater degree of shielding in the sample

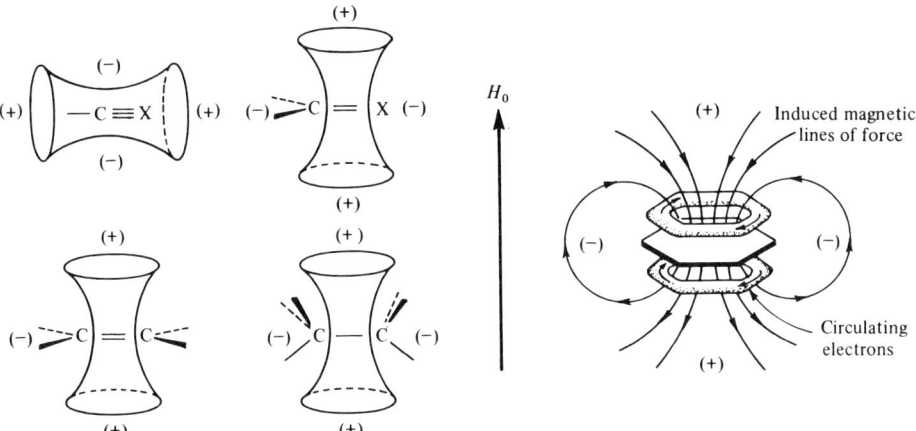

FIGURE 11-8 Shielding (+) and deshielding (−) zones in the neighborhood of triple, double, and single bonds to carbon and aromatic rings.

than in the reference. Another frequently used convention, but not officially approved, is the τ scale, in which $\tau = 10.0 - \delta$.

Recommended reference materials for other nuclei include CS_2 or TMS for ^{13}C, trichlorofluoromethane (CCl_3F) for ^{19}F, and phosphoric acid for ^{31}P. The numbers on the dimensionless (shift) scale upfield from the reference are designated positive.

Proton resonances from C–H bonds are located in the range from $\delta = 0.9$ to 1.5 when only aliphatic groups are substituents. A CH_3 group usually appears at $\delta = 0.9$ when the adjacent three bonds are methylene groups; CH_2 and CH protons are slightly further downfield in that order. An adjacent unsaturated bond shifts the resonant position of CH_3 to $\delta = 1.6-2.7$. An adjacent oxygen atom markedly shifts proton signals downfield to $\delta = 3.2-3.4$ for aliphatic entities and to $\delta = 3.6-3.9$ for aryl-O-CH situations. Many common groups produce special shielding effects because they allow circulation of electrons only in certain preferred directions within the molecule. Figure 11-8 shows shielding (+) and deshielding (−) zones in the neighborhood of triple, double, and single bonds to carbon. In C=C and C=O double bonds, the deshielding zone extends along the bond direction; even C–C bonds show some deshielding in this direction. This anisotropy of the magnetic susceptibility of chemical bonds means that the shielding or deshielding of a neighboring proton in the molecule is dependent on its distance from the bond and its orientation with respect to that bond. Aromatic rings exhibit a strong anisotropic effect. When such compounds are placed in a magnetic field, the six π-electrons circulate in two parallel doughnut-shaped orbits on each side of the ring. The resulting local magnetic field opposes H_0 in a cone-shaped zone of excess shielding extending along the hexad axis, but reinforces H_0 in a zone of deshielding extending from the edge of the ring. In aromatic compounds the deshielding zone is more commonly occupied; thus, protons on aromatic rings appear at much lower field ($\delta = 7-8$) than olefinic protons ($\delta = 5-6$). In acetylenes, the electron current circulates in such a way that the shielding zone extends along the bond direction, and acetylenic protons appear at high fields ($\delta = 1.6-3.0$). Table 11-2 gives some proton chemical shifts; Table 11-3 gives selected ^{13}C chemical shifts.

TABLE 11-2 Proton Chemical Shifts (Values are given on the officially approved δ scale; $\tau = 10.00 - \delta$.)

Substituent Group	Methyl Protons	Methylene Protons	Methine Proton
HC–C–CH$_2$	0.95	1.20	1.55
HC–C–NR$_2$	1.05	1.45	1.70
HC–C–C=C	1.00	1.35	1.70
HC–C–C=O	1.05	1.55	1.95
HC–C–NRAr	1.10	1.50	1.80
HC–C–NH(C=O)R	1.10	1.50	1.90
HC–C–(C=O)NR$_2$	1.10	1.50	1.80
HC–C–(C=O)Ar	1.15	1.55	1.90
HC–C–(C=O)OR	1.15	1.70	1.90
HC–C–Ar	1.15	1.55	1.80
HC–C–OH (and OR)	1.20	1.50	1.75
HC–C–C≡CR	1.20	1.50	1.80
HC–C–C≡N	1.25	1.65	2.00
HC–C–SR	1.25	1.60	1.90
HC–C–OAr	1.30	1.55	2.00
HC–C–O(C=O)R	1.30	1.60	1.80
HC–C–SH	1.30	1.60	1.65
HC–C–(S=O)R and –SO$_2$R	1.35	1.70	
HC–C–NR$_3^+$	1.40	1.75	2.05
HC–C–O(C=O)CF$_3$	1.40	1.65	
HC–C–Cl	1.55	1.80	1.95
HC–C–O(C=O)Ar	1.65	1.75	1.85
HC–C–Br	1.80	1.85	1.90
HC–CH$_2$	0.90	1.30	1.50
HC–C=C	1.60	2.05	
HC–C≡C	1.70	2.20	2.80
HC–(C=O)OR (and NR$_3$)	2.00	2.25	2.50
HC–SR	2.05	2.55	3.00
HC–O–O	2.10	2.30	2.55
HC–(C=O)R	2.10	2.35	2.65
HC–C≡N	2.15	2.45	2.90
HC–CHO	2.20	2.40	
HC–Ar (and NR$_2$)	2.25	2.45	2.85
HC–SSR	2.35	2.70	
HC–(C=O)Ar	2.40	2.70	3.40
HC–SAr	2.40		
HC–NRAr	2.60	3.10	3.60
HC–SO$_2$R and –(SO)R	2.60	3.05	
HC–Br	2.70	3.40	4.10
HC–NR$_3^+$	2.95	3.10	3.60
HC–NH(C=O)R	2.95	3.35	3.85
HC–Cl	3.05	3.45	4.05
HC–OH and –OR	3.20	3.40	3.60
HC–NH$_2$	3.50	3.75	4.05
HC–O(C=O)R	3.65	4.10	4.95
HC–OAr	3.80	4.00	4.60
HC–O(C=O)Ar	3.80	4.20	5.05

TABLE 11-2 Continued

Substituent Group	Methyl Protons	Methylene Protons	Methine Proton
HC–F	4.25	4.50	4.80
HC–NO$_2$	4.30	4.35	4.60
Cyclopropane		0.20	0.40
Cyclobutane		2.45	
Cyclopentane		1.65	
Cyclohexane		1.50	1.80
Cycloheptane		1.25	

Substituent Group	Proton Shift	Substituent Group	Proton Shift
HC≡CH	2.35	HO–C=O	10–12
HC≡CAr	2.90	HO–SO$_2$	11–12
HC≡C–C=C	2.75	HO–Ar	4.5–6.5
HAr	7.20	HO–R	0.5–4.5
HCO–O	8.1	HS–Ar	2.8–3.6
HCO–R	9.4–10.0	HS–R	1–2
HCO–Ar	9.7–10.5	HN–Ar	3–6
HO–N=C(oxime)	9–12	HN–R	0.5–5

R = alkyl group; Ar = aryl group.
SOURCE: J. A. Dean, Ed., *Lange's Handbook of Chemistry*, 12th ed., McGraw-Hill, Inc., New York, 1979, from which additional proton chemical shifts may be found.

TABLE 11-3 ^{13}C Chemical Shifts (Values given on the δ scale, relative to TMS.)

Substituent Group	Primary Carbon	Secondary Carbon	Tertiary Carbon	Quaternary Carbon
Alkanes				
C–C$\lessgtr$	–20 to 30	25 to 45	30 to 60	35 to 70
C–O	40 to 60	40 to 70	60 to 75	70 to 85
C–N	20 to 45	40 to 60	50 to 70	65 to 75
C–S	10 to 30	25 to 45	40 to 55	55 to 70
C–Halide	–37 to 35	–10 to 45	30 to 65	35 to 75
	(I) (Cl)	(I) (Cl)	(I) (Cl)	(I) (Cl)

Alkynes	70 to 100	Isocyanides	130 to 150
Alkenes	110 to 150	Carbonates	150 to 160
Aromatics	110 to 135	Oximes	155 to 165
C-substituted	125 to 145	Ureas	150 to 170
Heteroaromatics	115 to 140	Thioureas	165 to 185
C-α	135 to 155	Esters, Anhydrides	150 to 175
Cyanates	105 to 120	Amides	160 to 180
Isocyanates	115 to 135	Acids, Acyl chlorides	160 to 185
Thiocyanates	110 to 120	Aldehydes	175 to 205
Isothiocyanates	120 to 140	Ketones	175 to 225
Cyanides	110 to 130		

Processes giving rise to chemical exchange or conformational change which are complete in 1–0.001 sec may give rise to spectra which are time averages in comparison with those expected in terms of instantaneous molecular conformations. If the exchange rate is high in comparison with the frequency of the chemical shifts and spin–spin couplings, the local fields seen from the nucleus of the exchanging atom will be averaged out to result in a single line, somewhat broader than normal.

Spin–Spin Coupling

Nuclei can interact with each other to cause mutual splitting of the otherwise sharp resonance lines into multiplets, called *spin–spin coupling*. These multiplets arise because magnetic moments of nuclei interact with the strongly magnetic electrons in the intervening bonds. The strength of the coupling, denoted by J, is given by the spacing of the multiplets and is expressed in hertz. Proton–proton couplings (Table 11-4) are usually transmitted only through two or three bonds, although weak couplings are often transmitted further.

TABLE 11-4 Proton Spin Coupling Constants

Structure	J, Hz	Structure	J, Hz
$\diagdown$C$\diagup^H_H$	12–15	Cyclohexane H_a (a-a), H_e (a-e), (e-e)	8–10, 2–3, 2–3
CH–CH (free rotation)	6–8	Cyclopentane (*cis*)	4–6
CH–OH (no exchange) (–NH)	5	(*trans*)	4–6
		Cyclobutane (*cis*)	8
		(*trans*)	8
CH–C=O with H	1–3	Cyclopropane (*cis*)	9–11
		(*trans*)	6–8
		(*hetero*)	4–6
H_t, H_g (*gem*) C=C (*cis*) H_c, H (*trans*)	0–3, 6–14, 11–18	Benzene (*o*), (*m*), (*p*)	6–10, 1–3, 0–1
H_c, CH (*cis*) C=C (*trans*) H_t, H_g (*gem*)	0.5–3, 0.5–3, 4–10	Pyridine (2–3), (3–4), (2–4), (3–5), (2–5), (2–6)	5–6, 7–9, 1–2, 1–2, 0–1, 0–1
C=CH–CH=C	10–13	Pyrrole (1–2), (1–3), (2–3), (3–4), (2–4)	2–3, 2–3, 2–3, 3–4, 1–2
=CH–C=O with H	6		

TABLE 11-4 Continued

Structure		J, Hz	Structure		J, Hz
$-CH_2-C{\equiv}C-CH$		0–3		(2–5)	1–3
$CH-C{\equiv}CH$		0–3	$\begin{array}{c}\diagdown\\ C\diagup\\ \diagup\diagdown\end{array}\begin{array}{c}H\\ \\ F\end{array}$		45–52
$\begin{array}{c}HH\\ \diagdown\diagup\\ C{=}C\\ (ring)\end{array}$	(3-member) (4-member) (5-member) (6-member) (7-member)	0–2 2–4 5–7 6–9 10–13	$CH-CF$	(gauche) (trans)	0–12 10–45
furan (O, positions 2,3,4,5)	(2–3) (3–4) (2–4) (2–5)	1.8 3.5 0–1 1–2	$\begin{array}{c}H_tH_g\\ \diagdown\diagup\\ C{=}C\\ \diagup\diagdown\\ H_cF\end{array}$	(gem) (cis) (trans)	72–90 1–8 12–40
thiophene (S, positions 2,3,4,5)	(2–3) (3–4) (2–4) (2–5)	5–6 3.5–5.0 1.5 3.4	$\begin{array}{c}\diagdown\\ C{=}C\\ \diagup\diagdown\\ FCH_3\end{array}$		2–4
benzene–F, H	(o) (m) (p)	6–10 5–6 0–2	$\begin{array}{c}\diagdownCF\\ \diagup\\ C{=}C\\ \diagup\diagdown\\ H\end{array}$		0–6
benzene–CH$_3$, F	(o) (m) (p)	2.5 1.5 0			

In certain rigid structures of favorable geometry, coupling through four bonds may be reasonably large. When unsaturated systems occur between the protons, long-range coupling may be enhanced. In allylic systems, four-bond couplings reach a maximum of about 3 Hz when the angle between the plane containing the olefinic protons and the C–H bond of the allylic carbon atom is about 90°. In H–C–C=C–C–H systems, five-bond couplings of about 3 Hz are observed. In acetylenes, allenes, and cumulenes, observable couplings may be transmitted over many bonds, up to nine in polyacetylenes. In aromatic rings, couplings of protons in *ortho* positions (through three bonds) are 7–9 Hz, *meta* (four bonds) 2–3 Hz, and *para* (five bonds) 0.5–1.0 Hz.

Couplings depend also on geometry. The dihedral angle between planes determines the coupling of protons on adjacent carbon atoms (Fig. 11-9). Adjacent axial–axial protons, displaying a dihedral angle of 180°, are strongly coupled, whereas axial–equatorial and equatorial–equatorial protons are coupled only moderately. *Trans* and *cis* protons on olefinic double bonds show $J_{trans}/J_{cis} \simeq 2$, which can be useful in assigning structures of geometrical isomers.

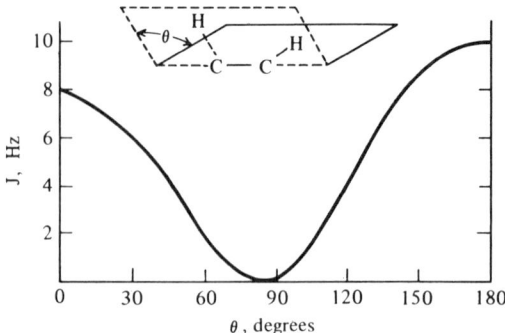

FIGURE 11-9 Dependence of the coupling constant J on the dihedral angle θ in the saturated system H–C–C–H. (By permission from M. Karplus and D. H. Anderson, *J. Chem. Phys.*, 30, 6 (1959); M. Karplus, *J. Chem. Phys.*, 30, 11 (1959).)

The number of lines in a multiplet is given by $2nI + 1$, where n is the number of nuclei producing the splitting. For protons, this becomes $n + 1$ lines. The relative intensity of each of the multiplets, as reflected in the integral curve, is proportional to the number of nuclei in the group. Intensities of the peaks within a multiplet are given by simple statistical considerations, and are proportional, therefore, to the coefficients of the binomial expansion. Thus, one neighboring proton splits the observed resonance to a doublet (1:1), two produce a triplet (1:2:1), three a quartet (1:3:3:1), four a quintet (1:4:6:4:1), and so on.

The magnitude of J is independent of the field strength, unlike the chemical shift. Thus, as H_0 increases, the multiplets move further apart but the spacing of the peaks within each multiplet remains the same. The ratio $J/\Delta\nu$, where $\Delta\nu$ is the chemical shift difference between the two coupled nuclei, is the critical parameter that determines the appearance of the spectrum. When $J/\Delta\nu$ is 0.1 or less, the spectrum consists of well-separated multiplets. As $J/\Delta\nu$ approaches unity, the spectrum begins to deviate noticeably from the simple first-order appearance. New peaks may appear and intensities are no longer binomial because some spin states that were degenerate when $J/\Delta\nu$ was small split because the magnetic field mixes states. Ultimately, when the chemical shift difference $\Delta\nu$ vanishes, the multiplet will collapse to a singlet. A strongly coupled system of three or more spins is difficult to unravel by inspection alone, although certain patterns become recognizable with experience. Use of 220- or 300-MHz NMR spectrometers is valuable. A higher field strength improves the ratio of chemical shift to coupling constant by causing the chemical shift to increase and spreading the NMR spectrum over a wider range (Fig. 11-10). It reduces complicating second-order spectrum effects and produces a sensitivity gain in addition.

Other nuclei with spins of $\frac{1}{2}$ will interact with protons (and each other) and cause observable spin–spin coupling. Without deliberate isotopic substitution, significant numbers of only fluorine and phosphorus occur naturally. In fact, the presence of one of these elements may be deduced from an otherwise unexplained coupling effect. Usually J is larger than for most proton–proton couplings. The direct coupling of ^{13}C–H is often noticeable as sidebands on a proton NMR spectrum; these sidebands will not vary as the sample spinning rate is changed as will the spinning sidebands. To a first approximation,

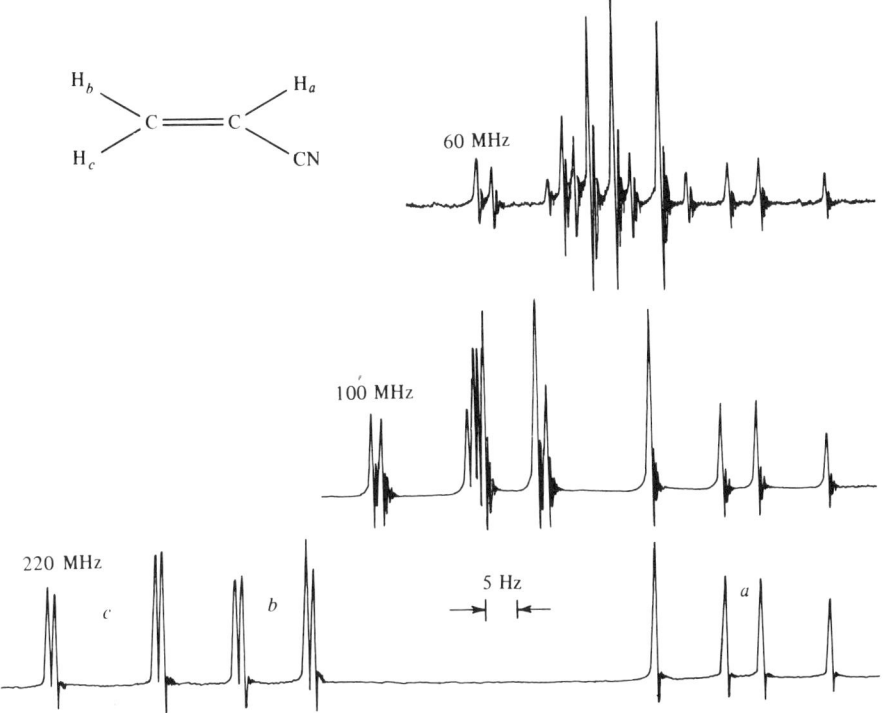

FIGURE 11-10 NMR spectra of acrylonitrile at 60, 100, and 220 MHz illustrate the primary advantage of operating at the highest attainable rf frequency and magnetic field. (Courtesy of Varian Associates.)

the magnitude of the ^{13}C–H coupling is proportional to percent s-character in the C–H bond.

When a proton is coupled to a nucleus which has a nonzero quadrupole moment, the latter provides efficient spin–lattice relaxation (T_1 is decreased) and this is usually sufficient to decouple, completely or partially, the spin–spin interaction with the proton. Coupling of protons with chlorine, bromine, or iodine nuclei is not observed. In the case of ^{14}N, the decoupling is usually only partially effective so that the 1:1:1 triplet resulting from ^{14}N–^{1}H coupling is normally broad and featureless, except for ammonium ion in strongly acidic media.

Integration

The area under an absorption band is proportional to the number of nuclei responsible for the absorption. A device for electronically integrating the absorption signal is a standard item on most commercial spectrometers. The integral is represented as a step function; the height of each step is proportional to the number of nuclei in that particular region of the spectrum. Accuracy is typically within ± 2%. For quantitative analysis, a known amount of a reference compound can be included with the sample. The NMR signal of the reference compound preferably should contain a strong singlet lying in a region of the NMR spectrum unoccupied by sample peaks. Exact phasing out of the dispersion signal is crucial in integration. From the two peak areas, A_{unk} and A_{std}, and the weight of the internal standard taken, W_{std}, the amount of the unknown present is calculated by

$$W_{unk} = W_{std} \times \frac{N_{std}}{N_{unk}} \times \frac{M_{unk}}{M_{std}} \times \frac{A_{unk}}{A_{std}} \qquad (11\text{-}8)$$

where N's are the numbers of protons in the groups giving rise to the absorption peaks, and M's are the molecular weight of the compounds.

Whenever the empirical formula is known, the total height (in any arbitrary units) divided by the number of protons yields the increment of height per proton. Lacking this information, but deducing the assignment of a particular absorption band, one can calculate the increment per proton from the difference in elevation for the assigned group divided by the number of protons in the particular group. Unfortunately, there is no way of handling overlapping bands.

Example 11-1

A NMR spectrum shows three single peaks located at −440, −300, and −120 Hz in a field of 60 MHz with TMS as reference. The integral heights are 4.2, 1.7, and 2.5 units, respectively. Without knowledge of the empirical formula, the integral heights bear the ratio 5:2:3 and, since no splitting is observed, this must mean a group with five protons (probably an aromatic ring from the chemical shift), a methylene group, and a methyl group, each not coupled with one another.

If one knew that the empirical formula was $C_9H_{10}O_2$, dividing the total integral height of 8.4 units by the number of protons gives 0.84 unit as the increment per proton.

11.5 ELUCIDATION OF PROTON NMR SPECTRA

Application of NMR to structure analysis is based primarily on empirical correlation of structure with observed chemical shifts and coupling constants. Extensive surveys have been published.[2-5] These tabulations can be used to predict the position of resonance lines for a postulated compound, and these predictions can be compared with the sample spectrum. Conversely, one searches compilations and published spectra to ascertain groups which might occur at the positions observed in the sample spectrum. One of the unique advantages of NMR is that spectra can often be interpreted without reference to data from structurally related compounds.

Brief vigorous shaking with a few drops of D_2O generally results in a complete exchange of labile protons and collapse of their absorption signal. Deuterium does not absorb in the proton spectral region. Because of the smaller value of μ/I for deuterium, its coupling to hydrogen is much smaller (about one-sixth) than the corresponding H–H coupling and does not appear in the proton spectrum.

Double Resonance (or Spin Decoupling)

Nuclear magnetic double resonance, or spin decoupling, is achieved by irradiating an ensemble of nuclei not only with a rf H_1 at resonance with the nuclei to be observed but additionally with a second, relatively strong, alternating rf field H_2, perpendicular to H_0 and at resonance with the nuclei to be decoupled. Decoupling experiments can be carried out to convert homonuclear ($^1H-^1H$) or heteronuclear ($^{19}F-^1H$, $^{13}C-^1H$) multiplets into singlets, or less complex multiplets. Decoupling is achieved when $\nu_1 \nu_2/hI$ is greater than J but ν_2 is still sufficiently low in power so that saturation is not approached. The spin and magnetic moment of the nucleus irradiated with ν_2 is quantized in a direction perpendicular to its coupling partner, the coupling is removed, and any multiplets involved collapse into a single peak. By reversing the roles of irradiated and observed nuclei, unequivocal identification of the spin-coupled nuclei is provided.

Experimentally, in the frequency-sweep technique, the second rf field ν_2 is varied until a characteristic change occurs in the lines of a particular multiplet. For example, if ν_2 is held at a fixed separation from the reference line (at $H_5 - H_{TMS}$ in Fig. 11-11), and the spectrum is swept by ν_1 while the magnetic field and locking frequency are held constant, signals from protons H_3 change from a pair of quartets to a pair of triplets (upper spectrum of Fig. 11-11) indicating that H_3 and H_5 have a small long-range coupling constant. Small splittings also disappear in the patterns of H_4 and H_6. In a single experiment the effect on all other resonances of irradiating a chosen resonance may be observed.

In the field-sweep technique of decoupling, the spectrum is swept while maintaining a fixed frequency difference $\Delta \nu$ between the observing ν_1 and irradiating ν_2 frequencies. Only those coupled resonances separated by the chosen frequency difference will be ob-

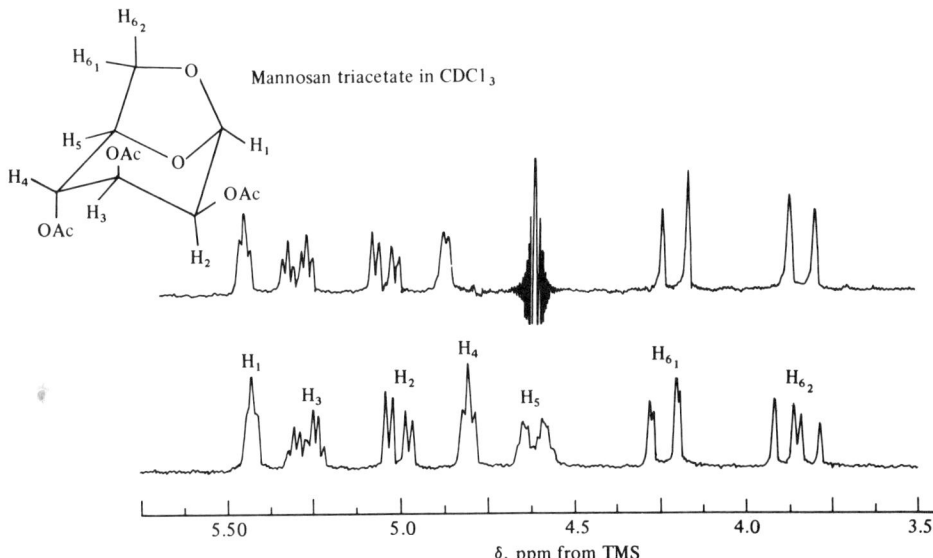

FIGURE 11-11 Frequency-swept spin decoupling at 100 MHz. (Courtesy of Varian Associates.)

served. For each critical frequency difference a separate decoupling experiment will be required.

In routine ^{13}C work usually all ^{13}C–^{1}H multiplets are decoupled for sensitivity and simplicity reasons. This is achieved when the decoupling field H_2 covers the range of all proton Larmor frequencies. This is at least 1 kHz at about 90 MHz in a magnetic field (H_0) of 23 kG. Decoupling fields with large frequency ranges can be realized either by a very large rf power (H_2), so that $\mu H_2/2\pi I = \Delta = 1$ kHz (for ^{1}H), or by application of broad band decoupling in which the coherent proton radio frequency is modulated with "white" noise. If the frequency spread of this noise is greater than the range of proton frequencies and high enough power is employed, the various C–H multiplets are collapsed, giving single lines for all carbons formerly coupled to protons. In the latter case, no detailed consideration is necessary to selection of the proton radio frequency.

Decoupling increases the sensitivity of NMR measurements because the intensities of all multiplet lines in a coupled spectrum are accumulated in one singlet in the decoupled spectrum. This technique is especially useful when working with the low abundance of natural ^{13}C in samples.

The Nuclear Overhauser Effect

The nuclear Overhauser effect observed in decoupled ^{13}C–^{1}H experiments arises from an intramolecular dipole–dipole relaxation mechanism. Although decoupling increases the sensitivity of NMR experiments because the intensities of all multiplet lines in a coupled spectrum are accumulated in one singlet signal in the decoupled spectrum, the intensity of the ^{13}C signal often increases much more than expected. In an ^{13}C–^{1}H decoupling experiment, the transitions of ^{1}H are irradiated while the resonance of ^{13}C nuclei are observed. Since the irradiating field is very strong, the homonuclear relaxation processes are not adequate to restore the equilibrium population of ^{1}H nuclei, and these nuclei transfer their energy to the ^{13}C nuclei via internuclear dipole–dipole interaction. The carbon nuclei receiving these transferred amounts of energy, behave as if they had been irradiated themselves and relax. Consequently, the population of the lower energy level of ^{13}C increases, and the intensity of the carbon signal is enhanced.

The maximum enhancement factor, f_C (H), mainly depends on the gyromagnetic ratios of the nuclei involved: f_C (H) = $[\mu/I]_H/2[\mu/I]_C$. Nuclear Overhauser effect sensitivity enhancements are attained in heteronuclear double resonance experiments when the nucleus with low μ/I is observed while the nucleus with high μ/I is decoupled. Negligible enhancement of proton NMR sensitivity would be expected from decoupling ^{13}C. This extra intensity enhancement is welcome when obtaining qualitative NMR spectra, as it saves a factor of $[f_C(H)]^2$ in time-averaging in attaining a desired signal-to-noise ratio. However, in integration of peak areas for quantitative work, the enhancement is a bane.

A further complication arises from the pulsed Fourier transform mode of detection required to achieve adequate sensitivity, which results in a dependence upon the ratio of the pulse repetition rate and the relaxation time, T_1, for each carbon nucleus. The nuclear Overhauser enhancement factors are made either all equal to zero by gating the decoupler on only during acquisition, and off during the delay period, or are all assumed to have equal value if the carbon atoms being studied are all protonated and restricted in their motions by being part of the framework of a fairly large molecule. However, if nonprotonated carbons must be integrated, the required delay times can be very long and the analysis time becomes prohibitive. In such cases, it often proves useful to add a relaxation agent such as chromium(III) acetyl acetonate which quenches the nuclear Overhauser effect for all carbons and shortens all T_1 values enough so that pulses can be repeated at about 2-sec intervals.

Spin Tickling

In *spin tickling* a weak irradiating field ν_2 is used, a field whose bandwidth is only slightly larger than the widths of individual lines. The effect of irradiating one line in the spectrum is to split other lines in the same spectrum coupled to it. Hidden lines in a complex region of the spectrum can be located by irradiating these lines one at a time while looking at the effect on other regions of the spectrum. Frequency-swept spin tickling is illustrated in Fig. 11-12. In this spectrum, one of the olefinic protons produces the set of four lines centered around 6.3δ. The proton coupling of 14 Hz is midway between the normal *cis-* and *trans-*couplings of 10 and 17 Hz, respectively. While "sitting on" the peak around 646 Hz from TMS, an external audio oscillator ν_2 is slowly varied so as to sweep ν_2 through the low-field pattern of signals from 7 to 9δ. A dip in the signal at 646 Hz occurs when ν_2 is 777.6 or 764.2 Hz larger than the frequency of the oscillator used to lock to the TMS signal. Likewise, the recorder pen is positioned on the peak at about 627 Hz and the splitting effects are noted when ν_2 is 737.5 and 723.7 Hz larger than the TMS lock frequency. Thus, the four hidden lines are located and their positions in the spectrum are indicated by the arrows. From the large spacings the average value of phos-

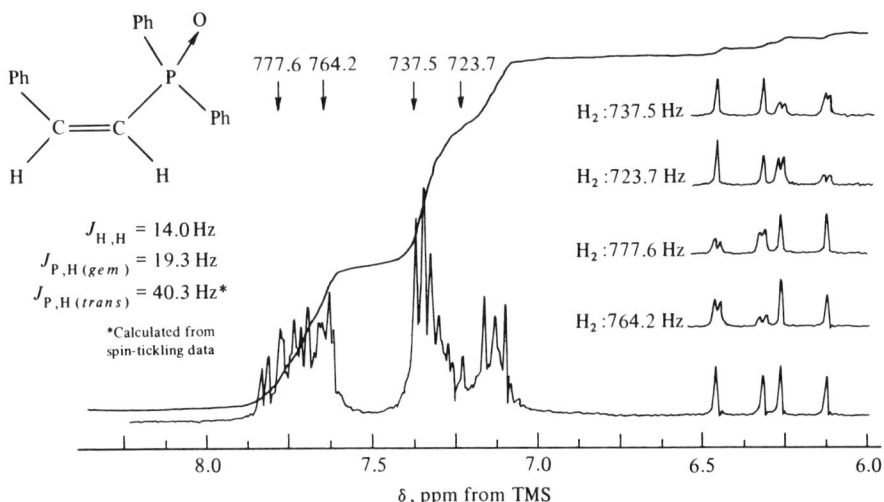

FIGURE 11-12 Frequency-swept spin tickling at 100 MHz. (Courtesy of Varian Associates.)

phorus coupling to this proton is 40.3 Hz, as compared with the *cis*-coupling constant of 13.5 Hz, and agrees with *trans*-phosphorus–hydrogen coupling.

Solvent Influence and Shift Reagents

Proton resonance peaks will be spread across a broader range of magnetic field strength by addition of a paramagnetic compound to the solution being studied. In Fig. 11-13 the complex pattern has been resolved (see the upper trace) by the addition of 50% benzene to the CCl_4 solution in which the NMR spectrum was initially taken. Benzene solvent molecules associate with electron-deficient sites in solute molecules. Because benzene is highly anisotropic, different protons in the solute experience shielding or deshielding depending on their orientation to the benzene ring. In this particular example, the ring protons are shielded (located above the benzene ring) and appear at higher field.

An isolated carbonyl group induces a solvent effect which changes sign near a plane drawn through the carbonyl carbon atom and perpendicular to the carbonyl group. The corresponding shift relative to pyridine changes sign near a plane drawn through the α-carbon atoms. Thus, in the case of an equatorial proton adjacent to the carbonyl group in a cyclohexanone ring, the shift will be small or zero in benzene and negative in pyridine, whereas for axial protons or methyl groups the shift will be positive in benzene and small or zero in pyridine.[6]

The most commonly used shift reagents are lanthanide fluorinated β-diketones. They function by acting as Lewis acids, forming a complex with the substance under analysis, which acts as a nucleophile. Induced shifts are attributed to a pseudocontact, or dipolar interaction between the shift reagent and the nucleophile. The most commonly used metal chelates are those of Eu(III) and Yb(III), which normally induce downfield shifts, and Pr(III), which induces upfield shifts. Shift reagents can give resolution of peaks (Fig.

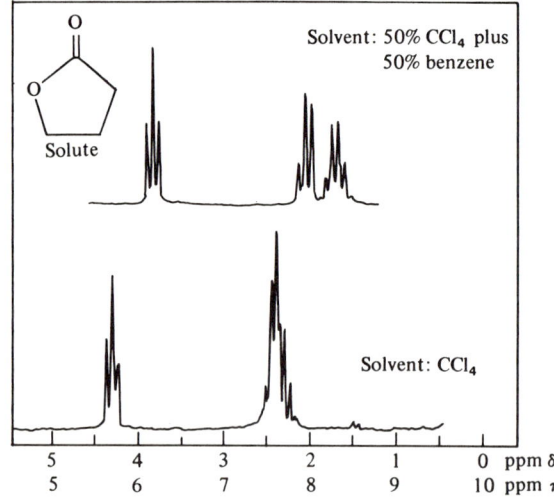

FIGURE 11-13 Solvent effect on an NMR spectrum.

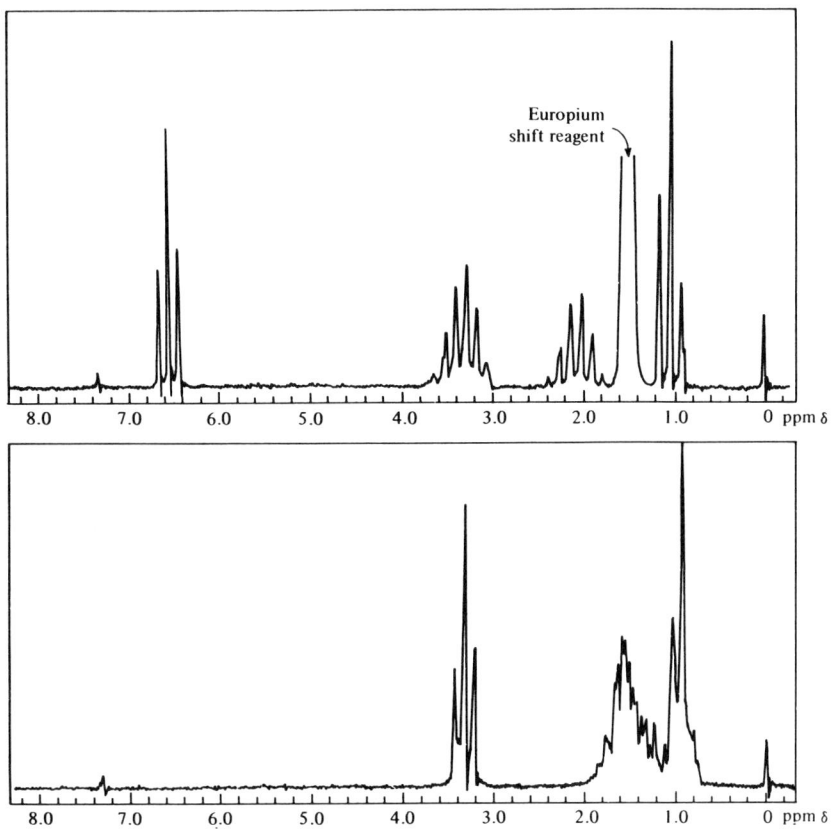

FIGURE 11-14 NMR spectra of di-*n*-butyl ether (1.0×10^{-4} moles) in 0.5 ml CCl_4 alone (*lower trace*), and with 5.0×10^{-5} moles of tris (1,1,1,2,2,3,3-heptafluoro-7, 7-dimethyl-4,6-octanedione) europium(III) added (*upper trace*).

11-14) comparable to the resolution achievable with 100- or even 220-MHz spectrometers. The fastest and easiest technique for obtaining induced shifts is to add a few milligrams of shift reagent directly to the nucleophile dissolved in solvent. Increments of shift reagent can be added until sufficient resolution is attained. There is essentially a linear dependence of the chemical shifts upon added shift reagent. The most frequently used solvents are chloroform and carbon tetrachloride. Once shift reagents have been used to resolve the NMR peaks of a compound, spin-decoupling experiments become possible.[7,8]

Elucidation of NMR Spectra

Example 11-2

Returning to the information given in Example 11-1, these structures are present:

⬡— —CH_2— —CH_3

From the position of the phenyl group and the lack of multiplet structure, the benzyl group is suggested. Placing an oxygen next to the phenyl ring would shift upfield the *ortho* and *para* protons; likewise, a carbonyl group would shift the *ortho* protons downfield. The position of the isolated methyl group suggests an adjacent double bond, either carbonyl or phenyl ring. The latter is impossible because a methylene group must be accommodated. Two structural candidates remain as possibilities: benzyl methyl ketone or benzyl acetate (if one can eliminate the presence of sulfur). In the former compound the methylene protons would be expected at δ 3.6; in the latter at δ 5.1. The observed position is δ 5.0, and the compound is benzyl acetate.

Example 11-3

The NMR spectrum shows a single peak at δ 6.83 (τ 3.17), a quadruplet at δ 4.27 (τ 5.73), and a triplet at δ 1.32 (τ 8.68). For the multiplets the coupling constant is about 7 Hz. The integrator readings are in the ratio 1:2:3, respectively. Empirical formula is $C_8H_{12}O_4$.

The upfield methyl group is split into a triplet by an adjacent methylene group which, in turn, is split into a quadruplet—the typical ethyl pattern. This is confirmed by the integrator readings. Assignment of the low-field group is not so simple. It is not quite in the location expected for a benzene ring, nor does it contain sufficient numbers of protons. However, an olefinic proton absorption is a possibility, although further shielding is indicated. Returning to the methylene group, its downfield position could be due to an adjacent –O–(C=O)– structure. If the olefinic proton were alongside the carbonyl group, its absorption position would be reasonable. Summarizing our present information, we have

$$\text{CH}_3-\text{CH}_2-\text{O}-\overset{\overset{\text{O}}{\|}}{\text{C}}-\text{CH}=$$

which is exactly one-half the empirical formula. The complete structure is either diethyl fumarate or diethyl maleate. Having gotten this far we would take a NMR

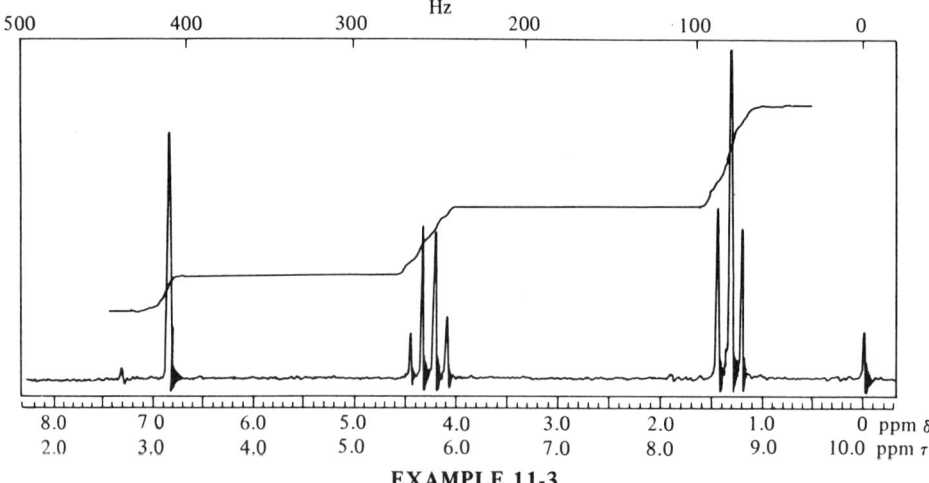

EXAMPLE 11-3

spectrum of each compound and would find that in diethyl maleate the low-field single peak occurs at δ 6.28 (τ 3.72). Coupling between the olefinic protons collapsed because $\delta_2 - \delta_1 = 0$.

Overlapping spectral features can be a source of difficulty in interpretation. The distribution of protons shown by the integration curve is helpful.

Example 11-4

The NMR spectrum for the compound with empirical formula $C_7H_{16}O$ shows a symmetrical heptet centered around δ 3.78 (τ 6.22) and an unsymmetrical doublet with individual peaks at δ 1.18 (τ 8.82) and δ 1.12 (τ 8.88). The integrated areas are 2 units for the heptet and 24 + 6 units for the doublet. The absence of peaks at low field indicates absence of aldehyde, unsaturation, and probably hydroxyl absorption.

The heptet spells out a probable isopropyl group whose methine proton is split by six methyl protons. Its field position indicates an adjacent oxygen. In turn, the methine proton splits the *gem*-methyl protons into a doublet. A logical presumption is an isopropyl ether group, the oxygen atom serving to isolate the methine proton from the remainder. Using the heptet proton as divisor, the integrator readings spell out nine unassigned protons in a single peak superimposed on the low-field wing of the methyl doublet in the isopropyl group. This could only mean three isolated methyl groups—a *tert*-butyl group. The compound is *t*-butyl isopropyl ether.

11.6 QUANTITATIVE ANALYSIS

Quantitative analysis by NMR deserves more attention. Accurate electronic integrators with reproducibility better than 2%, available even on low-cost instruments, offer great potential as a means of quality control. Provided that at least one resonance band from each component of a mixture is free from extensive overlap by other absorptions, NMR quantitative analysis should be possible.

The determination of the ethylene oxide chain length in commercial nonionic detergents provides an example. Approximately 10% solutions of each surfactant in CCl_4 are prepared, and the spectrum and several repeat integrals of each are run (Fig. 11-15). The ratio of the ethylene oxide proton signal to the cetyl chain signal is calculated and, since the number of protons in the cetyl chain is known, the calculation of the number of $[O-CH_2-CH_2-O]$ units is a matter of simple proportion. For each detergent the total analysis time is about 10 min; the alternative is a lengthy titration procedure.

Impure samples can be assayed using another compound as the internal standard. The NMR spectrum of

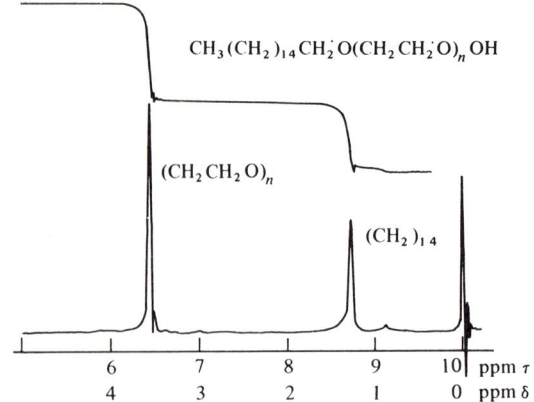

FIGURE 11-15 Ethylene oxide chain length of nonionic surfactants by quantitative NMR.

possesses a singlet peak due to the N–CH$_3$ protons at $\delta = 3.40$ and a doublet due to the –(C=O)–CH$_2$–S–P– group at $\delta = 4.3$. Benzyl benzoate, which gives a well-resolved peak at $\delta = 5.4$, is an excellent internal standard.

Products from the air oxidation of p-cymene yield a complex mixture. The NMR spectrum, with peak assignments indicated, is given in Fig. 11-16, which also shows the position of the internal standard line from benzyl benzoate.

The determination of water in liquid N$_2$O$_4$ in the less than 0.1 weight % range illustrates the usefulness of NMR in trace analysis and in a system for which calibration stan-

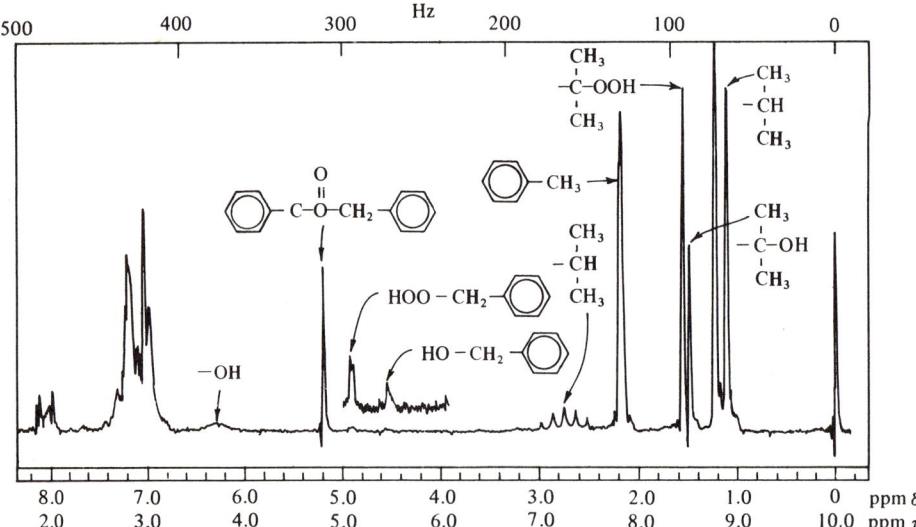

FIGURE 11-16 NMR spectrum of mixtures of products obtained from the oxidation of p-cymene, benzoyl benzoate added as internal standard. Integrals are not shown. (Courtesy of International Scientific Communications, Inc.)

dards are not available. At $-10°C$ the exchange of protons between dissolved H_2O, HNO_2, and HNO_3 molecules is rapid, so that a single NMR peak is observed for the exchanging protons. As little as 0.03% H_2O can be detected with a relative standard deviation of 7.5% based on a single spectral scan. Accuracy is improved by using the accumulated average of 50 scans, which can be run in 90 min. With this procedure, 0.01% H_2O (100 $\mu g/ml$) can be determined. Samples of N_2O_4 are weighed into a tarred NMR tube, and 30–50 μl of a benzene internal standard are added.

Residual H_2O in samples of high purity D_2O can be determined by the method of standard additions, using standard microvolumetric techniques. The resulting NMR integrals are plotted versus the weight percent of added H_2O and extrapolated to zero integral.

A well-known assay of a pharmaceutical formulation by NMR is that of aspirin, phenacetin, and caffeine mixtures.[9] The procedure takes about 20 min and does not require the separation of individual constituents. A separate analytical peak of known origin is present for each component; no calibration curves are required since the integrals for the protons giving rise to the various analytical peaks are constant and unaffected by solvent or solute interactions. For aspirin the sharp peak at about $\delta = 2.3$, which represents the ester methyl group, is used. For phenacetin (see illustration for Problem 11-21) the amide methyl group at $\delta = 2.1$ is preferable; the quartet at $\delta = 4.0$ is suitable although one of the methyl peaks of caffeine overlaps the quartet at $\delta = 3.9$, and a correction is necessary. For caffeine the two methyl resonances at $\delta = 3.4$ and 3.6 are used.

PROBLEMS

1. At $43°C$ the NMR spectrum of liquid acetylacetone shows a peak at δ 5.62 (37 units on the integrator) and a peak at δ 3.66 (19.5 units), plus additional peaks which do not concern us. Calculate the percent enol composition.

2. A hydrocarbon sample shows NMR bands over the interval δ 1.0–5.5. Benzophenone used as an internal standard shows NMR bands in the δ 6–7 region. The relative integrals were 228 and 184 units for 0.8023 g of benzophenone and 0.3055 g of sample. Calculate the percent hydrogen in the sample.

3. Phenol formaldehyde resins include a class prepared with excess phenol, called novolacs, which consist of phenolic nuclei linked by methylene bridges at positions *ortho* or *para* to the hydroxyl group. The integration of the spectrum gives the ratio of aromatic to methylene protons as 30 to 18. Calculate the average chain length and the average molecular weight.

4. Chlorination of *o*-cyanotoluene gave *o*-cyanobenzyl chloride in about 50% yield. Attempted fractionation of the washings and mother liquor gave a liquid whose proton magnetic resonance spectrum gave three singlets upfield from the aromatic ring signals as follows (integral units in parentheses): δ 2.52 (13), δ 4.72 (20), and δ 7.01 (10). (a) Assign the NMR signals. (b) From the signal intensities, deduce the relative molar proportions and the proportions by weight of the three constituents in the liquid mixture.

5. The NMR proton spectrum of the liquid diketene, $C_4H_4O_2$, shows two signals of equal intensity. What structure is consistent with this information?

6. A sample is believed from its mass spectrum to be either of the dicyanobutenes:

$$CH_3-CH=C\begin{matrix}CH_2CN\\ \\ CN\end{matrix} \qquad NC-CH=C\begin{matrix}CH_2CN\\ \\ CH_3\end{matrix}$$

$$\text{I} \qquad\qquad\qquad \text{II}$$

What characteristic in the NMR spectrum would identify each isomer?

7. The phosphorus resonance of phosphonic acid and phosphinic acid is reported to be a doublet in the former and a triplet in the latter compound. Write the structures of the two acids.

8. Addition of methyldichlorosilane to vinyl acetate gives an adduct whose likely structures are

$$(CH_3-Si(Cl)_2-CH_2-CH_2-O-CO-CH_3$$

or

$$CH_3-Si(Cl)_2-CH(CH_3)-O-CO-CH_3$$

The NMR spectrum shows two bands with clearly resolved triplet splitting. Which structure is supported by the NMR evidence?

9. On the basis of the two peaks of equal strength found in the NMR spectrum of the sodium salt of Fiest's acid in D_2O, which structure is correct?

$$\text{HOOC-C}\underset{\underset{CH_3}{|}}{\overset{\diagdown\;\;\diagup}{C}}\text{CH-COOH} \qquad \text{or} \qquad \text{HOOC-CH}\underset{\overset{\|}{CH_2}}{\overset{\diagdown\;\;\diagup}{C}}\text{CH-COOH}$$

10. For an impurity isolated from an important industrial chemical by gas–liquid chromatography, the empirical formula $C_5H_8O_4$ was established by elemental analysis, and mass spectrometry provided the molecular weight of 132. An infrared spectrum revealed a strong —C—O—C— absorption with no evidence of double bonds, carbonyl, or hydroxyl groups. From an examination of the proton NMR spectrum in the figure, which of the three structures is the correct one?

11. Isojasmone, a synthetic oil redolent of jasmine blossoms, consists of a mixture of the two isomers A and B. Isomer B has a greater odor value than isomer A. The NMR spectra and structures of the two compounds are shown in the figure. Compute the proportions of the two isomers after assigning the olefinic proton of each isomer to the proper spin–spin multiplet on the expanded trace.

NUCLEAR MAGNETIC RESONANCE SPECTROSCOPY 351

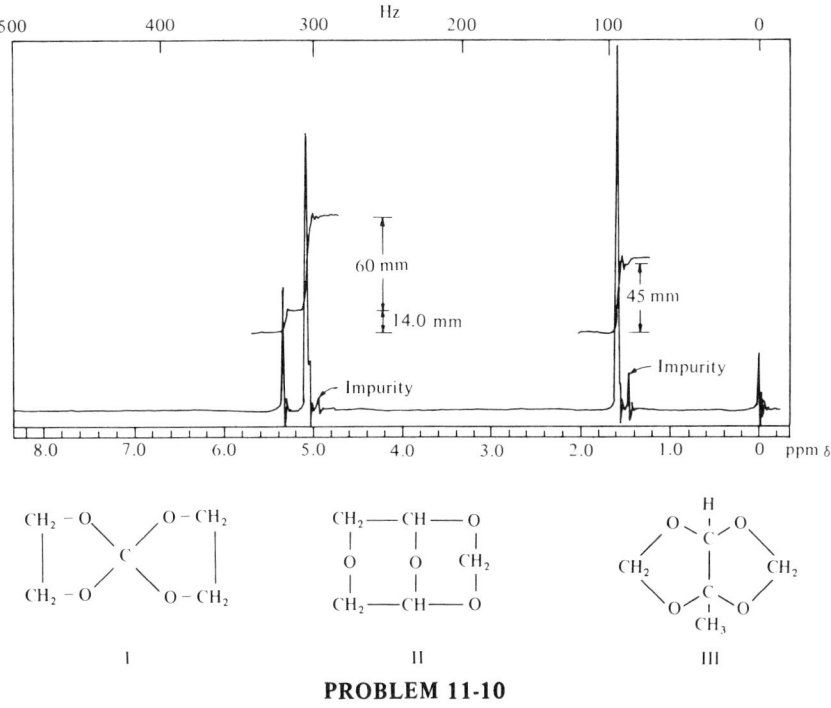

PROBLEM 11-10

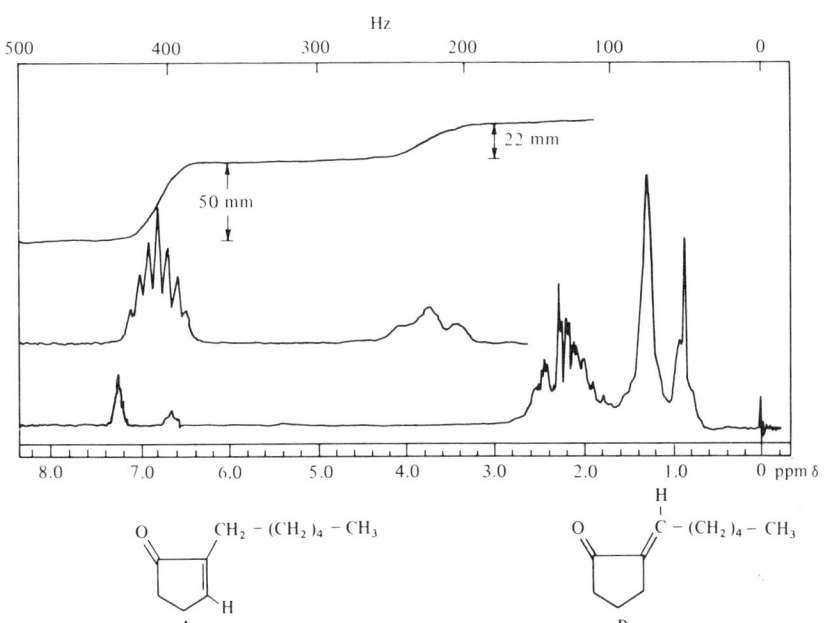

PROBLEM 11-11

12. Deduce the structure of the compound with the spectrum shown whose empirical formula is $C_{10}H_{11}NO_4$.

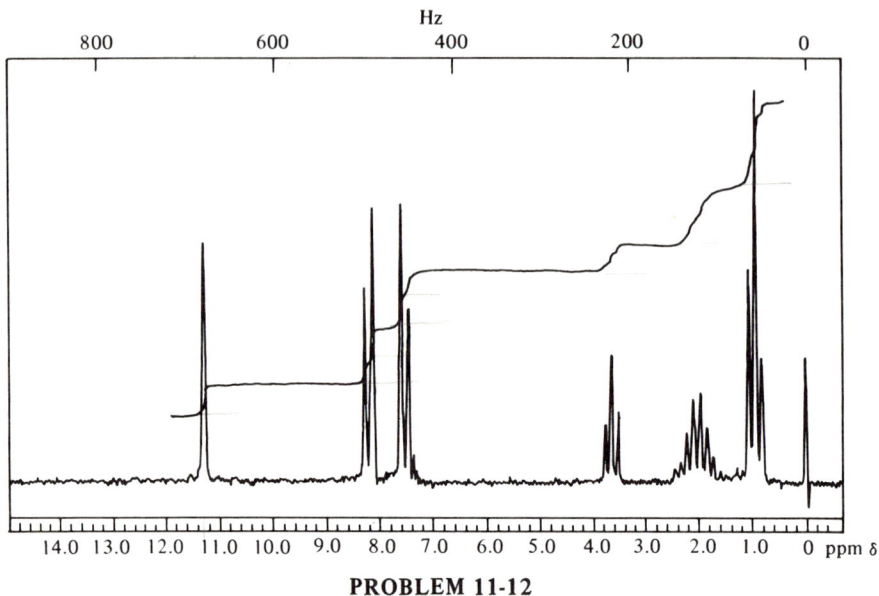

PROBLEM 11-12

13. The NMR spectrum contains a single peak at δ 3.58 and another single peak at δ 7.29. Integrated intensities are 8 and 20 units, respectively. From mass spectral information, the compound (mol. wt. 246) is known to contain two sulfur atoms. Deduce its structure.

14. The NMR spectrum contains single peaks at δ 7.27, δ 3.07, and δ 1.57. The empirical formula is $C_{10}H_{13}Cl$. Deduce the structure of the compound.

15. The molecular weight is 190; deduce the structure of the compound whose NMR spectrum is shown.

16. Overlapping multiplets always present a challenge in unraveling a NMR spectrum, such as the one for the compound $C_6H_{11}BrO_2$ shown. Deduce the structure. Ignore the small benzene peak at δ 7.32.

Nuclear Magnetic Resonance Spectroscopy 353

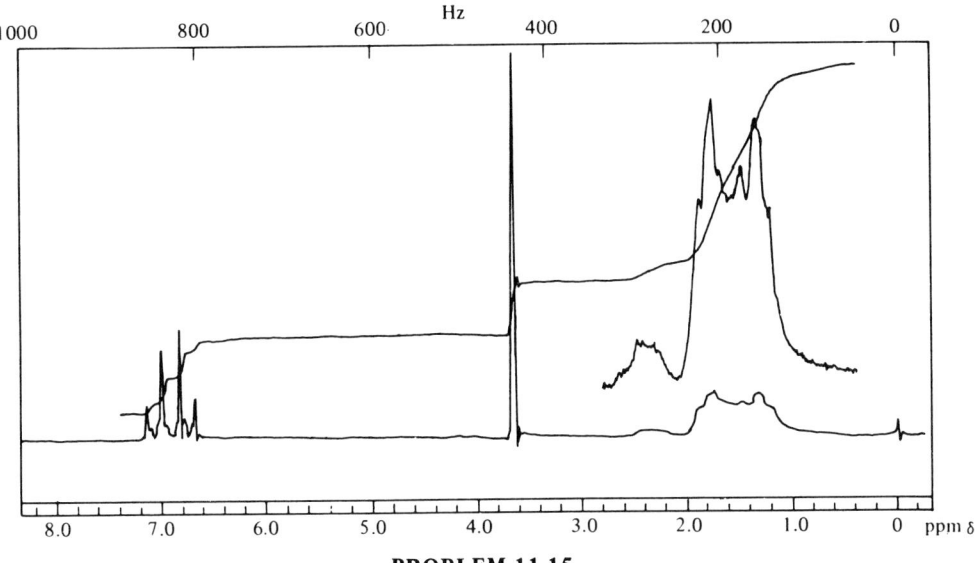

PROBLEM 11-15

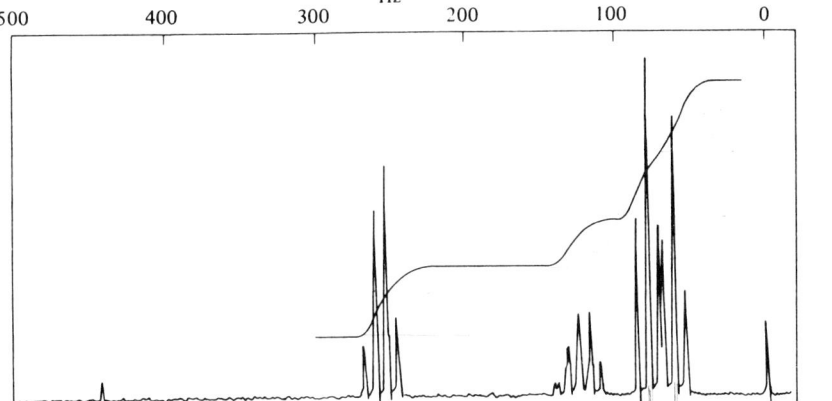

PROBLEM 11-16

17. The NMR spectrum of compound $C_4H_7ClO_2$, shown, also has one obvious overlapping spectral feature. Deduce the structure. The small peak at δ 7.32 is a benzene marker.

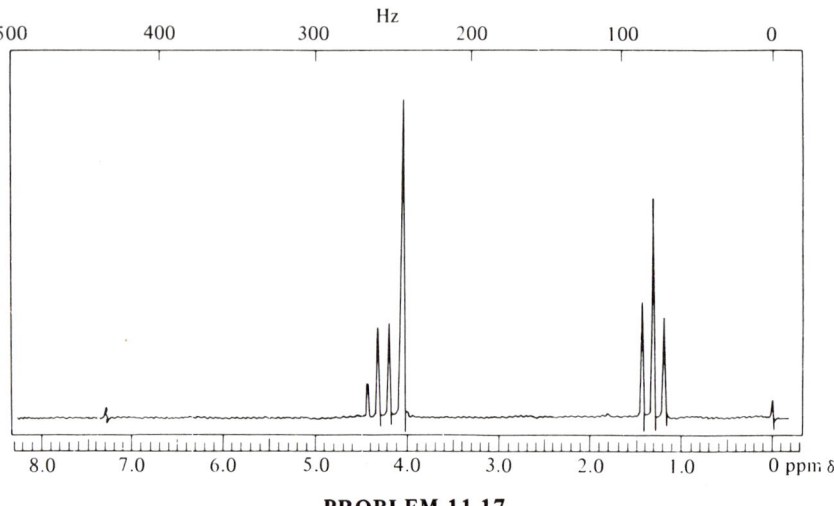

PROBLEM 11-17

18. Deduce the structure of the compound $C_8H_{14}O_4$ from the NMR spectrum shown. Be cognizant of the requirement for spin–spin coupling.

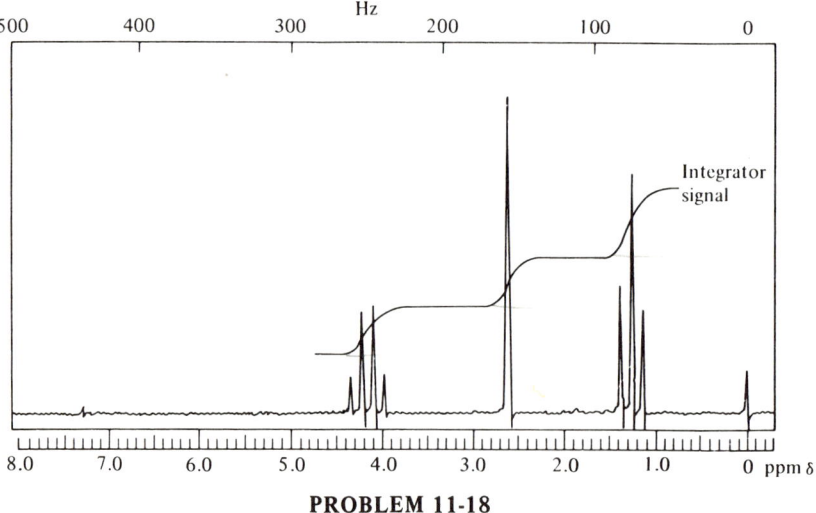

PROBLEM 11-18

19. The compound C_3H_4 has only one peak in its NMR spectrum located at δ 1.80. Can you ascertain its structure?

20. Chart the spin–spin couplings involved in the spectrum of methyl methacrylate shown. The lower trace is an expanded scale. The resonance position of external benzene was 418.4 Hz from TMS, the internal standard, in an rf field of 60 MHz.

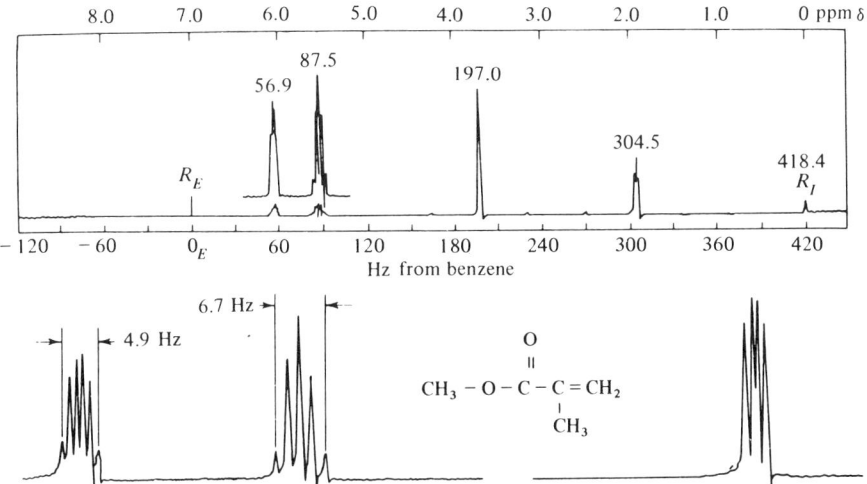

PROBLEM 11-20 The proton NMR spectrum of methyl methacrylate (liquid) taken on Varian model A-60 spectrometer. Internal standard was TMS; external standard was benzene.

21. The NMR spectrum of compound $C_{10}H_{13}NO_2$, isolated from a headache preparation, is shown. Write its structure.

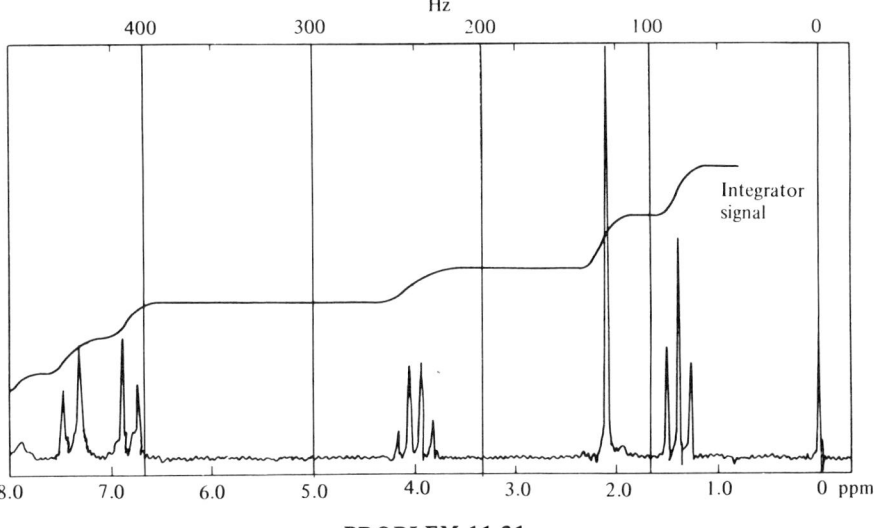

PROBLEM 11-21

22. Suggest a method for obtaining the resonance frequency of ^{11}B (19.3 MHz) on an NMR spectrometer equipped with a 5-MHz crystal oscillator when $H_0 = 14.09$ kG.

23. In the double resonance procedures, why cannot a sweep of the static magnetic field be used to generate the spectrum?

24. Estimate the nuclear Overhauser effect sensitivity enhancement for (a) decoupling protons from ^{13}C NMR spectra, and (b) decoupling ^{13}C from ^{1}H spectra.

BIBLIOGRAPHY

Farrar, T. C. and E. D. Becker, *Pulse and Fourier Transform NMR*, Academic, New York, 1971.
Jackman, L. M. and F. A. Cotton, Ed., *Dynamic Nuclear Magnetic Resonance Spectroscopy*, Academic, New York, 1975.
Jackman, L. M. and S. Sternhell, *Applications of Nuclear Magnetic Resonance Spectroscopy in Organic Chemistry*, Pergamon, New York, 1969.
Leyden, D. E. and R. H. Cox, *Analytical Applications of NMR*, Wiley, New York, 1977.
Mullen, K. and P. S. Pregosin, *Fourier Transform NMR Techniques: A Practical Approach*, Academic, New York, 1976.
Pople, J. A., W. G. Schneider, and H. J. Bernstein, *High-Resolution Nuclear Magnetic Resonance*, McGraw-Hill, New York, 1959.

LITERATURE CITED

1. Miknis, F. P., V. J. Bartuska, and G. E. Maciel, "Cross-polarization ^{13}C NMR with Magic-angle Spinning," *Am. Lab.*, p. 19 (November 1979).
2. *The Aldrich Library of NMR Spectra*, Aldrich Chemical Co., Milwaukee, Wisconsin, 1974.
3. Sadtler Research Laboratories, *Nuclear Magnetic Resonance Spectra*, Philadelphia, Pa., a continually updated subscription service.
4. Silverstein, R. M. and G. C. Bassler, *Spectrometric Identification of Organic Compounds*, 2d ed., Chap. 4, Wiley, New York, 1967.
5. Varian Associates, *High Resolution NMR Spectra Catalog*, Vol. 1, 1962; Vol. 2, 1963, Palo Alto, Calif.
6. Ronayne, J. and D. H. Williams, *J. Chem. Soc. Sect. B*, 535 (1967).
7. Kime, K. A. and R. E. Sievers, "A Practical Guide to Uses of Lanthanide NMR Shift Reagents," *Aldrichimica Acta*, **10**, 54 (1977).
8. Sievers, R. E., Ed., *Nuclear Magnetic Resonance Shift Reagents*, Academic, New York, 1973.
9. Hollis, D. P., *Anal. Chem.*, **35**, 1682 (1963).

CHAPTER 12

Electron Spin Resonance Spectroscopy

Electron spin resonance (ESR), or electron paramagnetic resonance (EPR), is a branch of absorption spectroscopy in which radiation of microwave frequency induces transitions between magnetic energy levels of electrons with unpaired spins. The magnetic energy splitting is created by a static magnetic field. Unpaired electrons, relatively unusual in occurrence, are present in odd molecules, free radicals, triplet electronic states, and transition metal and rare earth ions. There is much interest in the unpaired electrons of free radicals. These electrons are generally left after homolytic fission of a covalent bond, which is often produced by ultraviolet or gamma irradiation of the sample.

12.1 ELECTRON BEHAVIOR

Electron behavior in a magnetic field has been discussed in Chapter 1. Imposition of an external static magnetic field H_0 establishes two energy levels (Fig. 12-1). The difference in energy between the two levels is given by

$$\Delta E = \mu_e H_0 / M_s = h\nu \tag{12-1}$$

where μ_e is the electron magnetic moment and M_s is the angular momentum quantum number which can have values of $+\frac{1}{2}$ or $-\frac{1}{2}$. Substituting the values in Eq. 12-1, and rearranging

$$\nu = \frac{\mu_e H_0}{h M_s} = \frac{(9.285 \times 10^{-24} \text{ J T}^{-1}) H_0}{(6.626 \times 10^{-34} \text{ J sec})(\frac{1}{2})} = 2.803 \times 10^{10} \text{ sec}^{-1} \text{ T}^{-1} (H_0) \tag{12-2}$$

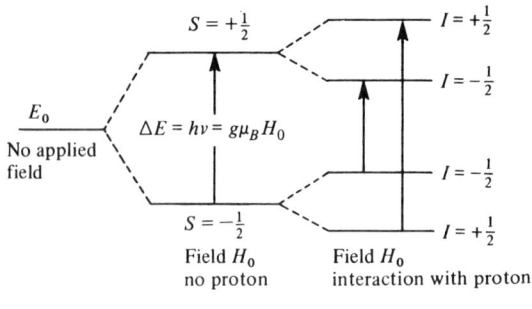

FIGURE 12-1 Electron energy levels. (a) Splitting of energy levels by a magnetic field H_0 and by interaction of unpaired electron with one proton. (b) Splitting of spectral line.

The SI unit* of magnetic field strength is a weber per meter squared, which is named a tesla (T). The conversion to gauss is 10,000 G ≡ 1 T. When the magnetic field is expressed in kilogauss, the resonance frequency for transitions from one spin state to the other for a free electron (in MHz) is given by $\nu = 2803 H_0$. Most ESR measurements in solution are made with an exciting frequency of 9500 MHz and a field strength of 3400 G.

For a single spin system, the excess of population of the lower energy level over that in the upper energy level is extremely small. It is governed by the Boltzmann distribution, similar to the situation in NMR (see Eq. 11-2). At room temperature and for a free electron, the relative ratio of the population of upper-to-lower energy levels is about 0.998. This much greater population difference is what gives ESR its superior sensitivity when compared with NMR. Even greater sensitivity can be obtained by working at low temperatures which was not always possible in NMR. Solids down to liquid helium temperatures have been studied by ESR.

12.2 ESR SPECTROMETER

The principal components of an ESR spectrometer are (1) a source of microwave radiation of constant frequency and variable amplitude; (2) a means of applying the microwave

*The International System of Units, Le Systéme International d'Unités, is officially abbreviated SI.

power to the sample—the microwave bridge; (3) a homogeneous and steady magnetic field to provide the magnetic field (spectroscopic) splitting; (4) an ac field superimposed on the steady field so as to sweep continuously through the resonance absorption of the sample; (5) a detector to measure the microwave power absorbed from the microwave field; and (6) an oscilloscope or a graphic x-y recorder. A simplified block diagram of an ESR spectrometer is shown in Fig. 12-2.

Source

Most ESR spectrometers employ radiation obtained from a klystron oscillator operating in the microwave X-band (3-cm wavelength) region. In a klystron the whole oscillating circuit is within the resonant cavity of the tube. A beam of electrons flows in pulses back and forth between the cathode and a reflector filament. Power may be withdrawn from the klystron through a *waveguide* by a loop of wire which couples with the oscillating magnetic field and sets up a corresponding field in the waveguide. A waveguide consists of hollow rectangular (copper or brass) tubing, 2.2×10 cm, with silver or gold plating inside to produce a highly conducting, flat surface. Reflection of microwave power back into the klystron is prevented by an isolator—a strip of ferrite material which passes microwaves in one direction only.

Components in the microwave assembly may be coupled together by irises or slots of various sizes. Waveguide elements are matched by using screws or stubs which can be positioned in the waveguide or across the coupling slits.

Sample Cavity

The sample, contained in a cylindrical quartz tube, is held in a cavity between the poles of the magnet. A standing wave is set up in the reflection cavity; the standing wave is composed of both magnetic and electric fields at right angles to each other. The cavity is analogous to a tuned circuit—a parallel LC combination. To minimize any influence of a high dielectric constant when such material is the sample, the sample tube is located in the cavity in a position of maximum rf magnetic field and minimum rf electric field. Tubing of 3–5 mm i.d. with a sample volume of 0.15–0.5 ml can be used with samples which do not possess a high dielectric constant. For samples with a high dielectric constant, flat cells with a thickness of about 0.25 mm, and sample volume of 0.05 ml, are often used. Rotatable cavities are used for studying anisotropic effects in single crystals and in solid samples.

Dual sample cavities are used for simultaneous observation of a sample and a reference material. Slots can be machined into the walls of the cavity to allow ultraviolet irradiation of the sample. The two cavity sections are separately modulated, one section by a 0.1-MHz field modulation and the other section by a 400-Hz field modulation. When the signals are fed through an appropriate 400-Hz amplifier and phase-sensitive detector, or a similar 0.1-MHz network, both resonances can be displayed on the recorder and superimposed. Both samples are literally in the same cavity. Thus, sources of error in intensity measurements are automatically compensated by comparing relative signal heights.

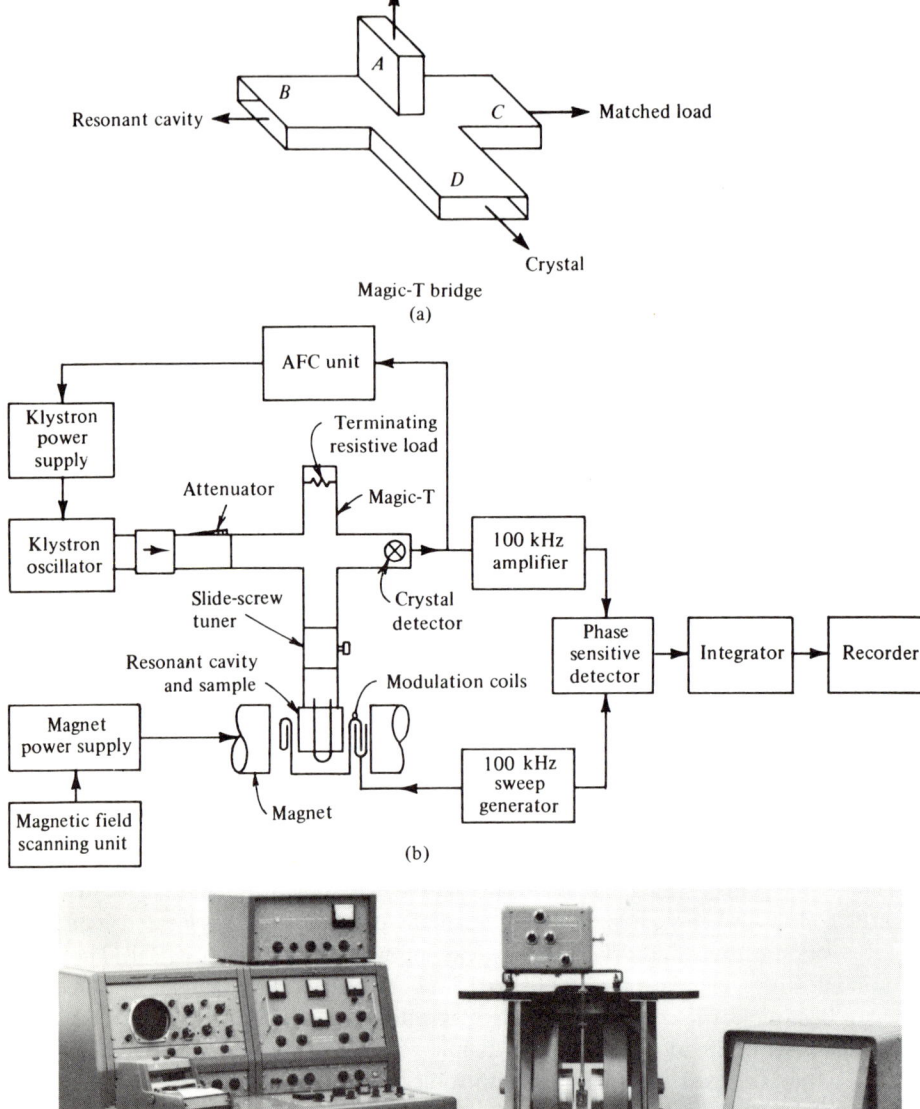

FIGURE 12-2 (a) Magic-T bridge, (b) schematic diagram of an ESR spectrometer, and (c) photograph of a commercial instrument. (Courtesy of Varian Associates.)

Magnet and Modulation Coils

An electromagnet capable of producing steady fields ranging from 50 to 5500 G is required to handle samples whose g-factor ranges from 1.5 to 6. The g-factor is given by

$$g = \mu_e/\mu_B M_s \qquad (12\text{-}3)$$

where g, called the spectroscopic splitting factor, has a value which is a function of the electron's environment (a value close to two for a free electron); μ_B is the Bohr magneton, a factor for converting angular momentum to magnetic moment. Because ESR spectrometers employ a sample resonant cavity to amplify the microwave signal, it is only feasible to vary the magnetic field. The cavity would not remain at resonance if the microwave frequency were varied. Also, it is difficult to vary the frequency of a klystron oscillator linearly and reproducibly. Field homogeneity and stability of one part in 10^6 are adequate for most ESR studies.

Microwave Bridge

The bridge, shown in Fig. 12-2, enables the microwave system to be operated as a balanced bridge with all the advantages of null methods in electrical circuits. The microwave bridge will not allow microwave power to pass in a straight line from one arm to the arm opposite. Power entering arm A will divide between arms B and C if the impedances of B and C are the same, so that no power will enter arm D. Under these conditions the bridge is said to be balanced. Arm C usually contains a balancing load. If the impedance of arm B (the sample cavity) changes because of some ESR resonance absorption by a sample in it, the bridge becomes unbalanced and some microwave power enters into arm D containing the detector—a semiconducting silicon–tungsten crystal which acts as a rectifier, converting the microwave power into direct current.

A set of coils mounted on the walls of the sample cavity and fed by a 0.1-MHz sweep generator provides modulation of the dc magnetic field at the sample position. As the main magnetic field is swept slowly through resonance over a period of several minutes, a dynamically recurring imbalance in the microwave bridge is detected and amplified for presentation on a recorder as the derivative of the microwave absorption spectrum against the magnetic field (Fig. 12-3). Because of instrumental considerations associated with the signal-to-noise ratio, ESR spectra are nearly always recorded as first-derivative spectra. For low-frequency modulation (400 Hz or less) the coils can be mounted outside the cavity and even on the magnet pole pieces. Higher modulation frequencies cannot penetrate metal effectively and the modulation coils must be mounted inside the sample cavity.

Sensitivity

The ultimate sensitivity at room temperature of practical X-band (9500 MHz) ESR spectrometers is often expressed as

$$N_{\min} = 1 \times 10^{11} \frac{\Delta H}{\sqrt{\tau}} \qquad (12\text{-}4)$$

where $N_{\min}$ is the minimum number of detectable spins per gauss, ΔH is the width between deflection points on the derivative absorption curve, and τ is the time constant of the detecting system which is inversely proportional to the bandwidth of the detection circuit. For a sample of very small dielectric loss, the minimum concentration is in the order of $10^{-9} M$ to barely see the line. For aqueous solutions, $10^{-7} M$ represents a reasonable lower limit. For structure determinations and quantitative analysis, the concentration should be about $10^{-6} M$. The limiting factor is generally the noise level in the spectrometer due to crystal noise and klystron noise. Working at higher magnetic field strengths gives a higher sensitivity. For example, at 35,000 MHz (K-band spectrometer) the sensitivity is 20 times greater.

12.3 ESR SPECTRA

In a homogeneous magnetic field an unpaired electron in an assemblage of other atoms, themselves possessing nuclear spins, can have a number of energy states. The possible energy states depend on the relative orientation of the magnetic moments of the unpaired electron, the closely associated nuclear spins, and the applied magnetic field. Contributions to the magnetic moment from orbital motion of the unpaired electron are not important for most organic free radicals.

Hyperfine Interaction

Since the radical electron is usually delocalized over the whole molecule or at least a large part of it, the unpaired electron comes into contact-interaction with many nuclei. Nuclei possessing a magnetic moment may interact and cause a further splitting of the electron resonance line. From the number and intensity distribution of the spectral lines, one can tell how many nuclei interact with the radical electron. The energies of a coupled level are given by

$$E = g\mu_B H_0 M_s + a_i h M_I \qquad (12\text{-}5)$$

where a_i is called the *hyperfine coupling constant* and M_I is the *spin quantum number* of the coupling nucleus. Ordinarily the hyperfine coupling constant is a small fraction of the electron splitting. On a spectrum it is the distance between associated peaks of a submultiplet, measured in gauss.

The selection rules for allowed ESR transitions are $\Delta M_I = 0$ and $\Delta M_s = \pm 1$. A single nucleus of spin $I = \frac{1}{2}$ will cause a splitting into two lines of equal intensity. Common nuclei with spin $\frac{1}{2}$ are ^{1}H, ^{19}F, ^{13}C, ^{15}N, and ^{31}P. Interaction with a single deuterium or nitrogen nucleus (^{2}H or ^{14}N, $I = 1$) will cause a splitting into three lines of equal intensity. Three nuclear orientations are permitted, one augmenting the external field, one diminishing it, and the other not changing it. In addition to the preceding nuclei, hyperfine splitting in polyatomic radicals by ^{6}Li, ^{7}Li, ^{10}B, ^{11}B, ^{23}Na, and ^{39}K have been reported, as have been splittings by the nuclei of transition elements, rare earths, and transuranic elements in compounds.

Hydrogen atoms that are attached to carbon atoms adjacent to the unpaired spin undergo interaction with the unpaired spin. This interaction is usually described as *hypercon-*

jugation and is a maximum when the carbon–hydrogen bond and orbital are coplanar. In general hyperfine splitting by hydrogen atoms is important only at the 1- or 2-position in an alkyl radical:

$$-CH_2-CH_2-CH_2-\overset{\cdot}{C}H-R$$
$$4321$$

Hyperfine splitting by β-hydrogens in acyclic radicals can be barely detected in most cases. In rigid molecules possessing a highly bridged structure, however, hydrogen atoms beta to the radical site undergo a strong interaction. If a double bond or a system of double bonds is followed by a single bond, as for example in methyl-substituted aromatic radicals, the double-bond character may be partially transferred to the single bond. In the ESR spectrum of 2-methyl-1,4-benzosemiquinone radical ion, the spin density is the same at the methyl protons as at the ring protons, and the spectrum consists of seven lines with an intensity ratio of 1:6:15:20:15:6:1, corresponding to the interaction with six equivalent protons. Sometimes the radical electron density penetrates even two C–C bonds, but with greatly diminished spin density. Other geometries can give rise to a long-range hyperfine splitting by hydrogen atoms. Any arrangement that places the back side of a carbon atom in close proximity to an orbital containing unpaired spin density will be expected to lead to interaction. A number of hyperfine splitting constants are gathered together in Table 12-1.

TABLE 12-1 Hyperfine Coupling Constants (in gauss)

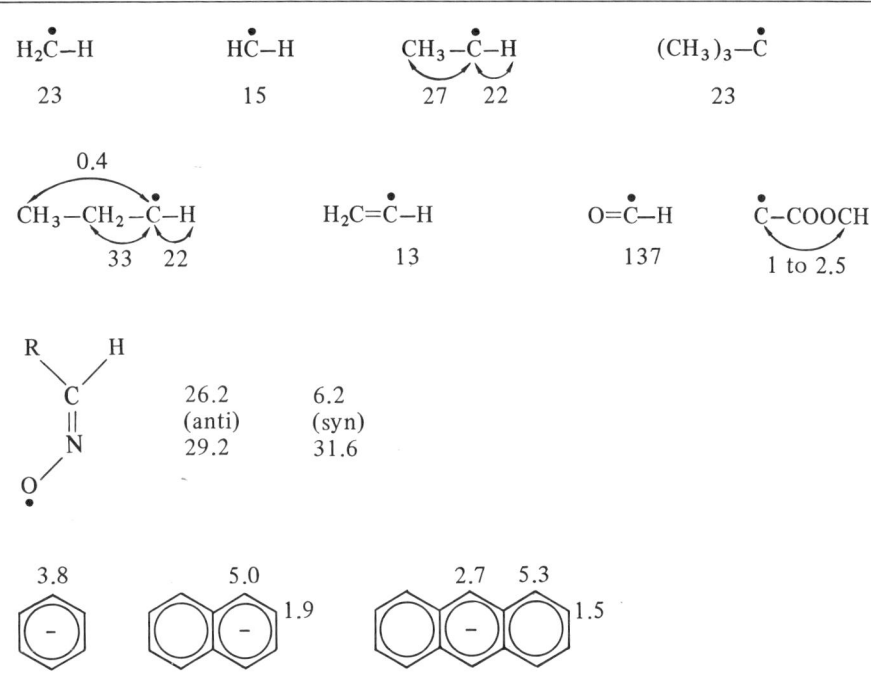

Example 12-1

Consider the free radical HO$_2$C–ĊH–COOH obtained by irradiating malonic acid. The methine proton is a charged spinning particle with a nuclear spin $\frac{1}{2}$; the proton magnetic moment is $+\frac{1}{2}$ or $-\frac{1}{2}$. The unpaired electron, represented by the "dot" over the carbon atom, will be affected by the magnetic field of the proton as well as that of the applied magnetic field H_0. Consequently, each electronic sublevel is further split into two nuclear sublevels that are equally separated, for a total of four levels (Fig. 12-1). Because the nuclear moment remains fixed during electronic transitions, only two transitions are found between these electronic states. Transi-

TABLE 12-1 Continued

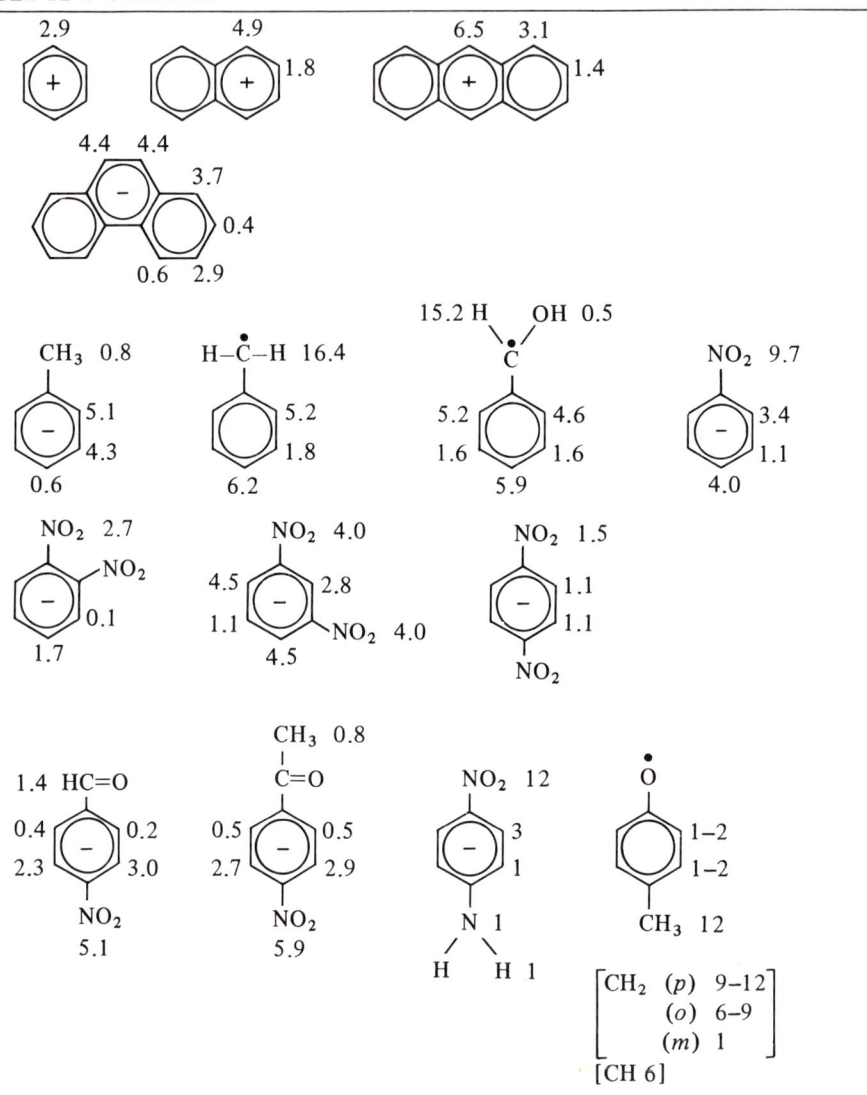

tions occur only between the states $M_s = +\frac{1}{2}$ or $M_s = -\frac{1}{2}$, conforming to the selection rules $\Delta M_s = \pm 1$ and $\Delta M_I = 0$. The result is a splitting of the original line into two absorption lines of equal intensity. The splittings from the carboxyl protons are too small to be detected.

If several magnetic nuclei are present, the situation is somewhat more complicated, because the electron experiences an interaction with each nucleus, and the spectrum is the result of a superposition of the hyperfine splittings for each nucleus. Several general types will be considered. When two equivalent nuclei are involved, the number of possible fields is reduced. Generally, $2nI + 1$ lines result from n equivalent nuclei; the relative intensity of these lines follow the coefficients of the binomial expansion. Two equivalent protons split each of the original electronic energy levels into three hyperfine levels in an intensity ratio 1:2:1. An illustration is the semiquinone formed on irradiation of 2,3-dichlorobenzoquinone and the triplet pattern from irradiated methanol (Fig. 12-3).

If the unpaired electron couples with nonequivalent protons, each proton will have its own coupling constant. In general, n nonequivalent protons will produce a spectrum with 2^n hyperfine lines.

The g-Factor

Hyperfine splitting is independent of the microwave frequency employed. This enables one to distinguish a nuclear hyperfine interaction from the effects of differences in g-factors. If, for instance, the separation of two peaks is observed to vary with the microwave frequency, it follows that they correspond to two transitions with different g-factors, not to interaction with a nuclear spin. The g-factor is a dimensionless constant and equal to 2.002319 for the unbound electron. The exact value of the g-factor reflects the chemical environment, particularly when heteroatoms are involved, because orbital angular momentum of the electron (orbital motion about a nucleus) can have an effect on the value of the transition $\Delta M_s = \pm 1$. In many organic free radicals, the g-value of the odd electron is close to that of a free electron because the electron available for spin resonance is usually near or at the periphery of the species with which it is associated. However, in metal ions, g-values are often greatly different from the free-electron value. There is a slight trend among organic radicals to higher values for free radicals containing oxygen or nitrogen, which are in turn lower than for radicals containing halogen or the peroxy group.

To measure the g-factor for free radicals, it is convenient to measure the field separation between the center of the unknown spectrum and that of a reference substance whose g-value is accurately known. A dual-sample cavity simplifies the measurement. The known may be a sample of finely powdered diphenylpicrylhydrazyl (DPPH), which is completely in the free-radical state, attached to the sample tube in a single cavity or in one chamber of a dual-cavity cell. Two signals will be observed simultaneously with a field separation of ΔH. The g-factor for the unknown is given by

$$g = g_{std}\left(1 - \frac{\Delta H}{H}\right) \qquad (12\text{-}6)$$

where H is the resonance frequency. ΔH is positive if the unknown has its center at a higher field.

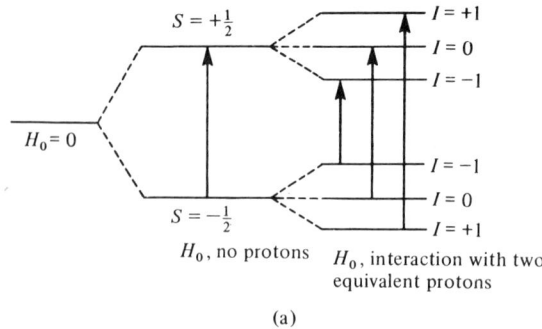
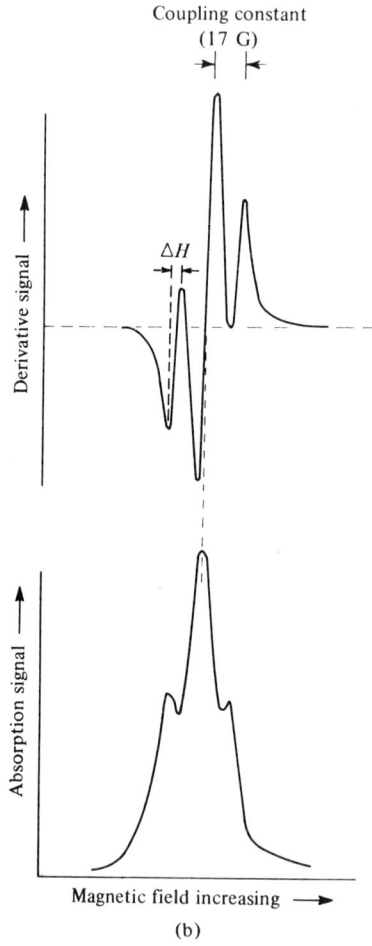

FIGURE 12-3 Irradiated methanol. (a) Interaction of an unpaired electron with two equivalent protons. (b) Derivative and absorption signals of the triplet spectrum.

Linewidths

ESR spectroscopy presents a time-averaged view of the geometry of a paramagnetic species. Because of the velocity of precession around the applied magnetic field of the magnetic moment connected with the unpaired electron spin, ESR spectra reflect a time-averaged period about 1000 times shorter than proton magnetic resonance. As the temperature is lowered the rate of conformational interconversion decreases and certain lines in the spectrum become broader. Those lines which broaden correspond to transitions for which conformational interconversion results in a change in spin state. Further lowering of temperature causes an increase in the percentage of time during which every molecule occupies a specific conformation. Eventually the spectrum approaches that of a blocked or frozen conformation (when the conformation lifetime is more than 10^{-6} sec). For sharp ESR lines, the peak-to-peak separation (ΔH) of a first-derivative spectrum should be about 0.1 G.

12.4 INTERPRETATION OF ESR SPECTRA

It should be apparent from the theoretical treatment that the interpretation of ESR spectra involves several parameters; the g-factor, the separation of the hyperfine lines and their relative intensities, the sample concentration, relaxation times, and linewidths. If the absorption lines are narrow and the spectrum fully resolved, the actual measurements are straightforward, provided that the field sweep rate is known. An ESR spectrum is assigned by perceiving the magnitudes of the coupling constants and correctly counting the lines. Frequently the measured spectrum will not contain all the lines expected because the g-factor and coupling values can be such that two lines come very close to one another and are not resolved. When many equivalent nuclei interact, relative peak heights become very large and it is difficult to see the smallest peaks. These smaller peaks are important because the outer portions of a spectrum are invariably the simplest, and the correct interpretation of these outer lines often provides the key to the unraveling of the more complex central parts. The best way of analyzing extremely complicated spectra is to introduce approximate coupling constants in an electronic computer and compare the computed spectra with the experimental ones.

Example 12-2

Figure 12-4 shows the spectrum of a single crystal of succinic acid, HOOC–CH_2–$\overset{\bullet}{C}H$–COOH, after gamma irradiation at 25°C. The coupling constants arise from the interaction of the unpaired electron with the protons on the carbon atoms. Reflecting back to the theoretical treatment in this chapter, a quartet with relative intensities 1:3:3:1 would indicate that the three protons couple equally. Obviously this is not the case. If the two methylene protons are equivalent, the spectrum would approximate a triplet (1:2:1) further split into a doublet by the CH proton. Again, this pattern is not observed. If the three protons couple unequally, an eight-

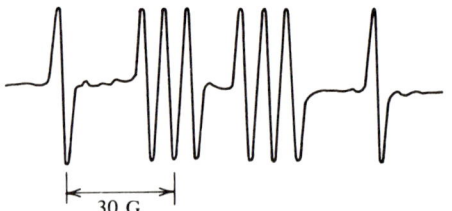

FIGURE 12-4 Spectrum of a single crystal of succinic acid after γ-irradiation at 25°C.

line pattern would be expected as follows: $2^3 = 8$, which is the pattern observed. A pair of dividers is very useful for ascertaining coupling constants and sorting out overlapping hyperfine patterns.

Example 12-3

Figure 12-5 shows the ESR spectrum obtained at room temperature from a single crystal of irradiated ammonium perchlorate. Hyperfine interaction is expected that involves the nitrogen nucleus of spin 1 and the proton nuclei of spin $\frac{1}{2}$. The spectrum probably arises from an unpaired electron spin strongly localized on an $\dot{N}H_3$ molecule. The 12-line pattern could arise from contact with the nitrogen nucleus– $2nI + 1 = 3$ lines, and contact with three equivalent protons– 4 lines, which splits each of the preceding three lines into a quartet of equally spaced lines with intensity ratios of 1:3:3:1. Verification is illustrated beneath the figure. With dividers and beginning at the left, we will ascertain the coupling constant from the left-most line and the first larger (more intense) line to its right (line 3). Note that this sepa-

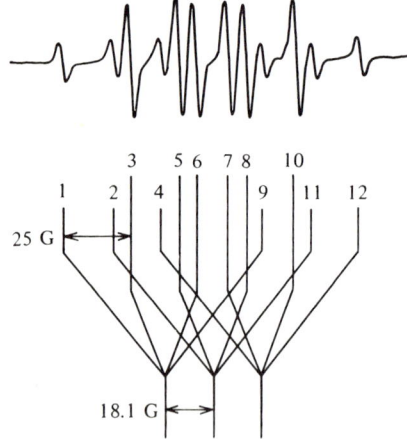

FIGURE 12-5 ESR spectrum of $\dot{N}H_3$ formed by X-ray irradiation of ammonium perchlorate crystals.

ration is found between the following pairs of lines: 1 and 3, 3 and 6, and 6 and 9, which together constitute one quartet. The second quartet involves lines 2, 5, 8, and 11; the remaining lines constitute the third quartet. Now from the center of each quartet one locates the individual members of the three-line pattern.

A three-line pattern, as in this example, and a triplet denote different environments. The former usually implies interaction with a single nucleus ($I = 1$), whereas the latter will involve interaction with two equivalent nuclei.

Example 12-4

The ESR spectrum of the ethyl radical trapped in argon at $4.2°K$ consists of a quartet of relatively sharp lines, each of which is further split into a triplet. The outer lines of the triplets are relatively broad. In liquid ethane at $-170°C$ the spectrum consists of 12 very sharp lines. A discussion of the origin of the two spectra and comments on the difference between the spectra follows.

The quartet is due to CH_3 and indicates that the unpaired electron interacts equally with all three hydrogen atoms and that, therefore, the CH_3 group freely rotates about the C–C bond. The triplet is caused by the CH_2 protons which produce a dipole magnetic anisotropy which cannot be averaged out by internal rotation. End-over-end rotation does not take place at $4.2°K$. At the higher temperature, rapid tumbling of radicals in solution cancels the anisotropic effects. In crystals powerful electrostatic fields are present. These crystalline fields may be sufficient to remove the degeneracy; that is, to split the spin levels, even in the absence of an external field. This gives rise to zero-field splitting. The ESR spectrum will be dependent upon the orientation of the crystalline field axis with respect to the applied laboratory magnetic field. These anisotropic interactions play important roles for radicals trapped in solids or highly viscous media. The hyperfine splitting for a radical that is not spherically symmetric depends on the orientation of the radical with respect to the magnetic field.

Example 12-5

A septet of triplets comprises the spectrum given by the semiquinone intermediate formed during the condensation of diacetyl (Fig. 12-6). The larger coupling constant is represented by the distance between the triplet centers; the smaller coupling constant is given by the separation between the hyperfine pattern of each triplet. Protons in two different environments are involved. The large septet with an intensity ratio of 1:6:15:20:15:6:1 must arise from six equivalent protons. Each of these lines is split again into a triplet by the weak coupling of two protons. The conclusion is that the condensation of diacetyl proceeds through a semiquinone intermediate whose structure must possess two methyl groups and two-ring protons. Knowledge of the magnitude of coupling constants is helpful (Table 12-1). Only relative hyperfine splittings can be obtained from a single sample. By placing a sample of known hyperfine splitting in one cavity of a dual-cavity sample holder, however, the unknown splitting can be accurately measured.

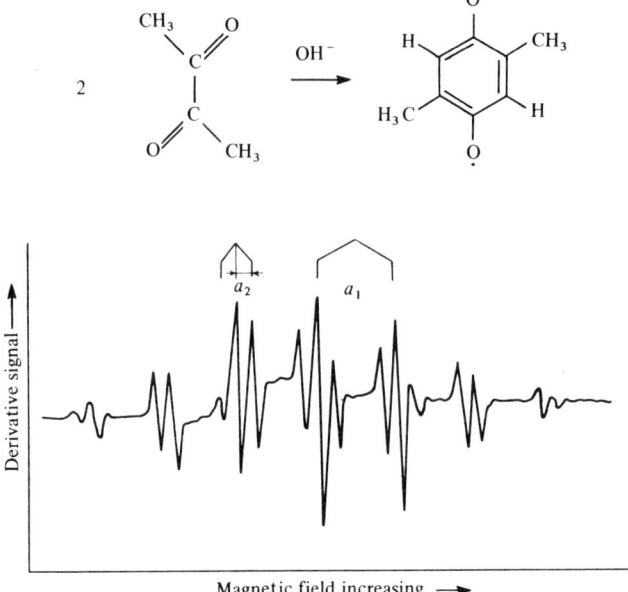

FIGURE 12-6 The septet of triplets in the ESR spectrum of semiquinone formed during the condensation of diacetyl.

Spin Label

Biological and chemical compounds which do not contain an unpaired electron can, nevertheless, be studied when they are chemically bonded to a stable free radical.[1,2] This radical, or *spin label*, produces a sharp and simple ESR spectrum that gives detailed information concerning the molecular environment of the label. Specific sites on a molecule can be tagged by choosing the appropriate spin label and reaction conditions. Five- or six-membered heterocyclic molecules incorporating a nitroxyl group whose nitrogen atom is bonded to a tertiary carbon are excellent spin-labeling compounds for biological macromolecules such as proteins and nucleic acids.

If the unpaired electron in a nitroxide radical is completely localized on the oxygen atom, there will be no isotropic coupling to the nitrogen nucleus. The observed coupling arises from the charged resonance structure.

where the unpaired electron is on the nitrogen atom. This charged structure is stabilized to a greater extent when the nitroxide is in a polar medium of high dielectric constant, such as water, than when it is in a nonpolar hydrocarbon environment. In a biological system, this solvent effect can be used to determine the polarity of the region surrounding a spin label.

12.5 ENDOR

Electron nuclear double resonance (ENDOR) is a method for improving the effective resolution of an ESR spectrum.[3] The sample is irradiated simultaneously with a microwave frequency suitable for electron resonance and a rf suitable for nuclear resonance. The rf is swept while one point of the ESR spectrum is observed under conditions of microwave saturation. The ENDOR display is ESR signal height as a function of the swept nuclear rf. For systems with short nuclear relaxation times, such as free radicals in solution, or with small nuclear magnetic moments, large rf fields are required. However, for systems with long relaxation times, such as solids, low power (5 W) equipment is sufficient.

The ENDOR technique is useful when a large variety of nuclear energy levels broaden the normal electron resonance line, masking the structure that contains important physical and chemical information. If the electron resonance can be held in a condition that is sensitive to nuclear transitions, the application of an rf field of the appropriate amplitude and frequency will expose the nuclear transitions responsible for the unresolved electron resonance.

12.6 ELDOR

In *electron double resonance* (ELDOR) the sample is irradiated simultaneously with two microwave frequencies. One of these is used to observe an ESR signal at some point of the spectrum, while the other is swept through other parts of the spectrum to display the ESR signal height as a function of the difference of the two microwave frequencies. The technique is useful for separating overlapping multiradical spectra and for studying various relaxation phenomena including chemical and spin exchange.

12.7 QUANTITATIVE ANALYSIS

The integrated intensity is usually related to the concentration of the paramagnetic species by comparison with a standard. The total area enclosed by either the absorption or dispersion signal is proportional to the number of unpaired electron spins in the sample. Comparison is made with a standard containing a known number of unpaired electrons and having the same line shape as the unknown (Gaussian or Lorentzian). A solid frequently used is 1,1-diphenyl-2'-picrylhydrazyl (DPPH) or solutions of peroxylamine disulfonate. DPPH contains 1.53×10^{21} unpaired spins per gram; substandards can be prepared by dilution with carbon black. Secondary standards include charred dextrose or synthetic ruby attached to the cavity. A dual-sample cavity is used to minimize difficulties with actual physical interchange of standard and sample.

Direct analytical applications of ESR are gradually being developed. The analysis of vanadium in petroleum products has proved to be a rapid and convenient method that covers the range 0.1–50 µg/ml. Vanadyl(IV) etioporphyrin dissolved in heavy oil distillate serves as a standard. A continuous process analyzer has been described in which the sample circulates continuously through the sample compartment. Manganese(II) ion can be determined in aqueous solutions over the range from 10^{-6} to $0.1 M$. Other ions that can be handled quantitatively by ESR include copper(II), chromium(III), gadolinium (III), iron(III), and titanium(III).

Polynuclear hydrocarbons, such as anthracene, perylene, dimethylanthracene, and naphthacene, have been determined after conversion to radical cations and adsorption onto the surface of an activated silica–alumina catalyst.

One problem inherent in the ESR method is that a relatively large fraction of the total area under the absorption (or first-derivative) curve tails off very slowly. The result is that overlap in the wings of a spectrum with peaks of another spectrum is more serious than immediately expected. Randolph[4] presents a critical evaluation of quantitative ESR methods.

PROBLEMS

1. For an ESR spectrometer that operates at a frequency of 35,000 MHz (K-band), calculate the static magnetic field required and the value of ΔE.

2. From the Heisenberg uncertainty principle, $\Delta E \Delta t \geqslant h/2$, where ΔE is the energy of the transition and Δt the time available for the measurement, estimate the range of Δt necessary to achieve reasonably sharp lines ($\Delta H \simeq 0.1$ G) for a radical with $g = 2$.

3. Which valency states of copper and silver will show a strong ESR signal?

4. The ESR spectra obtained during the enzymic oxidation by peroxidase–H_2O_2 of three substrates is shown. Match the spectrum with the free radical diagrammed below.

5. For each spectrum shown, identify the semiquinone from which it arises. The five compounds are p-semiquinone itself and the following derivatives: monochloro-, 2,3-dichloro-, trichloro-, and tetrachloro-. How does the splitting of each spectrum arise?

ELECTRON SPIN RESONANCE SPECTROSCOPY   373

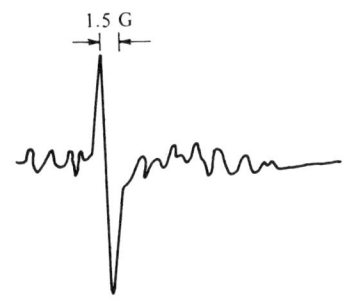

A. From reductic acid

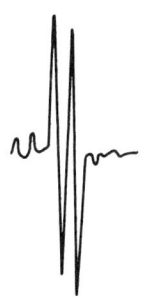

B. From dihydroxyfumaric acid

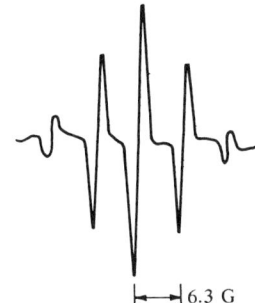

C. From ascorbic acid

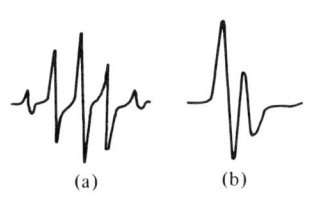

Spectrum 1   Spectrum 2   Spectrum 3

PROBLEM 12-4

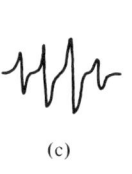 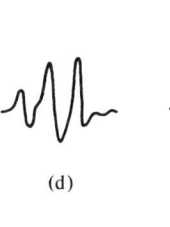 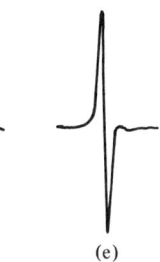

(a)   (b)   (c)   (d)   (e)

PROBLEM 12-5

374   CHAPTER 12

6. For each spectrum in the accompanying diagram, deduce the structure of the free radical and diagram the energy levels. (a) γ-Irradiation of dimethyl sulfone at 77°K, (b) alkali metal reduction of benzene in ethereal solution at −70°C, (c) γ-irradiation of *t*-butyl iodide at 77°K, (d) γ-irradiation of ethyl chloride at 77°K, (e) reaction of OH radical with glycollic acid, and (f) the general nature of the group in the 4-position of the semiquinone from 2,6-di-*t*-butyl phenol.

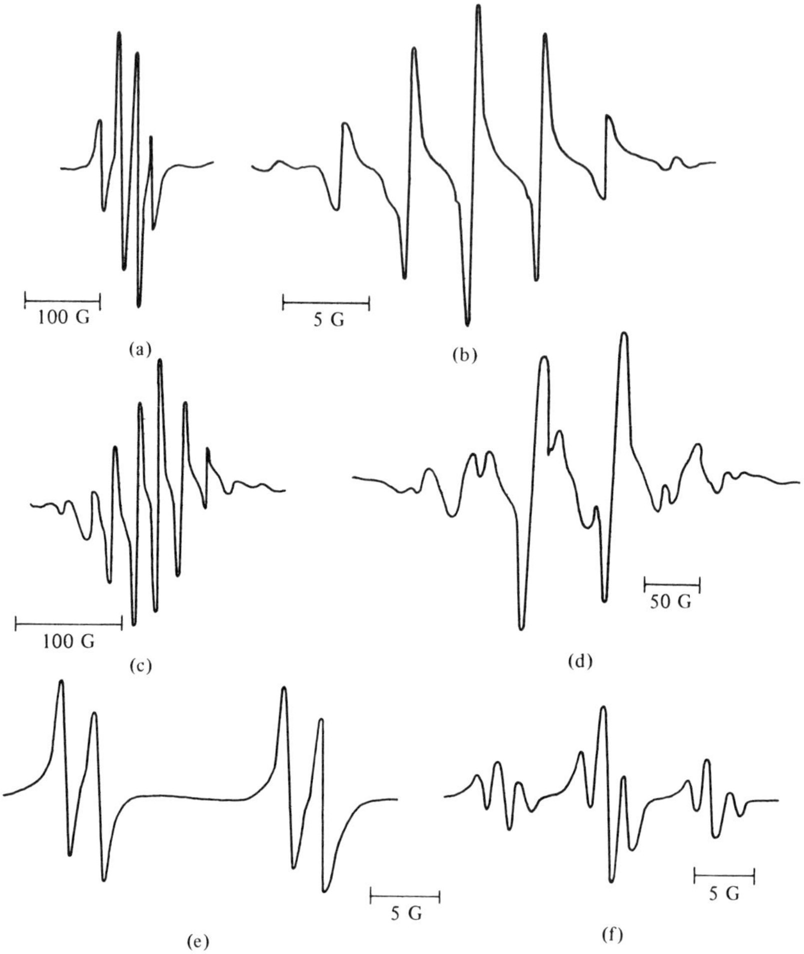

PROBLEM 12-6

7. Analyze the spectrum shown for the hypothetical radical H · X^+ ($I = 1$ for X).

PROBLEM 12-7

8. When malonic acid, $CH_2(COOH)_2$, is irradiated with X rays at room temperature, the spectrum appears to consist of a dominant doublet and a less intense, overlapping triplet. On standing for a few days, only a doublet remains. Determine the two products that are formed.

9. A single resonance line less than 1 G in width is observed at the position of the free-electron resonance in an aqueous solution of sodium dithionate(III) (hydrosulfite), $Na_2S_2O_4$. (a) What species is responsible? (b) Why is no hyperfine structure observed in the spectrum? (c) How would a sample enriched in ^{33}S verify the assignment?

10. The ion $(SO_3)_2NO^{2-}$ in an aqueous solution of the potassium salt gives the ESR spectrum shown in the figure: three sharp equally spaced lines (13-G separation) of equal intensity. After 50-fold amplification two additional lines, denoted by an asterisk, appear in the spectrum. Comment on the origin of the spectrum. [*Note:* ^{15}N ($I = \frac{1}{2}$) has a magnetic moment 0.6943 as large as that of ^{14}N.]

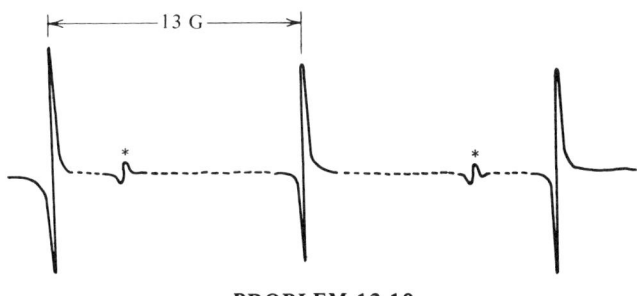

PROBLEM 12-10

11. When 2,6-di-*t*-butyl-4-methyl-4-*t*-butyl peroxycyclohexa-2,5-diene-1-one is decomposed at 130°C in xylene within the sample cavity of an ESR spectrometer, the spectrum shown is obtained. Deduce the principal radical intermediate.

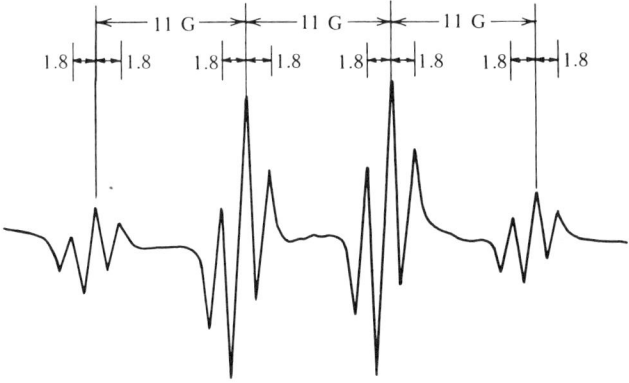

PROBLEM 12-11

12. When polyethylene is γ-irradiated in vacuum, a six-line spectrum is obtained with approximately a binomial intensity distribution and a splitting of 26 G. However, when stretched polyethylene samples are arranged with the direction of stretch perpendicular to the magnetic field, a spectrum of five doublets is obtained, with a doublet splitting of 13 G. (a) What is the most likely radical? (b) Comment on the coupling of the α- and β-protons.

13. Diagram the splitting of the energy levels giving rise to the ESR spectrum of the seminaphthoquinone shown.

PROBLEM 12-13

14. Assign the spectrum lines in the accompanying figure to the proper ring protons of naphthalene; the negative ion spectrum results from contact with alkali metal in an inert solvent.

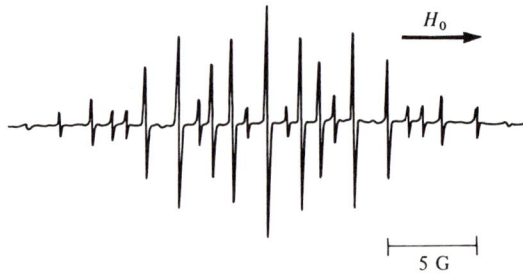

PROBLEM 12-14

15. Photolysis of a solution of hydrogen peroxide in isopropyl alcohol at 110°K leads to free radical formation and an ESR spectrum consisting of seven lines with a hyperfine splitting of 20 G, a linewidth of 10 G, and an approximate intensity distribution of 1:6:15:20:15:6:1. Which radical is responsible?

16. Match the individual structure of the three isomeric methylcyclohexanones with the ESR spectra shown. The radical anions were detected by exposing a dimethyl sulfoxide solution to air for 15–25 sec. The solution contained 0.1M potassium *tert*-butoxide and a particular methylcyclohexanone. (*Note:* An alkyl group confers conformation stability on a cyclohexene ring in terms of ESR frequencies.)

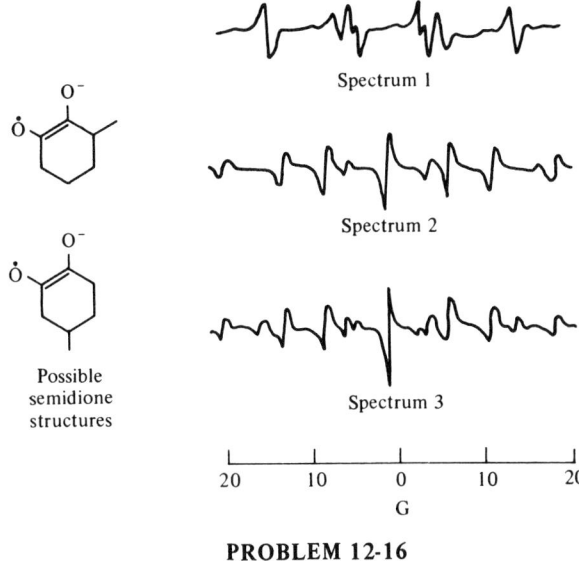

PROBLEM 12-16

17. What difference would be anticipated between the ESR spectra of the following aryloxy radicals: 2,6-di-*t*-butyl-4-methyl derivative (I) and 2,6-di-*t*-butyl-4-ethyl derivative (II).

18. Predict the ESR spectrum obtained from the aryloxy radicals formed from each of these compounds: (a) 2,6-di-*t*-butyl-4-hydroxymethylene phenol, (b) 2,4,6-tri-*t*-butyl phenol, (c) 2,6-dimethyl-4-*t*-butyl phenol, and (d) 2-methyl-4,6-di-*t*-butyl phenol.

19. For each of the following compounds, determine the structure of the free radical that is formed upon treatment with PbO_2: (a) 2,4-dimethyl-6-*t*-butyl phenol, spectrum shows a quartet of triplets; (b) 2-methyl-4-ethyl-6-*t*-butyl phenol, spectrum shows a quartet of triplets; (c) 2-ethyl-4-methyl-6-*t*-butyl phenol, spectrum shows a triplet of triplets; and (d) 2,4,6-trimethyl phenol, spectrum shows a septet of triplets.

BIBLIOGRAPHY

Alger, R. S., *Electron Paramagnetic Resonance, Techniques and Applications*, Wiley-Interscience, New York, 1968.

Ayscough, P. B., *Electron Spin Resonance in Chemistry*, Methuen, London, 1967.

Berliner, L. J., *Spin Labeling–Theory and Applications*, Academic, New York, 1976.

Dorio, M. M. and J. H. Freed, Eds., *Multiple Electron Resonance Spectroscopy*, Plenum, New York, 1979.

Ranby, B. and J. F. Rabek, *ESR Spectroscopy in Polymer Research*, Springer-Verlag, New York, 1977.

LITERATURE CITED

1. Griffith, O. H. and A. S. Waggonner, *Accounts Chem. Res.*, **2**, 17 (1969).
2. Hamilton, C. L. and H. M. McConnell, in *Structural Chemistry and Molecular Biology*, A. Rich and N. Davidson, Eds., W. H. Freeman, San Francisco, 1968.
3. Kedzie, R. W., *Am. Lab.*, **1** (12), 19 (December 1969).
4. Randolph, M. L., in *Biological Applications of Electron Spin Resonance*, H. M. Swartz, J. R. Bolton, and D. C. Borg, Eds., Chap. 3, Wiley, New York, 1972.

CHAPTER 13

Chemical Analysis of Surfaces

Most chapters in this book describe specific methods and their applications to many substances or use in many situations. Due to the recent interest in analysis of surfaces, this chapter will, in a sense, reverse the above procedure and describe several different methods of analyses applicable to qualitative and semiquantitative analyses of surfaces as an aid in surface characterization.

A surface is, theoretically, an infinitely thin layer separating two phases. The scanning electron microscope provides pictures of surface topography at a resolution of about 10 nm. Most analyses are concerned with the interface between a solid and a gas (usually air). Furthermore, in practice the layer of interest has a finite thickness, at least the thickness of an atom or a molecule. One might be interested in determining what is present in the surface layer and whether the surface layer is homogeneous in a direction parallel to the surface. Sometimes, however, there is a concentration gradient in a direction perpendicular to the surface that may extend from atomic dimensions to 10 nm or more. Surface analysis then might involve studying the differences in concentrations of elements at the surface and into the bulk layer by layer. On most technological surfaces the greatest surface compositional changes occur in approximately the first 20 monolayers.

Surface Spectroscopy

Surface spectroscopy involves probing a sample target with a flux of energetic particles and detecting characteristic particles emitted from the surface after interaction. The probe beam may be photons, electrons, or ions.

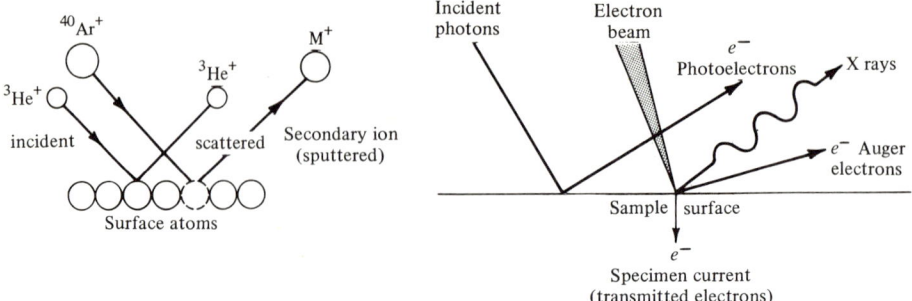

FIGURE 13-1 Energy interchange involved in surface analysis methods.

Surface analysis has many applications, for example, adsorption of contaminants on a surface, adsorption of reacting or interfering molecules on the surface of a catalyst, and segregation of components of the bulk material at a surface or at grain boundaries or imperfections. Corrosion products are identified at the surface where the corrosion is occurring. For conductors, semiconductors, and insulators, one can detect and identify impurities that critically affect bond integrity and the electrical properties. Evaluation of predeposition and cleaning is facilitated. Adhesion studies benefit from the determination of the composition of surface layers that influence adhesive properties of materials. Embrittlement of metals and alloys can be studied in conjunction with in-position fracture of specimens to detect and identify grain boundary impurities that cause embrittlement. Thin film composition can be correlated to electrical, transport, and emission characteristics of the film.

The energy exchange interactions that give rise to these signals may be effectively stimulated by a variety of energy sources (Fig. 13-1). Activation may be by electromagnetic radiation, as in electron spectroscopy for chemical analysis (ESCA), by a beam of incident ions, as in secondary ion mass spectrometry (SIMS) and ion scattering spectroscopy (ISS), or by an incident electron beam as in Auger electron spectroscopy (AES) or electron microprobe (X-ray fluorescence) analysis (discussed in Chapter 9).

Photons of high energy penetrate deeply into a solid; an X-ray beam of 1000 eV may penetrate up to 1000 nm. By contrast, low-energy electrons and low-energy ions have very short penetration depths on the order of 1–2 nm. However, when the ion energy is increased from 10^3 to 10^7 eV, the penetration depth increases from 2.0 to 10^4 nm. A comparison of relative depths of penetration and of analysis volumes for the several surface analysis methods to be discussed in this chapter are shown in Fig. 13-2. Thus, if one wishes to do surface spectroscopy, one should use either low-energy electrons or low-energy ions to limit the sampling depth to roughly 2.0 nm or less.

There are advantages and disadvantages offered by the various sources of incident energy. Excitation by electromagnetic radiation generally provides excellent chemical information but offers poor elemental sensitivity and spatial resolution. Electromagnetic radiation offers the additional advantage of being the least destructive to organic and other highly sensitive surfaces. The use of an incident ion beam generally provides the best elemental sensitivity, but it is limited in spatial resolution. It offers little chemical information and

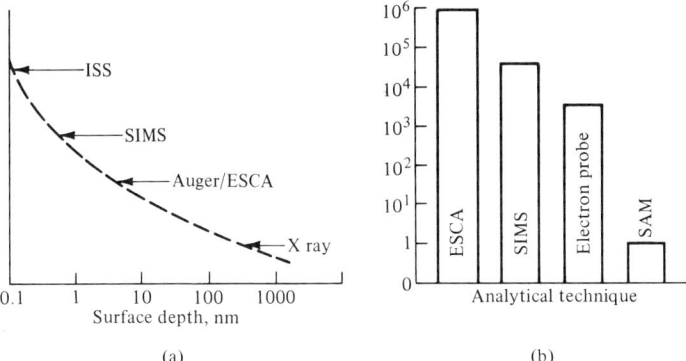

FIGURE 13-2 (a) Comparison of relative depths from which information originates for various surface analysis techniques, and (b) comparison of analysis volumes normalized to scanning Auger microprobe.

it can be used for destructive testing only. The electron beam, on the other hand, offers excellent spatial resolution for sample imaging and analysis. It offers a sensitivity much better than electromagnetic radiation-based techniques and it can be used for nondestructive analysis.

Solids are most conveniently handled as flat pieces. Pulverized material may be pressed into a disk, spread onto a metallic backing, such as abraded aluminum plate, or dusted onto conducting tape. Liquids may be handled if frozen on a cryostat finger.

Problems Unique to Surface Analysis

Samples must be transferred from their point of origin to the spectrometer without contacting air, which may alter the surface, giving erroneous results. This would be particularly critical when examining an active catalyst.

For most solid surfaces, adsorption of residual gases in a vacuum chamber occurs at an approximate rate of 1 monolayer/sec at 10^{-6} torr. A vacuum of 10^{-10} torr provides analysis times on the order of 30 min before about one-tenth of a monolayer of residual gas (more than enough to contaminate the surface) is adsorbed. In vacuums of only 10^{-5} to 10^{-6} torr, heavy contamination from pump oil is a problem. All the techniques to be discussed in this chapter have a shallow escape depth for the analytical signal and consequently place the most stringent requirements on the cleanliness of the residual vacuum in the vicinity of the sample surface. The presence of a surface contaminant will attenuate the analytical signal coming from the true sample and add the contaminant's characteristic spectrum to that of the real sample.

An even more severe problem is the possible disruption of a surface by the measurement techniques themselves. Little surface disruption occurs in ESCA and the measurements are characteristic of the surface. Charging effects can be compensated. AES is more destructive and is particularly bad for organic materials. Electrons are chemically active and thus can cause chemical effects. These effects will usually be most severe in insulators and least in conductors. Again the charging effect can be compensated. SIMS and ISS, techniques that use sputtering as the mechanism of obtaining surface information, will disrupt the surface. However, in the static SIMS mode the beam current density incident on the surface is kept at a low level, which corresponds to removal of about one monolayer in 10 hr. Thus, although SIMS causes long-term disruption, particularly in dynamic SIMS, short-term information (15 min) can be characteristic of the intact surface layer. Dynamic SIMS uses larger beam currents and does not attempt to preserve the outer layer of a surface. In fact, dynamic SIMS experiments scramble about 2.0–5.0 nm of the surface and, at best, give analyses averaged over several atomic layers.

Distribution of Surface Species

The composition of surfaces is not necessarily the same as the bulk. Frequently, in otherwise homogeneous solid systems, inhomogeneities occur as one approaches an interface. The distribution of surface species becomes an important parameter when considering chemical analysis of a surface. Surface analysis that is independent of the instrumental technique occurs only if the information depth of the technique is small relative to the concentration profile. A combination of the nature of the distribution of species and the actual measurement depth used can create vastly different ratios of measured components.

If the interaction volume being probed in a particular measurement is homogeneous and consists of a single surface phase, it can be straightforward to deduce elemental concentrations from the observed intensities. Real samples, however, can be expected to be inhomogeneous and it is a demanding task for the analyst to ensure that meaningful measurements are made.

Sputter-Etching, Depth Profiling, and Elemental Imaging or Mapping

The procedures used to obtain clean specimens include argon ion sputtering, scribing, and bake-out. The high vapor pressure of many reactive metals may prevent attainment of ultra-high vacuum through bake-out. Most spectra are obtained during argon ion sputtering. This technique produces relatively clean surfaces without baking the system. A sputtering rate of about 1.0 nm/min is considerably faster than the adsorption rate of active residual gases.

For many highly reactive metals the surface oxide layer may be too thick for convenient removal by sputtering. A more expedient method for producing a clean surface is to scribe the surface with a carbide tip. The base of the scratched groove should be about as wide as the probe beam diameter.

Depth profiling analysis can be done in conjunction with a noble gas ion beam for sputtering etching of a surface. This technique extends the spatial resolution provided in the plane of a surface to the depth dimension. The process involves bombardment of the specimen surface with a beam of argon or xenon ions to sputter etch the specimen surface while simultaneously bombarding the specimen with the electron beam (for AES) or with an X-ray beam (for ESCA). It is important that the sputter-etched area be much larger than the area of the electron or photon beams so that the region of analysis is in the center of a relatively flat-bottomed crater. This precludes analysis of the sides of the crater, where the depth resolution would be seriously impaired. In typical equipment the ion beam diameter is of the order of a few millimeters, while the electron or photon beam might be focused down to 25 μm. Typical etching rates are variable up to about 3.0 nm/min.

Raster/gating permits automatic scanning of the probe beam over a preselected surface area. The gating circuit can be used in conjunction with the rastering beam to restrict data collection to a smaller area than the scanned beam is covering. Elemental analysis can be secured by simply scanning across the specimen or by focusing on selected points.

13.1 ION SCATTERING SPECTROMETRY (ISS)

ISS involves observing a binary elastic collision between an incident noble gas ion of 300–3000 eV in energy and a surface atom or ion (Fig. 13-3). A fraction of the bombarding ions leave the surface after only a single binary elastic collision with surface atoms. Through conservation of momentum, an ion scattered at a particular angle retains a specific energy dependent only upon the mass of the surface atom and the energy of the primary beam ion. By scanning the energy of the scattered ion, the ion scattering spectrum is obtained. Knowing the mass, M_0, and the energy, E_0, of the primary ion beam, and the energy, E_i, of the particular ion scattered at 90°, the mass of the surface atom, M_i, is determined by the expression

$$\frac{E_i}{E_0} = \frac{M_i - M_0}{M_i + M_0} \tag{13-1}$$

Primary ions that penetrate past the top monolayer of exposed surface atoms do not retain the energy necessary to contribute a signal indicative of the elements present. Thus, the scattered signal is derived entirely from the top exposed layer of atoms on the specimen surface. ISS is sensitive to all elements with atomic number greater than that of the bombarding noble gas ion. The limit of detection is approximately 10^{-4} monolayer; the overall sensitivity is about 1% of the monolayer for most elements.

ISS has limited spatial capabilities; spatial resolution is about 100 μm. For elements of high atomic weight, sputtering with argon ions in place of helium ions improves the response. ISS uses several different noble gas ion species to maximize the element coverage

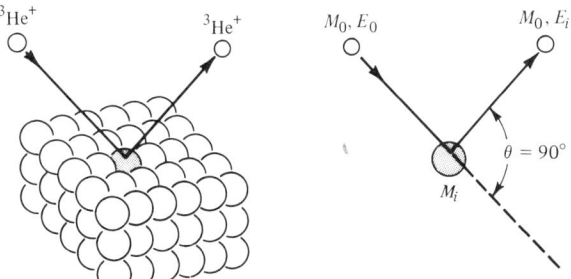

FIGURE 13-3 Elastic collision of noble gas ion and surface atom.

384   CHAPTER 13

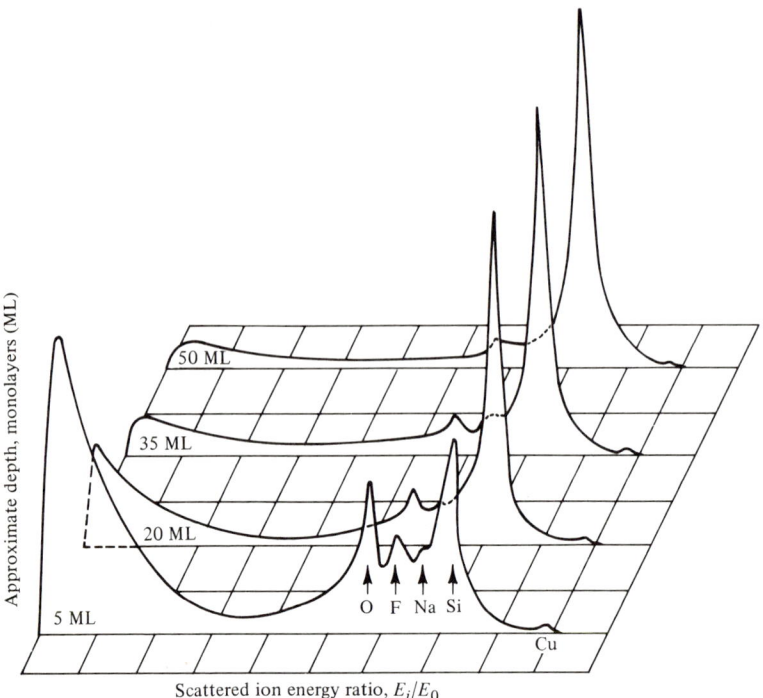

FIGURE 13-4 Depth profiling with ISS; analysis of a silicon wafer from 5 to 50 monolayers.

and mass resolution, since the scattering process is blind to masses below the probe ion's mass and the ability to resolve atoms of similar mass degrades as the difference between the probe ion and target masses increases.

Each particular element of mass M_i will cause scattering of the noble gas ion beam at a unique energy. For example, bombardment with $^3He^+$ with an incident energy of E_0, the calculated energy after 90° scattering is $0.684(E_i/E_0)$ from ^{16}O and $0.714(E_i/E_0)$ from ^{18}O. The intensity of a given scattering peak is proportional to the number of scattered ions, and thus is directly related to the amount of material present on the surface.

The probe ion beam can be used in a dual manner to both analyze and to clean the surface controllably. By successive recording of scattering spectra, using controlled beam current densities, a depth-profile composition analysis can be performed, as shown in Fig. 13-4. For $^3He^+$, the rate of removal generally ranges from 3 to 50 monolayers/hr, thus enabling a single monolayer to be carefully examined or quickly removed. When surface cleaning at a higher rate is desired, a more massive noble gas ion is used, generally argon. In this case, the removal rates are increased by approximately a factor of 10.

ISS is one technique that is sensitive to different isotopic masses. The technique finds use in studying catalytic reaction mechanisms, self-diffusion processes, adsorption–desorption phenomena, the interaction of air pollutants with solid surfaces, or any other reaction involving an exchange of the same atomic species.

Ion Scattering Spectrometer

The principal components of the ion scattering spectrometer are (Fig. 13-5): (1) a monoenergetic ion gun with excellent focusing properties and controllable current densities that generates the primary ion beam; (2) a high-resolution energy analyzer for measurement of the scattering ion energies which also defines a narrow scattering angle; (3) an ion detector that detects and amplifies the scattered ion signal; (4) a rotatable multiple sample holder that carefully defines the correct sample angle with respect to the primary ion beam and analyzer; and (5) an ion gun drive, controls, and associated instrumentation. In addition there is needed an ultra-high vacuum system with pressure monitoring equipment; a residual vacuum of 10^{-9} torr is obtained with ion and titanium sublimation pumping. Sample charging of insulating surfaces is alleviated with a charge neutralization flood gun.

Ion Source An electron impact ionization source provides a beam of noble gas ions, such as $^{3}He^{+}$, $^{4}He^{+}$, $^{20}Ne^{+}$, or $^{40}Ar^{+}$, with a low initial kinetic energy spread. The ion source consists of a cylindrical grid with an external filament. Ions formed by electron bombardment of the noble gas inside the grid are extracted axially from one end and focused on the target by an electrostatic aperture lens system. The ion gun produces a nominal 1-mm diameter ion beam throughout an operating range of 300–3000 eV. Beam diameter can be decreased in steps to a minimum of 100 μm. Such convenient beam selection permits

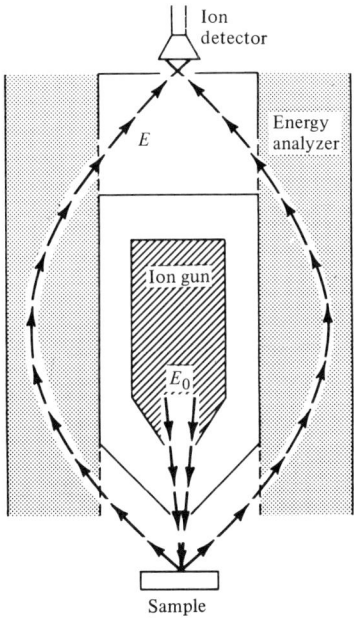

FIGURE 13-5 ISS optical system; ion gun is coaxial with cylindrical mirror analyzer (CMA).

the rapid transition from localized surface analysis with the small beam, to a large area, higher sensitivity analysis where spatial resolution is not critical. By focusing adjustments current densities from 1 to 50 $\mu A/cm^2$ are obtained.

Energy Analyzer The ion gun, which is used to form and direct the primary ion beam normal to the specimen surface, is located coaxially within a cylindrical mirror (energy) analyzer (CMA). The normal incident beam eliminates shadowing caused by surface topography and permits analysis of fine powders without resorting to the use of adhesives to hold them in place.

The CMA accepts those electrons which are emitted within a few degrees of $42.3°$; the acceptance solid angle of this device is about 10% of the total 2π solid angle. The principle of the CMA is based on the electrostatic focusing properties of two coaxial cylinders. With the specimen placed on axis, electrons are allowed to pass into the analyzer through an annular aperture cut in the inner cylinder. A deflecting potential is applied to the outer cylinder, which focuses electrons of a particular energy back onto the axis after they have passed through another annular aperture at exit. As no retardation is involved and only electrons in a very narrow range of energy (about 0.5% of the energy) are passed at any one deflecting voltage, the signal-to-noise characteristics are good. Spectral scan rates of 0.1 sec or faster are possible across the entire energy spectrum. This permits real-time oscilloscopic monitoring and signal averaging of spectral data.

Detector Ion detection is accomplished by use of a channel electron multiplier mounted directly behind the exit slit of the analyzer. The detector is enclosed in an electrostatic shield, electrically common with the analyzer. The detector is operated in a pulse counting mode. Its pulse output is shaped, amplified, and counted.

13.2 SECONDARY ION MASS SPECTROMETRY (SIMS)

In SIMS an energetic primary ion strikes the surface and releases secondary ions. If atomic, these secondary ions are analyzed and detected as such, while molecular ions can dissociate to give positive and negative ion mass spectra for the molecules present in the surface. The SIMS spectrum therefore consists of both those secondary ions which are stable to such dissociation and the ionic fragments of those that are not. SIMS spectra also incorporate features caused by ion–neutral association reactions occurring in the selvedge, the region of relatively high pressure just above the surface.

SIMS is a rapid, easily used technique that not only affords qualitative identification of all surface elements (including hydrogen), but also permits identification of isotopes and the structural elucidation of molecular compounds present on a surface and permits very high detection sensitivity (parts per million levels) with a minimum of sample volume (0.2–0.5 nm depth). In addition to depth composition information of about two atomic monolayers, SIMS can also obtain spatially resolved surface information, such as absorbed ion images, elemental line scans, and images of surface constituents. It is capable of unit mass resolution from hydrogen to its mass limit of 300 daltons. The positive SIMS spectra are extremely sensitive to elements on the left side of the periodic table while the negative SIMS spectra favor the right side. Signal-to-background ratios with SIMS spectra are improved, as compared with ISS, because of the inherent advantages that come with the state-of-the-art quadrupole mass spectrometer. Molecular SIMS shows the characteristic structural specificity of mass spectrometry which extends to isomer and isotopic distinction. Deleterious effects of the sampling ion beam are smaller for organic, as opposed to metallic surfaces, hence higher primary ion currents can be used without causing surface damage.

Given the existence of strong, well-defined bonds between specific atoms in organic compounds, the final bonding pattern revealed in the ionic compositions making up the SIMS spectrum can be related to the surface bonding pattern. The interpretation of SIMS spectra of organic compounds is facilitated by the large body of data available on the dissociations of gas phase organic ions. SIMS spectra incorporate features due to low-energy ion–molecule reactions and to unimolecular dissociations which parallel those observed in other forms of mass spectrometry. While parent ions such as $(M + Ag^+)$ and $(M - H)^-$ are not yet familiar ionic species, the dissociation reactions that they and other secondary ions undergo can be inferred from other forms of mass spectrometry.

While SIMS has been applied mainly to metals and semiconductors, it also has been used for analysis of glass, rare earth compounds, and minerals. One trend is the micro-area analysis by microprobe SIMS. Contributions are being made to important subjects such as analysis of precipitates, grain boundary segregations, and welded bonds, and to some serious problems such as temper brittleness, hot workability, hydrogen embrittlement, and hydrogen-induced cracking.

Ion yields in SIMS are very dependent on and sensitive to the chemical environment in which the ions are formed. The use of oxygen as a bombarding species gives significantly enhanced positive ion yields for electropositive elements in comparison to the yields obtained by using a nonreactive bombarding species such as $^{40}Ar^+$. If Ar^+ is used, positive ion yields can be enhanced by increasing the oxygen pressure in the vicinity of the sample surface.

The interactions of the sample and the primary ion beam are very complex. The matrix effects are extreme, causing variations of up to several orders of magnitude in measured elemental percentages. Generally the secondary ion yield is erratic until a reactive layer forms on the sample and a steady state is reached. The ion yield is then relatively constant, and quantitative analysis can be attempted. In any case, quantitative analysis of the first few monolayers of the sample is especially difficult.

SIMS Instrumentation

Instrumentation for SIMS ranges from simple plasma discharge source/quadrupole mass analyzers to sophisticated ion microprobe mass analyzers (IMMA) and ion microanalyzers. Whatever the level of sophistication, all instruments employ a source of primary ions, a mass analyzer, and a sensitive secondary ion detector, enclosed in a high vacuum or an ultra-high vacuum chamber. Since the ion beam used in SIMS fulfills the excitation as well as sputtering requirements of the technique, the ion source is a key part of the instrument and determines many of the instrumental capabilities.

Since the ISS technique also produces sputtered ions from the surface, it is convenient to simultaneously conduct SIMS. In the combined ISS/SIMS system, the incident beam of monoenergetic noble gas ions is directed vertically onto a properly positioned sample. The entrance aperture for the CMA is shown in Fig. 13-5. In the third orthogonal direction from the sample is the quadrupole mass analyzer whose output is the SIMS signal. A prefilter ion lens rejects the neutrals and high-energy ions. Generally, the energy range of the scattered ions is quite different from the energy range of the sputtered ions. This is beneficial for maximal signal-to-noise ratio in both techniques. The sputtered ions have low energies and interfere little with the scattered ion signal in ISS. Because of their low energy, however, the sputtered ions can be easily analyzed in a quadrupole mass analyzer (Chapter 19). The ability of the ISS/SIMS system to generate large amounts of useful information rapidly requires the use of data processing and storage to realize optimum analytical capabilities. These functions are provided through three separate units: a digital multiplexer, a data processor, and a tape drive data storage unit.

Ion Microprobe Mass Analyzer (IMMA) The IMMA type of instrument (Fig. 13-6) uses a microfocused primary ion beam from a duoplasmatron source to provide lateral microanalysis with high spatial resolution (2–10 μm) as well as surface analysis and in-depth profiling. The impinging ion beam is generated in a duoplasmatron source which is essentially a low-voltage, low-pressure, hot-cathode arc that is capable of producing either positively or negatively charged ions. This source employs a plasma with both electrostatic and magnetic constriction (hence the term *duo*) to produce a high brightness source

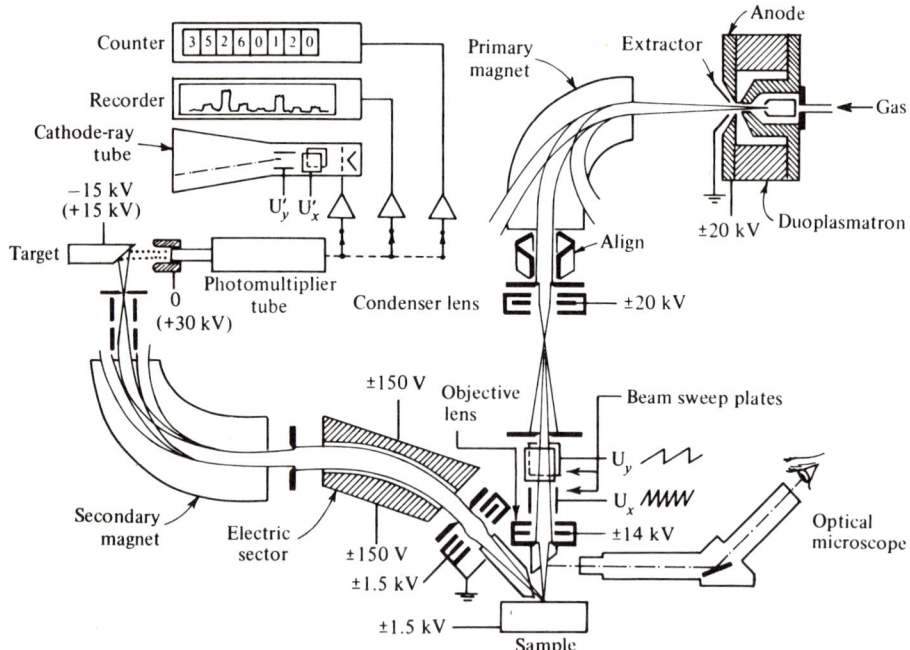

FIGURE 13-6 Schematic diagram of an ion microprobe mass analyzer. (Courtesy of Applied Research Laboratories.)

of inert or reactive gas ions (usually either O⁻, F⁻, or Ar⁺). Ions are extracted from the source through a hole in the anode and accelerated to energies ranging from 5 to 25 keV. Two electrostatic lenses then provide demagnification of the duoplasmatron source image to produce an ion beam diameter of 2–10 μm. The primary beam spot is held stationary for local analysis or rastered about the sample surface for secondary ion imaging or for producing a flat-bottomed crater for accurate depth profiling.

The duoplasmatron produces a variety of ionized species and, in some cases, ionized molecular fragments. Since it is desirable to have only one type of ion interact with the sample, the primary beam is passed through a mass analyzer placed at the entrance to the electrostatic lens column. The magnetic field is adjusted so that only the desired ion is deflected into the lenses. The composition of the primary beam can be checked by using a sample surface as an electrostatic mirror, thereby deflecting a fraction of the primary beam into the mass spectrometer. In IMMA the magnetic field is the wedge-type with plane but inclined pole pieces. In the less sophisticated ion microanalyzer, the source mass analyzer is omitted.

Sorting out the sputtered secondary sample ions by mass charge ratio is accomplished in the IMMA system by a two-stage mass spectrometer. The first-stage electrostatic sector sorts the ions on the basis of velocity and brings most of them to a similar speed. Next, the second-stage magnetic sector sorts the ions on the basis of mass charge ratio. Only those ions having a discrete selected ratio are passed through the exit slit to the detector. By varying the magnetic field strength, the entire mass range from 1 to 300 daltons can be scanned in 30 sec. An electrostatic lens increases the angular aperture of the detector.

13.3 AUGER EMISSION SPECTROSCOPY (AES)

AES measures electrons emitted from a surface, induced by electron bombardment. The first step is ionization of an inner atomic level by a primary electron. Once the atom is ionized it must relax by emitting either a photon, as discussed in Chapter 9, or an electron—the nonradiative Auger process. In most instances nature chooses the Auger process. For example, a *KLL* Auger transition means that the *K* level electron undergoes the initial ionization. An *L* level electron moves in to fill the *K* level vacancy and, at the same time, gives up the energy of that transition (*L* to *K*) to another *L* level electron which then becomes the ejected Auger electron. Other Auger electrons originate from *LMM* and *MNN* transitions. At this point the atom is doubly ionized. The energy of the ejected electron is a function only of the atomic energy levels involved in the Auger transition, and is thus characteristic of the atom from which it came. A threshold energy, related to the transition energy, exists and a primary energy of 5–6 times the Auger energy maximizes the sensitivity to that particular transition. All elements except hydrogen and helium, produce Auger peaks. Most elements have more than one intense Auger peak, so that a recording of the spectrum of energies of Auger electrons released from any surface, compared with the known spectra of pure elements, enables a chemical analysis to be made.

Because the X-ray and Auger emission processes are competitive, the relative sensitivities of these two techniques are complementary when considering the relative abundance of X rays and Auger electrons after ionization of a particular level. In the light elements ($Z < 30$), the Auger process dominates, which makes AES relatively more sensitive. For heavier elements the electron microprobe (Chapter 9) becomes more sensitive for transitions following ionization of inner shells. The sensitivity of AES is maintained, however, by utilizing Auger transitions between outer shells, for example *MNN*, where the Auger process dominates.

Although the penetration of the primary electron beam is several atomic layers, the Auger electrons are on the average of much lower energies. Electrons of such low energy must originate very close to the surface if they are to escape without being lost by inelastic scattering before reaching the surface. Typically, Auger electrons come from the first few atomic layers; the sampling depth is about 2.0 nm.

The sensitivity of the Auger technique is determined by the probability of the Auger transitions involved, the incident beam current and energy, and by the collection efficiency of the energy analyzer. With a high-sensitivity CMA, the detection limit for the elements varies between 0.1 and 1 atomic percent (or 10^{-3} of a monolayer). Because the electron beam can be focused to a small diameter (50.0 nm), it is possible to do spatial resolution on a sample. Operated in this mode, AES is usually referred to as the Auger microprobe. AES is traditionally run with high-intensity electron guns and low-resolution analyzers, resulting in fast analysis speed. Unfortunately, this limits one to primarily elemental information with little information on chemical bonding, such as one clearly obtains in the chemical shifts of ESCA or in the molecular fragments found in SIMS. For example, hydrocarbon compounds appear in AES only as a C peak, since H is not observed.

Figure 13-7 shows the Auger spectra of Ag, Cd, In, and Sb, in these cases the *MNN* transitions. The spectra are very similar with the only major difference being the shifts in energy from one element to the next. These shifts are of the order of 25 eV and, since the peak positions can be measured to an accuracy of ±1 eV, there is no ambiguity in identification of adjacent elements in the periodic chart. Auger spectra of all the elements lie between 50 and 1000 eV. The *KLL* transitions correspond quite nicely with tabulated X-ray energies. There are, however, some minor differences due to the energy Auger electrons must lose in escaping from the sample. The Auger lines are relatively broad because of the double uncertainty of the origins within a band of both the downward and upward electrons involved in the Auger process. Lines due to *LMM* transitions also tend to correlate with X-ray energies, but some lines are beginning to appear which the selection rules for photon emission forbid. The selection rules must be relaxed for the radiationless Auger process. As the atomic number increases, the Auger spectra become more complex and overlapping may occur.

The strength of AES lies in its ability to give both a qualitative and quantitative nondestructive analysis of the elements present in the immediate atomic layers from a very small area of a solid surface. When combined with a controlled removal of surface layers by ion sputtering, AES provides the means to solve some very important problems. To provide ion sputtering for surface cleaning and/or depth profiling, the sample chamber is backfilled with argon and an electron impact ion source used, as in ISS.

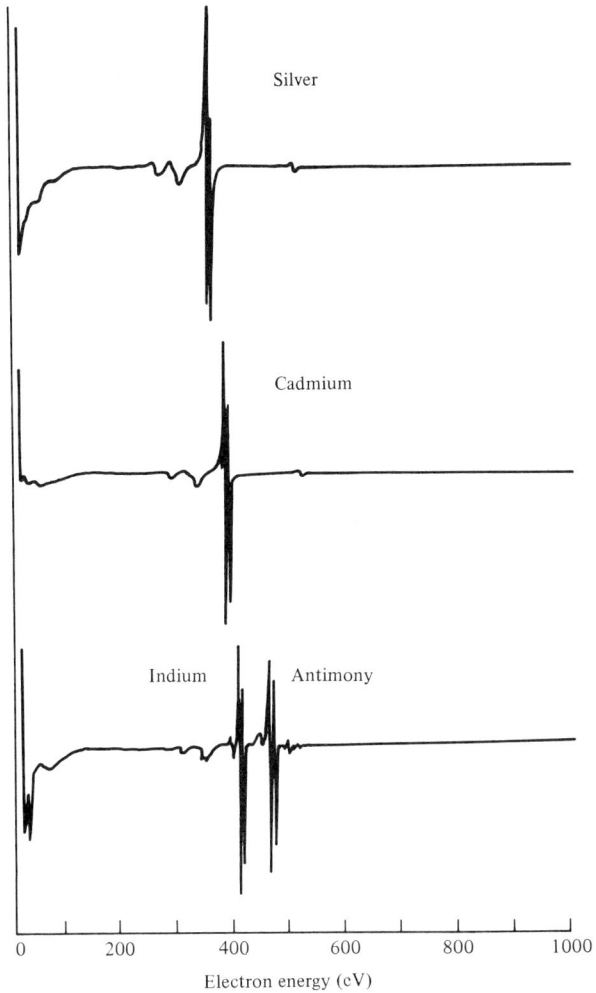

FIGURE 13-7 Auger spectra (differential form) from silver, cadmium, indium, and antimony; all are *MNN* transitions.

AES Instrumentation

The Auger spectrometer consists of an ultra-high vacuum chamber console, a sample carrousel and manipulator unit, and a combination electron gun/energy analyzer unit (Fig. 13-8). Auxiliary equipment often includes a grazing incidence electron gun and a sputter ion gun for cleaning surfaces and for profiling studies. Auger spectrometers are available as large-beam ($\sim$25 μm) depth-profiling instruments or as Auger microprobes with beam diameters of 5 μm. If an instrument is to be used as a high-sensitivity depth

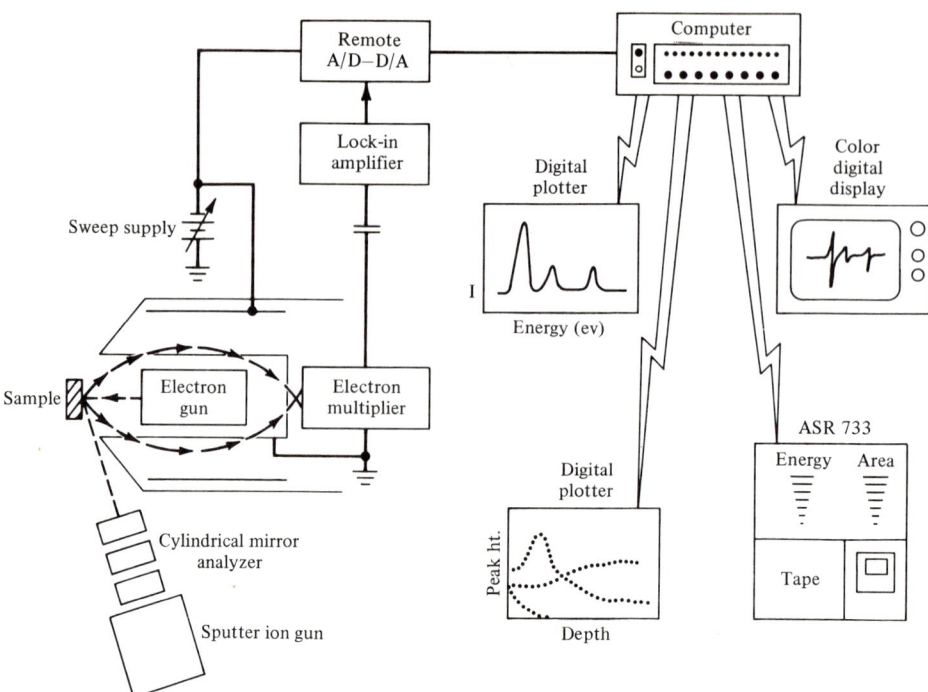

FIGURE 13-8 Schematic diagram of an Auger spectrometer with computer control.

profiler and also as a high lateral resolution microprobe, two electron guns may be required to realize optimum use of each operating mode.

The energy distribution of electrons emitted from the target, $N(E)$, is evaluated by scanning the negative voltage applied to the outer cylinder of the CMA. Thus, as the voltage applied to the outer cylinder is scanned, the secondary electron distribution is generated by the current output of the analyzer. Because the Auger peaks are superimposed on a rather large continuous background of secondary electrons, it has become popular to differentiate electronically the $N(E)$ function. This is accomplished by applying a small ac voltage on the dc energy selecting voltage and synchronously detecting the in-phase component of the output current of the electron multiplier with a lock-in amplifier. Unfortunately, the use of the lock-in amplifier in data acquisition places significant limits on sensitivity and quantification of the data. The limitations can be substantially reduced through the use of digital techniques. The digital storage of signal-averaged data with a high dynamic range makes it possible to use digital filters to remove high-frequency noise components from the signal. Signal averaging using multiple passes over a given energy window can be used to ensure that the statistical noise and the Auger peaks of interest present different spatial frequencies. This separation in frequency allows the postacquisition processing of these data to remove the high-frequency noise. In this way, one can substantially improve the elemental sensitivity without serious loss of energy resolution. Consequently, one now has a means for preserving the chemical shift information, while improving sensitivity for elements present in low concentration. In effect, this allows the choice between energy resolution (that is, oxidation state information as in ESCA) and elemental sensitivity to be made well after the data are acquired.

Quantitative Analysis in AES

The difficulties in the quantification of Auger data are illustrated in depth-profiling analysis. In this operation the change in peak shape of the differentiated spectra distorts the conventional peak-height estimates of concentration. Changes in peak shape can occur because of changes in oxidation state of the emitting atom. This is particularly severe in cases where bonding orbitals are involved in the Auger transitions. Peak-shape changes also may be induced by matrix effects, since the energy loss mechanisms and their effect on peak shape may be dominated by matrix elemental composition.

In order to provide accurate quantitative information within the normal quantitative limits of AES, it is necessary to use the total Auger current. This is expressed by the area under the curve for the peak of interest, with background subtracted. By using undifferentiated data and integrating the total number of Auger electrons in the peak, one is permitted a choice of the limits of integration. Now, chemical shifts may be observed, and a depth profile is realized without the artifacts.

Scanning Auger Microprobe (SAM)

The SAM employs a finely focused, scanning electron beam as the probe for AES analysis of a given surface. For a surface area, the SAM will provide an electron micrograph, an Auger image of selected elements on a CRT display for observation or photographing, and a depth-composition profile. First, the CRT image is used to determine specific points or an area of interest on the sample. From the micrograph obtained, points of interest or an area up to 200 μm can be chosen for compositional analysis.

The image mode is useful because position of elements are delineated in a few minutes. An x-y recording of a line scan across a selected region of the image for a given element will give a relative quantitative measure of its concentration.

The thin film analyzer mode of the SAM simultaneously sputter-etches a relatively large surface area (several millimeters in diameter) and multiplexes the Auger signals from a smaller area about 15 μm in diameter. Up to six selected Auger peaks can be recorded as the surface is etched. Depth resolution is about 10% of the total etched thickness. Elements are detected when present in quantities down to 0.1% of monolayer.

At best, the spatial resolving power of SAM is 500 nm with tungsten thermionic emitters as the electron source. The use of LaB_6 emitters extends the resolving power to 100–200 nm because of their high electron optical brightness. Field emitters, which provide even higher brightness, are expected to allow one to analyze smaller areas in less time when they come into use.

Many surface composition problems involve compositional inhomogeneity, both in-depth and across a material surface. Scanning Auger microscopy often provides a more detailed characterization of material that exhibits inhomogeneous composition. SAM offers high spatial resolution, in the form of SEM-type imaging combined with elemental mapping capabilities. For example, graphite present in spherical nodules indicates that the specimen of cast iron is ductile cast iron. Examination of the carbon Auger images shows carbon to be present in both the graphite nodules and, at lower concentration, in regions between nodules. Carbon is not evident in the circular regions around nodules. The iron Auger image shows iron to be present everywhere in the specimen surface except in the graphite regions.

Spatially resolved SIMS is especially useful when combined with SAM, as suggested by the study of an automotive ignition contact. A high-resolution absorbed current SAM image of the contact obtained first showed detailed surface topography. An absorbed ion current image was then obtained and compared with the SAM image to identify specific areas of interest for further study. High-sensitivity SIMS point analysis was performed at the center and the outer edge of the contact. The resulting spectra showed higher concentrations of calcium and iron in the center region of the contact, as indicated by the large increase in the Ca^+/K^+ and Fe^+/Ni^+ ratios, compared to the spectrum from the side of the contact.

13.4 ELECTRON SPECTROSCOPY FOR CHEMICAL ANALYSIS (ESCA)

ESCA is concerned with the measurement of core-electron binding energies. A molecule or atom is bombarded with a source of high-energy X rays which cause the emission from sample atoms of inner-shell electrons. All electrons whose binding energies are less than the energy of the exciting X rays are ejected. The kinetic energies, E_k, of these photoelectrons are then measured by an energy analyzer. The core-electron binding energies, E_b, relative to the Fermi level* can then be computed via the relationship

$$E_b = h\nu - E_k - \phi \tag{13-2}$$

where $h\nu$ is the energy of the exciting radiation and ϕ is the spectrometer work function, a constant for a given analyzer. Binding energies unambiguously define a specific atom. AES can be compared to ESCA because both originate from similar fundamental processes. In ESCA the ionizing source is an X-ray photon that ejects an inner-core electron, whereas in AES the electron ejection is caused by an impinging electron. The photoelectron escape depths are the same as those for Auger electrons of the same kinetic energy. The energy of the ejected electron (Auger electron or E_k) is thus characteristic of the atom involved and its chemical environment.

Although the X-ray photon may penetrate and excite photoelectrons to a depth of several hundred nanometers, only the photoelectrons from the outermost layers have any

---

*The quantum mechanical (chemical) zero of energy, called the vacuum level for gaseous systems, is the zero of energy determined by following the potential energy interaction of an electron as it is removed to infinity with respect to its parent atom. When the system being examined is a solid, the quantum-mechanical zero in energy is the Fermi level. The Fermi level differs from the vacuum level by the amount of energy that is needed to remove the electron, already outside its chemical environment, from its physical environment, that is, the surface. This extra energy for a material is proportional to the work function.

chance to escape from the material environment and to be eventually measured. Most ESCA measurements of solids generate useful information from only the outer 2.0 nm of the surface layer.

The required sample size is a microgram or less. The sampling area is approximately 1 cm^2. Applicability to the second row elements, including carbon, nitrogen, and oxygen, makes ESCA an important structural tool for organic materials. The detection limit depends on the particular element being measured, but will range from 1% of a monolayer for light elements to 0.1% of a monolayer for heavy elements.

To determine the absolute binding energy, the work function of the sample and of the spectrometer must be known. Unfortunately, the work function is not independent of the physical state of the material; neither is it easy to measure or calculate. The problem can be circumvented for many materials by employing a physical trick. If the material being examined is a conductor, it usually is possible to couple the conduction bands of the material with that for the spectrometer, which presumably is also a good conductor. The coupling is such that the Fermi levels of the material and the spectrometer merge. The value of ϕ may also be calculated relative to a reference element, such as carbon (1s) or gold spectral features.

Frequently, only a relative binding energy (chemical shift) is desired in chemical studies. Here it is sufficient that ϕ be constant. For example, the chemical shift of a metal oxide relative to the metal may be calculated from the measured kinetic energies:

$$\Delta E_{\text{oxide}} = E_{k(\text{metal})} - E_{k(\text{oxide})} \tag{13-3}$$

It is ΔE_{oxide}, the chemical shift, which gives chemical structure information.

Chemical Shift

The utility of ESCA for the chemist is the result of chemical shifts that are observed in electron binding energies. The binding energies of core electrons are affected by the valence electrons and therefore by the chemical environment of the atom. When the atomic arrangement surrounding the atom ejecting a photoelectron is changed, it alters the local (quantum) charge environment at that atomic site. This change, in turn, reflects itself as a variation in the binding energy of *all* the electrons of that atom. Thus, not only the valence electrons, but also the binding energies of the core electrons experience a characteristic shift. Such a shift is inherent to the chemical species producing the results and thus provides the capability of chemical analysis. In a simple sense, the shifts of the photoelectron lines in an ESCA spectrum reflect the increase in binding energy as the oxidation state of the atom becomes more positive. In general, any parameter, such as oxidation state, ligand electronegativity, or coordination, that affects the electron density about the atom is expected to result in a chemical shift in electron binding energy.

A major portion of the strength of ESCA as an analytical tool lies in the fact that chemical shifts can be observed for every element in the periodic chart except for hydrogen and helium. Magnitudes of chemical shifts will vary from element to element, and the sensitivity for a particular element will vary with the photoelectron cross section. In general, the ESCA chemical shifts lie in the range 0–1500 eV. For instance, the position

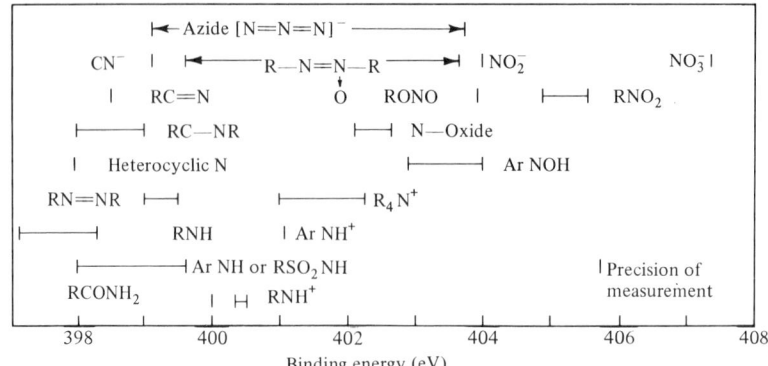

FIGURE 13-9 Correlation chart for nitrogen (1s) electron binding energies and organic functional groups.

of the nitrogen 1s peak at a binding energy of 398 eV correlates very well with nitrogen being in a formal oxidation state of −1. In order to make this assignment one must refer to a catalog of reference nitrogen spectra or to correlation charts, such as Fig. 13-9. One should be cautious about making structural assignments on the basis of small ESCA binding energy shifts since factors such as crystal potential energy differences and sample charging can cause apparent shifts of the order of magnitude of the observed chemical shifts.

In general, photoelectron peaks are narrower than the corresponding X-ray emission lines, and in most cases, vary from 1 to 3 eV (FWHM). Since the chemical shifts for a given element are of the order of 10 eV, ESCA is not a high-resolution method. The FWHM resolution of about 1.0 eV makes it possible to resolve electron energies that differ by about 0.5–1.0 eV, and thus to measure chemical shifts which are of interest to most chemists. For example, Fig. 13-10 shows the partial ESCA spectrum of Cu_2O, CuO, and metallic copper. One could clearly distinguish between metallic copper and CuO, but it would be difficult to decide between metallic copper and Cu_2O. However, by also observ-

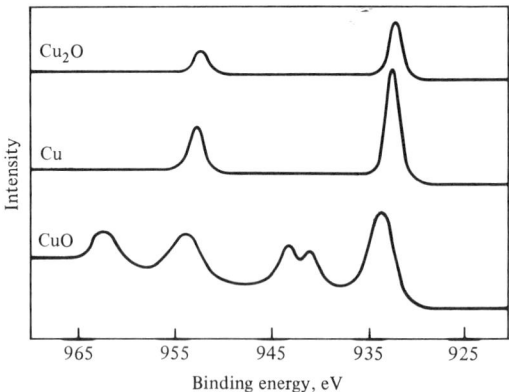

FIGURE 13-10 ESCA spectra of Cu_2O, CuO, and metallic copper.

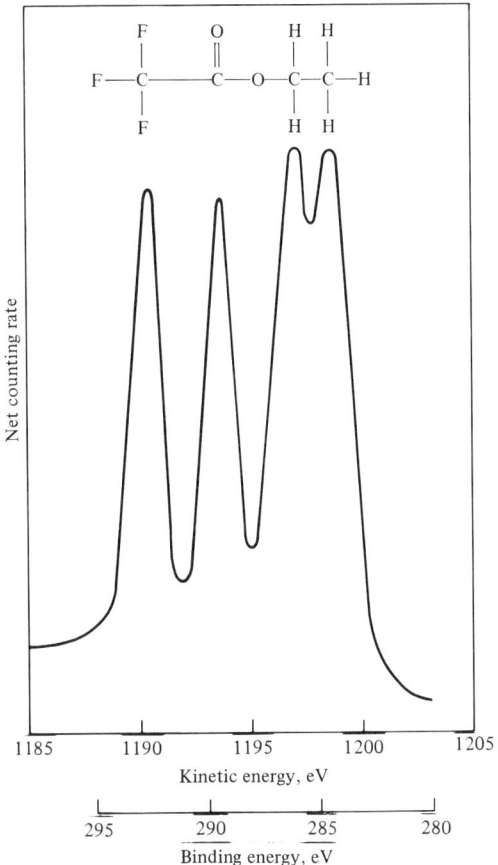

FIGURE 13-11 ESCA spectrum of ethyltrifluoroacetate.

ing the oxygen 1s spectrum (530.8 eV for Cu_2O and 530.1 eV for CuO), one can easily distinguish between metallic copper and its two oxidation states. The classic example of chemical shifts is the carbon 1s ESCA spectrum of ethyltrifluoroacetate shown in Fig. 13-11. Each carbon atom is located in a different chemical environment and the ESCA spectrum contains four distinct photoelectron lines. The trifluorocarbon yields the photoelectron line at highest binding energy since the fluorine atoms withdraw electron density from the carbon atom most efficiently. In this example, the relative positions of the photoelectron lines reflect the relative electronegativity of the various substituents; the respective carbon photoelectron lines actually appear above their peaks in the spectrum.

ESCA Instrumentation

Instrumentation for ESCA, shown in Fig. 13-12, involves a radiation source of sufficient energy to eject an electron from the sample. There must also be a device that collects the emitted electrons, counts them, and carefully measures their kinetic energy. A

398 CHAPTER 13

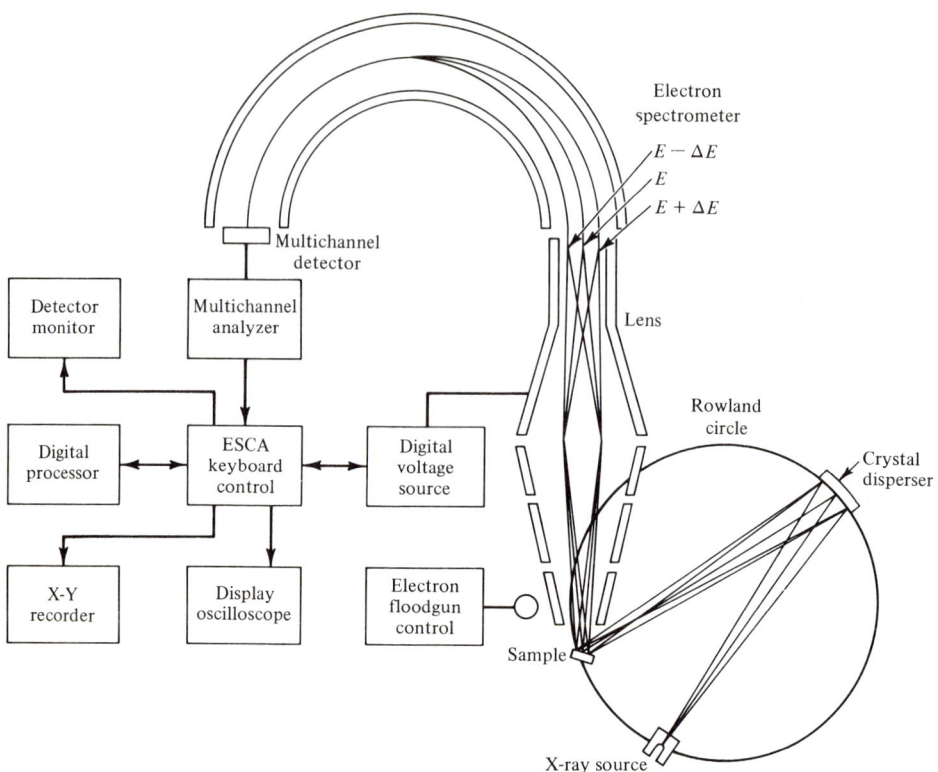

FIGURE 13-12 Schematic of an ESCA spectrometer using crystal dispersion to achieve X-ray monochromatization and a digital processor. Commercial version is the Hewlett-Packard model 5950B.

storage and display unit is usually included. Since it is necessary to ensure that the mean free path of the photoelectrons is large enough to allow them to traverse the distance from the sample to the detector without suffering energy loss, ESCA is a vacuum technique with a maximum operating pressure of about 5×10^{-6} torr.

ESCA spectra can be obtained on solids, liquids, and gases; the physical form of the sample is not important. However, the technique is a vacuum technique; therefore, low vapor pressure solids are most easily run. A solid sample need only be placed on a probe that is appropriately positioned relative to the X-ray beam and the spectrometer slit. Liquids cannot be run as such, but must be condensed onto a cryogenic probe and run in the condensed phase. Alternatively, liquids can be vaporized and run in the gaseous state. To obtain spectra on the gaseous sample, the spectrometer must be equipped with a differential pumping system to prevent the pressure in the analyzer from rising above 10^{-6} torr.

Source Soft X rays, such as Mg $K\alpha_{1,2}$ and Al $K\alpha_{1,2}$ with a FWHM of 0.75 and 0.95 eV, respectively, are usually employed for photoelectron excitation. It is important to have at least two alternate sources in order to distinguish photoelectron peaks from Auger peaks. When a different X-ray source is used, the photoelectron peaks shift in kinetic energy but the kinetic energies of Auger peaks remain constant and appear at the same energy position in the spectrum. A source of small linewidth is advantageous to have if an element exists in several oxidation states in a sample and it is desired to extract the binding energies of each of these oxidation states.

Two types of X-ray systems are used in commercial instruments. In one the sample is illuminated directly by the output of the source. This polychromatic type of source is quite simple and permits the use of different target materials and hence different energies of photoelectron excitation. The other system incorporates an X-ray monochromator to disperse the X radiation and thus provide monochromatic illumination of the sample surface. Spectral interferences and background are thereby reduced but X-ray intensities impinging on the sample surface are much lower.

The monochromatic source shown in Fig. 13-12 places the X-ray anode, a spherically bent crystal disperser, and the sample on a Rowland circle. Only the X radiation from the anode is reflected to the sample by the crystal. The source radiation will be free from the X-ray satellite structure (arising from $K\alpha_{3,4}$ and $K\beta$ lines) that plagues the photoelectron spectra generated with polychromatic beams. Elimination of the broad background bremsstrahlung radiation improves the signal-to-background ratio and sharply reduces X-radiation damage in the sample. For example, radiation-induced changes, such as reduction of Cu(II) to Cu(I), that occur in 20 sec with bremsstrahlung radiation, will only occur after 10 hr with the crystal-reflection method.

ESCA Electron Analyzers The electron analyzers used in ESCA instrumentation employ the double-focusing principle. An electrostatic field sorts the electrons. As the field is varied, electrons of appropriate kinetic energy are focused at the detector. Initially, in the instrument illustrated, the photoelectrons are channeled through a series of four interconnected electrostatic lenses. This method for photoelectron collection is referred to as dispersion compensation, with the four lenses acting as selective apertures. In addition, a retarding voltage is employed to bring the photoelectrons into focus so that the inherent linewidths of the X-ray photon can be removed. In addition, the lenses provide the mechanism for selecting and scanning the particular photoelectron energy of interest.

Detector Both continuous channel and discrete dynode electron multipliers are used to count the electrons. The continuous channel detector counts electrons with high efficiency to very low energies and, compared to discrete dynode multipliers, are more stable to atmospheric and other gases.

Scan and readout systems are either of a continuous or incremental type. In the continuous mode the focusing field is increased continuously as a function of time as the signal from the detector is simultaneously monitored by a rate meter. The focusing field and output from the rate meter are synchronized to allow accurate recording of spectra. In the continuous mode the energy region of interest can be scanned only once. This is somewhat of a handicap since no signal averaging can be performed. Signal averaging becomes necessary when the signal is weak or the quantity of sample is small.

The incremental scanning mode increases the field in a series of small steps, counting the signal during each increment. When the counting rate at each increment is plotted as a function of focusing field, a spectrum is produced. Instrumentation using the incremental scanning mode uses either a multichannel analyzer or a small dedicated computer to accumulate the data. In such systems, signal averaging can be achieved by performing repetitive scans over the energy region of interest. The counting rate in each increment is added to the preceding one. When the system contains a dedicated computer, both control of energy scan and data acquisition are under its control. Usually several energy regions can be scanned sequentially, with the computer storing the data until they are retrieved by the operator.

For display the digital nature of the results makes it convenient to use a point plotter. Spectra of two formats are generated. First step in an analysis is to run a 0–1000 eV survey scan that displays peaks from all major components of the sample. A scan time is automatically chosen by the computer that is optimum for the span and memory channels used. Next, each spectrum interval where peaks appear is examined separately in more detail giving the analytical data needed to identify an element and its relative amount.

Scanning ESCA

ESCA in its earlier years was always thought of as a low spatial resolution technique because the specimen is excited by flooding the surface with X rays. These X rays could not be readily focused. Thus, elemental images could not be generated in a manner analogous to that of SAM where the source of excitation, a scanning electron beam, is focused to a small spot. However, if a focused electron beam is used to bombard a thin foil of aluminum which has a thin specimen mounted on the side opposite the beam a localized source of Al Kα X rays is produced in the aluminum foil (Fig. 13-13). This causes a spatially localized source of photoelectrons to be created in the specimen (but also containing satellite structure and bremsstrahlung radiation). The result is an ESCA spectrum

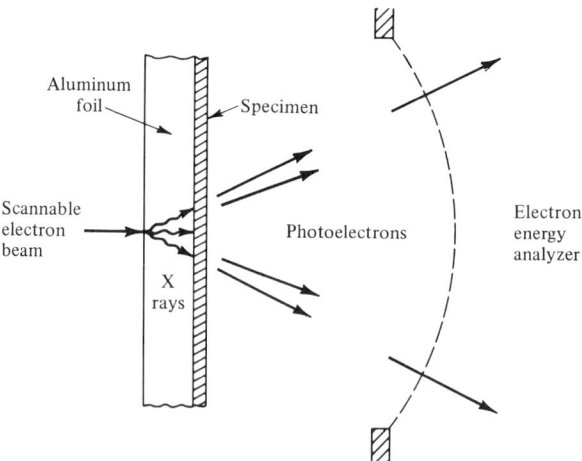

FIGURE 13-13 Schematic arrangement for spatially resolved ESCA.

from an area less than 20 μm in diameter. If a scanning electron beam is employed, two-dimensional photoelectron images can be obtained.

A practical example of the use of surface analysis involved the study of a metallographically polished cast iron specimen. An ESCA survey spectrum indicated the presence of Fe, O, and C as major components of the material. High-resolution spectra of Fe and C showed the presence of both oxidized and reduced iron species and of at least two carbon species. Detailed analysis of the carbon binding energies revealed the presence of a carbide and a species with a binding energy characteristic of graphite. These conclusions are consistent with the fact that cast iron contains precipitated graphite as well as phases rich in iron carbide. The oxidized iron species probably was due to surface oxidation of the specimen.

Quantitative Analysis

The intensity of a photoelectron line is proportional to not only the photoelectric cross section of a particular element, but also to the number of atoms of that particular element that are present in the sample. Analyses of mixtures are often accurate to ±2%. An example involves measuring the intensities of photoelectron lines in spectra obtained from mixtures of MoO_2 and MoO_3 at binding energies that correspond to each oxide. No instrumental technique had existed that was capable of performing this analysis. Even though the surface of MoO_2 was significantly contaminated with MoO_3, a linear calibration curve resulted. Mixtures of $PbO-PbO_2$, $Cr_2O_3-CrO_3$, and $As_2O_3-As_2O_5$ can also be measured quantitatively. Estimates of the total protein content of various grains can be made by measuring the intensities of the nitrogen and sulfur peaks.

BIBLIOGRAPHY

Barr, T. L., "Applications of ESCA in Industrial Research," *Am. Lab.*, p. 65 (November 1978); p. 40 (December 1978).

Betteridge, D. and A. D. Baker, "Analytical Potential of Photoelectron Spectroscopy," *Anal. Chem.*, **42**, 43A (1970).

Carlson, T. A., *Photoelectron Auger Spectroscopy*, Plenum, New York, 1975.

Czanderna, A. W., Ed., *Methods of Surface Analysis*, Elsevier, New York, 1975.

Davis, L. E., N. C. MacDonald, P. W. Palmberg, G. E. Riach, and R. E. Weber, *Handbook of Auger Electron Spectroscopy*, 2nd ed., Physical Electronics Industries, Eden Prairie, Minn., 1976.

Day, R. J., S. E. Unger, and R. G. Cooks, "Molecular Secondary Ion Mass Spectrometry," *Anal. Chem.*, **52**, 557A (1980).

Evans, C. A., Jr., "Secondary Ion Mass Analysis," *Anal. Chem.*, **44**, 67A (1972).

Evans, C. A., Jr., "Surface and Thin Film Compositional Analysis: Description and Comparison of Techniques," *Anal. Chem.*, **47**, 818A (1975).

Evans, C. A., Jr., "Surface and Thin Film Analysis: Instrumentation," *Anal. Chem.*, **47**, 855A (1975).

Harris, L. A., "Auger Electron Emission Analysis," *Anal. Chem.*, **40**, 24A (1968).

Heinrich, K. F. J. and D. E. Newbury, Eds., *Secondary Ion Mass Spectrometry,* NBS Special Publ. No. 427, GPO, Washington, D. C., 1975.

Hercules, D. M., "Electron Spectroscopy," *Anal. Chem.,* **42,** 20A (1970).

Hercules, D. M., "Challenges in Surface Analysis," *Anal. Chem.,* **50,** 734A (1978).

Kane, P. F. and G. B. Larrabee, Eds., *Characterization of Solid Surfaces,* Plenum, New York, 1974.

Karasek, F. W., "Developments in ISS/SIMS," *Research/Development,* p. 26 (January 1978).

Lee, L. H., Ed., *Characterization of Metal and Polymer Surfaces,* Vol. 1, Academic, New York, 1977.

Liebl, H., "Ion Microprobe Analyzers: History and Outlook," *Anal. Chem.,* **46,** 22A (1974).

Lucchesi, C. A. and J. E. Lester, "Electron Spectroscopy Instrumentation," *J. Chem. Educ.,* **50,** A205, A269 (1973).

Riach, G. E. and R. F. Goff, "Electrons or Ions," *Industrial Research,* p. 84 (June 1974).

Sparrow, G. R., "Ions Working for You," *Industrial Research,* p. 81 (September 1976).

Swartz, W. E., Jr., "X-Ray Photoelectron Spectroscopy," *Anal. Chem.,* **45,** 788A (1973).

CHAPTER 14

Refractometry and Interferometry; Polarimetry, Circular Dichroism, and Optical Rotatory Dispersion

14.1 THEORY

When a ray of light passes obliquely from one medium into another of different density, its direction is changed on passing through the surface. This is called *refraction*. If the second medium is optically denser than the first, the ray will become more nearly perpendicular to the dividing surface. The angle between the ray in the first medium and the perpendicular to the dividing surface is called the angle of incidence, i, whereas the corresponding angle in the second medium is called the angle of refraction, r. Sin i and sin r are directly proportional to the velocities of the light in the two media. The ratio sin i/sin r is called the *index of refraction*, η. If the incident ray is in the denser medium, η will be less than 1; if in the rarer, greater than 1. Commonly η is taken as greater than 1, the ray passing from the optically rarer medium (usually air) to the denser.

The index of refraction for two given media varies with the temperature and the wavelength of light and also with the pressure, if we are dealing with gases. If these factors are kept constant, the index of refraction is a characteristic constant for the particular medium and is used in identifying or determining the purity of substances and for determining the composition of homogeneous binary mixtures of known constituents.

The refractive index is theoretically referred to vacuum as the first medium, but the index referred to air differs from this by only 0.03% and, for convenience, is more commonly used. The refractive index of a transparent substance gradually decreases with increasing wavelength except at regions of absorption where the refractive index changes abruptly. The change of refraction with wavelength is known as *dispersion*. Because of dispersion,

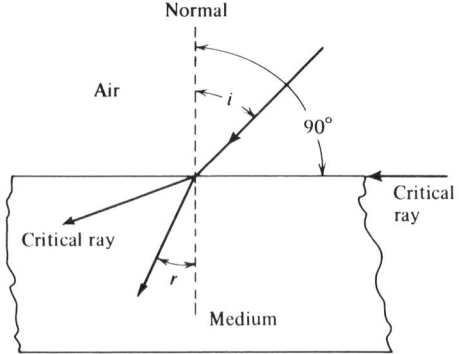

FIGURE 14-1 Angles of incidence (i) and refraction (r).

the wavelength must be specified when refractive indices are stated. The symbol η_D^{20} means the index of refraction for the D lines of sodium* measured at 20°C.

When the beam of light passes from a denser to a rarer medium, the angle r will be greater than the angle i. As angle i increases, the ratio sin i/sin r remaining constant, the angle r must also increase and remain greater than i. If angle i is increased to the value where r becomes 90°, the beam of light will no longer pass from the first medium to the second, but will travel through the first medium to the dividing surface and then pass along this surface, thus making an angle of 90° with the perpendicular to the surface (Fig. 14-1). This is called the *critical ray*. In Fig. 14-1, i and r would then be interchanged, and the direction of the arrows would be reversed. If i is smaller than this particular value, light will pass through the second medium; if greater, all light will be reflected from the surface back into the first medium. This furnishes the basis for the reference line used in several refractometers. Total reflection can occur only when light passes from the denser to the rarer medium.

The refractive index of a liquid varies with temperature and pressure, but the *specific refraction*, r_D,

$$r_D = \frac{\eta^2 - 1}{\eta^2 + 2} \frac{1}{\rho} \tag{14-1}$$

where ρ is the density, is independent of these variables. This relationship is known as the Lorentz and Lorenz equation. The molar refraction is equal to the specific refraction multiplied by the molecular weight. It is a more or less additive property of the groups or elements comprising the compound. Tables of atomic refractions are available in the literature; an abridged set of values is given in Table 14-1. Thus, specific refraction is valuable as a means for identification of a substance and as a criterion of its purity. In a homologous series of compounds the specific refraction of higher members generally increases fairly regularly with increasing length of the carbon chain.

*The yellow doublet at 589.0/589.6 nm.

TABLE 14-1 Atomic Refractions

Group	Mr_D	Group	Mr_D
H	1.100	Br	8.865
C	2.418	I	13.900
Double bond (C=C)	1.733	N (primary aliphatic amine)	2.322
Triple bond (C≡C)	2.398	N (*sec* aliphatic amine)	2.499
O (carbonyl)(C=O)	2.211	N (*tert* aliphatic amine)	2.840
O (hydroxyl)(O–H)	1.525	N (primary aromatic amine)	3.21
O (ether, ester)(C–O–)	1.643	N (*sec* aromatic amine)	3.59
S (thiocarbonyl)(C=S)	7.97	N (*tert* aromatic amine)	4.36
S (mercapto)(S–H)	7.69	N (amide)	2.65
F	1.0	–NO$_2$ group (aromatic)	7.30
Cl	5.967	–C≡N group	5.459

The values of refractive index for organic liquids range from about 1.2 to 1.8, those for organic solids from about 1.3 to 2.5.

Dispersion is also sometimes useful in the identification of compounds. Dispersion is often taken as the Abbé number, ν, defined as

$$\nu = \frac{\eta_D - 1}{\eta_F - \eta_C} \tag{14-2}$$

where η_F and η_C are the refractive indices for the F and C lines of hydrogen ($\lambda = 486.1$ nm and $\lambda = 656.3$ nm, respectively).

Example 14-1

The refractive index of acetic acid at 20°C is 1.3698, the density at 20°C is 1.049 g cm^{-3}, and the molecular weight is 60.0. From Eq. 14-1 the specific refraction is found:

$$r_D = \frac{[(1.3698)^2 - 1]}{[(1.3698)^2 + 2]} \frac{1}{1.049} = 0.2155 \text{ cm}^3 \text{ g}^{-1}$$

The molar refraction is

$$Mr_D = (60.0)(0.2155) = 12.93 \text{ cm}^3 \text{ mole}^{-1}$$

The molar refraction is a constitutive property depending upon the structural arrangement of the atoms within the molecule. From the atomic and group refractions in Table 14-1, the molar refraction of acetic acid can be computed and compared with the experimental value as follows:

ACETIC ACID			METHYL FORMATE		
2 carbons	=	4.836	2 carbons	=	4.836
4 hydrogens	=	4.400	4 hydrogens	=	4.400
1 carbonyl oxygen	=	2.211	1 carbonyl oxygen	=	2.211
1 hydroxyl oxygen	=	1.525	1 ester oxygen	=	1.643
		12.972			13.090

Methyl formate possesses the same empirical formula as acetic acid and differs only slightly in structure. This difference is apparent in the molar refraction values, although the difference only amounts to about 0.90%. To distinguish between the two compounds would require a precision of ±0.006 in refractive index and ±0.005 in density. For methyl formate, $\eta_D^{20} = 1.344$ and $d^{20} = 0.974$.

14.2 REFRACTOMETERS

Refractometers determine the index of refraction by measuring the position of the critical ray. A prism of glass is wet with a thin layer of the liquid to be measured or dipped into the liquid. In one instrument, the Abbé refractometer, the position of the critical ray, indicated by the border between the light and dark portions of the field seen in a telescope viewing the prism, is measured by rotating a telescope that is attached to a scale. Since the index of refraction of the prism is known, the scale can be graduated in index of refraction rather than angular degrees. In a second type of instrument, the immersion refractometer, the position of the critical ray emerging from the prism is read by a fixed telescope with a graduated scale mounted at one focal point. The range of the immersion refractometer is much less than the Abbé but the precision is much higher, ±0.000037 in η_D for the immersion instrument compared to ±0.0002 for the Abbé. In order to cover a wide range of index of refraction the immersion instrument uses several different prisms. The Abbé refractometer is shown schematically in Fig. 14-2 and the immersion refractometer in Fig. 14-3. In the Abbé a drop of the sample is placed between the two prisms while in the immersion instrument the prism is dipped directly into the sample contained in a small beaker.

Applications

The refractometer measures concentration more accurately than ordinary density measurements with a hydrometer. For example, with an immersion refractometer, assuming a sufficiently accurate temperature control of about 0.1°C, a scale reading of 0.02 division (which is about the best one can do in reading the instrument, estimating the nearest 0.01 division) corresponds to the following weight of substances per 100 ml: methyl alcohol, 24 mg; ethyl alcohol, 12 mg; ammonium chloride, 4 mg; perchloric acid, 10 mg.

If both density and refractive index are determined, it is possible to determine each of two components, such as methyl and ethyl alcohol, with a fair degree of accuracy if nothing else is present. It should be noted that both density and refractive index are measures of the total amount of substance in solution, no matter how many different ones there may be.

The immersion refractometer is especially useful in determining the concentration of aqueous and alcoholic solutions. Wagner[1] describes precautions to be used, such as constancy of temperature, rinsing the prism with water of the same temperature, wiping lightly, and allowing 2 min before reading.

Shippy and Barrows[2] showed how the index of refraction could be used to determine the composition of solutions of sodium chloride and potassium chloride. A curve was

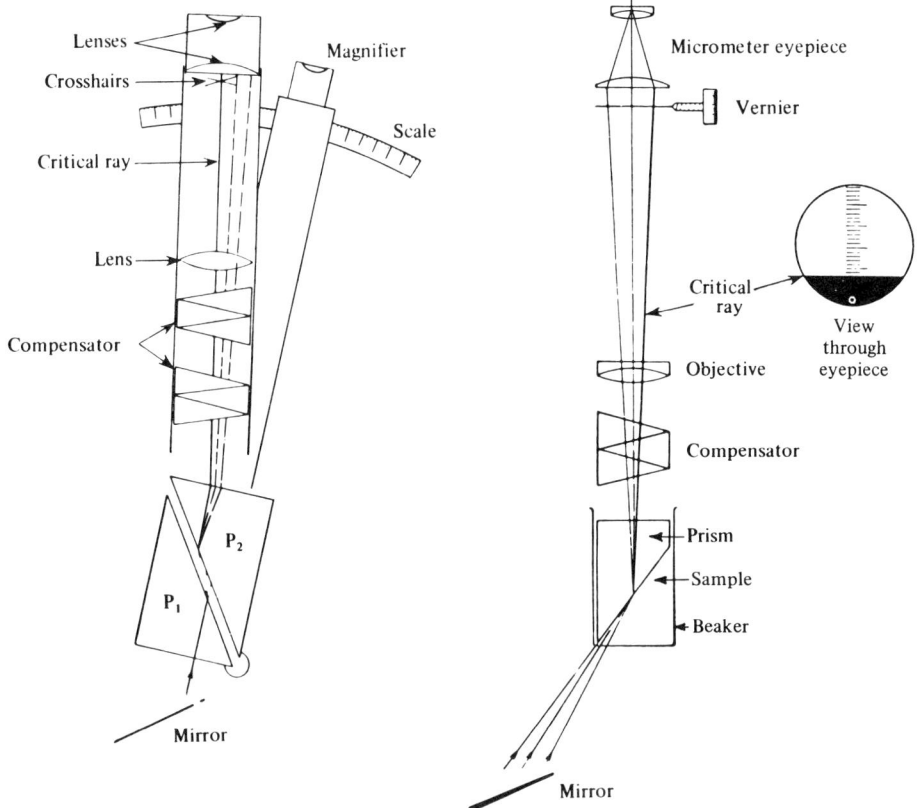

FIGURE 14-2 The Abbé refractometer.   **FIGURE 14-3** The immersion refractometer.

constructed by plotting percentage of sodium chloride against index of refraction. A fair degree of accuracy was attained.

In physiological chemistry the refractometer is very important. In only 2 ml of serum it can be used to determine nonalbuminous constituents, total globulins, insoluble globulins, albumens, and total albumen, with great accuracy. The action of ferments can be followed with the refractometer. The refractometer is also useful in controlling the analysis of commercial products, in identifying unknown substances, and in distinguishing substances of the same boiling point and compounds of the same nature, such as halogenated hydrocarbons.

Recording Refractometers

Instruments have been designed for continuous and automatic recording of refractive indices (or differences between a reference and specimen). These instruments utilize servomechanisms which track the position of a slit image or critical boundary, or operate a compensating mechanism to maintain a constant position of the image or boundary. Fig-

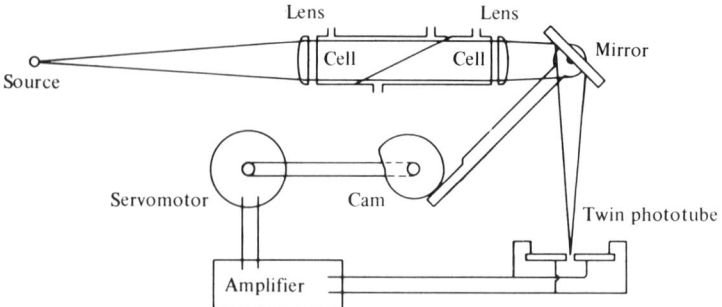

FIGURE 14-4 Schematic optical system of a recording refractometer.

ure 14-4 is a schematic representation of the latter type of instrument. Light from the source is defined by a slit and chopped by a rotating sector (not shown), then passes through the double-prism cell. The refracted beam then strikes a mirror (or beam-splitting arrangement) which focuses the light on twin detectors. Imbalance is removed by the servomotor which is geared to the mirror. The amount of rotation, which is proportional to the refractive index, is correlated with known standards. The instrument may be used to compare the refractive index of two process streams.

Refractometers are dependent upon the refraction of light when passing from one substance to another. The finest measurement of refractive index, however, is based upon the interference of light. In this method there is no diffraction. The light enters and leaves the solution at right angles. The interferometer, reduced to its simplest terms, may be represented by Fig. 14-5.

Parallel light passes through two small openings, R_1 and R_2 in Fig. 14-5. Since R_1O and R_2O are of equal length, the two beams arrive at O in phase and a bright spot results. At other points on the screen the lengths of the two beams are not the same. Thus at some point X_1 the two beams differ by half a wavelength. Interference of the two beams produces a dark spot here. At a point a little farther along, Y_1, the difference is one wavelength and a bright spot is formed. With monochromatic light, this succession of light and dark spots (maximum and minimum) continues indefinitely. Now, if a substance of slightly greater refractive index is placed at C, the optical length of the beam R_2O is increased by an amount Δb because the velocity of light through C is decreased. The mag-

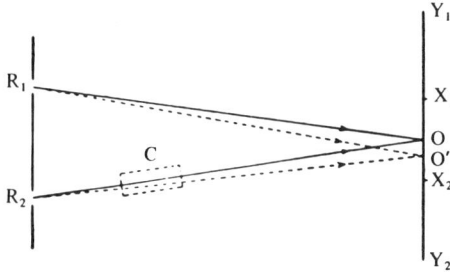

FIGURE 14-5 Optical principle of an interferometer.

nitude of this increase depends on the thickness of the sample and upon its refractive index, where

$$\Delta b = b(\eta - \eta_0) \tag{14-3}$$

b = thickness of sample
η = refractive index of sample
η_0 = refractive index of medium (air)

The velocities of light in two media are proportional to their indices of refraction.

The two beams will no longer arrive in phase at O, but at some other point O', which is now optically equally distant from R_1 and R_2. The entire band system will be shifted by this amount. For light of wavelength λ, the distance between O and O', measured in numbers of fringes, N (each made up of a dark and a light band), is

$$N = \frac{\Delta b}{\lambda} \tag{14-4}$$

$$N = \frac{b(\eta - \eta_0)}{\lambda} \tag{14-5}$$

If N is greater than 1, it is impossible to tell how many whole numbers of bands greater it is, because all bands are alike. This difficulty is avoided by using white instead of monochromatic light. Now the central band is the only one that is pure white. The bands on either side of this maximum of the first order are fringed with blue toward the center of the system and with red along the outer edge. This is due to the different wavelengths that make up white light. The next adjacent bands are even more highly fringed. After six or seven bands, the diffusion is so great that the rest of the field is again uniformly white. Thus, with substance at C in the path of one of the beams, by counting the number of bands which the central band has been shifted, we can determine the value of $b(\eta - \eta_0)$. Any one of these terms can then be calculated if the others are known. If two plates of equal thickness were placed in the two beams, the number of bands that the central band shifts would be a means of calculating the refractive index of one of the plates, provided that the value of the other one was known. The interferometer is, however, not used primarily for measuring index of refraction but for comparing and measuring concentrations of solutions and gases.

In one type of instrument, shown in Fig. 14-6(a), the optical length of the two beams is equalized by means of a glass plate in the path of each beam P_1 and P_2 at an angle of

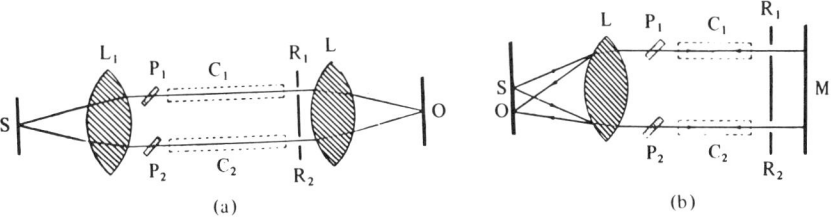

FIGURE 14-6 Optical schematic of interferometers: (a) extended version and (b) folded type.

about 45° to the beam, one plate being fixed and the other attached to a lever by which it can be rotated, thus increasing or decreasing its effective thickness. The movement is measured by a micrometer screw. This is turned until the central achromatic bands of the two systems correspond; that is, the optical path of the two beams is the same length. It is possible to match them to $\frac{1}{20}$ of a band, corresponding to a reading of about one scale division on the micrometer screw. This instrument was originally used for measuring changes in the composition of gases, and when the gas chambers are 1 m long, one scale division corresponds to a change in η_D of 0.000000015. It is capable, therefore, of measuring quantities of such substances as CO_2 and CH_4 present in air in amounts as low as 0.02%.

In a later type of portable gas-and-water interferometer, made by Zeiss and shown in Fig. 14-6(b), the light is reflected by the mirror M so that it passes twice through the chambers, and the bands are observed at the same end as the light source. In this way it is possible to obtain the same precision with half the length. The light is furnished by a small electric lamp and is focused on a slit. The interference bands are not as brilliant as those of the other instrument but are just as plain and as easily set. The chambers can be jacketed with an air thermostat or with water, and gases as well as liquids can be used. With the latter, the cells vary from 1 to 40 mm in length. The scale reading is proportional to the thickness of the liquid, so that the range and precision of the instrument may be varied by changing the length of the cell.

If water is put in one-half of the cell and a dilute solution of salt in the other, the number of scale divisions of displacement will be determined by the difference in refractive indices of the solution and the pure solvent. A calibration curve can be obtained by making up solutions of known concentration and comparing them with water. Plotting scale readings against concentration, a line is obtained which is almost straight. The deviation from a straight line is due, not to an inconstant variation of refractive index, but because a variation of 10 scale units at one end of the scale may increase the optical length of the beam more than it would at the other end of the scale. Stated in other words, the thickness of band varies with the scale division because of properties inherent in the lever arm action. The band thickness may vary as much as 10%. But since this variation is quite regular, a few points will be sufficient for calibration. The range of the 5-mm chamber is from η_D 1.33320 to 1.34010 for 3000 divisions using water as the comparison liquid. This corresponds to a range of 15.0–33.0 on the immersion refractometer. Assuming that the latter can be read to 0.05 division, the precision with this particular chamber is about 10 times that of the refractometer. With the 40-mm chamber it would be about 80 times as great, but the range would be $\frac{1}{8}$ as great. With this chamber one division corresponds to 1.5–3.0 mg of solute per liter for most aqueous solutions. The greatest differences of concentration which can be directly compared are therefore from 0.45 to 0.90%. The range of the measurement with the interferometer can be increased by comparing the solution against solutions having a known amount of solute present. This does not decrease the precision of the measurement. Thus any concentration of solute can be determined if a series of known solutions has been prepared so that each solution of the series differs by no more than 3000 scale divisions from the preceding one.

There are two procedures which one may follow when using the interferometer for analysis. First, it may be used as a direct-reading instrument, as just outlined. A calibration curve is constructed by making up a number of solutions and comparing them with water, preferably the same water as that used in making up the solutions. The readings are plotted against concentration and connected to make a smooth curve. When the unknown is compared with water, its concentration can be read from the curve.

In the other method, the interferometer is used as a zero-reading instrument. In this method, no previous calibration is necessary but an approximate knowledge of the concentration of the unknown solution is required. It is then compared with two solutions, one slightly more and one slightly less concentrated. This method is slightly more laborious for a single determination, but one gains in precision what one loses in convenience. This is so because only a limited portion of the scale is used (the solutions should not differ by more than 200 scale divisions). In addition to not requiring a previous calibration, this method is not subject to the error caused by the apparent shifting of the achromatic band of the interference system. When comparing a solution of a salt with water, it must be remembered that the central band is brought back to its zero position by turning a glass plate. The increase in the refractive index of the liquid is counterbalanced by decreasing the effective thickness of the compensator plate. Since the dispersion power of the solution differs from that of glass, the band system changes its appearance, so that after the concentration of salt has increased sufficiently (usually about 300 scale divisions), there is an apparent shift in the position of the achromatic band which, if not considered, would cause an error of one bandwidth (18 scale divisions). This shows the advantage of working over a very limited portion of the scale.

To secure a precision of ±0.000001 with the refractometer, the temperature must be regulated to 0.01°C. Since the interferometric method is a differential one, no special regulation of temperature is required in order to determine the difference in η_D of two solutions to 0.0000001. The sensitivity of the instrument, in terms of average parts of solute per million of solvent, is as follows for one scale division:

Refractometer (temperature to 0.01°C) 200–300
Interferometer (simple temperature control) 40-mm chamber 1.5–3.0

The interferometer is not entirely independent of temperature because there is a slight difference between the temperature coefficients of solution and solvent. Thus, a solution of potassium chloride giving a reading of 200 when compared with water at 25°C will give a reading of 202 when the two are compared at 20°C. With water solutions a variation of ±0.5°C is permissible even in very accurate work. It is *absolutely necessary,* however, that the *two chambers* should be at exactly the same temperature. With organic liquids accurate control of temperature is required.

Applications

The use of the interferometer in analyzing gases has already been mentioned. It has been used to determine the permeability of balloon fabrics to hydrogen and helium. Many applications are possible in the analysis of gases. Using a 1-m chamber, 0.02% of carbon dioxide or methane in air can be determined.

The interferometer has been used in the investigation of sea water to chart ocean currents and in the analysis of dilute solutions used for freezing-point determinations. It can be used to determine potassium and sodium in a mixture of their sulfates or chlorides with a precision of ±0.1 mg on a 50-mg sample. The mixture is dissolved in exactly 200 times its weight of water and compared with a standard solution of pure potassium sulfate dissolved in 200 times its weight of water. The reading will range from 430 to 0 as the composition of the mixture ranges from pure sodium sulfate to pure potassium sulfate, and a calibration curve is constructed.

The instrument has been used in water investigations, in measuring adsorption, in investigating colloidal solutions, sewage, fermented liquids, and milk and in biological problems such as measurement of serums, CO_2 in blood, ethyl alcohol in blood, ferment activity, and concentration of heavy water. It has been used to determine the end point in titrations and to follow the velocity of reactions. In acidimetric and precipitation reactions it gives as accurate results as good visual methods. It is necessary to plot the straight lines showing the change in reading with solution added; at the end point a sharp angle occurs. The interferometer is particularly valuable in measuring small changes in the composition of mixtures of two organic liquids as a result of preferential adsorption.

14.3 POLARIMETRY THEORY

Polarimetry, the measurement of the change of the direction of vibration of polarized light when it interacts with optically active materials, is one of the oldest of the instrumental procedures. Much of the work on the development of prisms and other devices for the production of polarized light was done in the early part of the nineteenth century.*
A small, rough polarimeter seems to be a simple piece of apparatus, but a precision polarimeter is an example of complicated optical equipment.

Ordinary, natural, unreflected light behaves as though it consisted of a large number of electromagnetic waves vibrating in all possible orientations around the direction of propagation. If, by some means, we sort out from the natural conglomeration only those rays vibrating in one particular plane, we say that we have *plane-polarized light*. Of course, since a light wave consists of an electric and a magnetic component vibrating at right angles to each other, the term "plane" may not be quite descriptive, but the ray can be considered planar if we restrict ourselves to noting the direction of the electrical component. *Circularly polarized light* represents a wave in which the electrical component (and therefore the magnetic component also) spirals around the direction of propagation of the ray, either clockwise ("right-handed" or dextrorotatory) or counterclockwise ("left-handed" or levorotatory).

---

*See, for example, J. B. Biot, *Mem. prem. classe Inst. France*, **13**, 218 (1812); W. Nicol, *Edinburgh New Phil. J.*, **6**, 83 (1828).

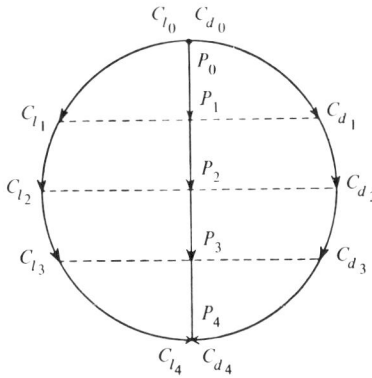

FIGURE 14-7 Representation of a plane-polarized ray as the sum of two circularly polarized rays.

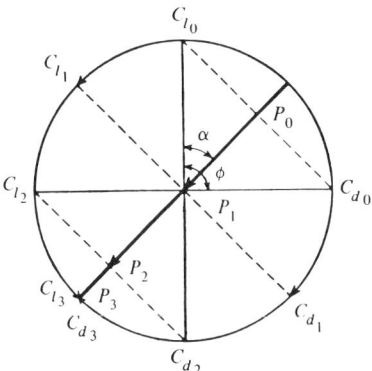

FIGURE 14-8 Rotation of the plane of polarized light due to slowing down of one of the circular components.

If one combines a polarimeter with a monochromator so that measurements of optical rotation can be made at various, known wavelengths, then the optical rotatory dispersion (ORD) can be determined. On the other hand, if a polarimeter or some other device is used to produce d and l circularly polarized light, a spectrophotometer can be used to measure the differing absorption by certain compounds at certain wavelengths and thus determine the circular dichroism (CD). Both the ORD and CD characteristics are useful in structural determinations of optically active compounds.

It is possible, and for many explanations of the interaction of light and matter it is quite enlightening, to represent a plane-polarized ray as the vector sum of two circularly polarized rays, one moving clockwise and one counterclockwise and with the same amplitude of vibration. It is obvious from Fig. 14-7 that at zero time the sum of C_{l_0} and C_{d_0}, the left and right circularly polarized rays, equals P_0, the plane-polarized ray. At the time when C_l is at C_{l_1}, C_d is at C_{d_1}, and the vector sum is P_1, and so on around the circle.

If, following the passage of the plane-polarized ray through some material, one of the circularly polarized components—say, the left circularly polarized ray—has been slowed down, then the resultant would be a plane-polarized ray rotated somewhat to the right from its original position. Figure 14-8 illustrates the case where the right-handed ray is 90° ahead of the left-handed ray. The plane-polarized resultant ray is rotated 45°, that is, 90°/2, from the original position. The rotation α is just one-half of the phase difference φ of the two circular components.

The index of refraction, η, represents the ratio of the velocity of a ray of light in a vacuum, c, to its velocity in a medium, v. That is,

$$\eta = \frac{c}{v} \tag{14-6}$$

If a substance showed different indices of refraction for the l and d components of a plane-polarized ray, then one beam would be slowed down on passage through the medium, and the plane of polarization of the ray would be rotated. Thus

$$\eta_l = \frac{c}{v_l} \quad \text{and} \quad \eta_d = \frac{c}{v_d} \quad (14\text{-}7)$$

$$\frac{v_l}{v_d} = \frac{\eta_d}{\eta_l} \quad (14\text{-}8)$$

If we let b represent the length of the column of material traversed by the ray; λ_0, the wavelength of the light; ν, the frequency of rotation (or vibration) of the light; and c, the velocity of light in a vacuum; then the difference in degrees, φ, between the two rays is given by Eq. 14-9:

$$\varphi = \frac{2\pi b \nu}{v_d} - \frac{2\pi b \nu}{v_l} \quad (14\text{-}9)$$

$$= \frac{2\pi b \nu c}{v_d c} - \frac{2\pi b \nu c}{v_l c} \quad (14\text{-}10)$$

$$= \frac{2\pi b \eta_d}{\lambda_0} - \frac{2\pi b \eta_l}{\lambda_0} \quad (14\text{-}11)$$

$$= \frac{2\pi b}{\lambda_0}(\eta_d - \eta_l) \quad (14\text{-}12)$$

or

$$\alpha = \frac{\pi b}{\lambda_0}(\eta_d - \eta_l) \quad (14\text{-}13)$$

and

$$\eta_d - \eta_l = \frac{\alpha \lambda_0}{\pi b} \quad (14\text{-}14)$$

A solution of 3.45 g of sucrose per 100 g of aqueous solution at 18°C shows a rotation α of 99.8° for light of 500.0 nm. The tube length b is 10 cm. Using Eq. 14-14, one calculates

$$\eta_d - \eta_l = \frac{99.8° \times 500.0 \times 10^{-7} \text{ cm}}{180° \times 10 \text{ cm}}$$

$$\eta_d - \eta_l = 2.77 \times 10^{-6}$$

Thus relatively small differences in the index of refraction for right and left circularly polarized light cause appreciable rotation of the plane of polarized light. The difference in indices of refraction for right and left circularly polarized light is known as *circular birefringence*.

Pasteur,[3] van't Hoff,[4] and Le Bel[5] worked out the principles which chemists now recognize as requirements in order that a given molecule possess "optical activity," that is, rotate the plane of polarized light. A compound is optically active in solution if its structure cannot be brought to coincide with that of its mirror image; that is, the compound does not possess a plane or a center of symmetry. If a tetrahedral carbon atom is substituted with four different groups, it is said to be asymmetric and would lead to optical activity unless a second similar asymmetrically substituted carbon atom is contained in the molecule. For example, mesotartaric acid and other meso compounds are not optically active. In the case of many nonplanar compounds, such as spiro compounds, allylenic compounds, and certain substituted biphenyl compounds, dissymmetrical structures with optical activity may result without an asymmetric carbon atom in the molecule. Likewise, optical activity is not limited to carbon compounds but may occur in any dissymmetrical three-dimensional compound.

Some substances show optical activity only in the crystalline, solid state. In noncubic crystals, there are at least two primary directions in the crystal that show different spatial arrangements of the atoms and thus different force fields. Radiation is transmitted at different velocities in different directions. Such crystals are known as anisotropic crystals and rotate the plane of polarized light.

Optical Rotatory Dispersion and Circular Dichroism Theory

Cotton[6] discovered an interesting connection between rotatory power and light absorption in optically active compounds. As one approaches certain optically active absorption bands in a compound from the long-wavelength side, the rotatory power at first increases strongly, then falls off and changes sign. This effect is known as the *Cotton effect*. Within the absorption band, the molar absorptivity for right- and left-hand circularly polarized light is different; that is, $(\epsilon_d - \epsilon_l) \neq 0$. This effect changes linearly polarized light into elliptically polarized light and is known as *circular dichroism*.

If one assumes that a substance near an absorption band absorbs left circularly polarized light, the l component, more strongly than the right circularly polarized light, the d component, that is, $\epsilon_l > \epsilon_d$, then the amplitude of the d component will be greater than the l. Furthermore, if one assumes that $\eta_d > \eta_l$, then the d component will be retarded more than the l component. The resulting elliptically polarized light is represented in Fig. 14-9.

The angle between the major axis of the ellipse and the plane of the original radiation is the angle of rotation, α. The ellipticity, that is, the angle whose tangent is the ratio of the minor axis of the ellipse, OB, to the major axis, OA, is known as θ. The molecular ellipticity, $[\theta]$, can be shown to be given by the relationship

$$[\theta] = 3305 \, (\epsilon_l - \epsilon_d) \tag{14-15}$$

Circular dichroism graphs are plots of $[\theta]$ against wavelength and might resemble the curve in Fig. 14-10.

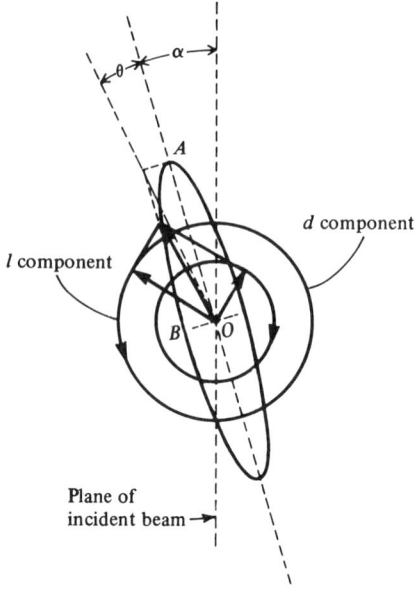

FIGURE 14-9 Elliptically polarized light produced when $\eta_d > \eta_l$ and $\epsilon_l > \epsilon_d$.

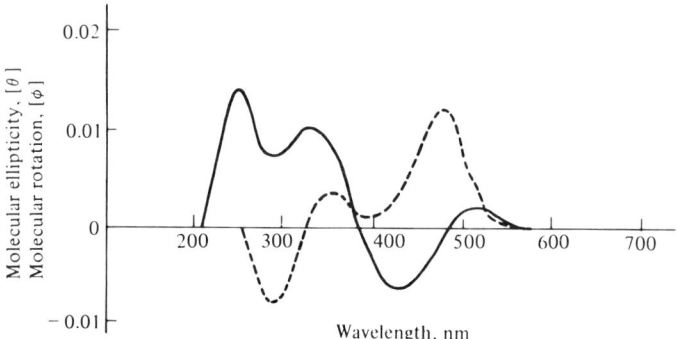

FIGURE 14-10 Circular dichroism, dotted line, and optical rotatory dispersion, solid line, for a hypothetical substance with absorption bands in the region 200–600 nm.

The specific rotation [α] defined below changes with wavelength, and the rate of change of specific rotation with wavelength is known as *optical rotatory dispersion*.[7] Drude[8] has shown that the specific rotation may be expressed as a function of wavelength by an equation with several terms:

$$[\alpha] = \frac{k_1}{\lambda^2 - \lambda_1^2} + \frac{k_2}{\lambda^2 - \lambda_2^2} + \frac{k_3}{\lambda^2 - \lambda_3^2} + \ldots \quad (14\text{-}16)$$

where λ is the wavelength of measurement and $k_1, k_2, k_3, \ldots$ are the constants that can be identified with the wavelengths of maximum absorption of the optically active absorption bands.

In a region far removed from an optically active absorption band, the dispersion is normal, and the equation of Drude can be simplified to Eq. 14-17.

$$[\alpha] = \frac{k}{\lambda^2 - \lambda_0^2} \quad (14\text{-}17)$$

where λ_0 is a constant representing the wavelength of the nearest optically active absorption band. When $\lambda \gg \lambda_0$, Eq. 14-17 may be reduced further to

$$[\alpha] = \frac{k}{\lambda^2} \quad (14\text{-}18)$$

A plot of molecular rotation, $[\phi]$, (see Eq. 14-25) versus wavelength for a hypothetical compound is shown in Fig. 14-10. Note that both molecular ellipticity and molecular rotation may be either positive or negative depending on the relationships between ϵ_d and ϵ_l or η_d and η_l.

Measurement of Optical Rotation

The rotation exhibited by an optically active substance depends on the thickness of the layer traversed by the light, the wavelength of the light used for the measurement, and the temperature. If the substance measured is a solution, then the concentration of the optically active material is also involved, and the nature of the solvent may also be important. There are certain substances that change their rotation with time. Some are substances that change from one structure to another with a different rotatory power and are said to show *mutarotation*. Mutarotation is common among the sugars. Other substances, owing to enolization within the molecules, may rotate so as to become symmetrical and thus lose their rotatory power. These substances are said to show *racemization*. Mutarotation and racemization are influenced not only by time but by pH, temperature, and other factors. In expressing the results of any polarimetric measurement, it is therefore very important to include all experimental conditions.

The results of polarimetric measurements are reduced to a set of standard conditions. The length employed as standard is 10 cm for liquids and 1 mm for solids. The standard wavelength is that of the green mercury line (546.1 nm), although the sodium doublet (589.0/589.6 nm) has been widely employed, especially in the older measurements. The standard temperature is 20°C.

Thus, if b is the layer thickness in decimeters, C is the concentration of solute in grams per 100 ml of solution, α is the observed rotation in degrees, and $[\alpha]$ is the specific rotation or rotation under standard conditions, then

$$[\alpha] = \frac{100\alpha}{bC} \quad (14\text{-}19)$$

The temperature of the measurement is indicated by a superscript and the wavelength by a subscript written after the brackets.

For a pure liquid, the concentration is unimportant, but temperature changes cause expansion and contraction of the liquid and a consequent change in the number of active molecules in the path of the light. For pure liquids unit density is assumed as standard, and the definition of specific rotation becomes

$$[\alpha] = \frac{\alpha}{b\rho} \qquad (14\text{-}20)$$

where ρ is density.

Temperature changes have several effects upon the rotation of a solution or liquid. An increase in temperature increases the length of the tube; it also decreases the density, thus reducing the number of molecules involved. It causes changes in the rotatory power of the molecules themselves due to association or dissociation, increased mobility of the atoms, and affects other properties. In general, the effect of temperature may be expressed by Eq. 14-21:

$$[\alpha]^t = [\alpha]^{20} + z(t-20) \qquad (14\text{-}21)$$

where z is the temperature coefficient of rotation and t is the temperature in degrees Celsius. Substances vary widely in their values of z.

Any liquid or solution, when placed in a magnetic field, rotates the plane of polarized light because of the effect of the magnetic field upon the motion of the electrons in the molecules. This effect was discovered by Faraday[9] and is known as the *Faraday effect*. The rotation χ is positive if in the same direction as the magnetizing current. For most substances χ is positive and is appreciable for many organic substances.

For analytical purposes, the chief interest in polarimetry is to determine the concentration of substances, although abundant correlation between rotation and chemical structure has been found.[10] The relationship between rotation and concentration of a solution is, unfortunately, not strictly linear, so that the specific rotation of a solution is not a constant. The concentration of the measurement should always be stated when $[\alpha]$ is given. The values of the specific rotation extrapolated to infinite dilution may be employed.

The relationship between $[\alpha]$ and concentration may usually be expressed by one of the three equations proposed by Biot.

$$[\alpha] = A + Bq \qquad (14\text{-}22)$$

$$[\alpha] = A + Bq + Cq^2 \qquad (14\text{-}23)$$

$$[\alpha] = A + \frac{Bq}{C+q} \qquad (14\text{-}24)$$

where q is the percentage of solvent in the solution and A, B, C are constants.

Equation 14-22 represents a straight line, Eq. 14-23 a parabola, and Eq. 14-24 a hyperbola. The constants A, B, and C are determined from several measurements at different concentrations.

Calculations of Polarimetry and Saccharimetry

The equations for the calculation of specific rotation, $[\alpha]$, from measurement of the angle of rotation have already been given (Eqs. 14-19 and 14-20). Often it is desired to calculate the molecular rotation $[\phi]$, which is given by the equation

$$[\phi] = \frac{[\alpha] \times \text{mol wt}}{100} \qquad (14\text{-}25)$$

The polarimeter is widely used as a saccharimeter in sugar analysis. In the determination of the concentration of sucrose in a substance containing no other optically active material except sucrose, it is convenient to take 75.2 g of unknown as sample and dissolve it in enough water to make 100 ml of solution. Then, with a 2-dm tube, the rotation in degrees is numerically equal to the concentration of sucrose in percent by weight. This follows from Eq. 14-19 using $[\alpha]_D^{20} = 66.5°$ for sucrose.

This sample, 75.2 g, is rather large, and consequently most modern saccharimeters are not graduated in degrees but in smaller divisions. Several types of graduation have been used. Most modern saccharimeters are graduated in the "International" scheme in which one division equals 0.3462°. It is obvious that one should check the normal weight of any saccharimeter, preferably by calibration with pure sucrose, before any unknowns are run.

Most raw sugar samples contain other optically active substances besides sucrose. When sucrose is heated with acid or with the enzyme invertase, it is "inverted" to form one molecule of fructose and one of glucose. Sucrose has a specific rotation of +66.5°, fructose has a specific rotation of –93°, and glucose +52.5°. Thus the specific rotation changes from +66.5° to (–93° + 52.5°)/2 = –20.2° upon inversion. By measuring the change in rotation upon inversion it is possible to determine sucrose in the presence of other optically active substances. The formula used for the determination of sucrose is known as the *Clerget formula*, Eq. 14-26. See Browne and Zerban[11] for a more complete discussion of this formula and of sugar analysis in general.

$$\text{percent sucrose} = \frac{100\,(a-h)}{144 - \dfrac{t}{2}} \times \frac{W}{w} \qquad (14\text{-}26)$$

where a is the rotation in Ventzke degrees of sucrose solution before inversion, h is the rotation in Ventzke degrees after hydrolysis, t is the temperature in degrees Celsius, W is the normal weight for saccharimeter employed, and w is the weight of sample taken per 100 ml of solution.

The Clerget factor, 144 in Eq. 14-26, varies slightly with concentration and with the details of the method of hydrolysis employed. The factor is derived by considering the experimentally determined Ventzke readings for inverted sugar, the dilution introduced by the addition of hydrochloric acid (10 ml per 100 ml of solution) necessary for the hydrolysis, and the fact that one molecule of water is consumed for each molecule of sucrose hydrolyzed.

14.4 APPLICATIONS OF OPTICAL ROTATORY DISPERSION AND CIRCULAR DICHROISM

The main applications of both optical rotatory dispersion and of circular dichroism lie in the area of structure determination of such optically active substances as amino acids, polypeptides and proteins, steroids, antibiotics, terpenes, and metal–ligand complexes. Many applications to date are empirical in nature and depend upon knowledge of the behavior of compounds similar to those under investigation. Some general rules are evident, however.

Aliphatic amino acids show a unique Cotton effect, the sign of which reflects the stereochemistry at the asymmetric center. α-Amino acids of levo configuration show a positive Cotton effect around 215 nm, whereas their dextro enantiomers display a negative Cotton effect. In polypeptides it is possible to estimate the percent of α-helix structure by measurements of optical rotatory dispersion.

In steroids, *cis* and *trans* ring junctions lead to different types of optical rotatory dispersion curves. In one form the specific rotation increases with decreasing wavelength until it reaches a peak and reverses (a positive curve), and in the other the specific rotation decreases with decreasing wavelength until it reaches a trough and reverses (a negative curve). The location of carbonyl groups in steroids also can often be narrowed down to a limited number of possibilities by noting the sign of the Cotton-effect curve and the wavelength and specific rotation of the peak or trough.

Theoretical studies of radiant energy absorption by chromophores asymmetrically surrounded in a molecule have led to the so-called "octant" rule. The octant rule states that the sign of the contribution that a given atom at point $P(x, y, z)$ makes to anomalous rotatory dispersion will vary as the simple product, $x \cdot y \cdot z$, of its coordinates. As an illustration, look at the carbonyl chromophore group in the cyclohexanone molecule. Represent the cyclohexanone molecule in the chair form with coordinates defined by the x-y, x-z, and y-z planes as shown in Fig. 14-11. The carbonyl group lies in the x-y plane and

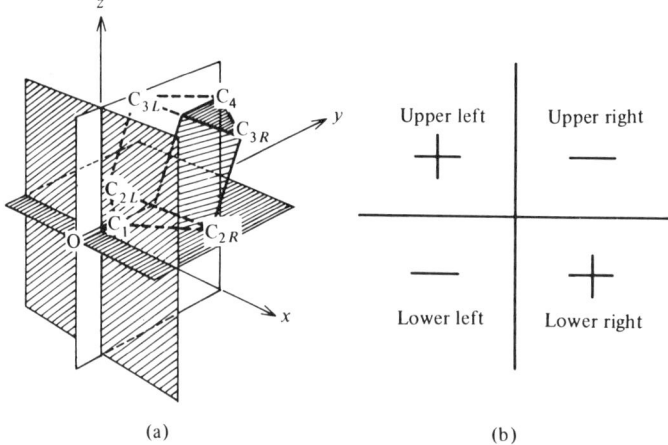

FIGURE 14-11 Illustration of the octant rule for cyclohexanone: (a) the octants and (b) contributions of groups in the four octants.

is bisected by the y-z plane. Atoms or substituent atoms can now be located in one of the eight octants defined by ±x, ±y, and ±z. Atoms or substituents in the +x, +y, +z; +x, –y, –z; –x, –y, –z, or –x, –y, +z octants have positive contributions to the Cotton effect. Atoms or substituents in the other four octants have negative effects and atoms or groups which lie on or near a plane have little or no effect.

14.5 THE POLARIMETER

The polarimeter consists of the following basic parts:

1. A light source
2. A polarizer
3. An analyzer
4. A graduated circle to measure the amount of rotation
5. Sample tubes

Except in the simplest instruments a half-shade device is also included. Some polarimeters may be equipped with photocells or other devices for measuring the intensity of light emerging from the instrument, although most polarimeters are designed for visual observation.

The most common light sources for polarimetry are sodium vapor lamps and mercury vapor lamps. The sodium lamp emits light of wavelengths 589.0 and 589.6 nm plus a little continuous background which can be largely eliminated with a filter of 7% potassium dichromate used in a 6-cm-thick layer. The mercury lamp emits light of several wavelengths, the prominent visual lines being at 435.8, 491.6, 546.1, 577.0, and 579.1 nm. The proper choice of filters will permit the isolation of each line. If a continuous light source can be employed, then ordinary sunlight or light from a tungsten filament lamp can be used.

The polarizer (and the analyzer) may be of several different types. One type consists of a crystal, usually calcite or quartz, cut diagonally at such an angle that one component of the light is totally reflected. The second component passes through the second half of the crystal and thus emerges, going in the same direction as the original beam (Fig. 14-12). The two halves of the prism are cemented together with a cement having an index of refraction as near as possible to 1.4865, which is the value of η_e, the extraordinary index of refraction for calcite. The index of refraction for calcite at right angles to the above-mentioned ray is $\eta_0 = 1.6584$.

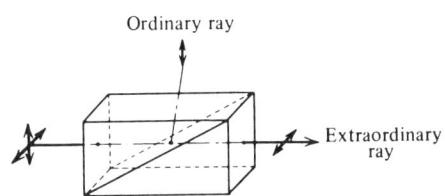

FIGURE 14-12 The Glan–Thompson prism.

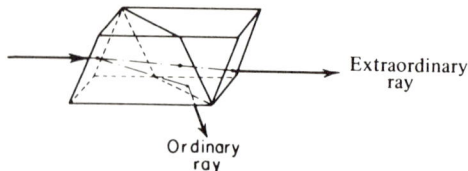

FIGURE 14-13 The Nicol prism.

Several different varieties of polarizing (and analyzing) prisms are known. They vary in the angles of the faces of the prism and of the cut diagonally through the prism. The Glan-Thompson prism (Fig. 14-12) and the Nicol prism (Fig. 14-13) are the most common. The Nicol prism requires smaller pieces of calcite and is cheaper but is not as good as the Glan-Thompson prism. The light emerging from a true Nicol prism is displaced from the original beam and will revolve in a circle as the prism is rotated. Again, two Nicols used together will not produce total extinction at any point for the whole field and thus will introduce uncertainty in the balance point. With either a Nicol or Glan-Thompson prism the entering light must be essentially parallel; otherwise some unpolarized light will be transmitted. Light must not, therefore, be concentrated on the prism by using a converging beam.

Light can also be polarized by reflection from a mirror at the proper angle—Brewster's angle. If light strikes a mirror at such an angle i that

$$\tan i = \eta \qquad (14\text{-}27)$$

where η is the refractive index of the mirror material, then only the component vibrating perpendicular to the plane of incidence (parallel to the mirror surface) will be reflected. Reflection is not used in modern polarimeters to produce polarized light. It is interesting to note, however, that light emerging from a monochromator is partially polarized, with the greatest intensity perpendicular to the exit slit. Thus if a monochromator precedes a polarimeter, the slit should be perpendicular to the direction of transmission of the polarizing prism.

A third method of producing polarized light is by Polaroid filters. Polaroid filters are composed of strongly dichroitic crystals oriented in a plastic material. These crystals strongly absorb light vibrating in one direction and only weakly absorb light vibrating in the perpendicular direction. Polaroids can never give 100% polarization; also, the light must lie in the region from about 500.0 to 680.0 nm. Polaroids are used, therefore, only on less expensive instruments.

When two prisms, a polarizer and an analyzer, are used together, the intensity of light transmitted through the combination is given by the law of Malus:

$$I = KI_0 \cos^2 \theta \qquad (14\text{-}28)$$

where I is the emerging intensity from analyzer, I_0 is the incident intensity on analyzer, θ is the angle between the directions of transmission of the two prisms, and K is the factor taking into account reflection and absorption losses in the analyzing prism; K is approximately equal to 1.

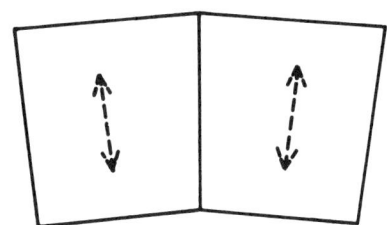

FIGURE 14-14 The Jellett–Cornu prism, end view.

The graduated circle is fitted with a vernier for more precise measurement of the angle through which the analyzing prism has been rotated. Special reading devices employing a pair of parallel index lines are used on the most precise instruments. A tangent screw with graduated drum allows the borders of an etched line on the main scale to be made to coincide with the two hairlines. With such a device readings can be made to $0.002°$.

The polarimeter tubes must have plane and parallel glass disks at the ends. The glass must be free from strain; otherwise the disks will produce a partial circular polarization of the light, and complete extinction of light will be impossible. Each tube should be tested by filling it with water, placing it between crossed prisms, and noting whether the dark field remains dark. The length of polarimeter tubes may be determined by measuring the rotation of a known, strongly rotating liquid or solution—for example, nicotine in ethyl alcohol—at a definite temperature.

In cheaper instruments one measures the position of the analyzer required to give a minimum intensity without the sample in the tubes, and again with the sample in the tubes. The difference in the two readings is the rotation caused by the introduction of the sample. The human eye, however, is much better at *matching* light intensities than is the human mind at *remembering* intensities, as it must if the minimum intensity is to be determined. Consequently, the more precise polarimeters make use of so-called half-shade devices which result in matching two half-fields for a balance point.

Many different half-shade devices are available. Each has its own advantages and disadvantages. The Jellett–Cornu prism (Fig. 14-14) is constructed by sawing a Glan–Thompson prism in two lengthwise, grinding one face down a little, and cementing the parts back together. When light passes through this polarizing prism, the two halves will produce polarized light beams tilted slightly with respect to each other. Rotation of the analyzer prism in front of such a polarizing prism would result in complete extinction, first of one-half of the field and then of the other half. At some intermediate position the two halves of the field would appear of equal brightness. This point is taken as the balance point. The Jellett–Cornu device does not allow variation of the half-shade angle, the angle between the two prisms. Variation of this angle is desirable, because large angles are necessary for precise balancing with weak light sources and small angles give more precise balancing with strong light sources.

The Lippich prism (Fig. 14-15) is a popular half-shade device. A small polarizing prism A precedes the large polarizing prism B. With such an arrangement one-half of the field can be rotated slightly with respect to the other half. The analyzer must be at some inter-

FIGURE 14-15 The Lippich prism, side view.

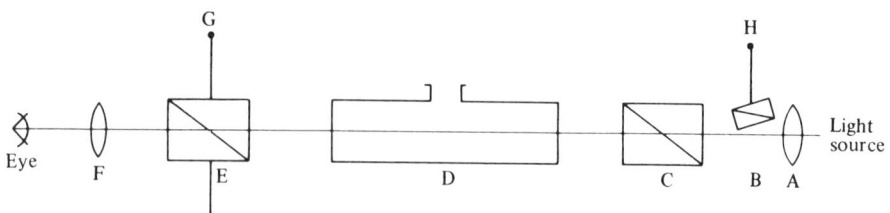

FIGURE 14-16 Optical arrangement of a polarimeter with a Lippich half-shade device; A, collimating lens; B, Lippich half-shade prism; C, polarizing prism; D, tube; E, analyzing prism; F, eyepiece; G, scale; H, lever to adjust half-shade angle.

mediate position in order to achieve equal illumination of both halves of the field. Sometimes two Lippich prisms are used, dividing the field into three parts that match at the balance point. The half-shade angle of any Lippich arrangement can be varied by rotating the Lippich prism.

A popular, inexpensive half-shade device is the Laurent half-wave plate. A thin plate of quartz cut parallel to its optic axis is placed over one-half of the field of the polarizer. The quartz plate is cut just thick enough that for a given wavelength of light (usually the yellow sodium doublet) the slow ray lags exactly one-half wavelength behind the fast ray. This results in a slight rotation of the light passing through the part of the polarizer covered by the plate. The amount of rotation can be varied by rotating the quartz plate with respect to the polarizing prism, and thus the half-shade angle is variable. The quartz plate is suitable, however, only for the wavelength for which it was constructed. At other wavelengths it becomes more difficult to find the balance point due to lack of contrast in the fields as the analyzer is rotated.

The optical arrangement of a precision polarimeter is shown in Fig. 14-16 and an example of a commercial instrument is shown in Fig. 14-17. The Rudolph instrument has a Lippich double field polarizer.

In the type of instruments described above, the rotation is measured by rotating the analyzer with respect to the polarizer. It is also possible to measure rotatory power by leaving the analyzer permanently crossed with respect to the polarizer and compensating any rotation caused by the sample with a piece of quartz which rotates light in the opposite direction to that of the sample. The design of such an instrument is shown in Fig. 14-18. Wedges I and II are made of levorotatory quartz and are ground to the same angle. Wedge I is stationary, but II is movable. Moving wedge II thus varies the thickness of the block of levorotatory quartz. Block III is made of dextrorotatory quartz and is of thickness equal to that of wedges I and II when II is in an intermediate position. Thus both positive and negative rotations of the sample can be compensated by moving the wedge in or out

FIGURE 14-17 Rudolph polarimeter. (Courtesy of O. C. Rudolph & Sons.)

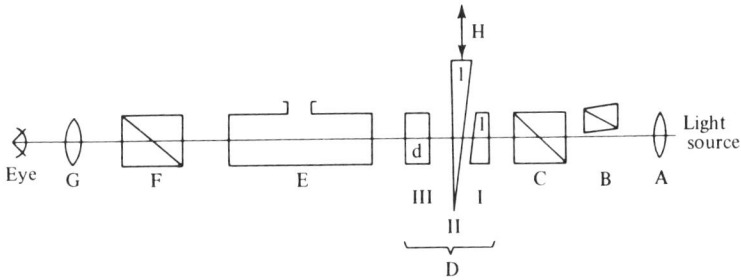

FIGURE 14-18 Quartz-wedge compensating polarimeter: A, collimating lens; B, Lippich half-shade prism; C, polarizing prism; D, quartz-wedge compensator; E, tube; F, analyzer prism (position fixed); G, eyepiece; H, scale and movement device for compensator.

from its intermediate position. Compensating polarimeters of the type described above are used largely for sugar analyses and are known as saccharimeters. Fortunately, the rotatory dispersion of quartz, sucrose, and a few other sugar solutions is very nearly the same. Thus if white light is used as a source, the quartz can compensate the rotation of the sugar solution at all wavelengths. If the dispersions of quartz and sugar were not the same, only light of one wavelength would be compensated completely, and the field would appear colored rather than dark.

Automatic Recording Spectropolarimeters

Automatic recording spectropolarimeters, necessary for the determination of optical rotatory dispersion, may be placed broadly in two classes: instruments that use a null-point method and instruments that use a ratio method. At present all commercially available instruments fall into the first class. They differ among themselves in how the null point is achieved. Two will be described in some detail.

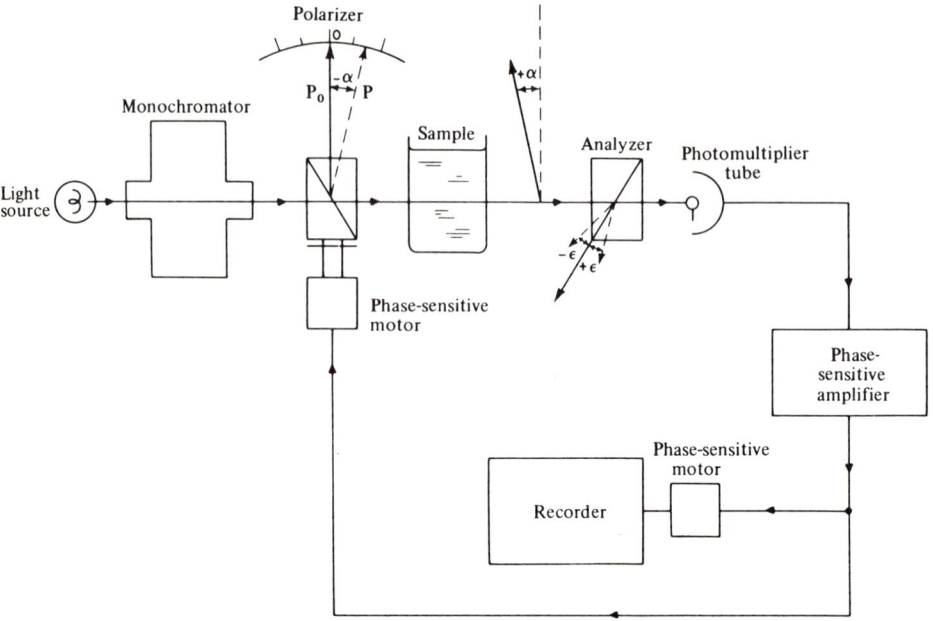

FIGURE 14-19 Schematic diagram of the Rudolph spectropolarimeter. The polarizer is rotated through an angle $(-\alpha)$ that is equal and opposite to the angle of rotation of the sample. The angle of scan, $\pm \epsilon$, is induced by a mechanical oscillator. P_0, initial position of polarizer axis. (Courtesy of O. C. Rudolph & Sons.)

The Rudolph spectropolarimeter works on the null-point principle, the null point being ascertained by an imposed mechanical oscillation ($\pm \epsilon$) of the analyzer, whose mean angular position is orthogonal with respect to the plane of polarization of the entering light beam (Fig. 14-19). At this point, angular changes of $+\epsilon$ and $-\epsilon$ produce the same current in the phototransducer system. The optical rotation α produced by the introduction of a sample between the polarizer and the analyzer is measured by the angular rotation $(-\alpha)$ of the polarizer that is required to restore the balance of the signal output. The analyzer prism is mechanically oscillated to produce a 20-Hz modulation of the light beam striking the photomultiplier tube. A portion of its signal is separated, by means of a chopper, into two signals, corresponding to the right and left oscillations of the analyzer. The difference between these two signals is fed into a null-point-seeking servo system, which drives the polarizer to a position that compensates for the rotation of the sample. The recording of the optical activity is obtained by means of a linkage between the angular position of the polarizer and an X-Y recording system.

The Cary spectropolarimeter is similar in principle to the Rudolph instrument, except that in the Cary instrument the mechanical oscillation of the analyzer is replaced by an oscillation brought on by a magneto-optical effect. To achieve this, a Faraday cell is placed ahead of the analyzer (Fig. 14-20). The Faraday cell consists of a silica cylinder surrounded by a coil. An alternating current (60 Hz) passes through the coil, thus cyclically displacing the plane of polarization of the beam. A motor energized by the amplified current from the photomultiplier moves the polarizer by means of a mechanical linkage.

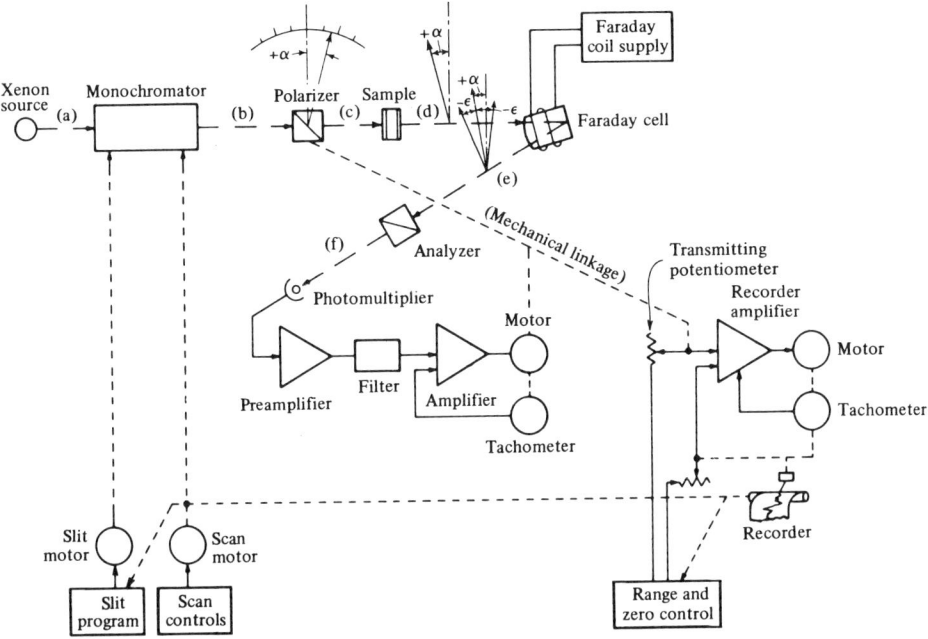

FIGURE 14-20 Schematic diagram of the Cary spectropolarimeter. The angle of scan, $\pm \epsilon$, is induced by a Faraday cell. (a) Undispersed, nonpolarized beam; (b) monochromatic nonpolarized beam; (c) monochromatic polarized beam; (d) beam "c" rotated by sample; (e) beam "d" cyclically displaced by Faraday cell; (f) component transmitted by analyzer. (Courtesy of Applied Physics Corp.)

14.6 INSTRUMENTS FOR CIRCULAR DICHROISM MEASUREMENT

An ordinary spectrophotometer can be adapted to measure circular dichroism. It is only necessary to provide some means of producing d and l circularly polarized radiation. For this purpose a plane-polarized beam can be passed through a quarter-wave plate. If the plate is rotated from $+45°$ to $-45°$, first d and then l circularly polarized light is produced. In order to cover a large wavelength region, several quarter-wavelength plates are necessary. Other devices can also be used to produce circularly polarized beams so that the relative absorption can be measured by the spectrophotometer.

PROBLEMS

1. A substance having the analysis C_3H_6O might be either acetone or allyl alcohol. Determine which of these two substances it is from the fact that the molar refraction is 16.97.

2. The refractive index of carbon tetrachloride at $20°C$ is 1.4573, and the density at $20°C$ is 1.595 g cm^{-3}. Calculate the molar refraction.

3. From the atomic refractions, calculate the refractive index at 20°C of nitrobenzene. The density is 1.210 g cm^{-3}.

4. Calculate the specific refraction and the molar refraction for each of these liquids:

Compound	n_D^{20}	d^{20}
Benzene	1.4979	0.879
Ethanol	1.3590	0.788
Ethyl acetate	1.3701	0.901
Toluene	1.4929	0.866
Nitrobenzene	1.5524	1.21
Water	1.3328	0.998

5. For D$_2$O, $n_D^{20} = 1.32830$, and for water, $n_D^{20} = 1.33280$. If the refractive index for a sample is 1.32980, calculate the percent D$_2$O present.

6. What is the refractive index of a mixture of 10 ml of benzene and 40 ml of nitrobenzene? See Problem 4 for necessary data.

7. A 120-ml sample of wine was distilled to remove all the alcohol; the distillate was diluted to a volume of 100 ml. The reading on the immersion refractometer at 25°C was 26.8. What was the percentage by volume of alcohol in the wine? Necessary tables are in J. A. Dean, Ed., *Lange's Handbook of Chemistry*, 11th ed., 10, p. 251, McGraw-Hill, New York, 1973.

8. One gram of an organic substance was dissolved in 50.00 ml of water. This solution, in a 20-cm tube, read +2.676° in a polarimeter, while distilled water in the same tube read +0.016°. Calculate the specific rotation of the substance.

9. Exactly 10 g of raw sugar was dissolved in water and made up to a volume of 100 ml. In a 20-cm tube at 25°C this solution read 12.648° in a polarimeter. After inversion, according to the directions of Experiment 14-2, the solution read −3.922°. Calculate the percentage of sucrose in the raw sugar.

10. Calculate the Brewster angle for borosilicate glass of index of refraction $\eta_D = 1.47$. Devise an experiment to check this result.

11. Approximately what fraction of light would be removed from a beam of light passing through two Polaroid filters set at 45° with respect to each other?

12. Substitute Beer's law into Eq. 14-15. Assume that initial beam intensity, P_0, for d radiation equals P_0 for l radiation and thus develop the equation for determining molecular ellipticity using a spectrophotometer.

13. Calculate the difference in indices of refraction for right and left circularly polarized light for a substance giving a rotation of −10.5° at 600.0 nm in a 10 g per 100 ml solution in a 20-cm polarimeter tube.

14. Draw curves like Fig. 14-8 for the situation where $\eta_l > \eta_d$ and $\epsilon_d > \epsilon_l$, and for the situation where $\eta_l > \eta_d$ and $\epsilon_l > \epsilon_d$.

15. Ten grams of a compound was dissolved in water and made up to a volume of 100 ml. What is the molecular ellipticity if the solution placed in a 10-cm tube showed spectrophotometer readings of 40% transmittance for d circularly polarized light and 42% for l circularly polarized light. The incident beams are equal in intensity.

BIBLIOGRAPHY

Bauer, N., K. Fajans, and S. Z. Lewin, in *Physical Methods of Organic Chemistry*, 3rd ed., A. Weissberger, Ed., Vol. 1, Part II, Chap. 28, Wiley-Interscience, New York, 1960.
Carroll, B. and I. Blei, *Science*, **142**, 200 (1963).
Degenhard, W. E., "Interferometry" in *Encyclopedia of Industrial Chemical Analysis*, F. D. Snell and C. L. Hilton, Eds., Vol. 2, pp. 334–346, Interscience, New York, 1966.
Degenhard, W. E., "Refractometry" in *Encyclopedia of Industrial Chemical Analysis*, F. D. Snell and C. L. Hilton, Eds., Vol. 3, pp. 392–407, Interscience, New York, 1966.
Djerassi, C., *Optical Rotatory Dispersion*, McGraw-Hill, New York, 1960.
Heller, W. and D. D. Fitto, in *Physical Methods of Organic Chemistry*, 3rd ed., A. Weissberger, Ed., Vol. I, Part II, Chap. 33, Wiley-Interscience, New York, 1960.
Lewin, S. Z. and N. Bauer, in *Treatise on Analytical Chemistry*, I. M. Kolthoff, and P. J. Elving, Eds., Vol. 6, Part I, Chap. 70, Wiley-Interscience, New York, 1965.
Maley, L. E., "Refractometers," *J. Chem. Educ.*, **45**, A467 (1968).
O'Brien, R. N., in *Techniques of Chemistry*, A. Weissberger and B. W. Rossiter, Eds., Vol. 1, Part 3A, Wiley-Interscience, New York, 1972.
Tilton, L. W. and J. K. Taylor, in *Physical Methods in Chemical Analysis*, W. C. Berl, Ed., Vol. 1, Academic, New York, 1950.
Velluz, L., M. Legrand, and M. Grosjean, *Optical Circular Dichroism*, Academic, New York, 1965.

LITERATURE CITED

1. Wagner, B., *Z. Angew. Chemie*, **33**, 262 (1920).
2. Shippy, B. A. and G. H. Barrows, *J. Am. Chem. Soc.*, **40**, 185 (1918).
3. Pasteur, L., *Ann. Chim. Phys.*, **24**, 442 (1848).
4. van't Hoff, J. H., *La chimie dans l'espace*, Rotterdam, 1874.
5. Le Bel, J. A., *Bull. Soc. Chim.*, **22**, 337 (1874).
6. Cotton, A., *Compt. Rend.*, **120**, 989, 1044 (1895); *Ann. Chim. Phys.*, **8**, 347 (1896).
7. Djerassi, C., *Optical Rotatory Dispersion*, McGraw-Hill, New York, 1960.
8. Drude, P., *The Theory of Optics*, Longmans, New York, 1929.
9. Faraday, M., *Phil. Mag.*, **28**, 294 (1846).
10. Djerassi, C., *Science*, **134**, 649 (1961).
11. Browne, C. A. and F. W. Zerban, *Physical and Chemical Methods of Sugar Analysis*, Wiley, New York, 1941.

CHAPTER 15

Chromatography — General Principles

The term chromatography embraces a family of closely related separation methods based on experiments described by Day[1] and Tswett[2] in 1897–1906. The feature distinguishing chromatography from most other physical and chemical methods of separation is that two mutually immiscible phases are brought into contact wherein one phase is stationary and the other mobile. The sample mixture, introduced into the mobile phase, undergoes a series of interactions (partitions) many times between the stationary and mobile phases as it is being carried through the system by the mobile phase. Interactions exploit differences in the physical or chemical properties of the components in the sample. These differences govern the rate of migration of the individual components under the influence of a mobile phase moving through a column containing the stationary phase. Separated components emerge in the order of increasing interaction with the stationary phase. The least retarded component elutes first, the most strongly retained material elutes last. Separation is obtained when one component is retarded sufficiently to prevent overlap with the zone of an adjacent solute as sample components elute from the column.

The column is the heart of a chromatograph and provides versatility in the types of analyses that can be obtained with a single instrument. This versatility, due to the wide choice of materials for the stationary and mobile phases, makes it possible to separate molecules that differ only slightly in their physical and chemical properties. It can also be a problem when it comes time to select an appropriate chromatographic method.

15.1 CLASSIFICATION OF CHROMATOGRAPHIC METHODS

The mobile phase can be a gas or a liquid, whereas the stationary phase can only be a liquid or a solid. The field of liquid column chromatography (LCC) embraces several distinct types of interaction. When the separation involves predominantly a simple partitioning between two immiscible liquid phases, one stationary and the other mobile, the process is called liquid–liquid (or partition) chromatography (LLC). When physical surface forces are mainly involved in the retentive ability of the stationary phase, the process is denoted liquid–solid (or adsorption) chromatography (LSC). Two other chromatographic methods differ somewhat in their mode of action. In ion-exchange chromatography (IEC), ionic components of the sample are separated by selective exchange with counterions of the stationary phase. The use of exclusion packings as the stationary phase brings about a classification of molecules based largely on molecular geometry and size. Exclusion chromatography (EC) is referred to as gel-permeation chromatography by polymer chemists and as gel filtration by biochemists. If the mobile phase is a gas, the methods are called gas–liquid chromatography (GLC) and gas–solid chromatography (GSC). Gas and liquid mobile phase methods will be treated individually. There are few fundamental reasons for separate treatment. However, differences in operating technique and equipment warrant individual chapters.

15.2 NATURE OF PARTITION FORCES

Broadly speaking, the distribution of a solute between two phases results from the balance of forces between solute molecules and the molecules of each phase. It reflects the relative attraction or repulsion that molecules or ions of the competing phases show for the solute and for themselves. These forces can be polar in nature, arising from permanent or induced electric fields associated with both solute and solvent (also adsorbent) molecules, or they can be due to London's dispersion forces, which depend on the relative masses of the solute and solvent molecules. In ion-exchange chromatography, the forces on the solute molecules will be substantially ionic in nature but will include polar and nonpolar forces as well. The relative polarity of solvents is manifested in their dielectric constants.

Dispersion Interaction

In a nonpolar liquid, such as CCl_4, London's dispersion interaction is the only force present between two molecules. This interaction is produced by rapidly varying ("instantaneous") dipoles formed between nuclei and electrons at zero-point motion of the molecule acting on the polarizability of other molecules to produce induced dipoles in phase. For instantaneous dipoles moving in phase, there would arise an attractive interaction between them owing to the movement in phase of their electronic systems. This effect can be additive over large numbers of molecules in a system. Dispersion forces are relatively weak. When the CCl_4 molecules approach each other so closely that their electron orbitals begin to overlap, the weak attraction changes to repulsion. As a consequence, the CCl_4 molecules exist in a disordered array relative to each other. A second nonpolar solute or solvent mixes in all proportions commensurate with its solubility. Neither kind of molecule has much attraction for its fellows and therefore does not oppose molecules of the other kind intermingling.

Dipole–Dipole Interaction

In a liquid composed of molecules with permanent dipoles, these molecules will be much more strongly attracted to one another than to those of a nonpolar molecule. This dipole–dipole interaction leads to association of liquids or of liquid–solute pairs. Only substances whose attraction for polar molecules of the solvent is about as strong as the attraction of solvent molecules for one another will be able to force the solvent molecules apart and mix with them.

Hydrogen Bond Interactions

Dipoles capable of forming hydrogen bonds are exceptional in their behavior. Although hydrogen can form only one covalent bond, the very small size of the hydrogen atom and its lack of inner closed electron orbitals makes it possible to form an additional bond with electron-rich atoms intermolecularly or intramolecularly. The actual strength of a hydrogen bond is dependent upon the geometry of particular combinations, the nature of neighboring atoms, resonance, and acid–base character.

Water has a particularly strong tendency to form intermolecular hydrogen bonds and can function both as an electron acceptor and an electron donor for dissolved substances. It exists in a relatively open, but highly hydrogen-bonded structure consisting of clusters of 4–5 water molecules. This energy-rich ordered array of water molecules must be broken down if solute molecules are to be accommodated. To break each hydrogen bond requires an expenditure of 4–6 kcal of energy. Consequently, the solubility of a solute in water is influenced strongly by its own ability to form hydrogen bonds with the water molecules or else to ionize, two means of supplying the energy required to disrupt the hydrogen bonds within water. No such network of hydrogen bonds exists in solvents such as ethers or CCl_4. Solvent molecules whose mutual attraction is enhanced by hydrogen bonding will resist penetration by nonpolar solute molecules. If these nonpolar molecules happen to be formed or be present, they will tend to be "squeezed out" of the more polar liquid phase and into the less polar phase when the two phases are in contact.

15.3 CHROMATOGRAPHIC BEHAVIOR OF SOLUTES

The chromatographic behavior of a solute can be described in numerous ways. For columnar chromatography, the retention volume (or corresponding retention time) and the partition ratio are the terms most frequently employed. By varying the stationary/mobile phase combinations and various operating parameters, the degree of retention can be varied from nearly total retention to a state of free migration. Optimizing performance implies a judicious combination of good separation of peak maxima, compactness of peaks, and high speed of elution (that is, short analysis time). Theory places these ideas on a systematic basis, relating them to the thermodynamics and kinetics of the chromatographic process.

Retention Behavior

Retention behavior reflects the distribution of a solute between the mobile and stationary phases. Consider Fig. 15-1 in which some substance is depicted that is well behaved in a chromatographic sense. The volume of mobile phase necessary to convey a solute band from the point of injection, through the column, and to the detector (to the apex of the solute peak) is defined as the retention volume, V_R. It may be obtained directly by measuring the retention time, t_R, and multiplying the latter by the volumetric flowrate, F_c, expressed as volume of mobile phase per unit time:

$$V_R = t_R F_c \qquad (15\text{-}1)$$

The volume flowrate, in terms of column parameters, is as follows:

$$F_c = (\pi d_c^2/4) \quad \epsilon_{tot} \quad (L/t_M) = V_{column}\epsilon_{tot}/t_M \qquad (15\text{-}2)$$

$$\underset{\substack{\text{cross} \\ \text{section} \\ \text{of empty} \\ \text{column}}}{} \quad \underset{\substack{\text{total} \\ \text{porosity}}}{} \quad \underset{\substack{\text{average} \\ \text{linear} \\ \text{velocity} \\ \text{of eluent}}}{}$$

where d_c is the column bore, L is the column length, V_{column} is the bed volume of the column, and $L/t_M = \bar{u}$ is the average linear velocity of the mobile phase as measured by the transit time of a nonretained solute, t_M. In interactive chromatography no material can elute prior to this time. When converted to volume, V_M (or V_0), it represents the dead space, void volume, or holdup volume of a column; it includes the effective volume contributions of the sample injector and detector. For solid packings the total porosity is 0.35–0.45, while for porous packings it ranges from 0.70 to 0.90.

The adjusted retention volume, V'_R, or time, t'_R, is given by

$$V'_R = V_R - V_M \qquad \text{or} \qquad t'_R = t_R - t_M \qquad (15\text{-}3)$$

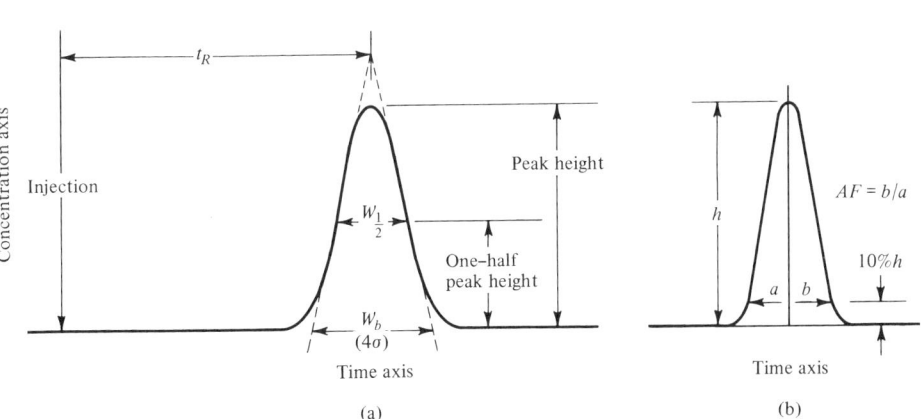

FIGURE 15-1 Evaluation of a chromatographic peak for (a) column efficiency and (b) peak asymmetry factor.

When the mobile phase is a gas, temperature and pressure must be specified, and retention volumes must be corrected for the compressibility of the gas since the gas moves more slowly near the inlet than at the exit of the column. The true retention volume is the integrated sum of gas volumes at the band positions along the column. The pressure-gradient correction (or compressibility) factor,* j, is expressed by

$$j = \frac{3[(P_i/P_0)^2 - 1]}{2[(P_i/P_0)^3 - 1]} \tag{15-4}$$

where P_i is the carrier gas pressure at the column inlet, and P_0 that at the outlet.

The specific retention volume, V_g^0, is the adjusted retention volume (or, in gas chromatography, the net retention volume after correction for the column pressure drop, namely, $V_N = jV_R'$) divided by the weight of liquid stationary phase, w_L (or weight of adsorbent phase, w_S) in grams. The data are converted to 0°C, unless otherwise specified by an appropriate superscript (V_g^T).

Partition Coefficient

When a solute enters a chromatographic system, it immediately distributes between the stationary and mobile phase. If, at some time during the movement of the solute down the column, the mobile phase flow is stopped, then the sample will be in equilibrium between the stationary and stopped mobile phase. The concentration in each phase as a function of distance down the column are mirror images but with a difference in magnitude. In this state the concentration in each phase is given by the partition coefficient

$$K = C_S/C_M \tag{15-5}$$

where C_S, C_M are the concentrations of solute in the stationary and mobile phases, respectively. For example, when $K = 1$, the solute is equally distributed between the two phases. The partition coefficient determines the average velocity of each solute zone; more specifically, the zone center as the mobile phase moves down the column.

At the appearance of a peak maximum at the column exit, one-half of the solute has eluted in the retention volume V_R, and half remains in either the volume of the mobile phase, or the volume of the stationary liquid phase, V_S. Thus,

$$V_R C_M = V_M C_M + V_S C_S \tag{15-6}$$

Rearranging and inserting the partition coefficient, we obtain a fundamental equation in chromatography:

$$V_R = V_M + KV_S \quad \text{or} \quad V_R - V_M = KV_S \tag{15-7}$$

It relates the retention volume of a component to the column dead volume, and the product of the partition coefficient and the volume of the stationary phase. This equation is correct for liquid partition columns, but for adsorption columns, V_S should be replaced by A_S, the surface area of the adsorbent.

---

*For a derivation of the pressure-gradient correction factor, see W. E. Harris and H. W. Habgood, *Programmed Temperature Gas Chromatography*, Wiley, New York, 1966, p. 49.

Partition Ratio

The solute partition ratio, k', is the most important quantity in column chromatography in that it relates the equilibrium distribution of the sample within the column to the thermodynamic properties of the column and to the temperature, as will be shown later. For a given set of operating parameters, k' is a measure of the time spent in the stationary phase relative to the time spent in the mobile phase. Also called the capacity factor, column capacity ratio, and mass distribution ratio, k' is defined as the ratio of the total amount of a solute in the stationary phase to the amount in the mobile phase at equilibrium:

$$k' = \frac{C_S V_S}{C_M V_M} = K \frac{V_S}{V_M} \tag{15-8}$$

Here the volumetric phase ratio, V_M/V_S, is often denoted by the symbol β; thus, $k' = K/\beta$. Stated another way, the partition ratio is the additional time (or volume) a solute band takes to elute over an unretained solute (for which $k' = 0$), divided by the elution time (or volume) of an unretained band:

$$k' = \frac{t_R - t_M}{t_M} = \frac{V_R - V_M}{V_M} \tag{15-9}$$

Rearranging Eq. 15-9, retention times are related to k' by the relation

$$t_R = t_M (1 + k') = \frac{L}{\bar{u}} (1 + k') \tag{15-10}$$

Values of k' higher than 8 waste valuable analytical time. Conversely, k' values less than unity are unfavorable due to potential interferences from nonretained peaks and early peaks perhaps of little or no analytical interest.

The fraction of time that a sample molecule spends in a particular phase will be very close to the fraction of all molecules of that solute which are instantaneously in the same phase. Thus, the average fraction of time spent by a solute molecule in the mobile phase will be

$$\frac{C_M V_M}{C_M V_M + C_S V_S} = \frac{1}{1 + k'} \tag{15-11}$$

Similarly, for the stationary phase

$$\frac{C_S V_S}{C_M V_M + C_S V_S} = \frac{k'}{1 + k'} \tag{15-12}$$

Equation 15-11 is sometimes denoted by **R**, called the retention ratio or retardation factor. It is the fraction of time spent by the solute in the mobile phase, that is,

$$\mathbf{R} = \frac{t_M}{t_M + t_S} = \frac{V_M}{V_R} = \frac{V_M}{V_M + K V_S} = \frac{\bar{u}}{t_R} \tag{15-13}$$

Relative Retention

The relative retention, α, of two solutes, where solute 1 precedes solute 2, is given by

$$\alpha = \frac{k'_2}{k'_1} = \frac{K_2}{K_1} = \frac{V'_{R,2}}{V'_{R,1}} = \frac{t'_{R,2}}{t'_{R,1}} = \frac{V^0_{g,2}}{V^0_{g,1}} \tag{15-14}$$

Relative retention depends upon two conditions: the nature of the two phases and the column temperature. The stationary phase is the most important selection and one should always attempt to choose as selective a phase as possible for the pair of solutes most difficult to separate. For a separation to be possible, the relative retention clearly must exceed unity. The useful ranges are from 1.05 to 2.0, with higher values wasting analytical time.

15.4 COLUMN EFFICIENCY AND RESOLUTION

Under operating conditions where the sorption isotherm is linear, that is, K and k' are independent of total solution concentration, and the solute has gone through the equivalent of 50 exchanges between the stationary and mobile phases, the resultant profile of a separated solute band closely approaches that given by a smooth Gaussian distribution curve. However, as solute zones pass through the column, they broaden and the concentration at peak maximum gradually falls. This broadening is important as it can ultimately affect resolution. The solute zones should be kept as narrow as possible by proper design of the experimental system.

Plate Height and Plate Number

The quantity that measures the column efficiency and is related to the peak width is called the plate height, sometimes called height equivalent of a theoretical plate, which has the dimension of distance. Plate height, H, has become the prime measure of peak dispersion in chromatography. It is determined for any solute from the elution record, as shown in Fig. 15-1, by applying Eq. 15-15:

$$H = \frac{L}{16} \left(\frac{W_b}{t'_R}\right)^2 \tag{15-15}$$

where L is the column length, W_b is the peak width at the base (the intersections of tangents to the inflection points with the base line), equal to 4σ in time units, and t'_R is the retention time corrected for transit time of an unabsorbed or nonretained solute. Often it is easier to measure the width at half the peak height (full width at half maximum), then

$$H = \frac{L}{5.54} \left(\frac{W_{1/2}}{t'_R}\right)^2 \tag{15-16}$$

The efficiency of a column can also be stated as a dimensionless quantity called the effective plate number, N_{eff}:

$$N_{\text{eff}} = \frac{L}{H} = 5.54 \left(\frac{t'_R}{W_{1/2}}\right)^2 \tag{15-17}$$

Column efficiency is often stated as the number of theoretical plates, with no correction for t_M made; then

$$n_{\text{theor}} = 5.54 \left(\frac{t_R}{W_{1/2}}\right)^2 \tag{15-18}$$

where N and n are related by

$$N = n\left(\frac{k'}{1+k'}\right)^2 \qquad (15\text{-}19)$$

The true separating power of the column is more accurately measured by the effective plate number, especially when comparing capillary with packed columns.

Plate height, rather than plate number, is a more meaningful measure of the column efficiency since it is independent (to a first approximation) of the column length. More importantly, from a theoretical point of view, plate height can be directly related to the experimental conditions and parameters. Any improvement in the compactness of peaks in chromatography is equivalent to a reduction in the plate height.

Peak Asymmetry

If k' is higher at lower concentrations, a situation often prevailing in adsorption chromatography, both gas and liquid, then the low concentration wings of bands will move more slowly than the high concentration parts, and as an initially symmetric band moves down the column it will become skewed and eventually tailed. The result is a band with a sharp front and a long, more or less exponential tail. When tailing is caused by nonlinearity of the partition isotherm, no amount of improvement in the kinetics of the process will improve the peak shape. Indeed, improvement of the kinetics will actually accentuate the skewness of the peak, a fact that may be used to diagnose whether skewness is due to thermodynamic or kinetic factors.

Asymmetrical peaks can also result from extra column effects, particularly injection problems. It can also arise from a poorly packed column. In fact, peak symmetry can be used as a criterion of column performance. A very simple comparison of the back half-width, b, of the peak at 10% of the peak height, divided by the corresponding front half-width, a, gives the asymmetry factor: $AF = b/a$. This is illustrated in Fig. 15-1. With a well-packed column an asymmetry factor between 0.90 and 1.10 should be achievable with standard solutes. A maximum of 1.3 is reasonable for large-scale commercial column production.

Resolution

As the solute zones migrate through a chromatographic column they always broaden. Resolution of components into discrete bands will occur only if the bands widen to a lesser extent than their maxima separate. In well-behaved chromatography where the peaks are Gaussian in shape, the resolution of two adjacent components is defined as the ratio of the peak separation to the mean peak width, as shown in Fig. 15-2, and expressed in Eq. 15-20:

$$R = \frac{t_{R,2} - t_{R,1}}{0.5\,(W_1 + W_2)} \qquad (15\text{-}20)$$

The resolution of adjacent peaks, if inadequate, can be improved in two independent ways: by increasing peak separation and by decreasing peak width. These two approaches

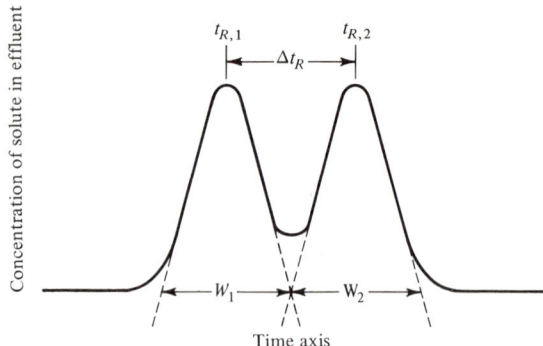

FIGURE 15-2 Definition of resolution.

to improving resolution are associated, respectively, with altering the thermodynamics and improving the kinetics of the chromatographic system.

The criterion for two peaks to be resolved will be arbitrary, but if the areas to be measured are to give reasonable quantitative accuracy, the peak maxima of the two components must be at least 4σ, that is, W_b or $2W_{1/2}$, apart. If so, then $R = 1.0$, which corresponds approximately to a 3% overlap of peak areas. A value of $R = 1.5$ represents essentially base line resolution (6σ) with only 0.2% overlap of peak areas. These criteria pertain to roughly equal solution concentrations. Increased resolution may be needed when a band from a major component is adjacent to a band of a minor constituent.

It is important to note that columns may possess adequate selectivity but exhibit poor efficiency, as depicted in Fig. 15-3 (*upper chromatogram*). Clearly, values of relative retention for the upper chromatogram are acceptable but either the column suffers from low efficiency or the mobile phase composition is inappropriate. On the other hand, the lower chromatogram exhibits excellent efficiency but poor selectivity; k' is too small.

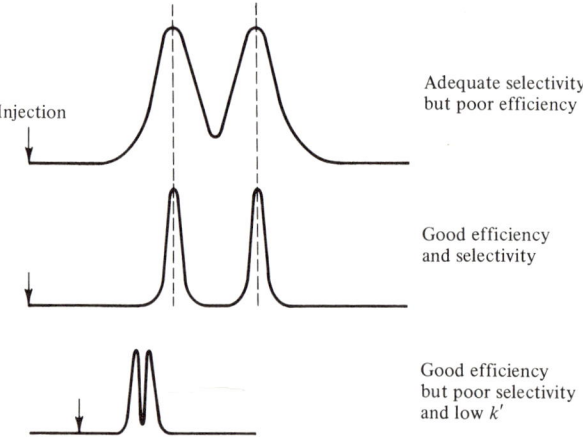

FIGURE 15-3 Selectivity, efficiency, and partition ratio for columns.

If approximately equal quantities of two solutes in adjacent bands are assumed, the resolution equation can be approximated in a more useful form as

$$R = \left(\frac{1}{4}\right)\left(\frac{\alpha - 1}{\alpha}\right)\left(\frac{k'}{1 + k'}\right)\left(\frac{L}{H}\right)^{1/2} \tag{15-21}$$

This version of the resolution equation can be divided into three parts: (1) a selectivity term dependent upon α, (2) a rate of migration term depending upon k' (here k' is taken variously as k'_2 or $\bar{k}'$), and (3) an efficiency term depending upon L and H (or theoretical plate number). Each term can be calculated directly from the recorded chromatogram. The first two factors are essentially thermodynamic, whereas the L/H term is mainly associated with the kinetic features of chromatography.

The selectivity term depends solely on the molecular forces between the solute and the two phases. If, after selection of the stationary phase, the separation is still a problem, then the relative retention is varied by changing the mobile phase composition in LLC, LSC, and IEC. Once the mobile phase, the nature of the stationary phase, and the temperature are chosen, the relative retention is fixed and cannot be changed. Because the first term in Eq. 15-21 is very sensitive to changes in the value of α, as shown in Table 15-1, it is desirable to select values of α within the range from about 1.05 to 2.0. An increase in α from 1.05 to 1.10, for example, improves resolution by a factor of 4 for the same L/H. Increases in resolution due to selectivity changes are most desirable since the necessity of higher column inlet pressures and longer analysis times will be avoided, as shown later.

When using the resolution equation to optimize a given separation, the k' term should be considered first. This term depends on the intermolecular interactions of the system as well as on the amount of the stationary phase or adsorbent surface area. As k' increases for a given column system, peaks become broader and it becomes more difficult to quantitate the more dilute peaks. In practice, the column loading of the stationary phase and the partition coefficient should be chosen such that the first eluted component of a given pair has a retention time that is greater than twice the passage time of a nonretained solute; that is, $t_R = 2t_M$. Maximum resolution in unit time will be obtained at $k' = 2$; the optimum

TABLE 15-1 Values Related to the Relative Retention

α	$\left(\dfrac{\alpha}{\alpha - 1}\right)^2$	N_{req} for $R = 1.5$ and $k' = 2$	L_{req}, meters for $H = 0.6$ mm
1.01	10,201	826,281	495
1.02	2601	210,681	126
1.03	1177	95,377	52
1.04	676	54,756	33
1.05	441	35,721	21
1.10	121	9801	5.8
1.15	58	4418	2.6
1.20	36	2916	1.7
1.25	25	2025	1.2
1.30	19	1514	1.0

range extends from 1 to 5. Although the resolution of two or three solutes of particular interest can usually be improved by adjustment of the composition of the mobile and stationary phases, this is not possible for a complex mixture of say 10 or 20 components, all of which are important. An effective solution to this problem is to change the band migration rates during the course of a separation by "k' programming." There are a number of ways to achieve this objective; they include temperature programming, primarily used in GLC; stepwise or gradient elution; flow programming; and coupled columns. Each technique will be discussed in subsequent chapters.

Finally, the L/H term is chosen to provide the maximum efficiency compatible with reasonable analysis time. Resolution may be improved by increasing the column length. However, higher inlet pressures may be required to maintain specified retention times. Longer analysis times are inevitable if the inlet pressure remains unchanged in LLC and packed GLC columns. And resolution only improves as the square root of the column length. An alternative method to increase resolution is by decreasing the plate height through improvement in the kinetic features of the column, perhaps by lowering the flow-rate of the mobile phase, but not lower than the minimum in the $H/\bar{u}$ graph, as discussed later. This allows for more efficient mass transfer of the sample during the sorption–desorption process, and results in modest increases in efficiency and hence small increases in resolution.

15.5 COLUMN PROCESSES AND BAND BROADENING

The various column processes that contribute to the peak variance, σ^2, and thereby broaden the solute band, will now be discussed. Band spreading theories in liquid and gas chromatography are nearly identical. The plate height is a function of thermodynamic and kinetic processes that take place in the column: transverse and longitudinal diffusion in the mobile phase, finite rate of equilibration of solute between the stationary and mobile phases (mass transfer), diffusion in the liquid stationary phase, and flow irregularities leading to convective mixing. Plate height is an effective way to express in simple terms the extent of band broadening, here stated in an abbreviated manner after the fashion of the van Deemter equation for GLC (see also Fig. 15-4):

$$H = A + B/\bar{u} + C_{\text{stationary}}\bar{u} + C_{\text{mobile}}\bar{u} \tag{15-22}$$

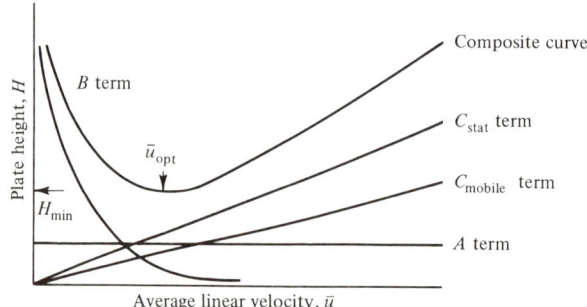

FIGURE 15-4 Typical $H/\bar{u}$ (van Deemter) curve for a gas chromatographic column.

The average linear velocity is used rather than the more easily measured flowrate. The velocity can be directly related to the speed of analysis, whereas the flowrate depends on the cross section of the column and the volume of the column occupied by the packing material. Experimentally the average linear velocity is determined by injecting a nonretained solute on the column being tested and measuring the time taken for it to pass through the column. Knowing the column length, $\bar{u} = L/t_M$.

Eddy Diffusion

The A-term, called "eddy" diffusion, in the van Deemter equation, results from the inhomogeneity of flow velocities and path lengths around the packing particles. $A = \lambda d_p$, where d_p is the particle diameter, and λ is an unspecified constant which is a function of the packing uniformity and the column geometry. Flow paths of unequal length must exist through any less than perfect packing. Some solute molecules of a single species will find themselves swept through the column close to the column wall where the density of packing is comparatively low, especially in small diameter columns, while other solute molecules will pass through the more tightly packed center of the column at a correspondingly lower velocity. In consequence, molecules following an easy path will elute ahead of those following a series of shorter and erratic streamlines, leading to a broadening of the elution band.

To minimize the A-term, the mean diameter of the particles constituting the chromatographic bed should be as small as possible consistent with obtaining a uniformly packed bed and with the pressure drop required to force the mobile phase through the packed bed. As the particle size decreases, the permeability of the column decreases, and higher inlet pressures are needed to drive the mobile phase. However, because of the higher efficiency, the column length can be decreased, thus decreasing the needed pressure drop. In GLC for open tubular columns the A-term is virtually zero.

Longitudinal Diffusion

The B-term in the van Deemter equation defines the effect of longitudinal, or axial, diffusion, that is, random molecular motion. It is expressed by $2\gamma D_M$, where γ is an obstructive (tortuosity) factor, which recognizes that longitudinal diffusion is hindered by the packing or bed structure, and D_M is the solute diffusion coefficient in the mobile phase. In coated capillary columns γ is unity, and in packed columns it has a value of about 0.6. High diffusion rates in the mobile phase cause solute bands to disperse axially along the column, particularly at low mobile phase velocities. This leads to peak broadening.

The contribution of longitudinal diffusion to the plate height is significant only at low mobile phase velocities. In LCC the ratio D_M/T_M should be used where T_M, the interparticle tortuosity factor, corrects the diffusion coefficient for the varying size and direction of the interstitial pores in the packing.

Mass Transfer

The $C_{\text{stationary}}$-term in the van Deemter equation results from resistance to mass transfer at the solute/stationary phase interface. It is proportional to d_f^2/D_S where d_f is the effective thickness of the stationary phase and D_S is the diffusion coefficient of the solute in the stationary phase. Slow mass transfer in the stationary phase means longer time spent in the phase for a solute molecule while perhaps other solutes are moving forward with the mobile phase. The faster the mobile phase moves through the column and the slower the rate of mass transfer, the broader will be the solute band which eventually elutes from the column. The rate of mass transfer can be improved by reducing the film thickness of the stationary phase, albeit with consequent loss of solute capacity, and so reducing the distance that a solute molecule must diffuse within the stationary phase. Whenever possible, nonviscous liquids should be chosen for the stationary phase so that D_S will not be unduly small.

The C_{mobile}-term represents radial mass transfer resistance. It is proportional to the square of the particle diameter of the packing, d_p^2, and inversely proportional to the diffusion coefficient, D_M, of the solute in the mobile phase.

In LCC, as contrasted with GLC, major differences will arise from the 10^4 decrease in the value of D_M and from the presence of stagnant pockets of mobile liquid phase trapped within the stationary phase. Molecules of solute will have time to diffuse randomly through any stagnant pools and will thereby be held back relative to the main band of sample. The velocity of the mobile phase will differ from point to point, owing to the perturbation caused by the support particles. Liquid streamlines near the particle boundaries move slowly, whereas streamlines near the center between particles move relatively more rapidly. Transfer of solute molecules by lateral diffusion to a different streamline is constantly occurring. Hence, the obstructive path of solute molecules is due both to diffusion from streamline to streamline, and to having to circumvent support particles. This molecular diffusion, coupled with uneven path lines (multipath effect), gives rise to a convective mixing, or coupled, term:

$$\frac{A}{1 + C_{\text{mobile}}/\bar{u}^{1/2}}$$

To minimize the effect of stagnant pools, the internal pore structure can be made impervious (that is, pellicular packings with a solid core), the overall diameter of the packing particles can be reduced, or supports can be chosen that possess very wide pores so that liquid flows easily in and out, or even through the pores. The mobile phase should possess a low viscosity so that diffusion of solutes in it is rapid.

The abbreviated equation for plate height in LCC includes four terms:

$$H = B/\bar{u} + \frac{A}{1 + C_{\text{mobile}}/\bar{u}^{1/2}} + C_{\text{mobile}} \bar{u}^{1/2} + C_{\text{stationary}} \bar{u} \qquad (15\text{-}23)$$

| longitudinal or axial diffusion | convective mixing | resistance to mass transfer in mobile phase | resistance to mass transfer in stationary phase |

These terms are shown graphically in Fig. 15-5.

For $H/\bar{u}$ curves from either Eq. 15-22 or 15-23, the limiting slope at high mobile phase velocities is due to the third and fourth terms. At very low velocities the plate height is

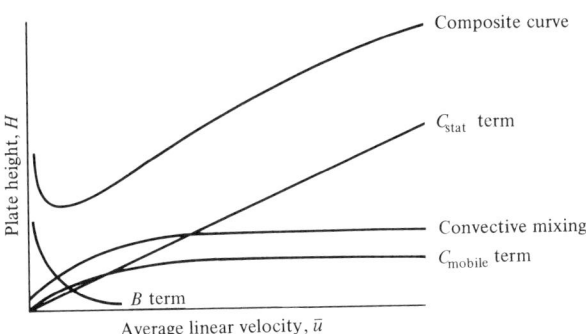

FIGURE 15-5 Typical $H/\bar{u}$ curve for a liquid chromatographic column.

controlled by the first term in the equation. Actually the region about the minimum in the $H/\bar{u}$ curves is quite broad for the usual GLC or LCC column, so that column efficiency may be traded for analysis speed.

15.6 REDUCED VARIABLES

The plate height/velocity equation may be simplified by casting it into a dimensionless or reduced form. Instead of measuring the plate height and velocity absolutely, they may be measured relative to the particle diameter of the packing and to the characteristic diffusion rate over a particle, respectively. Thus the reduced plate height, h, is

$$h = H/d_p \tag{15-24}$$

which simply states the number of particle diameters that constitute one plate. The reduced velocity, v, is given by

$$v = \bar{u}d_p/D_M = Ld_p/t_M D_M \tag{15-25}$$

which can be considered as the ratio of time required to displace solute molecules a distance equal to one particle diameter, to the time needed for the same displacement by molecular diffusion in the mobile phase. In other words the reduced velocity expresses the balance of mass transport by bulk flow and by diffusion or molecular motion across a single particle. The flow causes band dispersion, while the diffusion tends to limit this dispersive effect. Reduced plate height and reduced velocity are important parameters which permit comparison of columns with different packing sizes and mobile phases. One might also wish to use a reduced column length, ℓ, which is simply

$$\ell = L/d_p \tag{15-26}$$

In GLC it is usual to work at average linear gas velocities of 10 cm sec^{-1} and with particle diameters of about 200 μm. Solute molecules have D_M values around 0.1 cm^2 sec^{-1}. Consequently, these conditions give a reduced velocity of about unity and a reduced plate height close to two. By contrast, in LCC it is usual to work at average linear fluid velocities around 1 cm sec^{-1}, and with mobile phases in which solute molecules have

D_M values around 10^{-5} cm^2 sec^{-1}. Since particle diameters are commonly around 5–10 μm, the flow velocity should be chosen to give a reduced velocity in the range 3–10. For a "good" column in LCC the reduced plate height should be in the range 2–3.

15.7 TIME OF ANALYSIS AND RESOLUTION

Previous discussions have hinted at an important parameter of concern to the analyst, namely, the retention time required to perform a separation. This is the time needed to get the solute band through one plate, t_p, multiplied by the number of plates required, N_{req}, for the desired resolution. Thus, $t_R = N_{req} t_p$. Now t_p is given by the plate height divided by the band (or zone) velocity, $\bar{u}/(1 + k')$, where $1/(1 + k')$ is the fraction of time spent by the solute molecule in the mobile phase. Thus,

$$t_R = N_{req}(1 + k')(H/\bar{u}) \qquad (15\text{-}27)$$

Eliminating N_{req} between Eqs. 15-27 and 15-21, gives

$$t_R = 16R^2 \left(\frac{\alpha}{\alpha - 1}\right)^2 \left(\frac{(1 + k')^3}{(k')^2}\right) \left(\frac{H}{\bar{u}}\right) \qquad (15\text{-}28)$$

Differentiation of Eq. 15-28 with respect to k', placing all other variables into one constant, gives

$$\frac{dt_R}{dk'} = C\left[\frac{(k')^3 - 3k' - 2}{(k')^3}\right] \qquad (15\text{-}29)$$

Now t_R is a minimum when $k' = 2$, which is when $t_R = 3 t_M$. However, there is little increase in analysis time on varying k' between 1 and 5, provided k' has no effect on the other variables. Although a twofold increase in mobile phase velocity might be expected to halve analysis time, this is not strictly correct as H would increase. The ratio $H/\bar{u}$ can be obtained directly from the experimental plate height/velocity graph; it is the slope of the line drawn from the origin to a point on the graph. This slope decreases as the velocity increases although a point of diminishing return is reached at high velocities. Nevertheless, it is possible to use higher velocities, with a corresponding increase in column length, at least until the inlet pressure requirement becomes excessive. Although a longer column is needed, the analysis time will be less. These aspects will be considered in more detail in the subsequent chapters on GLC and LCC.

In view of the interrelationship of separation time and resolution, some chromatographers prefer to use plates per second to describe the performance of a chromatographic system. In GLC packed columns, N/t_R usually falls between 5 and 30 plates/sec, while for open tubular columns it varies from 50 to 100. In LCC it is possible to generate 25 or more plates per second.

Peak Capacity

While a precise determination of resolvable peaks will depend on the nature of solutes existing in a particular mixture, one can define and estimate a "peak capacity" which approximates the maximum number of peaks ideally spaced apart that can be separated on a given column. As we have seen, there is a lower limit to the retention volume, this being the column-free volume. However, except in exclusion chromatography, the upper limit is indefinite. Nevertheless, because the peaks from all chromatographic columns have a finite width as dictated by plate height, only a limited number of such peaks can crowd into the range accessible to them.

Following Giddings' treatment,[3] let us assume that the column possesses a fixed number of plates, equal for each solute. Thus, if the retention volume of the ith peak is $V_{R,i}$, its width ($W_b = 4\sigma = m\sigma$) in volume units is $4V_{R,i}/N^{1/2} = aV_{R,i}$, where a is a column constant, $m/N^{1/2}$. The next peak, with corresponding retention volume, $V_{R,i+1}$, and width, W_{i+1}, can therefore be no closer to the ith peak than the mean of the two widths, or $(a/2)(V_{R,i} + V_{R,i+1})$, if a minimum $m\sigma$ separation is required. Therefore, at closest spacing

$$V_{R,i+1} - V_{R,i} = (a/2)(V_{R,i} + V_{R,i+1}) \tag{15-30}$$

The ratio of retention volumes for adjacent peaks is

$$\frac{V_{R,i+1}}{V_{R,i}} = \frac{1 + a/2}{1 - a/2} \tag{15-31}$$

If the first peak elutes at volume V_1, the minimum elution volume of the nth and final peak, by comparison to V_1, will be given by an $(n-1)$-fold compounding of the ratio in Eq. 15-31. When logarithms are employed, the following expression for n results:

$$n = 1 + \frac{\ln V_n/V_1}{\ln[(1 + a/2)/(1 - a/2)]} \tag{15-32}$$

which gives the maximum number of peaks, spaced ideally, which can be resolved between volumes V_1 and V_n. In most practical situations, the plate number is sufficiently large and a sufficiently small, that the term in the denominator is approximately $\ln(1 + a)$. Furthermore, $\ln(1 + a) \simeq a$ when a is small. Thus, to a good approximation, Eq. 15-32 becomes

$$n = 1 + (1/a)\ln(V_n/V_1) = 1 + (N^{1/2}/m)\ln(V_n/V_1) \tag{15-33}$$

Additional peaks can be eluted in GC and LC, except exclusion, until time becomes excessive or until their dilution hinders proper detection. In GC it is practical to work with peaks whose retention volumes are 50 or so times that of the unretained peak. However, V_n/V_1 is often limited to 10, as would be reasonable in many gas and most liquid chromatographic systems. Table 15-2 shows some peak capacities for a given number of theoretical plates.

TABLE 15-2 Peak Capacities of Chromatographic Systems for a Given Number of Theoretical Plates

Theoretical Plates, N	Gas Chromatography $V_n/V_1 = 50$	Liquid Chromatography $V_n/V_1 = 10$	Exclusion Chromatography $V_n/V_1 = 2.3$
100	11	7	3
400	21	13	5
1000	33	20	7
2500	51	31	11
5000	69	41	14

15.8 QUANTITATIVE ANALYSIS

Detectors in chromatography are generally operated differentially, responding to either the concentration of solute or the mass flowrate. Those responding to the concentration yield a signal which is proportional to the solute concentration which traverses the detector. An elution peak results when the signal is plotted against time. For such detectors, the area under the peak is proportional to the mass of a component and inversely proportional to the flowrate of the mobile phase. It is crucially important that flow of mobile phase be kept constant for such detectors if quantitative analysis is to be carried out. In differential detectors that respond to mass flowrate, the peak area is directly proportional to the total mass and there is no dependency on flowrate of the mobile phase.

Peak Area Integration

In column chromatography the analog signal generated by the detector is graphically recorded in the form of the familiar chromatographic peaks. The area under these peaks can then be integrated in a variety of ways and the resulting data related to the composition of the samples being studied. Sometimes merely the peak height alone is measured.

Valuable discussions dealing with quantitative analysis will be found in the December 1967 issue of the *Journal of Gas Chromatography;* the entire issue was devoted to the subject. Errors in manual integration techniques are well handled in the paper by Ball, Harris, and Habgood.[4] Their conclusions: precision depends more on peak shape than on the particular manual method used. The relative error is large for extreme shapes.

Height Times Width at Half-Height Probably the most commonly used method for measuring chromatograms manually involves multiplying the actual peak height times the width at half-height. Operations involved are: drawing the base line of the peak, measuring the height from this base line, positioning the measuring scale parallel to the base line at one-half the height, and measuring the width of the peak at this position. The normal (zero signal) base line is not used because large deviations may be caused by tailing. The major factor affecting the precision of this method is the accuracy of measuring the peak width, particularly of narrow peaks.

Peak Height Measurement of peak height is inherently simple. The only operations involved in this type of measurement are drawing the base line and measuring the height. The precision is better than in measuring peak area, particularly of narrow peaks. Peak heights, however, are much more sensitive to determinate errors in chromatography, such as small changes in the technique of sample injection or in operating conditions. Peak height does not always remain directly proportional to sample size. As the latter increases there is a point at which the peak begins to broaden and no longer increases in height at the same rate.

Planimetry The peak is traced manually with a planimeter, a mechanical device which measures area by tracing the perimeter of the peak. The area is represented digitally on a dial. Indeterminate errors arise from placing the base line, tracing the peak outline, and

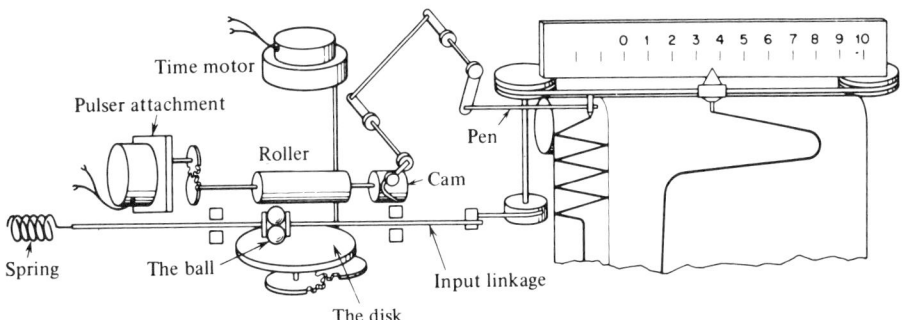

FIGURE 15-6 Ball-and-disk integrator *(schematic)*.

obtaining a reading. Planimetry is less precise than height–width integration for peaks of small area, but is comparable or somewhat better in precision for large or irregular areas.

Triangulation A variation of the height–width method of integration is the triangulation method in which tangents are drawn to the sides of the peak at the inflection points and the area of the triangle formed by these tangents and the base line is determined. The height is measured from the base line to the point where the tangents intersect. Uncertainties include the skill employed in constructing the traingle, the sharpness of the pencil used, and the shape of the peak. Triangulation has little to recommend it.

Ball-and-Disk Integrator The ball-and-disk integrator, shown schematically in Fig. 15-6, is an automatic mechanical type of integrator. A ball positioned on a rotating flat disk will rotate at a speed proportional to its distance from the center of rotation. The ball is positioned on the disk at a distance from the center in the same relationship as the position of the recorder pen to the base line of the chromatogram. If the disk is rotated at a constant speed (time), then the ball will rotate at a speed proportional to the position of the recorder pen from zero (provided the recorder zero and the center of the rotating disk have been exactly aligned). This speed is then transmitted to a roller through a second ball, which, by means of a "spiral in" and "spiral out" cam, actuates the integrator pen at a speed directly proportional to the position of the recorder pen. The drive between the disk and the ball is by traction through an oil film. Although this hydrostatic phenomenon is not clearly understood, the oil film acts similarly to an induction motor where slip is proportional to the driven load.

Reading the integrator trace is as follows (also refer to Fig. 15-7): First establish the desired chart time interval from the recorder pen trace of the chromatogram and project directly down to the integrator trace *(arrows)*. The value of an interval is obtained by counting the chart graduations crossed by the integrator trace. A full stroke of the "sawtooth" pattern in either direction represents 100 counts. Every horizontal division has a value of 10. Values less than 10 are estimated. In the example the interval for the main peak is 1083 counts. With care the pattern can be read within two counts. On some models the space between "blips" projecting slightly above the uppermost horizontal line is equivalent to 600 counts, making it possible to record up to 24,000 counts/in. of chart. The

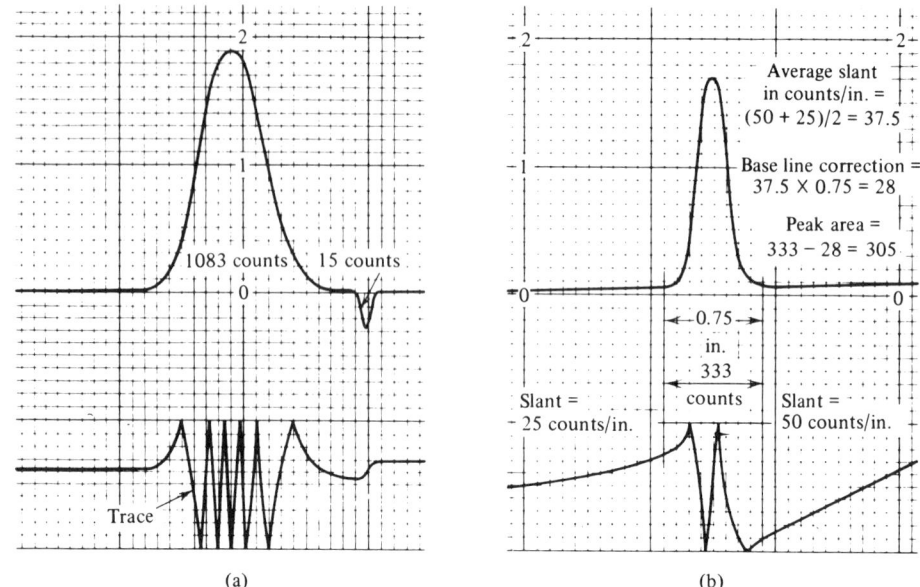

FIGURE 15-7 Estimation of peak areas with ball-and-disk integrator. (b) Method for handling base line correction.

pattern to the right in the illustration gives the method for estimating the base line correction when the peak base line does not coincide with the recorder base line.

Electronic Digital Integrator In this method the chromatographic input signal is fed into a voltage-to-frequency converter, which generates an output pulse rate proportional to the peak area. When the slope detector senses a peak, the pulses from the V-f converter are accumulated and counted. The key requirements for a digital integrator are wide linear range, high count rate/count capacity, and sensitive versatile peak detection logic circuits. Standard deviation at about 0.44% is the smallest of all methods discussed. A comparison of the precision of integration methods for a hydrocarbon mixture is shown in Table 15-3.

Computing Integrator Since 1967 on-line computer-based data systems have provided complete automation: automatic acquisition and reduction of data, storage of calculation methods and printout of analytical results (see Chapter 29). Initially the analog chromatographic signal is digitized by a hardware analog-to-digital converter. The software can then detect the presence of peaks, correct for base line drift, calculate areas and retention times, determine concentrations of components using stored calibration factors and gener-

TABLE 15-3 Comparison of Precision of Integration Methods

	Planimeter		Triangulation		$H \times W \times \frac{1}{2} H$		Cut & Weigh		Disk		Digital	
	Avg.	σ rel	Avg.	σ rel	Avg.	σ rel	Avg.	σ rel	Avg.	σ rel	Avg.	σ rel
Propane	0.04	20%	0.05	14%	0.04	12%	0.005	14%	0.04	18%	0.03	10%
i-Butane	4.84	1.65%	4.79	8.77%	4.52	3.76%	5.04	1.98%	4.57	0.44%	4.56	0.88%
n-Butane	14.01	5.64%	13.70	4.53%	13.56	1.62%	14.99	2.80%	14.05	1.00%	14.06	0.36%
Butene-1	18.52	3.29%	18.52	4.75%	18.66	3.05%	18.40	1.22%	18.63	2.46%	18.63	0.27%
i-Butene	8.12	5.67%	8.55	3.74%	8.56	2.22%	8.01	2.12%	8.07	2.60%	8.24	0.49%
t-Butene-2	20.18	6.49%	20.32	3.25%	20.07	2.04%	20.09	1.39%	20.21	1.14%	20.16	0.20%
cis-Butene-2	16.11	3.66%	15.83	1.58%	16.41	3.71%	15.85	1.45%	16.06	0.68%	15.95	0.38%
1,3-Butadiene	18.19	2.03%	18.25	1.81%	18.17	1.65%	17.58	1.19%	18.28	0.71%	18.27	0.49%
Time/Trace, min.	45–60		45–60		50–60		100–120		15–30		5–10	
Precision* σ rel	4.06%		4.06%		2.58%		1.74%		1.29%		0.44%	

*Excludes propane peak.
Courtesy of Spectra-Physics.

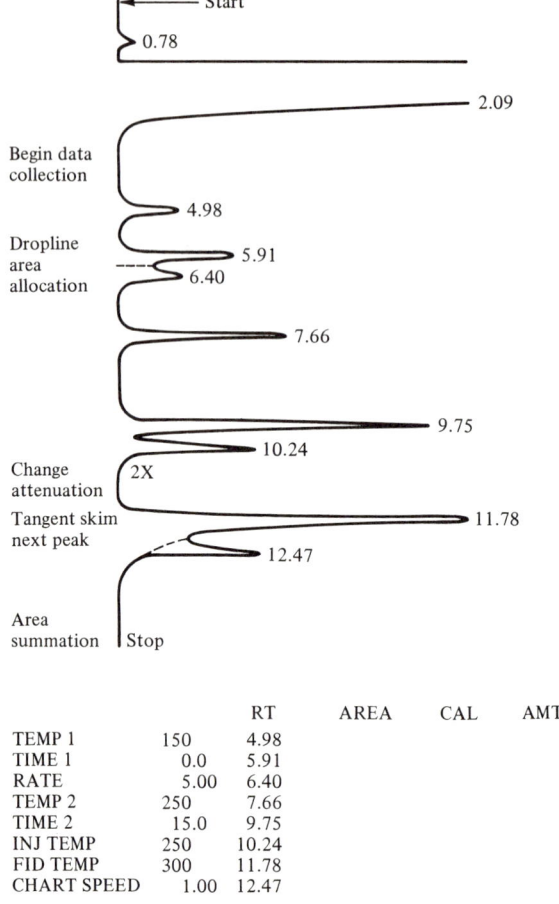

FIGURE 15-8 Partial printout of a chromatographic run and indicated computations that can be made by computer software programs.

ate a complete report of the analysis (Fig. 15-8). Peak area counts are accumulated when the signal leaves the base line. This departure from base line is usually determined by monitoring the slope of the signal. The retention time and signal heights of each peak maximum detected by the program are stored in memory. The termination of a component peak is established when the signal returns to base line. During isothermal runs the software can automatically increase slope sensitivity with time, thus ensuring the program's ability to detect both initial, sharp peaks and later low, flat peaks with equal precision. In the case of fused peaks, areas can be allocated to each component by dropping perpendiculars from valley points to the corrected base line (Fig. 15-9). A special algorithm, known as a tangent skim, is used to calculate the areas of isolated peaks appearing on the tail of a larger peak.

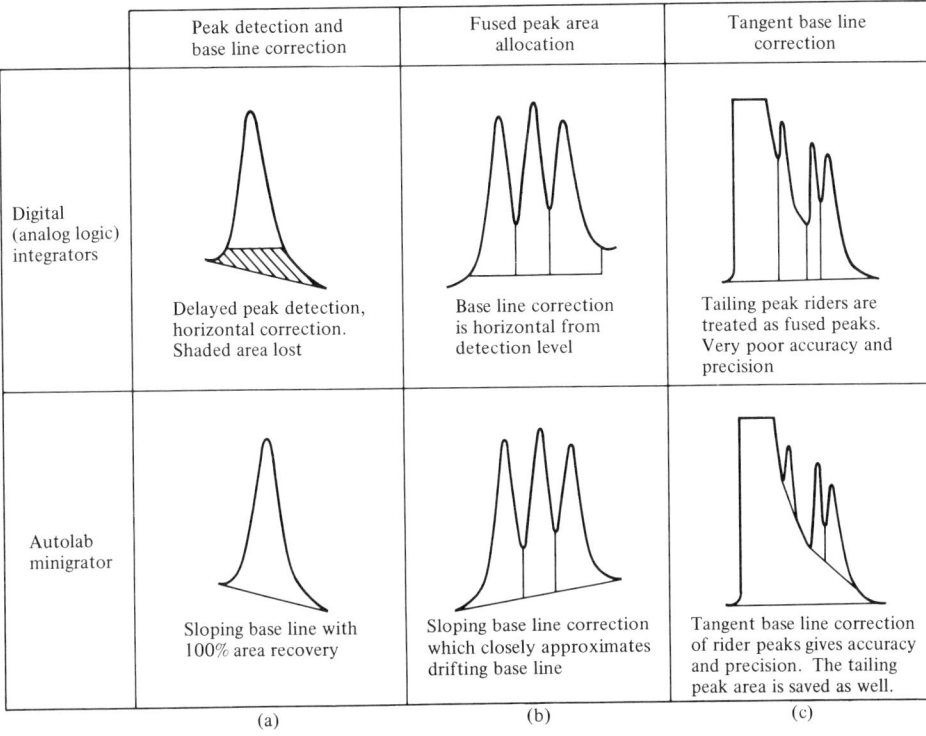

FIGURE 15-9 Computing integrator capabilities. (Courtesy of Spectra-Physics.)

Evaluation Methods

The three principal evaluation methods are (1) calibration by standards, (2) area normalization, and (3) internal standard. Each has its place, depending on the nature of the analysis.

Calibration by Standards When sample volume is known, calibration by standards is often used. It has the advantage that only the area of the peaks of interest need to be measured, but it does require that the same amount of sample be injected each time, and the necessary calibration standard(s) should be run under the same instrument operating conditions as the sample. The percent concentrations are obtained by ratioing the volume of each component of interest to the sample size. In practice, standard solutions of the component(s) of interest are prepared and injected into the chromatograph. Then an unknown $X = (\text{area})_X K$, where K is the proportionality constant (slope of the calibration curve), or the results may be interpreted graphically.

Relative response factors must be taken into account when converting area to volume, and when the response of a given detector differs for each molecular type or class of compounds. Response factors are best obtained by analyzing standard samples.

Area Normalization When it is known that the chromatogram represents the entire sample, that all components have been separated, and that each peak has been completely resolved, area normalization may be used for evaluation. To use this method, the area of each individual peak is measured, then divided by its response factor to give the peak's calculated area. Adding these together provides the total calculated area. The percent by volume for individual components is obtained by multiplying the individual calculated area by 100, and then dividing by the total calculated area. This method is particularly useful when analyzing complex mixtures.

Internal Standard The internal standard method permits the operating conditions to vary from sample to sample, and does not require repeatable sample injection. The internal standard has to be a component that can be completely resolved, is not present in the unknown mixture, and does not have any interference effects. A known quantity of this standard is chromatographed, and the area versus concentration plotted. A known amount of the standard is then added to the unknown mixture. Any variation in sample size will be immediately apparent by comparing the peak areas of the internal standard in different runs. A correction factor can then be used when determining the exact concentration of the other components.

Example 15-1

Assume there is 1.00 g of mixture to which is added 100-mg internal standard. Measurement of the resultant chromatogram shows four components (including the internal standard) with areas (in arbitrary units) as follows: $A_1 = 27$; $A_{std} = 80$; $A_2 = 20$; $A_3 = 70$; and area sum = 197. The amount of component 3 present in the sample is

$$W_3 = W_{std}(A_3/A_{std}) = 100 \text{ mg } (70/80) = 87.5 \text{ mg}$$

Percent component 3: $(0.0875 \text{ g}/1.000 \text{ g}) \, 100 = 8.75\%$

Note that component 3 represents less than 9% of the total sample, yet in peak area it appeared to be a major component. This indicates that a large part of the sample does not appear on the chromatogram as would be the case if the mixture included some inorganic salts.

In the foregoing example it was assumed that the internal standard and other components responded both to the column and to the detector in the same manner; that is, the internal standard was of the same homologous series as the components being measured. It is seldom possible for this assumption to be met, but the ratio of response factors (such as K_{std}/K_3) can be determined experimentally. This value would be applicable as long as the other requirements are met. When so, then $K_{std}/K_3 = (W_3 A_{std}/W_{std} A_3)$. The internal standard is most often used where a portion of the sample may not elute completely or may be lost in preliminary operations prior to the chromatographic step.

PROBLEMS

1. Consider a 50-cm column with a plate height of 1.5 mm that provides a theoretical plate number of 333 at a flowrate of 3 ml min^{-1}; $V_M = 1.0$ ml. (a) What is the solute

retention time and retention volume when k' is 1, 2, 5, and 10? (b) What is the base line peak width for each of the foregoing values of k'? (c) Estimate the peak capacity of the column for each of the listed values of k'.

2. Using the data in Problem 1, what is the zone velocity for each solute whose k' values are given? What is the retardation factor for each solute?

3. Chromatograms with a standard test mixture were obtained using porous alumina; assume the total porosity is 0.75. The inlet pressure was 22.5 atm for all columns. Operating conditions and retention times are tabulated below. (a) Calculate k' for each solute. (b) Calculate the average linear velocity and plate height for each column. (c) Calculate the reduced velocity, reduced plate height, and reduced column length for each test column. (d) Calculate the free column volume and flowrate for each column.

TEST SUBSTANCE, t_R, sec:	COLUMN 1	COLUMN 2	COLUMN 3
Nitrobenzene	538	182	91
Anisole	232	76	36
Biphenyl	168	56	26
Toluene	124	40	19
t_M, sec	104	34	16
L, cm	50.0	13.5	9.0
N	3200	4450	5000
d_p, μm	20	10	6.5
d_c, cm	0.2	0.5	0.5

4. Keeping all other parameters constant in the resolution equation, (a) graph the effect of k' on resolution and (b) graph the effect of α on resolution.

5. For a typical chromatographic separation giving just-resolved peaks ($R = 1.5$), assume that $N = 3600$, $\bar{k}' = 2$, and $\alpha = 1.15$. Sketch the effects of changing these parameters one at a time to (a) $N = 1600$, (b) $\bar{k}' = 0.8$, and (c) $\alpha = 1.10$.

6. To decrease the plate height and yet increase the resolution, what courses of action are available? What penalties may accrue for each approach?

BIBLIOGRAPHY

Ettre, L. S. and C. Horvath, "Foundations of Modern Liquid Chromatography," *Anal. Chem.*, **47**, 422A (1975).

Giddings, J. C., "Principles and Theory," *Dynamics of Chromatography*, Part I. Marcel Dekker, New York, 1965.

LITERATURE CITED

1. Day, D. T., *Proc. Am. Phil. Soc.*, **36**, 112 (1897); *Science*, **17**, 1007 (1903).
2. Tswett, M., *Ber. Dsch. Bot. Ges.*, **24**, 116, 384 (1906).
3. Giddings, J. C., *Anal. Chem.*, **39**, 1027 (1967).
4. Ball, D. L., W. E. Harris, and H. W. Habgood, *J. Gas Chromatogr.*, **5**, 613 (1967).

CHAPTER 16

Gas Chromatography

Gas–liquid chromatography (GLC) accomplishes a separation by partitioning solutes between a mobile gas phase and a stationary liquid phase held on a solid support. Gas–solid chromatography (GSC) employs a solid adsorbent as the stationary phase.

The sequence of a gas chromatographic separation is as follows. A sample containing the solutes is injected into a heating block where it is immediately vaporized and swept as a plug of vapor by the carrier gas stream into the column inlet. The solutes are adsorbed at the head of the column by the stationary phase and then desorbed by fresh carrier gas. This sorption–desorption process occurs repeatedly as the sample is moved toward the column outlet by the carrier (mobile phase) gas. Each solute will travel at its own rate through the column. Their bands will separate to a degree that is determined by the individual partition ratios and the extent of band spreading (see Chapter 15). The solutes are eluted sequentially in the increasing order of their partition ratios and enter a detector attached to the column exit. If a recorder is used, the signals appear on the chart as a plot of time versus the composition of the carrier gas stream. The time of emergence of a peak is characteristic for each component; the peak area is proportional to the concentration of the component in the mixture. Although gas chromatography is limited to volatile materials, about 15% of all organic compounds, the availability of column temperatures up to 450°C, pyrolytic techniques, and the possibility of converting nonvolatile materials into a volatile derivative extend somewhat the applicability of the method.

16.1 GAS CHROMATOGRAPHS

Basically, a gas chromatograph consists of six parts: (1) a supply of carrier gas in a high-pressure cylinder with attendant pressure regulators and flow meters, and a valve to intro-

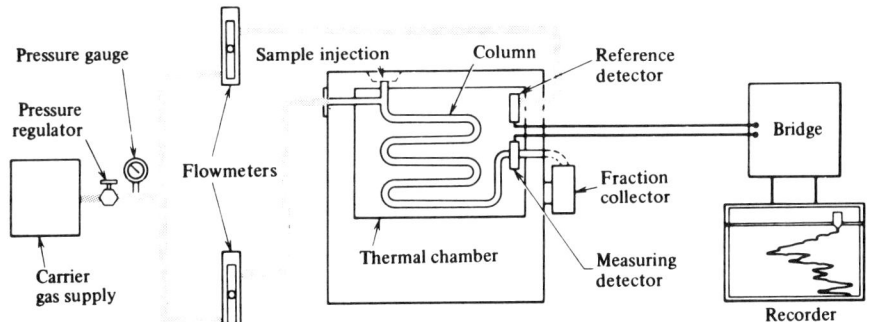

FIGURE 16-1 Schematic of a gas chromatograph.

duce extra make-up gas to some detectors, (2) a sample injection system, (3) the separation column, (4) the detector, (5) an electrometer and strip-chart recorder (and integrator perhaps), and (6) separate thermostated compartments for housing the column and the detector so as to regulate their temperature, or to program the column temperature. The components are shown schematically in Fig. 16-1.

Sample Injection System[1,2]

The most exacting problem in gas chromatography is presented by the sample injection system. The sample must be introduced as a vapor in the smallest possible volume and in a minimum of time without decomposition or fractionation occurring. Both quantity of sample introduced and the manner of introduction must be reproducible with a high degree of precision.

Liquid samples, 1–10 μl in volume, are usually injected by a microsyringe through a self-sealing silicone rubber septum onto a metal block that is heated at a fixed temperature. Here the sample is vaporized as a "plug" and carried into the column by the carrier gas stream (Fig. 16-2). Since the entire sample must undergo instantaneous vaporization to attain plug flow, the injection zone temperature must exceed the boiling points of all sample components. In a 1-μl sample, the individual solutes should be about 1% of the injected sample. For concentrated samples this implies dilution with a compatible solvent.

Although gas samples can be injected by a gas-tight syringe, the most accurate and precise method for gas samples uses a calibrated sample loop (0.5–10 ml) and a multiport rotary valve (Fig. 16-3). Liquid samples can be injected in the same manner using loops with smaller volumes, or with sliding plate valves.

Generally the smaller the sample used in gas chromatography (GC) the better the peak shape. Too large a sample overloads the column and an asymmetric peak with a trailing front results. The proper sample volume dissolves in the liquid stationary phase in a very short length of column, occupying just a few theoretical plates. A large sample will immediately spread over a much greater length, and a solute band can never be any narrower in time span than this distance. Also, a small sample produces a more dilute solution in the stationary phase which behaves more like an ideal solution in that the vapor pressure

456 CHAPTER 16

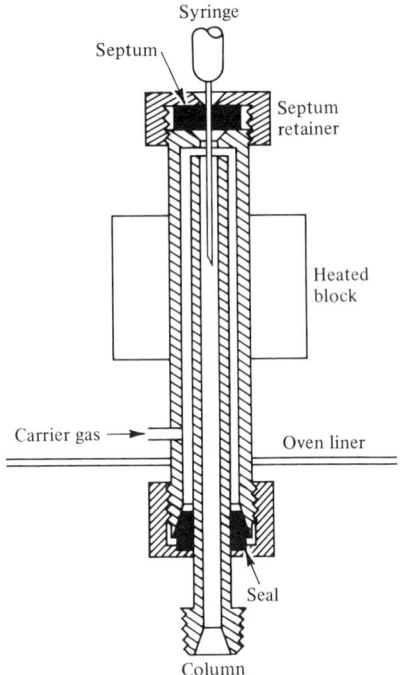

FIGURE 16-2 Schematic representation of a typical flash vaporizer injection port.

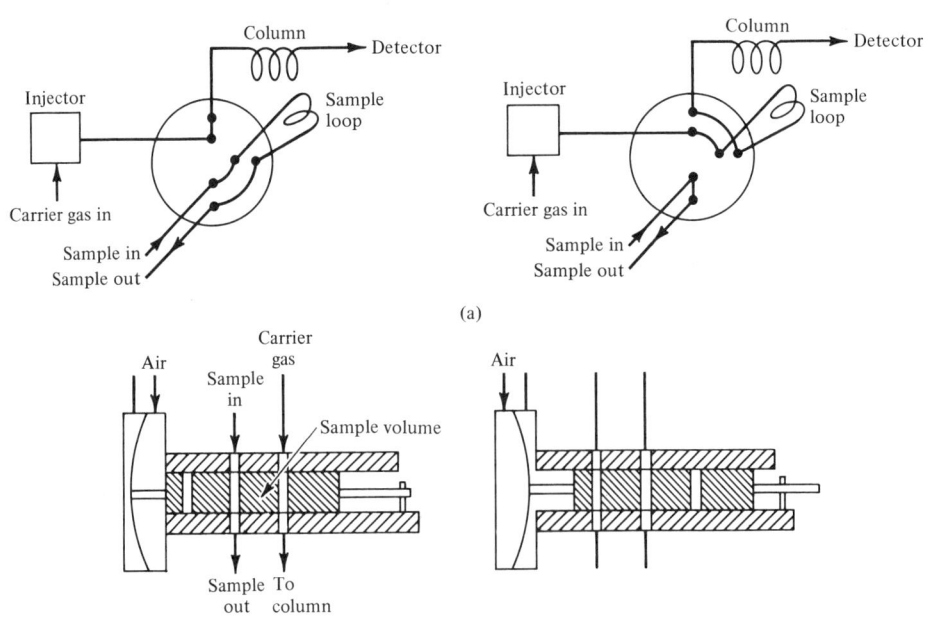

FIGURE 16-3 (a) Six-port rotary valve and (b) a sliding plate valve.

of the sample is linearly proportional to the concentration of sample in the partition liquid. Of course, the lower limit to sample size is set by the detector sensitivity.

Direct injection of dilute solutions onto a relatively low-temperature column is widely utilized in the trace analysis of environmental pollutants and components of physiological fluids. In this injection technique the less volatile trace sample components are "condensed" at the column inlet with minimum band spreading, while the more volatile (and large) solvent peak passes through the column with little retention. Following the solvent tail, the trace components of interest are eluted after a sudden or gradual elevation of the column temperature.

The automatic sampler duplicates manual sample measurement and injection. Sample vials are a glass, throwaway type with vapor-tight septum caps. The sampler flushes the syringe with new sample (or, by option, with a solvent) to remove traces of previous sample, pumps new sample to wet the syringe completely and eliminate bubbles, takes in a precisely measured (and adjustable) sample volume, and injects a reproducible "slug" into the gas chromatograph. There are two important differences between the automatic and manual method: (1) The automatic samplers are machine reproducible and consistently more precise than the skilled chromatographer; and (2) unattended operation releases the operator for other duties.

Volatile organic constituents of samples such as urine, human breath, and environmental air can be trapped on Tenax-GC contained in an 11-cm tube. Trapped samples can easily be stored or shipped to another site for analysis. Efficient desorption occurs with helium flow at 300°C. The desorbed volatiles are collected in a precolumn cooled by dry ice. The precolumn is then connected to the GC column, the dry ice removed, and the analysis started at room temperature. The precolumn contains the same liquid phase as the regular GC column.

When very dilute solutions are to be analyzed, the use of a concentration precolumn, such as a 2,6-diphenyl-p-phenylene oxide porous polymer, allows the quantitative transfer of up to 20 μl of sample. Solvent is removed and the sample is thermally desorbed into the first section of the regular capillary column held at a low temperature. A precolumn also prevents migration of nonvolatile (inorganic) impurities into the analytical column.

Reduction in sample volume is necessary when working with capillary columns. This is accomplished by an injector-splitter where typically a 1-μl sample is injected but only 0.01 μl enters the capillary; the remainder is vented through the split. The split ratio can vary between 1/20 and 1/1000. A typical capillary inlet splitter is shown in Fig. 16-4. It consists of an injector body with a glass insert that has an annular splitter at the end. At this point the vapors are split into two streams. One stream continues through the small internal-diameter glass-lined stainless steel tubing into the column. The other stream continues into the buffer volume and to vent. To ensure good mixing of sample before the split takes place, either the glass insert is packed with a bed of silanized glass beads or a series of baffles are used. The split region must be in a heated zone of constant temperature.

Pyrolysis is the accepted method for handling solids.[3] Operation based on the Curie principle utilizes filaments of certain alloys, which when subjected to intense rf energy, rapidly heat to a specific temperature unique to the alloy. Rise times to temperatures as high as 800°C are on the order of nanoseconds, but temperature control is difficult. The Pyroprobe enables the entire heating profile to be defined. A precision platinum element

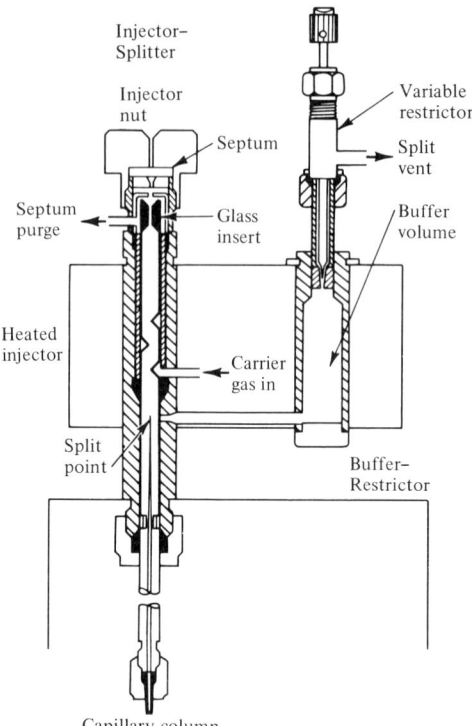

FIGURE 16-4 Capillary injector-splitter

serves as a temperature sensor, heater, and sample holder simultaneously. This element forms one leg of a conventional Wheatstone bridge circuit and the temperature setting control forms a balancing leg. By electronically programming the balancing leg, controlled linear heating rates are obtained. A ribbon probe is used for samples that can be dissolved or melted and deposited on the ribbon. The coil probe is used for material such as granular or fiber samples. The heating rate of the ribbon element can be controlled from 0.1°C/msec to 20°C/msec; this heats the ribbon to 600°C in 8 msec and to 1000°C in 17 msec. The coil element within a quartz tube takes 600 msec to heat to 600°C due to its higher mass.

Derivative Formation[4]

Derivatives play an important role in GC for the analysis of polar compounds such as fatty acids, steroids, many drugs, biological amines and phenols, or whenever polyfunctional groups of a compound are causing the peak to tail. Derivatives make a polar compound less polar, improve quantitation, and increase the volatility of high molecular weight compounds. Only a few typical examples will be discussed.

Silylating agents, such as *N,O-bis* (trimethylsilyl) acetamide, convert one active hydrogen (in $-OH$, $-COOH$, $-NH_2$, $=NH$, and $-SH$ groups) to a $-O-Si(CH_3)_3$ group. *N*-

Trimethylsilylimidazole in pyridine is preferred for carbohydrates since it gives a minimum of anomers and can be employed with aqueous solutions. When reaction is complete, a sample is taken from the solution with a syringe and injected into the gas chromatograph.

Amino acids, after drying, are converted to methyl ester hydrochlorides which are *trans*-esterified to butyl ester hydrochlorides. The latter are converted to volatile *N*-trifluoroacetyl *n*-butyl esters; subsequent chromatographic resolution into single peaks by temperature programming is easily accomplished.

For fatty acid analysis, methylation with BF_3 in methanol, followed by extraction of the formed methyl esters with petroleum ether before injection into the chromatograph is the usual derivatization method.

Chromatographic Columns

Two basic types of columns are in general use: the packed and the open tubular or capillary column. Packed columns are stainless steel, copper, or glass tubing, 1.6-, 3.2-, 6.4-, or 9.5-mm bore, and 3-m in length, that have been filled with a narrowly sieved, inert support that has been coated with a liquid phase for GLC. In GSC the filling is an adsorbent, such as silica gel, a bonded-phase support, or a molecular sieve.

Capillary columns[5-7] have an open unrestricted path for the carrier gas within the column. The wall-coated open tubular (WCOT) column is a long narrow-bore tubing (inside diameter about 0.25 mm) in which the inner wall is coated with the liquid stationary phase to about 1 μm in thickness. Lengths are typically 50–150 m. Columns constructed of silane-treated Pyrex glass have the most desirable features, in particular low catalytic activity. However, WCOT columns possess limited sample capacity. This limitation is overcome somewhat with the support-coated open tubular (SCOT) column[8] in which a porous layer is formed on the inside wall of the tubing. The porous layer can either be formed by chemical treatment of the inner wall of 0.5-mm bore tubing or deposited on the inside wall. In either case the advantageous open cross section and the unrestricted gas flow of open tubular columns is retained. The inert porous layer is impregnated with a liquid phase for GLC. With their increased surface area, resulting in increased sample capacity, SCOT columns need not be as long as WCOT columns, only about 16 m in length.

A good packed column will be nominally 1000–3000 plates/m; capillary columns range from 1000 to 4000 plates/m. Therefore, the total number of theoretical plates available for separation will be less than 10,000 plates for packed columns, but ranging up to 600,000 plates for capillary columns.

The sample capacity of capillary columns is determined principally by the thickness of the stationary phase on the column walls. Recently, high-efficiency columns have become available with liquid film thicknesses of 0.3–0.5 μm, substantially thicker than the 0.05–0.1 μm films previously used. The coating is a silicone gum similar in chemical structure to the usual oils. Thicker films increase sample capacity and coat the surface more uniformly. Even so, the stationary phase on capillary columns consists of droplets adhering to the surface which is never completely covered. Stationary phase films up to 2.0 μm can be used with 0.5-mm bore columns. The advantages of wide-bore columns include the ability to use direct injection techniques, to use nondestructive detectors, to use larger sample sizes for GC/mass spectrometry, and the large range of sample concentrations feasible. Disadvantages include the difficulty in eluting high-boiling-point compounds, lower column efficiency (only about 1000 plates/m), and longer analysis times.

Packed columns are usually formed into several coils and placed within the oven compartment. Capillary columns have tubing coiled into an open spiral, a basket-coil, or a flat pancake shape. Each coil should be separated by an air space to allow good circulation and to obtain maximum benefit when temperature programming is used.

Efficient capillary columns may well have no substitute for resolving mixtures with hundreds of components present because they are more effective and easier to use in developing separation methods. Capillary columns are more reproducible in terms of performance than packed columns; they are available from most suppliers, and they are individually tested and guaranteed on the basis of performance specifications. In contrast, the technology to make guaranteed packed columns has not been available.

Supports[9]

The purpose of the support is to provide an inert surface onto which the stationary liquid phase can be placed in a packed column. Only the diatomaceous earths and Teflon are used to any extent. The diatomaceous earth supports may be either firebrick derived materials, such as Chromosorb P and Gas Chrom R, or materials derived from filter aids, which include Chromosorb W, Anakrom ABS, and Gas Chrom Q. The latter supports have pore sizes about 9 μm, while the former material has 2-μm pores. Best particle size is 100/120 mesh (149–125 μm) for 2-mm-bore columns, and 80/100 mesh (177–149 μm) for 4-mm-bore columns. The effective internal diameter of the column should be at least eight times the diameter of the support particles.

Deactivation of all diatomaceous supports is necessary for most applications. Acid washing is effective in removing from the surface mineral impurities which can serve as adsorption sites. The acid-washed grade will perform quite well for the analysis of relatively nonpolar samples. However, the silanol (Si—OH) groups that cover the surface tend to adsorb solutes, particularly when the support is lightly loaded (<5%) or when nonpolar liquid phases are employed. These surface silanol groups can cause peak tailing through hydrogen bonding with polar samples. This problem increases in severity with the polarity of the sample and as the concentration of polar compounds is decreased. The problem can be minimized by treating the support with dimethyldichlorosilane which converts the silanol groups to silyl ethers, and then neutralizing the unused silylating agent with methanol. A third procedure for deactivation uses a polar stationary phase, one that contains functional groups such as an ester, a hydroxyl, or an amine group. These functional groups have strong hydrogen-bonding characteristics and tie up the active sites on the support surface. Even here, though, silanization can be useful. Stationary phase loading is limited to a maximum of 10% when using silanized supports.

Special supports are popular for particular applications. Very lightly loaded glass beads are used for very rapid analyses well below the boiling point of the sample components. The surface of the beads is roughened (texturized) to obtain better wetting and increased liquid phase capacity. Analysis of corrosive substances is performed with halocarbon supports coated with a halocarbon stationary phase. Sieved Teflon is the best available support.

Liquid Phases

The stationary liquid phase provides separation of the sample. In addition to having suitable selectivity, the liquid phase should have reasonable chemical and thermal stability. Upper temperature limits are included in Table 16-1. These should only be used as approximate guides because the true limit depends upon the type of detector used and the amount of column bleed tolerable for baseline stability. Operations should be kept at least 10–15°C below the upper temperature limit to prolong column life and to prevent rapid fouling of the detector. Column bleed is particularly disastrous when a gas chromatograph is directly coupled to a mass spectrometer. The column bleed material not only gives a large, obscuring background of ion peaks in the mass spectrometer, but can damage its ion source by contamination.

Bonding the partitioning compound to the column packing surface permanently anchors a film of oriented molecules to the surface[10,11] rather than mechanically coating the surface, and provides a liquid phase with no appreciable vapor pressure. Bonded phases are described in considerable detail in Chapter 18. For GLC the octadecyl alkane group, Carbowax 400, and β,β'-dioxypropionitrile among others have been bonded to the surface of a porous silica bead to produce a Si–O–C or a Si–O–Si linkage. Thermal stability ranges from 135°C to 160°C, a point at which the Si–O–Si bonds rupture or reactivity, such as hydrolysis with water vapor, apparently occurs. Inclusion of a carborane structure, which acts as an energy sink, periodically within the bonded group delays the destruction of the bonded molecule and enables the bonded phases to be used up to 500°C.

The *Kovats retention indices*[12] (R.I.) indicate where compounds will appear on a chromatogram with respect to *n*-alkanes injected with the sample. By definition, the R.I. for pentane is 500, for hexane is 600, for heptane is 700, and so on, regardless of the column used or the operating conditions, although the exact conditions and column must be specified, such as liquid loading, particular support used, and any pretreatment. For example, suppose that on a 20% squalane (a nonpolar) column at 100°C, the retention times for hexane, benzene, and octane were 15, 16, and 25 min, respectively. On a graph of ln t'_R of the alkanes versus their retention indices, a R.I. of 653 for benzene is read off the graph. The number 653 for benzene means that it elutes halfway between hexane and heptane on a logarithmic time scale. Thus, it is only necessary to run a standard set of alkanes on a particular column to determine retention times of compounds for which Kovats retention indices are known for that column and under the particular operating conditions. Repeating the experiment with a dinonylphthalate column, the R.I. for benzene is found to be 733, between heptane and octane, which implies that dinonylphthalate will retard benzene slightly more than will squalane; that is, dinonylphthalate is slightly more polar than squalane by $\Delta I = 80$ units.

The retention index system has many advantages. Given the R.I. values of a number of unknown compounds, homologous series are obvious. For example, if the R.I. of an unknown is 1260, the next higher homolog should appear at about 1360, the next lower homolog at about 1160. R.I. values are less influenced by column temperature than are relative retentions, even a 30°C difference enables valuable information to be obtained.

Classification of stationary phases by their ability to retard compounds, developed by Rohrschneider[13] and extended by McReynolds,[14] involves measurement of the R.I. for "index" compounds on a given column, as compared to those same compounds on a

TABLE 16-1 Stationary Phases in Gas Chromatography

Liquid Phase (similar phases; type)	Minimum/Maximum Temperature, °C	x'	y'	z'	u'	s'	Σ
For boiling point separation of broad molecular weight range of compounds:							
Squalane (2, 6, 10, 15, 19, 23-hexamethyltetracosane)	20/150	0	0	0	0	0	0
C-87 (24, 24-Diethyl-19, 29-dioctadecylheptatetracontane)*	30/280	21	10	3	12	25	71
OV-101 (DC 200, SE 30, SP-2100, Apiezons) (methyl silicone)	20/350	12	53	42	61	37	205
OV-73 (5.5% phenyl-substituted methyl silicone gum)	/350	16	55	44	65	42	222
Dexsil-300 (carborane/methyl silicone)	50/400	47	80	103	148	96	474
OV-7 (20% phenyl/80% methyl silicones)	20/350	69	113	111	171	128	592
For unsaturated hydrocarbons and other semipolar compounds:							
OV-17 (SP-2250, UCON LB550-X, dinonylphthalate) (50% phenyl silicone)	0/325	119	158	162	243	202	884
Tricresyl phosphate	20/125	176	321	250	374	299	1420
OV-215 (trifluoropropylmethylsilicone gum)	/275	149	240	363	478	315	1545
OV-225 (XE 60) (cyano propyl/phenyl/methyl silicone)	0/265	228	369	338	492	386	1813
For nitrogen compounds:							
Poly-A 103 (polyamide)	70/275	115	331	149	263	214	1072
Poly-A 135 (polyamide)	70/250	163	389	168	340	269	1329
XF-1150 (50% cyanoethyl/methyl silicone)	20/200	308	520	470	669	528	2495
For halogen compounds, freons, alkaloids; specifically retards compounds with keto groups:							
SP-2401 (OV-210, QF-1, FS-1265) (trifluoropropylsilicone)	20/275	144	233	355	468	305	1500
For alcohols, esters, ketones, and acetates:							
SP-2300 (Carbowax 20M, FFAP)	25/275	316	495	446	637	530	2424
SP-2310 (Silar 7CP)	25/275	440	637	605	840	670	3192
SP-2340 (Silar 10C, TCEPE)	25/275	520	757	659	942	800	3678
For fatty acid methyl esters:							
Neopentyl glycol succinate (HI-EFF-3BP)	50/230	272	469	386	539	474	2120
SP-2330 (DEGS, DEGA, HI-EFF-1BP, EGA)	25/275	490	725	630	913	778	3536
OV-275 (SP-2320) (cyanosilicone)	25/250	781	1006	885	1177	1089	4938
Kovats indices on squalane column for index compounds:		653	590	627	652	699	

*Kovats nonpolar, nonchiral, chemically pure, and characterized liquid phase; molecular weight:: 1222.37.

squalane column, to determine the degree to which each is retarded. The difference, ΔI, gives a measure of solute–solvent interaction due to all intermolecular forces other than London dispersion forces. The latter are the principal solute–solvent effects with squalane. Now the overall effects due to hydrogen bonding, dipole moment, acid–base properties, and molecular configuration can be expressed as

$$\Sigma \Delta I = ax' + by' + cz' + du' + es' \tag{16-1}$$

where $x' = \Delta I$ for benzene (intermolecular forces typical of aromatics and olefins), $y' = \Delta I$ for 1-butanol (electron attractors typical of alcohols, nitriles, acids, nitro-, and alkyl mono-, di-, and trichlorides), $z' = \Delta I$ for 2-pentanone (electron repellers typical of ketones, ethers, aldehydes, esters, epoxides, and dimethylamino derivatives), $u' = \Delta I$ for nitropropane (typical of nitro and nitrile derivatives), and $s' = \Delta I$ for pyridine (or dioxane). All the index compounds must be determined under identical conditions of column temperature and column loading. Rohrschneider/McReynolds constants, listed for selected liquid phases in Table 16-1, may be used to select the most appropriate stationary phase for a given separation and columns which have unique characteristics. For example, fluorosilicones are the only stationary phases for which the z' value is greater than the y' value, which indicates that this phase selectively retards ketones and therefore ketones would be eluted after alcohols. In general the constants are useful in selection of columns for separation of compounds that differ in functionality, such as alcohols from ketones, aromatics from aliphatics, and saturates from unsaturates. The system is of no value for separation of a homologous series of isomers. Another valuable use for these constants is in identifying similar stationary phases from the hundreds of partition phases which have been described in the literature.

Several examples should assist in understanding the method of column selection based on the Rohrschneider/McReynolds system. Columns for the separation of unsaturated and saturated fatty acid esters should be selected on the basis of high x' values alone; that is, a column that takes into account intermolecular forces, which is the only difference between the esters. Order of elution can be selected when two compounds of different classes have similar boiling points. To elute an alcohol first (from an ether), a column with a high z' value with respect to the y' value is needed, such as QF-1 or SP-2401. Just the reverse is needed, a high y' value relative to the z' value, if the ether is to be eluted ahead of the alcohol, perhaps a DEGS or SP-2340 stationary phase.

Another good rule to follow when selecting liquid phases is "like dissolves like." Unless the sample dissolves well in the liquid stationary phase little or no separation will occur. The gas phase is inert; separation occurs only in the liquid phase. The liquid phase can be selected on the basis of matching the polarity of the stationary phase to that of the sample components of interest. Lower temperatures will almost always increase the solubility and the selectivity. However, viscosity also increases as the temperature is lowered, and mass transfer may become too slow for practical separations. For preliminary screening separations one could begin with a stationary phase such as SE-30 (or OV-101 or SP-2100) for nonpolar samples or Carbowax 20M for polar samples, and analyze the sample by temperature programming the column from 35°C to the upper temperature limit of the column substrate at 2°C min^{-1}, and hold for 20 min at the upper limit. The resultant chromatogram will indicate whether a more polar phase is needed, if the concentration of the stationary phase is too low, and the temperature range in which the sample should be chromatographed. If the desired separation is not obtained, it is worthwhile trying the following sequence, again with temperature programming: 3% OV-17, 10% OV-225, 10% Carbowax 20M, and 3% SP-2401. These suggested loadings are for packed columns; the particular liquid phases are equally applicable for capillary columns. However, the choice of stationary phases is less critical for capillary columns because the total column efficiency is so high.

Ovens[15]

The column oven should be free from the influence of changing ambient temperatures and have a well-designed and adequate air flow system to maintain good temperature control of the column. In most designs the air is blown past the heating coils, then through the baffles that comprise the inner wall of the oven, past the column, and back to the blower to be reheated and recirculated. High-capacity, squirrel-cage blowers and carefully located baffles ensure uniform, rapid circulation of heated air within the oven and rapid cooling when desired. Most ovens are constructed of low-mass stainless steel to permit rapid heating and cooling cycles. For temperature programming it is desirable to have heat-up rates from $0.25°C$ min^{-1} to $10°C$ min^{-1}; the temperature steps at the bottom should be small for capillary columns. Heating from ambient to $400°C$ within 40 min, with cooling to $100°C$ in 3 min, are usual working requirements. Temperatures should be able to be maintained within $\pm 1°C$ for isothermal runs, and within $\pm 2°C$ of the desired temperature during programming.

16.2 DETECTORS[16, 17]

Located at the exit of the separation column, the detector senses the presence of the individual components as they leave the column. The detector output, after suitable amplification, is traced on a strip-chart recorder. The result is a chromatogram of concentration versus time. The temperature of the thermostatted detector compartment must be sufficiently high to prevent condensation of high-boiling sample vapors and stationary-phase bleed products, yet not high enough to cause decomposition of eluent material. The linear dynamic range of various GC detectors is shown in Fig. 16-5.

Thermal conductivity detectors were first and are still widely used; their simplicity is an advantage and they are nondestructive. For high-sensitivity analyses of organic compounds, the hydrogen flame ionization detector is used. Both types of detectors are essentially nondiscriminating. When samples contain so many constituents that the resulting chromatogram is a complex maze of peaks, the analyst is faced with the problem of identification and the elimination of interferences from nearby overlapping, or even obscuring peaks. Detectors which respond selectively or characteristically to some property of the eluted species offer one method of peak identification. Selective sample derivatization can be performed, enhancing utility of specific detectors. Compounds like steroids, biological amines, amino acids, and various drug metabolites are converted to perfluoro-derivatives prior to electron capture detection. Ketonic steroids can be derivatized to

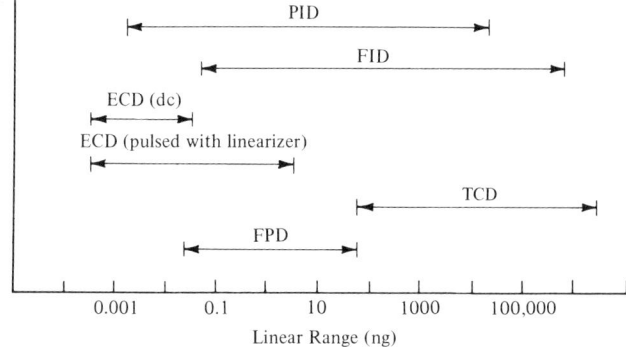

FIGURE 16-5 Linear dynamic ranges of gas chromatographic detectors: PID, photoionization; FID, flame ionization; ECD, electron capture; TCD, thermal conductivity; and FPD, flame photometric.

contain nitrogen, as methoxime derivatives, and be subsequently detected by the thermionic emission detector. A much higher degree of specific molecular identification can be achieved with on-line mass spectrometry or Fourier transform infrared spectrometry. The identification power of low-resolution mass spectrometers is enhanced significantly by the prior separation done in the chromatographic column. The power of combining the two methods should be obvious for the unambiguous identification of isomeric compounds. Specifically, whereas these compounds may yield similar mass spectra, their chromatographic mobility is quite different. Also, the combination of capillary GC with high-resolution mass spectrometry has scored impressive gains in sensitivity.

Thermal Conductivity Detector

In its classical design a thermal conductivity detector (TCD) is a cavity in a metal block with a tightly coiled filament (W, W-Re, or Au sheathed W) down the center and inlet and outlet tubes, as shown in Fig. 16-6. The filament is heated by a regulated dc current sup-

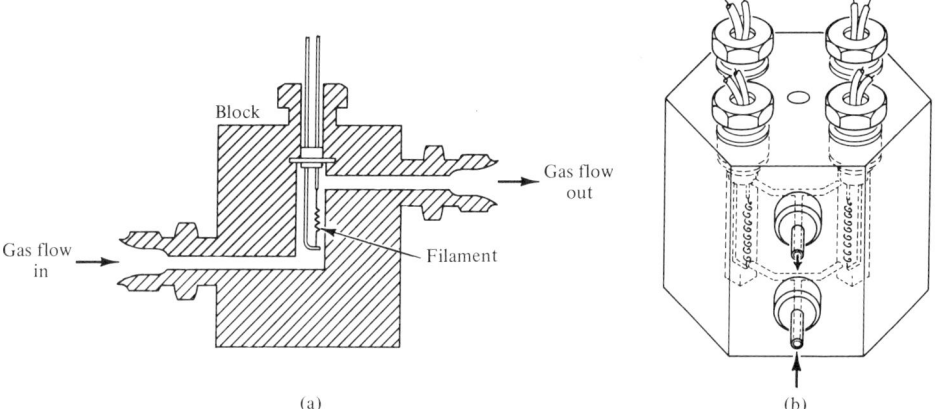

FIGURE 16-6 Thermal conductivity detector: (a) cross-sectional view and (b) four units mounted in a heat sink.

ply. Through changes in filament temperature, and therefore resistance, changes in the thermal conductance of the gas bathing the filament can be detected. The change in thermal conductance is proportional to the sample concentration momentarily in the detector. The thermal mass of the brass block serves as a heat sink.

The TCD system is sensitive to the cell geometry, the thermal conductivity, and the filament and cell temperatures. Therefore, if any of these change during a chromatographic run, the output is likely to have a drifting base line. The standard remedy to the drift problem is to use an additional similar cell as a reference and subtract the outputs. The cells are made two (or four) arms in a Wheatstone bridge circuit (Fig. 16-7). With the same carrier gas passing through all cavities, the network is balanced by means of the balancing potentiometers so the electrical output is zero. Perfect matching is never achieved. The primary limitation in the short term is that the temperatures of the cells do not track each other exactly. Long-term changes can be caused by oxidation of either filament. This destroys the balance of the electronics and changes sensitivity.

In analyses the column effluent is passed through one (or a pair) compartment; pure carrier gas is passed through a matched unit in the opposite arm of the bridge. Any appearance of a solute, or increase in its concentration, will decrease the rate of heat loss (except for hydrogen), increase the filament temperature since it retains more heat, and therefore increase the filament resistance. The net result will be an unbalance in the bridge circuit relative to the condition when only pure carrier gas is flowing through both compartments. The large difference in thermal conductivity between sample components and helium carrier gas gives a high output and good linearity.

The thermistor[18] is a metal oxide bead that exhibits 8000-Ω resistance at 25°C. With its negative temperature coefficient of resistance, it is highly efficient in the -20°C to 25°C range. A matched pair replaces the tungsten filaments; however, they are not interchangeable mechanically or electrically. The bridge is completed by a matched pair of 500-Ω, 3-W resistors. The choice between thermistor or filament is usually based on temperature considerations, thermistors for ambient or subambient, and filaments for higher temperatures.

Introduced in 1979, the modulated[19] TCD offers a novel design which is extremely stable and possesses sensitivity and speed to allow the detector to perform in capillary chromatography. The analytical and reference gases are alternately passed through a single detection cell containing a single straight-wire filament, rigidly mounted in a ceramic cartridge, actually a minute filament channel. By switching back and forth at 10 Hz a modulated intermediate signal is obtained with an amplitude proportional to the difference between the thermal conductivities of the gases. The detector is only sensitive to changes which occur at 10 Hz, and since there can be very little change in the cell's characteristics at that frequency, most signals other than thermal conductivity are invisible. A fluidic switching circuit controls the analytical and reference flows that enter at opposite ends of the filament channel. In the analytical mode a modulator gas flows into the analytical side of the system. This gas, flowing out to the vent, creates a back pressure of a few thousandths of a psi which forces most of the analytical column flow to go past the filament. At the right side the reference gas is forced, with the analytical gas, to flow out to the vent. After 1/20 sec, the modulator flow is switched from the analytical mode to the reference mode by a remotely located valve. The filament channel flow

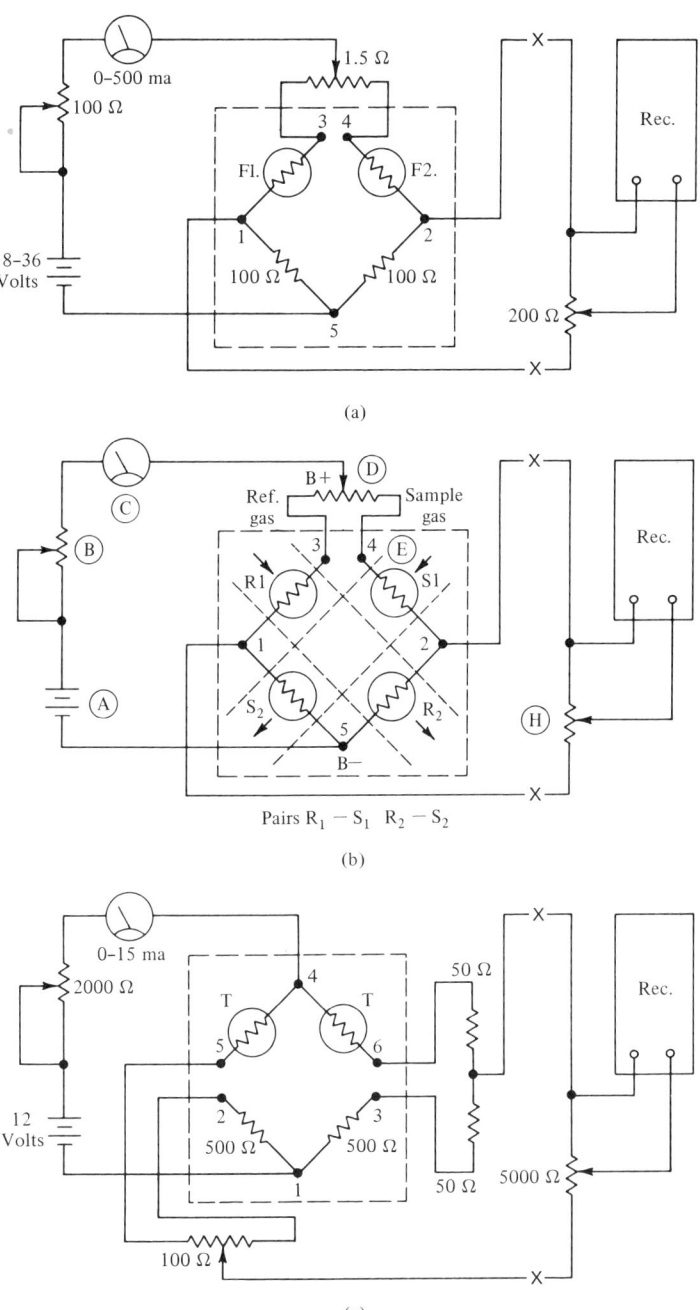

FIGURE 16-7 Circuitry for (a) thermal conductivity (two-filament) cells, (b) thermal conductivity (four-filament) cells, and (c) thermistor cells. (Courtesy of Gow-Mac Instrument Co.)

reverses and the reference flow ends up in the filament channel. This sequence is repeated 10 times/sec. The modulated intermediate signal is sensed by a synchronous demodulator which changes the signal to a smooth peak or base line (Fig. 16-8). There are two signals present in the modulated wave; one from the thermal conductivity and one from the carrier gas flow. The flowrate signal responds almost instantly when the modulator valve switches, but the thermal conductivity signal is delayed since it takes time for the gas to sweep the detection cavity. Therefore one can discriminate between these two signals and reject the unwanted flow signal. By synchronizing the electronics 90° out of phase with the flow signal, one obtains the maximum thermal conductivity with a much lower base line noise, drift, and wander due to flowrate variations.

Because of its simplicity, the TCD often is preferred for survey work and for moderate sensitivity work in all areas. It responds to all types of inorganic and organic compounds. It is nondestructive and therefore particularly suitable for fraction collection and preparative work. The cavity volume is 2.5 ml for detectors associated with 4-mm-bore columns. This is decreased to 30 μl in the microcell designed for use with SCOT columns, and even lower to 5 μl in the newest design of Hewlett-Packard. Sensitivity with the latter unit is about 0.3 ng; it can be used with SCOT or WCOT columns. The dynamic range of the detector is 10^7 with a linear range of 10^5. In the classical version of the TCD, coating and coking from column bleed, or dirty samples, not to be confused with oxidation or corrosion, will impair response.

Flame Ionization Detector

The flame ionization detector (FID) is currently one of the most popular detectors because of its high sensitivity, wide range, and great reliability. As shown in Fig. 16-9, column effluent enters the burner base through a millipore filter, is mixed with hydrogen gas, and the mixture burned at the tip of the jet with air or oxygen. Ions and free electrons are formed in the flame. These enter the gap between two electrodes, the flame jet and a collector, which may be parallel plates or in a cylindrical configuration mounted 0.5–1.0 cm above the flame tip. Across the two electrodes is imposed an applied potential about 400 V. This lowers the resistance across the gap and causes a current to flow. Normally an external bucking voltage is provided to balance the potential generated by the ions and free electrons generated in a pure hydrogen/air flame. This ensures that a net current flows only when ionized material enters the gap, thus enhancing the differential sensitivity of the detector. This current flow across an external resistor is sensed as a voltage drop, is amplified, and displayed on a recorder. When $-CH_2-$ groups are introduced into the flame, a complex process takes place in which positively charged carbon species and electrons are formed. The current is greatly increased.

The jet is enclosed in a chimney so that it is unaffected by drafts. The entire detector is housed in the oven or heated separately so water from the combustion does not condense. An ignitor coil and flame-out sensor is placed above the jet to reignite the flame should it become extinguished.

The FID responds only to substances that produce charged ions when burned in a hydrogen/air flame. In an organic compound the response is proportional to the number of oxidizable carbon atoms. For example, butane has twice as many carbon atoms as an

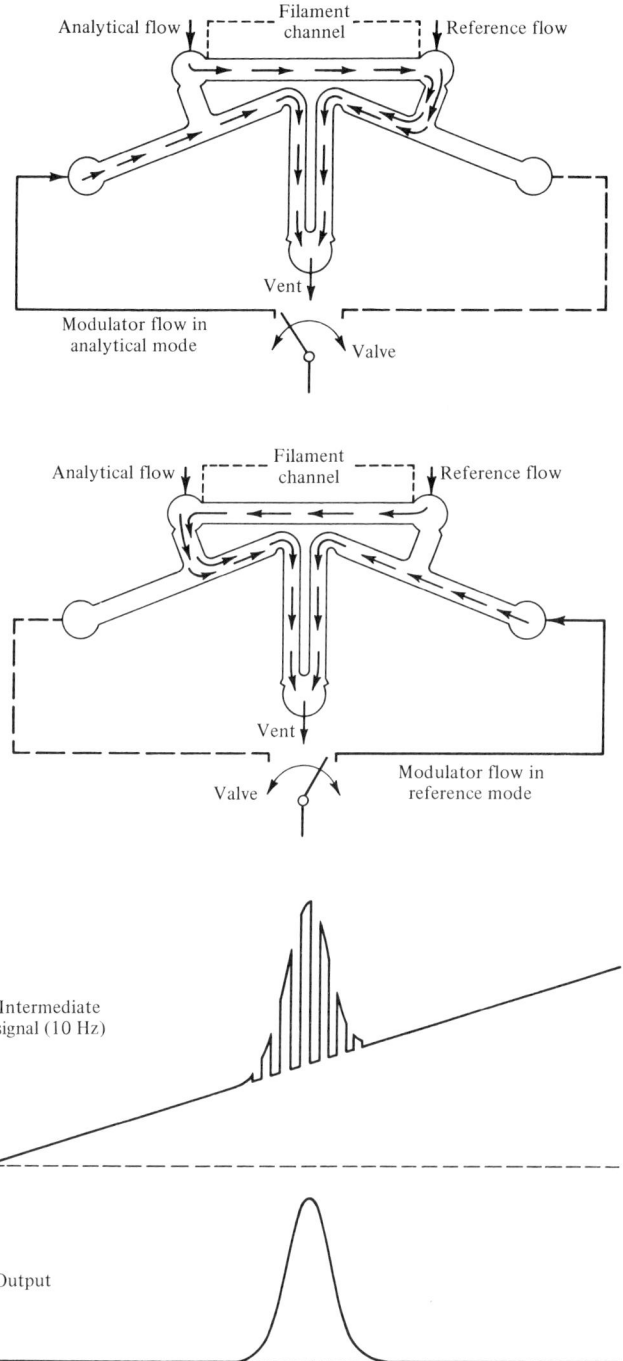

FIGURE 16-8 Modulated thermal conductivity detector. (Courtesy of Hewlett-Packard Co.)

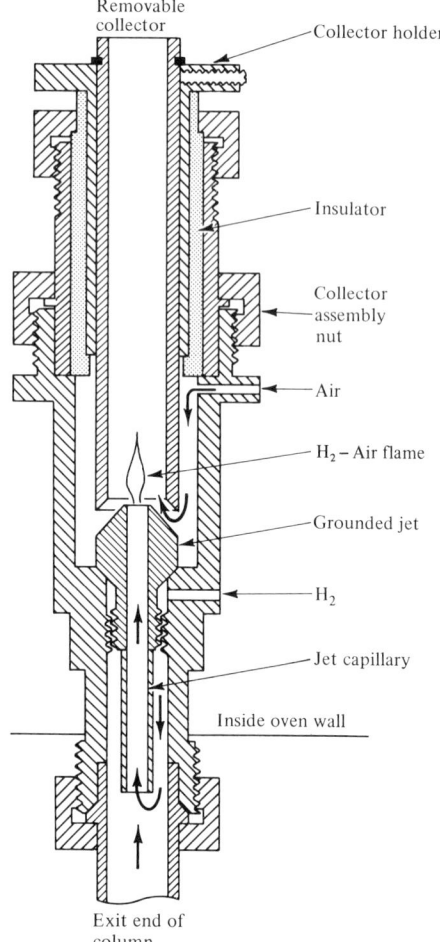

FIGURE 16-9 Cross section of a flame ionization detector. (Courtesy of Hewlett-Packard Co.)

equivalent volume of ethane. Within narrow limits, response of the FID to 1 mol of butane will be twice that to ethane. There is no response from fully oxidized carbons such as carbonyl or carboxyl groups (and thio analogs) and ethers, and response diminishes with increasing substitution of halogens, amines, and hydroxyl groups. The detector does not respond to inorganic compounds apart from those easily ionized in a hydrogen/air flame at 2100°C (see Table 5-2). Insensitivity to water and permanent gases (CO, CO_2, CS_2, SO_2, H_2S, NH_3, N_2O, NO, NO_2, SiF_4, and $SiCl_4$) is advantageous in analysis of moist samples and in air-pollution studies when small traces of organic materials have to be measured in these backgrounds. If desired, CO and CO_2 can easily be converted to CH_4 by reduction with hydrogen over a nickel catalyst and subsequently measured by the detector.

The FID is a mass flow detector. It thus depends directly on flowrate of carrier gas. It also varies in response with the applied voltage (until a plateau is reached), and with the temperature of the flame, which is a function of the hydrogen/air mixture ratio, as distinguished from the temperature of the detector housing. Detection limits are about 5 ng sec^{-1} for light hydrocarbon gases, increasing to 10 pg sec^{-1} for higher organic liquids and gases. Because of stationary phase bleeding at higher temperatures, the practical detection limit may often be only 1 ng sec^{-1}. Response is linear over seven orders of magnitude. Precise temperature control is not a requirement for this detector. The operating range is from 100°C to 420°C, an obvious advantage in programmed temperature applications. In hazardous areas the exhaust must be fitted with a flame-trap.

Thermionic Emission Detector

A diagram of a thermionic emission detector (TED), also called the alkali flame ionization detector, is shown in Fig. 16-10. It employs a fuel-poor hydrogen plasma.[20] This low-temperature source suppresses the normal flame ionization response of compounds not containing nitrogen or phosphorus, although the response to carbon is not entirely eliminated. A nonvolatile rubidium silicate bead, centered about 1.25 cm above the plasma jet, is electrically heated by a variable current supply to between 600°C and 800°C. This arrangement permits fine adjustment of the bead's temperature independent of the plasma as a source of thermal energy. With a very small hydrogen flow, the detector responds to both nitrogen and phosphorus compounds. Enlarging the plasma and changing the polarity between the plasma tip and collector, the detector responds only to phosphorus compounds. Compared with the FID, the TED is about 50 times more sensitive for nitrogen and about 500 times more sensitive for phosphorus. Compared with the flame photometric detector, it is about 100 times more sensitive for phosphorus. The minimum detectable limit (as caffeine) is 0.06 pg sec^{-1} for nitrogen. The selectivity against carbon, while

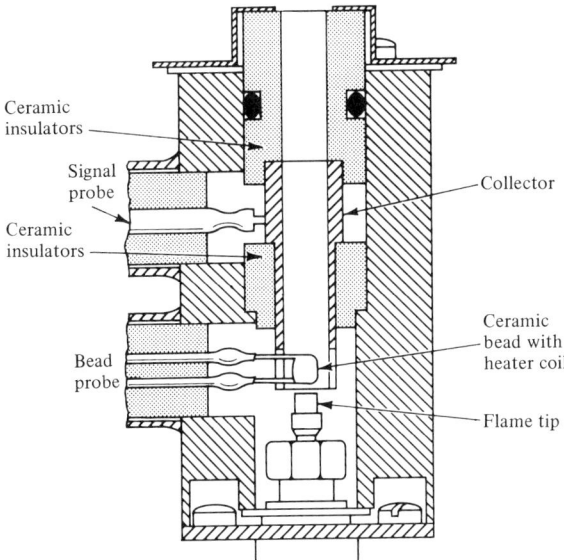

FIGURE 16-10 Thermionic emission detector. (Courtesy of Varian Associates.)

depending somewhat on the analytical conditions and the type of molecule, is always better than 1 in 5000.

Flame Photometric Detector

The flame photometric detector (FPD) is essentially a flame emission photometer.[21] The eluted species pass into a flame, usually a hydrogen-enriched, low-temperature plasma, inside a shielded jet which supplies sufficient energy first to produce atoms and simple molecular species, and then to excite them to a higher electronic state. The excited atoms and molecules subsequently return to their ground states with emission of characteristic atomic line or molecular band spectra which are measured by the photomultiplier tube. A narrow bandpass filter isolates the appropriate analytical wavelength range. A schematic diagram of the detector is shown in Fig. 16-11.

The most highly developed FPDs are selective for phosphorus and sulfur.[22] These elements are detected by monitoring band emissions from the molecular species HPO at 526 nm and S_2 at 394 nm. Detector response to phosphorus compounds is linear, whereas the response to compounds containing a single atom of sulfur is proportional to the square of the compound concentration. Quenching of the sulfur emission by other organic compounds present in the flame and extinguishing of the flame by solvent peaks are overcome with a dual-flame FPD. The lower flame causes partial combustion and decomposition of the solute molecules into relatively simple species. The upper flame then burns these and produces the emission which is detected with the photomultiplier tube. Although both flames produce the unique green or violet molecular emission, only the upper flame is viewed since this is less affected optically by the incoming sample components. Solvent peaks may extinguish the lower flame, but the upper one remains lit. After passage of the solvent, the lower flame is reignited by flashback from the upper one.

The FPD response to sulfur and phosphorus is about 10^4 times that elicited from hydrocarbons. Sensitivities are at the subnanogram level. Discrimination between sulfur

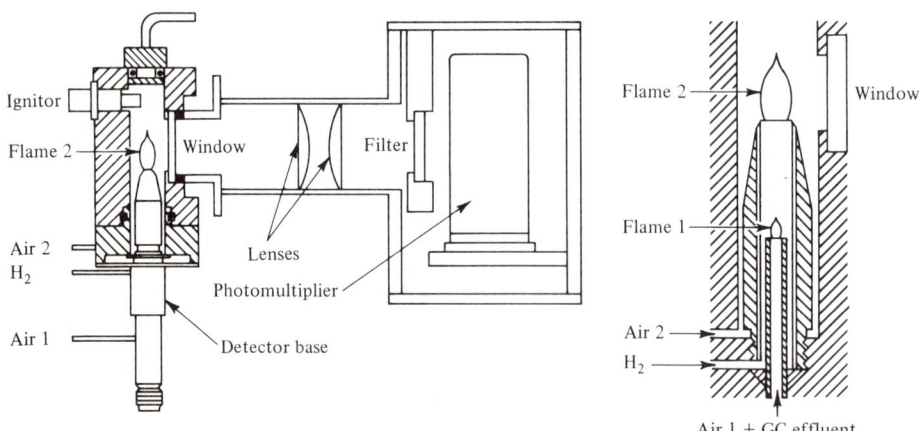

FIGURE 16-11 Flame photometric detector. (Courtesy of Varian Associates.)

and phosphorus is less impressive due to differential cross response, which arises because the band spectra of HPO and S_2 effectively overlap within the bandpass of the two filters. Sulfur-containing species may act as serious interferents in the detection of phosphorus compounds. It is necessary, therefore, to select a column which not only separates the sulfur compounds from each other, but also separates them from other sample components. This is not possible when both elements are present in the same compound.

The major field of application of GC/FPD systems has been in the determination of pesticides and pesticide residues containing sulfur and phosphorus. For such analyses, the high sensitivity and selectivity of FPD give it superiority over FID or electron capture detectors. The FPD has also been used to detect gaseous sulfur compounds such as mercaptans or thiophenes in gases, and H_2S, SO_2, and mercaptans in air pollution. A military use involves monitoring air for traces of nerve gases which contain phosphorus. Other uses include the detection of a variety of volatile metal salts and chelates, lead alkyls, and silylated compounds.

Electron Capture Detector

The electron capture detector[23] (ECD) has two electrodes with the column effluent passing between. One of the electrodes is treated with a radioisotope that emits high-energy electrons as it decays. These emitted electrons produce copious amounts of low-energy (thermal) secondary electrons in the GC carrier gas, all of which are collected by the other positively polarized electrode. Molecules that have an affinity for thermal electrons capture electrons as they pass between the electrodes and reduce this steady-state current, thus providing an electrical reproduction of the GC peak. Of the two general designs, the plane parallel and the concentric cell, the latter design is preferred (Fig. 16-12) since it is easier to construct a small, low dead-volume (0.3-ml) cell in this form. Radioactive sources include tritium adsorbed in titanium or scandium, and nickel-63 as a foil or plated on the interior of the cathode chamber. Tritium sources have a high specific activity, giving a large standing current and a high sensitivity, but the beta energy is so low that this source is extremely susceptible to contamination. The maximum working temperature is 225°C for Ti^3H and 325°C for Sc^3H foils. Nickel-63 is a higher-energy source and can be used up to 400°C. The tritium-type detector reaches a maximum sensitivity after a few days of operation and then constantly loses sensitivity which necessitates frequent recalibration and eventual foil replacement. The sensitivity of the nickel cell is approximately five times less than that of a tritium cell, but it remains constant and eventually surpasses the sensitivity of a tritium cell of the same age.

To prevent the formation of contact potentials and space charge effects, the detector voltage is applied as a sequence of narrow pulses with a duration and amplitude (1–3 μsec and 50 V, respectively) sufficient to collect the very mobile electrons but not the heavier, slower negative ions. During the interval between pulses (100–150 μsec) the electron concentration builds up inside the cell. When the pulse is applied, the concentration of electrons is essentially reduced to zero. This mode of operation has several advantages. During most of the time no field is applied; this enables the free electrons to reach thermal equilibrium with the gas molecules. The opportunity for electron capture

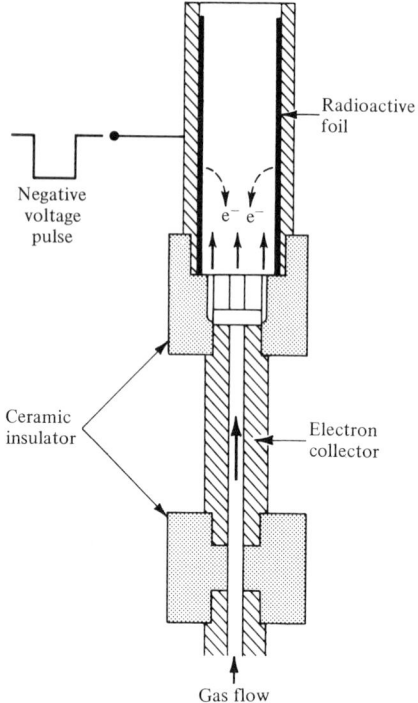

FIGURE 16-12 Electron capture detector. (Courtesy of Varian Associates.)

is maximized and stable response factors are thereby attained. Negative ion formation occurs in a region where positive ions are also present and recombination can efficiently take place. No collection of negative ions occurs. The formation of a space charge is prevented by the brevity of the pulse interval. The pulsed cell possesses a linearity of three or four decades, the actual range depends on which pulse interval is used and upon the radioisotope employed.

In another mode of pulsed operation the frequency of the pulse is automatically varied to maintain the collected current constant when a sample enters the detector. Most of these units operate with a base line frequency of a few hundred hertz, which increases to as high as 200 kHz with a sample component. The linear range is stated to be 10^5.

Argon mixed with 5-10% methane is the gas of choice for most pulsed ECDs. At the voltages employed, the electron velocity is ten times greater in this mixture than in nitrogen. Through inelastic collisions with methane, the metastables formed from argon are eliminated before they can cause undesirable sample ionization. A constant thermal equilibrium essential for reproducible electron capture, is maintained by these inelastic collisions. The argon/methane mixture is added as a makeup gas, introduced into the system after the column but prior to the ionization portion of the detector.

Efficient electron collection when nitrogen is the carrier gas can be achieved with a displaced coaxial cylindrical ECD cell. In addition to using a narrow pulsewidth of

0.64 μsec, the cell is constructed so that a 0.3-ml volume is formed by two closely spaced cylinders in which the carrier gas flow is counter to the electron flow.

All ECDs respond to electrophilic species which gives the detector its specificity. Polyhalogenated compounds, such as pesticides, give excellent response; detection is at the picogram level. The order of increasing response is F < Cl < Br < I. Other groups exhibiting good selectivity include anhydrides, peroxides, conjugated carbonyls, nitriles and nitrates, plus ozone and organometallics and sulfur-containing compounds. In addition to pesticides, applications involve halogenated anesthetics, polynuclear carcinogens, and sulfur hexafluoride tracer for following the course of pollution fumes from chimneys. The ECD responds exceptionally well to traces of oxygen; leak-free systems and oxygen-free carrier gases are a necessity. Response to traces of water vapor can cause unusable base lines so that molecular sieve traps are required prior to the flow controllers.

Helium Detector

The helium detector operates on the principle that metastable helium atoms (He*) will ionize all compounds which have lower ionization potentials than the excitation energy of He*. Such compounds will produce collectible ion particles when they collide with He*. A 250-mCi tritium foil is used to supply high-energy beta particles which promote ground-state helium atoms to He*. High field strengths of 4000–5000 V cm^{-1} are used to provide more efficient production of He* and to provide efficient collection of ionization products.

The helium detector must be used only in GSC because column bleed from liquid loaded columns will quench the He* before it can ionize sample molecules. However, the detector can handle the permanent gases (H_2, Ar, O_2, N_2, CH_4 and CO) eluting from a molecular sieve column; also the higher molecular weight compounds, such as CO_2, NO, N_2O, H_2S, SO_2, C_2H_4, and so on, eluting from a bonded phase packing. The part-per-billion sensitivity for the fixed gases makes the helium detector the most suitable detector for trace analysis of these gases. A minimum detectible level of 0.04 pg sec^{-1} and a linear range of over three orders of magnitude is possible.

Photoionization Detector [24]

The sensor (Fig. 16-13) contains a sealed interchangeable ultraviolet lamp that emits a selected energy line. Lamps with energies of 9.5, 10.0, 10.2, 10.9, and 11.7 eV are available. The absorption of ultraviolet light by a molecule leads to ionization. A chamber adjacent to the ultraviolet source contains a pair of electrodes. A positive potential applied to the accelerating electrode creates a field which drives ions formed by absorption of lamp energy to the collecting electrode where the current, proportional to concentration, is measured. The photoionization detector has a dynamic range of seven orders of magnitude, extending from 2 pg through 30 μg.

Compounds whose ionization potentials are lower than the lamp ionizing energy will respond. About one in a thousand molecules will be ionized. Molecules whose ionization potentials are up to 0.3 eV higher than a given lamp energy will also respond although

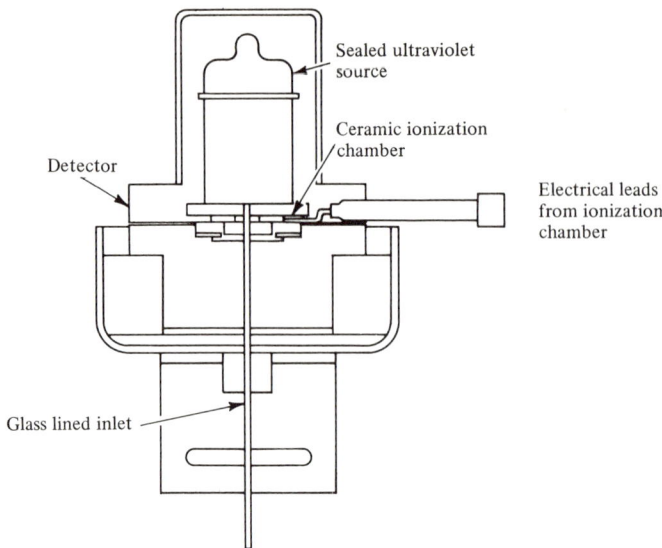

FIGURE 16-13   Photoionization detector. (Courtesy of HNU Systems, Inc.)

with a lower efficiency. This allows the detection of aliphatics (except CH_4), aromatics, ketones, aldehydes, esters, heterocyclics, amines, organic sulfur compounds, and some organometallics. The detector also responds to inorganics such as O_2, NH_3, H_2S, HI, ICl, Cl_2, I_2, and PH_3. For maximum selectivity in a given application, the lamp with the energy output just capable of photoionizing the species to be detected should be selected. When used with a 10.2-eV lamp, the photoionization detector does not respond to several commonly used solvents such as methanol, or to extraction solvents such as chloroform, dichloroethane, carbon tetrachloride, and acetonitrile.

Coulson Conductivity Detector

This detector operates by converting the analyte to an ionic species whose conductance is monitored in a dc conductivity cell from which the analyte ions are continuously removed so that the detector has a differential, rather than an integral, response. Selectivity depends on the specificity of the chemical reactions producing the conducting species. The detector offers considerable potential as a nitrogen selective detector. For nitrogen compounds the effluent from the gas chromatograph is mixed with hydrogen gas and hydrogenated over a nickel catalyst at 850°C in a quartz tube furnace. This results in the formation of ammonia from organic nitrogen, plus HCl from organic chlorides, H_2S from sulfur compounds, water from oxygen, and hydrogenated hydrocarbon fragments. Passage through a $Sr(OH)_2$ scrubber removes the acidic combustion products. Ammonia passes on through and into the conductivity cell. Detection limit is 100 pg as nitrogen. The conductivity cell has a solution circulating system which enables removal of the analyte ions in an ion-exchange column.

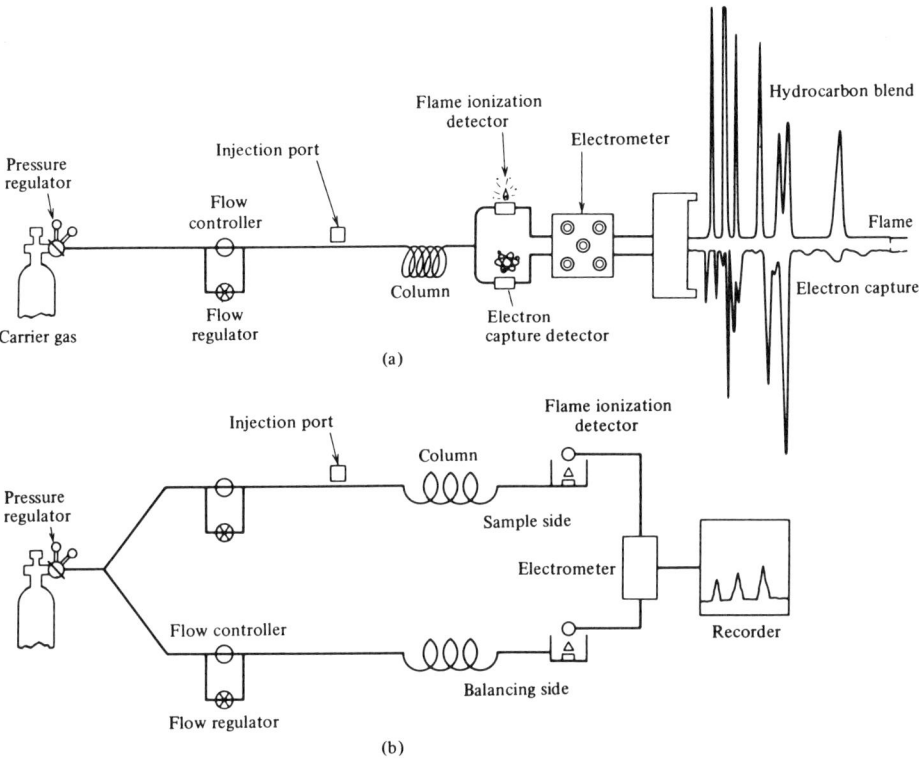

FIGURE 16-14 Dual-detector operation. (a) Two different detectors operated simultaneously and (b) differential operation with identical detectors.

Dual Detectors

Chromatography with dual-channel detection involves running a single sample through two dissimilar detectors simultaneously. Different detectors, each with its individual selectivity for the same compound, enable an operator to gain information as to the chemical composition of the sample and aids in distinguishing components comprising overlapping peaks. Each detector has its own amplifier. The amplified signal for each detector is fed to two single-pen recorders or to a dual-pen recorder where the simultaneous responses are recorded. An application is illustrated in Fig. 16-14. Unleaded gasoline shows a profile of hydrocarbons, detected by a flame ionization detector, and virtually no response from an electron capture detector. Leaded additives are detected only by the electron capture detector.

Identical dual detectors can be used in a differential mode of operation, particularly during temperature programming. The effluent from dual columns, one containing the sample and the other just a dummy, and each passing through identical detectors, eliminates the problem of a drifting base line or a rising base line as the upper temperature limit of the liquid substrate is reached or slightly exceeded in temperature programming. The signal from the reference column is used to cancel out a similar base line signal from the sample column. Peak area measurements can be made more accurately, and small peaks on a rising base line are not obscured.

16.3 OPTIMIZATION OF EXPERIMENTAL CONDITIONS

For packed columns, the plate height is given by the extended form of the van Deemter equation:

$$H = 2\lambda d_p + \frac{2\gamma D_M}{\bar{u}} + \frac{2}{3}\frac{k'}{(1+k')^2}\frac{d_f^2 \bar{u}}{D_S} + \frac{(k')^2 d_p^2 \bar{u}}{96 D_M(1+k')^2} \tag{16-2}$$

where λ is the packing factor (~ 0.5), γ is the obstruction or tortuosity factor (~ 0.7), d_p is the particle diameter, and d_f is the thickness of the stationary phase liquid droplets coated on the substrate. At the low reduced velocities employed in GLC, the coupled form of the A and C_{mobile} terms is not used. When heavily loaded columns are used, for example, 30% liquid coating, the third term on the right-hand side dominates among the C terms. This is caused by the slow diffusion in the stationary phase and the film thickness. Lower-loaded columns will be more efficient and have lower C values. However, as the liquid loadings become less, the C_{mobile} term becomes important (at approximately 5% loading). Good efficiency and short analysis times require small particle diameters. Particles of 100–200 μm represent the best compromise between efficiency and permeability, and require pressure drops of 3–4 atm for columns 2–3 m in length.

For open tubular columns in GLC, the plate height expression can be formulated as

$$H = \frac{2D_M}{\bar{u}} + \frac{1 + 6k' + 11(k')^2}{24(1+k')^2}\frac{r^2 \bar{u}}{D_M} + \frac{k'}{6(1+k')^2}\frac{d_f^2 \bar{u}}{D_S} \tag{16-3}$$

where r is the radius of the tubular column. Because film thicknesses are small, the second term in Eq. 16-3, involving the mobile-phase resistance to mass transfer, dominates at velocities above the optimum ($\bar{u}_{\text{opt}}$). Thus the tube diameter should be kept narrow. Secondarily plate height is related to k'. However, optimum k' values are determined on the basis of the resolution equation (Eq. 15-21).

Once the liquid phase has been selected, k' is inversely proportional to the phase ratio, or β value. For a packed column, β is about 15–20, for a SCOT column about 60–70, and for a WCOT column about 100–120. Thus k' will always be smaller on an open tubular column than on a packed column prepared with the same liquid phase and operated at the same temperature. For example, if k' is two on a packed column, the value at the same temperature would be 0.5–0.6 for a SCOT column and 0.3–0.4 for a WCOT column. In a column with a high β value, solutes with short retention times would be difficult to separate. The whole advantage of the high plate numbers given by open tubular columns may be lost if they are operated under conditions in which k' is fractional. For this reason, open tubular columns are operated at appreciably lower temperatures than are packed columns. The main advantage of open tubular columns is that the permeability is high, and long columns and/or high mobile-phase velocities can be used.

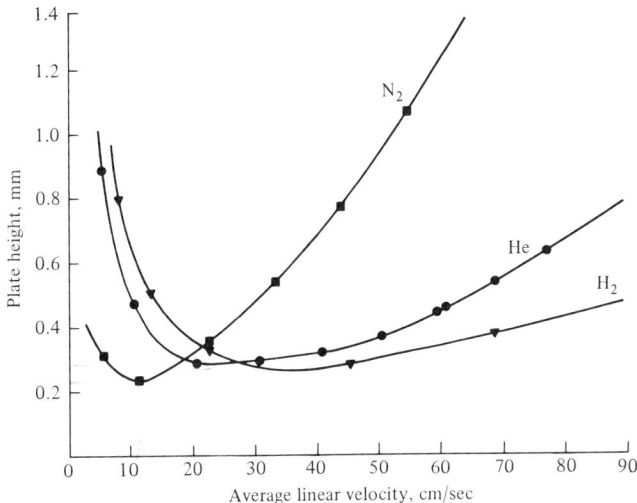

FIGURE 16-15 Van Deemter graph for various carrier gases flowing through a WCOT glass column coated with OV-101. Column temperature is 175°C.

Selection of carrier gas is an operator choice. A van Deemter plot for the three most common carrier gases is shown in Fig. 16-15. It is apparent from the graph that either helium or hydrogen should be used for most GLC work. Nitrogen achieves its optimum efficiency at a relatively low linear gas velocity and also loses efficiency much more rapidly as the velocity is increased.

As was shown in Fig. 15-4, the $H/\bar{u}$ relation is a hyperbola with an optimum velocity at which plate height is a minimum. That is, when $dH/d\bar{u} = 0$, $\bar{u}_{opt} = (B/C)^{1/2}$, with the C terms combined. The corresponding minimum plate height is: $H_{min} = A + 2(BC)^{1/2}$. The A term is absent with open tubular columns.

Temperature Dependence

The solute vapor pressure is described as a function of temperature by the Clausius–Clapeyron equation:

$$\ln p° = -\frac{\Delta \bar{H}_v}{RT} + C \tag{16-4}$$

where $\Delta \bar{H}_v$ is the molal heat of vaporization of bulk solute and C is a constant. When the solute activity coefficient is unity, the specific retention volume can be expressed as

$$V_g° = \frac{273R}{p° M w_L} \tag{16-5}$$

where M is the molecular weight of the stationary phase. Combining these two relations

$$\ln V_g° = \frac{273R}{M w_L} + \frac{\Delta \bar{H}_v}{RT} + C \tag{16-6}$$

Because the first term on the right-hand side is itself a constant, and $\Delta \bar{H}_v + \Delta \bar{H}_s = 0$ when the solute activity coefficient is unity, where $\Delta \bar{H}_s$ is the molal heat of solution,

$$\ln V_g^\circ = -\frac{\Delta \bar{H}_s}{RT} + C' \qquad (16\text{-}7)$$

Plots of $\ln V_g^\circ$ versus $1/T$ are linear and the slope yields $-\Delta \bar{H}_s/R$ (or $\Delta \bar{H}_v/R$). These plots have a positive slope for the molal heat of solution, generally -1 to -10 kcal mol^{-1}, is invariably negative, implying that heat is evolved. Among members of a homologous series, solutes will have the same value of V_g° at the column temperature corresponding to their boiling points.

An expression analogous to Eq. 16-7 but in terms of the partition coefficient gives

$$\ln K = \ln \frac{RT}{V_L} - \frac{\Delta \bar{H}_s}{RT} + C \qquad (16\text{-}8)$$

where V_L is the molar volume of stationary phase. Plots of $\ln K$ versus $1/T$ would therefore be expected to be curved, for the first term on the right-hand side of the equation is temperature dependent.

The relationship between relative retention and column temperature is

$$\ln \alpha = -(\Delta \bar{H}_2 - \Delta \bar{H}_1)/RT + \text{const} \qquad (16\text{-}9)$$

In most cases $(\Delta \bar{H}_2 - \Delta \bar{H}_1)$ will be greater than zero because the final component of an adjacent pair will have the higher molal heat of solution in the stationary phase. Thus the relative retention will decrease usually with increasing column temperature. Increasing the column temperature lowers the individual retention times on the order of 5%/°C.

The concentration of stationary phase and column temperature are related. With packed columns, increasing the amount of stationary phase will require a higher column temperature to obtain the analysis within the same period of time. However, even though two such analyses might be done within the same time, the column with the lower concentration of stationary phase operated at the lower temperature would give a chromatogram in which the early peaks would be eluted closer together than for a column with more stationary phase operated at a higher temperature. In isothermal analysis, and for mixtures up to approximately six carbons, a 15% concentration of stationary phase is useful. For six to twenty carbons, it is convenient to use 10% phase loading, and above these molecular weights the concentration should be reduced to 3–5%. The situation is similar with capillary columns. A SCOT column with a beta of 20–25 should be operated about 50°C higher than a standard WCOT column. Analogously, the temperature of analysis recommended for a SCOT column with a beta of 60–70 is about 20°C lower than for a SCOT column with a beta of 20–25, but about 30°C higher than for the corresponding WCOT column of similar beta value. Astute selection of column type and extent of loading can enable one to avoid subambient temperatures when working with a wide-range, low-boiling mixture of compounds.

To summarize, selection of column temperature for isothermal operation is a complex problem and usually one has to make a compromise, often dictated by the pair of solutes most difficult to separate. The simple solution to this problem is to change the band migration rates during the course of separation by temperature programming.

Time of Analysis[25]

The retention time for a solute with a known k' value was given by Eq. 15-28. The ratio $H/\bar{u}$ decreases as the carrier gas velocity increases, but a point of diminishing return is reached when the ascending part of the van Deemter curve starts to become linear. Nevertheless, it is always advantageous to employ gas velocities higher than $\bar{u}_{opt}$, with a corresponding increase in column length. Here open tubular columns prove valuable. The optimum practical gas velocity can be determined by the plot of N/t_R versus $\bar{u}$. Over the range of the first rise, and up to the maximum, an increase in column length can compensate for the loss of the overall column efficiency due to the higher gas velocity, but still keep the retention time shorter than the value corresponding to $\bar{u}_{opt}$.

Most modern gas chromatographs allow automatic resetting of the operating conditions in a short time, and automatic sampling systems are available. Instantaneous evaluation of the GC results with the help of computers is an option. Thus the time of a GC analysis should be reduced at least until it is compatible with the time of sample preparation.

Temperature Programming

Operation of the column at a constant temperature during a chromatographic run is sufficient for general applications, and for the separation of closely related compounds or samples whose components fall within a narrow boiling-point range. Complex multicomponent samples covering a wide temperature range cannot be satisfactorily chromatographed in a single isothermal run. A run at moderate column temperature will demonstrate good resolution of the low-boiling compounds, but requires a lengthy period for the elution of the high-boiling material. For the latter the retention time may be unusually long and the peaks so broad that they are indistinguishable from the base line. A run at higher temperature, while providing more rapid elution and better resolution of the higher-boiling materials, has the obvious disadvantage of poor or no resolution of the low-boiling components of the sample, as shown in Fig. 16-16.

Temperature programming combines the best results of runs at different temperatures. The sample is injected into the chromatographic system with the column temperature below the lowest-boiling component of the sample, preferably 90°C below. Then the column temperature is raised at some preselected heating rate. Earlier peaks, representing low-boiling components, emerge essentially as they would from an isothermal column operated at a relatively low temperature. As the column temperature increases, the higher-boiling components are forced through the column at an ever increasing rate.

Now the fraction of total solute found in the mobile phase is given by Eqs. 15-11 and 15-13. It is K that is responsible for the rather rapid increase in the amount of solute found in the vapor phase, as shown by Eq. 16-8. Since there is nearly always more solute in the liquid than in the gas phase, that is, the fraction $1/(1 + k')$ is much less than unity, it is usually a good approximation to ignore V_M in the denominator of Eq. 15-13. Remembering that $k' = K/\beta$, then from Eq. 16-8, ignoring the temperature dependency of the first term on the right-hand side,

$$\ln(1/k') = - \frac{\Delta \bar{H}_v}{RT} + C' \qquad (16\text{-}10)$$

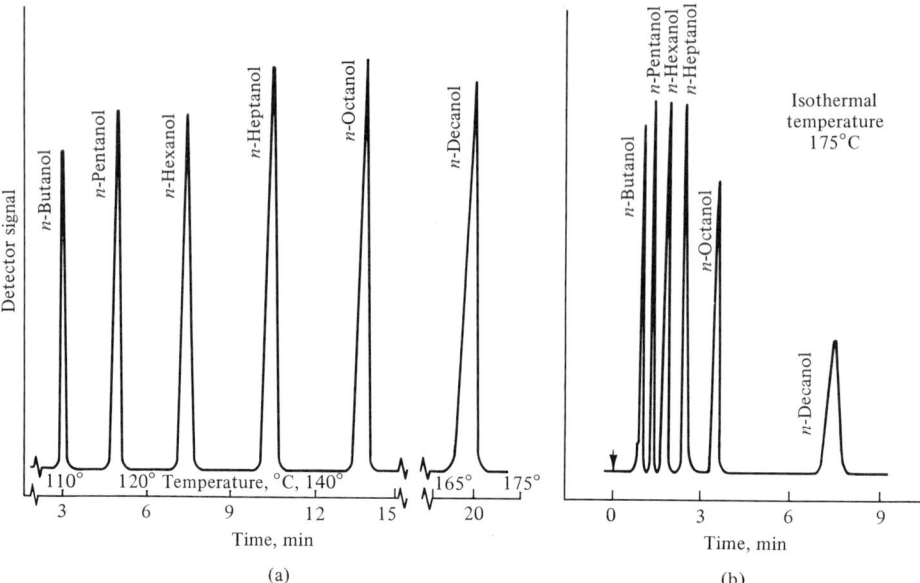

FIGURE 16-16 Chromatograms of an alcohol mixture. (a) Programmed temperature from 100°C to 175°C and (b) isothermal operation at 175°C.

Next it is important to determine the average increase in temperature needed to just halve the k' value, or double the fraction of total solute in the vapor phase. If k' is halved by increasing the temperature from T_1 to T_2, then from Eq. 16-10, the ratio of k' values is

$$\ln 2 = \frac{-\Delta \bar{H}_v/RT_2}{-\Delta H_v/RT_1} = \frac{\Delta \bar{H}_v}{RT} \frac{\Delta T}{T} \tag{16-11}$$

which gives

$$\Delta T = 0.693 RT^2/\Delta \bar{H}_v \tag{16-12}$$

where T is the geometric mean of the two temperatures and $\Delta T = T_2 - T_1$. Trouton's rule is approximately valid, so $\Delta \bar{H}_v/T_b \simeq 23$. Since the chromatographic process is operated near the solute boiling point, $\bar{T} \simeq T_b$. The ratio in Eq. 16-12 may also be approximated by 23. Substituting the typical operating temperature (T) into Eq. 16-12, we find the temperature increase (ΔT) that typically halves the k' value to be 21° at an operating temperature of 75°C, 24° at $T = 125°C$, and 30° at 225°C.

The significance of the k' value lies in the fact that the more solute one finds in the vapor state, the faster the peak migrates. The actual velocity of peak migration is a weighted average of the velocity of the mobile phase and the zero velocity of solute in the stationary phase. Thus, if $k' = 5$, this means that 1/5 of the solute molecules exist as vapor and possess a mean velocity $\bar{u}$. The average velocity of the solute zone is simply $\bar{u}/k'$; it may be considered as the migration rate relative to the inert carrier gas, t_M, for which $k' = 0$.

A step-function approximation can be used to follow the solute migration. Let us assume that the temperature increases in steps of 23°C to halve the value of k' each unit time. Furthermore, let us assume that $t_M = 1.0$ time units, and $k' = 60$. The isothermal adjusted retention time is then

$$t'_R = k'/t_M = 60/1 = 60$$

In unit time the fractional distance through the column length traveled by the band is $L/60 = 0.02L$. During the next step, when the effective $k' = 30$, and the column temperature is 23° higher,

$$t'_R = 30/1 = 30$$

and the fraction of column traversed by the band is $L/30 = 0.03L$; the total distance being $0.05L$. During succeeding steps:

$k' = 15$	$t'_R = 15$	$L/15 = 0.07L$	total distance: $0.12L$
$k' = 7.5$	$t'_R = 7.5$	$L/7.5 = 0.13L$	total distance: $0.25L$
$k' = 3.75$	$t'_R = 3.75$	$L/3.75 = 0.27L$	total distance: $0.52L$

Now only $0.48L$ remains for traversal so that when $k' = 1.88$ and $t'_R = 1.88$, the fractional distance that could be traversed in the last step would be $0.53L$. Consequently, the solute will emerge in $0.48/0.53$ or 0.90 time unit. The total adjusted retention time is 5.90 time units, and corresponds approximately to a temperature rise of $136°$.

Working backwards from the emergence of the zone peak, the solute will have migrated one-half the total column length in the last 23° interval. In the next-to-last interval the migration distance of the peak is half of the remaining distance, or $L/4$, and so on. Now the time required to migrate through the final interval before emergence is the interval length, $L/2$, divided by the peak velocity, $\bar{u}/k'$. Thus the migration time is $Lk'/2\bar{u} = t_M k'/2$. This time must equal the time needed to increase the temperature by 23°; it is equal to 23° divided by the programmed rate of temperature increase with respect to time, $r°$. Consequently, the interval migration time is $23°/r°$. Equating the two times, the mean $(\overline{1/k'})$ value of the last segment is

$$\overline{1/k'} = r°t_M/46 \qquad (16\text{-}13)$$

Since the elution occurs at the end of the final segment rather than in the middle where mean values are applicable; $(\overline{1/k'}) = [(1/k') + \frac{1}{2}(1/k')]/2 = \frac{3}{4}(1/k')$. Using this and Eq. 16-13, the terminal $\overline{1/k'}$ value is

$$\overline{1/k'} = r°t_M/35 \qquad (16\text{-}14)$$

for our particular example. To avoid significant losses in resolution, the heating rate should not exceed the relationship

$$r° \leq 35/t_M \bar{k}' \qquad (16\text{-}15)$$

If, as in the foregoing example, $t_M = 1$, and $\bar{k}' \simeq 2.5$ over the final interval, then the heating rate should be kept approximately 14° per time unit. The final temperature should be near the boiling point of the final solute, but not exceeding the maximum temperature limit of the stationary liquid phase.

As the temperature ramp proceeds, individual compounds will each automatically select their own ideal temperature in which to migrate and separate within the column. If the column temperature is increased linearly, the members of any homologous series will be eluted at approximately equally spaced intervals, as shown in Fig. 16-16, rather than proportional to log t_R.

16.4 GAS–SOLID CHROMATOGRAPHY

In GSC the columns are packed with interactive solids such as molecular sieves or porous polymers through which the carrier gas flows. Sample injection is by means of a gas sample loop or valve. Normal operation involves the use of columns 1–4-m length and temperatures in the range 40°C–90°C. A complete analysis can usually be obtained in a time equal to 1–2 min/component. Temperature programming is often useful.

Adsorbents

Inorganic molecular sieves are naturally occurring or synthetic zeolites which are comprised of interconnected cavities, or pores, of precisely uniform size. Only molecules smaller than these openings are able to enter. Four types of molecular sieves, varying in pore diameter and alkali metal content, are commercially available. Type 3A, a potassium aluminosilicate, with a pore diameter of 3 Å, will adsorb molecules such as water and ammonia. Type 4A, the sodium analog of Type 3A, adsorbs all molecules with critical diameters up to 4 Å; these include carbon dioxide, sulfur dioxide, hydrogen disulfide, ethane, ethylene, propylene, and ethanol. Type 5A, with calcium replacing part of the sodium content of Type 4A, has a pore diameter of 5 Å. It will separate straight-chain hydrocarbons (C_3 through C_{22}) from branched-chain and cyclic hydrocarbons. Type 13X, a sodium aluminosilicate of different crystalline structure from the preceding types, has a pore diameter of 10 Å.

The molecular sieves selectively separate molecules not only according to size and configuration, but according to polarity and degree of unsaturation as well. Thus polar unsaturated molecules are adsorbed in preference to nonpolar saturated molecules. The branched hydrocarbons are prevented from penetrating the small pore openings completely because of the larger cross section due to the side chain and will proceed through the column in a series of in-and-out movements over the particles.

Columns prepared with a carbon molecular sieve, Type B, with pore radius 5–15 Å, show very high efficiency and can be used for inorganic gas analysis and analysis of traces of some polar compounds with less than six carbon atoms per molecule. The Type B sieve is both extremely nonpolar and inert. Water is quickly eluted and well separated from methanol, ethanol, and formaldehyde. The Rohrschneider/McReynolds constants are negative: $x' = -103$, $y' = -84$, $z' = -62$, $u' = -193$, and $s' = -151$. For example, with this adsorbent benzene would elute before hexane; the adsorbent is less polar than squalane.

Porous polymer packings are analogous to the porous gels used in exclusion chromatography. Those made from copolymers of aromatic hydrocarbons provide column packings of low to moderate polarity. Polymers made from acrylic esters provide packings of moderate to high polarity. Most gases and low-boiling liquids can be separated on one or more of these polymers. Column temperature lies in the range 50°C–250°C.

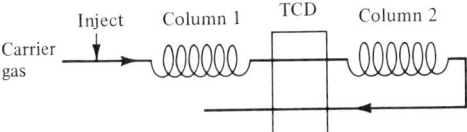

FIGURE 16-17 "Series-across-detector" arrangement.

Solid adsorbents, such as silica gel, alumina, and various forms of activated carbon, are used for some specific applications. The large retention of silica gel for CO_2, which elutes after ethane, is often useful in multicolumn systems. The almost irreversible adsorption of water is often a benefit. The most significant characteristic of alumina is its retention for unsaturated hydrocarbons. Graphitized thermal carbon black offers selectivity in the separation of molecules of different shapes, such as geometrical isomers. In addition, this support is used to separate high molecular weight compounds, since the minimal catalytic activity of the adsorbent permits operations at temperatures up to 500°C.

Multicolumn Systems

What makes GSC more complicated than GLC is that almost all gas mixtures contain components which are not separated on or will not pass through a particular column in a reasonable time. Backflushing substantially reduces the analysis time. For example, in trace level measurement of the hydrocarbons in natural gas, interest is usually centered on the C_1 to C_5 components. After the pentane peak has eluted from a porous polymer column at 110°C, the carrier gas flow is reversed so as to enter the column exit, and the C_6 and higher hydrocarbons are backflushed from the upper end of the separation column and, if desired, through the detector which is now positioned at the column entrance.

In the "series-across-detector" arrangement, shown in Fig. 16-17, one column is used before and a different column is used after the detector. Instead of two different column packings, the same packing may reside in both columns which are operated at different temperatures. Consider a mixture of H_2, O_2, N_2, and CO_2, to be separated on a polar porous polymer column followed by a column packed with molecular sieve 5A after the detector. After sample injection, H_2, O_2, and N_2 are detected in one arm of a TCD. These gases pass through the sieve column which separates O_2 and N_2, and further increases the separation of these from H_2, and into the other arm of the detector. Finally CO_2 elutes from the porous polymer and causes a signal in the first arm of the detector. Obviously the columns must be adjusted for length such that the components do not elute from the two columns at the same time.

In the two-column "series/bypass" technique, as shown in Fig. 16-18, a column switching valve enables column 2 to be bypassed by the carrier gas at selected times, so that certain components can be temporarily stored there while separations are made on column 1. Take for example a sample of sulfur-rich flue gases separated on column 1 (porous polymer) and column 2 (molecular sieve 5A), both at 97°C. Fifty seconds after sample injection, O_2, N_2, CH_4, and CO will have passed through column 1 and entered column 2 where they are stored by switching the carrier gas so as to pass only through column 1. Between a 1- and 3-min time interval, separate elution peaks for CO_2, C_2H_4,

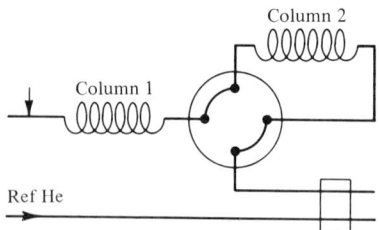

FIGURE 16-18 Series/bypass arrangement.

C_2H_6, and H_2S will appear. Then column 2 is switched into the gas stream to elute the stored components in the order enumerated over the next 3-min interval.

PROBLEMS

1. Often it is tempting to increase the loop volume in order to increase the amount of sample for trace analysis. What problems will arise if the loop volume were to be increased from 1 to 5 ml when columns with 2-mm bore are being used? Assume a flowrate of 30 ml min^{-1} at atmospheric pressure (outlet) and a column inlet pressure of 3 atm.

2. 1-Methylnaphthalene and 2-methylnaphthalene have adjusted retention volumes (also corrected for pressure gradient) of 1350 and 1200 ml, respectively, on a column with 3% liquid loading on Chromosorb P. On a column of glass beads with a liquid loading of 0.16%, 1-methylnaphthalene has an adjusted retention volume of 365 ml. What will be the adjusted retention volume of the 2-methylnaphthalene peak? Silicone oil 710 was liquid substrate in both columns.

3. The diffusion coefficient, D_M, for n-butane in helium at 25°C and 1 atm is 0.342 cm^2 sec^{-1}. In nitrogen the same constant has a value of 0.0960 cm^2 sec^{-1}. (a) Which terms in Eq. 16-2 are affected by this difference? (b) Sketch two graphs of H versus $\bar{u}$ showing how this difference will change the shape and location of $\bar{u}_{opt}$ and H_{min}. (c) Which carrier gas will allow faster mobile-phase velocities without significant loss of efficiency?

4. The following data were obtained from a column 360 cm in length that was packed with 30/50 mesh Chromosorb P and various amounts of hexadecane as the liquid substrate (whose density is 0.774 g ml^{-1}). Column operating temperature was 30°C. Chart speed was 61 cm hr^{-1}. Propane was the solute. Carrier gas was hydrogen, except for the final data set when nitrogen was employed.

	31% COLUMN LOADING; H$_2$ CARRIER GAS			23% COLUMN LOADING; H$_2$ CARRIER GAS	
$\bar{u}$, cm sec^{-1}	V'_R, mm CHART	W_b, mm CHART	$\bar{u}$, cm sec^{-1}	V'_R, mm CHART	W_b, mm CHART
1.00	335.4	47.9	1.43	157.4	19.2
1.65	205.0	25.2	2.81	78.0	8.1
3.22	108.1	12.7	4.15	54.1	5.2
6.16	57.3	7.7	5.43	40.6	4.0
8.80	38.5	5.8	6.64	32.8	3.3
13.28	25.0	4.5	12.26	18.1	2.1

	13% COLUMN LOADING; H$_2$ CARRIER GAS			23% COLUMN LOADING; N$_2$ CARRIER GAS	
$\bar{u}$, cm sec^{-1}	V'_R, mm CHART	W_b, mm CHART	$\bar{u}$, cm sec^{-1}	V'_R, mm CHART	W_b, mm CHART
2.49	60.1	6.3	0.71	307.2	27.2
4.78	31.2	2.7	1.38	158.9	12.3
6.86	21.3	1.8	2.72	82.4	6.2
10.51	14.5	1.3	6.12	36.2	3.2

(a) Graph the values of effective plate height versus the average linear velocity. (b) Estimate the B and C terms of the van Deemter equation. For the A term use the average particle diameter of the packing (0.045 cm). (c) Estimate the optimum velocity and the minimum plate height for each column. (d) Calculate the retention time for an unretained component at each linear velocity for each column. (e) Calculate k' for each column. (f) The three columns operated with hydrogen as carrier gas differed only in the amount of liquid stationary phase. How did their mobile-phase velocities differ? And why? (g) Through what range of mobile-phase velocities can 90% of the column efficiency be retained? (h) On the 23% column loading with H$_2$ carrier gas, a particular separation requires 1500 theoretical plates. What is the fastest mobile-phase velocity at which this can be achieved? How much time will be saved by running at the maximum flowrate instead of at the optimum velocity?

5. Tabulated are values of plate height versus average linear gas velocity for n-heptane ($k' = 7.25$) at 75°C on a 15-m SCOT column with 0.51-mm bore, and prepared with squalane liquid phase:

$\bar{u}$, cm sec^{-1}:	6.5	12.2	19	27	43
H, mm:	1.20	0.91	0.93	1.08	1.48

(a) For each data point, calculate t_R, t'_R, theoretical and effective plate numbers. (b) Graph H versus $\bar{u}$; calculate $\bar{u}_{opt}$ and H_{min}. (c) Graph N/t_R versus $\bar{u}$ and estimate the optimum practical gas velocity.

6. The performance of methylnaphthalenes on columns of 80/100 mesh Chromosorb P and on 200/230 mesh microbeads is given. All columns were 152 cm in length and 6.0 mm in diameter. Liquid substrate was silicone oil 710 [density = $1.000 - (0.0008 t_c)$].

Column Number		1	2	3	4
Solid Support			Chromosorb P		Glass Beads
Liquid loading, %(w/w)		30	10	3	0.16
Weight of liquid, g		5.55	1.44	0.40	0.060
Temperature, °C		182	142	100	90
Flowrate, ml min^{-1}		208	192	415	155
Average linear velocity, cm sec^{-1}		36.1	27.2	49.3	41.7
V_M, ml		15	18	22	9.4
V_R', ml	1-MeN	1020	870	1350	365
	2-MeN	910	780	1200	325
W_b, ml	1-MeN	92	72	127	32
	2-MeN	77	64	115	34

(a) For each column, calculate k', relative retention, and actual resolution obtained. (b) Compare the plate number required for $R = 1.0$ and 1.5 with the actual plate numbers obtained. (c) Calculate the volume of the stationary phase and the beta value for each column. (d) On four lines extending from 0 to 5 min, and placed one above the other, mark off the adjusted retention times for each solute on the four columns. Comment upon the analysis times.

7. The adjusted retention times for the hydrocarbons and alcohols are tabulated:

Solute	t_R', min	Solute	t_R', min
Toluene	0.297	Decane	1.226
Nonane	0.639	1-Octanol	1.863
1-Heptanol	0.932	Undecane	2.613
4-Octanol	1.051	Dodecane	5.024

(a) Calculate the Kovats retention index for each compound. (b) Predict the retention times for 1-nonanol and 1-decanol.

8. The following data were obtained on a 10-m WCOT column, 0.25-mm bore, coated with SP-2100, operated at 100°C. Helium carrier gas velocity was 37 cm sec^{-1}.

$W_{1/2}$, sec:	0.45	0.60	0.88	1.40	2.41	4.40
t_R', min:	0.067	0.88	1.27	2.03	3.48	6.22
Alkane:	$n\text{-}C_8$	$n\text{-}C_9$	$n\text{-}C_{10}$	$n\text{-}C_{11}$	$n\text{-}C_{12}$	$n\text{-}C_{13}$

(a) Calculate the retention time for a nonretained compound. (b) Graph log t_R' versus carbon number. Predict the value for $n\text{-}C_{14}$. (c) Calculate k' for each alkane. (d) Calculate both the theoretical plate number and the effective plate number for each alkane. Graph the results.

9. The separation of hydrocarbons was done on a column 1.5 m in length with a 2.3-mm bore, packed with n-C_8 alkane bonded to porous silica. Column temperature: 25°C. Detector: flame ionization. Carrier gas: nitrogen at 25 ml min^{-1}. Retention characteristics of selected peaks are:

COMPOUND	PEAK	t_R, min	W_b, min
Methane	1	0.39	
Propane	4	0.86	
n-Butane	7	1.75	
trans-Butene-2	10	3.12	
cis-Butene-2	11	3.43	0.31
n-Pentane	13	4.31	
2-Methylpentane	17	9.55	0.71

(a) Calculate the Kovats retention indices for all the compounds. (b) Estimate the position of n-hexane on the chromatogram. (c) What is the effective plate number for 2-methylpentane? Plate height? (d) What is the resolution between trans-butene-2 and cis-butene-2? (e) Calculate k' for cis-butene-2 and for 2-methylpentane. (f) To achieve a resolution of 1.5 for the separation of peaks 10 and 11, how many plates would be required? (g) Suggest three ways in which the additional plates could be achieved for part (f). Provide a firm number or basis for each suggested alteration in operating procedure. (h) What is the velocity of the carrier gas through the column?

10. The analysis of a benzene-cyclohexane pair at 65°C on a SCOT column prepared with squalane liquid phase (β = 75) was conducted (A) at the optimum average velocity (18 cm sec^{-1}) and (B) at the optimum practical velocity (60 cm sec^{-1}) for helium as carrier gas. α = 1.24. For the cyclohexane peak, k_2' = 1.41. Plate height was 0.56 mm and 1.08 mm, respectively. The desired resolution is 1.5. (a) Calculate the number of plates required and the theoretical plate number for each separation condition; also the required column length. (b) Note the ratio $H/\bar{u}$ for each condition. Comment on the column lengths required and on the analysis times needed. (c) What is t_M?

11. Adjusted retention volumes are given for a series of n-alcohols and n-acetates. Column temperature was 77°C. Helium flowrate was 89 ml min^{-1}.

	V_R', ml	
	15% CARBOWAX 400	15% NUJOL
Acetates		
Methyl	43.2	19.0
Ethyl	60.5	48.4
Propyl	106	120
Alcohols		
Methyl	57.5	6.3
Ethyl	110	14.1
Propyl	207	32.7
Butyl	408	74.1

(a) Prepare a graph of log retention volume against carbon number for each homologous series on each column. (b) Graph the log retention volume on one column substrate against the log retention volume on the other column for each family.

12. Specific retention volumes (in ml g^{-1}) for some chlorinated hydrocarbons and benzene at several column temperatures on three column substrates are listed.

TEMP., °C	CH_2Cl_2	$CHCl_3$	CCl_4	$CCl_2=CCl_2$	C_6H_5Cl	C_6H_6
			Paraffin			
74	29.5	74.3	141	568	677	136
97	17.0	41.5	76.5	273	323	74.3
125	8.08	19.6	34.5	105	124	
			Tricresyl phosphate			
74	56.1	137	99.8	359	826	133
97	30.3	68.5	52.7	170	372	69.4
125	13.9	29.6	24.7	69.5	143	
			Carbowax 4000			
74	69.6	138	55.8	165	548	89
97	31.9	59.3	27.4	78.7	235	43.3
125	13.5	24.2	13.5	31.1	83.6	

All columns were made from coated Chromosorb packed into columns 1.8 m in length and 6.4-mm bore. The paraffin column contained 4.58 g on 13.93-g support; the tricresyl phosphate column, 4.46 g on 13.71 g; and the Carbowax 4000 column, 4.38 g on 13.1 g. Densities of the liquid phases at 74°/4° in g cm^{-3}: paraffin, 0.768; tricresyl phosphate, 1.128; and Carbowax 4000, 1.081. (a) Compute the heat of vaporization for each solute on each of the stationary liquid phases. (b) Calculate the partition coefficient at each temperature for each solute on each column substrate. (c) Calculate the value of V_g at 150°C for the solutes on the Carbowax 4000 column. (d) What temperature change halves V_g on each column for each solute?

13. An isopropylbenzene peak has a retention time of 5.36 min at 200°C and 3.15 min at 225°C on a Carbopack C/0.1% SP-1000 column which has an efficiency of 2900 theoretical plates. What is the highest column temperature which can be used such that the peak width will not be less than 10 sec?

14. 1,2,3-Trimethylbenzene has an adjusted retention time of 21.3 min at 200°C and 13.3 min at 225°C. Will it be possible to elute this compound in less than 10 min? The liquid phase has a maximum operating temperature of 275°C.

15. Listed are the retention times for various compounds on a column of porous styrene-divinylbenzene copolymer, 360 cm in length and 3-mm bore, that contains 800 theoretical plates.

ADJUSTED RETENTION TIMES, MIN, FOR COLUMN TEMPERATURE, °C

COMPOUND	150°	200°
2-Propanol	1.11	0.47
1-Propanol	1.50	0.60
t-Butyl alcohol	1.73	0.60
Isobutyl alcohol	2.90	0.90
1-Butanol	3.50	1.00

(a) What column temperature would permit $R = 1.5$ for the most difficult pair to be separated? (b) At 150°C, how long a column would be required to achieve the same degree of separation, that is, $R = 1.5$? (c) What is the approximate difference in the $\Delta \bar{H}_v$ values of these two solutes?

16. Determine the time of analysis for each of these columns under the assumed conditions: (a) a packed column with a phase ratio of 20; (b) a WCOT column with a phase ratio of 120; and (c) a SCOT column with a phase ratio of 60. Squalane is the liquid substrate. Assume that $H = 0.60$ mm on each of these columns, and that the column temperature is 75°C. The partition ratio, k', of the second peak on the packed column is to be taken as 1, 2, 3, 6, 12, and 30 in the six comparative cases. Assume a relative retention ratio of 1.10 for the adjacent peaks and that the resolution desired is 1.5. The three columns are operated at the following average linear gas velocities with helium as carrier gas: packed column, 6 cm sec^{-1}; WCOT column, 12 cm sec^{-1}; and SCOT column, 18 cm sec^{-1}. Note the k' value for the minimum analysis time on each type of column. Graph the analysis time for each comparative case along sets of time scales.

17. The following information was obtained from three isothermal chromatograms of n-C_7 to n-C_{11} alkanes run on a column consisting of 10% SE-30 on Chromosorb W (30/60 mesh). Helium was the carrier gas. Column was 122 cm in length and the bore was 10 mm.

COLUMN TEMPERATURE

	120°C		140°C		160°C	
SOLUTE	t_R, min	W_b, min	t_R, min	W_b, min	t_R, min	W_b, min
Air peak	0.98		0.95		0.90	
n-C_7	1.72	0.24	1.49	0.18	1.22	0.14
n-C_8	2.28	0.31	1.84	0.25	1.43	0.20
n-C_9	3.26	0.42	2.46	0.34	1.78	0.25
n-C_{10}	5.03	0.65	3.46	0.43	2.30	0.33
n-C_{11}	8.02	0.96	5.07	0.60	3.15	0.43

(a) Devise a suitable temperature program for separating the n-alkanes using the step-function approximation method. Estimate the retention temperature for each alkane. What would be the maximum heating rate that should be employed? (b) Predict the retention temperatures for n-C_{12}, n-C_{13}, and n-C_{14} alkanes. (c) On separate graphs, one for each operating temperature, plot both the theoretical and effective plate numbers of the normal alkanes.

18. The following information was obtained from three isothermal chromatograms of the straight-chain alkyl acetates run on the same column described in Problem 17.

COLUMN TEMPERATURE

SOLUTE	80°C t_R, min	100°C t_R, min	120°C t_R, min
Air peak	1.00	0.95	0.90
Methyl	1.50	1.28	1.13
Ethyl	1.95	1.57	1.30
Propyl	2.95	2.08	1.60
Butyl	5.00	3.08	2.13
Pentyl	8.95	4.93	3.08

Devise a suitable temperature program for separating the *n*-alkyl acetates using the step-function approximation method. Estimate the retention temperature for each alkane. What would be the maximum heating rate that should be employed?

19. A chromatographic column, 45.7 m in length, is operated under two different average linear gas velocities: 28.5 and 49.4 cm sec^{-1}. The respective number of theoretical plates is 81,920 and 47,480. (a) What are the analysis times at each velocity for the methyl oleate peak ($k' = 4.71$)? (b) To what degree can the methol hexanoate ($k' = 0.12$) and the methyl octanoate ($k' = 0.18$) peaks be resolved on this column? How many theoretical plates would be required for $R = 1.5$ at these retention times? How long a column is required? (c) Repeat part (b) for methyl decanoate ($k' = 0.33$) and methyl dodecanoate ($k' = 0.63$).

20. The separation of C_1–C_4 formates and acetates on Poropak Q is made according to total carbon number. Among the formate esters, only *n*-alkyl members are present. Except for peak 1 (methanol), identify the number peaks in the chromatogram shown.

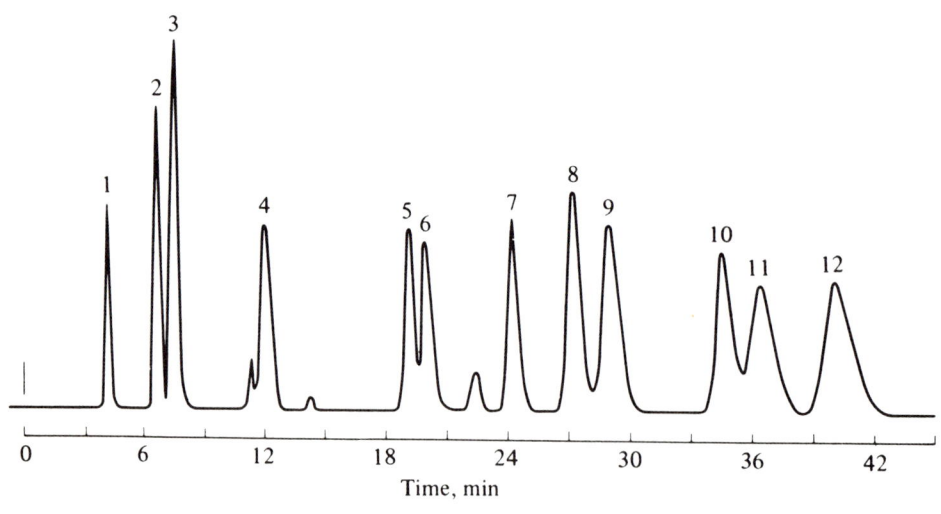

PROBLEM 16-20

21. The realtive response factors for *p*-xylene and toluene (relative to benzene assigned a value of unity) were found to be 0.570±0.0327 and 0.793±0.0178, respectively. Measurements were made by peak height. Unknown mixtures of these three solutes gave these peak heights. Calculate the percent composition of each sample.

	PEAK HEIGHT, mm		
SAMPLE	BENZENE	p-XYLENE	TOLUENE
1	98	87	86
2	136	82	63
3	148	51	97
4	52	48	81
5	85	35	42

22. A series of saturated aldehydes (total amount = 3.0 mg) were chromatographed on a column consisting of 3.3% silicone oil, operated at 180°C. Adjusted retention times, base widths, and peak heights were:

COMPONENT	C_5	C_6	C_7	C_8	C_9	C_{10}
t'_R, min	10.1	13.2	18.8	24.3	32.8	45.0
W_b, min	2.8	4.4	4.0	5.2	8.0	12.0
Peak height, mm	15.0	16.4	16.1	16.0	13.5	11.5

(a) Calculate the amount of each aldehyde in the sample. (b) The C_5/C_6 peaks are incompletely separated. If the column were to be lengthened, what should be the new length? (c) Focusing upon the column temperature, what should be the new operating temperature for $R = 1.5$? $\Delta H_s = 8.00$ kcal/mol for C_6 aldehyde.

BIBLIOGRAPHY

David, D. J., *Gas Chromatographic Detectors*, Wiley, New York, 1974.
Ettre, L. S. and W. H. McFadden, *Ancillary Techniques of Gas Chromatography*, Wiley-Interscience, New York, 1969.
Grob, R. L., Ed., *Modern Practice of Gas Chromatography*, Wiley, New York, 1977.
Harris, W. E. and H. W. Habgood, *Programmed Temperature Gas Chromatography*, Wiley, New York, 1966.
Laub, R. J. and R. L. Pecsok, *Physicochemical Applications of Gas Chromatography*, Wiley-Interscience, New York, 1978.

LITERATURE CITED

1. Hamilton, C. H., "Sampling Techniques," in *Instrumentation in Gas Chromatography*, J. Kruger, Ed., Centrex, Eindhoven, 1968.
2. Karasek, F. W., "Sampling Techniques in Chromatography," *Res/Dev.*, p. 54 (September 1973).
3. Wolfe, C., R. J. Levy, and J. Q. Walker, "Pyrolysis in Gas Chromatography," *Industrial Research*, p. 40 (January 1971).
4. Perry, J. A. and C. A. Feit, "Derivatization Techniques in Gas–Liquid Chromatography," in *GLC and HPLC Determination of Therapeutic Agents*, Part 1, K. Tsuji and W. Morozowich, Eds., Marcel Dekker, New York, 1978.

5. Rooney, T. A., L. H. Altmeyer, R. R. Freeman, and E. H. Zerenner, "Rapid GC Separations Using Short Glass Capillary Columns," *Am. Lab.*, p. 81 (February 1979).
6. Ettre, L. S., *Open Tubular Columns in Gas Chromatography*, Plenum, New York, 1965.
7. Novotny, M., "Contemporary Capillary Gas Chromatography," *Anal. Chem.*, **50**, 16A (1978).
8. Horvath, C., L. S. Ettre, and J. E. Purcell, "Support-coated Open Tubular Columns," *Am. Lab.*, p. 75 (August 1974).
9. Ottenstein, D. M., "Column Support Materials for Use in Gas Chromatography," *J. Gas Chromatogr.*, **1**, 11 (1963).
10. Dave, S. B., "A Comparison of the Chromatographic Properties of Porous Polymers," *J. Chromatogr. Sci.*, **7**, 389 (1969).
11. Grushka, E. and E. J. Kikta, Jr., "Chemically Bonded Stationary Phases in Chromatography," *Anal. Chem.*, **49**, 1005A (1977).
12. Ettre, L. S., "The Kovats Retention Index System," *Anal. Chem.*, **36**, 31A (1964).
13. Rohrschneider, L., *J. Chromatogr.*, **22**, 6 (1966); in *Advances in Chromatography*, Vol. IV, Marcel Dekker, New York, 1967.
14. McReynolds, W. O., *J. Chromatogr. Sci.*, **8**, 685 (1970).
15. Welsh, P., "Evaluation of Commercial Gas Chromatographic Ovens," *Am. Lab.*, p. 35 (May 1977).
16. Adlard, E. R., "A Review of Detectors for Gas Chromatography," Parts I and II, *CRC Critical Reviews in Analytical Chemistry*, **5**, 1, 13 (May 1975).
17. McCown, S. M. and C. M. Ernest, "Gas Chromatographic Detectors: A Syllabus of Errors," *Am. Lab.* p. 33 (May 1978).
18. Boucher, E. A., "Theory and Applications of Thermistors," *J. Chem. Educ.*, **44**, A935 (1967).
19. Craven, J. S. and D. E. Clouser, "A Fresh Design for Thermal Conductivity Detectors," Paper 408, 30th Pittsburgh Conference on Analytical Chemistry and Applied Spectroscopy, Cleveland, 1979.
20. Kolb, B., M. Auer, and P. Pospisil, "Reaction Mechanism in an Ionization Detector with Tunable Selectivity for Carbon, Nitrogen and Phosphorus," *J. Chromatogr. Sci.*, **15**, 53 (1977).
21. Natusch, D. F. S. and T. M. Thorpe, "Element Selective Detectors in Gas Chromatography," *Anal. Chem.*, **45**, 1184A (1973).
22. Patterson, P. L., R. L. Howe, and A. Abu-Shumays, "Dual-flame Photometric Detector for Sulfur and Phosphorus Compounds with Gas Chromatograph Effluents," *Anal. Chem.*, **50**, 339 (1978).
23. Burgett, C. A., "Electron Capture Detection," *Res/Dev.*, p. 28 (November 1974).
24. Driscoll, J. N., *et al.*, "Developments and Applications of the Photoionization Detector in Gas Chromatography," *Am. Lab.*, p. 137 (May 1978).
25. Ettre, L. S., "Factors Affecting the Speed of Gas Chromatographic Separations," *Am. Lab.*, p. 28 (December 1970).

CHAPTER 17

Liquid Column Chromatography: Instrumentation and Optimization

Only about 15% of the known compounds lend themselves to analysis by gas chromatography owing to insufficient volatility or thermal stability. Liquid column chromatography (LCC) does not have this limitation. The interchange or combination of solvents can provide special selectivity effects that are absent when the mobile phase is a gas. Ionic compounds, labile naturally occurring compounds, polymers, and high molecular weight polyfunctional compounds are conveniently analyzed by liquid column chromatography. While liquid flow in traditional liquid chromatography was achieved by gravity, modern LCC uses high-pressure pumps with relatively short, narrow-bore columns containing small particles of packing.

For isocratic operation, the general instrumentation for LCC incorporates these components (Fig. 17-1): (1) a solvent reservoir for the mobile phase; (2) a solvent pump (or pumps), equipped with a damping unit if a pulsating action results, to force the mobile phase through the chromatographic system; (3) a precolumn, except for bonded phases, and a guard column, the former to presaturate the mobile phase with the stationary phase and the latter to prevent contamination of the separation column; (4) a pressure gauge inserted close to the column to measure column inlet pressure; (5) a sampling or injection device to introduce the sample into the column; (6) the separation column; and (7) a detector with recorder readout or other data handling device. Many of these components will be discussed in some detail in this chapter.

17.1 SOLVENT DELIVERY SYSTEM[1]

Several features of the solvent delivery system must be considered: precise delivery of solvent over a relatively broad flow range; maximum pressure attainable; compatibility

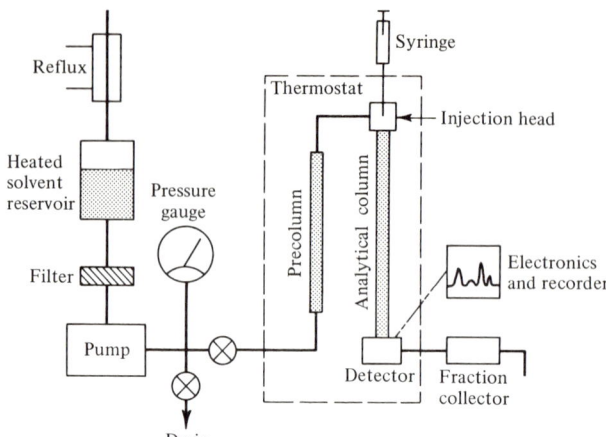

FIGURE 17-1 General instrumentation for liquid column chromatography.

with other components in the high-performance liquid chromatography (HPLC) system; compatibility with a wide choice of solvents; and noise level in the detector resulting from any pulsations. A degasser may be needed for removing dissolved air and other gases from the solvent. Final choice of pump will be interwoven with the type of separation column, the detector employed, whether isocratic or gradient elution is to be performed, the minimum detectability limit desired, precision in quantitation, and the cost of the packaged chromatograph. Three main types of pumps are used in LCC to propel the liquid mobile phase through the system: (1) displacement type which possesses a limited reservoir but offers nonpulsating flow; (2) the reciprocating type which has an unlimited reservoir but provides a pulsating flow; and (3) a pressure vessel which has a limited reservoir but gives a nonpulsating flow.

Reciprocating Pumps

Reciprocating pumps employ small-volume chambers with reciprocating pistons to work directly on the solvent, or diaphragms in which a hydraulic fluid transmits the pumping action to the solvent via a flexible diaphragm (Fig 17-2). Two check valves are synchronized with the piston or diaphragm drive to allow alternate filling and emptying of eluent from the chamber. Constant pressure is achieved with the out-of-phase relationship of the dual piston cycles; the displacement stroke of one piston is simultaneously matched to the filling stroke of the other. By overlapping the initial and final phases of both piston displacement strokes, the pressure drop caused by the slowing of one piston at the end of its stroke is compensated by the simultaneous engagement of the other piston. The net result of stroke overlap is constant pressure throughout the cycle for a solvent of maximum compressibility.

Although piston cycle overlap minimizes pressure variations, it does not eliminate pulsations in the flowrate arising from specific solvent compressibility. The residual pul-

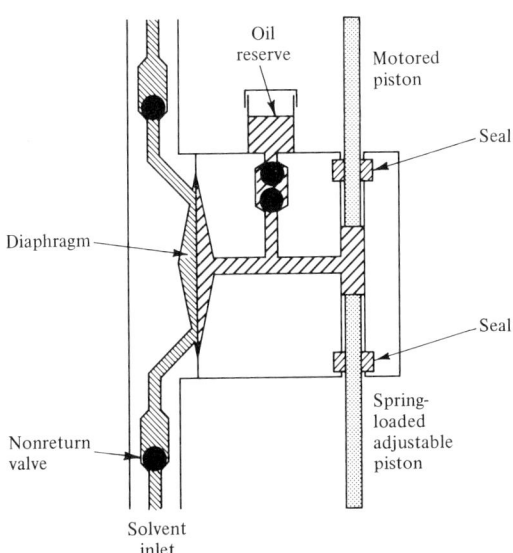

FIGURE 17-2 Reciprocating pump.

sations in each cycle occur because solvent drawn into the displacement chambers of the pump must be compressed to operational pressure prior to the onset of displacement. Therefore some type of damping must be employed. Damping methods include: (1) a triple-headed pump, (2) a tube with an air space, (3) a flexible bellows or tube, (4) a restrictor, or (5) incorporation of a correction based on a feedback monitoring of pressure and/or flowrate.

In a triple-headed pump, two heads are in different stages of filling as the third is pumping. Solvent flow is smoothed considerably. In the second and third damping methods, a gas (air space) or a flexible metal vessel takes up some of the pulsation energy, and when the pump refills, this energy is released to help smooth the pressure pulsations. This is similar to the action of a capacitor in electrical circuits. Simplest and as effective as most is the fourth method. A 25-cm length of 4-mm bore steel tubing, packed with 20-μm glass beads, is placed between the pump and the column. However, this type of damping device is only applicable to diaphragm reciprocating pumps which have a high stroke rate (about 100 min^{-1}). In one version of the fifth method, the pistons are driven by a cam with a specially designed profile. At constant rotation speed, this cam profile generates a precompression stroke sufficient for highly compressible solvents at maximum operating pressure. Differences between this primary setting and the actual compressibility of the operating solvent are monitored by an electronic drive control. Excessive precompression is corrected by decreasing the cam rotation speed during each precompression stroke. Automatic iterative adjustments of the cam speed program by the drive control logic produce the ideal compression compensation.

In the simplest single-head, sinusoidal-drive configuration, reciprocating pumps are relatively inexpensive and thus find frequent use in LCC. Two limitations of reciprocat-

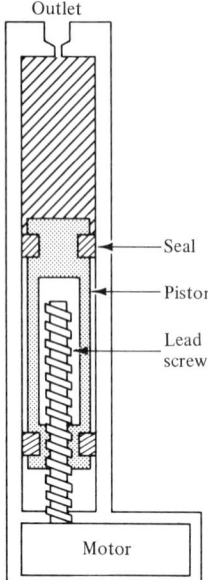

FIGURE 17-3 Syringe-type pump.

ing pumps are their inability to provide reproducible flow control over a range of solvent types without incorporation of flow monitoring devices or elaborate flow feedback circuitry, and their tendency to cavitate (with the formation of gas bubbles) with many useful solvents.

Syringe-Type Pumps

Syringe-type pumps work on the principle of positive solvent displacement by a piston mechanically driven at a constant rate in a piston chamber (250- to 500-ml capacity) with the generation of a pulseless flow with high-pressure capabilities (200–475 atm). A diagram is shown in Fig. 17-3. Tandem operation allows the use of multishaped forward and reverse gradients and simple flow programming. This type of pump provides the lowest dead volume pumping system and permits flushing solvents in any sequential manner simply by a push-button actuation of a purge valve.

Constant-Pressure Pumps

Constant-pressure pumps employ gas cylinder pressure either to drive or to regulate the pressure on the eluent, as shown in Fig. 17-4. The gas amplifier pump uses a large area piston to drive the eluent. Since the pressure on the eluent is proportional to the ratio of the area of the two pistons (usually between 30:1 and 50:1), a low-pressure gas source of 1–10 atm can be used to generate high liquid pressures of 1–400 atm. A valving arrangement permits rapid refill of the eluent chamber whose capacity is about 70 ml.

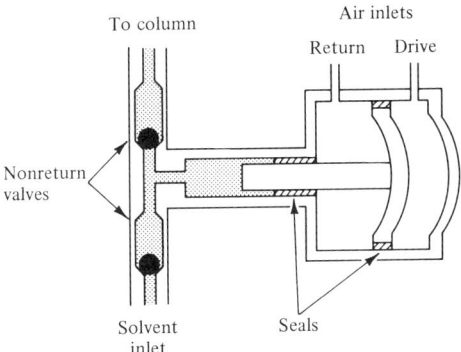

FIGURE 17-4 Constant-pressure pump.

Although pulseless, the flowrate and, hence, elution volumes can vary with changes in permeability of the column or viscosity of the solvent.

Applicability of Pumps

A pump should operate up to at least 200 atm, although 100 atm is adequate for much work. For analytical columns only moderate flowrates of approximately 1–2 ml min^{-1} need to be generated. The low noise levels generated by the syringe-type pump, the dual-head special-drive pump, and the gas amplifier and pressure-regulated pumps make these pumping systems most useful with columns of low capacity. Under preparative scale conditions involving wider diameter columns, pumps with high flowrates and eluent volume capabilities, such as multihead, sinusoidal-drive, and gas amplifier pumps, are required. With the large sample sizes that can be loaded on preparative columns, less sensitive detection is required. Consequently, pulsation noise is often a minor problem and inexpensive reciprocating pumps are usually adequate.

In LSC and LLC the permeability of columns will tend to be low, thus an important factor in selecting a pump is the upper pressure limit. The ease with which solvents may be changed is an important consideration in gradient elution or when scouting for the optimum solvent.

To control mobile-phase composition during gradient elution, the two liquids, contained in separate supply reservoirs, are supplied under pressure to a pair of proportioning valves, as shown in Fig. 17-5. Each valve is alternately opened to allow a prescribed amount of each liquid to flow into the mixing chamber of 0.5–2.0-ml capacity. To obtain a mobile phase of specific composition, one programs the ratio of valve A and valve B open times. If the ratio is constant, the mobile-phase composition is constant. If the ratio changes with time, a gradient is produced. Gradients may be generated from a number of basic gradient forms; nine are illustrated in Fig. 17-6. Time rate for addition of the secondary solvent may be selected for each of the basic gradient forms. Of the exponential gradient forms, four have fast rates of change at the beginning of the gradient with continually decreasing rates, and four have slow starting rates with continuously increasing rates of change. The interesting feature of the gradient (exponent 9), shown on the

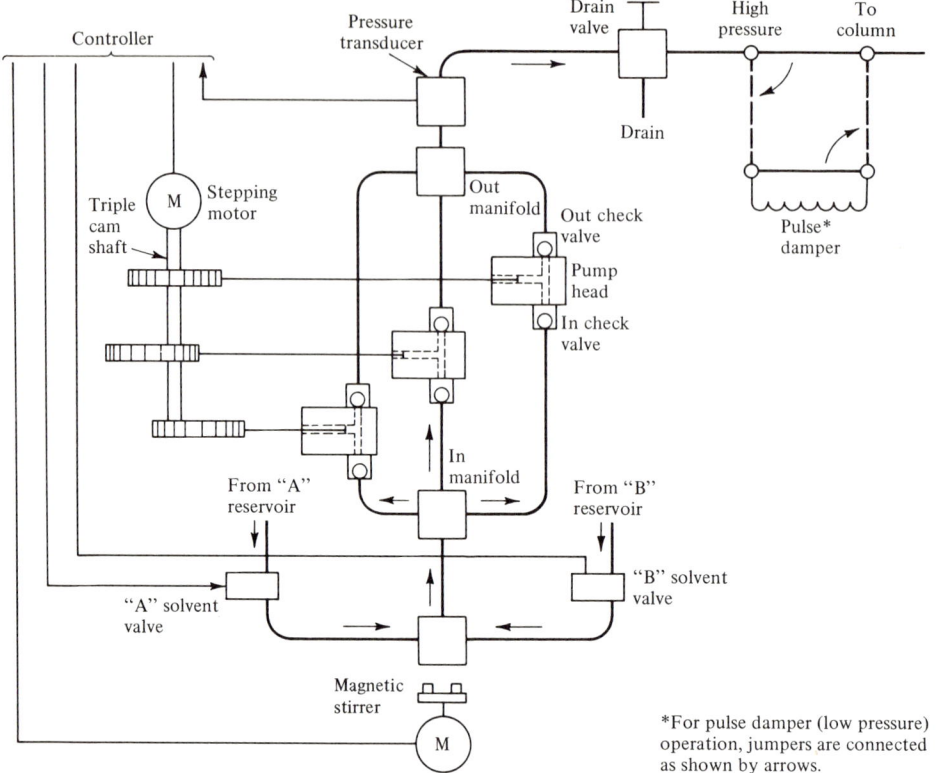

FIGURE 17-5 Gradient elution system schematic for mixing two solvents. (Drawing courtesy E. I. DuPont de Nemours, Inc.)

extreme right of Fig. 17-6, is that approximately 50% of the run is isocratic, holding at the initial composition. Thus, by using this gradient profile, it is possible to perform essentially isocratic separations followed by a built-in flush with the more powerful solvent in the mobile phase.

17.2 SAMPLE INTRODUCTION

The ideal sample introduction method should be able to insert reproducibly and conveniently a wide range of sample volumes into the pressurized column as a sharp plug (rectangular injection) with little loss in efficiency. The injection system should possess zero dead volume to prevent loss of resolution.

Syringe Injection

In the syringe-septum injection method, a small (10 μl) sample is introduced into the pressurized column with a high-pressure syringe through a self-sealing elastomer septum

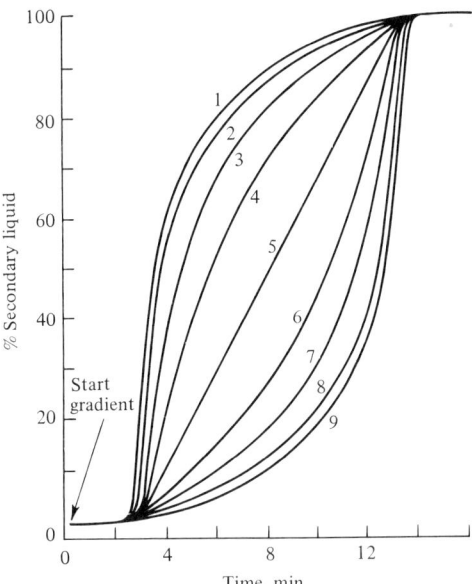

FIGURE 17-6 Basic gradient forms.

and directly on top of the column packing. The method is limited to low pressures (up to 34 atm) unless a double septa design is employed. The latter can withstand pressures up to 200 atm; at higher pressures there is sample leakage around the syringe plunger, difficulty in maintaining the septum leak proof, and difficulty in inserting the needle into the pressurized system. Advantages include: low initial cost, variable and small sample volumes can be handled, and low band spreading.

In stopped-flow injection, the pressure in the column is fully released while the pressure on the upstream side is maintained constant. The syringe deposits the sample at the head of the column and, after its withdrawal, the pressure to the system is resumed. A number of factors can lead to band spreading but the method is simple, inexpensive, and can be used up to very high pressures. For syringe-type and reciprocating pumps, flow can be brought to zero in the column and rapidly resumed by diverting the eluent stream by means of a three-way valve before the injector.

In the syringe-septumless system the needle of the high-pressure syringe is sealed to the injector by a Teflon collar. The needle is then opened to the pressurized system by a special valving arrangement and slid into the moving stream for sample injection. Reversing the procedure allows the syringe to be removed for filling. With compression fittings these units operate up to 575 atm, otherwise about 300 atm is the upper pressure limit.

Sampling Valves and Loops

Sampling valves allow the sample to be introduced reproducibly into a pressurized column (up to 475 atm) without signficant interruption of the flow. The technique is adaptable to automation. The sample is drawn into the valve core at atmospheric pressure.

When the core is pushed in with the slider system, the sample is pushed into the head of the column with minimal dilution.

The principle involved in loop injectors requires filling a small sample loop with sample via a syringe. By valve switching, the sample in the loop is displaced by diverting the flow of mobile phase through the loop and onto the column. The system is shown in Fig. 16-3 for GLC, but the arrangement applies to LCC also.

With either sliding valves or loops, the sample should be dissolved in portions of the mobile phase to eliminate unnecessary solvent peaks. Sample loop or core volumes are nominal but may be calibrated. Valve injection produces somewhat higher band spreading than syringe injection unless great care is taken in the coupling of the valve to the column.

General Considerations

The maximum sample volume (and sample concentration) should not exceed the linear capacity of the column. Linear capacity is defined as the weight of sample per gram of packing which causes a 10% reduction in the specific retention volume relative to the constant retention volume observed for smaller samples. Several experiments should be run with different solute concentrations to see whether or not the peak width, peak asymmetry, and retention time are constant. Generally, the maximum sample volume lies between 1 and 20 μl, but should be no greater than $0.5 t_R F/\sqrt{N} = 0.5 V_M (1+k')/\sqrt{N}$, a value which corresponds to an increase in plate height of about 2%. A sample volume of 10 μl is maximum for a 10-cm column with 5-mm bore and packed with particles possessing a porosity of 0.75 when the plate number is 10,000. Injection right on the top of the column packing, but not into it, is best for efficiency and long-term stability of the packing. Quite large injections can be accepted without affecting column performance. For injections of large amounts of solute, it is better to inject a larger volume of a more dilute sample than risk overloading at the top of the column by injecting a small volume of very concentrated solution.

17.3 SEPARATION COLUMN

Heavy wall glass or stainless steel tubing has generally been used to construct LCC columns in order to withstand high pressures. The tubing must have a smooth, precision-bore internal diameter to ensure that a well-packed column will not channel near the wall/packing interface because of wall irregularities. This would result in broader peaks and lower efficiency. Straight columns are preferred, operated in the vertical position, with the flow being directed either up or down through the packing. Connections to the column are made with low dead-volume fittings designed to eliminate stagnant pockets of mobile phase.

Columns with an internal diameter of 5 mm provide a good compromise between sample capacity, the amount of packing used and solvent required, and column efficiency. A 2.1-mm-bore packed column requires about five times the inlet pressure as the same length of 4.6-mm-bore column for the same flowrate. Of course, the linear velocity will not be the same. On the other hand, the larger bore column holds almost five times the amount of packing for the same length. Also, the larger bore column gives broader peaks which, however, minimize the effect of band spreading due to fittings, tubing, and detector dead volumes.

In a distinctly novel approach, the Waters radial compression module[2] applies hydraulic pressure (via glycerol contained within a plastic sleeve) to compress radially a flexible wall cartridge, 10 cm in length with an 8-mm bore, as shown in Fig. 17-7, and thereby produces a highly efficient column bed that is free of voids and channels. Separations are performed while the cartridge is under compression. When separations are complete, the cartridge

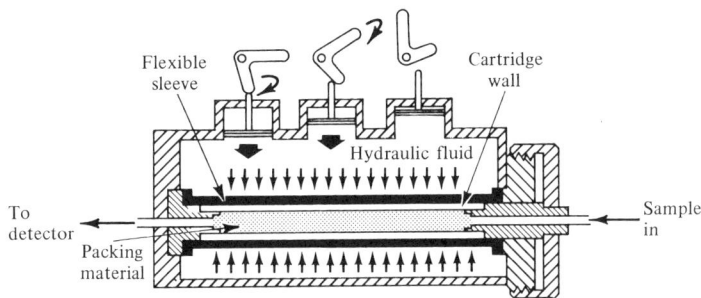

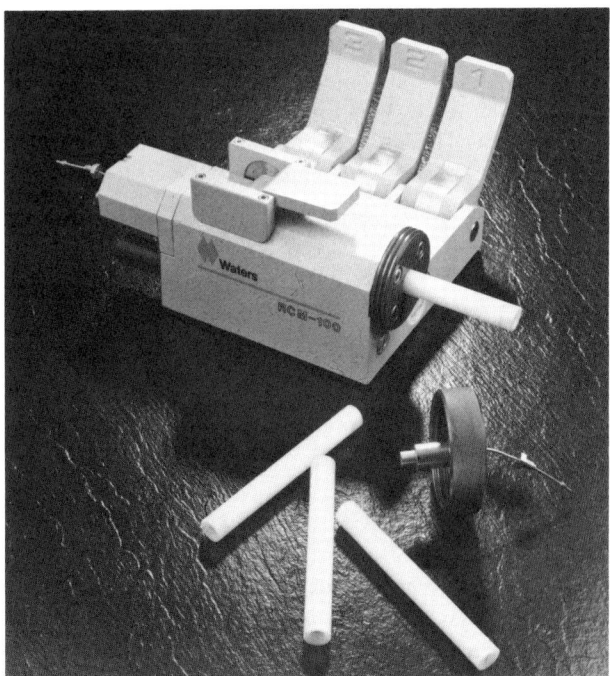

FIGURE 17-7 Radial compression module. The hydraulic pressure is applied by manually moving three pressure actuator levers in sequence. A heavy-duty rubber sleeve isolates the cartridge from the hydraulic fluid. (Courtesy of Waters Associates.)

can be decompressed, removed from the module, and reused repeatedly without losing efficiency. Low cartridge price, as compared with comparable steel columns, permits the dedication of a cartridge to each application. This maximizes the effective life of the cartridge and eliminates consumption of costly solvents for column purging. Column efficiency is 5000 plates per 10 cm, which is about twice the performance of a steel cartridge of the same dimensions and packed with the same material. Efficiency also decreases more slowly as the velocity of the mobile phase increases. Photomicrographs reveal the actual depressions formed on the flexible wall of the cartridge by the spheres of the packing material.

Current practice utilizes column packings that lie in the range from 3 to 7 μm in diameter; occasionally up to 10 μm or higher, especially for exclusion chromatography. These are much smaller diameters than were used in packings only a year or so ago. The rational will be obvious after reading the section on optimization. Each time the particle diameter is halved, the pressure drop required is raised by approximately a factor of 4. Often, however, the column length can be shortened significantly. Standard column lengths are 10-25 cm with totally porous particles, and ranging up to 50-100 cm for pellicular packings when the particle diameters exceed 10 μm.

Circulating air baths or water jackets are used to control the column temperature within 0.2°C over a range in temperature from 35°C to 140°C. The solvent is preheated with forced oven air, or a water bath, to the compartment temperature. Temperature control is especially important in liquid-liquid partition and ion-exchange chromatography which are more influenced by this parameter than adsorption or exclusion chromatography.

In the *coupled column* technique, the sample is placed upon a short precolumn. The fast-moving components are directed to a suitable analytical column; the slower-moving components are directed via a switching valve to a shorter analytical column on which the analysis is completed first. The valve is then switched to the first analytical column and the faster-moving components eluted.

Preceding the separation column should be a short 5-cm protection column to adsorb or filter out unwanted material. These units, called guard columns, are by design expendable and are periodically repacked, replaced, or reconditioned. Guard columns can decrease detector background, improve separation, decrease degradation of the sample, and extend column lifetimes by preventing contamination of the expensive separation column. If effectively matched and packed efficiently, guard columns have little effect on the plate height of the analytical column.

17.4 DETECTORS

Unfortunately there is, at present, no universal detector for LCC that possesses all the necessary attributes that are required for a completely versatile liquid chromatograph. Thus it is necessary to select a detector on the basis of the problem at hand and, in doing a variety of separations, it is expected that more than one detector will be needed.

Detectors for LCC can be placed into two categories. Bulk property detectors, typified by the refractive index monitor, compare an overall change in a physical property of the mobile phase with and without an eluting solute. Although universal, this type of detector tends to be relatively insensitive and requires good temperature control. The second class of detectors, the solute property detectors, respond to a physical property of the solute which is not exhibited by the pure mobile phase. Solute property detection is roughly 1000 times more sensitive, giving a detectable signal for a few nanograms of sample. Ultraviolet/visible absorption, fluorescence, and electrochemical detectors have achieved popularity in this category. Precolumn and postcolumn derivatization expand their applicability.

The response time of the detector is critical and should be at least 10 times smaller than the peak width in time units. Only then is the peak area not changed, although the corresponding reduction in the plate number is 16%. This condition can be written as $\tau \leqslant W/10 = 2t_R/5\sqrt{N}$. The time constant observed on many commercial detectors vary from 1 to 3 μsec, making the detector the weakest part of the chromatograph.

To estimate the detection limit using a particular detection system, the zone spreading must be known. This spreading includes the dilution factor caused by the injector, the column dead space, the connecting tubes, and the detector volume.[3] Any spreading caused by injection, transport through capillary connections, or detector volume is added to the random dispersion in the column. Symmetrical bands, such as chromatographic peaks, are additive as variances, that is,

$$\sigma_{tot}^2 = (\sigma_{inj}^2 + \sigma_{trans}^2 + \sigma_{det}^2) + \sigma_{column}^2 \qquad (17\text{-}1)$$

For the variances arising from the extra-column system to be neglected, they must collectively be less than $\sigma_{column}^2/2$ at the flow velocities needed for HPLC. Typically the dilution factor varies from 5 to 100, meaning that the concentration of the band peak in the detector is 1/5 to 1/100 the initial sample concentration before injection. A 10-fold greater concentration than this estimate would be needed for precise quantitation. When the system concentration is desired in terms of sample weight, the sample volume must be considered. Here is where the choice of column size enters the picture. Narrow-bore columns dilute small samples less than large-bore columns. To keep the increase in plate height resulting from remixing of the band in the detector cell smaller than 1%, the detector volume should be less than or equal to $0.1 t_R F_c/\sqrt{N}$. For example, the preceding condition is fulfilled with a 10-μl detector cell for a 10,000 plate peak with a flowrate of 0.5 ml min^{-1} if t_R is larger than 20 min. With the use of particle packings less than, or equal to, 5 μm, the detector volume will have to be 2–3 μl to ensure that the separations achieved in the column are not lost in the detector. The connecting tubing between the column and the detector must be kept to an absolute minimum. It should have a bore no more than 0.25 mm and be no more than 200 mm in length. Couplings should possess zero dead volume.

Ultimate detection limits in any detection system depend upon the availability of ultrapure solvents for mobile phases, or the ability to purify solvents in-house.

Photometric Detectors

An ultraviolet photometer operating at fixed wavelengths of 254 or 280 nm is one of the most widely used detectors for LCC. The advantages of this detector are its relatively low cost, the high sensitivity (nanogram level) achieved for many compounds of chemical and biological interest that absorb ultraviolet light, and its insensitivity to changes in temperature, flowrate, and mobil-phase composition. Compounds absorbing ultraviolet radiation include all substances having one or more double bonds and substances having unshared nonbonded electrons as, for example, all olefins, all aromatics, and compounds containing the carbonyl, thiocarbonyl, nitroso, and azo groups. The mobil-phase solvent should absorb only weakly or not at all. For solvents the ultraviolet cutoff is often considered to be the wavelength below which the solvent will absorb more than 1.0 absorbance unit in a 1-cm path cell. Water, methanol, and acetonitrile all allow operation in the far ultraviolet to at least 205 nm, and are commonly used as solvents for reversed-phase and ion-exchange chromatography. Hexane, pentane, isooctane with 2-propanol, acetonitrile, and tetrahydrofuran (cutoff at 212 nm) can be used for adsorption and exclusion chromatography.

The optical diagram of a fixed wavelength photometer is shown in Fig. 17-8. Two beams of light from a low-pressure mercury lamp pass through a collimating lens, the measuring and reference sample cells, a visible light-blocking filter, and finally fall upon dual CdS photodetectors. A filter placed after the lamp selects the 254-nm emission line of mercury or a specially developed phosphor convertor produces an emission band peaking at 280 nm. The low-pressure mercury lamp with approximately 90% of its radiation at 254 nm provides a very intense light source. Short-term noise levels are very low with this detector, usually less than 0.0001 absorbance unit. A feedback regulated power supply automatically compensates the bridge voltage for changes in lamp brightness and solvent absorption, thus providing an output voltage in the detector which is directly related to solute concentration. The photometer can operate in the single cell configuration with the analytical output of the photometer being referenced against air. However, it may be desirable to have not only a reference cell but also to have mobile phase flowing through the reference cell which is placed between the lamp and reference phototube. In this way the contribution to the absorbance from the mobile phase is negated; this is particularly desirable when working with gradients.

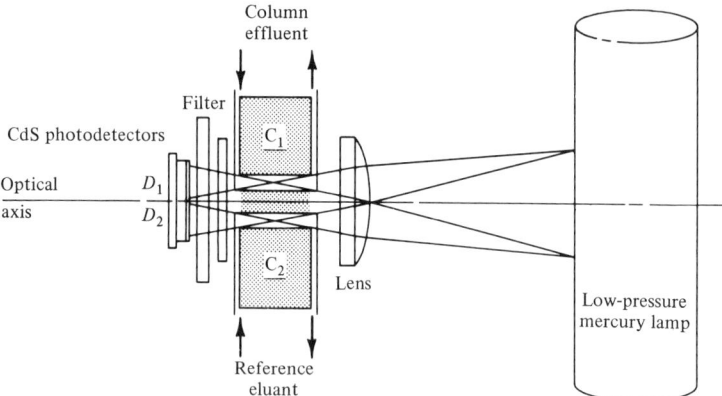

FIGURE 17-8   Fixed wavelength ultraviolet/visible photometric detector. (Courtesy of Laboratory Data Control.)

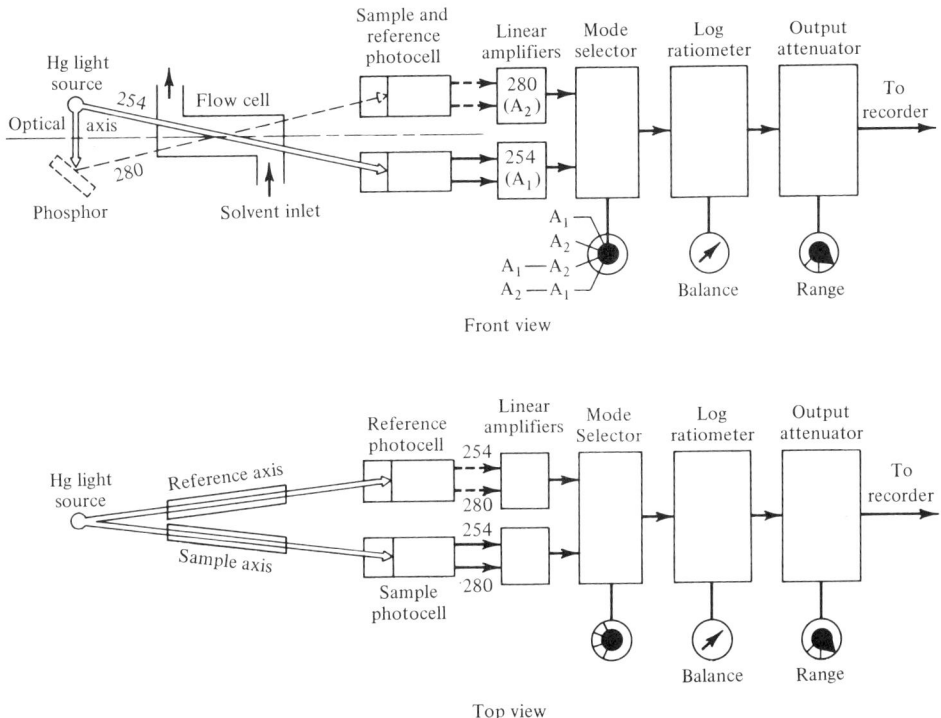

FIGURE 17-9 Dual-wavelength photometer. (Courtesy of Laboratory Data Control.)

In a dual-wavelength photometer the mercury lamp illuminates a screen partially coated with the phosphor convertor (or the coating is placed on the lamp envelope). The screen or coating acts as a composite light source. Each wavelength is separated into two beams, as shown in Fig. 17-9. The beams are completely separate in the flow cell.

The fixed wavelength ultraviolet photometer has the advantages of a very intense light source, and very low noise levels. The majority of compounds have some ultraviolet absorbance at 254 or 280 nm, even though the wavelengths do not coincide with the absorption maximum of the solutes. When compounds of protein or nucleic acid origin are being studied, the relative absorbance at 254 and 280 nm is often used for qualitatively classifying the chromatographic peaks of separated components. Proteins absorb more at 280 nm than at 254 nm; nucleic acid compounds do just the opposite.

The multiwavelength filter photometer uses the medium-pressure mercury lamp whose emission lines (and relative intensities) lie at 254 nm (150), 280 nm (9), 313 nm (81), 334 nm (7), and 365 nm (100). Narrow bandpass filters are used to select one of these emission lines. There may be cases in which it would be desirable to employ a broad bandpass filter which transmits light of several wavelengths. This type of filter would be particularly useful in preliminary investigations in which little or nothing is known about the ultraviolet absorbing characteristics of the sample.

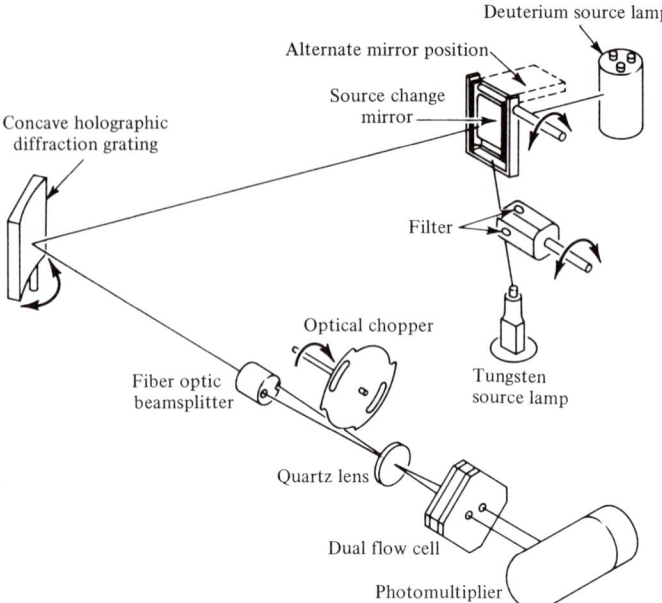

FIGURE 17-10 Optical schematic of a variable wavelength ultraviolet/visible spectrophotometer. (Courtesy of Laboratory Data Control.)

It would be ideal to use for each type of compound the wavelength most strongly absorbed. A spectrophotometer is the choice provided it can be efficiently coupled to a chromatographic column. One schematic diagram is shown in Fig. 17-10. By rotating the diffraction grating to the appropriate angle of incidence, the desired wavelength will be projected onto the fiber optic beamsplitter. Two separate beams of equal intensity are produced and focused onto the dual flow cells by a quartz lens. The advantages of the variable wavelength spectrophotometer are the availability of a wide range of wavelengths, the ability to change wavelengths without changing filters or lamps, and the ability to operate at wavelengths below 254 nm. The noise levels of the spectrophotometer are greater than the filter photometer since the wavelength is selected from the continuum of the deuterium or tungsten lamp through the use of a monochromator.

Typically, the flow cells for photometric detectors are 10 mm in length and 8–10 μl in volume. The path length should be as long as possible to enhance absorbance. If the sample cell is too short, a lens is required to concentrate the light energy into the cell space. Then, a change toward shorter wavelengths causes a change in focus. Internal scattering occurs, and nonlinear behavior may be encountered.

Several cell geometries are used: the Z-form, the H-form or split flow cells where eluent enters the center of the cell and flows outward in both directions, and a tapered cell. Light passes down the center limb in all designs. The H-design is claimed to be more suitable with respect to drift caused by flow variations of the mobile phase and would permit flow programming. The tapered cell construction allows better light transmission and offers greater stability and sensitivity.

A problem in absorbance detection is that of distinguishing between true sample output versus anomalies which occur due to refractive index phenomena of pure solvents within the flow cell. Deflected light which strikes the cell wall is absorbed and appears as an absorbance change on the detector. From the source lamp, light is emitted which, upon entering the cell and striking the liquid interface, is bent because of the refractive index difference between the quartz cell window and the solvent. Within the cell there are transients occurring due to changes in flow, solvent gradients, and sample components. The tapered cell geometry ensures that all of the emitted light which enters the cell will leave the cell if there is no true absorbance.

Absorbance ratioing is a fast, easy, and accurate method for measuring the purity of a chromatographic peak. These ratios, obtained at any two wavelengths under standard conditions, are absolutely specific for each compound and can be used to identify them and determine their impurity. As shown in Fig. 17-11, this method verifies that one is quantitating the correct peak and measures the purity of that peak prior to quantitation. By this method one can also monitor for the presence of unresolved impurities even when using gradient elution.

The limit of detection of photometric detectors to various compounds can be estimated if the approximate molar absorptivity is known at the operating wavelength and the noise level is known. Assuming a path length of 10 mm and a noise level less than 0.0004 absorbance unit, a concentration detection limit is: $2(\text{noise})/\epsilon b = 0.0008/\epsilon$ in mol cm^{-1} liter^{-1}. If ϵ is 10,000, the minimum detectable concentration is 80 nM liter^{-1} (or 40 ng ml^{-1} for a compound whose molecular weight is 500). If the system concentration is desired in terms of sample weight instead of concentration, the sample volume must be considered and the system dilution factor. For example, for a sample of 5 μl, a dilution factor of 20, and molecular weight of 500, the detection limit is:

$$\frac{(2)(0.0004)(20)(5 \times 10^{-6} \text{ liter})(500)}{(10{,}000 \text{ liter mol}^{-1} \text{ cm}^{-1})(1 \text{ cm})} = 4.0 \text{ ng}$$

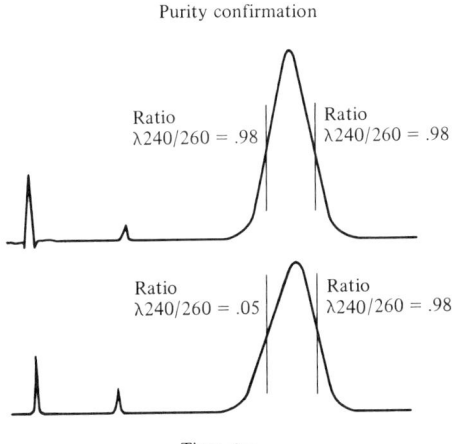

FIGURE 17-11 Absorbance ratioing data for a peak.

Thus the detection limit is about 1 ng at best and may be considerably larger when the operating wavelength lies to one side of the absorption maximum of the compound being monitored when using fixed wavelength photometers.

The response time of photometric detectors is 1 sec to reach 98% of full scale. They equilibrate instantly to solvent changes, as during gradient elution. The detector is subject to pressure effects, and a pulseless pump or a reciprocating pump with pulse damping must be used. Temperature affects the signal but to a lesser degree than with the refractive index monitor. The signal is severely impaired by the presence of air bubbles in the mobile phase which appear as spikes in the chromatogram.

Fluorometric Detector

Fluorescent detectors are either straight-through or 90° types. Due to stray light and subsequent light-scatter, the 90° type is recommended to minimize noise and to observe increased sensitivity. Increased selectivity for specific compounds may be obtained depending upon the excitation wavelength of the solutes. Fluorescence monitoring in tandem with ultraviolet monitoring may simplify a complex matrix problem.

Some instruments use a continuously variable monochromator with a deuterium or tungsten lamp which enables the operator to select excitation wavelengths. The detection system consists of an end-on photomultiplier tube with appropriate secondary filters to transmit the fluorescence signal and reject the excitation wavelength. The flow-through cell has an internal volume of 5 μl; its emission energy is intercepted by a half-sphere mirror and reflected onto the detector.

Less sophisticated instruments use appropriate primary and secondary filters, as shown in Fig. 17-12. The excitation lamp is a low-pressure, hot-cathode mercury lamp with a phosphor coating that emits near-ultraviolet radiation peaking around 360 nm. Visible light is blocked by filter F1 before entering the large diameter end of the cone condenser B. The highly reflecting internal surface of the condenser and its metal bifurcating plate direct the excitation light to cell chambers C1 and C2. A fluorescent coating on two

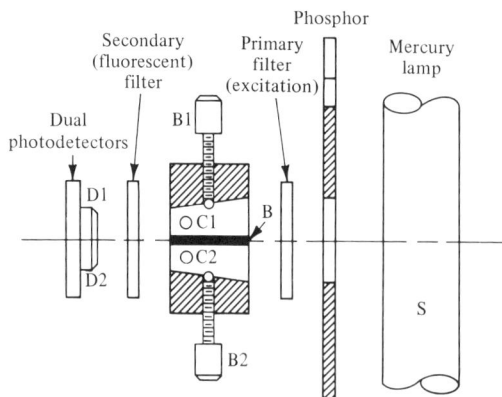

FIGURE 17-12 Optical diagram of a filter fluorometer. (Courtesy of Laboratory Data Control.)

adjusting screws B1 and B2 is provided to compensate for background solvent fluorescence. Emitted fluorescent light from the cell chambers passes through a sharp cutoff ultraviolet blocking filter F2 and impinges on the corresponding photosensitive elements of a dual photocell D. Cuvette volume is 13 μl.

Almost all fluorescing compounds absorb above 400 nm, and most of those absorbing below 300 nm, will have some absorption in the near ultraviolet and will, therefore, yield some fluorescent light although at less sensitivity. Typical fluorescing compounds would be polynuclear aromatics, steroids, plant pigments, vitamins, alkaloids, catecholamines, and systems using tagging reagents, such as fluorescamine or o-phthalaldehyde for amino acid analysis via postcolumn reactions. The solvents must be transparent to both the exciting ultraviolet energy and the fluorescence wavelengths. Sensitivity of the detector is approximately 1 ng ml^{-1} for strongly fluorescing compounds. It is usable with gradient elution. Fluorescence suffers from vulnerabilities such as turbidity and quenching.

Infrared Detectors

Organic compounds absorb energy at various wavelengths over the infrared spectrum. Many of these absorptions are directly related to the presence of specific functional groups, as outlined in Chapter 7. For example, the ester carbonyl absorbs at 5.78 μm. An infrared detector set at this wavelength is, for all practical purposes, a universal detector which responds to any organic compound exhibiting an ester carbonyl band.

The sensing unit for a fixed wavelength is a single-beam infrared spectrometer that employs an interference filter for wavelength selection. Cells consist of a stainless steel body with amalgam sealed windows. Path length of 1.5 or 3.0 mm, with internal volumes from 5 to 50 μl, enable detection limits to extend down to 5–10 ng ml^{-1}.

Infrared detectors are especially useful when the component of interest is not detected by ultraviolet/visible detectors or refractive index detectors, or when the analyst wishes to determine the functional group distribution. Care must be taken when selecting the solvent system and analytical wavelength, but once on line, the infrared detector is unequaled in the specificity of information it provides.

Differential Refractometers

Differential refractometers can operate on either one of two principles. The reflection type, governed by Fresnel's law, measures the intensity of the reflected light. The deflection type is governed by Snell's law and senses the deflection of the light beam. With either type the response depends on the difference between the refractive indices of the solute and the mobile liquid phase. The best refractive index monitors can discriminate to 10^{-7} unit, but since an average solute will only differ from the solvent by 0.1 unit, the overall sensitivity will be about 1 in 10^6, or at the microgram level. Response is affected by pressure fluctuations; thus either pulse damping or nonpulsating pump system is needed. Since the temperature coefficient for detector response is 0.0001 unit $°C^{-1}$, an efficient heat exchanger is required to preheat an incoming solvent and to control the cell temperature within 0.001°C. Use in gradient elution is restricted to a few pairs of liquids with virtually identical refractive indices. Response time is 2 sec to reach 99% final output reading.

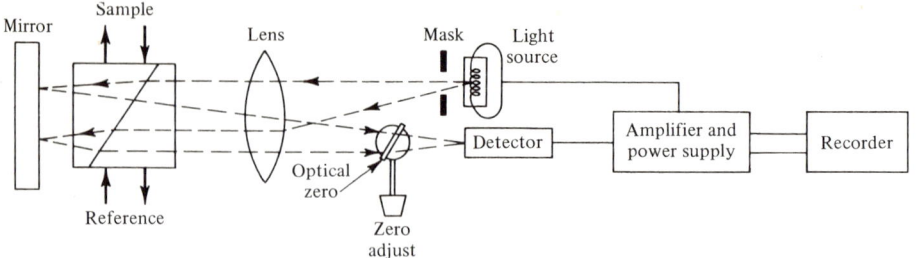

FIGURE 17-13 Deflection-type refractometer. (Courtesy of Waters Associates.)

The *deflection-type* of refractometer, illustrated in Fig. 17-13, measures the deflection of a beam of monochromatic light by a double prism as the eluent is passed through one half of it, the other half of the prism being filled with pure solvent. A beam of light from an incandescent lamp passes through an optical mask, which confines the beam within the region of the sample and reference cells. The lens collimates the light beam which passes through the cells, and is reflected back by the mirror through the cells again. The beam is then focused on a beamsplitter before passing to twin photodetectors. The reference and sample cell are separated by a diagonal glass divider. If the refractive index of the mobile phase changes due to the elution of a sample component, the beam from the sample cell is slightly deflected. As the beam changes location on the detector, an out-of-balance electrical signal is generated that is proportional to solute concentration. An optical flat, which deflects the beam from side to side, is used to adjust for zero output signal when identical material is in both prisms. One sample cell suffices for the entire refractive index range from 1.00 to 1.75 units. No alignment, other than resetting the base line reading to zero, is required when the solvent is changed. This type of refractometer is the easier of the two to use but has the larger cell volume (15–25 μl).

The *reflection-type* refractometer measures the change in percentage of reflected light at a glass–liquid interface as the refractive index of the liquid changes. In the optical path (Fig. 17-14) two collimated beams from the projector (light source, masks and lens)

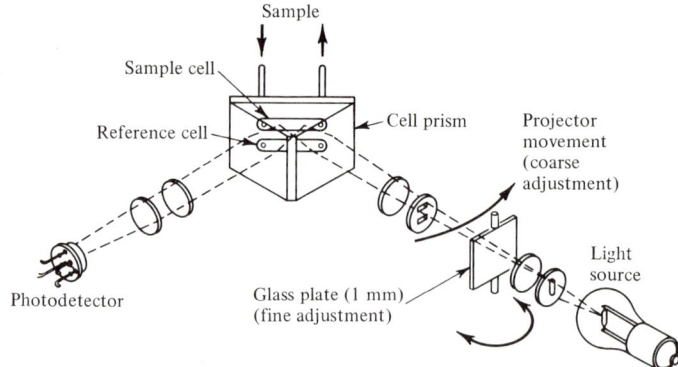

FIGURE 17-14 Optical diagram of reflection-type (Fresnel) refractometer. (Courtesy of Laboratory Data Control.)

illuminate the two liquid cells which are formed between the cell prism and a stainless steel, reflecting back-plate (finely ground to diffuse the light). The light passes through the flowing liquid in the cell and is reflected from the back-plate. This diffuse, reflected light appears as two spots of light which are imaged by lenses on dual photodetectors. Since the ratio of reflected light to transmitted light is a function of the refractive index of the two liquids, the illumination of the cell back-plate is a direct measure of the refractive index of the liquid (3 μl) in each chamber. When the angle of the incident light is correctly adjusted to slightly less than the critical angle for pure solvent passing through both chambers, equal intensities of light will strike the two detector elements. Rotation of the projector assembly permits coarse adjustment; fine adjustment is done with an optical flat rotated ±30° from normal. During the passage of a solute through one of the chambers, the refractive index will change and the fractions of reflected and transmitted light will alter, thus producing an imbalance in the signals of the detectors.

The Fresnel-type refractometer is not as sensitive or stable as the deflection-type refractometer. Interchangeable single units are required for the range from 1.31 to 1.45 (water, ethanol and hexane solvents) or 1.40 to 1.55 (chloroform, carbon tetrachloride, and benzene solvents). Dirt, bubbles, and films all exert a large effect on the signal with the reflection instrument, although bubbles should be excluded from the deflection type also.

Electrochemical Detectors[4]

The suitability of electrochemical detection ultimately depends on the voltametric characteristics of the molecules of interest in a polar aqueous or aqueous-alcohol mobile phase and at a suitable electrode surface. Most practical applications involve oxidizable components. Amperometric detectors, although their efficiency is often only 1–10%, are currently favored over coulometric detectors because they are more sensitive and less complex. Once the operating potential is chosen, chromatograms are obtained by plotting current as a function of time. Excellent current-to-voltage converters can be constructed from rugged integrated circuit operational amplifiers with metal-oxide-semiconductor/field-effect transistor (MOS/FET) inputs; consequently, electrochemical detection is quite inexpensive even for work at the pico-equivalent level and below where nanoampere currents are often encountered.

The active region of an amperometric detector consists either of a tubular electrode or a thin-layer cell of 1-ml volume. One design is shown in Fig. 17-15. Carbon paste forms a suitable electrode material. Because of low currents involved, it is possible to operate the cell in the classical two-electrode mode. Positioning an auxiliary third electrode downstream and opposite the working electrode reduces the uncompensated resistance to a negligible value even when employing mobile phases containing less than 10 mM of ionic material.

Typical compounds handled by electrochemical detection include aromatic amines and their derivatives (hydroxylamines, amides, quinoneimines), ascorbic acid, uric acid, cysteine and penicillamine, benzidine, folic acid, vitamin B_6, indoleacetic acid, 5-hydroxytryptamine, tetrahydroisoquinoline alkaloids, and phenothiazines.

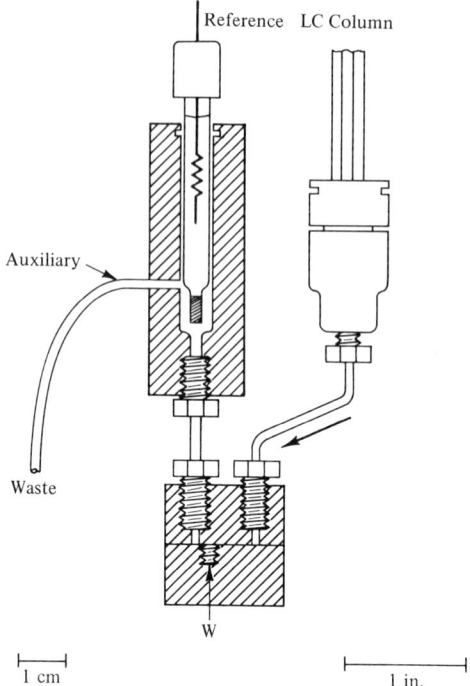

FIGURE 17-15 Thin-layer amperometric detector. (Courtesy of Bioanalytical Systems, Inc.)

Derivatization[5]

Workers in LCC are finding increasing value in pre- and postcolumn derivatization with respect to either selective and/or trace detection. Off-line derivatization is carried out almost exclusively before separation. The derivatized sample is then injected into the chromatographic system. On-line derivatization becomes part of the chromatographic system. It offers the largest measure of user convenience for a specific analysis or type of analysis. It does involve complicated preparation and in many cases the necessary instrument modifications must be considered permanent.

The most common derivatization techniques in HPLC are intended to enhance detectability by ultraviolet absorption or by fluorescence. Criteria that must be met include: only a single derivative should be formed from each parent compound; the reaction mixture should be free of by-products, including excess reagent, which can interfere with separation or detection; and the derivative should preserve differences in the parent compounds which allow separation.

In those cases where the detection sensitivity to the 254-nm detector in ultraviolet absorption is zero or very low, detection can be enhanced by attaching a chromophore with high absorption at 254 nm to the solute. This might be accomplished, for example, with the following reagents:

REAGENTS	REACTANTS
N-Succinimidyl p-nitrophenylacetate	Amines and amino acids
3,5-Dinitrobenzoyl chloride	Alcohols, amines, and phenols
p-Nitrobenzyloxyamine hydrochloride	Aldehydes and ketones
O-p-Nitrobenzyl-N,N'-diisopropylisourea	Carboxylic acids
p-Nitrobenzyl-N-n-propylamine HCl	Isocyanate monomers

The formation of fluorescent derivatives serves two purposes: it allows sensitive detection of otherwise nonfluorescent molecules; and it exploits the selectivity of fluorescence by allowing detection of all compounds with a particular functional group in a sample after derivatization. The number of reagents suitable for precolumn derivatization is limited:

REAGENTS	REACTANTS
4-Bromomethyl-7-methoxycoumarin	Carboxylic acids
7-Chloro-4-nitrobenzyl 2-oxa-1,3-diazole	Amines (primary and secondary) and thiols
1-Dimethylaminonaphthalene-5-sulfonyl chloride	Amines (primary) and phenols
1-Dimethylaminonaphthalene-5-sulfonyl hydrazine	Carbonyls

A major advantage of precolumn derivatization over postcolumn reactions is that the product need not have different detection properties from the reactant or analyte; different separation properties are adequate. Somewhat greater freedom is available in the selection of precolumn derivatives since long reaction times can be allowed. In postcolumn derivatization such times may degrade separation via an increase in bandwidth. However, in postcolumn work there is much less need to convert quantitatively each analyte into a single product, and the postcolumn reaction occurs in a constant environment.

17.5 OPTIMIZATION OF COLUMN PERFORMANCE[6-11]

With more parameters than equations interrelating them, the problem of optimization in liquid chromatography is difficult because an analytical problem will have several degrees of freedom. Nevertheless, workers now have a fairly good appreciation of the compromises that must be struck with respect to theoretical plates, time, pressure, and the central role of particle diameter on column performance. Depending upon the desires of the analyst or the nature of the sample, one might seek the lowest pressure drop and therefore probably the lowest cost equipment, the shortest analysis time, the shortest column length, or a large peak capacity. Naturally trade-offs are possible. It should be remembered that the time needed to prepare the sample for analysis, that is, preliminary operations such as sampling, extraction, derivatization, and cleanup, are often longer than the total analysis time in LCC.

Most columns now consist of fully porous particles in the 5–10-μm-diameter range, and come in lengths of 10–25 cm. Smaller particle sizes lead to faster analyses. Efficiency will diminish for columns with particles much less than 5 μm, since the velocity will be less than the optimum value. Kinetic processes of stationary/mobile phase transfer may also become limiting at very small (<3 μm) particle size.

Column efficiency is best given by stating the reduced plate height, the reduced velocity, and the column flow resistance parameter, ϕ, or column permeability k_o ($\phi = 1/k_o$). The former is given by

$$\phi = \Delta P d_p^2 t_M / \eta L^2 \tag{17-2}$$

where ΔP is the pressure drop across the column and η is the eluent viscosity. For a "well-packed" column, the reduced plate height should be between 3 and 4 for nonionic substances at reduced velocities approximately 10, and it should be possible to achieve $h = 10$ at $v = 100$. For 5-μm particles, $h = 3$–4 translates to about 9000 plates for $L = 15$ cm, and about 15,000 plates for $L = 25$ cm. The flow resistance parameter will be in the range 500 (pellicular packings) to 1200 (highly porous packings). A doubling of solvent viscosity in either isocratic or gradient elution means a doubling in the time required for a given separation, or a doubling in the pressure drop across the column to maintain the retention time unchanged, other factors being unaltered. The solvent viscosity should not exceed 0.4 cP (centipoise equivalent to 0.04 Pa sec or 0.04 pascal second).

In the earlier years of HPLC the conventional approach had been to use the column at the maximum pressure attainable with the equipment, which means the highest possible mobile-phase velocity. For a more difficult separation, a longer column was packed with the same material. The latter column has a lower mobile-phase velocity and a smaller plate height, and thus will be more efficient on the two counts. Alternatively, most difficult separations can usually be handled by reducing the mobile-phase velocity of the original column to achieve higher efficiency with the attendant penalty of a much longer retention time, and thus analysis time. Because high column pressure makes sample injection difficult and is a major factor in equipment price, one may wish to select optimum experimental conditions equivalent to operating the column at its optimum flow velocity. The column length and the particle size of the packing are chosen to obtain the desired performance in terms of retention time and plate number.

The complete equation for the dependence of the reduced plate height upon reduced velocity, as obtained by Knox,[8] is

$$h = B/v + Av^{0.33} + Cv \tag{17-3}$$

The B term equals 2γ; the obstructive parameter γ is 0.6 for solid-core or pellicular packings and close to unity for completely porous materials. Usually for a well-packed column the A term is about unity, but may rise to 2.9 or greater for poorly packed columns. For columns of porous particles, the C term is in the order of 0.05, decreasing to 0.003 for pellicular particles.

A logarithmic plot of Eq. 17-3 is shown in Fig. 17-16. At low reduced velocities the B term, arising from axial or longitudinal spreading, dominates. At high velocities the C term dominates and is responsible for the rise of h as the velocity increases. The C term contains the contribution from mass transport kinetics and includes the contribution

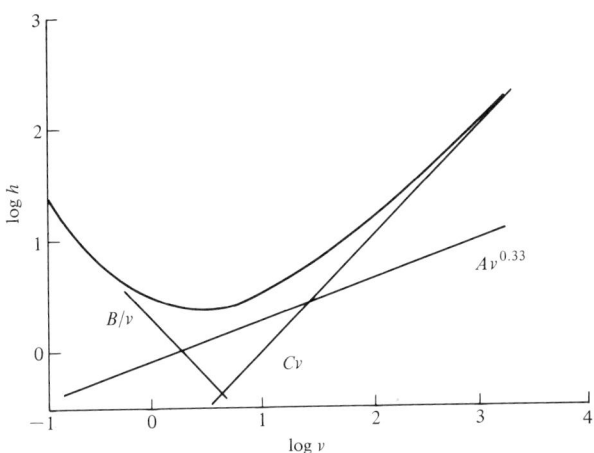

FIGURE 17-16 Logarithmic plot of reduced plate height, h, against reduced velocity, v (Knox equation), with $A = 1$, $B = 2$, $C = 0.1$.

from stagnant pockets of mobile phase. In the intermediate region of velocities, h shows a well-defined minimum at about $h = 2.5$ and at $v = 3$. In this region the A term, which arises from the flow-generated nonequilibrium in the mobile phase, dominates. Over the range of reduced velocity from about 1 to 8, the h/v curve is rather flat; there is less than 25% loss in performance when compared to the best achievable by working at $v = 3$.

Useful qualitative information can be obtained from the form of the h/v curve. If the curve shows a flat and rather high minimum, perhaps $h_{min} = 10$, then one suspects that the A term is high and that the column may be poorly packed. If the rising part of the h/v curve at higher velocities is steep, and $h = 10$ at perhaps $v = 20$, this implies a high value of the C term and poor mass transfer is suspected.

Optimization in Terms of Time and Cost

A realistic approach to optimization would be to search for conditions that give the resolution needed in a reasonably short analysis time at minimal cost and trouble. This implies a choice of working conditions that require only a moderate pressure. Desirable separation parameters selected will be $k' = 2$ and $N = 9000$, a situation that corresponds to a resolution of 1.5 and a relative retention of 1.10 (see Eq. 15-21). For water as eluent, $\eta = 10^{-3}$ Pa sec and $D_M \simeq 1 \times 10^{-5}$ cm^2 sec^{-1} for many solutes. The column is prepared from a highly porous packing, for which $\phi = 1000$.

The illustrative calculations will be performed for an inlet pressure of 20 atm and a column packing whose particle diameter is 10 μm. The basic equations required are:

$$N = \ell/h \tag{17-4}$$

$$\Delta P = \phi \eta \ell^2 / t_M = \phi \eta N^2 h^2 / t_M \tag{17-5}$$

$$v = \ell d_p^2 / t_M D_M = N h d_p^2 / t_M D_M \tag{17-6}$$

and the experimentally determined dependence of h upon v, Eq. 17-3. Eliminate t_M between Eqs. 17-5 and 17-6 to give

$$\Delta P = \phi \eta \ell v D_M / d_p^2 \tag{17-7}$$

Next eliminate the reduced column length between Eqs. 17-7 and 17-4:

$$hv = \Delta P d_p^2 / \phi \eta N D_M \tag{17-8}$$

The right-hand side of Eq. 17-8 can be evaluated for any ΔP and d_p, so

$$hv = \frac{(20 \text{ atm})(1.0145 \times 10^5 \text{ Pa atm}^{-1})(10 \times 10^{-4} \text{ cm})}{(1000)(10^{-3} \text{ Pa sec})(9000)(1 \times 10^{-5} \text{ cm}^2 \text{ sec}^{-1})} = 22.5$$

From Eq. 17-3, rearranged

$$hv = 22.5 = 2 + v^{1.33} + 0.05 v^2 \tag{17-9}$$

which can be solved for v, then h; these are $v = 8.38$ and $h = 2.69$. From Eqs. 17-4 and 17-5,

$$t_M = N^2 h^2 \phi \eta / \Delta P \tag{17-10}$$

$$= \frac{(9000)^2 (2.69)^2 (1000)(10^{-3} \text{ Pa sec})}{(20 \text{ atm})(1.0145 \times 10^5 \text{ Pa atm}^{-1})} = 289 \text{ sec}$$

Now

$$t_R = (1 + k') t_M \tag{17-11}$$
$$= (1 + 2)(289 \text{ sec}) = 865 \text{ sec (or 14.4 min)}$$

The required column length is

$$L = N h d_p \tag{17-12}$$
$$= (9000)(2.69)(10 \times 10^{-4} \text{ cm}) = 24.2 \text{ cm}$$

The values of t_R and L for various values of ΔP and d_p are given in Table 17-1.

From the data in Table 17-1, the required analysis time decreases steadily with the decrease in particle size. This explains why the recent trend to the use of smaller particles is paralleled by a decrease in the length of the columns used. However, while in theory one should use particles smaller than 5 µm, there arise severe practical difficulties in handling particles of such small diameters. Heavy demands are placed on the minimization of extra column effects, as previously discussed. Another problem can arise from thermal effects caused by frictional forces associated with flow in the small particle columns.

Column Operated to Achieve Preselected Retention Time

It is possible to obtain a given separation in a preselected time using quite different experimental conditions while maintaining the plate number constant. The calculations will be done for an analysis time, $t_R = 4$ min, and $N = 4000$. All other parameters selected

LIQUID COLUMN CHROMATOGRAPHY: INSTRUMENTATION AND OPTIMIZATION

TABLE 17-1 Values of t_M, t_R, and L for Various Values of ΔP and d_p

ΔP, atm		d_p, μm		
		5	7	10
20	t_M, min	3.5	3.6	4.8
	t_R, min	10.4	10.9	14.4
	L, cm	10.3	14.7	24.2
50	t_M, min	1.6	2.1	3.2
	t_R, min	4.7	6.4	9.6
	L, cm	10.9	17.8	31.2
100	t_M, min	1.0	1.6	2.5
	t_R, min	3.2	4.7	7.5
	L, cm	12.8	21.7	39.1

$k' = 2$, $N = 9000$, $\eta = 10^{-3}$ Pa sec, $D_M = 10^{-5}$ cm^2 sec^{-1}, $\phi = 1000$, $A = 1$, $C = 0.05$, $\gamma = 1$.

for the preceding example will be retained except a particle size of 5 μm will be employed. One proceeds by eliminating the reduced column length between Eqs. 17-4 and 17-6 to give

$$h/v = t_M D_M / N d_p^2 \qquad (17\text{-}13)$$

Remembering that $t_M = t_R/(1 + k')$, then $t_M = 80$ sec. Now from Eq. 17-13,

$$\frac{h}{v} = \frac{(80 \text{ sec})(1 \times 10^{-5} \text{ cm}^2 \text{ sec}^{-1})}{(4000)(5 \times 10^{-4} \text{ cm})} = 0.80$$

Since the dependence of h upon v is known from Eq. 17-3, divide by v, and

$$\frac{h}{v} = 2v^{-2} + v^{-0.67} + 0.05 = 0.80 \qquad (17\text{-}14)$$

Equation 17-14 may be solved for v, then h; the values are $v = 2.82$ and $h = 2.26$. The required column length, given by Eq. 17-12, is

$$L = (4000)(2.26)(5 \times 10^{-4} \text{ cm}) = 4.5 \text{ cm}$$

Lastly, the inlet pressure is calculated from Eq. 17-5:

$$\Delta P = \frac{(1000)(10^{-3} \text{ Pa sec})(4000)^2 (2.26)^2}{(80 \text{ sec})(1.0145 \times 10^5 \text{ Pa atm}^{-1})} = 10.1 \text{ atm (or 148 psi)}$$

Other column lengths and inlet pressures necessary to ensure an efficiency of 4000 plates and an analysis time of 4 min are shown in Table 17-2 as a function of particle diameter. Exactly the same chromatograms would be obtained with a pressure of 264 atm (or 3880 psi) when using 15-μm particles as contrasted with a pressure of only 10.1 atm (or 148 psi) for 5-μm particles. So, a threefold reduction in particle diameter results in a 26-fold reduction in the pressure drop at constant resolution and analysis time. This is achieved

TABLE 17-2 Length and Inlet Pressure Necessary to Ensure an Efficiency of 4000 Plates and an Analysis Time of 4 min, as a Function of Particle Diameter

Particle diameter, μm	Column length, cm	Inlet pressure, atm
3	3.5	14.9
4	3.8	11.2
5	4.5	10.1
7	6.9	11.9
10	14.6	26.4
12	25.6	56.0
15	69.5	264

$A = 1, C = 0.05, \gamma = 1, D_M = 10^{-5}$ cm^2 sec^{-1}, $k' = 2, \phi = 1000, \eta = 10^{-3}$ Pa sec.

through a 15-fold reduction in the column length. Although particles of 20-μm diameter were used until rather recently, it becomes impossible to operate with this size packing under the specified separation conditions used in this illustration.

If a longer retention time for nonretained materials is desired, even a few minutes, lower inlet pressures achieve the same performances. For a given resolution, accepting a slight increase in the retention time will allow work at a much lower pressure with a shorter column or a column packed with coarser particles. A larger reduction in analysis time is obtained by using small particles rather than high pressures. Yet one must be careful to remain at reduced velocities larger than the minimum in the h/v curve. Note in Table 17-2 that the inlet pressure passes through a minimum with 5-μm particles under the specified conditions.

Selecting a low viscosity solvent whenever possible without changing k' and relative retention is an easy way to substantially decrease the analysis time. Typically D_M is 0.3×10^{-5} cm^2 sec^{-1} in 1-propanol and 3×10^{-5} cm^2 sec^{-1} in hexane. The viscosities of these eluents at 20°C are 0.0020 and 0.00028 Pa sec (or N sec m^{-2}), respectively. As D_M increases in going from a viscous eluent like 1-propanol to a very fluid hexane, the optimum particle diameter will increase. This is because the larger diffusion coefficient allows molecules to diffuse over greater distances in a fixed time.

Solvent Programming (Gradient Elution)

No single isocratic solvent composition can elute all substances with good resolution in a reasonable time or in a measurable concentration. To adequately handle samples having both weakly adsorbing and strongly adsorbing substances, the rates of band migration must be changed during a chromatographic run by solvent programming, thereby gradually attaining the optimum k' value for each solute. Solvent programming involves changing the mobile phase either stepwise or continuously to increase the strength of the mobile phase as eluent during the separation. This technique is analogous to temperature programming in gas chromatography. The optimum gradient for a particular separation is selected by trial and error. For example, starting a method development scheme with too powerful an eluting solvent causes an almost immediate elution of the sample;

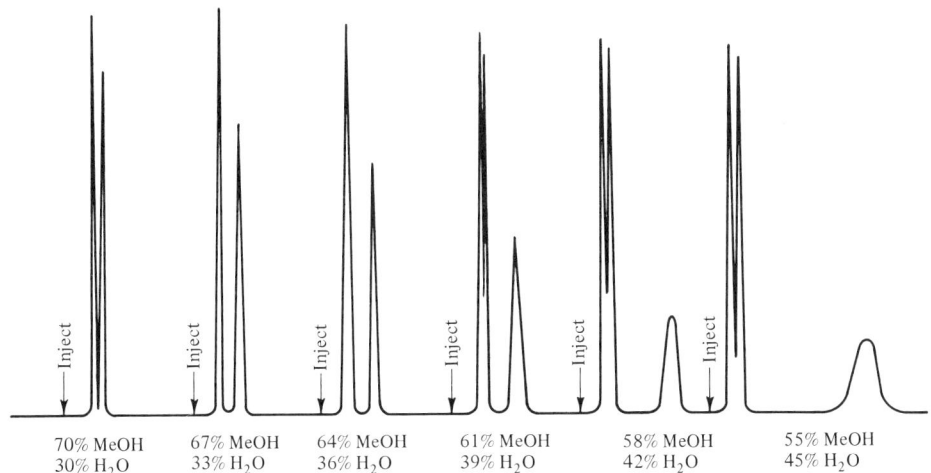

FIGURE 17-17 Stepwise elution with decreasing elution strength. Aqueous phase contained a phosphate buffer, pH 7.0. Sample was a mixture of dyes: amaranth, erythrosin, and sunset yellow. Column: alkyl nitrile bonded phase.

thus information about the chromatographic system is available very quickly. Successive samples are then injected with the eluting strength progressively decreased, often by stepwise changes in the mobile-phase composition, as shown in Fig. 17-17. Solvent programming information can be obtained in greater detail if an automated sampler is used to inject samples periodically while the solvent strength of the mobile phase is slowly, yet continuously, decreased (Fig. 17-6). The resultant chromatograms will show many trial separations. The early chromatograms reflect the fact that the samples are poorly retained on the column packing, as shown in Fig. 17-18. Some peaks may actually reverse their elution order as the concentration of the weaker eluent is increased. At the completion of the run, one determines from the gradient profile the approximate concentration of mobile phase required to elute the peak of interest. Set the programmer at this concentration and reinject the sample. The sample should now elute with close to the desired retention and only a slight change in the solvent ratio should be required to optimize the separation.

Stepwise elution is worth considering if the sample contains some fast-moving components followed by some slow-moving solutes. In this case the mobile phase is stepped from one isocratic composition to another with much stronger eluting power to effect a total separation within a reasonable time. Care must be exercised to ensure that the step does not bring about a displacement effect; that is, where the change in solvent composition causes two or more components to elute at the changing solvent front. In this case little or no separation will be achieved.

After solvent programming, the column must be regenerated. Regeneration can be achieved by pumping initial mobile phase through the column until the initial solvent activity is obtained, or retracing a gradient by decreasing the solvent strength as a function of time. Running a negative gradient permits a more rapid return to the initial conditions.

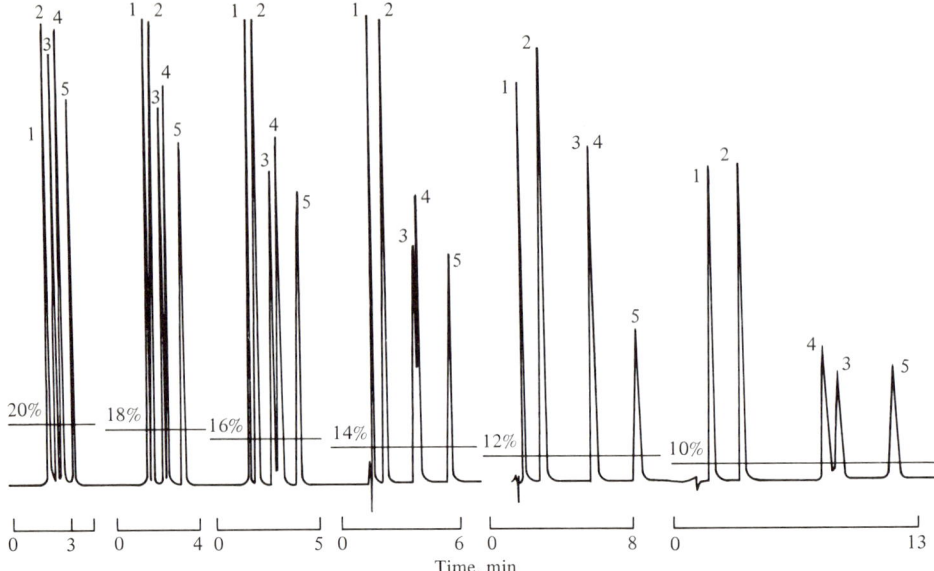

FIGURE 17-18 Isocratic separation of methyl xanthines on bonded phase C-8 column. Mobile phase is 20-mM phosphate buffer, pH 2.6, and percent acetonitrile marked on graph. Peaks: (1), theobromine; (2), theophylline; (3), hydroxypropyl theophylline; (4), caffeine; and (5), 8-chlorotheophylline.

Flow Programming

Flow programming is an alternative to gradient elution. Effective flow programming, up to 9 ml min^{-1}, can be done with the radial compression separation column described earlier, thus eliminating the need for one pump extra as required for gradient elution. Flow programming is not quite as good as solvent programming, but it is a very useful alternate for inexpensive gradients and a good technique when all one needs is a quick scouting run to check feasibility. The chromatograms in Fig. 17-19 demonstrates only a small loss in efficiency (resolution) when the flowrate is increased.

Recycle Technique

Recycle capability involves multiple cycling of partially resolved components through the same column. As shown in Fig. 17-20, the initial pass through the column system did not achieve base line separation of naphthalene and biphenyl. By recycling the sample three times through the column system, base line separation was achieved. The maximum number of recycle passes depends on the efficiency of the column and the extent to which the system will spread peaks since some band broadening occurs due to the small added dead volumes in the system. From the information obtained during the initial pass through the column, $R = 0.84$ and $N_{eff} = 64$ (first peak). Thus for $R = 1.5$, 204 effective plates are needed or roughly four times the column length employed. This implies that three recycles will be adequate, which indeed proved to be true. Recycle chromatography is much more practical than purchasing three additional column lengths with the attendant problems of pressure drop and solvent holdup. Obviously, recycle technique is not the method of choice when many samples have to be run, but when one does not want to invest a lot of time in methods development, recycle gives the operator a great deal of separating power for those difficult separations of adjacent peaks.

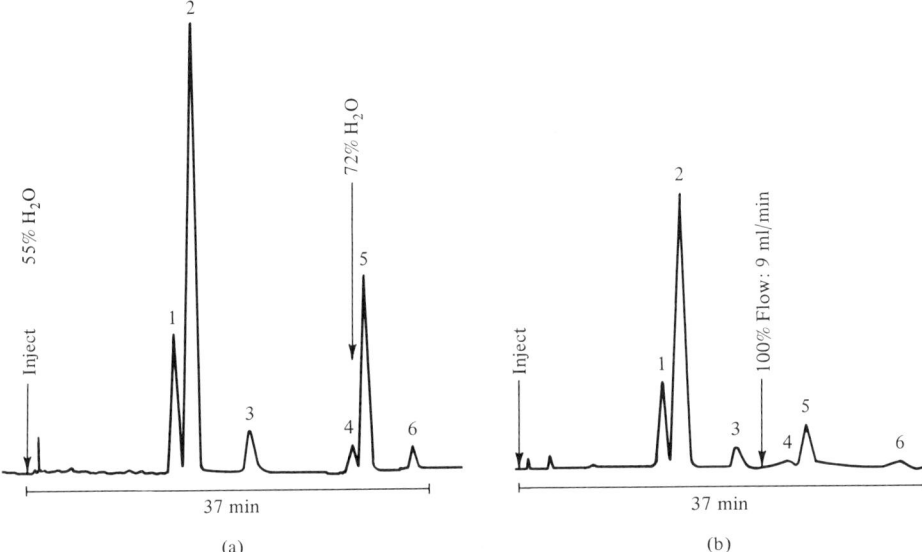

FIGURE 17-19 Comparison of (a) a gradient run to (b) a flow programming run. Gradient run: Acetonitrile/water, 55-72% water in 30 min at a flowrate of 4 ml/min. Programming run: Acetonitrile/water in a ratio of 56:44; flowrate increased from 3 ml/min to 9 ml/min in a 20-min period (indicated on the curve by 100% flow). Radial compression cartridge, 8 mm × 10 cm. Peaks: (1), cinerin II: (2), pyrethrin II; (3), jasmolin II; (4), cinerin I; (5), pyrethrin I; (6), jasmolin I. (Courtesy of Waters Associates, Inc.)

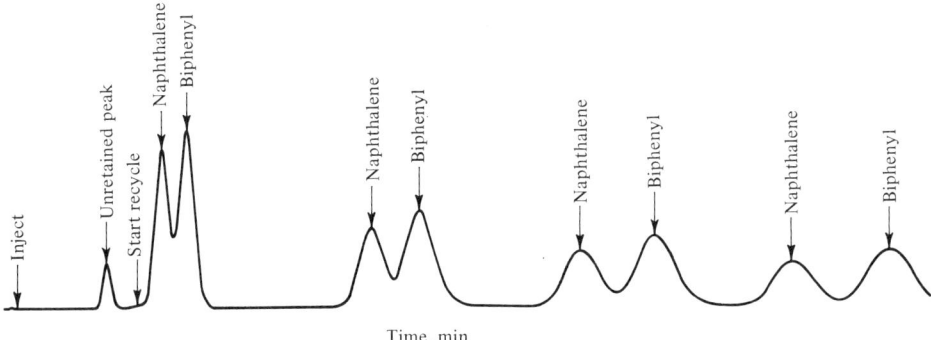

FIGURE 17-20 Recycle chromatographic separation of a mixture of naphthalene and biphenyl. Column: 2 mm × 122 cm packed with porous silica. Sample recycled three times through the column system. (Courtesy of Waters Associates, Inc.)

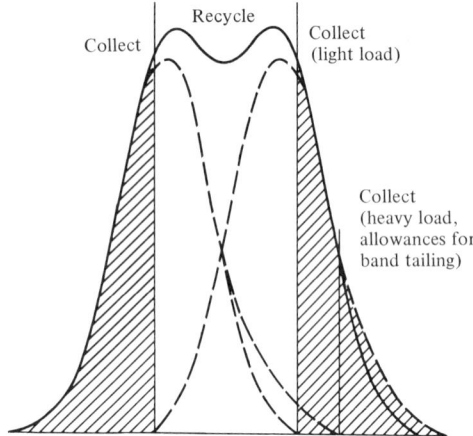

FIGURE 17-21 Shave-recycle technique showing the collection and recycle intervals. (Courtesy of Waters Associates, Inc.)

An easy answer to the purity question is to trap a given peak, then reinject it for recycle chromatography through a high-efficiency column. If the peak remains homogenous after multiple recycle passes, there is a very high degree of certainty that the material is pure. However, should the peak begin to develop "shoulders" after a few recycles, additional compounds become evident. Qualitative and quantitative errors in an analysis can thus be avoided.

Shave-recycle enables an operator to isolate milligram quantities of pure material on an analytical column; that is, one can purify sufficient material for NMR or mass spectrometry with a single analytical column because the loading capacity can be increased. The technique is illustrated in Fig. 17-21. Shave-recycle is useful in easy separations ($\alpha \geqslant 1.3$) when it is possible to heavily load a column until $R = 0.7$. Advantages with this technique include increased throughput, reduced separation time, lower solvent consumption, and minimum column investment.

Effect of Temperature

Column temperature during a chromatographic run has some influence. As shown in Fig. 17-22, the same materials are separated at different temperatures while the mobile-phase flowrate (1 ml min^{-1}) and composition (45% methanol:55% water) are held constant. The k' values decrease and the peaks become sharper as the temperature is increased. This observation is fairly general for most modes of LCC. An exception is ion-exchange chromatography in which selectivity changes are frequently observed with changes in temperature.

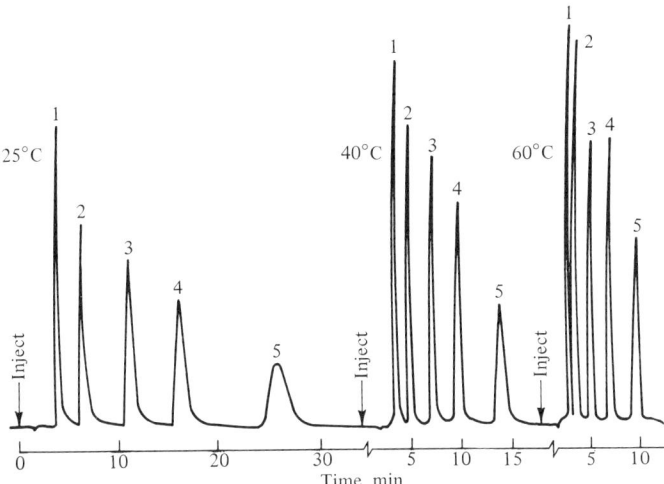

FIGURE 17-22 Influence of column temperature in liquid column chromatography. Mobile phase is 45% methanol/55% water. Peaks: (1), 9,10-anthraquinone; (2), 2-methyl-9,10-anthraquinone; (3), 2-ethyl-9,10-anthraquinone; (4), 1,4-dimethyl- 9,10-anthraquinone; and (5), 2-t-butyl-9,10-anthraquinone.

PROBLEMS

1. What problems may arise in use from a poorly packed LCC column?

2. Prepare a table containing a geometrically similar column set such that for each column the reduced length is 20,000. Do this for particle diameters: 1, 2, 3, 4, 5, 7, 10, 12, 15, and 20 μm.

3. Some optimization procedures in LCC originate from the idea of generating the maximum number of plates within the shortest possible time, that is, minimizing t_R/N. (a) Under what conditions can this be achieved? (b) Because of equipment limitations, what does this approach lead to in terms of t_R and ΔP?

4. Express the volume over which the base width of the peak elutes in terms of N, t_M, and k'.

5. For a column exhibiting a plate number of 5000 for a peak whose retention time is 3.0 min, what would be the reduction in plate number for a detector time constant equal to one-tenth of the peak width at base line (4σ)?

6. Suppose that the volume over which the base width of the peak elutes, V_w, plus the spreading due to the detector volume, V_{det}, is not to be more than 12% greater than V_w alone. Experiment shows that for a typical 8-μl photometer cell into which a 1-μl sample is directly injected by syringe or high-grade injection valve, V_{det} is about 30 μl.

(a) Calculate the minimum column volume which will allow these conditions to be met for an unretained solute from a column giving 10,000 plates and packed with 5-μm particles of totally porous silica whose porosity is 0.75. (b) If the column length is 10 cm, what is the minimum column diameter?

7. Estimate the maximum sample volume that should be injected into each of these columns whose porosity is 0.75. Column A: 500 mm by 2 mm, t_M = 104 sec, N = 3200. Column B: 135 mm by 5 mm, t_M = 34 sec, N = 4450. Column C: 90 mm by 5 mm, t_M = 16 sec, N = 5000.

8. Prepare a table giving optimum particle diameters for a column to be operated at a reduced velocity of 3.0. The reduced length is 20,000. Assume t_M values of 10, 30, 100, and 300 sec. Do this for three eluents: 1-propanol (D_M = 3 × 10^{-5} cm^2 sec^{-1}); water (D_M = 1.0 × 10^{-5} cm^2 sec^{-1}); and n-hexane (D_M = 3.0 × 10^{-5} cm^2 sec^{-1}).

9. Calculate the pressure drop required for the individual eluents at each of the values of t_M and d_p tabulated in answer to Problem 8. Do this for solid core particles (ϕ = 500) and for porous particles (ϕ = 1000). Viscosity at 25°C, in mN sec m^{-2}: 1-propanol, 2.004; water, 0.8903; and n-hexane, 0.313 (at 20°C).

10. Determine the pressure drop and column length required to maintain N = 5000 with t_M = 100 sec for these particle diameters: 2, 3, 4, 5, 7, 9, 10, 12, and 15 μm. Assume that ϕ = 1,000, η = 0.89 × 10^{-3} Nm^{-2} sec, and D_M = 1.0 × 10^{-5} cm^2 sec^{-1}.

11. Graph the relation between column inlet pressure as a function of column length to obtain a peak with 5000 plates (k' = 2) in 300 sec. On the same graph, but with the x-axis extending to the left, graph the column inlet pressure as a function of the packing particle diameter. Use the data previously calculated for Problem 10. Note the trade-off between L and d_p at constant inlet pressure.

12. Calculate the elution time for a nonretained material and column length required when the pressure drop is constant at 30 or 200 atm, and plate number is 5000, with these particle diameters: 3, 5, 7, and 10 μm. Assume that η = 10^{-3} N sec m^{-2}, D_M = 1 × 10^{-5} cm^2 sec^{-1}, and ϕ = 500.

13. Determine the efficiency and plate number of individual columns with these particle diameters: 3, 4, 5, 7, 10, 12, 15, and 20 μm. All columns are operated at the same pressure drop (20 atm) and elution time (t_M = 100 sec). Water is the eluent. Other parameters: η = 10^{-3} N sec m^{-2}, D_M = 1 × 10^{-5} cm^2 sec^{-1}, ϕ = 1000, ℓ = 20,000, A = 1, C = 0.05, and γ = 1.

14. Prepare a graph of reduced plate height versus reduced velocity from the results of Problem 13. Calculate the physical column length, L, and the average linear velocity, $\bar{u}$, for columns packed with particles 4, 5, and 10 μm in diameter.

15. What is the maximum number of plates which can be obtained from any particle size in a given elution time?

16. Determine the elution time and column length required when the pressure drop is maintained constant at 20 or 100 atm, the plate number is 5000, and the column is packed

with these particle diameters: 3, 4, 5, 7, 10, 15, and 20 μm. All other parameters the same as in Problem 13.

17. Derive an expression for the maximum number of plates that can be obtained for any particle size given a stated inlet pressure and eluent.

18. To achieve the desired resolution, a change in R from 0.8 to 1.25 is required. Assume that the conditions for the starting separation are a 30-cm column of 10-μm particles, reversed-phase LCC (30% acetonitrile:70% water as mobile phase) at 25°C, t_M = 80 sec, and sample molecular weights in the range 200–500. Explore each of the possibilities for changing resolution by the required amount: (a) a change in ΔP (or F_c), (b) a change in L (with ΔP constant). D_M = 0.56 $\times$ 10^{-5} cm^2 sec^{-1}.

19. In gradient elution two sets of conditions may be considered to exist: one to provide maximum resolution, $F_c t/V$ = 0.70; and the other to provide minimum analysis time, $F_c t/V$ = 0.85. Here t is the time interval for each solvent, and V is the volume of the mixing vessel. The programmer has a 1-min discrimination. Also, $F_c t$ = 2.5V_M, a condition that ensures that each solvent is used to develop a chromatogram over a k' range of 2.5. What should be the volume of the mixing vessel when employing a column 25 cm in length, 4.6-mm bore, and with a porosity of 0.72? Do this for flowrates of 1 and 2 ml min^{-1}.

20. The separation of adenosine mono-, di-, and triphosphate nucleotides (AMP, ADP, and ATP) was accomplished in a little over 3 min using 0.4M KH$_2$PO$_4$ (plus 3% methanol) and a 15-cm by 2-mm column, packed with 10-μm particles of silica to which was bonded a 3-aminopropyl siloxane phase. The mobile phase viscosity was 1.4 cP. Flowrate was 100 ml hr^{-1} at an inlet pressure of 2900 psi. Suggest improvements (with reasons) in the operating procedure.

BIBLIOGRAPHY

Ettre, L. S. and C. Horvath, "Foundations of Modern Liquid Chromatography," *Anal. Chem.*, **47**, 422A (1975).

Simpson, C. F., Ed., *Practical High Performance Liquid Chromatography*, Heyden, New York, 1976.

Snyder, L. R. and J. J. Kirkland, *Introduction to Modern Liquid Chromatography*, 2nd ed., Wiley-Interscience, New York, 1979.

LITERATURE CITED

1. Berry, L. and B. L. Karger, "Pumps and Injectors for Modern Liquid Chromatography," *Anal. Chem.*, **45**, 819A (1973).
2. Fallick, G. J., "New HPLC Column Technology," Papers 636, 637, 30th Pittsburgh Conference on Analytical Chemistry and Applied Spectroscopy, Cleveland, 1979; *Am. Lab.*, p. 87 (November 1979).

3. Karger, B. L., M. Martin, and G. Guiochon, "Role of Column Parameters and Injection Volume on Detection Limits in Liquid Chromatography," *Anal. Chem.*, **46**, 1640 (1974).
4. Kissinger, P. T., "Amperometric and Coulometric Detectors for High-Performance Liquid Chromatography," *Anal. Chem.*, **49**, 447A (1977).
5. Jupille, T. H., "Derivatization for Detectability in HPLC," *Am. Lab.*, p. 85 (May 1976).
6. Eon, C. and G. Guiochon, "The Pertinency of Pressure in Liquid Chromatography," *J. Chromatogr.*, **99**, 357 (1974).
7. Grushka, E., "Solute Band Spreading in Liquid Chromatography," *Anal. Chem.*, **46**, 510A (1974).
8. Knox, J. H., "Practical Aspects of LC Theory," *J. Chromatogr. Sci.*, **15**, 352 (1977).
9. Martin, M., G. Blu, C. Eon, and G. Guiochon, "Optimization of Column Design and Operating Parameters in High Speed Liquid Chromatography," *J. Chromatogr. Sci.*, **12**, 438 (1974).
10. Martin, M., C. Eon, and G. Guiochon, "Trends in Liquid Chromatography," *Res/Dev.* p. 24 (April 1975).
11. Snyder, L. R., "A Rapid Approach to Selecting the Best Experimental Conditions for High-Speed Liquid Column Chromatography. Small Particle Columns," *J. Chromatogr. Sci.*, **15**, 441 (1977).

CHAPTER 18

High-Performance Liquid Chromatography Methods

The choice of the correct liquid chromatographic system to use for a given mixture of solutes cannot be made with certainty and must be confirmed by experiment. If the likely chemical nature of the sample components is known, then the phase system can be chosen from the references in the literature of such mixtures. If nothing is known about the chemical nature of the sample, then knowledge about sample solubility will give some indication as to which chromatographic method to employ. Figure 18-1 provides a general guide.

If the sample is completely soluble in organic solvents and the molecular weight is less than 2000 for all the sample components, then a liquid–solid system can be used. Physical selectivity is dominant in adsorption chromatography. In this method the components are separated on the basis of their polarities.

LLC best separates homologs where the separation is based on chain length differences (or molecular weight selectivity). A successful separation is achieved by establishing the proper balance between attraction of the mobile-phase solvent and the stationary liquid phase for the sample. Most good separations are achieved by matching the polarity of the sample and stationary phase, and using a solvent which has a markedly different polarity.

Ion-exchange chromatography is the method of choice for compounds with ionic or ionizable functional groups. Thus, if the sample is insoluble in organic solvents but soluble in water, giving a solution that is not neutral, or is only soluble in dilute acid or alkali, then ion-exchange chromatography is a likely technique.

When it is known or suspected that the molecular weight exceeds approximately 2000 for some or all the sample components, then a separation utilizing exclusion chromatography is indicated. This method is based on the ability of controlled-porosity substrates

530 CHAPTER 18

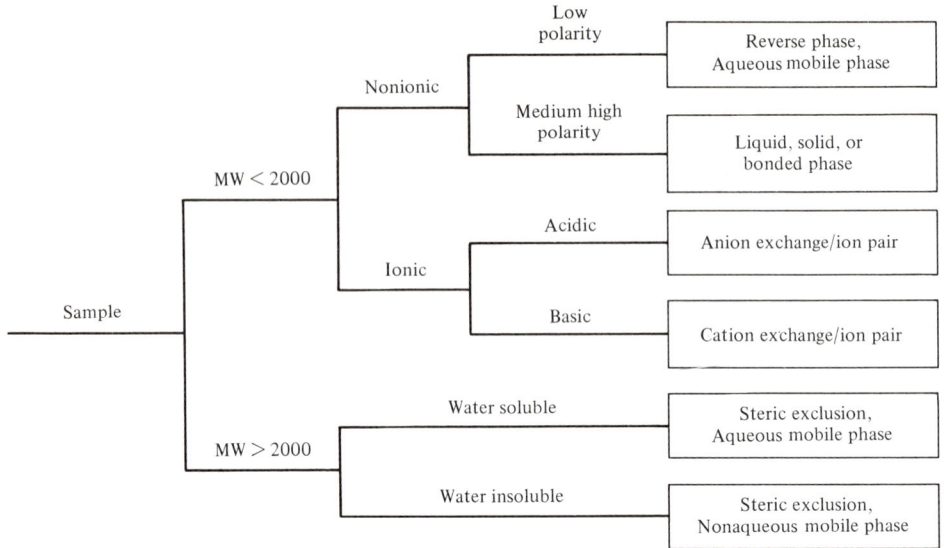

FIGURE 18-1 Guide to liquid chromatography mode selection. (Courtesy of Waters Associates, Inc.)

to sort and separate sample mixtures according to the size and shape of the components of the sample.

18.1 ADSORPTION CHROMATOGRAPHY

Adsorption HPLC, because of its versatility, is often the answer to a separation problem. If the sample is soluble in nonpolar or moderately polar solvents such as hexane, methylene chloride, chloroform, or diethyl ether, then adsorption chromatography is a likely choice. In liquid–solid adsorption chromatography (LSC), the mobile phase is a liquid while the stationary phase is either a totally porous particle or a porous layer bead—the pellicular packings, as shown in Fig. 18-2.

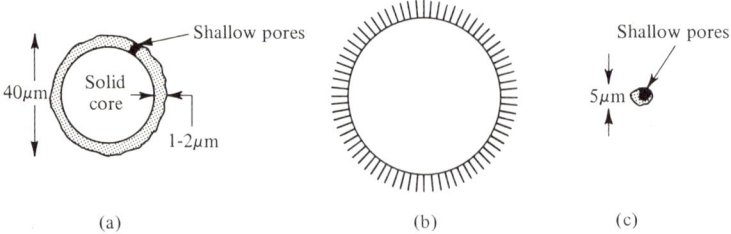

FIGURE 18-2 Stationary phases used in adsorption HPLC: (a) porous layer beads; (b) bonded phase, and (c) porous microparticle.

Porous Layer Beads

A pellicular or porous layer bead type of packing consists of a solid, spherical glass bead with an average particle diameter of 30–40 μm, and a thin, porous outer shell. The outer shell, typically 1–3 μm thick, may be a silica gel layer, a network of small spherical particles bonded to the solid core, or a bonded monomeric or polymeric organic phase. Surface areas of the porous layer beads range from 5 to 15 m^2 g^{-1}. Columns are easy to pack with these materials because of the dense core, but due to their small surface areas, porous layer beads suffer from limited sample capacity (approximately 0.1 mg g^{-1}). Thus they are less useful for preparative work or for use with the lower sensitivity detectors. However, relative to a porous packing of equivalent diameter, stationary phase mass transfer is greatly improved in a thin porous layer and, consequently, these packings exhibit good efficiency. Longer columns are possible because the pressure drop is lower due to the larger particle size of porous layer supports.

Porous Particles

The totally porous particle is generally a high surface area, active material such as silica gel or alumina. These are the most common phases in adsorption chromatography. Both silica gel and alumina have a high concentration of surface hydroxyl groups whose number and geometric arrangement determine the activity (retention) of the adsorbent. The slightly acidic silanol groups (Si–OH) in silica gel exist at the surface and extend into the internal channels of the pore structure. These hydroxyl groups interact with polar or unsaturated moieties by hydrogen bonding. During dehydration (activation) a certain proportion of these silanol groups will form siloxane linkages (Si–O–Si) between neighboring silicon atoms. Siloxane groups are very weak in their adsorptive properties.

For silicas, surface areas range from a low of 100 m^2 g^{-1} to a high of 860 m^2 g^{-1} with the average being 400 m^2 g^{-1}. Since pore diameter is inversely related to surface area, average pore diameter is the smallest (35 Å) for silicas with high surface area, and largest (330 Å) for silicas with low surface areas. The linear adsorption coefficient of a solute is independent of both these parameters provided the solute molecule is small enough to enter the pores unimpeded and provided the nature of the active surface sites is independent of the pore diameter. Silica gels comprise about 70% of the column packings in LSC.

Aluminas comprise about 20% of the solid adsorbents in use. Aluminas have low surface areas, larger average pore diameters, and higher packing densities than silicas. Aluminas do have some selectivity advantages, particularly for unsaturated hydrocarbons and halogen-containing compounds, and are useful for very basic compounds which may adsorb too strongly on the acidic silicas. In addition, alumina possesses greater stability at higher pHs where silica tends to dissolve.

The isocratic operation of a high surface area silica gel column will only be effective for the separation of solutes of narrow polarity range and for solutes having polarities close to that of the solvent. Columns packed with silica gel of low surface area will separate solute mixtures of wide polarity range. It follows that high surface area microparticulate adsorbents with high resolving power would be used to separate complex mixtures of solutes of similar polarity, whereas the low surface area pellicular packings with relatively low resolving power would be used for separating simple mixtures of wide polarity range.

A number of group-selective adsorbents have been prepared by impregnating an adsorbent with a material that will form a complex with a specific organic functional group. For example, silver nitrate-impregnated silica gel has been used for the separation of unsaturates.

With the production of commercial quantities of the microparticulates and the development of techniques to pack them, the microparticulate packings have now displaced the porous layer beads (pellicular packings) in popularity. Compared with porous layer beads, the porous microparticles offer the advantages of at least an order of magnitude in column efficiency, sample capacity, and speed of analysis. Exceptionally high efficiency adsorption columns, having up to 40,000 plates/m, are now available.

Adsorption Processes in LSC

The range of compound types that can be separated by LSC extends from very nonpolar hydrocarbons to very polar multifunctional compounds.[1] However, ionic compounds are best handled by ion-exchange or ion-pair chromatography. The selectivity is truly unique. The order of adsorption follows the general polarity scale for various classes of compounds, as outlined in Table 18-1. Generally the most polar group of a polyfunctional compound

TABLE 18-1 Solvent Strength Parameter, $e°$, and Physical Properties of Selected Solvents

Solvent	$e°(SiO_2)$	$e°(Al_2O_3)$	Viscosity (20°C) mN sec m^{-2}	Refractive Index (20°C)
Pentane	0.00	0.00	0.23	1.358
Hexane		0.00	0.313	1.375
Cyclohexane	−0.05	0.04	0.980	1.426
Carbon disulfide	0.14	0.15	0.363	1.628
Carbon tetrachloride	0.14	0.18	0.965	1.460
1-Chlorobutane		0.26	0.47	1.402
Diisopropyl ether		0.28	0.379	1.368
2-Chloropropane		0.29	0.335	1.378
Benzene	0.25	0.32	0.65	1.501
Diethyl ether	0.38	0.38	0.23	1.353
Chloroform	0.26	0.40	0.57	1.443
Methylene dichloride		0.42	0.44	1.425
Methyl isobutyl ketone		0.43		1.394
Tetrahydrofuran		0.45	0.55	1.407
Acetone	0.47	0.56	0.32	1.359
1,4-Dioxane	0.49	0.56	1.54	1.422
Ethyl acetate	0.38	0.58	0.45	1.370
1-Pentanol		0.61	4.1	1.410
Acetonitrile	0.50	0.65	0.375	1.344
1-Propanol		0.82	2.00(25°)	1.38
Methanol		0.95	0.60	1.329
Water		Large	1.00	1.333

Solvent strength parameter ($e°$) values from L.R. Snyder, *Principles of Adsorption Chromatography*, Marcel Dekker, New York, 1968.

governs its adsorption characteristics. For compounds of low to moderate polarity, adsorption chromatography often makes possible the separation of complex mixtures into classes of compounds of similar chemical functionality. The weak dispersive interactions between the adsorbent and hydrocarbon moieties provide little or no retention. Thus, LSC is less influenced by molecular weight differences and more by specific functional groups. LSC also excels in the separation of positional isomers. Consequently, the separation of compounds differing only in the degree or type of alkyl substitution, such as members of an homologous series, is usually poor by adsorption. However, even this shortcoming can be used to advantage if the isolation of all the members of a particular class of compounds is desired. For example, LSC can be used to isolate the polynuclear aromatics from a petroleum sample, or the triglycerides from a lipid extract.

The surface hydroxyl group is the predominating group on silica and alumina, and governs their adsorption characteristics. The number and topographical arrangement of those groups determine the activity of the solid adsorbent. A silica surface that has not been heated for long periods in excess of 400°C is covered, to a greater or lesser extent, by hydroxyl groups. These hydroxyl groups can be divided into three types: I, free hydroxyl; II, bound hydroxyl; and III, reactive hydroxyl.

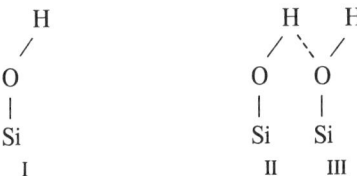

Some hydroxyl groups exist also in a hydrated (or solvated) form; namely, $Si-OH \cdot OH_2$. The surface of large-pore silicas consists predominately of free hydroxyls, while that of small-pore silicas contains reactive and bound hydroxyls. The activity of the different types of site increases in the order: bound < free < reactive hydroxyls. This means that the surface activity of small-pore silica particles is greater than that of wide-pore silica particles because of the greater concentration of reactive hydroxyls in the small-pore silica particles. However, the effect of adding water (or some polar solvent) to an activated silica surface is to deactivate the reactive hydroxyls of a small-pore silica first, leaving a surface of bound hydroxyls. Corresponding deactivation of a large-pore silica leaves a surface of free hydroxyls. Consequently, the surface activity of a heavily deactivated small-pore silica is less than a similar deactivated large-pore silica.

Existing models of the adsorption process assume that the adsorption sites are completely covered by either adsorbed solute molecules or solvent molecules. The surface hydroxyl groups interact with the functional groups of the solute (or solvent) molecules and, depending on the strength of this interaction, preferentially adsorb one molecule relative to another. Retention is governed mainly by interaction with the polar functional groups of the solute through electrostatic interactions involving permanent dipole or hydrogen bonding.

For LSC, a molecule held on the surface of the adsorbent will have a given potential energy due to the intermolecular forces holding it there, and a kinetic energy due to vibrational movement. When the kinetic energy exceeds the potential energy, the molecule will leave the surface and move back into the mobile phase. Conversely, if a molecule of solute in the mobile phase has a kinetic energy less than the potential energy, on striking the adsorbent surface the molecule will be adsorbed. Now the molecules held on the surface have a range of energies depending on their mass, shape, and temperature, and this range of energies will probably have a Boltzmann distribution. Furthermore, the adsorption sites themselves have a distribution of energies. Thus the condition for adsorption and desorption will depend not only on the probability of a molecule having less or more than a particular energy but also on its striking a respective site having a particular activity. Thus the resistance to mass transfer at the adsorbent/liquid interface will be composed of two components, one associated with the static liquid in the adsorbent pore and the other associated with the adsorbent surface at the pore walls.

The linear capacity of an adsorbent has been defined as the maximum weight of sample that can be applied to a gram of adsorbent before the adsorption coefficient falls more than 10% below its linear isotherm value. The linear capacity of more activated adsorbents is very low, less than 10^{-4} g g^{-1}. To control adsorbent activity, water is added to selectively cover or block the active sites on the surface. The addition of 1–2% water per 100 m^2 g^{-1} of the surface of a polar adsorbent, such as silica or alumina, increases the linear capacity from 5- to 100-fold. Where highly polar eluents or elevated temperatures are to be used, physically adsorbed water is likely to be lost. In these circumstances deactivation is best achieved using ethylene glycol or glycerol. In the packed column adsorbent water content is adjusted conveniently by modification of the mobile phase, or a polar modifier, such as 0.1–0.5% of an alcohol, is added to the nonpolar solvent. Finally, a standard solute or test sample, with k' about 1–5, is injected and its retention time measured. If too long or too short, further adjustments are made in the water content. If the retention time changes upon subsequent injections of test samples, the water contents of adsorbent and mobile phase are not in equilibrium.

Solvent Strength Parameter

Snyder[2] defines a solvent strength parameter $e°$ as the adsorption energy per unit area of standard adsorbent. Thus solvents can be rated according to their strength of adsorption. Table 18-1 is a list of common solvents used in adsorption chromatography in order of increasing $e°$ on alumina. The values are different, but the order is essentially the same on silica gel. In the older literature the listing was called the "eluotropic series"; it is also called the "polarity index." For a given solute and adsorbent, log k' varies linearly with $e°$.

For a given separation the initial solvent can be selected by matching the relative polarity of the solvent to that of the sample components. As a first approximation, a solvent is chosen to match the most polar functional group present in the sample; for example, alcohols for the hydroxyl group and amines for the amino group. From this first isocratic chromatogram the separation can be refined. If the k' values are too small (sample elutes too rapidly), then a weaker (less polar) solvent is substituted. Conversely, if the sample does not elute in a reasonable time because of high k' values, then a solvent or solvent blend with higher polarity is selected. Saunders[3] has formulated approximate rules for estimating the solvent strength required to give $k' = 3$ for a variety of polyfunctional compounds on various porous silicas.

Often it is more convenient to find an optimum $e°$ value by the use of binary solvent mixtures in order to utilize secondary solvent effects, namely, the contribution to the total solvent strength of the dispersion, dipole, proton acceptor, and proton donor forces. These secondary solvent effects which are due to solute–solvent localization effects may overshadow the solvent strength parameter. Solvent strength is held constant while the binary (or even ternary) solvent composition is varied. Saunders has published a graph (Fig. 18-3) showing mixed solvent strengths on silica gel. In Fig. 18-3, $e°$ is plotted across the top and various binary solvent compositions in each of the horizontal lines below it. Each line

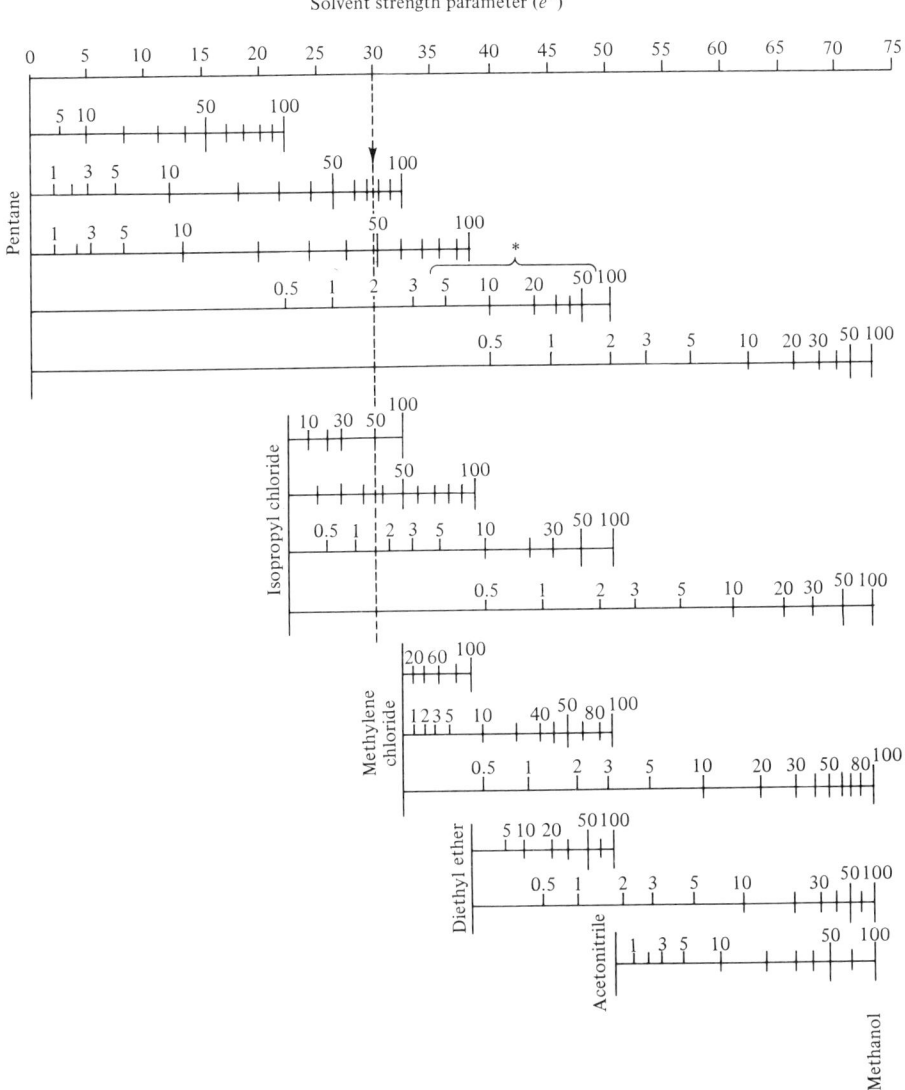

FIGURE 18-3 Mixed solvent strengths on silica gel. CCl_4 must be substituted for part of the pentane to achieve miscibility for mixtures indicated (*). (Reprinted with permission from D. L. Saunders, *Anal. Chem.*, **46**, 470 (1974). Copyright (1974) American Chemical Society.)

corresponds to a range from 0 to 100% by volume of that particular solvent in the binary pair. The first five lines represent mixtures of pentane with five other solvents: 2-chloropropane, dichloromethane, diethyl ether, acetonitrile, and methanol. The second series of lines correspond to binary mixtures of 2-chloropropane with solvents of higher $e°$, and so on for each series of lines. For any given solvent strength, Fig. 18-3 indicates several binary mixtures. For example, the dashed line in the illustration indicates solvent mixtures with $e° = 0.30$. One of these mixtures is 76% by volume of dichloromethane in pentane. Similarly, 49% diethyl ether in pentane, or 37% diethyl ether in 2-chloropropane, would also give $e° = 0.30$. Secondary solvation effects will be the greatest if the concentration of stronger component in the binary mixture is less than 5% by volume or greater than 50% by volume.

Selective localization of adsorbed solutes also affects retention. Basic solvents preferentially increase the retention of proton donor solutes as a result of the hydrogen bonding of solvent and solute molecules in the adsorbed phase.

Modifiers[4]

The addition of a small quantity of sodium phosphate or sodium acetate to the mobile phase will frequently sharpen peaks. Similarly, 1–2% of modifiers, such as tetrahydrofuran added to acetonitrile or methanol, will produce the same effect. The objective in the use of these additives is to reduce peak tailing.

Another technique is available in water-base mobile-phase systems. By using a relatively low pH buffer, ionization of the solute molecules is suppressed, thus increasing k' values. This is useful in situations where greater retention is required. Conversely, if the material is forced to ionize, thus reducing the amount of its nonpolar surface area, its degree of retention is less. The use of pH is an excellent way to control k' values for weakly ionizable materials in reversed phase chromatography.

18.2 LIQUID–LIQUID PARTITION CHROMATOGRAPHY

In liquid–liquid partition chromatography, partition of an organic solute occurs between a liquid mobile phase and an organic liquid adsorbed on, or chemically bonded onto, a support. Separations in LLC are very similar to solvent extraction. In fact, solvent extraction data can be used to predict LLC partition coefficients.[5]

In classical LLC the stationary liquid phase is uniformly coated on a support material. This can be accomplished as described for GLC supports. Coating a prepacked stationary phase is the preferred method for viscous phases. To avoid phase miscibility, the two partitioning liquids must differ greatly in polarity. The stationary phase should be a good solvent for the sample but a poor solvent for the mobile phase. This tends to limit classical LLC to compounds with low k' values. There are problems with mechanically coated phases. The support surface should not interact with the solute if the true partition separation is desired. Consequently, there is a minimum loading for each liquid phase on each type of support material. If below this minimum, exposed support sites will adsorb additional stationary phase from the presaturated mobile phase, or possibly adsorb solutes on exposed sites as in LSC. The coated phase may be stripped by strong solvents or gradually removed due to lack of mobile-phase presaturation. Under high pressures shearing forces tend to remove the stationary liquid. Column temperature must be carefully controlled within 0.5°C. These limitations, and the inability to use classical LLC with gradients, coupled with the advent of efficient, stable bonded phase packings, has discouraged use of conventional coated phases.

LLC packings can be divided into two types: normal phase and reverse phase. This division, based on relative polarities of the mobile and stationary phases, is somewhat arbitrary. *Normal phase* LLC uses a polar stationary phase (often hydrophilic) and a nonpolar mobile phase, whereas *reverse phase* LLC employs a nonpolar stationary phase (usually hydrophobic) and a polar mobile phase. In the reverse phase mode the compounds that are nonpolar selectively interact with the liquid phase and are retained more strongly than the ones which are polar. The elution order of the classes of compounds in Table 18-1 would be in reverse. Hydrocarbons would be retained more strongly than alcohols. Reverse phase chromatography will most likely provide optimum retention and selectivity when compounds have no hydrogen bonding groups or have a predominance of aliphatic or aromatic character. The reverse phase technique in its various forms—regular partition, ion-pair partition, and ion suppresson—is the most widely used mode in HPLC now that bonded phase supports have provided a suitable stationary phase.

Bonded Phase Supports[6-9]

Bonded phase supports possess functional groups incorporated on saturated hydrocarbon chains which are chemically bonded to the surface of a support, usually silica gel. Substrates include large porous silica gels, porous layer beads, and microparticles. The latter dominate although large porous particles find use occasionally for preparative work. Most of the problems encountered in conventional LLC are eliminated by bonding the static liquid to the packing.

Silicate ester bonded phases, or "brushes," are prepared by a one-to-one esterification reaction between a surface silanol group of the silica support and an alcohol such as 3-hydroxypropionitrile or Carbowax 400. Silicate esters lack hydrolytic and thermal stability, and they cannot be used with aqueous or alcoholic mobile phases.

The siloxane (Si–O–Si–C) type of bond has become the standard for commerical bonded phases. It is formed by reactions of di-(or tri-)chloro-organosilane or di-(or tri-) alkoxysilane with the surface silanol groups of fully hydroxylated silica gel. Equation 18-1 shows in simplified form a typical chlorosilane bonding reaction with a silanol group on the silica gel surface:

$$\equiv Si-OH + ClSi(CH_3)_2 R \rightarrow \equiv Si-O-Si(CH_3)_2 R + HCl \qquad (18\text{-}1)$$

For steric reasons, it is not possible for all silanol groups to react. The silica surface then will consist of a mixture of silanols and hydrophobic chains. More silanols can be removed by treatment with trimethylchlorosilane, a small silylating agent. Some silanols are inaccessible in micropores while others are blocked by bulky surface groups already bonded. By varying the nature of the organic portion of the bonding silane, the surface polarity of the bonded phase packing can be altered from an essentially hydrophobic one, consisting of a hydrocarbonaceous layer, to one of various functional groups. These bonded phases are attacked only in very acidic (pH < 2) or basic (pH > 9) environments, and are stable up to 80°C.

Although the functional groups that could be bonded is endless, a vast number of bonded stationary phases are not necessary because the mobile phase composition can be altered easily to change selectivity (see Fig. 18-4). The most popular bonded phase comprises a linear hydrocarbon, most commonly an octadecyl (C_{18}) alkyl chain, although C_8 and C_2 moieties are also available. The C_{18} sorbent is used to perform reversed phase separations on samples of low/medium polarity using methanol/water or acetonitrile/water mobile phases. With these partitioning phases selectivity is established by the hydrophobic functional groups (C_2, C_8, or C_{18}), but individual compounds are separated on the basis of their relative solubility in the mobile phase. Thus one can separate the polyaromatic hydrocarbons, phthalate esters, fatty acids, and other homologous series. For samples of moderate polarity, a C_8 sorbent, in conjunction with an acetonitrile/water mobile phase, is useful.

Bonded phases of medium polarity alkyl nitrile (the cyano column) can be substituted in separations where β, β'-oxypropionitrile was used as a stationary phase in classical LLC. In the normal mode, the nitrile phase is less retentive than silica gel in LSC, but displays similar selectivity. It does provide some selective interactions with compounds having double bonds. The use of tetrahydrofuran, acetonitrile, or methanol as mobile-phase components produce selectivity changes. Figure 18-4 shows three aromatic compounds

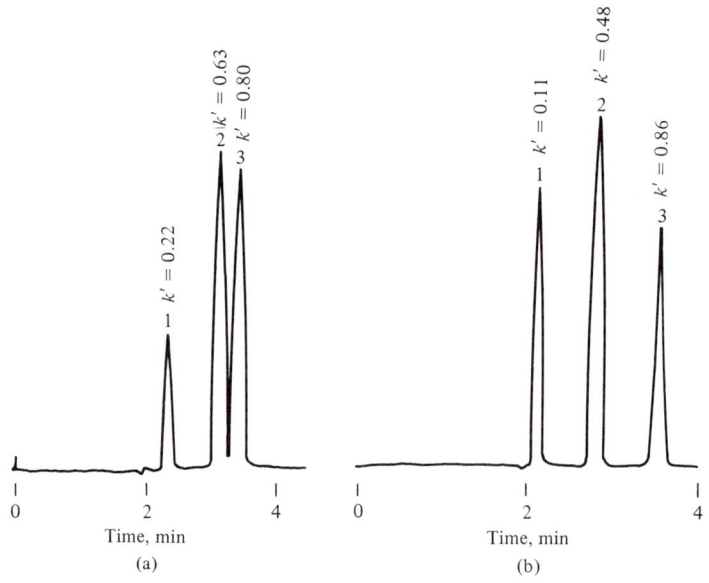

FIGURE 18-4 Selectivity changes through changes in mobile-phase components in normal phase liquid chromatography on nitrile bonded phase. (a) 25% isopropanol, 75% cyclohexane; (b) 25% tetrahydrofuran, 75% cyclohexane. Peaks: (1), anisole; (2), nitrobenzene, (3) o-dimethylphthalate.

separated on the cyano column in the normal phase mode using 2-propanol or tetrahydrofuran as the strong solvent component in the solvent mixture. Separations of oil-soluble vitamins, essential oils, nitrophenols, or more polar homologous series have been performed using alcohol/heptane mobile phases. In the reverse phase mode, the nitrile packing poorly retains the more polar compounds in a mixture. In fact, this packing shows less retention than the more hydrophobic C_{18} sorbent.

The polar aminoalkyl functional group imparts unusual chromatographic selectivity. It may function as either a Bronsted acid or base, depending on the solute, or interact with solutes by hydrogen bonding. In water/acetonitrile mobile phases, polar compounds such as carbohydrates and peptides can be separated; in acidic solution the functional group behaves as a weak anion exchanger.

The bonded phenylsilane group will also operate in the reverse phase mode. This functional group generally shows less selectivity than octadecylsilanes for nonpolar compounds, but more selective interactions with polar compounds. Retention times are intermediate between those found on a C_{18} and a cyano packing.

From a mass transfer point of view, a monomeric layer consisting of "bristles" of alkyl chains is generally preferred over a polymeric layer that can swell and undergo penetration by molecules, resulting in slow diffusion. In most cases, the actual surface topology of bonded phase materials on the molecular level is not known. It is unlikely that the C_8 or C_{18} hydrocarbonaceous portion would fully extend into the mobile phase. If wetted by the mobile phase, monomeric phases respond rapidly to changes in mobile-phase composition. Lack of wetting causes poor efficiency, with adsorption occurring at the sorbent/solvent interface in addition to the expected partition equilibrium.

For very low surface area silicas, polymerization of the reactants may be required to provide sufficient coverage and adequate sample capacity. For satisfactory chromatography with polymeric layers, the presence of even a small amount of polar solvent appears to be important. This polar solvent apparently penetrates into the polymeric layer, swelling it, and establishing a partitioning phase of solvated polymer.

The weak surface energies of the bonded alkyl phase permit not only rapid analyses but also rapid reequilibration when the mobile phase is altered, as in gradient elution or in solvent exploration for optimum resolution. Typically, only 5-10 column volumes are necessary for equilibrium to be reached versus often 50-100 or more column volumes for silica gel.

Reverse Phase Liquid Chromatography

As regularly practiced, reverse phase chromatography utilizes a hydrophobic bonded phase packing, usually possessing a C_{18} or C_8 functional group and a polar mobile phase, usually a partially or fully aqueous mobile phase (Fig. 18-5). Polar substances prefer the mobile phase and elute first. As the hydrophobic character of the solute increases, retention becomes longer. Generally, the lower the polarity of the mobile phase, the higher is its eluent strength. The eluent strength of the various solvents in reverse phase chromatography has been found to follow approximately the reverse order given in the eluotropic series (Table 18-1). Thus, water is the weakest eluent. Methanol and acetonitrile are the most popular organic solvents because they have relatively low ultraviolet cutoff points,

	Sample	Column packing	Mobile phase	
High POLARITY Low	Low/moderate polarity (soluble in aliphatic hydrocarbons)	Bonded C-18	Methanol/water	Low POLARITY High
	Moderate polarity (soluble in methyl ethyl ketone)	Bonded C-8	Acetonitrile/water	
	High polarity (soluble in lower alcohols)	Bonded C-2	1,4-Dioxane/water	

FIGURE 18-5 Column packing material and mobile phase solvents for reverse phase chromatography.

low viscosity, and are readily available with sufficient purity. In most applications neat methanol or acetonitrile are the strongest eluents employed in reverse phase mode. Eluents of intermediate strength are usually obtained by mixing one of these solvents with water or an aqueous buffer. Other water-miscible, nonultraviolet absorbing solvents, such as dioxane, tetrahydrofuran and other alcohols, have also been used.

The strong attractive forces between water molecules, arising from the three-dimensional intermolecular hydrogen bonded network, form a "structure" of the water (that is, a very high cohesive energy density) that must be distorted or disrupted when a solute is dissolved. Only highly polar or ionic solutes can interact with the water structure. Nonpolar solutes are "squeezed out" of the mobile phase and are relatively insoluble in it but bind with the hydrocarbon moieties of the stationary phase. The driving force for retention is *not* the favorable interaction of solute with stationary phase, but the effect of solvent in forcing the solute to the hydrocarbonaceous bonded layer.[10] As this phenomenon is opposed by polar group interaction of the solute with the mobile phase, hydrophobic retention deals mainly with nonpolar substances or the nonpolar portions of molecules. Hydrophobic selectivity arises then as a consequence of nonpolar size differences, that is, nonpolar surface areas, of solutes. Hydrophobic force reduction can be effected by adding any miscible organic solvent to water. The effect will be larger with less polar solvents and with greater concentration. Thus one speaks of the added organic modifier as the *stronger* solvent and water as the weaker solvent. Consequently, good resolution of homologs is observed with improved separation as the water content of the mobile phase is increased, for a given stationary phase.

In resolving functional group differences, the central role of the organic modifier has also become clear. Polar group selectivity is quite dependent on the organic solvent chosen. This appreciation has led to the use of ternary mobile phases consisting of water and two organic modifiers, as for example, methanol and tetrahydrofuran, for the precise control of selectivity of solutes particularly with different functionalities, as shown in Table 18-2. When the sample mixture contains components having a wide range of polarity, the peak capacity of the separation column often has to be increased by using gradient elution. In reversed phase mode, solvent gradients are generated by a continuous decrease in

TABLE 18-2   Retention Time (in min) of Different Functional Groups in Certain Ternary Solvent Mixtures

Solute	40% H_2O, 50% CH_3CN, Plus 10% C Solvent				
	CH_3OH	CH_3CN	THF	CH_2Cl_2	DMSO
ϕNH_2	2.9	2.9	2.4	2.8	3.3
ϕOH	3.1	2.9	2.7	2.9	3.3
ϕOCH_3	5.5	4.7	3.9	6.4	5.6
ϕ (benzene)	5.5	4.7	4.3	6.6	5.6
ϕCl	8.5	6.8	5.9	9.3	8.7

THF is tetrahydrofuran; DMSO is dimethyl sulfoxide.
From S.R. Bakalyar and R. McIlwrick, *Ind. Res.,* p. C4 (February 1978).

the polarity of the eluent during the separation; for example, by gradually increasing the organic solvent content in water/methanol or water/acetonitrile mixtures.

In many applications, particularly when ionogenic solutes are separated, the eluent pH has to be controlled by using a buffer. Adjusting the pH on the acid side will suppress ionization of acids, allowing the acids to be separated by reverse phase techniques. This approach is termed *ion suppression* and is useful for weak acids and bases in the pH 2-8 range. It is advisable to maintain a relatively high buffer concentration in order to facilitate a rapid establishment of the protonic equilibria and to avoid asymmetrical peaks or band splitting due to the slowness of the secondary equilibria involved in the chromatographic process. Phosphate buffer can be used in a wide pH range. At low pH perchloric acid is frequently employed. Use of an alkaline eluent is precluded because siliceous bonded phases are not stable in general.

The simplicity and reproducibility of reversed phase chromatography on bonded phases make this method particularly attractive in clinical chemistry. The quantitative analysis of drugs of abuse is increasingly carried out by this technique. In the pharmaceutical field the use of this technique has been increasing at the expense of adsorption chromatography on silica or alumina. Because the water content of the mobile phase can range from 100% to quite low percentages or none at all, a broad spectrum of biomolecules can be chromatographed; lipophilic or ionic, small or large. Water in the injection solvent will not in general affect retention, because of the presence of water already in the mobile phase. Trace enrichment is readily adaptable to reverse phase HPLC. One uses an injection solvent (often water) that is significantly weaker than that of the mobile phase. A preconcentration of the solute will occur at the top of the column as the mobile phase at the column entrance is converted to the weaker injection solvent. Gradient elution will subsequently separate the sample ingredients.

Strong acids and bases cannot be handled by ordinary liquid chromatographic techniques since the ion suppression method is limited by the instability of stationary bonded phases to the range pH 2.0-7.5. However, by forming an ion pair (a coulombic association species formed between two ions of opposite electrical charge) with a suitable counterion, ionic or ionizable compounds can be converted to electrically neutral compounds and therefore partitioned into the respective nonpolar phase. A large organic counterion added to the mobile phase forms a reversible ion-pair complex with the ionized sample; this complex behaves as an electrically neutral and nonpolar (lipophilic) compound. The extent to which the ionized sample and counterion form an ion-pair complex affects the degree to which the retention is increased. By adjusting the pH so that the sample is present in its ionic form, and choosing a strongly ionic counterion of opposite charge with a very lipophilic group attached, the situation can be represented by the following equilibrium:

$$\text{solute}^{\pm} + \text{counterion}^{\mp} \rightleftharpoons [\text{solute}^{\pm}, \text{counterion}^{\mp}]^{\circ} \text{ pair} \qquad (18\text{-}2)$$

The exact mechanism for ion-pair chromatography has not been clearly established. There are two fundamental proposed models.[11] The first postulates that the solute molecule forms an ion pair with the counterion in the mobile phase. This uncharged ion pair then partitions into the lipophilic stationary phase (Fig. 18-6a). The other mechanism postulates that the counterion partitions into the stationary phase, or is "loaded" onto the bonded reverse phase packing, with its ionic group oriented at the surface. This produces two possibilities for the material to be chromatographed. It can be attracted to the

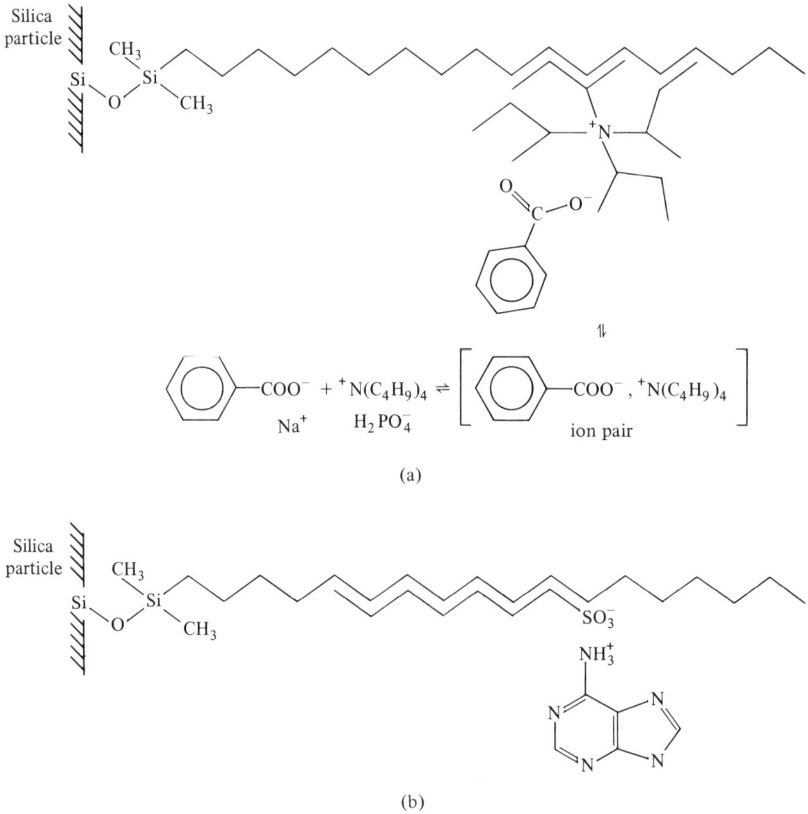

FIGURE 18-6 Ion-pair chromatography: (a) ion-pair partitions into the lipophilic stationary phase (bonded C-18) and (b) counterion (alkyl sulfonate) loaded onto stationary phase.

hydrocarbon portion in the usual reverse phase manner, or it can interact in an ion-exchange mode (Fig. 18-6b). Quite likely the true mechanism involves both postulates but is probably made more complex by adsorption and micelle formation. Whatever the mechanism, ion-pair chromatography allows for some unique separations otherwise not obtainable by either reverse phase or ion exchange.

The major property of ion pairs in an analytical context is their resultant ability to move from an aqueous environment to areas of lower dielectric. For hydrophobic ions the formation of the ion pair will most likely be in the aqueous bulk phase, followed by subsequent transfer. But for inorganic molecules and ionized organic molecules of lower molecular weight, ion-pair formation will probably occur in the interfacial or diffusion layers between the two phases where the dielectric constant will be far lower than that of the aqueous phase.

Reverse Phase Ion-Pair Partition

Reverse phase ion-pair partition (RP-IPP),[12] in which the pairing ion is located in the eluent, is the preferred method. A bonded stationary phase is used. Both counterion and buffer are added directly to the aqueous mobile phase. Elution follows in order of decreasing ion-pair polarity, typical for reverse phase separations.

There are many factors which enter into an RP-IPP separation. Retention may be controlled by the binding strength of the ion pair and the concentration of the counterion. The relative size of the lipophilic group on the counterion will affect the degree of retention obtained. For example, for basic samples, alkyl sulfonates adjusted to a pH of 3.5 are used often as the mobile phase counterion. When heptane sulfonic acid is used, as opposed to pentane sulfonic acid, retention for the ionic sample components is increased (Fig. 18-7). With alkyl chain lengths as high as 12-16 carbons, a significant amount of counterion is adsorbed onto a C_{18} or C_8 bonded phase. Alkyl sulfates (for example, dodecyl sulfate) behave similarly to alkyl sulfonic acids but yield different selectivities when used as the counterion. Even perchloric acid forms very strong ion pairs with a wide range of basic solutes. Increasing the concentration of counterion increases retention up to a limit which is usually set by the solubility of the counterion in the mobile phase, particularly when an organic modifier is employed in the system.

The control of the pH is a most important parameter. Retention is increased as pH maximizes concentration of the ionic form of the solutes. Maintaining a pH around 2.0 ensures that both strong and weak bases will be in their protonated ionic forms, and any weak acids present will be primarily in their nonionic forms. For acidic samples, a quaternary amine is recommended as the counterion. Phosphate buffer keeps the mobile phase at a pH of about 7.5. At this pH both strong and weak acids are in their ionic forms and weak bases are in their nonionic form.

A good first choice of bonded phase column would be a monolayer C_{18} or a C_8 reverse phase column. Start with a 0.01M concentration of shorter chain counterion, or if the counterion contains C_{10} or larger alkyl groups prepare a 0.005M solution. The most commonly employed solvent combinations are water/methanol and water/acetonitrile. Although use of acetonitrile offers better column efficiencies due to its lower viscosity, its usefulness is limited by the poor solubility of many ion-pair reagents. A

544 CHAPTER 18

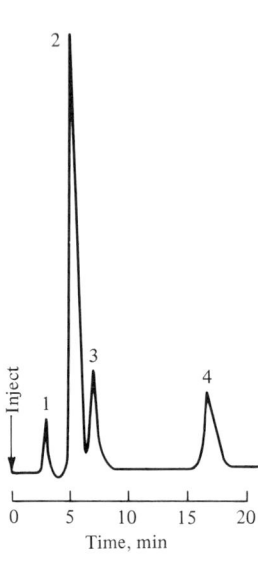
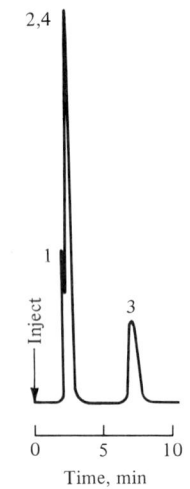
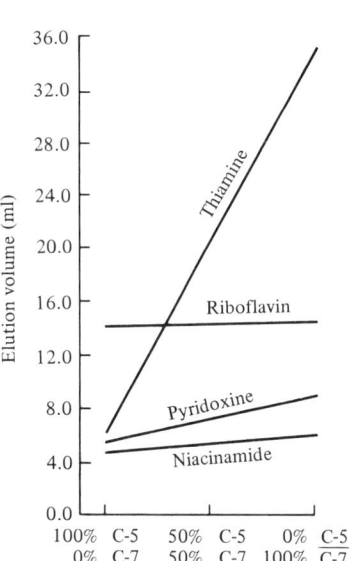
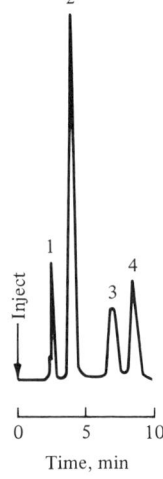

FIGURE 18-7 Separation of mixtures of ionic and nonionic compounds by reverse phase ion-pair chromatography. Peaks: (1), niacinamide; (2), pyridoxine; (3), riboflavin; and (4), thiamine. (Courtesy of Waters Associates, Inc.)

reasonable concentration of any buffer required is 0.001M–0.005M; buffer components should be selected with poor ion-pair properties but good solubilities.

Unlike conventional ion exchange, RP-IPP can separate nonionic and ionic compounds in the same sample. One should optimize the separation of the nonionic solutes first, then select and add a counterion to the mobile phase whereby the ionic solutes become retained. Take, for example, the separation of the water-soluble vitamins. Thiamine is strongly ionic, pyridoxine and niacinamide less so, and riboflavin is nonionic at pH 3.5. A two-step procedure is required. In the first step, the methanol/water ratio (polarity of the mobile phase) is adjusted to obtain good retention of the nonionic compound, riboflavin. In the second stage, an organic counterion is chosen and added to the eluent mobile phase to separate the three ionic compounds. The latter exhibit differences in retention with the alkyl chain length of the counterion. The extent of this effect depends upon the ease of ionization of the solutes. Thus, thiamine, a quaternary amine, shows the greatest sensitivity to a change in counterion. As shown in Fig. 18-7, the optimum separation is achieved with a 50/50 mixture of C_5/C_7 alkyl sulfonic acids. A mixture of counterions added to the mobile phase produces a retention proportional to the concentration of each counterion.

18.3 ION-EXCHANGE HPLC

In this technique the stationary phase is an ion exchanger which consists of two components: the polymer matrix and attached functional groups. The latter are permanently bonded ionic groups with their counterions of opposite charge. These counterions can be exchanged for an equivalent number of other ions of the same sign in the mobile phase. Ion-exchange chromatography is particularly adapted to the analysis of ionized or ionizable compounds via charge–charge interactions.

Ion-Exchange Packings

Some ion exchangers bear negatively charged groups and are used for exchanging cationic species. Others, designed for exchanging anionic species, are provided with positively charged groups. The most commonly used functional groups are the sulfonate type for cation exchange and the quaternary amine type for anion exchange (Fig. 18-8). Sulfonate exchangers are strongly acidic exchangers because they have the properties of strong acids when in the H-form. Likewise, quaternary ammonium exchangers are strongly basic because, when in the OH-form, their properties are similar to those of a strong base. Both functional groups are totally dissociated; therefore their exchange properties are independent of solution pH. That is, their exchange capacity, which is the number of functional groups available for exchange per mass (or volume) unit of exchanger, is constant and not subject to change with pH.

There are some exchangers whose functional groups have weak acidic or basic properties. Thus a carboxylate group carried by an exchanger permits the exchange of cationic species only when the pH is high enough to permit dissociation of the –COOH site. For the same reason a tertiary amine exchanger has exchanging properties only when in an

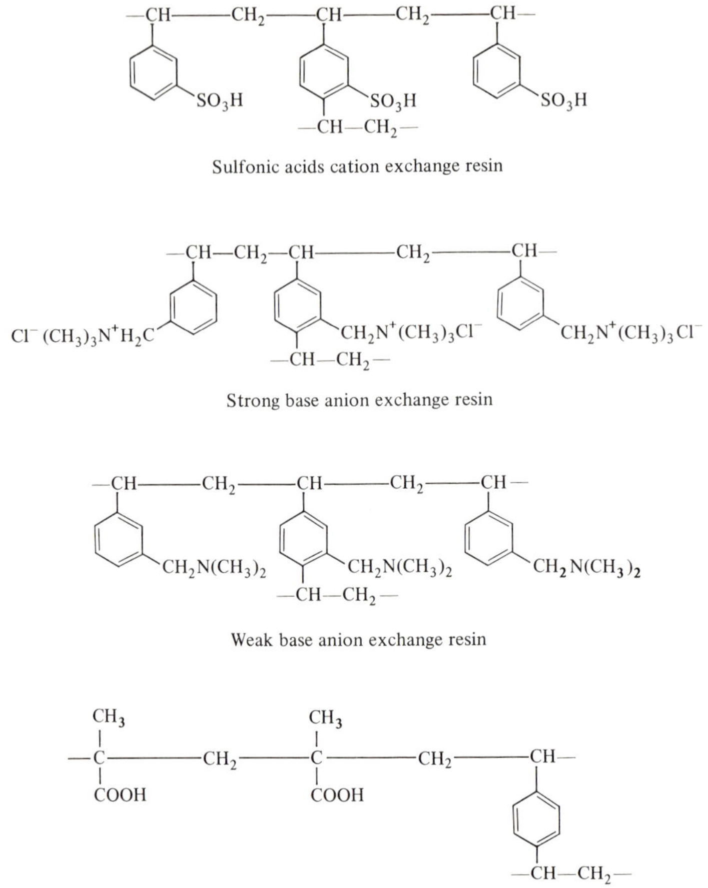

Sulfonic acids cation exchange resin

Strong base anion exchange resin

Weak base anion exchange resin

Carboxylic cation exchange resin

FIGURE 18-8 Functional groups of ion-exchange resins.

acidic medium, because its functional groups are then carrying a positive charge, a proton having been bound to the nitrogen atom.

In addition to their weak acid character, some functional groups have chelating properties towards some metallic cations, namely,

$$-N\begin{matrix} \nearrow CH_2COOH \\ \searrow CH_2COOH \end{matrix} \qquad\qquad -PO_3^{2-}$$

Aminodiacetate Phosphonate

These exchangers have considerable affinity for heavy metal cations and, to a lesser extent, for alkaline earth cations. The former type finds use as a column packing for ligand exchange.

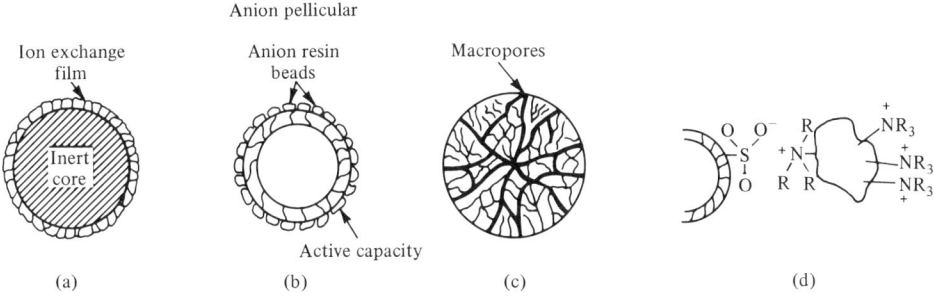

FIGURE 18-9 Structural types of ion-exchange resins: (a) pellicular with ion-exchange film; (b) superficially porous resin coated with exchanger beads; (c) macroreticular resin bead; and (d) surface sulfonated and bonded electrostatically with anion exchanger.

Ion-exchange HPLC may be performed on one of three types of packings: pellicular, phases bonded to silica microparticles, and classical cross-linked polystyrene-divinylbenzene resins of small particle diameter (Fig. 18-9). The pellicular type consists of a resin coating, about 1-2 μm in thickness, onto a glass bead whose diameter is 30-40 μm. Superficially porous resins are obtained by coating glass beads with a thin layer of silica microspheres (mean diameter, 0.2 μm) on which is coated or bonded an ion exchanger. This increases the interface between the resin and mobile phase. For either type packing the exchange capacity is low: 0.01-0.1 mequiv g^{-1}.

The resin may also be bonded to silica microparticles by means of silylation reactions or polymerized into the pores of a superficially porous silica gel. When preparing ion exchangers by silylation, a vinyl group is chosen for R_3 of $HOSiR_1 R_2 R_3$ which leads to a vinylated silica onto which styrene is then polymerized. Afterwards the bonded phase is sulfonated or treated with chloromethyl ether and subsequently trimethylamine (or hydroxyethyldimethylamine) to prepare the quaternary amine exchanger. Spherical particles are available with diameters 5-10 μm; exchange capacities are 0.5-2 mequiv g^{-1}.

Macroporous (or macroreticular) styrene-divinylbenzene polymers have large-size channels, in addition to micropores, which offer the ions easy access to the functional groups of the exchanger. Compared to the classical microreticular cross-linked resins, the macroporous beads (Fig. 18-9c) of small diameter do not swell or shrink appreciably with changes in ionic strength, or deform at high flow velocities. They are well suited to separations carried out in nonaqueous media although they have less selectivity than the microreticular gel resins as far as simple species are concerned.

The type of polystyrene-divinylbenzene exchanger used in ion chromatography is prepared by surface (only) sulfonation which leads to the formation of a thin surface shell of sulfonic acid groups. The surface of the anion resin is unique in that the active capacity of the surface-sulfonate resin is used to bond the very small anion resin particles electrostatically. This creates the aminated resin pictured in Fig. 18-9d which is quite stable and long-lived. Due to the proximity of all of the active sites to the eluent/resin interface, this type of exchanger possesses favorable mass transfer characteristics. The capacity is around 0.020 mequiv g^{-1} of starting copolymer.

Exchange Equilibrium

The primary process of ion-exchange chromatography involves adsorption/desorption of charged ionic materials in the mobile phase with a permanently charged (opposite sign) stationary phase. For example, in the case of an ionized resin, initially in the resin, H^+-form, in contact with a completely ionized solution containing K^+ ions, an equilibrium is set up:

$$\text{resin}, H^+ + K^+ \rightleftharpoons \text{resin}, K^+ + H^+ \tag{18-3}$$

which is characterized by the selectivity coefficient, $k_{K/H}$:

$$k_{K/H} = \frac{[K^+]_r [H^+]}{[H^+]_r [K^+]} \tag{18-4}$$

where the subscript r refers to the resin phase. Strictly, the selectivity coefficient is constant only if the activity coefficient ratios in the resin and in the solution phases are constant.

Affinity differences are essentially governed by the physical properties of the solvated ions. The resin phase will show a preference for (1) the ion of higher charge, (2) the ion with the smaller solvated radius, and (3) the ion which has greater polarizability. The solvated ionic radius limits the coulombic interaction between ions, and the polarizability of the ions determines the van der Waals' attraction. Together these factors control the total energy of interaction between oppositely charged species, and hence their tendency to exist as ion pairs in the resin matrix. Energy is needed to strip away the solvation shell surrounding ions with large hydrated radii, even though their crystallographic ionic radii may be less than the average pore opening in the resin matrix. This explains the position of the lithium ion among the alkali metal ions, and the fluoride ion among the halide ions.

The partitioning of each trace ion, K^+ in our example, between the resin phase and the solution phase can be described by means of a concentration distribution ratio, D_c:

$$(D_c)_K = \frac{[K^+]_r}{[K^+]} \tag{18-5}$$

Combining the equations for the selectivity coefficient and for D_c:

$$(D_c)_K = k_{K/H} \frac{[H^+]_r}{[H^+]} \tag{18-6}$$

Equation 18-6 reveals that the concentration distribution ratio for trace concentrations of an exchanging ion is independent of the respective solution concentration of that ion (neglecting any activity effects). Therefore, the uptake of each trace ion by the resin is directly proportional to its solution concentration. However, the concentration distribution ratios are inversely proportional to the solution concentration of the resin counterion, which is to be expected since the counterion competes with the trace ion for exchange sites in the resin phase.

In order to accomplish any separation of two cations (or two anions) from each other, it is necessary that one of these cations be taken up by the resin in distinct preference to the other. This is expressed by the separation factor (or relative retention), $\alpha_{K/Na}$, using K^+ and Na^+ as the example. Thus,

$$\alpha_{K/Na} = \frac{(D_c)_K}{(D_c)_{Na}} = \frac{k_{K/H}}{k_{Na/H}} = k_{K/Na} \qquad (18\text{-}7)$$

If the selectivity coefficient, $k_{K/Na}$, is unfavorable for the separation of K^+ from Na^+, no variation in the concentration of H^+ (the eluent) will improve the separation. The situation is entirely different if the exchange involves ions of different net charges. Now the separation factor does depend on the concentration. For example, the more dilute the counterion concentration in the eluent, the more selective the exchange becomes for polyvalent ions.

Applications

The efficiency of the separation of cations can be markedly influenced through the use of complexing agents in the eluent. For example, buffered solutions containing citrate or ethylenediaminetetraacetate (EDTA) will bring about separations because some cations are converted into neutral or negatively charged ionic species. The extent of conversion into such species will be a function of pH and metal formation constants.

To a substantial extent, practically every metal may be converted to a negatively charged complex ion through suitable masking systems. This fact, coupled with the greater selectivity of anion exchangers, makes anion exchange a logical tool for handling metals. The complexes, negatively charged, are adsorbed by the exchanger and eluted by changing the concentration of masking agent in the eluent sufficiently to cause dissociation of the metal complexes or a decrease in the fraction of the metal present as an anionic complex ion in solution. If interconversion of complexes is fast, control of ligand concentration affords a powerful tool for control of adsorbability, since ligand concentration controls the fraction of the metal as adsorbable complex. Extensive studies have been reported of metals which form chloride and fluoride complexes, just two examples of many such systems. In some fortunate cases one or a group of the metal ions in a mixture will not be transformed into a negatively charged species and will not be bound by the resin, whereas the remainder of the metal ions will be converted to anionic complexes and selectively adsorbed.

Cation exchangers containing counterions, such as Cu^{2+}, Ni^{2+}, or Zn^{2+}, show a unique preference for absorbing molecules that can act as ligands. Thus even though they are bound to an exchanger, these metals retain their ability to be the central atom of a coordination compound. Furthermore, the ligand associated with the coordinating metal attached to the resin (R^-) can be replaced by a different ligand, namely,

$$R^-[Cu(NH_3)_4^{2+}] + 4\text{ caffeine} \rightleftharpoons R^-[Cu(\text{caffeine})_4^{2+}] + 4NH_3 \qquad (18\text{-}8)$$

(Caffeine is 3,7-dihydro-1,3,7-trimethyl-1H-purine-2,6-dione.) No ion exchange takes place; the exchanger acts simply as a support for the coordinating metal ion. This process is termed *ligand exchange*. The chelating resins are ideally suited for ligand-exchange work. The chelating polymers have iminodiacetate functional groups attached to a styrene-divinylbenzene matrix. Divalent metal ions from the transition elements are tightly bound to such exchangers. Consequently, leakage of metal ions from the chelating resins by ordinary ion-exchange reactions with cationic materials in eluting solutions is held to a minimum.

Ligands may be separated from each other by ligand-exchange chromatography. Different ligands have different affinities for the coordinating metal attached to the exchanger. Hence, their migration rates down a column differ and thus separation occurs. For example, purine metabolites have been separated on a chelating resin loaded with copper(II) ion.[13] Eluent was $1M$ NH_3. The elution order (V'_R values in parentheses) was guanine (30 ml), xanthine (34 ml), adenine (44 ml), and hypoxanthine (65 ml). Uric acid was not retained. In the example chosen for Eq. 18-8, caffeine is strongly retained on the column, and is well separated from theophylline, theobromine, and most other compounds appearing in coffee, cola beverages, tea, commercial decongestants, and analgesic tablets.

Sugars, as with polyhydroxy compounds in general, form with borate ions a series of complexes with various stabilities which dissociate to various degrees. Disaccharides, having a lesser tendency to form complexes, can readily be separated from monosaccharides which form more stable complexes. Moreover, monosaccharides can be separated from each other within the classes of hexoses, pentoses, and tetroses with a borate buffer and pH gradient raising from pH 7 to 10.

The analytical separation of aldehydes and ketones, and the separation of these from alcohols, is based on the fact that the carbonyl group forms addition compounds with the hydrogen sulfite ion which is strongly adsorbed by anion-exchange resins. Alcohols do not form the sulfite-addition product and are therefore not adsorbed. Subsequently, the ketones are desorbed by hot water and the aldehydes eluted with a NaCl solution.

Acids can be separated according to their strengths on strong anion exchangers, with the weakest acids emerging first, either by elution with a strong acid or, since acid dissociation depends upon pH, by gradient elution with buffers of decreasing pH. Amino acids, which add protons to form cations in the pH range below their isoelectric points, can be separated on cation exchangers by gradient elution with buffers of increasing pH; here the most acidic components emerge first and the most basic last. Commercial automatic amino acid analyzers are available.

Advantage can be taken of weakly basic anion-exchange resins to separate acids of different strengths. Acids having a small dissociation constant are retained only to a slight extent and, if the dissociation constant of the acid is less than the base constant of the exchanger, virtually no retention occurs. In this manner, HCN, carbonic, silicic, and boric acids can be separated from phosphoric, sulfuric, and hydrochloric acids.

Ion Chromatography[14]

The limitation of conventional ion exchange is the limitation in detectors. Unless the ion exiting from the HPLC column has a directly measurable property, such as light absorption or radioactivity, some kind of chemical work-up is necessary for continuous effluent monitoring. Ion chromatography solves this detection problem by using an additional suppressor column downstream from the separator column. The suppressor column strips out or neutralizes the ions of the eluent and leaves only the species of interest passing through the conductance cell. In anion analysis, the anion ion-exchanger column is followed by a cation-exchange suppressor column. On entering the suppressor column, the eluting base is removed by the acidic resin:

$$\text{resin}-H^+ + Na^+ + OH^- \rightarrow \text{resin}-Na^+ + H_2O \qquad (18\text{-}9)$$

and the separating anions are converted to their acids:

$$\text{resin}-H^+ + M^+ + A^- \rightarrow \text{resin}-M^+ + H^+ + A^- \qquad (18\text{-}10)$$

which flow through the suppressor column and into the conductivity cell of the instrument where they are monitored. An analogous system exists for cation analysis where HCl is the eluent, a cation exchanger is in the separator column, and an anion-exchange resin is in the suppressor column. Additional schemes are outlined in Table 18-3 and shown in Fig. 18-10.

Obviously the suppressor column will become depleted and must be regenerated. Also the ratio V_B/V_A, volume of suppressor bed/volume of separator bed, must be kept as low as possible to minimize loss of resolution in the separator column. In practice, a ratio of ten or less has been found acceptable. The ratio C_B/C_A (specific capacity of suppressor bed/specific capacity of separator bed) must be as large as possible to maximize the number of ions that can be removed from the eluent stream before the capacity of the suppressor bed is saturated. In practice, these requirements are met by using conventional resins with a high degree of cross-linking in the suppressor column while using special low-capacity resins in the separator column. The latter are either surface-sulfonated styrene-divinylbenzene microbeads for cation exchange, or the latter bonded with microbeads of quaternary resin particles. The bonding is electrostatic. With only surface sites, ion exchange is rapid.

The choice of eluent revolves about balancing eluting power and eluent concentration with the selectivities of the sample ions in the separator resin. To ensure sufficient residence time in the separator column, one should select an eluting ion that has about the same affinity for the resin as does the sample ions. It is for this reason that the carbonate ion, or HCO_3^-, is sometimes a suitable alternate to the hydroxide ion. The difference in

TABLE 18-3 Schemes for Ion Chromatography

Separating Column	Eluent	Suppressor Column	Stripping Reaction
Resin-H^+	HCl	Resin-OH^-	Resin-OH^- + HCl $\rightarrow$ Resin-Cl^- + H_2O
Resin-Ag^+	$AgNO_3$	Resin-Cl^-	Resin-Cl^- + $AgNO_3$ $\rightarrow$ Resin-NO_3^- + AgCl
Resin-Cu^{2+}	$Cu(NO_3)_2$	Resin-amine	Resin-amine + $Cu(NO_3)_2$ $\rightarrow$ Resin-amine [$Cu(NO_3)_2$]
Resin-Ag^+	$AgNO_3$-HNO_3	(1) Resin-Cl^-	Resin-Cl^- + $AgNO_3$ $\rightarrow$ Resin-NO_3^- + AgCl
		(2) Resin-OH^-	Resin-OH^- + HNO_3 $\rightarrow$ Resin-NO_3^- + H_2O
Resin-OH^-	NaOH	Resin-H^+	Resin-H^+ + NaOH $\rightarrow$ Resin-Na^+ + H_2O
Resin-HCO_3^-	$NaHCO_3^-$	Resin-H^+	Resin-H^+ + $NaHCO_3^-$ $\rightarrow$ Resin-Na^+ + H_2O + CO_2

Reprinted with permission from H. Small, T.S. Stevens, and W.C. Bauman, *Anal. Chem.*, 47, 1801 (1975). Copyright (1975) American Chemical Society.

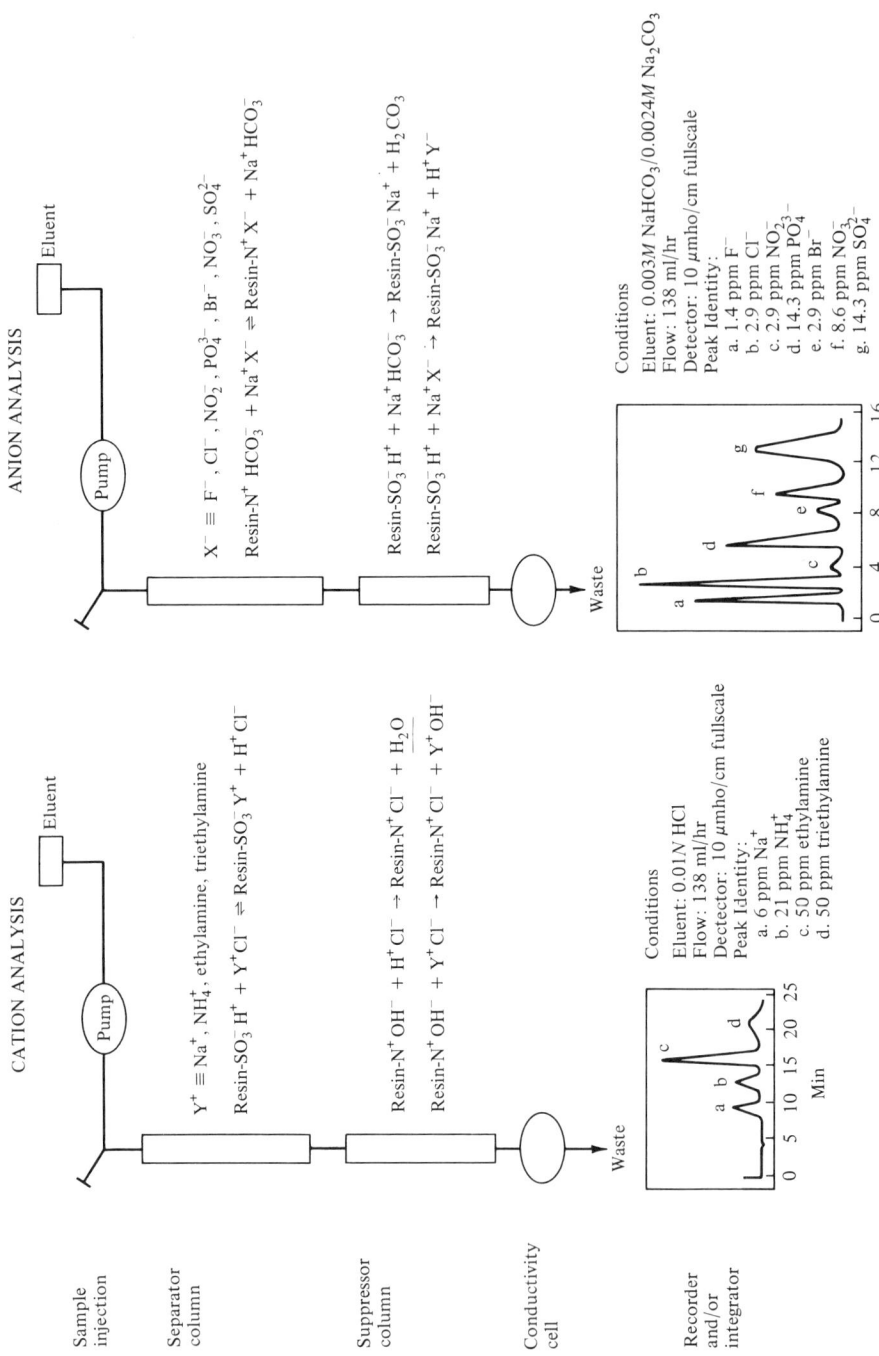

FIGURE 18-10 Ion chromatography (IC) flow scheme. (Courtesy of Dionex Corp.)

affinities between the eluent and sample ions must not be greater than a factor of 10. To decrease residence time one can increase the eluent concentration; to increase residence time, one can decrease the eluent concentration.

Matrices that have been analyzed include brine solutions, pulp and paper liquors, soil extracts, pond waters, plating baths, and industrial process, waste, and boiler waters. Among ions determined are: F^-, Cl^-, PO_4^{3-}, NO_3^-, CrO_4^{2-}, SO_4^{2-}, $C_2O_4^{2-}$, Na^+, K^+, NH_4^+, Mg^{2+}, and Ca^{2+}. Air samples and vapors from Schöniger combustions are trapped in an appropriate solvent and diluted as necessary.

18.4 EXCLUSION CHROMATOGRAPHY

Liquid exclusion chromatography, also called gel permeation chromatography (GPC), is a noninteractive mode of separation. Essentially a maze for molecules, the particles of the column packing have various size pores and pore networks, so that solute molecules are retained or excluded on the basis of their hydrodynamic volumes; that is, their size and shape. In so doing, the stationary phase can effect a separation according to molecular weight.

As the sample passes through the column, the solute molecules are sorted by the pores of the packing material. Very large molecules cannot enter many of the pores and will also penetrate (permeate) less into the comparatively open regions of the packing, and thus excluded, will travel mostly around the packing and elute at the velocity of the mobile phase. Very small molecules, diffusing into all or many of the pores have accessible to them both the flowing mobile solvent and the stationary solvent trapped within the pores. With a larger column volume at their disposal, small molecules exit the column last. Between these two extremes, intermediate-size molecules can permeate some passages but not others and, consequently suffer retardations in their progress down the column, exiting at intermediate times. In a sense, a column packed with porous particles can be regarded as being a column of variable path length; it is a short column for solute molecules of molecular dimensions greater than the average pore size of the column packing, whereas it is a long column for solute molecules smaller than the pore size of the packing.

Column Packings

Column packings are available in two types: semirigid, cross-linked polymer gels; and rigid, controlled-pore-size glasses or silicas. The semirigid materials swell slightly, and some care must be taken in their use because these materials are pressure-limited due to bed compressibility. Depending upon the solvent and hence degree of swelling, maximum pressure is limited to about 300 psi. The styrene-divinylbenzene cross-linked polymers allow fractionation within the molecular weight range from 100 to 5×10^8. Partially sulfonated polystyrene beads are compatible with aqueous systems; the nonsulfonated with nonaqueous systems. Packings are available in prepacked columns of 5- or 10-μm-diameter particles, and may be used up to 6000 psi. Another class of hydrophilic porous packing is prepared by suspension copolymerization of 2-hydroxyethyl methacrylate with ethylene dimethacrylate. The packing can withstand pressures up to 3000 psi and are usable with aqueous systems and with a variety of polar organic solvents.

Porous glasses and silicas cover a wide range of pore-size diameters. For example, one series has these pore diameters; the operating range in molecular weight units (daltons) is given also:

Pore-Size Diameter, Å	Operating Range, Daltons
40	1000–8000
100	1000–30,000
250	2500–125,000
550	11,000–350,000
1500	100,000–1,000,000
2500	200,000–1,500,000

These packings are chemically resistant at pH < 10, and can be used with aqueous and polar organic solvents. With nonpolar solvents it is desirable to deactivate the surface by silylation to avoid irreversible retention by polar solutes. Porous inorganic materials have distinct advantages over organic exclusion packings. After calibration, columns can be used routinely and indefinitely, with no possibility of sample contamination or biodegradation. The bed volume remains constant at high flowrates or high pressures. Thermal stability permits the use of high temperature.

The various pore sizes available permit separating small molecules with molecular weights of under 100 to large polymers and accompanying additives with molecular weights up to 5×10^8 daltons (Fig. 18-11). Separations in exclusion chromatography include analysis of molecular weight distributions of polydisperse polymer samples,

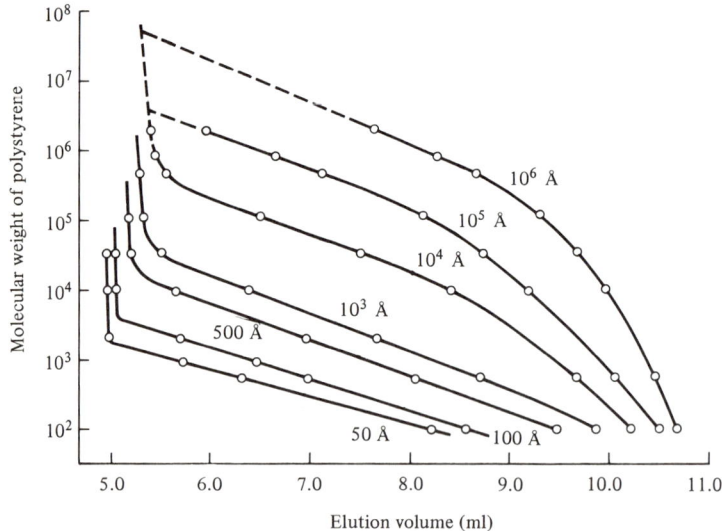

FIGURE 18-11 Separation range of various pore-size exclusion chromatographic packings.

preparative fractionation of polymers to obtain narrow molecular-weight-dispersion fractions, purification and analysis of biological materials, and studies of complex equilibria. Biological applications of aqueous exclusion chromatography involving proteins, amino acids, viruses, carbohydrates, and natural polymers such as lignin sulfonates have been reported in addition to separations of inorganic simple salts, macro-ions, and metal complexes. Steric exclusion may also serve as an exploratory or prefractionation technique.

Solvents

The time required for development of new methods is short since solvent selection is easier. Exclusion chromatography requires only a single solvent in which to dissolve and run the sample. In contrast to the other modes, usually all sample components elute between the excluded volume and the total permeation volume.

Retention Behavior

The essential behavior of a solute and the characteristics of porous column packings can be discussed in very simple terms. For a packed column of porous particles with a total bed volume, V_t:

$$V_t = V_M + V_S + V_g \qquad (18\text{-}11)$$

where V_M is the void volume of the mobile phase (that is, the unbound solvent in interstices between the solvent loaded porous particles), as estimated by elution of a totally excluded solute; V_S is the cumulative internal volume within the porous particles and available to a totally included solute or molecule of solvent (also called V_i); and V_g is the volume occupied by the matrix.

If it is assumed that the time taken for a solute molecule to diffuse into a pore is small with respect to the time which the molecule spends in the vicinity of the pore, then the separation process will be completely independent of diffusion processes. Under these conditions the retention (elution) volume, V_R, of a solute is the volume of effluent that flows from a column between the sample injection and its emergence in the effluent. That is,

$$V_R = V_M + KV_S \qquad (18\text{-}12)$$

or, rearranging as the distribution coefficient:

$$K = \frac{V_R - V_M}{V_S} \qquad (18\text{-}13)$$

which can be stated as the fraction of internal pore volume that is accessible to the solute. Totally excluded molecules will elute in one void volume; that is, $V_R = V_M$, and so $K = 0$. For small molecules which can enter all of the pores of the packing, $V_R = V_M + V_S$, and hence $K = 1$. Intermediate-size molecules elute between these two limits, and K

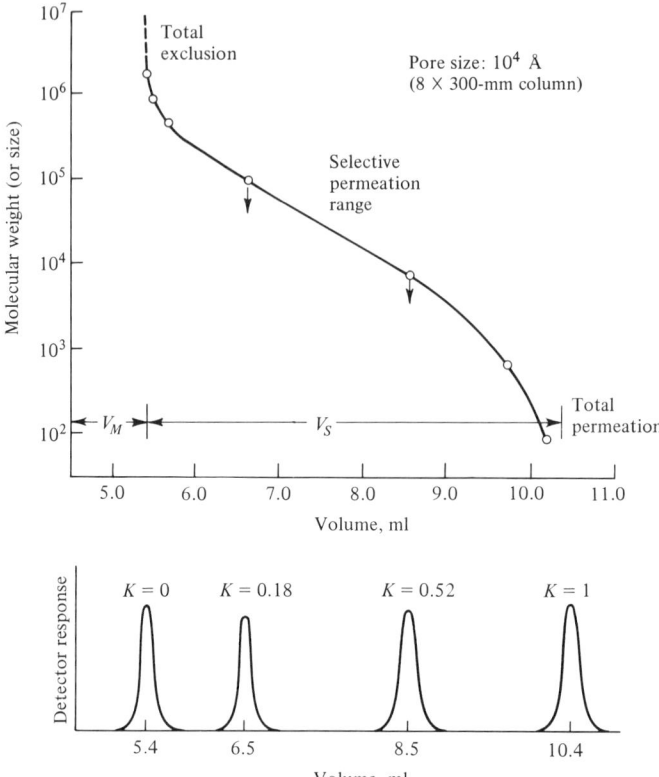

FIGURE 18-12 Retention behavior in exclusion chromatography.

ranges from 0 to 1. An elution graph is shown in Fig. 18-12. In the upper portion is shown the graph of logarithm of the molecular weight versus retention volume. It is a sigmoid-shaped curve, in which there is a linear range of effective permeation between the limiting values that correspond to exclusion ($K = 0$) and to total permeation ($K = 1$). Maximum elution volume is often only twice the column void volume.

Solute distribution in exclusion chromatography is governed mainly by the entropy change between phases (in contrast to other LC methods where enthalpy changes are substantial). Thus, K can be derived as

$$K \simeq e^{\Delta S^\circ/R} \qquad (18\text{-}14)$$

Furthermore, the temperature independence of peak retention is predicted by Eq. 18-14, and is substantiated by experimental observations.

Some workers might consider the small operating range of K to be a serious limitation. The number of peaks that can be resolved is limited (see Table 15-2). The total separation of very complex samples is exceedingly difficult. Also the separation of molecules of very similar size and isomers is generally precluded. In contrast, exclusion chromatography offers a number of advantages: (1) narrow bands for easy detection, (2) short separation times without gradient elution, (3) predictable separation times and values according to molecular size, (4) freedom from sample loss or reaction during separation, and (5) little problem of column deactivation.

Sources of Error

A possible source of error is sample interaction with the packing. Naturally, favorable interactions between sample molecules and substrate can lead to enhanced or total retention. This can occur when the sample adsorbs or partitions with the substrate. For example, siliceous substrates would tend to adsorb solutes with polar functional groups. In this regard, styragel (and related phases) could provide a more hospitable environment for polar solutes. To counteract these effects the surfaces of the exclusion packing may be chemically modified to reduce adsorptivity, or the column adsorptivity can be eliminated by proper solvent selection. Adsorptive sites can be reacted with an inert moiety, such as glycol ether or alkyl silyl groups; the inherent danger in this approach is that the chemically modified silica surface will now act as a partitioning medium. A second approach involves using a mobile phase which competes effectively for the adsorptive sites. Tetrahydrofuran is quite polar, and provides adequate deactivation for a wide range of polymers when using silica columns. An example of this problem occurs when polyethylene oxylated polyols are run on a silica column using methanol as a mobile phase; elution is in the reverse order expected for a size separation because adsorption predominates. If, however, a mobile phase containing propylene glycol and acetonitrile is used, the desired separation according to molecular size is observed.

In styragel/dimethylformamide systems, low molecular weight solutes elute in order of decreasing polarity. This partitioning is probably solvophobic in nature; that is, dipole-dipole interactions oppose the intrusion of apolar solutes in the medium and favor association of solute and gel.

When polyelectrolytes or simple salts are eluted in aqueous solutions, with or without added electrolyte in the mobile phase, several effects may occur which cause deviation from the normal steric exclusion mechanism. These effects arise from interaction of the solute with its ionic environment: ionic strength, free charges on the pore surface, and nature of the eluent.

Uncharged, sterically unhindered water-soluble polymers adopt random coil configurations in solutions. Polymers with ionizing sites, however, exhibit site-site repulsion that leads to chain extensions, and hence higher viscosities.

An ion exclusion effect arises when the pore surface contains charged groups of the same sign as the solute. Charge repulsion may cause limited pore penetration and therefore smaller elution volumes may be exhibited toward anionic species if the eluent is of insufficient ionic strength to neutralize ion-ion repulsion. This effect can be eliminated by employing $0.01 M$ NaCl in the eluent.

Ion inclusion can occur when a polydisperse polyelectrolyte or mixture of different sized salts is chromatographed using a porous packing at low ionic strength eluent. The presence of the larger sterically excluded molecules leads to the establishment of a Donnan equilibrium. This leads to a higher included concentration of the smaller charged molecules, and a higher elution volume than would be found in the absence of excluded charged molecules. This effect is also suppressed by addition of sufficient electrolyte to screen the charge on the excluded charged molecule.

Resolution

In exclusion chromatography a meaningful resolution expression can be derived by introducing the mathematical form of the exclusion curve (Fig. 18-11):

$$M = D_1 e^{-D_2 V} \quad \text{or} \quad \ln M = -D_2 V + \ln D_1 \tag{18-15}$$

where D_1 is a constant and D_2 is the slope of the curve. Substituting molecular weights M_1 and M_2 for each of two solutes that elute as peaks at volumes V_1 and V_2, then

$$V_2 - V_1 = \frac{1}{D_2} \ln \frac{M_1}{M_2} \tag{18-16}$$

and the resolution expression (Eq. 15-14) in volume units becomes

$$R = \frac{1}{4\sigma D_2} \ln \frac{M_1}{M_2} \tag{18-17}$$

Column performance in exclusion chromatography is properly measured by resolving materials that differ by a known ratio of molecular weights. It is not measured by plate number. The plate number only accounts for band spreading at total permeation and does not reflect the efficiency of the pores in the operating region of interest. The term $4\sigma D_2$, on the other hand, is a direct measure of both band spreading and calibration slope in the selective separating region of the column.

The plate number is given by

$$N = 16 \left[\frac{(V_M + V_S)}{4\sigma} \right]^2 \tag{18-18}$$

Typically, N is 10,000 and the total permeation volume $(V_M + V_S)$ is about 5 ml. Therefore, the base line width, $4\sigma = 0.20$ ml, and $\sigma = 50$ μl. This means that the elution volumes observed in high-performance exclusion chromatography are extremely small, and that minimization of dead volume is very important. Response time of detectors can become limiting.

The use of microparticles in exclusion chromatography has the same advantages as in the other modes of liquid chromatography. Small particle size promotes rapid mass transfer so that flowrates much higher than normally used in the classical mode with nonrigid organic gels may be employed to achieve separations in several minutes rather than several hours. The higher resolution obtained with microparticle packings allows separation of molecules much closer in molecular size, including small molecules. However, for larger molecules such as polymers, the smaller elution volumes obtained with these packings require more precise control over flowrate to get accurate molecular weight data.

When the probable molecular dimensions or weights of the sample components are known, selecting the particular pore size or exclusion limit of the packing is usually straightforward. For species in the 100-1000 range of molecular weights, differences on the order of 40-50 molecular weight units should exist between components for discrete resolution. Column lengths are determined by the magnitude of the differences; as size differences diminish, longer columns of a given exclusion-limit packing are required.

Proper column selection is a key to optimum results; the analyst desires to obtain maximum resolving power in the molecular weight range of interest with minimum band spreading. A quick preliminary run is made using a column set with one column packed with 1000-Å material and the second with 100 Å. Using a flowrate of 3 ml min^{-1}, the entire run requires about 4 min/sample. If most of the sample elutes near the exclusion limit, use a column packed with 1000-Å material. Elution of the majority of the material halfway between exclusion and total permeation volumes suggests a column packing of 500-Å pore size. Near total permeation indicates a 100-Å packing, and total permeation would indicate a 60-Å pore size. Should sample components elute over the entire range, a full column set of all four packings should be used.

Column Calibration

To determine molecular weights for monodisperse species or molecular weight averages and distributions for polydisperse systems, the exclusion columns must be calibrated. This is achieved by eluting appropriate calibration standards and monitoring the elution volume.[15]

Narrow dispersed standards of polystyrene, polytetrahydrofuran, and polyisopropene are available for use in organic solvents. They produce sharp and very well-characterized peaks. Polystyrene standards come in these nominal molecular weights: 600, 1000, 3000, 10^4, 3×10^4, 10^5, 3×10^5, 10^6, and 3×10^6. Samples of dextrans, polyethylene glycols, polystyrene sulfonates, and proteins are available for use in hydrophilic solvents.

Care must be exercised in correlating molecular weight/elution volume data for polymers having different chemical compositions. Differences in structure and solvent–polymer interactions lead to different hydrodynamic volumes for equivalent molecular weights. The hydrodynamic radius of the molecule is proportional to the logarithm of the product of molecular weight and intrinsic viscosity; that is, $\ln M[\eta]$. It is a valid parameter for linear polymers. To circumvent the problem of preparing narrow molecular-weight-dispersed fractions for each polymer of interest, the concept of universal calibration has been proposed. The Mark–Houwink equation, $[\eta] = KM^a$, describes the correlation between the intrinsic viscosity of a polymer and its molecular weight. A graph of $\ln [\eta]M$ versus the elution volume produces a unique relationship for determining molecular weights of polymers structurally different from those used for calibration of the column. Values of K and a have been tabulated for a large number of polymer-solvent systems. Thus,

$$\ln M_2 = \frac{1}{1+a_2} \ln \frac{K_1}{K_2} + \frac{1+a_1}{1+a_2} \ln M_1 \qquad (18\text{-}19)$$

For polymers, specific parameters of interest include: $\overline{M}_w$, the weight-average molecular weight; $\overline{M}_n$, the number-average molecular weight; and $\overline{M}_w/\overline{M}_n$, the dispersity. A simplified example will illustrate these parameters. Assume the chromatogram exhibits two peaks of equal concentration which arose from 1 g each of two components, one of molecular weight 150,000 and the other of molecular weight 50,000; and N is the number of molecules in 1 g of each molecular weight component. Then

$$\overline{M}_w = \frac{(1 \text{ g} \times 150{,}000) + (1 \text{ g} \times 50{,}000)}{1 \text{ g} + 1 \text{ g}} = 100{,}000$$

$$\overline{M}_n = \frac{(0.33 \times 150{,}000) + (1 \times 50{,}000)}{0.33 + 1.0} = 74{,}800$$

$$\frac{\overline{M}_w}{\overline{M}_n} = \frac{100{,}000}{74{,}800} = 1.34$$

$\overline{M}_w$ values are particularly sensitive to the amount of high molecular weight material present. An analogous situation exists for $\overline{M}_n$ values at the low molecular weight end of the distribution. The dispersivity is essentially a measure of the relative spread in molecular weights present in a polymer. An exclusion chromatogram provides a complete, detailed molecular weight dispersion of a polymer and an analysis of many of the additives that are generally found in finished plastic products. Direct comparison between chromatograms of two or more materials can quickly establish "good" versus "bad" (Fig. 18-13), "theirs" and "ours," "new" versus "old," or product stability such as virgin versus used lubricating oils (Fig. 18-14).

When the tertiary structure of a protein is destroyed, it becomes amenable to conventional calibration. This can be accomplished by using $6M$ guanidine hydrochloride, urea, or by complexing with sodium dodecyl sulfonate. For example, protein-sodium dodecyl sulfonate complexes yield a unique calibration curve for each pore diameter.

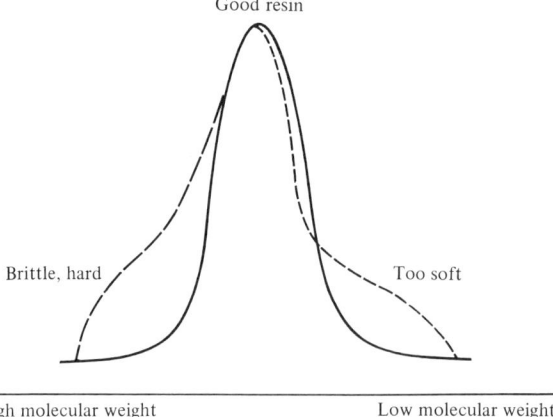

FIGURE 18-13 Exclusion chromatography employed in quality control to evaluate incoming resin. (Courtesy of Waters Associates, Inc.)

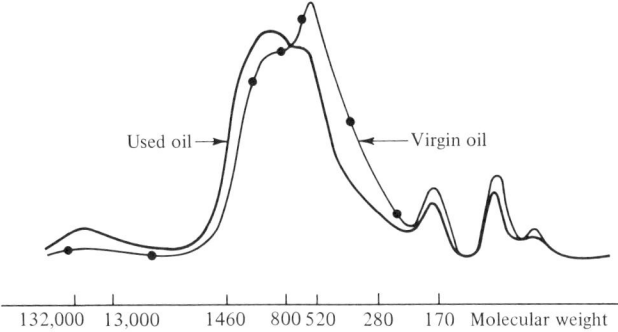

FIGURE 18-14 Exclusion chromatography employed to monitor changes in lubricating oil before serious damage occurs to precision machines. (Courtesy of Waters Associates, Inc.)

PROBLEMS

1. What are the major causes of tailing and memory effects in adsorption chromatography?

2. A particular LSC separation specifies that the mobile phase be 10% methylene chloride in hexane. Unfortunately the laboratory supply of methylene chloride is temporarily exhausted. What mobile phase might be substituted that has approximately the same strength?

3. A particular compound is eluted from a silica gel column too rapidly when dioxane is used as the mobile phase. Will methyl isobutyl ketone make the compound move faster or slower?

4. On a C_{18} bonded phase, a particular compound moves too slowly in a methanol/ acetonitrile (50:50) solvent. What adjustment should be made in the ratio of solvents?

5. During the separation of carbohydrates using a bonded aminoalkyl functional group, an increase in the water concentration of the acetonitrile/water mobile phase decreases retention. Is the bonded phase acting in the "normal" or "reverse" mode?

6. For each of the samples, develop a scheme for ion chromatography by suggesting the separating column, the eluent, and the suppressor column including the chemical stripping action. (a) Analysis of fruit juices for Na^+, K^+, and NH_4^+. (b) Separation of tetraethyl ammonium and tetra-n-butyl ammonium ions. (c) Determination of Cl^-, SO_4^{2-}, and PO_4^{3-} ions in municipal water supplies and sources.

7. What is the smallest number of samples that can be injected into an ion chromatograph before the suppressor column will become depleted and must be regenerated?

8. Some headache preparations contain aspirin and phenylpropanolamine. Suggest a LCC method for their separation employing ion suppression plus ion-pair reverse phase chromatography.

9. Linear alkyl benzene sulfonates are the major surfactants in household detergents at the present time. Therefore the detection of such substances, and their separation based upon the alkyl chain length, in environmental samples is desirable. Reverse phase techniques using a methanol/water solvent resulted in only two major peaks and no peaks for individual alkyl members. Success is achieved if the ion-pair technique is used. Predict the effect of counterion size on retention considering ammonium, tetramethylammonium, and tetrabutylammonium chlorides.

10. Amphiprotic compounds, such as the monofunctional amino acids, are difficult to chromatograph. (a) What two different approaches could be used if ion exchange is selected? (b) Reverse phase ion-pair chromatography also offers two approaches. What are they?

11. A cation-exchange resin column is saturated with copper(II) ions recovered from rinse waters from plating operations. It is desired to recover the copper and to convert the resin to the H^+ form for reuse. Normally this might be done by washing the column with $3M$ sulfuric acid, but $6M$ hydrochloric acid is found to be superior. Why?

12. An anion-exchange column is saturated with chromate ions recovered from an industrial operation. Regeneration of the column with sodium chloride solution is slow and consumes a large volume of reagent. (a) Why is sodium chloride unsatisfactory? (b) Suggest an alternate approach for regeneration of the resin bed.

13. In separate containers are exactly 1-g portions of a cation-exchange resin (H^+ form, 4.3 mequiv g^{-1} capacity) and 10 ml of $0.25M$ HCl. To one container is added solute A and to the other solute B, each in concentrations of exactly $10^{-4}M$. After equilibration 58.84% of A and 32.26% of B remain in the solution phase. (a) Calculate the weight concentration distribution ratio for each solute. (b) Express the separation factor (relative retention) of solute B relative to solute A.

14. Consider the separation of trace quantities of K^+ and Mg^{2+} on a column of pellicular cation-exchange resin. The selectivity coefficients, $k_{K/H}$ and $k_{Mg/H}$, are 2.28 and 1.15, respectively. The exchange capacity of the resin bed (1.92 ml) is 1.70 mequiv ml^{-1} when fully swollen. Estimate the individual concentration distribution ratios for these concentrations of hydrogen ion in the aqueous phase; also the relative retention of K/Mg: (a) $0.10M$, (b) $0.50M$, (c) $1.00M$, and (d) $3.00M$.

15. On a particular-size exclusion column, retention volumes for the individual peaks (molecular weight in parentheses) are: #1, 3.55 ml (8,100,000); #2, 4.45 ml (1,800,000); #3, 5.05 ml (500,000); #4, 5.81 ml (25,100); and #5, 5.93 ml (76). The base line width (4σ) of peak 5 is 0.47 ml. (a) What is the effective operational range, in log molecular weight units, of the column packing? (b) What is the value of V_M? Of V_S? (c) Indicate on the volume axis where $K = 0$ and $K = 1$. (d) Calculate the individual partition coefficients for the components exhibiting peaks #1–#4. (e) Calculate the resolution for peaks

#1 and #2. (f) Calculate the plate number from the benzene peak; express this also as plates per meter. (g) Given that the total column volume is 7.55 ml, estimate the "void" porosity and the "total" porosity of this particular packing.

16. Suggest a method for calculation of V_S in an exclusion chromatographic packing that contains few or no labile protons.

BIBLIOGRAPHY

Bakalyar, S.R., "Mobile Phases for High Performance Liquid Chromatography," *Am. Lab.*, p. 43 (June 1978).
Done, J.N., J.H. Knox, and J. Loheac, *Applications of High Speed Liquid Chromatography*, Wiley, New York, 1975.
Gruschka, E., Ed., *Bonded Stationary Phases in Chromatography*, Ann Arbor Science, Ann Arbor, Mich., 1974.
Helfferich, F., *Ion Exchange*, McGraw-Hill, New York, 1972.
Majors, R.E., "Recent Advances in High Performance LC Packings and Columns," *J. Chromatogr. Sci.*, **15**, 334 (1977).
Perry, S.G., R. Amos, and P.I. Brewer, *Practical Liquid Chromatography*, Plenum, New York, 1972.
Snyder, L.R., *Principles of Adsorption Chromatography*, Marcel Dekker, New York, 1968.
Snyder, L.R., and J.J. Kirkland, *Introduction to Modern Liquid Chromatography*, 2d ed., Wiley-Interscience, New York, 1979.
Yau, W.W., J.J. Kirkland, and D.D. Bly, *Modern Size Exclusion Liquid Chromatography*, Wiley-Interscience, New York, 1979.

LITERATURE CITED

1. Saunders, D.L., "Practical Aspects of Adsorption HPLC," *J. Chromatogr. Sci.*, **15**, 372 (1977).
2. Snyder, L.R., "Role of the Solvent in Liquid Solid Chromatography—a Review," *Anal. Chem.*, **46**, 1384 (1974).
3. Saunders, D.L., "Solvent Selection in Adsorption Liquid Chromatography," *Anal. Chem.*, **46**, 470 (1974).
4. Engelhardt, H., "The Role of Moderators in Liquid–Solid Chromatography," *J. Chromatogr. Sci.*, **15**, 380 (1977).
5. Locke, D.C. and D.E. Martire, "Theory of Solute Retention in Liquid–Liquid Chromatography," *Anal. Chem.*, **39**, 921 (1967).
6. Karger, B.L. and R.W. Giese, "Reversed Phase Liquid Chromatography and Its Applications to Biochemistry," *Anal. Chem.*, **50**, 1048A (1978).
7. Cox, G.B., "Practical Aspects of Bonded Phase Chromatography," *J. Chromatogr. Sci.*, **15**, 385 (1977).
8. Grushka, E. and E. J. Kikta, "Chemically Bonded Stationary Phases in Chromatography," *Anal. Chem.*, **49**, 1004A (1977).

9. Horvath, C. and W. Melander, "LC with Hydrocarbonaceous Bonded Phases; Theory and Practice of Reversed Phase Chromatography," *J. Chromatogr. Sci.,* **15**, 394 (1977).
10. Horvath, C. and W. Melander, "Reversed Phase Chromatography and the Hydrophobic Effect," *Am. Lab.,* p. 17 (October 1978).
11. Tomlinson, E., T. M. Jefferies, and C. M. Riley, "Ion-Pair High-Performance Liquid Chromatography," *J. Chromatogr.,* **159**, 315 (1978).
12. Gloor, R. and E. L. Johnson, "Practical Aspects of Reverse Phase Ion Pair Chromatography," *J. Chromatogr. Sci.,* **15**, 413 (1977).
13. Wolford, J. C., J. A. Dean, and G. Goldstein, "Separation of Oxypurines by Ligand-Exchange Chromatography and Determination of Caffeine in Beverages and Pharmaceuticals," *J. Chromatogr.,* **62**, 148 (1971).
14. Small, H., T. S. Stevens, and W. C. Bauman, "Novel Ion Exchange Chromatographic Method Using Conductimetric Detection," *Anal. Chem.,* **47**, 1801 (1975).
15. Abbott, S. D., "Size Exclusion Chromatography in the Characterization of Polymers," *Am. Lab.,* p. 41 (August 1977).

CHAPTER 19

Mass Spectrometry

The first mass spectrometer dates back to the work in England of J. J. Thompson in 1912 and of F. W. Aston in 1919, but the instrument that served as a model for more recent ones was constructed in 1932. The mass spectrometer produces charged particles consisting of the parent ion and ionic fragments of the original molecule, and sorts these ions according to their mass/charge ratio. The mass spectrum is a record of the numbers of different kinds of ions—the relative numbers of each are characteristic for every compound, including isomers.

The main advantages of mass spectrometry as an analytical technique are its increased sensitivity over other analytical techniques and its specificity in identifying unknowns or for confirming the presence of suspected compounds. The enhanced sensitivity results primarily from the action of the analyzer as a mass filter to reduce background interference and from the sensitive electron multipliers used for detection. Sample size requirements for solids and liquids range from a few milligrams to subnanogram quantities as long as the material can exist in the gaseous state at the temperature and pressure existing in the ion source. The excellent specificity results from characteristic fragmentation patterns, which can give information about molecular weight and molecular structure. In addition, a mass spectrometer is an essential adjunct to the use of stable isotopes in investigating reaction mechanisms and in tracer work. Also, mass spectrometry has contributed greatly to a more detailed understanding of kinetics and mechanisms of unimolecular decomposition of molecules.

19.1 COMPONENTS OF MASS SPECTROMETERS

There is no universal mass spectrometer. Certain designs and configurations lend themselves to the solution of specific problems better than do others. However a mass spectrometer can be divided into the following main parts: (1) sample inlet system(s); (2) ion source; (3) ion analyzer system; (4) ion detector; (5) spectrum recording system; (6) vacuum chamber and pumping system; and (7) electronic power and control system. Interfacing a chromatograph with a mass spectrometer is often desirable.

Inlet Sample System

To handle all types of material, different sample systems are required (Fig. 19-1). Introduction of gases merely involves transfer of the sample from a gas bulb into the metering volume. The latter is a small glass manifold of known volume (about 3 ml), coupled to a mercury manometer and attached by a port to the inlet manifold. A sample is metered in the standard volume and then expanded into a reservoir volume (perhaps 3 liters) immediately ahead of the sample "leak." The meter pressure ranges from 30 to 50 torr; after expansion the pressure ranges from 10^{-3} to 10^{-1} torr.

Liquids are introduced in various ways—by break-off devices (see Fig. 19-9) by touching a micropipet to a sintered glass disk under mercury or gallium, or by hypodermic needle injection through a silicone rubber dam. The low pressure in the reservoir draws in the liquid and vaporizes it instantly.

Heated inlet systems extend the usefulness of mass spectrometry to polar materials, which are prone to be adsorbed on the walls of the chambers at room temperature, and

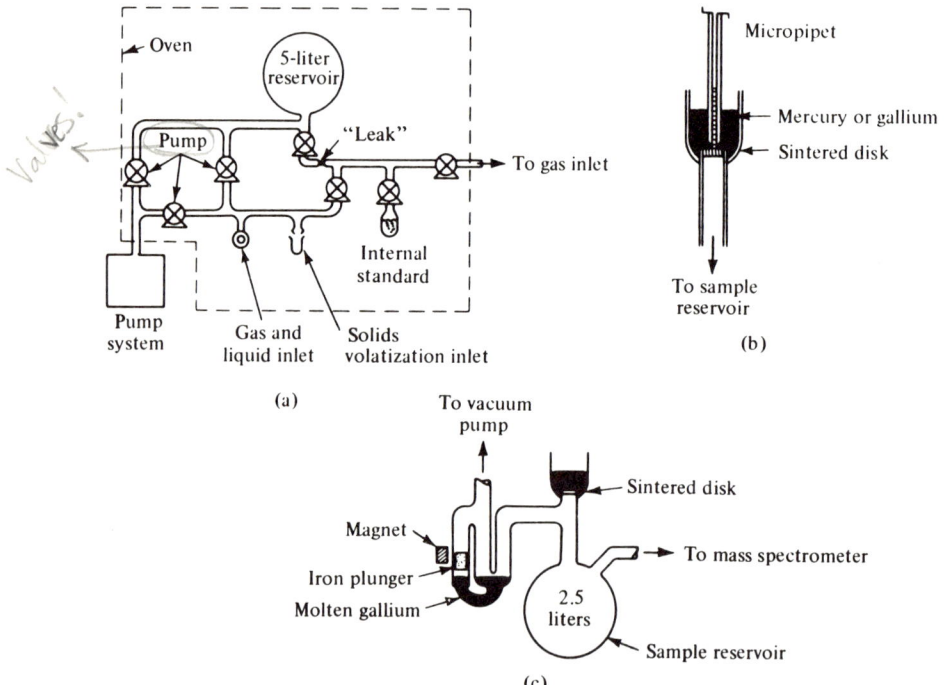

FIGURE 19-1 (a) Inlet sample system for a mass spectrometer. (b) Introduction of liquids through a sintered disk. (c) Magnetically actuated, gallium cutoff valve.

to less volatile compounds insofar as they possess a vapor pressure of the order of 0.02 torr at the temperature of the sample reservoir, usually 200°C. The temperature is limited by the materials of construction and by thermal degradation. Above 200°C most compounds containing oxygen or nitrogen are thermally decomposed. Solids melting below the reservoir temperature can be introduced directly. With a direct sample introduction probe, the sample is loaded into a short length of melting point capillary, placed in the well at the end of the probe, and inserted to within a few millimeters of the ion source through a vacuum lock that maintains the vacuum-tight arrangement. Then the sample temperature is raised until sufficient vapor pressure is indicated by the total ion current indicator or by appearance of a spectrum. Oftentimes a small amount of chemistry suffices to convert a compound, itself not volatile, into a derivative which still retains all the important structural features but has now sufficient vapor pressure. Magnetically actuated gallium cutoffs are employed as valves (Fig. 19-1).

From the sample reservoir the gases diffuse through a molecular leak into the ion source. The leak is a pinhole restriction (about 0.013–0.050 mm in diameter) in a gold foil. The preferred type of flow into the ion source depends on the purpose for which the instrument is intended. For analytical work, conditions for molecular flow are usually employed in which collisions between molecules and the walls are much more frequent than collisions between molecules. However, this type of leak is less desirable in instruments designed primarily for isotope work, since repeated measurements are made of the relative concentrations of two members of a mixture. In isotope studies viscous flow is preferred in which a gas molecule is more likely to collide with other gas molecules than with the surfaces of the container—thus there is no tendency for various components to flow differently from the others.

In continuous-monitoring inlet systems the sample must be admitted to the instrument at or near atmospheric pressure. Consequently, it is necessary to drop the pressure by a viscous flow system to a range in which molecular flow can be achieved by the use of leak perforations of reasonable size. In one system a pair of viscous leaks is arranged through which gas is drawn by an auxiliary mechanical pump; the leaks are so proportioned that the pressure intermediate between them is about 1 torr. The sample is admitted to the mass spectrometer through a perforated foil from the region of intermediate pressure between the two leaks.

Ionization Sources

The ion source is of primary importance and must be considered as the heart of the mass spectrometer. In fact the ion source might be regarded as a chemical reaction vessel, and each source must be chosen as appropriate to the sample. Ion sources have the dual function of producing ions without mass discrimination from the sample and accelerating them into the mass analyzer with only a small spread of kinetic energies prior to acceleration. Sources that produce a large spread of energies in the ion beam must be used with double-focusing mass analyzers to obtain sufficient resolution.

All source designs have several features in common. There must be an ionization chamber in which the actual ionization of the sample occurs and there must be an ion withdrawal and focusing system in which the ions are removed electrostatically from the chamber and are accelerated toward the mass analyzer. Several pairs of focusing elements and slits then control the direction, shape, and width of the ion beam. Generally the source will have its own separate high vacuum pump.

The ionization efficiency of a source must be high so that a large portion of the neutral sample particles presented will become ions to be analyzed and detected as a mass spectrum. High efficiency is particularly important for the analysis of nanogram quantities of sample material and trace impurities in solids. An ion beam current of 10^{-10} A is a desirable source output.

The positive ions formed in the ionization chamber are drawn out by a small electrostatic field between the large repeller plate (charged positive) behind them—the original entrance to the ion source which did not affect the molecules while they were yet unionized—and the first accelerating slit (charged negative) (see Fig. 19-2). A strong electrostatic field between the first and second accelerating slit of 400–4000 V accelerates the ions of masses m_1, m_2, m_3, and so on, to their final velocities. The ions emerge from the final accelerating slit as a collimated ribbon of ions with velocities and kinetic energies given by

$$eV = \tfrac{1}{2} mv_1^2 = \tfrac{1}{2} mv_2^2 = \tfrac{1}{2} mv_3^2 = \cdots \tag{19-1}$$

Combinations of ion sources offer dual capabilities within one source housing. Sometimes both electron-impact and chemical ionization spectra (or field ionization spectra) are needed for a given compound. The chemical ionization or field ionization spectrum might show a molecular ion and the fragmentation of an electron-impact spectrum might identify the class of compound. Calibration of a mass spectrometer must be done with the electron-impact mode; perfluoroalkanes so often used as mass markers would deactivate emitter wires used in field ionization or field desorption, and chemical ionization would provide few if any fragment ions.

Electron-Impact Ionization The electron-impact ion source is the most commonly used and highly developed ionization method. As shown in Fig. 19-2, once past the molecular leak, the neutral molecules find themselves in a chamber that is maintained at a pressure of 0.005 torr and at a temperature of 200 ± 0.25°C. Located perpendicular to the incoming gas stream is an electron gun. Electrons emitted from a glowing filament (rhe-

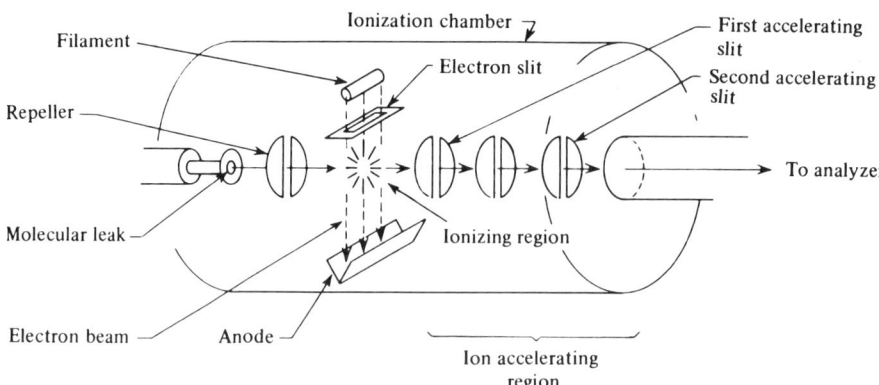

FIGURE 19-2 Electron-impact ion source and ion accelerating system.

nium, thoriated iridium, or carbonized tungsten) are drawn off by a pair of positively charged slits through which the electrons pass into the body of the chamber. An electric field maintained between these slits accelerates the electrons. The number of electrons is controlled by the filament temperature, whereas the energy of the electrons is controlled by filament potential. Ions are formed by the exchange of energy during the collision of the electron beam and sample molecules. This results in a Franck–Condon transition producing a molecular ion, which has an odd number of electrons and is usually in a high state of electronic and vibrational excitation.

The electric field can be varied from 6 to 100 V. A range from 6 to 14 V is employed in molecular weight determinations when it is desirable to avoid fragmentation. Also, little or no fragmentation is desirable when an analysis of a mixture of compounds is needed and the list of possible components is limited. A source operating at 70 V, the conventional operating potential, provides sufficient energy to ionize and cause the characteristic fragmentation of sample molecules. Minor fragmentation is helpful as confirmatory identification, while moderate fragmentation can provide positive identification of an unknown. At 70 V the appearance of the spectrum is nearly independent of the electric field, and reproducibility for quantitative work is thereby secured.

It is customary to form the ionizing electrons from the cathode into a tight helical beam by a small magnetic field, on the order of 100 G, which is confined within the ionization region.

Many compounds do not give a molecular ion in an electron-impact source because of the excess ionization energy imparted to the molecule during the ionization step. This can be considered as a disadvantage of this source.

Chemical Ionization Chemical ionization results from ion–molecule chemical interactions involving a small amount of sample with an extremely large amount of a reagent gas. A two-part process occurs. First, a reagent gas (methane in our example) is ionized by electron-impact ionization in the source (Eq. 19-2). The electron energy must be 200–500 V to ensure penetration of the ionization electrons into the active volume.

$$[\text{electron impact}] \quad CH_4 + e^- = CH_4^+ + 2e^- (CH_3^+, CH_2^+, \text{ and so on}) \quad (19\text{-}2)$$

Primary reagent ionization is followed by second-order processes in which the primary ion reacts with additional reagent gas molecules to produce a stabilized reagent ion plasma (Eq. 19-3).

$$[\text{secondary ions}] \quad CH_4^+ + CH_4 = CH_5^+ + CH_3 \quad \text{and} \quad (19\text{-}3)$$

$$CH_3^+ + CH_4 = C_2H_5^+ + H_2 \quad (19\text{-}4)$$

The second part of the chemical ionization process occurs when a reagent ion (CH_5^+ or $C_2H_5^+$) encounters a sample molecule (MH). Reagent ion and sample molecule may react via any of several modes:

$$[\text{proton exchange}] \quad CH_5^+ + MH = CH_4 + MH_2^+ \quad (19\text{-}5)$$

$$[\text{hydride abstraction}] \quad CH_3^+ + MH = CH_4 + M^+ \quad (19\text{-}6)$$

$$[\text{charge exchange}] \quad CH_4^+ + MH = CH_4 + MH^+ \quad (19\text{-}7)$$

The ion–molecule reaction is a much gentler process than electron-impact ionization, and the quasimolecule ion is an even-electron ion which is more stable than the odd-electron ion molecule produced during electron-impact ionization. In general, chemical ionization spectra are intermediate in their extent of fragmentation between the very simple field ionization spectra and the complex electron-impact spectra. The general absence of carbon–carbon cleavage reactions from the chemical ionization spectra means that they will provide little skeletal information.

Because of the relatively high source pressures, a large number of collisions occur between sample ions and neutral molecules. These collisions tend to remove any excess energy in the sample ions, thus stabilizing them. The small amount of fragmentation provides a sensitivity increase up to 100 times for the chemical ionization process because all of the ionization is concentrated in the molecular ion plus only a small number of fragment ions. Sensitivity is enhanced even more by high cross sections for the chemical ionization process and long ion residence times (10^{-3}–10^{-5} sec).

The basic physical requirements for an effective chemical ionization unit (Fig. 19-3) are a tightly enclosed source housing, a high-speed pumping system, and a differential pumping barrier between the source and analyzer regions. The source should include

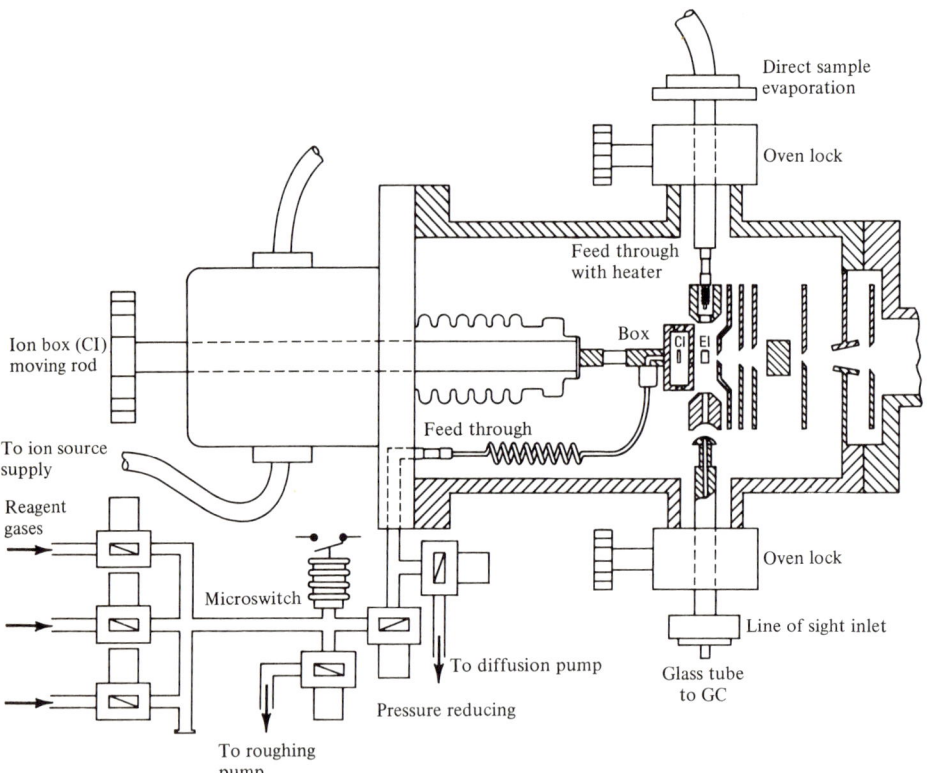

FIGURE 19-3 Combination chemical ionization/electron-impact ionization source. (Courtesy of Varian Associates.)

well-sealed inlet connections and small orifices for the entrance of the electron beam and the exit of the ion beam to the mass analyzer. This will allow pressures inside the source to reach 0.5-4.0 torr while the pressures outside the source are about four orders of magnitude smaller. With the unit illustrated in Fig. 19-3, electron-impact spectra can be obtained by simply turning off the reagent gas and admitting the sample at normal electron-impact pressures.

Different reagent gases can be specific for certain functional groups, and can provide excellent control of sample ion fragmentation. When the interest is primarily to determine the molecular weight of a compound or to confirm it as one of a small set of compounds, a low-energy reactant, such as $t\text{-}C_4H_9^+$ (from $i\text{-}C_4H_{10}$), is frequently used. An even weaker protonating agent like NH_4^+ generates $(M + H)^+$ and $(M + NH_4^+)$ ions, useful in characterizing polyhydroxy compounds like sugars. Deuterium oxide can be used to determine the presence of active hydrogen. Oxygen and hydrogen have found use as reagent gases in negative ion chemical ionization/mass spectrometry. Competition between localized chemical-ionization induced reactions at various sites in a molecule can produce structural information that is often absent from the electron-impact spectrum. With NO as reagent gas, all alcohols give ions at m/e $(M - 17)$, but only primary and secondary alcohols give ions at m/e $(M - 1)$ and $(M + 30 - 2)$ in addition to this.

Argon, helium, and nitrogen, as reagent gases, produce fragmentation patterns essentially identical to electron-impact ionization but with increased sensitivity. Helium, because of its use as a carrier gas in gas-liquid chromatography, has been used as a reagent gas for charge exchange reactions (illustrated by Eq. 19-7).

Field Ionization and Field Desorption The unique properties of a field ionization source arise from the behavior of chemical compounds under high potential fields. When a molecule is brought between two closely spaced electrodes in the presence of a high electric field (10^7-10^8 V cm^{-1}), it experiences an electrostatic force similar to that on the plates of a charged condenser. If the metal surface (anode) has the proper geometry, either a sharp tip or tips (or a thin wire) and under high vacuum (10^{-6} torr), this force can be sufficient to remove an electron from the molecule without imparting much excess energy. Resonance electron tunneling and field induced surface reactions are two proposed mechanisms by which molecules are converted to the molecular ion. The high potential gradient produced on the tip of a needle of an activated emitter in the presence of a ground plane will deform the atomic potential wall of a molecule that an electron sees to such an extent that the probability of tunneling under that wall becomes quite high. In the absence of electron collisons there is little or no fragmentation of the ion. Sensitivities are an order of magnitude below those of electron-impact ionization.

In construction a thin needle of a few micrometers diameter constitutes the anode which is located 1 mm away and immediately behind the exit slit of the ion chamber. The exit slit serves as the cathode and opposite field-forming member. The remainder of the source consists of the focusing slits common to all ion sources; in fact, the field ionization unit can be combined with the electron-impact source in a single assembly.

Field desorption is a variant of field ionization. In field desorption an activated field ionization emitter wire is dipped into or has deposited on it a solution of the sample under study. The emitter wire is then gradually heated with an electric current while it is at the high-voltage field-ionization conditions. Field desorption is a valuable technique for studying surface phenomena, such as adsorbed species and trapped samples, and the results of chemical reactions on surfaces. Due to the nonhomogeneity of the microneedle (or wire) size and directional orientation, the surface ions generated during the field desorption possess an energy distribution and thus spread over a small angle. Refocusing of the mass spectrometer after repetitive scans is necessary for good field desorption spectra.

Spark Source Ionization Samples are formed into two electrodes held in small movable vises that are encapsulated in an insulated, evacuated spark housing. The electrodes form part of the secondary circuit of an rf oscillator circuit (Fig. 19-4). An intermittent voltage sufficient to produce a spark (8–20 kV) is applied to vaporize and ionize part of the sample without heating the entire sample substantially. The rf voltage, produced by a Tesla coil circuit, builds up rapidly during the prebreakdown phase. At a critical voltage, determined by the microgeometry of the electrodes, breakdown is initiated, causing the current to increase rapidly and the voltage to fall, depending on the reactance of the spark circuit. Within 10 nsec the voltage decreases from several thousand volts to less than 100 volts. During the initial breakdown period, material (approximately 10^{-8} g) is evaporated from the electrodes and bombarded with high-energy electrons, producing highly charged ions (+10 or higher) with large kinetic energies. After the interelectrode voltage has fallen to a few volts, the discharge resembles a dc arc, with low charge states (primarily +1) and low kinetic energies. Typical pulse lengths are from 25 to 200 μsec with repetition rates of 30–1000 pulses/second.

Spark source mass spectrometry as a technique for the analysis of solids has grown in use enormously in recent years because of two inherent advantages. First, its detection sensitivity is high whether defined in terms of the low concentration of an impurity that can be detected in a matrix or in terms of the total amount of sample needed to detect impurities. The second outstanding advantage of the spark source as an analytical tool is its ability to atomize and ionize all elements with nearly equal sensitivity.

The rf spark can be used to analyze many different kinds of materials. With electrical conductors and semiconductors, the spark can be formed directly between electrodes of the sample. Techniques have been developed for sintering powders or for mixing insulators in powder form with suitable conductors, which after compacting into rod form are suit-

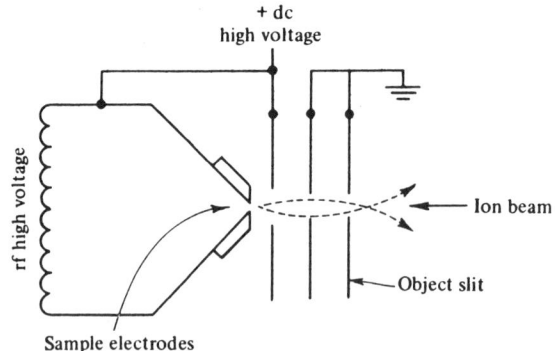

FIGURE 19-4 Spark source ionization.

able for sampling with the spark source. The preparation, cleaning, and handling of samples is critical because a monatomic layer of surface contamination is readily detected.

It is an erratic ion source and the ion current of a singly charged species fluctuates widely with time, necessitating an integrating detector, such as a photographic plate, or else recording the data only during a fixed time interval following each sparking. To maintain adequate resolution, a double-focusing mass analyzer must be used.

For quantitative analysis uniform sampling must be achieved. With many sparkings, as would be needed in trace analysis, uniformity is likely. At higher concentrations, fewer sparkings are required, and the condition of uniformity may not be met. In these cases, the ion beam may be chopped so that only a small percentage of the ions reaches the detector, and thereby the number of sparkings can be increased.

Thermal (Surface) Ionization The thermal or surface ionization source is useful for inorganic solid materials. Samples are coated on a tungsten ribbon filament and then heated until they evaporate. When an atom or molecule is evaporated from a surface (at approximately 2000°C) it has a certain probability of being evaporated as a positive ion. This probability is predictable and is a function of the ionization potential, E_i, of the sample and the work function, ϕ, of the filament material. The relationship for the ratio of ions, n^+, to neutral species, n^0, is given by the Langmuir–Saha equation:

$$\frac{n^+}{n^0} = \exp\left[\frac{e(\phi - E_i)}{kT}\right] \tag{19-8}$$

where e is the electronic charge.

This technique is appropriate for inorganic compounds that generally have low ionization potentials (3–6 eV). Surface ionization is especially useful in determining isotope ratios in inorganic compounds for geochemical applications or in studies of elements involved in nuclear chemistry. No ionization of the background gases in the mass spectrometer occurs. On the other hand, surface ionization is inefficient for organic compounds whose ionization potentials usually lie in the range from 7 to 16 eV. The energy spread of the ions is small; thus only a single-focusing mass spectrometer is needed.

Ion-Collection Systems

Resolved ion beams, after passage through a mass analyzer, sequentially strike a detector. Several types of detectors are available.

Faraday Cup Collector The Faraday cup collector provides a simple and effective means of monitoring ion current in the focal plane of the mass spectrometer. It consists of a cup with suitable suppressor electrodes and guard electrodes, as shown in Fig. 19-5. Currents as low as 10^{-15} A may be detected.

574 CHAPTER 19

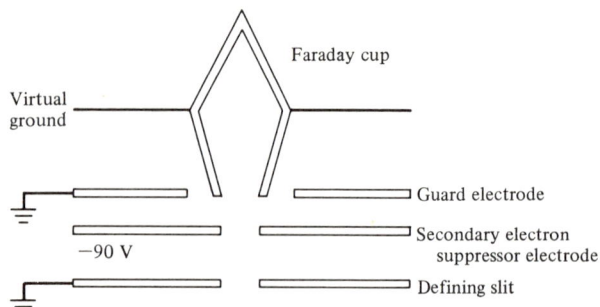

FIGURE 19-5 Schematic diagram of a Faraday cup collector.

Electron Multiplier For ion currents below 10^{-15} A, an electron multiplier is necessary (Fig. 19-6). The ion beam strikes the conversion dynode of a multistage ion multiplier tube. Typical ion multiplier tubes used in mass spectrometry have between 15 and 20 Cu–Be dynodes arranged in either venetian blind or box-and-grid fashion. Secondary electrons, emitted by the dynodes, are constrained by a magnetic field to follow circular

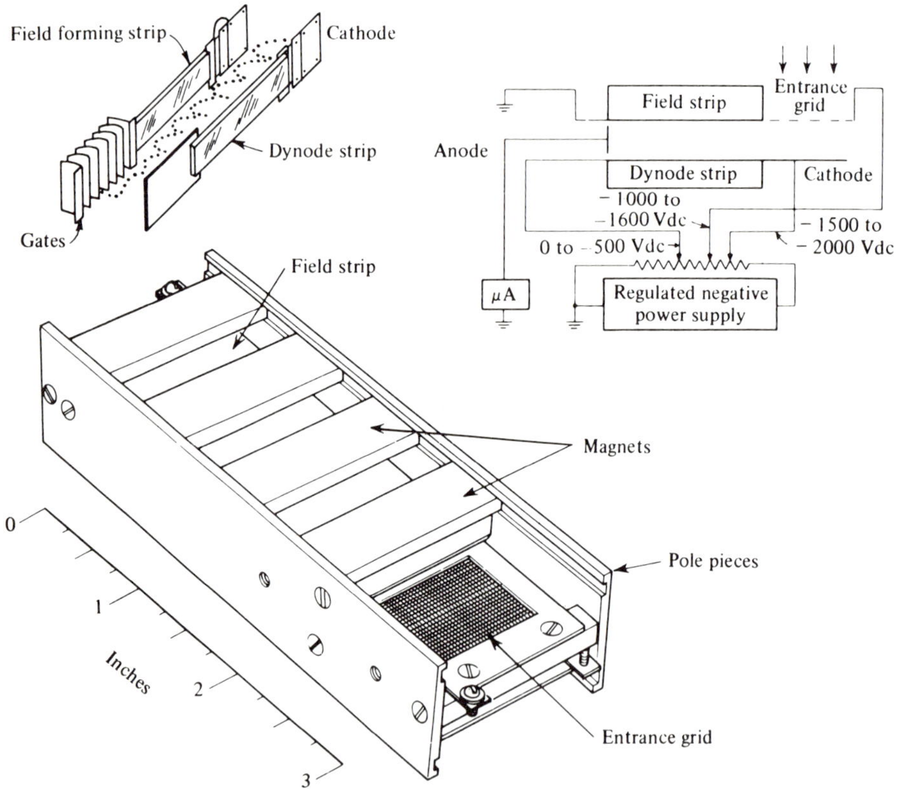

FIGURE 19-6 Electron multiplier phototube and typical electrical circuit for operating the tube. (Courtesy of Bendix Corp.)

paths, causing them to strike successive dynodes (or the same electrode from which they were emitted, but at a different point). Gain ranges from 10^5 to 10^7. The limiting factor is either the system noise level or the system background. The magnetic field is produced by a number of small permanent magnets.

Channel Electron Multiplier Array A channel plate is composed of a regular (usually hexagonal) close packed array of channels in a flat plate of semiconducting material. Typical pore diameters lie in the 10–25-μm range. The length-to-diameter ratio determines the gain characteristics of the device, with a ratio of 40 giving a gain of 10^3 electrons per initial ion. The plate is about 1 mm thick. The inside of each pore, or channel, is coated with a secondary electron emissive material; thus each channel constitutes an independent electron multiplier. To achieve higher gain, two plates can be operated in tandem, with the output of one plate forming the input of the next. A schematic of a channel electron multiplier array is shown in Fig. 19-7. Energetic particles enter the first channel plate where they collide with the wall and produce secondary electrons. For spark source mass spectrometry, the efficiency of this process is between 70 and 80%.

Photographic Plate A photographic plate can give greater resolution than an electrical detector. It is more cumbersome though. Photographic emulsions have been discussed in Chapter 6. Because the photographic plate is a time-integrating device, it can provide the highest sensitivity of any detector. A 36-to-1 mass range can be recorded simultaneously. Spectra from extremely small samples, samples with low vapor pressure, and ions of short life can be detected. All of these might be missed with an electrical detector. Photographic plate detection is used with most rf spark instruments.

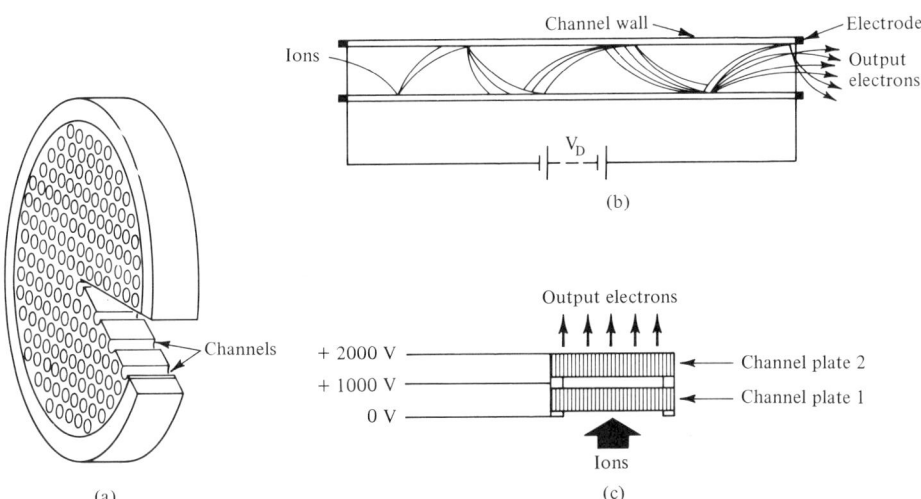

FIGURE 19-7 Multichannel electron multiplier arrays. (a) Schematic construction of a multichannel array. (b) Operation of electron amplification. (c) Schematic arrangement of the tandem type. Actually a slice angle (5°, 8°, or 13°) is selected to prevent a primary ion from passing through the channel and to prevent ionic feedback between two plates.

Data Handling

On the older mass spectrometers the readout display usually consisted of a direct writing-recording oscillograph with perhaps five galvanometers with relative sensitivities of 1, 3, 10, 30, and 100. This system is easy to operate, reliable, and of low initial cost. However, the resultant spectra often need considerable manual manipulation because each mass number must be counted, and the amplitudes measured and normalized before identification can be made.

More recently data have been digitized and collected on magnetic tape or stored in the memory of a computer for subsequent processing. This system permits rapid accumulation of the wealth of data generated. On request, the dedicated microcomputer will reconstruct the mass spectrum (or the chromatogram on GC/MS or LC/MS instruments). Cathode-ray tubes are frequently used to provide real time display of data acquisition.

Electrical detection of the mass spectrum allows operation of the ion-collection system in one of several modes. Scanning the mass spectrum across the detector is an extremely inefficient way to collect information and is seldom used. Peak switching allows the detector to view selected regions of the spectrum, bypassing those that contain no information or unwanted information. In this mode the magnetic field is varied to scan the spectral regions. Resolution generally suffers in peak switching because the slits must be widened to allow for slight inaccuracies in field settings while still viewing a portion of the mass spectrum that includes the mass number of interest. In the third mode the instrument simply monitors a single mass number to achieve low limits of detection by long-term integration of the signal.

Vacuum System

For the operation of a mass spectrometer, the ion source, the mass analyzer, and the detector must be kept under high vacuum conditions of 10^{-6}–10^{-7} torr. Both the speed at which the instrument can be operational after cleaning or opening to atmospheric pressure, and the efficiency of maintaining high vacuum are related to the capacity and speed of the vacuum system. Most systems use combination of oil diffusion pumps to maintain high vacuum together with backing rotary pumps to reduce the initial pressure to approximately 10^{-3} torr. However, oil diffusion pumps are being replaced more and more by turbomolecular pumps. Turbomolecular pumps contain no working fluid; the pumping effect is purely mechanical. Therefore background spectra are practically nonexistent, and even accidental venting does not create any problems.

For gas chromatographic/mass spectrometer systems, differential pumping is recommended. One pumping system is connected to the ion source part. Only a small hole is provided for the ion transmission between the ion source and the mass analyzer. A second pumping system keeps a very low pressure in the analyzer part of the mass spectrometer.

For spark source work, much higher vacuums are needed, and this usually requires the use of ion pumps or cryogenic pumping methods.

19.2 RESOLUTION

The ability to separate ions of different masses reaching the detector ranges from 1 part in 20 to better than 1 part in 30,000 on high-resolution instruments. Generally, resolving powers below 1000 are used for routine mixture analyses. The resolving power of a mass spectrometer is defined as the ratio $M/\Delta M$, where M and $M + \Delta M$ are the mass numbers of two neighboring peaks of equal intensity in the mass spectrum. For example, to distinguish oxygen of mass 31.9988 from sulfur of mass 32.06, a resolution of 533 is necessary. Similarly, to resolve the CH_4–O doublet at mass 100 fragment, $M/\Delta M \geqslant 3165$. On the other hand, a resolution of 1 part in 200 adequately distinguishes between mass 200 and mass 201.

The "valley definition" of the resolving power is based on the relative height of the valley formed between two overlapping peaks. A figure of 10% is commonly used, with each peak contributing 5% to the valley; that is, peaks with separation $\Delta M/M$ equal the peak width at 5% height points, as shown in Fig. 19-8. This definition is unduly pessimistic for distinguishing doublets. The doublet will generally be distinguishable if the two

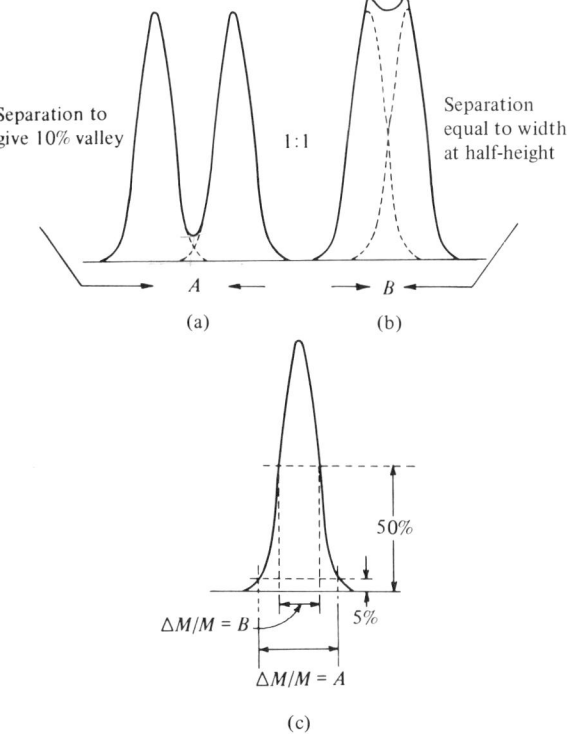

FIGURE 19-8 (a) Resolution equal to 10,000 on *10% valley definition*; that is, peaks with separation $\Delta M/M = A$ equals peak width at 5% height points (c). (b) Resolution based on *width at half-height* definition: that is, peaks drawn with separation $\Delta M/M = B$ equals peak width at 50% height points (c).

peaks are separated by their "width at half-height," provided their intensity ratio is not greater than 10 to 1.

Resolution is strongly influenced by the pressure in the spectrometer and is a function of slit widths, deflection radius, and homogeneity of the ion source. Variation of the source and collector slit widths is the usual method of changing the resolving power. Of course, a decrease in the slit widths to improve the resolving power results in a decrease in sensitivity.

19.3 MASS SPECTROMETERS

Magnetic-Deflection Mass Analyzer Systems

The magnetic-deflection-type system is simply a stable, controllable magnetic field which causes ions to be deflected along curved paths according to their mass-to-charge ratio. Ions formed in the source are accelerated by the electrostatic slits and diverted into circular paths by a magnetic field parallel to the slits and perpendicular to the ion beam. Ions of mass m and charge e, on passage through an accelerating electric field, attain a velocity v, which can be expressed in terms of the accelerating potential V and the kinetic energy of the individual ion as it leaves the electric field,

$$\tfrac{1}{2} mv^2 = eV \tag{19-9}$$

or, solving for the velocity term,

$$v = \sqrt{\frac{2eV}{m}} \tag{19-10}$$

Equating the centripetal and the centrifugal forces to which the ion beam is subjected on entering the uniform magnetic field H,

$$\frac{mv^2 r}{r} = Hev \tag{19-11}$$

The radius of curvature of the path a deflected ion follows is proportional to its momentum and inversely proportional to the strength of the magnetic field,

$$r = \frac{mv}{eH} \tag{19-12}$$

Substituting Eq. 19-12, the radius is also expressed by

$$r = \frac{1}{H}\sqrt{2V\frac{m}{e}} \tag{19-13}$$

Ions accelerated through a uniform electrostatic field and then deflected through a uniform magnetic field will therefore have different radii of curvature of their orbits. Only those ions which follow the path which coincides with the arc of the analyzer tube in the magnetic field are brought to a focus on the exit slit where the detector is located. Ions of other mass/charge (m/e) ratio strike the analyzer tube (which is grounded) at some point, are neutralized, and are pumped out of the system along with all other un-ionized molecules and uncharged fragments. Thus, the magnetic field classifies and segregates the ions into beams, each of a different m/e, where

$$\frac{m}{e} = \frac{H^2 r^2}{2V} \tag{19-14}$$

To obtain the mass spectrum, the accelerating voltage or the magnetic field strength is varied at a constant rate. Usually the accelerating voltage is varied and each m/e ion from light to heavy is successively swept past the detector slit at a known rate. A magnetic analyzer will also focus ions that are of the same mass and velocity but of different initial directions. Thus, this type of instrument resolves ions of different masses, and maximizes the resolved ion beam intensity by focusing. The Dempster (180°) design is shown in Fig. 19-9.

The rather bulky magnet required in the 180° design prompted the development of sector instruments in which the ion source, the collector slit, and the apex of the sector-shaped magnetic field are collinear, as shown in Fig. 19-10 for a Nier 60° sector instrument. Sector instruments are unique because the ion source and collector are completely removed from the magnet region. This isolation permits the use of unusual and diverse ion source constructions, and the use of conventional electron multipliers for ion detection. The ion transit time from the accelerator slits to the magnetic field is signifi-

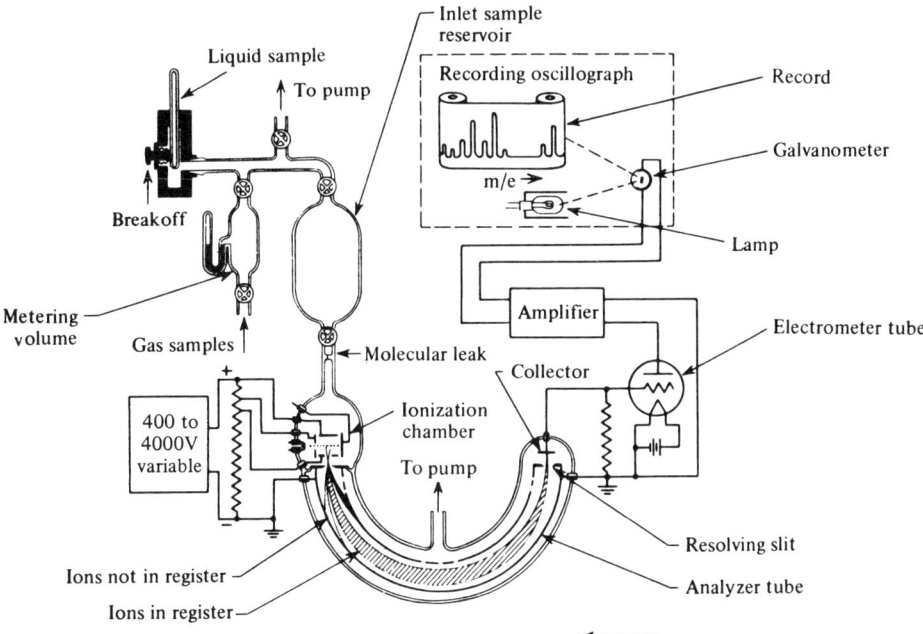

FIGURE 19-9 The Dempster (180°) magnetic-deflection analyzer system. (Courtesy of Consolidated Electrodynamics Corp.)

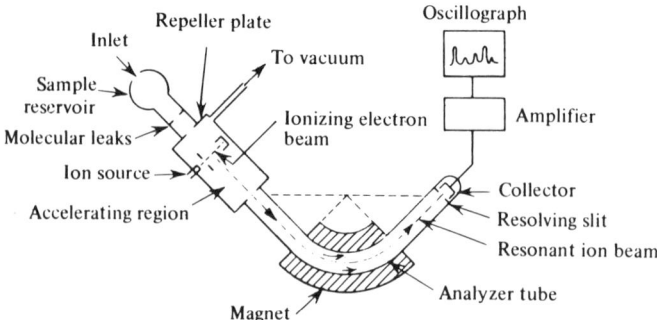

FIGURE 19-10 Schematic diagram of a Nier 60° sector mass spectrometer.

cantly longer in the sector-type instrument, and, consequently, any peaks resulting from metastable transition products are several times larger than the corresponding peaks in a 180°-type spectrometer.

With magnetic analyzer systems a resolution of 1 in 200 mass units can be obtained. Narrow slits, strict alignment of components, and additional signal amplification enables mass peaks in the range from 200 to 600 to be resolved—the ultimate limit for this class of spectrometer.

Instruments designed for process control use a permanent Alnico magnet with a field of 4000–6000 G. The overall range is restricted. For example, one commercial instrument has a range from 2 to 80 mass units and adequate resolution for separation of adjacent peaks up to about 35. Small, portable spectrometers with a radius of curvature of 5 cm can withstand temperatures of 450°C and operate in the ultrahigh vacuum range (10^{-15} torr).

Time-of-Flight Spectrometer

The essential principle of time-of-flight (TOF) mass spectrometry is that if ions of different mass are given the same kinetic energy, they will acquire different velocities and will therefore have a time of flight which is mass dependent. Sample molecules are ionized by electron impact. An electron beam is pulsed through the ionization region for 1 μsec at some preselected energy, typically 70 eV. Immediately following this ionization pulse, the first accelerating grid is given a pulse of negative charge which starts the positive ions through the stages of acceleration. The ion beam reaches drift energy, typically 2700 eV, in a distance of less than 2 cm. Each ion has a kinetic energy, eV, where e is its charge and V is the electric potential through which it has passed:

$$eV = \tfrac{1}{2} mv^2 \tag{19-15}$$

Because all ions have essentially the same energy at this point, their velocities will be inversely proportional to the square roots of their masses. The ions are now allowed to enter and move down a field-free region (Fig. 19-11), 1 m in length, with whatever velocity they may have acquired. The lighter ions speed on ahead while the heavier ions travel at lower velocities. Hence the original beam becomes separated into "wafers" of ions ac-

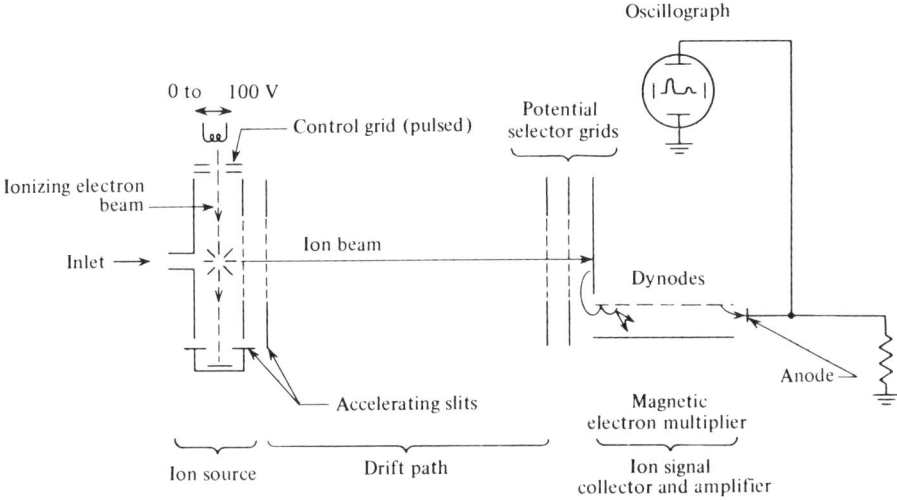

FIGURE 19-11 Schematic diagram of a time-of-flight mass spectrometer. (Courtesy of Bendix Corp.)

cording to their masses. The wafers of ions impact sequentially on the flat cathode of the ion detector. A cathode-ray oscilloscope is synchronized with the pulse repetition rate of the spectrometer. The transit time t (in microseconds) of ions through a distance L (in centimeters) is given by

$$t = L\sqrt{\left(\frac{m}{e}\right)\left(\frac{1}{2V}\right)} \quad \text{or} \quad \frac{m}{e} = \frac{2Vt^2}{L^2} \qquad (19\text{-}16)$$

A complete mass spectrum of a sample can be repeated 20,000 times in 1 sec; thus a complete mass spectrum is generated each 50 μsec. The instrument is excellent for kinetic studies of fast reactions and for direct analysis of effluent peaks from a gas chromatograph. Resolution is about 1 part in 400. A disadvantage is the possible overlap of pulse masses. Insertion of an energy selector grid before the detector limits ions that are allowed to reach the detector to only selected masses.

Quadrupole Mass Analyzer

A quadrupole field is formed by four electrically conducting, parallel rods (Fig. 19-12). In accordance with the theory of quadrupoles, the rods must have a hyperbolic cross section.[*] One diagonally opposite pair of rods is held at $+U_{dc}$ volts, and the other pair at $-U_{dc}$ volts. An rf oscillator supplies a signal to the first pair of rods that is $+V \cos \omega t$, and an rf signal retarded by 180° ($-V \cos \omega t$) to the second pair. The equipotential surfaces in the region between the four rods appear as oscillating hyperbolic potentials.

---

*Until recently, the manufacturing of rods with hyperbolic cross section not exceeding the very small tolerances (approximately 0.00025 cm) seemed to be too complicated and expensive. For this reason rods with circular cross sections were used as an approximation. These quadrupole mass analyzers required special ion optics to optimize the ion entrance into the analyzer system.

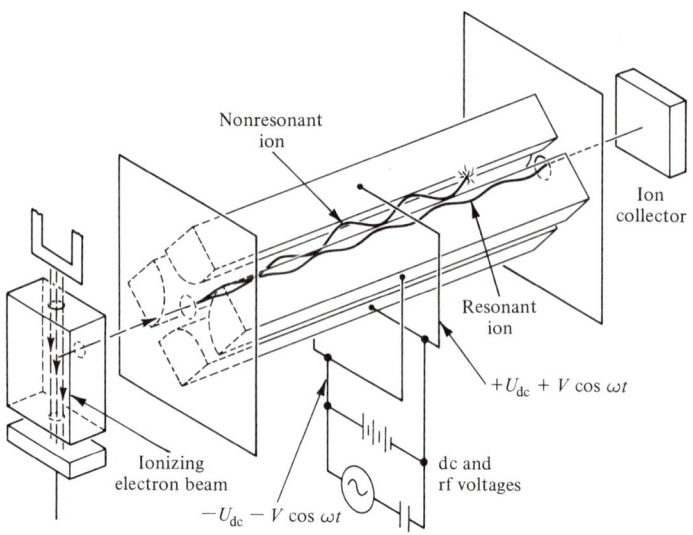

FIGURE 19-12 Quadrupole mass analyzer.

An ion injected down the longitudinal axis will undergo transverse motion in the plane perpendicular to the longitudinal axis in addition to its injection velocity down this axis. There are no field gradients along the device, so the ions travel in the axial direction. The dc electric fields tend to focus positive ions in the positive plane, and defocus them in the negative plane. When an alternating rf field is superimposed, an ion of light mass responds to the changes in the electric field without striking an electrode. As the resultant field becomes negative during part of the negative half-cycle of the alternating field, the positive ion will be accelerated toward the electrodes and will achieve a substantial velocity. The following positive half-cycle will have an even greater influence on the motion of the ion, causing it to reverse its direction and accelerate even more. The ion will exhibit oscillations with increasing amplitudes until it finally strikes on the electrodes. The lighter the ion in mass, the smaller the number of cycles before it is collected by the electrode. On the other hand, heavy positive ions will gradually drift toward the electrodes because they will not respond to any significant extent to the small repulsive force existing during part of the positive half-cycle of the alternating field. Only one m/e ratio can pass through the quadrupole mass analyzer and be detected for a given rf potential and frequency. An entire spectrum can be produced by varying the rf frequency, while the rf and dc potentials remain constant, or by varying the rf potential (and the dc potential simultaneously so that these ratios of potentials remain fixed) while the rf frequency is held constant. The sweep can be as rapid as 1000 mass units/sec. Modern units can deliver practically nondiscriminated spectra up to 500 mass units. An advantage of the quadrupole mass analyzer is that there are only a few potentials of the unit to be adjusted; the tuning can be done automatically by computer control.

The quadrupole analyzer is not restricted to detection of monoenergetic sources. Ions are accepted within a 60° cone around the axis. The quadrupole analyzer therefore does not require focusing slits, which results in higher sensitivity. The resolution is a function of the number of cycles an ion spends in the field. Increasing the rod length (usually 5–20 cm) increases the resolution and the capability to handle ions of higher energies. If the rf frequency is increased, the length of the analyzer can be reduced. Rod diameters are also a factor: increasing the rod diameter increases the sensitivity by a large factor, whereas decreasing the diameter increases the mass range.

The quadrupole mass analyzer is well suited for the registration of negative ions since the analyzer does not discriminate between the polarity of the ions. Simultaneous pulsed positive and negative ion chemical ionization/mass spectrometry is a new method in which both the positive and negative ions produced in a conventional chemical ionization source are alternately pulsed from the source, with the appropriate potentials, through a quadrupole analyzer to two electron multipliers, one for positive and one for negative ions. For collection of negative ions the first dynode of the multiplier must be supplied with a positive potential. The total accelerating voltage of the multiplier must be added up to the point of collection, and a floating preamplifier has to be provided. The advantage here is that some compounds have negative quasimolecular ion spectra that are 1–3 orders of magnitude more intense than their positive ion spectra.

Double-Focusing Mass Spectrometers

All the preceding instruments are termed single-focusing mass spectrometers. In a single-focusing instrument there is a lack of uniformity of ion energies, since the accelerating potential experienced by an ion depends upon where in the source it is formed. The resulting spread in ionic energies produces a spread in their radii of curvature in the magnetic field. The result is peak broadening and low to moderate resolution.

In a double-focusing mass spectrometer, an electrostatic deflection field is incorporated between the ion source and the mass analyzer. This geometry utilizes a cylindrical electrostatic analyzer as an energy filter which allows a band of energies to pass into the magnetic sector where the ions are mass analyzed. Ions are focused thereby both for velocity and direction. Focusing is accomplished by acceleration or deceleration of the ions as they enter the electrostatic field. Positive ions traveling more closely to the positive plate are slowed down, while those traveling more closely to the negative plate are accelerated.

The Mattauch–Herzog geometry, shown in Fig. 19-13, involves an electrostatic sector with 31°50′ angular deflection followed by a β slit which determines the energy bandpass of the energy filter. Following this slit, a monitor assembly is allowed to intercept a fixed fraction of the total ion beam for use in both the photographic plate and electron multiplier detection modes. The ions then enter a homogeneous magnetic field, are deflected through 90°, and come to a focus along a plane which is nearly parallel to the exit face of the magnet. The ions which form this focal image have been separated according to their m/e ratio, as dictated by

$$r = \frac{mv}{eH} \qquad (19\text{-}17)$$

$$v = \sqrt{\frac{2eV}{m}} \pm \Delta v \qquad (19\text{-}18)$$

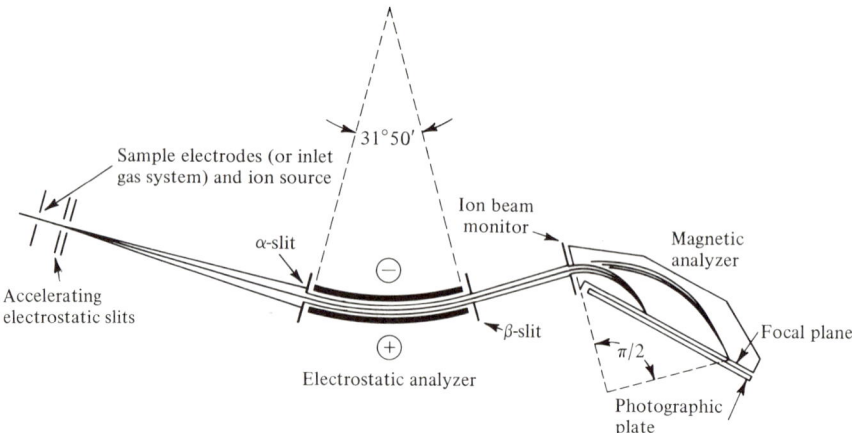

FIGURE 19-13 Schematic diagram of a double-focusing mass spectrometer showing basic Mattauch–Herzog geometry.

Equations 19-17 and 19-18 indicate that the mass spectral resolution will depend on the effect of the spread in the velocity, Δv, being small compared to the effects of the accelerating voltage and magnetic field strength. For this reason the accelerating voltage is kept high (20–25 kV) as is the magnetic field strength (15 kG). At these values, ions of mass 6–240 atomic mass units will be displayed along a focal plane 25 cm in length. The Mattauch–Herzog geometry lends itself to the use of image forming detectors, such as the photographic plate or channel electron multiplier arrays. All of the ions from m/e 6 to m/e 240 will be simultaneously recorded. Resolution is at least 20,000.

The Nier–Johnson geometry (Fig. 19-14) is another version of a double-focusing mass spectrometer. In this configuration, ions of only one m/e value are sharply in focus at

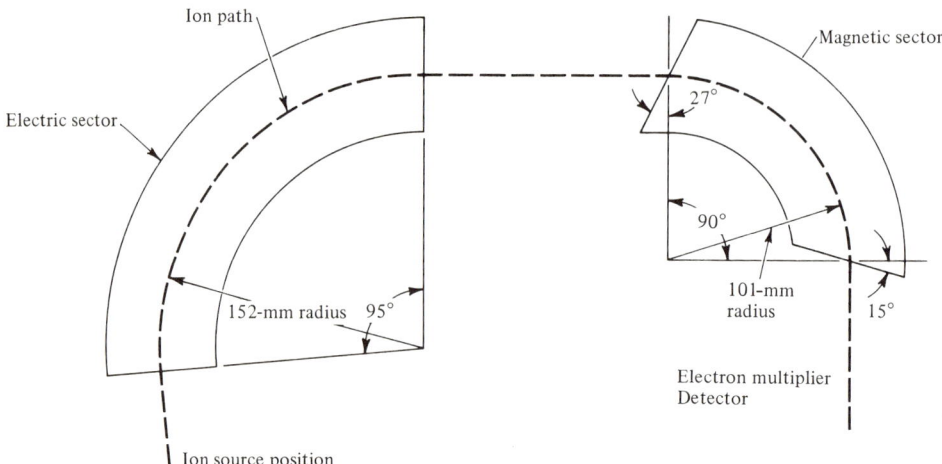

FIGURE 19-14 Schematic of a double-focusing mass spectrometer showing Nier–Johnson geometry.

any given combination of field strengths; hence this geometry is not suitable for photographic detection. On the other hand, mass spectrometers with this configuration now have provision for rapid scanning of the normal sample mass range (m/e 10 to m/e 300) in 1-3 sec.

19.4 INTERFACING CHROMATOGRAPHY AND MASS SPECTROMETRY

The aim of an interfacing arrangement is to operate both a chromatograph and a mass spectrometer without degrading the performance of either instrument. The problems of interfacing a chromatograph with a detection system depend upon the properties of the chromatographic phase, the properties of the sample, and the quantity flowrate of the mobile phase. Chromatography is an ideal separator, whereas mass spectrometry is excellent for identification. The problem is compatibility. In the combined chromatograph/mass spectrometer system, the mobile phase carries sample through the chromatographic column at a flowrate, pressure, and temperature dictated by the chemical nature of the sample. This flow must be transferred to a mass spectrometer constrained by practical design to operate under a flowrate, pressure, and temperature that may differ considerably from those of the chromatograph. Compromise must be minimal.

Gas Chromatograph/Mass Spectrometer Interface

The gas chromatograph/mass spectrometer (GC/MS) interface has several functions. It must attenuate the carrier gas, normally at a pressure drop of one atmosphere at the exit of the column to around 10^{-6} torr in the mass spectrometer. It must divert excess carrier gas into an auxiliary exhaust system. These functions should not introduce any broadening of the gas chromatographic peak.

When favorable instrumental parameters eliminate the need for enrichment, the pressure drop can easily be implemented either with a simple needle valve or by a capillary restrictor. Involved in any interface design are dead volume considerations and flow balance. Each problem may occur with either a needle valve or a capillary restrictor, but the dead volume problem is more likely encountered with a needle valve, whereas improper flow balance is more probable when using an inflexible capillary restrictor. With a capillary restrictor optimum operation occurs only at one flowrate, which depends on the dimensions of the restrictor.

An answer to the difficulty of a single flowrate is to use a variable restrictor in the form of a needle valve. One schematic configuration is shown in Fig. 19-15. The conductance of valve 1 can be adjusted so that the quantity of gas flow $Q_1 = Q_3$ or, if Q_1 is greater than the allowed value of Q_3, valve 1 is set for Q_3 and the excess carrier gas is vented through valve 2. With such flexibility it is easy to maintain a pressure of one atmosphere if desired, and the system can be used at any reasonable flowrate from 0.1 to 25 ml/min.

The degree of enrichment that may be desired in the GC/MS interface will depend upon the type of column and the particular mass spectrometer coupled with the gas chromatograph. The lower flow employed in most capillary GC columns greatly facilitates GC/MS

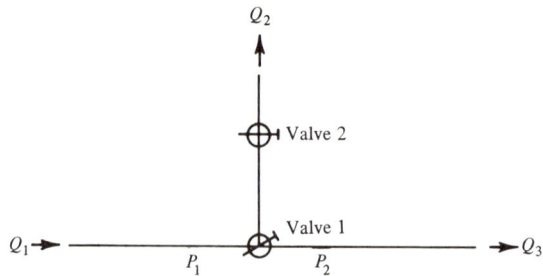

FIGURE 19-15 Schematic of a needle valve GC/MS interface.

interfacing. Effluent from a typical capillary column can be led directly into most modern instruments designed for GC/MS work. Large-bore capillary columns or micropacked columns usually operate with a flow around 5 ml/min. Depending upon the type of mass spectrometer, they should be interfaced either directly through a restrictor or with a small degree of enrichment through a separator.

Flowrates in packed columns are 1–25 ml/min, whereas optimum values are 0.1–2 ml/min in a mass spectrometer. This suggests an enrichment factor of 10–20 should suffice for most GC/MS work. The indicated enrichment can be attained in a single stage separator. A packed column can be coupled directly to a well-pumped chemical ionization system without any enrichment; it is common to use the reagent gas (for example, CH_4) as a carrier gas in the GC column.

Enrichment Devices (Separators)

Almost all GC/MS interface systems contain an enrichment device. Sample utilization is only 10% or less if a simple restrictor with a bypass valve is employed. Although this permits analyses in the 1–10-ng range, one cannot afford to ignore 90% of the sample, particularly on low level samples.

Effusion Separator or Effluent Splitter   Based on the concept of enrichment by diffusion, the effluent splitter relies on the fact that carrier gas molecules usually are much lighter than those of the sample and therefore can be removed preferentially by vacuum pumps through an effusion chamber. Effluent from the gas chromatograph passes through a pressure restrictor into a tube constructed of ultrafine porosity sintered glass whose average pore size is about 10^{-4} cm. The sintered glass tube is surrounded by a vacuum chamber pumped with a rotary forepump. Entrance to the mass spectrometer occurs through another capillary restrictor with an i.d. of about 0.02 cm.

Enrichment occurs if the pressure within the sintered glass tube is maintained in the range 0.5–5 torr. At this pressure the molecular mean free path is about 10^{-3} cm, which is greater than the diameter of the sintered pores. As a result, molecular flow occurs and each molecular species effuses through the pores independently of the presence of other gaseous molecules. The lighter helium carrier gas permeates the effusion barrier in preference to the heavier organic molecules. Enrichment typically is around five- to sixfold, and the yield around 27%.

Viscous flow occurs through the capillary to the mass spectrometer because the internal diameter of 0.02 cm is greater than the mean free path of 10^{-3} cm. All molecular species are thus carried to the mass spectrometer with the same conductance. Obviously, if the pressure in the sintered tube is too high, molecular flow, hence enrichment, will not occur through the pores. Conversely, if the pressure is too low, viscous flow will not occur through the restrictor to the mass spectrometer, with the result that molecular flow occurs in both directions and there is no net enrichment.

A typical porous barrier is shown in Fig. 19-16(a). Separators of this type are most effective operating at flowrates from 1 to 50 ml/min, and at temperatures as high as 350°C to prevent heavy materials from condensing in the pores or on the porous surfaces. The interior surfaces of the separator should be silanized to prevent peak tailing.

Jet/Orifice Separator A precisely aligned, supersonic jet/orifice system is effective in removing the carrier gas by the effusion principle. Effluent from the gas chromatograph is throttled through a fine orifice [Fig. 19-16(b)], where it rapidly expands into a vacuum chamber. During this expansion, the faster diffusion rate of helium results in higher sample concentration in the core of the gas stream which is directed toward a second jet or orifice aligned with the first jet. Alignment and relative spacing of the expansion and collector orifices are very critical and small deviations result in much lower efficiency. This type of separator must be optimized for the flowrate of one particular column; the distance between jets must be changed for use at other flowrates. Yields are about 25% under normal operating conditions. Two jet/orifice assemblies may be used in series. A two-stage separator gives a yield of 40%; it is useful for a low-capacity mass spectrometer. A single stage jet/orifice separator is primarily used with packed columns with gas flowrates of 30–50 ml/min. Peak distortion due to the separator is slight.

Membrane Separator The membrane separator takes advantage of large differences in permeability between most organic molecules and the carrier gas when confronted by a membrane. Effluent from a gas chromatograph enters a cavity which is separated from the mass spectrometer vacuum system by a dimethyl silicone rubber membrane, usually about 0.025–0.040 mm thick. Helium, or any permanent gas, has a low permeability and is not adsorbed by the membrane, whereas the organic molecules are adsorbed and pass through the membrane and directly into the high vacuum of the mass spectrometer system. The permeability of the membrane is determined by solubility and diffusion rate. For helium the product of these two quantities is very small even though diffusion for such a small molecule as helium is very rapid. Enrichment values in the range of 10- to 20- fold are commonly attained, and the yield may be in the range 30–90%. Major problems with this type of separator are the temperature limits (80–220°C) and temperature optimization. The upper limit is serious and cuts out a segment of GC/MS work. Each compound has an optimum temperature that will give best compromise of solubility and diffusivity for membrane enrichment. Consequently, sample discrimination occurs. There is also a time lag of about 0.1 sec while the sample molecules pass through the membrane. Polar compounds and high-boiling compounds tend to be partially adsorbed on the silicone membrane, resulting in tailing of these materials into the mass spectrometer.

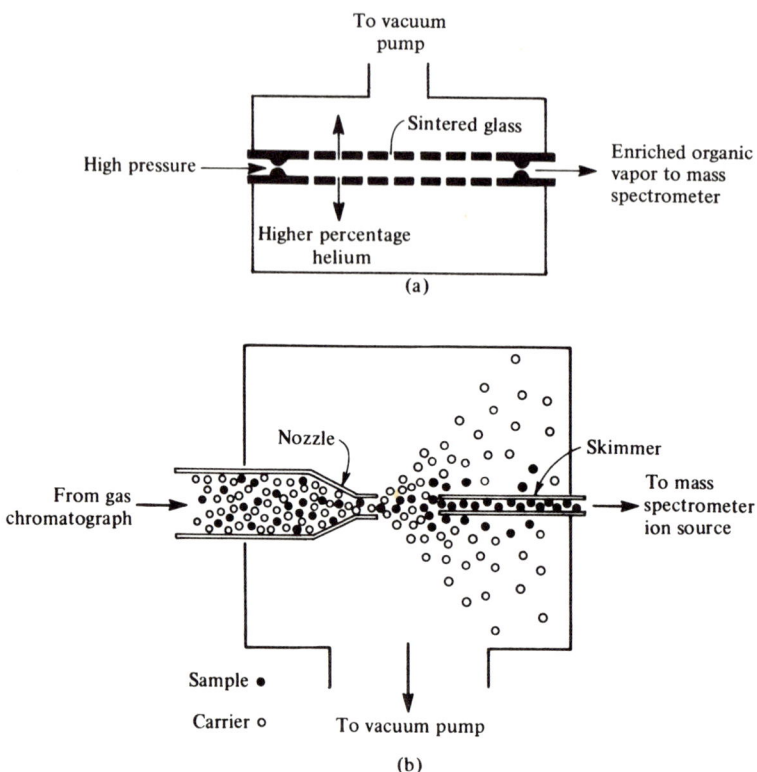
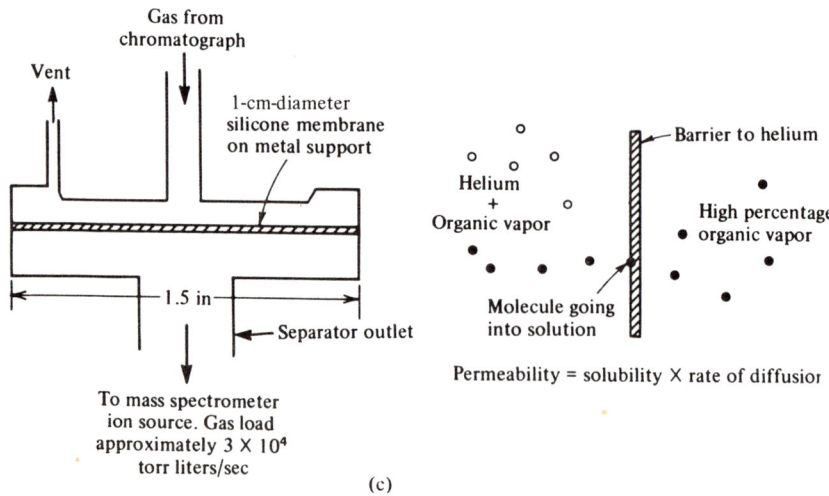

FIGURE 19-16 Enrichment devices: (a) Porous barrier separator or effluent splitter, (b) jet/orifice separator, and (c) molecular separator using a permeable membrane.

GC/MS Instruments

Two types of mass spectrometers are mainly used for GC/MS work: magnetic sector mass spectrometers and quadrupole mass filters. There seems to be no clear preference for one or the other type of instrument. Advantages and disadvantages are well balanced between the two types of instruments. Fast peak switching is important for qualitative and quantitative multi-ion selection analyses. With sector field instruments mass differences of up to 30 or 40% between the selected peaks can be covered by switching the accelerating voltage (which is 800–8000 V). For larger differences the quadrupole mass filter is preferable. The linear mass scale of the quadrupole instrument is ideally suited for digital control, a feature which is important in connection with routine analysis with high sample throughput and for unattended operation. Another advantage of the mass filter is the low ion-accelerating voltage (10–20 V).

For GC/MS usually low resolution of mass spectra is required; only nominal masses need be measured. The molecular peak and the fragment peaks form a typical pattern which can be interpreted by experience or compared against reference spectra in a library. However, if an unknown spectrum cannot be identified by the comparison method, precise masses are very helpful to elucidate the spectrum. This requires a medium- or high-resolution mass spectrometer which is the domain of magnetic sector field instruments.

A typical bench-top GC/MS system is shown in Fig. 19-17. The ion source, quadrupole mass analyzer, and detector units are all located within a concentric diffusion pump and, in turn, surrounded by a spun-steel cylinder. This is surrounded by a cavity containing vapors from the boiling pump fluid.

Liquid Chromatography/Mass Spectromety Interface

Liquid chromatography coupled with mass spectrometry (LC/MS) is in an early stage of development but it has enormous potential. LC/MS is particularly attractive because HPLC can handle nonvolatile and thermally sensitive compounds. The interface is the key. For LC/MS an enrichment factor of around 10^5 is usually required. Such an extreme demand on the interface illustrates one of the important differences between LC/MS and GC/MS. Two commercial units are available.

The design shown in Fig. 19-18 consists of a moving belt-loop that traverses a two-stage-vacuum lock. Effluent from the liquid chromatograph is dropped onto the belt and passes under an infrared heater, which helps to evaporate the solvent. The sample is conveyed into the source housing when the pressure is reduced to about 10^{-5} torr or lower. A small heater vaporizes the sample in the ion chamber close to the electron beam. Another, more powerful heater positioned near the exit of the ion chamber removes any trace of the sample that remains. The belt consists of a polyimide film that can be heated to 400°C. The interface can be used with magnetic sector or quadrupole instruments, and in either the electron-impact or chemical ionization modes. The maximum effluent flow-rates range from 0.4 ml/min for water to 2 ml/min for more volatile organic solvents such as hexane.

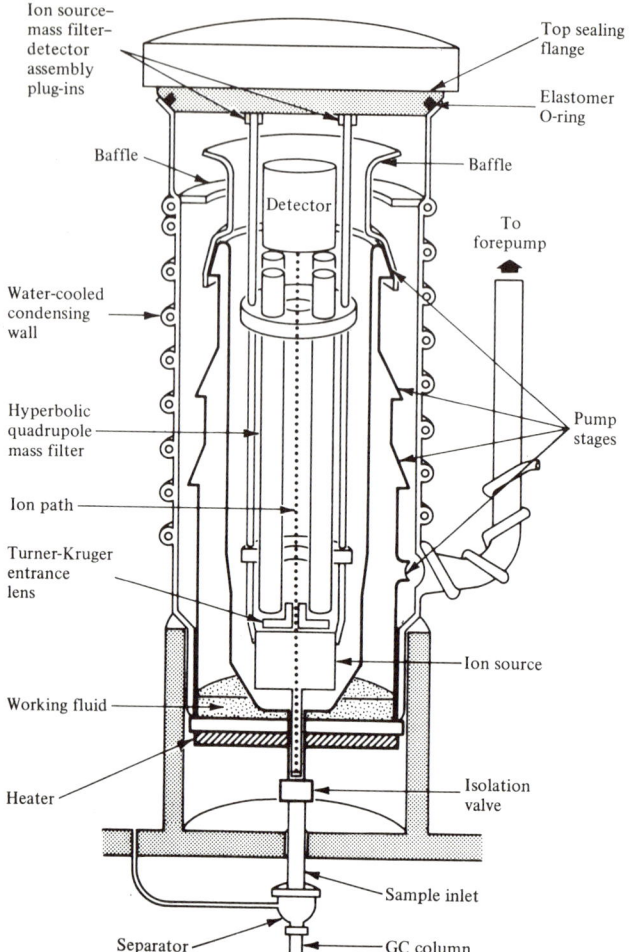

FIGURE 19-17 Schematic of the vacuum system and the quadrupole analyzer assembly of a GC/MS instrument. (Courtesy of Hewlett-Packard Co.)

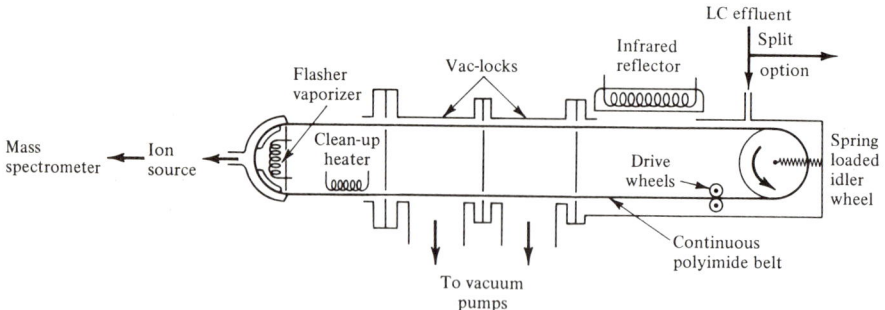

FIGURE 19-18 Schematic diagram of a moving belt LC/MS interface. (Courtesy of Finnigan Corp.)

A second design, shown in Fig. 19-19, splits the LC effluent and injects into the chemical ionization source directly only that small amount of effluent allowed by the flow capacity of the mass spectrometer. The solvent serves as the reactant gas for chemical ionization. The interface consists of a replaceable diaphragm with a tiny orifice, normally 5–15 μm, the exact value being selected according to the solvent viscosity. A cryogenic trap in the source region supplements the normal chemical ionization pumping system.

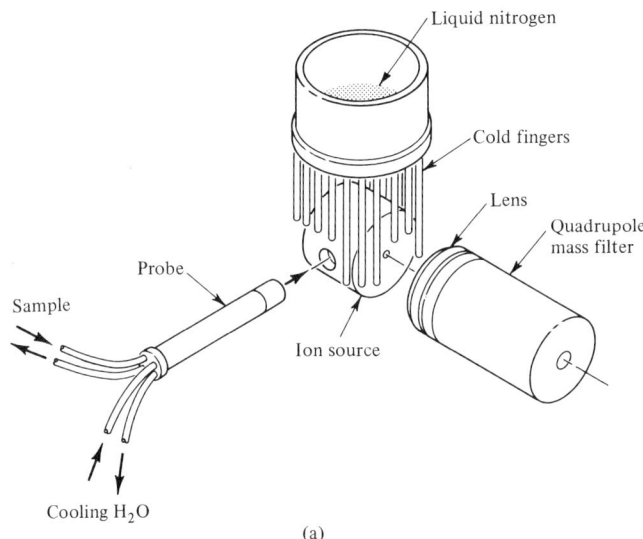

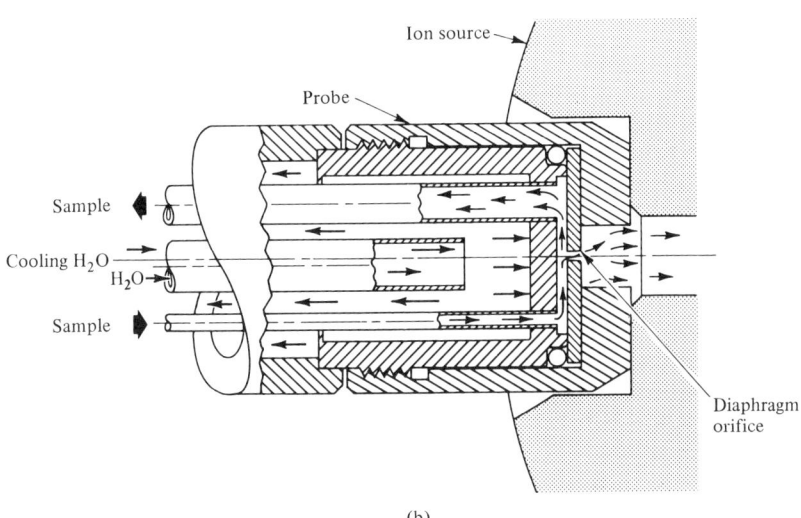

FIGURE 19-19 (a) Schematic of the direct introduction LC/MS interface using an orifice in a diaphragm. (b) Detail of the interface unit. (Courtesy of Hewlett-Packard Co.)

19.5 QUANTITATIVE ANALYSIS OF MIXTURES

The system employed in quantitative analysis by mass spectrometry is basically the same as that employed in infrared or ultraviolet absorption spectrometry. Spectra are recorded for each component. Consequently, samples of each compound must be available in a fairly pure state. From inspection of the individual mass spectra known or suspected to be present in a mixture, analysis peaks are selected on the basis of both intensity and freedom from interference by the presence of another component. If possible, monocomponent peaks (perhaps parent-ion peaks) are selected. The sensitivity is usually given in terms of the height of the analysis peak per unit pressure—obtained by dividing the peak height for the analysis peak by the pressure of the pure compound in the sample reservoir.

Calculation of sample compositions is simplified if the components of the mixture give spectra with at least one peak whose height is due entirely to the presence of one component. The height of the monocomponent peak is measured and divided by the appropriate sensitivity factor to give its partial pressure. Division then by the total pressure in the sample reservoir at the time of analysis yields the mole percent of the particular component.

If the mixture has no monocomponent peaks, simultaneous linear equations are then set up from the coefficients (percent of base peak) at each analysis peak, one equation for each compound in the mixture with n terms (unless one or more terms are zero) when n components are in the mixture. For the analysis of a butanol mixture, the equations at four masses are set up from coefficients listed in Table 19-1 for the particular instrument:

$$90.58x_1 + 1.47x_2 + 1.02x_3 + 2.46x_4 = M_{56} = 126.7$$
$$0.26x_1 + 100.00x_2 + 17.78x_3 + 4.98x_4 = M_{59} = 301.5$$
$$6.59x_1 + 0.59x_2 + 100.00x_3 + 5.03x_4 = M_{45} = 322.6$$
$$0.79x_1 + 0 + 0.29x_3 + 9.06x_4 = M_{74} = 14.8$$

To achieve greater speed in computation, the matrix of coefficients is inverted, yielding a set of equations in terms of each unknown and the analytic masses:

$$x_1 = 110.70M_{56} - 1.625M_{59} - 0.7442M_{45} - 28.77M_{74} \quad (n\text{-butyl})$$
$$x_2 = 1.39M_{56} + 100.08M_{59} - 17.67M_{45} - 45.53M_{74} \quad (t\text{-butyl})$$
$$x_3 = -6.83M_{56} - 0.489M_{59} + 100.31M_{45} - 53.56M_{74} \quad (sec\text{-butyl})$$
$$x_4 = -9.39M_{56} + 0.157M_{59} - 3.17M_{45} + 1108.0M_{74} \quad (\text{isobutyl})$$

Peaks from the mixture spectrum are substituted into the inverse matrix equations, yielding the number of divisions of base peak due to each component. Division by the appropriate sensitivity factor (Table 19-2) yields the partial pressure of each component. Each partial pressure is divided by the total computed pressure, yielding mole percent. The sum of the partial pressures determined in this way should equal the total sample pressure. A discrepancy would indicate an unsuspected component or a change in operating sensitivity.

An outstanding feature of mass spectrometric analysis is the large number of components that can be handled without need for fractionation or concentration. Mixtures containing up to as many as 30 components can be analyzed, and quantities of material as low as 0.001 mol% can be detected in hydrocarbon mixtures. Calculations are usually

TABLE 19-1 Mass Spectral Data (Relative Intensities) for the Butyl Alcohols

	PERCENT OF BASE PEAK (ITALIC)			
m/e	n-Butyl	sec-Butyl	t-Butyl	Isobutyl
15	8.39	6.80	13.30	7.47
18	2.18	0.23	0.49	2.05
27	50.89	15.87	9.87	42.20
28	16.19	2.98	1.67	5.94
29	29.90	13.94	12.65	21.17
31	*100.00*	20.31	35.53	63.10
33	8.50			53.40
39	15.63	3.36	7.70	19.03
41	61.57	10.13	20.82	55.68
42	32.36	1.64	3.32	60.46
43	61.36	9.83	14.45	*100.00*
45	6.59	*100.00*	0.59	5.03
55	12.29	2.06	1.55	4.35
56	90.58	1.02	1.47	2.46
57	6.68	2.74	9.02	3.89
59	0.26	17.78	*100.00*	4.98
60		0.64	3.26	0.57
74	0.79	0.29		9.06

SOURCE: A. P. Gifford, S. M. Rock, and D. J. Comaford, *Anal Chem.*, **21**, 1026 (1949).

TABLE 19-2 Analysis of a Mixture of Butyl Alcohols[a]

Component	Value of x	Sensitivity, Divisions/ 10^{-3} torr	Partial Pressures, 10^{-3} torr	Mol%
n-Butyl	$x_1 = 12{,}871$	1151	11.18	24.4
t-Butyl	$x_2 = 23{,}976$	2093	11.46	25.0
sec-Butyl	$x_3 = 30{,}555$	2698	11.33	24.8
Isobutyl	$x_4 = 14{,}234$	1205	11.81	25.8

SOURCE: A. P. Gifford, S. M. Rock, and D. J. Comaford, *Anal Chem.*, **21**, 1026 (1949).
[a]Mass peaks used: 45, 56, 59, and 74.

carried out on high-speed automatic computers. More complex mixtures, covering a wide boiling range, may require a rough or simple distillation before analysis. Precision normally falls within the range of ±0.05 to 1.0 mol%.

19.6 USE OF STABLE ISOTOPES

Stable isotopes can be used to "tag" compounds and thus serve as tracers to determine the ultimate fate of the compound in chemical or biological reactions. The mass spectrum displays amounts of the added isotope in the fragment ions as well as in the parent ion. Thus, the position of the tracer isotope in the molecule can often be determined without laborious chemical degradation techniques.

A number of stable isotopes in sufficiently concentrated form are available, including practically all of the isotopes of the lighter elements: H, B, C, N, O, S, and Cl. These isotopes complement the relatively larger number of radioactive isotopes.

The isotope-dilution method, described in Chapter 10 for radioactive isotopes, can be employed equally well with stable isotopes to determine the amount of substance present in a complex mixture. It is only necessary to know the ratio of isotopes present in the added sample of the substance, the ratio present in the final sample isolated from the mixture, and the weight of the added sample. It is not necessary that the test substance be separated quantitatively from the mixture; only a few milligrams of pure substance are necessary.

Isotope-Ratio Spectrometer

A less expensive adaptation of the usual mass spectrometer, the isotope-ratio mass spectrometer, has been made available for work in these fields. In the modified instrument the ion currents from two ion beams—for example, the ion beams from $^{12}CO_2$ and $^{13}CO_2$—are collected simultaneously by means of a double exit slit and are amplified simultaneously by two separate amplifiers. The larger of the two amplified currents is then attenuated by the operator, with a set of decade resistors, until it will exactly balance the smaller current from the other amplifier. The ratio of the two currents is determined from the resistance required. This is a null method and practically eliminates the effect of other variables in the system. A tracer material can be detected even after great dilution. Medical researchers study body functions using isotopically labeled tracers. Precise age dating is based on the rate of decay of radioactive nuclides. If the decay rates of ^{238}U to ^{206}Pb, ^{40}K to ^{40}Ar, and ^{87}Rb to ^{87}Sr are known, and the ratio of the isotopes of one of these pairs is measured, then the age of minerals and rocks can be determined.

19.7 LEAK DETECTION

The helium mass spectrometer, widely used to detect leaks, consists of a magnetic deflection-type mass spectrometer that is tuned to maximum sensitivity in the presence of helium, and is coupled to a vacuum system. This is an inexpensive, compact, portable instrument. For determination of exact leak locations (within 1.0 mm), the exterior of the evacuated test piece is carefully sprayed with a hand-held helium jet. When the jet passes over a leak, a small amount of helium enters the object, passes into the spectrometer, and is detected. Sensitivity is as low as 2×10^{-11} atm ml/sec of helium. To determine total leakage, an evacuated test piece is placed within a helium-filled hood connected to the leak detector. If leakage is present, the sum total of helium admitted into the object travels through the system and is measured. When objects contain helium under pressure or can be pressurized with helium, a sampling probe "sniffer" is passed over the test piece. Escaping helium is drawn into the instrument and the exact location of the leak determined.

19.8 CORRELATION OF MASS SPECTRA WITH MOLECULAR STRUCTURE

When bombarded by electrons, every substance ionizes and fragments uniquely. A molecule may simply lose an electron or it may fragment into two smaller units, an ionized fragment and a neutral particle, the sum of whose masses equals their precursor. A molecular or parent ion is generally observed in considerable intensity when the gaseous molecules are bombarded with electrons of energy just sufficient to cause ionization, but not bond breakage, which equals about 8–14 eV for most organic molecules. As the electron energy is increased further, the molecular ion can be formed with excess energy in its electronic and vibrational degrees of freedom. Because redistribution of energy between the bonds takes place rapidly, all of the bonds are affected simultaneously. As soon as the excess energy over the ground-state energy possessed by the molecular ion becomes equal to the dissociation energy of some particular bond, the appropriate fragment ion can be formed. Although the peak intensities are extremely sensitive to ionizing voltage at low values of the ionizing voltage, the relative peak intensitites become fairly constant once the ionization voltage exceeds 50 eV. At higher ionizing energies the total production of ions is higher, but the net effect of higher overall intensity and the resultant severe fragmentation is an increase in relative intensity of the fragment peaks at the expense of the parent peak.

Molecular Identification

In identification of a compound the most important single item of information is the molecular weight. The mass spectrometer is unique among analytical methods in being able to provide this information very accurately. At ionizing voltages ranging from 9 to 14 V it can be assumed that no ions heavier than the molecular ion will be formed and, therefore, the mass of the heaviest ion, exclusive of isotopic contributions, gives the nominal molecular weight.

Restriction on the number of possible molecular formulas can be achieved by study of the relative abundance of natural isotopes for different elements (Table 19-3) at masses 1 and 2 or more units larger than the parent ion. Observed values are compared with those calculated for all possible combinations of the naturally occurring heavy isotopes of the elements. For a compound $C_w H_x O_z N_y$, a simple formula allows one to calculate the percent of the heavy isotopic contributions from a monoisotopic peak, P_M, to the P_{M+1} peak:

$$100 \frac{P_{M+1}}{P_M} = 0.015x + 1.11w + 0.37y + 0.037z$$

and to the P_{M+2} peak:

$$100 \frac{P_{M+2}}{P_M} = 0.20z + 0.006w(w-1) + 0.004wy + 0.0002wx$$

TABLE 19-3(a) Abundances of Some Polyisotopic Elements (%)

Element	% Abundance	Element	% Abundance	Element	% Abundance
^{1}H	99.985	^{16}O	99.76	^{33}S	0.76
^{2}H	0.015	^{17}O	0.037	^{34}S	4.22
^{12}C	98.892	^{18}O	0.204	^{35}Cl	75.53
^{13}C	1.108	^{28}Si	92.18	^{37}Cl	24.47
^{14}N	99.63	^{29}Si	4.71	^{79}Br	50.52
^{15}N	0.37	^{30}Si	3.12	^{81}Br	49.48

(b) Selected Isotope Masses

Element	Mass	Element	Mass
^{1}H	1.0078	^{31}P	30.9738
^{12}C	12.0000	^{32}S	31.9721
^{14}N	14.0031	^{35}Cl	34.9689
^{16}O	15.9949	^{56}Fe	55.9349
^{19}F	18.9984	^{79}Br	78.9184
^{28}Si	27.9769	^{127}I	126.9047

SOURCE: *Lange's Handbook of Chemistry*, J. A. Dean, Ed., 12th ed., McGraw-Hill Book Co., New York, 1979.

Tables of abundance factors have been calculated by Beynon[1] for all combinations of C, H, N, and O up to mass 500. Table 19-4 illustrates the spectral peak contributions at nominal masses 135 and 136 for a few of the compounds having a parent peak (or possibly a fragment peak) of mass 134. Once the empirical formula is established with reasonable assurance, hypothetical molecular formulas are written. One can utilize the entries in the formula indices of Beilstein and Chemical Abstracts.

Compounds containing chlorine, bromine, sulfur, or silicon are usually apparent from the prominent peaks at masses P_{M+2}, P_{M+4}, and so on. The abundance of heavy isotopes is treated in terms of the binomial expansion $(a + b)^m$, where a is the relative abundance of light isotopes, b is the relative abundance of heavy isotopes, and m is the number of atoms of the particular element present in the molecule. When two elements are present, the binomial expansion $(a + b)^m (c + d)^n$ is used.

TABLE 19-4 Heavy-Isotope Contributions to Parent Peak of Mass 134 (%)

Empirical Formula	P_{M+1} Peak	P_{M+2} Peak
$C_5H_{10}O_4$	5.72	0.94
$C_5H_{14}N_2O_2$	6.47	0.58
$C_6H_{14}O_3$	6.83	0.80
$C_8H_6O_2$	8.82	0.74
$C_9H_{10}O$	9.93	0.64
$C_9H_{12}N$	10.30	0.48
$C_{10}H_{14}$	11.03	0.55

If the mass of the parent ion is measured with a high-resolution mass spectrometer, the number of possible empirical formulas can be still further restricted. Because the masses of the elements are not exactly integral multiples of a unit mass, a sufficiently accurate mass measurement alone enables the elemental composition of the ion to be determined. For combinations of C, H, N, and O, the relationship is

$$\frac{\text{exact mass difference from nearest integral mass} + 0.0051z - 0.0031y}{0.0078} = \text{number of H's}$$

For example, a crystalline solid containing only C, H, and O, gave the mass 134.0368 for the molecular ion. Thus,

$$\frac{0.0368 + 0.0051z}{0.0078} = 6 \text{ H's when } z = 2 \text{ oxygen atoms}$$

and the empirical formula is $C_8H_6O_2$. One substitutes integral numbers for z (oxygen) and y (nitrogen) until the divisor becomes an integral multiple of the numerator within 0.0002 mass units.

Two general rules aid in writing formulas. If the molecular weight of a C, H, N, and O compound is even, so is the number of hydrogen atoms it contains; if the molecular weight is divisible by four, the number of hydrogen atoms is also divisible by this number. When nitrogen is known to be present in any compound of C, H, O, As, P, S, Si, and the halogens that have an odd molecular weight, the number of nitrogen atoms must be odd. Once the exact molecular formula has been decided, the sum total of the number of rings and double bonds can be determined by the formula

$$R = \tfrac{1}{2}(2w - x + y + 2)$$

when covalent bonds make up the molecular structure. From the formula $C_8H_6O_2$, $R = 6$; one strong possibility is a phenyl ring (4) plus two additional double bonds.

Metastable Peaks

A one-step decomposition process may be indicated by an appropriate *metastable peak* in the mass spectrum. Metastable peaks arise from ions which decompose in the field-free region after they are accelerated out of the ion source but before entering the analyzer. Their lifetime is about 10^{-6} sec. A metastable ion transition takes the general form: Original ion → daughter ion + neutral fragment. The metastable peak m^* will appear as a weak, diffuse peak, usually at a nonintegral mass, given by:

$$m^* = \frac{(\text{mass of daughter ion})^2}{\text{mass of original ion}}$$

In a spectrum which is linear with respect to m/e values, the distance of m^* below the daughter ion will be of similar magnitude to the distance of the daughter ion below the original ion. The foregoing relationship holds only for ions decomposing in a small portion of the accelerating region and will be more frequently observed with 60° and 90° sector instruments where a field-free region exists after the accelerating slits and before the magnetic analyzer. Of course, the absence of a metastable peak from the spectrum does not preclude a particular decomposition.

Example 19-1

A compound of molecular formula $C_9H_{12}S$ gave a mass spectrum with the parent peak at mass 152 (45%) and fragment ion peaks at masses 137 (7%), 110 (100%), 77 (7%), 66 (11%), 65 (8%), and 43 (12%). Metastable peaks were located at 123, 79.6, and 54.1 These correspond to the transitions:

$$\begin{array}{cc} m^* & \\ 123 & 152^+ \rightarrow 137^+ + 15 \\ 79.6 & 152^+ \rightarrow 110^+ + 42 \\ 54.1 & 110^+ \rightarrow 77^+ + 33 \end{array}$$

which suggests the elimination of CH_3, $CH_3-CH=CH_2$, and HS, respectively. The loss of HS from the base peak also suggests that —SH was connected to a phenyl ring (mass 77). Furthermore, the base peak at 110 is the result of a rearrangement because it is an even-mass ion originating from an even-mass molecular ion. This establishes the presence of a mass 43 group attached to a mass 109 entity. The initial loss of a methyl group strongly suggests an isopropyl group, although this should be confirmed from reference spectra or from a NMR spectrum. Finally, the strength of the molecular peak hints at a stable molecule. The compound is isopropylphenylthioether.

Mass Spectra

The mass spectrum of a compound contains the masses of the ion fragments and the relative abundance of these ions plus often the parent ion. The uniqueness of the molecular fragmentation aids in identification. Because sufficient molecules are present and dissociated for the probability law to hold, the dissociation fragments will always occur in the same relative abundance for a particular compound. The mass spectrum becomes a "fingerprint" for each compound, as no two molecules will be fragmented and ionized in exactly the same manner on electron bombardment. There are sufficient differences in these molecular fingerprints to permit identification of different molecules in complex mixtures. To a considerable extent the breakdown pattern can be predicted. Conversely, the size and structure of the molecule can often be reconstructed from the fragment ions in the spectrum of a pure compound. For example, Table 19-5 indicates the relative abundance of the significant fragments produced from three isomeric octanes. In the structural formulas the asterisk indicates the bond which is broken in the most probable process of scission, and the plus sign indicates the next most probable process. Favored sites for bond rupture in the molecule parallel chemical bond lability. The mass 114 corresponds to the parent ion formed by the loss of a single electron from the parent compound; mass 99 corresponds to the loss of a methyl group plus an electron; mass 71, to the loss of a propyl group plus an electron; mass 57, to the loss of a butyl group; and mass 43, to the loss of a pentyl group.

It is usual practice in reporting mass spectra to normalize the data by assigning the most intense peak (the so-called *base peak*) a value of 100; other peaks are reported as percentages of the base peak.

When working from a spectrum it is advisable to tabulate the prominent ion peaks, starting with the highest mass, and also to record the group lost to give these ion peaks. All possible molecular structures are listed, employing a file of common fragment ions

TABLE 19-5 Mass Spectral Pattern of Trimethylpentanes

	RELATIVE ABUNDANCES, %		
	2,3,3-Trimethylpentane	2,2,4-Trimethylpentane	2,3,4-Trimethylpentane
Mass/Charge Ratio	C C + | C–C * C + C–C | C	C C | | C–C * C+C–C | C	C C | | C–C * C * C–C | C
114	0.1	0.02	0.3
99	3	5	0.1
71	1	1	40
57	70	80	9
43	15	20	50

SOURCE: H. W. Washburn, H. F. Wiley, S. M. Rock, and C. E. Berry, *Ind. Eng. Chem. Anal. Ed.,* **17**, 75 (1945).

encountered in mass spectra. Finally, one attempts to predict the mass spectral features from available correlation data and to check these features against the actual spectrum. Usually only one bond is cleaved; in succeeding fragmentations a new bond is formed for each additional bond that is broken. When fragmentation is accompanied by formation of a new bond as well as by breaking an existing bond, a *rearrangement* process is said to have occurred. The migrating atom will almost exclusively be hydrogen: six-membered cyclic transition states are most important although alternative ring sizes also operate.

Some general features of the mass spectra of compounds can be predicted from general rules for fragmentation patterns:

1. Cleavage is favored at branched carbon atoms: tertiary > secondary > primary, with the positive charge tending to stay with the branched carbon (carbonium ion).
2. Double bonds favor cleavage beta to the bond (but see Note 8).
3. A substance having a strong parent peak often contains a ring, and the more stable the ring the stronger the parent peak.
4. Ring compounds usually contain peaks at mass numbers characteristic of the ring.
5. Saturated rings lose side chains at the alpha carbon. The peak corresponding to the loss of two ring atoms is much larger than for the loss of one ring atom.
6. In alkyl-substituted ring compounds, cleavage is most probable at the bond beta to the ring if the ring has a double bond next to the side chain.
7. A hetero-atom will induce cleavage at the bond beta to it.
8. Compounds containing a $>$C=O group tend to break at this group, with the positive charge remaining with the carbonyl portion.

The presence of Cl, Br, S, and Si is easy to deduce from the unusual isotopic abundance patterns of these elements. These and other elements, such as P, F, and I, are also detectable from the unusual mass differences that they produce between some fragment ions in the spectrum.

PROBLEMS

1. For a field strength of 2400 G in 180° magnetic-deflection spectrometer, what electrostatic voltage range suffices for scanning from mass 18 to mass 200? The radius of curvature of the 180° analyzer tube is 12.7 cm.

2. For a drift length of 100 cm in a time-of-flight mass spectrometer, what is the difference in arrival time between ions of $m/e = 44$ and $m/e = 43$ when the accelerating voltage is 2800 V?

3. The spectrum of cetyl palmitate (M.W. 480) was recorded on a linear scale in the presence of $C_{10}F_{19}$ (M.W. 480.970). On the spectrum the molecular ion peak of cetyl palmitate was separated from the $M + 1$ peak by 200.5 nm, and from the perfluorocarbon molecular ion peak by 95.3 mm. What is the precise mass of the molecular ion of cetyl palmitate?

4. The parent peak spectrum of tridecylbenzene (260.2504), phenyl undecylketone (260.2140), 1,2-dimethyl-4-benzoyl napthalene (260.1201), and 2,2-naphthyl benzothiophene (260.0922) would require what resolution for quantitative analysis based on the parent peak?

5. What resolving power is needed to separate (a) the $CH_2N-C_2H_4$ doublet at mass 200: (b) the N_2-CO doublet at mass 150; and (c) the CH_2-N doublet at mass 200?

6. A peptide was admitted to a high-resolution mass spectrometer and the parent peak mass was measured relative to the parent peak in the spectrum of dibromobenzene (236.8638). The measured ratio of unknown mass/reference mass was 1.001197 ± 0.000002. Compute the exact weight of the peptide and deduce the molecular formula.

7. From the following exact molecular weights, estimate the empirical formulas assuming only C, H, O, or N is present unless otherwise indicated: (a) 164.0473, (b) 120.0575, (c) 180.0939, (d) 94.0531, (e) 109.0528, (f) 190.9540 (contains Cl), (g) 181.0891, (h) 334.0873 (contains S), and (i) 177.0426.

8. (a) In the high-resolution spectrum of methionine, a quartet of peaks appear at nominal mass 88: 88.0220, 88.0345, 88.0335, and 88.0267. Deduce the fragment ion responsible for each peak. (b) Methionine also gives a doublet at nominal mass 75. One line corresponds to $C_2H_4NO_2$; the other has m/e 75.0267. Outline the process leading to the fragment $C_2H_4NO_2$. Deduce the fragment ion of m/e 75.0267.

9. What is the probable composition of a molecule of mass 142 whose P_{M+1} peak is 1.1% of the parent peak?

10. Deduce the number and type of halogen atoms present in a molecule from the abundance of heavy isotopes and the intensity ratios given in the following table:

	P_M	P_{M+2}	P_{M+4}	P_{M+6}	P_{M+8}
Compound A	30	29	10	1	—
Compound B	13	30	19	6	1
Compound C	5	20	30	19	5
Compound D	23	30	7	—	—
Compound E	18	30	14	2	—

11. From the isotopic abundance information, what can be deduced concerning the empirical formula of each of the following compounds?

m/e	% OF BASE PEAK	m/e	% OF BASE PEAK	m/e	% OF BASE PEAK
90(P)	100	89(P)	17.12	206(P)	25.90
91	5.61	90	0.58	207	3.24
92	4.69	91	5.36	208	2.48
		92	0.17		

m/e	% OF BASE PEAK	m/e	% OF BASE PEAK	m/e	% OF BASE PEAK
230(P)	1.10	140(P)	14.8	151(P)	100
232	2.12	141	1.40	152	10.4
234	1.06	142	0.85	153	32.1
				154	2.9

12. The significant portion of the mass spectral data is given for the individual alcohols. Select appropriate analytical masses and write a series of four equations in terms of divisions of base peak due to each of the four alcohols.

	% OF BASE PEAK				UNKNOWN MIXTURES		
m/e	METHYL	ETHYL	n-PROPYL	ISOPROPYL	A	B	C
15	35.48	9.44	3.77	10.70			
19	0.29	3.13	0.90	6.51			
27	—	21.62	15.20	15.50			
29	58.80	21.24	14.14	9.49			
31	100	100	100	5.75			
32	68.03(P)	1.14	2.25	—	600	600	2350
33	0.98	—	—	—			
39		—	4.00	5.52	4800	3000	3000
43		7.45	3.18	16.76			
45		37.33	4.39	100			
46		16.23(P)	—	—	1000	1100	698
59			9.61	3.58	4000	2300	5000
60			6.36(P)	0.44(P)			
Sensitivity, 8.76 divisions/10^{-3} torr		17.98	26.51	23.47			

13. The mixture peaks for three unknown mixtures of alcohols are shown in Problem 12. For each mixture compute the mole percent of each alcohol.

14. A material containing only C, H, and O, and in the form of leaflets melting at 40°C, possesses a rather simple mass spectrum with the parent peak at m/e 184 (10%), the base peak at m/e 91, and small peaks at m/e 77 and 65. Metastable peaks appear at m/e 45.0 and 46.5. Deduce the structure of the compound.

15. The mass spectrum possesses a strong parent peak at m/e 122 (35%) plus peaks at m/e 92 (65%), m/e 91 (100%), and m/e 65 (15%). In addition there are metastable peaks at 46.5 and 69.4 mass units. Deduce the compound's structure.

16. A solid, melting at 33°C, has the following mass spectrum: Parent ion at m/e 200 with P_{M+1} exhibiting a strength equal to 10.55% of the parent peak, and P_{M+2}, 5.77% of the parent peak. The major ion peaks occur at these mass units: 172 (18%), 155 (61%), 108 (10%), 107 (11%), 92 (30%), 91 (100%), and 65 (30%). Metastable peaks appear at 46.5, 53.5, 67.9, 106.3, 121.7, and 147.9 mass units. Deduce the structure of the compound insofar as possible.

17. Deduce the structural formula for each of these compounds from the mass spectral data:

	$C_4H_8O_2$		$C_4H_6O_2$
m/e	% OF BASE PEAK	m/e	% OF BASE PEAK
27	39.3	15	27.7
29	19.8	26	22.4
39	14.8	27	68.1
41	23.7	29	13.0
42	24.7	42	11.8
43	22.3	55	100
45	19.1	58	8.4
60	100	59	5.2
73	27.1	85	12.3
88(P)	1.6	86(P)	2.1

	C		D
m/e	% OF BASE PEAK	m/e	% OF BASE PEAK
29	18	63	22
39	23	64	20
51	29	65	33
65	18	92	82
78	50	93	18
91	100	120	100
105	41	121	34
134(P)	57.4	152(P)	45.0
135	5.80	153	4.1
136	0.41	154	0.4

18. Deduce the complete structural formula of the compound from the mass spectrum in the figure.

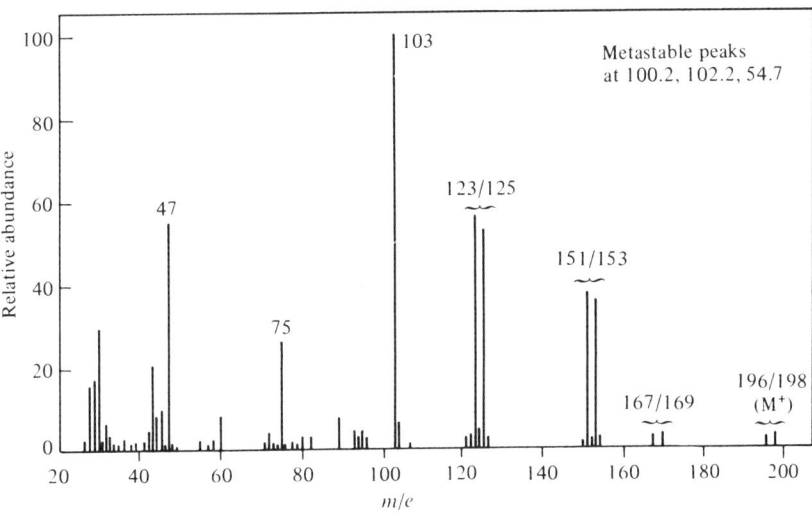

PROBLEM 19-18

19. Deduce the structural formula of the compound (bp 74°C) whose mass spectrum is shown.

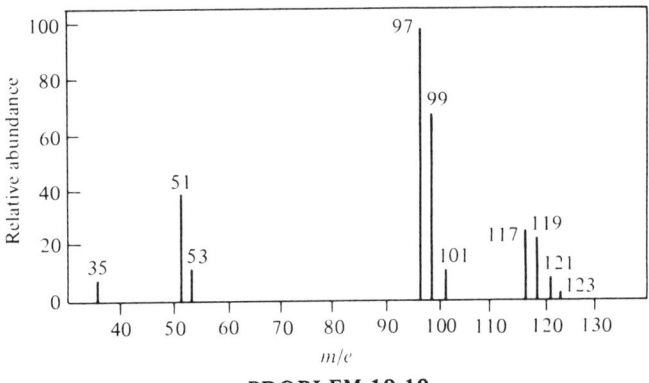

PROBLEM 19-19

20. A low melting solid with molecular formula $C_5H_8O_4$ gave the mass spectrum shown. Deduce the structural formula of the compound.

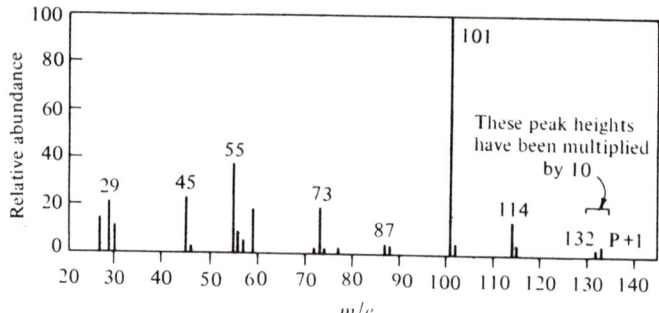

PROBLEM 19-20

BIBLIOGRAPHY

Ahern, A. J., "Spark Source Mass Spectrometric Analysis of Solids," in *Trace Characterization*, W. W. Meinke and B. F. Scribner, Eds., National Bureau of Standards Monograph 100, Washington, D. C., 1967.

Anbar, M. and W. H. Aberth, "Field Ionization Mass Spectrometry: A Tool for the Analytical Chemist," *Anal. Chem.*, **46**, 59A (1974).

Arpino, P. J. and G. Guiochon, "LC/MS Coupling," *Anal. Chem.*, **51**, 683A (1979).

Brown, R., M. L. Jacobs, and H. E. Taylor, "Spark Source Mass Spectrometry," *Am. Lab.*, p. 29 (November 1972).

Brunnee, C., G. Kappus, H. Rache, E. U. Seiler, and B. Windel, "A Computerized Mass Spectrometer for Isotopic Analyses of Solids, *Am. Lab.*, p. 141 (February 1978).

Budzikiewicz, H., C. Djerassi, and D. H. Williams, *Mass Spectrometry of Organic Compounds*, Holden-Day, San Francisco, 1967.

Campbell, I. M., "Radiogas Chromatography/Mass Spectrometry," *Anal. Chem.*, **51**, 1012A (1979).

Chait, E. M., "Ionization Sources in Mass Spectrometry," *Anal. Chem.*, **44**, 77A (1972).

Feser, K. and W. Kogler, "The Quadrupole Mass Filter for GC/MS Applications," *J. Chromatogr. Sci.*, **17**, 57 (1979).

Gochman, N., L. J. Bowie, and D. N. Bailey, "Specialized Gas Chromatography–Mass Spectrometry Systems for Clinical Chemistry," *Anal. Chem.*, **51**, 525A (1979).

McFadden, W. H., "Interfacing Chromatography and Mass Spectrometry," *J. Chromatogr. Sci.*, **17**, 2 (1979).

McLafferty, F. W., *Interpretation of Mass Spectra*, 2nd ed., Benjamin, Menlo Park, Calif., 1973.

Middleditch, B. S., Ed., *Practical Mass Spectrometry*, Plenum, New York, 1979.

Milberg, R. M. and J. C. Cook, Jr., "Design Considerations of MS Sources: EI, CI, FI, FD, and API," *J. Chromatogr. Sci.*, **17**, 17 (1979).

Munson, B., "Chemical Ionization Mass Spectrometry," *Anal. Chem.*, **43**, 28A (1971); **49**, 772A (1977).

Reynolds, W. D., "Field Desorption Mass Spectrometry," *Anal. Chem.*, **51**, 283A (1979).
Wilkins, C. L., "Fourier Transform Mass Spectrometry," *Anal. Chem.*, **50**, 493A (1978).
Yost, R. A., and C. G. Enke, "Triple Quadrupole Mass Spectrometry," *Anal. Chem.*, **51**, 1251A (1979).

LITERATURE CITED

1. Beynon, J. H. and A. E. Williams, *Mass and Abundance Tables for Use in Mass Spectrometry*, Elsevier, Amsterdam, 1963.

CHAPTER 20

Thermal Analysis

Thermal analysis includes a group of techniques in which a physical property of a substance is measured as a function of temperature while the substance is subjected to a controlled temperature program. A complete modern thermal analysis instrument measures temperatures of transitions, weight losses in materials, energies of transitions, dimensional changes, modulus, and viscoelastic properties. Current applications include environmental measurements, product reliability, compositional analysis, stability, chemical reactions, and dynamic properties.

Thermometric titrimetry, also discussed in this chapter, involves changes in solution temperature that are plotted as a function of time or volume of titrant.

20.1 DIFFERENTIAL THERMAL ANALYSIS AND DIFFERENTIAL SCANNING CALORIMETRY

In *differential thermal analysis (DTA)* the temperature of a sample and a thermally inert reference material are measured as a function of temperature (usually sample temperature). Any transition which the sample undergoes will result in liberation or absorption of energy by the sample with a corresponding deviation of its temperature from that of the reference. This differential temperature (ΔT) versus the programmed temperature (T) at which the whole system is being changed tells the analyst the temperature of transitions and whether the transition is exothermic or endothermic.

Closely related to DTA is *differential scanning calorimetry (DSC)*. In this method the sample and reference material are also subjected to a closely controlled programmed temperature. In the event that a transition occurs in the sample, however, thermal energy is added to or subtracted from the sample or reference containers in order to maintain both sample and reference at the same temperature. Because this energy input is precisely equivalent in magnitude to the energy absorbed or evolved in the particular transition, a recording of this balancing energy yields a direct calorimetric measurement of the transition energy.

The information obtained from DTA and DSC techniques, coupled with thermomechanical analysis, X-ray diffraction patterns, and chemical analysis of residues and any evolved gases, provides a quantitative and qualitative estimation of solid-state reactions. Comparison of data can be made from successive runs utilizing various environmental conditions and pressures.

Instrumentation

Typical DTA equipment is illustrated in Fig. 20-1. The furnace contains a sample block with identical and symmetrically located chambers. Each chamber contains a centered thermocouple. The sample is placed in one chamber and a reference material, such as α-Al_2O_3, is placed in the other chamber. The furnace and sample block temperature are then increased at a linear rate, most often 5° to 12°C/min, either by increasing the voltage through the heater element by a motor-driven variable transformer or by a thermo-

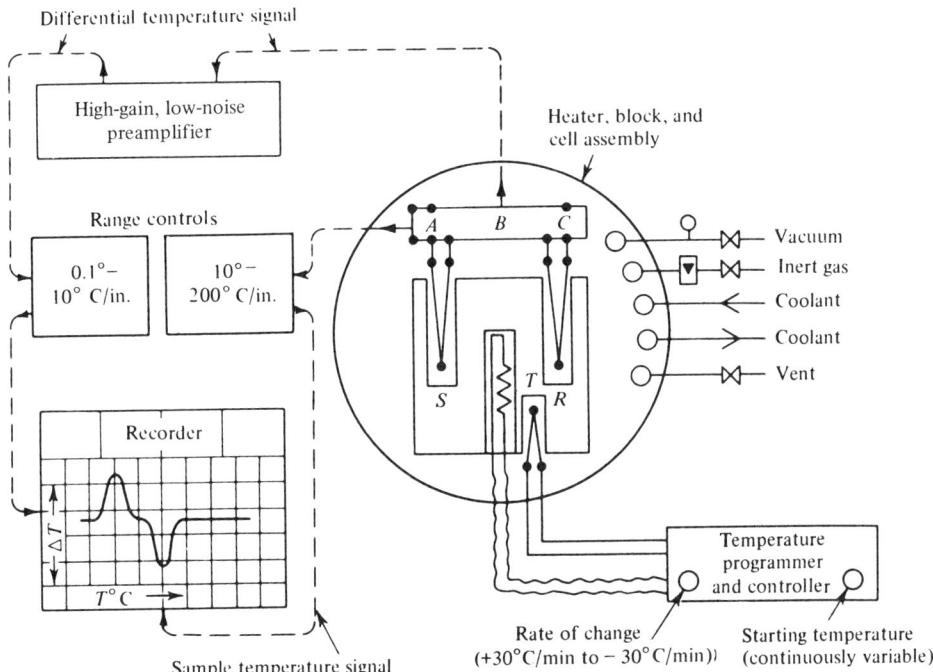

FIGURE 20-1 Schematic diagram of the DuPont differential thermal analyzer. (Courtesy of E. I. DuPont de Nemours, Inc.)

couple-actuated feedback type of controller. The difference in temperature between sample and reference (S, R) thermocouples, connected in series-opposition, is continuously measured. After amplification (about 1000 times) by a high-gain, low-noise, dc amplifier for the microvolt-level signals, the difference signal is recorded on the y-axis of a millivolt recorder. The temperature of the furnace is measured by a separate thermocouple which is connected to the x-axis of the recorder, frequently through a reference ice junction or room-temperature compensator. Because the thermocouple is placed directly in the sample, or attached to the sample container, the DTA technique provides the highest thermometric accuracy of all the thermal methods. The area under the output curve, however, is not necessarily proportional to the amount of energy transferred in or out of the sample. If maximum calorimetric accuracy is desired, as in DSC, the sample and reference thermocouples are removed from direct contact with the sample (see Fig. 20-2). The temperature range is between $-190°$ and $1600°C$. Sample sizes range from 0.1 to 100 mg.

Quantitative heats of transition require area integration to determine the total amount of energy transferred into or out of the sample. Precise measurement requires enlargement of the thermogram for accurate area measurements and a time-base display rather than the temperature-base thermogram. Also, precise quantitative heat capacity measurements are typically made at high sensitivity settings. Microprocessors avoid the necessity for running several thermograms. A disk memory retains all occurrences at maximum sensitivity. An expanded thermogram can be replotted automatically over a temperature range selected by the operator. Temperature expansion to $0.2°C/cm$ is possible. Combined with high calorimetric sensitivity of 0.01 mW/cm, this allows virtually unlimited expansion of

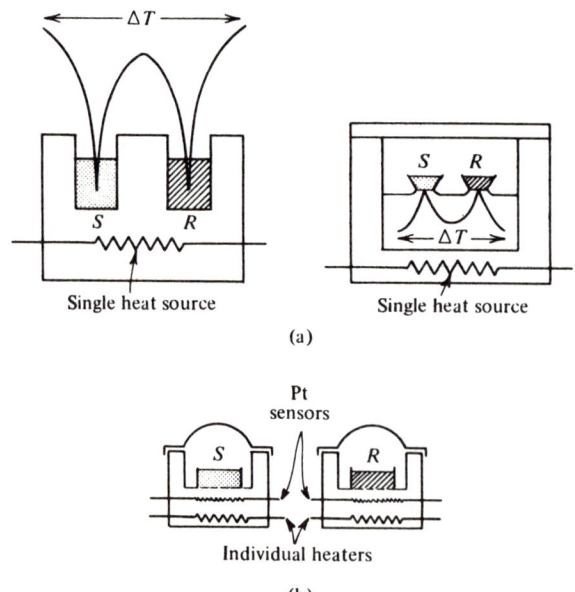

FIGURE 20-2 Arrangement of temperature sensors in (a) DTA and (b) DSC.

thermal occurrences. For maximum accuracy, a base line can be run and subtracted from the sample thermogram to determine heat capacity. Heats of transitions are calculated by a microcomputer in a data analyzer without time-base recordings and laborious data computations.

20.2 THERMOGRAVIMETRY

Thermogravimetry (TG) provides the analyst with a quantitative measurement of any weight change associated with a transition. For example, TG can directly record the loss in weight with time or temperature due to dehydration or decomposition. Thermogravimetric curves are characteristic for a given compound or system because of the unique sequence of physicochemical reactions which occur over definite temperature ranges and at rates that are a function of the molecular structure. Changes in weight are a result of the rupture and/or formation of various physical and chemical bonds at elevated temperatures that lead to the evolution of volatile products or the formation of heavier reaction products. From such curves data are obtained concerning the thermodynamics and kinetics of the various chemical reactions, reaction mechanisms, and the intermediate and final reaction products. The usual temperature range is from ambient to 1200°C with inert or reactive atmospheres.

The derivative in TG is often used to pinpoint completion of weight-loss steps or to increase resolution of overlapping weight-loss occurrences.

Instrumentation

For TG the sample is continuously weighed as it is heated to elevated temperatures. Samples are placed in a crucible or shallow dish that is attached to an automatic-recording balance. The automatic null-type balance incorporates a sensing element which detects a deviation of the balance beam from its null position. One transducer is a pair of photocells, a slotted flag connected to the balance arm, and a lamp (Fig. 20-3). Once an initial balance has been established, any changes in sample weight cause the balance to rotate. This moves the flag so that the light falling on each photocell is no longer equal. The resulting nonzero signal is amplified and fed back as a current to a taut-band torque motor (the pivot-point of the balance) to restore the balance to equilibrium. This current is proportional to the weight change and is recorded on the y-axis of the recorder. Changes in mass can also be detected by contraction or elongation of a precision helical spring whose movement is detected by the movement of an attached core in a linear variable differential transformer (Fig. 20-4). With either type balance the sample container is mounted inside a quartz or Pyrex housing which is located inside the furnace. Furnace temperature is continuously monitored by a thermocouple whose signal is applied to the x-axis of the recorder. Linear heating rates from 5° to 10°C/min are generally employed. In differential thermogravimetry the actual measurement signal is derived from a solid-state resistance-capacitance circuit which uses the direct output of the electrical weight-change signal from the thermobalance for the primary signal input. The resulting output is the derivative, $\Delta w/\Delta t$, which is used in kinetic interpretations. Sample sizes range from 1 to 300 mg, and sensitivities down to a few micrograms of weight change are common.

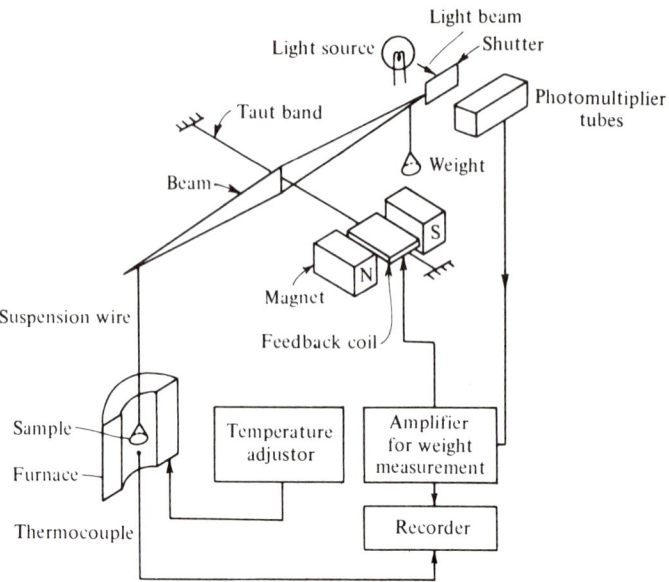

FIGURE 20-3 Schematic diagram of TG equipment with optical sensor. (Courtesy of Shimadzu Seisakusho, Ltd.)

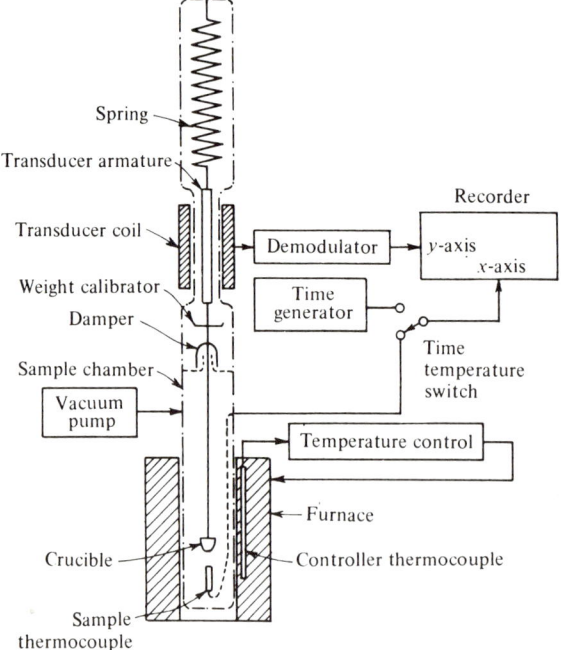

FIGURE 20-4 Modular diagram of TG equipment with spring and transducer coil as sensor. (Courtesy of American Instrument Co.)

20.3 METHODOLOGY OF DSC (OR DTA) AND TG

The weight-change curve for calcium oxalate monohydrate is shown in Fig. 20-5. Water is evolved beginning slightly above 100°C. At about 250°C a break is obtained in the curve at the stoichiometry corresponding to that of the anhydrous salt. Further heating gives definite weight plateaus for the carbonate (from 500° to 600°C) and finally the oxide (above about 870°C). Exact locations of the weight plateaus are dependent on the heating rate (a slower heating rate will shift values to lower temperatures) and the ambient atmosphere around the sample particles. The curve is quantitative in that calculations can be made to determine the stoichiometry of the compound at any given temperature.

Thermal analysis will be affected by the experimental conditions. Deviations caused by instrumental factors include furnace atmosphere, size and shape of the furnace and sample holder, sample holder material and its resistance to corrosive attack, wire and bead size of the thermocouple junction, heating rate, speed and response of the recording equipment, and location of the thermocouples in the sample and reference chambers. Another set of factors influencing results depends on the sample characteristics; these include layer thickness, particle size, packing density, amount of sample, thermal conductivity of sample material, heat capacity, ease with which gaseous effluents can escape, and the atmosphere surrounding the sample. Details concerning these factors should be sought in the treatises cited in the Bibliography.

Thermogravimetry, a valuable tool in its own right, is perhaps most useful when it complements differential thermal analysis studies. Virtually all weight-change processes absorb or release energy and are thus measurable by DTA or DSC, but not all energy-change processes are accompanied by changes in weight. This difference in the two techniques enables a clear distinction to be made between physical and chemical changes when the samples are subjected to both DSC (or DTA) and TG tests.

In general, each substance will give a DSC or DTA curve whose number, shape, and position of the various endothermic and exothermic features serve as a means of qualitative identification of the substance. When an endothermic change occurs, the sample temperature lags behind the reference temperature because of the heat in the sample.

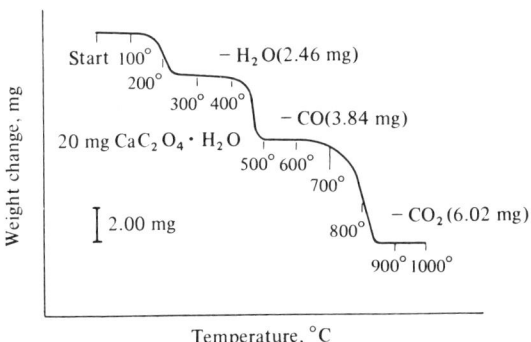

FIGURE 20-5 Thermogravimetric evaluation of calcium oxalate monohydrate; heating rate 6°C/min.

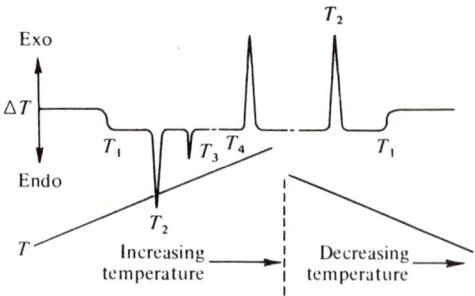

FIGURE 20-6 DTA curve of a hypothetical substance contrived to illustrate the discussion.

The initiation point for a phase change or chemical reaction is the point at which the curve first deviates from the base line. When the transition is complete, thermal diffusion brings the sample back to equilibrium quickly. The peak (or minimum) temperature indicates the temperature at which the reaction is completed. When the break is not sharp, a reproducible point may be obtained by drawing one line tangent to the base line and another tangent to the initial slope of the curve.

Various behaviors that may be deduced from a DSC curve are shown in Fig. 20-6. The heat capacity at any point is proportional to the displacement from the blank base line. A broad endotherm indicates a slow change in heat capacity A "second-order or glass" transition, observed as a base line shift (T_1), denotes a decrease in order within the system. This is the temperature at which a polymer changes from a brittle, glasslike material to a tough, resilient material. The lower the glass transition temperature, the lower the temperature at which the polymer is useful in applications, such as adhesives or impact-resistant structures. In a thermoset, a high glass transition temperature may indicate incomplete cure of the resin; in a thermoplastic, a high glass transition temperature may indicate use of the wrong plasticizer or incomplete reaction in the formation of the polymer itself. Endotherms generally represent physical rather than chemical changes. Sharp endotherms (T_3) are indicative of crystalline rearrangements, fusions, or solid-state transitions for relatively pure materials. Broader endotherms (T_2) cover behavior ranging from dehydration, temperature-dependent phase behaviors, to melting of polymers Exothermic behavior (without decomposition) is associated with the decrease in enthalpy of a phase or chemical system. Narrow exotherms usually indicate crystallization (ordering) of a metastable system, whether it be supercooled organic, inorganic, amorphous polymer, or liquid, or annealing of stored energy resulting from mechanical stress. Broad exotherms denote chemical reactions, polymerization, or curing of thermosetting resins. Exotherms with decomposition can be either narrow or broad depending on kinetics of the behavior. Explosives and propellants are sharpest, and "unzipping" of polyvinylchloride is rapid, while oxidative combustion and decomposition are generally broad.

On cooling one would expect the reverse of features observed on the heating cycle (see Fig. 20-6). Since T_4 is not found to recur on cooling, the reaction is obviously nonreversible (perhaps a pyrolytic decomposition). Instead of taking the system up to T_4, start the cooling cycle before that temperature. As it cools, the substance is seen to lose its transition peak at T_3. Judging from the area under the T_2 peak, the transition energy of T_3 has been added to T_2. This indicates a metastable condition at T_3, the retained energy being released in one large step at a lower temperature. Further along the cooling curve, the glass transition at T_1 falls properly into place to complete the cycle.

Unravelling the significance of curves is not always straightforward. A reference library of curves of interest to a particular laboratory is vital. Temperature data on commercial products or melting points for pure substances reported in the literature are of little value when comparison with a dynamically scanned thermal profile is desired. Complementary techniques are valuable. Establishing whether a gaseous product is evolved at a corresponding DTA or DSC transition, and its identification, often assists in elucidating the decomposition route. Gas chromatographs and mass spectrometers[1] can be coupled to thermal analysis equipment for repetitive analysis of gaseous decomposition products. Analysis of the evolved gases by chemical means is also possible. Thermal decompositions in inert, oxidative, or special atmospheres provide clues through changes in the curves or displacement of features.

Example 20-1

The TG and DTA curves of manganese phosphinate monohydrate are shown in Fig. 20-7. The weight-loss data (TG curve) from a 200-mg sample run under vacuum and with the analysis of effluent gases showed the loss of one mole of water at 150°C, one mole of phosphine at 360°C, and the slow loss of another mole of water starting around 800°C. In comparison with the DTA curve, two major peaks remain unidentified: the large exotherm at 590°C and the endotherm at 1180°C, plus several smaller thermic features. Thermogravimetric data obtained from runs performed under vacuum and in a nitrogen atmosphere failed to show any loss associated with these peaks. Each sample was measured for its real density; the resulting data are shown on the DTA curve. Undoubtedly the sharp DTA exotherm at 590°C represents a phase change. The relatively small endotherm starting above 900°C must represent a recrystallization exotherm following the elimination of water which is superimposed on the latter endotherm. The peak at 1180°C is due to melting. With this information the thermal decomposition reactions and phase changes are:

$$Mn(PH_2O_2)_2 \cdot H_2O(s) \rightarrow Mn(PH_2O_2)_2(s) + H_2O(g)$$
$$Mn(PH_2O_2)_2(s) \rightarrow MnHPO_4(s) + PH_3(g)$$
$$\alpha\text{-}MnHPO_4(s) \rightarrow \beta\text{-}MnHPO_4(s)$$
$$2MnHPO_4(s) \rightarrow Mn_2P_2O_7(s) + H_2O(g) \text{ (and recrystallization)}$$
$$Mn_2P_2O_7(s) \rightarrow Mn_2P_2O_7(l)$$

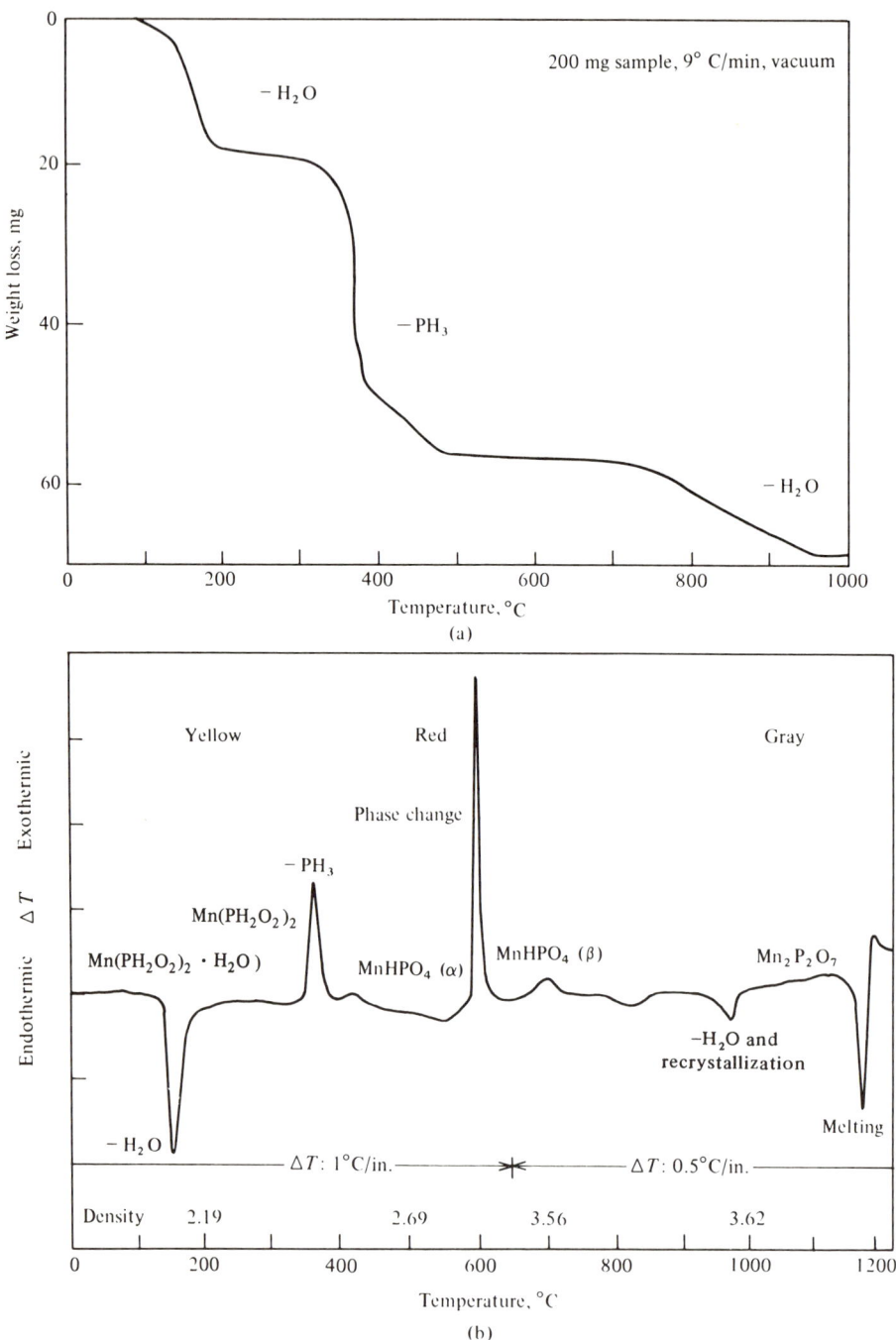

FIGURE 20-7 (a) TG and (b) DTA curves for Mn $(PH_2O_2)_2 \cdot H_2O$. (Courtesy of American Instrument Co.)

Example 20-2

The curves for calcium acetate monohydrate (Fig. 20-8) illustrate the influence of different atmospheres. The first endothermal peak is unaffected by change in atmosphere. The weight-loss data indicate loss of water and thus conversion of the monohydrate to the anhydrous salt. The next feature on the DTA curve is an endotherm in CO_2 and Ar, but an exotherm in an O_2 atmosphere; the weight loss corresponds to one mole each of CO_2 and CO. In an oxygen atmosphere the highly exothermic reaction must be the oxidation of carbon monoxide. The final stage is the decomposition of calcium carbonate to calcium oxide, which is a function of the partial pressure of CO_2 in contact with the sample and consequently is shifted to higher temperatures in the CO_2 atmosphere.

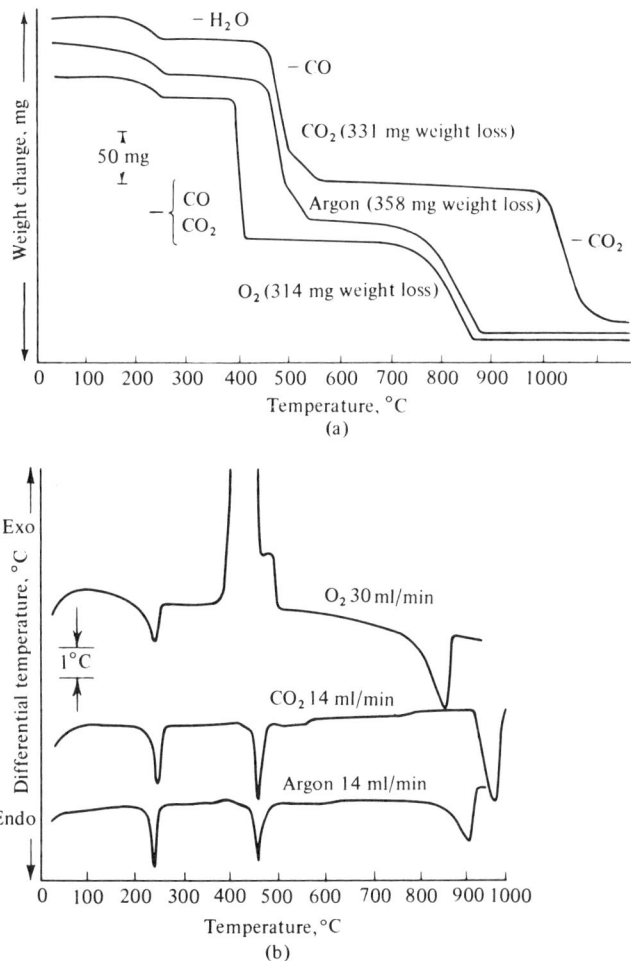

FIGURE 20-8 Curves for calcium acetate monohydrate. (a) TG, 6°C/min and (b) DTA, 12°C/min. Particular atmosphere above the solid phase is indicated for each curve. (Courtesy of American Instrument Co.)

Thermal studies with polymers can predict a product's performance in use; that is, its stiffness, toughness, or stability. Melting-point, phase-transition, pyrolysis, and curing temperatures can be accurately measured. Once a polymer has been broadly classified by other methods, curves often can be used to establish, by comparison with known reference materials, the degree of polymerization, the thermal history of the sample, crystal perfection and orientation, the effect of different coreactants and catalysts, the percentage of crystalline polymer, and the extent of chain branching. For example, curves for a low molecular weight, nonlinear, branched-chain polymer will show a continuous series of rather broad and low-melting endotherms, whereas a high molecular weight, stereo-regular, linear polymer will reveal a single narrow and higher-melting endotherm. If a polymer has been incompletely cured, the heating cycle may reveal an exotherm at a temperature close to the one employed for the polymerization reaction. An exotherm occurring just below the melting temperature can indicate "cold crystallization," which results if a sample is quenched quickly after being melted. On reheating, crystallites form rapidly and exothermically just prior to remelting of the polymer. Annealing temperatures are similarly revealed as exotherms.

If the molecular weight or density of a polymer has been established by appropriate (often lengthy) methods, subsequent determination of its melt temperature (a 15-min process) can be related to molecular weight or density. Product quality can be maintained subsequently by simply examining curves of polymer materials to obtain molecular weights or densities from an appropriate calibrated graph.

Instead of using the traditional method of preparing a derivative from the organic sample and a reagent, the sample can be heated with a specific reagent at a programmed heating rate in a selected atmosphere. The DTA or DSC curve will show the derivative-forming reaction, the physical transitions of the sample or reagent (whichever is in excess), and the physical transitions of the intermediates and final products. When one reactant is volatile and in excess, a rerun will usually show only the derivative characteristics.[2]

The area of exotherms or endotherms can be used to calculate the heat of the reaction or the heat of a phase transition. Suitable calibration is necessary with DTA equipment, but the values are given directly with DSC instruments.

20.4 THERMOMECHANICAL ANALYSIS

Thermomechanical analysis (TMA) provides measurements of penetration, expansion, contraction, and extension of materials as a function of temperature. Typical apparatus, diagrammed in Fig. 20-9, consists of a probe connected mechanically to the core of a linear variable differential transformer (LVDT). The core is coupled to the sample by means of a quartz probe containing a thermocouple for measurement of sample temperature. Any movement of the sample is translated into a movement of the transformer core and results in an output that is proportional to the displacement of the probe, and whose sign is indicative of the direction of movement. The temperature range is from that of liquid nitrogen to 850°C.

In the penetration and expansion modes, the sample rests on a quartz stage surrounded by the furnace. Under no load, expansion with temperature is observed. Calculation of the thermal coefficient of linear expansion may be made directly from the slope of the resulting curve. A weight tray attached to the upper end of the probe allows a predetermined force to be applied to the sample to study variations under load. Probes of small tip diameter and a loaded weight tray are used when the sensitive detection of softening temperatures, heat distortion temperatures, and glass transitions are of interest. Larger tip diameters and zero loading are used in the expansion mode when coefficients of expansion and dimensional changes due to stress relief are the objects of investigation. Sam-

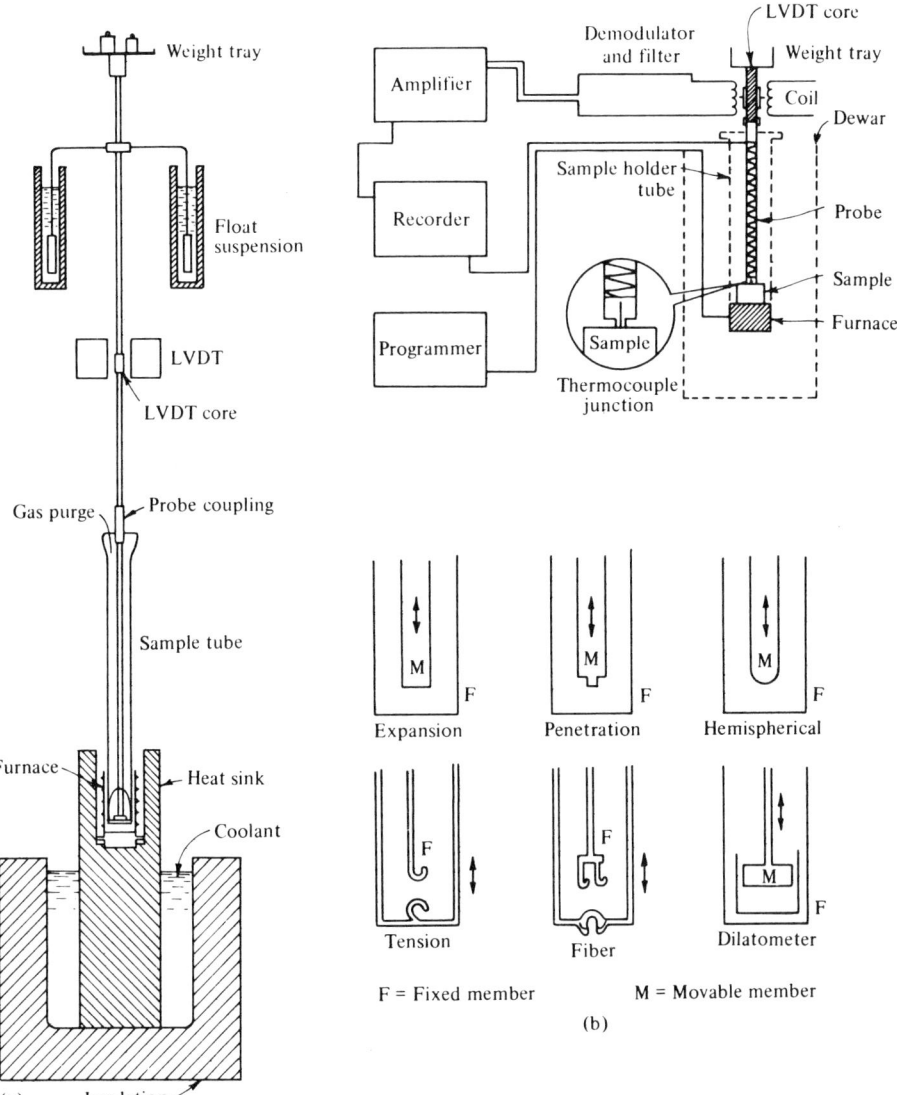

FIGURE 20-9 (a) Thermomechanical analyzer. (Courtesy of Perkin-Elmer Corp.) (b) Probe configurations. (Courtesy of E. I. DuPont de Nemours, Inc.)

ple sizes may range from a 2.54-μm coating to a 1.3-cm-thick solid. Sensitivities down to a few microinches are observable.

For measurement of samples in tension, sample stage and probe are replaced by a sample holder system consisting of a stationary and movable hook constructed of fused silica. This permits extension measurements on films and fibers. Holes about 0.6 cm apart are punched into injection-molded pieces or solution-cast or extruded films; also a fiber fused into a loop can be used for this test. The double-hook probe is designed to grasp a pair of aluminum spheres that are crimped onto either end of a fiber sample. Measurements made with these probes can be related to the tensile modulus of a sample.

Volume expansion characteristics of samples are measured by placing the sample in a quartz cylinder fitted with a flat-tipped quartz probe in a cylinder-piston arrangement. Sample volume changes are translated into linear motion of the piston.

Example 20-3

The expansion and heat capacity behavior of herring oil is shown in Fig. 20-10. The expansion characteristics show changes at $-108°$, $-80°$, $-47°$, $-27°$, and $-15°C$, at which point the material is apparently fluid. These volume changes confirm the changes in heat capacity measured by DSC at $-106°$, $-51°$, $-26°$, and $-13°C$, and emphasize the need for using more than one mode of thermal analysis to illustrate the thermal response of a system.

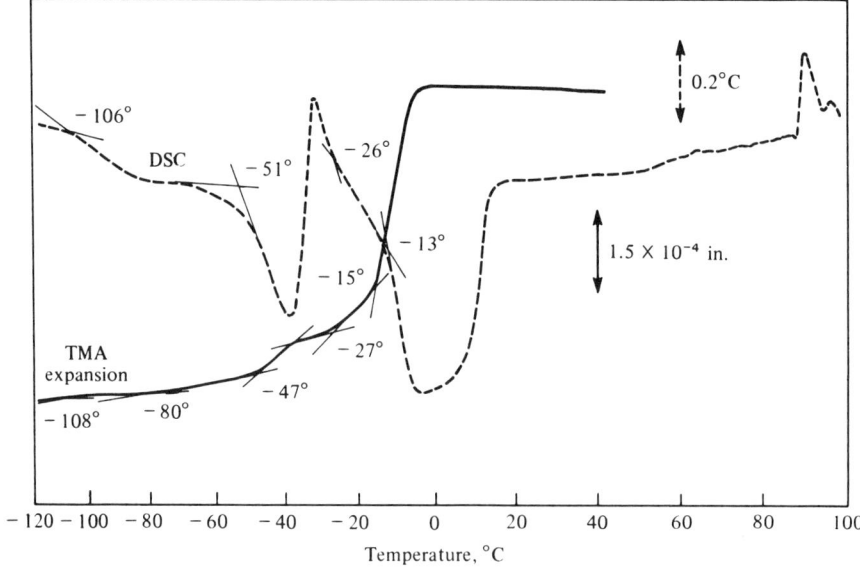

FIGURE 20-10 Thermomechanical (expansion) behavior and differential scanning calorimetry of herring oil.

20.5 DYNAMIC MECHANICAL ANALYSIS

Dynamic mechanical analysis is the most sensitive thermal analytical technique for detecting transitions associated with movement of polymer chains. The technique involves measuring the resonant frequency and mechanical damping of a material forced to flex at a selected amplitude. Mechanical damping is the amount of energy dissipated by the sample as it oscillates, while the resonant frequency defines Young's (elastic) modulus or stiffness. Loss modulus and the ratio of loss modulus to elastic modulus can be calculated from the raw frequency and/or damping data. In general, modulus and frequency, as well as damping, change more dramatically than heat capacity or thermal expansion during secondary transitions. For example, dynamic mechanical analysis is helpful in determining the effectiveness of reinforcing agents and fillers used in thermoset resins.

20.6 THERMOMETRIC TITRIMETRY

Thermometric titrimetry and *titration calorimetry* are techniques in which the temperature of a system is measured as a function of titrant added. The resultant temperature–volume curve is similar to other linear titration curves.

Instrumentation

The equipment (Fig. 20-11) consists basically of a motor-driven automatic buret, an adiabatic titration chamber, a thermistor and Wheatstone bridge circuit, and a strip-chart recorder. To minimize heat transfer between the solution and its surroundings, the titrations are performed under as near adiabatic conditions as possible in an insulated beaker or Dewar flask of 100- to 250-ml capacity that is closed with a stopper provided with holes for the buret tip, a glass stirrer, and the thermistor. The titrant is delivered at flow-rates of 0.1–1.0 ml/min. To obviate volume corrections and to minimize temperature

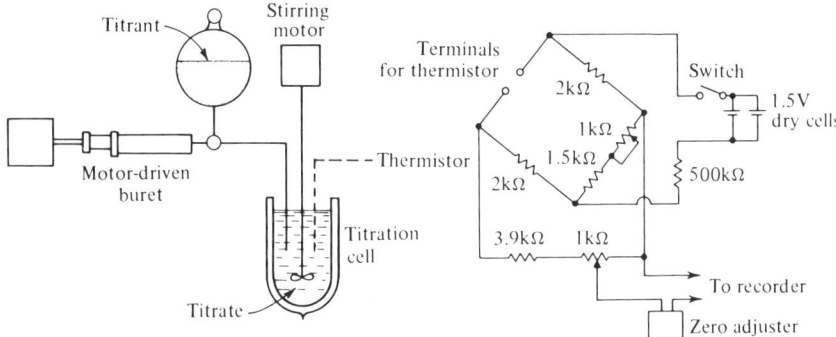

FIGURE 20-11 Schematic titration assembly and bridge circuit for conducting thermometric titrations. (After H. W. Linde, L. B. Rogers, and D. N. Hume, *Anal. Chem.*, 25, 494 (1953). Courtesy of American Chemical Society.)

variations between the titrant and sample, the titrant concentration is usually 100 times larger than that of the reactant. Amounts of sample are selected so that a volume of titrant not exceeding 1–3 ml is required.

Because temperature changes in the course of a titration range between 0.01° and 0.2°C, the accuracy of the temperature measurement must be about 10^{-4}°C. For a thermistor having a resistance of 2 kΩ and a sensitivity of -0.04 Ω Ω^{-1} deg^{-1} Celsius in the 25°C temperature range, a change of 0.01°C corresponds to an imbalance potential of 0.157 mV. Temperatures of titrant and sample should be within 0.2°C before a titration is begun. A small heating element, located inside the titration vessel, can be used to warm the sample to the temperature of the titrant or as a calibrating device when estimating heats of reaction or mixing.

In a differential thermometric apparatus, temperature-sensing elements are placed in both the sample and blank (pure solvent plus titrant) solutions. Sensitivity is improved and extraneous heat effects, such as stirring and heats of dilution, are minimized.[3]

Methodology

Contrary to potentiometric titrations of various types that depend solely on equilibrium constants and, hence, the free energy of the reaction, $\Delta G°$, or

$$-\Delta G° = RT \ln K \qquad (20\text{-}1)$$

thermometric titrations depend only on the heat of the reaction, ΔH, or

$$\Delta H = \Delta G + T\Delta S \qquad (20\text{-}2)$$

Thus, a thermometric titration may be feasible when all "free energy" methods fail. This point is clearly shown in Fig. 20-12, where a comparison is given of the potentiometric

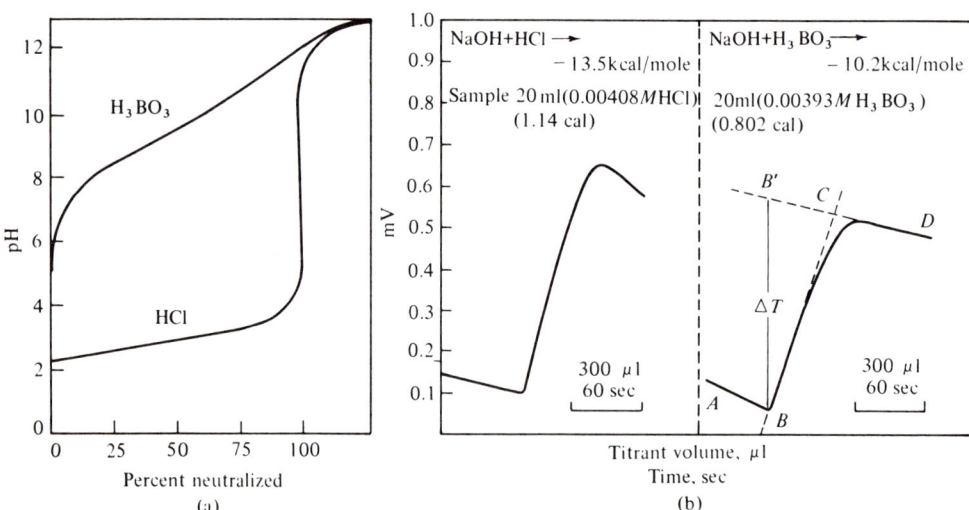

FIGURE 20-12 (a) Potentiometric and (b) thermometric titration curves for hydrochloric and boric acids with 0.2610M sodium hydroxide.

and thermometric titration curves for HCl and H_3BO_3. In contrast to the potentiometric curve, the thermometric titration curve has a well-defined end point for the weak acid. The change in temperature of the titration is dependent on the heat of reaction of the system, according to the equation

$$\Delta T = \frac{N\Delta H}{Q} \tag{20-3}$$

where N represents the number of moles of water formed by neutralization, ΔH is the molar heat of neutralization, and Q is the heat capacity of the system. In practice, ΔH and Q are constant throughout the titration so that ΔT is proportional to N.[4]

On the thermometric titration curve shown in Fig. 20-12, point A is where the temperature readings were begun, and line AB is a trace of the temperature of the solution before the addition of titrant. If the line AB shows a marked slope, it is an indication of excessive heat transfer between the solution and its surroundings. At point B the addition of titrant was begun; line BC shows the gradual evolution of the heat of reaction. Point C is the end point. Line CD may slope downward or upward. The linear portions of the curve are extrapolated to give the initial and equivalence points, and the distance between them is measured along the volume (or time) axis of the chart to ascertain the volume of titrant consumed in the reaction. The vertical line BB' is the temperature difference (ΔT) used to evaluate enthalpies (Eq. 20-3).

Applications

Applications of thermometric titrimetry include the determination of the concentration of an unknown substance, determination of reaction stoichiometry, and the determination of the thermodynamic quantities: ΔG, ΔH, and ΔS. The first application is perhaps the most useful to the analytical chemist. Precision and accuracy of measurements depend largely on enthalpy of the reaction involved, and range from 0.2 to 2%. About $0.0001M$ is the lowest limit of concentration that can be successfully titrated in the more favorable cases.

All acids with $K_a \geqslant 10^{-10}$ can be titrated thermometrically in $0.01M$ solution with a precision of 1% if the heat of neutralization is 13 ± 3 kcal/mole. The extension to acids too weak to titrate potentiometrically is clearly demonstrated by the curves in Fig. 20-12. Good end points are obtained for other weak acids and bases, even in emulsions and thick slurries.

Nonaqueous systems are well suited for thermometric titrations, although attention must be paid to the heat of mixing of solvents and dilution. The lower specific heat of many organic solvents introduces a favorable sensitivity factor. Under strictly anhydrous conditions, even diphenylamine, urea, acetamide, and acetanilide are readily titratable with perchloric acid in glacial acetic acid.[5] Lewis bases, such as dioxane, morpholine, pyridine, and tetrahydrofuran, have been titrated with the Lewis acid $SnCl_4$ in the solvents CCl_4, benzene, and nitrobenzene.[6]

Thermometric titrations are very useful in titrating acetic anhydride in acetic acid-sulfuric acid acetylating baths, water in concentrated acids by titration with fuming acids, and free anhydrides in fuming acids. In fact, methods based on heats of reaction offer one of the few approaches to the analysis of concentrated solutions of these materials.[7]

Good results can be obtained in precipitation and ion-combination reactions such as the halides with silver or mercury(II), and cations with EDTA and oxalate. Silver titration of halides has been done at elevated temperatures in molten salts.[8]

When the titration reaction is appreciably incomplete in the vicinity of the equivalence point, actual titration curves exhibit curvature from which equilibrium constants and corresponding free energies can be calculated. The temperature rise that occurs during an exothermic reaction can be used to determine constituents. For example, benzene has been determined rapidly and with good precision in the presence of cyclohexane by measuring the heat of nitration when a standard nitrating acid mixture is added to the sample; the temperature rise is a direct function of the benzene present. In a similar manner, heats of reaction have been used to estimate the heats of successive steps in the formation of metal–ammine complexes,[9] the heats of chelation,[10,11] and heats of reaction in fused salts under virtually isothermal conditions.[8]

PROBLEMS

1. Formulate the solid-state reaction of sodium bicarbonate when heated. It decomposes between 100° and 225°C with the evolution of water and carbon dioxide. The combined loss of water and carbon dioxide totaled 36.6% by weight, whereas the weight loss due to carbon dioxide alone was found to be 25.4%.

2. A definite relation exists between decomposition temperature of $CaCO_3$ and equilibrium partial pressure of CO_2. A series of thermograms were obtained with a dynamic flow of CO_2 in the pressure range from 40 to 600 torr. Since pure CO_2 was used, the partial pressure is equivalent to the system pressure. Estimate the heat of dissociation from the following data:

Initial decomposition temperature, °C	926	895	840	802	759	749
Pressure CO_2, torr	600	400	200	100	50	40

3. Ascertain the glass transition of a polycarbonate resin from the following heat capacity measurements:

TEMPERATURE RANGE, °K	SPECIFIC HEAT	TEMPERATURE RANGE, °K	SPECIFIC HEAT
400.0–402.5	0.345	412.5–415.0	0.373
402.5–405.0	0.349	415.0–417.5	0.385
405.0–407.5	0.355	417.5–420.0	0.417
407.5–410.0	0.361	420.0–422.5	0.449
410.0–412.5	0.367		

4. The heat of fusion of a mixture of AgCl–AgBr can be used for the analysis of Cl–Br mixtures because these ions form ideal solid solutions in all proportions. ΔH_{fusion} (in cal/g) was found to be 12.1 for pure AgBr and 22.0 for pure AgCl. What weight percent of AgCl is present in a mixture which has the following values of heat of fusion: (a) 14.4, (b) 20.0, (c) 16.9, (d) 16.05, and (e) 19.6?

5. On a DSC curve obtained with 10.2 mg of dotriacontane and an external 12.1-mg standard of indium whose ΔH_{fusion} = 6.8 cal/g, the following areas, as measured by a planimeter, were obtained: chain rotation (65°C), 158 units; fusion of dotriacontane (72°C), 439 units; and fusion of indium, 93 units. Calculate the transition energies of dotriacontane.

6. A mixture of 95% Ar and 5% O_2 was passed through a DTA oven which was heated at 10°C/min. A sample of 1.000 g of UO_2 registered an exotherm at 360°C whose peak area was 25.6 cm². When a current of 2.1 A at 3.6 V was passed for 30 sec, a calibration peak of 15.6 cm² was obtained. Determine the energy liberated in the reaction: $3UO_2 + O_2 \rightarrow U_3O_8$.

7. In the accompanying figure curve A is the weight-loss thermogram from pure $CaCO_3$, curve C shows a similar trace from $MgCO_3$, and curve B is the thermogram of a limestone sample. (a) Derive an expression for the direct quantitative analysis of CaO and MgO. (b) Write equations for the solid-state decomposition of $MgCO_3$. (c) Calculate the percent CaO and MgO in the limestone sample.

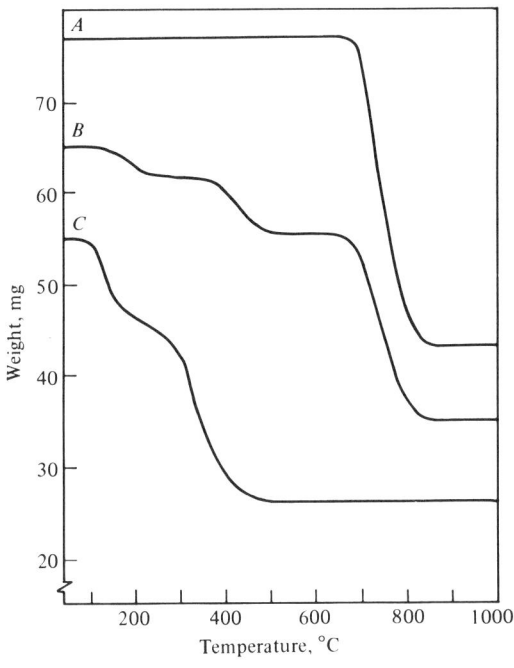

PROBLEM 20-7

8. The decomposition reactions of a 100-mg sample of nickel oxalate dihydrate vary in different atmospheres. In both flowing and stationary air, successive weight losses of 19 and 39 mg were observed. However, in flowing CO_2 and in flowing N_2 the successive weight losses were 19 and 49 mg. The same temperature program was employed in the four runs. Write the decomposition reactions.

624 CHAPTER 20

9. In fused lithium nitrate–potassium nitrate at 158°C, the shape of the titration curve obtained in 8.6×10^{-4} molal solution of potassium chloride with 1.40 molal silver nitrate showed that precipitation at the equivalence point was about 20% incomplete. In contrast, precipitation of 1.17×10^{-2} molal solution of KCl was 98.5% complete. Estimate the molal solubility product of AgCl in the eutectic salt melt.

10. Demonstrate, by discrete ionization and neutralization steps and suitable calculations, why an acid as weak as boric acid gives an adequate thermometric titration curve. For boric acid neutralization, ΔS is -31.1 e.u.

11. Estimate the values of ΔH and sketch the hypothetical titration curve for a mixture of calcium and magnesium ions titrated with EDTA. Thermodynamic characteristics at 25°C of chelation equilibria with EDTA are:

Cation	$pK_{STABILITY}$	$\Delta S°$, e.u.
Ca^{2+}	-11.0	+31
Mg^{2+}	-9.1	+60

12. A simultaneous DTA–TG curve for manganese(II) carbonate in a porous crucible is shown (solid lines). (a) What are the transitions involved at each peak on the DTA trace, and what are the products at each TG plateau? (b) Another laboratory, using a controlled atmosphere with 13 atm CO_2, obtained the curves shown (dashed line). Why is the initial oxide different?

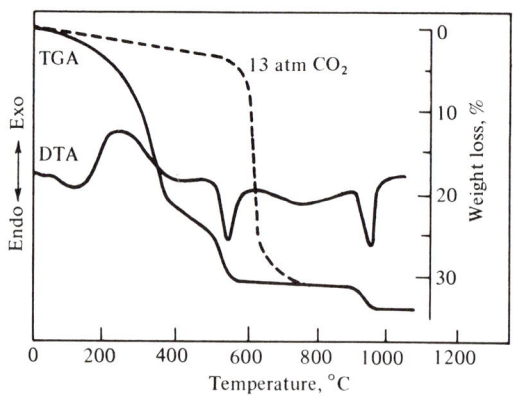

PROBLEM 20-12

13. From the thermomechanical penetration curve shown, deduce the nature of the two transitions.

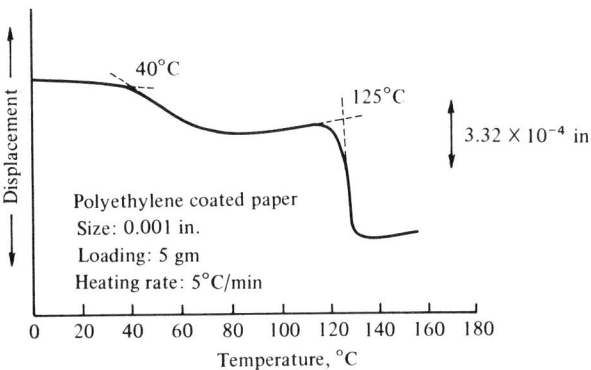

PROBLEM 20-13

14. Three successive runs on the same sample of a fiberglass mat impregnated with an uncured epoxy resin are shown. The scan in run A was stopped at 90°C, and the sample cooled and rerun (run B). Run C is the sample from run B after cooling. Discuss the features observed in the DSC scans.

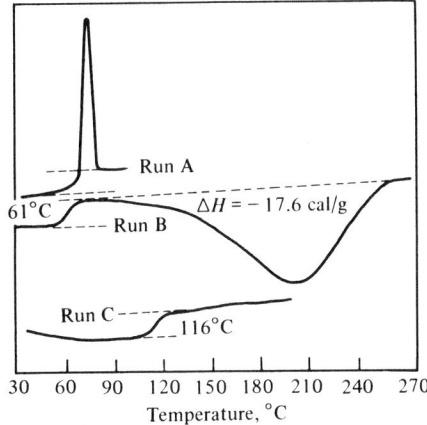

PROBLEM 20-14

626 CHAPTER 20

15. The figure shows the micro DTA curves of the two-phase system involving picryl-chloride and hexamethylbenzene; construct the phase diagram.

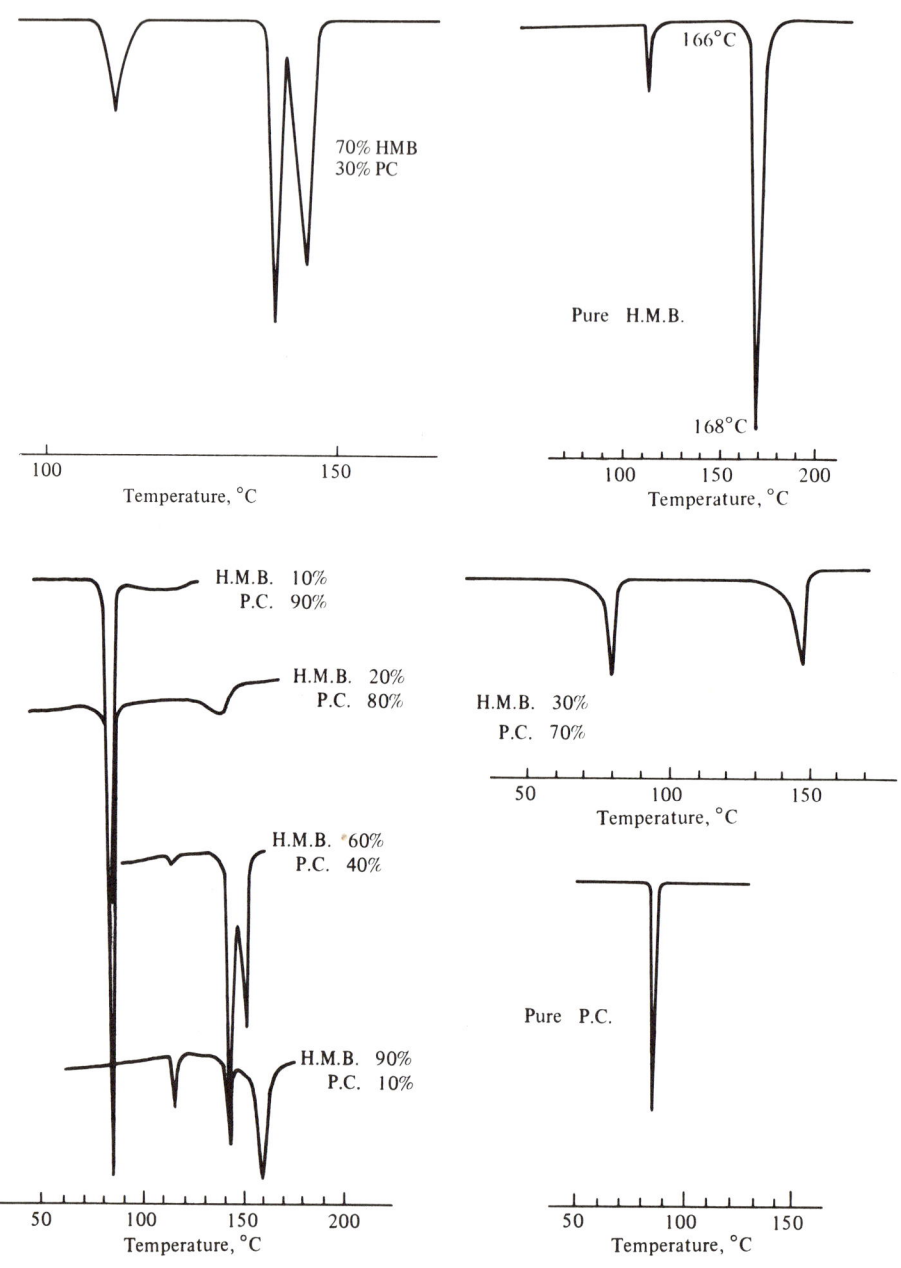

PROBLEM 20-15

BIBLIOGRAPHY

Blazek, A., *Thermal Analysis,* Van Nostrand-Reinhold, London, 1974.
Levy, P. F., R. L. Blaine, P. S. Gill, and J. D. Lear, "Thermal Analysis: Advances in Instrumentation,"*Am. Lab.*, p. 79 (June 1979).
Mackenzie, R. C., Ed., *Differential Thermal Analysis,* Academic, London, 1970.
Schwenker, R. F., Jr. and P. D. Garn, Eds., *Thermal Analysis,* Vol. 1: *Instrumentation, Organic Materials and Polymers,* 1969; Vol. 2: *Inorganic Materials and Physical Chemistry,* 1968, Academic, New York.
Vaughan, H. P. and J. P. Elder, "Advances in Quantitative Differential Thermal Analysis," *Am. Lab.*, p. 53 (January 1974).
Wendlandt, W. W., *Thermal Methods of Analysis,* 2nd ed., Wiley-Interscience, New York, 1974.
Wendlandt, W. W. and L. W. Collins, *Benchmark Papers in Analytical Chemistry,* Vol. 2, *Thermal Analysis,* Dowden, Ross, and Hutchinson, Stroudsburg, Pa., 1976.
Wendlandt, W. W. and L. W. Collins, *Thermal Analysis,* Wiley, Chichester, England, 1977.

LITERATURE CITED

1. Gohlke, R. S. and H. G. Langer, *Anal. Chem.,* **37** (10), 25A (1965); **37**, 433 (1965).
2. Chiu, J., *Anal. Chem.,* **34**, 1841 (1962).
3. Tyson, B. C., W. H. McCurdy, and C. E. Bricker, *Anal. Chem.,* **32**, 1640 (1961).
4. Jordan, J., *Record of Chem. Prog.,* **19**, 193 (1958).
5. Keily, H. J. and D. N. Hume, *Anal. Chem.,* **36**, 543 (1964).
6. Cioffi, F. J. and S. T. Zenchelsky, *J. Phys. Chem.,* **67**, 357 (1963).
7. Somiya, T., *J. Soc. Chem. Ind. (Japan),* **32**, 306, 490 (1929).
8. Jordan, J., J. Meir, E. J. Billingham, and J. Pendergrast, *Anal. Chem.,* **32**, 651 (1960).
9. Poulsen, I. and J. Bjerrum, *Acta Chem. Scand.,* **9**, 1407 (1955).
10. Jordan, J. and T. G. Alleman, *Anal. Chem.,* **29**, 9 (1957).
11. Jordan, J., J. Meir, E. J. Billingham, and J. Pendergrast, *Anal. Chem.,* **31**, 1439 (1959).

CHAPTER 21

Introduction to Electrometric Methods of Analysis

Electroanalytical chemistry encompasses a wide variety of techniques, each based upon a particular phenomenon occurring within an electrochemical cell. Each basic electrical measurement—current, resistance, and voltage—has been used alone or in combination for analytical purposes. If these electrical properties are measured as a function of time, many other techniques are possible. Table 21-1 contains a brief classification of electrochemical methods arranged according to the quantity measured.

Electrometric methods can also be grouped into two categories: *steady-state* or equilibrium method (nonpolarized), *transient* or dynamic (polarized or diffusion-controlled) methods. In the first category, equilibrium is effectively assured by vigorously stirring the solution or rotating the electrode or both so that concentration gradients at the electode are completely or nearly completely eliminated. In such cases, the electrode potential is related to concentration by the Nernst equation, 21-6. In transient methods, both the electrode and solution are static, in which case, after electrolysis begins, the concentration gradients at the electrodes are time dependent or diffusion controlled. This category includes such methods as chronopotentiometry, chronoamperometry, and polarography (voltammetry). When either impressed potential or current are controlled, the dependent variable is a function of both concentration and time. If data are recorded as current versus potential, time still enters the experiment because the value of the current observed for a given potential depends upon the elapsed time between application of the potential and the measurement.

The general relationships between current, potential, and composition of an electroactive system are depicted in the three-dimensional representation of Fig. 21-1. The various types of two-dimensional variations observed in the diverse techniques can be

TABLE 21-1 A Summary of Electrometric Methods of Analysis

Quantity Measured	Variable Controlled	Name or Description of Method
E versus volume of reactant	$i = 0$	Potentiometric titrations
	i	Potentiometric titration at constant current
E	$i = 0$	1. Measurement of ionic activities or concentrations
		2. Null-point potentiometry
E versus time	i	Chronopotentiometry
Weight of separated phase	i or E	Electrogravimetry
$1/R$ (conductance)	E	Concentration measured after calibration with known mixtures
$1/R$ (conductance) versus volume of reagent	E	Conductometric titrations
i versus E	Concentration	1. Polarography or voltammetry
		2. Stripping voltammetry
i versus volume of reagent	E	Amperometric titrations
Coulombs (current × time)	E	Coulometry at controlled electrode potential
	i	1. Coulometry at constant current
		2. Stripping analysis

visualized from the intersection with the solid surface of a plane perpendicular to a particular axis. For example, at any particular concentration a polarographic curve can be obtained by changing the voltage and observing the current. Amperometric titration curves are obtained by observing the current at a constant voltage. A plane corresponding to zero current intersects the surface to yield the usual potentiometric titration curve.

Electrometric methods are characterized by a high degree of sensitivity, selectivity, and accuracy. Highly refined ways of making electrical measurements permit reliable deter-

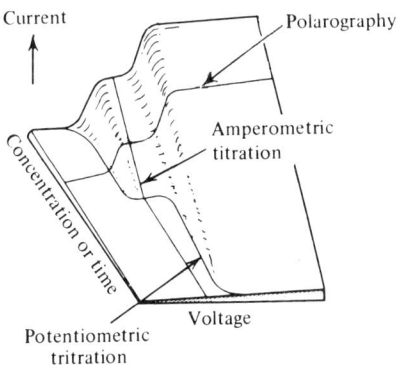

FIGURE 21-1 Three-dimensional representation of the relationship between current, potential, and composition of a system. (After C. N. Reilley, W. D. Cooke, and N. H. Furman, *Anal. Chem.*, 23, 1226 (1951). Courtesy of American Chemical Society.)

minations at the submicroampere and microvolt range. This means that analytical sensitivity, limited largely by noise considerations, approaches and even exceeds the 10^{-10} molar level. Because fractions of a drop can be analyzed, reliable analyses at the subnanogram range are possible. Electrochemical selectivity minimizes preseparations, providing a further advantage. Remote control, in-line installations, and automation are possible with these techniques.

21.1 TYPES OF ELECTROCHEMICAL CELLS

There are two types of electrochemical cells: galvanic (or voltaic) and electrolytic. A *galvanic cell* consists of two electrodes and one or more solutions (that is, two half-cells) and is capable of spontaneously converting chemical energy more or less completely into electrical energy and supplying this energy to an external source. In these cells, a chemical reaction involving an oxidation at one electrode and a reduction at the other electrode occurs. The electrons evolved in the oxidation step are transferred at the electrode surface, pass through the external circuit, and then return to the other electrode where reduction takes place. When one of the chemical components responsible for these reactions is depleted, the cell is no longer capable of supplying electrical energy to an external source and the cell is "dead."

If electrical energy is supplied from an external source, the cell through which the current is forced to flow is called an *electrolytic cell.* Electrochemical changes are produced at the electrode/solution interfaces, and concentration changes are produced in the bulk of the system. During electrolysis a galvanic cell is built up from the products that accumulate at the electrodes. If the external current is turned off, the products tend to produce current in the opposite direction.

At the exact point where the galvanic emf is opposed by an equal applied emf, no current flows through the cell in either direction. In this condition of null balance, the potential generated at the interface of an indicator electrode will reflect the composition of the solution phase, provided that the indicator electrode is selected so that its potential is sensitive to the desired component in the solution phase. However, it is not possible to measure the potential of a single electrode, because any electrical contact between the bulk of the solution and an external circuit is itself another electrode/solution interface. It is only possible to measure the potential of one electrode relative to another— the reference electrode. Before we proceed, the matter of electrode potentials requires clarification.

21.2 ELECTRODE POTENTIALS

Two major factors determine the electrode potential relative to another electrode. One factor is the electrolytic solution pressure of the element, which is the tendency of an active element to send its ions into solution. At a given temperature this is a characteristic constant for a stable form of an element, but it varies if the electrode is strained mechanically or if a metastable crystalline form of the metal is present. The second factor is the activity of the dissolved ions of the element, which in turn varies with concentration at constant temperature.

Following the convention adopted by the International Union of Pure and Applied Chemistry at Stockholm in 1953,* electrode potentials are regarded as the emf of cells formed by the combination of an individual half-cell with a standard hydrogen electrode, any liquid junction potential which arises being set at zero. Thus, when the emf of each half-cell is mentioned, what is actually implied is the emf of the cell:

$$\text{Pt}, \text{H}_2(1 \text{ atm}) \mid \text{H}^+ (m = 1.228) \parallel M^{n+}(a = 1), M°$$

standard hydrogen electrode   individual half-cell
            liquid junction

The emf is divided into two contributory electrode potentials, $E°_{\text{H}^+, \text{H}_2}$ and $E°_{M^{n+}, M°}$, the cell emf being their difference. If all the substances participating in the reversible operation of the cell at a particular temperature are in their standard states, the free energy change of the cell reaction

$$\frac{n}{2} \text{H}_2(g) + M^{n+} = n\text{H}^+ + M° \qquad (21\text{-}1)$$

will have its standard value $\Delta G°$, and the emf of the cell will be the standard cell emf, $E°_{\text{cell}}$. These are related by the expression

$$\Delta G° = -nFE°_{\text{cell}} \qquad (21\text{-}2)$$

in which F is the value of the faraday. When the cell reaction is a spontaneous one, ΔG is negative; this requires the cell emf to be positive. The convention universally adopted is that the standard potential of the hydrogen electrode shall be taken as zero at all temperatures, thus setting up the standard hydrogen electrode (SHE) scale of electrode potentials. Returning to the expression for the emf of the cell, the electrode potential of the metal half-cell is equal in sign and magnitude to the electrical potential of the metallic conducting lead on the right when that of the similar lead on the left is taken as zero. The expression implies further that a reaction, as shown in Eq. 21-1, occurs when positive electricity flows through the cell from left to right or, equivalently, when electrons flow through the cell from right to left. If this is the direction of the current when the cell is short-circuited, the emf of the half-cell (a reduction) will be positive, the reaction will proceed spontaneously (a galvanic cell), and the free energy change will be negative. Standard electrode potentials for a number of selected half-cell reactions are given in Appendix A.

The individual half-cell on the right is written to represent a metal/metal ion electrode reaction. It could equally well have been written to represent another gas/ion system or ion/ion system; each of these reactions takes place at an inert metal electrode, such as gold or platinum.

---

*A complete account of the several sign conventions and the IUPAC recommendations can be found in a paper by T. S. Licht and A. J. deBéthune in *J. Chem. Educ.*, 34, 433 (1957).

When the electromotive forces of the half-cells

$$\text{Zn} \mid \text{Zn}^{2+}$$
$$\text{Ag, AgCl}(s) \mid \text{Cl}^-$$
$$\text{Pt} \mid \text{Fe}^{2+}, \text{Fe}^{3+}$$

are intended, these being oxidation reactions, the reactions implied are

$\text{Zn} \mid \text{Zn}^{2+} \parallel \text{H}^+ \mid \text{H}_2(g), \text{Pt}$	$\text{Zn}° + 2\text{H}^+ \rightarrow \text{Zn}^{2+} + \text{H}_2(g)$ (21-3)
$\text{Ag, AgCl}(s) \mid \text{Cl}^- \parallel \text{H}^+ \mid \text{H}_2(g), \text{Pt}$	$\text{Ag}° + \text{Cl}^- + \text{H}^+ \rightarrow \text{AgCl}(s) + \frac{1}{2}\text{H}_2(g)$ (21-4)
$\text{Pt} \mid \text{Fe}^{2+}, \text{Fe}^{3+} \parallel \text{H}^+ \mid \text{H}_2(g), \text{Pt}$	$\text{Fe}^{2+} + \text{H}^+ \rightarrow \text{Fe}^{3+} + \frac{1}{2}\text{H}_2(g)$ (21-5)

These electromotive forces should not be called electrode potentials, although they may be denoted oxidation potentials.

To lessen the confusion that has arisen over the years concerning the two terms—electrode potential, an observed, invariant physical quantity, and the emf of a half-reaction, which may be defined as a reduction reaction or as an oxidation reaction—it should be remembered that the two terms are distinctly different. The sign of the emf of a half-reaction depends on the direction in which the reaction is written. Only when written as a reduction reaction will the sign of the emf of the half-reaction correspond to the sign of the electrode potential.

If direct electrical measurements prove impractical, the position of a couple in the standard electromotive series may be determined by thermochemical measurements, from equilibrium studies, or from kinetic experiments that show whether the half-cell is oxidizing or reducing relative to couples of known potentials.

Effect of Concentration on Electrode Potentials

The potential E of any electrode is given by the generalized form of the Nernst equation

$$E = E° - \frac{RT}{nF} \ln \frac{a_{\text{red}}}{a_{\text{ox}}} = E° - \frac{2.3026RT}{nF} \log \frac{a_{\text{red}}}{a_{\text{ox}}} \qquad (21\text{-}6)$$

where $E°$ is the standard electrode potential, R is the molar gas constant ($8.314 \text{ J mol}^{-1}\text{K}^{-1}$), T is the absolute temperature, n is the number of electrons transferred in the electrode reaction, F is the faraday, and a_{ox} and a_{red} are the activities of the oxidized and reduced forms, respectively, of the electrode action. If concentrations are substituted for activities, common logarithms for natural logarithms, and numerical values inserted for the constants, assuming the temperature to be 25°C, the Nernst equation becomes

$$E = E° - \frac{0.05915}{n} \log \frac{[\text{red}]}{[\text{ox}]} \qquad (21\text{-}7)$$

A change of one unit in the logarithmic term changes the value of E by $59.15/n$ mV. For many analytical purposes, a system is considered quantitatively converted when 0.1% or less of the original electroactive species remains. For a metallic ion–metal system, such as the $Ag^+/Ag^°$ system,

$$E = E° + 0.0591 \log [Ag^+] \quad (21\text{-}8)$$

the value of the electrode need shift by only $3 \times 0.0591 = 0.177$ V, or in general, by $3 \times 0.0591/n$ V for a quantitative conversion. On the other hand, for an ion–ion system, such as Fe^{3+}/Fe^{2+},

$$E = E° - 0.0591 \log \frac{[Fe^{2+}]}{[Fe^{3+}]} \quad (21\text{-}9)$$

the shift would depend on the original concentration of both ions.

Effect of Complex Formation on Electrode Potentials

The effect of reagents that can react with one or both participants of an electrode process will be examined next. The simplest case involves a single ionic species formed over a range of concentrations of complexing agent. A typical example is the silver ion/silver metal couple in the presence of aqueous ammonia, where the $Ag(NH_3)_2^+$ complex ion constitutes the major ionic species in the solution phase.

The formation of the silver diammine complex is represented by the equilibrium

$$Ag^+ + 2NH_3 \rightleftharpoons Ag(NH_3)_2^+ \quad (21\text{-}10)$$

for which the formation constant is written as

$$K_f = \frac{[Ag(NH_3)_2^+]}{[Ag^+][NH_3]^2} = 6 \times 10^8 \quad (21\text{-}11)$$

For the half-reaction involving the silver ion/silver couple,

$$Ag^+ + e^- = Ag^° \quad (21\text{-}12)$$

the Nernst equation is expressed by Eq. 21-8. Combining Eqs. 21-11 with 21-8 yields the potential of a silver electrode in aqueous ammonia systems,

$$E = E° + 0.0591 \log \frac{1}{K_f[NH_3]^2} + 0.0591 \log [Ag(NH_3)_2^+] \quad (21\text{-}13)$$

The shift in electrode potential caused by the complexing agent is contained in the second term of Eq. 21-13.

For a couple involving two oxidation states of a metal in solution, such as the aquo-cobalt species,

$$Co^{3+} + e^- = Co^{2+}; \quad E° = 1.84 \text{ V} \tag{21-14}$$

in the presence of aqueous ammonia, both the cobalt(II) hexammine and the cobalt(III) hexammine species predominate. The respective formation constants are

$$\frac{[Co(NH_3)_6^{2+}]}{[Co^{2+}][NH_3]^6} = K'_f = 10^5 \tag{21-15}$$

$$\frac{[Co(NH_3)_6^{3+}]}{[Co^{3+}][NH_3]^6} = K''_f = 10^{34}$$

Substitution of these values into the Nernst equation for the cobalt system gives

$$E = E° + 0.0591 \log \frac{K'_f}{K''_f} + 0.0591 \log \frac{[Co(NH_3)_6^{3+}]}{[Co(NH_3)_6^{2+}]} \tag{21-16}$$

Here the shift in potential is a function of the ratio of the formation constants for each electroactive species. Generally, the higher oxidation state will form the more stable complex and, if it does, the shift in electrode potential will be in the negative direction.

21.3 ELECTROCHEMICAL CELLS

For a complete electrochemical cell from which negligible current is drawn, the emf is given by

$$E_{cell} = E_{ind} + E_{ref} + E_j \tag{21-17}$$

where E_{ind}, E_{ref}, and E_j are the indicator electrode, the reference electrode, and the liquid junction potentials, respectively. The indicator electrode is the sensing or probe electrode, the reference electrode is independent of the sample solution composition, and the liquid junction is an interface between dissimilar solutions. In a properly designed system, E_{ref} is a constant and E_j is either constant or negligible. When these conditions are realized, the indicator electrode can supply information about ion activities.

21.4 REFERENCE ELECTRODES

Ideally, the reference electrode is of known and constant potential with negligible variation in the liquid-junction potential from one test or calibration solution to another. Reference electrodes contain these components (Fig. 21-2): (1) the actual reference internal half-cell, usually either silver chloride/silver or calomel; (2) the salt-bridge electrolyte; and (3) a small channel in the tip of the electrode through which the salt-bridge electrolyte flows very slowly and electrical contact is made with the other components of the electrochemical cell. The potential of the internal half-cell must not be significantly altered if a small current passes through it (approximately 10^{-8} A or less); the resistance of the entire electrode must not be too great; the electrode should be easily assembled,

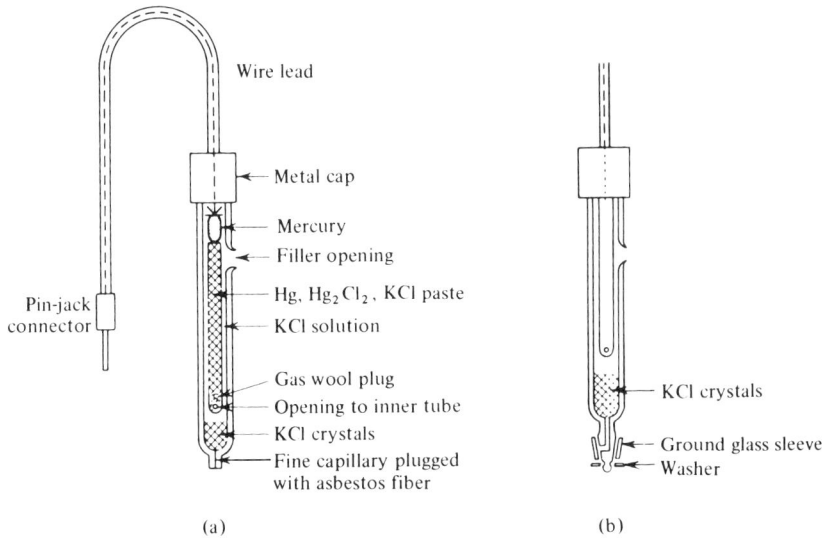

FIGURE 21-2 Calomel electrodes: (a) fiber-type and (b) sleeve-type.

and the components should be stable in contact with the atmosphere and at the operating temperature. The standard potential of the reference electrode includes the liquid-junction potential of the electrochemical cell

$$\text{Pt} \mid \text{H}_2(g), \text{H}^+(a = 1) \parallel \text{KCl solution, reference half-cell}$$

Potentials of the common reference electrodes are given in Table 21-2; a more complete compilation appears in *Lange's Handbook of Chemistry,* 12th ed., 1979.

Hydrogen-Gas Electrode

The hydrogen-gas electrode consists essentially of a piece of clean platinum foil, coated with a thin layer of finely divided platinum to hasten establishment of the electrical potential. The platinum is capable of making the reaction

$$2\text{H}^+ + 2e^- = \text{H}_2(g)$$

at the platinum–solution interface proceed reversibly. The electrode is immersed in the solution under investigation or a known reference standard and electrolytic hydrogen gas (99.8% purity adequate) at 1 atm pressure is bubbled through the solution and over the electrode in such a way that the electrode surface and the adjacent solution will be saturated with the gas at all times. Electrode life is 7-20 days after which its response becomes sluggish.[1] One construction is illustrated in Fig. 21-3.

The essential purpose of a hydrogen-gas electrode in an analytical laboratory is to check the accuracy of other reference and indicator electrodes, in particular the errors of electrode combinations. However, the hydrogen-gas electrode is also used to test the magnitude of liquid-junction potentials and the accuracy and stability of reference standard

TABLE 21-2 Potentials of Reference Electrodes in Volts as a Function of Temperature (Liquid-Junction Potential Included)

Temperature, °C	0.1M KCl Calomel[a]	Sat'd. KCl Calomel[a]	1.0M KCl Ag/AgCl[b]
0	0.3367	0.25918	0.23655
5			0.23413
10	0.3362	0.25387	0.23142
15	0.3361	0.2511	0.22857
20	0.3358	0.24775	0.22557
25	0.3356	0.24453	0.22234
30	0.3354	0.24118	0.21904
35	0.3351	0.2376	0.21565
38	0.3350	0.2355	
40	0.3345	0.23449	0.21208
45			0.20835
50	0.3315	0.22737	0.20449
55			0.20056
60	0.3248	0.2235	0.19649
70			0.18782
80		0.2083	0.1787
90			0.1695

[a] R. G. Bates et al., J. Res Natl. Bur. Std., **45**, 418 (1950).
[b] R. G. Bates and V. E. Bower, J. Res. Natl. Bur. Std., **53**, 283 (1954).

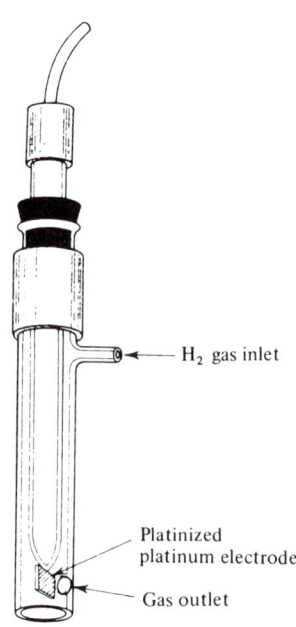

FIGURE 21-3 Hydrogen-gas electrode assembly.

solutions. It should be recalled that the hydrogen-gas electrode is the primary standard against which all other electrode potentials are measured.

Calomel Electrodes

Calomel electrodes comprise a nonattackable element, such as platinum, in contact with mercury, mercury(I) chloride (calomel), and a neutral solution of potassium chloride of known concentration and saturated with calomel. The half-cell may be represented by

$$\text{Hg} \mid \text{Hg}_2\text{Cl}_{2 \text{ sat'd}}, \text{KCl}(xM)$$

where x represents the molar concentration of potassium chloride in the solution. The saturated calomel electrode (SCE), in which the solution is saturated with potassium chloride ($4.2M$), is commonly used because it is easy to prepare and maintain. For accurate work the $0.1M$ or $1.0M$ electrodes are preferred because they reach their equilibrium potentials more quickly and their potential depends less on temperature than does the saturated type.

Construction of some commercial versions of calomel electrodes is illustrated in Fig. 21-2. A typical one consists of a tube 5–15 cm in length and 0.5–1.0 cm in diameter. The mercury–mercury(I) chloride paste is contained in an inner tube connected to the saturated potassium chloride solution in the outer tube by means of an asbestos fiber or ground glass seal in the end of the outer tubing. An electrode such as this has a relatively high resistance (2000–3000 Ω) and very limited current-carrying capacity before exhibiting severe polarization.

The saturated calomel electrode exhibits a perceptible hysteresis following temperature changes, due in part to the time required for solubility equilibrium to be established. Those designed for measurements at elevated temperatures have a large reservoir for potassium chloride crystals. Calomel electrodes become unstable at temperatures above 80°C and should be replaced with silver/silver chloride electrodes.

In measurements in which any chloride ion contamination must be avoided, the mercury(I) sulfate and potassium sulfate electrode may be used.

Silver/Silver Chloride Electrodes

The silver/silver chloride electrode consists of metallic silver (wire, rod, or gauze) coated with a layer of silver chloride and immersed in a chloride solution of known concentration that is also saturated with silver chloride. The cell formed is

$$\text{Ag} \mid \text{AgCl}_{\text{sat'd}}, \text{KCl}(xM)$$

It is a small compact electrode and can be used in any orientation. Electrode potentials are known up to 275°C.

Preparation of the silver chloride coating can be more difficult than fabrication of a calomel electrode. Silver chloride is appreciably soluble in concentrated chloride solution, necessitating the addition of solid silver chloride to assure saturation in the bridge solution, yet entailing the risk that silver chloride may precipitate at the liquid junction when it is in contact with a solution of low chloride-ion content.

In nonaqueous titration studies this electrode occupied a preeminent position for many years, although the calomel electrode can and has been employed in virtually all types of solvent systems. Reproducibility of results vary from ±10 to ±20 mV in the more aqueous solvent mixtures, to ±50 mV in the nearly anhydrous media. Special salt bridges are often necessary.

Salt Bridges and Liquid Junctions

Connection between a separate reference and indicator electrode (or anode and cathode in electrolytic cells) is usually by a junction which allows the passage of ions but does not permit the solutions to mix. Various styles of electrolyte junction have been designed: a ground glass plug or tapered sleeve, a wick of asbestos fiber sealed into glass, an agar bridge rendered conductive by an electrolyte, a porous glass plug, a dual-junction glass rivet, and a flowing junction involving a palladium annulus or capillary drip. No single type can be used in all situations.

At the boundary between two dissimilar solutions or solids, there is always a fairly high resistance that involves an appreciable ohmic drop. A junction potential is always set up. It results from the fact that the mobilities of positive and negative ions diffusing across the boundary are unequal. Because of this difference, one side of the boundary accumulates an excess of positive ions. The junction potential adds to or subtracts from the potential of the reference electrode, depending on which side of the boundary becomes positive. In making electrode measurements it is very important that this potential be the same when the reference electrode is in the standardizing solution and in the sample solution; otherwise, the change in liquid-junction potential will appear as an error in the measured electrode potential.

The junction potential is less when the ions of the electrolyte have nearly the same mobilities (see Table 26-1). In general a filling solution should be within 5% of being equitransferent. The ionic strength of the electrolyte bridge solution should be at least 5–10 times greater than the maximum ionic strength expected in the sample and standardizing solutions. The electrolytic filling solution need not be limited to only one salt. One equitransferent solution is a mixture of K^+, Na^+, NO_3^-, and a small amount of Cl^- in the appropriate ratios to satisfy the equitransferent condition. It gives a lower junction potential than saturated KCl in dilute (less than $10^{-3} M$) ionic strength samples. Above $0.5M$ (or $\mu = 0.5$), KCl is preferable. For trace Cl^- determination, saturated KNO_3 should be used; likewise, saturated K_2SO_4 for trace nitrate determination, and $5M$ lithium trichloroacetate for trace K^+ determination and solutions containing ClO_4^-.

When a saturated solution comprises one side of a boundary, the junction potential amounts to 1–2 mV under many aqueous solution conditions unless the second solution is strongly acidic or alkaline or the difference in concentration between the two sides of the junction is very large. A junction between $0.1M$ HCl and $0.01M$ HCl has a potential of 40 mV with the dilute side positive. For KCl, however, at the same concentrations, the junction potential is 1.0 mV with the dilute side negative because the mobility of K^+ is slightly less than that of Cl^-. Approximate liquid-junction potentials are given by Milazzo[2] for a number of boundary systems.

The salt-bridge electrolyte forms a continuous electrical conduction path from the internal cell to the lower tip of the electrode. It also serves to protect the internal half-cell from contamination that would cause changes in the reference electrode potential. The leakage rate of the bridge solution should be low, but satisfactory performance depends upon continuous, unimpeded, positive flow of the filling solution through the junction. The asbestos fiber and palladium annulus junction provide a small electrolyte flow, only 0.1–0.01 ml per day. Day-to-day stability of the junction potential is 2 mV for the asbestos fiber junction, 0.2 mV for the palladium annulus, and 0.06 mV for the ground glass sleeve. The asbestos fiber junction is good for general use, although it tends to clog in some media, especially colloids and suspensions. The palladium annulus is recommended for microtitrations and pH measurements for clinical solutions and applications under high pressure or vacuum. The ground glass sleeve is better for precipitation titrations, titrations in nonaqueous solvent systems, and the handling of colloids and suspensions. Double-junction, sleeve-type salt bridges overcome problems with leakage of undesirable ions into the sample solution or compatibility of filling and sample solutions (for example, nonaqueous systems). The double-junction reference electrode contains inner and outer filling solution chambers connected by a ceramic plug. The inner chamber is inside the hollow body of the electrode, and the outer chamber is formed by the body of the electrode and the outer sleeve. Both chambers can be drained and refilled quickly. This type electrode must be used if the filling solution of the single-junction electrode contains the ion being measured or an ion that interferes with the electrode response to the ion being measured. The double-junction electrode must also be used if the filling solution of the single junction contains any ion that complexes or precipitates the ion being measured or that forms a precipitate with any ion in the sample.

BIBLIOGRAPHY

Clark, W. M., *Oxidation-Reduction Potentials of Organic Systems,* Williams and Wilkins, Baltimore, 1960.
Ives, D. J. G. and G. J. Janz, *Reference Electrodes,* Academic, New York, 1961.
Latimer, W. M., *The Oxidation States of the Elements and their Potentials in Aqueous Solutions,* 2nd ed., Prentice-Hall, Englewood Cliffs, N.J., 1952.
Lingane, J. J., *Electroanalytical Chemistry,* 2nd ed., Wiley-Interscience, New York, 1958.

LITERATURE CITED

1. Perley, G. A., *J. Electrochem. Soc.,* **92,** 485 (1948).
2. Milazzo, G., *Electrochemistry,* W. Schwabl, Trans., p. 98, Springer-Verlag, Vienna, 1952.

CHAPTER 22

pH and Ion Selective Potentiometry

Potentiometric methods embrace two major types of analyses: the direct measurement of an electrode potential from which the activity (or concentration) of an active ion may be derived, the topic of this chapter; and the changes in the electromotive force brought about through the addition of a titrant, which will be discussed in Chapter 23. The field of analytical potentiometry is experiencing a renewal of substantial activity, sparked by the development of novel ion-selective electrodes which are taking their place alongside the historical pH glass electrode. The electrodes considered in this chapter fall into the following major categories:

1. Glass electrodes
2. Solid-state and precipitate electrodes
3. Liquid–liquid membrane electrodes
4. Enzyme and gas-sensing electrodes

All seem to involve an ion-exchange process in the potential-determining mechanism. Many yield potentials, within appropriate concentration limits, that can be adequately described by the classical Nernst equation or its expanded modifications.

Ion-selective electrodes measure ion activities, the thermodynamically effective free ion concentration, not concentrations. Activity measurements are valuable because the activities of ions determine rates of reactions and chemical equilibria. For example, ion activities are important parameters in predicting corrosion rates, extent of precipitation, formation of complexes, degree of acidity, solution conductivities, effectiveness of metal pickling baths and electroplating bath solutions, and physiological effects of ions in biological fluids. In dilute solutions, ion activity usually approaches the concentration.

Ion-selective electrodes have several advantages over conventional methods of determining ionic concentration. They do not affect the solution being studied; they are portable, not too expensive, and they are suitable both for direct determination of ion activities and as sensors in titrations.

22.1 GLASS-MEMBRANE ELECTRODES

The various types of ion-selective glass electrodes available are all members of a continuum of glass electrodes. To obtain desired response characteristics, a particular composition must be selected. Three subtypes of glass electrodes and their selectivity characteristics can be summarized as follows:

pH type, selectivity order: $H^+ \ggg Na^+ > K^+, Rb^+, Cs^+ \ldots \gg Ca^{2+}$

cation-sensitive type, selectivity order: $H^+ > K^+ > Na^+ > NH_4^+, Li^+ \ldots \gg Ca^{2+}$

sodium-sensitive type, selectivity order: $Ag^+ > H^+ > Na^+ \gg K^+, Li^+ \ldots \gg Ca^{2+}$

The second two subtypes also display considerable response to such univalent cations as Tl^+, Cu^+, and $R_nH_{4-n}N^+$, but are primarily responsive to univalent cations and generally are quite unresponsive to anions.

Not only the degree of electrode selectivity, but even the selectivity order can be changed with appropriate adjustment of the glass composition. As a general rule, cation selectivity (over hydrogen ion) can be achieved by adding elements whose coordination numbers are greater than their oxidation numbers [as in the substitution of aluminum(III) for silicon(IV)] to alkali metal–silicate glasses (20% Na_2O–10% CaO–70% SiO_2). Such charge-deficient elements apparently leave the glass with an excess of negatively charged ion sites which attract cations having the proper charge-to-size ratio. Glasses containing less than about 1% Al_2O_3 yield good pH electrodes with little metal-ion response (see the section "Glass Electrodes for pH Measurement"). Glasses having a composition about 27% Na_2O–5% Al_2O_3–68% SiO_2 show a general cation response. Glasses of the composition 11% Na_2O–18% Al_2O_3–71% SiO_2 are highly sodium-selective with respect to other alkali metal ions. More complex glasses containing additives frequently yield electrodes with superior mechanical and electrical properties.

At present, according to Eisenman,[1] glass electrodes are the preferred electrodes for H^+, Na^+, Ag^+, and Li^+, because of their high specificity for these ions and their excellent stability characteristics. In addition, because these electrodes function well in organic solvents they can be used in nonaqueous media, as well as in the presence of lipid-soluble or surface-active molecules. Electrode response is relatively indifferent to the type of anion present, unless it chemically attacks the glass or reacts with the cation.

The glass electrode comprises a thin-walled bulb of cation-responsive glass sealed to a stem of noncation-responsive, high-resistance glass. In this manner the cation response is confined entirely to the area of the special glass membrane, eliminating any variance caused by the depth of immersion. A typical electrode is illustrated in Fig. 22-1. If intended for use outside a shielded electrode compartment, the stem of the electrode is shielded and a shielded lead is provided which should be grounded.

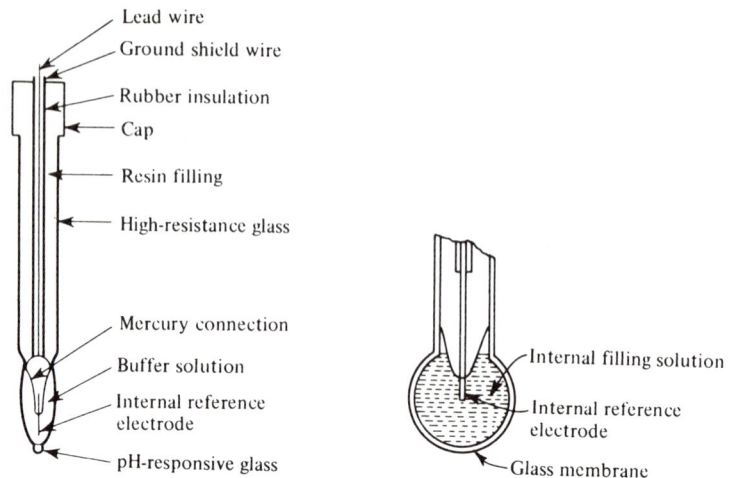

FIGURE 22-1 Construction of a glass membrane, pH-responsive electrode.

Both surfaces of the glass membrane are cation-responsive. Changes in the electrical potential of the outer membrane surface are measured by means of an external reference electrode and its associated salt bridge. An electrolyte of high buffer capacity and suitable chloride concentration fills the inside of the glass membrane; into this electrolyte dips an inner reference electrode. The complete electrochemical cell is

internal reference electrode	internal electrolyte	glass membrane	standard or unknown solution	external reference electrode

When immersed in an electrolyte, the glass electrode must be viewed as a structured continuum consisting of "dry" glass between a "swollen" or hydrated gel layer at the electrode–solution interface. A cross section through a glass membrane might be represented as

internal solution	hydrated gel layer	dry glass layer	hydrated gel layer	external solution

The dry glass layer constitutes the bulk of the membrane's thickness, about 50 μm; the hydrated layers vary in depth from 5 to 100 nm, depending upon the hygroscopicity of the glass. When the dry glass electrode is first immersed in an aqueous medium, the formation of the hydrated (external) layer causes some swelling of the membrane. Thereafter, a constant dissolution of the hydrated layer takes place with accompanying further hydration of additional dry glass so as to maintain the thickness of the hydrated layer at some steady-state value. The rate of dissolution of the hydrated layer depends on the composition of the glass and also on the nature of the sample solution. The dissolution rate largely determines the practical lifetime of the electrode; lifetimes vary from a few weeks to several years.

The mechanism by which cations affect the potential of the glass membrane is not well understood.[2] According to the presently accepted concept, it is an ion-exchange process in the gel layer of the glass membrane producing a phase-boundary potential that determines the cation response of the electrode. Only the gel layer of the glass directly participates in the equilibrium. Cationic exchange occurs only in the external part of the gel layer, which does in fact act as a semipermeable membrane to cations. The inner regions of the glass have little effect on the potential formation. The concept of an actual penetration through the glass membrane by substantial amounts of cations has been definitely disproved.

22.2 SOLID-STATE SENSORS

The nonglass, solid-state sensors replace the glass membrane with an ionically conducting membrane. Solid-state electrodes are of two main types, homogeneous and heterogeneous. Homogeneous electrodes may consist of a single crystal or pressed pellets of precipitates. They must not have internal cracks or other leakages or have too high a resistance. The LaF_3 electrode for fluoride determination is an example of the homogeneous type. Heterogeneous electrodes are formed by incorporating solid material in a supporting medium such as silicone rubber, polyvinyl chloride or other polymeric materials. They are usually simpler to prepare than homogeneous electrodes.

In the fluoride electrode, the active membrane is a single crystal of LaF_3 doped with europium(II) to lower its electrical resistance and facilitate ionic charge transport. The LaF_3 crystal, sealed into the end of a rigid plastic tube, is in contact with the internal and external solutions (Fig. 22-2). Typically, the internal solution is 0.1M each in NaF and NaCl; the fluoride ion activity controls the potential of the inner surface of the LaF_3 membrane, and the chloride ion activity fixes the potential of the internal Ag/AgCl wire reference electrode. The electrochemical cell incorporating the LaF_3 membrane electrode is

Ag | AgCl(s), Cl$^-$(0.1M), F$^-$(0.1M) | LaF_3 crystal | test solution ‖ reference electrode

It obeys a Nernst-type relation of the form

$$E = \text{constant} + \frac{RT}{F} \ln \frac{[F^-]_{int}}{[F^-]_{ext}} \quad (22\text{-}1)$$

which, because the $[F^-]_{int}$ is constant, simplifies to

$$E = \text{constant} + 0.05916\, pF \quad (22\text{-}2)$$

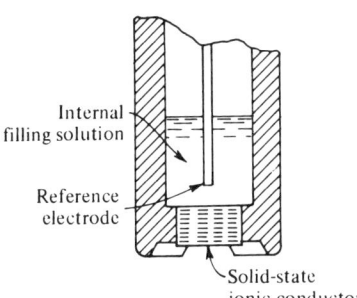

FIGURE 22-2 Cross-sectional view of solid-state sensor. (Courtesy of Orion Research, Inc.)

at 25°C. The activity calibration curve shows that the electrode follows Nernstian behavior to fluoride concentrations as low as $10^{-5} M$ and a useful, although non-Nernstian, response to at least $10^{-6} M$ fluoride ion. The fluoride electrode also responds to hydroxide ion concentration and, theoretically, a second term involving [OH$^-$] concentrations should be added to Eqs. 22-1 and 22-2. In practice the hydroxide ion concentration is kept constant with buffer solutions. One buffer which is recommended for fluoride ion determinations is known as TISAB (Total Ionic Strength Adjustment Buffer). It consists of $0.25 M$ acetic acid, $0.75 M$ sodium acetate, $1 M$ sodium chloride, and 1 mM sodium citrate (for masking Al^{3+} and Fe^{3+} which interfere by complexing F$^-$), and, as its name suggests, it controls the overall ionic strength as well as the pH.

The sulfide ion electrode has as its active element a polycrystalline Ag$_2$S membrane. If this membrane is altered from pure Ag$_2$S by dispersing within it another metal sulfide, such as CuS, CdS, or PbS, one obtains the corresponding metal-selective electrode. These electrodes transport charge by the movement of silver ions, but their potential is determined indirectly by the availability of the S^{2-} ion which, in turn, is fixed by the activity of the silver ion or divalent cation in contact with the membrane. Mixed crystals of AgX–Ag$_2$S compose the anion-selective electrodes of chloride, bromide, iodide, and thiocyanate, respectively. The solubility of the divalent metal sulfide, or the silver halide, must be greater than that of Ag$_2$S. Any of these electrodes also function as a silver-selective electrode.

Cast pellets of silver halides can also serve as the active membrane material for the respective halide-selective electrode. Selectivity is basically a function of the silver halide solubility.

The fine particles of a sparingly soluble salt can be immobilized in a coherent silicone rubber matrix[3] to form heterogeneous electrodes. The resulting membranes, prepared by the cold catalyzed polymerization of silicone rubber monomer, mixed with at least 50 weight % of the appropriate precipitate (for example, silver chloride for chloride response), have good mechanical properties and give reproducible potentials. The membrane is cemented to a glass body. Because the silicone rubber matrix has an immense resistance, the embedded precipitate must provide electrical conductivity across the membrane and this requires the precipitate particles to be in contact with each other.

A crystal membrane can be a highly selective device. Conduction in the crystalline phase proceeds by a lattice defect mechanism in which a mobile ion (usually the lattice ion with the smallest ionic radius and the smallest charge) adjacent to a vacancy defect moves into the vacancy. A vacancy for a particular ion is ideally tailored with respect to size, shape, and charge distribution to admit only the mobile ion; all other ions are unable to move and cannot contribute to the conduction process. For example, in the fluoride electrode the fluoride ion alone transports the electrical charge within the crystal, whereas in the mixed crystal–Ag$_2$S electrodes the mobile ion is the silver ion. Because no foreign ions can enter the crystalline phase, these electrodes behave as a Nernstian device within limits of concentration. Especially at lower concentrations the Nernst behavior fails due to solubility of the crystalline component. Interferences can occur, but they arise from chemical reactions at the crystal–solution interface.

The main advantage of solid-state sensors over silver metal–silver halide electrodes of the second type is their insensitivity to redox interferences and surface poisoning. In the silver metal–silver halide electrodes, the potential is established via the couple Ag/Ag^+, which of course is sensitive to oxidants and reductants.

Solid-state electrodes have a typical life of 1–2 years in the laboratory and 1–3 months when used continuously at elevated temperatures or in continuously flowing systems containing abrasive materials. Although their operating temperature range for continuous use is 0° to 80°C, they can be used intermittently in the 80° to 100°C range. When the activities in solution approach the solubility of the membrane material, the potential value converges to a limit which gives the lower detection limit. For example, the silver halide electrodes exhibit useful pH ranges up to 5, 6, and 7, respectively, for chloride, bromide, and iodide ions.

A new solid-state ion-sensitive electrode has been discovered by Bergveld,[4] who removed the metal gate from a metal oxide semiconductor field-effect transistor (MOSFET) and exposed the silicon oxide gate insulator to a solution. In the unsaturated region the drain current is proportional to the magnitude of the voltage applied to the gate metal and in the ion-sensitive transistor without the metal gate the drain current may become proportional to $(RT/nF) \ln a_1$, where a_1 is the activity of an ion in the solution. An interesting application of OSFETs (a MOSFET without the metal gate, M) is described by Moss, Janata, and Johnson[5] who replaced the metal gate with a valinomycin/plasticizer/polyvinylchloride film and thus created an OSFET sensitive to K^+ ions.

22.3 LIQUID-MEMBRANE ELECTRODES

Liquid-membrane electrodes can be classified into two types. In one the ion exchanger is drawn into a porous diaphragm which encloses the electrode body and thus separates the inner solution of the electrode from the test solution. In the second type, the ion exchanger functions as a plasticizer in a polymeric film which is sealed or otherwise fastened to the electrode body and separates the inner and test solutions. Construction of the latter type resembles, in many ways, a solid-state electrode.

The liquid-membrane electrode, using a liquid ion exchanger and a membrane, consists of a double concentric tube arrangement in which the inner tube contains the aqueous reference solution and internal reference electrode (Fig. 22-3). The outer compartment contains the organic liquid ion-exchanger reservoir which occupies the pores (100 μm) of a hydrophobic membrane. This membrane, a fluorocarbon body, is replaceable. The electrode can be taken apart to change the membrane or the ion exchanger, thus altering the electrode's ion selectivity. For example, the calcium-selective electrode uses the calcium salt of bis(2-ethylhexyl) phosphoric acid (d2EHP) dissolved in various straight-chain alcohols, or didecylphosphoric acid dissolved in di-n-octylphenyl phosphonate, as the liquid ion exchanger. The neutral, undissociated molecules of $Ca(d2EHP)_2$ diffuse easily in the solvent-saturated pores of the membrane but are insoluble in water. At the membrane–sample interface, these molecules can exchange their calcium ions for those in the sample solution. In this way the d2EHP groups can transport calcium ions back and forth across the membrane until equilibrium is established. The internal aqueous filling solution consists of a fixed concentration of calcium and chloride ions. This solution is in contact

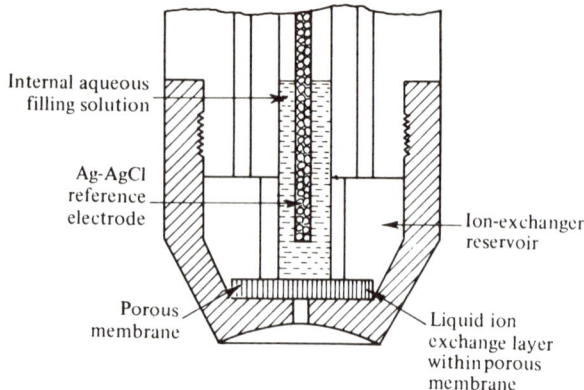

FIGURE 22-3 Construction of liquid ion-exchange electrode. (Courtesy of Orion Research Inc.)

with the ion-exchanger reservoir (via the porous membrane) and with the inner Ag/AgCl reference electrode.

Site groups of the R–S–CH$_2$–COO$^-$ type, in which sulfur and the carboxylate groups are in position to form a chelate ring with cations, show good selectivity for copper(II) and lead(II) ions. For the nitrate and fluoroborate electrodes, a substituted nickel(II)-1, 10-phenanthroline ion-pair site group is used. The corresponding iron(III) ion association complex is used in the perchlorate-selective electrode. With liquid ion exchangers the potential selectivity is largely dominated by their equilibrium ion-exchange selectivity toward cations or anions. Electrodes of this type will be only moderately selective for the ion of interest. However, considerable latitude exists for the development of numerous systems responsive either to cations or anions of varying charge.

Neutral extraction membranes are low dielectric liquids. A typical example is decane containing 10^{-4} to $10^{-7} M$ of an extractant such as valinomycin, nonactin, or monactin. These molecules are devoid of charged groups but contain an arrangement of ring oxygens energetically suitable through ion–dipole interaction to replace the hydration shell around cations. Thus, these lipid-soluble molecules are able to dissolve cations in organic solvents, forming mobile charged complexes with the cations therein, and in this way provide a mechanism for cation permeation across such normally insulating media. Neutral carriers are one of the ways used by biological membranes to discriminate among cations; and by using neutral carriers it may be possible to duplicate the exquisite selectivities characteristic of living cells in an artificial electrode. Valinomycin membranes show great potassium selectivity, about 3800 times that of sodium and much better than that observed (30:1) with the best available potassium-sensitive glass electrode. The actin-base membrane electrode is about 4 times more responsive to NH$_4^+$ than to K$^+$. Even more unusual is the 18,000 to 1 selectivity of the valinomycin membrane for potassium with respect to hydrogen ion; this means that the electrode is usable in strongly acidic media.

Liquid-membrane electrodes must be recharged every 1–3 months to replace the liquid ion exchanger and renew the porous membrane. It is desirable to have the electrode self-cleaning, and to effect this, the internal liquids of the electrode are put under mild hydrostatic pressure. The internal fluids then tend to flow out, rather than the sample fluid flowing into the electrode. This loss and the slight water-solubility of the exchanger limit the sensitivity of these electrodes to low concentrations of cations. The electrode will show a Nernstian response until the activity of the test ion is within a factor of 100 of the solubility of the liquid ion exchanger. Response time is about 30 sec in pure solutions; however, it is longer in mixtures because of the time required for the internal-membrane diffusion potential to reach steady state via adjustment in the liquid ion-exchanger concentration profile. In the presence of interfering ions, the ion-exchange process must establish new ion activities in the surface and this is not necessarily an instantaneous process. The temperature of operation must be kept within limits specified by the manufacturer, generally between $0°$ and $50°C$, so that water does not permeate the membrane and membrane liquids do not bleed excessively into the aqueous solution.

22.4 GAS-SENSING AND ENZYME ELECTRODES

Although the ammonia electrode is a gas-sensing electrode, it is used just as if it were a selective-ion electrode. Its construction is shown in Fig. 22-4. Dissolved ammonia from the sample diffuses through a gas-permeable fluorocarbon membrane until a reversible equilibrium is established between the ammonia level of the sample and the internal filling solution. Hydroxide ions are formed in the internal filling solution by the reaction of ammonia with water: $NH_3 + H_2O = NH_4^+ + OH^-$. The hydroxide level of the internal filling solution is measured by the internal sensing element and is directly proportional to the level of ammonia in the sample. Samples and standards are adjusted to a fixed pH, or to $pH > 11$. The ammonia electrode, unlike glass and other ammonium ion electrodes, is almost totally free from interferences, although volatile amines may interfere. Anions,

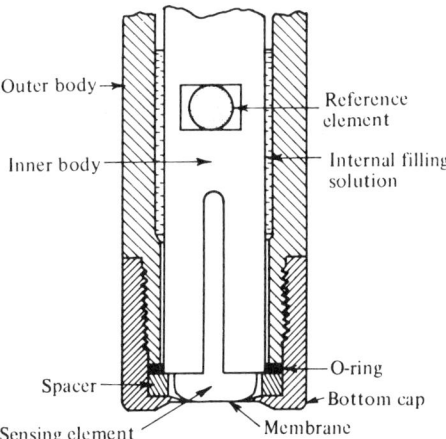

FIGURE 22-4 Construction of the ammonia gas-sensing electrode. (Courtesy of Orion Research, Inc.)

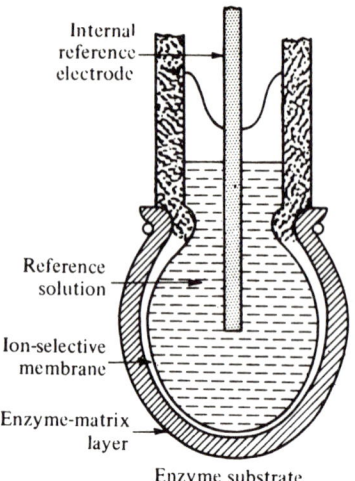

FIGURE 22-5 Enzyme electrode.

cations, and common gases, such as carbon dioxide, sulfide, cyanide, and sulfur dioxide do not interfere. Sensitivity extends from 10^{-6} to $1M$, that is, from 0.017 to 17,000 μg/ml.

The basic arrangement of the enzyme electrode is shown in Fig. 22-5. The enzyme is immobilized in a gel layer which coats a conventional cation-sensitive glass electrode. For the ammonium ion-selective electrode, the enzyme urease is fixed in a layer of acrylamide gel held in place around the glass electrode bulb by porous nylon netting or a thin cellophane film. The urease acts specifically upon urea in the sample solution to yield ammonium ions which diffuse through the gel layer and are sensed by the electrode. The resulting potential is proportional to the substrate (urea) concentration in the sample solution. There are thousands of enzyme–substrate combinations that would yield products measurable with ion-selective electrodes. With some modifications, this electrode system can be reversed; that is, the substrate would surround the glass membrane, resulting in an enzyme-sensing electrode.

22.5 INTERFERENCES

Selective-ion electrodes are subject to two types of interferences: method interference and electrode interference. Method interferences occur when some characteristic of the sample prevents the probe from sensing the ion of interest. For example, a fluoride electrode can detect only fluoride ions. In acid solution, however, fluoride forms complexes with the hydrogen ion and is thereby masked from the fluoride detector.

Electrode interferences arise when the electrode responds to ions in the sample solution other than the ion being measured. The selectivity coefficient is an index of the ability of an electrode to measure a particular ion in the presence of another ion. In liquid- and glass-membrane electrodes, interference occurs when an interfering ion passes into the membrane just as does the ion being measured. As an example, barium ions will be transported across the liquid membrane of the calcium electrode if the activity of barium ion in the sample is sufficiently high and the activity of calcium ion is very low. Under these conditions the electrode potential will depend to some extent on the barium ion activity. The electrode response incorrectly attributed to the test ion's activity a, whose ionic charge is z, is related to the selectivity coefficient k_i by the equation

$$a = k_i a_i^{(z/z_i)} \tag{22-3}$$

where a_i is the background level of the interference whose ionic charge is z_i. Thus a selectivity coefficient of 0.01 means that the response attributable to Ca^{2+} alone at $a_{Ca^{2+}}$ would be matched in the absence of calcium by barium at an activity 100 times larger.

The interference mechanism that occurs in crystal electrodes is quite different and requires a different method for expressing the selectivity coefficient. Surface reactions can convert one of the components of the solid membrane to a second insoluble compound. As a result, the membrane loses sensitivity to the ion being measured. For example, thiocyanate ion can interfere with bromide ion measurements if the reaction

$$SCN^- + AgBr(s) \rightarrow AgSCN(s) + Br^- \tag{22-4}$$

takes place, which it will begin to do if the ratio of the thiocyanate ion activity to the bromide ion activity exceeds the value given by the ratio of the solubility products of silver thiocyanate to silver bromide, or

$$\frac{1}{k_i} = \frac{a_{SCN}}{a_{Br}} = \frac{1.00 \times 10^{-12}}{5.0 \times 10^{-13}} = 2.0 \tag{22-5}$$

Failure to give a Nernstian response also occurs in dilute solutions approximating the solubility of the membrane material; this is the limiting value of the cell emf regardless of dilution.

Solid electrodes can also give problems if they are used in samples containing a species that forms a very stable complex with one of the component ions of the crystal. For example, citrate ion forms with lanthanum ion a very stable citrate complex, which has the effect of increasing the solubility of the membrane and thereby increasing the lower limit of detection of the fluoride ion. More drastic is the effect of cyanide ion in contact with a mixed silver halide/silver sulfide membrane. In this case the reaction proceeds virtually to completion with the consumption of silver halide from the membrane. However, if an appropriate diffusion barrier is placed on the membrane surface, then a steady state is quickly established in which the cyanide level at the crystal surface is virtually zero, whereas the iodide (or other halide) level is very nearly one-half the sample cyanide concentration. The silver ion activity at the membrane surface is fixed by the iodide level via the solubility product of silver iodide. As a result the electrode depends in an almost Nernstian manner on the sample cyanide ion concentration. It has been found that the porous silver sulfide matrix left behind as the silver iodide is consumed provides an almost ideal diffusion barrier; thus an interference can be turned to an advantage. The cyanide electrode has a lifetime of several months of continuous operation.

22.6 ION-ACTIVITY EVALUATION METHODS

The direct-measurement technique requires a single potentiometric measurement on the sample solution. The sample's millivolt reading from an expanded scale pH/mV meter is compared to a previously prepared calibration curve, or to the sample concentration or activity read directly from the meter scale of a calibrated specific ion meter. Calibration procedures must use solutions of known activity or concentration, depending on which parameter is required. Below 10^{-3}–$10^{-4} M$ the two are practically indistinguishable. Direct-measurement techniques are useful where samples are essentially pure solutions of the ion sensed, or have a relatively high and constant total ionic strength. To swamp effects caused by variations in total ionic strength, the ionic strength of the sample solution and the calibrating solutions may be adjusted by adding a high level of a noninterfering ion, for example, $1M$ KNO_3, when making halide determinations. Naturally the ion sensed must possess a large selectivity coefficient relative to possible interfering ions.

An approach, similar to the operational definition of pH values, can be applied to the problem of measuring the activities of other ions in solution. Based on procedures set forth by Bates and Alfenaar,[6] standard reference values for pNa and pCl in sodium chloride solutions, pCa in calcium chloride solutions, and pF in sodium fluoride solutions are suggested in Table 22-1. The uncertainty in activity of an ion from an emf measurement of ±0.2 mV amounts to approximately 0.75% for a monovalent ion and 1.5% for a divalent ion. Because of their freedom from junction potentials, cation-sensitive glass electrodes can yield a higher degree of precision with respect to emf measurements.

Potential sources of errors to be considered in ion-activity evaluation methods include response time, temperature coefficient and drift of the electrode. Solid-state sensors usually have response times in the 10–100 msec range, although Ag_2S responds in about 1 msec. Liquid exchanger or neutral carrier sensors may have response times of several seconds or even minutes.

TABLE 22-1 Suggested Reference Standard Values, pI,[a] at 25°C

Material	Molality, mole kg^{-1}	pNa	pCa	pCl	pF
NaCl	0.001	3.015	—	3.015	—
	0.01	2.044	—	2.044	—
	0.1	1.108	—	1.110	—
	1.0	0.160	—	0.204	—
NaF	0.001	3.015	—	—	3.015
	0.01	2.044	—	—	2.048
	0.1	1.108	—	—	1.124
CaCl$_2$	0.000333	—	3.537	3.191	—
	0.00333	—	2.653	2.220	—
	0.0333	—	1.887	1.286	—
	0.333	—	1.105	0.381	—

[a] pI = log [I].

The temperature coefficients are generally predictable from the general Nernst equation. Silver sulfide electrodes have a coefficient of -0.40 mV/°C in $AgNO_3$ but $+0.05$ mV/°C in $0.1M$ Na_2S and $0.1M$ NaOH.

The drift of ion-selective electrodes varies greatly. For liquid membranes a figure of 2 mV per day is not unusual and thus one should recalibrate electrodes daily.

In direct-measurement methods the relative determination error, at 25°C, is given by

$$\% \text{ error} = 100 \Delta c_1 / C_1 \approx 4/z/\Delta E \tag{22-6}$$

where ΔE is the error in millivolts of the potential measurement and z is the charge of the ion to be determined. The accuracy can be increased by use of the *known increment* (or *decrement*) method.

In the *known increment* (or *decrement*) methods, another designation for the method of additions, the concentration of a specific ion sample is estimated by observing the change in electrode potential when a known incremental (or decremental) change is made in concentration of the ion in the sample. This approach requires neither preparation of a calibration curve nor calibration of logarithmic scales with standard solutions. The ion is added to the test solution in a known amount that changes the total concentration by a known amount ΔC, but does not change the total ionic strength appreciably or the fraction of the total concentration that is free. Thus, the initial reading is taken on a sample C_1, and the electrode response is

$$E_1 = \epsilon + S \log(\gamma_1 C_1) \tag{22-7}$$

After the incremental addition

$$E_2 = \epsilon + S \log[\gamma_1 (C_1 + \Delta C)] \tag{22-8}$$

Combining equations,

$$\Delta E = E_2 - E_1 = S \log\left(\frac{C_1 + \Delta C}{C_1}\right) \tag{22-9}$$

where S is the Nernstian factor or emf/pC slope. Known addition and subtraction methods are particularly suitable for samples with high unknown total ionic strength. Where the species being measured is especially unstable, known subtraction is preferred over known addition. Computers can be used to control the addition of a standard solution, read the potential after each addition, and even calculate the end point by fitting the data by the method of least squares to a plot like Eq. 22-9 for several additions of standard solution.[7]

Ion-selective electrodes can be used to monitor and record concentrations in flowing solutions; however, high flowrates may distort the membranes, especially liquid membranes, and cause errors in the measurement. Furthermore, the response time of the electrode limits the flowrate, especially if large changes of concentration are occurring in the stream.

22.7 THE MEASUREMENT OF pH

The pH scale is a series of numbers that express the degree of acidity (or alkalinity) of a solution, as contrasted with the total quantity of acid or base in some material as found by an alkalimetric (or acidimetric) titration. As defined by Sørensen, who introduced the term,

$$\text{pH} = -\log [\text{H}^+] \qquad (22\text{-}10)$$

Note that what is involved is the negative logarithm of the hydrogen-ion concentration expressed in molarity. However, it is the activity of the hydrogen ion that is formally consistent with the thermodynamics of the pH electromotive cell, and the activity definition is

$$\text{paH} = -\log a_{\text{H}^+} \qquad (22\text{-}11)$$

Now $[\text{H}^+]$ and $f_\pm[\text{H}^+]$ are often the most useful units for expressing the acidity of aqueous solutions, where $f_\pm$ is the mean ionic activity coefficient. Unfortunately, the established experimental pH method cannot furnish either of these quantities. Consequently, the term pH is merely a mathematical symbol of convenience, widely accepted, but devoid of exact thermodynamic validity. For those interested in a detailed treatment of the historical development of the concept of pH, the treatise by Clark and the more recent work by Bates, both cited in the Bibliography, should be consulted.

The acidity of a solution will depend upon several factors: (a) the chemical nature of the acid, as expressed by the degree of dissociation (or association of a base), pK_a; (b) the relative concentrations of acid and its conjugate base, and the total ionic strength of the solution; and (c) the temperature of the solution as it affects the dissociation of water and the dissociation of the acid. For acids which are not dissociated completely, the expression for the pH of the solution is

$$\text{pH} = pK_a + \log \frac{[\text{A}^-] + [\text{H}^+]}{[\text{HA}] - [\text{H}^+]} \qquad (22\text{-}12)$$

Except when the hydrogen-ion concentration is comparable to the concentrations of HA or A^-, the expression can be simplified to

$$\text{pH} = pK_a + \log \frac{[\text{A}^-]}{[\text{HA}]} \qquad (22\text{-}13)$$

Buffer Solutions

A *buffer* may be defined as a solution which maintains a nearly constant pH value despite the addition of substantial quantities of acid or base. Generally it consists of a mixture of an incompletely dissociated acid and its conjugate base. In selecting a particular buffer, three characteristics should be considered: the buffer value β, the dilution value $\text{pH}_{1/2}$, and the change of pH with change in temperature, $\Delta \text{pH}/\Delta T$.

The Van Slyke buffer value β indicates the resistance of a buffer to change in pH upon addition of an acid or base. It is defined as

$$\beta = \frac{\Delta B}{\Delta \text{pH}}$$

where ΔB is an increment of completely dissociated base (or acid) in gram-equivalents per liter which produces a change of ΔpH in the solution. In the selection of a buffer system, pK_a should be as close as possible to the desired pH. Under this condition the ratio $[A^-]/[HA]$ in Eq. 22-13 is close to unity, and the buffer value will be large. For high buffering capacity, the concentrations of the buffering components should be high, yet consistent with considerations of ionic strength of the medium and the concomitant effect upon pH measurements.

The pH of the buffer solution should also be relatively insensitive to changes in the total concentration of the buffer components at a fixed ratio of $[A^-]/[HA]$. The dilution value is defined as the change of pH that results from a 1:1 dilution of the solution with pure water.

Solutions of specified composition for many pH values are compiled in the treatises by Clark and Bates. In the Clark and Lubs series, which spans the range from pH 1.0 to 10.2, the ingredients are phthalic acid ($pK_1 = 2.90$, $pK_2 = 5.41$), potassium dihydrogen phosphate ($pK_2 = 7.13$), and boric acid ($pK = 9.24$), which are combined in suitable proportions with hydrochloric acid or sodium hydroxide. MacIlvaine's standard buffer spans the range from pH 2.2 to 8.0 and involves mixing citric acid ($pK_1 = 3.09$, $pK_2 = 4.75$, $pK_3 = 5.50$) and potassium dihydrogen phosphate solutions in certain proportions. In the last example, combining several acids of varying strength, but whose pK_a values differ by less than two units, with the respective conjugate bases provides a universal buffer solution which covers a wider range of pH values than any single system, yet a solution which exhibits considerable buffering capacity over the entire useful range. Buffer tablets, available from chemical supply houses, eliminate the preparation, storage, and mixing of buffering ingredients, and need only be dissolved in the specified volume of pure water to obtain the pH value specified on the container.

Operational Definition of pH

The pH value is defined for an aqueous solution in an operational manner, according to the Bates–Guggenheim convention, as

$$\text{pH} = \text{pH}_s + \frac{E - E_s}{2.302 \, RT/F} \qquad (22\text{-}14)$$

In this definition, T is the temperature in degrees Kelvin, and E and E_s are, respectively, the emf of an electrochemical cell of the usual design

| electrode reversible to hydrogen ions | unknown or standard (s) buffer solution | salt bridge | reference electrode |

which contains first the "unknown" solution, and secondly, a standard reference solution of known pH, namely, pH_s.

The NBS pH standards were assigned pH_s values from measurements of the emf of cells containing hydrogen gas and silver-silver chloride electrodes (that is, without a liquid junction):

$$Pt \mid H_2(1\text{ atm});\ H^+\ Cl^-(\text{plus}\ K^+\ Cl^-);\ AgCl(s) \mid Ag$$

by the equation

$$E = E° - 0.000198\,T\ \log f_{H^+} f_{Cl^-} m_{H^+} m_{Cl^-} \qquad (22\text{-}15)$$

where $E°$ is the standard potential of the cell.[6,7] Upon rearranging Eq. 22-15 in terms of the acidity function, $p\,(a_{H^+} f_{Cl^-})$

$$p(a_{H^+} f_{Cl^-}) = -\log f_{H^+} f_{Cl^-} m_{H^+} = \frac{E - E°}{0.000198\,T} + \log m_{Cl^-} \qquad (22\text{-}16)$$

The pH_s of the chloride-free buffer solution is computed from the equation

$$pH_s = p(a_{H^+} f_{Cl^-})° + \log f°_{Cl^-} \qquad (22\text{-}17)$$

where $p(a_{H^+} f_{Cl^-})°$ is the value obtained by evaluation of $p(a_{H^+} f_{Cl^-})$ at several concentrations of chloride, and extrapolation to zero chloride concentration. The activitiy coefficient of chloride ion is given by the equation

$$-\log f°_{Cl^-} = \frac{A\sqrt{\mu}}{1 + 1.5\sqrt{\mu}} \qquad (22\text{-}18)$$

where μ is the ionic strength, which should be maintained equal to, or less than, 0.1, and A is a parameter of the Debye-Hückel theory having a different value at each temperature. The recommended values of pH_s are summarized in Table 22-2. The total uncertainty in pH_s, exclusive of any liquid junction potentials introduced during calibration of pH equipment, is estimated as 0.005 pH unit (0°-60°C) and 0.008 pH unit (60°-95°C). The necessity of estimating the individual activity coefficients of chloride ion in each reference solution deprives the pH_s value of exact fundamental meaning. Nevertheless, the operational definition of pH, chosen in part for its reasonableness but largely for its utility, agrees as closely as possible with the mathematical concepts evolved from the present state of solution theory.

Interpretation of Measured pH

The operational definition of pH emphasizes that the determination of pH is essentially a determination of a difference of emf as recorded in a pH cell containing first a reference buffer and then a test solution. The definition demands only that the electrode potential of the reference electrode remains constant while measurements of E and E_s are being made. Unfortunately, the definition makes no allowances for the presence of a liquid-junction potential or a change in the value of the junction potential when the reference standard is replaced by an unknown solution. Hopefully, the liquid-junction potential will remain constant from one measurement to another, and its value is combined with the

TABLE 22-2 National Bureau of Standards Reference pH_s Buffer Solutions

Temp., °C	Secondary Standard, 0.05M K Tetroxalate	KH Tartrate (Sat'd. at 25°C)	0.05M KH_2 Citrate	0.05M KH Phthalate	0.025M each KH_2PO_4 Na_2HPO_4	0.008695M KH_2PO_4 0.03043M Na_2HPO_4	0.01M $Na_2B_4O_7$	0.025M each $NaHCO_3$ Na_2CO_3	Secondary Standard, $Ca(OH)_2$ (Satd. at 25°C)
0	1.666		3.863	4.003	6.984	7.534	9.464	10.317	13.423
5	1.668		3.840	3.999	6.951	7.500	9.395	10.245	13.207
10	1.670		3.820	3.998	6.923	7.472	9.332	10.179	13.003
15	1.672		3.802	3.999	6.900	7.448	9.276	10.118	12.810
20	1.675		3.788	4.002	6.881	7.429	9.225	10.062	12.627
25	1.679	3.557	3.776	4.008	6.865	7.413	9.180	10.012	12.454
30	1.683	3.552	3.766	4.015	6.853	7.400	9.139	9.966	12.289
35	1.688	3.549	3.759	4.024	6.844	7.389	9.102	9.925	12.133
40	1.694	3.547	3.753	4.035	6.838	7.380	9.068	9.889	11.984
50	1.707	3.549	3.749	4.060	6.833	7.367	9.011	9.828	11.705
60	1.723	3.560		4.091	6.836		8.962		11.449
70	1.743	3.580		4.126	6.845		8.921		
80	1.766	3.609		4.164	6.859		8.885		
90	1.792	3.650		4.205	6.877		8.850		
95	1.806	3.674		4.227	6.886		8.833		
Buffer value, β	0.070	0.027	0.034	0.016	0.029	0.016	0.020	0.029	0.09
Dilution value, $\Delta pH_{1/2}$	+0.186	+0.049	+0.052	+0.052	+0.080	+0.07	+0.01	+0.079	−0.28

SOURCE R.G. Bates, *J. Res. Natl. Bur. Std.*, 66A, 179–183 (1962); B.R. Staples and R.G. Bates, ibid., 73A, 37 (1969).
NOTE: Numbers given are "conventional" pH values. Properties of these buffer solutions are included at the foot of each column.

value of the reference electrode. However, at pH values less than 2 or greater than 12, and for ionic strengths greater than 0.1, the reproducibility of the liquid-junction potential is seriously impaired and errors as large as several tenths of a pH unit can result. To detect any serious impairment of the response of the measuring device and electrode assembly outside the pH range 2–12, the tetroxalate solution and the calcium hydroxide solution are included among the pH reference buffers, but are designated secondary standards.

For pH measurements with an accuracy of 0.01 to 0.1 pH unit, the limiting factor is often the electrochemical system; that is, the characteristics of the electrodes and the solution in which they are immersed. Another source of error is due to temperature, for not only does the proportionality factor between cell emf and pH vary with temperature, but dissociation equilibria and junction potentials also have significant temperature coefficients. For accuracy of ±0.01 pH unit, the temperature should be known to ±2°C. Ideal solutions are those with compositions that match closely the primary standards of reference. Specifically, they are aqueous solutions of buffers and simple salts, of ionic strengths between 0.01 and 0.1, with only low concentrations of nonelectrolytes. In industrial processes, fortunately, a highly accurate knowledge of the pH of a solution is seldom required. Neither is it necessary to know exactly what a particular pH value means. It is sufficient to know that at a certain stage in an industrial process a particular pH value is maintained.

22.8 GLASS ELECTRODES FOR pH MEASUREMENT

Typical high-quality pH-sensitive glass membranes are chiefly lithium silicates with lanthanum and barium ions added. These ions act as lattice "tighteners" to retard silicate hydrolysis and lessen alkali ion, chiefly sodium ion, mobility. Lithium ions are the bulk mobile charge carriers under an applied electric field. After the membrane is soaked in water, the surface layer is depleted of Li^+, which is replaced by H^+. Virtually all surface silicate anion sites are neutralized by H^+ ions. Content of H^+ decreases in a complex way with increasing distance into the membrane, while Li^+ content increases in such a way that the sum of positive ions, charge carriers, and other cations, balance the presumed uniform fixed-site concentration.

The activity of water in the solution appears to play an important role in the development of the pH response of the glass membrane. If the ionic strength is extremely high, or if a nonaqueous solvent is present, the measured potential deviates from the expected value. A direct relationship between hygroscopicity of the glass and pH response has been demonstrated. All glass electrodes must be conditioned for a time by soaking in water or in a dilute buffer solution, even though they may be used subsequently in media that are only partly aqueous.

The glass electrode displays an amazing versatility. Involving no electron exchange, it is the only hydrogen ion electrode uninfluenced by oxidizing and reducing agents. Nor is it disturbed by common electrode poisons. However, the glass membrane reaches equilibrium with the test solution slowly (response time is normally several seconds) and the surface of the glass is easily contaminated by adsorbed ions and particulate matter which delay the attainment of equilibrium between electrode and solution. The high electrical resistance (5–500 MΩ) necessitates measuring circuits with high input impedance, and is accompanied by a large temperature coefficient of resistance which changes exponentially with temperature. Chemical durability and electrical resistance are linked together. Electrodes durable against chemical attack at elevated temperatures have excessive electrical resistances when the temperature is lowered. Conversely, electrodes that are robust at low temperatures will corrode rapidly in solutions at high temperatures. Consequently, electrodes are designed specifically for certain ranges of temperature and for certain ranges of pH. Frequently a general-purpose electrode is useful from $-5°$ to $60°C$ in acids and dilute alkalis with a negligible error to a pH of 11. In more alkaline solutions the observed values of pH are too small and must be corrected from nomographs supplied by the manufacturer (see Problem 22-1). The positive alkaline error is due to the partial exchange of cations, other than hydrogen ion, between the glass membrane and the solution. In general, the error will be large when the test solution contains a univalent cation in common with the glass membrane. For measurements above $60°$–$80°C$, a special glass is used that will withstand $100°C$ continuously, with intermittent use up to $130°C$; however, below $35°C$ these glasses are sluggish in response.

The negative acid error is due to the change in the activity of the water in the gel layer. The gradual dissolution of the outermost layer of glass may also account for the error in strongly acid solution by preventing formation of a steady-state potential.

A glass electrode exhibits a reasonably rapid response to rapid and wide changes of pH in buffered solutions. However, valid readings are obtained more slowly in poorly buffered or unbuffered solutions, particularly when changing to these from buffered solutions, as after standardization. The electrodes should be thoroughly washed with distilled water after each measurement and then rinsed with several portions of the next test solution before making the final reading. Poorly buffered solutions should be vigorously stirred during measurement, otherwise the stagnant layer of solution at the glass–solution interface tends toward the composition of the particular kind of pH-responsive glass. Suspensions and colloidal material should be wiped from the glass surface with a soft tissue.

Commercial glass electrodes are fabricated in a wide variety of sizes and shapes and for many special applications. Syringe and capillary electrodes (Fig. 22-6) require only 1 or 2 drops of solution, even as little as 1 mm^3 volume in ultramicro work, whereas others will penetrate soft solids or are designed for measurements on smooth surfaces. The normal-size electrode operates with a volume of solution from 1 to 5 ml. Polyalcohols added to the solution in the reference electrode, and mercury inside the glass membrane to make direct contact with the glass, permit measurements of semifrozen materials at $-30°C$.

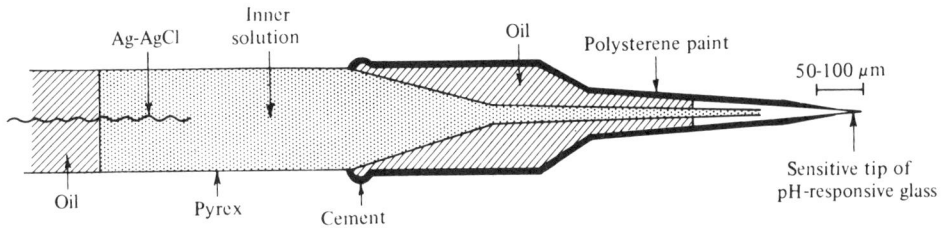

FIGURE 22-6 Cross-sectional view of a glass microelectrode.

Glass electrodes of special construction are available for operation under pressure conditions. Combination glass indicator and reference electrodes as a single unit are shown in Fig. 22-7. The outer cylinder contains the electrolyte for the salt bridge of the reference electrode and surrounds the usual glass-electrode assembly except for the pH-sensitive bulb.

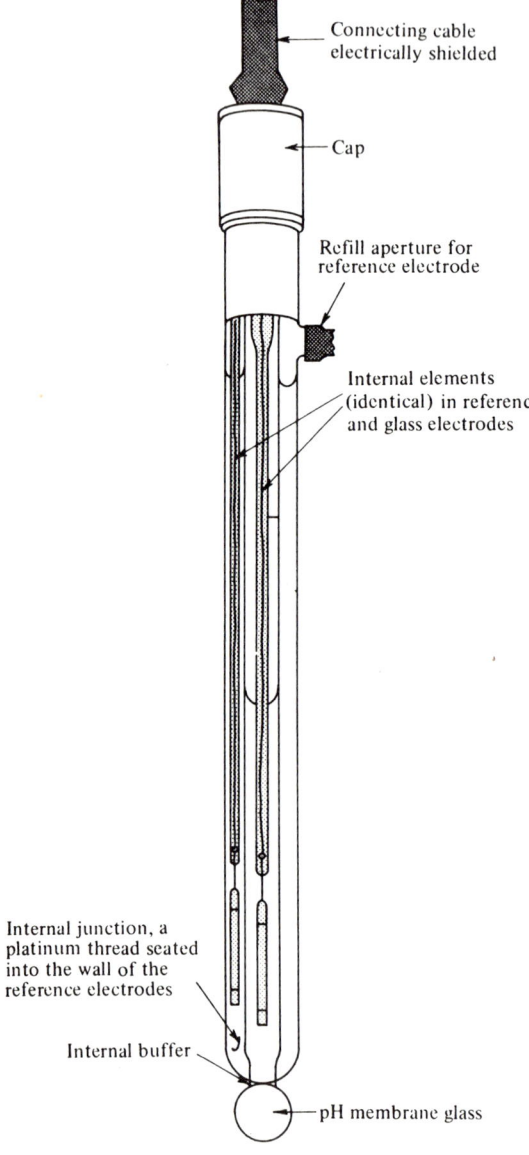

FIGURE 22-7 Combination pH/reference electrode. (Courtesy of Sargent-Welch.)

22.9 ELECTROMETRIC MEASUREMENT OF pH AND pI

To achieve a reproducibility of ±0.005 pH or pI unit, the assigned limit of certainty of many reference buffer and ion-activity standards and including unavoidable variations in liquid-junction potentials, an instrument is needed that will be sensitive and reproducible to at least 0.2 mV. Negligible current must be drawn during measurement if changes in the ion concentration at the electrode surface are to be avoided, and no error is to arise from the voltage drop across the inherent resistance of the electrochemical cell. With glass, solid-state, and membrane electrodes the current drawn should be 10^{-12} A or less. This restricts the choice of instrument to a high-impedance electronic voltmeter.

Null-Balance pH Meters

A schematic diagram of the first pH meter to be marketed is shown in Fig. 22-8. It incorporates an ordinary potentiometric circuit with a null-balance amplifier circuit. The role of each of the adjustments and circuit components perhaps can best be appreciated in terms of the sequence of operations involved in the standardization and calibration of the instrument.

1. With the ganged switch in position 1, the grid of the input stage is connected to the contactor on slidewire R_1. The contactor is moved until the meter needle stands in its center position. Note that R_1 is the center portion of a potentiometer circuit which

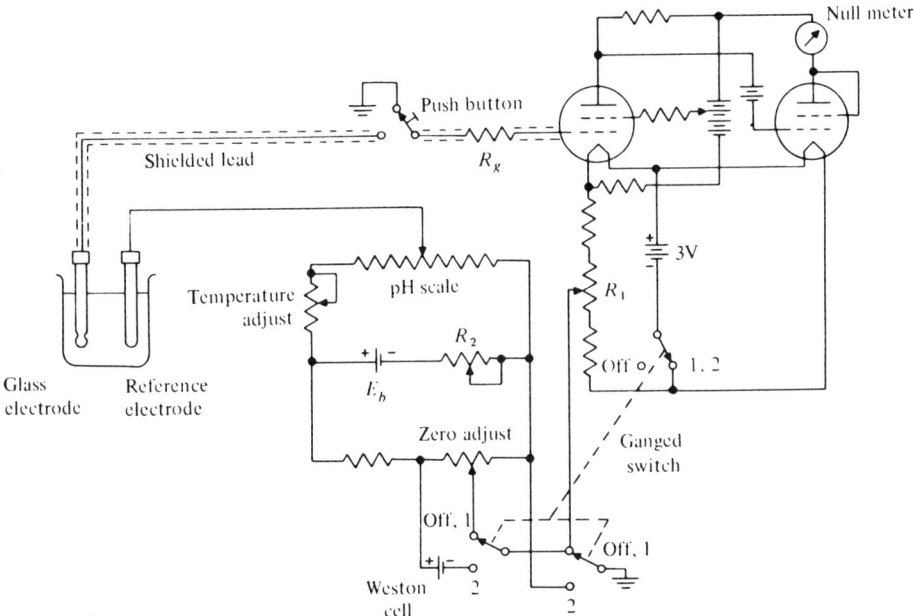

FIGURE 22-8 Schematic circuit diagram of the Beckman Model G pH meter, a null-balance amplifier circuit. (Courtesy of Beckman Instruments, Inc.)

involves the same dry cells that supply the current through the filaments of the tubes. This operation selects an arbitrary "null" position on the meter, analogous to the mechanical adjustment of a D'Arsonval galvanometer, and compensates for aging of the batteries and changes in tube characteristics.

2. The rheostat marked *temperature adjust*, which shunts the *pH scale* slidewire, is set to the value of the solution temperature, thereby altering the denominator of the second term in Eq. 22-14. With the ganged switch in position 2, one side of the Weston standard cell is attached to the cathode of the input tube (through the contactor on R_1) and the grid of this tube is attached to the pH slidewire. Actually the Weston cell is in opposition to the potentiometer involving the pH slidewire, battery E_b, and rheostat R_2, with the amplifier serving as the current-measuring galvanometer. The rheostat R_2 is adjusted until the meter returns to its center position. Now the slidewire scale is standardized in the correct number of millivolts per inscribed pH unit.

3. In the final step, the electrode assembly is immersed in a pH reference buffer and the pH slidewire is set at the value of the pH standard. With the ganged switch in position 1, a button is depressed which connects the glass electrode to the grid of the input tube. All this time the reference electrode has been connected to the contactor on the pH slidewire. Now the *zero adjust* rheostat (also called the *asymmetry control*) is adjusted until the meter returns to its center position. The inscribed pH scale is thereby brought into juxtaposition with the actual pH value by the zero adjust resistance, which serves to lengthen one end of the pH scale while shortening by an equal amount the other end. Any changes in the asymmetry potential of the glass electrode are also compensated.

The slidewire in these instruments is also marked off in units of 100 mV and can be used in potentiometric methods other than pH measurements. The limiting factor in measurements with null-detector circuits is the slidewire accuracy, which is generally at least 0.1%. Because these circuits are subject to zero drift and require circuit readjustments frequently, they are unsuited to long-time unattended operation.

Direct-Reading pH and pI Meters

Meters with indicating scales in pH values are calibrated in voltage units for a glass–reference electrode pair on the basis of the relationship for the emf of a pH cell, described by

$$E = k - KT \text{(pH)} \tag{22-19}$$

the equation of a straight line with slope $-KT$ and a zero intercept of k. The electrode and its reference electrode must possess an isothermal point at 0 V. This is accomplished by using as the internal solution in the glass electrode a buffer whose pH change with temperature exactly compensates the temperature changes of the internal and external reference electrodes. The proper emf/pH slope involves adjusting the KT factor (actually $2.3026RT/F$) to 59.16 mV per pH unit at 25°C by means of a slope control to rotate the emf/pH slope about the isothermal point. The temperature compensator, which is reserved to correct the slope for the actual temperature of the sample, varies the instrument definition of a

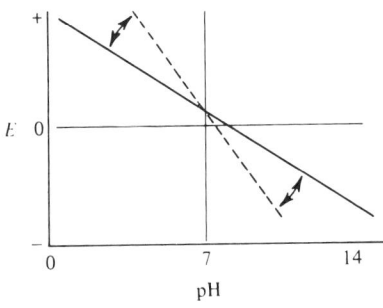 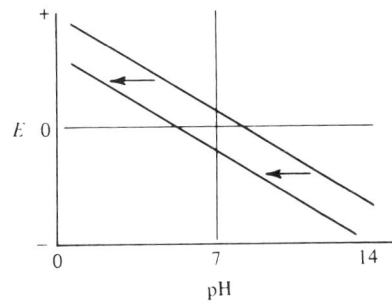

FIGURE 22-9 Operation of the (a) *slope* and (b) *intercept* controls shown schematically.

pH unit from 54.20 mV at 0° to 66.10 mV at 60°C. Finally the meter scale is brought into juxtaposition with the pH value of a standard reference buffer whose pH value most closely approximates that of the test solution. The intercept (asymmetry) control effectively shifts the response curve laterally until it passes through the isothermal point. Figure 22-9 schematically illustrates these operations. A check with a second reference pH buffer, such that the two buffers bracket the pH of the test solution, should be done to ensure proper functioning of the cell assembly and measurement equipment and to verify conformity of the pH response with the theoretical Nernst slope. In effect, then, the pH value of the test solution is determined by interpolation.

A direct-reading instrument has few manipulative steps and is adaptable to continuous recording or control of industrial operations or processes. Temperature compensation can be provided by causing the feedback current to flow through a temperature-sensitive resistor (thermistor) immersed in the test solution. A useful feature found in some meters is expanded ranges covering 0.5, 1, or 2 pH units over the full meter scale, thus permitting expanded range readings to 0.001 pH unit.

For ion-activity measurements, readout is required in terms of activity. This requires that the meter be equipped not only with slope and intercept calibrating controls but also with a logarithmic scale and, in the event of a divalent ion, the means for halving the slope ($n = 2$). Alternatively, an expanded millivolt scale and a calibration curve on semilog paper may be used. A scale with infinity at the center allows concentration to be read directly when using the known addition and subtraction techniques with selective ion electrodes.

PROBLEMS

1. The sodium ion correction nomograph for general-purpose glass electrodes manufactured by Beckman Instruments, Inc., is shown in the figure. (a) Estimate the corrected pH for these measurements:

Solution number	A	B	C	D
pH scale reading	13.50	11.25	12.00	12.10
Temperature, °C	25	30	25	40
[Na^+]	0.05	0.2	0.1	0.02

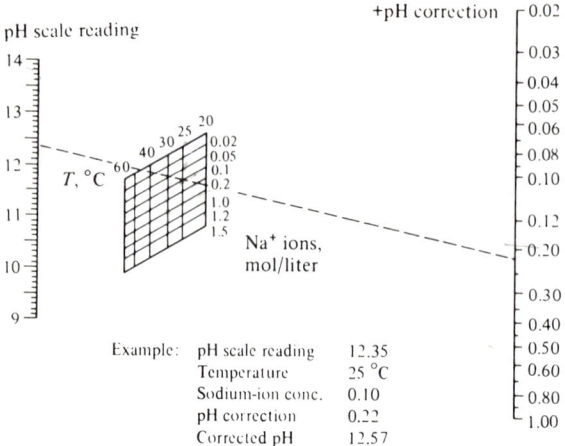

PROBLEM 22-1

(b) Below what pH value is the pH correction less than 0.02 pH unit at [Na⁺] = 0.2? At [Na⁺] = 1.0? (c) To render negligible the pH correction at pH 10.9, below what value must the sodium ion concentration be maintained?

2. Why is it necessary to protect carefully a potassium acid phthalate buffer solution from contamination with acids or alkalis?

3. Would the error in pH_s be significant if potassium acid tartrate solution were not completely saturated?

4. Why are alkaline reference buffer solutions subject to larger errors due to temperature changes than are acid buffers?

5. Why does the pH_s value of borax buffer solutions change so considerably with temperature? (*Hint*: Consider polymeric species of boric acid.)

6. Although no nitrite-selective electrode is available, suggest an indirect method to measure nitrite ion activity.

7. A fluoride solid-state electrode has a selectivity coefficient of 0.10 relative to hydroxide ion. At $10^{-2}M$ fluoride concentration, what hydroxide ion concentration could be tolerated?

8. What should be the lower pH limit when using the fluoride electrode if a 1% error is to be tolerated, and samples and standards are not adjusted to the same pH value?

9. Estimate the ratios at which the following ions can be present without impairing the response of the solid-state bromide electrode: chloride, iodide, hydroxide, and cyanide ions.

10. If calcium ion activity is to be measured with a liquid-membrane electrode in samples containing up to $0.7M$ sodium ion, estimate the minimum level of calcium which

can be measured under these conditions. Assume a minimum level of 5% interference. The selectivity coefficient for sodium ion is 1.0×10^{-4}.

11. What is the maximum concentration of interfering anions that can be tolerated for a 1% interference level when measuring $10^{-5} M$ BF_4^- with a fluoroborate liquid ion-exchange membrane electrode? The interfering anions and their selectivity coefficients are: OH^-, 10^{-3}; I^-, 20; NO_3^-, 0.1; HCO_3^-, 4×10^{-3}; SO_4^{2-}, 1×10^{-3}.

12. What is the total sulfide concentration in 100 ml of sample that gives a potential reading of −845 mV before the addition of 1 ml of $0.1 M$ $AgNO_3$ and a reading −839 mV after the addition?

BIBLIOGRAPHY

Bates, R. G., *Determination of pH: Theory and Practice*, 2nd ed., Wiley, New York, 1973.
Britton, H. T. S., *Hydrogen Ions*, 4th ed., D. Van Nostrand, New York, 1956.
Clark, W. M., *The Determination of Hydrogen Ions*, 3rd ed., Williams and Wilkins, Baltimore, 1928.
Durst, R. A., Ed., *Ion-Selective Electrodes*, National Bureau of Standards Spec. Publ. 314, Washington, D.C., 1969.
Eisenman, G., R. G. Bates, G. Mattock, and W. M. Friedman, *The Glass Electrode*, Wiley-Interscience Reprint, New York.
Kolthoff, I. M., *Acid-Base Indicators*, 4th ed., Macmillan, New York, 1937.
Koryta, J., "Ion-Selective Electrodes," Cambridge Monographs in Physical Chemistry, No. 2, Cambridge University, New York, 1975.
Lakshminarayanaiah, N., *Membrane Electrodes*, Academic, New York, 1976.
Lingane, J. J., *Electroanalytical Chemistry*, 2nd ed., Wiley-Interscience, New York, 1958.
Midgley, D. and K. Torrance, *Potentiometric Water Analysis*, Wiley, New York, 1979.
Orion Research, Inc., *Analytical Methods Guide*, 7th ed., Orion Research, Inc., Cambridge, Mass., May 1975.
Weissberger, A. and B. W. Rossiter, Eds., *Physical Methods of Chemistry*, Vol. 1, Part IIA, Wiley-Interscience, New York, 1971.

LITERATURE CITED

1. Eisenman, G., in *Ion-Selective Electrodes*, R. A. Durst, Ed., Natl. Bur. Std. Publ. 314, Washington, D. C., 1969.
2. Durst, R. A., *J. Chem. Educ.*, **44**, 175 (1967); see also Koryta, J., "Ion-Selective Electrodes" listed in Bibliography.
3. Pungor, E., *Anal. Chem.*, **39** (13), 28A (1967).
4. Bergveld, P., *I.E.E.E. Trans.*, BME-19, 342 (1972).
5. Moss, S. D., J. Janata, and C. C. Johnson, *Anal. Chem.*, **47**, 2238 (1975).
6. Bates, R. G. and M. Alfenaar, "Activity Standards for Ion-Selective Electrodes," in *Ion-Selective Electrodes*, R. A. Durst, Ed., Natl. Bur. Std. Spec. Publ. 314, Washington, D. C., 1969.
7. Ariano, J. M. and W. F. Gutnecht, *Anal. Chem.*, **48**, 281 (1976).

CHAPTER 23

Potentiometric Titrations

The measurement of the potential of an appropriate indicator electrode has been used for many years as a method of detecting the equivalence point in a variety of titrations. When a potentiometric titration is being performed, interest is focused upon changes in the emf of the electrochemical cell as a titrant of precisely known concentration is added to a solution of the test element. The method can be applied to any titrimetric reaction for which an indicator electrode is available to follow the activity of at least one of the substances involved. Reproducible equilibrium is of little concern here. Requirements for reference electrodes are greatly relaxed; it is only necessary that the response of one member of a pair of electrodes be substantially greater or faster than that of the other. In addition to the establishment of the equivalence point of a reaction, further information about the sample and its reactions may be obtained by the complete recording of a potentiometric titration curve.

The chief advantages of potentiometric titrations are applicability to turbid, fluorescent, opaque, or colored solutions, or when suitable visual indicators are unavailable or inapplicable. A succession of equivalence points in the titration of mixtures can be followed. In contrast to the direct potentiometric methods discussed in Chapter 22, potentiometric titrations generally offer an increase in accuracy and precision at the cost of increased time and difficulty. Accuracy is increased because measured potentials are used to detect rapid changes in activity that occur at the equivalence point of the titration, and this rate of emf change is usually considerably greater than the response slope which limits precision in direct potentiometry. Furthermore, it is the change in emf versus titrant volume rather than the absolute value of the emf which is of interest. Thus, the influences of liquid junction potentials and activity coefficients have little or no effect.

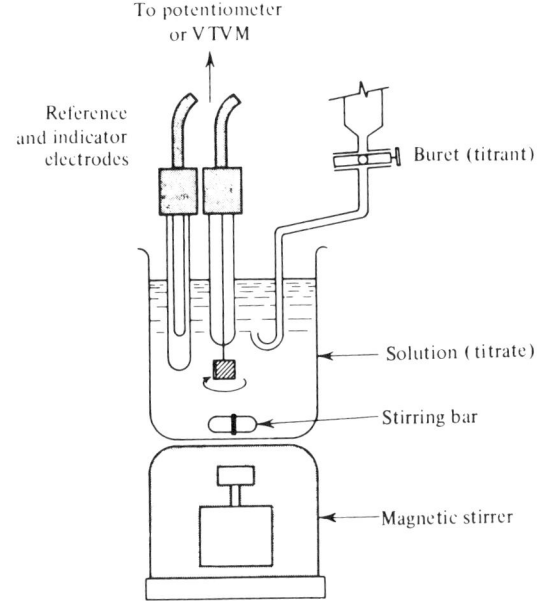

FIGURE 23-1 Equipment for potentiometric titrations.

Equipment needed to carry out a classical potentiometric titration is illustrated in Fig. 23.1. In certain instances simplifications in equipment are possible.

23.1 CLASSIFICATION OF INDICATOR ELECTRODES

Electrodes of the First Kind

Electrodes of the first kind are reversible with respect to the ions of the metal phase. The electrode is a piece of metal in contact with a solution of its ions, for example, silver dipping into a silver nitrate solution. One interface is involved. For the half-cell

$$Ag^+ + e^- = Ag \qquad (23\text{-}1)$$

the Nernst expression is

$$E = 0.799 - 0.0591 \log \frac{1}{[Ag^+]} \qquad (23\text{-}2)$$

By convention, the activity of the pure massive metal (or any solid phase) is taken as unity.

The metal must be thermodynamically stable with respect to air oxidation, especially at low ion activities. In neutral solutions, suitable electrodes are restricted to Hg_2^{2+}/Hg and Ag^+/Ag. If oxygen is removed from the solution by deaeration, other electrodes become feasible: Cu^{2+}/Cu, Bi^{3+}/Bi, Pb^{2+}/Pb, Cd^{2+}/Cd, Sn^{2+}/Sn, Tl^+/Tl, and Zn^{2+}/Zn.

Simple amalgam electrodes are also electrodes of the first kind. For zinc, the reaction at the electrode is

$$Zn^{2+} + Hg + 2e^- = Zn(Hg) \tag{23-3}$$

Electrodes of the Second Kind

Electrodes of the second kind involve two interfaces, such as metal coated with a layer of one of its sparingly soluble salts. The underlying electrode must be reversible. Consider a silver wire coated with a thin deposit of silver chloride. At the Ag/AgCl solution interface the electrochemical equilibrium is

$$AgCl(s) + e^- = Ag + Cl^- \tag{23-4}$$

In addition, there is a chemical equilibrium

$$AgCl(s) = Ag^+ + Cl^-; \quad K_{sp} = 1.8 \times 10^{-10} \tag{23-5}$$

Combining these two equations, we arrive at the Nernst expression for Eq. 23-4:

$$E = 0.799 + 0.0591 \log K_{sp} - 0.0591 \log [Cl^-] \tag{23-6}$$

This simplifies to

$$E = 0.222 - 0.0591 \log [Cl^-] \tag{23-7}$$

Electrodes of the second kind can be used for the direct determination of the activity of either the metal ion or the anion in the coating and also as an indicator electrode to follow titrations involving either. Limitations on these electrodes are severe. They can be used only over a range of anion activities such that the solution remains saturated with respect to the metal coating. Interferences from other anions can occur if they too form an insoluble salt with the cation of the underlying electrode.

Electrodes of the Third Kind

Reilley and co-workers[1,2] showed how to use an electrode of known reversibility to measure activities of ions for which no electrode of the first kind exists. They used a small mercury electrode (or gold amalgam wire) in contact with a solution containing metal ions to be titrated with a chelon Y, such as EDTA. A small added quantity of mercury(II) chelonate, HgY^{2-}, saturated the solution and established the half-cell:

$$Hg \mid HgY^{2-}; \; MY^{(n-4)+}; \; M^{n+}$$

where the electrode potential is given by

$$E = E° + \frac{0.0591}{2} \log \frac{[M^{n+}] [HgY^{2-}]}{[MY^{(n-4)+}]} \tag{23-8}$$

Because a fixed amount of HgY^{2-} is present, the potential is dependent upon the ratio $[M^{n+}]/[MY^{(n-4)+}]$. The species HgY^{2-} must be considerably more stable than $MY^{(n-4)+}$.

Redox Electrode

The redox electrode, usually gold, platinum, or carbon, immersed in a solution containing both the oxidized and reduced states of a homogeneous and reversible oxidation-reduction system, develops a potential proportional to the ratio of the two oxidation states. The only role of the redox electrode is to provide or accept electrons. An example is platinum in contact with a solution of iron(III) and iron(II) ions. For the half-reaction

$$Fe^{3+} + e^- = Fe^{2+}; \quad E° = 0.771 \text{ V} \tag{23-9}$$

the Nernst expression is

$$E = 0.771 - \frac{0.0591}{1} \log \frac{[Fe^{2+}]}{[Fe^{3+}]} \tag{23-10}$$

Platinum electrodes are unsuitable for work with solutions containing powerful reducing agents, such as chromium(II), titanium(III), and vanadium(II) ions, because platinum catalyzes the reduction of hydrogen ion by these reductants at the platinum surface. Consequently, the interfacial electrode potential will not reflect the changes in the composition of the solution. In these cases a small pool of mercury can be used as the electrode because of the high overpotential associated with the deposition of hydrogen gas on a mercury surface.

In many redox titrations, the inert electrode is not reversible for one of the half-reactions, as in the case for thiosulfate in iodometric titrations. However, if the nonreversible system attains chemical equilibrium quickly with a reversible system (for example, with I_2, I^-), the latter will serve as the potential-determining half-reaction. When chemical equilibrium is attained more slowly, mixed potentials may be involved; these bear no simple relationship to the activities of the reacting species although a stable equilibrium potential is rapidly attained. Sometimes the inert redox electrode behaves more or less like an oxygen electrode toward dissolved oxygen. If properly preconditioned, it may exhibit a memory for particular systems.

Ion-Selective Electrodes

Ion-selective electrodes (Chapter 22) can also be used as indicating electrodes. Their usefulness can be extended to the determination of species for which ion-selective electrodes are not available by using the appropriate selective ion as the titrant. An example is the determination of lithium by precipitation of LiF in alcohol, the potentiometric titration being followed with the fluoride electrode.

23.2 LOCATION OF THE EQUIVALENCE POINT

The critical problem in a titration is the recognition of the point at which the quantities of reacting species are present in equivalent amounts—the *equivalence point*. The

TABLE 23-1 Potentiometric Titration Data in Vicinity of End Point[a]

Volume of Titrant, ml	Cell emf, pH units	$\Delta pH/\Delta V$	$\Delta^2 pH/\Delta V^2$
19.50	6.46		
		1.30	
19.60	6.59		+1.00
		1.40	
19.70	6.73		+2.00
		1.60	
19.80	6.89		+9.35
		3.00	
19.85	7.04		+52.0
		5.60	
19.90	7.32		+124.0
		11.80	
19.95	7.91		−136.0
		5.00	
20.00	8.16		−24.0
		3.80	
20.05	8.35		−5.00
		3.40	
20.10	8.52		
		1.60	
20.20	8.68		

[a] Volume at end point = 19.90 + 0.050[124/(124 + 136)] = 19.92 ml.
pH at end point = 7.32 + 0.59[124/(124 + 136)] = 7.60.

titration curve can be followed point by point, plotting as ordinate successive values of the cell emf versus the corresponding volume (or milliequivalents) of titrant added as abscissa. Additions of titrant should be the smallest accurately measured increments that provide an adequate density of points across the pH (or emf) range. Typical data are gathered in Table 23-1. Over most of the titration range the cell emf varies gradually, but near the end point the cell emf changes very abruptly as the logarithm of the concentration(s) undergoes a rapid variation. The resulting titration curve will resemble Fig. 23-2a. The problem in general is to detect this sharp change in cell emf that occurs in the vicinity of the equivalence point. The equivalence point may be calculated, as will be outlined later. Usually the analyst must be content with finding a reproducible point, as close as possible to the equivalence point, at which the titration can be considered complete—the *end point*. By inspection the end point can be located from the inflection point of the titration curve: the point which corresponds to the maximum rate of change of cell emf per unit volume of titrant added. Distinctness of the end point increases as the reaction involved becomes more nearly quantitative. Once the cell emf has been established for a given titration, it can be used to indicate subsequent end points for the same chemical reaction.

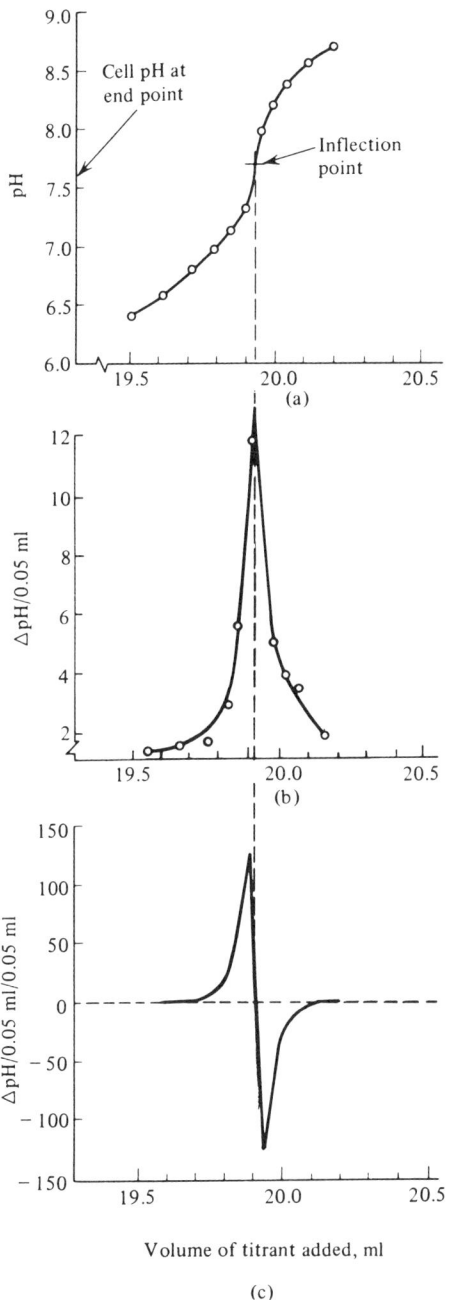

FIGURE 23-2 Titration curves for data in Table 23-1. (a) Experimental titration curve, (b) first derivative curve, and (c) second derivative curve.

In the immediate vicinity of the equivalence point the concentration of the original reactant becomes very small, and it usually becomes impossible for the ion or ions to control the electrode potential. The cell emf will become unstable and indefinite because the indicating electrode is no longer poised; that is, it is not bathed with sufficient quantities of each electroactive species of the desired redox couple. If the electroactive species are not too dilute, a drop or two of titrant will suffice to carry the titration through the equivalence point and into the region stabilized by the electroactive species of the titrant. However, solutions more dilute than $10^{-3} M$ generally do not give satisfactory titration end points unless special procedures are employed.

An end point may be located more precisely by plotting successive values of the rate of change of cell emf versus each increment of titrant in the vicinity of the inflection point. Increments need not be equal but should not be too large or too small. The position of the maximum on the first derivative curve, Fig. 23-2b, corresponds to the inflection point on the normal titration curve. Once the end point volume is known, the corresponding cell emf at the end point can be obtained from the original titration curve. The end point can be even more precisely located from the second derivative curve, Fig. 23-2c, which is obtained by plotting the cell emf-volume acceleration versus the volume of titrant added. At the end point the second derivative becomes numerically equal to zero as the value of the ordinate rapidly changes from a positive to a negative number. Although either of these methods of selecting the end point is too laborious to do manually for each titration, they become feasible with appropriate electronic circuits (Fig. 23-10).

Oftentimes, tabulation of titration data will suffice to locate the end point by interpolation without the necessity of constructing derivative curves. In Table 23-1, the first two columns are original data. The third and fourth columns are the calculated values which would be used to plot the first and second derivative titration curves, respectively. A simple mathematical method for arriving at the pH (or emf) and volume of titrant at the end point is also outlined.

Particularly in acid–base titrations, a titration of a solution prepared like the sample solution, omitting only the sample itself (that is, a blank), should also be run. In Fig. 23-3, curve B is a conventional titration curve with volume of titrant plotted against pH. Only the inflection at pH 8.0 is immediately apparent. The titration of the blank, curve C, is shown below. When the blank curve is subtracted volume-wise from the sample curve, the resulting curve A shows clearly the presence of two more inflections at pH values of 2.0 and 10.5. The preparation of blank curves is also desirable when handling relatively dilute solutions, when a minor constituent is suspected in the sample, or when impurities are present in the solvent.

Gran[3] proposed that plots of $\Delta ml/\Delta E$ versus ml of titrant be made to locate the equivalence point. Such plots are linear just before and after the end point, having a V-shape. Special graph paper is available on which the electrode potential (in mV) is plotted on the vertical (antilogarithmic) axis versus volume of titrant added on the horizontal linear axis. Only four or five points are necessary to define each segment of the curve; the intersection (extrapolated) of the two segments locates the end point. In some titrations the lines become curved some distance from the end point.

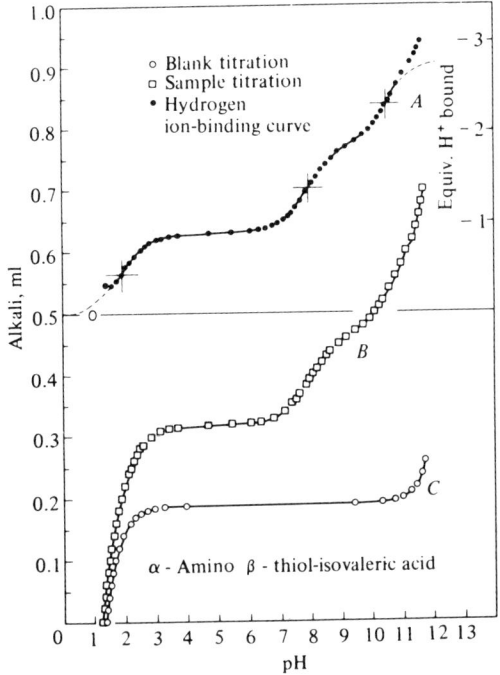

FIGURE 23-3 Titration curves for α-amino-β-thiolisovaleric acid. (T. V. Parke and W. W. Davis, *Anal. Chem.*, 26, 642 (1954). Courtesy of American Chemical Society.)

Use of Two Indicating Electrodes

Variation of the difference in electrode potential of two indicating electrodes can be followed sometimes during a titration. Generally only changes in the cell emf, but not the actual value, will be provided, but this is sufficient for many titration purposes. Elimination of the usual reference electrode and its attendant salt bridge eliminates leakage of the bridge electrolyte and minimizes the liquid-junction potential, which is desirable when solutions possessing high resistance (true for many nonaqueous systems) are involved.

A simple indicating electrode for reference purposes consists of a platinum (or other type) electrode inserted into the delivery tip from the buret, the buret electrode,[4] or inside a capillary tube which contains a small portion of the original solution, the shielded electrode.[5] These electrodes, pictured in Fig. 23-4, assume a definite though not predictable potential and, because each is in contact with a solution of unchanging composition, maintain a constant potential which may be used for reference throughout the titration. Paired with a second indicating electrode dipping into the main solution, the usual S-shape titration curve is obtained. A platinum electrode serves as reference for most types of titrations. An antimony electrode in the buret paired with an antimony indicating electrode is useful for a strongly basic titrant such as sodium aminoethoxide in basic nonaqueous solvents.

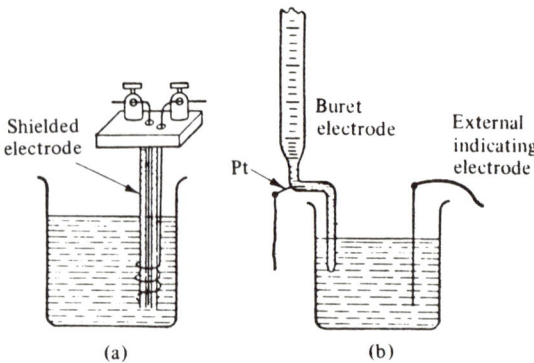

FIGURE 23-4 Systems consisting of two indicating electrodes. (a) Shielded capillary reference electrode and (b) buret electrode.

The glass electrode may be employed as an electrode of reference in a medium of fixed, or buffered, pH or rather high hydrogen ion concentration. Paired with a silver electrode, the combination finds use in argentometric titrations because leakage from the salt-filling solution of a reference electrode is eliminated. In redox titrations, the glass electrode can be paired with platinum.

A graphite or tungsten electrode placed in the main solution serves well as a reference electrode, although these elements tend to reduce the amplitude of the change of cell emf by reason of their own response curves. This is insignificant in the case of a graphite/platinum electrode pair in neutralization reactions, and the tungsten/platinum pair in many redox systems. With two fast chemical half-reactions, a differential-shaped titration curve is obtained, but when one reaction is slow in establishing its equilibrium at the graphite or tungsten surface, a distorted S-shaped curve is obtained. An informative study of the tungsten electrode is available.[6]

An interesting system is provided by the pair: platinum and platinum/10% rhodium electrodes. Each acts as an indicating electrode of the second kind in neutralization reactions, presumably because of a thin layer of platinum oxide on each surface which renders each electrode responsive to hydrogen ions, in addition to the usual response to redox systems exhibited by an "inert" electrode. Whereas the platinum electrode responds rapidly to changes in solution composition, the platinum/10% rhodium electrode lags ever so slightly. Consequently, if the titrant is added rapidly and uniformly, the pair will exhibit a maximum difference in response at the equivalence point when the logarithmic term in the Nernst expression is changing most rapidly. A first-derivative curve is obtained. This pair of electrodes proves useful with automatic differential titrators.

Constant Current Potentiometry

Differential electrolytic potentiometry involves the observation of potentials across two indicator electrodes in a stirred solution during the passage of a minute, highly stabilized current. The apparatus required is a pair of wire electrodes, a resistor, a battery, and a meter (Fig. 23-5).

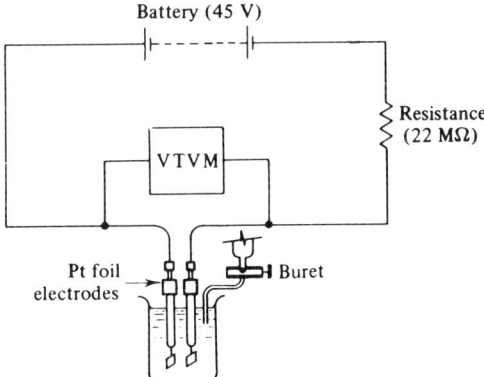

FIGURE 23-5 Equipment for conducting titrations at constant electrolysis current (approximately 2 μA with values of voltage and resistance indicated).

The electrode chosen must be appropriate to the reaction involved. Glass electrodes are not suitable. The electrode area should be kept small, 0.5 cm^2 or less, and should be of equal area for reversible reactions. For irreversible reactions there is some advantage in making the inactive electrode smaller. The current density required for optimum differentiation depends on the equilibrium constant of the reaction and especially on the concentration of the titrant. For a 0.1N titrant it is about 1 μA cm^{-2}, and it decreases with decreasing titrant concentration. Good stabilization of the current is required. The ballast load, the product of the ballast resistance in ohms, and the source voltage in volts, should exceed 10^9 and preferably 10^{10}. At low values, potentials become unsteady and erratic, and current fluctuations also occur. At very high values, response times increase and the Johnson noise in the ballast resistor increases.

The shape of the titration curve can be predicted from current–voltage curves.[6] For example, consider the titration of copper(II) with EDTA.

$$Cu^{2+} + H_2Y^{2-} = CuY^{2-} + 2H^+ \qquad (23\text{-}11)$$

wherein the indicator electrodes are polarized with the small current indicated by the horizontal dashed line in Fig. 23-6, which contains the pertinent current–voltage curves. The initial potentials adopted by the cathode and anode will be the values at the intersections of the horizontal current line with the current–voltage curves; namely, E and F, respectively. The cell emf is then the difference, $F - E$. Curves 1 through 3 represent the Cu^{2+}/Cu$^\circ$ system at the beginning of the titration, when the titration is 91% complete, and when it is 99% complete. The difference in potential between each succeeding curve (from E to D, and from D to C) is 29.6 mV, as would be expected from the Nernst expression. The place where each curve intersects the zero-current axis (points $E, D,$ and C) is the "null potential," as measured by the usual zero-current technique. However, in the vicinity of the equivalence point the mass transfer characteristics of copper(II) ions up to the cathode surface must be considered. A concentration gradient is established and the current carried through the cell by copper(II) ions becomes limited (curves 4 through 6). As soon as the current-carrying ability (the limiting current plateau) of the copper system

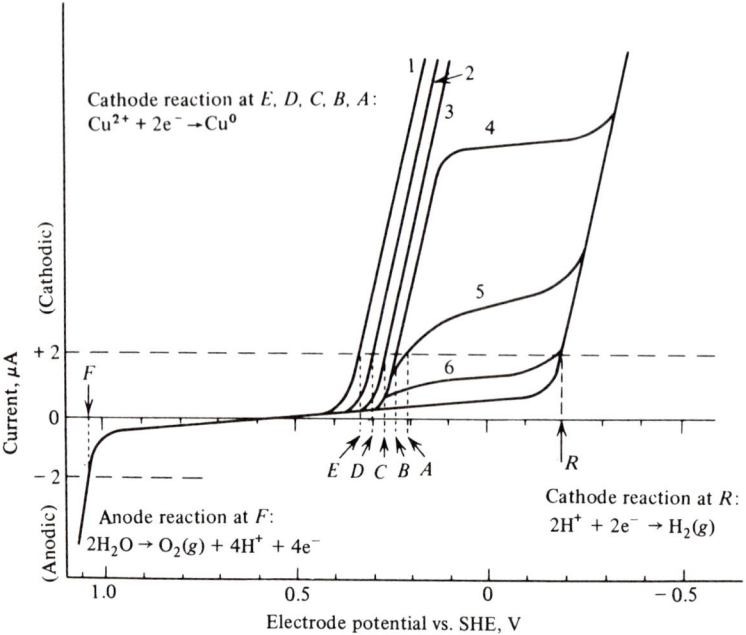

FIGURE 23-6 Schematic current–voltage curves of copper(II)/(0) system in the presence of hydrogen ions. The potential scale is approximately real; the current scale is arbitrary.

falls below the electrolysis current being forced through the circuit, the potential of the cathode shifts quickly from point A to a value set by any other redox system which is able to poise the electrode, as, for example, the reduction of hydrogen ions at point R. All this time the other indicating electrode has maintained a constant potential because the anode reaction at its surface is the oxidation of water (EDTA is not electroactive). The appearance of the titration curve is shown by curve 1 in Fig. 23-7.

When the titrant also possesses a set of current–voltage curves which lie within the boundary conditions imposed by the solvent and supporting electrolyte, the titration curve resembles curve 2 in Fig. 23-7. Such a titration would be iron(II) and cerium(IV). Prior to the equivalence point the cathode potential is established by the Fe(III)/Fe(II) system. After the equivalence point the cathode potential becomes stabilized and the anode becomes the indicating electrode for the Ce(IV)/Ce(III) system. Titration curves take the form of the first derivative of the zero-current indicator electrode curve. The anode curve is displaced along the volume axis so that its potential rise occurs before that of the zero-current electrode, while the cathode curve is displaced in the opposite direction.

The vertical displacement of the two curves, the differential potential, then traces a sharp peak. The magnitudes of the anode lead and cathode lag are proportional to the differentiating current density. With electrodes of equal area, reversible reactions, and equal diffusion coefficients, the lead and lag are equal and a symmetrical differential curve is obtained. However, the end point and equivalence point usually do not coincide because of imperfection in reversibility of the electrode processes. Also electrolysis is

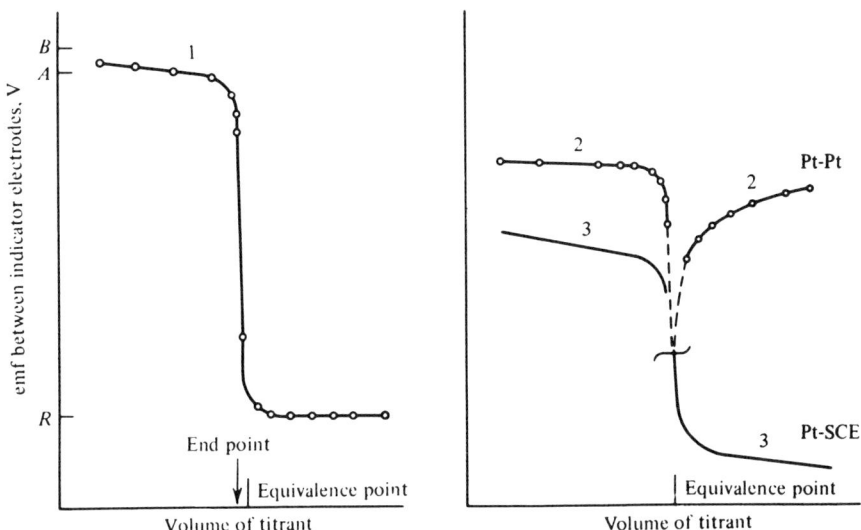

FIGURE 23-7 Titration curves using constant electrolysis current. Curve 1: Copper(II) titrated with EDTA (data from Fig. 23-6); Curve 2: Iron(II) titrated with cerium(IV) at finite current flow; and Curve 3: Iron(II) titrated with cerium(IV) under zero-current condition.

occurring continuously at both electrodes; often the products are removed from the solution; for example, the copper(II) ions in the first example would be reduced to metallic copper at the cathode. The titration error will be a function of the electrolysis current and the time taken for the titration. Furthermore, the end point will be premature by an amount that is a function of the magnitude of the electrolysis current in comparison with the concentration of unreacted test element, which gives rise to the limiting current of identical value. The titration error can be minimized by setting the electrolysis current at as small a value as possible and providing rapid stirring to improve the rate of mass transfer to (and away from) the electrode surfaces.

A major advantage of titrations at constant current is that only one electroactive system needs to be present, either the titrant or the test element. In fact, no advantage accrues in use of the method for two electroactive systems which establish steady potentials rapidly. For the latter case the largest potential change occurs prematurely with polarized electrodes, whereas it occurs exactly at 100% of the equivalence-point volume when no electrolysis current is flowing.

Automatic Titrators

When performed manually so as to give a detailed titration curve or merely to locate precisely an end point, a potentiometric titration is a tedious and time-consuming operation. For routine analyses the method does not have the speed and simplicity of comparable procedures employing visual indicators. Automatic equipment for performing, and if desired, recording, the titration curve in its entirety provides a logical solution to the problem, albeit at some capital outlay. An automatic titrator enables an operator to

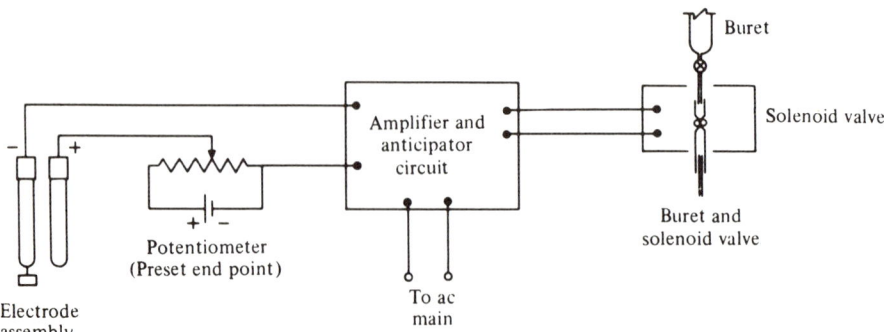

FIGURE 23-8 Automatic titrator and its schematic circuit diagram.

perform other tasks while the instrument delivers the requisite titrant and stops the delivery at a preset end point or, perhaps, continues beyond the end point when the entire curve is traced. Maximum benefit may be obtained from the equipment, particularly with slow reactions. The addition of the next increment will be delayed until the measured electrode potential falls below the value selected for the end point. The instrument will continue to repeat the final stages of the titration until a stable end point is obtained.

The basic features of commercial automatic titrators are similar. In the delivery unit, with no current passing through the solenoids, a short length of flexible tubing is squeezed shut in some manner. With the instrument set up and the buret level read, a switch is pressed to start the titration. The solenoid is energized, the pressure on the tubing is released, and titrant is allowed to flow through the delivery tip. The titration proceeds at a fast rate until a predetermined distance from the end point, when the anticipation control automatically slows the delivery of titrant. At the end point the delivery is stopped. The anticipation control is the key to highly precise automatic operation. It is set to anticipate the end point (preset on some instruments) by a chosen number of pH units or millivolts. The decreased rate of delivery precludes overstepping the preset end point while permitting a rapid delivery of titrant during the initial stages of the titration.

A schematic circuit diagram of an automatic titrator is shown in Fig. 23-8. The control unit includes a calibrated potentiometer, a null-sensing amplifier, and an anticipator circuit. To operate, the potentiometer is set at the pH or potential expected at the end point, the electrode assembly is immersed in the sample solution, and the operating switch is depressed. The difference signal arising between the cell emf and the preset voltage on the potentiometer is amplified, and the output from the amplifier energizes the solenoid valve, or relay, in the delivery unit. As the end point is approached, the difference signal diminishes. When the two signals are matched, the delivery of titrant is stopped. If, upon additional mixing, the cell emf falls below the preset voltage, the controller relay will cause the delivery unit to dispense more titrant. This cycling repeats until a stable end point is reached without ever overshooting the end point.

An automatic recording titrator (Fig. 23-9) plots the complete titration curve. It is started and stopped manually. In this type of titrator the difference signal arises between

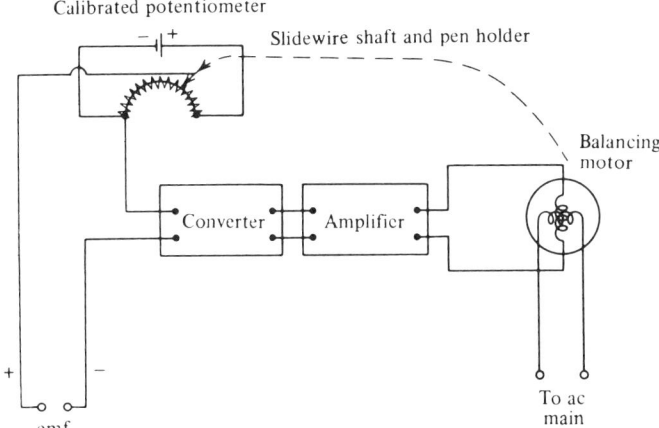

FIGURE 23-9 Schematic circuit diagram of an automatic recording titrator.

the cell emf and an adjustable voltage from a calibrated potentiometer whose slidewire contact is positioned by the same motor that drives the recorder pen. The difference signal, always very small, is converted to alternating current by a converter and then amplified. The output of the amplifier energizes one winding of the two-phase motor; the other winding is permanently connected to the 110-V main. Thus, the servosystem involves actuating the motor which drives the contactor on the slidewire of the potentiometer in a direction to match the cell emf and to preserve the null-balance. The pen traces the change in balancing voltage, and the corresponding cell emf, on the chart to provide a permanent record. No previous knowledge of the end point is required, and reevaluations can be made at a later date. This feature is a distinct advantage where completely unknown systems are run and where there may be successive inflection points. The chart-drive motor and syringe-delivery or constant buret-delivery unit must be synchronized to ensure a constant delivery rate throughout the entire operation.

In place of conventional burets, the delivery units can be designed around various types of syringes. These are operated by a motor-driven micrometer screw which actuates the plunger. Syringe-delivery units offer protection to the titrant from atmospheric oxidation, contamination, and loss of volatile solvent. To ensure rapid signal response, the delivery tip is placed close to the indicating electrode and in front, with respect to the direction of stirring, so that the indicating electrode is bathed by solution at a more advanced stage of titration.

A fully automatic unit will accept serially samples placed in a turntable. After each titration the turntable rotates, indexes the next sample solution beneath the electrode holder, lowers the electrode assembly, delivery tip, and stirring rod into the beaker, and actuates the titration switch to perform the next titration. Each time, the syringe is refilled with titrant and a printer prints out the amount of titrant delivered. This type of automatic instrument is ideal for performing multiple analyses in which the fundamental analytical procedure remains fixed over a period of time, as in a quality control situation.

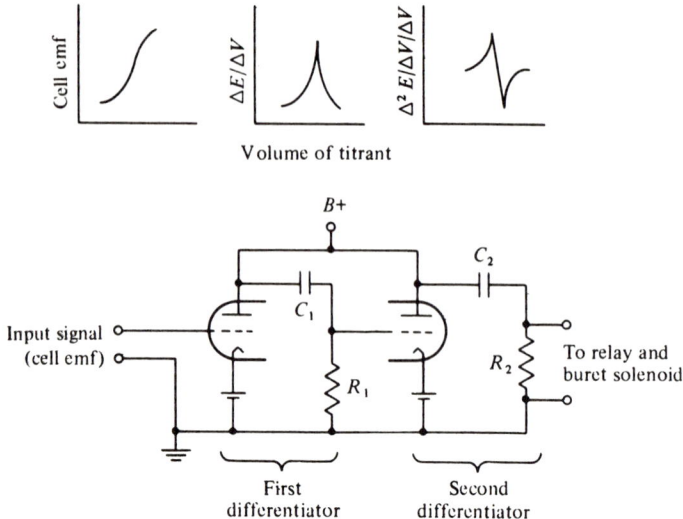

FIGURE 23-10 Schematic diagram of a (Sargent–Malmstadt) derivative-autotitrator. (Courtesy of Sargent–Welch Co.)

The equivalent of plotting a second derivative curve is accomplished automatically in another type of autotitrator,[7] whose schematic diagram is shown in Fig. 23-10. The cell emf is fed directly to the control grid of a conventional amplifier. The amplified voltage is differentiated by a resistance–capacitance differentiator, $R_1 C_1$, and the output is closely proportional to the first derivative curve. Repeating the operation once again produces an output proportional to the second derivative of the titration curve, and a voltage ideally suited to trigger a relay system which closes the buret-delivery unit, or terminates an external electrolysis signal, at the inflection point of the titration curve. All types of reactions are applicable if they possess a suitable reaction rate and the concentrations lie within 0.1–0.01N. The derivative-autotitrator is definitely not applicable to titrations based on very slow reactions, whether these are the electrode reactions themselves, the fundamental chemical reaction, or intermediate secondary reactions. The method is invalidated when acceleration of cell emf becomes a function of some rate of change other than that of the rate of delivery of titrant, which is 1–6 ml/min. Too slow a delivery rate means too small a second-derivative signal, which may not actuate the thyratron relay. The relay must be set to reject spurious fluctuations to prevent a premature cutoff of the delivery unit; consequently, the actuating signal at the true end point must exceed any operating fluctuations.

Automatic titrators can also be devised so that they will measure the volume of titrant required to maintain the indicator electrode at a constant potential. The volume added is plotted automatically versus time and becomes useful in kinetic studies. Since many enzymatic reactions result in the release or consumption of hydrogen ions, a measure of the amount of acid or base required to maintain a constant pH versus time gives a measure of the enzyme activity.

Sensitivity

The sensitivity of a potentiometric titration is limited by the accuracy of the measurement of electrode potentials at low concentrations. Below $10^{-5}N$, the residual current interferes with zero-current potentiometry. Similarly, the current in polarized titrations cannot be fixed at less than the residual current which is the order of the limiting current for a $10^{-5}N$ solution. A $10^{-2}N$ solution can therefore be titrated with an accuracy of 0.1%, but a $10^{-3}N$ solution can be titrated with an accuracy of only 1%. Other titration methods are needed for solutions more dilute than $10^{-3}N$, the limiting concentration in potentiometric titration methods.

23.3 NULL-POINT POTENTIOMETRY

In principle, null-point potentiometry is a concentration cell technique that compares the solution to be analyzed with a solution of known composition.[8] The usual procedure consists of adjusting the composition of one of the half-cell solutions to match the other as evidenced by zero cell potential when measured between identical indicator electrodes selective for the species being determined. The dependence upon an absolute measurement of emf is eliminated. Very small volumes of solution, ranging from 5 to 100 μl, may be handled.

For a cell of this type, the emf is given by

$$E_{cell} = \frac{RT}{nF} \ln \frac{\gamma_1 C_1}{\gamma_2 C_2} + E_j \qquad (23\text{-}12)$$

where the subscripts 1 and 2 indicate the test and variable known solutions. If a sufficiently large excess of an inert electrolyte is used in both half-cells, E_j will become negligible. Also, the activity coefficients of the ionic species (γ_1, γ_2) in the two half-cells will be approximately equal due to the constant high ionic strength maintained by the inert electrolyte.

The null-point technique can be applied in two ways. In the *dead-stop* method, diluent or a standard solution of the species of interest is added until the null point is reached; the test element concentration is evaluated from the known concentration in the variable half-cell. In the *linear-interpolation titration* technique, aliquots of the diluent or standard solution are added incrementally, and the emf measurements are made prior to and after the null point. These emf data are then plotted (E_{cell} vs. log C_2); at the $E_{cell} = 0$ intercept, $C_1 = C_2$. Ideally the slope is 59.16/n mV per decade change in concentration; the slope serves as a check on the electrode behavior.

23.4 CLASSES OF POTENTIOMETRIC TITRATIONS

The principles governing the major types of potentiometric titrations will be treated only briefly here since most beginning textbooks of analytical chemistry cover the subject in much detail and, for more detailed discussions, there are excellent sections in the books by Laitinen, Lingane, Kolthoff and Elving, and Charlot, Badoz-Lambling, and Tremillon.

Oxidation–Reduction Reactions

Oxidation–reduction reactions can be followed by an inert indicator electrode. The electrode assumes a potential proportional to the logarithm of the concentration ratio of the two oxidation states of the reactant or the titrant, whichever is capable of properly poising the electrode. Let us assume that the reactant is the principal system; for example, the iron(III)/(II) system in the titration with cerium(IV). At the start of the titration the minute amount of one oxidation form iron(III) leaves the system without a definite electrode potential. However, as soon as a drop or two of cerium(IV) has been added, the concentration ratio of iron(III)/(II) assumes a definite value and, likewise, the electrode potential of the indicator electrode. During the major portion of the titration the electrode potential changes gradually. Only as the equivalence point is approached does the concentration ratio change rapidly again. Past the equivalence point, the indicator electrode ceases to be affected by the iron(III)/(II) system and will assume a potential dictated by the cerium(IV)/(III) system. For various ratios of iron(III)/(II) system, the corresponding electrode potential is

Ratio, Fe(III)/Fe(II)	10^{-3}	10^{-2}	10^{-1}	1	10	100	1000
Electrode potential, E	0.594	0.653	0.712	0.771	0.830	0.889	0.948

At the equivalence point in an oxidation–reduction reaction

$$a_{ox_1} + b_{red_2} = a_{red_1} + b_{ox_2} \tag{23-13}$$

the electrode potential is the weighted mean of the standard electrode potentials of reactant and titrant,

$$E_{\text{equiv pt}} = \frac{bE_1^\circ + aE_2^\circ}{a + b} \tag{23-14}$$

When $a = b$, the titration curve is symmetrical around the equivalence point, but when $a \neq b$, the titration curve will be markedly asymmetrical and the point of inflection will not coincide with the equivalence point. The difference will depend upon the ratio a/b. If $a > b$, the inflection point will occur when excess oxidant$_1$ is present in solution; that is, before the equivalence point. The inverse is true when $b > a$. These errors are, however, generally insignificant when compared to other errors of titrations, such as slowness of attainment of equilibrium potential, slowness of reaction, or inexact stoichiometry.

In general, the reduction potential of a metal complex, or of a metal ion in equilibrium with a complexing agent, is decreased by complex formation. Three effects are involved[9]: The *coordinating effect* is the combination of a metal ion with an electron donor. The *charge effect* is simply the charge on the resulting complex. The *electronic effect* relates to the degree of stability of the electron configuration in the metal complex. The first two always tend to increase the tendency toward oxidation to a higher valence state; the third effect may work in either direction.

Besides the straightforward redox titrations described above, there are many others that can be made to proceed quantitatively by varying the chemical nature of one or more of the reactants. For example, cobalt salts may be titrated potentiometrically by potassium ferricyanide in the presence of a high concentration of ammonium citrate and aqueous ammonia.[10] The ammonia converts the cobalt aquo ions to the cobalt hexammine complex ions and, since the $Co(NH_3)_6^{3+}$ species is much more stable than the $Co(NH_3)_6^{2+}$ species, the oxidation potential of the cobalt system shifts from +1.84 to about +0.13 V, permitting quantitative oxidation with ferricyanide ($E° = +0.36$ V) if the temperature is lowered to 0°-5°C and oxygen is excluded. Likewise, manganese(II) ion can be titrated to a manganese(III) complex ion by permanganate ions in the presence of diphosphate(V) ions and at a pH of 6 to 7 because diphosphate(V) forms a very stable complex $Mn(H_2P_2O_7)_3^{3-}$ with Mn^{3+}. The pH is important here as with other reactions involving the hydrogen or hydroxyl ions.

Ion Combination Reactions

Reactions in this category involve the formation of a sparingly soluble compound or a slightly dissociated material. Argentometric titrations are illustrative. For the chloride, and similar precipitation systems, the titration curve can be calculated from Eq. 23-7 or equivalent equations. At the equivalence point,

$$[Ag^+] = [Cl^-] = \sqrt{K_{sp}} \tag{23-15}$$

Typical halide titration curves are shown in Fig. 23-11.

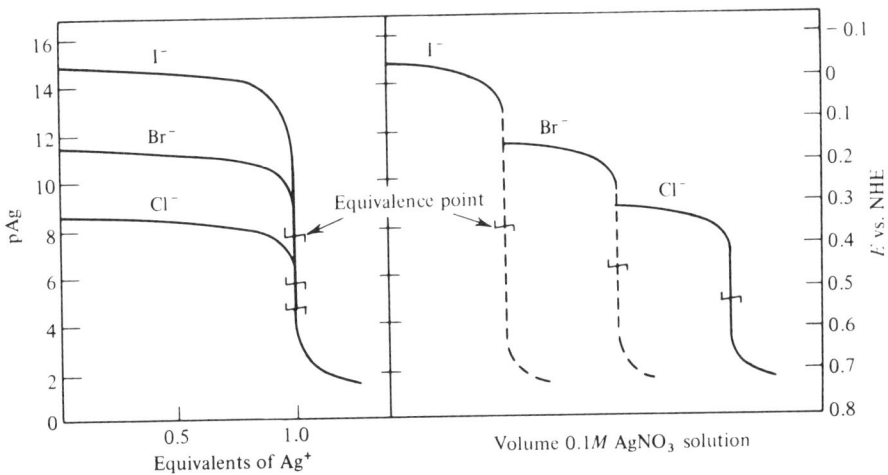

FIGURE 23-11 Theoretical titration curves of halide ions (0.1M each) with silver nitrate solution and a silver indicator electrode. The dashed segments are the separate curves for iodide and bromide ions.

For an ion combination reaction involving a soluble complex, such as

$$Ag^+ + 2CN^- = Ag(CN)_2^- \tag{23-16}$$

the electrode potential is given by the expression

$$E = 0.799 + 0.0591 \log \frac{1}{K_f[CN^-]^2} + 0.0591 \log [Ag(CN)_2^-] \tag{23-17}$$

where K_f is the formation constant of the dicyanoargenate(I) ion. At the equivalence point

$$[Ag^+] = \tfrac{1}{2} [CN^-] = \sqrt[3]{\frac{[Ag(CN)_2^-]}{4K_f}} \tag{23-18}$$

The magnitude of the inflection point on the titration curve depends upon the degree of insolubility of a precipitate, or the extent of dissociation of a complex. Successive titrations are feasible when one compound is markedly less soluble, or dissociated, than another; that is, $K_1/K_2 \geqslant 10^6$ and $K_1 > 10^8$, where the constants symbolize solubility product constants or instability constants ($K_{\text{instab}} \equiv 1/K_f$). Application of potentiometric titration methods to precipitation reactions is limited by factors adversely affecting the character of precipitates, and applicability to this category is restricted by unavailability of indicator electrodes.

Ion-selective electrodes can be utilized for many complexometric titrations. Best results can be obtained with reagents such as EDTA which form only one complex with the metal ion and, consequently, give a single end-point break. An important type of complexometric titration involves the titration of a solution containing two metal ions, both of which form a complex with a reagent L. If an electrode is available that senses the ion M_1, which forms the more stable complex, then a titration curve similar to Fig. 23-12 results. If only M_1 were present in the sample, curve A would be obtained. If M_2 is also present, curve B results. In this case, after all the M_1 has been complexed by L at the first end point, the concentration of L increases up to a point where the second complex involving M_2 begins to form. This type of titration is very useful, as it allows an electrode sensitive to M_1 to be used to titrate any metal M_2 that forms a less stable complex with a given reagent than does M_1. In practice, the sample containing M_2 is spiked with an indicator concentration of the complex M_1L (10^{-5} to $10^{-4} M$), which need not be accurately measured. The initial solution is therefore equivalent to the solution in Fig. 23-12 at a point corresponding to the first end point. The rest of the titration with L proceeds exactly as described.

In order to establish the most suitable pH values for titrations involving H^+ or OH^- ions it is convenient to measure the pH as a function of electrode potential for the various types of ions which may be present before, during, and after the reaction. From the curves so obtained the range of suitable pH values can be determined.

Acid–Base Reactions

Titrations of acids and their conjugate bases can be broken down conveniently into several categories, including consideration of nonaqueous solvent systems. Indicating electrodes are discussed in Chapter 22.

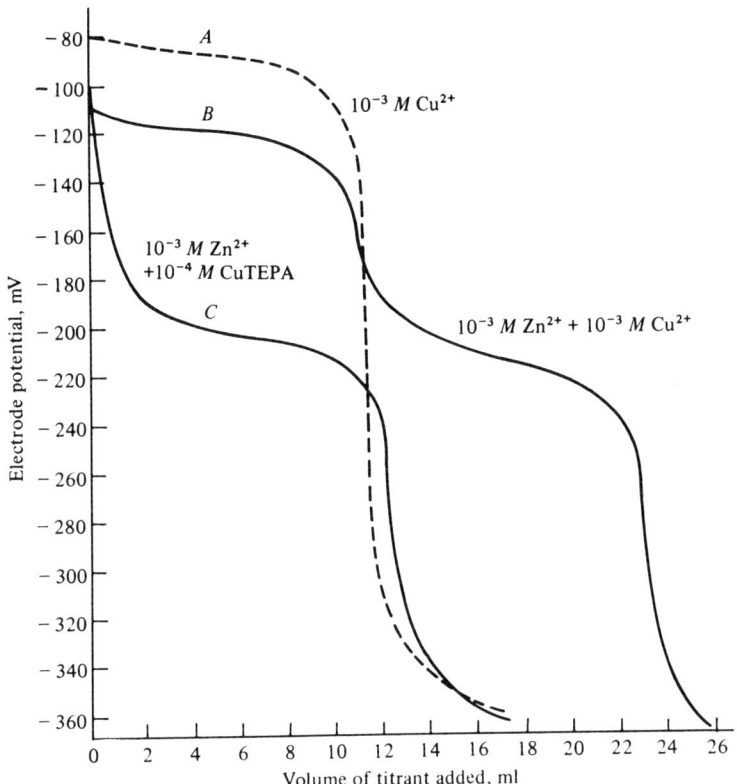

FIGURE 23-12 Complexometric titration using a copper-selective electrode and tetraethylene pentamine (TEPA) titrant at pH 8 (curve *A*) and at pH 10 (curves *B* and *C*). (Courtesy of Orion Research, Inc.)

In the titration of a completely dissociated acid or base, the pH at the equivalence point is that of pure water (in the absence of dissolved CO_2), namely, 7. For a reaction to be complete within 0.1%, the initial concentration must not be less than $10^{-4} N$.

For an incompletely dissociated acid, the hydrogen-ion concentration at the equivalence point is given by the expression

$$[H^+] = \sqrt{\frac{K_w K_a}{C_{salt}}} \qquad (23\text{-}19)$$

As the acid becomes progressively weaker, the distinctness of the inflection point diminishes, and the pH at the equivalence point shifts to higher values. Feasibility of a particular titration is determined by the product, $K_a[HA]$. For an uncertainty of 0.1% or less, and in aqueous solution, the product should exceed 10^{-8}, assuming that the titrant is completely dissociated and 0.1N in strength.

The accuracy with which two successive equivalence points may be located will depend on the absolute and relative strengths of the two acid groups and their concentrations. Difficulty is encountered in locating the break in the titration curve at a ratio of the first to the second dissociation constant of 100, and the ratio must be greater than 10^5 to 1 to give a sharp inflection point. However, by using transparent masks with theoretical curves, it is possible to estimate pK_a values at the extremes of the pH scale where only a portion of the complete theoretical curve can be distinguished (see Fig. 23-3). Groups whose pK_a values differ by as little as 1 pK unit can be resolved, provided at least half of each hydrogen ion-binding curve is free from overlap and permits matching with the masks.

Acid–Base Titrations in Nonaqueous Solvents

Many acids or bases that are too weak for determination in water become susceptible to titration in appropriate nonaqueous solvents. The resolution of mixtures, particularly of dibasic acids, may be improved.

The major considerations in the choice of a solvent for acidimetric reactions are its acidity and basicity, its dielectric constant, and the physical solubility of a solute. Acidity is important because it determines to a large extent whether or not a weak acid can be titrated in the presence of a relatively high concentration of solvent molecules. Phenol, for example, cannot be titrated as an acid in aqueous solution because water is too acid and present in too high a concentration to permit the phenolate ion to be formed stoichiometrically by titration with a base. In other words the intrinsic basic strength of the phenolate and hydroxide ions are not sufficiently different for the reaction

$$\phi\text{--OH} + \text{OH}^- \rightleftharpoons \phi\text{--O}^- + H_2O \tag{23-20}$$

to proceed quantitatively to completion. In less acidic solvents, such as dimethylformamide or pyridine, this titration can be carried out readily with a stronger basic titrant, the alkoxide ion,

$$\phi\text{--OH} + \text{RO}^- \rightarrow \phi\text{--O}^- + \text{ROH} \tag{23-21}$$

The solvent must not be strongly basic if resolution of the strong and moderately strong acids is to be achieved, because of the "leveling effect" of a basic solvent on the stronger acids. In ethylenediamine, sulfonic and carboxylic acids are both leveled to the ammonium type ion, thus,

$$\left.\begin{array}{l}\text{RSO}_3\text{H}\\\text{RCOOH}\end{array}\right\} + H_2N\text{--}C_2H_2\text{--}NH_2 \rightarrow \left.\begin{array}{l}\text{RSO}_3^-\\\text{RCOO}^-\end{array}\right\} + H_2N\text{--}C_2H_2\text{--}NH_3^+ \tag{23-22}$$

whereas in dimethylformamide only the sulfonic acid is leveled. An ideal solvent for the titration of an acidic mixture should be sufficiently weak in acidity to permit titration of the most weakly acid component and sufficiently weak in basicity to permit resolution of the strongest components.

Acids can be classified as uncharged, positively charged, or negatively charged. The members of any one class may vary in relative strength to a certain extent as the dielectric constant of the solvent is changed, but in general they behave in a similar manner. However, acids of different charge type change greatly in relative strength as the solvent is changed. The positively charged acids, such as the ammonium ion, become stronger relative to an uncharged acid, such as acetic acid, as the dielectric constant is reduced.

The protolysis reaction involving the solvent, SH,

$$NH_4^+ + SH \rightleftharpoons SH_2^+ + NH_3 \tag{23-23}$$

does not result in the formation of new charged species as does the reaction for an uncharged acid such as acetic acid,

$$HOAc + SH \rightleftharpoons SH_2^+ + OAc^- \tag{23-24}$$

Negatively charged acids, such as the hydrogen succinate ion, tend to become weaker relative to an uncharged acid.

The apparent strength of an acid in a solvent may be expressed empirically in terms of its midpoint potential or half-neutralization point. The difference in the midpoint potentials of two acids in the same solvent can serve as a measure of the resolution which can be achieved. It should be roughly 200-300 mV, depending on the slope of the plateaus in the titration curves.

Solvents may be divided into several classes. *Amphiprotic* solvents are those that possess both acidic and basic properties. They undergo self-dissociation, or autoprotolysis, that is,

$$SH + SH \rightleftharpoons SH_2^+ + S^- \tag{23-25}$$

to produce a solvonium ion SH_2^+ and a solvate ion S^-. Representative amphiprotic solvents include water, the lower alcohols, and glacial acetic acid. The product of the ion concentrations gives the autoprotolysis constant

$$K_{auto} = [SH_2^+][S^-] \tag{23-26}$$

It is 14 in water, varies from 15 to 19 in alcohols, and is about 14 in glacial acetic acid. Amphiprotic solvents are also classified as protogenic solvents if they are more acidic than water (for example, acetic acid) or protophylic solvents if they are more basic than water (for example, ethylenediamine, dimethylformamide, ketones, and others).

Aprotic solvents have no acidic or basic properties, and if their dielectric constant is low, they have low ionizing power. Aprotic solvents include aromatic and aliphatic hydrocarbons, carbon tetrachloride, and methyl isobutyl ketone.

When a solute is dissolved in an amphiprotic solvent, the position of equilibrium depends on the relative acidic or basic strengths of the solute and solvonium ion (or solvate). The position of the autoprotolysis ranges, relative to the intrinsic strength of index acids, is indicated schematically in Fig. 23-13. Although a base must have a dissociation constant greater than about 10^{-8} for successful titration in water, the accurate determination of compounds with basic dissociation constants (in water) of 10^{-12} is possible with a more acidic solvent such as glacial acetic acid. Amino acids yield sharp end points because the carboxylic acid group is swamped, thus removing the zwitterion equilibrium. Commonly, the titrant is a solution of perchloric acid in dioxane or glacial acetic acid which has been standardized with either potassium acid phthalate (phthalic acid is the product) or sodium carbonate.

In an analogous fashion, basic solvents enhance the properties of weak acids. Phenols and carboxylic acids produce distinctive end points in butylamine or ethylenediamine,

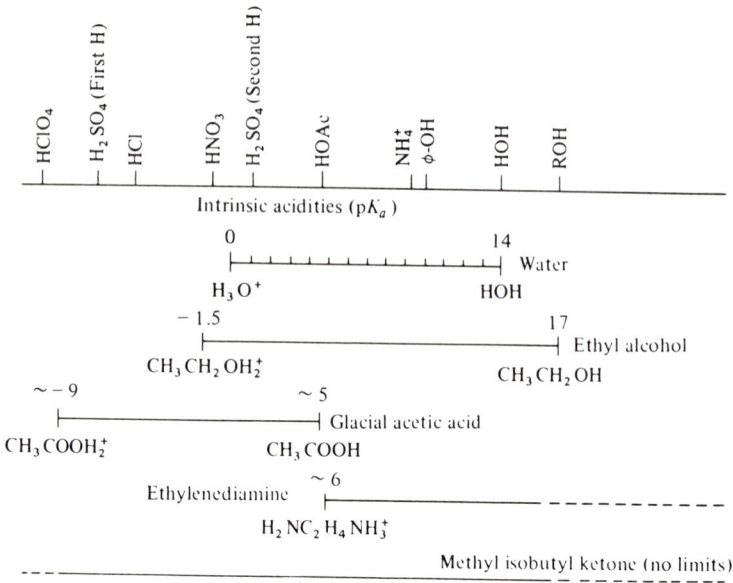

FIGURE 23-13 Schematic representation of autoprotolysis ranges of selected solvents, in relation to the intrinsic strength of certain index acids. Influence of dielectric constant is not included.

and these plus sulfonic acids can be differentiated from each other in dimethylformamide. The titrant is usually sodium aminoethoxide or a quaternary base.

The effect of the dielectric constant coupled with the decrease in solvent basicity is well illustrated for the titration of oxalic acid, succinic acid, or sulfuric acid singly in isopropyl alcohol. In each instance two well-resolved end points are present, whereas in water a single inflection is obtained for sulfuric and succinic acid and only a poorly defined inflection indicates the existence of the first replaceable hydrogen in oxalic acid. Similarly, solutions containing both perchloric and acetic acid exhibit two distinct inflections, and the degree of resolution is comparable with that of the dibasic acids. The reason is the difference in charge type between perchloric and acetic acid (or more correctly the solvonium ion and acetic acid), which makes the change in relative strength so large. Likewise, dibasic acids are a mixture of a positively charged acid (the solvonium ion) or an uncharged acid (succinic acid, first hydrogen) and a negatively charged acid (HSO_4^-, $HC_2O_4^-$, and so on). In addition, the weaker basicity of isopropyl alcohol lessens the extent of "leveling" of the first hydrogen of sulfuric acid and the perchloric acid, which increases the difference between the midpoint potentials of the titration curves.

Aprotic solvents have certain advantages over the other classes of solvents. By not interacting with dissolved solutes, no leveling action is exerted. Except for the influence of the dielectric constant, each solute will exhibit its intrinsic acidic or basic strength. Having no autoprotolysis limits, the range of applicability is limited only by the strength of the acid or base titrant. The former is usually perchloric acid or p-toluene sulfonic acid, the latter a quaternary ammonium base. If suitably spaced, a series of successive end points can be achieved in a solvent such as methyl isobutyl ketone (Fig. 23-14).

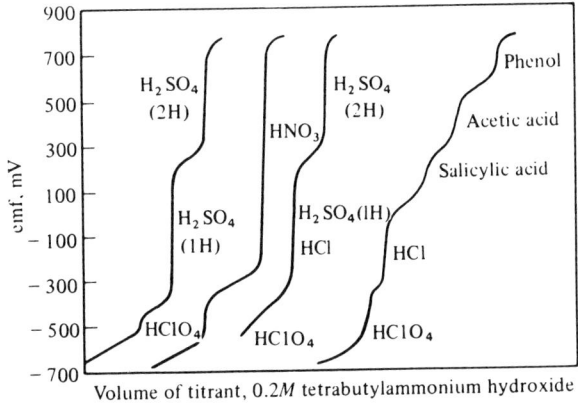

FIGURE 23-14 Resolution of acid mixtures in methyl isobutyl ketone. Glass–calomel electrodes. (D. B. Bruss and G. E. A. Wyld, *Anal. Chem.*, 29, 232 (1957). Courtesy of American Chemical Society.)

The electrode systems vary with the solvent employed. The glass–calomel electrode system is suitable where the solvent is either acetonitrile, an alcohol, or a ketone, or for differentiating titrations in dimethylformamide, provided that the titrant consists of either potassium hydroxide or alkoxide or tetraalkyl-ammonium hydroxide (that is, no sodium compounds). It is advisable to replace the aqueous salt bridge of the calomel electrode by either a saturated solution of potassium chloride in methanol or N-tetraalkyl-ammonium chloride solution. A glass electrode does not function as an indicator electrode in the more strongly basic solvents if the titrant contains sodium compounds. In this situation a pair of antimony electrodes forms a satisfactory combination, one dipping into the titrant and the other into the solution. A platinum electrode sealed into the buret and a glass indicator electrode give stable and reproducible potentials when solutions in methyl isobutyl ketone are titrated with quaternary ammonium hydroxide in benzene-methanol. The chloranil indicator electrode has been used in glacial acetic acid. Solvents with low dielectric constant exhibit such high internal resistance that it is difficult to find electrodes which function satisfactorily. When the dielectric constant is 5 or less, potentiometric methods are unsuitable. A review of potentiometric electrode systems in nonaqueous titrimetry has been published.[11] Brief surveys of titrations in nonaqueous media are available and should be consulted for operational details.[12,13]

Other Nonaqueous Titrations

Other types of reactions are also feasible in nonaqueous solvents, particularly when solubility is a strong consideration. In petroleum products, the titration of hydrogen sulfide and mercaptans, either singly or in combination, is done in a 1:1 mixture of methanol-benzene (plus dissolved sodium acetate) with a methanolic solution of silver nitrate. The electrode system consists of a silver indicator electrode coupled with a calomel reference electrode connected to the sample solution by means of an agar bridge containing 3% potassium nitrate in the gel.

The Karl Fischer method for the determination of water is an excellent example of a well-known redox reaction conducted in methanol solution.[14] Iron(II) perchlorate is useful as a reductant in glacial acetic acid.

PROBLEMS

1. Sketch the titration curves you would expect from the titration of each of the following aqueous systems: (a) $0.1M$ solution of H_3PO_4 titrated with $0.2M$ NaOH, (b) a solution $0.05M$ in Na_3PO_4 and $0.1M$ in Na_2HPO_4 titrated with $0.1M$ HCl, (c) a $0.1M$ solution of ammonia titrated with $0.1M$ HCl.

2. Construct the complete curve for the titration of 50 ml of $0.1M$ titanium(III) chloride with a $0.1M$ solution of methylene blue ($E° = 0.52$) in $1M$ HCl.

3. Construct the titration curve for the titration of the solution resulting from passage of $0.01M$ vanadium solution through an amalgamated zinc reductor; titrant is $0.1M$ cerium(IV).

4. Sketch the titration curves you would expect from the titration of each of the following nonaqueous systems. Express the ordinate values in units of 0.059 V. (a) A mixture of an alkyl sulfonic acid ($pK_a = -7$) and an alkyl carboxylic acid ($pK_a = 4$) dissolved in methyl isobutyl ketone and titrated with tetra-n-butylammonium hydroxide. (b) A solution of aniline in glacial acetic acid titrated with $HClO_4$. (c) A mixture of HCl and acetic acid dissolved in isopropyl alcohol and titrated with sodium isopropoxide. (d) A mixture of acetic acid and phenol in n-butylamine titrated with sodium aminoethoxide.

5. Calculate the potential at the equivalence point in the potentiometric titration of each of these systems; assume the reference electrode is saturated calomel: (a) Titration of tin(II) with cerium(IV) ions; (b) titration of uranium(IV) with iron(III); (c) titration of VO^{2+} with cerium(IV) in $1M$ H_2SO_4; (d) titration of arsenic(III) with bromate in $5M$ HCl.

6. From the information in Fig. 23-11, estimate the error in the location of the end point, as contrasted with the equivalence point, for (a) iodide and (b) bromide when all three halides are present in mixtures. Disregard any error attributable to mixed salt formation.

7. The following pH readings were obtained for corresponding volumes of $0.100N$ NaOH in the potentiometric titration of a weak monobasic acid:

0.00 ml = 2.90	14.00 ml = 6.60	16.00 ml = 10.61
1.00 ml = 4.00	15.00 ml = 7.04	17.00 ml = 11.30
2.00 ml = 4.50	15.50 ml = 7.70	18.00 ml = 11.60
4.00 ml = 5.05	15.60 ml = 8.24	20.00 ml = 11.96
7.00 ml = 5.47	15.70 ml = 9.43	24.00 ml = 12.39
10.00 ml = 5.85	15.80 ml = 10.03	28.00 ml = 12.57
12.00 ml = 6.11		

(a) Plot the above values of pH against milliliters of NaOH solution. (b) What is the pH value at the equivalence point? (c) What volume of NaOH corresponds to the equivalence point? (d) What is the ionization constant of the acid?

8. In the titration of iron(II) using the differential potentiometric method with a constant electrolysis current, the following readings were obtained for corresponding volumes of $0.1 M$ cerium(IV):

2.00 ml = 50 mV	9.60 ml = 410 mV	10.75 ml = 400 mV
4.00 ml = 50 mV	9.80 ml = 740 mV	11.00 ml = 365 mV
6.00 ml = 50 mV	10.00 ml = 705 mV	12.00 ml = 300 mV
8.00 ml = 100 mV	10.25 ml = 515 mV	14.00 ml = 250 mV
9.00 ml = 155 mV	10.50 ml = 460 mV	16.00 ml = 205 mV
9.40 ml = 205 mV.		

(a) Plot the millivolt readings against the volume of cerium solution. (b) What volume of cerium corresponds to the end point?

BIBLIOGRAPHY

Beckett, A. H. and E. H. Tinley, *Titrations in Non-Aqueous Solvents*, 3rd ed., British Drug Houses, Poole, England, 1962.
Bockris, J. O'M. and B. E. Conway, *Modern Aspects of Electrochemistry*, Vols. 9–11, Plenum, New York, 1974-5.
Browning, D. R., Ed., *Electrometric Methods*, McGraw-Hill, Maidenhead, England, 1969.
Charlot, G., J. Badoz-Lambling, and G. Tremillion, *Electrochemical Reactions*, American Elsevier, New York, 1962.
Gyenes, I., *Titration in Nonaqueous Media*, Van Nostrand Reinhold, New York, 1967.
Kolthoff, I. M. and P. J. Elving, Eds., *Treatise on Analytical Chemistry,* Chaps. 11–14 and 16, Vol. I, Part I, Wiley-Interscience, New York, 1959.
Lagowski, J. J., Ed., *The Chemistry of Nonaqueous Solutions*, Academic, New York, 1976-8.
Laitinen, H. A. and W. E. Harris, *Chemical Analysis*, 2nd ed., McGraw-Hill, New York, 1975.
Lingane, J. J., *Electroanalytical Chemistry*, 2nd ed., Wiley-Interscience, New York, 1958.
Ringbom, A., *Complexation in Analytical Chemistry*, Wiley, New York, 1963.
Rossotti, H., *Chemical Applications of Potentiometry*, Van Nostrand Reinhold, New York, 1969.
Sawyer, D. T. and J. L. Roberts, Jr., *Experimental Electrochemistry for Chemists*, Wiley-Interscience, New York, 1974.
Streuli, C. A., "Titrimetry: Acid–Base Titrations in Nonaqueous Solvents," in *Treatise on Analytical Chemistry*, Vol. II, Part I, I.M. Kolthoff and P.J. Elving, Eds., Wiley-Interscience, New York, 1975.

LITERATURE CITED

1. Reilley, C. N. and R. W. Schmid, *Anal. Chem.*, **30**, 947 (1958).
2. Reilley, C. N., R. W. Schmid, and D. W. Lamson, *Anal. Chem.*, **30**, 953 (1958).

3. Gran, G., *Acta Chem. Scand.*, **4**, 559 (1950); *Analyst*, **77**, 661 (1952).
4. Willard, H. H. and A. W. Boldyreff, *J. Am. Chem. Soc.*, **51**, 471 (1929).
5. Müller, E., *Z. Physik. Chem.*, **135**, 102 (1928).
6. Kolthoff, I. M., *Anal. Chem.*, **26**, 1685 (1954).
7. Malmstadt, H. V. and E. R. Fett, *Anal. Chem.*, **26**, 1348 (1954).
8. Malmstadt, H. V. and J. D. Winefordner, *Anal. Chim. Acta*, **20**, 283 (1959).
9. Laitinen, H. A., *Chemical Analysis*, McGraw-Hill, New York, 1960.
10. Chirnside, R. C., H. J. Cluley, and P. M. C. Proffitt, *Analyst*, **72**, 354 (1947).
11. Stock, J. T. and W. C. Purdy, *Chem. Rev.*, **57**, 1159 (1957).
12. Beckett, A. H. and E. H. Tinley, *Titrations in Non-Aqueous Solvents*, 3rd ed., British Drug Houses, Poole, England, 1962.
13. Fritz, J. S., *Acid–Base Titrations in Nonaqueous Solvents*, G. Frederick Smith Chemical Co., Columbus, Ohio, 1952; also Allyn and Bacon, Boston, 1973.
14. Mitchell, J. and D. M. Smith, *Chemical Analysis*, 2nd ed., Part III, Wiley, New York, 1977.

CHAPTER 24

Voltammetry, Polarography, and Related Techniques

The polarographic method of analysis is based on the current–voltage curves arising at a microelectrode when diffusion is the rate-determining step in the electrochemical reaction. The development of polarography, commencing with the work of Heyrovsky in 1922, marked a significant advance in electrochemical methodology because it introduced the element of selectivity through control of electrode potential, an element which was largely lacking in the older electrochemical methods of potentiometry and conductimetry. However, the fundamental dc polarographic technique suffered from a number of difficulties that made it less than ideal for routine analytical purposes and made the results obtained somewhat difficult to interpret. With the advent of low-cost, fast, stable, operational amplifiers in the early 1960s, various problems began to be overcome. Investigations of such techniques as ac polarography, pulse polarography, and derivative pulse polarography demonstrated the utility and desirability of the *new* polarography for fingerprint purposes and analytical applications. Today multipurpose instruments in an all-electronic form provide sensitivity to the parts-per-billion level for many electroactive substances. All the techniques discussed in this chapter are united in their applicability to electrodes other than the standard dropping mercury electrode employed in classical polarography; these include glassy carbon and wax-impregnated graphite electrodes that permit voltammetry in the anodic region. The reemergence of voltammetry is timely, since it has great sensitivity to environmentally important elements such as lead and cadmium. It is being used in forensic work to determine drugs and arsenic in urine.

24.1 CURRENT–VOLTAGE RELATIONSHIPS

An electrode is considered to be *polarized* when it adopts a potential impressed upon it with little or no change of the current. Take, for example, a platinum electrode dipping into a solution of copper(II) ions which is also $0.1M$ in sulfuric acid. When short-circuited with a calomel reference electrode, the platinum electrode will assume the potential of the calomel electrode with no flow of current. The platinum electrode is polarized, and it will remain polarized until an emf is impressed across the two electrodes that is sufficient to exceed the decomposition potential of the copper(II) ions. When the impressed voltage does exceed the decomposition potential, copper will deposit on the platinum. Until this potential is attained, there is no reversible electrode reaction. After some copper has plated out, the electrode becomes depolarized and its potential is determined by the Nernst equation,

$$E = E° + \frac{0.0591}{2} \log [Cu^{2+}] \qquad (24\text{-}1)$$

As long as the electrode is ideally depolarized, passage of current does not cause the potential to deviate from its reversible value.

Figure 24-1 gives idealized current–voltage curves of copper(II) ion solutions with varying concentrations. Starting out with a well-stirred solution of $0.05M$ copper(II) ions and impressing a voltage across the platinum electrode and reference electrode, the current–voltage curve will be traced by curve OAB. No current will be observed to flow until the applied emf exceeds the decomposition potential of a solution of $0.05M$ copper(II) ions. At point A, copper commences to plate out on the electrode and current starts to flow. As the voltage is increased further, the current increases linearly in accordance with Ohm's law. On the other hand, if the voltage is reduced from B to A, the current will diminish

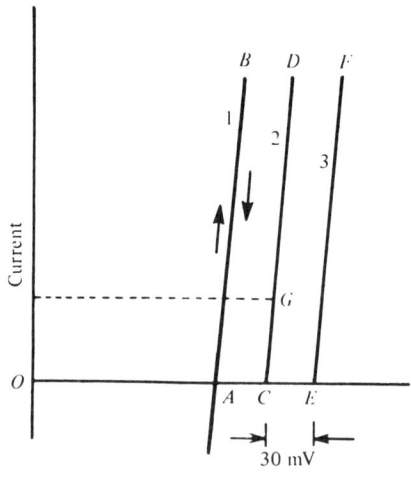

FIGURE 24-1 Current–voltage curves without polarization. Curve 1: $0.05M$ copper; curve 2: $0.005M$ copper; curve 3: $0.0005M$ copper.

gradually to zero as the deposit of copper dissolves from the platinum electrode. Similarly, for smaller concentrations of copper the current–voltage traces will be given by curves *OCD* and *OEF*. In each case, the decomposition potential will be shifted along the voltage axis to a more negative value by 29.5 mV per 10^{-1} change in concentration.

The picture changes if the experiment is repeated without stirring the solution and with a microelectrode; that is, an electrode with a small area of contact with the test solution. Consider a concentration of 0.005 M copper(II) ions at an impressed emf corresponding to *D* in Fig. 24-1. If the voltage is held constant, the current will start to flow at a value corresponding to *D* but will decrease rapidly to some value *G* as the concentration of copper(II) ions at the microelectrode surface becomes depleted by deposition. The current, represented by *G*, is characteristic of the rate at which fresh copper(II) ions are supplied to the microelectrode. The microelectrode is polarized under these conditions.

In general, there are three mass-transfer processes by which a reacting species may be brought to an electrode surface. These are (1) diffusion under the influence of a concentration gradient; (2) migration of charged ions in an electric field; and (3) convection due to motion of the solution or the electrode. In voltammetry the effect of migration is usually eliminated by adding a 50- or 100-fold excess of an inert "supporting electrolyte." The ions of this electrolyte migrate to relieve the electric fields but do not undergo an electrochemical reaction at the electrode. A potassium salt is often employed. Because the potassium ions cannot be discharged at the cathode until the impressed voltage becomes rather large, large numbers of them remain as a cloud around the cathode. This positively charged cloud restricts the potential gradient to a region so very close to the electrode surface that there is no longer an electrostatic attraction operative to attract other reducible ions from the bulk of the solution. Convection can be minimized by using unstirred vibration-free solutions. Under such conditions, the limiting current is controlled solely by diffusion of the reacting species through the concentration gradient adjacent to the electrode. The latter gradient arises because of the relative slowness of diffusion.

According to Fick's law, the net rate of diffusion of a species to a unit area of electrode surface *A* at any time *t* is proportional to the magnitude of the concentration gradient, that is,

$$\text{flux} = -D \left(\frac{dC}{dx} \right)_{x=0} = \frac{-D(C_{\text{bulk}} - C_0)}{\delta} \tag{24-2}$$

where D is the diffusion coefficient of the species and δ is the thickness of the hypothetical diffusion layer about the microelectrode (Fig. 24-2). As the region around the microelectrode becomes depleted of electroactive species, that is, as C_0 approaches zero, the rate of diffusion becomes proportional to the concentration in the bulk of the solution, C_{bulk}.

When equilibrium is established at the microelectrode, the rate of discharge of the ions will be equal to the rate of diffusion to the electrode. If i is the faradaic current, then the rate of discharge of ions is equal to i/nFA, where n is the number of electrons involved in the discharge process and F, the faraday, is the quantity of electricity carried by one

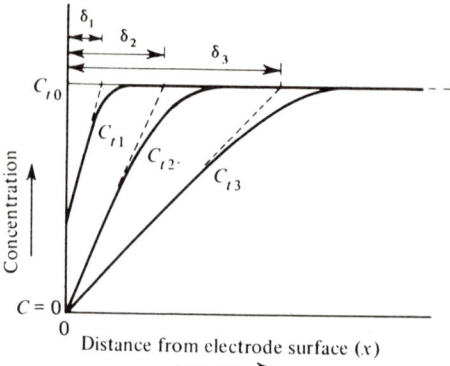

FIGURE 24-2 Concentration profiles for reacting species at various times after the start of electrolysis. $C_{t0} = C_{\text{bulk}}$; $t_3 > t_2 > t_1$; t_0 = before electrolysis; δ_i refers to thickness of the Nernst diffusion layer at time t_i.

equivalent of electroactive species. Expressing the flux in terms of electrical current density, i/A, the diffusion-limited current is given by

$$i_{\text{lim}} = \frac{nFADC_{\text{bulk}}}{\delta} \qquad (24\text{-}3)$$

showing that the limiting current is proportional to concentration and inversely proportional to the thickness of the diffusion layer. A diffusion-limited current decreases with time because of the buildup of the diffusion layer.

In its simplest form, classical or conventional polarography involves applying a linearly varying dc potential between two electrodes, one small and easily polarizable and the other large and relatively immune to polarization. The current between these electrodes is recorded as a function of the applied potential. A characteristic steplike current–voltage curve is obtained for each electroactive species in the solution (see Fig. 24-4). The potential at the midpoint of the rising portion of each step, the half-wave potential, $E_{1/2}$, is characteristic of the particular active species causing the transition in the solvent system. In addition, the difference in current i_d between the base line before the step and the plateau after it is proportional to the concentration of the species in question.

Current–voltage curves with several types of microelectrodes are shown in Fig. 24-3. For the curves shown, the voltage scanning rate was between 2 and 8 mV/sec. Curve A was obtained with a stationary gold wire microelectrode. The electrochemical reaction begins gradually and a hump-shaped current–voltage curve is observed. The height of the hump is proportional to the square root of the scanning rate. Curve B was recorded using a rotating platinum microelectrode. The electrode is rotated rapidly through the solution so that the diffusion layer cannot increase in thickness, but remains constant and very thin. A steady current is obtained immediately. Rotating microelectrodes find considerable use in amperometric titrations. Curve C was obtained with a dropping mercury electrode. With this type of microelectrode the growth of the drops offsets the effect of a widening diffusion layer. The current oscillates between a near-zero value, just after a drop falls, to a maximum value, just before the next drop falls. Usually no attempt is made to follow or

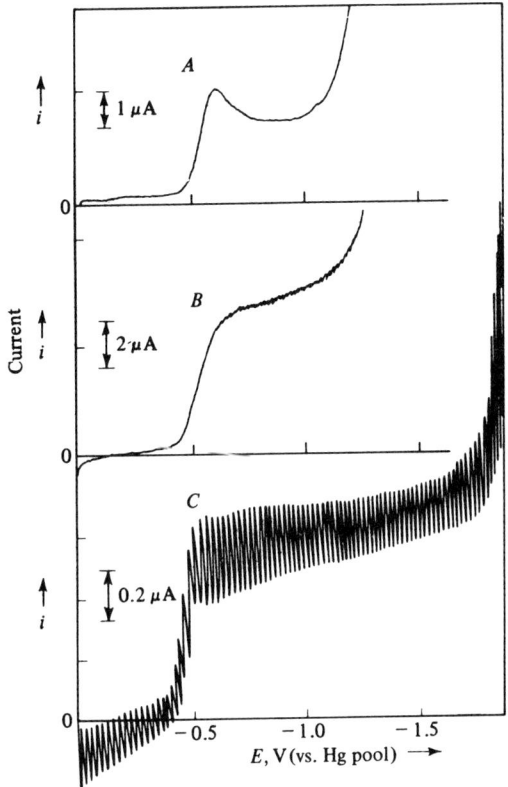

FIGURE 24-3 Qualitative comparison of current–voltage curves recorded for $10^{-4}M$ Tl^+ in $1M$ KCl with various types of electrodes. Curve A, gold wire electrode (mercury coated), scanning rate = 500 mV/min; curve B, rotating platinum wire electrode (600 rpm); curve C, dropping mercury electrode.

to record the entire current excursion, but rather a somewhat damped system is used to indicate the average current flowing during the life of the drop.

24.2 CHARACTERISTICS OF THE DROPPING MERCURY ELECTRODE

The most commonly used type of microelectrode is the *dropping mercury electrode*, the electrode of classical polarography. This small polarizable electrode is produced by passing a stream of mercury through a very fine bore (0.05–0.08 mm i.d.) glass capillary. A steady flow of droplets issues from the capillary at a rate of one drop every 3–5 sec. A dropping mercury electrode has several advantages: (1) Its surface area is reproducible with any given capillary. (2) The constant renewal of the electrode surface eliminates passivity or poisoning effects. (3) The high overpotential of hydrogen on mercury renders the electrode useful for electroactive species whose reduction potential is considerably more negative than the reversible potential of hydrogen discharge. (4) Mercury forms amalgams with many metals and thereby lowers their reduction potential. (5) The diffusion current assumes a steady value immediately and is reproducible.

The dropping mercury electrode is useful over the range +0.3 to −2.8 V versus the SCE. At potentials more positive than 0.3 V, mercury dissolves and gives an anodic wave. The most positive potentials may be attained in the presence of noncomplexing anions that form soluble mercury (I and II) salts, for example, nitrate or perchlorate ions. Anions that form insoluble mercury salts or stable complexes shift the anodic dissolution potential to more negative values. At potentials more negative than −1.2 V, visible hydrogen evolution occurs in $1M$ HCl solutions, and at −2 V the usual supporting electrolytes of alkali salts begin to discharge. The most negative potentials may be attained in solutions in which a quaternary ammonium hydroxide is used as supporting electrolyte. With tetra-n-butyl-ammonium hydroxide the limit is −2.7 V.

The Charging or Capacitative Current

Even when no reducible species is present in solution, an appreciable current must flow to charge the double layer capacitance at the surface of each growing drop up to the new applied potential. At the solution–electrode interface, a separation of charge takes place which makes the interface look like a large capacitor to the external circuitry. Current is required to charge this capacitor, in addition to the current required by any reacting species. Because the electrode surface repeatedly grows to a maximum, then suddenly falls to zero as the drop detaches, the current flowing in the system fluctuates in the same fashion. More importantly, this charging current appears as a surge at the beginning of each drop when a new capacitor must be charged. The magnitude of this surge increases with applied potential, because the capacitor must be charged to a higher potential, thereby producing a highly sloping base line.* This charging current is the principal factor limiting the sensitivity of polarography and its accuracy at low concentrations. At concentrations of electroactive species of $10^{-3}M$ or greater, the charging current is negligible compared to the faradaic current and may be ignored. At concentrations of $10^{-4}M$ the charging current is an appreciable fraction of the total current and a correction must be made for it. At concentrations around $10^{-5}M$, the charging current is usually larger than the faradaic current and the precision of the polarographic determination depends principally on how precisely the contribution of the charging current can be estimated, compensated, or eliminated at the potential at which the diffusion current is measured.

Diffusion Current

The theoretical equation, which has become known as the Ilkovic equation, for the faradaic diffusion current is†

$$(i_d)_{av} = 607 \, nCD^{1/2} m^{2/3} t^{1/6} \tag{24-4}$$

---

*The capacity of a mercury surface in dilute aqueous chloride or nitrate solution is about 44 μF/cm² on the positive side of the electrocapillary maximum and about half this value on the negative side.
†With instantaneous current, the constant is 708.

where $(i_d)_{av}$ is the average current (in microamperes) flowing during the life of a drop, n is the number of equivalents per mole of the electrode reactant, D is the diffusion coefficient of the electroactive substance in square centimeters per second, C is the concentration of the electroactive material in millimoles per liter, m is the mass rate of flow of mercury through the capillary in milligrams per second, and t is the drop time in seconds. For a typical dropping mercury electrode with the following characteristics: $m = 2$ mg/sec, $t = 4$ sec, and taking D as 1×10^{-5} cm^2/sec, the response would be $i_d/nC = 3.8$ μA/mequiv/liter.

A more dimensionally correct modification of the diffusion equation, which recognizes that the diffusion is toward a curved surface, is

$$i_d = 607\, nCD^{1/2} m^{2/3} t^{1/6} \left(1 + 34\, \frac{D^{1/2} t^{1/6}}{m^{1/3}}\right) \qquad (24\text{-}5)$$

For typical values of D, m, and t, the last term is roughly 5–10% of the first term. Equation 24-5 should be used whenever maximum precision is desired for comparing results from different capillaries and for calculating the diffusion coefficient. For analytical applications of polarography, the original Ilkovic equation is adequate and much more convenient; its errors tend to cancel in practical use.

Factors Affecting the Diffusion Current

The Ilkovic equation indicates that the diffusion current should increase directly with the sixth root of the drop lifetime. Galvanometers usually employed have a 3- to 6-sec time period and therefore are unable to follow the periodic growth and fall of the current with each individual drop. The sawtoothed waves actually observed (Fig. 24-3c) correspond to the oscillations about the true average current to which the Ilkovic equation refers. In measuring the diffusion current, therefore, one should measure the average of the oscillations.

To obtain the true diffusion current of a substance, a correction must be made for the residual current i_r. The most reliable method for making this correction is to evaluate in a separate polarogram the residual current of the supporting electrolyte alone. The value of the residual current at any particular potential of the dropping electrode is then subtracted from the total current observed. In practice, an adequate correction can be obtained by extrapolating the residual current portion of the polarogram immediately preceding the rising part of the polarogram, and taking as the diffusion current the difference between this extrapolated line and the current–voltage plateau. Both methods are illustrated in Fig. 24-4. Because the slope of the charging current curve is not linear with changing potential and because a change in drop time affects the charging current and the faradaic current differently, estimation of residual currents by extrapolation techniques are inaccurate and questionable at low concentrations. Even with introduction of linear compensation for residual current and a "curve follower" to permit subtraction of the residual current run on a blank solution, the lower limit with sophisticated instrumentation and an ideal reductant lies in the concentration range of 0.2 to $2 \times 10^{-6} M$.

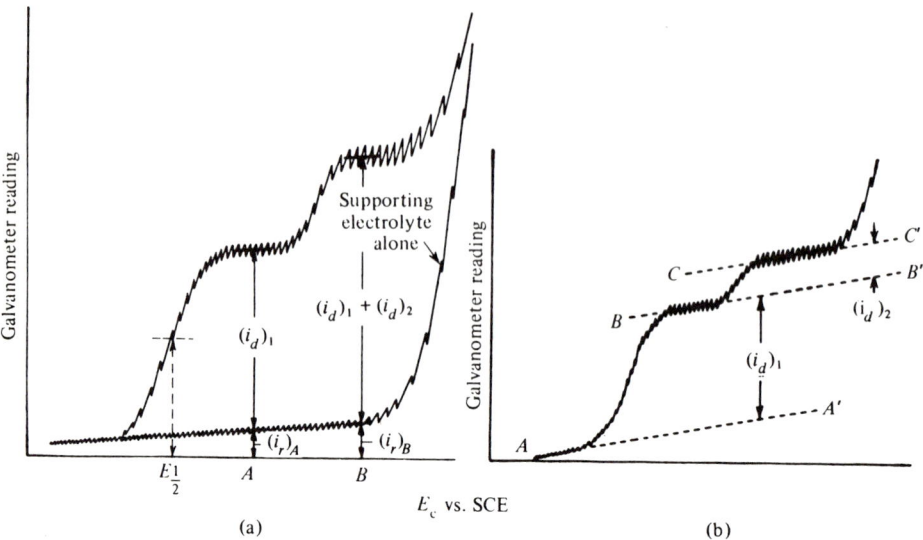

FIGURE 24-4 Measuring a diffusion current: (a) exact method and (b) extrapolation method.

The Ilkovic equation points out two facts: (1) The observed diffusion current is directly proportional to the concentration of electroactive material. This relationship is the foundation of quantitative polarography. (2) The diffusion current is proportional to the product $m^{2/3} t^{1/6}$. The quantities m and t depend on the dimensions of the dropping capillary microelectrode and on the pressure exerted on the capillary orifice due to the height of the mercury column attached to the electrode. An increase in pressure will not alter the size of the individual drops, which is a function of the capillary bore, but it will increase the number of drops forming in a given time and consequently the total electrode area exposed to the solution. A mercury reservoir of large area is customarily attached to the mercury column to prevent any change in height of the column during a series of analyses.

The drop time varies as a function of the emf impressed across the polarographic cell. Actually the drop time follows very closely the electrocapillary curve of mercury,* as shown in Fig. 24-5. As the emf is increased, the drop time first increases, then passes through a maximum at about −0.52 V, and decreases rapidly with increasing negative cathodic potential. The product $m^{2/3} t^{1/6}$ is less affected because it is influenced by the sixth root of t only, and for practical purposes, may be assumed constant over the range of cathode potential from zero to −1.0 V. At more negative potentials, however, its decrease is more rapid and must be taken into account.

The influence of temperature on the diffusion current is quite marked, particularly as the diffusion coefficient of many ions changes 1–2% per degree in the vicinity of 25.0°C, the standard temperature chosen for polarographic work. This implies that the temperature of the solution in the polarographic cell must be controlled to within 0.5°C.

---

*The electrocapillary curve expresses the relation between the potential of mercury and the surface tension at a mercury–electrolyte solution interface.

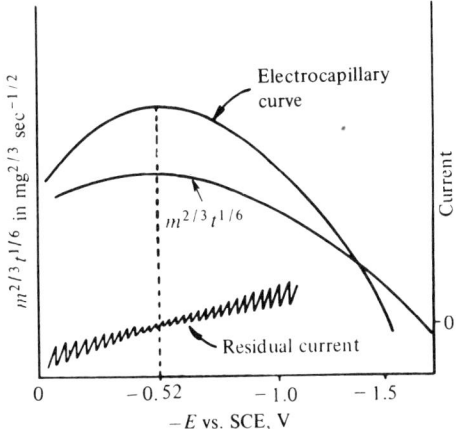

FIGURE 24-5 Comparison of the capillary characteristics and the electrocapillary curve for mercury with increasing negative potential. The magnitude of the residual current is shown on the lower curve.

Gelatin or some other maximum suppressor has a very pronounced effect on the critical drop time below which the Ilkovic equation fails. Without gelatin the Ilkovic equation fails with drop times less than 4 or 5 sec. As gelatin is added, the critical drop time decreases to the neighborhood of 1.5 sec. At faster drop rates there is appreciable stirring of the solution and a significant variation in the thickness of the diffusion layer which produces an abnormally large current. With drop times between 2 and 5 sec, 0.005–0.01% of gelatin present, and 0.5M or larger concentration of supporting electrolyte, the diffusion current will be directly proportional to concentration.

The nature and viscosity of the solvent medium also influence the diffusion current. The diffusion coefficient varies inversely with the viscosity coefficient of the solution. Ionic species vary in size and consequently in their rate of diffusion, depending on whether they are present as aquo complexes or some other type. The effect of complex formation is shown by the data in Appendix B. In some cases the nature of the complex species determines whether or not a satisfactory polarographic wave will be obtained. With tin(IV) ions, for example, no reduction is obtained in nitrate or perchlorate media in which only an aquo complex exists, whereas well-defined waves are found in chloride solutions in which the predominent species is $SnCl_6^{2-}$.

Polarographic Maxima

Current–voltage curves obtained with the dropping mercury electrode are frequently distorted by more or less pronounced maxima. These maxima vary in shape from sharp peaks to rounded humps. In all cases the current rises sharply, but instead of developing into a normal diffusion current, it increases abnormally until a critical value is reached and then rapidly decreases to the normal diffusion-current plateau. No exact explanation has been proposed. Maxima are especially prevalent when the decomposition potential is considerably removed from the electrocapillary zero of mercury.

Whatever the cause, maxima must be eliminated in order to obtain the true diffusion-current plateau. They can usually be suppressed by surface-active agents. Gelatin is often used, but the amount present in the solution must be carefully controlled and should lie between 0.005 and 0.01%. Less is useless, and more will suppress the diffusion current. Agar and methyl cellulose are also employed. Generally the proper amount of suppressor is added to every polarographic solution during the preparative step as a precautionary measure.

24.3 THE HALF-WAVE POTENTIAL

The electroactive material in polarography is characterized by its half-wave potential, $E_{1/2}$. This is the potential at the point of inflection of the current–voltage curve, one-half the distance between the residual current and the final limiting current plateau, as shown in Fig. 24-4. The significance of the half-wave potential is demonstrated by an oxidation–reduction system

$$\text{ox} + ne^- \rightleftharpoons \text{red} \tag{24-6}$$

The reversible potential of the system as it exists at the electrode–solution interface will be recorded on the polarogram. This electrochemical equilibrium may be represented as follows:

$$E = E° - \frac{0.0591}{n} \log \frac{[\text{red}]_i}{[\text{ox}]_i} \tag{24-7}$$

where the subscripts denote concentrations at the electrode–solution interface.

Assume, for example, that the solution at the electrode surface consists entirely of the oxidized form before the commencement of the current–voltage scan. As soon as the applied emf is made large enough to reduce some of the oxidant, the concentration of oxidant at the electrode surface begins to decrease. Some ions will move in from the bulk of the solution as the concentration gradient builds up between the electrode surface and the bulk of the solution. The observed current depends upon the rate of diffusion established by the concentration gradient

$$i = K([\text{ox}] - [\text{ox}]_i) D_{\text{ox}}^{1/2} \tag{24-8}$$

where K includes capillary characteristics and other terms from the Ilkovic equation. When the current attains the limiting value represented by the diffusion-current plateau, the concentration of oxidant at the electrode–solution interface will be essentially zero, and

$$i_d = K[\text{ox}] D_{\text{ox}}^{1/2} \tag{24-9}$$

Solving Eq. 24-8 for $[\text{ox}]_i$, and then combining with Eq. 24-9,

$$[\text{ox}]_i = \frac{i_d - i}{K D_{\text{ox}}^{1/2}} \tag{24-10}$$

For metals that form amalgams with the dropping electrode, the concentration of metal amalgam at the surface of the drop is directly proportional to the current on the current–voltage curve, and generally the concentration of reductant formed is proportional to the observed current, so

$$i = K \, [\text{red}]_i \, D_{\text{red}}^{1/2} \tag{24-11}$$

Solving for $[\text{red}]_i$ and substituting the result into Eq. 24-7 along with Eq. 24-10, the potential of an oxidation–reduction system can be expressed as

$$E = E° - \frac{0.0591}{n} \log \frac{i}{i_d - i} + \frac{0.0591}{n} \log \left(\frac{D_{\text{red}}}{D_{\text{ox}}} \right)^{1/2} \tag{24-12}$$

By definition, the half-wave potential is the point where

$$i = \frac{i_d}{2}$$

At this point the first logarithmic term becomes zero and,

$$E_{1/2} = E° + \frac{0.0591}{n} \log \left(\frac{D_{\text{red}}}{D_{\text{ox}}} \right)^{1/2} \tag{24-13}$$

This equation represents the current–potential relation for a reversible voltammetric wave. The half-wave potential is a quantity characteristic of a given electroactive species in a given medium. It is usually very close to the value of the standard potential of the couple since the remaining logarithmic term is constant and often nearly zero.

If a solution contains an electroactive species in which both the reduced and oxidized forms are soluble ions, the reaction may be represented as

$$\text{ox}^{a+} + ne^- \rightleftharpoons \text{red}^{(a-n)+} \tag{24-14}$$

The dropping mercury electrode then should behave as an inert electrode just as would a platinum electrode. If both the oxidation of the reduced form and the reduction of the oxidized form are rapid and reversible and if both the oxidized and reduced forms are initially present in the solution, then a polarographic wave would look like Fig. 24-6.

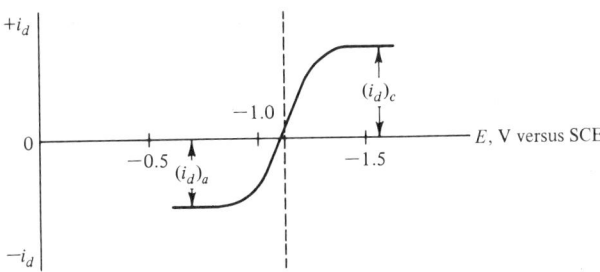

FIGURE 24-6 Idealized polarographic curve for reversible system, $\text{Ox}^{a+} + ne^- \rightleftharpoons \text{red}^{(a-n)+}$, $(i_d)_a$ = anodic limiting current and $(i_d)_c$ = cathodic limiting current.

The anodic current $(i_d)_a$ arises when the reduced form is being oxidized at the electrode and vice versa. Under these conditions the current–potential relation will assume the form

$$E = E_{1/2} + \frac{0.0591}{n} \log \frac{(i_d)_c - i}{i - (i_d)_a} \qquad (24\text{-}15)$$

and $E_{1/2}$ will be independent of concentration and equal or close to $E°$ for the reaction when

$$i = \frac{(i_d)_c + (i_d)_a}{2}$$

There is no characteristic half-wave potential for a reversible reaction that yields an insoluble product on the electrode surface, such as the deposition of silver on a platinum microelectrode. The potential at $i = 0.5 i_d$ is dependent on the bulk concentration of the metal and the stirring rate.

Reversible and Irreversible Reactions

For the current–potential relations described above, the rate for the electron transfer at the electrode surface was assumed to proceed fast enough to maintain the surface concentrations of reactants and products very close to their equilibruim values. When this is true, the reaction is termed *reversible*. If the energy of activation for the electron transfer reaction is large, and the rate of electron transfer correspondingly slow, the concentrations at the electrode surface will not be equilibrium values and the Nernst equation is not applicable. Voltammetric reactions of this type are termed *irreversible*. To determine whether an electrode reaction is reversible, one should prepare a plot of E versus $\log i/(i_d - i)$. For a reversible electrode reaction, a straight line should result whose slope is $0.0592/n$. Furthermore, $E_{1/2}$ for the cathodic reduction should coincide with $E_{1/2}$ in the anodic oxidation reaction, after a correction is made for the iR drop across the electrode–solution interface. When the electrode reaction is irreversible, a straight line may still result but the slope of the log plot will differ from the theoretical value.

In essence, there is no such thing as a perfectly reversible electrode reaction. In a practical sense a reaction is termed reversible if, within the limits of error of experimental measurement, its behavior follows the Nernst equation under the given experimental conditions. Consequently, a reaction that appears to be reversible under one set of experimental conditions may not behave reversibly under a different set of experimental conditions. For example, the reaction may become irreversible if it is required to proceed at a much faster pace, or if it is carried out in the presence of a masking agent which forms a complex with unfavorable electron transfer reaction-rate characteristics.

24.4 INSTRUMENTATION

Conventional polarography uses a dropping mercury electrode as the microelectrode and a layer of mercury as the counter electrode. A typical electrical circuit and polarographic cell are shown in Fig. 24-7. A suitable slidewire is a linear 10-turn Helipot potentiometer. The resistance of the potentiometer should not exceed 100 Ω in order that the current flowing through the cell always remains a negligible fraction of the current flowing

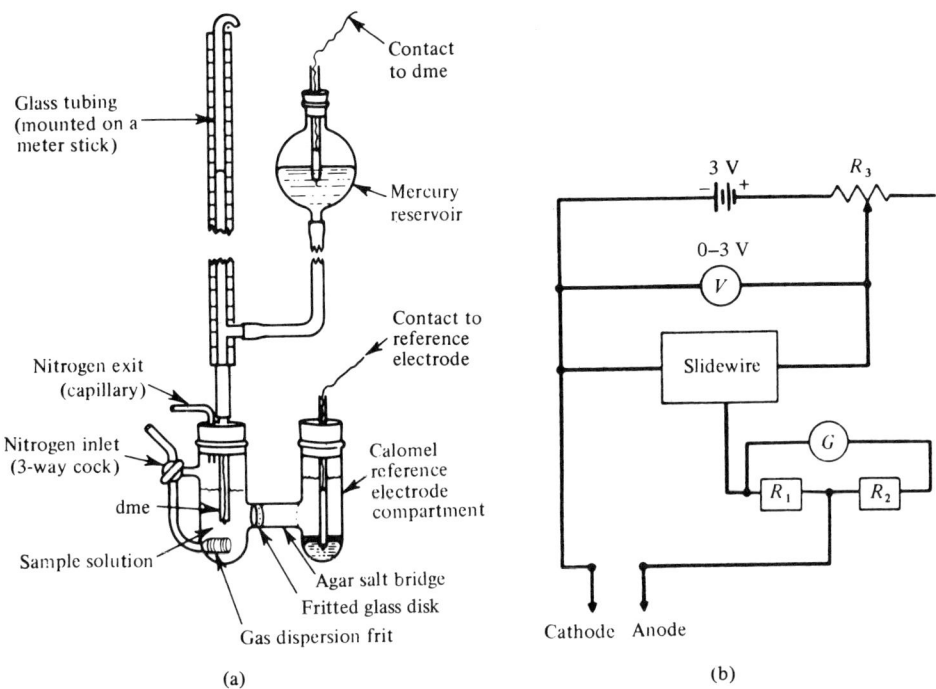

FIGURE 24-7 (a) H-cell and reservoir arrangement for polarographic cell. (b) Basic circuit for obtaining classical dc polarograms.

through the slidewire and does not affect the iR drop at any point along the slidewire. The potential drop across the slidewire is adjusted to any desired value by inserting two dry cells (3.0 V) and regulating the voltage by means of a radio potentiometer (about 500 Ω) placed in series. The voltage span selected is indicated by the voltmeter, V. To measure the current a galvanometer with sensitivity of 0.005 μA/mm scale deflection is satisfactory. The shunt for the galvanometer can be an Ayrton shunt or two 1000-Ω resistance boxes connected as shown. Keeping the total resistance at some fixed value, for example, $R_1 + R_2$ = 1000 Ω, the sensitivity of the galvanometer circuit will be given by $S_g R_1 /(R_1 + R_2)$, where S_g is the sensitivity of the galvanometer alone.

The circuit shown in Fig. 24-7 can be made the basis of a simple automatic recording polarograph if the slidewire is driven at 1 rpm with a synchronous motor, and if the galvanometer is removed and a recording millivolt potentiometer connected across R_1. Capacitive current and drop-induced fluctuations limit the use of simple battery-potentiometer polarographs to systems in which test element concentrations are high enough to yield polarograms undisturbed by these difficulties.

A mercury pool at the bottom of the polarographic cell is the simplest form of a counterelectrode. Because the layer of mercury has a large area and the current is generally very small, the concentration overpotential at this electrode is negligible and its potential may be regarded as constant. In chloride solutions it maintains approximately the potential of the calomel electrode of the particular chloride-ion concentration. While convenient, the mercury pool never possesses a definite, known potential. In the absence of chloride ions or another depolarizing electrolyte, and particularly in nonaqueous solutions, the potential does not attain a steady value. Moreover, in the presence of substances capable of forming complexes with mercury ions, the dissolution potential of mercury will be shifted to more negative reduction potentials, thus compressing the useful range of the dropping mercury electrode. In order to eliminate the possibility of unknown or nonreproducible anode potentials, it is necessary to replace the mercury pool with a separate reference (external) electrode connected to the polarographic cell through a salt bridge. The SCE is commonly employed, and it is almost universal practice in polarography to express half-wave potentials with reference to this electrode.

At potentials where mercury oxidizes, the dropping mercury electrode can be replaced by various electrode materials including pyrolytic graphite, carbon paste, wax-impregnated carbon rods, boron carbide, and platinum.

Drop Detachment

Mercury drop synchronization is essential for reproducible measurements in single-sweep polarography and desirable in other types of polarography. The detachment of a drop of mercury from the capillary is accomplished by an electromechanical drop dislodger. One type moves the capillary away from the drop at a fixed time interval; others deliver a sharp knock to the capillary. At the same moment a trigger signal is sent to the time-base. Each drop is permitted to grow until its area changes the least, usually after the first 1.5–2.0 sec of drop life. Then the current in the cell is sampled just prior to the dislodgement of the drop. The sampled current is stored in a memory and read out to the recorder until the next measurement is taken. In this way the recorder plots a curve representing the peak current flowing during each drop's life cycle, and is devoid of drop-induced fluctuations. The EG&G Princeton Applied Research Corporation markets a Model 303 Static Mercury Drop Electrode which, by valving the mercury flow, first allows a drop to form, then stops the flow to maintain a static drop while measurements are being made, then finally dislodges the drop and starts a new cycle.

Three-Electrode Potentiostat

A characteristic of modern polarographic instrumentation is potentiostatic control of the working electrode potential. In classical polarography the dc ramp voltage was applied across the entire cell, rather than across the working electrode–solution interface. Current flowing through a polarographic cell of high resistance causes an appreciable voltage drop. Thus, the potential at the dropping mercury electrode differs from that at the other end of the polarographic cell. This causes several deleterious effects: (1) The half-wave potential (or peak potential in derivative methods) is shifted to more negative values. (2) The total reduction current is less. (3) The distortion of the shape of the wave and slope of the polarographic step is often severe. Despite these shortcomings, it is desirable to be able to use nonaqueous solvents of high resistance and aqueous electrolytes of millimolar concentrations. Three-electrode potentiostatic control makes this possible.

A third electrode, a reference electrode of constant potential, can be considered a probe that is positioned as close as possible to the mercury drop, and can sense the potential at that point. It is connected to the polarograph (point C in Fig. 24-8) through a circuit which

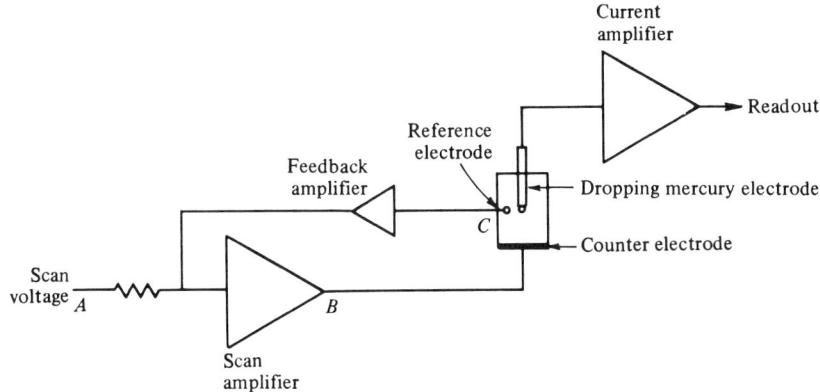

FIGURE 24-8 Schematic diagram of a three-electrode potentiostat.

draws essentially no current; the input impedance is in excess of 10^{11} Ω. If the voltage sensed by the reference electrode is less than the dc ramp voltage provided to the scan amplifier, the feedback to the scan amplifier from the operational amplifier control loop will provide a corrective voltage that will change the applied emf enough to compensate for the resistance of the cell and electrolyte. Thus, the voltage measured at point C should always be the same as that applied to point A and, if momentarily it is not, the voltage at point B will automatically increase to maintain point C equal to point A. Present-day analog potentiostats operate in the same manner but they are extremely fast, with a rise time on the order of microseconds when driving a resistive load. A set control potential can be maintained to within better than 1 mV for long periods, and the reference electrode is loaded only to the extent of picoamperes or less.

A modern polarograph, the EG&G Princeton Applied Research Corporation Model 174A Polarographic Analyzer, is shown in Figs. 24-9 and 24-10. The instrument is very

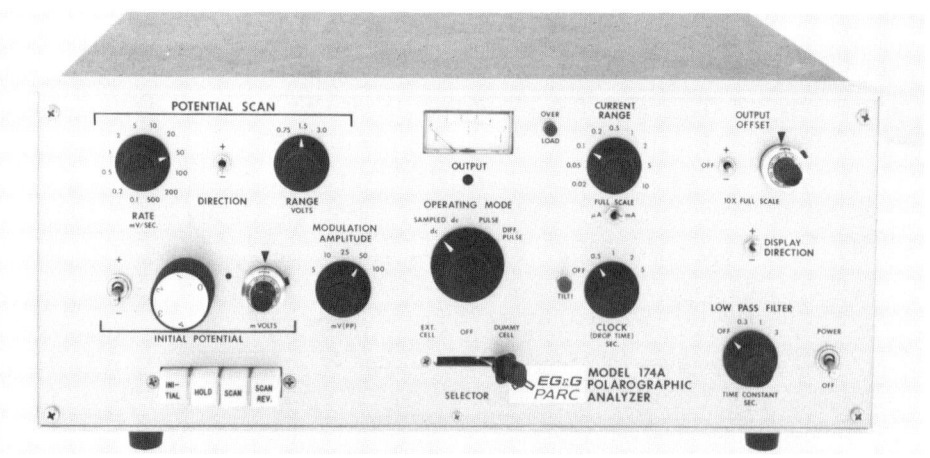

FIGURE 24-9 EG&G Princeton Applied Research Corporation Model 174A Polarographic Analyzer. (Courtesy of EG&G Princeton Applied Research Corp.)

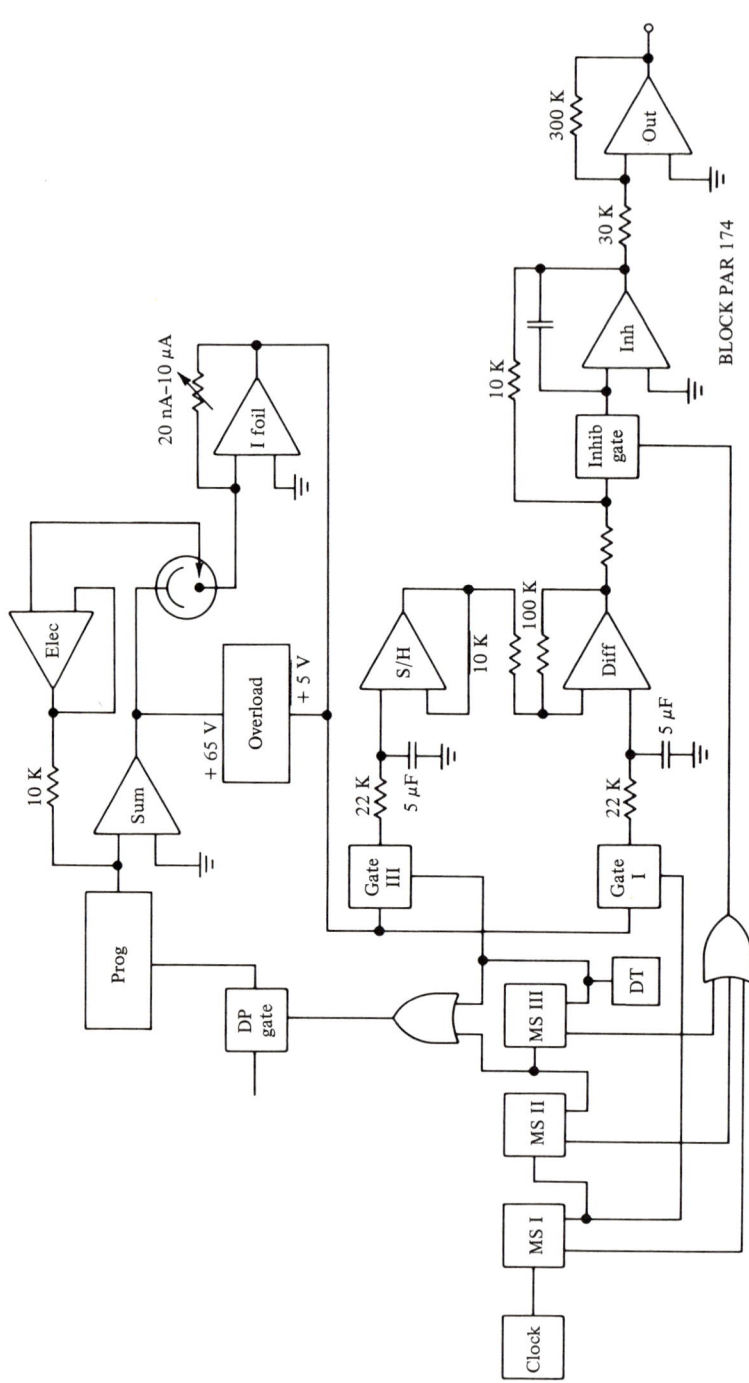

FIGURE 24-10 Block diagram of EG&G PARC Model 174A Polarographic Analyzer. (Courtesy of EG&G Princeton Applied Research Corp.)

versatile and is useful for dc polarography, sampled dc polarography, differential and integral pulse polarography, direct and differential pulse stripping voltammetry, linear potential sweep voltammetry and, with attachments, phase-sensitive ac polarography. It has an initial potential range up to ±5 V, a scan rate from 0.1 mV/sec to 5 V/sec, and a scan range of 0.75, 1.5, or 3.0 V, either positive or negative starting at the initial potential. The current range is from 0.02 μA to 10 mA full scale. It has available 5, 10, 25, 50, or 100 mV pulses for differential pulse polarography, a clock with 0.5, 1, 2, and 5 sec positions for timing pulse applications and control of a static mercury drop electrode or a drop timer, and an output offset of up to 1000% of full-scale output to offset large base line signals.

The EG&G Princeton Applied Research Corporation also makes a simpler Model 364 Polarographic Analyzer for routine analyses at minimum cost or for student use and an elaborate Model 374 Microprocessor-Based Polarographic Analyzer with a built-in microprocessor which controls all aspects of the analysis including instrument setup, sample deaeration, data recording, background subtraction, concentration calculation, and chart labeling.

24.5 MODERN VOLTAMMETRIC TECHNIQUES

For the remainder of this chapter it will be assumed that all the instrumentation discussed possesses potentiostatic capabilities. Only those techniques that have found the greatest degree of analytical application will be treated. Commercial instrumentation for all these techniques is available, and all techniques are applicable to electrodes other than the standard dropping mercury electrode.

In any electrochemical system, an equilibrated diffusion layer is established between the electrode and the bulk of the solution. Once the reaction potentials have been reached, the concentration of the electroactive species of interest changes from the bulk concentration value at positions far removed from the electrode surface to essentially zero at the electrode surface. When the classical slow scan rates are employed, the slope of the concentration gradient within the diffusion layer is determined primarily by the rate of depletion of the electroactive species at the electrode surface. It varies from essentially zero at potentials significantly more positive than the reduction potential, to a value governed by the concentration and diffusion coefficient of the electroactive species at potentials well past the reduction potential. However, when the polarographic method uses rapid changes in potential, either because the scan rate is fast or because a pulse modulation of some type is employed, the slope of the concentration gradient at any particular potential will be greater than in the slow scan instance. The bulk concentration will be closer to the electrode surface, the number of electroactive particles arriving at the surface per unit time will be greater, and larger current signals will result.

Derivative Polarography

Derivative methods of polarographic analysis depend on differentiation of the current–voltage curve while it is being recorded. The resulting polarogram consists of a series of peaks superimposed upon the background. Measurement of a peak rather than a plateau is inherently more sensitive, particularly when the nonlinearity of the base line and the oscillations accompanying drop formation are considered. An elegant experimental answer to the problem of derivative polarography is the apparatus of Kelley et al.[1] in which a three-electrode system using potentiostatic control of the dropping electrode potential is employed to yield first derivatives of theoretical slopes. A special RC filter network is used to smooth the curves. For the case of general interest, that is, negligible solution resistance and appreciable residual current,

$$\left[\frac{d(\Sigma i)}{dt}\right]_{max} = -\left(\frac{dE}{dt}\right)\left(\frac{ni_d}{4RT/F - di_r/dE}\right) \quad (24\text{-}16)$$

where Σi is the total current (faradaic plus residual), $-dE/dt$ is the rate of potential scan, and di_r/dE is the slope of the residual current curve at the half-wave potential. According to Eq. 24-16 the sensitivity of derivative polarography (about $10^{-7}M$) is proportional to the scan rate up to the maximum of about $40/n$ mV/min. As the scan rate increases, the potential at the electrode–solution interface, through the capacitive nature of the interface, penetrates deeper into the solution and causes a larger number of ions to be reduced. The maximum scan rate at which undistorted peaks may be observed is set by time lags introduced by the RC filter network. Important also is the slope of the residual current curve, which varies with the resistance of the solution. The main advantage of this technique is the minimization of the effect of diffusion currents due to the presence of more easily reducible constituents. Polarographic waves differing in half-wave potential by $90/n$ mV are completely resolved, whereas smaller differences lead to peaks that are sufficiently separated to be used analytically (Fig. 24-11). Because the derivative signal eturns to the base line, each wave can be recorded at its maximum sensitivity. The peak potential is typically $28/n$ mV more negative than the half-wave potential reported for classical polarography.

Pulse Polarography[2]

Pulse polarography takes advantage of the fact that, following a sudden change in applied potential, the capacitative current surge decays much more rapidly than does the faradaic current. In this technique a small amplitude voltage pulse, in addition to the linearly increasing dc ramp (about 1 mV/sec) normally used for dc polarography, is applied to the polarographic cell. As each mercury drop forms it is allowed to grow for a period of time, perhaps 1.9 sec at the dc ramp potential, after which a sudden voltage pulse of perhaps 50-msec duration is applied. The pulse is synchronized with the maximum growth of the mercury drop, if a dropping mercury electrode is used. The current is measured 40 msec after the application of the pulse, to allow time for the charging current to decay to a very low value. The capacitative current actually decays exponentially at a rate governed by the magnitude of the capacitance and series resistance of the system. During this time interval the faradaic current also decays somewhat but does not reach the diffusion-controlled level because the concentration gradient at the instant of current measurement is considerably larger. Each succeeding drop is polarized with a

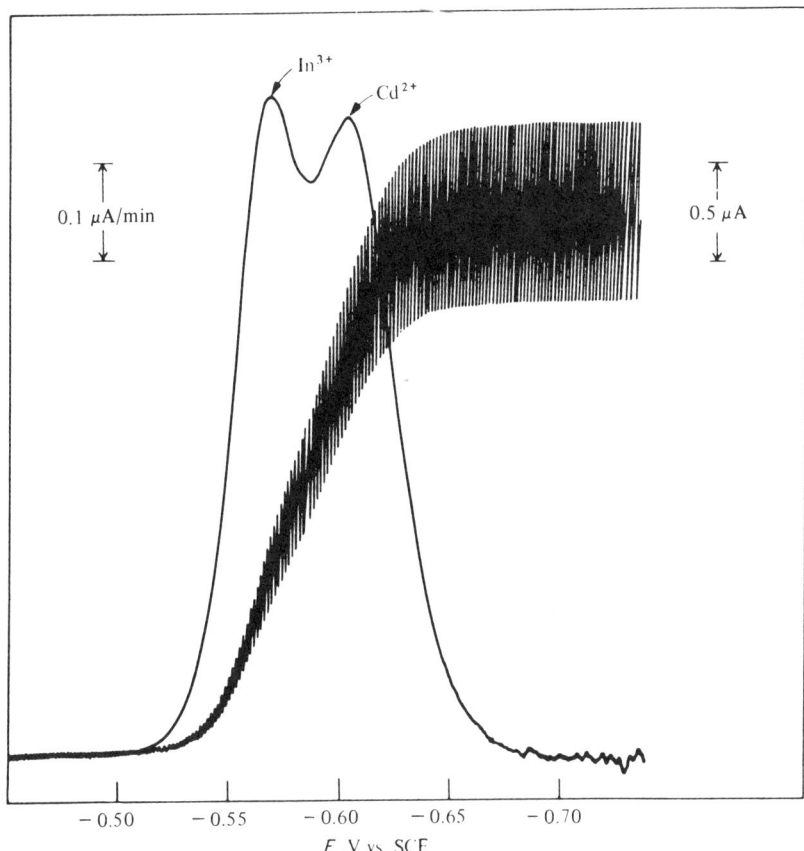

FIGURE 24-11 Comparison of resolution of a regular and a derivative polarogram recorded with a solution containing $1 \times 10^{-4} M$ In^{3+} and $2 \times 10^{-4} M$ Cd^{2+} in $0.1 M$ KCl.

somewhat larger pulse. This method gives a current–voltage curve similar to that obtained in dc polarography except for the cancellation of the capacitive components. The measured signal is the faradaic current that flows at the pulse potential minus any faradaic current flowing due to the fixed dc potential (Fig. 24-12). The limiting current is given by the Cottrell equation,

$$i_{\lim} = nFCA \sqrt{\frac{D}{\pi t}} \qquad (24\text{-}17)$$

In comparison with classical dc polarography, the sensitivity is about 6.5 times better. A much larger gain in real sensitivity is achieved through the virtual elimination of the charging current.

Derivative pulse polarography employs two current sampling intervals of equal time periods. The first sample period occurs just before the pulse application; the second sample period occurs at the end of the pulse. The current samples are stored on memory capaci-

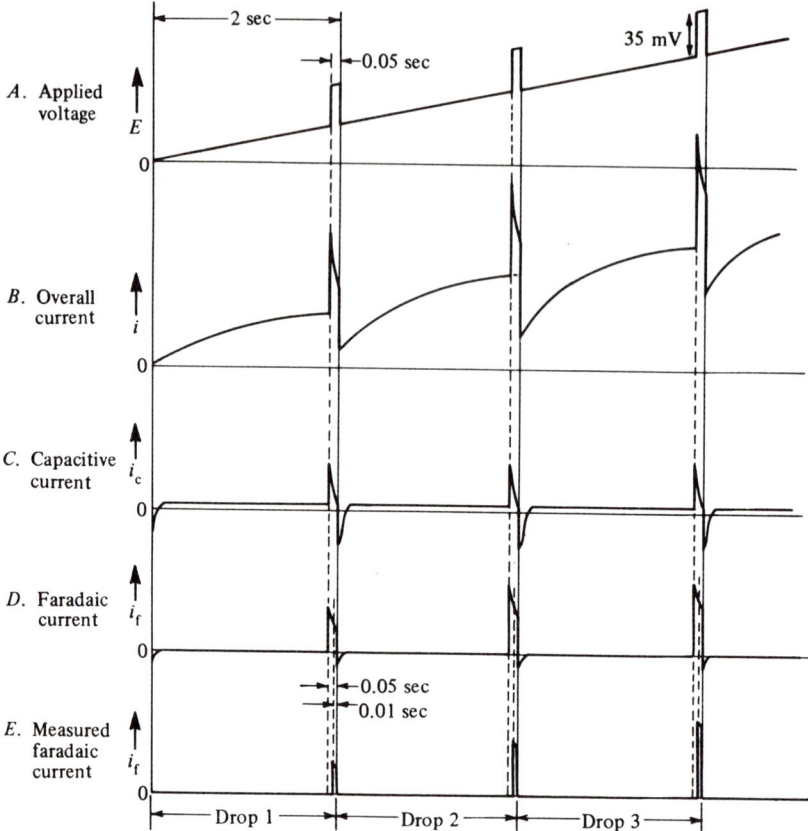

FIGURE 24-12 Current and voltage diagrams illustrating the operation of a pulse polarograph. Curve A, a linearly increasing scan voltage upon which a 35-mV pulse is superimposed during the last 50 msec of the 2-sec drop time; B, the overall current flowing through the cell as a result of the applied voltage; C, the capacitive component of cell current; D, the faradaic component of the cell current measured above the "dc" background; and E, the net current signal measured during the last 10 msec of the life of the drop after the capacitive current has decayed to near zero.

tors and the difference is displayed on a recorder operated in a display-and-hold manner. Differences in the two stored samples occur only in the region of the half-wave potential where the current is changing rapidly with the potential. The voltammogram recorded in this way is a peak-shaped incremental derivative of a conventional dc polarogram. Because the applied pulse is held for an appreciable length of time, the system response is not strongly dependent on the electrode kinetics, at least not to the extent expected of ac polarography. This implies that pulse polarographic techniques may be readily used for organic and other electrochemically irreversible systems. For pulses of small amplitude, ΔE, the peak current is given by

$$\Delta i_{max} = \frac{n^2 F^2}{4RT} AC(\Delta E) \sqrt{\frac{D}{\pi t}} \qquad (24\text{-}18)$$

whereas for pulses of large amplitude, it is given by

$$\Delta i_{max} = nFAC\sqrt{\frac{D}{\pi t}}\left(\frac{\sigma - 1}{\sigma + 1}\right) \qquad (24\text{-}19)$$

where $\sigma = \exp[(\Delta E)nF/2RT]$. In the limit, for $\Delta E \gg RT/nF$, $(\sigma -1)/(\sigma + 1)$ approaches unity and i_{max} becomes the limiting current as given by the Cottrell equation. The use of a large pulse amplitude allows one to obtain large signals from extremely dilute solutions, albeit with some distortion in the curve shape, because the pulse amplitude is an appreciable portion of the total polarographic wave. In principle the derivative mode is less sensitive than the normal mode, but the resolution is better, about $50/n$ mV. It is capable of detecting $10^{-8} M$, a 100-fold increase over classical polarography and a 10-fold increase over derivative dc polarography.

ac Polarography[3]

In the ac technique a potential periodic in time, such as a sine wave, and of relatively small amplitude (1–35 mV) and low frequency (10–60 Hz), is superimposed upon the slow linear voltage sweep of dc polarography. This sinusoidal variation in potential is similar to that employed in cyclic voltammetry with the exception that the potential excursions are of much smaller magnitude, typically 10 mV peak-to-peak. The direct-current component of the total current is blocked out and only the rectified and damped alternating component is displayed as a function of dc potential. By looking only at the alternating portion of the current that flows and detecting its amplitude, one is in effect looking at the difference in current that flows between the minimum and maximum applied potentials during the modulation period. The current is sampled just before the mercury drop is dislodged. A peak output signal with its maximum amplitude at the half-wave potential, rather than a step-shaped wave, is produced, because the reversible wave attains its maximum slope at the half-wave potential; therefore, a given ac signal causes the periodic changes in concentration of the electroactive species to be maximal at this potential. At other dc potentials along the wave, the sinusoidal potential variation causes less of a perturbation of the dc surface concentration and the ac current is correspondingly less. An important characteristic of ac polarography is that it responds only to reversible electrode reactions. This limits its applicability, but in some cases can be advantageous in avoiding interferences.

Detection of only the ac component allows one to separate the faradaic and capacitative currents because of the phase difference between them. By employing a phase-sensitive lock-in amplifier, one can select either the faradaic (phase shifted 45° from the applied potential) or the capacitative current (phase shifted 90°), while rejecting the other. The capacitative current is important in studies of kinetics and adsorption. The maximum height of the faradaic alternating current is given by

$$\Delta i_{max} = \frac{n2F^2 AVC\omega^{1/2} D_{ox}^{1/2}}{4RT} \qquad (24\text{-}20)$$

where A is the electrode area, V is the amplitude of the voltage signal, and ω is the angular frequency. Because even a moderate cell resistance serves to mix the phases, successful application of the phase discriminator requires very low cell resistances.

Another successful approach to the separation of faradaic and capacitive components of the cell current is through the measurement of second (or higher) harmonics of the alternating current. The method is based on the fact that the capacitive current varies essentially linearly with voltage, whereas the faradaic process varies nonlinearly. Thus, although the first derivative of the capacitive current may be appreciable, the second derivative will be near zero. Whereas the base line current may be 75% of the peak current in the case of normal ac polarography, it is only about 5% or less when the second harmonics are recorded. The second harmonic, because its signal crosses through zero at the half-wave potential, is very useful for resolution problems, especially in complex mixtures.

Cyclic or Fast Linear Sweep Voltammetry

Cyclic voltammetry involves the measurement of current–voltage curves under diffusion-controlled, mass transfer conditions at a stationary electrode, utilizing symmetrical triangular scan rates ranging from a few millivolts per second to hundreds of volts per second. The triangle returns at the same speed and permits the display of a complete polarogram with cathodic (reduction) and anodic (oxidation) waveforms one above the other (Fig. 24-13). Two seconds or less is required to record a complete polarogram following deaeration of the solution. This technique yields information about reaction reversibilities and also offers a very rapid means of analysis for suitable systems. The method is particularly valuable for the investigation of stepwise reactions, and in many cases direct investigation of reactive intermediates is possible. By varying the scan rate, systems exhibiting a wide range of rate constants can be studied, and transient species with half-lives of the order of milliseconds are readily detected. The method can be applied to stationary electrodes as well as to a single mercury drop, and to reactions for which stripping analysis is inapplicable due to highly irreversible electrode processes or the formation of solution-soluble reaction products.

In cyclic voltammetry the typical concentration gradient prevails at potentials more positive than the reduction potential. Once this point has been passed, however, the rate of variation of potential is too rapid for diffusional processes to maintain equilibrium

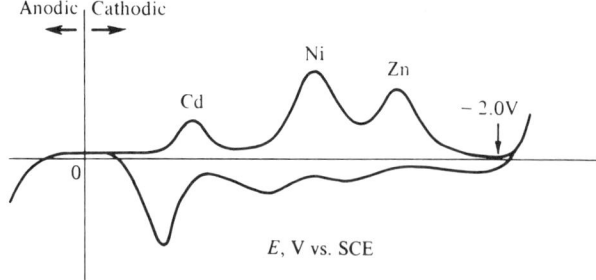

FIGURE 24-13 Cyclic voltammetric plot of Cd, Ni, and Zn in 0.1M KCl. The initial scan in the cathodic direction from zero to −2.0 V causes the reduction of Cd, Ni, and Zn. The reverse scan from −2.0 V to zero causes the oxidation of the metals and produces a peak moving in a downward direction whose height is related to both concentration and reversibility of the reaction.

with the bulk of the solution. More and more material is consumed and the diffusion layer extends further and further into the solution. Unlike the case with dropping electrodes, this process is not periodically reversed by the stirring associated with drop fall, so that the signal decay continues and a peak-shaped curve is obtained. Furthermore, in the course of the cathodic variation in potential, the reduced form of the reactant is produced in the vicinity of the electrode while a depletion of the oxidized form occurs. Given sufficient time, the reduced form would diffuse into the bulk of the solution, but the potential is taken back to the initial value at a rate such that some of the reduced form is still present at the electrode surface and undergoes a process of oxidation back to the form of the couple initially present in the solution. A maximum occurs at $(E_{dc} - E_{1/2})n =$ 0.0285 V. Thus the peak potential is

$$E_{peak} = E_{1/2} - 0.0285 \tag{24-21}$$

for a reduction reaction. For an oxidation reaction the sign of the numerical term is reversed. On the reverse scan, the position of the peak depends on the switching potential. As this potential moves more negative, the position of the anodic peak becomes constant at $29.5/n$ mV anodic of the half-wave potential. With the switching potential more than $100/n$ mV cathodic of the reduction peak, the separation of the two peaks will be $59/n$ mV and independent of the rate of potential scan. This is a commonly used criterion of reversibility. Reversibility can also be ascertained by plotting $(E_{cath} - E_{anod})$ as a function of $\sqrt{\text{velocity}}$, which should be a straight line if reversible.

To do qualitative analysis with cyclic voltammetry, one observes four characteristics of an electroactive substance: peak potential, wave slope, reversibility, and the effect of changing the supporting electrolyte. To illustrate: An unknown sample in $0.1M$ KCl exhibits peaks at −0.40 (slope 28 mV) and at −0.60 V (slope 25 mV); on the reverse scan there is an anodic peak at −0.60 V. In considering the first peak at −0.40 V, likely choices would be lead(II) and thallium(I). Lead has a slope of 28 mV for a two-electron reduction, whereas thallium(I) has a slope of 56 mV. The wave at −0.60 V could be cadmium(II), chromium(III), indium(III), or europium(III). Cadmium has a slope of 28 mV and is reversible, indium has a slope of 19 mV and is reversible, chromium is not reversible, and europium has a peak potential near −0.7 V with a slope of 64 mV. The tentative conclusion is that the sample contains lead and cadmium. Further validation would involve the use of a different electrolyte.

Chronopotentiometry

Chronopotentiometry is based on the observation of the change in potential of a working electrode as a function of time during electrolysis. Usually the electrolysis is performed with a constant current in a quiescent solution. Ultimately, the exhaustion of electroactive substance at the surface of the electrode causes a more or less rapid change of potential. The time necessary for the potential to go from one level to another is measured. This transition time is a measure of the rate at which the concentration of the electroactive species at the electrode surface is reduced to the point where it can no longer sustain the

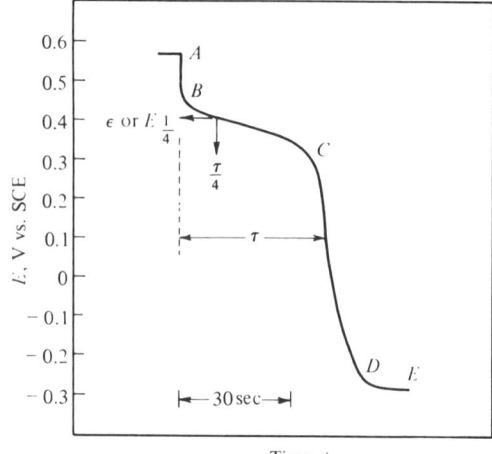

FIGURE 24-14 Chronopotentiogram for the reduction of iron(III) at a platinum cathode. The concentration of iron(III) was 0.0250M, the supporting electrolyte solution was 1M HCl, and the temperature was 25°C. A cylindrical wire electrode of area 0.281 cm^2 and radius 0.0255 cm was used, and the constant current was 2.79 × 10^{-4} A.

required current. A schematic chronopotentiogram is shown in Fig. 24-14. For a symmetrical, reversible reaction the potential of the electrode as a function of the time t obeys the relationship (at 25°C):

$$E = \epsilon - \frac{0.0591}{n} \log \frac{t^{1/2}}{\tau^{1/2} - t^{1/2}} \qquad (24\text{-}22)$$

where τ is the transition time, and ϵ, which is identical with the polarographic half-wave potential, is very nearly equal to the standard potential of the electrode reaction. The potential becomes equal to ϵ when $t = \tau/4$. With linear diffusion to a plane electrode and when diffusion is the controlling factor, the transition time (in seconds) is given by the Sand equation

$$\tau^{1/2} = \frac{\pi^{1/2} \, nFAD^{1/2} C}{2i} \qquad (24\text{-}23)$$

where i is the constant current (in amperes) imposed on the system. The constant-current density is selected so that the transition time will be less than about 30 sec, as otherwise the strong convective stirring produced by the density gradient at the electrode disrupts the diffusion layer at unshielded electrodes. The Sand equation is seldom used directly because D usually is not known with sufficient accuracy in a variety of supporting electrolytes. Instead the electrode is calibrated empirically under a given set of conditions with known concentrations of the test element. Lingane[4] has discussed the analytical aspects of chronopotentiometry.

When several electroactive species are present in the solution, the chronopotentiogram will show a succession of transition times. However, transition times are nonadditive. After the first inflection, when reaction of the second or subsequent substance begins, the first (or prior) substance still continues to diffuse to, and react at, the electrode. Consequently, the current results from both reactions at points on the second wave. With a constant total current, the current from the second reaction is smaller than it would be in the absence of the first substance, and the transition time is increased.

In the two-stage reduction or oxidation of a single substance, only the two n-values differ. Since $C_1 = C_2$, and $D_1 = D_2$, it follows that the ratio of the individual transition times is

$$\frac{\tau_2}{\tau_1} = \frac{2n_2}{n_1} + \left(\frac{n_2}{n_1}\right)^2 \tag{24-24}$$

where n_1 and n_2 are the separate number of electrons for each stage.

Both chronopotentiometry and dc polarography are based on similar principles. Because chronopotentiometry is more sensitive to the rates of the various steps that may be involved in an overall electrode process, chronopotentiograms often tend to be less well defined than polarograms, especially at small concentrations. In general, chronopotentiometry is disappointing at concentrations below about $5 \times 10^{-4} M$. It is a powerful tool for the study of electrode processes at higher concentration. Thin-layer chronopotentiometry is most important when run under conditions of quantitative electrolysis; it is discussed in the section on coulometry (Chapter 25).

Stripping Analysis[5]

The technique of stripping analysis, also called linear-potential sweep stripping chronoamperometry, involves two steps: A concentration or pre-electrolysis step in which the desired component is deposited cathodically or anodically, followed by a reverse electrolysis in which the component is determined. In the anodic variant of stripping analysis, the metal (or metals) concerned is reduced at a controlled potential for a definite time under fixed conditions of geometry and stirring. The working electrode may be a hanging mercury drop or a mercury film on a wax-impregnated graphite electrode, or in some cases an inert solid electrode such as platinum or carbon. The final anodic dissolution, or stripping process, involves a linear anodic scan in which the metal is oxidized. The resulting stripping voltammogram shows peaks, the heights (or areas) of which are generally proportional to the concentrations of the corresponding electroactive metal ions, and the potentials of which have the same qualitative interpretation as their half-wave potentials in polarography. Standards are carried through identical pre-electrolysis and stripping steps. A typical anodic scan is shown in Fig. 24-15 for a 0.100-ml sample of whole blood prepared for analysis by digestion with perchloric acid and brought to 5.0-ml volume with sodium acetate. As little as $10^{-8} M$ cadmium (corresponding to about 10^{-6} weight %) has been determined with a precision of ±3% using 15 min of pre-electrolysis and linear scan rate of 21 mV/sec. Because the standard addition method is usually used for evaluating the unknown concentration, the reproducibility will be the same as the precision. By extending the pre-electrolysis time to 60 min, the sensitivity is extended to $10^{-9} M$ but deviations will be about 10–20% at this concentration level.

Cathodic stripping analysis consists of forming an insoluble layer at the electrode surface during an anodic pre-electrolysis, and stripping it off by reverse electrolysis. Unlike

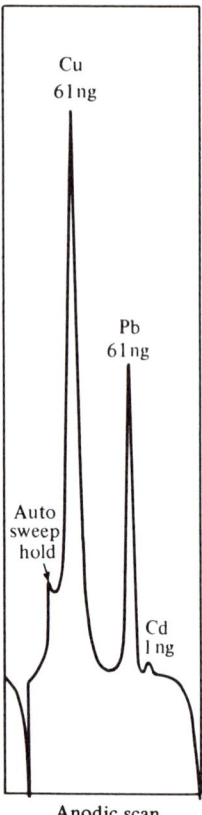

FIGURE 24-15 Stripping analysis: anodic scan. Experimental data: plating time, 30 min; plating potential, −1.0 V; sweep rate, 60 mV/sec; chart speed, 12.7 cm/min; blood sample, 0.2 ml; current range, 200 μA full scale; stripping time, ∼ 16 sec. (Courtesy of Environmental Sciences Associates, Inc.)

anodic stripping analysis, the deposition potential depends on the concentration to be determined and shifts about $60/n$ mV to more positive values for each order-of-magnitude decrease in the anion concentration. It is not possible to carry the deposition to completion because of the limitation imposed by finite solubility of the layer; usually salts of mercury(I) precipitate. Such a method has been used for determination of the halides, tungstate, and molybdate. Lower concentration limits are about $5 \times 10^{-6} M$.

Complete exhaustion of the desired component in the solution during the concentration step is not necessary and not even desirable, as long as this step is carried out under reproducible conditions. Thus an enormous saving in time is effected with only a moderate sacrifice in sensitivity. The plating potential should surpass the half-wave potential by some hundreds of millivolts until the current reaches its limiting value. Then after a short rest period to allow the amalgam concentration to become homogeneous and to ensure that convection in the solution has ceased (perhaps 30–60 sec), the anodic redissolution procedure is initiated. If desired, the electrolyte may be changed to one better suited for the stripping process. Using the hanging mercury drop electrode, the peak current during anodic scan is proportional to the square root of the scan rate; with thin-film mercury electrodes, a direct proportionality is observed. Compared with other highly sensitive electroanalytical methods, stripping analysis gives comparable performance at lower cost.

Kryger and Jagner[6] force metals into and out of a mercury film on a glassy carbon electrode several times, rapidly, averaging the signals obtained. They then remove the metals completely and obtain a background signal by again averaging several pulses. The precision obtained by this method is comparatively good.

24.6 APPLICATIONS

Determinations of inorganic or organic species that are either molecular or ionic can be performed if they undergo oxidation or reduction at a mercury electrode in the region of potential bounded at the positive limit by the potential of oxidation of mercury in the medium employed, and at the negative limit by the potential at which the supporting electrolyte or the solvent is reduced. If materials other than mercury are used for the microelectrode, the positive limit can be extended considerably and will be limited only by the oxidation of the solvent, the anion present, or the electrode material. Nonaqueous solvents can be used for organic substances that are insoluble in water, the only limitation being that if the resistance of the medium is high, the polarographic wave may be severely distorted. A precision of ±3% to ±5% is readily attainable at concentrations between 10^{-2} and $10^{-4} M$. With additional effort and care to maintain the experimental conditions constant, a precision approaching ±1% can be obtained.

Reversible organic systems are confined largely to quinones and a few other functional systems such as the phenylene diamines, which resemble quinones in forming resonating systems. However, polarography offers the possibility of characterizing oxidation–reduction properties of numerous irreversible systems. These systems involve a step with a high activation energy, and the half-wave potential is a function of the rate constant of the electrode process. Among structural factors which affect half-wave potentials are the nature of the electroactive group; that is, the group where cleavage or bond formation occurs during electrolysis, stereochemistry, and the nature of substituents. Also important is the molecular frame to which the electroactive group is attached and the groups situated in the immediate vicinity of the electroactive group.

The types of bonds that can be reduced at a dropping mercury electrode are enumerated in Table 24-1. Often the presence of a single group is insufficient to bring the wave of the substance studied to an accessible potential range. For these compounds it is necessary for the molecule to contain, in addition to the electroactive group, another activating group that affects the electron distribution in the substrate and shifts the half-wave potential of the reduction wave. In a system of conjugated multiple bonds, the whole system of conjugated bonds represents a single electroactive group. Both inductive and resonance effects can ease the reduction of groups attached to aromatic systems or to double bonds. For example, the reduction potential of the disulfide group linked to a phenyl group is –0.5 V versus SCE and to an alkyl group is –1.25 V. As the number of condensed aromatic rings increases, the reduction is made easier. Single C–X (where X is halogen)

TABLE 24-1 Reducible Organic Functional Groups

$>C=O$	Ketone	$-C{\equiv}N$	Nitrile	$-NO_2$	Nitro
$-CHO$	Aldehyde	$-N=N-$	Azo	$-NO$	Nitroso
$>C=C<$	Alkene	$-NO=N-$	Azoxy	$-NHOH$	Hydroxylamine
$\phi-C{\equiv}C-$	Aryl alkyne	$-O-O-$	Peroxy	$-ONO$	Nitrite
$>C=N-$	Azomethine	$-S-S-$	Disulfide	$-ONO_2$	Nitrate

Also dibromides, aryl halide, alpha-halogenated ketone or aryl methane, conjugated alkenes and ketones, polynuclear aromatic ring systems, and heterocyclic double bond.

bonds are usually reduced at more negative potentials than C=C–C–X, but at more positive potentials than C–C=C–X. The ease of reduction of the C–X bond increases with the increasing polarizability of the halogen, that is, C–F $<$ C–Cl $<$ C–Br $<$ C–I. One should be aware of the extent to which organic functional groups can be converted to an active polarographic group because such conversion can markedly extend the method. A number of examples are enumerated in Table 24-2. These are only a few of the many facets of organic polarography. The subject is thoroughly discussed by Zuman[7].

Oxygen Determinations

Several compact portable units are available for the determination of dissolved oxygen. The oxygen-sensing probe is an electrolytic cell with gold (or platinum) cathode separated from a tubular silver anode by an epoxy casting. The anode is electrically connected to the cathode by electrolytic gel, and the entire chemical system is isolated from the environment by a thin gas-permeable membrane (often Teflon). A potential of approximately 0.8 V (from a solid-state power supply) is applied between the electrodes. The oxygen in the sample diffuses through the membrane and is reduced at the cathode with the formation of

TABLE 24-2 Organic Functional Group Analysis of Nonpolarographic Active Groups

Functional Group	Reagent	Active Polarographic Group
Carbonyl	Girard T and D	Azomethine
	Semicarbazide	Carbazide
	Hydroxylamine	Hydroxylamine
Primary amine	Piperonal	Azomethine
	CS_2	Dithiocarbonate (anodic)
	$Cu_3(PO_4)_2$ suspension	Copper(II) amine
Secondary amine	HNO_2	Nitrosoamine
Alcohols	Chromic acid	Aldehyde
1,2-Diols	Periodic acid	Aldehyde
Carboxyl	(Transform to thiouronium salts)	$-SH$ (anodic)
Phenyl	Nitration	$-NO_2$

the oxidation product, silver oxide, at the silver anode. The resultant current is proportional to the amount of oxygen reduced. To counteract temperature effects, a thermistor is built into the sensor. The analyzer unit operates over the range from 0.2 to 50 ppm of dissolved oxygen. Gases that reduce at −0.8 V will interfere; these include the halogens and SO_2. H_2S contaminates the electrodes.

24.7 EVALUATION METHODS

Direct Comparison

The direct comparison method calls for recording the current−voltage curves of a standard solution of the test ion under the same conditions as the unknown. Then, using the Ilkovic equation in the simplified form, the diffusion current quotient, i_d/C, can be computed. When divided into the height of the unknown wave, it yields the concentration of test ion in the unknown. The unknown will be most accurately determined when the concentration of the comparison standard is about the same as that of the unknown, particularly if a nonlinear relation exists or is suspected to exist between wave height and concentration. The quantity of standard can be estimated by remembering that the diffusion current of simple ions in neutral or acid solution is about 4 μA per milliequivalent of reducible ion.

Relative measurements of this type do not demand knowledge of the exact capillary characteristics, only that they remain constant during the comparison. Likewise, temperature need not be controlled at any fixed value, merely maintained the same for all solutions. Immersion of the solution cells in a large container of water is adequate. However, it is important that the composition of the supporting electrolyte and the amount of maximum suppressor added be identical for the unknown and the comparison standard.

Standard Addition

If a single analysis is to be performed, it is possible to dispense with the preparation of a known solution that is the exact duplicate of the test solution. The polarogram of the unknown solution is recorded, then a known volume of a standard solution of the test ion is added and the polarogram repeated. From the increase in the diffusion current, the original concentration can be computed by interpolation. For the unknown solution,

$$i_d = KC_x = h \qquad (24\text{-}25)$$

and after the addition of v ml of a standard solution, whose concentration of test ion is C_3, to V ml of unknown,

$$KC_x\left(\frac{V}{V+v}\right) + KC_s\left(\frac{v}{V+v}\right) = H \qquad (24\text{-}26)$$

Solving for the concentration of the unknown,

$$C_x = \frac{-vC_s h}{hV - H(V+v)} \qquad (24\text{-}27)$$

For the maximum precision the amount of standard solution added should be sufficient to about double the original wave height (see also Chapter 29).

Internal Standard Method

The internal standard method, also called the "pilot ion" method, is based on the fact that the relative wave heights of two electroactive substances in a particular supporting electrolyte are constant for equal concentrations and independent of capillary characteristics. Even small temperature differences between analyses can be tolerated. In practice, one element is used to standardize the dropping assembly and all the other diffusion-current constants for other elements are measured relative to this same ion. Thus, it is only necessary to add a known concentration of the reference ion to an unknown; and from the wave heights of the unknown and the reference ion, the concentration of the unknown can be computed

$$C_x = \frac{i_x (I_d)_s}{i_s (I_d)_x} \cdot C_s \qquad (24\text{-}28)$$

This method simplifies work when different capillary systems must be used. Only a single standard solution is required for a series of test substances once the internal standard ratio, $(I_d)_s/(I_d)_x$, has been established. Whenever the nature of the concentration of the supporting electrolyte is altered in any manner, the ratio must be determined anew. The method has only limited application because only a small number of ions give sufficiently well-defined waves for use as internal standards. In multicomponent mixtures there may not be sufficient difference among existing half-wave potentials to introduce another wave.

24.8 AMPEROMETRIC TITRATION METHODS

The polarographic method can be used as the basis of an electrometric titration method comparable with the potentiometric, the conductometric, and the photometric methods. In this case the voltage applied across the indicator electrode and reference electrode is kept constant, and the current passing through the cell is measured and plotted against the volume of reagent added. Hence the name *amperometric titration*.

The current is measured, in general, on a diffusion-current region of a current–voltage curve. On such a region the current is independent of the potential of the indicator electrode because of an extreme state of concentration polarization at the electrode. Because at the electrode surface the concentration of material undergoing electrode reaction is maintained at a value practically equal to zero, the current is limited by the supply of fresh material to the electrode surface by diffusion. The rate of diffusion, and hence the current, is proportional to the concentration of diffusing substance in the bulk of the solution.

This technique can best be described by an example—the titration of a reducible substance, lead ion, with a nonreducible reagent, sulfate ion. A polarogram of a solution containing lead ions is represented by curve A in Fig. 24-16. If the voltage is held at any value on the diffusion-current plateau, the current will be represented by i_0. The titrant exhibits

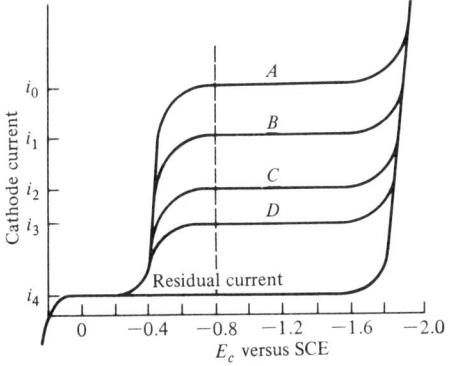
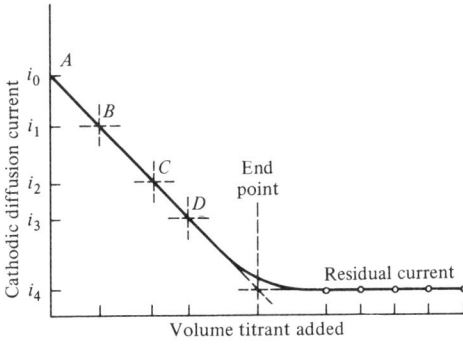

FIGURE 24-16 Successive current–voltage curves of lead ion made after increments of sulfate ion were added.

FIGURE 24-17 Amperometric titration curve for the reaction of lead ions with sulfate ions. See Fig. 24-16 for corresponding current–voltage curves. Performed at −0.8 V vs. SCE.

no diffusion current at the applied emf. Increments of titrant remove some of the electroactive lead ions. As the concentration of lead ions decreases, the current decreases to i_1, i_2, i_3, and finally i_r, at which point the lead ions have completely reacted and the only current flowing is a residual current characteristic of the supporting electrolyte.

If successive values of the diffusion current are plotted against the volume of titrant added, the result is a straight line which levels off at the end point (Fig. 24-17). The intersection of the extrapolated branches of the titration curve gives the end point.

When both titrant and unknown give diffusion currents at the applied voltage chosen, the current will drop to the end point, then increase again to give a V-shaped titration curve, seen in Fig. 24-18. If the original material does not react electrolytically, but the titrant does, a horizontal line, rising at the end point, results. This is shown in Fig. 24-19.

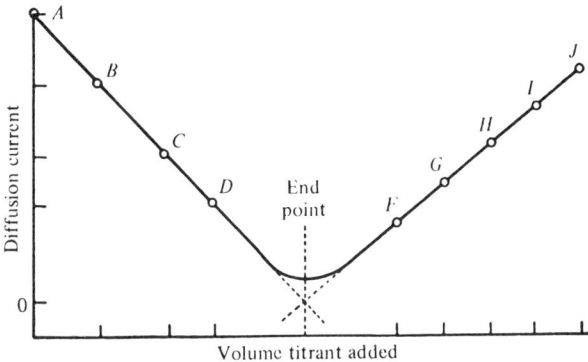

FIGURE 24-18 Type of amperometric titration curve when both reactant and titrant give diffusion currents; for example, the titration of lead ions with dichromate(VI) ions at −0.8 V vs. SCE. (See also Fig. 24-20).

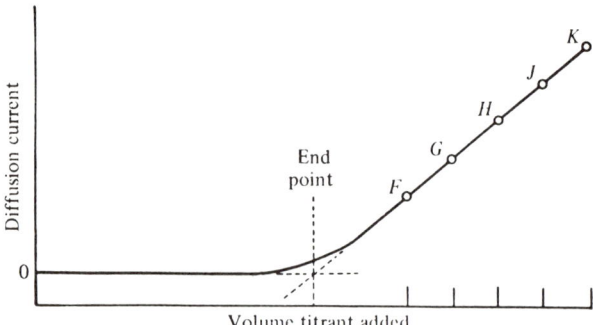

FIGURE 24-19 Type of amperometric titration curve when only titrant gives a diffusion current; for example, the titration of lead ions with dichromate(VI) ions performed at 0.0 V vs. SCE in an acetate buffer of pH 4.2. (See also Fig. 24-20.)

Methodology

If the optimum value needed to maintain the titration voltage is not known, the polarograms are determined for the materials involved, an appropriate voltage is selected, and the titration is carried out. As sometimes happens, a choice between two applied emf values can be made. In the titration of lead with dichromate, the titration can be conducted by choosing as the voltage a value E_1 at which dichromate ions are reduced, but not lead ions. The current–voltage curves are shown in Fig. 24-20. The titration curve is a horizontal line, rising at the end point, resembling Fig. 24-19. By shifting the cathode potential to E_2, both dichromate and lead ions are reduced. The current will drop to the end point, then increase again to give a V-shaped titration curve resembling Fig. 24-18.

In practice, the reversed L-shaped type of curve, illustrated by Fig. 24-19, is preferred. Because the titrant in this case produces no current, it can be added continuously at a moderate rate until the end point is passed. This will be noted by a permanent increase in the diffusion current. Then three or four additional readings, taken after successive increments of excess titrant have been added, will establish the rising branch of the curve.

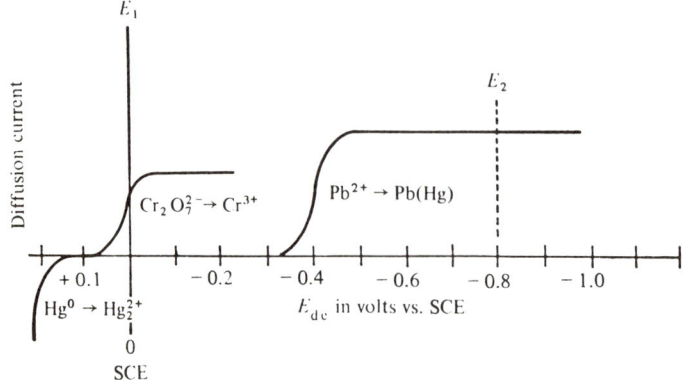

FIGURE 24-20 Current–voltage curves of dichromate(VI) and lead ions shown schematically.

Strictly speaking, a correction for dilution is necessary to attain a linear relation between current and volume of titrant, but by working with a reagent which is tenfold more concentrated than the solution being titrated, the correction becomes negligibly small. Incompleteness of reaction in the vicinity of the end point usually will not detract from the results provided reaction equilibrium is attained rapidly during the titration. Points can be selected between 0 and 50% and 150 and 200% of the end-point volume for the construction of the two branches of the titration curve. In these regions the common ion effect will repress dissociation and solubility of precipitates.

Apparatus

The equipment for conducting amperometric titrations is simple. Although it may be the same as for polarography, several simplifications are possible. The potential of the indicator electrode need only be selected within 0.1 V if it lies on a limiting current region of a current–voltage curve. Often the potential of a reference electrode will lie in the permissible range, so that it is necessary only to short-circuit the indicator electrode through a suitable current-measuring instrument to a reference electrode of relatively large area.

No thermostat is necessary. The temperature of a solution will seldom vary appreciably during the short time, 10 min or less, necessary to conduct a titration.

The indicator electrode may be a dropping mercury electrode or a rotating metal microelectrode. The latter is simple to construct. It consists of a short length of wire, usually platinum, protruding 5–10 mm from the wall of a piece of glass tubing. The latter is bent at right angles a short distance from the end of the stem so as to sweep an area of the solution with the wire. It is illustrated in Fig. 24-21. The electrode is mounted in the shaft of a motor and rotated at a constant speed of about 600 rpm. By using a rotating electrode, the diffusion layer thickness is decreased, thereby increasing the sensitivity and the rate of attainment of a steady diffusion state. The limiting current may be up to 18 or 20 times larger than that with a dropping electrode; it is proportional to the $\frac{1}{3}$ power of the number of revolutions per minute above 200 rpm. A stationary electrode with a magnetic stirrer to pass the solution by the electrode exhibits the same response as a rotated electrode.

The larger currents attained with a rotating electrode allow correspondingly smaller concentrations to be measured without loss of accuracy. Also the absence of drops disengaging themselves at regular intervals eliminates the charging current observed with a dropping electrode, which in turn permits the use of ordinary rugged microammeters. However, many systems whose oxidation potentials or reduction potentials lie in the range of the platinum microelectrode do not give limiting currents with a rotating electrode. Where the discharge of hydrogen interferes, a dropping mercury electrode, with its larger value of hydrogen overpotential, must be used.

The removal of oxygen is generally mandatory over most of the useful range of the dropping electrode. Nitrogen or hydrogen gas must be bubbled through the solution preceding the titration and for a minute or two after the addition of each increment of titrant. The oxidation of mercury limits the anodic range of the dropping electrode, but when applicable, the rotating electrode extends the useful range to about +0.9 V versus the SCE, at which point the oxidation of water to oxygen commences.

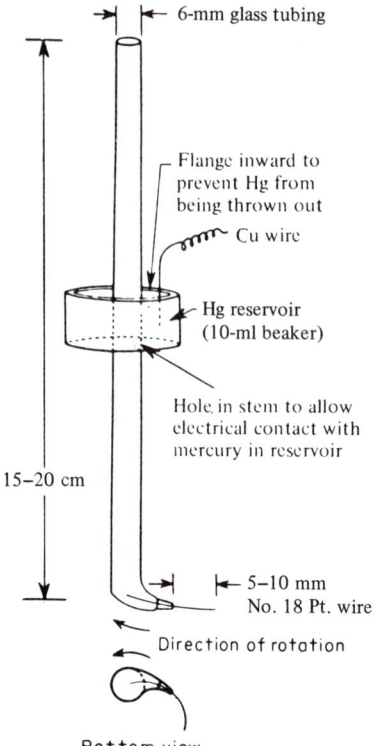

FIGURE 24-21   Rotating platinum microelectrode.

Successive Titrations

Iodide, bromide, and chloride can be successively titrated in mixtures with silver, using the rotating electrode.[8] In a 0.1–0.3N solution of ammonia only silver iodide will precipitate when a silver solution is added. The indicator reaction is the reduction of the complex diammine silver ion. Consequently, the potential of the rotating electrode must be made negative enough to plate out silver, but must not be negative enough to give an appreciable current due to the reduction of dissolved oxygen. The range of permissible potential is strictly limited, as is evident from an examination of the current–voltage curves of diammine silver reduction and oxygen reduction using a silver-plated microelectrode in an ammoniacal solution (Fig. 24-22). Fortunately the mercury/mercuric iodide/potassium iodide reference electrode happens to lie in the permissible range (–0.23 V vs. SCE) and can be short-circuited through the current-measuring device to the rotating electrode. During the titration of iodide the current remains constant at zero, or nearly so, until the iodide ions are consumed, and then rises. After three or four points have been recorded past the end point, the solution is acidified to make it 0.8N in nitric acid. Immediately the silver ions added in excess and now released from the ammine complex combine with the bromide ions and precipitate as silver bromide, and the current drops to zero.

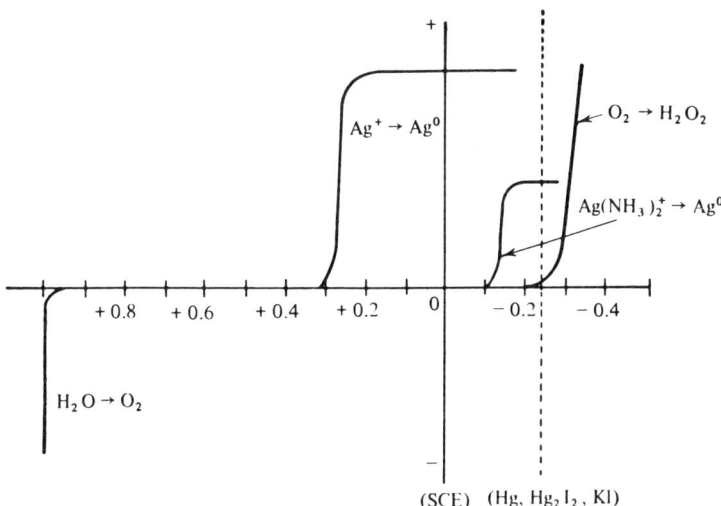

FIGURE 24-22 Schematic current–voltage curves of silver and oxygen obtained with a rotating microelectrode.

The titration of bromide and chloride is carried out at a less negative potential, for in these titrations the indicator reaction is the deposition of silver from aquosilver ions. Because the potential of the saturated calomel electrode lies in the limiting current region, it also may be short-circuited to the rotating electrode. Chloride does not interfere with the titration of bromide because silver chloride particles cause a cathodic current even in the presence of a large excess of chloride. Therefore a second rise in the current indicates the end point of the bromide titration. A chloride end point can be obtained by adding gelatin, which suppresses the current due to silver chloride, and continuing the titration until the current again rises after the chloride end point. A composite of these titration curves is shown schematically in Fig. 24-23.

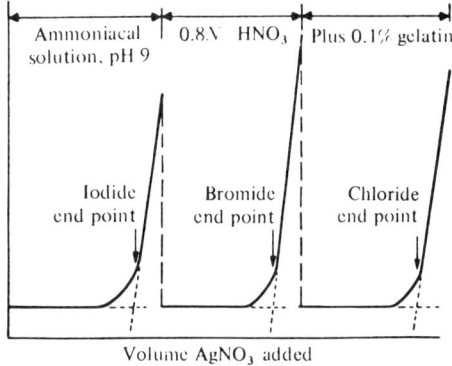

FIGURE 24-23 Composite of the consecutive titration curves for a mixture of iodide, bromide, and chloride ions.

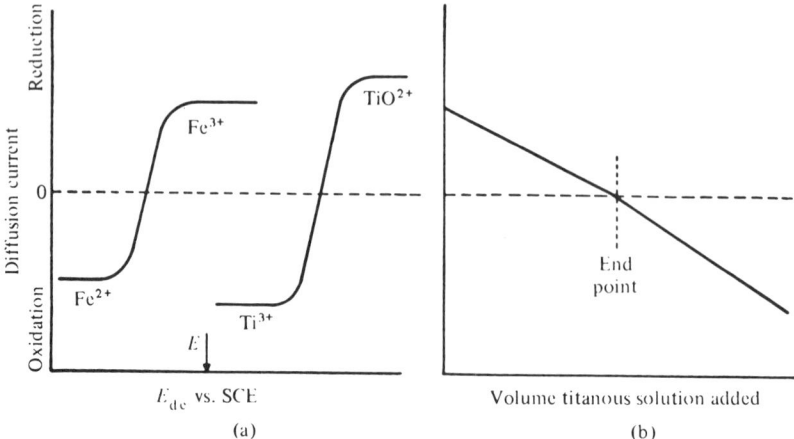

FIGURE 24-24 Current–voltage curves of (a) iron(II/III) and titanium(IV/III) systems and (b) amperometric titration system for iron(III) titrated with titanium(III) solution at E_{dc} = point E on the graph.

Titrations to Zero Current

With systems in which both the oxidant and reductant yield a diffusion current, the titration curve obtained is of the type shown in Fig. 24-24. In such systems as, for example, the titration of iron(III) with titanium(III), a voltage E is impressed upon the indicator electrode, so that the diffusion current for the reduction of iron(III) is set up at the start of the titration. As the iron(III) concentration is decreased linearly, the current decreases in a similar fashion and reaches zero at the end point. When the end point is passed, a diffusion current caused by the oxidation of the titanium(III) is set up. A change in slope caused by the difference in diffusion coefficients is usually evident as the lines cross the zero axis. Actually the zero axis is the value of the residual current for the supporting electrolyte and usually will be different from zero.

Titrations of this type without any chemical reaction are also possible.[9] Copper(II) and tin(II) in a tartrate medium at pH 4 possess the current–voltage curves schematically represented in Fig. 24-25. At an applied potential midway between the half-wave potential of the cathodic copper(II to 0) wave and the anodic tin(II to IV) wave, the diffusion currents of both waves are fully developed. Titrating with a solution of copper (II), the anodic diffusion current of the tin(II) wave will be compensated by the increasing cathodic current of the copper(II) wave. The net diffusion current will be zero at the end point.

24.9 TWO INDICATOR ELECTRODES

In a modification of the usual or classical amperometric system, two similar platinum electrodes can be immersed in the titration cell. A small and constant voltage is applied to these electrodes as in the classical method. For the method to be applicable, the only requirement is that a reversible oxidation–reduction system be present either before or after the end point.

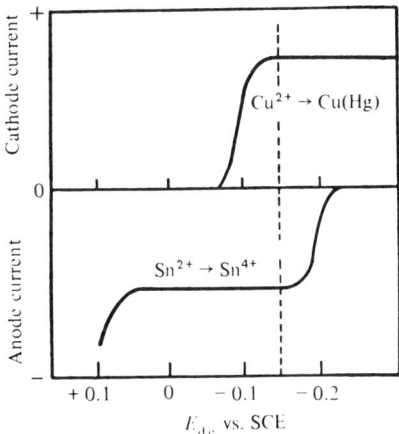

FIGURE 24-25 Current–voltage curves for cathodic reduction of copper(II) ions and anodic oxidation of tin(II) ions, both in tartrate medium at pH = 4.

In a titration with two indicator electrodes, and when the reactant involves a reversible system, a small amount of electrolysis takes place. The amount of oxidized form reduced at the cathode is equal to that formed by oxidation of the reduced form at the anode. Both electrodes are depolarized until either the oxidized or the reduced member of the system has been consumed by a titrant. After the end point, only one electrode remains depolarized if the titrant does not involve a reversible system. The solution at this juncture resembles a one-electrode method connected to a depolarized (reference) electrode. Current flows until the end point. At and after the end point the current is zero or close to zero.

The method was introduced years ago under the name "dead-stop end point."[10] The reverse of this type of end point, and the more desirable in practice, might be called "kick off" and resembles a reversed L-shaped amperometric curve. When both the system titrated and the reagent are reversible oxidation–reduction systems, the current is zero or close to zero only at the end point, and a V-shaped titration curve results.

Three regions appear in a titration curve when two indicator electrodes are employed.[11] Take for example the iodine–iodide system being titrated with thiosulfate. If a considerable quantity of iodide is in solution, the system will be well poised and the current will maintain a steady value (Region 1, solid line, on Fig. 24-26). As the titration progresses and the concentration of iodine gets smaller, the concentration overpotential (polarization) at the cathode begins to play a role, and the current tends to become diffusion controlled. Now the current tends to vary in proportion to the concentration of iodine remaining in the bulk of the solution. The characteristics of the line giving the change of current from the point where the system is well poised to the vicinity of the end point is represented as Region 2. Near the end point the line becomes straight, as in a titration with one indicator electrode at constant applied emf. No current flows after the end point, because the thiosulfate–tetrathionate system is not a reversible couple and insufficient emf is applied to cause the oxidation of iodide ions at the anode and the discharge of hydrogen (or dissolved oxygen) at the cathode.

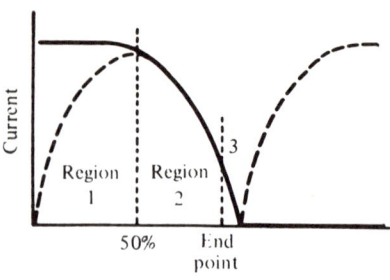

FIGURE 24-26 Amperometric titration lines using two indicator electrodes. The dashed line, in Region 1, is followed when the reactant is poorly poised initially and also after the end point, when the titrant also forms a reversible oxidation–reduction system.

If the system involving the reactant is poorly poised, as it is when no iodide ions are initially present in the solution, the current at the start of the titration will be essentially zero. The current will rise as iodide ions are formed and will reach a maximum when the degree of completion of the titration is 50%. This portion of the curve is represented by the dashed lines in Region 1 of Fig. 24-26. The remainder of the titration curve will follow the descending branch.

Both one- and two-indicator electrode methods become identical if the applied emf in the two-indicator method is made large enough to yield a diffusion-controlled current early in the titration. Whereas some workers recommend only sufficient applied emf to balance the back emf, approximately 20 mV in the iodine titration, the diffusion current is not completely developed until the applied emf is 100 mV or greater, as shown in Fig. 24-27. With larger values of applied emf the value for the iR term remains negligibly small over a larger region, and thus the titration line remains straight over a longer distance in the vicinity of the end point. In effect, this places the system under a diffusion-controlled condition well in advance of the end point. A second effect is to increase the current sensitivity, but this can be varied at will by varying the size of the electrodes and the speed of stirring.

Comparison With Other Titration Methods

Several advantages of amperometric methods are immediately apparent. The equipment is simple. Because it is a relative method, there are fewer disturbing variables, as contrasted with polarography. Electrode characteristics are unimportant. There is no need to determine the capillary characteristics of a dropping electrode. Use of rotating electrodes is possible, and, in fact, desirable when applicable. The lack of current oscillations when using rotating electrodes makes it possible to use rugged microammeters for measurement of the current. Accuracy is higher than in polarography because each branch of the titration curve effectively is the average of the recorded points. An error in the end point is primarily determined by the accuracy of the titrant delivery.

The method possesses greater sensitivity than conductometric and potentiometric titrations. In fact, amperometric methods are best for determining traces with good precision. Concentrations from 0.1 to 0.0001M, and even in favorable cases to $10^{-6}M$, can be measured with ease and accuracy.

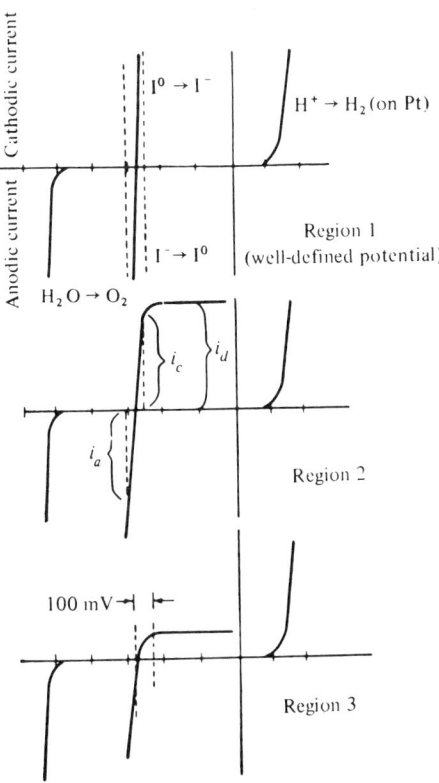

FIGURE 24-27 Schematic current–voltage curves for two indicator electrodes corresponding to regions on Fig. 24-26; iodine titrated with thiosulfate with excess iodide present.

Applications of amperometric methods are more general than classical potentiometric methods and polarography. Many systems do not possess a measurable equilibrium potential but can be electrolyzed under an applied emf. However, even if one reactant is not oxidizable or reducible, the titration can be conducted by utilizing the oxidation–reduction characteristics of the other reactant. This method is one of the few generally applicable to precipitation reactions.

PROBLEMS

1. Compare the polarographic evaluation methods with respect to (a) applicability to routine analyses of similar type samples, (b) applicability to routine analyses of samples of widely varying composition but for the analysis of a few constituents by polarography, and (c) applicability to occasional analyses of a wide variety of samples and for a wide variety of constituents.

2. Discuss the factors that may contribute to the observed polarographic limiting current of a single ion in addition to the diffusion current. How may all the undesirable factors be eliminated?

3. Polarographic curves resemble potentiometric titration curves. When might polarography yield useful data not obtainable by potentiometric methods?

4. The data below were obtained in 25°C with cadmium ion in a supporting electrolyte composed of 0.1M potassium chloride and 0.005% gelatin. Galvanometer deflections were measured at −1.00 V versus SCE. The galvanometer sensitivity was stated as 0.0055 μA/ mm; the current multiplier was set at 50. $t = 2.47$ sec; $m = 3.30$ mg sec^{-1}.

Cd^{2+}, mM	i_d, mm
0.00	4.5
0.20	11.0
0.50	21.0
1.00	37.5
1.50	54.0
2.00	70.5
2.50	86.5

Plot the calibration curve on a sheet of graph paper with the ordinate the diffusion current in microamperes (or arbitrary scale divisions) and the abscissa the corresponding cadmium concentration. From the calibration curve determine the concentration of an unknown solution, prepared similarly, which has a diffusion current of 39.5 mm after correction for the residual current.

5. Exactly 25.0 ml of the unknown solution, which gave a diffusion current of 39.5 mm in Problem 4, has been transferred to the polarographic cell. To this solution is added exactly 5.0 ml of 0.0120M cadmium solution. The corrected diffusion current is now 88 mm. Calculate the concentration of the unknown cadmium solution.

6. At the time the unknown cadmium solution was prepared in Problem 4 a second solution was also prepared, identical in all respects except that it also contained zinc ions, 10.0 ml of 0.0100M zinc in a total volume of 100 ml. The corrected diffusion current was 32 mm for the zinc wave. Calculate the concentration of the unknown cadmium solution by the internal standard method. Obtain the necessary diffusion current constants from Appendix B.

7. An unknown amount of copper(II) ions produces 12.3 μA on a dc polarogram. By adding 0.100 ml of $1.00 \times 10^{-3}M$ copper(II) ions to the original volume of 5.00 ml, the new current is 28.2 μA. Calculate the original amount of copper.

8. From each of the following sets of transition times obtained from chronopotentiograms, deduce the successive electron reactions: (a) $\tau_2 = 3\tau_1$ for the reduction of oxygen at a mercury cathode; (b) $\tau_2 = 8\tau_1$ for the reduction of uranium; and (c) $\tau_2 = 35\tau_1$ for the oxidation of iodide ion ($<0.0025M$) at a platinum anode in dilute sulfuric acid.

9. To what extent will the charging current decay under the following operating conditions in normal pulse polarography? $R = 1000\ \Omega$ (the electrolytic resistance of the solution); C (the double layer capacitance) $= 20\ \mu F/cm^2$ for a dropping mercury electrode of area $0.03\ cm^2$; measurements made 40 msec after the application of each pulse.

10. The polarogram of a sample solution is shown. Before the solution was made up to volume in $0.1M$ KCl, sufficient cadmium was added to bring its concentration to 0.03 mM. Identify and estimate the concentrations of the other two metals present in the nonferrous alloy sample.

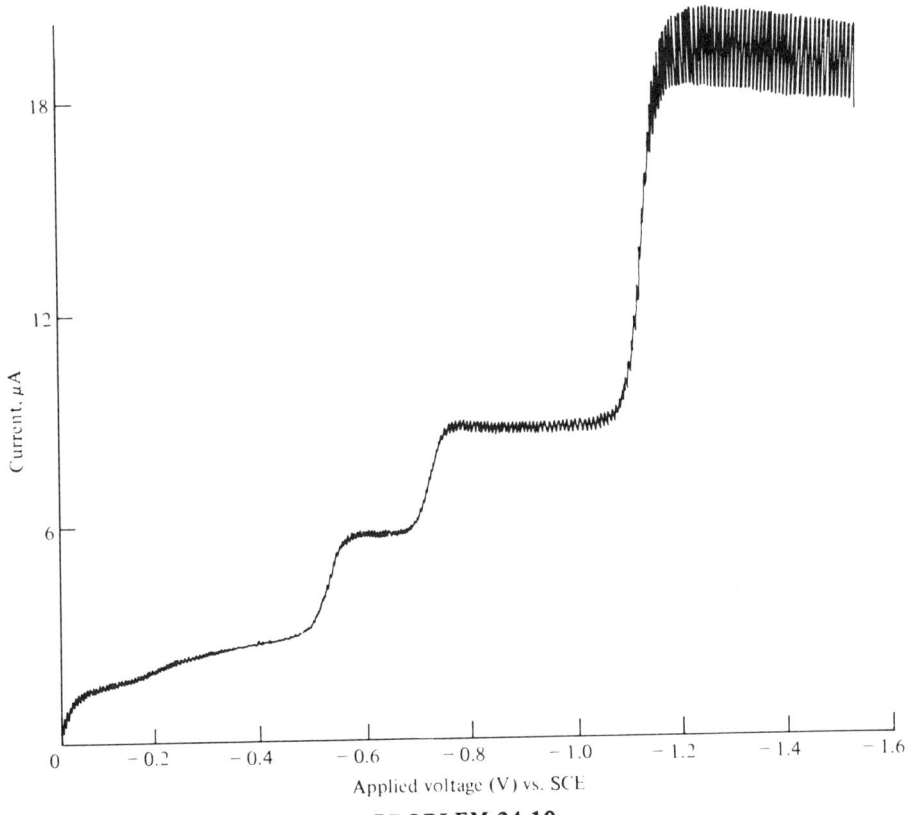

PROBLEM 24-10

11. On the conventional dc polarogram shown, identify the five different metals present in the 0.1M KCl solution.

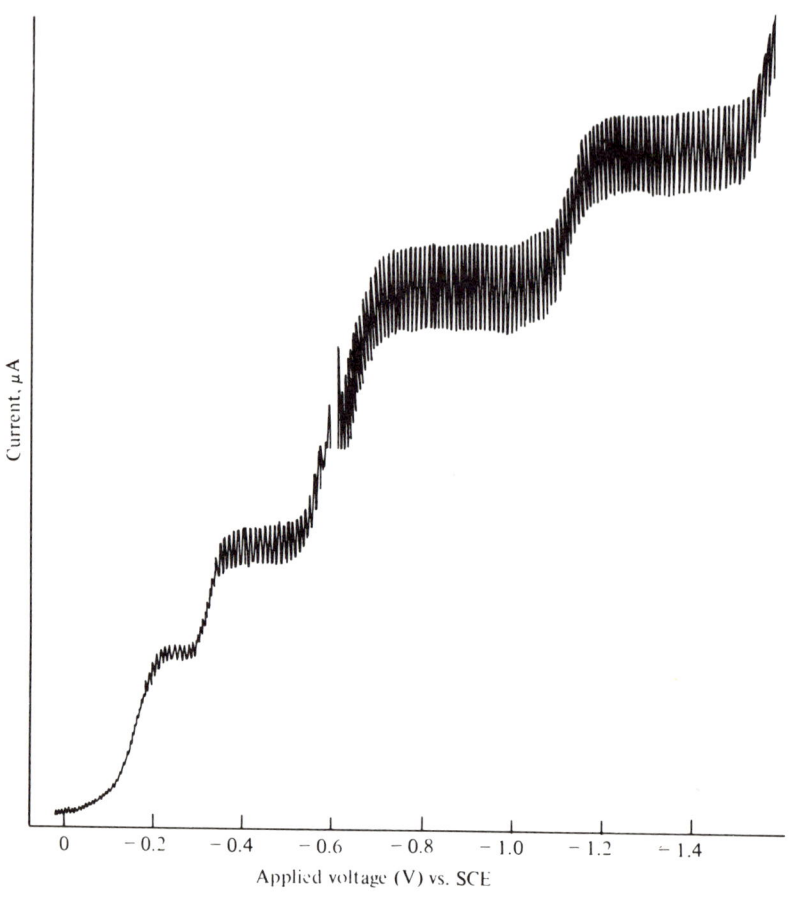

PROBLEM 24-11

12. Identify the six metals present in the derivative polarogram shown which was obtained in 0.1M HCl.

13. The polarogram of a sample is shown. Sufficient cadmium was added to the sample before dilution to volume in 0.1M HCl to bring its concentration to 2.00 × $10^{-4}M$. Calculate the concentration of the other two metals present.

14. Compare polarographic determinations with amperometric titrations with respect to (a) relative accuracy, (b) permissible range of applied potential usable, and (c) applicability of each method in regard to types of ions and range of concentrations.

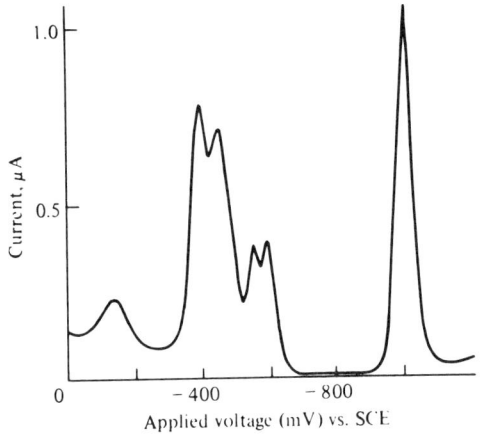
PROBLEM 24-12

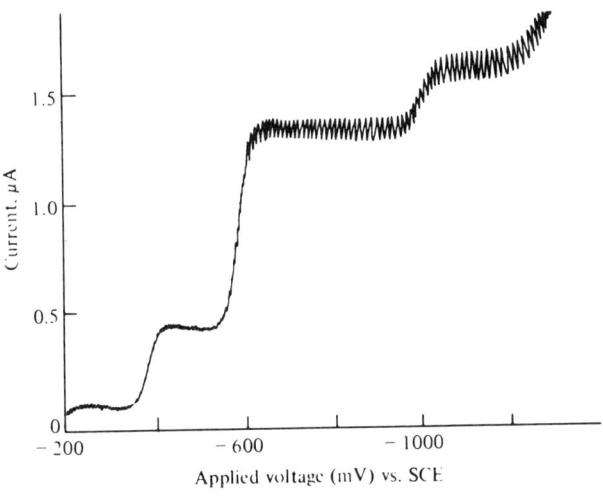
PROBLEM 24-13

15. Cupric ions form a precipitate with alpha-benzoinoxime in an ammoniacal solution. The $Cu(NH_3)_4^{2+}$ present in a supporting electrolyte consists of $0.05M$ ammonia and $0.1M$ ammonium chloride is reduced stepwise, giving polarographic waves at -0.2 and -0.5 V versus SCE. Alpha-benzoinoxime gives a polarographic wave with $E_{1/2} = -1.6$ V versus SCE. Deduce the shape of the amperometric titration curve that will be obtained at applied voltage of -0.8 V versus SCE, and also the sketch of the titration curve obtained at -1.7 V. Which potential would be preferred under normal circumstances? When nickel and zinc ions are also present in the titrating solution, which potential would be preferred?

16. Contrast amperometric titration methods with potentiometric titration methods.

BIBLIOGRAPHY

Adams, R. N., *Electrochemistry at Solid Electrodes,* Marcel Dekker, New York, 1969.
Bond, A. M., *Modern Polarographic Methods in Analytical Chemistry,* Marcel Dekker, New York, 1980.
Copeland, T. R. and R. K. Skogerbee, "Anodic Stripping Voltammetry," *Anal Chem.,* **46,** 1257A (1974).
Dolezal, J. and J. Zyka, in *Standard Methods of Chemical Analysis,* 6th ed., F. J. Welcher, Ed., Vol. IIIA, Chap. 20, D. Van Nostrand, New York, 1966.
Ellis, W. D., "Anodic Stripping Voltammetry," *J. Chem. Educ.,* **50,** A131 (1973).
Evans, D. H., "Voltammetry: Doing Chemistry with Electrodes," *Acc. Chem. Res.,* **10,** 313 (1977).
Fisher, D. J., "Advances in Instrumentation for dc Polarography and Coulometry," *Adv. Anal. Chem. Instrum.,* **10,** 1 (1974).
Flato, J. B., "The Renaissance in Polarographic and Voltammetric Analysis," *Anal. Chem.,* **44(11),** 75A (1972).
Flato, J. B., "Two New Electrochemical Instruments," *Am. Lab.,* **1** (2), p. 10 (February 1969).
Heyrovsky, J. and J. Kuta, *Principles of Polarography,* Academic, New York, 1965.
Jain, R. K, H. C. Gauer, and B. J. Welch, "Chronopotentiometry: A Review of Theoretical Principles," *J. Electroanal. Chem., Interfacial Electrochem.,* **79,** 211 (1977).
Kolthoff, I. M. and J. J. Lingane, *Polarography,* 2nd ed., Wiley-Interscience, New York, 1952, in 2 volumes.
Lingane, J. J., *Electroanalytical Chemistry,* 2nd ed., Wiley-Interscience, New York, 1958.
Meites, L., *Polarographic Techniques,* 2nd ed., Wiley-Interscience, New York, 1965.
Milner, G. W. C., *The Principles and Application of Polarography,* Wiley, New York, 1957.
Müller, O. H., in *Physical Methods of Chemistry,* A. Weissberger and B. W. Rossiter, Eds., Vol. I, Part IIA, Chap. 5, Wiley-Interscience, New York, 1971.
Schaap, W. B., in *Standard Methods of Chemical Analysis,* 6th ed., F. J. Welcher, Ed., Vol. 3, Part A, Chap. 19, Van Nostrand Reinhold, New York, 1966.
Schmidt, H. and M. von Stackelberg, *Modern Polarographic Methods,* Academic, New York, 1963.
Stock, J. T., *Amperometric Titrations,* Wiley, New York, 1965.
Vydra, F., K. Stulik, and E. Julakova, *Electrochemical Stripping Analysis,* Wiley, New York, 1976.
Zeeman, P., "Techniques, Advantages and Limitations of Organic Electroanalytical Procedures," *Electrochim. Acta,* **21,** 687 (1976).
Zeeman, P. and I. M. Kolthoff, *Progress in Polarography,* Wiley-Interscience, New York, Vol. I, 1962; Vol. II, 1962; Vol. 3, with L. Meites, 1972.

LITERATURE CITED

1. Kelley, M. T., H. C. Jones, and D. J. Fisher, *Anal. Chem.,* **31,** 1475 (1959).
2. Burge, D. E., *J. Chem. Educ.,* **47,** A81 (1970).
3. Smith, D. E., in *Electroanalytical Chemistry,* A. J. Bard, Ed., Vol. 1, Chap. 1, Marcel Dekker, New York, 1966.

4. Lingane, J. J., *Analyst,* **91**, 1 (1966).
5. Barendrecht, E., "Stripping Voltammetry," in *Electroanalytical Chemistry,* A. J. Bard, Ed., Vol. 2, p. 53, Marcel Dekker, New York, 1967.
6. Kryger, L. and D. Jagner, *Anal. Chem. Acta,* **78**, 251 (1975).
7. Zuman, P., *Chem. Eng. News,* p. 94 (March 18, 1968); *Substituent Effects in Organic Polarography,* Plenum, New York, 1967.
8. Laitinen, H. A., W. P. Jennings, and T. D. Parks, *Ind. Eng. Chem., Anal. Ed.,* **18**, 355, 358 (1946).
9. Lingane, J. J., *J. Am. Chem. Soc.,* **65**, 866 (1943).
10. Foulk, C. W. and A. T. Bawden, *J. Am. Chem. Soc.,* **48**, 2045 (1926).
11. Kolthoff, I. M., *Anal. Chem.,* **26**, 1685 (1954).

CHAPTER 25

Electrogravimetry and Coulometry

25.1 ELECTROSEPARATIONS

Electroseparation is electrolysis in which a quantitative reaction or, at the very least, an appreciable amount of reaction of analytical interest takes place at an electrode. Electrooxidation or reduction may be accomplished. Electroreductions for analytical purposes have long been carried out with the current kept more or less constant by adjusting the voltage applied to the cell. Probably the most widely used type of electroreduction involves a mercury cathode to which a constant potential is applied. Although selectivity is poor, this method finds extensive use for removal of interfering elements. Better selectivity can be achieved by controlling the potential of the working electrode with a potentiostat such that constituents may be sequentially removed from a solution either by reduction or oxidation. Electroorganic synthesis is a useful tool for synthetic organic chemists.

Other uses of electroseparations include the dissolution of refractory metals; the metal sample is made the anode and a platinum, tantalum, or graphite cathode is employed. Electrolytic methods are useful in the preparation of very pure inorganic substances or in the preparation of unusual oxidation states of metals. As a preliminary step in the preparation of samples for spectroscopic techniques, elements are deposited on a suitable electrode material and determined by emission spectroscopy or by spark source mass spectrometry.

25.2 BASIC PRINCIPLES

In a typical electrogravimetric cell a pair of relatively large platinum electrodes is immersed in an electrolyte and a voltage is applied across them. At first, as the applied volt-

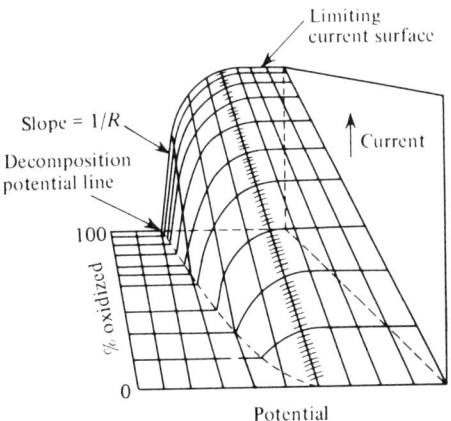

FIGURE 25-1 Three-dimensional representation of a reversible oxidation–reduction system as it applies to electroanalysis. (W. H. Reinmuth, *J. Chem. Educ.*, 38, 149 (1961). Courtesy of Journal of Chemical Education.)

age is gradually increased, virtually no current (except a small residual current) is observed to flow through the cell. However, as a particular point on the voltage axis is reached (Fig. 25-1), a noticeable reaction will be observed and a current begins to flow through the cell. This particular value of applied voltage is called the *decomposition potential* with respect to the particular electrode reaction. As soon as the decomposition potential is exceeded, continuous electrolysis of the solution is sustained. With further increase in applied voltage the current increases linearly in accordance with Ohm's law. Ideally, the slope of the curve is the reciprocal of the resistance between the terminals of the electrolytic cell. Actually, the current is soon limited by the rate of mass transfer of electroactive material to the electrode surface. The mass transfer coefficient depends on the rate of stirring, cell and electrode geometries, diffusion coefficient of the electroactive species, and the like. As the concentration of electroactive species in solution decreases, so also does the limiting current plateau. The decomposition potential simultaneously shifts to more cathodic (if reducible species are involved) or anodic (if oxidizable species are involved) values.

To determine what voltage must be applied to a cell to cause electrolysis, it is necessary to know first what reactions will occur at the two electrodes. If these reactions are known, it is possible to calculate the potential of each electrode and thereby determine the emf of the galvanic cell which exerts its potential in opposition to the applied voltage. For example, in the electrolysis of 0.100M $CuSO_4$ in 0.5M H_2SO_4 with platinum electrodes, the reaction at the cathode will be

$$Cu^{2+} + 2e^- \rightarrow Cu^\circ \tag{25-1}$$

and at the anode

$$2H_2O \rightarrow O_2(g) + 4H^+ + 4e^- \tag{25-2}$$

Using the Nernst equation to calculate the potential of the copper electrode,

$$E = 0.337 + 0.0296 \log(0.100) = 0.307 \text{ V} \tag{25-3}$$

For the other electrode at which oxygen is evolved,

$$E = 1.229 + \frac{0.0591}{4} \log \frac{[O_2][H^+]^4}{[H_2O]^2} \tag{25-4}$$

This expression can be simplified by realizing that the term $[O_2]$ is the same as the partial pressure of oxygen gas in the atmosphere (approximately 0.21), and that the concentration of water is essentially constant. Then,

$$E = 1.229 + 0.0148 \log p_{O_2} [H^+]^4 \tag{25-5}$$

In $1N$ H_2SO_4, the potential is about 1.22 V.

The cell emf is the algebraic difference of the electrode potentials of the two half-cells comprising the electrolytic cell. The spontaneous reaction which will occur in this cell is

$$2Cu^\circ + O_2(g) + 4H^+ = 2Cu^{2+} + 2H_2O \tag{25-6}$$

This corresponds to subtracting Eq. 25-3 from 25-5; therefore, the emf of the galvanic cell is $1.22-0.31 = 0.91$ V. Correspondingly, the emf of the electrolytic cell is -0.91 V. When the applied voltage is exactly 0.91 V, no current will flow through the cell in either direction, which is, of course, the principle of potentiometric measurements. As soon as the applied voltage is in excess of the galvanic cell emf, the iR drop, and the sum of any overpotential effects,

$$E_{\text{applied}} = E_{\text{cell}} + iR + \omega_{\text{anod}} + \omega_{\text{cath}} \tag{25-7}$$

current is forced to flow through the electrolytic cell and the reaction expressed by Eq. 25-6 proceeds from right to left. The maximum value of the current is governed by the cell resistance and mass transfer conditions.

Completeness of Depositions

The emf that is applied to an electrolytic cell must be sufficient to ensure the removal of the desired electroactive species to an extent adequate for the purpose of the experiment. Take the deposition of silver, for example. As the silver deposits, the concentration of silver ions in solution decreases, and, according to the Nernst equation,

$$E = 0.80 + 0.0591 \log [Ag^+] \tag{25-8}$$

the potential at which silver deposits becomes more negative. The cathode potential changes very nearly 59 mV for each tenfold decrease in the concentration of the silver ion remaining undeposited (and, in general, $59/n$ mV for metallic ions). Assuming that the anode process proceeds at a constant level and that the iR drop is essentially constant, the decomposition potential of the solution should vary by a like amount.

If only a very minute amount of a metal is to be deposited on platinum, perhaps a fraction of a microgram, then the amount of deposit may not be enough to form a monolayer of atoms, and hence the activity of the metal phase cannot be assumed to be constant. Rogers and Stehney[1] discuss the theory of depositions in such situations.

Overpotentials

When a current flows across an electrode-solution interface, it is normally found that the electrode potential changes from the reversible value it possesses before the passage of current. The difference between the measured potential (or cell emf) and its reversible value is the *overpotential* (also called overvoltage). The electrodes are said to be polarized. Both cathodic and anodic processes exhibit overpotential, and it is affected by many factors. When an anodic process shows an overpotential effect, the applied potential necessary to cause electrolysis will always be a more positive value than the calculated potential and, for cathodic processes, overpotential causes the applied potential to be more negative than the calculated value. Although overpotential phenomena complicate the calculation of the applied voltage necessary for electrolysis to occur, its effect makes feasible certain separations that would not be expected from standard electrode potentials.

Various types of overpotential may be distinguished. In some electrode processes a film of oxide or some other substance forms on the electrode surface and sets up a resistance to the passage of current across it. What may be termed an ohmic-pseudo-overpotential is also observed when the capillary tip used in measuring the potential of an electrode is at an appreciable distance from the electrode surface. The latter effect only becomes appreciable at high current densities or low concentrations.

A second type of overpotential is due to concentration changes in the vicinity of the electrode and is consequently referred to as *concentration overpotential*. Whenever a finite current flows across an electrode-solution interface, the concentration of the electrode surface is somewhat altered from its concentration in the bulk of the solution. The reason is that the species at the electrode surface is not replenished at a rate commensurate with the current demand. Consequently, the potential of the cathode will exhibit a more negative value as the voltage applied to the electrolytic cell is increased. Generally, the metal-ion concentration at the electrode interface is only 1% of the bulk concentration; thus, the concentration overpotential is approximately $0.118/n$ V. In practice, concentration overpotential is minimized by the use of electrodes with large surface areas and by keeping electrolysis currents small, although the latter stipulation is not conducive to rapid electrolysis. The rate of mass transfer is aided by mechanical stirring and increase in temperature of the solution.

Although concentration changes are probably the most important source of overpotential accompanying the deposition of a metal, small overpotentials arise from other causes. As a general rule, the overpotential of a metal upon itself is not large (about 0.01 V) at low current densities. This is not true, however, for such hard metals as cobalt, nickel, iron, chromium, and molybdenum. In an ammoniacal solution the deposition potential of copper on a platinum electrode is considerably lower than on a silver electrode. Consequently, in the separation of silver from copper in an ammoniacal solution, if the silver is plated on a platinum electrode and if toward the end of the electrolysis the level of the solution is raised, copper will plate on the bare platinum surface exposed.

Complexation may affect the overpotential in either direction, because the rate of exchange of electrons between the electrode and complex species may be greater or less than the rate of exchange with the aquated ion. For example, aquonickel ions show an overpotential of about 0.6 V at a mercury surface, whereas complexes of nickel with thiocyanate or pyridine, although actually shifting the equilibrium electrode potential more

TABLE 25-1 Hydrogen Overpotential on Various Cathodes

ELECTROLYTE IS $1M$ H_2SO_4. OVERPOTENTIAL GIVEN IN VOLTS.

Cathode	First Visible Gas Bubbles	Current Density	
		0.01 A cm^{-2}	0.1 A cm^{-2}
Antimony	0.23	0.4	—
Bismuth	0.39	0.4	—
Cadmium	0.39	~0.4[a]	—
		—	1.2
Copper	0.19	0.4	0.8
Gold	0.017	0.4	1.0
Lead	0.40	0.4	1.2
Mercury	0.80	1.2	1.3
Platinum (bright)	~0	0.09	0.16
Silver	0.097	0.3	0.9
Tin	0.40	0.5	1.2
Zinc	0.48	0.7[b]	—

SOURCE: J.J. Lingane, *Electroanalytical Chemistry*, 2nd ed., p. 209 Wiley-Interscience, New York, 1958. Reproduced by permission.

[a] $0.005M$ H_2SO_4.
[b] $0.01M$ $Zn(C_2H_3O_2)_2$.

negative, show a decrease in overpotential which more than compensates for the shift in the equilibrium potential.

The evolution of gases at an electrode is usually associated with an overpotential significantly larger than concentration overpotentials. It is particularly marked in the evolution of hydrogen and oxygen, and is called the *gas overpotential*. Some values of the overpotential of hydrogen on various surfaces are given in Table 25-1. The anodic overpotential of oxygen on smooth platinum in acid solutions is approximately 0.4 V.

Gas overpotential depends on several factors: (a) *Electrode material*. At a given current density, overpotential for many metal surfaces seems to decrease roughly in a parallel manner to the thermionic work function of the electrode material. (b) *Current density*. An increase in current density invariably increases the overpotential up to a limiting value. (c) *Electrode condition*. Whether smooth or rough, bright or platinized, the overpotential at a given current density decreases if the electrode surface is roughened. This is due partly, if not entirely, to an increase in the effective area and the consequent decrease in the actual current density. (d) *Temperature*. As the temperature is raised, overpotential diminishes. For most electrodes the temperature coefficient is about 2 mV/°C. (e) *The pH*. At low current densities the overpotential is independent of the pH. At high acid concentrations and with some metals there appears to be some dependence on pH.

The hydrogen overpotential is large on metals such as bismuth, cadmium, lead, tin, zinc, and especially mercury. With a mercury cathode, a number of useful separations become possible, as will be discussed in a later section. Overpotential of hydrogen on cadmium and zinc is important in electrolysis because it permits their determination in an aqueous solution.

Processes at the Anode

The behavior at an anode is, in general, analogous to that at a cathode. The process associated with the smallest oxidation potential, whether it be dissolution of the metallic anode to form cations or the discharge of anions, will take place first. Subsequent anodic processes will follow in order of increasing oxidation potential. The discharge of anions involves a consideration of sulfide, halide, and hydroxyl ions only. For other anions, the hydroxyl ion, derived from the water at the surface of the anode, or water itself, will be preferentially discharged, leading to the evolution of oxygen. Consequently, for solutions of metal salts other than halides or sulfides, the decomposition potential will depend primarily upon the metal ion. Equations 25-4 and 25-5 will express the anodic reaction.

Under suitable conditions, PbO_2, MnO_2, and Tl_2O_3 can be deposited at the anode and thereby separated from nearly all other metallic ions. Halide ions can be deposited on a silver anode—selectively, if the anode potential is controlled.

25.3 EQUIPMENT FOR ELECTROLYTIC SEPARATIONS[2]

In order to make electrolytic separations, it is necessary to have a source of direct current, an adjustable resistance, a cell for electrolysis, including the electrodes, and usually some means for stirring the solution. In order to measure the current and applied voltage, an appropriate ammeter and voltmeter are needed. The schematic arrangement of the equipment is shown in Fig. 25-2. The direct current is most conveniently supplied from storage batteries because they give a steady voltage. However, compact commercial

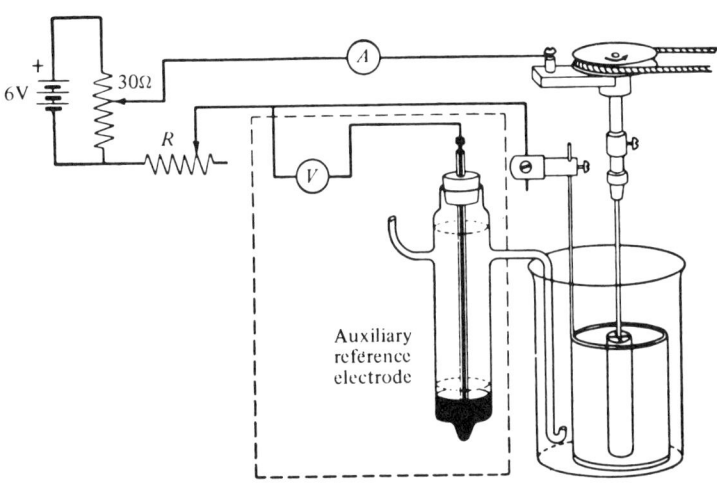

FIGURE 25-2 Equipment for electrodeposition. Enclosed within the dashed lines is the additional equipment required for measuring the electrode potential.

power supplies are available that operate from alternating current to supply the direct current. A schematic diagram of one unit is shown in Fig. 25-3. A fixed transformer steps the voltage down to 6 or 10 V, and the current is then passed through a selenium rectifier bridge of the full- or half-wave type, and finally through a filter circuit. The latter, a combination of an inductance or choke and a capacitor, converts the pulse of raw dc output from the rectifiers into a more or less smooth flow of direct current. A filter circuit which leaves the ripple remaining at 1% or less is satisfactory. The variable autotransformer is used to control the voltage applied to the stepdown transformer, and thereby the rectified output voltage.

The electrolysis cell is frequently a tall-form beaker, covered with a split watch glass to exclude dirt and to minimize loss of solution through spray during the electrolysis. The cathode is generally a cylinder (perhaps corrugated) of platinum gauze or a perforated platinum foil. Anodes may be a coiled wire, a platinum paddle, or a second gauze electrode smaller in diameter than the cathode. A gauze construction presents the largest surface area consistent with adequate mechanical strength.

After the electrolysis is complete, the deposited metal must be removed from the solution without contaminating the solution if further analyses are to be made on the solution, and without loss of the deposited metal if this deposit is to be weighed or analyzed. If the deposit has been made on a platinum electrode and is to be weighed, the electrode must be washed thoroughly as it is removed from the solution. Furthermore, because of the voltaic cell which is present and which would cause dissolution of the deposited metal if the applied voltage were interrupted, the electrode should be washed without breaking the electric current. This is best done by slowly lowering the electrolysis cell from the electrodes while washing the electrodes with a stream of water from a wash bottle. The electrodes are then rinsed with alcohol or acetone prior to drying them at an elevated temperature. The weight of the deposit is obtained by weighing the electrode before and after deposition.

25.4 ELECTROGRAVIMETRY

Constant-Current Electrolysis

Electrolysis has long been carried out with the current kept more or less constant with time by adjusting the voltage applied to the cell. The technique can be represented by the intersection of the system surface with a plane parallel to the base, or zero-current plane, of Fig. 25-1. No control is exerted over the cathode potential; rather, a predetermined current is forced through the electrolytic cells regardless of mass transport conditions. The electrochemical process with the most positive reduction potential will occur first at the cathode, then the next most positive electrochemical process, and so on. Thus, if a current is passed through a solution containing cupric, hydrogen, and zinc ions, copper will be deposited first at the cathode. As the copper deposits, the reduction potential of the cupric ions becomes more negative, requiring periodic changes in the applied emf to more negative values as the electrolysis proceeds. More significantly, the rate at which cupric ions can be brought to the electrode surface will eventually fall under the rate re-

ELECTROGRAVIMETRY AND COULOMETRY 743

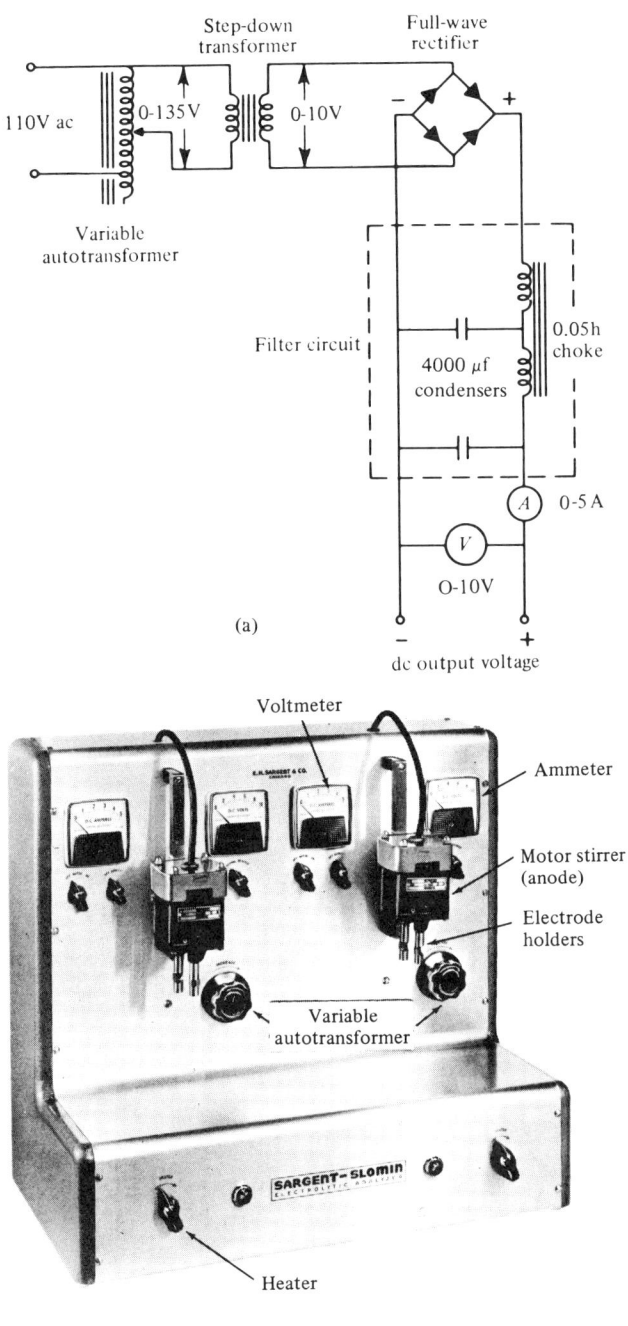

FIGURE 25-3 Direct-current power supply for electrodeposition. (a) The electrical circuit and (b) exterior view showing the controls. (Courtesy of Sargent-Welch.)

quired by the current forced through the cell; that is, the reduction of cupric ions alone will not hold the current at the desired level. Further increases in the applied emf will then result in a rapid change of cathode potential to a point where it equals that of hydrogen ions, and liberation of hydrogen gas begins. From this point on an increasing fraction of the current is devoted to the evolution of hydrogen, although the cathode potential will become relatively stable at a level fixed by the electrode potential and the overpotential for the evolution of hydrogen gas. This second process would be intolerable were it to involve another metal deposit; even so, the continual evolution of gas at the electrode is unsuited for adherent deposits.

Since, in the foregoing example, the hydrogen-ion concentration remains virtually constant in a solution during the evolution of hydrogen at the cathode and oxygen at the anode, the potential of the cathode cannot become sufficiently negative for the deposition of the zinc ions to commence. It should be evident, then, that metallic ions with a positive reduction potential may be separated, without external control of the cathode potential, from metallic ions having negative reduction potentials. However, for this separation to be successful, the hydrogen overpotential on the cathode plus the reversible reduction potential of the hydrogen ions must be less than the negative reduction potential of any of the metallic ions that are to remain in solution. For example, cupric ions in a solution containing $1M$ hydrogen ions may be separated from all metallic ions whose reduction potentials are more negative than about -0.4 V, the hydrogen overpotential on a copper electrode for relatively large current densities. Additional selectivity can be achieved through use of masking agents or potential buffers, or control of pH.

Example 25-1

Under what conditions would it be possible to initiate the deposition of zinc onto a copper-clad electrode from a solution that is $0.01M$ in zinc ions? Also, what conditions are necessary for quantitative removal of zinc?

Answer From the Nernst equation, the deposition potential for zinc is

$$E = -0.76 + 0.03 \log (0.01) = -0.82 \text{ V}$$

and this value increases to -0.94 V when 0.01% remains in solution. Turning to the expression for the evolution of hydrogen, we can evaluate the minimum pH necessary to allow the deposition of zinc to commence, assuming the overpotential of hydrogen on copper to be 0.4 V:

$$E = 0.0 + 0.059 \log [H^+] + (-0.4) = -0.82$$

from which the pH is calculated to be 7. Although it might be expected that the pH would have to be raised to about 8.5 to remove the zinc quantitatively, this is true only if the amount of zinc is insufficient to coat completely the electrode surface exposed to the solution. As soon as the electrode becomes coated with zinc metal, the overpotential of hydrogen rises to the value on a zinc surface, and, consequently, the deposition of zinc proceeds to completion at pH 7 approximately. In practice, an ammoniacal buffer is employed, partly to take advantage of the superior deposit from zinc ammine ions.

Separations with Controlled Electrode Potentials

To carry out the electrolytic separation of two metals whose deposition potentials differ by an adequate amount, yet which lie on the same side of hydrogen, provision must be made to control the electrode potential. This is achieved by introducing an auxiliary reference electrode and placing the tip of the salt bridge adjacent to the working electrode. The potential of the working electrode is determined by measuring the emf of the cell established by the electrode and the reference electrode, from which the potential of the working electrode (herein assumed to be the cathode) can be calculated:

$$E_{cath} = E_{cell} - E_{ref} \tag{25-9}$$

A potentiometer or vacuum-tube voltmeter serves to measure the cell emf. The extra items of equipment, enclosed within the dashed lines of Fig. 25-2, are the only changes required in the conventional apparatus for electrodeposition by the constant-current method.

With the isolation of the term: $E_{cath} + \omega_c$, from the cell emf, controlled-potential electrolysis may cleanly separate two elements, put an element into a particular oxidation state, or synthesize an organic compound. The potential of the cathode (or anode in oxidation reactions) is controlled so that it never becomes sufficiently negative to allow the deposition of the next element. In practice, the controlled potential should be $0.118/n$ V more negative than the final equilibrium electrode potential to correspond to the 100-fold difference that usually prevails in concentration between the electrode surface and the bulk concentration. This control of the electrode potential is achieved by adjusting the voltage applied to the electrolysis cell—manually with a battery and a variable resistor, by manual adjustment of the autotransformer at the input of commercial electroanalyzers, or electronically with a potentiostat. The current steadily decreases as the metallic ions are removed, but the maximum current permissible is used at all times and thus the electrolysis proceeds at its maximum rate. The electrolysis is discontinued when the current has fallen to a constant low value, usually 10 or 20 mA. The intersection of the system surface with a plane parallel to the zero-potential plane in Fig. 25-1 shows the course of the current and the percent reduced (or oxidized) during the electrolysis.

A variety of electronic and electromechanical circuits are available which are capable of decreasing or increasing automatically the applied emf in order to maintain a constant electrode potential. The schematic circuit of one instrument is shown in Fig. 25-4. It can be used in conjunction with the power supply shown in Fig. 25-3. In use, the electrode potential is set at any desired value from 0 to 3 V versus the reference electrode. The reference half-cell is then balanced against the working electrode through a potentiometer. When the voltage of the working electrode changes from the preset potential, a current flows in the potentiometer circuit. This current, amplified by a dc amplifier, activates one of two relays that control the reversible motor that adjusts the contactor of the autotransformer (Fig. 25-3). If the electrode potential is low, one of the relays operates the autotransformer to increase the electrode potential. If it is high, the other relay turns the motor in reverse to decrease it until the difference signal has been reduced to zero. Completely electronic circuits are also available (see Fig. 25-14).

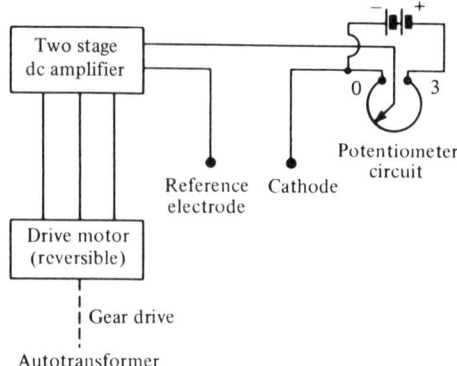

FIGURE 25-4 Schematic circuit diagram of an automatic potentiostat.

An approximate value of the limiting electrode potential can be calculated from the Nernst equation, but lack of knowledge concerning the overpotential term for a system severely limits its usefulness. A more reliable method involves the determination of the limiting potential empirically from current–potential curves. The current–potential curve is determined for each reaction under exactly the same conditions that will prevail in the actual analysis. The potential of the working electrode is increased in regular increments by increasing the voltage applied to the cell. The current is observed at each value of the electrode potential. To minimize any change in the concentration around the electrode, the cell circuit should be closed only long enough to secure the current measurement. Schematic current-cathode potential curves for the reduction of copper(II) are shown in Fig. 23-6. Ordinarily polarograms obtained with the dropping mercury electrode serve excellently to define the conditions for electrolysis with a large mercury cathode. Usually these will be about 0.1–0.15 V more negative than the polarographic half-wave potential.

Separations with controlled electrode potentials are very satisfactorily done with a mercury cathode. By including a coulometer in series with the electrolysis cell, it is comparatively simple to perform a series of separations and analyses without replacing the mercury cathode between successive separations and, perhaps, without weighing any electrodes.

Halides may be determined by electrolyzing their solutions between a platinum cathode and a silver anode. By controlling the anode potential, the gain in weight of the anode is equal to the weight of the particular halide in the original solution.

The use of controlled electrode potentials is now quite prevalent in preparative organic chemistry. If a certain organic compound can undergo a series of reductions (or oxidations), each at a definite potential, it is then possible to reduce the starting material selectively and efficiently to some desired compound by controlling the potential of the cathode during the reduction. Because this procedure for organic compounds produces essentially only the desired product, it is much more economical than most reductions or oxidations performed chemically, where side reactions producing undesired products usually occur.

Constant-Voltage Electrolysis

If we consider all the terms embodied in Eq. 25-7 for the separation of copper and lead ions at the cathode, with oxygen liberated at the anode, the copper is practically completely deposited at an applied voltage of 1.91 V:

$$E_{applied} = (1.23 + 0.40) - (0.21 - 0.01) + 0.50 \qquad (25\text{-}10)$$

where the ohmic drop is assumed to be 0.50 V. The lead will not begin to deposit until the applied voltage exceeds 2.28 V. Consequently, the separation of copper from lead can be accomplished by an electrolysis at any constant applied voltage between these two values.

This technique assumes that the ohmic drop and the overpotential terms in Eq. 25-7 remain constant. However, as a consequence of the cell reaction, the copper concentration diminishes and the hydrogen-ion concentration increases. This results in both E_{anod} and E_{cath} becoming more negative. From the Nernst equation the theoretical cathode potential will change from 0.31 to 0.19 V if the corresponding copper(II) concentration changes from 0.1 to $10^{-5} M$. On the other hand, the anode potential will have decreased only slightly. The greater conductance of the hydrogen ion decreases the cell resistance and affects the iR term. Finally, the overpotential terms are dependent on the current density and cannot be expected to remain constant when the current tends to vary. These variations render it difficult to control the cathode potential within as close limits as is possible by the controlled-potential method.

The electrolysis process can be represented by the intersection of the system surface in Fig. 25-1 with a slanted plane. When the current is zero, the cell emf is equal to the applied voltage. During the electrolysis, as the cathode potential becomes more negative, the copper(II) concentration becomes depleted to the point where the supply of ions at the cathode surface is insufficient to meet the current demand. When this occurs, the current must of necessity decrease, ultimately falling to zero. Thus, the deposition of lead does not occur, but this advantage is achieved at the cost of diminished current at any time. Hence a longer time is required for complete electrolysis in comparison with the controlled-potential method.

Composition of the Electrolyte

It is clearly possible to separate one metal from another if the respective deposition potentials are sufficiently far apart and, in constant-current electrolysis, if one potential is more negative than that required for the evolution of hydrogen. When two metals have similar discharge potentials, sometimes the electrolyte composition can be altered sufficiently for separation to be possible. By addition of a masking agent which forms a complex with one of the metal ions, the discharge potential for this ion usually becomes more negative. Take, for example, an alkaline solution of copper and bismuth ions to which cyanide ions are added. Copper(II) ions are reduced to the monovalent state and form the tricyanocuprate(I) ion:

$$Cu^+ + 3CN^- \rightleftharpoons Cu(CN)_3^{2-}; \quad K_f = 10^{27} \qquad (25\text{-}11)$$

In this environment the discharge potential of the copper complex is −1.05 V when 1M cyanide ion is present. By contrast, the addition of cyanide hardly affects the deposition potential for bismuth, and if tartrate ions are present to keep the bismuth in solution, quantitative separation from copper is possible.

The temperature dependence of a series of homologous metal complexes sometimes provides a means for separating discharge potentials. For example, in the case of nickel and zinc in ammoniacal solution, the deposition potentials are similar at 20°C, being −0.90 and −1.14 V, respectively, but differ markedly at 90°C, now being −0.60 and −1.05 V. In general, the less dissociated the complex at room temperature, the greater the change in the degree of dissociation as the temperature rises.

Potential Buffers

Suitable oxidation-reduction systems that are preferentially reduced at the cathode, or oxidized at the anode, may be employed to limit and maintain a constant potential at the electrode. The name *potential buffer* is applied to this type of electrolyte because of its functional resemblance to pH buffers. For example, the uranium(III)/(IV) system has been used to prevent the cathode potential from exceeding approximately −0.5 V, the reduction potential of the uranium system. As the cathode potential attains this value, there is increased competition of the uranium(IV) reduction reaction with the deposition process. The reduced uranium(III) ions are reoxidized at the anode and remain in the electrolyte to continue the cyclic process. Eventually the entire current flow through the electrolytic cell is consumed in the reduction of uranium(IV) at the cathode and the oxidation of uranium(III) at the anode. One application of the uranium couple has been to prevent the deposition of chromium and manganese at a mercury cathode while permitting copper, tin, lead, and nickel to deposit in the normal manner.

Nitrate ions have long been employed in the constant-current deposition of copper and lead dioxide to prevent the formation of metallic lead at the cathode. Nitrate ions are less easily reduced than copper ions, and sufficiently so that the copper can be quantitatively removed; yet nitrate ions are more easily reduced than lead ions. At the anode, on the other hand, lead(II) ions are oxidized to lead dioxide more easily than hydroxyl ions to oxygen, thus,

$$Pb^{2+} + 2H_2O \rightarrow PbO_2(s) + 4H^+ + 2e^- \tag{25-12}$$

for which the electrode potential is 1.46 V, whereas the liberation of oxygen occurs at about 1.70 V. Chloride ions must be absent for several reasons: Chloride ions are oxidized more easily ($E° = 1.36$ V), the overpotential of chlorine gas is less than that for the deposition of PbO_2, and formation of $PbCl_3^-$ is appreciable.

The successful deposition of copper from chlorocuprate(I) ions, and the prevention of the competing oxidation of these ions to copper(II) at the anode, are due to the buffering action of hydrazine. The oxidation of hydrazine,

$$N_2H_5^+ \rightarrow N_2(g) + 5H^+ + 4e^- \tag{25-13}$$

for which the electrode potential is 0.17 V, takes place in preference to chlorocuprate(I) ions,

$$CuCl_3^{2-} \rightarrow Cu^{2+} + 3Cl^- + e^- \tag{25-14}$$

for which the electrode potential is 0.51 V. As long as hydrazine is present in excess, the anodic oxidation of chloride ions and copper(I) will be prevented. Without any hydrazine, the anode potential remains at a relatively high value between the chlorocuprate(I)–copper and the water–oxygen couples.

Physical Characteristics of Metal Deposits

It is important to conduct the electrolysis under conditions which ensure that the deposit is pure, adherent to the electrode, and quantitative. This requires consideration of the factors that influence the nature of the deposit: current density, the chemical nature of the ion in solution, that is, complexed or as an aquated ion, the rate of stirring, the temperature, and the presence of depolarizers to minimize the evolution of gases.

The optimum conditions for achieving the best deposit vary from one metal to another. Adherence to the electrode is the most important physical characteristic of a deposit. Generally a smooth deposit and adherence are congruent. Flaky, spongy, or powdery deposits adhere only loosely to an electrode. Simultaneous evolution of a gas is often detrimental. Continual evolution of bubbles on the electrode surface disturbs the orderly growth of the crystal structure of a metal deposit, and porous and spongy deposits may be obtained. Once under way, these continue because the current density is high at points on an otherwise smooth surface, and this is conducive to irregular, treelike growths. The discharge of hydrogen frequently causes the film of solution in the vicinity of the cathode to become alkaline, with the consequent formation of hydrous oxides or basic salts.

The chemical nature of the ion in solution often has an important influence on the physical form of the deposited metal. For example, a pure, bright, and adherent deposit of copper can be obtained by electrolyzing a nitric acid solution of Cu^{2+} ions. By contrast, a coarse, treelike deposit of silver is obtained under similar conditions. If a suitable deposit of silver is to be obtained, the electrolysis must be carried out from a solution in which the silver ions are complexed, as $Ag(CN)_2^-$. Similarly, the best deposits of iron are obtained from an oxalate complex and those of nickel from an ammonia complex. Halide ions facilitate the deposition of some metals, probably because the overpotential is lower for metal halide ions than for aquated metal ions. Complex ions also exhibit what is known as "throwing power" to a considerable degree—that is, the property of a solution by virtue of which a relatively uniform deposit of metal may be obtained on irregular surfaces.

The time required for electrodeposition is shortened if the electrode is rotated or the electrolyte stirred vigorously. As an adjunct to the normal diffusion process and mass transfer conditions, it lowers the concentration overpotential and enables a higher current density to be employed without deleterious results.

Increase in temperature has two effects which oppose each other. On one hand, diffusion is favored. On the other hand, hydrogen overpotential is decreased and the stability of many complex ions is decreased.

Factors Governing Current

If a reaction is 100% current efficient, the passage of one faraday of electricity, 96,487 coulombs, will cause the reaction of one equivalent weight of substance. The relationship between the weight in grams, W, the number of coulombs, Q, the molecular weight, M, the number of faradays involved in the reaction of one mole, n, and the value of the faraday in coulombs, F, is given by the relationship

$$W = \frac{QM}{Fn} \tag{25-15}$$

If the current, i, remains constant as in constant-current coulometry, then the total number of coulombs is given by the product of the current times the time; thus

$$Q = it \tag{25-16}$$

If, as in controlled-potential coulometry, the current changes continuously, then Q is given by the integration of time versus current as in Eq. 25-17.

$$Q = \int_0^\infty i\,dt \tag{25-17}$$

Most controlled-potential coulometric titrations are carried out under conditions in which the current is diffusion controlled and the relationship between current, i_t, at any time, t, and concentration, C_t, is given by the equation

$$i_t = \frac{nFADC_t}{\delta} \tag{25-18}$$

where A is the area of the electrode, D is the diffusion coefficient, and δ is the thickness of the Nernst diffusion layer. From Faraday's law, the rate of change of concentration with time is given by the relationship

$$\frac{dC_t}{dt} = -\frac{i_t}{nFV} \tag{25-19}$$

where V is the volume of the solution. Substitution[3] of Eq. 25-18 into 25-19 and integration yields the following results for concentration as a function of initial concentration, C_0, and current as a function of initial current, i_0,

$$C_t = C_0 e^{-kt} \tag{25-20}$$

$$i_t = i_0 e^{-kt} \quad \text{or} \quad 2.3 \log\left(\frac{i_0}{i_t}\right) = kt \tag{25-21}$$

where $k = DA/V\delta$.

The number of coulombs, Q_t, passing up to time t is given by integration.

$$Q_t = \int_0^t i\,dt = \int_0^t i_0 e^{-kt}\,dt = \frac{i_0}{k} - \frac{i_t}{k} \tag{25-22}$$

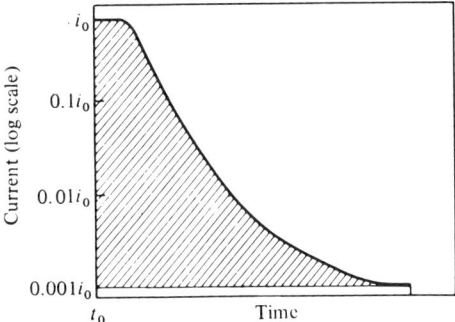

FIGURE 25-5 Typical current–time relationship in controlled-potential coulometry. Shaded area indicates graphical method of integration.

Such a relationship is useful in estimating the total number of coulombs required for complete reaction before the reaction is actually completed. Q_t is read at several values of t, preferably in the range of 90-99% completion, and then these values of Q_t are plotted versus i_t. Q_∞ is determined by extrapolation of the straight line so obtained to the coulomb axis. The limiting value of Q is, obviously, i_0/k.

If the extrapolation method is not used, then the electrolysis is usually carried out until i_t has diminished to 0.1% or less of its original value or until the current becomes equal to the residual current as measured on a sample of supporting electrolyte alone.

The behavior of current versus time in an actual controlled-potential electrolysis is represented in Fig. 25-5. The initial horizontal portion of the curve arises because most coulometers have an upper limit to the current they can furnish and therefore may not be able to carry out the electrolysis at a rate sufficiently high to reach the limiting current set by Eq. 25-18 until the concentration, C, is reduced somewhat. The tail at the end of the curve is due to residual current in the supporting electrolyte.

Example 25-2

If a constant current of 10.00 mA passes through a chloride solution for 200 sec, what weight of chloride reacts with the silver anode?

Answer The net charge involved is

$$Q = it = (10 \times 10^{-3} \text{ A})(200 \text{ sec}) = 2.00 \text{ coulombs}$$

or

$$\frac{2.00}{96,487} = 2.075 \times 10^{-5} \text{ equivalents of chloride ion}$$

Since $n = 1$ for the reaction

$$Ag^\circ + Cl^- \rightarrow AgCl + e^-$$

2.075×10^{-5} equivalents equal 2.075×10^{-5} mole of chloride ion. In weight, the amount of chloride ion, now present as AgCl, is

$$(2.075 \times 10^{-5})(35.45) = 0.735 \times 10^{-3} \text{ g (or 0.735 mg)}$$

The Mercury Cathode

Cathodes comprising a pool of mercury or an amalgamated platinum or brass gauze electrode warrant special consideration. The mercury cathode is not generally used to determine any of the metals plated out because of the difficulties involved in weighing and drying the mercury before and after the determination. However, it is one of the most useful aids for the removal of certain base metals, even in considerable quantities, that interfere in the determination of elements high in the electromotive series. Two factors set mercury apart from other electrode materials. Many of the metals depositing on mercury can form an alloy (amalgam) with the mercury. Owing to the alloy formation the deposition potentials of these metals on mercury are displaced from their normal value in the positive direction with respect to reduction potentials. Their deposition is also aided by the fact that the hydrogen overpotential on mercury is particularly large. As a result, the deposition from a fairly acid solution is possible for such metals as iron, nickel, chromium, zinc, and even manganese under certain conditions.

In its simplest form the mercury cathode cell consists of a shallow pool of mercury covering the bottom of a beaker. Electrical contact to the mercury is made by a glass-enclosed platinum wire, either immersed in the mercury pool or sealed through the base of the container. The cell designed by Melaven,[4] shown in Fig. 25-6, is a slightly more refined form and is in common use. The cathode consists of 35-50 ml of pure mercury in a modified separatory funnel. The apparatus has a conical base fitted with a three-way stopcock. One arm of the stopcock is connected to a leveling bulb that controls the level of the mercury in the cell; the other permits removal of the electrolyte. With a beaker, this removal is accomplished by siphoning. The anode is a platinum wire in the form of a spiral. Agitation is accomplished by a mechanical stirrer or a stream of air.

Simple, unitized cells have been designed.[5] A sturdy, compact, self-contained immersion electrode is shown in Fig. 25-7. This cell is a glass dish about 30 mm in diameter by

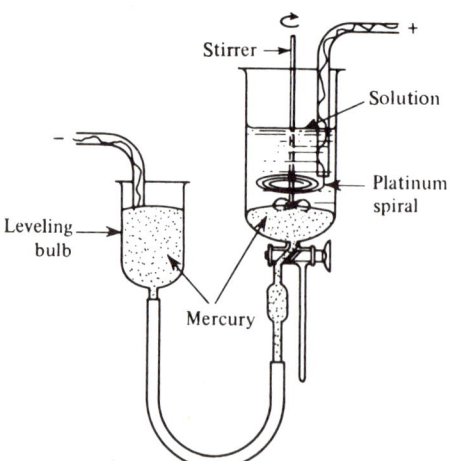

FIGURE 25-6 Mercury cathode cell. (Reprinted with permission from A. D. Melaven, *Ind. Eng. Chem., Anal. Ed.*, 2, 180 (1930). Copyright 1930 American Chemical Society.)

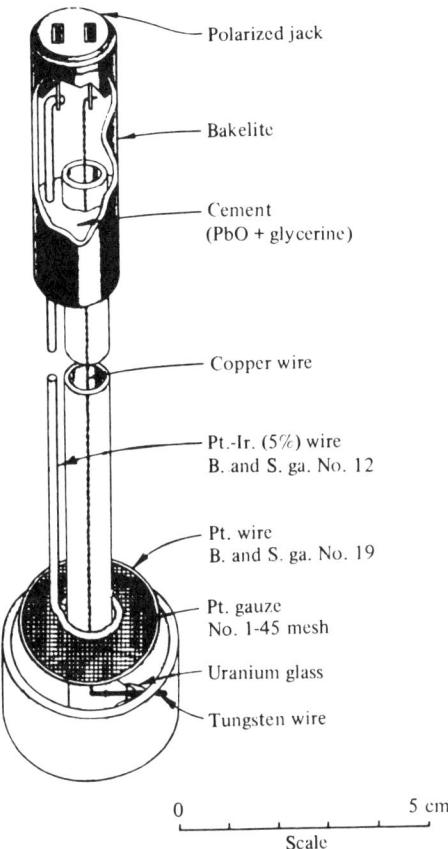

FIGURE 25-7 Unitized mercury cathode cell. (Reprinted with permission from H. O. Johnson, J. R. Weaver, and L. Lykken, *Ind. Eng. Chem. Anal. Ed.*, 19, 481 (1947). Copyright 1947 American Chemical Society.)

15 mm high, from the side of which extends a glass tube carrying the wire for electrical contact. A flat, spiral anode completes the cell. The unitized electrode is easily removed from the electrolyte and washed with a stream of wash solution quickly enough to prevent appreciable dissolution of the deposited metals. The consumption of mercury is a minimum, usually 5 ml per electrolysis, and the simplicity with which duplicate assemblies are interchanged encourages frequent substitution of fresh mercury. This increases the efficiency of a separation and decreases the time for electrolysis. Finally, the difficulty from loss of mercury in handling and from dispersion during electrolysis is minimized.

Specially designed mercury cathode equipment patterned after the preceding designs is available commercially and can be used in conjunction with ordinary electrolysis apparatus.

Vigorous stirring materially shortens the electrolysis. Agitation can be accomplished by any type of mechanical stirrer or by a stream of air. Rapid countercurrent stirring of

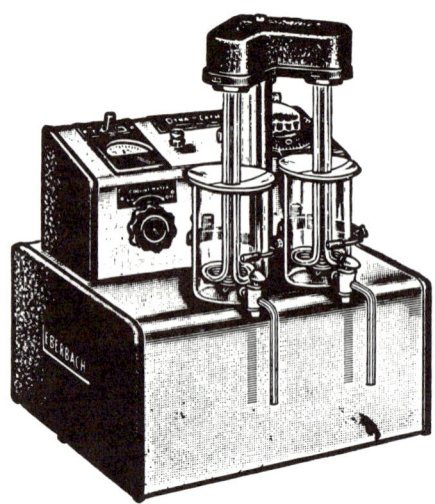

FIGURE 25-8 The Dyna-Cath, a commercial mercury cathode instrument. (Courtesy of Eberbach & Son.)

the mercury and the electrolyte at the deposition interface favors a more efficient deposition and constantly exposes fresh mercury to the electrolyte. This can be provided by a magnetic stirring bar floating on the mercury surface, or by letting the impeller blades be only partially immersed in the mercury. A commercial unit has been devised in which a magnetic circuit provides the stirring, the electrolyte and the mercury becoming the two independent rotors of a dc motor.[6] In addition, the magnetic field immediately removes deposited ferromagnetic materials from the mercury-solution interface and retains them beneath the surface of the mercury. The instrument is pictured in Fig. 25-7.

The electrolyte is usually a 0.1–0.5M solution of sulfuric acid or perchloric acid. Nitric and hydrochloric acids are avoided; the reduction of nitrate lowers the current efficiency for the reaction of interest, and the anode may be attacked in the presence of chloride.

A current density of 0.1–0.2 A cm^{-2} is common, but substantially higher current densities have been used in cells with appropriate cooling devices to remove heat developed by the resistance of the electrolyte, as is done in the Dyna-Cath (Fig. 25-8). The amount of metal removed is proportional to the current and the area of the mercury surface.

In most cases the constant-current technique serves excellently as the separation method for the elements shown in Table 25-2. Most uses of the mercury cathode concern the removal of an interfering element or elements before the determination of a substance that remains in the electrolyte. In this respect it has been extensively applied to facilitate the determination of aluminum, titanium, vanadium, and magnesium in a wide variety of materials. The element most commonly deposited is iron. The mercury cathode has also been used to effect the reduction of an element or compound to a lower oxidation state in solution. Other applications are enumerated in review articles.[7,8]

TABLE 25-2 Electrolysis with a Mercury Cathode in 0.3N Sulfuric Acid Solution

H																	He
Li	Be											B	C	N	O	F	Ne
Na	Mg											Al	Si	P	S	Cl	Ar
K	Ca	Sc	Ti	V	Cr	Mn	Fe	Co	Ni	Cu	Zn	Ga	Ge	As	Se	Br	Kr
Rb	Sr	Y	Zr	Nb	Mo	Tc	Ru	Rh	Pd	Ag	Cd	In	Sn	Sb	Te	I	Xe
Cs	Ba	La	Hf	Ta	W	Re	Os	Ir	Pt	Au	Hg	Tl	Pb	Bi	Po	At	Rn
Fr	Ra	Ac															
		Ce	Pr	Nd	Pm	Sm	Eu	Gd	Tb	Dy	Ho	Er	Tm	Yb	Lu		
		Th	Pa	U	Np	Pu	Am	Cm	Bk	Cf	Es	Fm	Md	No	Lw		

SOURCE: G. E. F. Lundell and J. I. Hoffman, *Outlines of Methods of Chemical Analysis*, p. 94 John Wiley & Sons, Inc., New York, 1938.
NOTE: On the periodic chart of the atoms the theoretical separation possibilities have been indicated for the mercury cathode. The elements enclosed by solid line (———) are quantitatively deposited in the mercury. Those surrounded by a dotted line (· · · ·) are quantitatively separated from the electrolyte but not quantitatively deposited in the mercury. Elements enclosed by a wavy line (∼∼∼) are incompletely separated from the electrolyte.

Internal Electrolysis

Internal electrolysis is the term applied by Sand[9] to electrogravimetric analyses which employ an attackable anode. The latter is connected directly to the cathode. In reality the arrangement is nothing but a short-circuited galvanic cell. It is convenient in some applications because the electrolysis proceeds spontaneously without the application of an external voltage and the choice of an attackable anode limits the cathode potential without elaborate instrumentation or the operator's attention. However, the driving force—that is, the difference between the potential of the system plating at the cathode and the dissolution of the anode—is small, and in consequence, the cell resistance is a critical factor in determining the rate of metal deposition. The application of the method is restricted to small amounts of material if the time of electrolysis is not to be excessively long.

The selection of an anode is made with a knowledge of the reversible potentials of the various metal ion–metal couples. A typical application is the removal of small amounts of copper and bismuth from pig lead.[10] Because the reduction potential of lead is sufficiently far apart from the reduction potentials of the copper and bismuth systems, the anodes can be constructed from helices of pure lead wire. The arrangement of equipment is shown in Fig. 25-9. Dual anodes are often used to provide a larger electrode area. These are inserted within a porous membrane (alundum shell) in order to isolate them from the sample and forestall any direct plating on the lead itself. A platinum gauze electrode is placed between the anode compartments. The electrolysis is begun by short-circuiting the cathode to the anode.

To keep the ohmic resistance small, the anode solution must have a high concentration of electrolyte. In addition it must contain a higher concentration of the ions formed from

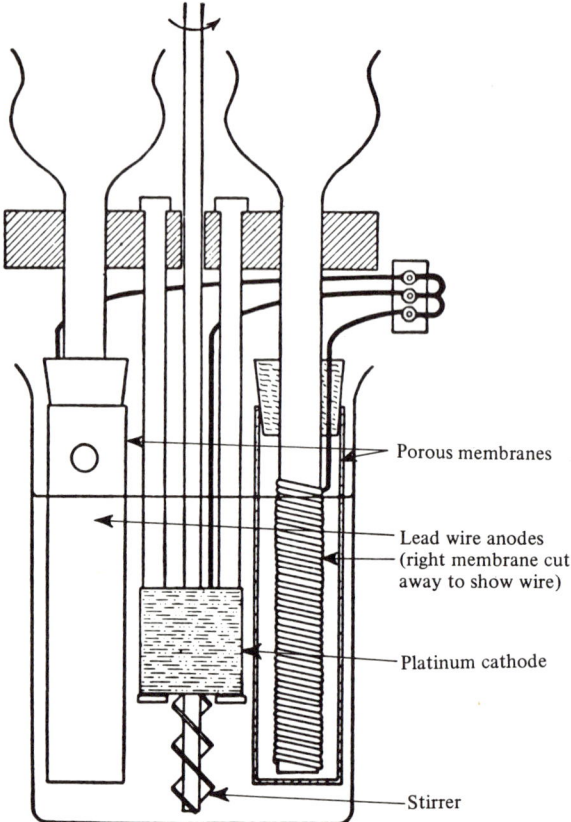

FIGURE 25-9 Apparatus for internal electrolysis.

the dissolution of the anode (that is, lead ions in the example) than does the catholyte containing the dissolved sample. The anode reaction will be the dissolution of the lead; the cathode reaction the deposition of copper. The cell can be represented as

$$-Pb° \mid Pb^{2+} \parallel Cu^{2+} \mid Cu°(Pt) +$$

Because the cell operates spontaneously, the cathode is the positive electrode. As the cathode and anode are short-circuited, the only dissipation of energy is in the form of the ohmic resistance, which in turn limits the maximum current flow through the cell:

$$iR = -E_{Pb^{2+},Pb°} + E_{Cu^{2+},Cu°} + \frac{0.0591}{2} \log \frac{[Cu^{2+}]}{[Pb^{2+}]} \qquad (25\text{-}23)$$

For an anolyte solution that is $1M$ in Pb^{2+} ion, and inserting standard electrode potentials, Eq. 25-23 becomes

$$iR = 0.22 + 0.0296 \log [Cu^{2+}]. \qquad (25\text{-}24)$$

With this arrangement the electrode potential of the cathode cannot exceed −0.12 V. Only those metal ions will deposit whose electrode potentials are more positive than this value. In the example taken, as the electrolysis progresses, the concentration of copper(II) ions diminishes and the electrode potential of the cathode becomes more negative until it becomes equal to the anode potential (or the decomposition potential of another substance is exceeded). At no time will the decomposition potential of lead at the cathode be exceeded. There is no danger of lead contamination due to the concentration-overpotential factor because the rate of cathodic deposition is controlled by the rate of anodic dissolution.

The anode need not always be constructed of the material that constitutes the matrix of the sample. For selective reduction of several trace constituents in zinc, for example, four separate samples would be dissolved for the separation of traces of silver, copper, lead, and cadmium. In the first, an attackable anode of copper would permit the complete removal of silver, but control the cathode potential below the deposition potentials of the others. Similarly, a lead anode would make it possible to remove silver plus copper; a cadmium electrode would remove silver, copper, and lead; and with a zinc anode, all four elements would be removed.

The amount of deposit is generally limited to quantities not exceeding 25 mg. Although larger quantities have been handled, the deposit is apt to be spongy and some of the metal ions may diffuse to the anode during the longer time required for complete electrolysis. Little attention is required during an analysis except to flush the anolyte compartments once or twice. Halide solutions may be employed without removing the halide ion and without adding an anodic depolarizer. Average running time is 30 min/sample.

25.5 ELECTROGRAPHY

The electrographic method,[11] developed by Glazunov,[12] and Fritz,[13] is a useful microanalytical tool for accurately identifying and determining substances. This method consists in anodically dissolving a minute amount of the test substance onto a piece of bibulous paper or, for more accurate rendition, gelatin-coated paper which has been soaked in a suitable electrolyte. The test sheet is held under pressure between the sample surface, the anode, and a suitable cathode surface. The latter may be a flat square electrode for flat surfaces, a long narrow electrode for use on metal ribbons, or sponge rubber covered with aluminum foil for uneven sample surfaces. The unit is connected to a battery of dry cells and a current is allowed to flow for several seconds. A general laboratory circuit is shown in Fig. 25-10. While the current is flowing, ions leave the surface of the specimen

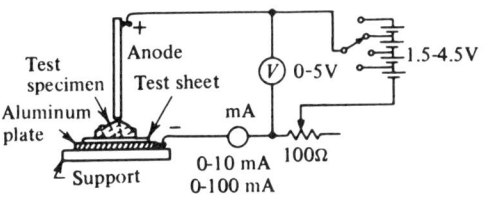

FIGURE 25-10 Schematic arrangement of equipment and electrical circuit for electrographic analysis.

and migrate into the permeable test sheet. Their presence can be made manifest, if they are colorless, by treating the test sheet with selective reagents. Distinctive identifying colors result and appear in an exact chemical and physical image of the surface.

The magnitude of the current required and its duration can be approximated from the second law of Faraday

$$it = \frac{96{,}487 A d n}{W} \qquad (25\text{-}25)$$

where i is the current in amperes, t is the duration of the current in seconds, A is the area of the specimen surface in centimeters squared, W is the atomic weight of the element dissolving and forming n equivalents per gram atom, d is the minimum weight of material needed for detection by the method chosen in grams per square centimeter. In general, 50 μg of most metals will produce brilliantly colored products when the reaction is confined to an area of 1 cm^2. These conditions would require a current of 15 mA and an exposure time of 10 sec. For multicomponent samples the current may be carried almost entirely by ions of highest mobility, with the result that the pattern will be underexposed with respect to the poorly conducting constituents. This difficulty may be partially remedied by moistening the specimen with a mineral acid instead of a neutral electrolyte and by eluting the interfering ions prior to spot testing.

The test sheet may be moistened with only a neutral electrolyte, such as sodium nitrate or sodium chloride, or it may be impregnated with a reagent for the metal or metals to be detected, such as potassium ferricyanide for iron and ammonium sulfide for copper and silver. With a neutral electrolyte the print must be further developed for immersion in a developing reagent that forms a reaction product of distinctive color. Individual patterns can be secured by developing successive prints with different selective reagents, for example, α-benzoinoxime for copper, dimethylglyoxime for nickel, and α-nitroso-β-naphthol for cobalt from a sulfide mineral surface which has been electrographed with an ammonia solution. The test sheet should be fine-grained and held in close contact with the surface to be tested. By this method, in contrast with contact printing, lateral diffusion or "bleeding" is minimized. Prints are sharp and permit many fine features to be detected.

The electrographic method is applicable only to materials that are conductors of the electric current. It can be applied for the inspection of lacquer coating and of plated metals for pinholes and cracks in their surface. It can be used for many alloy identifications, such as the differentiation of lead-containing brass from ordinary brass, nickel in steel, and the distribution of metal constituents within an alloy. In the biological field the method is applicable to the localization of those constituents which are normally present within the tissue in an ionic state. One important advantage of this method, besides those of simplicity and rapidity, is the fact that so little of the sample is consumed; the sample remains essentially unaltered. Portable field kits have found extensive use in inspection and sorting work, in the laboratory as well as in the stockroom and in mineralogical field work.

Analogous to the anodic oxidation transfer is the cathodic reduction of certain anions of tarnish or corrosion films on metals. These are often tied up as basic insoluble salts and are not detectable in simple contact printing. Electrolytic reduction will free these ions.

By controlling the time, pressure, and current in a series of transfers, quantitative determinations can be made by comparing the color intensity of the pattern of an unknown sample with a series of patterns produced by known amounts of the metallic ion.

25.6 ELECTROLYTIC PURIFICATION

Electrolysis is of great practical value in the commercial refining of certain metals. Purification of copper by electrolytic refining is a large-scale operation. The relatively impure copper metal (about 99% copper) is made the anode in a sulfuric acid bath, and the copper is electrolytically dissolved, forming copper sulfate solution. The copper ions are redeposited at the cathode to form the purified copper metal (about 99.99%). Total impurities are decreased by a factor of 100. The high efficiency of the method depends on the fact that most of the impurities present at this particular stage in the refining of metallic copper are not soluble in the electrolyte.

In the refining of gold, impure gold is made the anode in a hydrochloric acid bath, and the purified gold is electroplated out at the cathode. The small amounts of silver impurity will form insoluble silver chloride. However, when too much silver is present, as in the reclaiming of jewelry gold, a nitric acid bath is used. The gold falls to the bottom of the bath as purified metal powder, and the silver is deposited at the cathode.

25.7 COULOMETRIC METHODS

Coulometric methods of analysis measure the quantity of electricity, that is, the number of coulombs, required to carry out a chemical reaction. Reactions may be carried out either directly by oxidation or reduction at the proper electrode (primary coulometric analysis), or indirectly by quantitative reaction in the solution with a primary reactant produced at one of the electrodes (secondary coulometric analysis). In any case, the fundamental requirement of coulometric analysis is that only one overall reaction must occur and that the electrode reaction used for the determination proceed with 100% current efficiency.

Coulometric methods eliminate the need for burets and balances, and the preparation, storage, and standardization of standard solutions. These methods can be automated readily. In a sense the electron becomes the primary standard. Coulometric methods can be used to produce reagents in solution that would otherwise be difficult to employ: volatile reactants such as chlorine, bromine, or iodine, or unstable reactants such as titanium(III), chromium(II), copper(I), or silver(II). The method is particularly useful and accurate in the range from milligram quantities down to microgram quantities and, therefore, in trace analyses. In addition, coulometric techniques are especially adaptable to remote control and operation and are presently very useful in the analyses of radioactive materials.

There are two general techniques used in coulometry. One, the *controlled-potential method*, maintains a constant electrode potential by continuously monitoring the potential of the working electrode as compared to a reference electrode. The current is adjusted continuously to maintain the desired potential. The other method, known as *constant-current coulometry*, maintains a constant current throughout the reaction period. In this method an excess of a redox buffer substance must be added so that the potential does not rise to a value that will cause some unwanted reaction to take place. Furthermore, the product of electrolysis of the redox buffer must react quantitatively with the substance to be determined; that is, it serves as an intermediate in the reaction. Examples of both types of coulometry will be discussed later in this chapter.

Instruments Used in Constant-Current Coulometry

Constant-current procedures require only a knowledge of the current and elapsed time to determine the number of coulombs. Since both current and time can be measured with high accuracy and with relatively simple equipment, this method of coulometry is both accurate and simple.

The schematic diagram of a coulometric setup for constant-current methods is illustrated in Fig. 25-11. The major problem is adequate stabilization of the constant-current supply in the range 1–200 mA. A true constant-current source as described below is, of course, desirable, but fairly constant current can be obtained from batteries with a series-regulating resistance. Either standard radio B batteries with voltages of 45–300 V or a line-

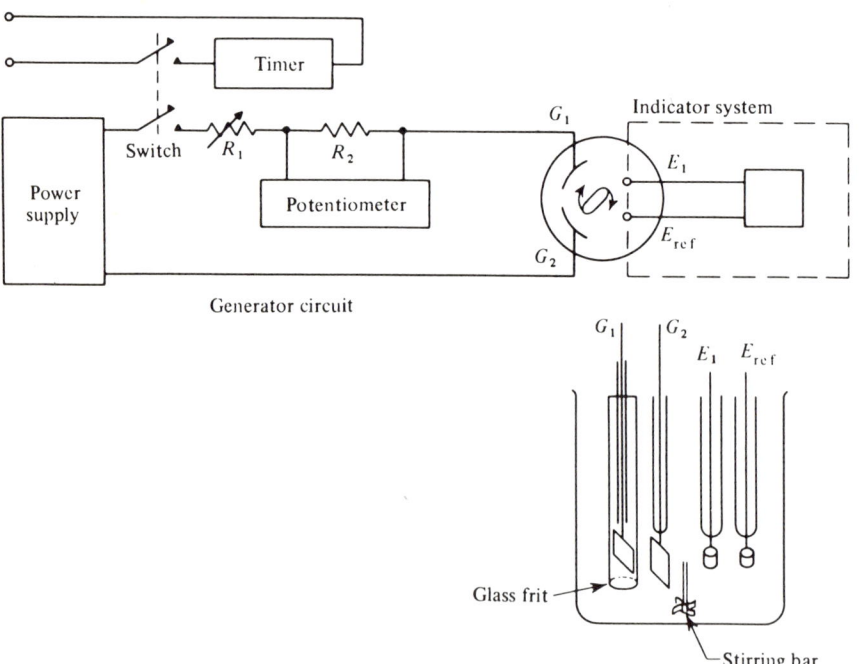

FIGURE 25-11 Schematic of equipment (and titration vessel) for constant-current coulometry. R_1 is the series (ballast) resistor; R_2 is the precision resistor; G_1 and G_2 are the generator electrodes (one isolated behind a porous frit barrier); and E_1 and E_{ref} are the electrodes for the end-point detector system.

operated, constant-voltage power supply with large series resistors (ballast resistor) can be used to maintain a constant current. If the voltage and the resistor are sufficiently large, changes in the resistance of the cell and the cell potential will have little effect on the current. This resistor is also varied to adjust the cell current to the desired level. Usually a current is selected that allows the electrolysis to be completed within 10–200 sec. To maintain the series resistance in thermal equilibrium and minimize adjustments of the cell current, it is advisable to employ a switching arrangement whereby the electrolytic cell is replaced by a dummy resistance (high-wattage type, approximately 20 Ω) during the intervals between analyses.

The current can be indicated approximately by a calibrated milliammeter, and measured precisely by means of the voltage drop across a precision resistor incorporated directly in series with the electrolytic cell. The voltage drop across the resistor can be measured very precisely with a manual or a recording potentiometer when the voltage drop is about 1 V. Under these conditions the error in the current measurement is about 0.002%.

Time measurements are normally made with a precision electric stopclock. A single switch control actuates both the timer and the electrolysis current. Times accurate to 0.01 sec are possible with modern electric chronometers.

One commercial source of equipment (Sargent-Welch) that maintains a constant current in any of several selected ranges is shown in Fig. 25-12. It comprises basically a power supply of the conventional ac rectifier and filter type to provide a dc voltage with a maximum of about 300 V, sufficient for work with high-resistance electrolytes. Current that is drawn from this power supply by connection to the cell electrode system passes through a series-regulating tube and a precision resistor, which is one of several selected by a current-selector switch. The size of this resistor is so chosen that, at the specified current level, an iR drop or voltage is developed which is equal to the potential of a standard cadmium cell. Any instantaneous error or difference resulting from a change in current due to line voltage or cell resistance variation is converted to an ac signal, amplified many times, and reconverted to a dc signal which is applied to the series regulator. Other commercial units are available from Fisher Scientific Co., Allied Electronics, Ltd., and A.E.I. (Woolwich), Ltd.

Generating electrodes must be of sufficient area to permit a low enough current density to keep electrode polarization within the limits necessary for 100% current efficiency. Currents normally employed require a substantial electrode surface area (10 cm^2 or larger), and often utilize a half-cylinder of sheet or gauze platinum. The nonworking electrode must be isolated in most cases by a salt bridge and frit barrier. The latter arrangement, however, increases the internal resistance of the cell and entails larger energy losses, so that a potentiostat may have difficulty in stabilizing the potential properly.

Instruments Used in Controlled-Potential Coulometry

In controlled-potential methods, the current is continuously changing and some sort of integrating device, a coulometer, is needed. In addition, a potentiostat is necessary to control the potential of the working electrode at the desired value.

Chemical Coulometers The electrolytic method for determining the current-time integral employs a standard chemical coulometer in series with the electrolysis cell. With electrogravimetric coulometers, the change in weight of one electrode is a measure of the charge transferred. The coulometer is simply a second electrolytic cell in which an electrochemical reaction is known to proceed with 100% current efficiency, as, for example, the

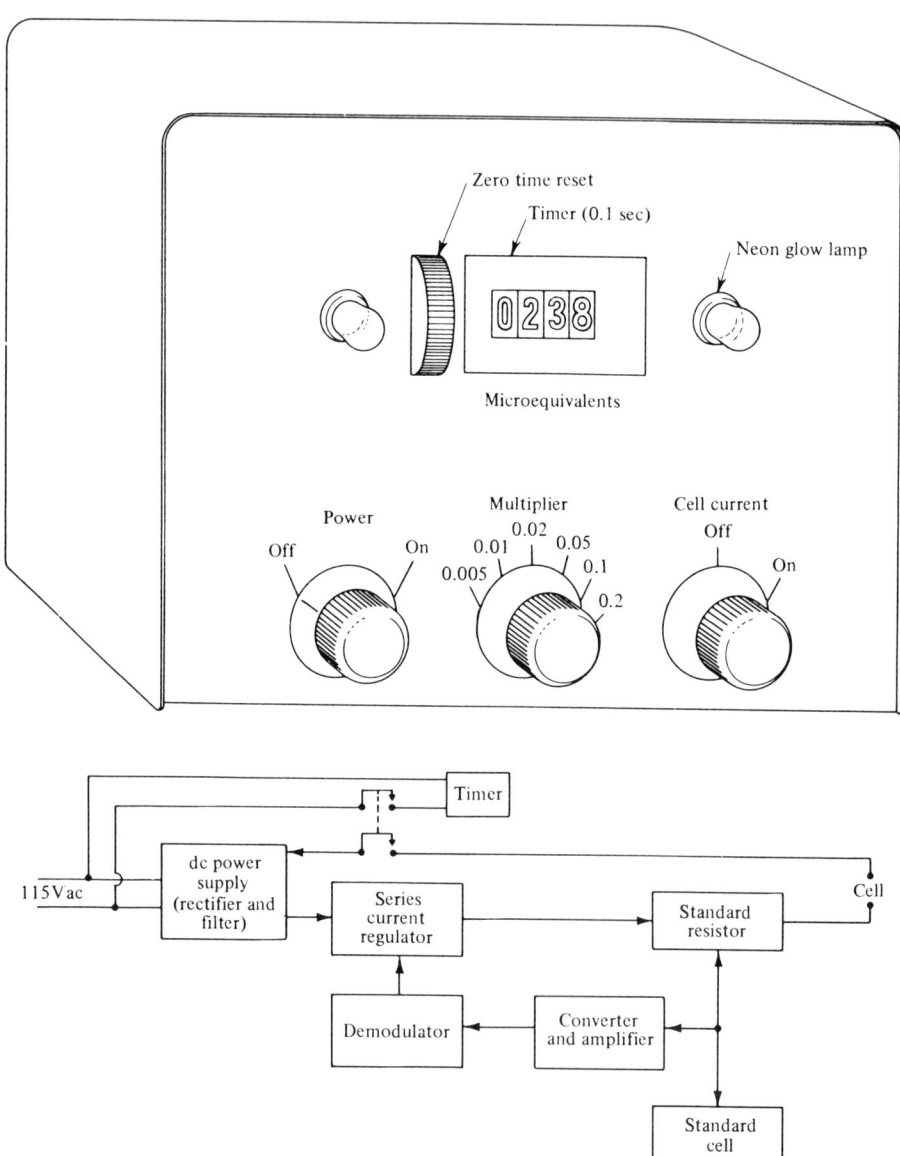

FIGURE 25-12 Sargent coulometric current source block diagram (and front panel). (Courtesy of Sargent-Welch.)

deposition of silver or copper. A modern version of the electrogravimetric coulometer is the coulometric coulometer.[14,15] After completion of the coulometric step, the coulometer is included in another circuit, by means of which a perfectly constant current is passed through the coulometer in the opposite direction. In this way the electrodes are returned to their original condition. The time required for this, multiplied by the current, gives the number of coulombs consumed in the actual determination. For example, if the reaction involved reduction of copper(II) to metallic copper, the deposit is redissolved anodically and the end point is indicated by a sharp change in electrode potential from the value for copper to that for the discharge of oxygen (see "Electrolytic Stripping"). This method is particularly suited for analyses on a microscale.

Lingane has described two gas coulometers. In one the total volume of hydrogen and oxygen liberated in the electrolysis of an aqueous solution of $0.5M$ potassium sulfate is collected in a thermostatted gas-measuring buret.[16] Its lower limit of accuracy is 10 Q. For the range from 5 to 20 Q, $0.1M$ hydrazine sulfate is used as electrolyte. Nitrogen is evolved at the anode and hydrogen at the cathode. The net coulometer reaction is

$$N_2H_5^+ = N_2(g) + 2H_2(g) + H^+ \qquad (25\text{-}26)$$

The total volume of nitrogen and hydrogen evolved is measured in a 5-ml buret.[17] A typical arrangement of the gas coulometer is shown in Fig. 25-13. In using a gas coulometer, one must adjust liquid levels before, during, and after the electrolysis, but the coulometer thereby furnishes a semicontinuous indication of the progress of the reaction. After correction of the gas volume to standard conditions of pressure and temperature, 16,810 ml

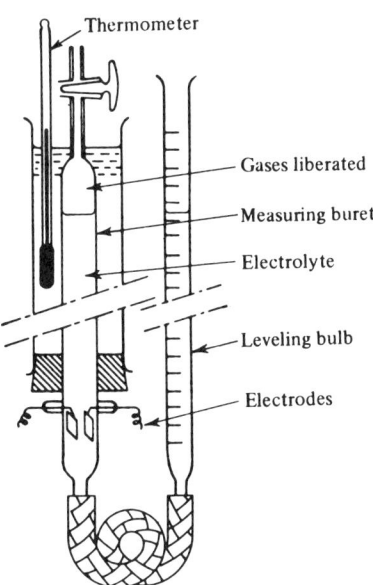

FIGURE 25-13 Gas coulometer. (Reprinted with permission from J. J. Lingane, *J. Am. Chem. Soc.*, 67, 1916 (1945). Copyright 1945 American Chemical Society.)

of gas corresponds to one faraday (96,485Q) theoretically; the actual volume per coulomb is 0.1739 ml at standard conditions.

Electromechanical Coulometers A simple electromechanical integrator is constructed by connecting the ends of a series resistor (in the current circuit) to one of the coil windings of a low-inertia integrating motor—essentially a dc motor built for 0.25–1.5 V operation in which friction and heat losses have been reduced to a minimum. The speed of shaft rotation is a linear function of the applied voltage—derived from the current flowing through the series resistor. Rotation of the armature shaft is followed by a mechanical counter.[18]

Electronic Coulometers Electronic integration, although requiring complicated equipment, enables one to integrate even small charge transfers and is extremely accurate. The main limitation (0.01%) with electronic equipment arises from background signal. Voltage-to-frequency converters measure the voltage drop over a standard resistor and feed the output to a scaler, from which the current–time integral is obtained as a number of counts. For example, an input signal of 1 V may be converted to an output signal of 10,000 counts/sec. The operational amplifier–capacitor integrator is similar to those used in analog computers. Response time to current changes is as fast as 10 μsec. Usable range extends from 10 μA to 10 mA or greater. A typical circuit is included in Fig. 25-14.

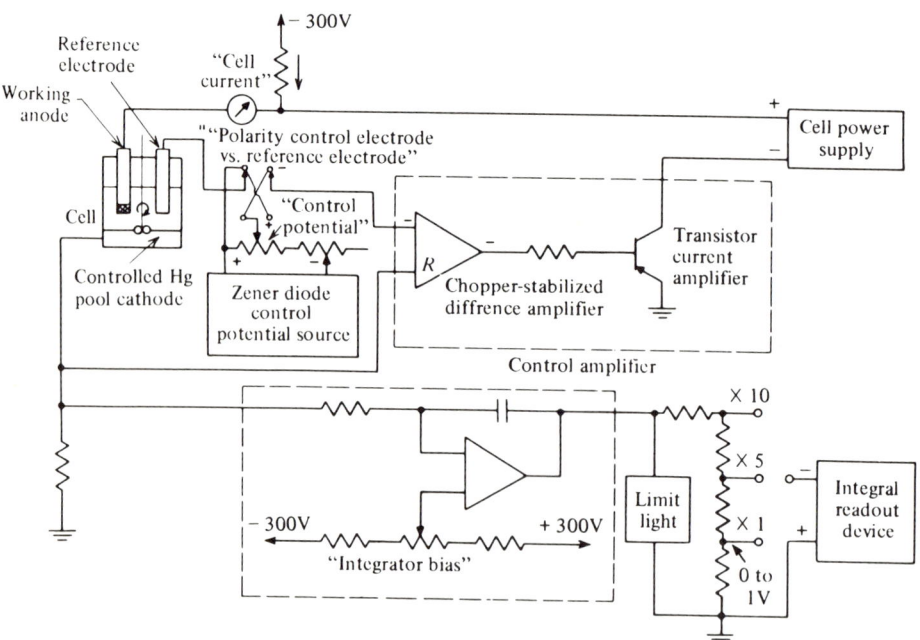

FIGURE 25-14 Electronic controlled-potential coulometric titrator. Block diagram, switched for reduction. (M. T. Kelley, H. C. Jones, and D. J. Fisher, *Anal. Chem.*, **31**, 488 (1959). Courtesy of American Chemical Society.)

In controlled-potential coulometry, four instrumental units are involved: a coulometer, a dc current supply, a potentiostat, and an electrolytic cell. The test material is reduced (or oxidized) directly at the working electrode, and the charge transfer during this process is integrated by a coulometer. In order that only the desired reaction may take place, the potential of the working electrode is controlled within 1–5 mV of the limiting electrode potential with the aid of a potentiostat. As the desired constituent reacts at the working electrode, the current decreases from a relatively large value at the beginning to essentially zero at the completion of the reaction (Fig. 25-5). Potentiostats have been mentioned previously (see Fig. 25-3). An electronic potentiostat is included in the complete instrument described below.

In the apparatus of Kelley, Jones, and Fisher the potential of the working electrode is controlled by a stabilized difference amplifier combined with a transistor current amplifier.[19] The electrolysis current is integrated by a stabilized amplifier and the integral is read out as a voltage. The block diagram, switched for reduction, is shown in Fig. 25-14. The command signal to the control amplifier is the algebraic sum of the control potential from the control-potential source and the potential of the controlled electrode with respect to the solution, as seen through the reference electrode. The control potential is a selected fraction of the constant potential across a silicon voltage (Zener) diode. The current integrator is an analog computer circuit.

Titrations are made in an inert atmosphere provided by a nitrogen blanket. For controlled anode-potential oxidations, such as those of iodide or iron(II), platinum electrodes are used, and the cathode is isolated by a salt bridge and frit barrier. For controlled cathode-potential reductions, such as those of uranium(VI) or copper(II), the anode is a platinum wire that is isolated by a sulfuric acid salt bridge and frit barrier, and the cathode is a mercury pool. Thorough agitation of the mercury–solution interface is necessary to obtain a high initial current by providing adequate mixing of the solution to replenish ions depleted by electrolysis. A rotated platinum cell with fast sparging characteristics, low sample volumes (2 ml), and high electrolysis rates has also been developed.[20] A standard reference electrode, positioned as close to the working electrode as possible, monitors the potential of the working electrode. The titration is terminated when the current drops to a predetermined fraction of the initial current or to the residual current from the supporting electrolyte.

Applications of Controlled-Potential Coulometry

To apply controlled-potential coulometry, current–potential diagrams must be available for the oxidation–reduction system to be determined and also for any other system capable of reaction at the working electrode. As discussed earlier, in any sample system where two or more ions are capable of oxidation (or reduction) at a working electrode, that which requires the least free energy for transformation will determine the electrode process. For this to be compatible with the requirement of 100% current efficiency in generation, it is necessary to control the potential of the working (generating) electrode within specific limits.

These limits can be understood better from an example.[21] Consider a mixture of antimony(V) and antimony(III) in a supporting electrolyte containing $6M$ HCl plus $0.4M$

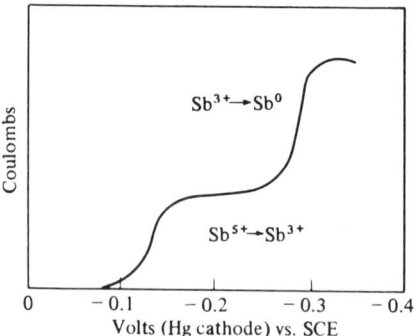

FIGURE 25-15 Electrolytic reduction of antimony(V) by two-step process in 6M HCl plus 0.4M tartaric acid. (L. B. Dunlap and W. D. Shults, *Anal. Chem.*, 34, 499 (1962). Courtesy of American Chemical Society.)

tartaric acid (Fig. 25-15). Plots of Q versus cathode potential show plateaus centered at −0.21 and −0.35 V versus SCE. First, one prereduces the supporting electrolyte at −0.35 V. Then the sample is introduced and the system deaerated. Finally the reduction is started, at −0.21 V for $Sb^{5+} \rightarrow Sb^{3+}$, followed at −0.35 V for $Sb^{3+} \rightarrow Sb°$. Initially, electrolysis proceeds at a constant rate (initial current is i_0) until the potential of the working electrode reaches the limited value, in this case −0.21 V versus SCE. At this point the potentiostat takes over, and the current through the cell gradually decreases until all antimony(V) has been reduced to antimony(III). This pattern is repeated at −0.35 V.

Current–potential diagrams (also denoted "coulograms") are obtained by plotting current against cathode–reference electrode potential (rather than cathode–anode potential which would include the large and variable iR drop in the cell). The necessary data can be obtained by setting the potentiostat to one cathode–reference electrode potential after another in sequence, allowing only enough time at each voltage setting for the current indicator to balance. Alternatively, the reduction (or oxidation) is performed in the usual manner except that periodically throughout the electrolysis the potential is adjusted to a value that causes cessation of current flow. The net charge transferred up to this point and the electrode potential are noted and the electrolysis then continued. Curves plotted from a series of points establish optimum electrode potentials because they relate extent of reaction with electrode potentials under actual titration conditions and electrode material.

By controlling the potential of the electrode at a suitable value, it is possible to reduce a metal completely to a lower valency state, and then, by controlling at a more positive potential, the metal can be oxidized quantitatively to a higher valency state on allowing the current to reach its background value. For example, at −0.15 V with a mercury electrode, reduction of uranium(IV to III) and chromium(III to II) occur simultaneously. If a pre-electrolysis is carried out at −0.55 V, only uranium(III) is oxidized. Although the reaction does not occur with 100% current efficiency, it is complete. When all uranium(III) has been removed from the solution, chromium is determined by oxidation to chromium(III) with a 100% yield at −0.15 V.

Indirect methods are possible. In the determination of plutonium in the presence of iron, the first step is the reduction of plutonium(VI) to plutonium(III) and partial reduction of iron(III) to iron(II) at a platinum electrode in a sulfuric acid electrolyte. When this is followed by oxidation of the mixture to plutonium(IV) and iron(III), the net reaction is the reduction of plutonium(VI) to plutonium(IV). Interference caused by the presence of uranium is thereby avoided.[22] Also, the reaction which occurs at a mercury electrode,

$$Hg° + Y^{4-} = HgY^{2-} + 2e^- \qquad (25\text{-}27)$$

may be used to follow a number of electrochemical reactions wherein a metal, M^{n+}, is not electroactive

$$M^{n+} + HgY^{2-} + 2e^- = Hg° + MY^{(n-4)+} \qquad (25\text{-}28)$$

An excess of HgY^{2-} is added to the solution of M^{n+} (Y^{4-} is the symbol for the anion of EDTA). The current is limited by the diffusion of M^{n+} to the electrode, and becomes zero at the end point.

Controlled-potential methods are finding ever-increasing numbers of applications, especially since the advent of the modern, electronic coulometric titrators. A recent example[23] is the determination of nitrite, which is difficult by ordinary chemical means. The nitrite is oxidized directly to nitrate at a platinum electrode in pH 4.7 acetate buffer solution. Errors of about 0.05% were recorded for the determination of 1 mg of NO_2^-. Methods for a few organic substances, particularly halogen and nitro compounds, have also been developed. Controlled-potential methods can be used in preparative work to oxidize or reduce compounds at controlled potentials. They can also be used to determine the probable result of redox reactions because, if the weight of material and the coulombs required in reaction are known, then n, the number of electrons required per mole, can be computed.

Controlled-potential coulometry suffers from the disadvantages of requiring relatively long electrolysis times and expensive equipment, although it proceeds virtually unattended with automatic coulometers. However, direct indication of optimum conditions for successive reactions are easily obtained. No indicator electrode system is necessary, since the magnitude of the final current is sufficient indication of the degree of completion of the reaction. Although the concentration limits vary for each individual case, the upper limit is about 2 milliequivalents and the lower limit is about 0.05 microequivalent. The latter limit is largely set by the magnitude of the residual current and the many factors that affect it.

Constant-Current Coulometry

In constant-current coulometric methods some external means must be used to determine the end point of the reaction because, at constant current, the potential will rise at the completion of each reaction to a value which will permit some other reaction to take place and thus maintain the current.

Detection of the End Point Various methods are used for detection of the end point. It can be found by means of normal colored indicators, provided the indicator itself is not electroactive, or by instrumental methods—potentiometry, amperometry, and photometry. No correction for volume changes is necessary when plotting the results if internal generation of titrant is employed. Potentiometric and photometric indication find use in acid-base and redox titrations, while amperometric procedures are applicable to redox and precipitation reactions and, in particular, for these systems as the solutions become more dilute.

Electrolytic Stripping The removal (stripping) of deposits has been used to measure the thickness of plated metals and of corrosion or tarnish films. In the case of oxide tarnish on the surface of metallic copper, the specimen is made the cathode and the copper oxide is reduced slowly with a small, but constant, known current to metallic copper. When the oxide film has been quantitatively reduced, the potential of the cathode changes rapidly to the discharge potential of hydrogen. The equivalence point is taken as the point of inflection of the voltage–time curve, as illustrated in Fig. 25-16. From the known current i, expressed in milliamperes, and the elapsed time t, in seconds, the film thickness T, in angstrom units, can be calculated from the known film area A in centimeters squared, and the film density ρ according to the equation

$$T = \frac{10^5 Mit}{AnF\rho} \qquad (25\text{-}29)$$

where M is the gram-molecular weight of the oxide comprising the film. From mixed films of oxide and sulfide on a metal, two inflection points are obtained. Similar methods have been described for the determination of the relative amounts of tin(IV) and tin(II) oxides on a tinplate surface.

Analogous anodic dissolution is used to determine the successive coatings on a metal surface. Iron is sometimes clad with a tin undercoating for adhesion and a copper–tin surface layer for protection from corrosion. The two coatings will exhibit individual potential breaks. In a similar manner, the thickness of chromium plate on iron, copper, or nickel, and of zinc or nickel plate on either copper or iron can be determined. The method may

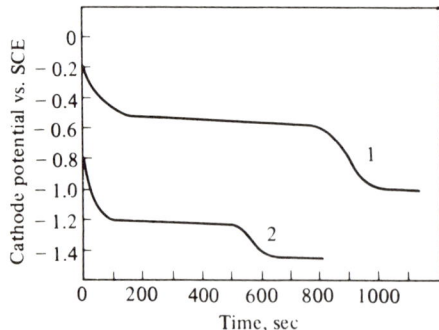

FIGURE 25-16 Cathodic reduction of tarnish films on copper: curve 1, copper(I) oxide; curve 2, copper(I) sulfide.

also be used whenever the substance to be determined can be deposited beforehand so as to adhere to a solid electrode or to form an amalgam with a mercury electrode. Sensitivity is high. Accuracy is limited by the residual current and by the fact that the last traces of deposit do not dissolve uniformly from the surface.

Primary Coulometric Titrations In primary coulometric titrations at constant current the substance to be determined reacts directly at the electrode. Consquently, no other substance should be able to be electrolyzed at the working electrode until much higher potentials are attained, usually at least 0.5 V from the desired value. Since the potential of the working electrode is not controlled, this class of titrations is limited generally to reactants which are nondiffusible.

One major area of application involves the electrode material itself participating in an anodic process as, for example, the reaction of mercaptans, sulfhydryl groups, and ionic halide ions with silver ions generated at a silver anode. For chloride samples the initial reaction may be

$$Cl^- + Ag^\circ \rightarrow AgCl(s) + e^- \qquad (25\text{-}30)$$

followed by

$$Ag^\circ \rightarrow Ag^+ + e^- \qquad (25\text{-}31)$$

as soon as the limiting current (supply of chloride ions to the anode) has become smaller than the current forced through the electrolytic cell. At this point the silver ion generated anodically diffuses into the solution, and precipitation occurs with the chloride ions left in solution. Of course, the result of the two reactions is identical. The end point of the titration is ascertained amperometrically. Commercial titrators for biological and industrial samples based on this method are available (Aminco-Cotlove, Buchler Instruments). Combustion by the oxygen flask method precedes the titration step for nonionic halides in organic compounds. Mercaptan samples are dissolved in a mixture of aqueous methanol and benzene to which aqueous ammonia and ammonium nitrate are added to buffer the solution and to supply sufficient electrolyte to lower the solution resistance.

Secondary Coulometric Titrations In secondary coulometric titrations an active intermediate is first generated quantitatively by the electrode process, and this then reacts directly with the substance to be determined. The standard potential of the auxiliary system has to lie between the potential of the system to which the substance to be determined belongs and the potential at which the supporting electrolyte or a second electroactive system undergoes an electrode reaction.

A knowledge of current–potential curves aids the analyst in choosing the auxiliary system. Current–potential curves for systems pertinent to the coulometric determination of

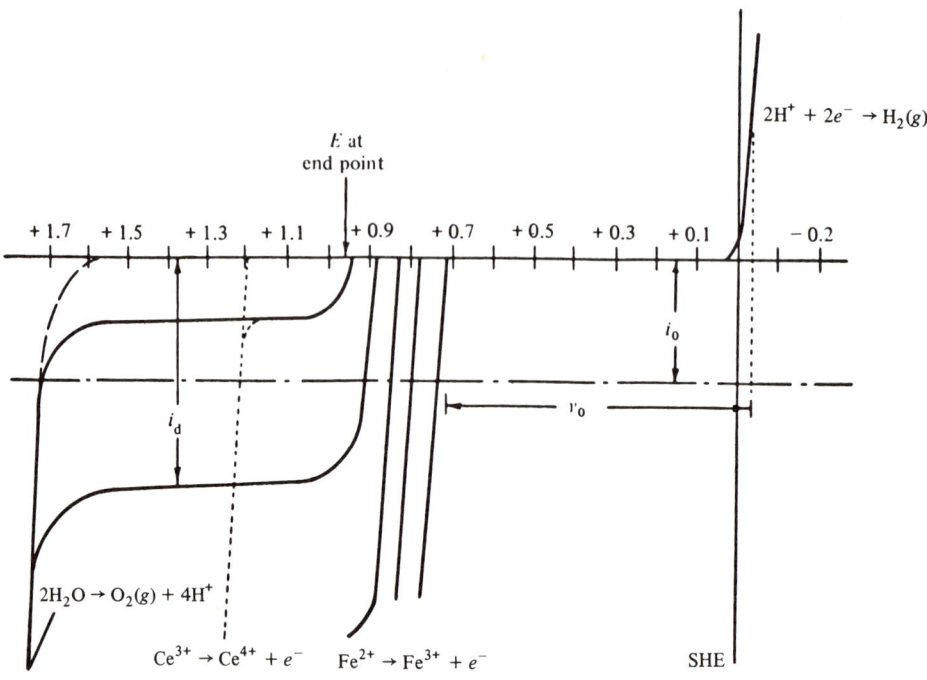

FIGURE 25-17 Current–potential curves pertinent to the coulometric titration of iron(II) with cerium(III) as the auxiliary system.

iron(II) at constant current are illustrated in Fig. 25-17. To complete the titration within a reasonable period of time, usually 10–200 sec, a finite current must be selected, say i_0. The necessary applied emf will result in a voltage drop across the cell given by V_0 for the initial concentration of iron(II) present in the solution. At the beginning, iron(II) is oxidized directly at the anode

$$Fe^{2+} \rightarrow Fe^{3+} + e^- \tag{25-32}$$

As the concentration of iron(II) decreases with the progress of the oxidation, the current will tend to decrease. However, since the current is being maintained constant, the voltage must be increased continually until ultimately the decomposition potential of water is exceeded. If i_0 is selected sufficiently small to delay the other anodic oxidation, the time required for a determination will become too long for practical consideration. One alternative, of course, is to control the anode potential at a value below the decomposition potential of the undesired reactant, as was discussed in an earlier section.

Interposition of an auxiliary system between the potentials at which iron(II) and water are oxidized is the basis of secondary coulometric titrations. No interfering electrode reaction can occur if the potential of the working electrode, in this case the anode, is prevented from reaching the value that would occasion initiating the decomposition of water. Limitation of such potential drift is achieved by having a precursor of the secondary titrating agent present in relatively high concentration. In our example a large excess of cerium(III) is added to the solution. Now as soon as the limiting current of iron(II) falls below the value of the current forced through the cell, that is, $i_0 > i_{\text{limiting}}$, the cerium(III) commences to undergo oxidation at the anode in increasing amounts until it may be the preponderant anode reactant. Since the cerium(IV) formed reacts instantly and stoichiometrically with the iron(II),

$$Ce^{4+} + Fe^{2+} \rightarrow Ce^{3+} + Fe^{3+} \qquad (25\text{-}33)$$

the total current ultimately employed in attaining the oxidation of iron(II) is the same as would have been required for the direct oxidation. Because there is a relatively inexhaustible supply of cerium(III), the anode potential is stabilized at a value less than the decomposition potential of water. The end point is signaled by the first persistence of excess cerium(IV) in the solution and may be detected the ordinary way with a platinum–reference electrode pair, or photometrically at a wavelength at which cerium(IV) absorbs strongly.

Errors due to impurities in the supporting electrolyte or in the auxiliary substance can be avoided by performing a pretitration, then performing the sample titration, or a succession of titrations, in the same supporting electrolyte.

Exploitation of titrants that for one reason or another are difficult to use in conventional titrimetry are among the virtues of secondary coulometric methods. Electrolytic generation of hydroxyl ion has some advantages over conventional methods. Very small amounts of titrant can be prepared, and in a carbonate-free condition. To analyze dilute acid solutions, such as would result from adsorption of acidic gases, the cathode reaction

$$2H_2O + 2e^- \rightarrow H_2(g) + 2OH^- \qquad (25\text{-}34)$$

generates the hydroxyl ion. Of course, in the initial stages it is also possible for the hydrogen ion to react directly at the cathode,

$$2H_3O^+ + 2e^- \rightarrow 2H_2O + H_2(g) \qquad (25\text{-}35)$$

but in the vicinity of the end point, the secondary generation predominates. The anode reaction must also be considered. If a platinum anode is used, it must be isolated in a separate compartment, for hydrogen ions would be liberated at its surface. Alternatively, a silver anode may be used within the electrolytic cell in the presence of bromide ions, for then the anode reaction is

$$Ag^\circ + Br^- \rightarrow AgBr(s) + e^- \qquad (25\text{-}36)$$

and the silver and bromide ions are fixed as a coating of silver bromide on the electrode surface.

Halogens generated internally, and particularly bromine, have found widespread application, especially in organic analysis. In contrast with certain difficulties encountered in the use of bromate–bromide mixtures by conventional volumetric procedures, coulometry is much simpler.[24] Bromates are not soluble in many organic solvents, and many organic samples are not soluble in water. However, sodium and lithium bromides are quite soluble in various organic solvents in which brominations can be conducted.

The complexing ability of EDTA has been exploited in the coulometric titration of metal ions. The method depends on the reduction of the mercury(II) or cadmium chelate of EDTA and the titration, by the anion of EDTA that is released, of the metal ion to be determined. If the direct reaction of metal with EDTA is too slow, excess of the EDTA anion is generated and then the excess is back-titrated by cadmium generated at a cadmium–amalgam electrode.

Dual intermediates can be used whenever the substance to be titrated does not react rapidly with the auxiliary system or, at least, when the reaction rate is not as rapid as the generation rate. An excess of titrant is generated and permitted to react for the necessary time. Then the polarity of the working electrode is reversed and a back-titration is conducted with a second auxiliary system. In this manner an excess of bromine can be titrated with electrically generated trichlorocuprate(I) ion,

$$Cu^{2+} + 3Cl^- + e^- \rightarrow CuCl_3^{2-} \tag{25-37}$$

$$Br_2 \text{ (in excess)} + 2CuCl_3^{2-} \rightarrow 2Cu^{2+} + 2Br^- + 6Cl^- \tag{25-38}$$

External Generation Internal generation methods possess limitations. Often conditions conducive to optimum generation of reactant and to rapid reaction with the substance to be titrated are not compatible. Or the sample may contain two or more substances that are capable of undergoing reactions at the electrode and that are not different sufficiently in electrode potential to permit use of an auxiliary system. For example, the titration of acids by electrically generated hydroxyl ion is precluded in the presence of certain other reducible substances. These limitations are circumvented when the reagent is generated in an electrolytic cell that is isolated from the solution to be titrated, and the desired electrolytic product is allowed to flow via a capillary tube into the test solution.

Cross-sectional views of a double-arm[25] and a single-arm generator cell[26] are illustrated in Figs. 25-18 and 25-19. The supporting electrolyte is fed continuously from a reservoir into the top of the generator cell. The incoming solution is then divided at the T-joint, in the two-arm design, so that about equal quantities flow through each of the arms of the

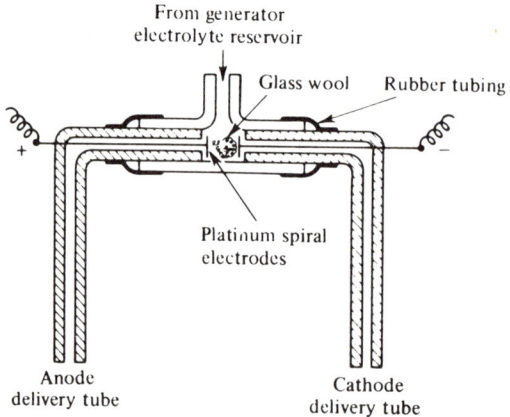

FIGURE 25-18 Double-arm electrolytic cell for external generation of titrant. (D. D. DeFord, J. N. Pitts, and C. J. Johns, *Anal. Chem.*, 23, 938 (1951). Courtesy of American Chemical Society.)

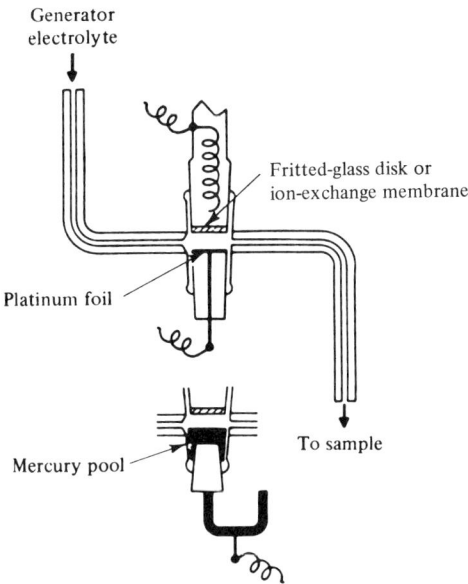

FIGURE 25-19 Single-arm generator cell with working electrode either of platinum or a mercury pool. (J. N. Pitts *et al., Anal. Chem.,* 26, 628 (1954). Courtesy of American Chemical Society.)

cell. Platinum electrodes are sealed on either side of the T-joint. The products of electrolysis are swept along by the flow of solution through the arms and emerge from the delivery tips on either side. A beaker containing the sample to be titrated is placed beneath the appropriate delivery tip. Thus, determinations performed with external generation of titrant hardly differ from normal volumetric methods in essentials; the only difference is that the titer is referred to unit of time and not unit of volume. Naturally, one-half of the liquid continuously discharged is conducted to waste. When a solution of Na_2SO_4 is supplied, H_2SO_4 is formed at the anode and NaOH at the cathode. For bromination, a solution of KBr is used.

The single-arm generator cell is useful for the generation of reagents in those cases in which mixing of the cathode and anode electrolysis products can be tolerated. The working electrode can be made of platinum or it can be a mercury pool. The other electrode compartment is isolated by a frit barrier. The flow of supporting electrolyte is usually 6 ml/min, or larger.

The titration of azo dyes with titanium(III) illustrates the advantage of external generation. At room temperature the rate of reaction of titanium(III) with the dye is slow. Yet on raising the temperature, hydrolysis of titanium(IV) and bubble formation at the electrode surface lead to low current efficiencies. However, if the titanium(III) is generated at room temperature and then delivered to the hot dye solution, optimum conditions prevail for each step. A mercury-pool cathode or an amalgamated working electrode is used to take advantage of the favorable hydrogen overpotential on mercury.

Precision Constant-Current Coulometry

Constant-current methods with careful attention to eliminating or decreasing all sources of error can be extremely precise. Diehl and co-workers[27] have recently determined the value of the faraday by neutralization of carefully purified 4-aminopyridine by hydrogen ions generated and coulometrically measured by the reaction

$$N_2H_5^+ \rightarrow N_2(g) + 5H^+ + 4e^- \tag{25-39}$$

They obtained for the faraday a value of 96,486.57 coulombs in close agreement with the value of 96,486.72[28] reported by the National Bureau of Standards from the anodic dissolution of silver. It has been proposed that, when applicable, the conformance of a given material to its theoretical electrochemical equivalent should define the absolute purity. The faraday,* then, would in fact be the primary standard.

PROBLEMS

1. At what value should the cathode potential be controlled if one desires to separate silver from a 0.005M solution of Cu^{2+} ions? If the initial silver concentration is 0.05M, how long should the deposition take, assuming $\delta = 2 \times 10^{-3}$ cm; $D = 7 \times 10^{-5}$ cm^2 sec^{-1}; V = 200 ml; and A = 150 cm^2?

2. Copper and nickel can be separated by a constant-current procedure provided the pH is carefully controlled. Calculate the minimum pH necessary to initiate the deposition of nickel from a 0.005M solution. Assume that the current is 1A, area = 150 cm^2, and remember that the electrode will be covered with the copper deposit.

3. Under the conditions of Problem 2, how completely will the copper have been removed up to the point when hydrogen gas is initially liberated?

4. What is the initial cathode potential when cadmium deposits form a solution 0.01M in cadmium?

5. What weights of each of the following would be deposited by 3378 coulombs? (a) Cu° from Cu^{2+}, (b) PbO_2 from Pb^{2+}, (c) Cl^- as AgCl at a silver anode, (d) Sn° from $SnCl_4$.

6. A solution is initially 0.01M in silver ion and 0.5M in copper(II) ions. (a) What cathode potential is needed theoretically for the complete deposition of silver? (b) What cathode potential may be required considering concentration polarization? (c) How much silver remains in the solution when the cathode potential has been brought to 0.45 V vs. SHE?

7. Under a given set of electrolysis conditions, 0.500 g of silver was deposited at the cathode and oxygen liberated simultaneously at the anode. Calculate the number of millimoles of hydrogen ion added to the solution. If the solution volume were 200 ml, what would be the change in pH, assuming initially that the solution was neutral and unbuffered?

---

*96,484.56 (27) C mol^{-1} is the recommended value for the Faraday constant (uncertainty ±0.27).

8. In an electrolytic determination of bromide ion from 100 ml of solution, the silver anode, after electrolysis was completed, was found to have gained 0.8735 g. (a) Calculate the molarity of bromide in the original solution. (b) Calculate the potential of the silver electrode at the beginning of the electrolysis, assuming the solubility product of AgBr is 4×10^{-13}.

9. A solution which is $0.01M$ in zinc sulfate, and buffered at pH 4 with an acetate buffer, is to be electrolyzed using a copper-clad cathode. If the overpotential of hydrogen on copper is 0.75 V at the current density to be used, and that of oxygen on the platinum anode is 0.50 V: (a) Calculate the decomposition potential of the solution, assuming that the iR drop is 0.5 V. (b) Will the decomposition potential change as the electrolysis proceeds? (c) How much zinc will remain in solution at the point when hydrogen gas begins to be liberated?

10. At a current density of 0.01 A cm^{-2} the overpotential of hydrogen gas on cadmium is 0.4 V. Would it be possible to deposit cadmium quantitatively in a solution buffered at pH = 2?

11. The cathode potential is controlled 0.05 V less negative than the value at which tin would be deposited from a $0.005M$ solution. (a) Calculate the molarity of copper ions remaining in a sulfate solution. (b) Estimate the quantity of undeposited copper in a solution $1.0M$ in HCl and containing hydrazine hydrochloride.

12. If lead is used as the soluble anode in the internal electrolysis of a lead solution containing a small amount of copper, what would be the final concentration of copper in solution if the lead concentration is $0.2M$?

13. By means of suitable calculations, show why zinc can be successfully plated onto a copper-clad electrode from a solution buffered at pH = 10.5, whereas the deposition would not occur without concurrent evolution of hydrogen gas if smooth platinum electrodes were substituted. Assume a current density equivalent to the appearance of first visible gas bubbles.

14. Suggest an electrographic method for detecting the presence of copper filings on an ax blade suspected of being used to cut telephone cables. The method should be adaptable to courtroom demonstration before a jury of laymen.

15. Suggest a field method for the detection of fool's gold (FeS_2).

16. For a typical laboratory deposition of 0.200 g copper onto a platinum electrode (area 160 cm^2) from 200 ml of $0.5M$ tartrate solution adjusted to pH 4.5 and containing hydrazine, calculate the time required to reduce the copper concentration (a) to 1% of its original value and (b) to 0.1%. Starting at an initial value of 2.6 A, the current decreased to 1.3 A after 2 min, to 0.65 A after 4 min, and to 0.33 A after 6 min.

17. In Problem 16, assuming that the diffusion coefficient of the tartrate complex is approximately 2×10^{-5} cm^2 sec^{-1}, estimate the thickness of the diffusion layer about the electrode.

18. A solution is $0.1M$ in cadmium ion and $0.01M$ in hydrogen ion. (a) What is the difference between the two cathode potentials? (b) To this solution is added sufficient ammonia to convert the cadmium to $Cd(NH_3)_4^{2+}$ and to make the solution $0.57M$ in free NH_3 and $0.10M$ in NH_4^+. Assuming that the volume of the solution is unchanged, what is the difference between the two cathode potentials?

19. In an ammoniacal solution, $0.10M$ in both NH_3 and NH_4^+, would the deposition of copper be complete if the cathode potential were limited at -0.40 V vs. SCE?

20. Outline a procedure for the successive determination of lead, cadmium, and zinc in metallurgical materials, such as flue dust from zinc refineries, composed chiefly of the oxides of these three metals.

21. During the determination of silver in silver solder, the $Cu(NH_3)_4^{2+}$ complex undergoes reduction to the $Cu(NH_3)_2^+$ complex. What difficulty is thereby introduced into the electrodeposition of silver from the silver ammine complex?

22. For the determination of cyanide ion, Baker and Morrison [*Anal. Chem.*, **27**, 1306 (1955)] employed a silver and platinum electrode pair, connected together through a microammeter, immersed in $1M$ NaOH solution to which the cyanide sample is added. (a) Write the cell representing the reaction and (b) the electrode reactions involved. (c) Why does a reaction, nevertheless, proceed spontaneously?

23. During the deposition of copper from a chloride medium and in the presence of hydrazine hydrochloride as anodic depolarizer, the following current readings were obtained for corresponding times:

3.00 A = 1.00 min	1.2 A = 7.0 min	0.075 A = 13.0 min
3.00 A = 2.00 min	0.8 A = 8.0 min	0.052 A = 14.0 min
2.85 A = 3.00 min	0.50 A = 9.0 min	0.036 A = 15.0 min
2.70 A = 4.00 min	0.30 A = 10.0 min	0.027 A = 16.0 min
2.2 A = 5.0 min	0.18 A = 11.0 min	0.020 A = 17.0 min
1.8 A = 6.0 min	0.12 A = 12.0 min	0.016 A = 18.0 min
		0.016 A = 20.0 min

(a) Graph the results on semilog paper. (b) From the descending slope of the graph, determine the value of the constant k (see Eq. 25-21). (c) Estimate the time required to reduce the copper concentration to 0.1% of its original value [after all the residual copper(II) ions are reduced to the chlorocuprate(I) ion].

24. The initial current is 90.0 mA and decreases exponentially with $k = 0.0058$ sec^{-1}; the titration time is 714 sec. How many milligrams of uranium(VI) are reduced to uranium(IV)?

25. When an integrating motor was calibrated, these results were obtained:

CURRENT, mA	SHUNT RESISTANCE, Ω	TIME, SEC	COUNTS, N
10.02	2220	600	9102
20.03	1110	600	9180
30.00	770	600	9458
50.00	475	600	9773

Calculate the microequivalents per count.

26. The calibration factor of an integrating motor is 0.00267 microequivalents per count. Calculate the normality of an acid solution, 10.0 ml of which produced 40.72 counts during a titration.

27. These results were obtained during the titration of three successive 1.00-ml aliquots of As_2O_3 solution with electrically generated iodine at pH 8 and using amperometric indication of the end point. Graph the results and determine the normality of the As_2O_3 solution. The microequivalents of iodine generated are followed by the amperometric signal in microamperes: Pretitration–0.00 microequivalents = 0.4 μA; 5.10 = 0.7; 9.90 = 1.3; and 15.0 = 1.7. First aliquot–15.0 = 0.4; 50.0 = 0.4; 100.0 = 0.4; 149.5 = 1.3; 154.5 = 2.0; 160.0 = 2.6; 164.9 = 3.0. Second aliquot–164.9 = 0.4; 200.0 = 0.4; 250.0 = 0.4; 273.8 = 1.0; 277.4 = 2.2; 280.8 = 2.7; 286.1 = 3.2. Third aliquot–286.1 = 0.4; 350.0 = 0.4; 400.0 = 0.4; 402.0 = 0.8; 406.3 = 2.6; 411.0 = 3.1; 416.1 = 3.8.

28. In coulometric titrations, a milliampere-second corresponds to how many grams of (a) hydroxyl ions, (b) antimony(III to V), (c) chloride ions, (d) copper(II to 0), (e) arsenious oxide(III to V)?

29. A 0.5M K_2SO_4 solution in a gas coulometer gave 22.33 ml of hydrogen plus oxygen at the end of an electrolysis; temperature was 24.0°C in the water jacket and the pressure was 740 mm. The partial pressure of water over the electrolyte is 22 mm at 24°C. How many coulombs were involved in the electrolysis?

30. Calculate the concentration of acid in a 10.0-ml aliquot that required a generation time of 165 sec for the appearance of the pink color of phenolphthalein. The voltage drop across a 100-Ω resistor was 0.849 V.

31. Sketch the current–potential curves that would pertain to each of these coulometric systems: (a) The titration of acids with electrically generated hydroxyl ion in a potassium bromide electrolyte and using a silver anode. (b) The generation of excess bromine in a potassium bromide electrolyte, followed by the generation of copper(I) to react with the unused bromine. (c) The titration of zinc with generated ferrocyanide ions.

32. In the coulometric determination of permanganate ion by generating iron(II) from iron(III), the permanganate was all reduced to manganese(II) by a constant current of 2.50 mA acting for 10.37 min. Calculate the molarity of the permanganate if the initial volume was 25.00 ml.

33. In an electrolytic determination of bromide from 100.0 ml of solution, the quantity of electricity, as read on a mechanical current–time integrator, was 105.2Q. Calculate the weight of bromide ions in the original solution. Calculate the potential of the silver electrode that should be employed throughout the electrolysis. K_{sp} of AgBr = 4 × 10^{-13}.

34. The following measurements were made in a coulometric titration of arsenic(III) ions with generated bromine.

> Generation time: 132.6 sec
> Calibrated resistance: 100 Ω
> iR drop across resistance: 0.620 V

Calculate the amount of arsenic present in the sample.

35. Using generating currents of 1–10 mA, and corresponding titration times of about 300–100 sec, what range in weights of mercaptans may be present in a solution volume of 50 ml? The reaction is $Ag° + RSH \rightarrow AgSR + H^+ + e^-$.

36. Problem 25-36 shows (courtesy of B. W. Conroy and O. Memis, NUMEC, Apollo, Pa.) the coulograms of iron(III), vanadium(V), and manganese(VII), for their reduction in 1M phosphate medium at pH 2 with a platinum cathode. Outline a procedure for the determination of each element in a mixture of the others by controlling the cathode potential.

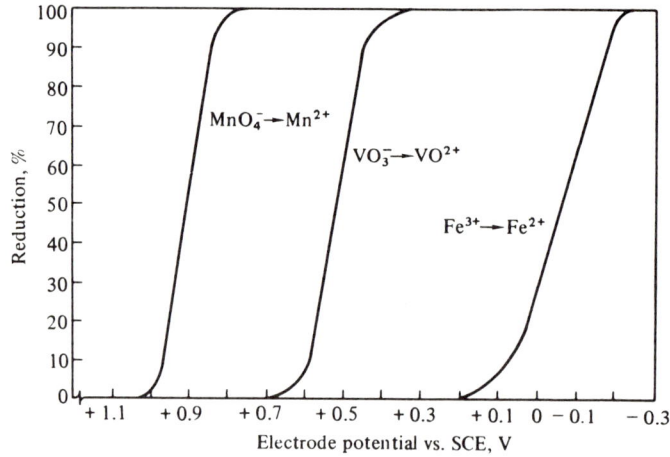

PROBLEM 25-36

37. Assuming that the coulograms for iron and vanadium, shown in Problem 25-36, are reversible, outline a constant-current procedure for the determination of iron(II) in the presence of vanadium(IV).

38. From the information contained in Problem 25-36, outline a procedure for the determination of the amounts of vanadium(V) and vanadium(IV) in a mixture containing the two oxidation states.

39. How long should a constant current of 100.0 mA be passed through a solution to prepare 100 ml of a solution of 0.0100M Ni^{2+} using an anode of pure nickel?

40. Determine the equivalent weight of an organic acid if 0.0400 g in alcohol–water mixture required a constant current of 50 mA for 500 sec to generate sufficient hydroxyl ion to reach a phenolphthalein end point.

BIBLIOGRAPHY

Abresch, K. and I. Classen, *Die coulometrische Analyse*, Verlag Chemie, Weinheim, 1961.
Adams, R. N., *Electrochemistry at Solid Electrodes*, Marcel Dekker, New York, 1969.
Bard, A. J., Ed., *Electroanalytical Chemistry, A Series of Advances,* Marcel Dekker, New York, Vol. 1, 1966; Vol. 2, 1967; Vol. 3, 1969; Vol. 4, 1970; Vol. 5, 1971; Vol. 6, 1973; Vol. 7, 1974; Vol. 8, 1975; Vol. 9, 1976; Vol. 10, 1977; Vol. 11, 1979.
Charlot, G., J. Badoz-Lambling, and B. Tremillon, *Electrochemical Reactions*, American Elsevier, New York, 1962.
DeFord, D. D. and J. W. Miller, in *Treatise on Analytical Chemistry*, I. M. Kolthoff and P. J. Elving, Eds., Vol. 4, Part I, Chap. 49, Wiley-Interscience, New York, 1963.
Eberson, L. E. and N. L. Weinberg, "Electroorganic Synthesis," *Chem. Eng. News*, p. 40 (Jan. 25, 1971).
Kies, H. L., *J. Electronanal. Chem.*, **4**, 257 (1962).
Lewis, D. T., *Analyst*, **86**, 494 (1961).
Lingane, J. J., *Electroanalytical Chemistry*, 2nd ed., Wiley-Interscience, New York, 1958.
Milner, G. W. C. and G. Phillips, *Coulometry in Analytical Chemistry*, Pergamon, London, 1968.
Rechnitz, G. A., *Controlled Potential Analysis*, Pergamon, London, 1963.
Shain, I., in *Treatise on Analytical Chemistry*, I. M. Kolthoff and P. J. Elving, Eds., Vol. 4, Part I, Chap. 50, Wiley-Interscience, New York, 1963.
Shanefield, D., "Electrolysis as a Purification Tool," *Ann. N.Y. Acad. Sci.*, **137**, 135 (1966).
Tutundzic, P. S., *Anal. Chim. Acta*, **18**, 60 (1958).
Wawzonek, S., "Synthetic Electroorganic Chemistry," *Science*, **155**, 39 (1967).

LITERATURE CITED

1. Rogers, L. B. and A. F. Stehney, *J. Electrochem. Soc.*, **95**, 25 (1949).
2. Lott, P. F., *J. Chem. Educ.*, **42**, A261 (1965).
3. Lingane, J. J., *Anal. Chim. Acta*, **2**, 591 (1948).
4. Melaven, A. D., *Ind. Eng. Chem., Anal. Ed.*, **2**, 180 (1930).
5. Johnson, H. O., J. R. Weaver, and L. Lykken, *Anal. Chem.*, **19**, 481 (1950).
6. Center, E. J., R. C. Overbeck, and D. L. Chase, *Anal. Chem.*, **23**, 1134 (1951).
7. Maxwell, J. A. and R. P. Graham, *Chem. Rev.*, **46**, 471 (1950).
8. Page, J. A., J. A. Maxwell, and R. P. Graham, *Analyst*, **87**, 245 (1962).
9. Sand, H. J. S., *Analyst*, **55**, 309 (1930).
10. Clark, B. L., L. A. Wooten, and C. L. Luke, *Trans. Electrochem. Soc.*, **76**, 63 (1939).
11. Hermance, H. W. and H. V. Wadlow, "Electrography and Electrospot Testing," in *Standard Methods of Chemical Analysis*, 6th ed., F. J. Welcher, Ed., Vol. 3, Part A, pp. 500–520, D. Van Nostrand, New York, 1966.

12. Glazunov, A., *Chim. Ind.*, Special Number, p. 425 (Feb. 1929).
13. Fritz, H., *Z. Anal. Chem.*, **78**, 418 (1929).
14. Ehlers, V. B. and J. W. Sease, *Anal. Chem.*, **26**, 513 (1954).
15. Smith, S. W. and J. K. Taylor, *J. Res. Natl. Bur. Std. (U.S.)*, **63C**, 65 (1959).
16. Lingane, J. J., *J. Am. Chem. Soc.*, **67**, 1916 (1945).
17. Page, J. A. and J. J. Lingane, *Anal. Chim. Acta*, **16**, 175 (1957).
18. Bett, N., W. Nock, and G. Morris, *Analyst*, **79**, 607 (1954).
19. Kelley, M. T., H. C. Jones, and D. J. Fisher, *Anal. Chem.*, **31**, 489 (1959).
20. Clem, R. G., *Anal. Chem.*, **43**, 1853 (1971).
21. Dunlap, L. B. and W. D. Shults, *Anal. Chem.*, **34**, 499 (1962).
22. Shults, W. D., *Anal. Chem.*, **33**, 15 (1961).
23. Harrar, J. E., *Anal. Chem.*, **43**, 143 (1971).
24. Swift, E. H. *et al.*, *Anal. Chem.*, **19**, 197 (1947); **22**, 332 (1950); **24**, 1195 (1952); **25**, 591 (1953); and *J. Am. Chem. Soc.*, **70**, 1047 (1948); **71**, 1457, 2717 (1949).
25. DeFord, D. D., J. N. Pitts, and C. J. Johns, *Anal. Chem.*, **23**, 938 (1951).
26. Pitts, J. N., D. D. DeFord, T. W. Martin, and E. A. Schmall, *Anal. Chem.*, **26**, 628 (1954).
27. Diehl, H., *Anal. Chem.*, **51**, 318A (1979).
28. Taylor, B. N., W. H. Parker, and D. N. Langenberg, *Rev. Mod. Phys.*, **41**, 403 (1969).

CHAPTER 26

Conductance Methods

Conductance measurements were among the first to be used for determining solubility products, dissociation constants, and other properties of electrolyte solutions. Conductance is an additive property of a solution depending on all the ions present. Solution conductance measurements, therefore, are nonspecific. This nonspecificity restricts the quantitative analytical use of this technique to situations where only a single electrolyte is present or where the total ionic species needs to be ascertained. In these special situations, however, conductance measurements are capable of extreme sensitivity.

The possibility of using conductance to locate end points in titrations was also recognized early in the development of instrumental methods. Changes in the slope of conductance versus titrant volume occur because ionic mobilities vary and also because of the formation of insoluble or nonionized materials. Accordingly, conductometric indication of end points is possible. A related technique, high-frequency conductometric titration, was developed in recent years. High-frequency measurements permit the determination of changes in conductance, or dielectric constant, without the introduction of electrodes into direct contact with the solution.

26.1 ELECTROLYTIC CONDUCTIVITY

Electrolytic conductivity is a measure of the ability of a solution to carry an electric current. Solutions of electrolytes conduct an electric current by the migration of ions under the influence of an electric field. Like a metallic conductor, they obey Ohm's law. Exceptions to this law occur only under abnormal conditions; for example, very high voltages or high-frequency currents. Thus, for an applied electromotive force E, maintained constant but at a value that exceeds the decomposition voltage of the electrolyte, the current i flowing between the electrodes immersed in the electrolyte will vary inversely with the resistance of the electrolytic solution R. The reciprocal of the resistance $1/R$ is called the *conductance*, and is expressed in reciprocal ohms, or mhos.

The standard unit of conductance is specific conductance κ, which is defined as the reciprocal of the resistance in ohms of a 1-cm cube of liquid at a specified temperature. The units of specific conductance are the reciprocal ohm cm (or mho/cm). The observed conductance of a solution depends inversely on the distance d between the electrodes and directly upon their area A,

$$\frac{1}{R} = \kappa \frac{A}{d} \tag{26-1}$$

The electrical conductance of a solution is a summation of contributions from all the ions present. It depends upon the number of ions per unit volume of the solution and upon the velocities with which these ions move under the influence of the applied electromotive force. As a solution of an electrolyte is diluted, the specific conductance in Eq. 26-1 will decrease. Fewer ions to carry the electric current are present in each cubic centimeter of solution. However, in order to express the ability of individual ions to conduct, a function called the *equivalent conductance* is employed. It may be derived from Eq. 26-1, where A is equal to the area of two large parallel electrodes set 1 cm apart and holding between them a solution containing one equivalent of solute. If C_s is the concentration of the solution in gram equivalents per liter, then the volume of solution in cubic centimeters per equivalent is equal to $1000/C_s$, so that Eq. 26-1 becomes

$$\Lambda = 1000 \frac{\kappa}{C_s} \tag{26-2}$$

At infinite dilution the ions theoretically are independent of each other and each ion contributes its part to the total conductance, thus

$$\Lambda_\infty = \Sigma(\lambda_+) + \Sigma(\lambda_-) \tag{26-3}$$

where λ_+ and λ_- are the ionic conductances of cations and anions, respectively, at infinite dilution. Values for the limiting ionic conductances for selected ions in water at 25°C are given in Table 26-1. The ionic conductance is a definite constant for each ion in a given solvent, its value depending only on the temperature. Since these are actually equivalent conductances, symbols such as $\frac{1}{2}$ Ba^{2+} are sometimes employed. At finite concentrations interionic forces generally lower the ionic mobilities.

Example 26-1

The equivalent conductance at infinite dilution of H_2SO_4 is

$$\Lambda_\infty = 350 + 80 = 430 \; \Omega^{-1} \; cm^2$$

The molar conductance is given by

$$(2)(350) + (2)(80) = 860 \; \Omega^{-1} \; cm^2$$

TABLE 26-1 Limiting Equivalent Ionic Conductances in Aqueous Solution at 25°C

Cations	λ_+	Anions	λ_-
H^+	350	OH^-	198
Li^+	39	F^-	54
Na^+	50	Cl^-	76
K^+	74	Br^-	78
NH_4^+	73	I^-	77
Ag^+	62	NO_3^-	71
Mg^{2+}	53	IO_4^-	55
Ca^{2+}	60	HCO_3^-	45
Sr^{2+}	59	Formate	55
Ba^{2+}	64	Acetate	41
Zn^{2+}	53	Propionate	36
Hg^{2+}	53	Butyrate	33
Cu^{2+}	55	Benzoate	32
Pb^{2+}	71	Picrate	30
Co^{2+}	53	SCN^-	66
Fe^{2+}	54	SO_4^{2-}	80
Fe^{3+}	68	CO_3^{2-}	72
La^{3+}	70	$C_2O_4^{2-}$	74
Ce^{3+}	70	CrO_4^{2-}	85
$CH_3NH_3^+$	58	PO_4^{3-}	69
$N(Et)_4^+$	33	$Fe(CN)_6^{3-}$	101
$N(Bu)_4^+$	19	$Fe(CN)_6^{4-}$	111

SOURCE: J. A. Dean, Ed., *Lange's Handbook of Chemistry*, 12th ed., Table 6-7, p. 6-34, McGraw-Hill Book Company, New York, 1979.
NOTE: All values rounded to nearest unit.

The conductivity of solutions is quite temperature dependent. An increase of temperature invariably results in an increase of ionic conductance, and for most ions this amounts to about 2% per degree. For precise work, conductance cells must be immersed in a constant-temperature bath. It is customary to select 25°C for measurements in the United States, although generally 18°C is preferred in Europe. For relative measurements, as in titrations, the conductance cell need only attain thermal equilibrium with its surroundings before proceeding with conductance measurements.

26.2 MEASUREMENT OF ELECTROLYTIC CONDUCTANCE

Electrolytic conductance measurements usually involve determination of the resistance of a segment of solution between two parallel electrodes by means of Ohm's law. These electrodes are platinum metal that has been coated with a deposit of platinum black to increase the surface area and reduce the polarization resistance. Some of the more important phenomena associated with the application of a voltage between electrodes immersed in a liquid electrolyte are indicated in Fig. 26-1 for an idealized system. To eliminate the effects of processes associated with the electrodes, such as those discussed in

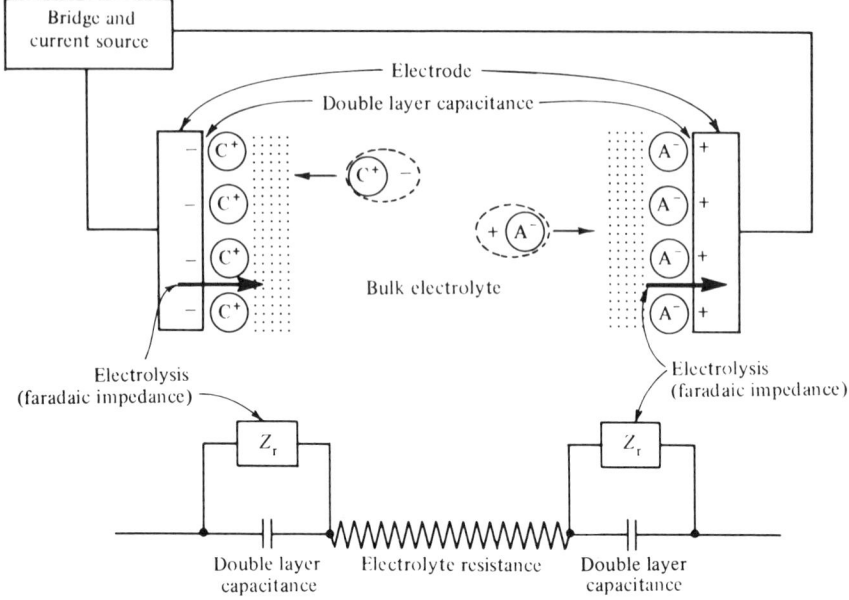

FIGURE 26-1 Electrolytic conductance cell—a simplified representation of the double layer at the electrodes, faradaic processes, and migration of ions through the bulk electrolyte. (J. Braunstein and G.D. Robbins, *J. Chem. Educ.*, 48, 52 (1971). Reproduced by permission.)

Chapter 25, measurements are made with an alternating current at 60, 100, 1000, or 3000 Hz. Some variation of the Wheatstone bridge is generally employed. The evaluation involves a comparative procedure. A conductance cell is calibrated by determining its cell constant, using a solution of known conductivity as will be described later.

Generally the bridge circuit must contain not only resistance, but also capacitance (or inductance) to balance the capacitive effects in the conductance cell. The latter arises from the electric double layer at the electrode–electrolyte interface at applied voltages below the decomposition voltage, and from the frequency-dependent resistance (impedance) associated with the faradaic processes at voltages above the decomposition voltage. For resistances of less than 10^4 Ω the model of a conductance cell as the electrolyte resistance in series with the double layer capacitance is a reasonable physical approximation.[1] The magnitude of impedance at 1000 Hz is of the order of

$$\frac{1}{2\pi fC} = 1.6 \text{ to } 16 \text{ }\Omega \qquad (26\text{-}4)$$

for capacitance values of 10–100 μF/cm² of electrode surface. Introduction of a variable capacitance (or inductance) into the bridge circuit permits compensation of the phase shift between current and voltage caused by the capacitance in the electrolytic cell. Parallel *RC* balancing arms are used more frequently than a series arrangement because smaller capacitance values are needed, and small capacitances can be obtained with higher accuracy

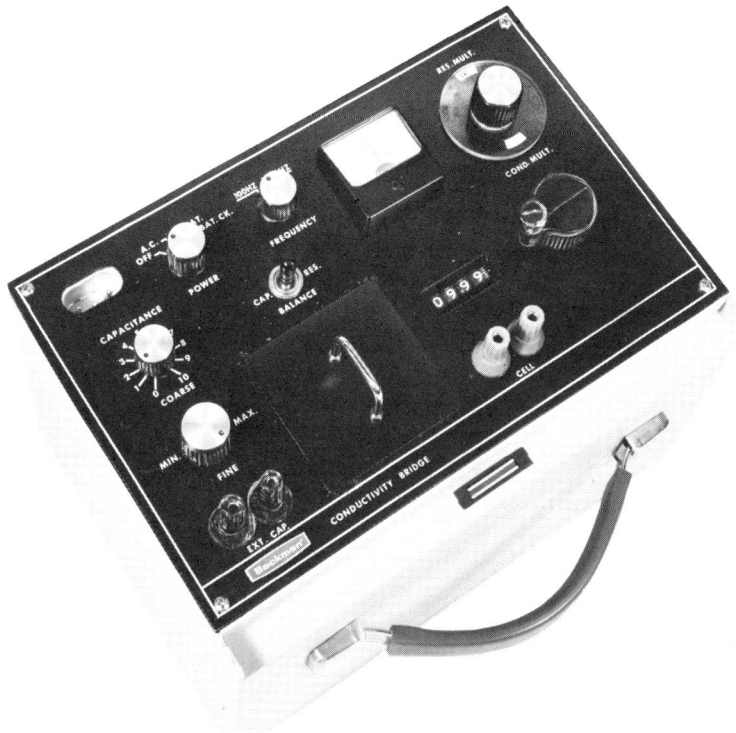

FIGURE 26-2 Laboratory and field conductivity bridge. (Courtesy of Beckman Instruments, Inc.)

and less frequency dependence than large ones. For example, the parallel capacitance required to compensate a series capacitance of 100 μf at 1000 Hz with a resistance of 1000 Ω is only 300 pF.

Instrumentation

A typical commercial conductivity bridge, shown in Fig. 26-2, is designed to measure electrolytic conductance in micromhos and resistance in ohms. The reading for either is indicated directly on a digital readout dial. A built-in generator provides bridge current at frequencies of 100 and 1000 Hz. Generally the lower frequency should be used when the measured resistance is high, and the higher frequency when the measured resistance is low. The instrument is balanced with the aid of a phase-sensitive detector and a null meter. Inexpensive process control instruments use an electron-ray (magic-eye) tube. With precision-class instruments, oscilloscopes are used for balance indication. The null condition is indicated as a straight horizontal line on the oscilloscope, whereas resistive and reactive imbalance are shown, respectively, by tilting of the line and by widening to an ellipse.

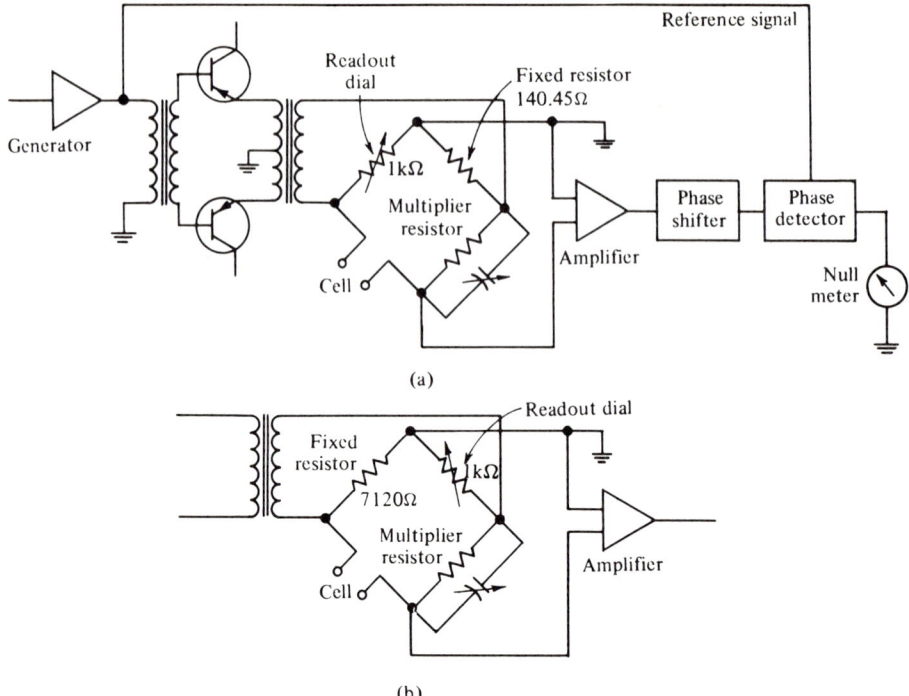

FIGURE 26-3 Simplified schematic of a conductivity bridge: (a) resistance mode and (b) conductance model. (Courtesy of Beckman Instruments, Inc.)

The operation of the instrument can be understood from Fig. 26-3a. The generator supplies a sinusoidal drive voltage to the bridge arm as well as a reference voltage for the phase detector. In the resistance mode, the bridge is an equal-arm bridge with cell and decade resistance in adjacent arms. At balance the decade resistance equals the cell resistance. The bridge is balanced by adjusting the readout dial resistor, a 10-turn potentiometer coupled to a mechanical counter which constitutes the readout device. The range of the instrument is changed by switching in different multiplier resistors for each range. The cell constant of the conductivity cell should be selected to maintain the measured resistance between 100 Ω and 1.1 MΩ.

The small output signal of the bridge at balance is amplified and applied to a phase-shifting circuit where, in the resistive position, the signal is shifted slightly to make up for any phase change that occurs in the amplifier and transformers. In the capacitive position the phase is shifted 90° so that the detector will be sensitive only to capacitive imbalances in the bridge circuit. To facilitate reactive balance, a continuously variable 8–200-pF capacitor, plus 10 steps of 200-pF capacitance, is placed in the arm of the bridge adjacent to the cell. The output of this circuit is then applied to the detector where it is added to the reference signal and the resultant signal is read on the meter.

Figure 26-3b shows the bridge in the conductance mode. Now the decade resistance is in the arm opposite the cell. All other circuits remain the same and perform the same functions as they do when measuring resistance.

TABLE 26-2 Specific Conductances of Potassium Chloride Solutions

Grams KCl/kg of Solution	κ IN Ω^{-1} cm^{-1}	
	18°C	25°C
71.1352	0.09784	0.11134
7.4191	0.01117	0.01286
0.7453[a]	0.001221	0.001409

SOURCE: G. Jones and B. C. Bradshaw, *J. Am. Chem. Soc.*, 55, 1780 (1933).
[a]Virtually 0.0100M.

Conductance Cells

In the design of conductance cells for precision measurements a number of factors must be taken into consideration. However, for many purposes two parallel sheets of platinum fixed in position by sealing the connecting tubes into the sides of the measuring cell are adequate. Also satisfactory are two sheet-platinum electrodes or wands immersed in the solution and held by ordinary clamps. These arrangements make the measured resistance independent of sample volume and proximity to surface.

There are practical limits of measured electrolytic resistance for any desired accuracy and sensitivity. The optimum appears to be in the vicinity of 500–10,000 Ω when errors are to be ±0.1%. In solutions of low conductance, the electrode area A should be large and the plates spaced (ℓ) close together; for highly conducting solutions, the area should be small and the electrodes far apart. The platinum electrodes are almost always lightly plated with platinum black to reduce the polarizing effect of the passage of current between the electrodes.

For a given cell with fixed electrodes, the ratio ℓ/A is a constant, called the cell constant Θ. It follows that

$$\kappa = \frac{1}{R}\left(\frac{\ell}{A}\right) = \frac{\Theta}{R} \tag{26-5}$$

For conductance measurements a cell is calibrated by measuring R when the cell contains a standard solution of known specific conductance, and Θ is then computed by means of Eq. 26-5. The electrolyte almost invariably used for this purpose is potassium chloride. Values of the specific conductance of potassium chloride solutions are given in Table 26-2. For conductometric titrations the absolute conductance need not be known, merely relative conductances as the titration progresses.

Example 26-2

When a certain conductance cell was filled with 0.0100M solution of KCl, it had a resistance of 161.8 Ω at 25°C, and when filled with 0.005M NaOH it had a resistance of 190 Ω.

The cell constant is

$$\Theta = (0.001409)(161.8) = 0.2280 \text{ cm}^{-1}$$

The specific conductance of the sodium hydroxide solution is

$$\kappa = \frac{\Theta}{R} = \frac{0.2280}{190} = 0.00120 \text{ } \Omega^{-1} \text{ cm}^{-1}$$

and the equivalent conductance is

$$\Lambda = \frac{(1000)(0.00120)}{0.005} = 240 \text{ cm}^2 \text{ equiv}^{-1} \text{ } \Omega^{-1}$$

Various types of conductivity cells are commercially available. The dip cell (Fig. 26-4) is the simplest to use whenever the liquid to be tested is in an open container. It is merely immersed in the solution to a depth sufficient to cover the electrodes and the vent holes. Liquid volumes of 5 ml or less suffice for small-diameter dip

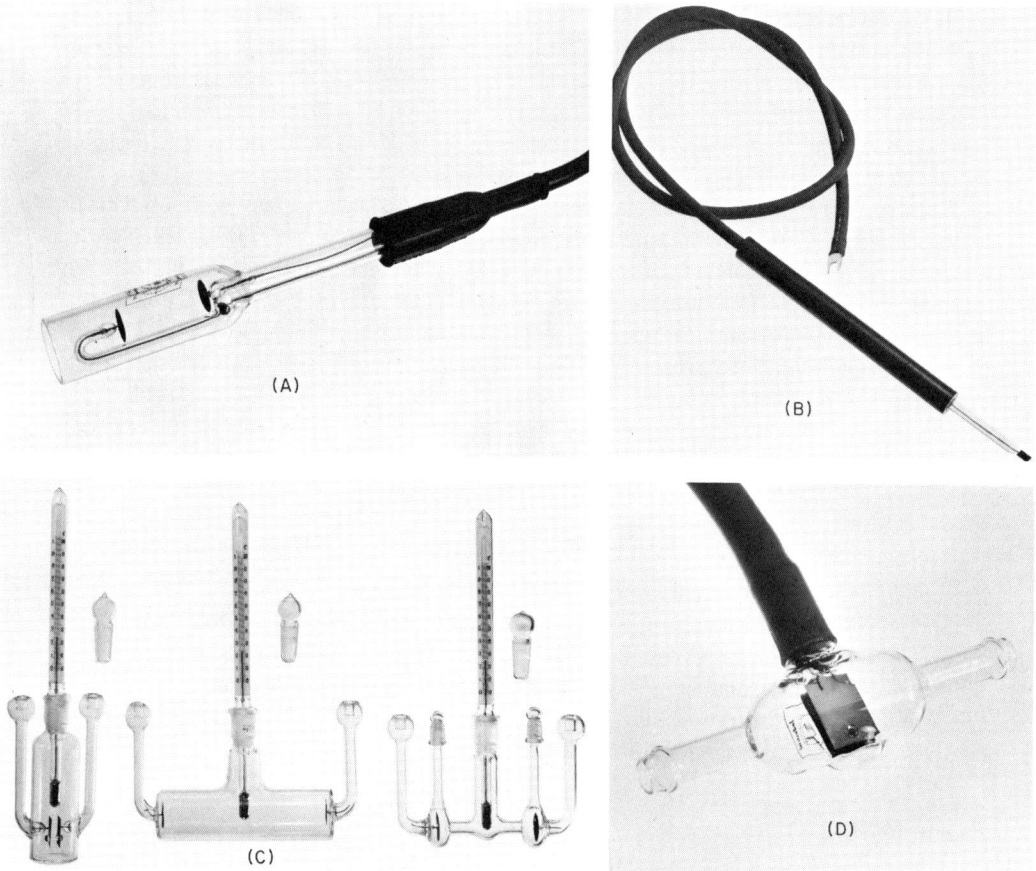

FIGURE 26-4 Conductivity cells. (A) Dip-type cell for medium-conductance solutions; cell constants from 0.5 to 2. (B) Individual wands for conductometric titrations. (C) Fill-type cell for laboratory work to contain the sample under test and to be immersed in a temperature bath. (D) Flow-through cell; cell constants from 0.01 to 0.2. (Courtesy of Beckman Instruments, Inc.)

cells. A pair of individual platinum electrodes on glass wands is useful in conductance titrations. Epoxy cells are used for high-temperature work in corrosive solutions except concentrated oxidizing acids. Pipet cells permit measurements with small volumes of solution, as little as 0.01 ml in some designs.

Temperature Compensation

Conductivity varies with temperature as well as with electrolyte concentration. The temperature coefficient of conductance of electrolyte solutions in water is almost always positive and of a magnitude from about 0.5-3% per degree Celsius. A practical means of providing temperature compensation is to introduce into the bridge circuit a resistive element that will change with temperature at the same rate as the solution under test. In different forms, this temperature compensator arm of the bridge can be a rheostat calibrated in temperature and requiring manual adjustment or a thermistor and fixed resistive network in thermal contact with the test solution, to provide automatic compensation. Regardless of the means employed, accurate compensation for temperatures changes requires that the temperature coefficient of resistance of the compensator match that of the test solution.

26.3 DIRECT CONCENTRATION DETERMINATIONS

Although the electrical conductance of a solution is a general property and is not specific for any particular ion, a number of analyses can be made by means of a measurement of conductivity. In general, the success of a measurement depends upon relating the property of the sample that is to be estimated to the conductance of some highly conducting ion. For example, free caustic remaining in scrubbing-tower solutions can be estimated by observing the decrease in conductance of the solution. This is possible even in the presence of the salts formed upon neutralization of the alkali, because the conductance of the hydroxyl ion is approximately fivefold greater than that of any other anion. Similarly, the unusually high conductance of the hydrogen ion permits an estimation of the free acid content in acid pickling baths. The changing conductivity resulting from the adsorption of gaseous combustion products in suitable solutions is frequently employed for the determination of carbon, hydrogen, oxygen, and sulfur individually, in organic and in inorganic compounds. The change in conductivity is always measured in relation to an identical solution not in contact with the combustion products. On the other hand, when checking the purity of distilled or deionized water, steam distillates, rinse waters, boiler waters, or in regeneration of ion exchanges, it is the total salt content which is sought.

For all these purposes very compact and inexpensive conductance bridges are available with scales calibrated directly in pounds per gallon, parts per million, grains per gallon, or percent. For example, instruments can be supplied for direct indication such as 1-12 lb Na_2CO_3 per 100 gal., 0-40 parts per thousand salinity, 0.4-10% H_2SO_4, 0.4-12% NaOH, and 96-99.5% H_2SO_4. These units are intended for industrial monitoring as well as for following an industrial process.

Example 26-3

The scale of a conductivity bridge is inscribed from 0.005 to 2.0% H_2SO_4 in approximately a logarithmic manner. For these solutions the specific conductance ranges from about 0.00044 to 0.176 Ω^{-1} cm^{-1}. What range of resistances are involved and what cell constant is compatible?

Answer The resistance values will range from

$$R = \frac{\Theta}{0.00044} = 22{,}800$$

to

$$R = \frac{\Theta}{0.176} = 56.9$$

A suitable cell constant is 10.0 cm^{-1}; the resistance ranging from 57 to 22,800 Ω. A cell constant of 20 cm^{-1} would also be suitable. A smaller cell constant would provide too low a resistance for the stronger acid solutions.

A cell with a constant 10 cm^{-1} would have electrodes of moderate area and some distance apart, perhaps electrodes 0.5 cm^2 in area and spaced 5 cm apart.

Direct-reading conductivity meters are also available. These instruments apply a stabilized ac voltage to the conductivity cell and a series resistor, rectify the voltage drop across the series resistor, and measure the resultant dc signal. As long as the resistor in series with the cell is smaller in resistance, the dc signal will be directly related to the cell resistance. Continuous indication is provided on a linear meter scale.

26.4 CONDUCTOMETRIC TITRATIONS

In this method the variation of the electrical conductivity of a solution during the course of a titration is followed. It is not necessary to know the actual specific conductance of the solution; any quantity proportional to it is satisfactory. This may result in considerable simplification of equipment. The titrant is introduced by means of a buret, and the conductance readings corresponding to various increments of titrant are plotted against the latter. Figure 26-5 illustrates the conductometric titration of hydrochloric acid with sodium hydroxide. As seen from Eq. 26-3, the measured conductance is a linear function of the concentration of ions present. In the example, the falling branch represents the conductance of the hydrochloric acid still present in the solution, together with that of the sodium chloride already formed. The rising branch represents the conductance of the excess base present after neutralization, together with that of the sodium chloride. Since the variation of conductance is linear, it is sufficient to obtain six or eight readings, covering the range before and after the end point, and draw two straight lines through them. The point of intersection of the two branches gives the end point.

If the reaction is not quantitative, there is curvature in the vicinity of the end point. Hydrolysis, dissociation of the reaction product, or appreciable solubility in the case of precipitation reactions will give rise to this type of curvature. At a sufficient distance on

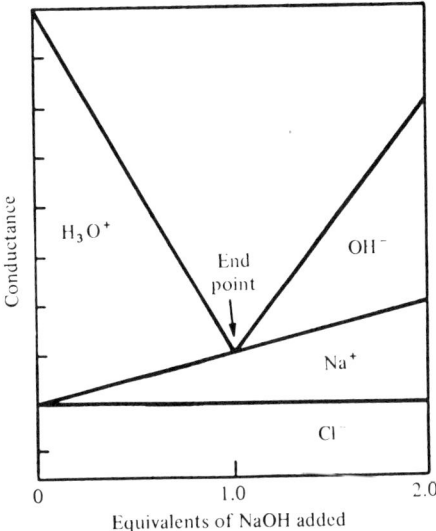

FIGURE 26-5 Titration of hydrochloric acid with sodium hydroxide.

either side away from the end point, from 0 to 50% and between 150 and 200% of the equivalent volume of titrant, sufficient common ion is present to repress these effects, and the branches are straight lines. By extrapolating these portions of the lines, the position of the end point can be determined.

The acuteness of the angle at the point of intersection of the two branches will be a function of the individual ionic conductances of the reactants. In Fig. 26-5, the falling branch was steep because it involved the replacement of hydrogen ion ($\lambda_+ = 350$) by sodium ions ($\lambda_+ = 50$), and a large difference exists between the two conductances. Similarly, the rising branch on this curve is relatively steep also, but not as steep as the falling branch, because the conductance of the hydroxyl ion ($\lambda_- = 198$) is considerably smaller than the corresponding value for the hydrogen ion.

The titrant should be at least ten times as concentrated as the solution being titrated in order to keep the volume change small. If necessary, a correction may be applied:

$$\left(\frac{1}{R}\right)_{actual} = \left(\frac{V+v}{V}\right)\left(\frac{1}{R}\right)_{obs} \qquad (26\text{-}6)$$

where V is the initial volume and v is the volume of titrant added up to the particular conductance reading.

In principle, all types of reactions can be employed. The method can be used with very dilute solutions, about $0.0001M$. On the other hand, because every ion present contributes to the electrolytic conductivity, large amounts of extraneous electrolytes should be absent. In the presence of large amounts of such electrolytes, the change in conductance accompanying a reaction would be a very small part of the total conductance and would be difficult to measure with accuracy. For this reason oxidation–reduction reactions usually cannot be performed because the solutions are generally well buffered or strongly acidic.

Acid–Base Titrations

The conductometric titration of an acid with a base, each completely dissociated, is illustrated by Fig. 26-5 for the titration of hydrochloric acid with sodium hydroxide:

$$(H^+ + Cl^-) + (Na^+ + OH^-) \rightarrow (Na^+ + Cl^-) + H_2O \tag{26-7}$$

The highly conducting hydrogen ions initially present in the solution are replaced by sodium ions having a much smaller ionic conductance, while the concentration of chloride ions remains constant except for the small dilution by the titrant. The conductance of the solution at any point on the descending branch of the titration curve is given by the expression

$$\frac{1}{R} = \frac{1}{1000\Theta}(C_H\lambda_H + C_{Na}\lambda_{Na} + C_{Cl}\lambda_{Cl}) \tag{26-8}$$

In terms of the initial concentration of hydrochloric acid C_i and the fraction of the acid titrated f,

$$C_H = C_i(1-f), \quad C_{Na} = C_i f, \quad \text{and} \quad C_{Cl} = C_i$$

Substituting these values into Eq. 26-8,

$$\frac{1}{R} = \frac{C_i}{1000\Theta}[\lambda_H + \lambda_{Cl} + f(\lambda_{Na} - \lambda_H)] \tag{26-9}$$

As a result of the term within the parentheses in Eq. 26-9, the conductance of the solution diminishes up to the equivalence point. Beyond the equivalence point the conductance increases in direct proportion to the excess base added.

Example 26-4

Let us assume that one is titrating 100 ml of 0.01N HCl solution with 0.1N NaOH in a cell whose constant is 1.0 cm^{-1}. Under these conditions the initial conductance is

$$\frac{1}{R} = \frac{0.01}{(1000)(1)}(350 + 76) = 0.00426 \text{ } \Omega^{-1}$$

In this dilute solution, little error is introduced by assuming that the actual ionic conductances are essentially those at infinite dilution when dealing with completely ionized materials.

When the titration is 0.9 complete,

$$\frac{1}{R} = \frac{0.01}{(1000)(1)}[350 + 76 + (0.9)(50 - 350)]$$

$$= 0.00156 \text{ } \Omega^{-1}$$

and, correcting for the dilution caused by the titrant,

$$\frac{1}{R} = \frac{(100+9)}{100}(0.00156) = 0.00170 \text{ } \Omega^{-1}$$

At the equivalence point, there exists a solution of sodium chloride whose conductance is

$$\frac{1}{R} = \frac{0.01}{(1000)(1)} (76 + 50) \left(\frac{110}{100}\right) = 0.00139 \ \Omega^{-1}$$

When the equivalence point has been exceeded by 10%

$$\frac{1}{R} = \frac{0.01}{(1000)(1)} [126 + (0.1)(248)] \left(\frac{111}{100}\right) = 0.00167 \ \Omega^{-1}$$

If the strong acid is titrated with a weak base, for example, an aqueous solution of ammonia, the first part of the conductance titration curve, representing the removal of hydrogen ion and replacement by ammonium ion, will be very similar to the descending branch of Fig. 26-5, because most cations have similar ionic conductances. After the equivalence point is passed, however, the conductance will remain almost constant, because a solution of ammonia has a very small conductance compared with that of ammonium chloride (Fig. 26-6).

$$(H^+ + Cl^-) + NH_3 \rightarrow NH_4^+ + Cl^- \tag{26-10}$$

Titrations of weak acids or weak bases are somewhat more difficult. Acetic acid, for example, is present partly in the form of H^+ and CH_3COO^-, but largely as nonionized molecules. The proportions of each are regulated by the law of mass action. Initially the solution has a low conductance. As neutralization proceeds, the common ion formed, that is, the acetate ion, represses the dissociation of the acetic acid so that an initial fall in conductance may occur. With further addition of sodium hydroxide the conductance of the sodium and acetate ions soon exceeds that of the acetic acid which they replace, and so the curve passes through a minimum and thereafter the conductance of the solution increases. The shape of the initial portion of these conductance curves will vary with the strength of the weak acid and its concentration, as indicated in Figs. 26-6 and 26-7.

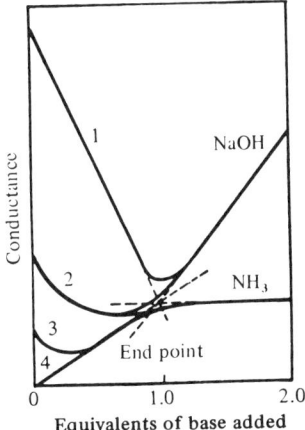

FIGURE 26-6 Conductometric titration curves of various acids by sodium hydroxide and by aqueous ammonia. The numbered curves are (1) HCl, (2) acid, pK_a of 3, (3) acid, pK_a of 5, and (4) acids, $pK_a > 7$.

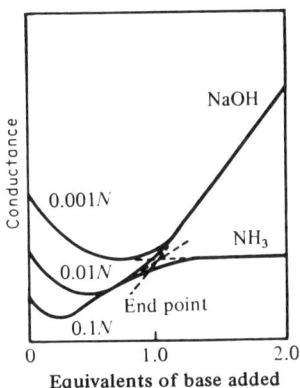

FIGURE 26-7 Conductometric titration curves of acetic acid ($pK_a = 4.8$) at various concentrations.

When a weak acid is titrated with a weak base, the initial portion of the conductance titration curve follows the pattern described above. Beyond the equivalence point, there is no change in the conductance because of the very small conductance of the excess free base. The intersection of the two branches is sharper than for a corresponding titration of a weak acid with a strong base.

Pronounced hydrolysis in the vicinity of the equivalence point makes it necessary to select the experimental points for the construction of the two branches considerably removed from the equivalence point. It will be observed in Fig. 26-7 that for dilute solutions of weak acids no linear region is obtained preceding the equivalence point, and therefore no reliable point of intersection seems possible. Addition of ethanol or other solvent with a smaller dielectric constant than water often reduces the dissociation of the weak acid sufficiently to yield a region of linear conductance preceding the equivalence point. When this does not suffice, two other experimental modifications may aid in the location of the equivalence point. In one procedure, titrations are conducted with duplicate portions of test solution, using an aqueous solution of ammonia as one titrant and a strong base as the other titrant, both titrants equivalent in normality. The conductance curves preceding the equivalence point are practically identical in shape, but beyond the equivalence point the curves will have quite different slopes. After superimposing the foreportions, the intersection of the two branches beyond the equivalence point establishes the end point, as shown in Fig. 26-7. In the second procedure, sufficient ammonia is added to neutralize about 80% of the weak acid; then the titration is carried out with standard sodium hydroxide. If the conductance were plotted during the addition of the aqueous solution of ammonia, the curve would resemble those discussed. As long as any of the original acid is present the curve will continue to parallel the shape shown in Fig. 26-6, but when all the acid has been consumed, the ammonium ion formed commences to react with the hydroxyl ion, thus

$$(NH_4^+ + CH_3COO^-) + (Na^+ + OH^-) \rightarrow NH_3 + H_2O + Na^+ + CH_3COO^- \qquad (26\text{-}11)$$

and the conductance falls owing to the replacement of the ammonium ion ($\lambda_+ = 73$) by the sodium ion ($\lambda_+ = 50$). When the replacement is complete, the conductance abruptly increases, as shown in Fig. 26-8, and is then parallel to the corresponding part of Fig. 26-5 after the equivalence point.

The determination of a very weak acid or a very weak base is merely an extension of the titration of weak acids or weak bases. Take, for example, the titration of boric acid ($pK_a = 9.2$) with sodium hydroxide, as shown in Fig. 26-6. Initially the conductance is very small, since boric acid is dissociated to a negligible extent in aqueous solution. During the titration the reaction

$$HBO_2 + (Na^+ + OH^-) \rightarrow Na^+ + BO_2^- + H_2O \qquad (26\text{-}12)$$

occurs, and the conductance increases linearly with the formation of borate and the addition of sodium ions. Beyond the equivalence point, further addition of sodium hydroxide introduces hydroxyl ions and there is a further increase of conductance. The inflection point will not be particularly sharp; in fact, it is often impossible to locate accurately the intersection of the two branches. This is true when the acid becomes so weak that extensive hydrolysis occurs throughout the reaction and the nonlinear portion extends from

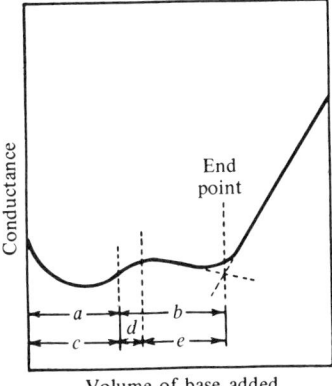

FIGURE 26-8 Conductometric titration of a weak acid employing preliminary addition of aqueous ammonia followed by sodium hydroxide: (a) volume of aqueous ammonia added; (b) volume of NaOH added up to the end point; (c) amount of acid neutralized by the NH_3; (d) amount of acid neutralized by the NaOH; and (e) displacement of NH_3 from NH_4^+ formed in (c).

the end point to the initial point. Only if the product of the ionization constant and the acid concentration exceeds 10^{-11} can the titration be performed.

One of the valuable features of the conductance method of titration is that it permits the analysis of a mixture of a strong and a weak acid in one titration. Figure 26-9 illustrates the neutralization of oxalic acid ($pK_1 = 1.2$; $pK_2 = 4.3$), representative of a strong acid and a moderately strong acid present in equivalent amounts. The initial decrease in conductance is due to the removal of hydrogen ions supplied by the relatively complete dissociation of $H_2C_2O_4$. This is followed by an increase in conductance as the weak acid, $HC_2O_4^-$, is consumed and replaced by $C_2O_4^{2-}$ ions and the cation of the titrant. A rounded section joins these two branches because the pK_a values are too close to each other and consequently the neutralization of the second acid begins while that of the first is being completed. When the neutralization of the second acid is complete, there is an increase in conductance when sodium hydroxide is the titrant due to the sodium and hydroxyl

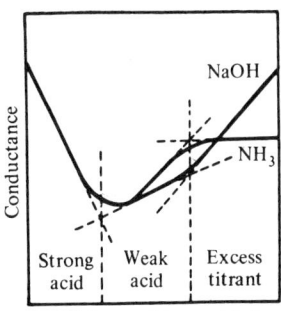

FIGURE 26-9 Conductometric titration of a mixture of a strong acid and a weak acid. Example: oxalic acid.

ions. Substitution of ammonia as the titrant results in a greater rise in conductance on the middle portion of the titration curve and little further change in conductance after the second end point. The first point of intersection gives the amount of strong acid in the mixture, and the difference between the first and second is equivalent to the amount of weak acid. Practical applications include the determination of mineral acids in vinegar and the titration of sulfonic acid groups followed by either carboxylic or phenolic groups in mixtures of organic acids.

The conductance method is also useful in the titration of the conjugate base of a weakly ionized acid, and vice versa. For example, organic acid salts such as acetates, benzoates, and nicotinates can be titrated with a standard solution of a completely ionized acid. As long as the ionization constant of the displaced acid or base divided by the original salt concentration does not exceed approximately 5×10^{-3}, the displaced acid will not contribute to the total conductance. The descending branch in the middle portion of Fig. 26-8 illustrates the nature of the titration curve that will be obtained.

Determinations by Precipitations and Through Formation of Complexes

Mercuric nitrate and perchlorate have found use as reagents for complexometric reactions. These salts exist almost entirely in the form of free ions. If a solution of a cyanide is added, the reaction

$$(Hg^{2+} + 2ClO_4^-) + 2(K^+ + CN^-) \rightarrow Hg(CN)_2 + 2(K^+ + ClO_4^-) \qquad (26\text{-}13)$$

occurs. Before the equivalence point, 1 mercuric ion is replaced by 2 potassium ions. The conductance varies only slightly. Beyond the end point the addition of potassium and cyanide ions causes the conductance to increase. In this class of reactions the slopes of the branches of the curve are determined both by the change in the ionic conductances of the ions present and by any change in the total number of electrical charges carried by the ions in solution. For these reasons, an acetate salt is preferable when titrating an anion, and a lithium salt when titrating a cation.

Even in favorable cases, results obtained through electrical conductivity tend to be less accurate than in acid–base systems. In precipitations, all the factors influencing the formation of the precipitate and the nature of the product must be considered just as in ordinary gravimetric methods. A slow rate of precipitation, excessive solubility of the insoluble materials, and all types of adsorption or occlusion difficulties make it difficult if not impossible to locate the equivalence point with any degree of accuracy.

26.5   MEASUREMENT OF DIELECTRIC CONSTANT

When a nonconducting liquid is placed between the metal plates of Fig. 26-10, the system behaves as two capacitors, C_g and C_s, in series. The capacitance C_s will vary from a fixed value C_0, determined by the geometry of the cell when filled with air, to a value D dependent upon the dielectric constant of the sample; that is,

$$C_g = C_0 D \qquad (26\text{-}14)$$

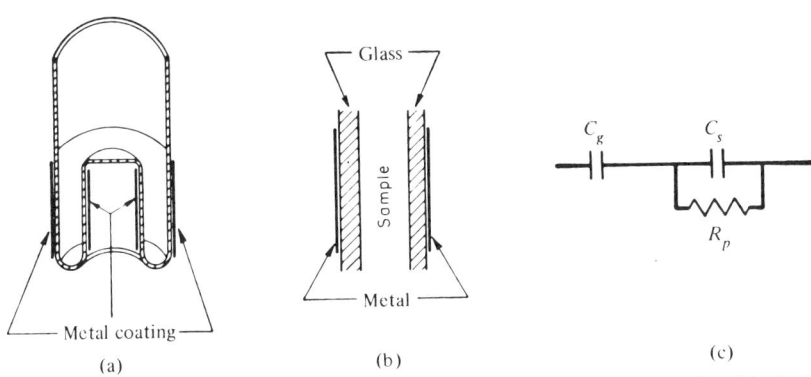

FIGURE 26-10 (a) Isometric vertical cross section of a typical cell; (b) functional equivalent of the cell; (c) equivalent circuit of cell and solution.

The value C_0 depends upon the thickness of the glass walls, the effective plate area, and the dielectric constant of the container itself. The capacitance of the cell is not a linear function of the dielectric constant, but will approach the value of C_g as a limit.

The capacitance of the cell can be determined by a capacitance bridge which resembles a Wheatstone bridge except that two arms of the bridge have capacitances rather than resistances and alternating current; often high frequency, rather than direct current is used. The unknown cell forms one arm of the bridge and a calibrated variable capacitor the other.

Other methods of measuring capacitance are available. One is the Sargent Oscillometer which measures the output frequency of two identical oscillator circuits, one contains the sample cell as part of the oscillator capacitance, while the other serves merely as a reference unit. The outputs of the two units are fed into a mixer unit which measures the difference in frequencies (beat frequency). A variable precision condenser in the unit containing the sample cell is altered to compensate for any capacitance change in the sample and thus to reduce the beat frequency to zero. This instrument can also be used for high-frequency titrations, but since these titrations are not widely used in analytical chemistry, they will not be further described in this book.

Table 26-3 lists the dielectric constants of some common materials. The high value for water renders it possible to determine extremely small amounts of water in organic liquids or moisture in granular or powdered substances. For many years a dielectric-type of moisture meter for use with grains, cereals, and other powdered substances has been available

TABLE 26-3 Dielectric Constants

Formamide	109.5	Acetic anhydride	20.7	Acetic acid		6.15
Water	81	1-Propanol	20.1	Ethyl acetate		6.02
Formic acid	58.5	2-Propanol	18.3	1-Butylamine		5.3
Acetonitrile	37.5	1-Butanol	17.1	Chloroform		4.806
Nitrobenzene	34.8	Ethylenediamine	14.2	Benzene		2.379
Methanol	32.6	Benzyl alcohol	13.1	Carbon tetrachloride		2.238
Ethanol	24.3	Phenol	9.78	1,4-Dioxane		2.209
Acetone	20.7	Aniline	6.89			

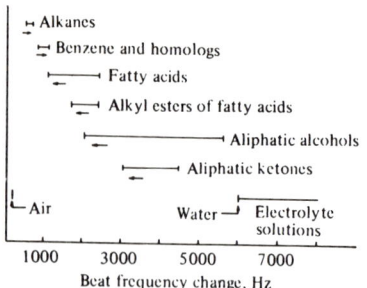

FIGURE 26-11 Frequency change induced by organic compounds. Arrow indicates direction of increasing molecular weight. (P. W. West, T. S. Burkhalter, and L. Broussard, *Anal. Chem.*, 22, 469 (1950). Courtesy of American Chemical Society.)

commercially. The test material is uniformly packed between two parallel plates which serve as the measuring condenser.

High-frequency methods have been applied to the discrimination of organic mixtures. Organic compounds lie in rather definite groups in the range between the dielectric constant of air and water, as shown in Fig. 26-11.

PROBLEMS

1. An aqueous 20% HCl solution has a specific conductance of about 0.85 Ω^{-1} cm^{-1} at 25°C. What is the measured resistance with a cell of constant (a) 100, (b) 20, (c) 10, (d) 1, (e) 0.2 cm^{-1}? Are these resistance values feasible to measure with standard conductivity bridges?

2. A cell constant of 20.0 cm^{-1} is recommended for a commercial conductivity bridge designed to span the range from 1 to 18% HCl. The corresponding conductance ranges from 0.0630 to about 0.750 Ω^{-1}. What range of resistance values are involved?

3. A meter scale is to be inscribed from 2 to 1000 ppm Na$_2$SO$_4$, and the midpoint of the logarithmic scale shall correspond to 40 ppm. Suggest a compatible set of instrument parameters; that is, resistance range and cell constant.

4. Similar to Problem 3, individual instruments are to be designed for each of these systems. Compute the resistance range and a compatible cell constant. Use handbooks to locate necessary conductance values, and assume average distilled water has a specific conductance of 2 × 10^{-6} Ω^{-1} cm^{-1}. (a) 0–5% HCl; (b) 0.5–5% NH$_3$; (c) 0–60 ppm sodium formate; (d) 0–40 ppm salinity (as NaCl); (e) 96–99.5% H$_2$SO$_4$; (f) 0.1–10% CrO$_3$.

5. The equivalent conductance of a 0.002414N acetic acid solution is found to be 32.22 at 25°C. Calculate the degree of dissociation of acetic acid at this concentration, and calculate the ionization constant.

6. The specific conductance at 25°C of a saturated solution of barium sulfate was 4.58 × 10^{-6} Ω^{-1} cm^{-1}, and that of the water used was 1.52 × 10^{-6}. What is the solubility of BaSO$_4$ at 25°C in moles per liter and in grams per liter? Calculate the solubility product.

7. The solubility product of silver iodate at 25°C is 3.1 × 10⁻⁸. What would be the resistance of a saturated solution of silver iodate, measured with a cell whose cell constant was 0.2 cm⁻¹? (Neglect the solvent correction.)

8. A very dilute solution of NaOH (100 ml) is titrated with 1.00N HCl. The following resistance readings (in ohms) were obtained at the indicated buret readings: 0.00 ml, 3175; 1.00 ml, 3850; 2.00 ml, 4900; 3.00 ml, 6500; 4.00 ml, 5080; 5.00 ml, 3495; 6.00 ml, 2733. Determine the normality of the solution and the weight of NaOH present.

9. In the titration of 100 ml of a dilute solution of acetic acid with 0.500N aqueous ammonia, the following resistance readings (in ohms) were obtained at the indicated buret readings: 8.00 ml, 750; 9.00 ml, 680; 10.00 ml, 620; 11.00 ml, 570; 12.00 ml, 530; 13.00 ml, 508; 15.00 ml, 515; 17.00 ml, 521. What is the normality of the acetic acid solution?

10. In the titration of 100 ml of H_2SO_4 in glacial acetic acid with 0.500M sodium acetate in the same solvent, the following specific conductance (× 10⁶) data were obtained at the indicated buret readings:

0.50 ml = 2.95	3.50 ml = 4.78	7.00 ml = 3.20
1.00 ml = 3.30	4.00 ml = 4.73	7.50 ml = 3.20
1.50 ml = 3.65	4.50 ml = 4.40	8.00 ml = 3.47
2.00 ml = 4.00	5.00 ml = 4.04	8.50 ml = 3.82
2.50 ml = 4.35	5.50 ml = 3.76	9.00 ml = 4.18
3.00 ml = 4.65	6.00 ml = 3.43	9.50 ml = 4.50

What is the molarity of the sulfuric acid solution?

11. The following relative conductance readings were obtained during the titration of a mixture containing an aliphatic carboxylic acid and an aromatic sulfonic acid. The titrant was 0.200N NH_3. Readings have been corrected for titrant volume.

0.00 ml = 2.01	3.20 ml = 1.19	5.00 ml = 1.51
1.00 ml = 1.75	3.50 ml = 1.26	6.00 ml = 1.52
2.00 ml = 1.47	4.00 ml = 1.41	8.00 ml = 1.53
2.50 ml = 1.33	4.20 ml = 1.47	
3.00 ml = 1.19	4.50 ml = 1.51	

Calculate the number of equivalents of each acid present in the mixture.

12. Using the equivalent conductance values obtained from Table 26-1, sketch the general form of the titration curve in each of the following cases: (a) titration of $Ba(OH)_2$ with HCl; (b) titration of NH_4Cl with NaOH; (c) titration of silver nitrate with potassium chloride; (d) titration of silver acetate with lithium chloride; (e) titration of sodium acetate with HCl; (f) titration of a mixture of a sulfonic acid and a carboxylic acid with NaOH; (g) titration of $KH_3(C_2O_4)_2$ with NH_3.

13. A commercial liquor contains nicotinic acid, ammonium nicotinate, and nicotinamide. Devise a conductometric titration for the determination of the free acid and the ammonium salt.

14. Exactly 50 ml of a 0.001N solution of HCl is titrated with 0.01N KOH. Calculate the conductance and resistance observed after the addition of 0, 25, 50, 90, 100, 110, 150, 175, and 200% of the equivalent amount of titrant. Assume the cell constant is 0.2 cm^{-1}. Plot the results.

15. Exactly 100 ml of a 0.1N solution of NH_4NO_3 is titrated with 1.0N KOH. As in Problem 14, calculate the conductance observed after the addition of the stated increments of titrant. Assume the cell constant is 0.5 cm^{-1}. Plot the results.

16. Ammonia in gas streams has been determined by bubbling the gas through exactly 100 ml of 0.0400M boric acid at a rate of 10 liters/min and measuring the change in conductance. The following data were obtained for a series of standard solutions of ammonia added to boric acid:

NH_3 ADDED, MEQUIV	CONDUCTANCE, Ω^{-1}, ($\times 10^4$)	
	HBO_2	$HBO_2 + NH_3$
48.55	0.55	45.55
0.715	0.55	10.65
0.0715	0.55	1.61
0.0071	0.55	0.66

Plot the calibration curve. Calculate the ammonia concentration present in each of these gas streams for which these conductance readings ($\times 10^4$) were obtained: sample A, 12.75; sample B, 6.35; sample C, 1.15; sample D, 3.45.

17. In Problem 16 the cell constant was 0.069 cm^{-1}. Assuming that the equivalent conductance of the borate ion at infinite dilution is about 30 cm^2 equivalent^{-1} ohm^{-1}, and that the specific conductance of the water employed was 2×10^{-6} Ω^{-1} cm^{-1}, estimate the degree of ionization of boric acid at the concentration employed and calculate the ionization constant.

BIBLIOGRAPHY

Britton, H. T. S., "Conductometric Analysis" in *Physical Methods in Chemical Analysis*, W. G. Berl, Ed., Vol. II, Academic, New York, 1951.
Braunstein, J. and G. D. Robbins, "Electrolytic Conductance Measurements and Capacitive Balance," *J. Chem. Educ.*, **48**, 52 (1971).
Lingane, J. J., *Electroanalytical Chemistry*, 2nd ed., Wiley-Interscience, New York, 1958.
Ross, J. W., "Conductometric Titrations," in *Standard Methods of Chemical Analysis*, 6th ed., F. J. Welcher, Ed., Vol. 3, Part A, Van Nostrand Reinhold, New York, 1966.
Shedlovsky, T. and L. Shedlovsky, in *Physical Methods of Chemistry*, 4th ed., A. Weissberger and B. W. Rossiter, Eds., Vol. I, Part 2A, Chap. III, Wiley-Interscience, New York, 1971.

LITERATURE CITED

1. Braunstein, J. and G. D. Robbins, *J. Chem. Educ.*, **48**, 52 (1971).

CHAPTER 27

Electronics: Fundamentals of Solid-State Devices

This chapter provides an overview of the basic principles and circuit components of the semiconductor technology employed in instrumentation. No attempt is made to review ac and dc circuits or older electronic devices, such as vacuum tubes.

27.1 BASIC FUNCTIONS OF INSTRUMENTATION

The purpose of chemical instrumentation is to obtain information from the substance being analyzed. In moving from its origin through the instrument to the output, the desired information, a chemical or physical quantity, undergoes a number of transformations. The number and complexity of these transformations is determined by the quality and quantity of data to be acquired from the sample under analysis.

Every analytical instrument may be divided into three basic components: an input transducer, an electronic signal modifier, and an output transducer. Input transducers, also known as detectors, transform the physical or chemical quantity of interest into an electrical signal. The signal modifiers perform necessary and desirable operations, such as amplification and filtering, on the signal from the input transducer. Finally, the output transducer converts the modified electrical signal into physical information which can be read and interpreted by the analyst.

Input Transducers

The majority of input transducers are analog devices that measure continuous physical properties (Table 27-1). These devices respond to input from their environment to produce

TABLE 27-1 Input Transducers

Physical Quantity	Input Transducer	Electrical Output
Concentration of electroactive species	Polarographic cell	Current
Ion activity in solution	Specific ion electrode	Voltage
Light intensity	Phototube	Current
	Photodiode	Resistance
Temperature	Thermistor	Resistance
	Thermocouple	Voltage

continuous electrical outputs. If the measured physical phenomenon is noncontinuous, the detector is designed to give pulse outputs, as in the case of high-energy radiation detectors which give pulse outputs when struck by gamma rays. The quality and capabilities of the detector untimately limit overall instrument performance.

Signal Transformation Modules

A signal transformation module takes information from the detector output, electrically converts this information into a more meaningful form, and presents it to the output transducer (Table 27-2). The type of detector and the final form of the output information will determine the electronic composition of the module. Module components range from a single simple resistor for current-to-voltage converison, to a complex microcomputer possessing a variety of signal transformation capabilities.

Output Transducers

The final instrument component, the output transducer, converts the modified electrical information into a form which can be utilized to retrieve the desired chemical information. This information may be displayed or recorded in either analog or digital form by a number of devices. Some commonly used output devices are shown in Table 27-3.

Modular Microelectronic Components

In the recent past an understanding of the operation of an instrument required a detailed knowledge of its component circuits. The advent of microelectronics has made a

TABLE 27-2 Electrical Signal Transformations

Amplification	Digital to analog
Analog to digital	Filtering
Attenuation	Integration
Comparison	Rectifying
Counting	Summing
Current to voltage	Voltage to current
Differentiation	Voltage to frequency

TABLE 27-3 Output Transducers

Alphanumeric printers	Oscilloscopes
Analog meters	Recorders, strip charts (y-t)
Computers	Recorders, x-y
Digital meters	Tape cassettes

large variety of electronic functions available in integrated circuit form. The integrated circuit (IC) package, known as a chip, becomes an electrical black box with input and output terminals. No detailed knowledge of the inner workings of the chip is necessary; rather its operation can be understood in terms of input and output signals. Each IC chip is considered to be a signal modifier. If a single chip cannot provide the desired signal transformation, several chips may be connected to obtain it.

Advances in technology have permitted increased numbers of circuit components to be packed into a single chip. Functions, such as counting, signal generation, modulation, and filtering, which formerly required many separate components, are now available on medium-scale integration (MSI) or large-scale integration (LSI) chips. A single LSI chip may contain the circuits required for a complex signal transformation. Instruments containing LSI and MSI chips are easier to understand, to operate, and to maintain than those containing larger numbers of discrete components.

Another trend in instrument design is the replacement of hardware components with software (computer programs) operated by microcomputers fabricated on LSI chips. Many signal transformations formerly accomplished by hardwired ICs are executed by the computer programs that interact directly with instruments. Replacement of conventional circuits with microcomputers has resulted in greater reliability, improved performance, expanded applications, and reduced maintenance costs (Chapter 30).

27.2 SEMICONDUCTOR COMPONENTS

Contemporary electrical signal modifiers have their origins in the transistor, a small low-power amplifier developed over 30 years ago to replace the large power-hungry vacuum tube. To understand the general concepts of transistor operation, it is necessary to consider briefly the properties of semiconductor materials.

Semiconductor devices are made by introducing controlled numbers of "impurity atoms" into a crystal of semiconductor material (silicon or germanium) by a process known as doping. The impurity atoms usually belong to either group V or group III of the periodic chart and possess atomic radii that allow them to replace individual silicon or germanium atoms without disruption of the crystalline structure. If a group V element, such as phosphorus, is introduced into a pure semiconductor crystal, the resulting material will contain an excess of mobile carrier electrons and is known as an n-type semiconductor.

Conversely, doping with boron, a group III element, yields a p-type semiconductor that has a deficiency of carrier electrons. Each electron deficiency is known as a hole, which is not really a particle but merely the absence of an electron at a position where one would normally be found in a pure lattice of silicon atoms. A hole possesses a

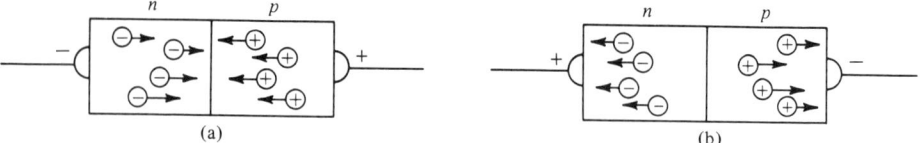

FIGURE 27-1 Schematic representation of diode operation.

positive electric charge and can therefore carry electric current. Holes move through semiconductor material in much the same way that bubbles move through a liquid medium and almost as rapidly as a carrier electron.

Diodes

A diode is the simplest semiconductor or device. It consists of a junction between n-type and p-type semiconductor regions in a single crystal. A positive voltage applied to the p-type region and a negative voltage applied to the n-type region causes a countercurrent of holes and electrons to flow (Fig. 27-1a). The holes are repelled by the positive voltage applied to the p-terminal and attracted to the negatively charged n-region. Conversely, electrons in the n-region are repelled by the voltage applied to the n-terminal and attracted to the p-region. The large current carried by electrons and holes traveling across the p-n junction is called the forward diode current. If the polarities of the p- and n-regions are reversed (Fig. 27-1b), the number of carriers (holes plus electrons) crossing the p-n junction is reduced practically to zero and no current flows through the diode. In reality a small "reverse" current is observed under these conditions due to the flow of "minority" carriers across the junction. Minority carriers consist of the very low concentrations of electrons found in p-regions and of holes found in n-regions.

A diode, unlike a transistor, is not capable of amplification and, therefore, cannot be classified as an active circuit component. However, it is unlike other passive components. Resistors, some capacitors, and inductors are symmetrical devices which have the same effect on a signal regardless of the signal polarity and the direction of current flow entering the device. The most significant property of the diode is its asymmetry, whereby it presents a high resistance to a signal of one polarity and a low resistance to a signal of opposite polarity.

The characteristics of a semiconductor diode are shown in Fig. 27-2. Note the change in scale for both axes on either side of zero. The forward-biased current is appreciable, whereas the reversed-biased current remains small until the Zener limit, Z (Fig. 27.2), is reached. At the Zener limit a breakdown in the normal semiconductor mechanism occurs and a large current flows. The voltage at the Zener limit is a property of the specific type of diode under consideration. As a result, the diode can be used as a switching device to control the flow of current as a function of the polarity and also as a voltage regulator to control the magnitude of the applied voltage.

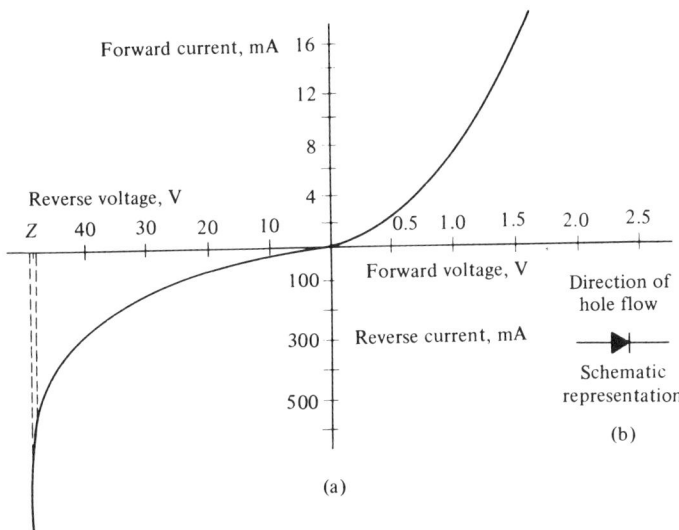

FIGURE 27-2 Characteristics of a semiconductor diode.

Bipolar Transistors

The earliest transistor to be fabricated was the bipolar or junction transistor constructed in 1948 by Bardeen, Brattain, and Shockley of Bell Telephone Laboratories. It consisted of two *p-n* junctions formed by sandwiching a thin slice of one kind of semiconducting material between two thicker sections of another semiconducting material (Fig. 27-3). Bipolar transistors are classified either *n-p-n* or *p-n-p*, depending upon the sequence of doped regions. One end of the transistor is the emitter, the other end the collector, and the thin middle section is the base.

Transistor performance depends upon the voltages applied to the three semiconductor regions. The operation of a bipolar transistor may be viewed as that of two diodes joined back to back. When the emitter of an *n-p-n* transistor is forward biased (the voltage applied to the emitter is more negative than that applied to the base) electrons flow from the emitter to the base. Because the base is so thin, most of these electrons (95–99%) pass

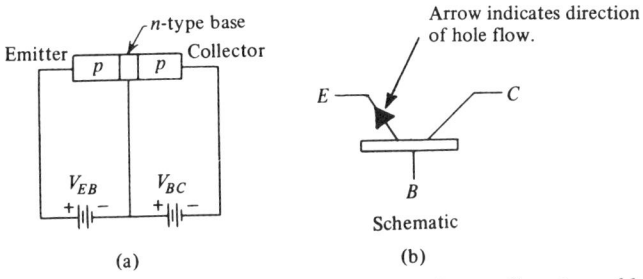

FIGURE 27-3 Transistor, *n-p-n* type. Arrow indicates direction of hole flow.

on into the collector region. The electrons which do not pass through the base combine with holes in this *p*-region, to form the base current. The collector is reverse biased with respect to the base region (the voltage applied to the collector is positive with respect to the voltage of the base).

Under these conditions the current is nearly the same in the collector and emitter circuits. As shown in Fig. 27-2, a small change of voltage across the emitter–base *n-p* junction causes a large change to occur in the current across the base–collector *p-n* junction. This base–collector current increase requires a large voltage change across the *p-n* junction. Now it must be remembered that the emitter current flows in a circuit with low resistance and the collector current flows in a high-resistance circuit. If a signal is present in the emitter circuit and a load resistance of 10 kΩ is added to the collector circuit, a high-voltage signal will appear across the load resistance. Since the currents in the two circuits are approximately equal, but the output voltage is much larger than the input voltage, power or voltage amplification has been achieved.

A similar explanation can be given for the operation of *p-n-p* transistors wherein the holes are considered to be the current carriers. Both types of devices *n-p-n* and *p-n-p*, are known as bipolar transistors since charge carriers of both polarities participate in their operation.

Transistors perform two basic roles in electronics: switching and amplification. In non-linear digital circuits the transistor typically functions as a high-speed electronic switch. The state of the switch, open or closed, depends upon the voltages applied across the two *p-n* junctions. In linear circuits, transistors operate as power amplifiers to increase the output current or voltage proportionally to the input.

The amplification obtained from a bipolar transistor depends upon how it is connected into the amplifier circuit. Figure 27-4 shows the three types of transistor amplifiers, the common base (described in preceding section), the common emitter, and the common collector. The impedance characteristics are important considerations when transferring power from one electrical module to another. In order to achieve the maximum power transfer, the impedances of the two modules must be matched (the output impedance of one must approximately equal the input impedance of the other). The common collector circuit is used primarily as an impedance matcher.

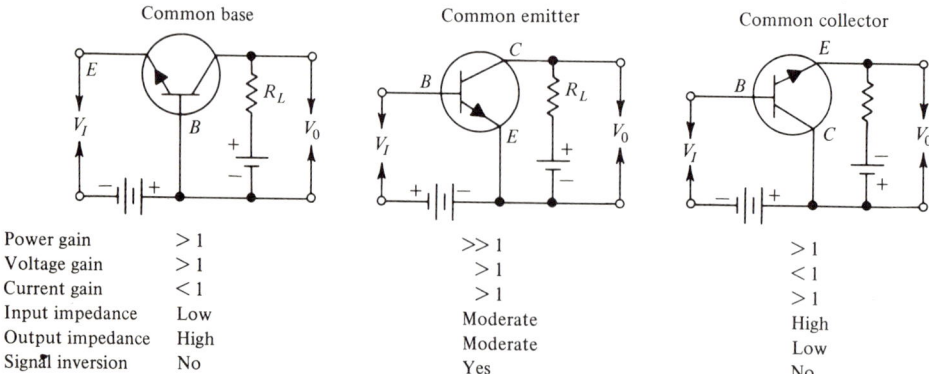

FIGURE 27-4 Transistor amplifier circuit characteristics.

The specific values of a transistor's operating parameters are determined by the geometry of the doped semiconductor regions. Large voltage or current gains are common. The most important dimension in a transistor is the thickness of the base. It must be as thin as possible for at least two reasons: (1) to allow virtually all the electrons injected by the emitter to cross over to the collector, and (2) to reduce the time necessary for an electron to move through the base. The latter function increases the switching speed of the transistor.

Field-Effect Transistors

The second basic type of transistor, the field-effect transistor (FET) was conceived in the early 1930s but was not fabricated in quantity until 30 years later. Although field-effect transistors employ n-type and p-type semiconductor materials, they differ markedly from bipolar devices in both operation and construction. The metal oxide semiconductor field-effect transistor (MOSFET) is the most commonly used type of FET in microelectronic circuits. The term MOS refers to the three materials used in the FETs constuction; metals, oxides, and semiconductors. A typical MOSFET consists of two island electrodes of n-type semiconductor material in a body of p-type material (Fig. 27-5). One island is known as the source and the other as the drain. On the surface of the body, between the source and the drain, a thin layer of silicon dioxide, an excellent insulator, is deposited. Finally a metal electrode is deposited over the oxide layer to form the third electrode known as the gate.

If a voltage is applied between the source and drain, no current will flow in the channel between the two n-islands since one of the n-p junctions will always be reverse biased. If the gate is positively charged with respect to the source, electrons will enter from the source onto the surface of the channel immediately under the gate. This region is inverted from a p-type region to an n-type as the electrons become the majority carriers. At this point there are no p-n junctions between the source and drain; consequently, a small current carried by the electrons in the channel region begins to flow. As the positive voltage applied to the gate increases, the conductive channel widens. The current flowing from source to drain increases as a function of the electric field generated by the applied gate voltage. This type of transistor is known as an enhancement-mode n-MOSFET.

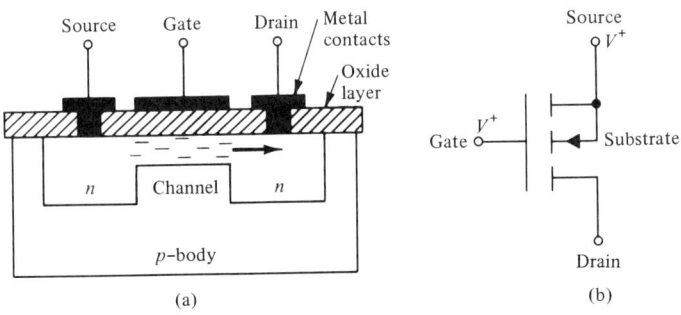

FIGURE 27-5 Enhancement-mode n-MOSFET. (a) n-Carriers conducting with positive gate voltage; (b) schematic.

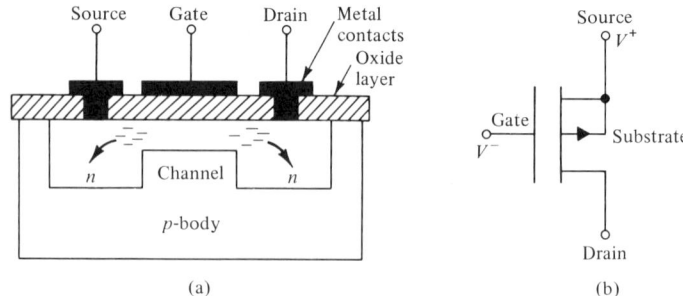

FIGURE 27-6 Depletion-mode *n*-MOSFET. (a) *n*-Carriers are driven out of conducting channel by positive gate voltage; (b) schematic.

The depletion-mode *n*-MOSFET (Fig. 27-6) differs slightly in design and operation from the enhancement mode *n*-MOSFET. The channel between *n*-type islands is constructed from *n*-type material. Since a conduction channel is already present, current flows from the source to the drain with no voltage applied to the gate. A negative voltage applied to the gate drives electrons from the channel and thus reduces the current flow.

In general terms, MOSFETs with conduction that can be turned on by applying voltage to the gate are called enhancement-mode devices, while those with current that can be turned off by application of gate voltage are known as depletion-mode devices. Altogether there are four types of MOS transistors: *n*-channel and *p*-channel types for both enhancement and depletion modes of operation.

Complimentary MOS (CMOS) devices include both *n*-MOS and *p*-MOS enhancement-type transistors on a single chip of silicon (Fig. 27-7). They offer the advantages of reduced power consumption and excellent noise immunity as compared with separate *n*- and *p*-MOS transistors.

Like the bipolar transistor, the MOSFET is capable of both switching and amplification. The MOSFET gain is usually measured in terms of a voltage ratio while the bipolar gain is given as a current ratio. It should also be noted that whereas both electrons and holes participate in the base current of a bipolar transistor, only one kind of carrier is present in the channel of a MOSFET.

Important dimensions in the fabrication of MOS transistors are the thickness of the oxide layer under the gate electrode and the distance between the source and drain. The sensitivity

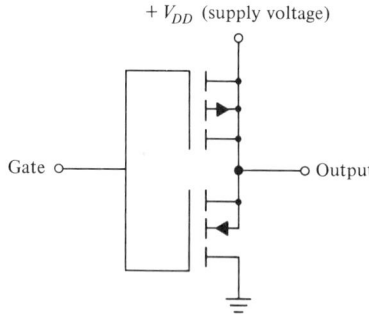

FIGURE 27-7 Schematic of CMOS amplifier.

of the transistor's response to gate voltage varies inversely as the thickness of the oxide layer. The difference in fabrication methods allows roughly four times as many MOS transistors to be fitted on a given area as bipolar ones. However, the bipolar transistors retain the major advantage of higher speed of response over all types of MOS transistors.

Integrated Circuits

Transistors may be employed as discrete units or as components of microelectronic circuits. The advent of microelectronics has not affected the functions of the basic components; namely, transistors, resistors, capacitors, and so on. The major difference is that all these components are available as an electrical functional unit fabricated on a single, small IC chip. Many problems of circuit design are taken care of within the IC, thus simplifying the design, operation, and maintenance of instrumentation.

An IC may be classified by method of construction and mode of operation. The two most important modes of fabrication of ICs are monolithic and thin film. Monolithic ICs have all circuit components and their interconnections on a single thin wafer of semiconductor material called a substrate. In the less commonly used thin-film ICs, circuit components are deposited in thin layers on the substrate. Transistors are not usually found as components of thin-film ICs. The hybrid IC incorporates features of both monolithic and thin-film construction.

ICs may function in a linear or nonlinear manner. The output of a linear IC is directly proportional to the input. Linear IC applications include many types of amplification, modulation, and voltage regulation. The operational amplifier is the single most important type of linear IC. Nonlinear ICs include all digital ICs and other circuits where there is not a linear relationship between the input and output signals. Digital ICs, the most important type of nonlinear ICs, usually employ some form of bistable (on/off) operation. These ICs are common in computer circuits and in other digital applications such as counters, calculators, and digital data communication equipment.

ICs must be placed in a protective housing and be provided connections to the outside world There are three methods of packaging ICs in containers (Fig. 27-8): the TO-5 glass

FIGURE 27-8 Types of IC packages. (Courtesy of Texas Instruments.)

metal can, the ceramic flat pack, and the dual-in-line ceramic or plastic flat packs known as dual-in-line packages (DIPs). The popular, less expensive plastic DIP packages can have 14, 16, 18, 24, or 40 connecting pins. A minimum of two pins is always required for connecting the IC to the power supply. The remaining connections are available for use as connecting terminals for input and output signals.

27.3 OPERATIONAL AMPLIFIERS

The first operational amplifier (op amp) circuits were composed of vacuum-tube amplifiers and other necessary elements such as diodes, resistors, and capacitors. Discrete transistors later replaced tubes and finally entire operational amplifier circuits became available on IC chips. About one-third of all ICs are op amps: over 2000 types are commercially available. These facts attest to the importance and diversity of op amp applications.

In addition to the obvious disparity in size, there are several other important differences between an IC op amp and earlier versions. An IC op amp consumes little power and operates at relatively low voltages (approximately ±20 V) with power supplies which need not be highly regulated. The IC op amps have very low drift with both temperature and time and can withstand short circuits on the output without damage. The major disadvantages of the IC op amps have been the limited operational voltage range (approximately ±20 V), the relatively low output current (< 10 mA), and low input impedance (< 1 MΩ). New generations of IC op amps have overcome the problems of low output current and low impedance. For high-voltage outputs, discrete transistors must still be employed.

Operational Amplifier Characteristics

The desirable characteristics of op amps are derived from two properties inherent in these devices: high-gain dc amplification and provision for external feedback (returning a fraction of the output signal to the input). This combination of high gain and feedback allows the output to be made independent of the internal parameters of the op amps. Thus, the output signal can be made to depend primarily on the components in the external feedback circuit. In addition, properties of high input impedance, low output impedance, and good response to high-frequency input signals are required for op amps to function properly.

Basic Operational Amplifier Circuits

Schematically the op amp is represented by a triangle with inputs and outputs (Fig. 27-9a). Although the noninverted input is most often connected to ground (Fig. 27-9b), it may be connected to an active signal input for certain applications. The three terminal representation of Fig. 27-9b is often abbreviated to the two terminal version (Fig. 27-9c); just the inverted input, V_i, and output, V_o, are shown. The gain ($G = V_o/V_i$) measured in this configuration is known as open-loop gain.

When the op amp is used for instrumental applications of signal transformation, it is normally employed in the closed-loop configuration (Fig. 27-10) where the impedances

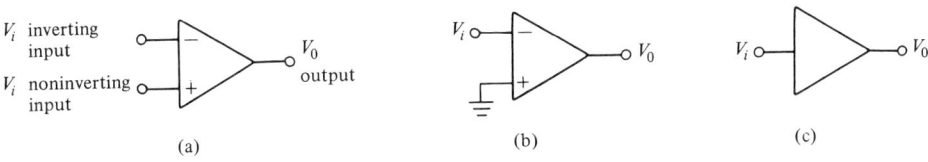

FIGURE 27-9 Operational amplifier representations.

Z_1 and Z_2 generally consist of resistors and/or capacitors external to the op amp. Since the input impedance of the op amp itself is large, i_a is practically zero and can be neglected. Thus $i_1 = i_2$.

Furthermore, normal operation of the operational amplifier requires that the two inputs be at essentially the same voltage, $V_d = (V_+ - V_-) \simeq 0$. The allowable difference at the inputs will then depend upon the gain, G, and the limiting output voltage V_o. With a differential input, $V_o = G(V_+ - V_-)$, having a minimum of approximately 10 V and the gain being very large, the input difference, $(V_+ - V_-)$ must be close to zero.

Under these conditions (Fig. 27-10)

$$i_1 = \frac{1}{Z_1}(V_i - V_d) = \frac{V_i}{Z_1} = i_2 = \frac{1}{Z_2}(V_d - V_o) = \frac{-V_o}{Z_2} \qquad (27\text{-}1)$$

and therefore

$$V_o = -\frac{Z_2}{Z_1} V_i \qquad (27\text{-}2)$$

The op amp is the basis for many signal modifying circuits. Table 27-4 illustrates some applications indicating how the composition of Z_1 and Z_2 determines the relationship between the input and output voltage signals. These circuits were initially developed as components for analog computers, but have since found extensive use in instrumentation.

Important Operational Amplifier Parameters

Regardless of the function of an op amp, its performance and reliability are determined by a number of parameters that appear as specifications for a particular device. The more important of these are:

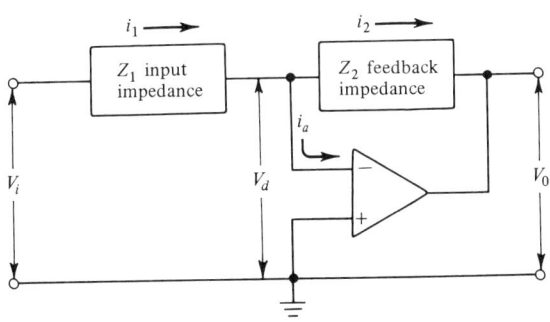

FIGURE 27-10 Operational amplifier with external components.

Input offset voltage: the voltage that must be applied across the input terminals to drive the output voltage to zero.

Input offset current: the current at the inputs necessary to zero the output voltage (this current usually taken as the difference in the currents at the two inputs).

Input bias current: average of the two currents required to obtain a zero output voltage.

Slew rate: the rate that the output voltage changes to the maximum saturated value (V/μsec).

Drift: the gradual change in offset voltage, bias current, and output voltage as a function of time and temperature.

TABLE 27-4   Basic Operational Amplifier Applications

Function	Impedance Z_1	Impedance Z_2	Relationship of V_o to V_i
Summation	V_1 — R_1; V_2 — R_2; V_3 — R_3	R_f	$V_o = -R_f \left(\dfrac{V_1}{R_1} + \dfrac{V_2}{R_2} + \dfrac{V_3}{R_3} \right)$
Integration	R_1	C	$V_o = \dfrac{-1}{RC} \int V_1 \, dt = \dfrac{-1}{C} \int i_{in} \, dt$
Differentiation	C_1	C_2 in parallel with R_f	$V_o = -R_1 C_1 \dfrac{dV_{in}}{dt}$ for $C_2 \ll C_1$
Voltage amplification	R_1	R_f	$V_o = -\dfrac{R_f}{R_1} V_{in}$
Current-to-voltage converter	None	R_f	$V_o = -R_f I_{in}$

Z_1 and Z_2 refer to Fig. 27.9. V_o is output voltage. V_i is input voltage. t is time.

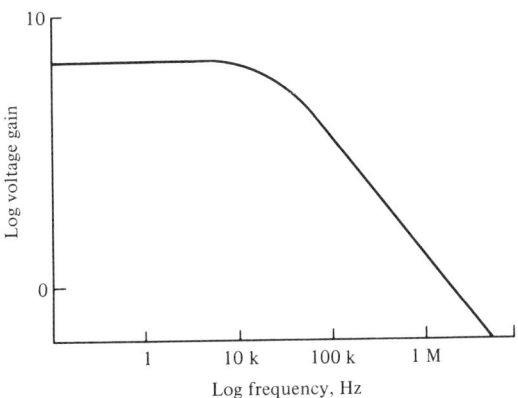

FIGURE 27-11 Bode plot of operational amplifier response.

Unity gain frequency bandwidth: frequency response of the operational amplifier given as a Bode plot (Fig. 27-11) of open-loop voltage gain versus frequency. It is an important parameter when ac signals are involved.

27.4 DIGITAL INTEGRATED CIRCUITS

Digital circuits were the first type of integrated circuit to be produced. They currently comprise the majority of nonlinear ICs and have found major applications in instrumentation for control and data processing. Logic functions generated by ICs are used to control instrument operation. For example, any one of three conditions may terminate a titration: depressing the off button, reaching the equivalence point on the titration curve, or emptying the buret. An IC containing the appropriate circuits can be used to implement this logic in the laboratory. Initially data processing performed by internal instrumental ICs was limited to relatively simple functions, such as counting and preparation of data for digital displays and computer interfaces. More complex data processing was done by computers external to the instrument.

Both the control and data processing functions have been expanded by designing instruments containing microprocessor ICs. The increasing number of "smart" instruments indicates the operational advantages and economic practicality of replacing conventional circuits and mechanisms with microprocessors. Chapters 29 and 30 cover microprocessor applications in more detail.

Digital ICs, like ordinary light switches, are binary devices existing in only two states—off or on. Their inputs and outputs can therefore have one of two voltage levels. Logic low (off) is about at ground, while logic high (on) is a few volts positive. The term "1" usually refers to logic high and the term "0" refers to logic low.

Logic Gates and Expressions

An understanding of digital circuits is possible using Boolean (logical) algebra. This algebra is used with digital circuitry in much the same manner as ordinary algebra is

TABLE 27-5   Basic Logic Functions

Function	Symbol and Boolean Expression	Truth Tables
AND	$A, B \rightarrow T$ $T = A \cdot B$	$\begin{array}{ccc} A & B & T \\ 0 & 0 & 0 \\ 1 & 0 & 0 \\ 0 & 1 & 0 \\ 1 & 1 & 1 \end{array}$
OR	$A, B \rightarrow T$ $T = A + B$	$\begin{array}{ccc} A & B & T \\ 0 & 0 & 0 \\ 1 & 0 & 1 \\ 0 & 1 & 1 \\ 1 & 1 & 1 \end{array}$
NOT	$A \rightarrow T$ $T = \overline{A}$	$\begin{array}{cc} A & T \\ 1 & 0 \\ 0 & 1 \end{array}$

employed in the analysis of analog circuits. A major difference is that variables in Boolean expressions can assume only one of two possible values.

There are three basic logic functions: AND, OR, and NOT. Each of these is represented by a logic symbol and its operation can be summarized by a truth table. Truth tables give output values as functions of various combinations of input values. The logic symbols and truth tables are given for the three basic functions in Table 27-5.

The symbols ⟆ and ⟆ represent the AND and OR functions, respectively, with inputs at A, B, and output at T. A dot, symbol $\cdot$, between two input values means they are ANDed together; a plus, symbol +, between values means the values are ORed with each other. The NOT function, a simple logic inversion, is symbolized by a bar over the negated expression, $\overline{A}$. The inversion symbol (Table 27-5) consists of two components, a triangle representing a current amplifier (a driver) and a circle at the output indicating the logic inversion (1 to 0 or 0 to 1) of the normal output.

Two of the most commonly used IC gates, the NAND (NOT AND) and the NOR (NOT OR) are shown in Table 27-6. When the input logic conditions are satisfied, the unique output is low instead of high. These devices are therefore said to have active level low outputs. The fact that more circuits are designed with NAND and NOR gates than with AND and OR gates can be explained by looking at the design of real ICs. Logic operations are performed first and then amplification returns logic levels to their specified voltage values. Since this amplification always results in logic inversion, real IC gate outputs are active level low unless additional circuit components are used to perform another inversion.

TABLE 27-6 NAND and NOR Logic Gates

Function	Symbol	Truth Table
NAND	$T = \overline{A \cdot B}$	A B T 0 0 1 1 0 1 0 1 1 1 1 0
NOR	$T = \overline{A + B}$	A B T 0 0 1 1 0 0 0 1 0 1 1 0

Logic Identities

A number of the important rules for Boolean logic combinations are given below:

1. Double inversion $(\overline{\overline{A}}) = A$
2. Distribution $A \cdot (B + C) = A \cdot B + A \cdot C$
3. Association $A + (B + C) = (A + B) + C$
4. DeMorgan's theorems $\quad (\overline{A + B}) = \overline{A} \cdot \overline{B}$
$\overline{A \cdot B} = \overline{A} + \overline{B}$

The first three rules are similar to those of ordinary algebra. DeMorgan's theorems are important as they show a useful relationship between AND and OR functions. These theorems are used in circuit design to simplify logic networks (Fig. 27-12). Here logic identities prove that the given configuration of three NOR gates can be reduced to one AND gate.

An important feature of Boolean logic networks is that any logic function may be constructed using only NOR gates or only NAND gates. The ability to assemble logic networks containing hundreds of identical gates greatly simplifies design, production, and maintenance. MSI and LSI circuits have greatly reduced the use of individual small-scale NAND and NOR gate circuits in instrumental design.

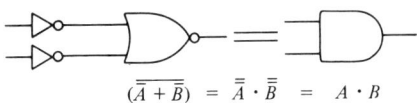

$(\overline{\overline{A} + \overline{B}}) = \overline{\overline{A}} \cdot \overline{\overline{B}} = A \cdot B$

FIGURE 27-12 Use of a deMorgan theorem to simplify a logic network.

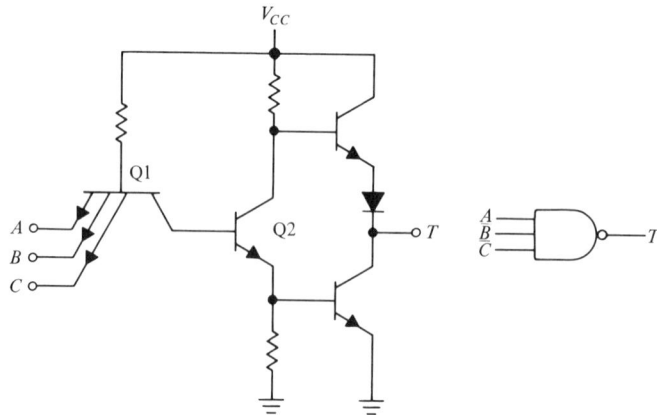

FIGURE 27-13 Three-input TTL NAND gate.

Logic Families

Gates (not to be confused with the gate electrode of an MOS transistor) are the fundamental units of electronic logic circuits. At the heart of every gate circuit is at least one active element which acts as a switch to open and close the output depending upon the input conditions. A number of logic families have emerged as new developments occurred in electronic circuit technology. The first logic families, transistor-resistor logic (TRL) and diode-transistor logic (DTL), were designed with discrete components which included transistors, resistors, and diodes. A typical TRL gate consisted of one transistor and five resistors, while the DTL equivalent contained a transistor, three resistors, and five diodes. The original ICs used resistor-transistor logic (RTL), employing more transistors and fewer resistors than earlier TRL circuits. This logic family represented an improvement in reduced power consumption and higher switching speed over its predecessors.

The next logic family to evolve remains the most common form of bipolar logic. It is known as transistor-transistor (TTL) logic. The control element in TTL gates is the multiple-emitter transistor which has no discrete transistor equivalent. This transistor, Q1 in Fig. 27-13, controls a single switching transistor, Q2, which in turn drives a network of amplifying transistors. TTL circuits provide greater output power, have wider manufacturing tolerances and greater noise immunity than RTL circuits. Integrated circuits containing hundreds of TTL gates are now common. Since all TTL ICs use the same power supply voltage, there are few problems in constructing circuits using TTL chips as components.

The majority of available and useful TTL IC devices employ a 7400 or a 5400 numbering system. The 7400 devices are lower in cost with a useful operating temperature range of 0° to 70°C. The 5400 devices are more expensive, meeting military operating temperature specifications of −55 to +125°C. Most important non-7400 TTL devices now have 7400 number equivalents. Figure 27-14a illustrates a 7402 IC containing four, dual input, NOR gates in a 14-pin dual in-line package (DIP), while Fig. 27-14b shows a 7420 with two 4-input NOR gates. These chips contain a small number of logic gates

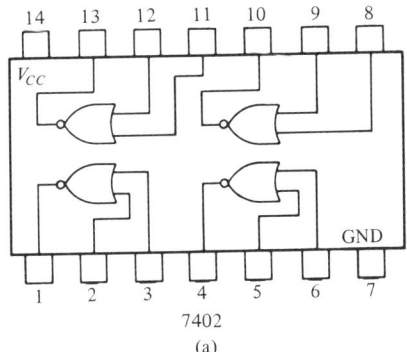 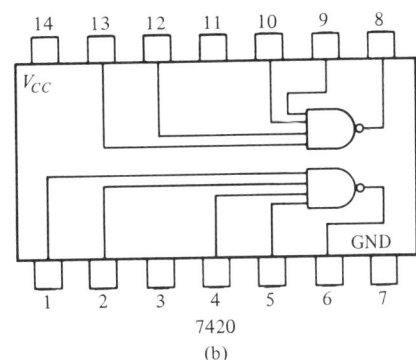

FIGURE 27-14 7400 Series TTL NOR gate ICs. (a) Quad 2-input NOR gates; (b) two 4-input NOR gates. (Courtesy of Signetics Corp.)

with inputs and outputs of individual gates available to the outside world. They represent small-scale integration (SSI).

There are a number of TTL subfamilies such as low power, high power, and Schottky, that interchange speed, power, and additional complexity for special uses. As a general rule, increasing the power consumption of an IC increases its switching speed and vice versa.

The implementation of logic functions using MOS circuits has evolved in a less complicated manner. The first MOSFET gates involved p-channel devices since they were the easiest to fabricate. They are also the slowest since holes have a lower mobility than electrons. The n-channel MOS circuits have largely replaced p-MOS devices in instrumentation where high performance is required. Circuit configuration using n-MOS transistors as the switching components may be classified according to the type of auxiliary load element used in the circuit. Figure 27-15a shows a dual enhancement-mode n-MOS transistor logic switch (Q1 and Q2) with a conventional resistor as the loading element.

The popular CMOS version of a NAND gate (Fig. 27-15b) uses the same switching elements, Q1 and Q2, with enhancement-mode p-MOS transistors, Q3 and Q4, as loading elements. This CMOS circuit consumes less power than the n-MOS circuit. CMOS devices are available in the same packages as TTL, with DIP types being the most popular. The 74C CMOS series of ICs are pin per pin compatible with the 7400 TTL series of ICs.

While the gap between high packing densities of CMOS devices and the fast speeds of bipolar circuits is being narrowed by continuing advances in materials science and fabrication processes, it is nevertheless substantial. Several new technologies are attempting to fill this gap. One is integrated injection logic (IIL), which compresses a complete logic circuit composed of two bipolar transistors into a single small unit. As a result IIL circuits have higher packing densities than normal TTL circuits and are faster than MOS circuits, approaching TTL speeds.

Flip-Flops

Gates are one of the two fundamental building blocks used in the design of logic circuits. Flip-flops, or bistable multivibrators, comprise the other type of basic circuit component.

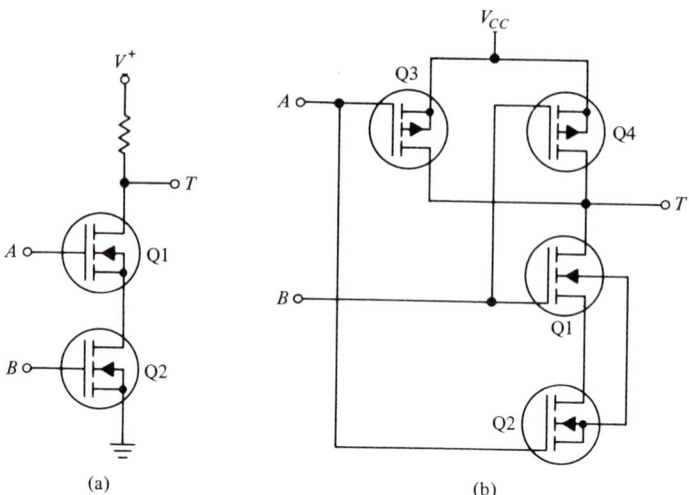

FIGURE 27-15 MOS NAND gates. (a) Dual-input MOS NAND gate; (b) dual-input CMOS NAND gate.

Circuits which perform the functions of counting, storing, and manipulating digital data are composed primarily of flip-flops.

The simplest bistable multivibrator is the reset–set (RS) flip-flop shown in Fig. 27-16. The truth table for this flip-flop indicates that a low (0) value at the S input sets the Q output to 1, and a zero value at the R input clears the Q output to 0. Simultaneous highs (1) on both inputs leave Q unchanged, while simultaneous lows (0) at the inputs give an ambiguous result. The device functions as a simple memory, remembering which input was most recently at logic state zero.

The clocked or gate RS flip-flop, Fig. 27-17, can respond to input logic signals only when the T (clock) input is high. The truth table indicates the logic of this flip-flop is reversed compared to the simple RS flip-flop. Figure 27-18a shows a simplified representation of the gated RS flip-flop.

The D flip-flop (Fig. 27-18c) is a useful modification of the gated RS flip-flop. This device, known as a data latch, stores the logic signal at the input during the last time interval that the clock input was high (1). The Q output cannot change state when the T input is low (0).

The JK flip-flop (Figure 27-18b) represents the ultimate in simple flip-flop devices. A clock pulse at T is necessary for the device to respond to the inputs at J and K. If J and K are both at logic level 1, the complementary outputs Q and $\overline{Q}$ will change state with

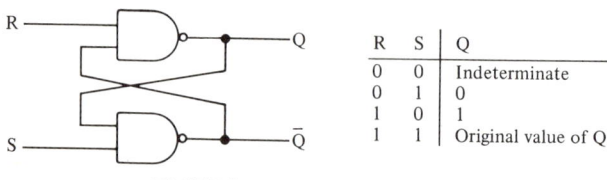

FIGURE 27-16 RS flip-flop.

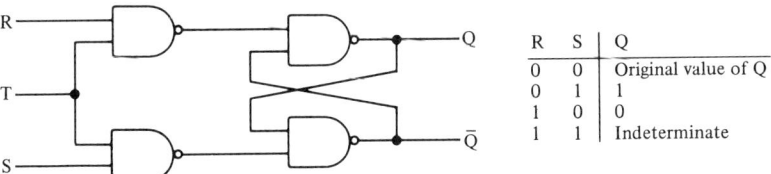

FIGURE 27-17 Gated RS flip-flop.

each successive clock pulse. A logic low at the J input prevents the Q output from going high while a logic low at the K input prevents the Q output from going low. The set and clear inputs override *all* other functions of the flip-flop. A zero at the clear input, clears Q low while a zero at the set input sets Q high.

Connecting Digital IC Packages

The output of one TTL or MOS IC, may be connected to the inputs of one or more IC packages. Connecting ICs from different logic families or subfamilies often requires that components be inserted between IC outputs and inputs in order to achieve signal level compatibility. The current driving capability of a digital IC is called its fan-out while its input requirements are referred to as fan-in. The average TTL gate output can drive the inputs of ten other gates, that is, the fan-out is ten. Most TTL gate inputs have a fan-in of one. Two special-purpose ICs are often used to increase the current driving capability above normal: buffers that have outputs with expanded fan-out and drivers that provide increased current levels.

A final widely used technique of connecting ICs is that of tri-stating. Tri-state devices have a third input state in addition to the usual 1 and 0 logic levels: This third state is equivalent to an open circuit or disconnect. It is controlled by the logic level applied to the output enable control (Fig. 27-19a). A logic 1 on the enable input will allow the results of the NAND operation to appear at the output of the gate, while a logic 0 at the enable input will result in an open circuit at the gate output. Tri-stating does not affect the normal function of the gate or the signal levels appearing at the inputs and outputs.

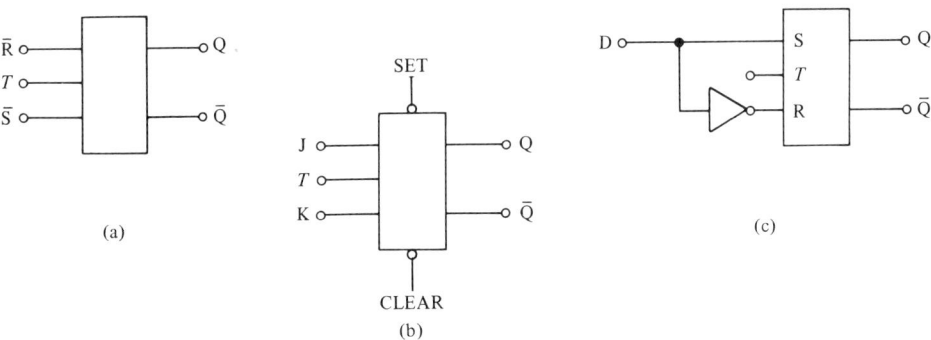

FIGURE 27-18 Flip-flop representations. (a) Gated RS flip-flop; (b) JK flip-flop; (c) D flip-flop.

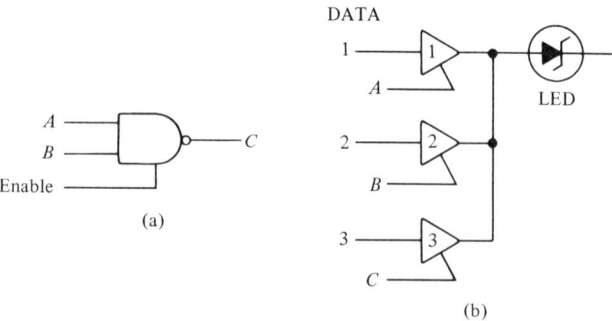

FIGURE 27-19 Tri-state devices. (a) Tri-state 2-input NAND gate; (b) application of tri-state drivers.

The utility of tri-state devices is illustrated in Fig. 27-19b. The status (logic state) of each driver input could be determined by sequentially connecting each driver output to the light emitting diode (LED) display using A, B, and C tri-state output enable controls. Tri-stating is widely employed for transmitting digital data from different sources over shared lines to a centralized signal modifier.

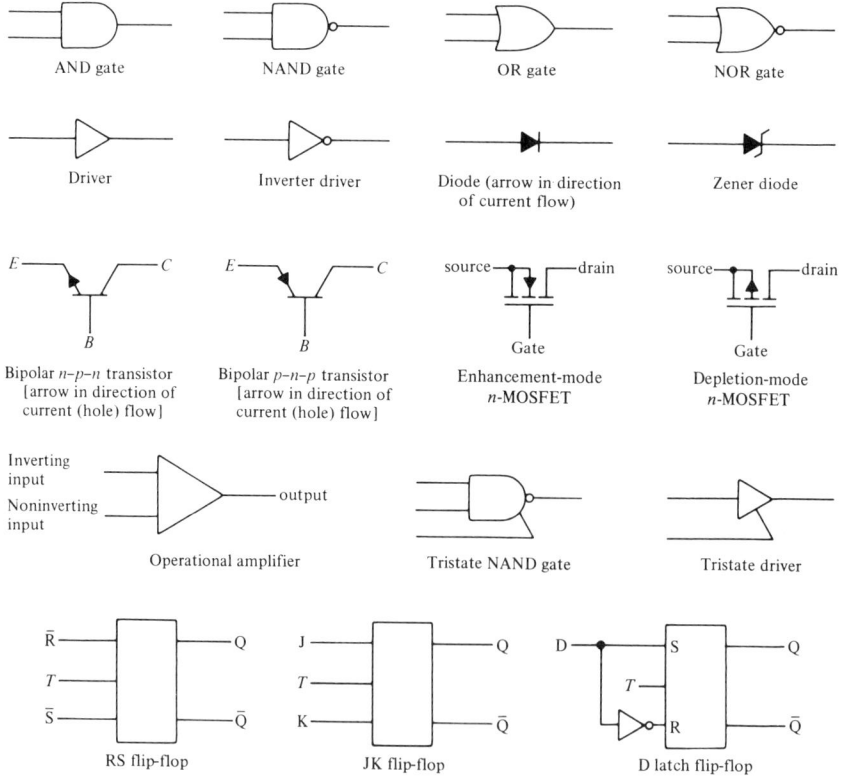

Notation for common solid-state components.

BIBLIOGRAPHY

Brophy, J. J., *Basic Electronics for Scientists,* 3rd ed., McGraw-Hill, New York, 1977.

Diefenderfer, A. J., *Principles of Electronic Instrumentation,* 2nd ed., Saunders, Philadelphia, 1979.

Lancaster, D., *MOS Cookbook,* Howard Sams, Indianapolis, 1976.

Lancaster, D., *TTL Cookbook,* Howard Sams, Indianapolis, 1974.

Lion, K. S., *Elements of Electrical and Electronic Instrumentation,* McGraw-Hill, New York, 1975.

Malmstadt, H. V., C. G. Enke, and S. R. Crouch, *Instrumentation for Scientists,* Benjamin, Reading, Mass., 1973.

Meindl, J. D., "Microelectronic Circuit Elements," *Sci. Am.,* **237** (3), 70 (Sept. 1977).

Noyce, R. N., "Microelectronics," *Sci. Am.,* **237** (3), 62 (Sept. 1977).

Weber, L. J. and D. I. McLean, *Electrical Measurement Systems for Biological and Physical Scientists,* Addison-Wesley, Reading, Mass., 1975.

CHAPTER 28

Electronics: Commonly Used Signal Modifying Circuits

This chapter introduces a number of circuits that are commonly used to modify signals from input transducers. No attempt has been made to treat auxiliary components such as power supplies.

28.1 DEVELOPMENT OF INTEGRATED CIRCUITS

Since the production of the first planar transistor in 1959, the number of elements in advanced integrated circuits has doubled each year. Circuits containing over 2^{18} (262,144) elements on a single chip are now available. This increase in packing density, still far from the limits imposed by the laws of physics, is predicted to continue for IC technology. Table 28-1 presents the chronological development of IC technology, classifies ICs according to component packing density, and indicates typical applications for each kind of IC.

The cost of executing a given electronic operation has declined as the space necessary to perform the operation has decreased. A NAND gate costing approximately ten dollars in 1961 is currently priced at less than ten cents. Op amp prices have likewise decreased by a corresponding factor during the same time period.

There are a number of reasons for this price decrease. Mass production of a complex circuit contained in a single high-density IC is less expensive than production of an equivalent circuit from a number of interconnected, lower-density IC packages. The smaller number of external interconnections reduces labor and materials costs. Interconnections of the IC within the package are more reliable than solder or connectors outside the package, thus reducing maintenance costs. Fewer external interconnections also mean that less intermediate testing is necessary during production.

TABLE 28-1 Increasing Component Packing Densities

Data	Components per IC	Circuit-Type	Application
1960	1	Discrete elements	Transistors
1962	10–64	Small-scale integration (SSI)	Gates, flip-flops, op amps
1964	64–1024	Medium-scale integration (MSI)	Counters, voltage-to-frequency converters Multiplexers
1969	1024–100,000	Large-scale integration (LSI)	Microprocessors Communications circuits
1974	100,000	Very large-scale integration (VLSI)	Microcomputers

Reducing the number of IC packages contained in an instrument results in less power consumption. Savings are therefore possible in power transformers, cooling fans, support racks, and cabinets. Instruments incorporating high-density MSI and LSI packages in their design are smaller, have less rigorous power requirements, and require less control of operating environments than instruments which do not contain high-density ICs.

28.2 DIGITAL MSI CIRCUITS

A large number of digital data transformations have been achieved using medium-scale integration (MSI). Two important groups of digital MSI circuits are those composed solely of logic gates and those containing flip-flops. Multiplexers/demultiplexers and decoders are included in the former group, while counters and registers belong to the latter.

Multiplexers/Demultiplexers and Decoders

Multiplexers are often referred to as selectors because they function in a manner similar to mechanical selector switches. A 4-input multiplexer "selects" one of the four inputs to appear at the output (Fig. 28-1). Data on the multiplexer input lines (D1–D2 in Fig. 28-1b) may be switched individually to the output using digital control switches S1 through S4. Signals on the input data lines may be either digital or analog depending upon the design of the multiplexer chip. Addition of gating logic within the chip allows the number of input lines to be increased and the number of control lines to be reduced. Figure 28-2a gives the block diagram of an 8-input multiplexer with three control lines. A 3-bit binary code (Fig. 28-2b) at inputs S1, S2, and S3 determines which data input signal will appear at the output.

Two time-dependent factors must be considered when multiplexing. If the switching frequency of the multiplexer is too high, some input data will not be transferred to the

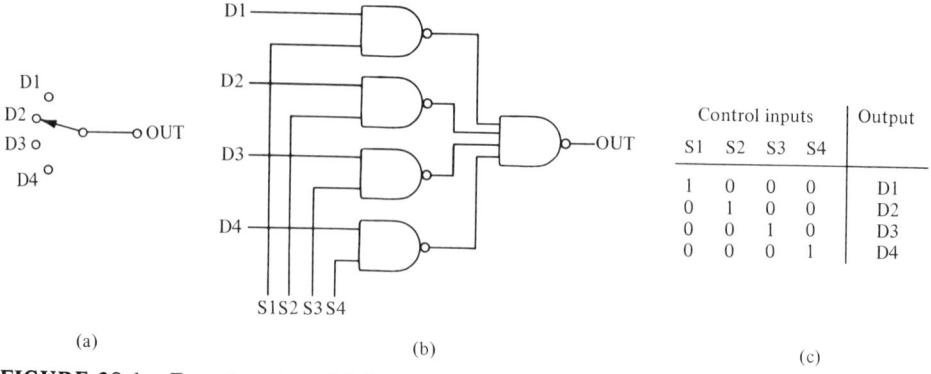

FIGURE 28-1 Four-input multiplexer. (a) Mechanical switch; (b) NAND gate, (c) truth table for NAND multiplexer.

output. According to the Nyquist sampling theorem, an input signal must be sampled at a rate twice that of the highest-frequency component of the signal. For example, if the maximum frequency component of an input signal is 200 Hz, the switching frequency for this channel (input line) should be 400 Hz (channel sampled every 2.5 msec) in order to prevent any loss of data.

The other time-dependent multiplexing consideration is the settling time. Since switching is not an instantaneous process, a finite amount of time is required for the multiplexer's output to reach the value of the input once the channel has been selected. Semiconductor switching devices have shorter settling times and thus higher switching frequencies than electromechanical (reed relays) and mechanical switches. In general, the switching times of TTL family ICs are faster than those of either MOS or CMOS ICs (Chapter 27). Another major limitation of semiconductor switching devices is the requirement of two power supplies, usually +15 and -15 V. Signal inputs and outputs may not exceed these limits without damaging the device.

The opposite of a multiplexer is a demultiplexer or decoder. A binary decoder produces a unique logic level output for each combination of binary inputs (Fig. 28-3). It is a

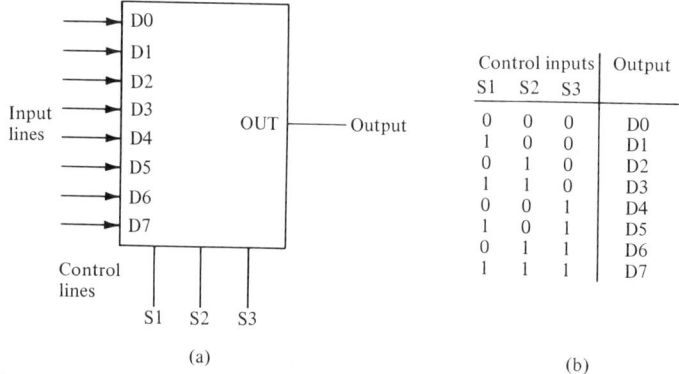

FIGURE 28-2 Eight-input multiplexer. (a) Block diagram; (b) truth table.

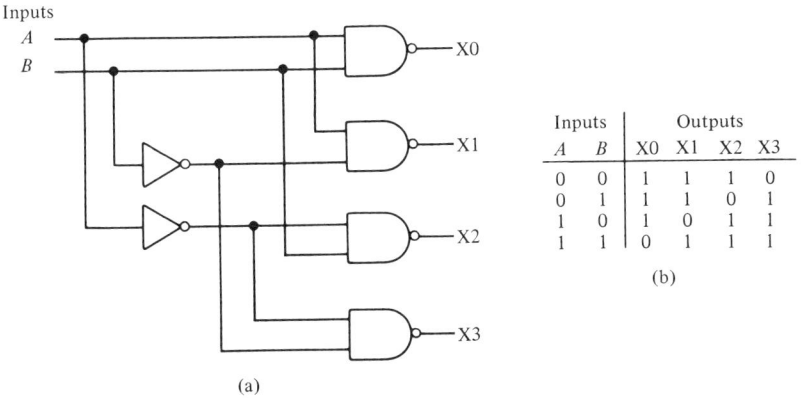

FIGURE 28-3 Two-bit binary decoder. (a) Logic circuit; (b) truth table.

common practice in digital IC circuit design to make the unique ouput a logic level zero. A demultiplexer may be viewed as a multiplexer with its input and output functions reversed (Fig. 28-4). Multiplexers/demultiplexers and decoders are used to implement logic functions in control devices and to route data signals through instruments and communication equipment.

Counters and Registers

Flip-flops (Chapter 27) are grouped together functionally as either registers or counters. A register can store data on the outputs of its component flip-flops. It is a simple device consisting of a group of interconnected flip-flops. Counters can perform more complex functions than registers, for they contain both flip-flops and logic gates. Applications for counters include controlling sequences of operations and recording the number of events as a function of time. The output of a counter's flip-flops at any given time is called the state of the counter. The sequence of states can be simple, such as recording a series of consecutive pulses, or more complex, as for example, counting up or down from a preset initial value.

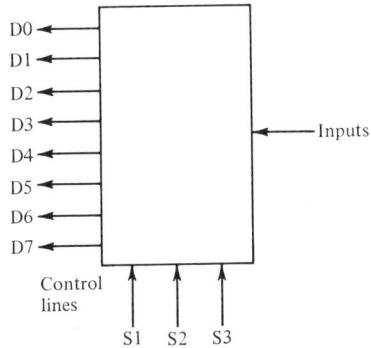

FIGURE 28-4 Eight-output demultiplexer.

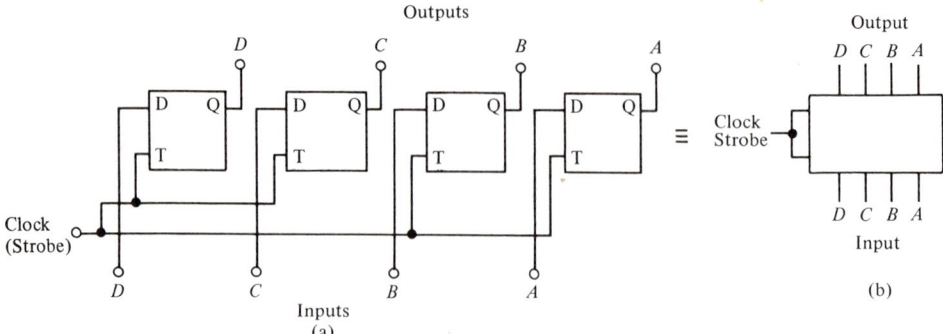

FIGURE 28-5 (a) 7475 Quad D-Latch. (b) Schematic.

Two common types of registers are the storage register (Fig. 28-5) and the shift register (Fig. 28-6). A logic level one on the strobe (clock) input of the register (Fig. 28-5) causes the Q output of each flip-flop to assume the logic level of its respective input. When the strobe line is cleared to zero, the Q output values are held (latched) in the state existing immediately prior to the one-to-zero level strobe transition. As long as the strobe input remains low, the outputs may not change with the inputs. The operation of this register may be described as parallel data in/parallel data out, since all the flip-flops change state simultaneously, that is, synchronous action.

The 4-bit shift register (Fig. 28-6) allows a sequence of digital pulses on a single input line (serial data) to be rippled through the register. The data are moved sequentially right to left from one flip-flop to the next by a succession of clock pulses. Data cannot be shifted within the register without a clock pulse. After four clock pulses, the original input bit is on the D output. The next clock pulse pushes this data bit out of the register. The parallel outputs indicate the state of the flip-flops after each clock pulse. This register is employed as a buffer to delay and store serial data and as a serial-to-parallel data converter. Eight-, 12-, and 16-bit shift registers are commonly used in instrumental circuits.

A simple binary counter can be constructed by connecting the output of one flip-flop to the input of the next (Fig. 28-7). The first flip-flop will toggle (change its Q output logic level) with every logic level one input. The Q output of the second flip-flop will toggle on every other input pulse, the third on every fourth pulse, and so on. A binary count at the Q outputs (A through D) results from the "toggling" traveling through the

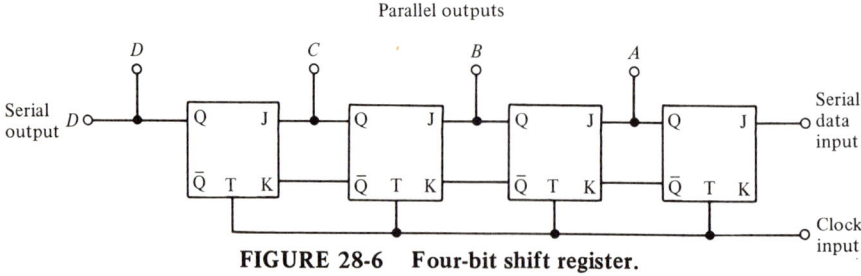

FIGURE 28-6 Four-bit shift register.

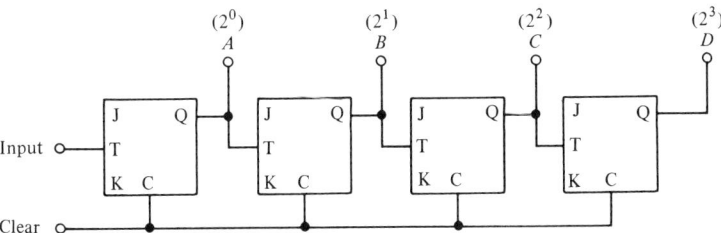

FIGURE 28-7   Simple modulo-16 counter.

series of flip-flops. This method of rippling pulses through a series of sequential flip-flops is known as asynchronous counting.

All the counter outputs can be reset to zero by one of two methods. A logic level low (zero) on the clear line (Fig. 28-7) will simultaneously clear all the outputs of the JK flip-flops to zero. The outputs are also cleared to zero after every 16 consecutive logic one input pulses have been registered by the counter. Thus, this particular counter is said to have a modulus of 16.

The counter rate of a ripple counter is limited by its propagation delay time. This time is the sum of the delay times of the individual flip-flops comprising the counter. Delay times of individual flip-flops, that is, time between a change in clock input and the resultant change in Q output, are typically 80 nsec per flip-flop. Thus the total propagation delay of a modulus 16 counter containing four flip-flops would be 0.32 μsec which limits the counting rates to 3.0 MHz.

The propagation delay is reduced by the synchronous binary counter which allows all four flip-flops to change state simultaneously. The input pulse is applied simultaneously to the clock inputs of all the component flip-flops. Logic circuits constructed from flip-flops and auxiliary gates prevent transitions at Q outputs until the appropriate count is reached. The maximum propagation time for the synchronous counter is therefore the delay time of a single flip-flop. Thus the maximum counting rate of a synchronous modulo-16 counter is four times that of the equivalent asynchronous counter.

Binary coded decimal (BCD) counters are used in instruments where the convenience of readout in decimal data is desired. Four interconnected flip-flops (Fig. 28-8) form one stage of an asynchronous BCD counter. This flip-flop combination represents one digit of a decimal number. Table 28-2 shows the state of each flip-flop Q output as a function of

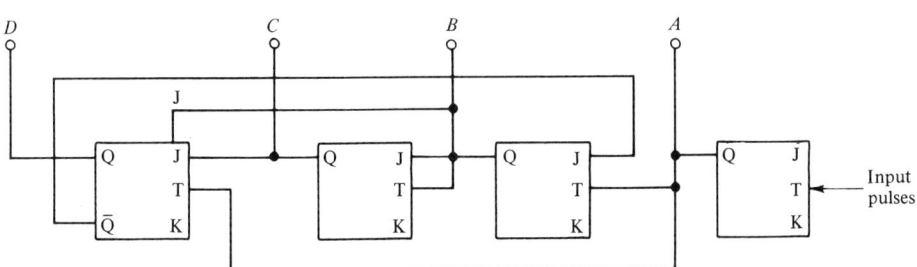

FIGURE 28-8   Asynchronous BCD counter.

TABLE 28-2 BCD Counter Outputs as a Function of Count Number

Input Count	BCD COUNTER OUTPUTS			
	D	C	B	A
0	0	0	0	0
1	0	0	0	1
2	0	0	1	0
3	0	0	1	1
4	0	1	0	0
5	0	1	0	1
6	0	1	1	0
7	0	1	1	1
8	1	0	0	0
9	1	0	0	1

the input count. Each digit requires one stage of four flip-flops—thus four stages could represent any number from 0 to 9999. A commonly used IC containing a single stage is the 7490 decade counter.

Counters and latches may be combined to perform the function of obtaining the total count at a given time without interrupting the counting process, for example, a digital wristwatch. Two ICs (Fig. 28-9), a 7490 decade (BCD) counter, and a 7475 quad D latch,

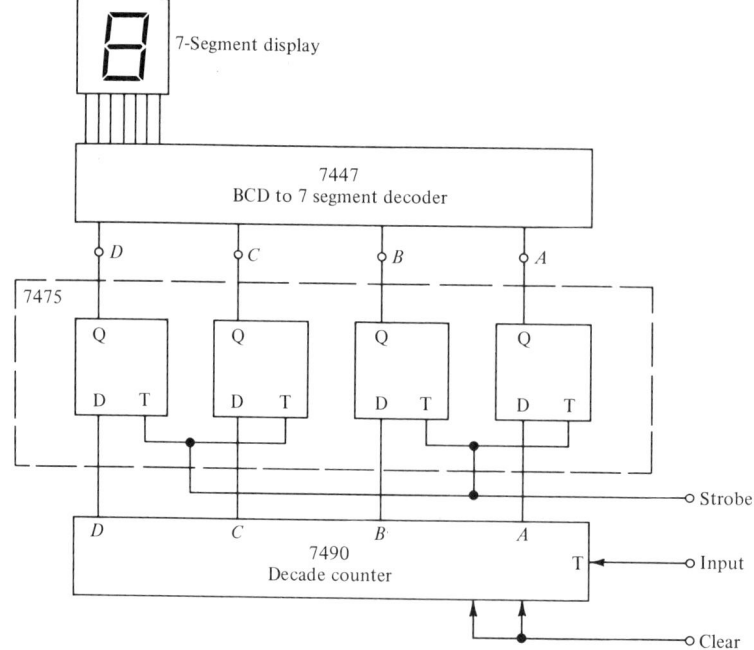

FIGURE 28-9 Latched BCD counter and 7-segmer display.

can be connected to obtain the desired counting function. A high (logic 1) on the strobe line causes each D flip-flop to assume the state of its respective counter flip-flop. Therefore the number in the counter at the time of the strobe pulse is stored (latched) on the outputs of the latch while the counter continues counting. The four flip-flops of the 7475 latch operating synchronously form a data register.

The data on the outputs of the 7475 latch may be the input to a decoder/driver IC producing a readout display. The 7447 BCD to 7-segment decoder/driver is connected into the total circuit. The decoder transforms the BCD input into output to light the proper light emitting diode (LED) segments of the display. BCD counters may be connected (cascaded) together to produce any number of decimal digits.

28.3 ANALOG MSI CIRCUITS

Since operational amplifiers are the key components of most analog instrument circuits, these circuits may be viewed as applications and extensions of the basic op amp configurations discussed in Chapter 27. Applications presented in this chapter were selected as examples of commonly used signal modifiers. They represent only a small fraction of the many practical analog instrumental circuits. These circuits are available in MSI packages. Combinations of these circuits also appear on LSI chips.

Measurement of Current and Voltage

Obtaining precise measurements of currents and voltages from input transducers is one of the most important functions of op amps. As feedback stabilized amplifiers they can increase the magnitude of input signals so that these signals may be more precisely registered by signal modifiers and output transducers.

A combination of the basic op amp summing and scaling circuits is used for current measurement (Fig. 28-10). An adjustable resistance, R, controls the degree of amplification (the sensitivity). Provision is also made for adding a signed current from a controlled source to the input current at summing point P. This current, known as either the summing or bucking current, is used to adjust the level of the base line, such as the dark current correction for photomultiplier output currents.

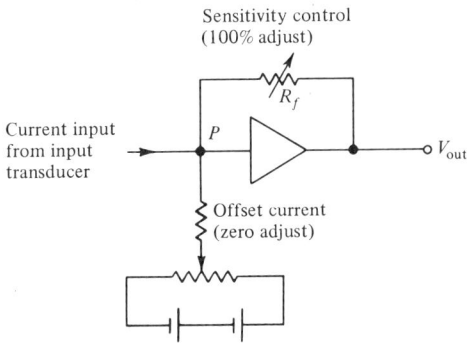

FIGURE 28-10 Current amplifier.

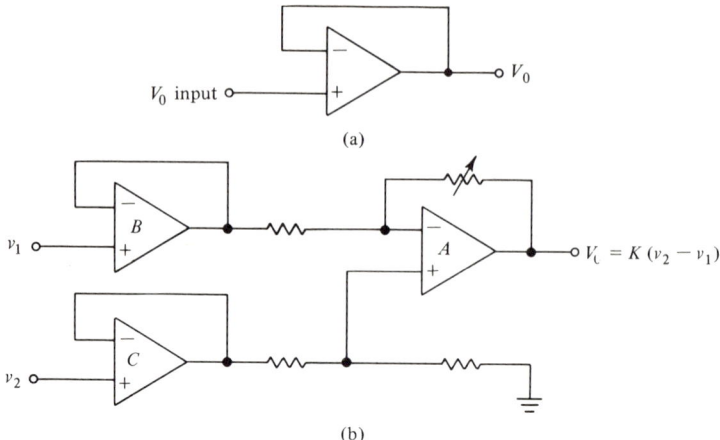

FIGURE 28-11 Voltage followers. (a) Simple voltage follower; (b) instrument amplifier.

If the voltage drop across two points in a circuit is to be precisely determined, no current should flow between these points during the measurement. Voltage followers (Fig. 28-11a) are used to isolate voltage measurement circuits in which no current flow is desirable from signal modifying circuits where current flow is necessary. These devices transfer voltage signals at the input to the voltage output without causing a current flow in the input circuit. The voltage follower op amp circuit has a gain of unity and provides a noninverted output voltage. The basic op amp properties of high input impedance and low output impedance are responsible for the isolating (buffering) action of the voltage follower. This circuit is used to obtain accurate electrochemical cell potentials in the absence of electrolytic current flow.

Voltage followers A and B (Fig. 28-11b) are attached to the inputs of a difference amplifier C to obtain a circuit known as an instrumental amplifier. This differential amplifier circuit is used to retrieve millivolts of analog data from volts of common-mode interference (signals of equal amplitude appearing simultaneously at the two inputs). It isolates the inputs from the outputs and thus protects the amplifier from high-voltage inputs and the device being measured from current leakage. Noise levels are also reduced. The gain, usually from 1 to 1000, may be adjusted by a single variable resistor R (Fig. 28-11b). Properties of low drift, excellent linearity, and good noise rejection make the instrumental amplifier a natural choice for extracting and amplifying low level signals in the presence of high common-mode-noise voltages. They are widely used as transducer amplifiers for thermocouples, current shunts, and specific ion electrode meters.

Sample-and-Hold Amplifiers

Another combination of op amps available on a single monolithic IC is the sample-and-hold (S/H) amplifier. It can be considered as a pair of voltage followers connected through an electronic switch (Fig. 28-12). The device has an analog input, a control input, and an analog output and is always in one of two operating modes: the sample mode in which the output tracks the input, or the hold mode in which the output retains the value of the

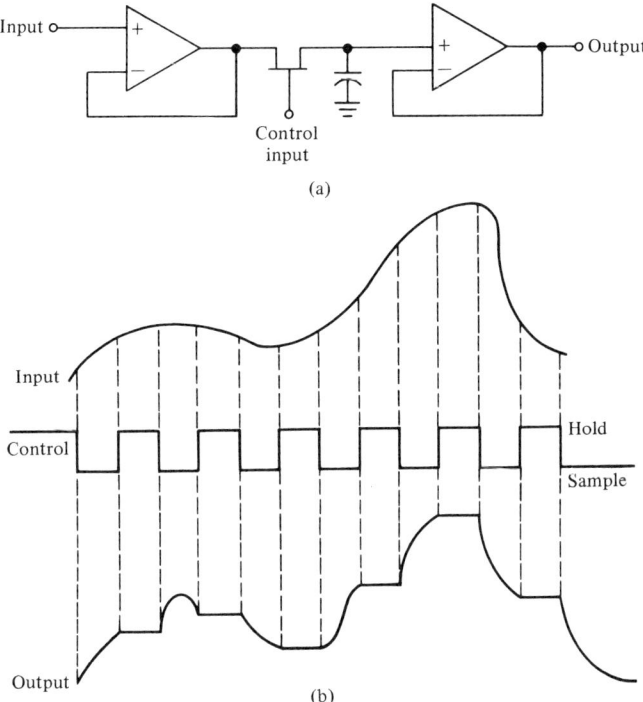

FIGURE 28-12 (a) Sample-and-hold amplifier. (b) S/H output as a function of input.

input signal at the time of the mode change (Fig. 28-12b). This circuit behaves as an analog switch which can sample the instantaneous input voltage and retain it as a constant dc level.

The S/H amplifier is similar to the track-and-hold amplifier, the two devices differing only in relative amounts of time spent in the sample-and-hold modes. If the sampling time is long compared to the holding time, the device is known as a track and hold. On the other hand, if the sampling time lasts only for the relatively brief period of time necessary to fully charge the capacitor C (Fig. 28-12a) the device becomes an S/H amplifier.

The S/H device can perform op amp and sampling functions, thus eliminating the need for separate sampling and scaling circuits. The S/H may be wired so as to extend its applications. If the control switch is always closed, the S/H functions as a conventional amplifier with excellent operating parameters. The control switch allows two or more S/H circuits to be multiplexed together since an open switch effectively disconnects the output of the first op amp (Fig. 28-12a).

Control of Current and Voltage

Circuits for the precise control of current and voltage may be constructed utilizing op amps. The potentiostat, a circuit providing a constant, accurately known voltage, is illustrated in Fig. 28-13a. The output voltage, which is equal to the reference voltage, V_{ref}, remains constant while allowing a considerable variation in the current output of the op

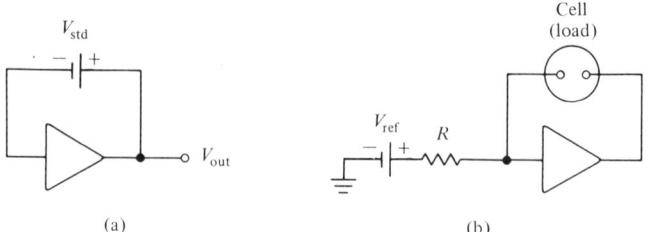

FIGURE 28-13 Voltage and current controllers. (a) Constant-voltage source (potentiostat); (b) constant-current source (amperostat).

amp. This circuit could be used to provide a constant potential across a varying load resistance; for example, the voltage control of a polarograph.

An amperostat or constant-current source (Fig. 28-13b) assures that the current through the load, such as an electrolysis cell, is equal to that in the input circuit, V_{ref}/R. Neither the current in the feedback circuit nor the voltage across the load may exceed the maximum output values of the op amp. The current in the electrolysis cell will remain constant as voltage across the load changes. This circuit, as well as the potentiostat circuit, are used extensively in electrochemical instrumentation.

Active Filters (Tuned Amplifiers)

Sometimes it is desirable to amplify a signal occurring at a given frequency or range of frequencies while suppressing signals of all other frequencies (Chapter 29). Op amps may be used to make active bandpass filter circuits (Fig. 28-14a). The output of this circuit (Fig. 28-14b) shows a maximum gain, A_0, at a frequency, f_0, over a frequency range Δf. (Δf is defined by the maximum allowable gain, usually 3.0 dB below A_0.) Values of circuit components and the operational parameters determine the magnitudes of A_0, f_0, and Δf. Amplifiers tuned to the frequency of a mechanical chopper can be used to improve signal-to-noise ratios in spectrometers.

At frequencies below 10 kHz, active filters offer a distinct advantage over passive filters (circuits containing only passive elements such as resistors, capacitors, and inductors). In order to operate at these low frequencies, passive filters require large inductors which are expensive, bulky, and sluggish in their response to input signals.

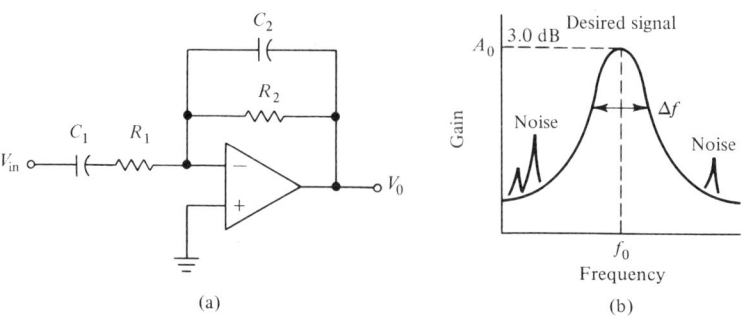

FIGURE 28-14 Active filter. (a) Circuit; (b) response curve.

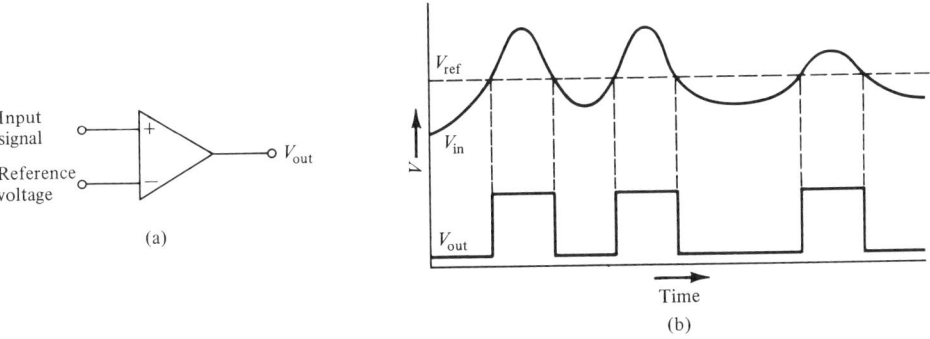

FIGURE 28-15 Voltage comparator. (a) Schematic; (b) voltage curve shapes.

28.4 COMPOSITE CIRCUITS

Analog and digital components are often combined into a single MSI circuit to obtain a desired signal transformation. Analog data may be used to control digital outputs while digital inputs may be used to generate analog signals. These circuits are often found in the interfaces between instruments and computers.

Voltage Comparators

Basic Level Testor A simple but important circuit is the voltage comparator (Fig. 28-15). The relative magnitude of two analog inputs controls the digital output of a difference amplifier. Voltage comparators can be designed to give a voltage level equivalent to a TTL logic 1 at the output when the voltage on the noninverting (+) input exceeds the voltage on the inverting (−) input. When the inverting input voltage exceeds the noninverting voltage, the output voltage drops to a level corresponding to TTL logic zero. If a constant reference voltage is applied to one input, the device can be used as a voltage level detector. Comparators are widely used as components of pulse counters and signal modifiers.

Schmitt Trigger If the voltages on the two inputs of an op amp voltage comparator are equal, then the output will change state randomly in response to the noise present on both inputs. The Schmitt trigger circuit (Fig. 28-16) is used to eliminate problems

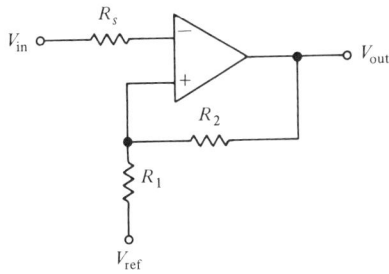

FIGURE 28-16 Schmitt trigger circuit.

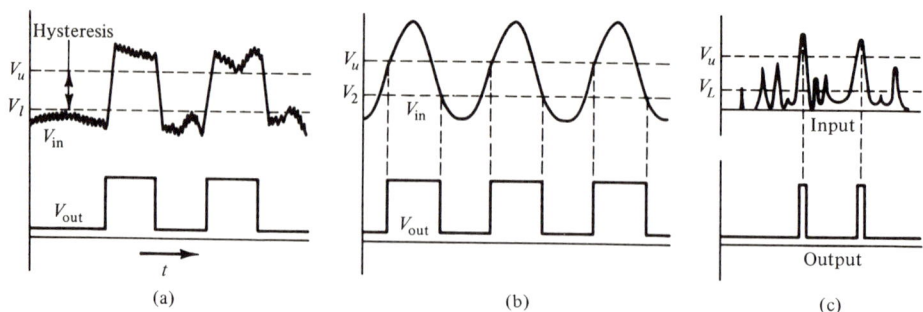

FIGURE 28-17 Applications of the Schmitt trigger. (a) Noise reduction; (b) square-wave generator; (c) pulse counting.

associated with noisy input signals. Two voltage levels, V_u and V_l, are produced by V_{ref}, R_1, and R_2 of the circuit. The output of the trigger changes state from 1 to 0 when $V_{in} > V_u$ and from 1 to 0 when $V_{in} < V_l$ (Fig. 28-17a). The voltage difference, $V_u - V_l$, and known as the hysteresis voltage, should be larger than the magnitudes of expected noise levels.

The major advantage of the Schmitt trigger comparator is its noise immunity which is equal to the voltage width of the hystersis. Once this device changes state, small noise pulses will not cause reswitching. The advantage of the comparator is its ability to narrowly define a single-voltage comparator level (Fig. 28-18). In addition to elimination of noise, the applications of a Schmitt trigger include square-wave generators (Fig. 28-17b) and pulse counters (Fig. 28-17c).

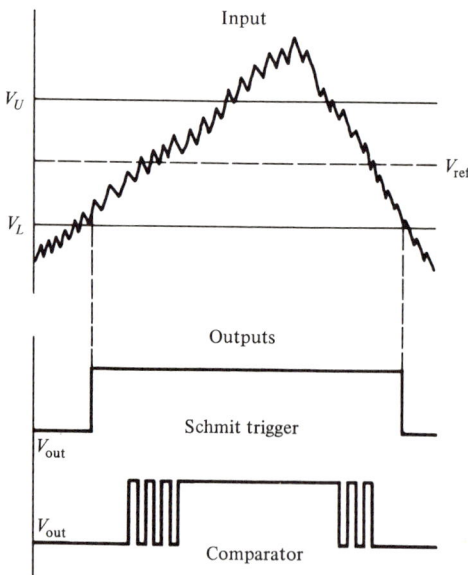

FIGURE 28-18 Schmitt trigger output vs. simple comparator output.

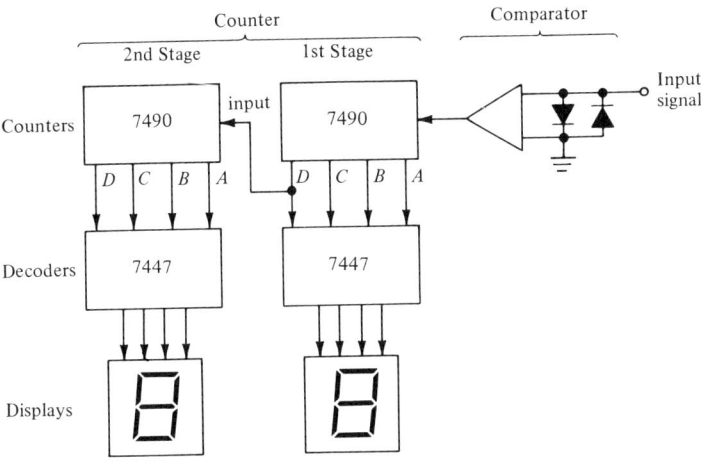

FIGURE 28-19 Pulse counter.

Pulse Counter The pulse counter (Fig. 28-19) uses a comparator or Schmitt trigger to detect pulses with voltage magnitudes greater than a preset voltage level. The output of the comparator is then counted and displayed using digital ICs. Diodes at the inputs protect the comparator from high-voltage signals. This signal modifier could be a part of a photon counting instrument. If the output of the photomultiplier detector is connected to the noninverting input of the comparator, the photon count can be read on the digital display.

Digital-to-Analog Converters

Analog output can be produced from digital input using a summing op amp circuit (Table 27-3). For example, a 3-bit digital-to-analog converter (DAC) (Fig. 28-20) can be

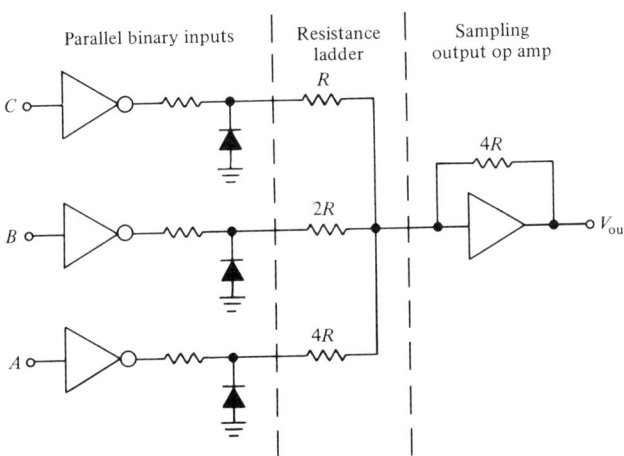

FIGURE 28-20 Three-bit digital-to-analog converter.

TABLE 28-3 DAC Output as a Function of Input

	Binary Data Word		
C	B	A	V_{out}-Volts*
0	0	0	0.0
0	0	1	0.5
0	1	0	1.0
0	1	1	1.5
1	0	0	2.0
1	0	1	2.5
1	1	0	3.0
1	1	1	3.5

*The logic 1 state is assumed to produce a signal that is ~ 0.5 V due to diode limiting. The feedback resistor of the output op amp multiplies the output by 4.

made by applying the total current from a resistor ladder to the input of an op amp. The drivers and diodes accurately set the voltage levels applied to the resistor ladder.

Three digital data bits are simultaneously applied to inputs A, B, and C, producing an output voltage V_0 given by the following equation:

$$V_0 = 8V_{in}\left(\frac{A}{4} + \frac{B}{2} + \frac{C}{1}\right) \quad (28\text{-}1)$$

where V_{in} is the voltage level of logic 1 state from any of the binary input bits A, B, or C. Note that the relative values of the resistors determine that A will be the least significant bit and C the most significant bit. The operational amplifier sums the input currents and converts them to a scaled output voltage (Table 28-3). The range of the scale is determined by the value of the feedback resistor. The number of input bits may be increased by expanding the weighted resistance ladder. DACs containing from 4 to 16 bits are common.

A monolithic 4-bit DAC contained on a single chip (Fig. 28-21) latches a 4-bit data word onto the chip inputs. When the strobe line undergoes a 1 to 0 transition, the latch

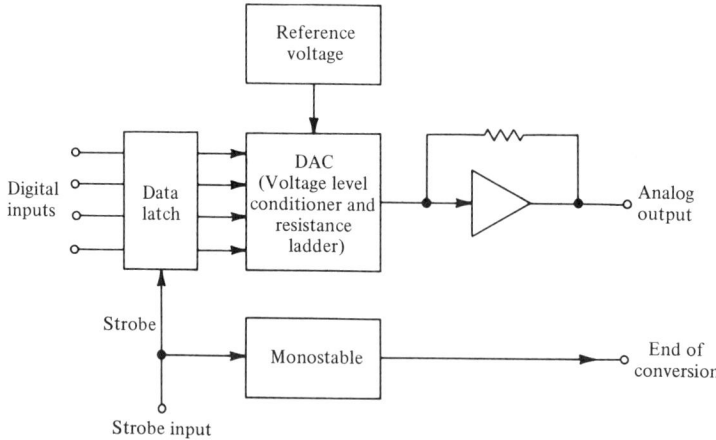

FIGURE 28-21 Complete 4-bit DAC with control inputs.

presents a new data word to the DAC inputs. The strobe pulse also triggers a monostable which delays the signal before it changes the logic state of the end of conversion pulse. This delay allows the DAC to perform its conversion, after which the end of conversion output indicates that the analog output has reached a stable value (settled). The DAC is then ready to make another conversion.

Important DAC parameters are resolution, accuracy, settling time, and analog outputs associated with the least significant and most significant bits (LSB and MSB). The resolution is determined by the number of input bits the converter will handle; for example, a 10-bit DAC has 2^{10} or 1024 output levels and therefore a resolution of 1 part in 1024. Accuracy is a measure of the deviation of the analog output voltage from its predicted value for a given input combination. The reference voltage is critical in determining accuracy. The settling time is the time required for a new, stable output value to be registered in response to a change in digital inputs.

The conveniently packaged 10-bit DAC (Fig. 28-22) unit is double buffered which means that there are two independent 10-bit registers within the device, the DAC register and the input buffer registers. Double buffering allows the set of digital data bits in the DAC register to be converted to analog output while a separate set of data bits is applied to the input buffer.

DACs are used to operate any device that requires an analog input signal and that is interfaced to a digital source such as a computer. Applications include: display of digital data stored in computer memory on analog peripherals, such as plotters and graphics

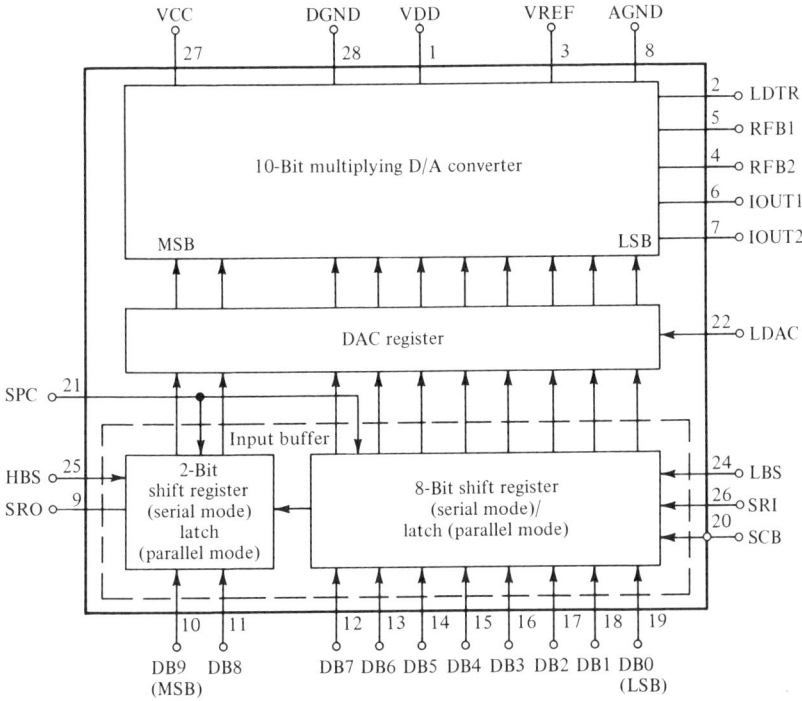

FIGURE 28-22 Functional diagram of the AD7522 chip. (Courtesy of Analog Devices.)

terminals; and control of chemical instrumentation using computers to generate analog voltage signals, such as ramps and waves.

Analog-to-Digital Converters

An analog-to-digital converter (ADC) produces a digital representation of an analog input signal. Since most input transducers initially produce analog signals, ADCs find many applications in digital instrumentation. The most important characteristics of ADCs are speed and accuracy of conversion. These parameters are determined by the particular method utilized for the analog-to-digital conversion which, like digital-to-analog conversions, may be achieved by several methods.

The voltage comparator is the heart of any ADC. The analog voltage to be converted is compared with a variable reference voltage. When the two voltages are equal, the comparator changes state and a digital output is registered. ADCs vary in the methods used to generate variable reference voltages and in the components used to register output data.

Counter Converters The simplest ADC is the counter converter (Fig. 28-23) which uses a binary up-counter latched to a DAC in order to produce a linear voltage ramp at the reference input of the converter. When the ramp voltage equals the analog input voltage the comparator changes state, stopping the counter and allowing the digital outputs of the counter to be recorded by an output device. The size of the count is proportional to the magnitude of the analog input voltage. For example, if the 4-bit ADC (Fig. 28-23) is set to cover analog inputs ranging from 0 to 1.0 V and a voltage of 0.40 V is placed on

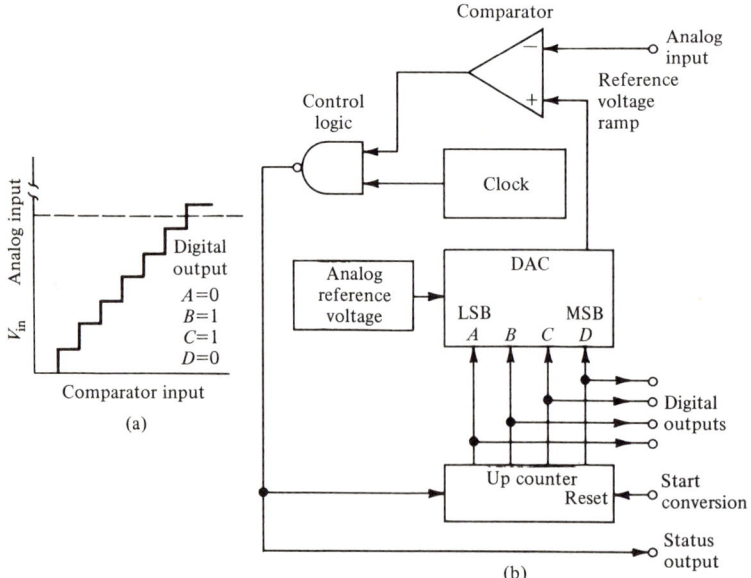

FIGURE 28-23 Four-bit counter ADC. (a) Example input and output; (b) composite circuit diagram.

TABLE 28-4 Example of a Successive Approximation by a 4-Bit ADC

Approximation	Answer	Logic State	Weighted Base 10 Value
# ⩾ 8 ?	Yes	1×2^3 (MSB)	8
# ⩾ 12 ?	Yes	1×2^2	4
# ⩾ 14 ?	No	0×2^1	0
# = 13 ?	Yes	1×2^0 (LSB)	1
			13

the input, the digital output will be 0110 corresponding to a value of 7/16 times the full-scale output. When the digital outputs have been recorded, the counter is reset to zero by a start conversion pulse and the process is repeated. The conversion time is directly proportional to the magnitude of the analog input voltage.

Successive Approximation Counter The popular successive approximation converter replaces the simple up-counter with a digital pattern generator and more complex control logic. This converter uses a series of logical guesses (approximations) to determine the digital equivalent of the analog input. Table 28-4 illustrates the sequence of approximations that a 4-bit ADC might use to digitize an analog input signal. A 4-bit binary number results from the series of approximations, the most significant bit corresponding to the result of the first approximation, the next bit matching the result of the second approximation, and so on. Thus the magnitude of the analog signal in this example yields a binary 1101_2 corresponding to 13_{10}. This means that the analog input voltage is 13/16 of the full-scale voltage of this DAC.

Successive approximation counters require N approximations for an N-bit conversion. Ten approximations would therefore be required to determine the value of an analog input to an accuracy of 1 part in 1024 (2^{10}). The previously described counter ADC could require up to $2N$ trials to produce an N-bit digital number. Successive approximation converters are widely used because they have fast conversion times which are independent of the magnitude of the analog input signal. The input signal for both the counter and successive approximation converters are usually in the high level voltage range, 0–10 V maximum.

Voltage-to-Frequency Converters In situations requiring conversion of small analog input signals (millivolt magnitudes), low-level ADCs are used. These converters are much slower (by factors of 100–1000) than the successive approximation converters, and trade speed for the ability to represent accurately low level input signals. These low level ADCs integrate the analog signal, then compare it to a reference voltage to produce the digital output.

The voltage-to-frequency converter (Fig. 28-24) is a simple converter for transforming the analog input voltage into a series of digital pulses whose frequency is proportional to the magnitude of the input. The circuit, composed of resistor R, capacitor C, and op amp A, is used to produce an integral of the input voltage at the inverting input of the com-

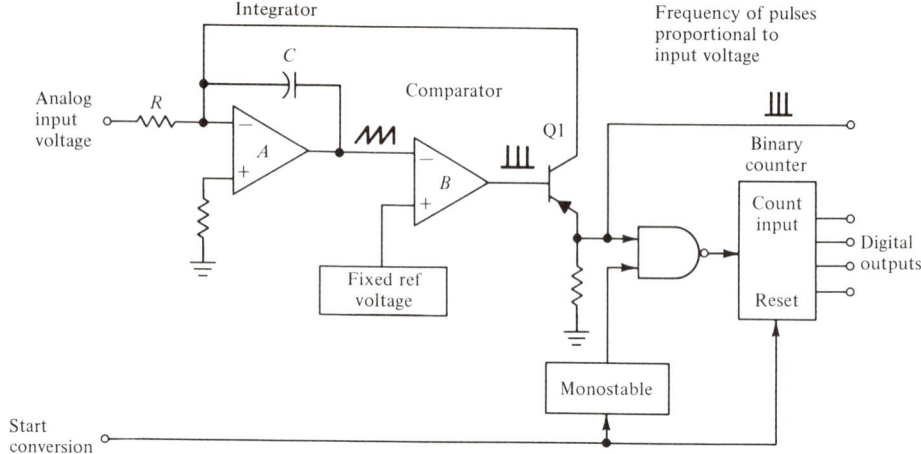

FIGURE 28-24 Four-bit voltage-to-frequency converter.

parator (operational amplifier B). When the voltage of the integral output exceeds the value of the reference voltage, the comparator changes state, which turns on transistor Q1 and allows the capacitor to discharge. When the output of op amp A reaches ground, op amp B again changes state, turning off the transistor and allowing C to recharge again. Thus a series of pulses is generated at the output of op amp B. The rate at which the integrator op amp A charges is a function of the input analog voltage signal. The larger the input voltage, the faster the integrator ramp and, therefore, the higher the frequency of output pulses.

The output pulses go to the input of a binary counter which counts pulses for a period of time determined by the monostable. The start-of-conversion input pulse resets the counter to zero and initiates the counting time interval by triggering the monostable. The monostable output, connected to one input of gate G, controls the counting time for pulses.

Dual Slope Converters Dual slope ADCs (Fig. 28-25) increase the accuracy of conversion over that of voltage-to-frequency converters by measuring a single integral rather than a series of integrals. An analog input voltage, V_1 (Fig. 28-26), is applied to the input of the integrator for a fixed period of time, t. At the end of this time a constant reference voltage, V_{ref}, of opposite polarity to V_{in} is applied to the input of the integrator by switch S_1. The counter is also cleared to zero at the end of time period t. It then begins counting up, recording the time (t_1) required for the reference voltage ramp (slope = V_{ref}/RC) to reach zero. When the comparator senses this zero crossing, the counter is halted, its output registered, and the analog circuitry is reset by switches S_1 and S_2. A larger analog input voltage V_2 (Fig. 28-26) requires a proportionally longer time, t_2, to cross the zero axis.

ELECTRONICS: COMMONLY USED SIGNAL MODIFYING CIRCUITS

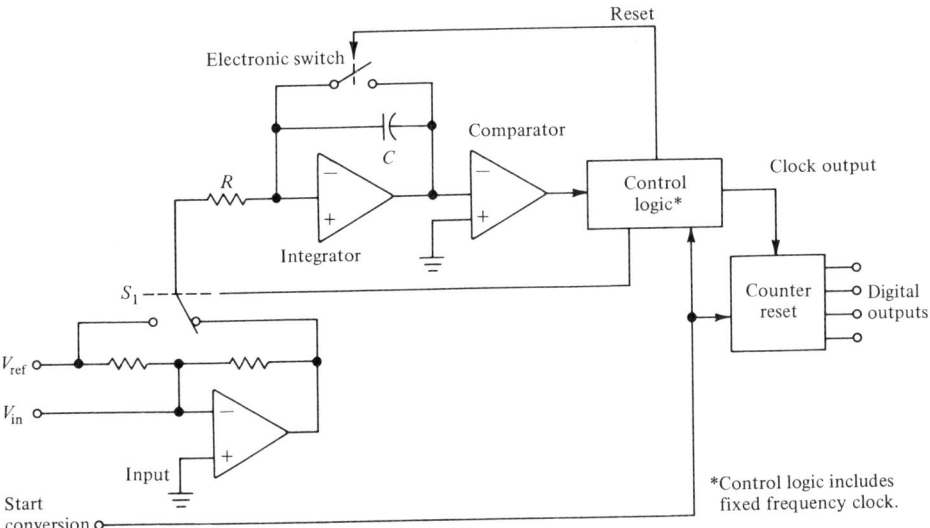

FIGURE 28-25 Dual slope ADC.

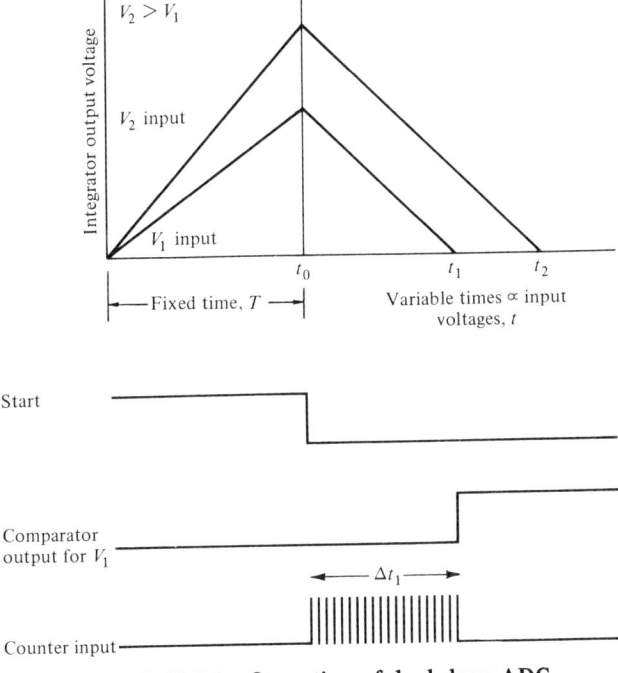

FIGURE 28-26 Operation of dual slope ADC.

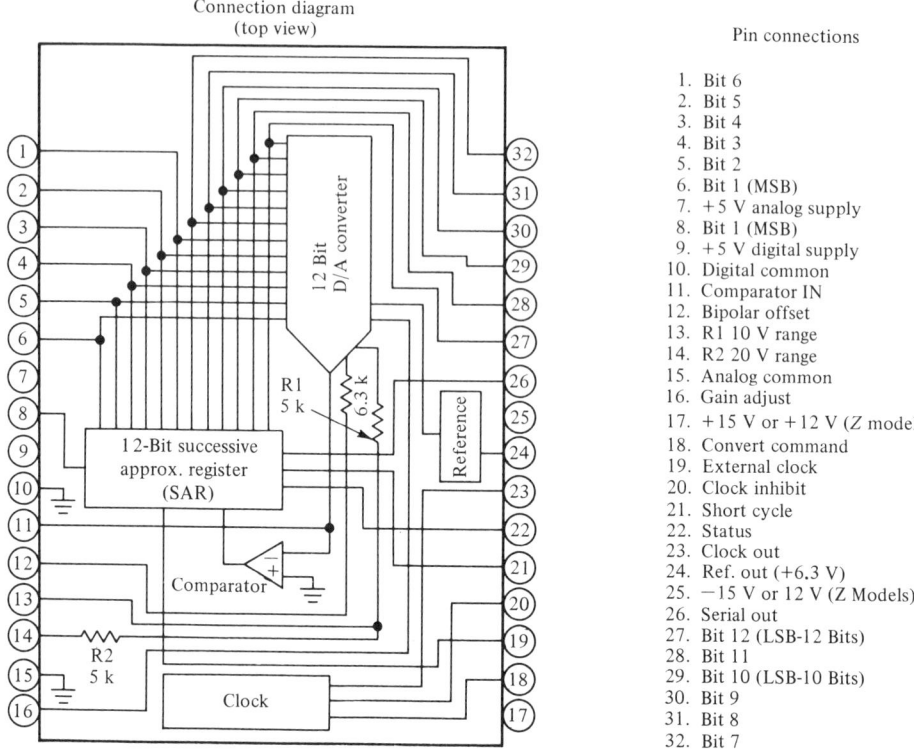

FIGURE 28-27 Twelve-bit ADC contained in a 32-pin ceramic package. (Courtesy of Burr-Brown.)

Since the total charge gained by the capacitor C with V_{in} applied to the integrator is equal to the charge lost by the capacitor with V_{ref} applied, then

$$\frac{V_{in}t}{R} = \frac{V_{ref}t_1}{R} \tag{28-2}$$

$$t_1 = \frac{t}{V_{ref}} V_{in} \tag{28-3}$$

If t and V_{ref} are constants for a given ADC, then the output of the counter (t_1) will be a binary representation of the analog input voltage (V_{in}). The conversion time must be greater than $2t$ and, therefore, the conversion frequency is limited to less than $(2t)^{-1}$ conversions per second.

ADCs, like DACs, are available in handy IC packages. The functional diagram of a totally self-contained 12-bit successive approximation ADC is illustrated in Fig. 28-27.

MSI Systems Circuits contained in MSI chips are combined to produce additional signal modification. For example, DACs and ADCs usually require auxiliary ICs to properly condition input or output signals. A typical data acquisition system (Fig. 28-28) uses

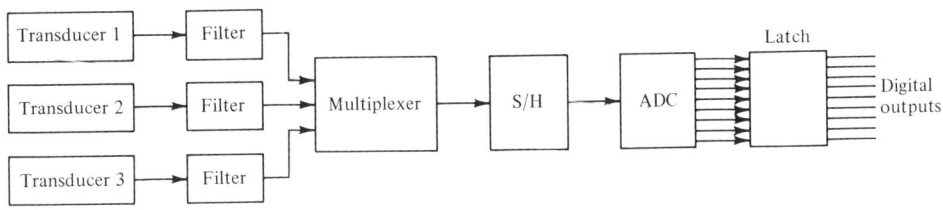

FIGURE 28-28 Configuration of a typical data acquisition system.

active filters, multiplexers, S/H, and data latch MSI circuits. S/H circuits are often used in conjunction with a variety of ADCs to sample the analog input voltage, V_1, and hold it constant during the time required for conversion. In situations requiring the conversion of analog signals from a number of different sources, several S/H's may be multiplexed to a single ADC.

28.5 LARGE-SCALE INTEGRATED CIRCUITS

As the number of components on an IC chip approaches a thousand, the integration classification changes from medium- to large-scale integration (LSI). Single LSI chips, representing combinations of SSI and MSI, are capable of providing a large number of signal transformations. The quality of low-cost LSI circuits is now satisfactory for the most demanding data transformations. LSI techniques can handle almost all digital electronic functions and many analog functions at a fraction of the cost of other approaches. Furthermore, these LSI techniques have allowed innovations in equipment design that heretofore would have been too expensive to market commercially.

LSI comes in many forms: programmable random logic, nonprogrammable functional logic, memories, microprocessors, and chips combining logic, memory, and analog circuits. Microprocessors and memories are discussed in Chapter 30. The LSI circuits discussed in the next section of this chapter represent combinations of some previously discussed MSI circuits.

Hybrid Data Acquisition System

The complete data acquisition system shown in Fig. 28-29 is contained in a single 80-pin LSI chip. The chip contains all components necessary to multiplex and convert analog signals from a minimum of 0 to a maximum of 5 V into equivalent digital outputs. Sampling rates range from 18 kHz (12-bit resolution) to 40 kHz (8-bit resolution). A low-drift instrumentation amplifier allows a selection of gains from 2 to 1000. The chip can be configured to accept either 8-channel differential or 16-channel single-ended analog signals. Tri-state outputs are provided for easy interface to microcomputer systems.

The system components include: (1) an analog multiplexer with data channel select latch, (2) instrumental amplifier with programmable gain, (3) an S/H circuit, (4) a 12-bit ADC, both parallel and serial outputs, (5) tri-state output buffers, (6) a monostable delay timer, and (7) an adjustable clock. The chip requires 15- and 5-V power supplies.

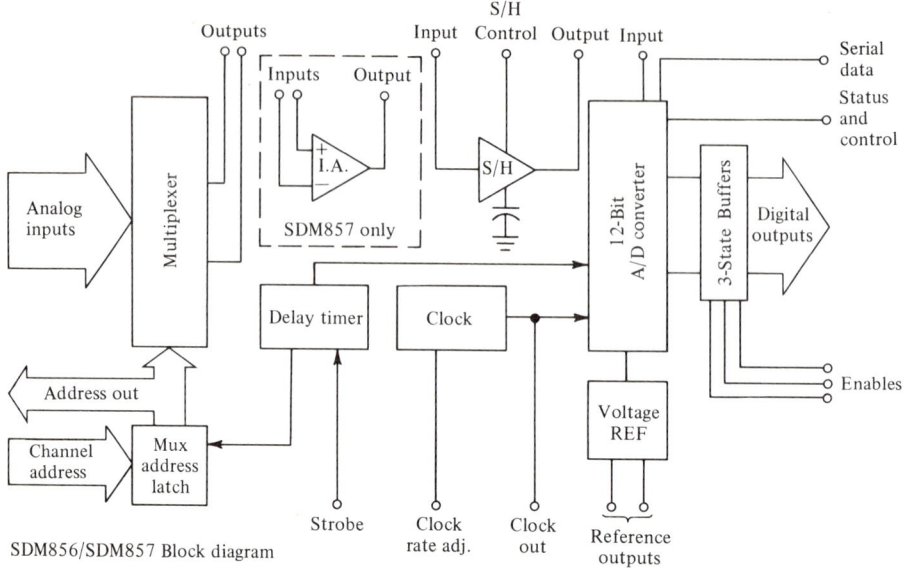

FIGURE 28-29 Hybrid data acquisition system. (Courtesy of Burr-Brown.)

This system provides the complete signal conditioning required for interfacing an input transducer directly to a computer. The analog input signal channel is selected by the multiplexer, the signal amplified by the instrumentation amplifier and transformed to a digital output using the S/H coupled to the dual slope ADC. Tri-state outputs allow the data to be conveniently transferred to the data input of a computer (Fig. 28-30). Digital control signals from the computer can be used to select the analog input channel, to set the gain of the instrumental amplifier, and to transfer data from the ADC output to the computer.

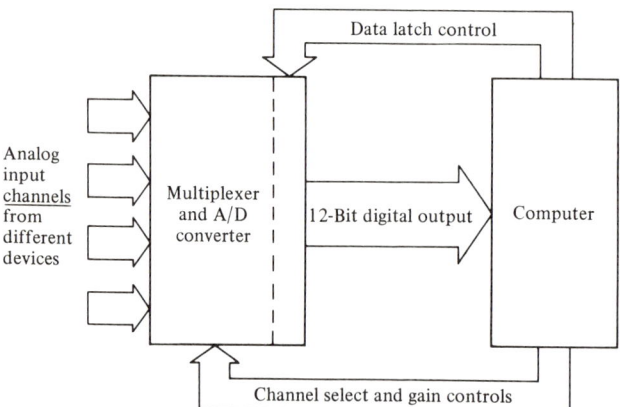

FIGURE 28-30 Hybrid data acquisition system as a computer interface.

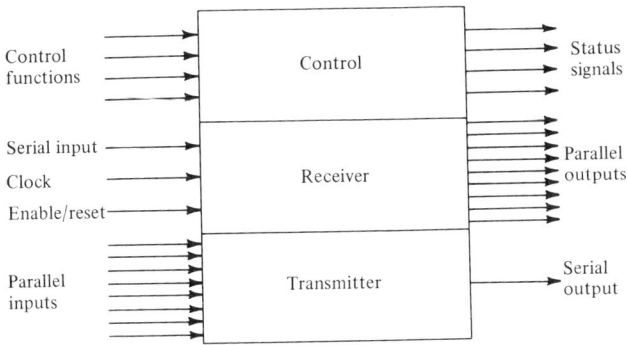

FIGURE 28-31 UART block diagram.

Universal Asynchronous Receiver/Transmitter

The LSI Universal Asynchronous Receiver/Transmitter (UART) provides a programmable digital communications center on a single 40-pin chip (Fig. 28-31). The transmitter register accepts bit parallel data and produces a bit serial output. The receiver register transforms bit serial input to bit parallel output. The two registers may operate independently or in series (full- or half-duplex, respectively). Other selectable operating parameters include the number of data bits in the transmitted code, presence or absence of parity bits for error checking, and the transmission rate (bits/sec). Transmitter buffer empty and receiver buffer full are two of the available status signals.

UARTs are used to link instruments with output devices, controllers, or computers at remote locations (Chapter 30). The digital serial data are commonly transmitted over a pair of twisted wires. This inexpensive chip has greatly facilitated applications requiring digital data tansmission, such as remote monitoring and feedback control.

BIBLIOGRAPHY

Altman, L., Ed., *Large Scale Integration*, McGraw-Hill, New York, 1976.
Blakeslee, T. R., *Digital Design with Standard MSI and LSI*, Wiley, New York, 1975.
Connelly, J. A., *Analog Integrated Circuits*, Wiley, New York, 1975.
Garrett, P. H., *Analog Systems for Microprocessors and Minicomputers*, Reston, Reston, Va., 1978.
Jung, W., *I. C. Op-Amp Cookbook,* H. W. Sams, Indianapolis, Ind. (1975).
Kalvoda, R., *Operational Amplifiers in Chemical Instrumentation*, Wiley, New York, 1975.
Oldham, W. G., "The Large-Scale Integration of Microelectronic Circuits," *Sci. Am.*, **237** (30), 110 (Sept. 1977).
Switzer, W. L., "Asynchronous Serial Computer Interfaces," *Anal. Chem.*, **48**, 1003A (1976).
Titus, J. A., C. A. Titus, P. R. Rony, and D. G. Larsen, *The Bugbook VII*, E and L Instruments, Derby, Conn., 1978.
Vassos, B. H. and G. Ewing, *Analog and Digital Electronics for Scientists*, Wiley, New York, 1980.

CHAPTER 29

Data Handling

29.1 INTRODUCTION

Although the ability to separate significant data-containing signals from meaningless noise has always been a desirable property of any instrument, it has become imperative with the demand for increasingly sensitive measurements. The amount of noise present in an instrument system determines the smallest concentration of analyte that can be accurately measured and also fixes the precision of measurement at larger concentrations. Noise reduction (or signal enhancement) is a primary consideration in obtaining useful data from measurements involving either weak signal sources, such as carbon-13 nuclear magnetic resonance spectroscopy, or trace amounts of analyte, such as polarography.

The two principal methods of achieving signal enhancement are: (1) the use of electronic hardware devices, such as filters, or equivalent computer software algorithms to process signals from the measurement as they pass through the instrument and (2) post-measurement mathematical treatment of data. Among the more useful of the post-measurement methods are statistical techniques. In addition to signal enhancement, these techniques aid in identification of sources of error and determination of precision, and also provide a method for an objective comparison of results. This chapter will present some common noise reduction techniques and briefly review important statistical methods typically employed in the treatment of instrumental data.

29.2 SIGNAL-TO-NOISE RATIO

As concentrations decrease to trace levels or as signal sources become weak, the problem of distinguishing signals from noise becomes increasingly difficult. The ability of an instrument to discriminate between signals and noise is usually expressed as a signal-to-noise (S/N) ratio where:

$$\frac{S}{N} = \frac{\text{average signal amplitude}}{\text{average noise amplitude}}$$

in the case of dc signals. An increase in the S/N ratio indicates a reduction in noise and thus a more desirable measurement. Once the physical or chemical quantity of interest has been converted to an electrical signal by an appropriate transducer, the S/N ratio can only be improved by decreasing the value of the noise. Since each increase in the magnitude of the signal will be accompanied by a proportional increase in the value of the noise, amplification cannot improve the S/N ratio.

29.3 SENSITIVITY AND DETECTION LIMIT

A number of parameters, including S/N ratio, affect the sensitivity of a particular instrumental method. Physical and chemical properties of the analyte, response of input transducer to the analyte, and the composition of the sample matrix are some of the more important factors which determine sensitivity. Sensitivity is defined as the ratio of the change in the instrument response (I_o, output signal) with a corresponding change in the stimulus (C, concentration of analyte):

$$S = \frac{dI_o}{dC} \tag{29-1}$$

Slopes of calibration curves (Figs. 29-1 and 29-2) are used to determine the sensitivity values. It is usually desirable to maximize the sensitivity value unless one wishes to extend the instrument's range of response without dilution of the sample.

Figure 29-1 indicates a linear response (constant sensitivity) over the entire range of measured concentrations for both substances A and B. From the slopes of the curves the sensitivity of the method is much greater for substance B than for substance A. A nonlinear response (Fig. 29-2) indicates a constantly changing value for sensitivity as a func-

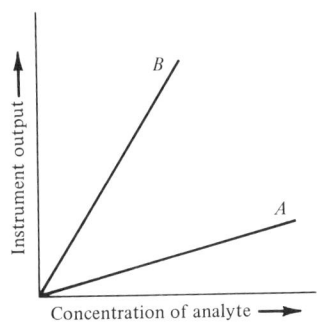

FIGURE 29-1 Linear response.

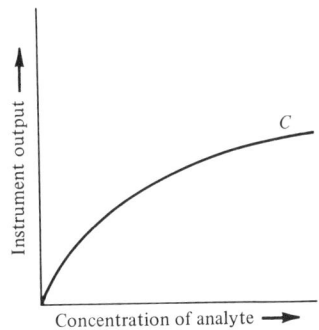

FIGURE 29-2 Nonlinear response.

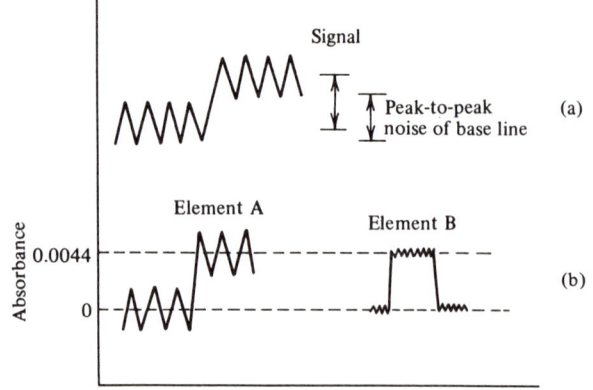

FIGURE 29-3 Detection limit for $S/N = 2$.

tion of concentration. Measurements of substance C become less sensitive with increasing concentration. Sensitivity may also be expressed as the concentration of analyte required to cause a given instrument response. For example, in atomic absorption spectroscopy, sensitivity is expressed as concentration in $\mu g/ml$ of analyte that produces an absorbance of 0.0043 absorbance unit (1.0% absorption). When comparing different techniques or instruments, one should be alert to the procedures used by the practitioners to arrive at sensitivity values.

As the concentration of the analyte approaches zero, the signal disappears into the noise, and the detection limit is reached. A quantitative definition of the detection limit is that concentration of analyte which produces an output signal twice the root mean square of the background noise (a signal-to-noise ratio of 2, Fig. 29-3a); it may also be expressed as peak-to-peak noise of base line. Thus reduction of noise levels in instrumentation yields lower detection limits. Detection limits are generally defined at a 95% confidence level ($\bar{x} \pm 2\sigma$), which means if a concentration at the detection limit were determined many times, it could be distinguished from zero in at least 95 out of 100 measurements. Concentration values smaller than the detection limit may be qualitatively detected but will be quantitatively unreliable.

A comparison of sensitivity and detection limit is illustrated in Fig. 29-3b. It shows the results of atomic absorption analysis of solutions containing equal concentrations of metals A and B. The sensitivities of the instrument system are identical for both metals. In the case of metal B, little noise is present and thus the detection limit for this metal is considerably smaller than that of metal A. In the analysis of B, if reserve amplification is available, the signal amplitude can be increased (with a corresponding increase in noise) until the noise level equal to that observed in the analysis of A is reached. At this point, the sensitivity of B would be much greater than that of A.

29.4 SOURCES OF NOISE

It is important for the analyst using a particular instrumental method to be conscious of the sources of noise and the instrument components used to minimize this noise. Noise

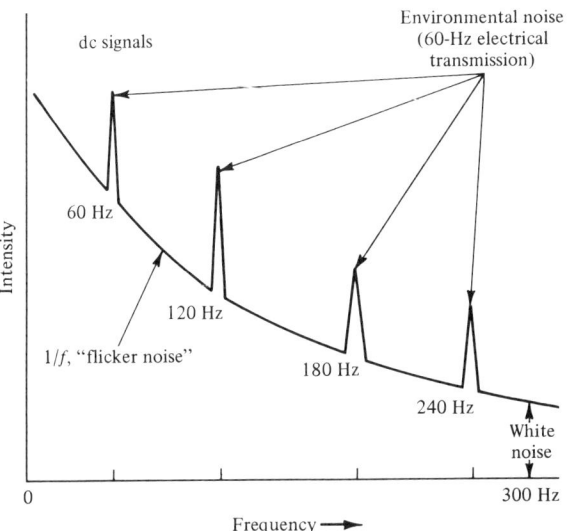

FIGURE 29-4 Kinds of noise.

enters a measurement system from environmental sources external to the measurement system or it appears as a result of fundamental, intrinsic properties of the system (Fig. 29-4). It is usually possible to identify the sources of environmental noise and either to eliminate or to avoid them. Such is not the case with fundamental noise because it arises from the discontinuous nature of matter and energy. Thus fundamental noise ultimately limits the sensitivity of every instrumental measurement.

The major kinds of fundamental noise associated with solid-state electronic devices are thermal, shot, and flicker.

Thermal Noise

Noise originating from the thermally induced motions in charge carriers is known as thermal noise. It exists even in the absence of current flow and is represented by the formula

$$V_{av} = \sqrt{4kTR\Delta f} \tag{29-2}$$

where V_{av} is average voltage due to thermal noise, k is the Boltzmann constant, T is the absolute temperature, R is the resistance of the electronic device, and Δf is the bandwidth of measurement frequencies. Since thermal noise is independent of the absolute values of frequencies, it is also known as "white noise."

Methods for reducing thermal noise are suggested by Eq. 29-2. Sensitive radiation detectors are often cooled to minimize this noise. Narrowing the frequency bandwidth of the detector is another method of reducing thermal noise, provided the frequencies important to the measurement of interest are not excluded. For example, if data-containing signals in the region between 10 and 20 kHz have a S/N ratio of 10 with a detector of $\Delta f = 1$ MHz, reducing the detector bandwidth by a factor of 100 to $\Delta f = 10$ kHz will increase the S/N

ratio by a factor of $\sqrt{100}$ with no loss of required measurement frequencies. Thermal noise is sometimes referred to as Nyquist noise after the physicist who derived Eq. 29-2 or Johnson noise commemorating the engineer who first measured it.

Shot Noise

The magnitude of shot noise is usually much smaller than that of thermal noise and therefore can often be ignored. This kind of noise originates from the movement of charge carriers as they cross n-p junctions or arrive at electrode surfaces. Because these motions involve individual carriers, variations of current due to shot noise are random. Like thermal noise, shot noise is proportional to the square root of the measurement bandwidth, Δf, and is also classified as "white noise." Thus shot noise is also minimized by reducing the bandwidth, Δf.

Flicker Noise

The third kind of fundamental noise, flicker noise, is often observed in the presence of dc current. Although the physical origins of this noise are not well understood, it can be represented by the following empirical equation:

$$V_{av} = \sqrt{K \frac{I^2}{f}} \tag{29-3}$$

where K is a constant depending on factors such as resistor materials and geometry, I is the dc current, and f is the frequency. Flicker noise predominates in measurements from 0 Hz (dc) up to about 300 Hz; it is due primarily to the contribution of the $1/f$ term. Although all solid-state devices are subject to flicker noise, field-effect transistors (FETs) seem to be affected less than bipolar devices. Flicker noise in amplifier systems is commonly referred to as drift. In sensitive measurements flicker noise may be eliminated by avoiding the use of low frequencies (including dc).

Environmental Noise

Environmental noise involves the transfer of energy from the surroundings to the measurement system and typically occurs at specific frequencies or a relatively narrow frequency of bandwidths. Two of the most common sources of environmental noise are the electric and magnetic fields produced by 60-Hz electrical transmission lines. This noise occurs not only at 60 Hz but also at frequencies corresponding to the harmonics (120, 180, 240, ... Hz). Other sources of environmental noise include reflected radiant energy, mechanical vibration, and electrical interaction between different instruments. Reduction or elimination of this kind of noise involves the shielding of circuits and wires utilized in signal transmission from external sources of energy. Proper grounding of all instruments and transmission of signals at frequencies well removed from those of environmental noise are specific examples of techniques for minimizing this noise.

DATA HANDLING   851

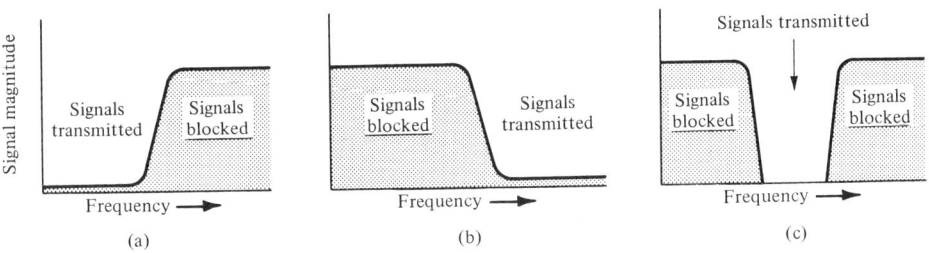

FIGURE 29-5   Filter types. (a) Low-pass, (b) high-pass, and (c) active or bandpass.

29.5   HARDWARE COMPONENTS

Although it is possible to ignore noise in situations in which the magnitude of the noise is small relative to that of the desired signal, an increasing number of instrumental methods require the extraction of information from noisy backgrounds. Noise reduction can be achieved through the use of electronic hardware components. An instrument user should become familiar with the commonly used noise reduction techniques in order to understand the function of noise reducing circuit components. This knowledge should result in more effective use of instruments to obtain precise measurements.

Filters

Although amplitude and phase relationship of input and output signals can be used to discriminate between meaningful signals and noise, frequency is the property most commonly used. As shown in the previous section, "white" noise can be reduced by narrowing the range of measured frequencies; environmental noise can be eliminated by proper frequency selection. Three kinds of electronic filters (Fig. 29-5) are used to select the band of measured frequencies; low-pass filters that allow passage of all signals below a predetermined cutoff frequency; high-pass filters that transmit all frequencies above a given cutoff point; and finally bandpass filters that combine the properties of the previous two filters to pass only a narrow band of frequencies (Fig. 29-5). The simplest filters are composed of passive circuit elements (Fig. 29-6) with the transmitted frequencies determined by values of the individual circuit components. Bandpass filters can be designed

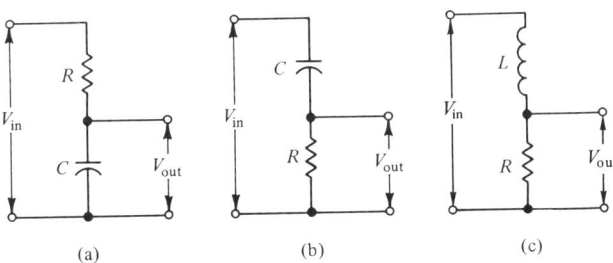

FIGURE 29-6   Passive filters.   (a) Low-pass RC filter; (b) high-pass LC filter; (c) low-pass LR filter.

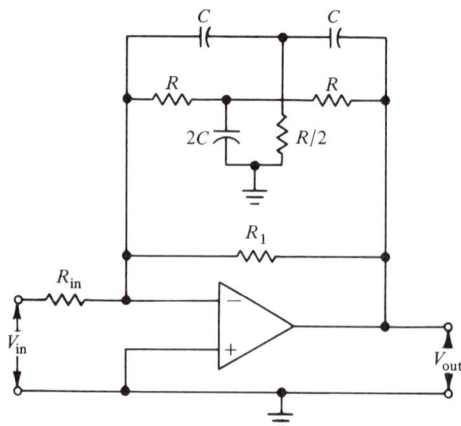

FIGURE 29-7 Twin-T bandpass filter in the feedback loop of an operational amplifier.

into op amp circuits. One such configuration involves a twin-T filter network in the feedback loop of an op amp and is shown in Fig. 29-7. The center band frequency is determined from

$$f_0 = \frac{1}{2\pi RC} \qquad (29\text{-}4)$$

Integrators

Integration of dc signals for precisely limited time periods is a powerful method for reducing white noise. The coherent (nonrandom) signal adds directly with respect to the integration time, while the random noise adds as the square root of the integration time; therefore, the S/N ratio improves as the square root of the integration time. Although a simple RC filter (Fig. 29-6a) can be used to integrate signals, an operational amplifier with a capacitor in the feedback loop usually serves as a hardware integrator (Fig. 29-7). Analog-to-digital converters such as voltage-to-frequency or dual slope devices have built-in S/N enhancement as a result of the integration techniques used in the signal conversion circuits.

Modulators

If the signal and noise cannot be separated by filtering, it is often advantageous to shift the signal of interest away from the noise frequency. To accomplish this, the signal is first transposed onto a carrier wave having a desirable frequency, then transmitted to an amplifier tuned to the frequency of the carrier signal and, finally, the original signal is recovered from the carrier wave. The first process is known as modulation; the final one as demodulation. Modulation/demodulation techniques can be used to process a signal in a region of minimum noise and also to discriminate between signal and noise on the basis of the signal's unique modulation configuration relative to the random pattern of the noise. This technique can be used, for example, to relocate signals away from dc where flicker

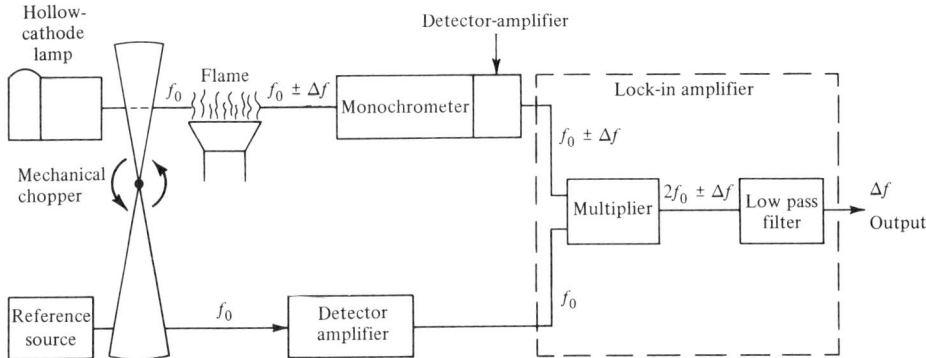

FIGURE 29-8 Application of a lock-in amplifier to an atomic absorption spectrophotometer.

noise is at its maximum. Any property of the carrier wave can be modulated by signals impressed upon it. Common examples include amplitude and frequency modulation used in radio broadcasting and the "chopping" of spectrometer signals.

Lock-In or Phase-Sensitive Amplifiers

Even when the signal is processed in a relatively noise-free environment, some noise will always be passed because of the bandwidth necessary to transmit the signal and the difficulty of obtaining and holding a match between signal frequencies and the filter bandpass. The lock-in or phase-sensitive amplifier offers a solution to these problems. Using a combination of signal frequency and phase relationships, it discriminates between both flicker and white noise. The functional components of a lock-in amplifier include a modulator (chopper), a multiplier, and a low-pass filter (Fig. 29-8).

The data-containing signal at frequency, f, is superimposed onto the carrier-wave frequency, f_0, to produce a modulated signal, $f_0 \pm \Delta f$, that is then transmitted to an electronic device known as a multiplier. Simultaneously, a reference signal, modulated at the same frequency as the carrier signal and held in a constant phase relationship with the carrier wave, is sent to the multiplier. Under these conditions of identical frequencies and constant phase relationship, the multiplier can synchronously demodulate the combination of carrier and reference signals to yield a waveform at $2f_0 \pm \Delta f$, where Δf is the frequency of the desired information. Since Δf is usually low-frequency data, it can be extracted from $2f_0$ by a low-pass filter. The bandwidth (of transmitted frequencies) can be adjusted by varying the RC time constant of the low-pass filter. These operations result in the transformation of the original spectrum of information frequencies, Δf, centered about the carrier frequency, f_0, to the same spectrum containing Δf centered at dc (0 Hz). As long as the carrier and reference waveforms occur at the same frequency and have a constant phase relationship (zero phase difference in the example of Fig. 29-9), then the desired information minus the noise will appear in the transformed spectrum. This method of noise reduction is limited to data-containing signals that are periodic or can be modulated in such a way as to be made periodic. When this is not possible, as in

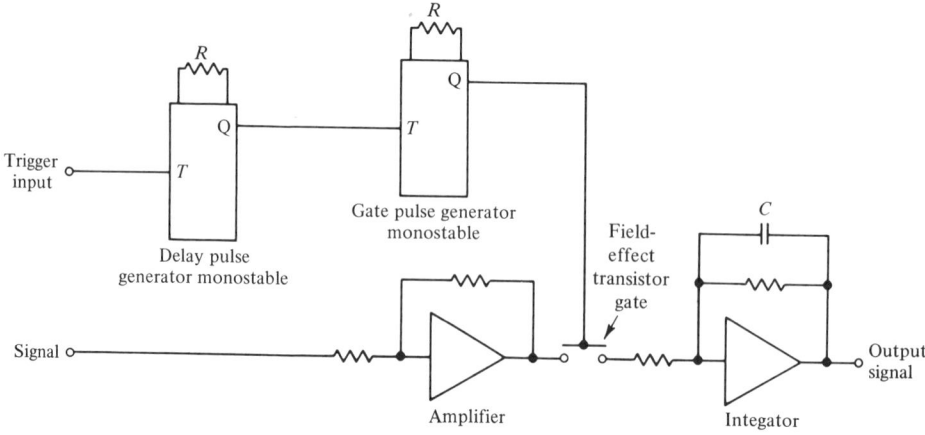

FIGURE 29-9 Simple boxcar integrator circuit.

the case of rapidly changing signals, other signal enhancement techniques must be employed. Since the final low-pass filtering step is centered at dc, some flicker noise may still persist in lock-in amplifier systems.

Phase-sensitive detection is often used in spectrophotometers to achieve increased S/N. In atomic absorption spectroscopy, for example, the major sources of noise are the light source (hollow-cathode lamp) and the flame. Simultaneously chopping a reference light beam and the hollow-cathode light beam (Fig. 29-8) produces two signals having identical frequencies and a constant phase difference (zero in this example). Random noise in the flame, detector, and amplifier are minimized in the output of the lock-in amplifier. Noise originating in the hollow-cathode source will not be removed from the final signal because the lamp input is not modulated by the chopper. To remove the lamp noise, the lamp power supply must also be modulated to produce a periodic signal.

Boxcar Integrators

The boxcar integrator provides a relatively simple method of signal enhancement for repetitive signals by periodically sampling the same portion of a signal for a fixed period of time and then averaging the samples using a low-pass RC filter. This triggerable, gated integrator (Fig. 29-9) is a versatile measurement device. It provides S/N enhancement for the portion of the signal that is sampled. This technique has found wide application in instruments requiring pulsed signal detection. It is best used for S/N reduction in repetitive signals, although it can be used for more complex variable input waveforms.

When compared to the average value of a single pulse, boxcar integration gives S/N enhancement equal to the square root of the number of pulses integrated. Since noise accumulates during the sampling time, further improvement in S/N results from the shortened total sampling time of the boxcar method as compared to the time required to average a single pulse. As in the case of the phase-lock amplifier, the sampling frequency should be carefully selected to avoid interference from environmental noise frequencies and their harmonics.

29.6 SOFTWARE SIGNAL ENHANCEMENT[1]

Many of the hardware noise reduction circuits presented in this chapter can be effectively implemented with computer software algorithms. The current availability of mini- and microcomputers make this an attractive alternative to hardware techniques. Operations, such as filtering, linearization, and attenuation, which were formerly accomplished by hardware devices, can now also be achieved using stored computer programs. Software operations offer advantages of flexibility and diversity. For example, a variety of software filters can be implemented by changing programs, while considerable effort is usually required to change hardware filters. Software methods also offer the additional advantage of postmeasurement treatment of data. Nevertheless, in situations where the computer cannot execute the required function at a satisfactory rate or when the programs and data storage require excessive memory, hardware techniques are necessary.

The minimum hardware necessary to implement software signal processing functions consists of signal conditioning and data conversion circuits as well as the computer components. This hardware is currently available as large-scale integrated (LSI) chips (Chapter 28). Input analog signals are first conditioned to meet the requirements of the analog-to-digital conversion unit. The data are then converted to digital form and processed by the appropriate computer programs. Finally, the processed data may be transformed back to the analog form employing a digital-to-analog converter.

Digital Filtering Technique[2-4]

Three of the most commonly used software signal enhancement techniques are boxcar averaging, ensemble averaging, and weighted digital filtering.

Boxcar Averaging In this method a group of closely spaced digital data points depicting a slowly changing analog signal are replaced by a single point representing the average of the group (Fig. 29-10). Since this technique is best suited for applications in which the analog signal changes slowly with time, boxcar averaging can often be implemented in real time (averaging occurs simultaneously with the acquisition of data). In this mode of operation, one group (boxcar) of points can be acquired and averaged before the next boxcar of data arrives. Improvement in S/N for this method can be calculated by the following equation:

$$S/N = \sqrt{n}(S/N)_0 \qquad (29\text{-}5)$$

where $(S/N)_0$ is the signal-to-noise ratio of the untreated data and n is the number of points averaged in each boxcar.

Ensemble Averaging This technique complements the boxcar method because it can be applied to signals that are changing rapidly. The results of n repeated sets of measurements of the same phenomenon are added and the resultant sum is divided by n to obtain an average scan. If each set of measurements is recorded in the same way, the data contained in the measurements will sum coherently, while the random noise should average to zero. To the extent that the n scans represent a normal statistical distribution, the resulting S/N will be

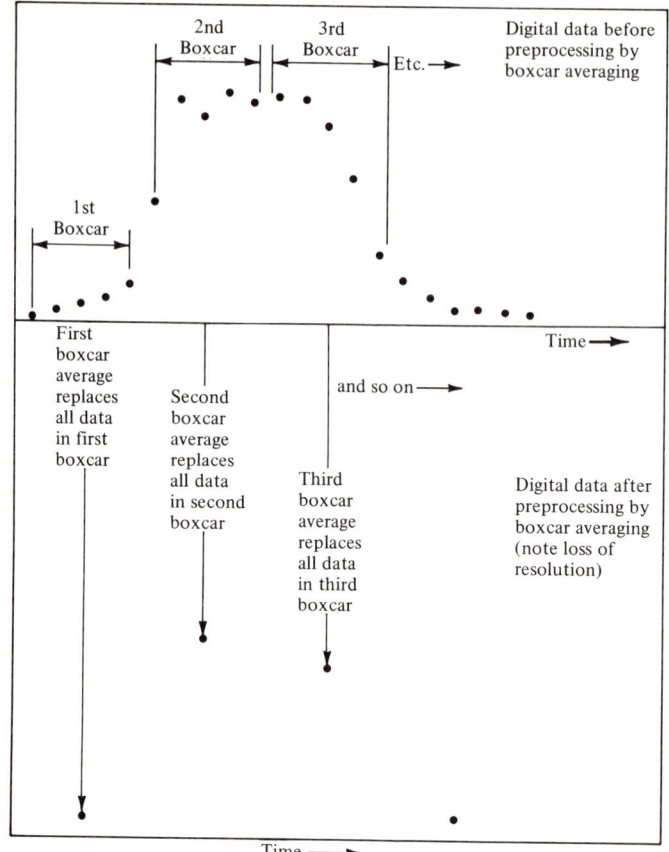

FIGURE 29-10 Digital data before and after preprocessing by boxcar averaging. (After G. F. Dulaney, *Anal. Chem.*, 47, 27A (1975). Courtesy of the American Chemical Society.)

improved by a factor of $\sqrt{n}$ over that of a single scan. As in any signal enhancement technique, the data sampling rate for each set of measurements should be twice that of the highest-frequency component found in the set so that data are not lost. If the sampling frequency significantly exceeds this minimum frequency, no additional data are transferred and, in fact, the noise increases due to the increased frequency bandwidth at the faster sampling rates. Sampling rates corresponding to the fundamental and harmonics of known environmental noise frequencies should be avoided.

Computer managed ensemble averaging has been used to extract small signals from background noise in instrumental methods such as carbon-13 nuclear magnetic resonance and electron spin resonance. The principal liability of the method is the time required to obtain significant improvement in the S/N ratio—100 scans required to improve S/N by an order of magnitude.

Weighted Digital Filtering In digital filtering each of the data points to be averaged contributes equally to the calculation of the average. Assigning different weights to points as a function of their position relative to the central point can produce more realistic filtering. Adjustable filtering parameters include the mathematical smoothing function, the number of points and their position relative to the central point in the moving average, and the number of times the data are processed by the smoothing function. Although this signal enhancement technique offers optimum flexibility in the choice of filter algorithms, the possibility of signal distortion is also great. The amount of time involved in weighted digital filtering usually requires that the method be applied after all the data have been acquired; in other words extensive software digital filtering is not usually performed in real time.

Combining the three software signal enhancement techniques discussed thus far can produce an algorithm which is more useful than any individual technique. For example, an 8-point boxcar average coupled with an ensemble average of nine scans and a single post-run application of a 7-point filter would give an S/N enhancement of 22.4. Over 500 scans would have to be averaged to achieve the same result. Even if the ensemble and boxcar techniques were combined, 64 scans would be necessary. Since the repetitive scans required for ensemble averaging usually take minutes while the digital filter execution time is measured in seconds, the combination of methods effects a substantial savings of time.

A summary of signal enhancement techniques is given in Table 29-1. The central processor unit (CPU) time referred to in the table is the amount of computer time required to perform the calculations for a given process.

Fourier Transformations[5,6] The mathematical operations known as Fourier transformations (FT) provide a powerful method of S/N enhancement. Applications of this technique in instrumental analysis usually fall into one of two general categories. The first involves the use of FT to produce spectroscopic methods which are much faster than conventional frequency-domain methods. Data are rapidly collected in the time domain and then converted by FT to the conventional frequency domain. Since the time required for a single scan is greatly reduced, the time required for ensemble averaging is also decreased. Thus, the efficiency of ensemble averaging is improved using FT spectroscopy. Second, transformations of conventional signals may be multiplied by appropriate conditioning functions to achieve digital filtering and other useful signal modifications. A second mathematical operation, an inverse Fourier transform, is required to restore the conditioned signal to its original form. Although a computer software algorithm, known as a fast Fourier transform (FFT), has reduced the execution time by orders of magnitude over previous transform algorithms, software FFT signal conditioning techniques remain slower than hardware methods.

Two methods of data representation are the frequency–amplitude function, $F(\nu)$, and the less common time–amplitude function, $f(t)$. Both functions contain the same physical information but are expressed in different formats. The functions, known as a Fourier transform pair, are related by the following transformation equations:

$$F(\nu) = \int_{-\infty}^{\infty} f(t) e^{-i(2\pi)\nu t} \, dt \qquad (29\text{-}6)$$

$$f(t) = \int_{-\infty}^{\infty} F(\nu) e^{i(2\pi)\nu t} \, 2\pi \, d\nu \qquad (29\text{-}7)$$

TABLE 29-1 Data Treatment by Filtering, Smoothing, and Averaging[a]

Technique	Function	S/N Improvement	Time Required	Advantages	Disadvantages
Boxcar averaging (software)	Low pass filtering	Proportional to: (number of samples in box)$^{1/2}$	Number of PTS (number of samples × T$_{conv}$) CPU-time, Small	Fast. Useful in real time.	Signal must slew slowly with respect to sampling rate. Resolution lowered. Some phase distortion.
Ensemble averaging	S/N ratio enhancement	Proportional to: (number of scans)$^{1/2}$	Number of PTS × T$_{conv}$ × number of scans CPU-time, Small	Useful even when S/N < 1 averages all random components regardless of f. Negligible phase distortion.	Signal must be stable. Repetitive. Noise must be random.
Unweighted digital filter	Low pass filtering	Proportional to: (number of PTS-in-window)$^{1/2}$	Postrun CPU time. Large	Capable of better resolution than boxcar. Minimal phase distortion.	
Weighted digital filter	Low pass, high pass, or bandpass filtering	Proportional to: (number of PTS-in-window)$^{1/2}$	Postrun CPU time. Very large	Any filtering imaginable can be implemented.	Slow. Filter must have appropriate shape and width or distortion will occur.
Analog filter hardware	Low pass, high pass, or bandpass filtering	Depends on components	Small	Fast	Possible phase and amplitude distortion.

[a] After D. Binkley and R. Dessy, *J. Chem. Ed.*, **56**, 152 (1979). (Courtesy of the Journal of Chemical Education.) PTS is the number of points; T$_{conv}$ is the conversion time; and CPU is the central processing unit.

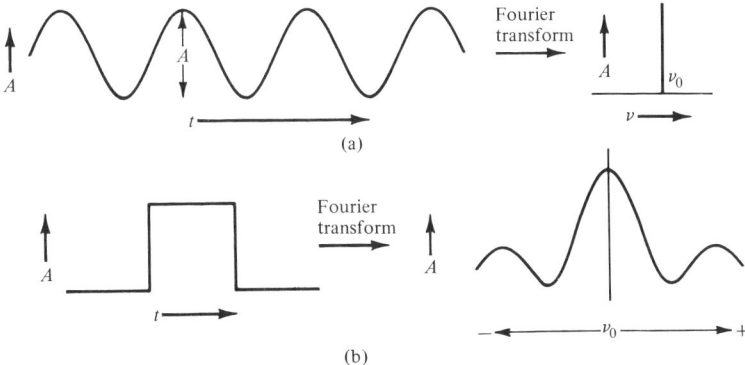

FIGURE 29-11 Simple Fourier transform pairs. (a) Cosine function and (b) square-wave function.

If, for example, $f(t) = A \cos 2\pi\nu_0 t$, then $F(\nu)$ is a single line at ν_0 with an amplitude A (Fig. 29-11a). If $f(t)$ is a square wave, then $F(\nu)$ broadens into the form shown in Fig. 29-11b. Addition of three component waveforms, each occurring at a different frequency, produces the $f(t)$ pattern shown in Fig. 29-12. The corresponding Fourier transformation, $F(\nu)$, consists of three lines located at the frequencies of the original waveforms. In this example, $f(t)$ represents the data produced by an optical interferometer or pulsed nuclear magnetic resonance (NMR) signal source and $F(\nu)$ is the conventional frequency spectrum.

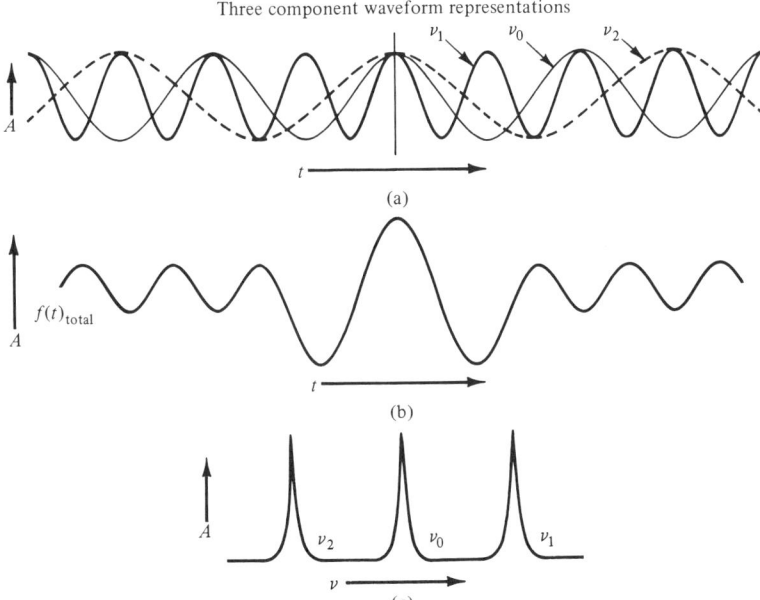

FIGURE 29-12 Fourier representation of data. (a) Individual $f(t)$'s, (b) resultant $f(t)$, and (c) Fourier transform to $F(\nu)$.

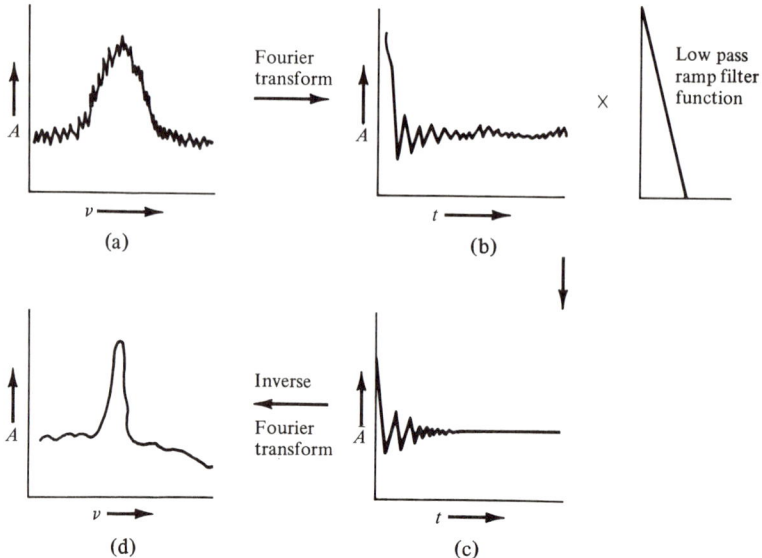

FIGURE 29-13 Application of Fourier transforms in digital filtering.

In Fourier transform spectroscopy the data are rapidly generated in the time-domain [$f(t)$] form by either an interferometer (in the case of optical spectroscopy) or a pulsed signal. The resulting data are in the form of superimposed waves and include all the frequencies of the spectral range of the instrument. In order for the computer to have sufficient data to perform the necessary Fourier transformation, sampling of the analog data must occur at a rate twice that of its highest-frequency component. Computer implementation of the transformation of the digitized data from time domain $f(t)$ to frequency domain $F(\nu)$ is carried out by means of a summation over a finite number of points, N,

$$F(\nu_j) = \sum_{k=1}^{N} f(t_k) e^{-i(2\pi)\nu_j t_k}; \quad j = 0, 1, 2, \ldots, M \qquad (29\text{-}8)$$

where N = total number of data points and ν_j, the jth component frequency out of a total of M frequencies. The results, $F(\nu)$, appear as a conventional frequency spectrum. Signal-to-noise enhancement is achieved by using the time saved by rapid scanning to make multiple scans that are then treated by signal averaging techniques. For example, in carbon-13 NMR spectroscopy, approximately 1000 scans are required to extract the weak carbon-13 signal from the noise. The FT technique requires 15 min to complete the scans while the normal continuous-wave method would require 60 days to obtain the same information.

The second application of FT is signal conditioning illustrated by digital filtering. Data contained in an amplitude–frequency spectrum (Fig. 29-13a) may be filtered by transforming the data to an amplitude–time spectrum using a Fourier transformation (Fig. 29-13b). The resulting waveform is then multiplied by an appropriate mathematical filter function to obtain the desired frequency response (Fig. 29-13c). Finally an inverse Fourier transformation regenerates the filtered amplitude–frequency spectrum (Fig. 29-13d). This method allows a variety of filter functions to be implemented using the appropriate software. Many of these filters would be impossible to implement using hardware devices. The major limitation of the software method is its slow speed relative to hardware techniques. In numerous situations where the speed of signal conditioning is not the limiting factor, the flexibility of software methods provides a great advantage over hardware techniques.

29.7 EVALUATION OF RESULTS

Total control of experimental variables is usually difficult and often impossible. Sampling methods, analysts' techniques, and instrument responses are each potential sources of error. Statistical methods provide a means for objectively evaluating the source and amount of error in analytical methods. The often used phrase "within experimental error" is meaningless if the magnitude of this error is not defined through use of statistical techniques.

Types of Errors

In order to obtain reliable results from an analytical method, sources of error must be identified and either eliminated or minimized. Errors may be classified as one of three major types—random, systematic, or gross.

Since the intrinsically uncertain nature of the measurement technique is the source of random error, this kind of error occurs in every analysis. Thermal, shot, and flicker noise, discussed earlier in this chapter, are sources of random error. The magnitude of a random error is usually small and can be minimized by filtering methods (either hardware or software).

The second kind of error, systematic or procedural error, causes results to deviate from the expected values in a constant manner. Sources include incorrect calibration of an instrument, insufficient purity of reagents, and improper operation of an instrument. Errors of this kind may often be identified and eliminated by modification of the analytical procedure.

Gross errors are distinguished from the previous two kinds of errors by their irregular nature and large magnitude. The occurrence of a single gross error in a series of repetitive analyses can greatly affect the accuracy of the final result. Carelessness in analytical technique, improper recording of intermediate data and results, and errors in calculations are common sources of gross error.

Expression of Error

Error may be expressed in absolute terms as the difference between an analytical result, x, and the known true value, μ:

$$d = (\mu - x) \tag{29-9}$$

When this difference is expressed as an unsigned number, it is known as an absolute error. Because the absolute error represents a difference between the result and the true value, it must be expressed in the same units of these quantities. An absolute error has no significance when separated from the result or true value. For example, an absolute error of 5.1

µg/ml may be acceptable in an analysis of a sample containing 511-µg/ml lead but unacceptable in a sample containing 1.7-µg/ml lead.

The relative error, E_{rel} is used to determine the accuracy of measurement and is typically expressed as the percentage of the known true value:

$$E_{rel} = \frac{d}{\mu}; \quad \%E_{rel} = \frac{d}{\mu} \times 100 \tag{29-10}$$

Since the relative error is a dimensionless number, it can be used to determine the accuracy of results as well as to compare the accuracies of results expressed in different units.

The following example illustrates the use of absolute and relative error for comparison of results. Analysis of lead and zinc in a sample yield the following results: Pb = 653 µg/ml, d_{Pb} = 4.3 µg/ml; and Zn = 4.5 µg/ml, d_{Zn} = 0.15 µg/ml. Therefore,

$$\%E_{Pb} = \frac{4.3}{653} \times 100 = 0.65\% \quad \text{and} \quad \%E_{Zn} = \frac{0.15}{4.5} \times 100 = 3.3\%$$

It is obvious that the lead determination gave the more accurate result, since it contains the smaller relative error, even though the absolute error was much greater than that of the zinc determination.

Precision and Accuracy

Accuracy may be defined as the agreement of a measurement with the known true value for the quantity being measured. Precision is concerned with the ability to reproduce the same values for a set of parallel observations. These terms are contrasted in Fig. 29-14. The shots in Fig. 29-14a are both accurate and precise, while the shots in Fig. 29-14b are precise but not accurate. The situation in Fig. 29-14b usually indicates the presence of a systematic error, in this case, perhaps poor alignment of the sighting mechanism on the rifle. While the accuracy of a measurement is determined by many factors, the precision is often limited by noise alone.

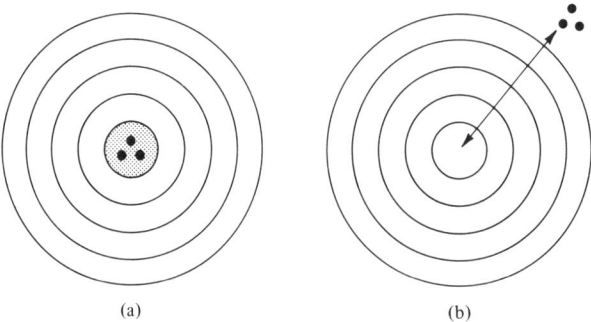

FIGURE 29-14 Accuracy vs. precision. (a) Shots accurate and precise; (b) shots precise only.

Precision and Significant Figures

Evaluation of an analytical method to discover the source and magnitude of errors requires careful acquisition and processing of data as well as application of appropriate statistical methods. Initial data must be reported with a precision which is indicated by the number of significant figures. Subsequent operations and calculations involving these data must preserve the correct number of figures so that the results will give a true indication of both the accuracy and precision of the analysis. Moreover, unless the proper number of significant figures are maintained, the results of any statistical treatments are meaningless.

Significant figures usually have economic importance—the more significant figures reported in a measurement, the more costly the analysis. Factors contributing to increased cost include more expensive instrumentation, higher-quality reagents, and increased expenditure of time and effort in performing the analysis. Once experimental observations have been recorded using the appropriate number of significant figures, care must be taken to ensure that the results of subsequent mathematical operations correctly represent the precision of the original measurements. It is impossible to increase the precision of experimental measurements by simple arithmetic operations. Likewise, care should be taken to assure that the number of significant figures is not improperly reduced in performing these simple operations. Readily available calculators and computers present the temptation to ignore the rules for handling significant figures and thus to produce results that nullify the experimental methods used to obtain the original data.

If the limit of error of a measurement is known, it should be represented as follows: 3946 ± 11. If the error is not stated, the last digit of the measured value is assumed to be uncertain and therefore determines the number of significant figures. Thus the value 3946 contains four significant figures and may represent a value between 3945.5 and 3946.5. The accepted rule for preserving the correct number of significant figures in multiplication and division is that the answer should contain *only* as many significant figures as the smallest number of figures in any factor. For example, the product of 493.15 times 32 is 15,780.8, but the above rule requires the answer be rounded off to 16,000. This value should be unambiguously reported as 1.6×10^4.

In addition and subtraction, the answer should be rounded off to the first column that has an uncertain digit. For example, 487.3751 g, the sum of 394.5 g, 91.47 g, and 1.4051 becomes 487.4 g according to this rule.

Statistical Methods and Their Applications[7,8]

In reporting the result of a set of identical (parallel) measurements, the mean and variance (usually reported in the form of its square root, the standard deviation) are fundamental parameters that form the basis for many statistical tests. If the number of parallel observations is greater than 30, these tests are assumed to be based on a Gaussian distribution, while if the number of measurements is 30 or less, a *t*-distribution is assumed and tests are based on Student's equations.

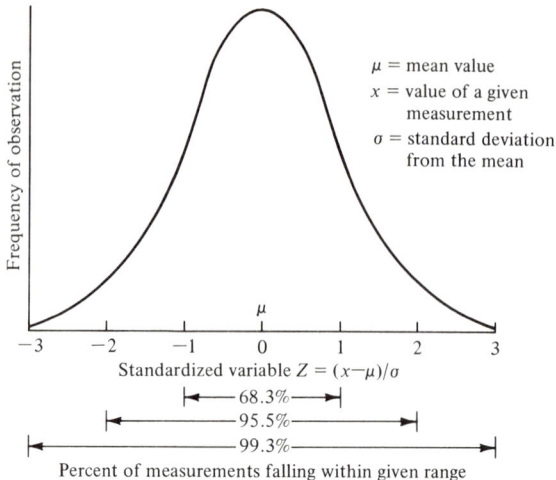

FIGURE 29-15   Gaussian distribution.

The precision of an analytical method is usually indicated by its confidence limit (Fig. 29-15); that is, the range of values about the mean that includes a specified percentage of the total observations. Precision increases with decreasing value of standard deviation. Other tests that are routinely employed include:

(1) The Q-test for outlying measurement (detects gross errors).
(2) The t-test for significance in the difference between two means (useful in comparing results from different instruments, methods, or samples of supposedly the same composition).
(3) The t-test for accuracy of results (for comparing the results of several methods against a known true value).

In addition to individual tests for significance of results, control charts are useful monitoring tools in laboratories where large amounts of data from a given method are generated over extended periods of time. The results of individual samples falling outside established confidence limits may be rejected. Trends in results may also be detected and corrective measures taken.

Although these and other statistical methods cannot solve all the problems involved in error analysis, they do provide the analyst with a systematic, objective approach to these problems. Statistical analysis of data should determine when differences among results have exceeded reasonable limits and have become large enough to imply the existence of real variations in sample compositions or analytical techniques.

29.8   ACCURACY AND INSTRUMENT CALIBRATION

Proper calibration (standardization) of instruments is essential in obtaining accurate analyses. The choice of a calibration technique is affected by the instrumental method, instrument response, interferences present in the sample matrix, and the number of samples to be analyzed. Three of the most commonly used calibration techniques are the analytical or working curve, the method of standard additions, and the internal standard method.

Analytical Curve

In the analytical (working) curve technique, a series of standard solutions containing known concentrations of the analyte is prepared. These solutions should cover the concentration range of interest and have a matrix composition as similar to that of the sample solutions as possible. A blank solution containing only the solvent matrix is also analyzed and the net readings—standard solution minus blank (background)—are plotted versus the concentrations of the standard solutions to obtain the working calibration curve (Fig. 29-16). If a nonlinear plot results, as is often the case, electronic hardware or computer software can be used to compensate for the curvature and produce an output that is a linear function of concentration. The number of standard solutions analyzed in nonlinear regions should be increased to maintain accurate analysis of unknown samples.

Linearity may also be achieved in some analyses by varying instrumental parameters. In the case of spectrophotometric analysis, changing the wavelength used to obtain absorption readings may produce a more linear working curve. It is of utmost importance to record all instrumental parameters used in obtaining data for the calibration curve because small variations in these parameters can affect the slope of the calibration curve. The calibration curve must be checked periodically using solutions of known concentration to detect any changes in instrument response.

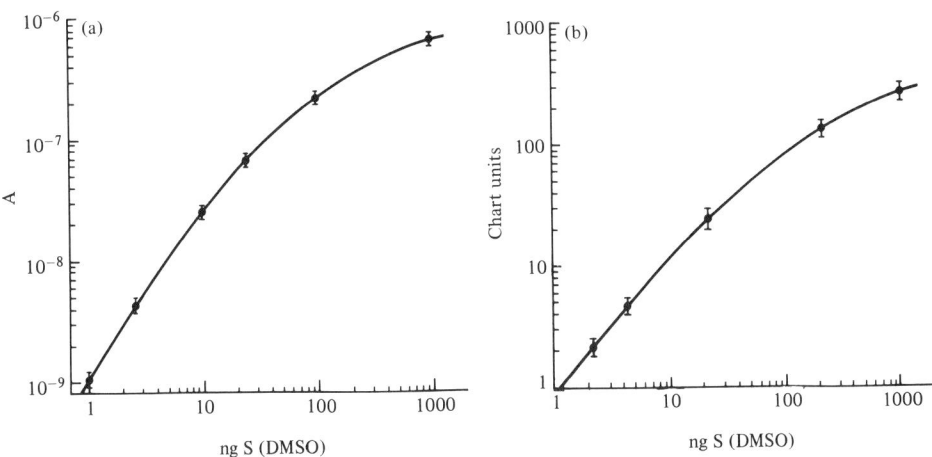

FIGURE 29-16 Calibration curves for the determination of dimethylsulfoxide in aqueous solutions by (a) flame photometric detector and (b) flame ionization detector. (After M. O. Andrea, *Anal. Chem.*, 52, 152 (1980). Courtesy of the American Chemical Society.)

Method of Standard Additions[9]

When it is impossible to suppress interferences from matrix elements, the standard addition method may be used. The instrument response must be a linear function of analyte concentration and yield no response at zero concentration of the analyte in order for this calibration technique to be employed. A small amount of analyte solution of known concentration is added to a portion of a previously analyzed sample and the analysis repeated using the same reagents, instrument parameters, and procedures. If an instrumental reading, R_x, is obtained from a sample solution of unknown concentration, x, and a reading R_1 from the sample solution to which a known concentration, a, of analyte has been added, then x can be calculated from the relation

$$\frac{x}{x+a} = \frac{R_x}{R_1} \tag{29-11}$$

Readings must be corrected for any background signal. It is always advisable to check the result with at least one other standard addition. Additions of analyte equal to twice and to one-half the amount of analyte in the original sample are optimum statistically. All solutions should be diluted to the same final volume so any interferent in the sample matrix will have an identical effect on each solution. Sufficient time must elapse between addition of the standard and actual analysis to allow equilibration of added standard with any matrix interferents.

A graphic solution using the standard addition method is shown in Fig. 29-17. The concentration scale (x-axis) is determined by the concentrations of analyte added to the sample solutions and thus the unknown concentration is given by the point at which the extrapolated line intersects the concentration axis.

The method of standard addition is widely used in electroanalytical chemistry to obtain results which are more accurate than those obtained using calibration curves. Since the unknown and standard solutions are measured under identical conditions, matrix-sensitive voltammetric techniques such as anodic stripping voltammetry rely almost exclusively on standard additions for quantitative results. Atomic absorption and flame emission spectrophotometry employ this method with complex sample matrices where viscosity, surface tension, flame effects, and other properties of the sample solution cannot

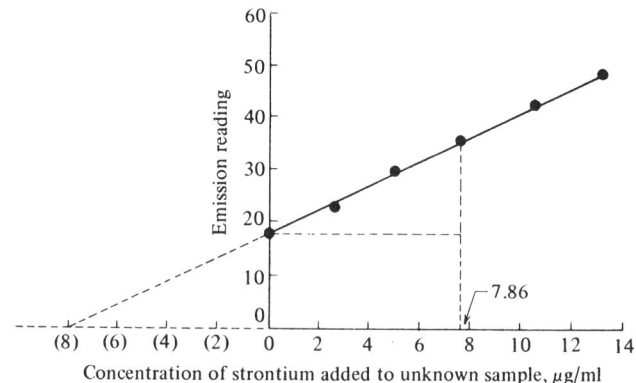

FIGURE 29-17 Graphic representation of the standard addition method of evaluation.

be accurately reproduced in calibration solutions. Results from standard additions can also provide a systematic means of identifying sources of error in analyses, such as depletion of test reagents, a defective instrument, or incorrect standard solutions.

Method of Internal Standard

In the internal standard method, a fixed quantity of a pure substance is added to samples and standards alike. The response of the analyte and internal standard, each corrected for background, are determined and the ratio of the two responses is calculated. If the parameters affecting the measured responses are well controlled, the response of the internal standard line will not change, since the concentration of the internal standard is fixed. If, however, one or more of the parameters affecting the measured responses varies, the internal standard and analyte responses should be affected equally. Thus the response ratio (analyte to internal standard) depends only on the analyte concentration. A plot of this response ratio as a function of analyte concentration yields a calibration curve.

The internal standard should be a substance similar to the analyte with an easily measured signal that does not interfere with the response of the analyte. In addition, the response of the internal standard should not be affected by other components of the sample. The concentration of the internal standard should be of the same order of magnitude as that of the analyte in order to minimize error in calculating the ratio. This method is used extensively in gas chromatographic analysis and to a lesser extent in infrared and emission spectroscopy.

BIBLIOGRAPHY

Currie, L. A., "Sources of Error and the Approach to Accuracy in Analytical Chemistry," Chap. 4, *Treatise on Analytical Chemistry,* 2nd ed., I. M. Kolthoff and P. J. Elving, Eds., Vol. 1, Part I, Wiley-Interscience, New York, 1978.

Eckschlager, K., *Errors, Measurements and Results in Chemical Analysis,* Van Nostrand Reinhold, London, 1969.

Mandel, John, "Accuracy and Precision: Evaluation and Interpretation of Analytical Results," Chap. 5, *Treatise on Analytical Chemistry,* 2nd ed., I. M. Kolthoff and P. J. Elving, Eds., Vol. 1, Part I, Wiley-Interscience, New York, 1978.

LITERATURE CITED

1. Cooper, J. W., *Anal. Chem.,* **50**, 801A (1978).
2. Savistzky, A. and M. J. E. Golay, *Anal. Chem.,* **36**, 1627 (1964).
3. Dulaney, G. F., *Anal. Chem.,* **47**, 25A (1975).
4. Binkley, D. and R. Dessy, *J. Chem. Educ.,* **56**, 148 (1979).
5. Becker, E. D. and T. C. Farrar, *Science,* **178**, 361 (1972).
6. Marshal, A. G. and A. W. Comisarow, *Anal. Chem.,* **37**, 491A (1975).
7. Youden, W. J., *Statistical Methods for Chemists,* Wiley, New York, 1951.
8. Bauer, E. L., *A Statistical Manual for Chemists,* 2nd ed., Academic, New York, 1971.
9. Bader, M., *J. Chem. Educ.,* **57**, 703 (1980).

CHAPTER 30

Computer-Aided Analysis

Computers have become basic laboratory tools that aid the chemist in performing required analyses. This chapter is an introduction to computers and their applications as components of analytical instrument systems. Concepts and "jargon" important in using laboratory computers are presented. Although minicomputers are discussed because of their impact on present and future instrumentation, major emphasis is given to microcomputers. These small devices have dispersed the capabilities of the computer to almost every type of laboratory instrumentation and have also facilitated communication between instruments and larger computers. The influence of microcomputers on analytical methods and instrumental design is just beginning to be felt.

30.1   INTRODUCTION

Digital computers have become integral components of modern methods of analysis. Applications of these devices to analytical instrumentation have increased with advances in computer technology. Initially computers were used to automate conventional calculations and existing instruments. Later new measurement methods were developed which were possible only through the use of computerized instrumentation and high-speed data processing techniques. Most recently, microcomputers are influencing instrument design as well as analytical methods. To understand the role of a computer in a specific instrumental method, it is necessary to consider the interactions among instrument, computer, and analyst. Three important combinations of interactions are off-line, on-line, and in-line (Figs. 30-1–30-3).[1]

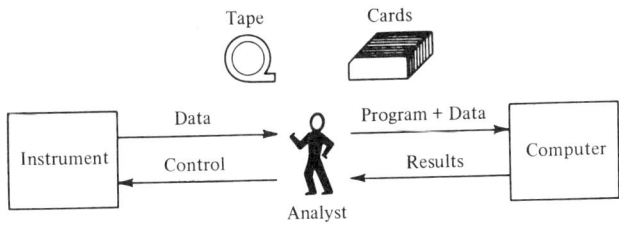

FIGURE 30-1 Off-line computer configuration.

Off-Line

Initial applications of digital computers to instrumentation utilized the off-line configuration (Fig. 30-1). In this configuration computer programs for processing instrument output data are written in an analyst-oriented language, such as FORTRAN or BASIC. Data are collected from the instrument, transferred to a suitable input medium (magnetic tape or punched cards), and then submitted with the appropriate program as a job to the computer. Jobs are processed sequentially by the computer operating in batch mode; the final results appear at an output device such as a line printer. Time required to obtain results depends upon the length and priority of the job as well as the total number of jobs to be processed by the computer.

Although small computers can be operated off-line, this configuration is generally implemented with large computers in situations requiring complex calculations, manipulation of sizable amounts of data, or a combination of the two. Efficient use of larger computers restricts their operation from being interrupted by an individual analyst or instrument for the purpose of data input or control. Because there is no direct communication between the computer and the instrument, data must be transferred indirectly under the control of the analyst. Thus, the computer running in an off-line environment cannot respond to the instantaneous needs of a specific instrument or analyst, but rather processes jobs in the order submitted.

On-Line

By the late 1960s advances in electronics resulted in the compact, moderately priced minicomputer that could be operated on-line with instrumentation (Fig. 30-2). In this configuration the computer is linked directly to instruments through an electronic interface. A single minicomputer dedicated to one or more instruments can perform specific tasks.

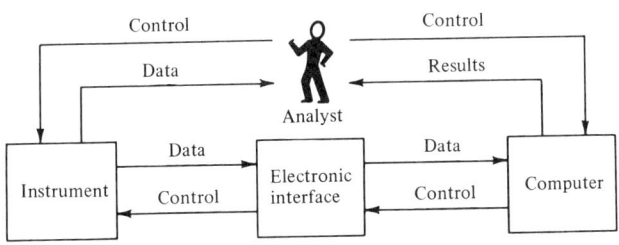

FIGURE 30-2 On-line computer configuration.

These include the acquisition and processing of data as well as instrument control functions. The analyst interacts with both the computer and the instrument to obtain and process data, to control instrument operation, and to retrieve results. Analyst–instrument interaction can be reduced by increasing the number of instrument control inputs and data outputs interfaced to the computer.

An on-line computer directly interfaced to an instrument can operate in a real-time mode. In this mode, the computer can respond instantaneously to data acquired from the instrument. Computations, control functions, and output of information occur rapidly enough to improve the dynamic operation of the instrument.

A variety of analytical instrumentation has been interfaced to on-line computers. Gas chromatography employs this technique for rapidly reducing large amounts of data taken from one or more instruments to concise, accurate results. Dedicated minicomputers have decreased the time necessary to obtain results of X-ray diffraction studies of absolute compound configurations from months to hours. Another beneficiary is high-resolution mass spectrometry. Fourier transform nuclear magnetic resonance and infrared spectroscopy, methods that require rapid execution of complex mathematical transformation functions, would be impossible without the aid of on-line computers.

In-Line

In the 1970s microcomputers became available as low-priced integrated circuit chips. The combination of reduced size and decreased price resulted in the replacement of minicomputers with microcomputers for some on-line applications. More importantly, however, instrument design was modified to include microcomputers as internal components. When the computer becomes an integral, dedicated part of the packaged instrument, the configuration is known as in-line (Fig. 30-3).

Although microcomputers were originally designed to replace hardwired, logic, control circuits in relatively simple machines, such as cash registers, microwave ovens, and sewing machines, later generations of these devices are capable of data processing tasks in addition to instrument control. In-line microcomputers are responsible for the intelligence of "smart" analytical instruments. Built-in microcomputers were introduced initially into chromatographs (gas and liquid) and spectrophotometers (infrared, visible, and ultraviolet) where precise control of instrument parameters and the ability to perform repetitive analyses are required.

In-line microcomputers can be controlled by programs stored in read-only memory (ROM). These programs are placed in the computer by the manufacturer and cannot be altered by the analyst. It is, however, possible to substitute or upgrade prestored programs (firmware) by changing ROM chips, thus avoiding extensive hardware modifications. The

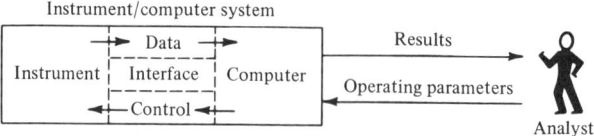

FIGURE 30-3 In-line computer configuration.

instrument designer may alter data handling functions, modify input and output formats, and even redefine instrument controls as new instrument applications emerge.

Advantages resulting from the replacement of conventional instrumental electronic hardware with microcomputers include:

(1) increased reliability (decreased maintenance) due to substitution of electrical and mechanical components with sequences of programmed instructions;
(2) more complete and reliable analysis by prompting the analyst to enter and check all variables;
(3) improved accuracy of results with automatic, periodic instrument calibration;
(4) easier troubleshooting and maintenance by built-in diagnostic tests that check the functions of instrument components;
(5) improved precision of results using digital signal processing;
(6) ease of communication with other devices and computers external to the instrument.

In summary, an in-line microcomputer running in real time can prompt the analyst for instrumental input parameters, supervise the operation of the instrument, process data, and report analytical results in the desired format.

30.2 COMPUTER ORGANIZATION—HARDWARE

Although digital computers differ widely in data processing rates, memory sizes, and word lengths, every computer comprises five basic functional units connected by signal pathways known as buses. These units are the central processing unit, the arithmetic-logical unit, input, output, and memory. The heart of a computer is the central processing unit (CPU) that contains the control unit (CU) and the arithmetic-logic (ALU) unit (Fig. 30-4).

Arithmetic-Logic Unit

The ALU performs the arithmetic and logic operations on data presented to it. Data are processed in the form of binary words, each word containing a specified number of binary bits. A bit consists of either a one or a zero. All operations are, therefore, performed using the principles of Boolean algebra (Chapter 27). Arithmetic operations include addition and subtraction; logic operations involve "ANDing," "ORing," and shifting all the bits of a word to the left or right. Although the number of basic operations performed by any computer is limited, the rapid execution of a series of these instructions, known as an algorithm, can produce many useful functions. Common algorithms include such functions as multiplication, integration, and manipulation of data arrays.

Control Unit

The CU is responsible for coordinating the operation of the entire computer system. Specifically, it generates and manages the control signals necessary to synchronize the flow of data on all buses with the operation of the functional units. Another essential role of

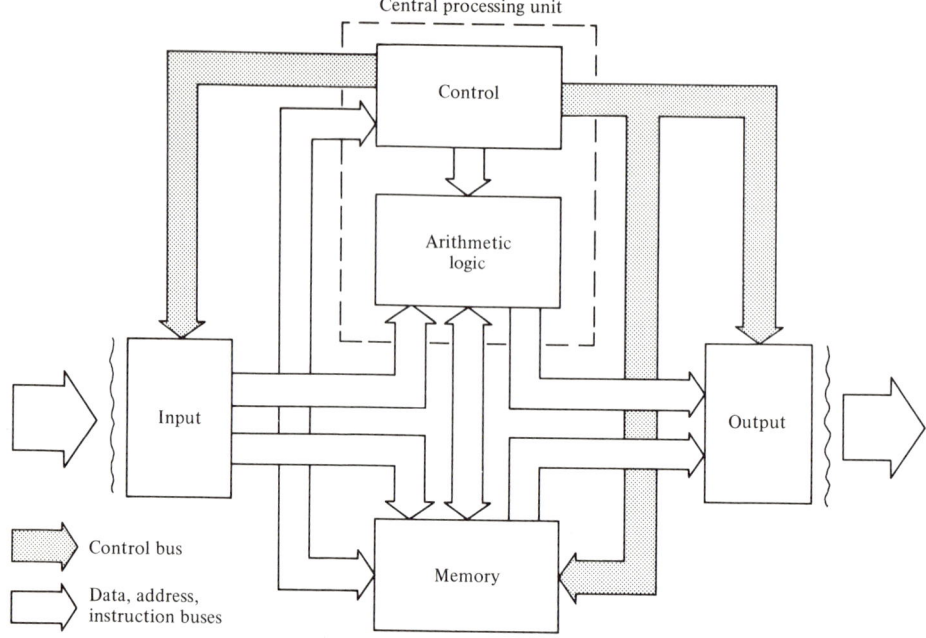

FIGURE 30-4 Basic computer organization.

the control unit is that of fetching, decoding, and executing successive instructions (a program) stored in the memory unit.

Central Processing Unit

The control unit is usually physically linked to the arithmetic-logic unit that it controls; the combination of CU and ALU is known as the CPU. Two critical parameters in evaluating CPU operation are the minimum time required to carry out specific instructions and the number of bits in a computer word. In general, both processing rates and the number of bits per word increase in going from micro to mini to the larger computers. A typical microcomputer has a word length ranging from 4 to 16 bits while in large computers, words vary between 32 and 64 bits. The data processing rate of a microcomputer may be from 10^{-2} to 10^{-6} times slower than the larger machines. It should be noted here that advances in solid-state electronic technology are increasing the processing speeds of all computers. Future microcomputers may have the speed of today's large computers.

Memory Units

The memory unit is used by the CPU for rapid storage and recall of information. It is linked to the CPU by address lines, data lines, and control lines. Address lines carry required binary information to locate specific parts of the memory, data lines carry information between memory locations and the CPU, and control lines direct the sequence of data transfers. Electronic memories currently available include ferrite core, bipolar, metal oxide semiconductor, magnetic bubble, and charge-coupled devices. Important memory characteristics are storage capacity, cost per bit of storage, reliability, and access time. Access time is the time required to read or write data at any storage location. Storage capacity is typically expressed as the number of storage locations, the number of bits in a given location matching the computer word length. Thus 1024 locations (known as 1K of memory) of 16-bit words would contain twice as many bits as 1K of 8-bit words. The cost per bit is roughly related to the complexity of the total memory; the more auxiliary electronic components required, the higher the price. In general, faster memories are more expensive than slower ones. Memory access time is important because it often limits the overall operating speed of the computer.

Memories can be classified as volatile or nonvolatile; volatile memories lose their data when power is removed while nonvolatile memories do not. There are memories that allow transfer of information in both directions between the CPU and memory storage locations, read/write memories, and ROM that permit transfer in only one direction—from memory to the CPU. When the access time of a read/write memory is independent of the position of the storage location in memory, the memory is known as random access memory (RAM). Although ferrite code memory can be classified as nonvolatile RAM, the terms RAM and ROM are usually applied to MOS- and CMOS-type memories; RAM is volatile and ROM nonvolatile. Once a program is entered into ROM it may never be altered by any means. An erasable programmable ROM, EPROM, may be erased and reprogrammed. Both ROM and EPROM programming require special hardware.

The memory types discussed thus far are known as internal memories since they are linked to the CPU by internal buses for rapid access to information. Information may also be stored external to the computer through use of a variety of peripheral devices and then transferred to the computer under the control of the CPU. External memory devices include magnetic disks, magnetic tape, paper tape, and punched cards.

Two types of information are contained in memory: instructions and data. Sequences of instructions directing the computer to perform specific tasks are known as programs. Under the supervision of the CPU, each successive instruction is fetched from a memory location, deposited in a special register of the CPU where it is decoded and finally executed by the computer. Data required by the program are processed by the ALU. Although data may be transferred to-and-from the computer in a number of formats, it is ultimately processed by the CPU as binary information.

Input/Output Units

The remaining two modules, the input and output (I/O) units, provide the computer with its links to the outside world and are thus important considerations in interfacing instrumentation. Input units supply data to the ALU or, in some situations, directly to the internal memory. External sources of data include keyboards, instruments, and external memory devices. Output units transfer data from the ALU or internal memory to external devices, such as light emitting diodes (LEDS), printers, video terminals, plotters, external memory units, and control devices (stepper motors, relay switches, and so on).

Thus a single computer control unit may supervise the operations of many individual peripheral I/O devices.

Buses

Buses connecting the component parts of a computer system are grouped according to the type of information they transmit. Buses linking the components of the CPU are known as internal buses, while those joining the CPU to memory, peripheral I/O devices, or other computers are designated as external buses. The specific number of transmission lines in a given kind of bus is determined by the computer architecture and the function of the bus. Since rapid communication among computer components is required, digital signals are transmitted simultaneously in parallel over the lines of a particular bus (Fig. 30-5).

The number of individual lines comprising an internal data bus is determined by the length of the word (in bits) processed by the computer. For example, a minicomputer capable of processing 16-bit words requires a data bus containing 16 parallel lines. Both the origin and the destination of information transmitted on the data bus are specified by signals on the address bus lines. Information may be sent to, or received from, an ALU register, a memory location, or any peripheral I/O device. An address bus with 16 lines can directly address 2^{16} (65,536) registers, locations, or devices. The timing of information transfers on the data bus is synchronized by signals appearing on the control bus that carries status and control information to-and-from the CPU. A typical control bus contains ten separate status and control lines.

External buses may be either parallel or serial. Parallel buses allow high-speed communication and require less complicated interface hardware. Many external devices may be connected to a single parallel bus. Serial buses require fewer lines (two for each device) but are slower in transferring data and require more complex interface hardware. Serial buses usually connect communications terminals and remote instruments to computers. Standardization of bus structures and protocols (the programmed response of the computer to specific data inputs) greatly simplifies the task of interfacing peripherals and instruments to computers. Two standard buses have gained wide acceptance among computer and instrument manufacturers; the IEEE Standard 488-1975 parallel bus and the RS-232 serial

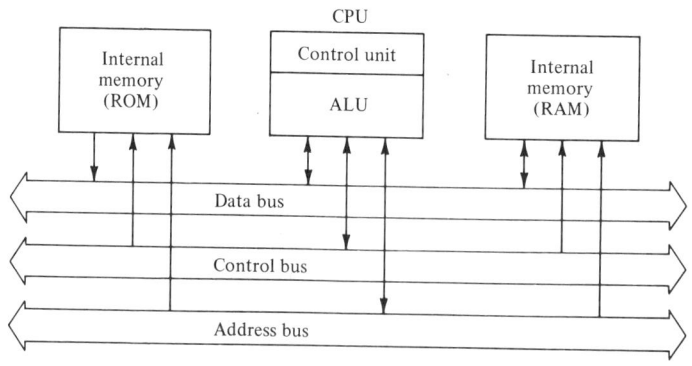

FIGURE 30-5 Internal bus connections.

bus.[2] Serial-to-parallel converters (registers) enable several serial devices to be attached to a parallel bus (see Chapter 28).

30.3 COMPUTER ORGANIZATION—SOFTWARE

While computer hardware costs have decreased at a rate of approximately 30% each year, software costs have increased by 15%. Therefore, a discussion of software development is necessary to complement the preceding discussion of hardware.

Programming can be carried out at various levels; from the lowest which is machine-oriented binary code to the highest which is machine independent. A comparison of programming a simple arithmetic addition at three different levels is given in Table 30-1.

Machine Language

The central processing unit of any computer responds to a set of binary coded instructions. The number of different instructions in the set and the binary codes for specific instructions depend on the internal architecture of the computer which in turn varies with different manufacturers and models. Sequences of these binary coded instructions are known as machine language programs. All higher level programs must ultimately be translated into series of machine language instructions, known as object programs. Writing low level machine language (object) programs is a tedious, time-consuming task and should be avoided whenever possible.

Assembly Language

The next highest level of programming is assembly language. It is composed of a set of mnemonics (groups of letters and numbers) representing machine language instructions; one assembly instruction for each machine instruction (Table 30-1). Programming at the assembly language level is easier and faster than machine language programming. A program known as an assembler is necessary to convert programs written in assembly language to machine language programs. Assembly language programs can be written, translated to

TABLE 30-1 Comparison of Programming an Add Instruction

	INSTRUCTION LANGUAGE	
High Level (FORTRAN or BASIC)	Assembly (Mnemonic)	Machine (Binary Code)
D = B + C	MOV A, M	01111110
	INR M	00110100
	ADD M	10000110
	INR M	00110100
	MOV M, A	01110111

object code, and executed on a single computer, provided an assembler program is resident in the computer's memory. Sufficient memory area must also be available for storage and execution of the user's program. In many situations involving minicomputers and microcomputers with limited memory space, cross-assembler programs that run on larger, faster computer systems are used. These cross assemblers produce machine language programs that can be transferred to smaller computers for execution. The instruction sets of both assemblers and cross assemblers are machine dependent (instructions for one kind of computer cannot be utilized to program a different kind of computer).

High-Level Languages

Languages utilized at the highest level of programming (FORTRAN, BASIC, PL/1, and so on) are the most convenient and concise for the human programmer. The detailed arrangement of machine language instructions is the responsibility of the translator program. High level languages are usually algebraic in nature with each line of the program, known as source code, producing several lines of object code (Table 30-1). Transportability (machine independence) is another factor contributing to the convenience of high level programs. The same source coded program may be run on different kinds of computers if a suitable translator program is available for each computer. It is important to note that while high level languages are usually machine independent, translator programs are not.

The most recent, highest level of computer software utilizes the concept of structured programming.[3,4] This concept involves the disciplined formulation and construction of algorithms in a systematic manner based on levels of abstraction. Dealing with problems at a high level of abstraction permits a top-down approach in which the tasks found on each level are broken down into individual subroutines. Each subroutine is written as a separate unit, starting with the most fundamental commands. These commands are combined into higher and higher level statements. Subroutines can be tested as units starting from the bottom up, a procedure which simplifies program debugging. Thus a single program is visualized as a collection of interacting subroutines.

PASCAL and FORTH are examples of languages designed to facilitate structured programming. PASCAL is better suited to run on larger computer systems, while FORTH runs well on smaller systems. FORTH was developed for programming tasks associated with the operation of radio telescopes. It provides the programmer with a flexible language for implementing the tasks associated with the operation and control of laboratory and process control instrumentation. Operator–instrument interaction is facilitated by the use of easily understood commands. These commands are developed by the programmer for a specific instrument system. High level languages such as PASCAL and FORTH should aid in reducing software development costs.

Translators

There are two kinds of high level translator programs: compilers and interpreters. Both translators produce machine language object code but differ in their method of operation. A compiler translates the entire high level program into object code. The resulting object program may not be executed until translation is complete. Programs may be compiled on one computer and executed on another. Programs are typically compiled in an off-line configuration; they may then be executed in off-line, on-line, or in-line configurations. Since compilers do not optimize the number of instructions required to perform a given task, object programs generated by a compiler are from two to five times longer than object programs produced from assembly language code written by skilled programmers.

Interpreters translate high level programs line-by-line and execute each series of resulting machine language instructions before interpreting another line of source code. Programs must be interpreted and executed by the same computer. Because a program cannot be interpreted by one computer and executed by another, the interpreter software must be resident in the memory of the computer executing the program. Interpreters (for example, BASIC) are typically used with on-line computers that respond to the programmer in real time.

Both interpreter and compiler programs suffer several disadvantages; both require large amounts of memory and neither is usually capable of directly generating instructions to handle I/O data associated with devices external to the computer. These I/O instructions are typically written in assembly language code. Once an object program in machine language has been prepared by a compiler, it may be stored and quickly recalled for execution at any time. In contrast, the use of an interpreter requires that the high level language source program must be reinterpreted and executed, one line at a time, *each* time the program is run. No permanent object program is available for storage. Thus the use of an interpreter slows program execution and can become a serious limitation in laboratory applications requiring high-speed data acquisition and manipulation. Interpreters, however, allow maximum flexibility in operator–computer interaction, including error and caution messages during program preparation and execution. In general, if memory costs are of little importance and relatively slow operation is acceptable, an interpreter language is the best choice for programming smaller computers. However, the method of programming in the future may be the use of efficient compilers that generate compact object programs.

Programming Aids

Auxiliary programs known as editors and debuggers expedite program development at all language levels. An editor allocates an area of memory for use as a scratch pad where program instructions can be written, appended, inserted, deleted, and so on. Certain kinds of errors, such as incorrect use of symbols, are detected by the editor program. The edited source program can then be assembled or compiled. A source program may be transferred to scratch pad memory and modified using the editor. If an interpreter program is used, each line of the source program can be edited before it is translated into object code.

Debugger software is used to detect and correct errors discovered during the execution of the object program. Features of debuggers include the ability to halt execution of an object program after any specified instruction (breakpoint), to display contents of both CPU registers and memory locations, and the ability to alter contents of memory locations. Diagnostic error messages can be generated while the program is running under the control of the debugger. Figure 30-6 outlines the steps involved in program development using auxiliary programming software.

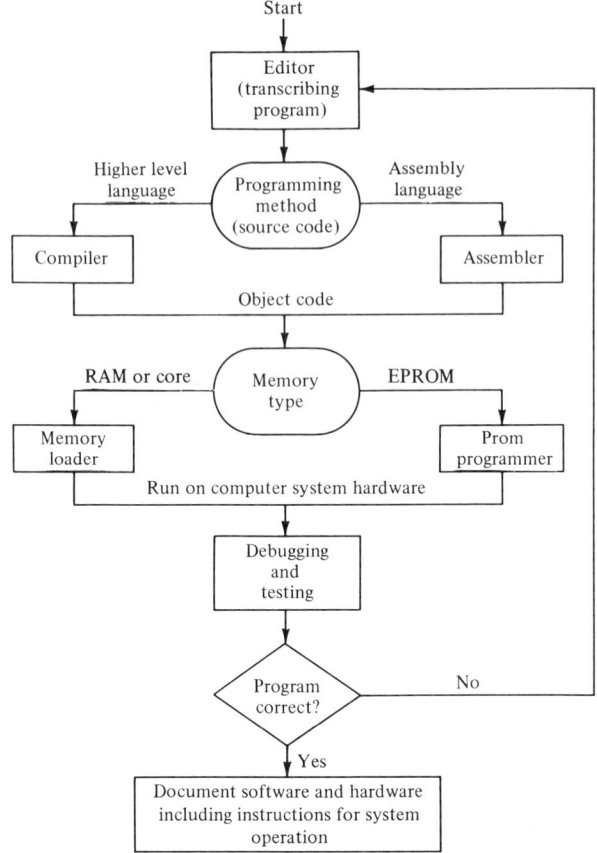

FIGURE 30-6   Software program development.

Operating System Software

If a computer system is to perform a variety of tasks such as program development, I/O control of peripheral devices, and job sequencing, an operating system software package is required. A set of programs, sometimes called an executive, is required to direct the overall operation of the computer and necessary peripherals. It may be written to allow the computer to operate in batch, real-time, or time-shared environments. Whenever possible, operating system software, as well as auxiliary programming software, is purchased directly from the computer manufacturer or commercial software vendor to minimize development costs.

Summary of Computer Software

In summary software may be divided into three major groups:

1. Applications software that performs a specific task such as preparing a given report, controlling an instrument, or solving a specific set of equations. This type of software

is most often written in the highest level language that can be supported by the computer system.
2. Auxiliary programming software that assists the programmer in preparing applications software. This includes assembly language, higher level compilers and interpreters, editors, and debugging programs.
3. Operating system software that controls the total operation of the computer system, handling the sequence of jobs, inputs, outputs, and so on.

30.4 IMPLEMENTATION—SOFTWARE VERSUS HARDWARE

Applications programs can be efficiently constructed using appropriate subroutines (subprograms whose functions are required repeatedly during the execution of the main program). An analogy can be drawn between the functions of these software subroutines and hardware integrated circuit chips. Just as groups of chips can be interconnected to build an operating unit, software subroutines may also be joined into a single program capable of operating a given instrument system. In addition, software programs can replace hardware at the functional level. The availability of inexpensive microcomputers makes it possible to program an algorithm (a sequence of instructions) that implements the same function performed by hardware, for example, logic control circuits, timers, and signal filtering (Chapter 29). Major advantages realized from replacement of hardware with software are increased flexibility and reliability. However, the slower speed of software driven devices can be an insurmountable disadvantage in applications requiring rapid response times.

Programming aids (editors, debuggers, and so on) facilitate changes in software. Modular program design allows subroutines to be added or removed as necessary for the computer system to accomplish the desired tasks. Modification of software is generally easier than redesigning hardware. New applications can be rapidly developed and implemented. Standardized computer hardware can also be used for a wide range of instrument applications. Experience gained in computerizing one kind of instrument may be applied to the automation of other instruments. For example, the same in-line microprocessor has been used by at least one instrument manufacturer in the design of "smart" gas chromatographs, atomic absorption spectrometers, and thermal analyzers. This standardization of hardware results in decreased costs of both hardware and software as well as easier maintenance of these instruments.

30.5 DATA REPRESENTATION

One of the major problems to be solved in any computer application is that of data transformation. Input data from peripheral devices and instruments must be converted to binary information required for the operation of all digital computers. Likewise, data transmitted from the computer to instrument or device outputs must be understood by the device or analyst. Converters that perform the necessary data transformations are comprised of either hardware and software, or a combination of both. They are commercially available in the form of standard devices or software packages.

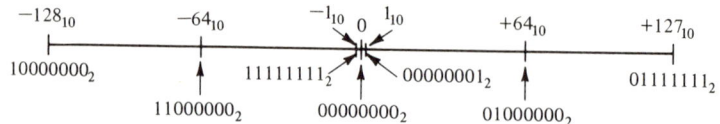

FIGURE 30-7 Representation of signed numbers by an 8-bit computer.

Binary Numbers

Binary information can represent both fixed point (integer) and floating point numbers, alphanumeric characters, and computer instructions. The computer's interpretation of a group of binary bits depends on the architecture of the computer and the context in which the information appears.

In unsigned integer arithmetic, a computer with an n-bit word can represent 2^n in positive integers; for example, an 8-bit computer can represent numbers from 0 to 256_{10}.* However, many computers utilize "two's complement arithmetic," which requires that both positive and negative numbers be represented. In fixed point notation the most significant bit is used to represent the sign of the number; 0 for positive integers and 1 for negative numbers. An 8-bit computer could represent signed numbers from -128_{10} through 0 to $+127_{10}$ (Fig. 30-7).

Floating point notation is used to expand the range of numbers that can be represented by a computer. For example, if an 8-bit computer can represent only 256_{10} integers directly, how can it handle the number $150,137_{10}$? Floating point notation solves the problem by representing the mantissa (number part) and the exponent (to some base) with two or more computer words both containing binary numbers. An 8-bit computer might represent the number 451.9×10^{-3} as shown in Fig. 30-8. The range of numbers represented by this computer would run from $-32,000 \times 10^{-256}$ to $+32,000 \times 10^{+255}$. Floating point subroutines are necessary to perform arithmetic and logic operations on floating point data. These routines are available as either hardware or software components of the computer system with the hardware having the usual advantage in speed over the software. In general, larger computers with 16, 32, or 64 bits per word, process floating point data more efficiently than those with smaller word lengths.

Although binary numbers consisting of many digits are handled easily by computers, they are tedious to the human programmer. The digits of binary numbers are therefore bunched together in groups of three or four, octal or hexadecimal notation, respectively. Table 30-2 illustrates examples of the relationship among these three notations which are commonly used for coding machine language programs.

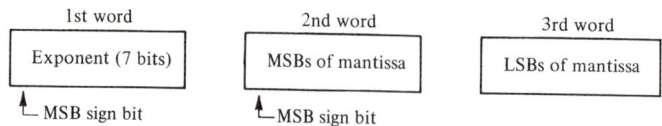

FIGURE 30-8 Three word representation of floating point numbers in an 8-bit computer. MSB, most significant bit; LSB, least significant bit.

---

*The subscript denotes the base of the number. For example, 256_{10} means 256 to the base 10, while 256_8 indicates 256 to the base 8.

TABLE 30-2 Binary Number Equivalents

Binary	Octal	Hexadecimal	Decimal	BCD
1	1	1	1	0001
10	2	2	2	0010
11	3	3	3	0011
111	7	7	7	0111
1000	10	8	8	1000
1001	11	9	9	1001
1010	12	A	10	00010000
1011	13	B	11	00010001
1100	14	C	12	00010010
1101	15	D	13	00010011
1110	16	E	14	00010100
1111	17	F	15	00010101
10000	20	10	16	00010110
10001	21	11	17	00010111
11011	33	1B	27	00100111

Data represented in binary coded decimal (BCD) format (Chapter 28) is often generated by instruments with digital output and can be easily interfaced to the data bus of a computer. Since each BCD digit requires 4 bits (see Table 30-2), the computer must be programmed to operate either on existing BCD data or to convert the data to pure binary numbers for subsequent operations.

Alphanumeric Code

In order to represent input and output data in a format that is easily understood by the analyst, it is necessary to code the alphanumeric characters and punctuation marks in binary format. If each alphanumeric character is represented by 7 bits (128 possible combinations), then all 52 letters (upper and lower case), 10 numerals, and 22 additional special symbols can easily be represented. The most commonly used code is the ASC II (American National Standard Code for Information Interchange).

30.6 COMPUTERIZED INSTRUMENT SYSTEMS

The ideal computerized instrument system joins the instrument directly to the computer and places the combination at the command of the analyst. Although large computers can be directly linked to instrumentation, minicomputers and microprocessors are used in the majority of applications. It is sometimes desirable to form a network consisting of a number of small computers that are dedicated to specific instruments; a host computer oversees the operation of all component computers and peripheral output devices.

The number of components of a computerized instrument system depends upon the tasks to be performed. In addition to the instrument–computer interface, a variety of standard peripheral devices are available to facilitate functions such as analyst–computer interaction, external storage of data and programs, and display of analytical results (Fig.

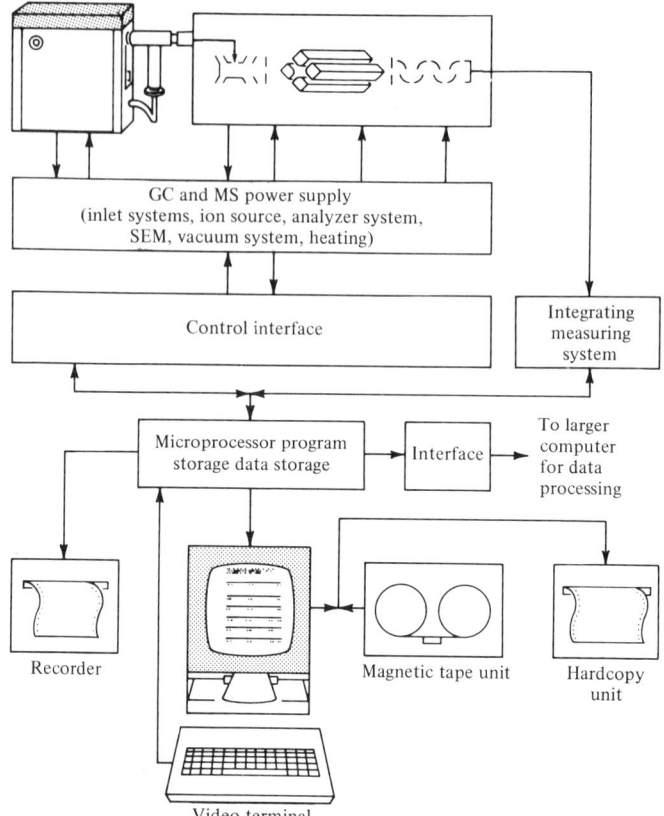

FIGURE 30-9 Microprocessor-controlled gas chromatography/mass spectrometry system. (Courtesy of Varian Associates.)

30-9). Typical I/O devices include keyboards and printers, paper tape and card reader/punches, magnetic tape transports, disk drives, plotters, and video terminals. Each device is interfaced to the computer's I/O buses with appropriate electronic hardware. Software, known as drivers, resident in computer memory, controls these standard interfaces and allows data to be transferred to and from peripheral devices. Since the cost of peripheral devices often constitutes the largest expense in the purchase of a computer system, effective utilization of the devices should be achieved in instrument computerization.

Computer–Instrument Interface

The focal point of a computerized instrument system is the interface between the computer's buses and the instrument's signal lines (Fig. 30-10). If the instrument I/O data are in a form compatible with the signal levels of the computer's I/O buses, the hardware components of the interface are simple (latches, multiplexers, and so on). An increasing number of instrument manufacturers are providing computer-compatible I/O signals. If the instrument's signals are not compatible (analog signals or different digital

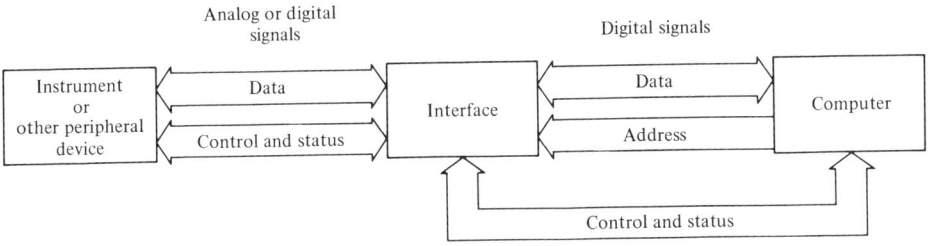

FIGURE 30-10 Functions of a computer interface.

logic levels) with those of the computer bus, the interface must contain more complex hardware components. These components include signal conditioning devices such as analog-to-digital converters, attenuators, and voltage level shifters in addition to the simple components previously mentioned. The hybrid data acquisition system (Chapter 28) is an example of a multifunctional interface.

Information on the data bus is transmitted through the interface in both directions under the management of signals on the control bus (Fig. 30-11). When several devices are multiplexed to a single computer, the controller uses signals on the address bus to select the device involved in the data transfer. When the hardwired buses have been connected to the interface, software programs must be written to operate the total system. The programmer must consider the hardwired address of each device, software methods of generating control signals, and software response to control signals coming from the instrument. Design of the hardware interface must therefore be coordinated with software development.

Instructions that enable the computer to communicate with instruments and other peripheral devices attached to the external computer buses are known as I/O instructions. An I/O instruction contains the address code of the particular device to be serviced by the computer. Execution of the I/O instruction by the computer causes the address code to appear on the address bus lines simultaneously with an I/O pulse on the appropriate control bus line. This combination of signals is used to direct the flow of data to or from the device specified by the address code (Fig. 30-11).

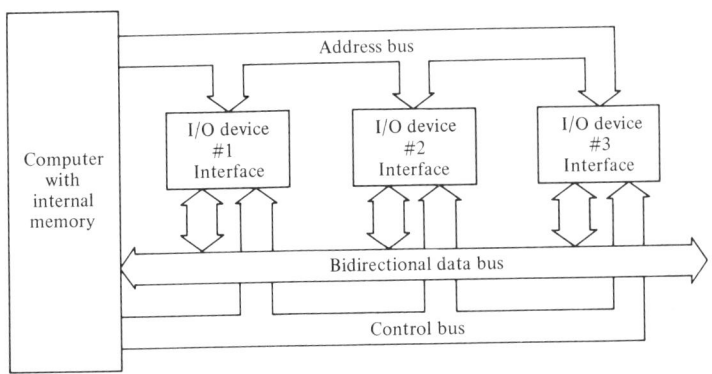

FIGURE 30-11 External bus structure.

Software Control of the Interface

A software programming strategy is required to control devices interfaced to a computer bus. Three strategies are employed: polling (programmed I/O), interrupt, and direct memory access (DMA).

Polling or programmed I/O is the simplest method to implement. After connecting the interface of each device to the three computer buses and assigning each device an address code, a program known as a polling loop is written to question sequentially each device to determine if it requires service (Fig. 30-12). The computer will periodically ask a device: "Do you require service?" The device responds with a "yes" or "no." A positive response causes the computer program to jump to a subroutine called a device handler that services the responding device. Whenever a negative response is received, the computer proceeds to poll the next device in sequence.

In actual practice, polling is implemented by the computer through a logic test of the status (1 or 0) of a specified data bit (flag) from the device. A typical service action is the transfer of a word or block of data to or from the device. The process of the computer's questioning a device and receiving information in return is called handshaking. A service routine for a given device may elicit a number of handshaking exchanges to ensure proper transfer of data. Two advantages of the polling method are minimal interface hardware and straightforward, simple software. The major disadvantage of this strategy is its demand on the computer's time. Each time a polling loop is entered, the status of each device in the loop must be checked even though it may not require service. If the time spent in the polling loop is objectionable for a given application, one of the other programming strategies may be implemented; if not, polling is the simplest technique to use.

In situations where computer response time to interfaced devices is critical or where it is undesirable to have the computer spend a large fraction of its time in polling loops, the

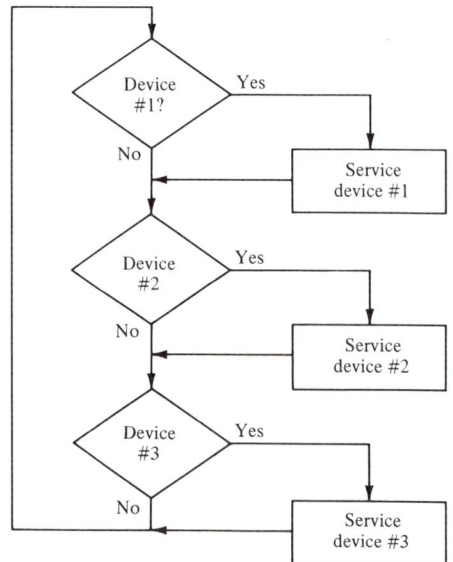

FIGURE 30-12 Polling method of input/output for three devices.

interrupt driven technique may be a suitable alternative. Using the method, each interfaced device is given the ability to initiate a request for service from the computer. The interrupt line on the control bus of the computer is used for this purpose. When an appropriate signal from a device is applied to the interrupt line, the computer breaks away from its current task, determines the device requesting service, and performs the specified service routine. It then, upon completion of this routine, returns to the task it was executing when the interrupt signal was received. If several devices initiate interrupt signals simultaneously, the computer services them in order of preassigned priorities (priority interrupts).

Interrupts provide fast response to I/O devices and are therefore required in real-time systems that must provide the fastest possible response time to external conditions. This technique of I/O handling also allows efficient use of computer time by scheduling the computer to work on less important (background) tasks while waiting for interrupts. The major disadvantages of this method are extra interface hardware, time required to service devices, and more complex software.

The final technique, direct memory access (DMA), can be implemented on many computers by adding DMA hardware controllers. Upon receiving the correct interrupt signal from an I/O device, the DMA hardware controls the transfer of data directly between the device's interface and the computer's memory at a much faster rate than the previous two methods. Thus less computer time is required for servicing I/O devices. DMA is typically used with fast I/O devices such as multichannel analyzers, disks, and video terminals. This method usually requires additional expense and adds significantly to the complexity of the computer system. However, once implemented, DMA requires little computer time to achieve data transfer.

30.7 MICROCOMPUTER INTERFACING[5]

The widespread use of microcomputers in analytical instrumentation requires a more detailed discussion of these devices. Although the general hardware and software components of microcomputers are the same as those of larger computers, there are significant differences in architecture and jargon. The distinction between microprocessor and microcomputer should be clearly defined at the outset. A microprocessor includes the central processing unit (arithmetic and control units) plus several general registers used for temporary storage communications with I/O devices and memory addressing (Fig. 30-13).

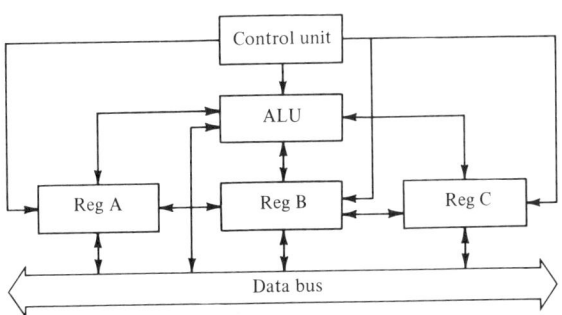

FIGURE 30-13 Basic microprocessor.

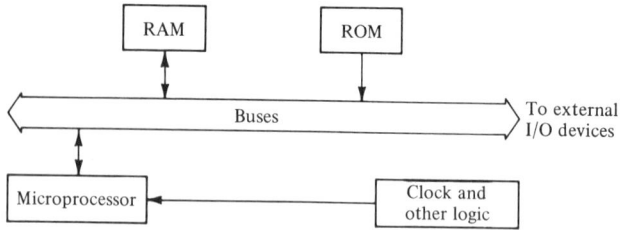

FIGURE 30-14 Parts of the microcomputer.

A microcomputer is produced when memory and external circuitry for control of processor timing and bus communications are added to the microprocessor (Fig. 30-14). Initially microcomputers were constructed from a number of component LSI chips including a microprocessor chip. Single very large-scale integrated circuits (VLSI) containing an entire microcomputer are now available on a single chip. Metal oxide semiconductors (MOS) are typically used in microcomputer memories, RAM for temporary data storage, and ROM for permanent program storage. Although microcomputers are similar to minicomputers in their general mode of operation, they are usually smaller in size and have slower operating speeds. However, developments in solid-state technology and computer architecture are erasing these distinctions. As in the case of larger computers, a number of I/O devices controlled by a single microcomputer forms a system (Fig. 30-9).

Microcomputers have replaced both hardwired circuits and general purpose minicomputers in numerous applications because their cost–performance characteristics are superior to either of the other approaches. Use of microcomputer-based control and logic software ensures flexibility in system modification and future expansion. ROM firmware provides the desirable fixed-function, dedicated characteristics of a hardwired system.

In-line microcomputers have become standard components of current instrumentation (Fig. 30-15). This development in instrument design is evidenced by references to microcomputers in almost every chapter of this text. The remainder of this chapter will be devoted to a more detailed description of microcomputer applications. These applications were chosen to illustrate different strategies for using microcomputers in instruments.

Evolution of Microprocessor Instrumentation[6]

The development of the LECO CS-144 carbon/sulfur determinator provides a clear, uncomplicated example of the transition from a conventional, hardware system to a hardware system with an on-line microcomputer and finally to an in-line microcomputer. The determinations of carbon and sulfur in a sample require the following operations: obtaining and recording the weights of a tarred and sample containing crucible; heating the sample for a given period of time; obtaining the infrared absorbance of CO_2 and SO_2 gases at predetermined frequencies; and using calibration factors obtained from analyses of standardized samples for calculation of the percentages of carbon and sulfur in the sample.

In the conventional system (Fig. 30-16), the hardware modules (shown in block form) operate together to provide the necessary data manipulations and timing functions required for the determinations. Calibration is accomplished by setting the potentiometer

COMPUTER-AIDED ANALYSIS   887

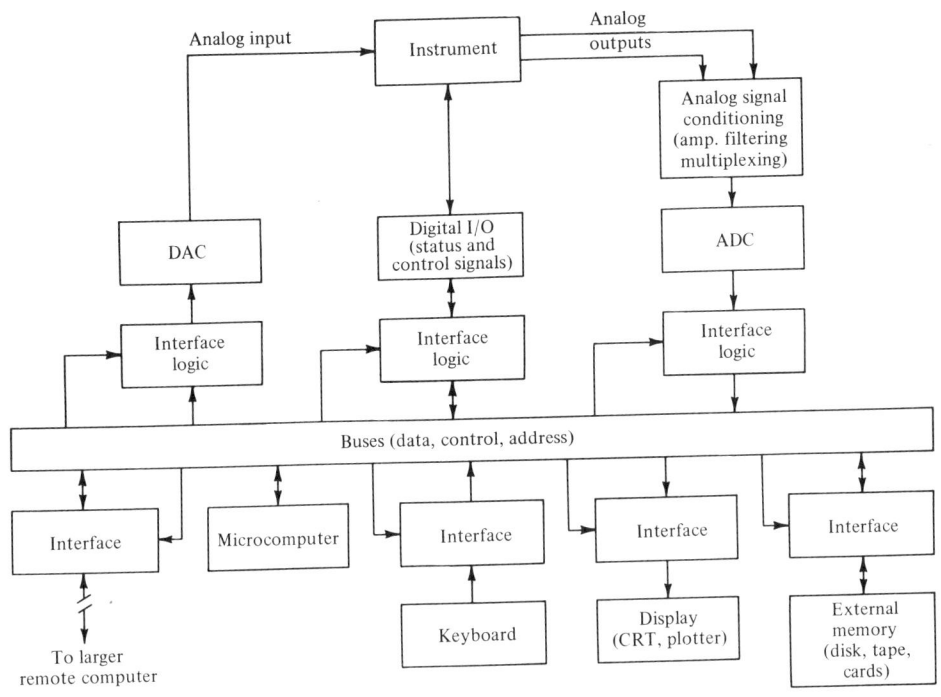

FIGURE 30-15 Microcomputer instrument system.

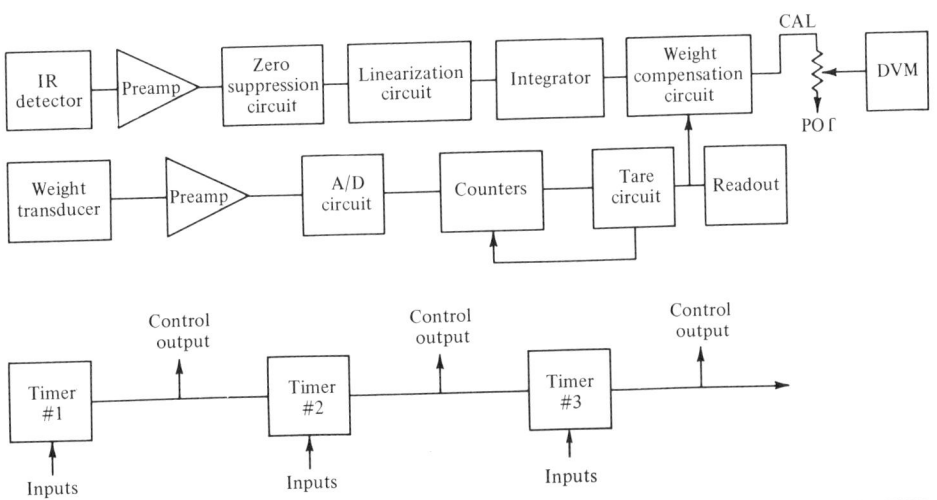

FIGURE 30-16 Carbon/sulfur determinator, conventional system. (Courtesy of LECO Corp.)

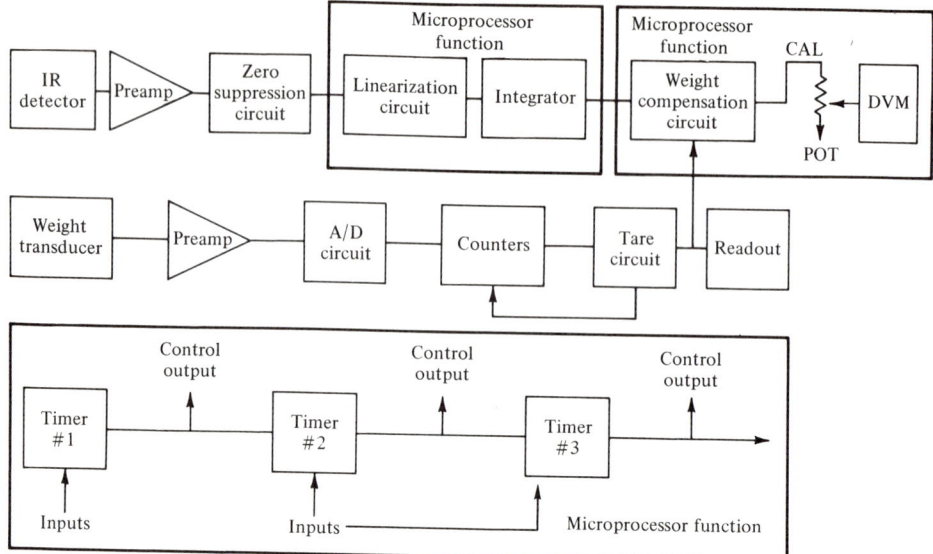

FIGURE 30-17 Conventional system with microprocessor add on. (Courtesy of LECO Corp.)

(POT) to obtain a reading on the digital voltmeter (DVM) corresponding to the known percentage of carbon or sulfur for a standard sample.

Addition of an on-line microcomputer to the conventional circuit design allowed the function of the hardware modules within the larger blocks (Fig. 30-17) to be replaced by software operations. Thus the microcomputer became responsible for the timing required for sample volatilization, acquisition and processing of infrared data, the calculation of the percentages of carbon and sulfur in the sample, and the output of the results to the display and printer.

In the final stage of development, the instrument system was redesigned around the central in-line microcomputer (Fig. 30-18). Analog inputs from the infrared detector and balance are conditioned by the preamplifiers and then transformed to digital form by the analog-to-digital converter. Other inputs include digital signals from the control keyboard

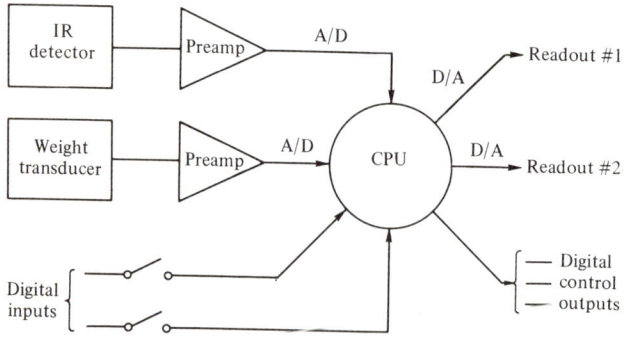

FIGURE 30-18 Total design microprocessor system. (Courtesy LECO Corp.)

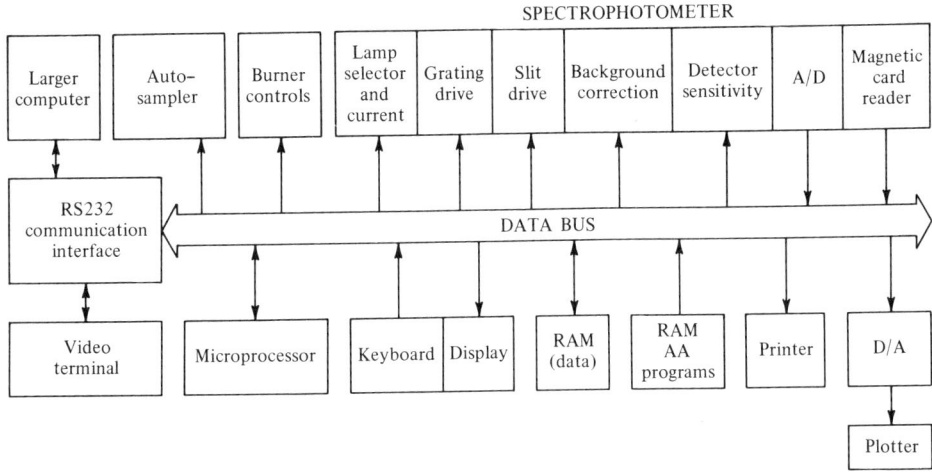

FIGURE 30-19 In-line microprocessor-controlled atomic absorption spectrophotometer (address and control buses omitted for simplicity).

and instrument control components. Outputs include the digital display and a printer as well as signals required for timing functions. The in-line microcomputer gives this instrument desirable characteristics of simplified operation and easy maintenance.

An In-Line Microcomputer-Controlled Atomic Absorption Spectrometer[7,8]

Incorporation of in-line microcomputers into the design of atomic absorption (AA) spectrophotometers minimizes human error and effort, while maximizing performance and analytical throughput (Fig. 30-19). Until the appearance of computerized instruments the rate of multielement sample analyses was limited by the time required for the analyst to change lamps and readjust operating parameters. The accuracy and precision of results were limited by the narrow dynamic quantitative range of the instrument. Microcomputer data processing results in a linearized working range of up to four orders of magnitude. This means that chromium at a concentration of 90 μg/ml can be determined with the same calibration and under the same conditions used to detect 0.01 μg/ml. The Perkin-Elmer Model 5000 atomic absorption spectrometer can handle up to 50 samples and can determine six different elements automatically. All control knobs are replaced by keyboards consisting of numeric and functional keys that allow entry of operating parameters such as wavelength, slitwidth, and flowrates of fuel and oxidant (Figs. 30-20 and 30-21). Data from the analysis of samples as well as the operational parameters are continually monitored and displayed on the control panel. Entering a set of parameters for a given analysis is a simple task. For example, inputing the desired wavelength from the keyboard and depressing the PEAK function key causes the grating in the monochrometer to move to the position appropriate for the selected wavelength and to automatically center on the peak of the line from the lamp. The analyst selects the desired lamp on the multilamp turret by entering the position number and touching the LAMP # key. The lamp

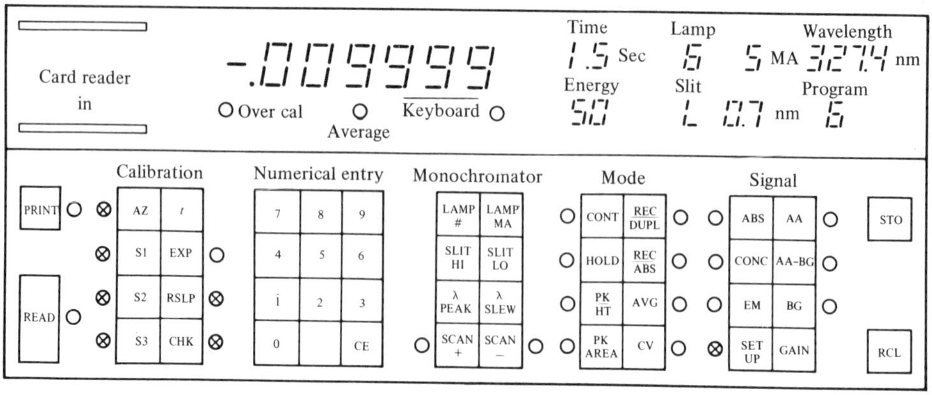

FIGURE 30-20 Keyboard/display panel of an atomic absorption spectrometer. (Courtesy of Perkin-Elmer Corp.)

current in milliamperes is then entered and the LAMP MA key depressed. When the automatic burner control unit is activated, normal flows for an air/acetylene flame are immediately available. The FLAME ON/OFF button ignites the flame; gas flow adjustments are entered through the keyboard (Fig. 30-21). Digital flow control offers more presision in resetting a previously optimized gas flow than does manually operated needle valves or pressure restrictor systems.

A set of optimized operating parameters for the analysis of an element (known as a method) can be entered into the RAM memory from the keyboard. At any subsequent time these parameters can be recalled and used to prepare the instrument for the analysis of the specified element. Since the amount of RAM memory available for storage is both limited and volatile, permanent storage methods are available on magnetic cards or a tape cassette. Up to six method programs may be read from magnetic cards and stored in mem-

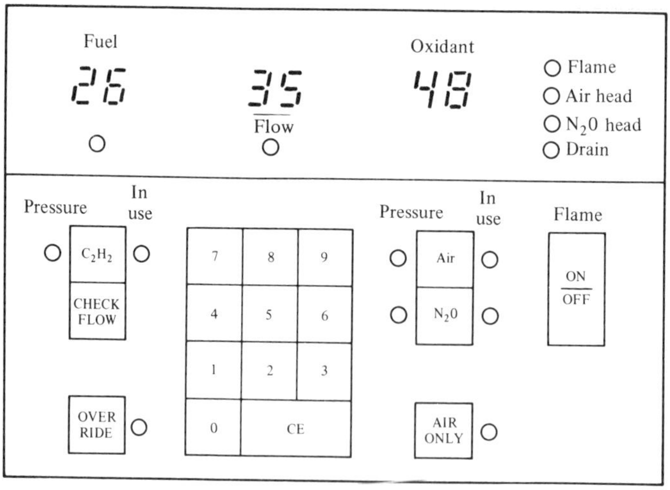

FIGURE 30-21 Control keyboard/display of flame parameters. (Courtesy Perkin-Elmer Corp.)

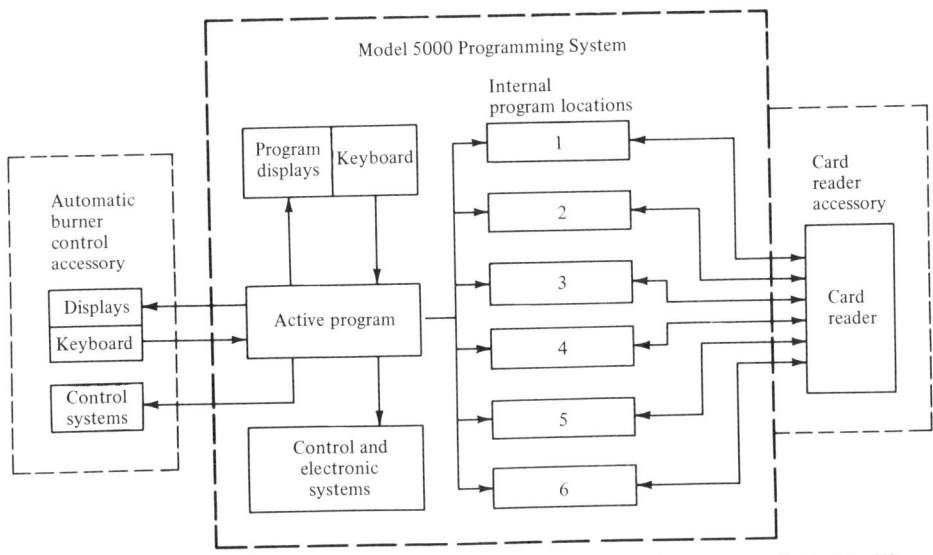

FIGURE 30-22 Atomic absorption programming system. (Courtesy of Perkin-Elmer Corp.)

ory for sequential execution by the instrument system (Fig. 30-22). Background correction is controlled by the microcomputer which selects the continuum source appropriate for the desired wavelength. Mechanical and electrical components required for operation of the optical system have been reduced by use of microcomputer software.

The instrument provides integrated readings in units of absorbance, concentration, or emission intensity. These readings can be updated continuously or held on the instrument display. Integration times are keyboard selectable from 0.2 to 60 sec. Peak height or integrated peak area can be measured with electrothermal (flameless) furnaces. The instrument can be used for flame emission as well as atomic absorption spectrometry.

Calibration of the instrument to read directly in concentration units requires only two steps. First, the operator enters the concentrations of the standards. The standards are then analyzed in the same sequence. Zero absorbance is set by aspirating a blank containing only solvent and depressing the ZERO button. If the calibration curve is known to be linear, only one standard is required. For nonlinear working curves, two or three standards are used, depending upon the degree of nonlinearity. An algorithm stored in ROM memory then linearizes the instrument's response with respect to sample concentration over a specified range.

To recalibrate in the middle of a run, a standard is rerun and the RESLOPE key is depressed. This causes the instrument to be recalibrated with the same program used for the original calibration. The analyst can use a fixed expansion rather than diluted standards by entering the desired expansion value (from 0.01 to 100) on the keyboard and depressing the EXP button.

Analytical accuracy can be improved by the averaging feature of the instrument. By entering the number of readings to be averaged and pressing the AVG key, the average is displayed on the readout. Lastly, the coefficient of variance and the standard deviation can be obtained by depressing appropriate keyboard buttons.

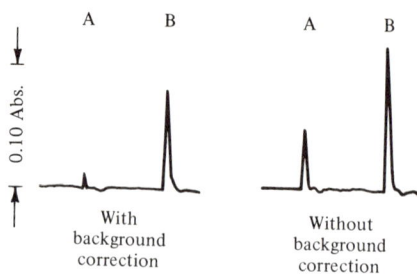

A. 0.1% solution of Ca as the chloride
B. Solution A + 0.02 µg/ml Ba

FIGURE 30-23 Determination of barium in calcium chloride with a graphite furnace using computerized background correction. (After R. D. Edigor, *Am. Lab.*, p. 75 (February 1978). Courtesy of International Scientific Communications, Inc.)

After replacement of the burner head with the graphite furnace head and the furnace control unit that contains a separate microcomputer, low level samples can be analyzed. Furnace operation is controlled by the following programmed parameters: time intervals necessary to obtain and hold temperatures for dry, ash, and atomize cycles; temperature for each cycle; times at which sample data are collected; time when AUTO ZERO control activated; and the flowrate of furnace inert gas stream. A base line feature makes it possible to store the difference between an electronic zero and a blank reading, thus permitting the true blank to be subtracted automatically. The use of this feature is demonstrated by the difficult analysis of barium in calcium chloride, using the graphite furnace (Fig. 30-23). Without the background correction the calcium matrix would have caused an erroneously high value of barium. As in the case of the flame spectrometer, method programs can be recorded on magnetic cards and reentered for subsequent analyses.

The computerized spectrophotometer is programmable not only from the keyboard and magnetic cards but also from an external terminal through a two-way RS232 communications interface. This standard interface permits external devices such as teletypes, CRT terminals, and computers to activate instrument functions. Using this capability, a tape cassette can store a large number of analytical methods for transfer later to the instrument. Data can be taken directly from the instrument's A/D converter at the chopping frequency and sent through the RS232 interface to an external computer for specialized data processing.

30.8 COMPUTER CONTROLLED LABORATORY AUTOMATION SYSTEMS

It is often desirable to implement a laboratory automation system that is capable of servicing a number of instruments, with a centralized computer. A commonly occurring example is a group of chromatographs interfaced to a single computer. Several approaches are available for achieving this automation. The individual components of the system, that is, the computer, auxiliary peripheral devices, interface hardware, and transmission equipment, can be purchased, all necessary software written, and the entire system assembled. In a second alternative, the computer with operating system software and standard external buses can be purchased as a system.

The host computer coordinates the operation of the entire system and provides a selection of data reduction methods, such as quantitation of sample components and deconvolution of overlapping peaks. Operating programs are stored in ROM while data reduction parameters and digital values are stored in RAM with optional disk storage available. Entry and display of alphanumeric information are provided by a keyboard and video monitor (CRT) located near the host computer. Instruments can then be connected to the external buses through hardware interfaces and appropriate applications programs developed or purchased. An example of such a system is Digital Equipment Corporation's 11/03 laboratory computer system.

In situations where minimal in-house computer expertise is available, turnkey laboratory computer systems offer the better alternative. The purchase price of a turnkey system should include all the hardware and software necessary to operate a group of instruments for the purpose of producing specified analytical results. An operator-oriented dialog allows the analyst to interact with the system software to develop methods for instrument operation, data processing and storage, and report generation. Included in the system's software are functions such as integration, background corrections, filtering, and differentiation that can easily be incorporated into the methods by the analyst.

Two commercially available turnkey systems representing different philosophies of implementation have been chosen as examples. The Hewlett-Packard 3354 Series Laboratory Automation System (Fig. 30-24) is a minicomputer-based general purpose system capable of servicing up to 30 different kinds of instruments. A network of microcomputers forms the Spectra-Physics 4000 Chromatography Data System (Fig. 30-25) that is capable of handling up to 16 chromatographs.

Minicomputer Lab Automation System

The heart of the Hewlett-Packard 3354 system is a 64K (16-bit) byte minicomputer linked to a 15-Mbyte disk. A variety of standard peripheral devices, such as hard copy and video terminals, can be supported by the system. Up to 30 different instruments may be connected to the system using remote hardware interface modules configured in a maximum of two transmission loops (Fig. 30-24). These loops consist of a twisted pair of wires and provide a digital communication link between the computer and any instrument in the loops for the purpose of acquiring data and issuing instructions. The signal from a given instrument is continuously monitored and integrated by an A/D converter in the analog module before it is digitized and serially transmitted to the computer. Digital modules permit data to be sent directly from an instrument with digital outputs to the computer. These modules also allow transmission of digital control signals from the computer to the instrument. Communication is greatly simplified by this approach which uses a twisted pair of wires to serially transmit digital signals. The system is flexible enough to allow the analyst to develop a method in a high level, easy-to-use language (LAB BASIC), while the system is simultaneously performing tasks associated with other instruments connected to the system.

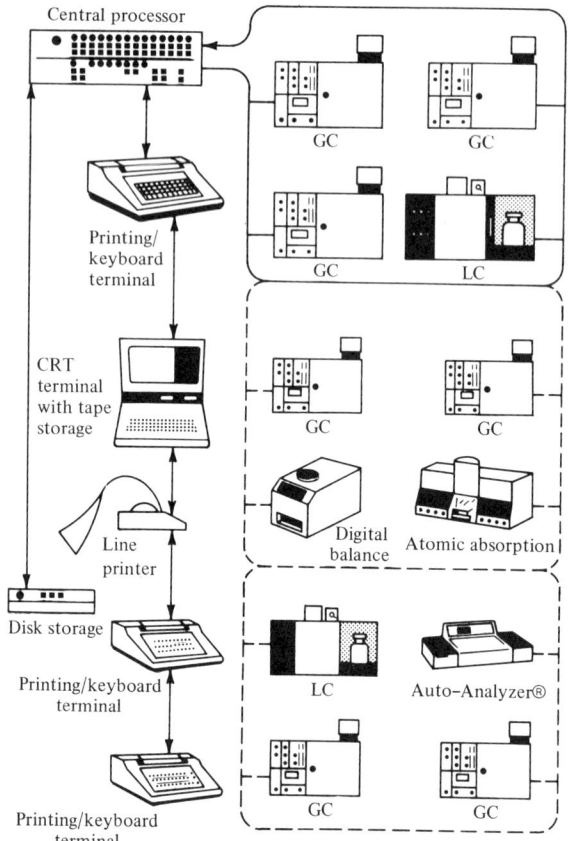

FIGURE 30-24 Minicomputer lab automation system. (Courtesy of Hewlett-Packard Co.)

Microcomputer Network System for Chromatographic Analysis

The SP4000 developed by Spectra Physics utilizes a network of hierarchial microcomputers to produce a multichannel chromatographic data system that can automate up to 16 chromatographs. The tasks of control, sampling, integration, digitization, and communication with the host computer are performed by subsidiary microcomputers contained in the data interface of each chromatograph. Data from each chromatograph are processed individually by the host computer, allowing a variety of applications to be handled simultaneously. Processed data can then be transmitted through the data interfaces to printer/plotters at the locations of the individual chromatographs.

Each data interface contains a microcomputer with both ROM and RAM. Signals from the chromatographic detector are continuously integrated, digitized, and sent to the host computer using serial data transmission. In addition, the operation of the chromatograph can be controlled from either the front panels of the individual data interface modules or by method programs resident in the host computer. Displays on the interface module allow the progress of the analysis to be monitored.

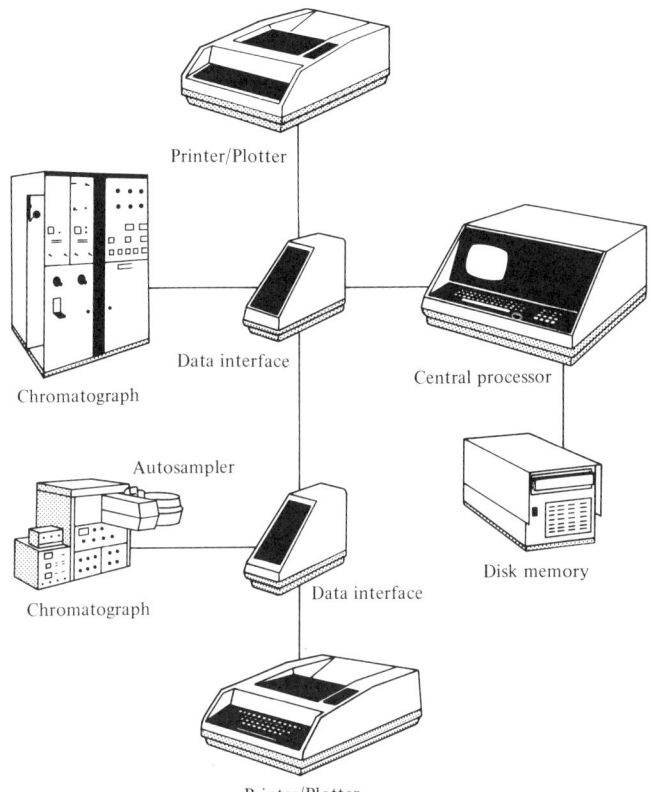

FIGURE 30-25 Microcomputer network system for chromatographic analysis. (Courtesy of Spectra Physics Corp.)

Summary

A large number of options exist for applying the capabilities of computers to chemical analyses. Selection of a computer system depends on the laboratory environment, the desired flexibility, and the cost effectiveness in obtaining the required analytical results. In choosing a method for implementing any computer application, the symbol TTL has a special significance, indicating the quantities of time, talent, and "loot." At the minimum one of these quantities must be present in excess; the larger the values of all three, the easier implementation becomes.

BIBLIOGRAPHY

Bauman, F., J. Hendrickson, and D. Wallace, *Chromatographia,* **7** (9), 530 (1974).
Bibbero, R. J., *Microprocessors in Instruments and Control,* Wiley, Philadelphia, 1979.
Carrick, A., *Computers and Instrumentation,* Heyden, Philadelphia, 1980.
Dessy, R. E., P. Janse-VanVurren, and J. A. Titus, *Anal. Chem.,* **46** (11), 917A (1974); **46** (12), 1055A (1974).

Frazer, J. W. and F. W. Kunz, Eds., *Computerized Laboratory Systems,* American Society for Testing Materials, Philadelphia, 1974.

Perone, S. P. and D. O. Jones, *Digital Computers in Scientific Instrumentation,* McGraw-Hill, New York, 1973.

Soucik, B., *Microprocessors and Microcomputers,* Wiley, New York, 1976.

LITERATURE CITED

1. Ratzloff, K. L., *Am. Lab.*, p. 17 (February 1978).
2. Zaks, R., *Microprocessors, from Chips to Systems,* Sybex, Berkeley, Calif., p. 296, 1977.
3. Davis, H., *Instruments and Control Systems,* p. 53 (June 1979).
4. Dessey, R. E. and M. K. Starling, *Am. Lab.,* p. 21 (February 1980).
5. Toong, H., *Sci. Am.*, **237**, 146 (1977).
6. Sitek, G. J. and R. B. Edwards, *Abstracts of the Pittsburg Conference,* p. 319 (March 1979).
7. Ediger, R. D., *Am. Lab.*, p. 68 (February 1978).
8. Barnard, T. W., *Anal. Chem.*, **51**, 1172A (1979).

CHAPTER 31

Process Instruments and Automatic Analysis

31.1 INTRODUCTION

This final chapter will discuss automatic analyzers and automated instrument systems. It is divided into two parts; laboratory instruments and on-line process analyzers. The number of methods available in each area is growing at a rapid rate, catalyzed by advances in sensor (detector) and microcomputer technologies. No attempt has been made to provide exhaustive coverage of these areas. Instead, examples have been selected to illustrate some of the important techniques used to solve applications involving large sample throughputs and continuous analysis.

At the outset it is necessary to distinguish between the characteristics of automatic and automated devices. According to the current definitions of the International Union of Pure and Applied Chemistry (IUPAC),[1] both devices are designed to replace, refine, extend, or supplement human effort and facilities in the performance of a given process. The unique feature of automated devices is the feedback mechanism which allows at least one operation associated with the device to be controlled without human intervention. For example, an automatic photometer might continuously monitor the absorbance of a given component in a process stream, generating some type of alarm if the absorbance exceeds a preset value. By contrast, an automated photometer system could transmit absorbance values to a control unit that adjusts process parameters (temperature, amount of additional reagent, and so on) to maintain the concentration of the measured component within preset limits. In spite of this fundamental difference, the terms automatic and automated are often interchanged.

Development of Process Control and Laboratory Analyzers

For the past 50 years both automatic and automated instruments have been used to monitor and control process stream environments. Nonselective properties of the process stream, such as density, viscosity, or conductivity, have long been used to monitor and to regulate process conditions, such as temperature, pressure, or flowrate. Selective determination of chemical components in process streams was originally performed on grab samples which were subsequently analyzed in control laboratories by off-line techniques with obvious shortcomings in time, economy, and human error. Increasing surveillance and control of the individual steps in industrial processes required rapid, selective analyses. In the 1950s and 1960s a variety of continuous analyzers with the capability for selective determination of chemical composition began to emerge. The more successful of these analyzers utilized refractometry, gas chromatography, potentiometry, and infrared spectroscopy. These analyzers provided dynamic rather than historic determination of the composition of starting materials, intermediates, products, and contaminants. Data from these analyzers, fed back to controllers, permitted better regulation of stream composition than earlier nonselective methods did. Most recently, process control analyzers have become the sensors for large computer-based process control systems.

The developments in process stream analyzers have been paralleled by the appearance of automatic laboratory instruments such as titrators, elemental analyzers, and continuous flow analyzers. Increased demand for medical testing has provided the impetus for the development of sophisticated automatic clinical analyzers.

During the 1970s progress in solid-state electronics and in computer technology greatly influenced the design of both process stream and laboratory analyzers. Instrument size has been reduced while reliability has increased. These analyzers possess one or more of the following capabilities: routine analysis and monitoring; acquisition, processing, and storage of data; on-line process control; and generation of data necessary for records and reports. Application of this technology to industrial process control has been slower than application to laboratory analyzers for two reasons: expense involved in replacing existing equipment and the apprehension concerning the reliability of this new equipment.

Automation Strategy

Automated (or automatic) analyses are usually implemented by one of two methods—continuous or discrete (batch). In a continuous flow analysis, samples are analyzed from a flowing stream with any necessary operations, such as filtering and reagent addition, performed prior to the measurement of chemical and physical properties. Actual determinations are made using flow-through sample cells. In discrete analysis, samples are placed in individual containers for the duration of the analysis. Preliminary operations of dilution, reagent addition, and mixing are performed on each sample at different locations within the analyzer. Each treated sample is then presented in sequence to the sensing device. A batch of samples is usually preloaded for processing by this technique. While both techniques are used in process control and laboratory analyses, continuous analysis is more often utilized in automated process control applications, because of its faster response time.

Not every process or analytical method lends itself to automation. Analyses involving gaseous or liquid samples have most often been successfully automated, while applications involving solid samples have generally proved most difficult. In fact, even though a wide variety of methods exist for automation, there will most likely always remain a hard core of complex chemical analyses that will defy automation or be too costly to automate.

31.2 INDUSTRIAL PROCESS ANALYZERS

Process control analyzers must have properties which in many instances differ markedly from those of analogous laboratory instruments. Rapid analysis is essential if the results are to be useful in high-speed processes. Analyzers must be simple in design and easy to maintain. This simplification is often achieved at the expense of analytical versatility. Gas chromatography has been widely used for process stream analysis partially because these analyzers are composed of simple, easily maintainable components. Ability to withstand harsh plant environments (mechanical shocks and vibrations, dust, and sometimes weather) is another important consideration.

Process control analyzers often compromise selectivity in order to increase sensitivity to changes in chemical composition of the process stream. Nondispersive infrared analyzers are more sensitive toward a specific component in many situations, but interfering components cause greater problems than they do in dispersive laboratory spectrophotometers. Decreasing the size of the analyzers facilitates installation and therefore reduces cost. Electrical components must be contained in explosion-proof housing for safe operation in areas where combustion hazards may occur. Reliability is of primary importance in an environment where downtime is expensive. An analyzer costing several thousand dollars cannot be permitted to delay a process involving millions of dollars in materials, equipment, and labor.

In deciding between continuous or discrete analysis streams, rapid and more sensitive analysis favor the continuous technique while increased selectivity can usually be achieved by processing individual samples in batches. In addition, continuous analyzers are usually simpler to design, operate, and maintain. The features of continuous analysis are illustrated schematically in Fig. 31-1. A sample is obtained at a controlled rate from the main process stream or other source of material. When necessary, provision is made for preparing the initial sample for analysis by further operations, such as the addition of reagents or filtration. The actual measurement is made by an appropriate sensor and the results of the determination indicated in terms of the concentration of the desired sample component. This information may be directly displayed by a readout device such as a meter or a strip chart recorder, or it may be transmitted to a computer for further processing and storage.

One class of primary sensors consists of devices that measure physical variables, such as temperature, pressure, fluid flowrate, and liquid level. The other broad category of primary sensors, often referred to as on-stream or on-line analyzers, determine chemical composition. In many cases they achieve this by measuring some physical property quantitatively related to chemical composition, such as the absorption of infrared, ultra-

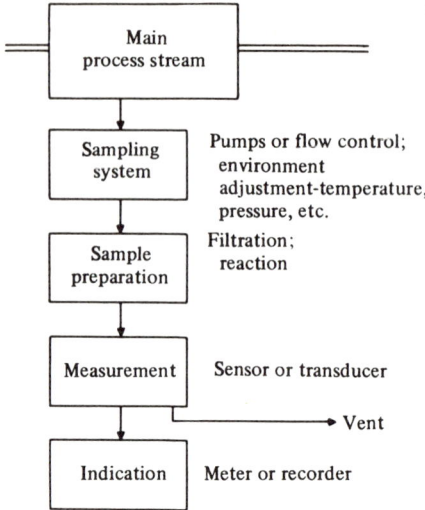

FIGURE 31-1 Features of continuous analysis.

violet or visible radiation, thermal or electrical conductivity, dielectric constant, paramagnetic susceptibility, density, or refractive index.

Sampling Problems

Most significant among the factors regarding the development of on-stream analyzers for chemical process control is the difficulty in obtaining samples and in preparing them properly for analysis. Sampling may often represent as much as 90% of the total analysis problem. Chemical process streams may be hot and under pressure, supersaturated with respect to certain dissolved salts, highly corrosive, laden with fibrous or particulate matter, or even contain radioactive materials. Even when acceptable samples can be obtained, they then may need to be cooled, filtered, or otherwise processed before analysis. All of these operations must be performed either on a continuous or on an automatic semicontinuous basis. Transport lag must be eliminated and dead volume reduced if events within the analyzer are to be representative of the process stream being monitored. Special bypass pumping devices are often needed to keep fresh sample rapidly supplied to the input of the analyzer. Extreme care is required in continuous analysis of trace quantities. Leakage of a sample stream at a fitting or connection can cause serious back diffusion of the atmosphere into the stream; examples include dissolved oxygen in power plant condensate streams or the analysis of water in dry gas streams.

Corrosive samples or materials under extremes of pressure or temperature may restrict the latitude in design of analyzers, especially in the area of sample cells, for various optical methods of analysis. Gas handling components should always include a small protective filter, preceded by a major filter if the gas contains suspended matter requiring removal. If a gas sample has a water vapor concentration high enough to cause condensation within the analyzer, or if moisture is an interferent, stream drying equipment must be installed.

If several separate process streams are reasonably identical in major constituents and the concentrations of the desired components do not vary over too large a range, all streams can be connected through solenoid-operated, three-way valves to a common manifold that leads to the measurement cell. Each stream may be directed to the cell in a sequence determined either by hardwired electronic control logic or by a software program resident in a computer interfaced to the control valves. This method of sampling can provide rapid analysis for several streams; the exact number depends on the required sampling rate.

In summary, process control instrumentation must provide quick, reliable results when operated in a harsh environment by semiskilled technicians. Thus, in many cases, the design of these instruments is quite different from those found in the controlled environment of a general purpose analytical laboratory.

31.3 METHODS BASED ON BULK PROPERTIES

A number of instrumental methods have been utilized to continuously measure bulk properties of process streams. These properties include pH, density, viscosity, conductance, capacitance, and combustibility. Selected examples will serve to illustrate these types of analyses.

In a binary system when the difference between the value of a given physical property of the components is large, the composition of the system may be determined easily by monitoring this property. A calibration curve of the property's signal versus composition yields the correlation. For example, water in many organic materials is readily determined because water has a dielectric constant of 80, whereas most organic materials have values between 1 and 10. Thus, the moisture content of the paper web can be monitored during manufacture at speeds of up to 3000 ft/min, as shown in Fig 31-2. The measuring head is an electrical capacitor which uses the paper web as a part of the measuring circuit. The dielectric constant of dry paper is about 3.

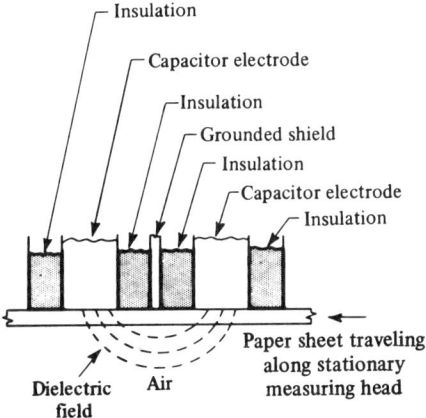

FIGURE 31-2 Schematic diagram of measuring head for control of moisture in paper web. (Courtesy of Foxboro Co.)

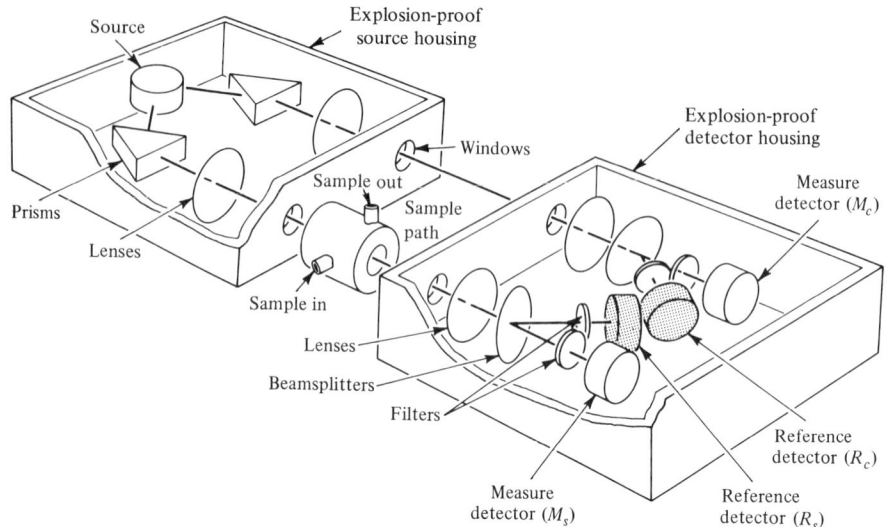

FIGURE 31-3 Dual path–dual frequency photometric process analyzer. (Courtesy of Anacon, Inc.)

In many process streams a pseudobinary situation exists in which the components of the stream fall into two responsive groups. If the signal difference between groups is large compared with the differences among individual components of each group, a successful "group-type" analysis is possible. Group analysis for aromatics, diolefins, ketones, and aldehydes in process streams is accomplished with ultraviolet instruments employing a suitable source and filters. The utilization of conductance measurements to monitor the ionic content of water has found important applications in boiler operations, cooling-tower losses, and in the manufacturing of pulp and paper.

Simple, rugged, continuous flow photometric analyzers are used to monitor either liquid or gaseous process streams. The dual path–dual frequency analyzer (Fig. 31-3) provides a high level of stability with minimum noise and maintenance. This analyzer measures either visible or ultraviolet radiation transmitted through the sample at two frequencies. The analyte intensity, M_s, is selected to be the wavelength that is most strongly absorbed by the component of interest in the sample. The reference wavelength is chosen for maximum intensity, R_s, by the sample. It is used to compensate for nonspectral variations in the optical system. In addition to the sample beam, a compensation beam is used to monitor variations in the intensity of the source at both the analyte, M_c, and reference, R_c, frequencies (Fig. 31-3). Thus four separate detectors produce electrical outputs M_s, R_s, M_c, R_c, each of which is proportional to the energy focused on the detector. The concentration of the component of interest in the sample, C, is given by Beer's law

$$C = \log_{10} \frac{R_s \times M_c}{R_c \times M_s} \tag{31-1}$$

Examination of Eq. 31-1 confirms that neither sample cell contamination nor variation in source intensity affect the value of C. Typical determinations include Cl_2, SO_2, or H_2S in stack gas; aromatics in hydrocarbons, dienes, polyenes, and polyacetylenes; and aromatic hydrocarbons, phenols, or inorganic (copper, cobalt, and nickel) salts in aqueous solutions.

A refractometer is the instrument of choice whenever the sample to be analyzed is a simple binary mixture. If the range of compositions is broad, analyzers utilizing density measurements are applicable. But when the concentration range is narrow and an analysis of the liquid phase of a suspension or slurry is required, refractometry is again the method of choice. The critical angle refractometer (Chapter 13) measures the index of refraction due to the interface between the prism and the process stream and therefore requires no penetration of the stream by the light from the source. This analyzer performs equally well on process streams that are dark or turbid as on those liquid streams that are clear and transparent. In addition, the measurement is not affected by large amounts of solid matter or gas bubbles contained in the streams. Streams that are viscous or must be held at high temperatures or pressures in order to prevent unwanted reactions may be analyzed directly by the critical angle refractometer. The probe sensing head (Fig. 31-4) may be easily inserted into any stream by threading it into a standard fitting in the process vessel, tank, or pipe line. Fiber optics are used to guide the light from the source to the prism. An automatic washing jet is provided to prevent film buildup on the prism surface. A thermistor probe inserted into the stream monitors the temperature and provides the data necessary to correct for temperature changes. It may also automatically correct the re-

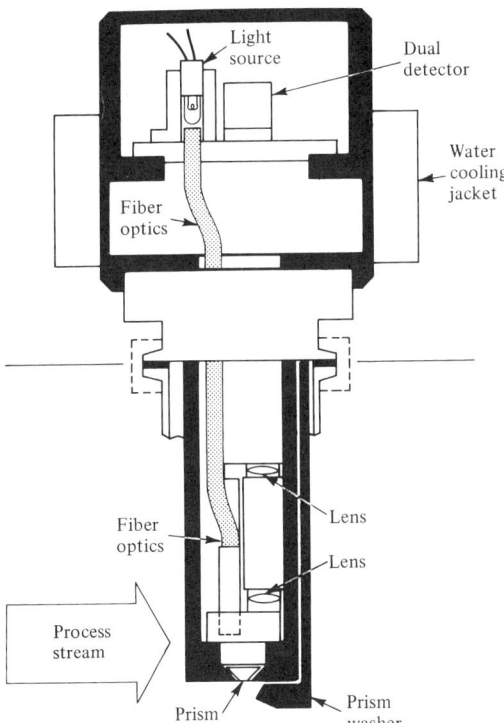

FIGURE 31-4 Probe head for critical angle process control refractometer. (Courtesy of Anacon, Inc.)

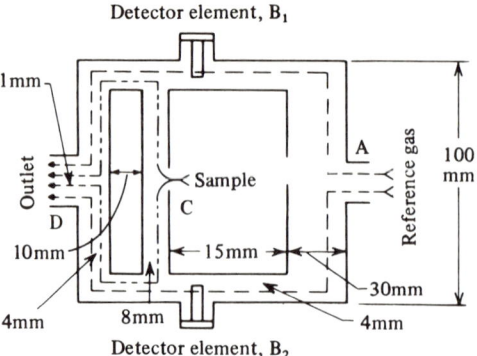

FIGURE 31-5 Nerheim gas-density balance. Schematic view as mounted in a vertical plane. (Courtesy of Gow-Mac Instrument Co.)

fractive index for variations in temperature if linked to a computer with appropriate software.

Gas Density

Although gas-density detectors can be used to measure concentration changes in the effluent from gas chromatography columns, they have also found applications in continuous, direct monitoring of process gas streams. These detectors can continuously measure the density or average molecular weight of binary or multicomponent gas streams.

At constant temperature and pressure, one mole of any pure, ideal gas will occupy the same volume as one mole of any other ideal gas. Consequently, the density of an ideal gas is a direct linear function of the molecular weight of that gas. Although this is strictly true only for ideal gases, nearly all real gases behave as ideal gases at temperatures near room temperature and pressures near atmospheric. Two designs of the gas-density detector are in commercial use: the original Martin design and the more recent Nerheim design. Both function on the same basic operational principles but differ in sensing elements, configuration, and simplicity.

The Nerheim configuration, Fig. 31-5, illustrates the principle of operation: With the conduit network mounted vertically, the reference (carrier) gas enters at A, splits into two streams, and exits at D. Two flowmeters, B_1, B_2, are installed, one in each stream, and are wired into a Wheatstone bridge. When the flow is balanced, the detector elements, which form a matched pair, are equally cooled and the bridge is balanced, thus giving a base line (zero) trace. The detector elements may be either hot wires or thermistors, depending upon the desired operating temperature. These are connected via an electrical bridge to a recording potentiometer. The sample gas (or effluent from a chromatographic column) enters at C, splits into two streams, mixes with the reference gas in the horizontal conduits, and exits at D. The sample gas never comes into contact with the detector element, thus avoiding problems caused by corrosion or carbonization.

If the sample gas is of the same density as the reference, there will be no unbalance of reference streams or of the detector elements. When the sample gas carries transient trace impurities which are heavier than the reference gas, the density of the heavier molecules will cause a net downward flow, partially obstructing the flow $A-B_2-D$, with a temperature rise of element B_2, and permitting a corresponding increase in the flow $A-B_1-D$, with a temperature decrease of element B_1. In a similar manner, lighter molecules will cause a net upward flow and the reverse will be true, namely, a temperature rise of element B_1 and decrease of element B_2 with a signal of opposite polarity from the first case. Bridge unbalance is linear over a broad range and directly related to the gas-density difference. Calibration for individual components is eliminated because the response depends on a predictable relationship, the difference in molecular weight of component, and reference gas:

$$\Delta \rho = k \ \frac{n_s(M_s - M_r)P}{RT} \qquad (31\text{-}2)$$

where k is a constant whose value depends on cell geometry, viscosity, the flow-measuring system, and the thermal conductivity of the gases; M_s and M_r are the molecular weights of the sample and reference gases; and n_s is the mole fraction of component in the sample.

The sensitivity of the gas-density detector with thermistor sensor is comparable to that of a thermistor-type of thermal-conductivity cell and requires no amplification at room temperature. The hot-wire sensor, which has one-sixth the sensitivity of thermistors at 25°C, may require low level amplification at 100°C. The low noise level of the detectors permits effective use to 300°C, although the sensitivity decreases rapidly with increasing temperature for thermistors, being at 250° one-fifth that at 50°C. The Nerheim design has an effective sample volume of approximately 5 ml.

31.4   INFRARED PROCESS ANALYZERS

Infrared instruments have been widely used in process stream analyses. Infrared analyzers fall into three categories: dispersive, nondispersive, and bandpass optical filter.

A dispersive infrared instrument is patterned after a double-beam spectrometer (Chapter 17). Radiation at two *fixed* wavelengths passes through a cell containing the process stream to provide a continuous measurement of the absorption ratio. The two wavelengths are chosen so that the component of interest in the process sample stream absorbs at one wavelength. At the other wavelength, the sample either does not absorb or else exhibits a small, constant absorption. The ratio of absorbance readings is converted directly into concentration of the component of interest and recorded. This type of instrument can handle liquid systems as well as gas streams, and has the ability to analyze quite complex mixtures. Tunable laser diodes, fabricated from lead–salt semiconductors, have been used as monochromatic infrared sources for trace gas monitoring. The detector may be the radiation thermocouple used in laboratory instruments. Dispersion infrared analyzers are used in process applications when the necessary selectivity cannot be achieved by nondispersive methods, in cases involving liquid process streams, and in analyses at wavelengths longer than 10 μm (less than 1000 cm^{-1}).

The popular method for selective analysis of gaseous process streams is nondispersive infrared absorption. These rugged, reliable nondispersive analyzers are commonly employed as sensing components of automated process-control loops. Since these instru-

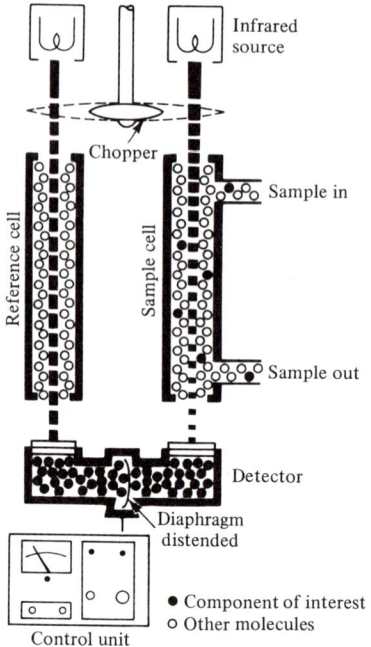

FIGURE 31-6 Positive-filter nondispersive process-stream infrared analyzer. (Courtesy of Beckman Instrument Co.)

ments do not require monochromatic radiation, their design and operation is simpler than that of dispersive instruments. Excellent sensitivity is achieved due to the strong signal power resulting from the fact that all of the radiation from the infrared source passes through the sample.

In practice, the nondispersive analyzer is really a filter photometer with the filter usually composed of the vapors of the component of interest. Interference filters constructed of nongaseous materials have also been employed successfully in the design of these compact analyzers. An analyzer containing an interference filter has the selectivity approaching that of a dispersive instrument while maintaining the simplicity of a colorimeter.

In the positive-filter type of analyzer (Fig. 31-6), the radiation from two matched energy sources is passed through a chopper, which blocks both sources simultaneously with equal on–off periods. This is done to minimize variations in signal due to fluctuations in ambient temperature and source intensity. One beam of radiation then passes through the sample cell, which is a highly polished tube that conducts radiant energy by multiple reflection. The gaseous sample from the process stream flows continuously through this cell. The other beam of radiation passes through a similar cell containing a nonabsorbing reference gas, such as nitrogen. Both beams of radiation then pass into the detector, a sealed compartment filled with the pure form of the gas being analyzed. A nonabsorbing pressurizing gas may be added to the detector to optimize its sensitivity. The detector is divided into two halves separated by a flexible diaphragm. When some of the component of interest appears in the sample stream, the sample side of the detector receives less radiant energy and the diaphragm expands outward from the reference side into the sample side. The diaphragm is positioned so that it forms a capacitor plate in an electrical circuit. Thus the difference in radiant energy received by the two compartments of the detector is transformed into an electrical signal, which is proportional to the concentration of the component of interest in the sample stream.

If, for example, the detector of this nondispersive analyzer is filled with carbon dioxide, the carbon dioxide absorbs infrared radiation at its characteristic wavelengths, principally 4.2 μm (Fig. 31-6). If carbon monoxide is present in the sample stream, it will not interfere with the analysis of carbon dioxide because its absorption bands do not overlap those of carbon dioxide. Thus, an analyzer with a detector filled with carbon dioxide detects primarily carbon dioxide and will not be affected by the presence of other components, such as carbon monoxide, unless these components are present at high concentrations or the component absorption bands overlap those of carbon dioxide. Where interference is a problem, standard optical filters, gas cell filters, and/or window material can be placed in both radiation beams to eliminate or to reduce the interference to an insignificant level. While these filtering techniques increase the selectivity of the analysis, they may also reduce the sensitivity of the analyzer toward the component of interest.

A dedicated ambient CO-monitoring system, designed around a nondispersive analyzer, is shown in Fig. 31-7. During sample analysis, the sample cell continuously receives ambient air containing CO, while the flow-through reference cell receives background air containing CO which has been treated to convert the CO to CO_2. Since both the sample and reference cells contain the same amounts of water vapor and nearly the same amount of CO_2, the resulting signal from the pneumatic detector is proportional to the CO in the sample cell. The autozero and span adjustment module ensures automatic periodic standardization of the system. Once every 4 hr the module introduces CO-free air to the sample and reference cells, thus allowing the system's output signal to be reset to zero. A reference CO span gas is run through the sample cell once every 24-hr period to recali-

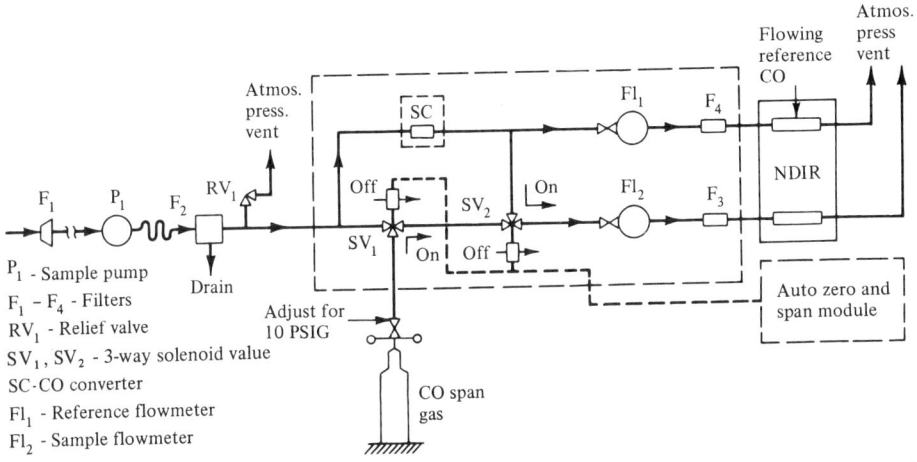

FIGURE 31-7 Ambient CO-monitoring system. (Courtesy of Beckman Instrument Co.)

brate the system's readout. The pump, P, draws an ambient air sample and supplies a constant stream to the analyzer. Particulates are removed by filters and the sample is cooled to remove condensible liquids.

Although this type of nondispersive instrument is both sensitive and selective, it suffers from a major limitation—a separate analyzer is required to monitor each component of interest. In many applications it is desirable to monitor several components of a process stream or of ambient air simultaneously. In order to perform these analyses at increased sensitivities and with a minimum of instrumentation, the design of the nondispersive photometer is modified. The pneumatic detector is replaced by a more sensitive broadband, solid-state detector. The functions of the chopper, gas-filled reference cell, and the component in the detector compartments are combined in a filter wheel. This wheel consists of pairs of gas-filled filters, alternately positioned in the single optical beam of the infrared source (Fig. 31-8). One cell is filled with pure sample of the component of interest, the next with nitrogen. When a filter containing a given component of interest is in the optical path, the amount of this component present in the sample cell has little effect on the detector since the component gas present in the filter has already absorbed most of the radiation. This filter is called the reference cell. When the filter cell containing the nitrogen is rotated into the optical beam, little energy is absorbed by the filter cell (known as the sensitive cell) and thus the amount of absorption depends upon the concentration of the component in the flow-through sample cell. If e_r is the detector output signal when the reference filter is placed in the optical path and e_s is the output when the filter containing the component of interest is in the optical path, then the ratio e_r/e_s is proportional to the concentration of the component of interest.

The first electronic circuitry developed for this type of nondispersive analyzer was entirely analog. Current instruments employ digital circuitry with microcomputer

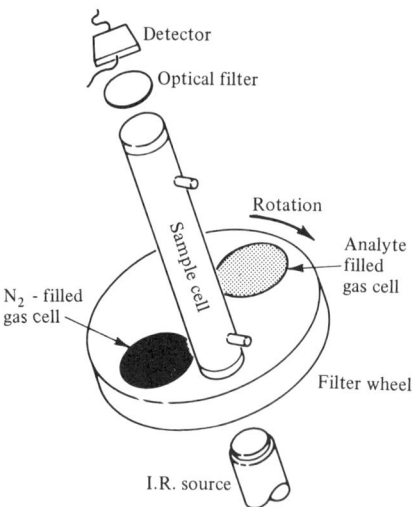

FIGURE 31-8 Filter wheel NDIR. (After R. J. Bibbero, *Microprocessor in Instruments and Control*, p. 180, Wiley, New York, 1977.)

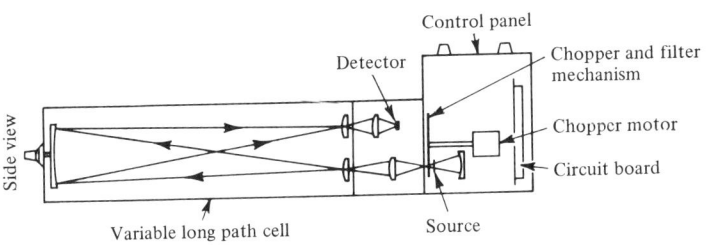

FIGURE 31-9 Single-beam infrared analyzer with variable path gas cell. (Courtesy of Foxboro Analytical Co.)

control. Replacement of analog hardware with microcomputer software leads to easier maintenance, increased reliability, and increased flexibility in both instrument design and operation (Chapter 30).

The versatile, portable Wilks miniature infrared analyzer (MIRAN) produced by Foxboro Analytical replaces the pairs of gas-filled cells with a variable filter wheel. A single component may be monitored by setting the wheel at the appropriate wavelength or the analyzer can scan through discrete portions of the spectrum to measure the concentrations of a number of different sample components. A number of different sampling accessories are available including variable path gas cells (Fig. 31-9), liquid flow-through multiple internal reflection (MIR) cells, and automatic film sample handler. More sophisticated models perform multicomponent analyses using a microprocessor to control analytical parameters, perform data reduction, and generate reports. The MIRAN analyzer is also the heart of a multipoint, multicomponent environmental air-monitoring system which can analyze samples in up to 24 locations for as many as 10 components. The system produces data in a form acceptable to the Occupational Safety and Health Agency (OSHA) and the Environmental Protection Agency (EPA).

Components frequently determined by infrared analyzers in industrial processes and in ambient air-monitoring stations are listed in Table 31-1; these include vapors of substances that are liquids at room temperatures. For example, if the measurement of carbon dioxide in industrial processes is considered, one finds a large number of applications, such as the control of combustion, the manufacture of cement, and the production of ethylene oxide, phthalic anhydride, and ammonia. By contrast, these infrared instruments cannot selectively measure similar substances, such as butane in propane.

31.5 OXYGEN ANALYZERS

The determination of oxygen is important in monitoring processes as varied as respiration (blood gas) and combustion (industrial flue gas). Energy conservation and environmental emission standards have forced fuel-consuming industries to improve combustion performance. Reliable determination of the oxygen content in flue gas is a good indicator of combustion efficiency.

Methods used to measure oxygen may be classified as either physical or chemical. Physical methods use the paramagnetic property of oxygen and thermal conductivity as

TABLE 31-1 Applications of Infrared Spectroscopy to Process Streams[a]

Gas	Analytical Wavelength, μm	Minimum Detectable Concentration (ppm) 20-m Path	Maximum Concentration 1.0 Absorbance (ppm or %) 0.75-m Path
Cyclohexane	3.4	0.04	4000
CO_2	4.25	0.08	7500
N_2O	4.5	0.03	2100
CO	4.65	1.2	8.3%
COS	4.85	0.02	1100
NO	5.3	1.5	8.2%
CH_3COCH_3	5.75	0.1	4000
SO_2	7.4	0.1	3000
Vinyl acetate	8.2	0.06	1000
Dioxane	8.8	0.2	2100
CH_3CH_2OH	9.4	0.4	5000
NH_3	10.75	2.2	2.1%
Freon 11	11.8	0.06	300
CCl_4	12.6	0.05	250
CH_2Cl_2	13.3	0.4	1900

SOURCE: Courtesy of Wilks Scientific Corp.
[a]Data for MIRAN (miniature infrared analyzers) dispersive gas analyzer with variable path gas cell (0.75 to 20 m).

the basis for quantitative determinations. Chemical methods include potentiometry and catalytic combustion. The choice of a particular method is determined by whether the measurement is for oxygen in gas samples or dissolved oxygen in liquids. The presence of interfering substances and the required limits of detection also are important factors in selecting a method. Several of these methods are briefly described in the following sections.

Magnetic Susceptibility

Oxygen, nitric oxide, and nitrogen dioxide are unique among the ordinary gases in that they are paramagnetic; that is, they are attracted into a magnetic field. Most gases are slightly diamagnetic and repelled out of a magnetic field. Oxygen is several times more paramagnetic than nitric oxide or nitrogen dioxide. The values of the volume susceptibilities are 146.6, 65.2, and 4.3×10^{-9}, respectively. Advantage is taken of this property of oxygen in gaseous oxygen analyzers.

Gaseous oxygen is measured on the basis of change in magnetic force acting on a test body suspended in a nonuniform magnetic field when the test body is surrounded by the gas sample. The Beckman paramagnetic oxygen analyzer (Fig. 31-10) incorporates a small glass dumbbell suspended on a taut quartz fiber in a nonuniform magnetic field. When no oxygen is present, the magnetic forces exactly balance the torque of the fiber. However, when a sample containing oxygen is drawn into the chamber surrounding the dumbbell, the magnetic force is altered, causing the dumbbell spheres to rotate away

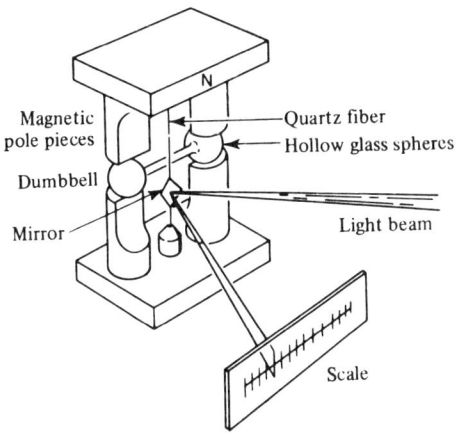

FIGURE 31-10 Schematic of Beckman paramagnetic oxygen analyzer. (Courtesy of Beckman Instruments, Inc.)

from the region of maximum magnetic flux density. The degree of rotation is proportional to the partial pressure of oxygen in the sample. A small mirror attached to, and rotating with, the dumbbell throws a beam of light on a translucent scale of the instrument. The scale is calibrated in concentration of oxygen. Instruments are capable of sampling static or flowing gas samples free of solids or liquids. With a span of 5% oxygen full scale, an accuracy of 0.05% oxygen can be achieved. Standard cell volume is 8-10 ml, although for static samples a 3-ml cell volume is possible. Response time is about 10 sec.

The instrument shown in Fig. 31-11 utilizes the magnetic properties of oxygen along with thermal conductivity for the measurement of oxygen in a gas. The gas sample is

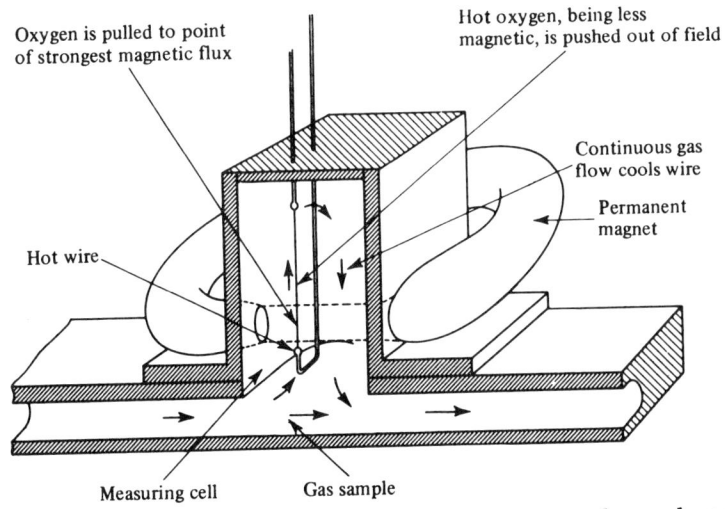

FIGURE 31-11 Schematic of Hays Magno-Therm oxygen recorder: schematic operation of analyzing cell. (Courtesy of Hays Corp.)

passed across the bottom of a gas cell containing an electrically heated hot wire. A strong magnetic flux from a permanent magnet is directed across the wire. The oxygen is pulled into the region around the hot wire by the magnetic flux and is heated by the wire. Oxygen loses its magnetic susceptibility in inverse proportion to the square of the absolute temperature, and therefore the heated, relatively nonparamagnetic gas is continually displaced by the cooler, more paramagnetic oxygen moving in from below. A flow of gas proportional to the amount of oxygen present is set up around the hot wire. The hot wire is cooled, and its resistance is thereby decreased. The resistance of the wire in the analysis cell is compared with the resistance of a similar wire in a comparison cell by means of a Wheatstone-bridge circuit. The comparison cell contains the sample of gas as does the measuring cell but does not have a magnetic flux around the wire. Thus all variables except the cooling due to the oxygen present are canceled. The zero setting of the instrument is checked by swinging the magnet away from the measuring cell without interrupting the gas flow. The overall accuracy of the instrument is claimed to be ±0.25% oxygen (up to 20%) and ±2.5% of range up to 100%.

Electroanalytical Methods

The Hersch galvanic cell provides an electrolytic method for the determination of low concentrations of oxygen in gas streams. The gas stream passes through an electrolytic cell, which consists of a silver cathode and an anode of active lead or cadmium. The electrodes are separated by a porous tube saturated with an electrolyte of potassium hydroxide. Oxygen in the gas sample is reduced to hydroxyl ions at the silver cathode. The metallic lead or cadmium in turn is oxidized to plumbate(II) ions or cadmium hydroxide. The magnitude of the cell current is a measure of the oxygen in the sample; a sensitivity of 1 ppm is attainable. Acidic substances are removed in advance by scrubbing the gas stream with an alkali hydroxide. Calibration is achieved by periodically generating known amounts of oxygen in a separate electrolysis cell.

Oxygen in flue gases may be monitored by probe-type analyzers that are inserted directly into high-temperature flue streams. Zirconium(IV) oxide (ZrO_2), stabilized by traces of materials such as CaO or Yb_2O_3, has a crystal structure containing cation vacancies or holes. At temperatures of around 1500°C, oxygen anions tend to migrate into the holes, causing the vacancy (hole) to be displaced. If platinum electrodes are placed on opposite sides of the ZrO_2 lattice and the temperature of the lattice is controlled at a high temperature, the voltage difference between the electrodes will indicate the difference in partial pressure of oxygen between the two sides (Fig. 31-12). If the oxygen partial pressure is controlled on the reference side, a millivolt signal proportional to the difference between the oxygen pressures is generated. At 816°C the magnitude of the signal is given by the Nernst equation

$$E = 0.054 \log \frac{P_r}{P_s} \tag{31-3}$$

where P_r and P_s are the partial pressures of oxygen on the reference and sample sides, respectively. An output voltage of 54 mV/decade of oxygen pressure difference is obtained at 816°C.

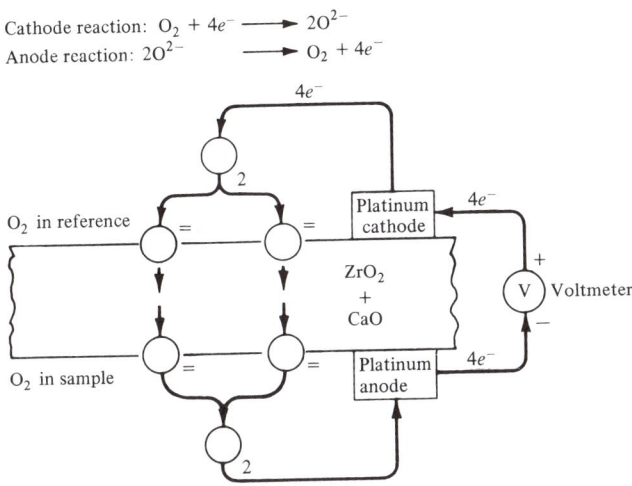

FIGURE 31-12 Zirconia oxygen analyzer. (Courtesy of Milton Roy Co., Hayes-Republic Division.)

This oxygen sensor can be operated at or above flue gas temperatures and provides a strong output signal specific to the oxygen content of the gas. No sample preparation is required. The sensor is rugged enough for service in the severe environment of a hot flue and can be repaired easily in the field. Sensor calibration can be performed without removing the probe from the flue stream. The temperature of the sensor must be controlled to minimize changes and gradients.

A unique method of sampling liquid streams for the measurement of dissolved oxygen is shown in Fig. 31-13. A sample is induced to flow past the oxygen electrode by injecting air into the immersion tube above the electrode. The resultant air/liquid mixture in the immersion tube is less dense than that of the sample stream. Thus the incoming liquid from the stream will force the liquid already in the immersion tube up the tube to a point of hydrostatic equilibrium. Since the immersion tube outlet is below the point of hydrostatic equilibrium, the incoming sample flow is maintained as long as air injection is continued. The sample stream is drawn through a filter and past the oxygen electrode to continuously present it with fresh sample. The injected air does not affect the measurement because it is injected above the sensing point.

If the filter becomes clogged, a back-flushing method is available. The immersion tube discharge is closed. If the air continues to flow into the immersion tube, it displaces the liquid which must flow back through the filter, dislodging trapped debris. After all the liquid has been flushed out of the immersion tube, the electrode is surrounded by damp air which provides a 100% saturated sample for calibration purposes.

31.6 ON-LINE POTENTIOMETRIC ANALYZERS

Potentiometric, single component analyzers that give continuous, real-time analyses are available for a variety of inorganic ions and dissolved gases. These systems use rugged specific ion electrodes to obtain laboratory accuracy in the environment of an industrial plant. The design of these analyzers is simple and straightforward (Fig. 31-14). In this system the sample is first filtered to prevent clogging of the electrode surfaces and then mixed with a reagent to minimize interferences. A unique bypass filter uses the flow of the sample stream to keep the filter surface clean. After mixing, the sample/reagent solution is pumped through a thermostated electrode assembly for measurement. Sample color and turbidity do not affect accuracy.

The electrode signals are transmitted to the electronic unit where their voltage difference is amplified. Both voltage and current outputs are provided for remote reading and control. A panel meter gives continuous readings, while a strip chart recorder gives a

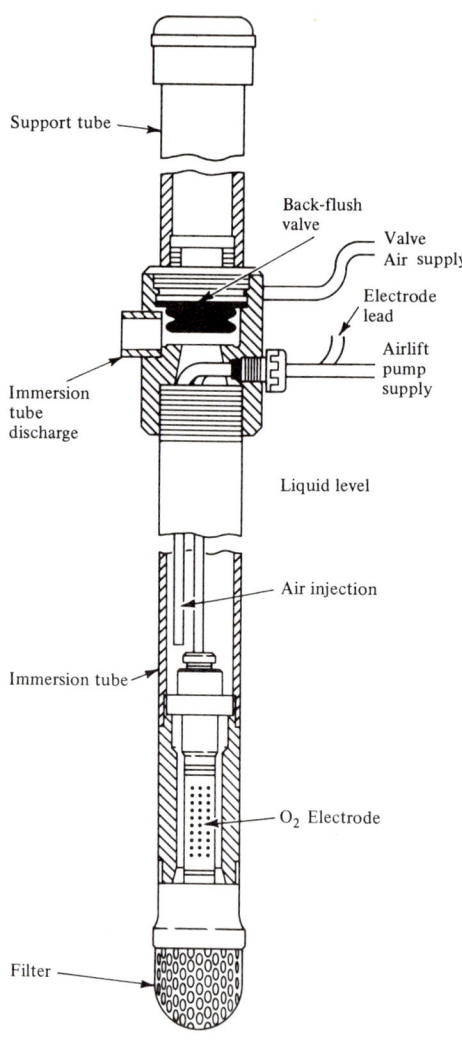

FIGURE 31-13 Dissolved oxygen electrode system. (Courtesy of Ionics, Inc.)

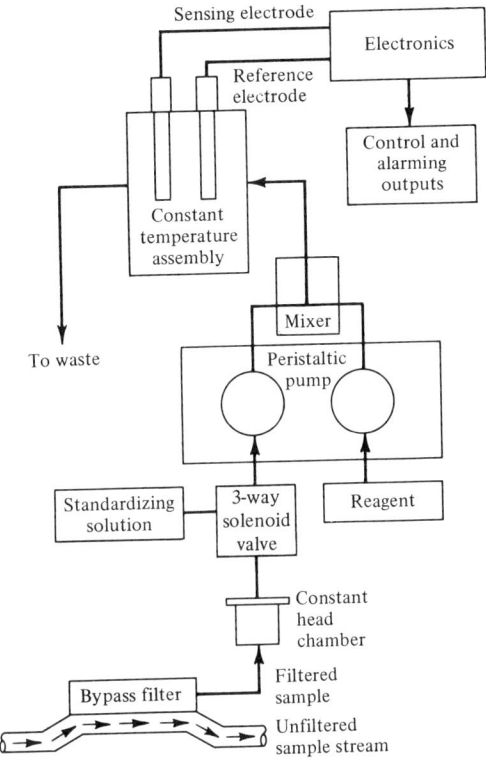

FIGURE 31-14 On-line potentiometric analyzer. (Courtesy of Orion Industrial.)

permanent record of the sample stream concentration. The results may be utilized for local alarm process and control or sent to a host computer for further analysis and process control. Applications include quality control measurements of chloride ion in the manufacture of ascorbic acid, monitoring cyanide in effluent from electroplating operations, and control of water fluoridation.

A continuous potentiometric chlorine analyzer is shown in Fig. 31-15. There are no pumps or other moving parts in the sampling system. After the pH of the solution is adjusted within the passive diffuser, the chlorine in the sample stream reacts with solid PbI_2 in the reaction chamber to produce iodine. The iodine now enters the electrode chamber and is sensed by an iodine-sensitive electrode. Correct operation of the analyzer is periodically checked by switching the three-way valve to the check position to obtain a stream of zero-chlorine water from the carbon purifier. This analyzer is used to monitor chlorine levels at critical points in water treatment plants (Fig. 31-16).

31.7 PROCESS GAS CHROMATOGRAPHY[2]

The first process gas chromatograph was put into service in 1954. Since that time these chromatographs have become the most widely used process analyzers in the petrochemical and refining industries. While the similarities between laboratory gas chromatographs (Chapter 16) and their process counterparts are apparent, differences important to process applications are less obvious. These differences include the physical appearance of the chromatograph, techniques used to acquire data, and most importantly, the way the analytical results are used.

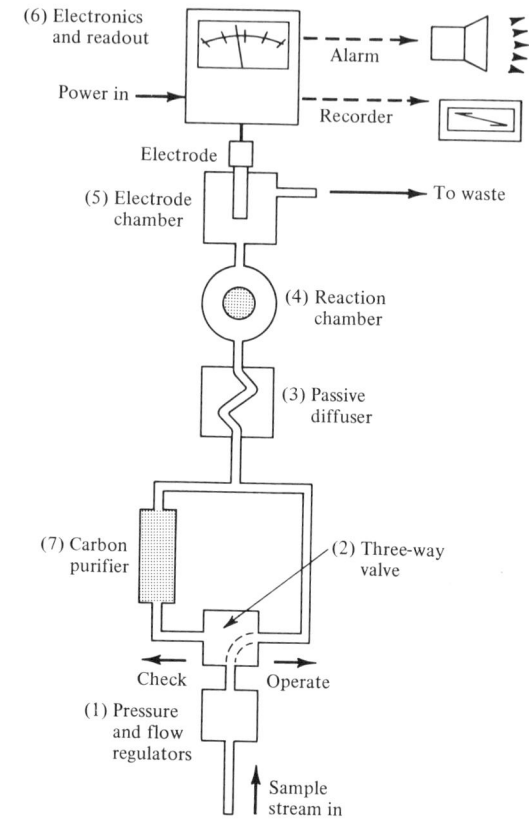

FIGURE 31-15 Continuous chlorine analyzer. (Courtesy of Orion Research, Inc.)

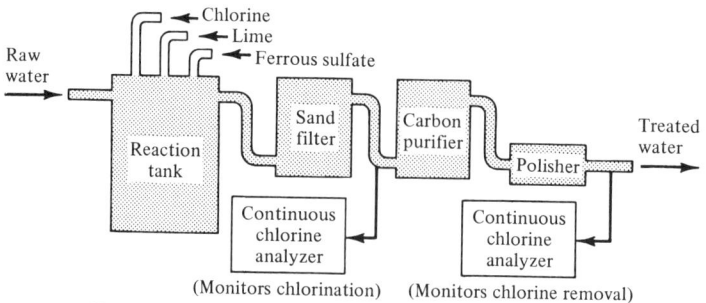

FIGURE 31-16 Typical water treatment system.

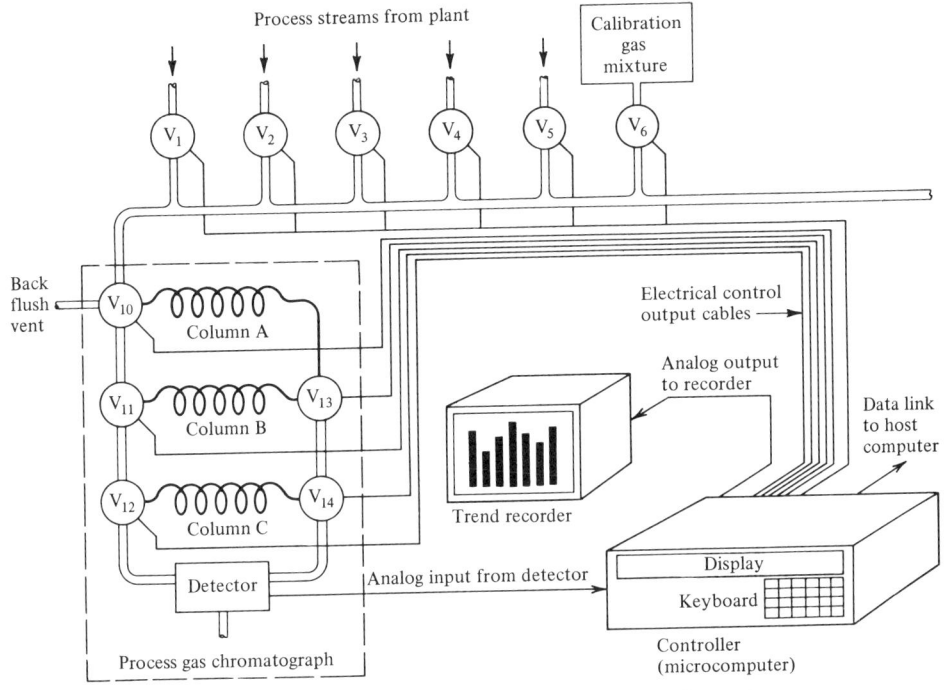

FIGURE 31-17 Process control gas chromatography system.

On-line analyzers are installed for the purpose of providing analytical results with a response time comparable to the changes in the process being monitored. This information may then be used to take appropriate control measures. The process gas chromatograph has been designed to accomplish this objective by operating continuously on-line and automatically analyzing one or more components of process streams in a cyclic, repetitive manner. A single analyzer typically monitors several streams (Fig. 31-17).

Differences Between Process and Laboratory Chromatographs

Among the characteristics unique to process gas chromatography is the method of sample handling. The sample is transferred directly from the sampled stream to the chromatographic column(s) with no human intervention. Sample valves and connecting lines can be maintained at elevated temperatures, thus allowing samples containing large amounts of water vapor or other condensibles to be transferred to the analyzer without altering sample composition. Many sampling situations that are difficult or impossible using laboratory chromatographs are routine for process control analyzers. The inlet septum, common to laboratory chromatographs, is replaced by a sample valve. This valve is a critical component in the process chromatograph, for it must repetitively and reproducibly transfer a precisely measured volume of the sample into the carrier gas, which flows through the analyzer columns. If the sample is in liquid form, the liquid sample valve must deliver a measured volume of sample to a vaporizer where it is first volatilized and then passed to the carrier gas. In addition to handling samples from one or more process streams, the valve system must possess a means for periodically providing a sample of known composition from a reference tank for the purpose of calibration.

As in laboratory chromatography, column parameters are optimized to provide separation of all components of interest in the shortest possible time. Analysis time is critical and must be minimized for process control applications. All components of the sample must be quantitatively accounted for and removed from the column(s) during each cycle. Residual components will either accumulate and change the properties of the column or they will elute during a later cycle and interfere with the analysis. The system should be as simple as possible to operate, adjust, and maintain. Personnel assigned to process analyzers usually do not have the training comparable to the personnel assigned to laboratory chromatographs. Process gas chromatographs are used in environments where downtime is prohibitive.

Because of the requirements of process analysis, column technology has developed in different directions from that used in laboratory practice. Isothermal, multicolumn analysis methods have been emphasized while little or no use has been made of techniques, such as temperature programming and derivatization, employed in laboratory separations. A large number of designs using valves to switch the sample among several columns have been developed. One widely used configuration is the stripper (Fig. 31-18), which consists of two columns in series: a stripper (or precut) and an analysis column. Provision is made for backflushing the stripper column through a separate vent in order to reject unwanted components while simultaneously providing carrier gas to the analysis column. The desired components are further separated on the analysis column while the stripper is being backflushed. This configuration can be used to remove moisture,

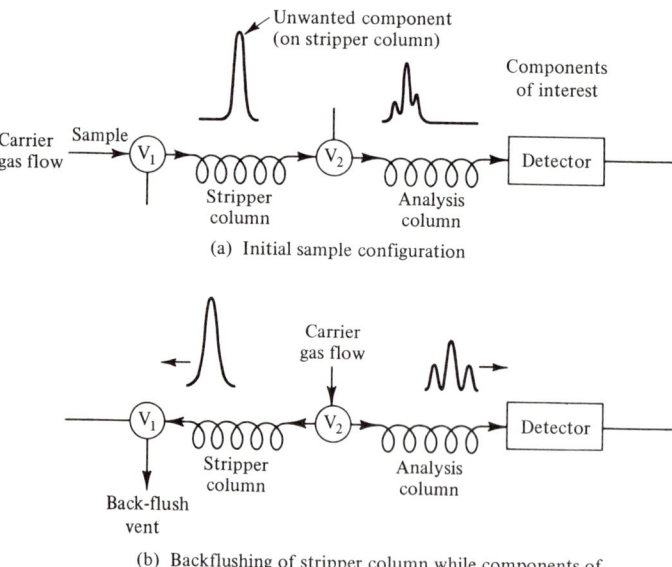

FIGURE 31-18 Use of a stripper column in process control gas chromatography.

an impurity often found in hydrocarbon samples. Use of a stripper column ensures that unknown components of higher molecular weight are quantitatively removed from the system during each cycle and will not appear unexpectedly during a subsequent analysis.

The thermal conductivity detector is the primary detector for process gas chromatography. It is reliable, simple, easy to maintain, inexpensive, and universal in its response to components. Increased demands for monitoring and process control require higher sensitivity and greater selectivity inherent in ionization detectors. The flame ionization detector has proved to be the most practical alternative to thermal conductivity detectors.

Process Chromatography Systems

Application of on-line digital computers have revolutionized methods utilized in both laboratory and process gas chromatography. In addition to providing real-time analysis of chromatograms and postrun calculation of results, the computer can control the operation of the process analyzer system by controlling both stream sampling and column switching valves. This computer control can improve the efficiency and reliability of the system as well as reduce maintenance costs. Initially large minicomputer systems were used to control as many as 30–40 on-line analyzers. This approach was useful in applications, such as ethylene production, which utilize large numbers of chromatographs. Smaller plants requiring only a few analyzers had difficulty justifying the large initial capital investment costs of the minicomputer system. This problem has been solved as microcomputers have begun to replace minicomputers. Individual chromatographic analyzers containing microcomputers have the full capabilities of the minicomputer at drastically reduced costs. A complete chromatographic analyzer system with microcomputer controller is shown in Fig. 31-17. Tasks performed by the computer include: program control (timing of operations in analysis), peak detection, base line correction, resolution of fused peaks, postrun calculations, calibration, scaling of analog outputs, and serial communications with display devices and the host computer. In addition to these basic tasks, the computer can perform maintenance tests, such as monitoring oven temperature, detector currents, and control the sampling system.

Applications of process chromatographs frequently encountered are for open- or closed-loop process control and environmental monitoring. In open-loop control the operator makes adjustments to the process conditions based on the chromatographic results. In closed-loop control the analyzer data are sent either to hardware electronic controllers or to a process control computer, which controls the process automatically. Monitoring of harmful components, such as vinyl chloride, in the atmosphere of plants that either produce or use them, is required by OSHA regulations. A single analyzer can obtain samples from a number of locations throughout the plant, pump them to the chromatograph equipped with a flame ionization detector, and record concentrations as small as 0.1 ppm.

31.8 CONTINUOUS ON-LINE PROCESS CONTROL

Automated control has become a major factor in the efficient operation of large-scale chemical processes. Production plants are facing difficulties due to rapidly rising costs of energy and raw materials, environmental regulations, and varying personnel attitudes. Sensitive control systems are required to reduce the margin of operational errors. Present control specifications require these systems to be highly flexible, yet easy to operate and maintain.

In process control applications, a continuous analyzer is attached to a sampling line and thereafter automatically and continuously obtains a signal that is proportional to the instantaneous concentration of a selected component in the flowing stream. The information provided is then automatically used to set the process environment controllers and to take any corrective action necessary to control the process. Thus, continuous stream analyzers take over the function of the control laboratory with an increase in speed and efficiency.

A number of steps are involved in setting up on-stream control facilities. The analyst, in close collaboration with the project engineer, must determine the analytical task or tasks to be done in order to follow a process as effectively as possible. The number of constituents monitored and the number and location of checkpoints must be decided. Never deluge the operator with a mass of information, much of which may be useless. The guiding principle should be to provide the operator with the least amount of data needed to produce the desired product. Economic considerations and the manpower requirements for installation, calibration, and maintenance must not be overlooked. Different analytical methods may be desirable at each checkpoint. As process control improves, production costs decrease; however, analysis costs rise. The selection of an optimum analytical method depends on minimizing the sum total of these costs. Leemans[3] discusses the selection procedure in detail.

Design Features

When an analyzer is selected for a given problem, the first step is to make certain that the particular analysis can be made by the instrument and that it has sufficient sensitivity to determine the component of interest in the range of concentrations expected. Although similar in operating principles to their laboratory cousins, process stream analyzers differ in a number of important respects. Moving the automatic instrument sensors from the laboratory to the plant confronts instrument designers with major problems. Design criteria must incorporate these features: (1) reliability, (2) operational simplicity, (3) readout as foolproof as possible, (4) ease of maintenance, and (5) flexibility for future growth.

First and foremost is reliability. Instrument downtime represents plant process downtime, or operation without control, which is very costly. Hence, long-term stability and reliability are essential characteristics. Availability of modular plug-in-type construction shortens necessary repairs and goes a long way toward making the instrumentation as reliable as any of the links in the process. Operational simplicity implies a minimum of controls and infrequent attention by the operator. Preferably once every shift, a cursory check is run. Thorough overhaul and inspection is carried out only during normal process downtime.

Readout and control functions must be made as foolproof as possible. Digital readout devices are utilized extensively. The environment in which the analyzer is used differs from the relative calm of the laboratory. Analyzers must withstand wide ambient temperature fluctuations and heavy vibrations, and they must not create explosion hazards. Often the units are completely sealed so as to operate independent of outside conditions and to withstand the onslaught of monkey-wrench mechanics.

Closed Loop Control

An automated control system typically consists of the sensing element with an amplifier, the controller, and a final control element. These elements react upon each other and form the closed loop as illustrated in Fig. 31-19. All elements are of equal importance.

Automated process control systems depend on deviations from a predetermined set-point signal for their operation. The controller acts on the difference between the set-point signal and the signal from the sensing element by sending an output signal.

Automated process control systems operate as follows. The signal from the sensing element is compared with the set-point signal. Any difference causes the controller to send an output signal to the final control element which produces a correction in the process. The corrective cycle terminates when the modified signal from the sensing element equals the set-point signal.

A time lag is involved in each of the steps of this feedback control loop; this time lag is composed of the time lapses before corrective action can be initiated and the dead time (time delay between corrective action and the corresponding signal change detected by the sensor).[4]

The combination of current computer technology with on-stream sensors can provide the ultimate in flexible, precise control of processes. In order to properly use this technology, the process under consideration must be studied to understand the reactions taking place, to determine how flows respond dynamically to changes in valve position and to work out the mathematics required to control the process. Equations are written to express the desired amount of influence from a given variable. In predictive control, the desired correction is made not on the basis of set points previously determined by the

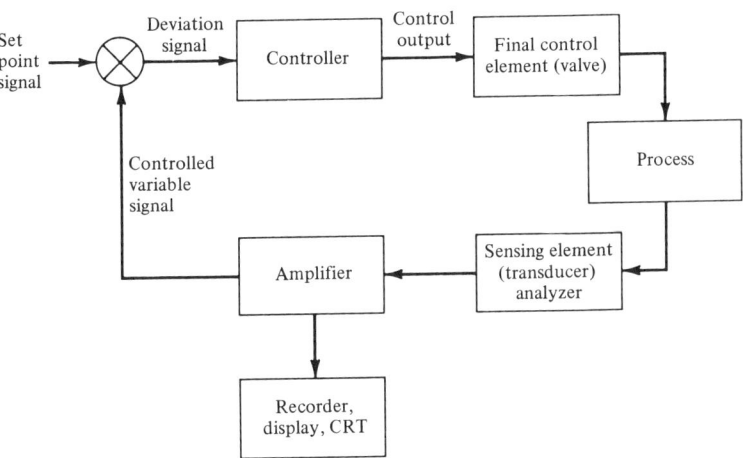

FIGURE 31-19 Simple control loop.

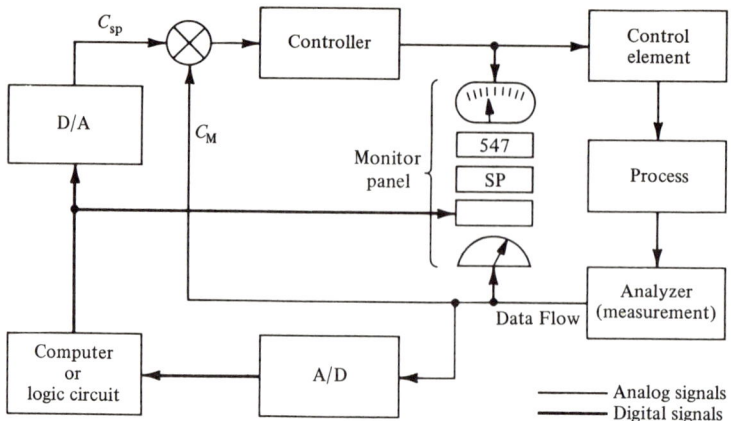

FIGURE 31-20 Supervisory computer control.

operator or process engineer, but on the basis of continuing calculations made by the computer. When a computer completes the closed control loop (Fig. 31-20), and should the operating conditions suddenly vary, the computer can rapidly make the necessary calculations so that several of the set points are properly adjusted to optimize the yield of the desired product. Computers can also be used to evaluate measurement results and to search for useful correlations not usually discernible by operators or engineers.

Methods of Process Control[5]

Classification of control methods has become difficult because each method has borrowed from the others, thus creating hybrid systems. The following general definitions are used to describe control systems.

Analog Control   Each process loop is controlled by a dedicated electronic or pneumatic control system (Fig. 31-19). Each controller is tuned and adjusted manually by operators.

Supervisory Control   This is another analog control system but with provision for allowing set points to be adjusted automatically by a supervisory computer (Fig. 31-20). This method is also known as digitally directed analog control.

Direct Digital Control (DDC)   This is a system in which all control outputs are computer generated on the basis of sensor inputs to the computer. No additional controllers exist between the computer and the process (Fig. 31-21).

Hierarchial Distributed Control (HDC)   Control of the overall process is divided into several areas, each area managed by its own control system. The individual systems communicate with each other and also with a larger central computer, which oversees the entire operation (Fig. 31-22). This approach allows the right amount of computing power to be used when and where it is needed and also provides the redundancy necessary for industrial process control.

The direct digital signal control (DDC) computers perform the primary process control functions, and communicate with the higher level host computers. These host or supervisory computers calculate the set points necessary for optimizing the desired products.

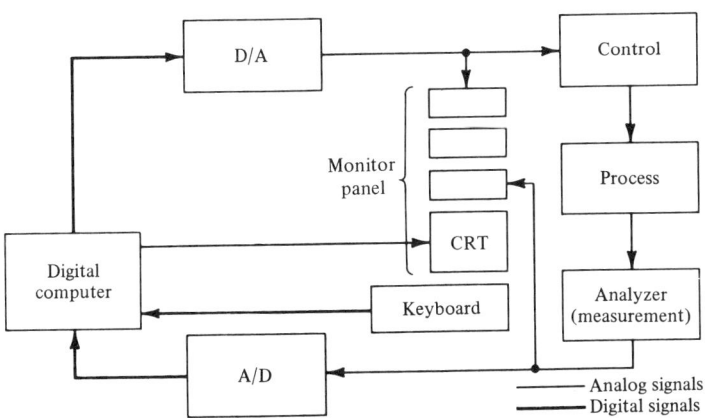

FIGURE 31-21 Direct digital control (DDC).

They also prepare data and transfer it to the central computer. In addition they can provide redundancy necessary in process control applications. This larger central computer, usually at a location remote to the plant, oversees the operation of several plants by setting production schedules and coordinating the shipment of materials from one plant to another.

Process Control System Applications

A typical use of continuous analyzers in ethylene purification is shown in Fig. 31-23. Purity and recovery are monitored at five checkpoints. Other stream components during ethylene manufacture and initial purification might be any mixture of these: CH_4, C_2H_6, C_3H_8, C_2H_2, CO, CO_2, and N_2. In the final purification, impurities are likely to be CH_4, C_2H_6, and C_2H_2, whose total concentration will be less than 2%. Instrument readability (1% of scale) should correspond to 0.05% ethylene in the manufacture, 0.02% in the initial purification, and 0.005% in the final purification. Long-term stability (8 hr or longer) should be within 2% of scale.

On-stream analyzers specifically designed for water treatment of cooling-tower waters are based on conductivity and pH measurements. A sample of recirculating water flows to an analyzer, where the conductivity and pH are continuously measured and recorded. The pH measurement provides a control signal for the addition of pH control chemicals. The signal for complete water replacement is based on the conductivity measurements, which are directly related to the total amount of dissolved solids in the system. In addition, the makeup water flowrate is measured and recorded, and provides a control signal for feed of corrosion inhibitor, biocide, and so forth.

31.9 AUTOMATIC CHEMICAL ANALYZERS[6]

The increased demand for medical services beginning in the late 1950s has resulted in growing work loads for clinical laboratories. Furthermore, new, specialized tests, such as

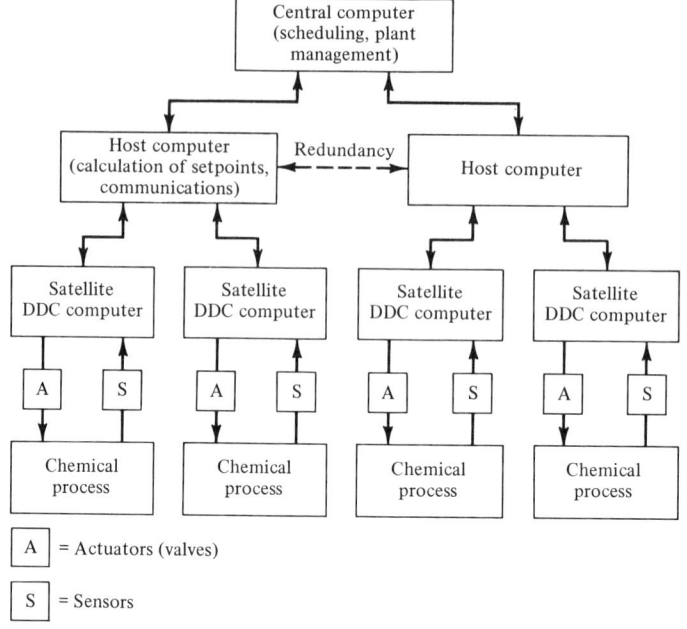

FIGURE 31-22 Multilevel, hierarchial distributed control.

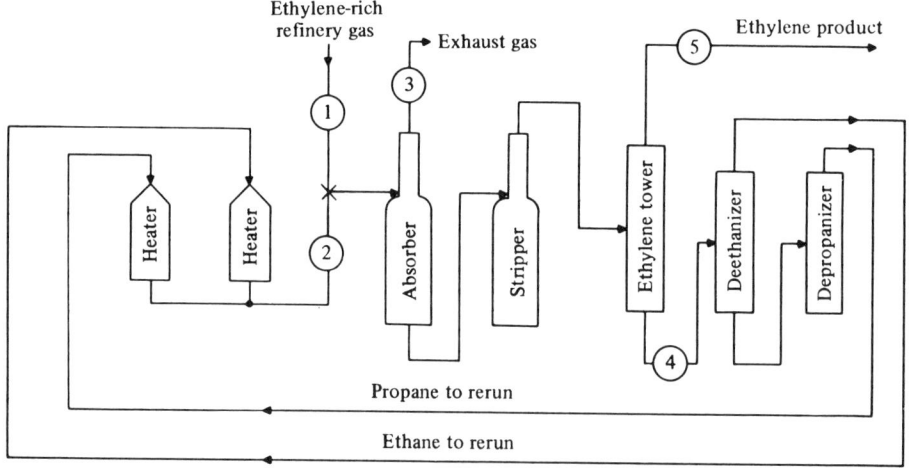

FIGURE 31-23 Use of continuous infrared analyzers in ethylene purification: (1) ethylene analysis on feed stock for accounting purposes; (2) ethylene analysis beyond crackers; (3) ethylene analysis of absorber off-gas for absorber efficiency; (4) ethylene analysis in ethylene-tower bottoms for fractionation-tower efficiency; (5) end-point analysis for ethylene purity. (After P. A. Wilks, Jr., *Chem. Eng. Progr.*, 51, 358 (1955).)

analyses of serum enzymes and protein fractions, have been added to the list of routine determinations. This increase both in the number of samples and in the number of available tests hastened the development of automatic chemical analyzers. Since the introduction of automatic instruments into clinical laboratories, these techniques have also been applied successfully to other areas such as environmental, pharmaceutical, food, and agricultural analyses, where large numbers of samples must be quickly analyzed.

Almost any repetitive analytical determination can be automated. Steps common to most analyses are: (1) sampling, (2) any required separation of sample components, (3) addition of one or more required reagents and subsequent mixing, (4) detection of analyte of interest by an appropriate sensor, and (5) collection, storage, and presentation of analytical data. Current automatic analyzers combine these steps to assay simultaneously up to 20 components in a 0.5-ml serum sample and can produce up to 3000 separate assays/hr.

An important difference between automatic and manual systems is that reactions do not have to be carried to completion in the automatic analyzer. Since conditions in a given automatic analyzer are unvarying and known, standard samples are subjected to the same treatment as samples of unknown concentration. Answers are supplied for most analyses from 10 to 100 times faster than manually operated methods; that is, in minutes rather than hours, or even days. Furthermore, results are not affected by operator fatigue or momentary inattentiveness.

Automatic clinical analyzers may be classified as one of three types based on their method of operation: discrete, continuous flow, or centrifugal.

Discrete Analyzers

This analyzer handles each sample as a separate entity and usually only one assay is made per sample. In some systems both the sample and required reagents are metered into discrete reaction vessels, test tubes, or specially designed cells (Fig. 31-24a). Other

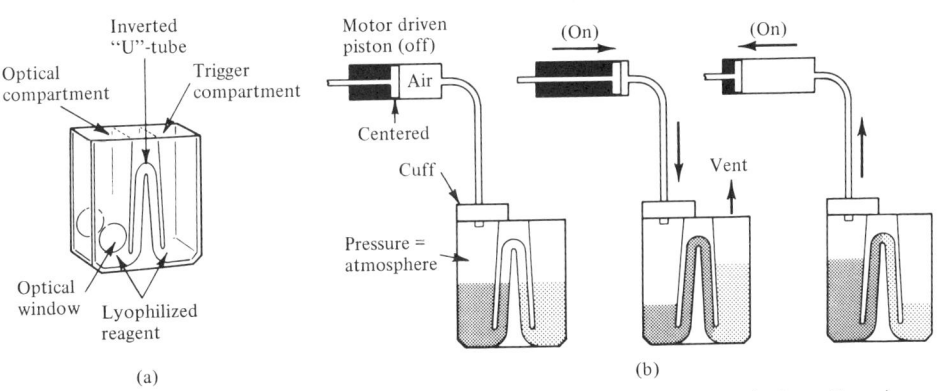

FIGURE 31-24 STAC cell and mixing procedure. (Courtesy of Technicon Corp.)

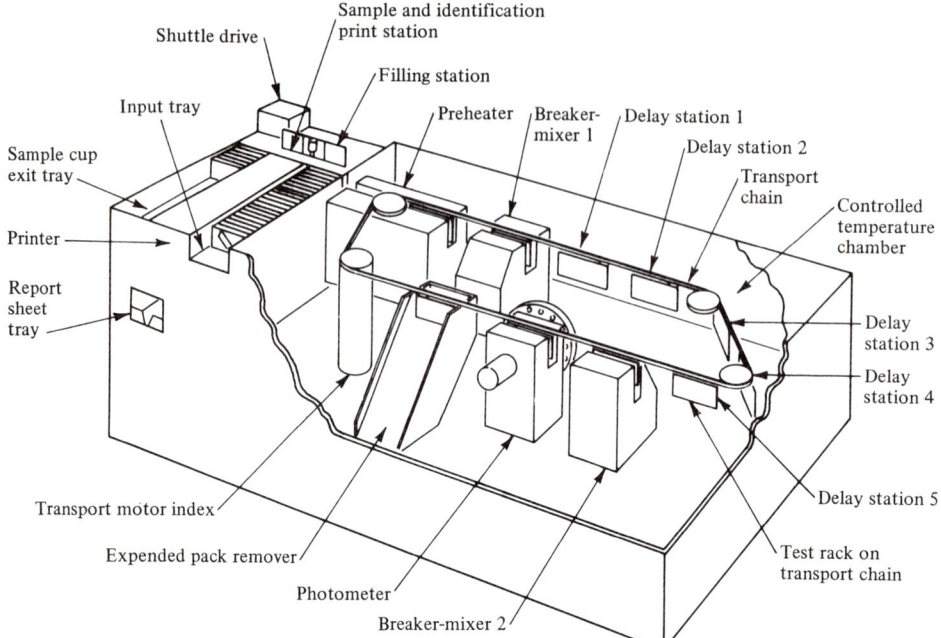

FIGURE 31-25 Automatic clinical analyzer (ACA). (Courtesy of E. I. DuPont de Nemours.)

systems employ cuvette cells containing prepacked quantities of the reagent required for a given sample; it is necessary to add only the sample. In both cases mixing and analysis take place in the test tube, sample cell, or cuvette. These containers are placed on a conveyor belt or turntable and the mixture is agitated (Fig. 31-24b). The mixture in the container may be passed through a water bath for temperature control (Fig. 31-25). Additional reagents may be added prior to measurement at the sensor station. Some systems include a component designed to wash the reaction vessels and recycle them back into the analysis train. Commercial analyzers of this type are generally capable of running a variety of different tests (Fig. 31-25).

The Technicon Single Test Analyzer (STAC) (Fig. 31-26) uses a computer to monitor and control the operation of the analyzer module (Fig. 31-27), and to acquire and process the desired analytical data. Built-in detection assists the operator in diagnosing system malfunctions. Flexibility in programming mixing times and photometer reading patterns allow individual enzyme determinations to be optimized to provide the best method at the lowest cost.

Discrete analyzers allow established manual techniques to be directly emulated by automatic operations. Only the test requested is performed on a given sample. This selectivity is not possible with continuous flow analyzers. The discrete analyzer uses smaller amounts of expensive reagents. A disadvantage associated with the use of discrete analyzers is their mechanical complexity compared with that of the other two types of

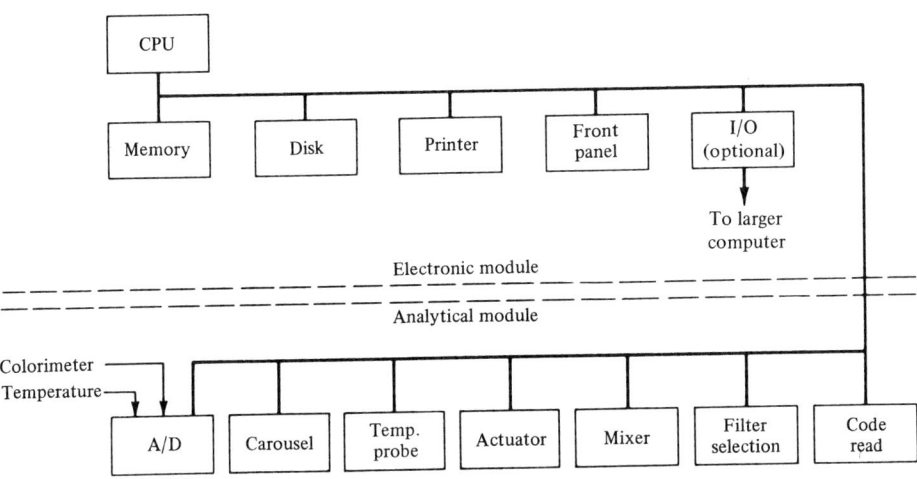

FIGURE 31-26 STAC analyzer unit. (Courtesy of Technicon Corp.)

analyzers. The use of prepackaged reagents in the sample containers adds to the expense of this method.

Continuous Flow Analyzers

In this second type of automatic analyzer, the samples are treated in a continuously flowing system. The analytical system of the AutoAnalyzer consists of a series of individual modules, each a separate component performing one specific function in a pro-

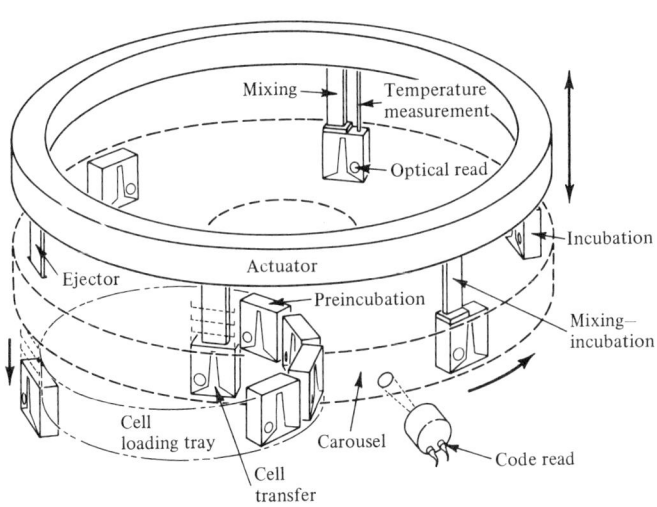

FIGURE 31-27 STAC computer-analyzer system. (Courtesy of Technicon Corp.)

grammed sequence (Fig. 31-28a). Modules can be interchanged and rearranged for different analytical methods. All aspects of sampling, separation, reagent addition, mixing, temperature control, sensing, readout, and recording are totally automatic. A sampling capillary (Fig. 31-28b) dips into each sample on the sampling tray, aspirates its contents for a timed interval and then feeds them into the analyzer. Between samples the capillary is raised to aspirate air for another timed interval. This step is followed by aspiration of an intersample liquid wash solution and another air aspiration before the next sample is introduced into the system.

The heart of the analyzer is the proportionating pump. It can deliver 12 or more separate fluids (reagents, diluents, air, and so on) simultaneously while varying their flowrates in any ratio from 1:1 to 79:1. The pump consists of two parallel stainless-steel roller chains with spaced roller thwarts that press continuously against a spring-loaded platen (Fig. 31-28c). Variable flowrates are obtained by using flexible tubing of different inside diameters.

At each point in the system where two liquids come together, there is a mixing coil of sufficient length to provide the required mixing time. In clinical analyzers, dialyzable analytes diffuse through a dialyzer membrane into a second segmented stream (Fig. 31-28a). This mixture is allowed to react as it passes through a thermostated reaction coil. After the sample has reacted, the analyte is quantitated by a suitable sensing device, such as a narrow bandpass ultraviolet/visible spectrometer, a fluorometer (for special cases requiring increased selectivity or sensitivity), a flame emission photometer (for Na and K), or ion selective electrodes. The results are printed on a continuous strip chart or are sent to a computer interfaced to the analyzer. In the latter case, these results are stored in a computer memory and may be accessed by one or more devices: a video terminal for surveying the data quickly, a printer which generates a report in the required format, or a larger computer for further processing and storage in a larger data file. Calibration of the system is checked periodically by interspersing standards of known concentration.

The performance of a continuous flow analyzer is determined from the shape of the signal produced by the sensor. Three principal parameters influence the quality of this signal: sample dispersion, mixing, and flow stability. Sample dispersion is minimized by the segmenting air bubbles and wash solutions, which serve as physical barriers to contamination. The bubbles also act as wiping agents in the tubular systems. Excessive sample dispersion leads to the overlap of adjacent samples. This effect can be countered by increasing the intersample wash time at the expense of either sampling time (volume) or total sample throughput (sampling time not reduced).

Complete mixing of the sample and all added reagents must occur in each sample segment. Incomplete mixing results in noisy, unstable detector output. Important mixing parameters include liquid viscosity, density, flowrate, internal diameter of tubing, helix coil diameter, and length of sample segment in the tube. Flow stability results in constant proportioning of sample to reagents for all segments from one sampling cycle to the next. An incorrectly proportioned sample stream also appears as a noisy detector output. Variation in liquid or air rates, compression of intersample air bubbles, a short sample segment, and blockages in the sampling probe or manifold tubes can result in flow instability in a continuous flow analyzer.

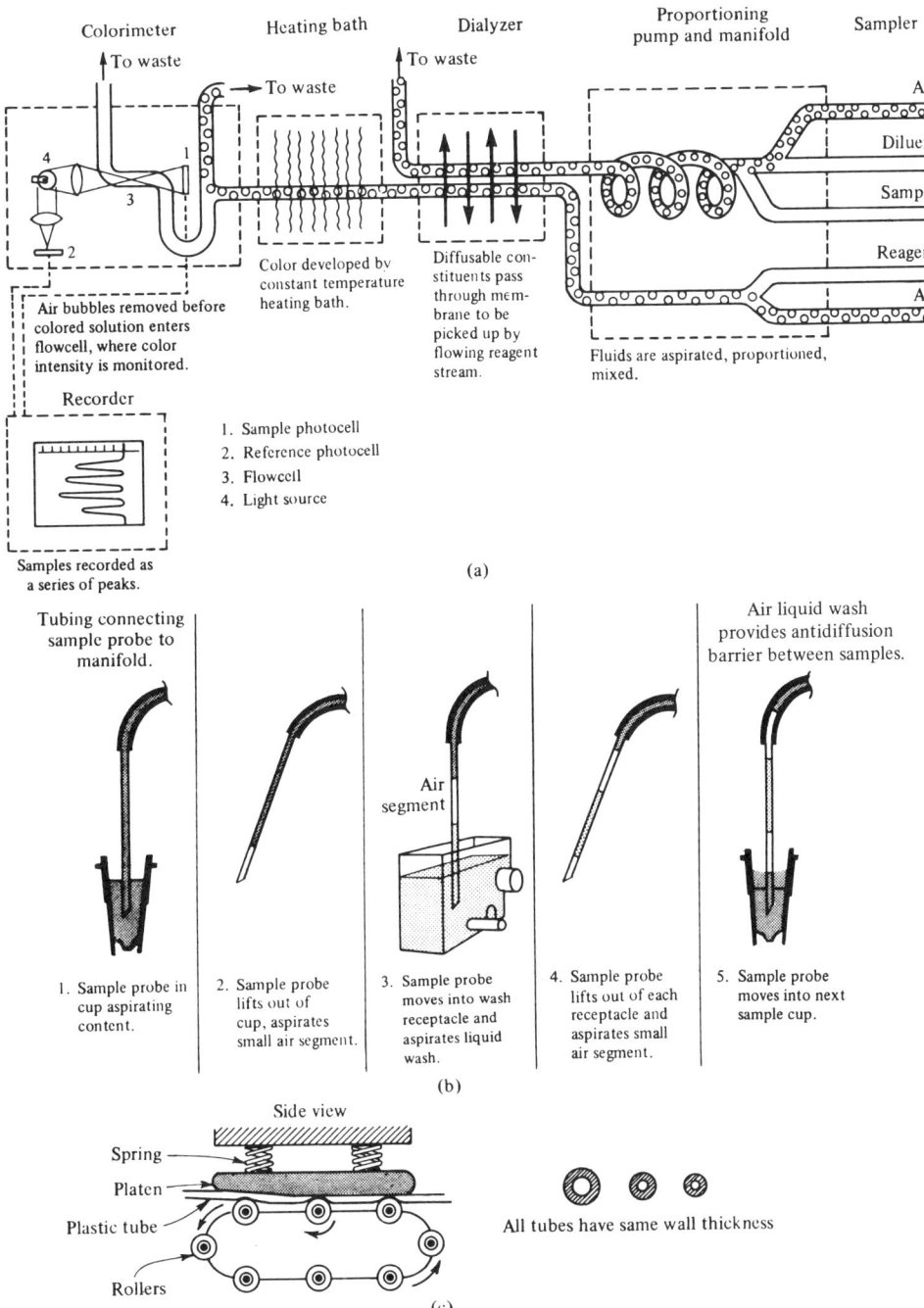

FIGURE 31-28 (a) Typical single channel flow schematic, (b) details of sampler, and (c) proportioning pump of the AutoAnalyzer. (Courtesy of Technicon Corp.)

The addition of computers to continuous flow analyzer systems has increased the accuracy and precision of analyses as well as producing higher sampling rates. Monitoring and analysis of detector output signals by the computer allow the results caused by the following errors to be flagged: malfunction in sample or reagent dispensor, insufficient sample or reagent, incomplete mixing, and carryover from a high-concentration sample to an adjacent low-concentration sample. When the detector output curves do not meet preselected "normal" criteria, the computer flags the assays for rerunning. At the same time the computer analyzers record abnormalities and alert the operator as to what corrective action is necessary. The Technicon SMAC/SDM high-speed computer-controlled biochemical analyzer and disk-based data processing system allows assay results to be stored on magnetic disk (64 analyses for each of 1500 patients) and also be transferred electronically to remote locations. Twenty different chemistries may be selected from 26 available methods. Only a 0.450-ml sample is required for the selected assays.

Multichannel flow analyzers offer advantages to laboratories with high volumes of multiassay samples. A typical physician's request for analysis of a single serum sample might include sodium, potassium, chloride, carbon dioxide, glucose, urea nitrogen, albumin, and total protein. One computer controlled analyzer can run approximately 240 complete analyses (1920 individual tests) per hour. By contrast, for laboratories with lighter workloads or irregular periods of activity, the discrete analyzer offers advantages of quick startup and ability to specify only certain types of tests.

Centrifugal Analyzers

In both discrete and continuous flow analyzers, the samples are presented serially to the detector with each sample spending a fixed amount of time in the detector. In situations, such as kinetic assays of serum enzymes, where it is necessary to monitor the detector output for extended time periods, the sample throughput (samples/hour) becomes a function of the time that each sample spends in the detector. The use of discrete or continuous flow analyzers requires a trade-off between the time that the sample is exposed to the detector and the rate of sample throughput; either exposure times must be shortened or throughput rates must be reduced. Centrifugal analyzers offer an alternative approach by effectively multiplexing the detector by rapidly rotating a disk containing samples past a single detector. Sample exposure time is increased without decreasing sample throughput since several samples are being monitored during a single extended exposure time.

In this type of analyzer, centrifugal force provides the means for mixing the sample and reagent and for transferring this reaction mixture to the cuvette. Figure 31-29 shows the cross section of the sample disk and detector compartment for one instrument. In use, samples and reagents are placed in cups arranged in concentric circles in a transfer disk. The cups in the various circles are aligned radially and arranged so that samples and reagents move through a transfer cavity when the rotor accelerates. After mixing, each sample solution is pulled into a cuvette where its transmittance or luminescence is measured. As the sample disk rotates, the cuvettes pass sequentially past the detector. The reading for each sample is referenced to a reagent blank contained in one of the cuvettes. Thus the analyzer functions as a multichannel double-beam spectrophotometer that pro-

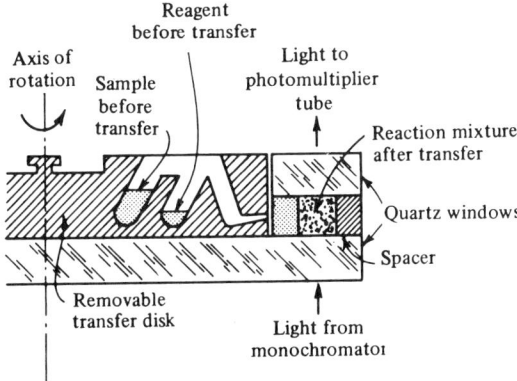

FIGURE 31-29 Cross section of sample disk and centrifugal analyzer section.

duces a complete set of sample readings with each revolution of the disk (100 msec per set at 600 rpm). Typically measurements from eight consecutive rotations are averaged for each sample. For extended kinetic determinations, sets of eight measurements are averaged over longer time periods (seconds or minutes) and these values displayed, printed, or stored for use in obtaining the final assay.

Computers built into the instrument control the operation of the analyzer and functions associated with the acquisition, averaging, manipulation, display, and storage of data (Fig. 31-30). The sample disk may be configured to run the same test on many different samples or different assays on the same sample. After the samples have been analyzed, the disk compartments can be cleaned automatically. The disks are usually loaded with samples and reagents outside the centrifugal analyzer. This may be done manually or by an automatic dispensing module. Sample volumes run from 5 to 100 μl for each assay.

Important advantages of this approach include virtually simultaneous measurement of samples and standards under the same conditions, optimization of analysis time for solutions in which sample-reagent reactions are slow, simple mechanical design, and the ability to change procedure easily by changing reagents in the disks and the monochromator setting of the photometric detector. The major disadvantage of centrifugal analyzers is their batch mode of operation, which necessitates reloading of disks with both reagents and samples for each group of samples analyzed.

31.10 AUTOMATIC ELEMENTAL ANALYZERS

Present instrumentation for carbon, hydrogen, and nitrogen (CHN) analyzers is based upon one of two general procedures. One involves the separation of carbon dioxide, nitrogen, and water by a gas-liquid chromatographic column. The other involves separation by means of specific absorbants for water and for carbon dioxide with the resulting change in composition of the gas mixture being measured. Thermal conductivity is the

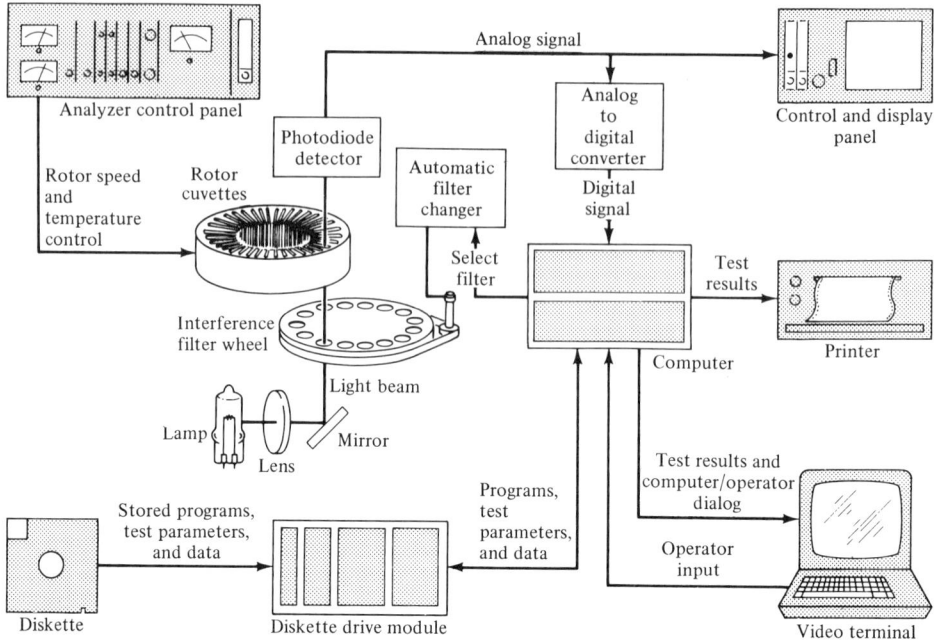

FIGURE 31-30   Centrifugal analyzer system. (Courtesy of American Instrument Co.)

detection method in both techniques. Results are calculated by analyzing standard samples and an occasional blank.

The sample (0.1-3 mg range) is either burned under static conditions in a pure oxygen atmosphere at 900°C, whereby the sample boat can subsequently be removed for weighing any residue, or mixed with cobalt(III) oxide [or a mixture of the oxides of manganese(IV) and tungsten(VI)] to provide the oxygen and heated to 900°C. Combustion converts the carbon in the sample to carbon dioxide and carbon monoxide, and hydrogen to water. Nitrogen is released as the free gas, along with some oxides. A stream of helium carries the combustion gases into a reaction furnace operated at 750°C. Here a chemical change completes the simultaneous oxidation and reduction of the sample gases. Hot copper reduces the nitrogen oxides to nitrogen and removes the oxygen. Copper oxide converts the carbon monoxide to carbon dioxide. If needed, a magnesium oxide layer in the middle of the furnace removes fluorine, and a silver-wool plug at the exit removes chlorine, iodine, and bromine, and also any sulfur or phosphorus compounds resulting from the combustion of the sample. In CH analyzers, oxides of nitrogen are removed a little later in the train with manganese dioxide.

Separation of the combustion gases by gas chromatography involves the following sequence: The gases pass through a charge of calcium carbide where water vapor is converted to acetylene. A nitrogen cold-trap freezes the sample gases and isolates them in a loop of tubing. A valve seals off the combustion train, which is then ready for another sample. The chromatographic stage is begun by lowering the cold-trap and heating the injection loop. Another stream of dry helium gas carries the gases (as a plug) into the chromatographic column where the three gases—N_2, CO_2, and C_2H_2—are completely

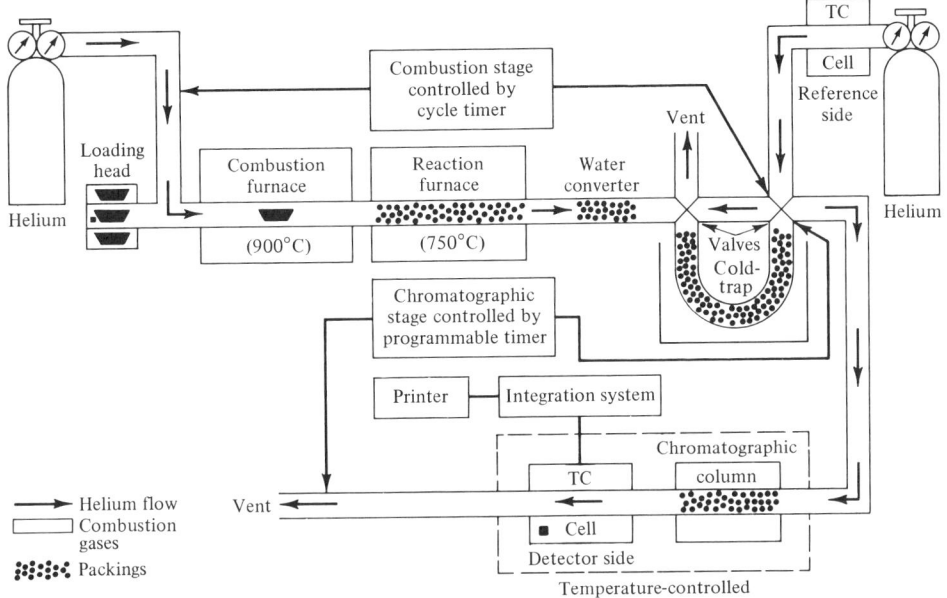

FIGURE 31-31 Schematic diagram of the Fisher Carbon-Hydrogen-Nitrogen Analyzer. (Courtesy of Fisher Scientific Co.)

separated. Most organic samples can be completely burned in 10–12 min, and the chromatographic separation requires another 10 min. Figure 31-31 gives the schematic diagram of a CHN analyzer of this type.

The other general procedure involves separation by means of specific adsorbants for water and carbon dioxide. Three pairs of thermal conductivity cells (glass-coated platinum filaments) are used in series for detection: one pair each for water, carbon dioxide, and nitrogen (Fig. 31-32). A magnesium perchlorate trap between the first pair of cells absorbs any water from the gas mixture before it enters the second pair of cells. The differential signal, measured before and after the trap, is proportional to the amount of hydrogen (measured as water removed) in the sample. Similarly a soda–asbestos trap between the second and third pair of cells results in a signal proportional to the carbon (carbon dioxide) in the sample. The last pair of cells detects nitrogen by comparing the helium–nitrogen mixture with pure helium.

For oxygen analysis, the combustion furnace is replaced by a quartz pyrolysis tube containing platinized carbon. The reduction furnace is replaced by a tube containing copper oxide. The operating temperatures are the same as for CHN analysis but the oxygen supply is shut off. Samples are handled exactly as they are in the CHN analysis mode with the sample initially heated in a helium atmosphere so that any oxygen in the sample forms carbon monoxide. This carbon monoxide is converted to carbon dioxide in the copper oxide tube, which is then detected and measured in precisely the same manner used for the carbon analysis.

Sulfur analysis can be performed if the combustion furnace is replaced with a tube containing tungsten(VI) oxide packing plus a dehydrating reagent. The water trap is re-

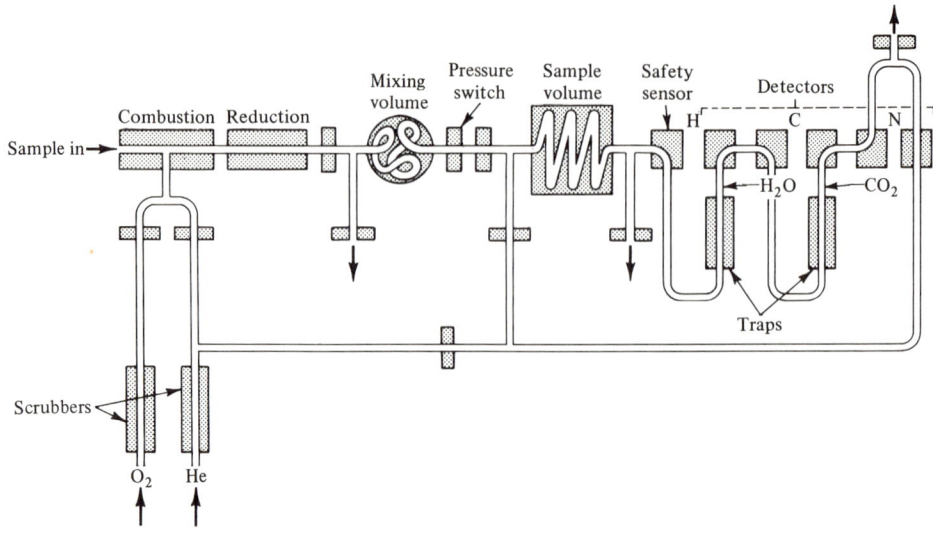

FIGURE 31-32 Flow schematic of the Perkin-Elmer Model 240B Elemental Analyzer. (Courtesy of Perkin-Elmer Corp.)

moved from the bridge area and replaced with a trap containing silver oxide to adsorb the SO_2 produced in the combustion tube.

In the first generation of CHN analyzers, sample throughput was limited by the semiautomatic design, which required each individual sample to be placed into the analyzer by the operator. Second-generation elemental analyzers resolved this limitation by adding automatic sampling and data handling modules to produce a fully automatic analyzer system. The Perkin-Elmer Model 240 analyzer (Fig. 31-33) can automatically feed and compute up to 60 samples without operator attention. Sample containers are loaded into a disk-shaped magazine, automatically fed through the analyzer, and removed at the end of each analysis. The data handling system includes an autobalance with calculator-compatible binary code decimal (BCD) output, and an electronic programmable calculator with printer. A functional keyboard allows easy operator interaction with the system.

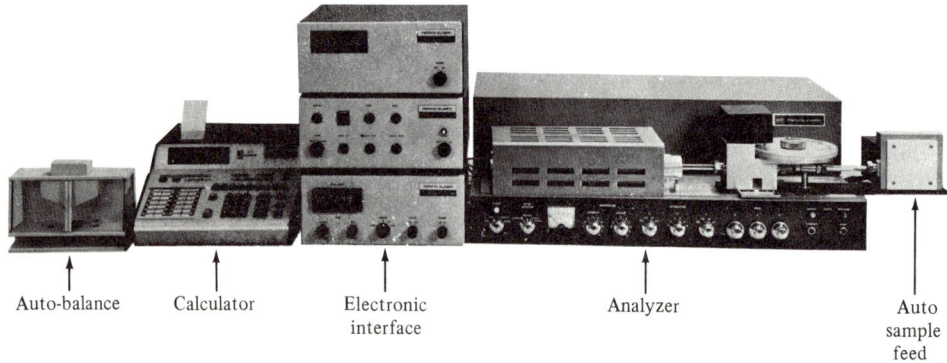

FIGURE 31-33 Automatic elemental analyzer. (Courtesy of Perkin-Elmer Corp.)

The typical analytical report for each sample includes sample identification number, data, sample weight, sample gas responses, weight percentages, and simplest formula. This automation results in increased sample throughput, reduction of human error, and flexibility of analyzer operation and data processing to meet a variety of applications.

PROBLEMS

1. Describe a simple *automatic* instrument for monitoring the pH in the production of disodium phosphate from soda ash and phosphoric acid. Explain how this instrument could be used as part of an *automated* system for the continuous production of disodium phosphate.

2. In the separation of aromatics from saturates, using refractive index to follow a process, what precision can be expected? The aromatics have a refractive index of about 1.50, and the saturates about 1.40.

3. What advantages do nondispersive infrared analyzers offer over dispersive instruments? What disadvantages are associated with nondispersive infrared analyzers?

4. Explain the major differences between process chromatographs and laboratory chromatographs.

5. Suggest a method for handling the following on-stream process situations: (a) fractionating-tower control in the separation of cyclohexane and *n*-hexane; (b) fractionating column for the determination of butane in isobutane; (c) control of the end point in making shortenings and margarine using materials such as soybean and cotton-seed oils; (d) changes in the concentration of individual solutions of inorganic salts, HCl, or H_2SO_4.

6. Design a method for monitoring boiler stack gas in pulp mills for soda, both to aid in minimizing soda losses and to help control air pollution. Keep in mind that light transmission or scattering are handicapped by rigid sample-handling requirements and high emission rates. In addition, highly conductive dissolved gases that are also present offset the advantages of measuring the electrical conductivity in scrubbed gas samples.

7. Design an on-line analyzer to close the loop in a hot process lime-soda softener.

8. Compare the three types of automatic chemical analyzers with respect to the mode of operation, advantages, and disadvantages.

9. Discuss several limitations inherent in the gas chromatographic system of analysis employed in some CHN analyzers.

10. A CHN analyzer was calibrated by burning 1.657 mg of dimethylglyoxime. The total signal for C = 17,500 units, H = 1062 units, and N = 4128 units. The unknown compound, a 2.021-mg sample, gave these signals: C = 40,760 units, H = 1078 units, and N = 3195 units. Calculate the percent carbon, hydrogen, and nitrogen.

BIBLIOGRAPHY

Considine, D. M., Ed., *Process Instruments and Controls Handbook,* 2nd ed., McGraw-Hill, New York, 1974.

Frant, M. S. and R. T. Oliver, *Anal. Chem.,* **52,** 1252A (1980).

Johnson, C. D., *Process Control Instrument Technology,* Wiley, New York, 1977.

Lewis, C. D., "Continuous Automatic Instrumentation for Process Applications," Chap. 105, in *Treatise on Analytical Chemistry,* I. M. Kolthoff and P. J. Elving, Eds., Vol. 10, Part 1, Wiley-Interscience, New York, 1972.

Snyder, L. R., J. Levine, R. Stoy, and A. Conetta, *Anal. Chem.,* **48,** 942A (1976).

Young, D. S., *Automation in Fundamentals of Clinical Chemistry,* N. W. Tietz, Ed., Chap. 4, Saunders, Philadelphia, 1976.

LITERATURE CITED

1. Irving, H. M., H. Freiser, and T. S. West, *Compendium of Analytical Nomenclature,* Pergamon, New York, 1978.
2. Villabobas, R., *Anal. Chem.,* **47,** 983A (1975).
3. Leemans, F. A., *Anal. Chem.,* **43,** 36A (1971).
4. Hall, G. A., S. P. Higgins, Jr., R. H. Kennedy, and J. M. Nelson, "Principles of Automatic Control," Chap. 18, in *Process Instruments and Control Handbook,* 2nd ed., D. M., Considine, Ed., McGraw-Hill, New York, 1974.
5. Merritt, R., *Instruments and Control Systems,* p. 23 (Nov. 1978).
6. Snyder, L. R., J. Levine, R. Stoy, and A. Conetta, *Anal. Chem.,* **48,** 942A (1976).

Experiments

Experiment 2-1   Determination of Spectral Response of Photosensitive Detectors

A direct-reading spectrometer will provide everything required. The following directions apply specifically to a Bausch & Lomb Spectronic 20 spectrometer.

Adjust the instrument, photocell dark, with the zero control in the normal manner. Insert into the sample holder a test tube filled with water. Rotate the 100% control until the meter needle reads near midscale. Now rotate the wavelength control until a maximum transmittance reading is achieved (readjusting the 100% control if necessary, to keep the meter needle on scale).

At the wavelength of maximum transmittance, carefully balance the instrument at 100% transmittance. Without changing any controls, henceforth, change the wavelength in intervals of 20 nm and record the corresponding transmittance reading (actually the relative response reading).

Repeat the series of operation with other phototubes. Plot the results on graph paper, plotting wavelength as abscissa. These results represent the overall response of the spectrometer–phototube, emissivity of light source, and intensity diffracted by the grating, each of which is a function of wavelength. Since the main difference lies for the most part with the tungsten source, it would be possible to estimate the relative response of the phototube to light of constant intensity by correcting each observed transmittance reading for the relative radiant power of the tungsten lamp.

Experiment 2-2 Determination of the Effective Slitwidth of Spectrometers

Equipment required is a scanning spectrophotometer equipped with an atomic line source, such as a mercury lamp or a flame emission spectrometer. It is desirable that the spectrometer possess an adjustable slit aperture.

Adjust the slit aperture to a given width. Adjust the dark current (background) signal on the recorder to a suitable initial value. Scan across the atomic emission feature(s). Multiplets, such as the sodium doublet at 589.0 and 589.6 nm, are ideal. Employ a series of slit openings, readjusting the initial position of the pen and the amplification of the signal so as to contain the recorded spectrum on the chart for each aperture.

Tabulate the full width at half maximum signal for each spectral line. For doublets note when the individual lines begin to merge at base line.

Correlate the experimental results with calculated values of bandpass and basewidth of the slit function (Eqs. 2-7 and 2-8).

Experiment 3-1 Determination of Iron with 1,10-Phenanthroline

Standard iron solution Dissolve 0.7022 g of reagent grade $(NH_4)_2 Fe(SO_4)_2 \cdot 6H_2O$ in 100 ml of distilled water, add 3 ml of $18M$ H_2SO_4, and dilute to 1 liter. One milliliter contains 0.100 mg of iron(II).

1,10-Phenanthroline Dissolve 0.25 g of the monohydrate in 100 ml of water, warming if necessary.

Hydroxylamine hydrochloride, 10% (w/v) Dissolve 10 g in 100 ml of water.

Sodium acetate, $2M$ Dissolve 17 g of sodium acetate in 100 ml of water.

Procedure Take an aliquot portion of the unknown solution containing 0.1–0.5 mg of iron and transfer it to a 100-ml volumetric flask. Add 5 ml each of the hydroxylamine hydrochloride and 1,10-phenantroline solutions, plus the volume of sodium acetate solution required to bring the pH to 4.0. Dilute to the mark, mix well, and measure the absorbance after 10 min in the region 460–520 nm. Calculate the molar absorptivity.

The volume of sodium acetate required to bring the pH to 4.0 is determined by the use of a similar sample aliquot, plus the hydroxylamine hydrochloride, with the aid of a pH meter, or a few drops of bromophenol blue indicator.

Experiment 3-2 Determination of the Dissociation Constant of Indicators

Indicator solution, 0.04% (w/v) Dissolve 0.1 g of the indicator (sulfonphthalein dyes) in water, adding 1–3 ml of $0.1M$ NaOH, if necessary. Dilute to 250 ml.

Buffer solutions Any standard buffer series may be used. It is desirable to maintain the ionic strength constant. In the following directions, $0.1M$ sodium acetate solution is employed. Dissolve 2.30 g of sodium acetate in water and dilute to 250 ml.

Procedure Transfer exactly 10.0 ml of 0.1M sodium acetate solution to a 100-ml volumetric flask. Add exactly 5.00 ml of 0.04% bromocresol green solution and dilute to the mark with distilled water. Mix thoroughly. Determine the absorption spectrum of the sodium acetate solution of bromocresol green (the "alkaline" color).

Pour the contents of the cuvette and the volumetric flask into a 250-ml beaker. Add precisely 2.00 ml of an acetic acid solution (0.25M in acetic acid and 0.1M in KCl). Mix well with a stirring rod, and measure the pH with a pH meter. Measure the absorbance at the wavelength of maximum absorbance of the "alkaline" form of the indicator. Repeat this procedure for additional 2.00-ml increments of acetic acid solution. Determine the absorption spectrum for the buffer solution which is equimolar in acetate and acetic acid. After five such measurements have been made, add 1.0 ml of 3M HCl and again determine the entire spectrum of the "acid" form of the indicator.

Correct all absorbance readings for the effect of dilution by multiplying each observed value by the factor $(100 + V)/100$, where V is the volume of acetic acid (and HCl) added. Plot the three absorption spectra on a single sheet of graph paper.

Another curve is plotted from the absorbance readings at the wavelength of maximum absorbance vs. the pH value for each buffer solution. Draw a smooth line through the points. The "acid" and "alkaline" solutions are assumed to represent the limiting values of absorbance. Determine the value of pK_a by insertion of the appropriate absorbance readings into the expression for the acidic dissociation constant, and averaging the results.

Other indicators in the sulfonphthalein series may be studied. These include bromophenol blue, bromocresol purple, bromothymol blue, phenol red, and cresol red. Other compounds that are suitable include vanillin (ϵ = 25,900), $4 \times 10^{-5}M$ at 347 nm; p-bromophenol, 1 mg/100 ml in hexane, at 244 nm; and o-chlorotoluic acid, $2 \times 10^{-5}M$, at 280 nm.

Use at least five buffer solutions whose pH values lie within the range p$K_{ind} \pm 1$. Also prepare the fully acid and fully alkaline solutions. Figure 3-2 illustrates some typical results.

A study of ionic strength effects merely involves assigning a fixed ionic strength to each of several individuals working with the same indicator. See R. W. Ramette, *J. Chem. Education*, **40**, 252 (1963).

Experiment 3-3 Simultaneous Determination of Binary Mixtures

Each student will be assigned a pair of solutes and the type of sample to be analyzed. Suggested systems include the following:

1. Manganese, as permanganate, 0.0004–0.00008M, plus chromium as dichromate, 0.0004–0.0017M, in 0.5M H_2SO_4. Analytical wavelengths are 440 and 525 nm.
2. Titanium, 0.8–8.0 mg of TiO_2 per 100 ml, plus vanadium, 1.0–20 mg as V per 100 ml, as their peroxide complexes. Add sufficient H_2SO_4 to adjust the final acidity to 1.5–3.5N. Add 10 ml of 3% hydrogen peroxide solution and dilute the solution to 100 ml in a volumetric flask. Measure the absorbance at 400 nm (titanium) and 460 nm (vanadium).

Standard vanadium solution Dissolve 2.2963 g NH_4VO_3 in 100 ml of water plus 10 ml of HNO_3; dilute to exactly 1 liter. One milliliter contains 1.00 mg of vanadium.

Standard titanium solution Fuse 0.2500 g of TiO_2 with 3–4 g of potassium pyrosulfate in a platinum or a porcelain crucible. Dissolve the melt in 50 ml of hot $4N$ H_2SO_4 and dilute to 250 ml with the same acid. One milliliter contains 1.00 mg of TiO_2.

Procedure Measure the absorbance of 3–5 standard solutions of each solute, whose concentrations span the interval suggested, at both analytical wavelengths. Compute the molar absorptivities from each individual measurement and determine the average value of the molar absorptivity and its associated deviation at the 95% confidence level.

Test the assumption that each solute contributes additively to the total absorbance at each analytical wavelength. Combine known aliquots of the individual solutions prepared for the preceding step, and measure the absorbance at each analytical wavelength. Calculate the amount of each solute present.

Determine the concentration of each solute in an unknown sample and report the results.

Experiment 3-4 High-Absorbance Differential Spectrophotometry

Each student will be assigned a single solute or colorimetric system. The range of concentrations to be studied and the particular spectrophotometer to be used will be designated.

Prepare a series of solutions whose absorbance readings extend from 0.2 to 2.0 (or approximately three times the maximum strength usually recommended) and separated from each other by about 0.2 absorbance unit.

Proceed in the ordinary way to set the instrument scale reading to zero transmittance when the photocell is darkened, and to 100 scale divisions (zero absorbance) with pure solvent in the sample container. Measure the absorbance and transmittance of each solution. Prepare a graph of absorbance vs. concentration.

Remove the solvent from its cuvette and substitute in it some of the solution which had the 0.2-absorbance value. Measure the transmittance to ascertain the ratio T_2/T_1. If not unity, this preliminary value can be used to correct subsequent readings for difference in path length of the two cuvettes: $-\log(T_2/T_1) = ab_1 C[\beta - 1]$, where $\beta = b_2/b_1$. Place the comparison standard (that is, solution with 0.2-absorbance value originally) in the light beam, with the reading scale set at 100 scale units, and balance the instrument by increasing the slit opening or by increasing the gain control. Measure the absorbance for the more concentrated solutions. Plot the results on the same graph.

Repeat the procedure with progressively more concentrated solutions being used as the comparison standard for setting the reading scale at 100 until it becomes impossible to establish the balance of the instrument.

Experiment 3-5 Photometric Titrations

Operating technique Set the spectrophotometer to the analytical wavelength or insert the proper filter into the photometer. Adjust the dark current to zero. Place the sample to be titrated into the light beam and set the instrument to read zero absorbance (if the reactant is colorless) or to some other starting value (if the reactant is colored) that lies within the range of linearity of the instrument. If the initial absorbance reading is too large, readjust the concentration of reactant.

Place the buret so that the tip extends into the solution. Turn on the stirrer and adjust the stirring rate so that the vortex does not obstruct the light path. Commence the titration by adding an increment from the buret, waiting until the absorbance reading is constant, and recording that value. More increments are then added and the absorbance noted. For exploratory work, 0.2- to 0.5-ml increments are desirable when using a 10-ml buret. Once the shape of the titration curve is known, it is usually necessary to take only three to four points on each side of the end point.

A plot of absorbance vs. milliliters of titrant is then made and the best straight lines are drawn between points taken well before and after the equivalence point. The intersection of the linear segments is taken as the end point. Correct all absorbance readings for the change in volume at each point on the titration curve.

Suggested Systems

Acid-base titrations Mixture of acetic acid with p-nitrophenol, each $10^{-3} M$, titrated with $0.1 M$ NaOH at 400-420 nm.

A solution of any strong acid, 10^{-3} to $10^{-4} M$, titrated with $0.01 M$ NaOH in the presence of thymol blue ($pK_a = 9.0$) at 615 nm.

Oxidation-reduction titrations A solution of arsenic(III), $10^{-3} M$, with ceric sulfate, $0.01 M$, at 320 nm, with a trace ($10^{-5} M$) of osmium tetroxide as catalyst.

Complexometric titrations Copper(II), $0.04 M$, with $0.1 M$ EDTA at pH 4.0 and a wavelength of 745 nm.

Calcium plus copper(II), each $0.02-0.06 M$, with $0.1 M$ EDTA in a $1 M$ ammonia-ammonium chloride buffer adjusted to pH 9.0 ± 0.2 and at a wavelength of 745 nm.

EDTA, $0.002 M$, titrated with $0.02 M$ zinc solution in an approximately $1 M$ ammonia-ammonium chloride buffer adjusted to pH 9 and with 0.002 g of Superchrome Black TS indicator present. Wavelength is 550 nm.

Iron(III), $0.05 M$, titrated with $0.1 M$ EDTA in an acetic acid solution ($1.0 M$) adjusted to pH 1.7-2.3 with HCl. Salicylic acid (4 ml of a 1% solution is added per 100-ml total volume) serves as indicator; the analytical wavelength of the iron-salicylic acid complex is at 525 nm.

Experiment 3-6 Composition of Complexes: Mole-Ratio Method, Method of Continuous Variations, and Slope-Ratio Method

A. Mole-Ratio Method

In this method the concentration of metal ion is held fixed and the concentration of the reagent is increased stepwise. On the graph of absorbance vs. moles of reagent added, the intersection of the extrapolated linear segments determines the ratio: moles of reagent/ moles of metal.

Procedure Transfer exactly 4 ml of the standard metal solution to 11 separate 50-ml volumetric flasks. Add any appropriate buffer solution and other necessary reductants. Add exactly 4, 6, 8, 10, 12, 14, 16, 18, 20, 22, and 24 ml of the standard reagent solution to each flask. Dilute to the mark with distilled water and mix. After any suggested waiting period, measure the absorbance at the designated wavelength.

B. Method of Continuous Variations

In this method the sum of the molar concentrations of the two reactants is kept constant as their ratio is varied. The abscissa of the extrapolated peak will correspond to the ratio present in the complex.

Procedure Transfer exactly 0, 2, 4, 6, 8, 10, 12, 14, 16, 18, and 20 ml of the metal solution to separate 50-ml volumetric flasks. Add 5 ml of the appropriate buffer solution and any necessary reductants. To the above flasks, arranged in serial order, add these amounts of the reagent solution, also in serial order: 20, 18, 16, 14, 12, 10, 8, 6, 4, 2, and 0 ml. Dilute to the mark and mix. After the suggested waiting period, measure the absorbance at the designated wavelength.

C. Slope-Ratio Method

In this method two series of solutions are prepared. In the first series various amounts of metal ion are added to a large excess of the reagent, while in the second series different quantities of reagent are added to a large excess of metal ion. The absorbance of the solutions in each series is measured and plotted vs. the concentration of the variable component. The combining ratio of the components in the complex is equal to the ratio of the slopes of the two straight lines.

Procedure Transfer exactly 30 ml of the standard metal solution to five separate 50-ml volumetric flasks. Add appropriate amounts of buffer solution and any necessary reductant. Add 1, 2, 3, 4, and 5 ml of the reagent solution, respectively, to the flasks, dilute to volume, and mix. After any waiting period, measure the absorbance at the designated wavelength. In the second series, use 30 ml of the reagent solution, and add 1, 2, 3, 4, and 5 ml, respectively, of the metal solution.

Suggested System

Iron(II) ammonium sulfate, $0.0005M$; 1,10-phenanthroline, $0.0005M$; acetic acid–sodium acetate buffer (total acetate = $0.1M$), pH 5.0 (use 10 ml for each solution); hydroxylamine hydrochloride, 5% (w/v), use 1 ml for each solution; measure at 510 nm after standing 10 min; $\epsilon = 12{,}000$.

Experiment 4-1 Determination of Quinine

Standard solution of quinine Dissolve 0.100 g of quinine bisulfate in 0.1N H_2SO_4 solution and dilute to 1 liter with additional acid solution. Dilute 10.0 ml of the foregoing solution to 1 liter with 0.1N acid solution. The resulting solution contains 0.00100 mg/ml of quinine bisulfate.

Dilute sulfuric acid, 0.1N Add 3 ml of concentrated sulfuric acid to 100 ml of water and dilute to 1 liter in a graduated cylinder.

Procedure Pipet 10.0, 25.0, 35.0, and 50.0 ml of the dilute standard quinine solution into a set of 100-ml volumetric flasks. Dilute to the mark with 0.1N H_2SO_4 solution. Treat the unknown sample similarly. Measure the fluorescence and prepare a calibration curve.

The fluorescence of quinine is constant in the concentration interval from 0.01 to 0.2N H_2SO_4. Proper primary and secondary filters, or wavelengths for excitation and fluorescence measurement, can be ascertained from the illustration below.

Experiment 4-2 Determination (Simplified) of Riboflavin

Standard riboflavin solution Dissolve 10.0 mg of riboflavin in 1 liter of 1% acetic acid solution. Keep the solution in a cool, dark place. This solution contains 10.0 μg/ml of riboflavin.

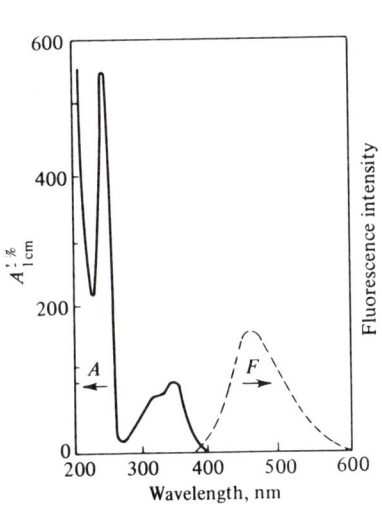

EXPERIMENT 4-1 Absorption and fluorescence spectra of quinine bisulfate in 0.1N H_2SO_4.

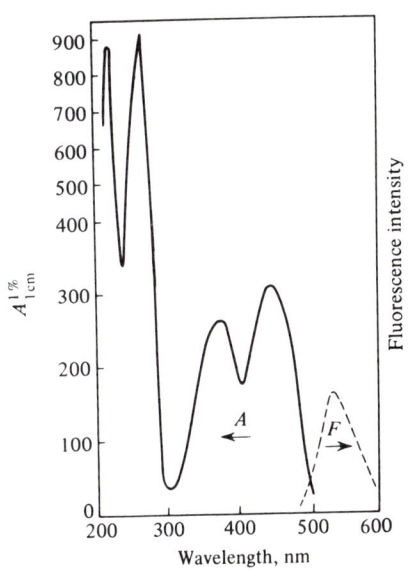

EXPERIMENT 4-2 Absorption and fluorescence spectra of riboflavin in water.

Procedure Prepare a series of standard riboflavin solutions, the strongest of which does not contain more than 1.0 μg/ml of riboflavin. Prepare a calibration curve. Appropriate excitation and fluorescence wavelengths can be deduced from the spectra in the illustration on p. 943.

Experiment 5-1 Determination of Sodium and Potassium

Standard solutions Dissolve 2.5420 g of sodium chloride in deionized water and dilute to 1 liter. Dissolve 1.9070 g of potassium chloride or 2.586 g of potassium nitrate in deionized water and dilute to 1 liter. Each solution contains 1000 μg/ml of the respective cation.

Weigh out 2.473 g of lithium carbonate and transfer to a 1-liter volumetric flask. Add approximately 300 ml of deionized water and then add slowly 15 ml of concentrated hydrochloric acid. After the CO_2 has been released, dilute the solution to 1 liter. This solution contains 1000 μg/ml of Li_2O. Most samples of lithium salts are contaminated with considerable sodium and some potassium, and therefore the same batch of stock lithium solution should be used in preparing the working standards used in the internal-standard method.

Working standards Prepare a set of six standards that contain 5, 10, 25, 50, 75, and 100 μg/ml of potassium (or sodium). These solutions will be used in the direct-intensity method.

Prepare a set of seven standards that contain 0, 5, 10, 25, 50, 75, and 100 μg/ml of potassium (or sodium), with each solution also containing 100 μg/ml of Li_2O. This series will be used in the internal-standard method.

Procedure Determine the calibration curve for potassium using the direct-intensity method. Use deionized water and the strongest standard solution to adjust the reading scale to zero and 100 divisions (or full scale). Without changing the instrument controls, determine the emission reading for the other concentrations. These directions presuppose that a single-beam flame photometer is available. If not, use the internal-standard method.

When using the internal-standard method, set the instrument to zero with the standard containing 100 μg/ml of Li_2O but with potassium absent. Be sure any reading scale knob is positioned at the zero mark. Next, set the 100-division (full-scale) reading with the strongest standard aspirating; again be sure the reading dial control is positioned at the 100 mark. Recheck each reading once again. Without changing the instrument controls, determine the emission reading for the other concentrations.

At the instructor's discretion, unknown samples may be compared by either or both procedures.

Plot emission intensity vs. concentration of potassium from the direct-intensity data on rectilinear graph paper and also on log–log graph paper. Mark the regions where the calibration curve is linear and the log–log plot has a slope of unity.

Experiment 5-2   Atomic Absorption Spectrometry

Standard calcium solution   Place 2.4973 g of $CaCO_3$ in a volumetric flask with 300 ml of water. Carefully add 10-ml HCl; after the CO_2 is released by swirling, dilute to volume. One milliliter contains 1.000 mg of calcium.

Standard iron solution   Dissolve 1.000-g iron wire in 20 ml of $5F$ HCl; dilute to volume. One milliliter contains 1.000 mg of iron.

1. Investigate designated lamp lines for absorption sensitivity by constructing calibration curves and noting slopes and range of linear response in concentration units. For calcium, use the line at 422.7 nm. For iron, the most sensitive line is at 248.3 nm; less sensitive lines occur at 251.1 and 372.0 nm. See Appendix E.
2. For the most sensitive absorption line of an element, investigate the effect of different slit openings on the nature of the calibration curve. Calculate the detection limit (for which noise data will be needed) and the sensitivity for each set of conditions.
3. For the most sensitive absorption line of an element, investigate the effect of fuel/oxidant ratio on the absorption signal. Increase the acetylene flow until the plume of the flame becomes distinctly yellowish (fuel-rich flame). Also decrease the acetylene flow until the flame becomes "hard;" that is, no light blue zone remains above the inner green–blue zone. A "normal" flame will have several millimeters of a light blue zone above the inner green–blue cone.
4. Vary the observation height in the flame at which absorption is observed by adjusting the burner relative to the optical path of the spectrometer.
5. Extend the usable scale range to lower concentrations of analyte by using scale expansion (2X, 5X, ...) until the noise becomes a limiting factor in determining the detection limit.
6. Extend the usable scale to higher concentrations by using a finite standard in place of the blank to make the zero setting. This illustrates base line offset. Compare this method (in the case of iron) with the alternate choice that involves selecting a less sensitive absorption line for your measurements.
7. For an analyte such as calcium, add some diammonium hydrogen phosphate (perhaps 1000 µg/ml for a 1–30-µg/ml series of calcium standards) and note the absorption readings. On a duplicate set of standards containing the phosphate, also add sufficient lanthanum to make its concentration about 5000 µg/ml. Note the absorption readings with lanthanum present.
8. For an element such as potassium, investigate the effect of an ionization suppressor by the addition of a fixed amount of (a) cesium or (b) sodium to a series of potassium standards.
9. Evaluate an unknown by the method of standard addition. Samples of copper oxide or iron oxide, used in sophomore analytical courses, make suitable unknowns. First, obtain an estimate of the apparent analyte concentration from comparison with an appropriate calibration curve. Then transfer a sufficient size aliquot to a 100-ml volumetric flask so that after dilution the absorption signal lies in the range 20–25% absorption. Two additional and equal aliquots are transferred into separate 100-ml

volumetric flasks. To one flask add an amount of standard analyte that is estimated to increase the absorbance 1.5 times; to the other flask add an amount estimated to increase the absorbance two times. Measure all three solutions and graph or calculate the amount of analyte present in the unknown sample.

Experiment 9-1   Identification of Substances by the Powder Diffraction Method

1. A very small sample of a crystalline material will be furnished. The sample must be very finely powdered (finer than 200 mesh) if it is not already so.
2. Prepare a small-diameter, thin-walled melting-point tube (diameter 0.7 mm or less) and fill it with the powdered sample. Place the tube in the chuck at the center of the powder camera and fix it in place with a drop of wax. Line the sample tube up so that it rotates without wobbling. Alternatively, the sample may be coated on the outside of a very fine glass rod, using collodion or Vaseline or other noncrystalline material to stick it on.
3. Take the camera into the darkroom and load it with film of the proper size. A punch is available for perforating the film so that the collimator and beam trap can be inserted into the camera. Insert the collimator and beam trap and place the cover on the camera.
4. Place the camera on the X-ray unit. An exposure of 20–30 min is recommended for a first trial. If this results in too light or too dark an exposure, estimate the required time for another exposure and repeat the whole procedure.
5. Remove the film in the darkroom. Develop for 4 min in X-ray developer and fix for twice the time required for the film to clear. Wash thoroughly and dry.
6. Measure the distance between lines of the film, using the film-measuring device. From the radius of the camera calculate the interplanar spacings creating the observed lines.

    If the distance between corresponding arcs of the same cone of diffracted rays—for example, the distance between A and B of Fig. 9-29—is measured and called S, then:

    $$\frac{S}{R} = 4\theta_{rad}$$

    where R = radius of camera and θ_{rad} is the angle of incidence, measured in radians. The angle, θ_{deg}, measured in degrees, is then:

    $$\theta_{deg} = \frac{S}{4R} \times 57.295$$

    Equation 9-3 can then be used to calculate the spacing d, using the values of λ for $K\alpha_1$ from Table 9-3. The Geological Survey has prepared sets of tables which give the d spacing directly if copper or iron target X-ray tubes are used and if the angle 2θ is calculated. These tables are very convenient to use.
7. For precise measurements of the camera radius and for compensation for film shrinkage during development, a pattern of sodium chloride should be taken. The main spacings in the sodium chloride pattern are 2.821, 1.99, 1.63, and 1.260 Å.
8. Estimate the relative intensity of the lines produced on the film. Refer to the A.S.T.M. X-ray tables of compounds to identify the unknown. Turn in all calculations, films, and the identification of the unknown to the instructor.

Experiment 10-1 Counter Geometry

The counting rate observed with any detector varies with the distance and the subtended angle from the detector window to the source. Because Geiger and proportional counters are usually mounted on a shelf arrangement, it is necessary to know the percentage of the activity recorded by the counter for each of the various shelf positions.

Place a radioactive standard with a known rate of disintegration on the uppermost shelf and record the activity. Repeat with the standard placed on each of the remaining shelf positions. Correct the observed activity for background and for absorption losses in the air path and window (see Experiment 10-2). Calculate the percentage of disintegrations registered by the counter for each shelf position.

Experiment 10-2 Absorption Curve for Beta or Gamma Emitters

Set the operating voltage of the counter at the predetermined value, and measure the background counting rate.

Place a sample of rather high activity in a fixed position; shelf 2, numbering downward, is convenient. Measure the counting rate. Add absorbers of known value in a position approximately midway between the source and the counter (uppermost shelf is convenient). Measure the counting rate for successive thicknesses of absorbers until the background counting rate or a constant counting rate is attained. With each absorber count for 5 min or until 5000 counts have been accumulated. The absorption curve for beta emitters is obtained with aluminum foil and massive metal of various thicknesses. The absorption curve for gamma radiation and high-energy beta emitters is studied by positioning lead absorbers of varying thicknesses between the counter and source. It is preferable to use radioisotopes that are single beta or gamma emitters.

Plot the results on semilog graph paper—the net activity as counts per minute, corrected for background and dead-time loss, on the ordinate logarithmic scale vs. the absorber thickness in milligrams per square centimeter on the abscissa. Extrapolate the plot to zero absorber thickness. Consider the absorption of the sample itself, any mounting cover (Scotch tape, 10 mg cm^{-2}), and the air path (1 mg cm^{-2}/cm) between the sample and counter plus the window of the counter (marked on counter). Often these total approximately 50 mg cm^{-2}. From the activity at zero absorber thickness and that at 50 mg cm^{-2}, estimate the correction to be added to each measured counting rate.

Estimate by visual extrapolation the range of the beta particles having the maximum energy. From Eq. 10-1 calculate the maximum energy of the beta particles, or use the curves given in Friedlander and Kennedy's book for the energy of the gamma radiation and the maximum energy of the beta particles. From the absorption curve, determine the half-thickness in aluminum (or in lead) for the particular emitter.

The self-absorption of the beta-radiation of ^{63}Ni in metallic nickel sources as a demonstration experiment has been described. [W. J. Gelsema, L. Donk, J. H. T. F. P. v. Enckevort, and H. A. Blijleven, *J. Chem. Educ.*, **46**, 528 (1969).]

Experiment 10-3 Back-Scattering of Beta Particles

Use a source that is mounted on a piece of cellophane (not over 1.5 mg cm^{-2}) that is cut to fit over a 2-cm hole in a sample holder. Place the source on the bottom of the holder with Scotch tape.

Measure the counting rate of the source with no added backing. Repeat with successive pieces of plastic (5.0 mg cm^{-2}) taped carefully to the back of the sample holder. Reposition the holder carefully each time to ensure reproducibility with respect to the counter.

Measure the counting rate of the source with various thicknesses of aluminum taped underneath the sample holder. Use thicknesses up to at least 0.4 range for E_{max}. Repeat with thick pieces of various metals taped to the back of the holder—aluminum, copper, silver, platinum, and lead of sufficient thickness to ensure saturation.

Plot the counting rate vs. the thickness (in mg cm^{-2}) of added backing material. Separate curves will be obtained for the plastic and aluminum. Show where the saturation thickness is effectively achieved with aluminum.

Plot the percentage increase in net counting rate against the atomic number of the backing material for saturation backing. At the instructor's discretion, determine the atomic number of an unknown material by using it in a back-scattering experiment.

Experiment 10-4 Identification of a Radionuclide from Its Decay Curve

The decay of a short-lived radionuclide is followed by measuring the activity at appropriate intervals of time, always placing the sample in the same position relative to the counter each time a counting rate is determined. Suitable elements include manganese-56, antimony-122, bromine-80, iodine-128, and indium-116. Barium-137m ($t_{1/2}$ = 2.554 min) can be separated from its cesium-137 parent by elution from Dowex 50W-x4 (20–50 mesh) in the potassium form at pH 11 with 0.05 to 1.0% (w/v) EDTA. [R. L. Hayes and W. R. Butler, Jr., *J. Chem. Educ.*, 37, 590 (1960).]

Longer-lived radionuclides can also be used, but it will necessitate measurements over a period of days or weeks.

Plot the net activity (corrected for background) against time on semilog paper. Determine the half-life over several time intervals and report the mean value. An unknown radioisotope (from among a restricted list) can also be identified from its absorption curve.

When a mixture of activities is present, a resolution of the decay curve is performed by the method of successive differences. First extrapolate the linear portion (the "tail" of the decay curve) to zero time. Subtract the activity associated with the longer-lived component from the gross decay curve to obtain the activity of the shorter-lived component at different time intervals and consequently its decay curve. If the gross residual decay curve is now linear, the half-life of the shorter-lived isotope can be computed; if not linear, the extrapolation process is repeated once more.

Experiment 10-5 Isotope Dilution Analysis

Experiments involving isotope dilution necessarily depend on the radioisotope and sample available. Directions for the use of iron-55/59 will illustrate the possibilities.

Prepare a solution of ferric chloride with a known concentration of iron and containing a small amount of iron-55/59. Add 1 ml of this solution, containing approximately 15 mg of iron, to a centrifuge tube which contains about 5 ml of water. Precipitate the iron with aqueous ammonia, centrifuge, wash with water, and centrifuge. Transfer a portion of the precipitate to a weighed metal planchet and dry under an infrared lamp. Weigh the sample and then count with an aluminum absorber (at least 200 mg cm^{-2}) above it. Record the weight (W_1) and the activity (A_1).

Next, take 1 ml of an iron solution of unknown concentration and add to it 200 μl of the known solution. Mix the solution thoroughly and treat the sample in the same manner as the standard. Record the final weight (W) and counting rate (A). Compute the amount of iron in the unknown solution by means of Eq. 10-18.

Relative isotope dilution analysis using samples of infinite thickness and involving calcium-45 has been described. [C. B. Johnston, G. W. Drake, and W. E. Wentworth, *J. Chem. Educ.*, **46**, 284 (1969).]

Experiment 10-6 Solubility of Precipitates

A 0.250-g sample of potassium iodide is spiked with 5 μCi of iodine-131 and then precipitated with an excess of lead ion. The precipitate is filtered, washed, and suspended in 100 ml of water (or an aqueous solution with a slight excess of either common ion). After filtration the activity of the filtrate is measured and the solubility of the lead iodide calculated.

An analogous experiment could involve measurement of the solubility of strontium sulfate, using about 0.200 g of potassium sulfate that has been spiked with 50 μCi of sulfur-35. Other experiments should suggest themselves.

Experiment 14-1 Determination of the Specific Rotation of a Substance

Weigh out from 10 to 20 g of the optically active substance assigned by the instructor. Dissolve in water and dilute to the mark in a 50-ml volumetric flask.

Determine the zero reading of the polarimeter with a tube filled with distilled water. Average several readings for precision.

Fill a 200-mm tube with the solution under investigation. Note the temperature of the solution and determine the rotation. Average several readings for precision. Calculate the specific rotation from the data, using Eq. 14-20.

Dilute 10.00 ml of solution to 25 ml in a volumetric flask. Repeat the determination of [α] on this solution. Repeat the dilution and determination of [α] once again. Plot specific rotation [α] against concentration.

Experiment 14-2 Determination of Concentration of Sucrose Solutions

This is a simplified experiment designed to illustrate the use of the polarimeter for sucrose analysis, employing inversion of the sucrose. For procedures employed in actual determinations of sucrose in commercial products, see the books by Browne and Zerban, the A.O.A.C., and Bates.

Procedure A solution of sucrose in water will be furnished as the unknown. Measure the rotation, α, at room temperature, t, in a 200-mm tube.

To 50 ml of solution in a flask, add exactly 5 ml of concentrated hydrochloric acid. Insert a thermometer and place the flask in a water bath. Heat the bath slowly at such a rate that the temperature of the sugar solution reaches 68°C in 15 min. Cool the flask quickly. Read the rotation in a 200-mm tube when the temperature has returned to its original value, t. Multiply the polarimeter reading on the invert sugar by $\frac{11}{10}$ to allow for the dilution by the acid. This rotation value equals α' in the formula below. Calculate grams of sucrose per 100 ml of solution, C, by the formula

$$C = \frac{\alpha - \alpha'}{1.9175 - 0.0066t}$$

Experiment 16-1 Parameters in the van Deemter Equation

Adjust the oven temperature to a suitable value for the column packing and the solutes assigned for study. Inject one μl of the solute mixture. Vary the helium flowrate over a wide interval so as to secure at least one, and preferably two, sets of data at sufficiently low mobile phase velocities such that the B-term can be ascertained. Three or four flow velocities greater than the optimum should provide an adequate number of data points for graphing the van Deemter equation. Record the mesh size of the column packing.

Calculate the adjusted retention time for each solute at each of the mobile phase velocities (cm sec^{-1}) from the carrier gas flowrate and column internal cross section. The air peak, or the peak of a nonretained solute (perhaps a low-boiling compound), serves as a measure of the column dead volume, t_M. Use the pressure-gradient correction on all data.

Calculate the plate number, both theoretical and effective, for each solute. From the stated column length, calculate the plate height. Plot plate height vs. mobile phase velocity to obtain the graph of the van Deemter equation. Estimate the A-, B-, and C-terms graphically and by regression analysis.

Regraph the plate height/velocity curve as H/H_{min} vs. $\bar{u}/\bar{u}_{opt}$. Tabulate the $H/\bar{u}$ terms at various velocities. Select a working velocity and column length adequate for the separation of your solute mixture.

Calculate the reduced plate height and reduced velocity for your system.

Experiment 16-2 Effect of Temperature on Retention Behavior

Adjust the carrier gas flowrate to the optimum value or slightly higher. Inject a sample mixture (n-alkyl acetates or n-alkanes are suitable) of members of a homologous family.

Secure chromatograms at successive temperatures commencing at 60°C and increasing the temperature in increments of 20°C until the maximum temperature recommended for your column packing has been attained or your assigned sample is insufficiently resolved.

Tabulate the retention time for each solute at each column temperature. Graph the logarithm of the adjusted retention time versus the carbon number for members of the homologous series.

Calculate the partition ratio, k', at each isothermal column temperature for each solute. What step interval of temperature halves the partition ratio? Estimate the optimum heating rate that should be used in a programmed temperature run.

Prepare a graph of adjusted retention time (on base-ten logarithm) vs. the reciprocal of the column absolute temperature for each solute. Note the retention value that corresponds to the boiling point for each solute in the homologous series.

Experiment 16-3 Temperature Programming

Using the isothermal column retention times from Experiment 16-2, design a temperature program for your mixture. Involved will be a selection of the initial and final column temperatures as well as the optimum heating rate.

Using your selected parameters, carry out a programmed temperature run. Compare the results with expectations; make necessary adjustments and rerun the temperature program.

Experiment 16-4 Kovats Retention Indices and McReynolds Constants

Inject a mixture along with at least two *n*-alkanes included that straddle the other compounds in terms of retention times. Calculate the Kovats R. I. values for each solute in the mixture.

Inject a sample of the individual pure compounds that serve as index compounds for the McReynolds constants. Inclusion of appropriate *n*-alkanes with the index compounds will provide a reasonable correction for column temperature differences. Estimate the values of the McReynolds constants for your particular column.

Experiment 16-5 Quantitative Analysis Using Relative Response Factors

Inject a sample containing equal volumes of benzene, toluene, and ethylbenzene. On the chromatogram measure the peak height. Assign the benzene (first) peak a response factor of unity. Assign the other peaks a response factor based on the their ratio of peak height to that of the benzene peak. Run several repetitive chromatograms to minimize the uncertainty in the relative response factors.

Chromatogram the unknown mixture assigned to you and report the volume percent of each component.

Experiment 20-1 Thermometric Titrations

Assemble the equipment shown in Fig. 20-11. Motor-driven burets, 5 ml in capacity, with delivery rate of about 0.01 ml/sec, are suitable. With efficient stirring, the titrant addition rate may approach 10 ml/min.

The sample is placed in a suitable-size polyethylene cup mounted in Styrofoam plastic and suspended in a rigid framework. With the sample container in position, each container should enclose a temperature-sensing element, a capillary buret tip, and the glass rod of a motor-driven stirrer, all of which are installed in holes drilled through the cover of the container.

Suitable thermistors will have characteristics similar to these specifications: cold resistance at 25°C, 2000–4000 Ω; temperature coefficient at 25°C, $-4.0\%/°C$; thermal time constant in still water, about 1 sec.

Adjust the bucking voltage (zero adjust) until a suitable base line is obtained on the recorder. Start the chart drive and the buret addition about 10 sec later. About 10 sec after the titration is complete, as indicated by a change in slope of the chart recording, stop the addition of titrant.

Suggested systems include $0.01–0.1M$ solutions of any acid or base; titrant concentration should be 10- to 100-fold greater than the concentration of the system being titrated. Extrapolate the linear segments of the plot to ascertain the starting and ending points.

Experiment 22-1 Selective Ion Electrodes

The object of this experiment is to evaluate the selectivity constant, actually a range of values, $K_{M/N}$, of a selective ion electrode, wherein

$$E = E° + \frac{0.0591}{z} \log [a_{M^{z+}} + K_{M/N}(a_{N^{n+}})^{z/n}]$$

where M^{z+} is the cation toward which the selective ion electrode is primarily responsive and N^{n+} is the interferent cation; z and n are the respective valences.

A. Separate Solution Methods

For illustrative purposes, the primary cation will be assumed to be sodium and a sodium glass electrode is being employed for its measurement. Plot the emf response to various activities and concentrations (0.1 to $10^{-5}M$) of sodium in a sodium nitrate salt solution. Add a drop of concentrated aqueous ammonia to each test solution; the pH must be 7 or greater to avoid interference from hydrogen ion.

Separately plot the emf response of the interferent cation over a concentration range that takes into account the selectivity constant; namely, 0.001 for K^+, 100 for H^+, and 350 for Ag^+.

When the primary and interferent activities are equal, and $z = n$,

$$\log K_{M/N} = (E_K - E_{Na})/0.0591z$$

assuming potassium is the interferent cation. Estimate $K_{M/N}$ for several $(E_K - E_{Na})$ values. The selectivity constant varies somewhat with the activities of sodium and interferent.

When $E_{Na} = E_K$, $\log a_{Na^+} = \log a_{K^+} K_{M/N}$, or $a_{Na^+}/a_{K^+} = K_{M/N}$. Again make several estimates of $K_{M/N}$.

B. Mixed Solution Methods

Measure the emf response in solutions containing a fixed amount of interferent cation and varying concentrations of that ion for which the electrode was designed. Plot the data. The intercept of the extrapolated (Nernstian) emf response segment of the graph with horizontal segment (total interference and therefore constant emf response) defines a particular intercept activity (a_M) for the primary cation. Then $K_{M/N} = a_M/a_N$.

The concentration of interferent can be varied with a constant level of primary cation. The interferent activity is found from the intercept of the two segments of the emf-response curve. Then $K_{M/N} = a_{M^{z+}}/(a_{N^+})^z$. This evaluation method is extensively used for hydrogen-ion interference studies.

Experiment 23-1 Acid–Base Titrations

A. Aqueous Systems

Each student will be assigned a titration system; the sample size and titrant strength will be specified. Suggested systems are:

1. Acetic acid, 0.1N, with 0.1N NaOH.
2. Sodium carbonate, 0.05M, with 0.1N HCl.
3. Phosphoric acid, 0.05M, with 0.1N NaOH.
4. Boric acid, 0.1N, in the presence of mannitol (4 g/50 ml volume) with 0.1N NaOH.

B. Glacial Acetic Acid

Perchloric acid, 0.1N, is prepared by adding slowly with stirring 8.5 ml of 72% perchloric acid to 900 ml of glacial acetic acid (or purified dioxane) followed by 30 ml of acetic anhydride. Allow the mixture to stand for 24 hr before use. Standardize against primary-grade potassium hydrogen phthalate.

Suggested basic solutes:

1. Potassium hydrogen phthalate. Dissolve approximately 0.1 g (weighed accurately) in 25 ml of glacial acetic acid with gentle boiling to effect solution. Insert a pair of small glass and calomel electrodes. Stir the solution and titrate with 0.1M perchloric acid solution, using a 10-ml buret.
2. Sodium carbonate. Dissolve approximately 0.05 g of the anhydrous salt, weighed accurately.
3. Aniline, or chloroaniline. Dissolve about 0.1 ml, weighed accurately by difference from a small beaker.

C. Methanol or Ethanol

Hydrochloric acid, 0.1N, is prepared by adding 9.0 ml of reagent-grade hydrochloric acid to 1 liter of absolute methanol. It is standardized against sodium carbonate as follows: Dissolve approximately 0.1 g of freshly dried sodium carbonate (weighed accurately) in the smallest possible amount of glacial acetic acid, taking care to avoid loss during the effervescence. Evaporate to dryness and dissolve the residue in 20 ml of methanol. Insert a glass–calomel electrode system. The calomel electrode should be the sleeve-type salt bridge or some type of salt bridge with a relatively large area of solution contact. Titrate with 0.1N hydrochloric acid.

Potassium hydroxide, 0.1M, is prepared by dissolving approximately 6 g of pellets in a small volume of methanol in the absence of air. Dilute to 1 liter with additional methanol and protect the solution from absorption of carbon dioxide. Standardize against the 0.1M hydrochloric acid solution.

Suggested titration systems:

1. Mixture of hydrochloric acid and acetic acid. Place in a 150-ml beaker 10 ml of 0.1M hydrochloric acid in methanol and 0.1 ml of glacial acetic acid. Dilute to 50 ml with methanol. Titrate with 0.1M potassium hydroxide in methanol.

   Repeat the titration with the inclusion of 10 ml of water in the solvent mixture. Note the change in position of the titration curve for hydrochloric acid and the indistinctness of the first inflection point.

2. Oxalic acid. Dissolve 0.2 g (weighed accurately) in 50 ml of isopropyl alcohol and titrate with 0.1M potassium hydroxide solution. Be sure to titrate both replaceable hydrogen ions.

Experiment 23-2 Oxidation–Reduction Titrations

A. Determination of Cobalt with Ferricyanide

Into a 250-ml beaker place 10 ml of 0.05M potassium ferricyanide solution, 10 ml of 5% ammonium citrate solution, and 100 ml of 5N aqueous ammonia solution. Immerse the beaker in crushed ice (or add crushed ice directly) to lower the temperature to 3°–5°C. Insert a smooth platinum-foil electrode and a calomel or other reference electrode. Stir the solution and titrate with 0.1N cobaltous solution (prepared from uneffloresced crystals of the reagent-grade cobalt sulfate heptahydrate). The cell emf falls from about 300 mV initially to about 100 mV in the vicinity of the end point when the reference electrode is a SCE.

An unknown cobalt solution may be determined by adding an aliquot to an excess of standard potassium ferricyanide solution, and titrating the unused ferricyanide solution with standard cobalt solution. For samples that may be high in iron, use 30% ammonium citrate solution.

B. Determination of Manganese with Permanganate

Place 150 ml of a 0.27M sodium pyrophosphate solution (freshly prepared) in a 400-ml beaker, and adjust the pH to 6–7 by the addition of concentrated sulfuric acid from a graduated pipet. Add 25 ml of 0.05M manganese solution and, if necessary, readjust the pH. Insert a bright platinum electrode and a calomel electrode. Stir the solution and titrate with 0.01M potassium permanganate solution. Remember that the permanganate ion undergoes only a 4-electron change to the 3+ state. If the vanadium content exceeds the manganese content in a steel sample, adjust the pH to 3.4–4.0.

C. Other Suggested Systems

Titrate an iron(II) sulfate solution, 0.05–0.1M, with 0.1M cerium(IV), 0.02M potassium permanganate, or 0.0167M potassium dichromate in a solution that is approximately 1M in sulfuric acid. Repeat the titration with the addition of 5 ml of 85% phosphoric acid to complex the iron(III) as it is produced.

Experiment 23-3   Argentometric Titration of Halides

A. Determination of a Single Halide

Place 25 ml of a 0.1M sodium chloride solution in a 250-ml beaker and dilute to about 100 ml with water. Insert a bright silver wire plus a reference electrode with a fiber-type connection or with an intermediate salt bridge containing a nonhalide filling solution. Titrate with 0.1M silver nitrate solution.

Repeat the titration with a 0.1M potassium iodide or a 0.1M potassium bromide solution. Note the differences in cell emf at the midpoint of each titration curve; compare with the calculated values.

Repeat the titration with a 0.1M potassium iodide solution in an aqueous solution that is 1M in ammonia. Note the difference in the position of the titration curve after the equivalence point.

B. Mixture of Halides

Place 10 ml of a 0.1M sodium chloride solution and 15 ml of a 0.1M potassium iodide solution in a 250-ml beaker and dilute to about 100 ml with water. Titrate with 0.1M silver nitrate solution.

Repeat the titration with this change: titrate the iodide ion in an ammoniacal solution, and after the end point is reached, acidify the solution with nitric acid and titrate the chloride ion.

C. Determination of Iodide with Redox Indicating System

Place 25 ml of a 0.1M potassium iodide solution in a 250-ml beaker and dilute to about 100 ml with water. Add about 0.5 ml of a saturated solution of iodine in alcohol, freshly prepared. Insert a bright platinum electrode and a calomel reference electrode. Titrate with 0.1M silver nitrate solution.

Experiment 23-4 Complexometric Titrations with EDTA

Mercury indicator electrode Lightly amalgamate a gold-wire electrode with mercury, or use a mercury pool emanating from a J-tube or other small cuplike container. Wash the mercury with dilute nitric acid and then rinse thoroughly with distilled water. After each titration, rinse the mercury thoroughly.

Procedure Place 25 ml of the metal-ion solution (approximately $0.05M$) in a 250-ml beaker, add 25 ml of the appropriate buffer solution (about $0.5M$ in each component), and 1 drop of $0.0025M$ mercury(II) EDTA solution. Insert the amalgam electrode and a calomel reference electrode. Titrate with $0.05M$ disodium dihydrogen ethylenediaminetetraacetate. A suitable selective ion electrode may be used in place of the amalgam electrode.

Conditions for the titration of selected metal ions are:

At pH 2 in chloroacetic acid system—thorium, mercury, or bismuth.
At pH 4.7 in a sodium acetate–acetic acid buffer—copper, zinc.
At pH 10 in an ammonium chloride–aqueous ammonia buffer—cobalt.

Experiment 23-5 Determination of Zinc with Ferrocyanide.

Place 25 ml of $0.1M$ zinc chloride solution in a 250-ml beaker. Add approximately 10 g of ammonium chloride and 1 ml of $0.001M$ potassium ferricyanide solution. Insert a bright platinum electrode and calomel reference electrode. Titrate with a $0.067M$ potassium ferrocyanide solution.

Experiment 23-6 Titration with Constant Electrolysis Current

Connect one terminal of a 45-V dry cell through a $22\text{-M}\Omega$ resistor to a platinum-foil electrode, and connect the other terminal of the dry cell directly to a second platinum-foil electrode. Also connect a high-impedance VTVM across the terminals of the platinum electrodes. The circuit should resemble Fig. 23-5.

Place 10 ml of a $0.1M$ iron(II) sulfate solution in a 150-ml beaker. Add 40 ml of water and 3 ml of concentrated sulfuric acid. Insert the pair of platinum electrodes. Titrate with $0.1M$ cerium(IV) solution.

Repeat the titration with $0.0167M$ potassium dichromate solution as the titrant.

If time is available, the titration can be performed with various values of the electrolysis current flowing through the solution other than the 2 μA suggested in the directions for assembly of the circuit components.

Experiment 24-1 Classical dc Polarography

Deaerate all solutions by bubbling a stream of nitrogen gas through the solution for 10 min (for 1–2 min if a filter stick is used to disperse the nitrogen).

1. Record the polarogram of a 0.2M KCl solution between 0 and −1.5 V vs. SCE. The current will be very small until a potential is reached where either the solvent or the supporting electrolyte is reduced. Note the variation of capacitance current with applied emf, and the potential at which there is little or no capacitance current.
2. Record the polarogram of a 0.001M CdCl$_2$ solution, also containing 0.2M KCl plus sufficient Triton X-100 to make its concentration 0.005%. In the vicinity of the half-wave potential of cadmium, determine m in the Ilkovic equation by collecting 20 drops of mercury for a known length of time (the drop time for individual drops should also be measured). The drops are collected in a plastic weighing bottle under the surface of the solution. Remove the solution with a dropper, wash with distilled water, dry with methanol, and weigh.

   Estimate the diffusion coefficient of cadmium in your system from the known values of n, C, m, i_d, and t. Plot the base-ten logarithm of $(i_d - i)/i$ against the applied emf for the cadmium polarogram. Select about six readings on each side of the half-wave potential. Evaluate the slope of the plot and calculate the value of n and the half-wave potential (uncorrected).
3. Record the polarograms of a series of solutions containing from 0.001 to 0.005M CdCl$_2$. Plot the value of the diffusion current against concentration of cadmium.
4. *Cathodic, Anodic, and Mixed Current–Voltage Curves* To a series of four 100-ml volumetric flasks add 50 ml of 1M sodium citrate and sufficient Triton X-100 to make its final concentration 0.005%. Dilute the first flask to the mark with distilled water. To the second flask add 10.0 ml of 0.0100M iron(III) ammonium sulfate solution and dilute to the mark. To the third flask add 10.0 ml of 0.0100M iron(II) ammonium sulfate solution and dilute to the mark. To the fourth flask add 5.0 ml each of the iron(III) and iron(II) solutions, and dilute to the mark. Each solution must be adjusted to pH 5.6 (use a pH meter) before diluting to the mark.

   Record the polarograms from +0.2 to −0.5 V versus SCE. To record the anodic portions of the polarograms, adjust the recorder pen initially to the midpoint of its scale when only the supporting electrolyte is recorded. Compare the half-wave potentials of the three solutions containing iron. From the small variation in individual values of the half-wave potential, estimate the iR drop within the electrolysis cell. Finally estimate the resistance of the solution from the known diffusion current observed.
5. Record the polarogram of a mixture containing 0.00100M each of CdCl$_2$, NiCl$_2$, and ZnCl$_2$ in a supporting electrolyte that is 1M in both NH$_3$ and NH$_4^+$.

Repeat the polarograms for successive 10-fold dilutions of each metal in the same supporting electrolyte until the diffusion current plateaus become indistinguishable from the preceding wave or the supporting electrolyte.

Estimate the approximate detection limits for cadmium, nickel, and zinc when determined by dc polarography. Use cadmium as the reference ion, and compute the ratio of diffusion constants of Ni/Cd and Zn/Cd. Compare with the values listed in Appendix B.

Experiment 24-2 Analysis of a Copper-Base Alloy

Lead, tin, nickel, and zinc are to be determined by the standard addition method. The major portion of the copper is removed from the dissolved sample, and lead is determined in one aliquot of the residual solution and tin in a second aliquot. Nickel and zinc are determined on a second sample from which copper, tin, and lead have been removed. Suitable solutions will be available if copper and tin plus lead were deposited electrolytically at controlled cathode potentials. Synthetic solutions that simulate the metal contents of the alloy may be employed if the electrolytic separations were not carried out. In the latter event a pseudoalloy solution is prepared that contains 4–20 mg each of lead and tin, 2–4 mg of nickel, and 4–100 mg of zinc, in a total volume of 100 ml.

Lead Transfer a 25-ml aliquot of the synthetic solution to a 100-ml volumetric flask. Add 4.8 g of sodium hydroxide pellets, 2.5 ml of 0.2% gelatin, and dilute to the mark. Mix well. Transfer exactly 25 ml of the solution to the electrolysis cell. Run the current–voltage curve from 0.6 to 0.9 V negative with respect to a SCE (approximately 0.5–0.8 V negative with respect to a mercury pool in contact with $1M$ sodium hydroxide solution).

Estimate the concentration of lead in the 25-ml aliquot from the fact that the diffusion current is about 4 μA per milliequivalent of lead. Evaluate the amount of lead present by the standard addition method.

Tin Transfer a 25-ml aliquot of the synthetic solution to a 100-ml volumetric flask. Add 21 g of ammonium chloride, 6.6 ml of $12M$ hydrochloric acid, and dilute to about 90 ml. Shake until all the salt has dissolved, warming if necessary. Add 2.5 ml of 0.2% gelatin solution and dilute to the mark. Mix well. Transfer exactly 25 ml of the solution to the electrolysis cell. Run the current–voltage curve from 0 to 0.7 V negative with respect to a SCE.

Estimate the amount of tin in the 25-ml aliquot from the diffusion current of the second wave. Subtract the diffusion current previously found for lead alone, because the half-wave potential of lead is virtually coincident with that of the second tin wave in the $4M$ ammonium chloride plus $1M$ hydrochloric acid–supporting electrolyte. Evaluate the amount of tin present by the standard addition method.

To correct the diffusion current of the second tin wave for the contribution of the coincident lead wave, multiply the diffusion current of lead found in the sodium hydroxide–supporting electrolyte by 1.035 and subtract the result from the total diffusion current due to tin plus lead. The factor 1.035 is the ratio of the diffusion current constants of lead in the ammonium chloride–hydrochloric acid medium and the sodium hydroxide medium, namely 3.52/3.40.

Nickel and Zinc Transfer a 25-ml aliquot of the synthetic solution from which tin and lead are absent to a 100-ml volumetric flask. Add 25 ml of a supporting electrolyte stock solution, which contains 43 g of ammonium chloride and 270 ml of concentrated aqueous ammonia, made up to 1 liter. Add 1 g of sodium sulfite, 2.5 ml of 0.2% gelatin solution, and dilute to the mark. Mix well and allow to stand 10 min to let the sulfite react with the dissolved oxygen. Transfer exactly 25 ml of the solution to the electrolysis cell. Run the current–voltage curves from 0.8 to 1.6 V negative with respect to the SCE.

If the height of the first wave due to nickel ions is much smaller than that of the second wave due to zinc ions, run a second current–voltage curve from 0.8 to 1.3 V negative with respect to the SCE at an increased galvanometer sensitivity.

Estimate the amounts of nickel and zinc in the 25-ml aliquot from the respective diffusion currents, and evaluate the amount of nickel and zinc present by the standard addition method.

Notes When working with the alkaline–supporting electrolytes, one observes a large diffusion current that begins at zero applied emf. Proceed as follows: Adjust the applied emf to 0.6 V negative to the reference electrode and return the galvanometer index to zero by means of the diffusion-current compensator (or bias control).

If the residual solutions from the electrodeposition experiments are employed, transfer the residual solutions to 250-ml volumetric flasks and dilute to the mark. Use 50-ml aliquots in each of the preceding steps.

Experiment 24-3 Pulse Polarography

Pulse polarography, because of its inherent greater sensitivity, is likely to yield from 2.5 to 5 times greater output current than would be obtained running a dc polarogram with the same solution.

Deaerate a solution which is $1.00 \times 10^{-4} M$ each in cadmium, nickel, and zinc ions, and containing a supporting electrolyte that is $1M$ each in NH_3 and NH_4^+. Use a scan rate of either 2 or 5 mV/sec over the interval from -0.2 to -1.5 V vs. SCE. In general, an initial potential must be used that is more positive than the half-wave potential of the first reacting species in the solution. The selected scan range should be great enough to bracket all half-wave potentials of interest.

Repeat the pulse polarogram with progressively more dilute analyte concentrations.

Experiment 24-4 Differential Pulse Polarography

Deaerate a solution, which is $1.00 \times 10^{-4} M$ each in cadmium, nickel, and zinc ions, and containing a supporting electrolyte that is $1M$ each in NH_3 and NH_4^+. Operation in differential pulse polarography is the same as in pulse polarography with the addition of a selected modulation amplitude. The choice of modulation amplitude is one of trading off sensitivity for resolution. Within the limitation that the modulation amplitude be less than the width of the overall rising portion of the step function, the larger the pulse amplitude, the greater the sensitivity and the greater the output peak. Small pulse amplitudes give better resolution and more faithful reproduction of the peak. For your system, use modulation amplitudes of 10, 25, and 50 mV on successive differential pulse polarograms. Compare results.

960 EXPERIMENTS

Experiment 24-5 dc Anodic Stripping Voltammetry

Deaerate a $1.00 \times 10^{-3} M$ solution of cadmium in $1 M$ KCl. Use a mercury-coated silver wire as the cathode. Electrolyze the solution at -0.8 V vs. SCE for exactly 1.00 min. Use 25 ml of solution. Set the scan range from -0.8 to -0.4 V vs. SCE and scan anodically at 10 mV/sec or greater to strip the cadmium.

Repeat the plating and stripping cycles with $1.00 \times 10^{-4} M$ cadmium solution, and progressively more dilute solutions. Plot the peak height against the cadmium concentration.

Run a series of plating cycles using a plating time of 2 or 3 min, and then strip as before.

The electrolysis time should be selected so that the solution is depleted by at most about 2%. Under these conditions the quantity of material present in the mercury film is proportional to the electrolysis time and the concentration of the solution. The electrolysis potential is maintained a few hundred millivolts more negative than the half-wave potential of the material to be determined.

Experiment 24-6 Linear Sweep and Cyclic Voltammetry

A three-electrode system is used. The working electrode will be a platinum disk of known area (you will need to measure it), some type of reference electrode, and a platinum-wire electrode for a counter electrode. Clean the platinum working electrode in hot, 30% nitric acid. Rinse in distilled water.

Place the electrode system into the solution to be investigated. Allow the system to remain quiescent for at least 3 min with the cell in "standby." There must be absolutely no vibrations during the quiescent period.

Set the initial potential to a value 0.2 V before the half-wave potential of the system to be studied. Set the scan rate to 2 V/min and the scan direction to the appropriate sign (anodic or cathodic). To record the voltammogram, follow this sequence of operations:

1. Recorder pen placed on chart paper.
2. Standby setting changed to cell connected.
3. Sweep generator set to scan position.
4. When the potential is 0.2 V past the peak potential, E_p, change the scan direction to the opposite sign.
5. When the scan voltage returns to the initial potential, raise the recorder pen and change the cell control to standby.

The samples that should be recorded are:

1. 100 ml of $1.00 M$ KCl
2. 100 ml of $1.00 M$ KCl plus exactly 1.00 ml of $0.100 M$ of the assigned ion. $K_4[Fe(CN)_6]$ is ideal for anodic scans, as is $K_3[Fe(CN)_6]$ for cathodic scans. Record the cyclic voltammogram of this solution for various scan rates.
3. At the optimum scan rate, run various concentrations of the assigned ion by successively adding 1.00-ml increments of the stock solution.

From the data accumulated in the experiment, calculate the diffusion constant at each of the scan rates and compare with the literature value. Use the equation

$$i_p = 268 n^{3/2} D^{1/2} A V^{1/2} C$$

where D is in cm²/sec, A in cm², V in V/sec, and C in mM/ml.

Prepare a table with these column headings, at the different scan rates:

$$i_p/V^{1/2}; (i_p)_{\text{anodic}}/(i_p)_{\text{cathodic}}; E_p; E_{p/2}; E_p - E_{p/2}; \text{and } i_p/ACV^{1/2}$$

Compare your data to the theory:

$$E_p = E_{1/2} - 0.0285/n \text{ and } E_{p/2} = E_{1/2} + 0.028/n$$

which implies that E_p should be independent of the scan rate and that

$$E_p - E_{p/2} = 0.0565/n$$

The ratio $i_p/V^{1/2}$ should be constant. Note the dependence of i_p upon concentration. Ideally the reverse peak current for your system should be equal to the forward peak current. How ideal is your ratio $(i_p)_a/(i_p)_c$?

Experiment 24-7 Amperometric Titration of Lead with Dichromate (VI)

1. Transfer 25 ml of 0.02M lead nitrate solution to a polarographic cell. Add 25 ml of a supporting electrolyte, which is approximately 0.1M in potassium nitrate, 0.17M in acetic acid, and 0.06M in sodium acetate. The pH should be about 4.2. (The weights of the three ingredients in the supporting electrolyte are 10, 10, and 5 g per liter, respectively.) Add 2.5 ml of 0.2% gelatin solution. Remove dissolved oxygen.
2. Determine the current–voltage curve of lead from 0 to 1.0 V negative, using a dropping mercury electrode and an external saturated calomel electrode of large area (> 10 cm²).
3. Plot the current–voltage curve on a sheet of graph paper.
4. Apply 0 V across the electrodes; that is, short the dropping electrode directly through galvanometer and shunt. Titrate with 0.05M potassium dichromate solution from a 10-ml buret. Take readings every 0.5 ml until the galvanometer registers a definite deflection, and then take readings every 0.25 ml until several points have been obtained beyond the equivalence point.
5. Run a current-voltage curve on the solution containing excess potassium dichromate. Plot the curve on the sheet of graph paper used in step 3.
6. Repeat step 4 with the cathode at 1.0 V negative with respect to the SCE.
7. Repeat steps 4 and 6 with 0.002M lead nitrate solution and 0.005M potassium dichromate solution.
8. On a second graph, plot the two pairs of curves of current vs. volume of titrant for steps 4, 6, and 7.

Note See Fig. 24-20 for the current–voltage curves of lead and chromium, and Figs. 24-18 and 24-19 for the general shape of the titration curves.

Experiment 24-8 Amperometric Titration of Arsenic(III) with Bromate

1. Transfer 25 ml of 0.001N arsenious oxide to the polarographic cell. Add 5 ml of 12M hydrochloric acid and 20 ml of 0.125M potassium bromide (15 g/liter). Center a button-type platinum electrode in the cell (a 150-ml beaker) and insert the arm of the salt bridge from a saturated calomel electrode whose surface area exceeds 10 cm^2. Insert a stirring bar and adjust the stirring rate between 200 and 600 rpm without producing a vortex.
2. Apply 0.2 V positive to the rotating platinum electrode (which is the anode in this titration). Titrate with 0.05N potassium bromate solution from a 10-ml buret. At first the galvanometer will be deflected only slightly, and 1-ml increments may be added. When a definite galvanometer deflection occurs, decrease the size of the increments to 0.25 ml and secure several readings after the equivalence point.
3. Run the current–voltage curve of bromine from 0.3 V negative to 0.9 V positive vs. the SCE. The excess bromate reacts with the bromide ions in the presence of hydrogen ions to form free bromine, the electroactive species undergoing reduction at the rotating platinum electrode.
4. Repeat step 2 with duplicate samples to ascertain the precision attainable by this titration method.

Experiment 24-9 Use of Two-Indicator Electrode System (Dead-Stop Method) for Amperometric Titrations

1. Transfer 5 ml of 0.01N iodine solution to a 150-ml beaker. Add 0.1 g of potassium iodide and dilute to about 60 ml.
2. Apply 0.1 V across two similar platinum-wire (or foil) electrodes. Pass the solution by the electrodes with a magnetic stirrer. Titrate with a 0.01N sodium thiosulfate solution from a 10-ml buret. Take readings every 0.5 ml until the galvanometer index remains constant through 4 or 5 additions. Plot the results on graph paper.

In place of a conventional instrument, connect a battery of 1.5 V through 0.1-MΩ and 7000-Ω resistors in series. Connect two similar platinum-wire electrodes to the terminals of the 7000-Ω resistor. In series with the cathode, insert a galvanometer with a sensitivity of about 0.1 μA/mm.

The titrant may be sodium arsenite; if it is, the pH of the solution must be adjusted between 4 and 9 with a suitable buffer.

Experiment 25-1 Controlled Cathode Separations

A. Determination of Copper in Brass

Transfer a 0.5-g sample of brass (weighed accurately) to a 200-ml tall-form beaker. Dissolve the brass in 10 ml of concentrated HCl plus 5 ml of water to which concentrated nitric acid is added dropwise until dissolution is complete. In an alternative method, add

10 ml of 1:1 HCl and, in small portions, 5 ml of 30% hydrogen peroxide. Boil the solution to expel the oxides of nitrogen and chlorine (or excess hydrogen peroxide).

Dilute the solution to 25 ml, and add 4 g of hydrazine hydrochloride. Heat the solution to 95°C and maintain it at this temperature until the dark green color changes to a light olive-green, indicating considerable reduction to the chlorocuprate(I) ion.

Place the beaker in position in the electrolysis apparatus, and add water until the electrodes are covered completely. Place the tip of the salt bridge from the reference electrode on the outside and near the middle of the cathode.

Turn on the electrolysis current and adjust the applied emf to maintain the cathode potential at −0.35 V vs. SCE. The initial current should range from 2 to 4 A. Copper may not deposit for several minutes or until all the copper(II) ions have been reduced to chlorocuprate(I) ions and, indeed, the reference electrode may at first be negative to the cathode. Copper will commence to plate out when the cathode potential is about −0.2 V. As the deposition of copper proceeds, continuously adjust the applied emf to maintain the cathode potential at the limiting value. Continue the electrolysis until the current decreases to 10-20 mA.

B. Determination of Tin Plus Lead in Brass

To the solution from which copper has been removed, add 1 g of hydrazine hydrochloride, and insert a copper-clad electrode (weighed accurately). Adjust the applied emf to −0.60 V vs. SCE initially, but after the electrode becomes coated with a deposit of tin and lead, raise the applied emf to −0.70 V. Continue the electrolysis until the current decreases to a constant value and remains steady for 10 min. Before removing the electrodes, carefully neutralize the solution to pH 5-6 with aqueous ammonia.

Experiment 25-2   *Separation of Copper by Internal Electrolysis*

Assemble the apparatus as shown in Fig. 25-9. For anodes, wind pure lead wire, 10 to 12 gauge, around the stem of a thistle tube which has been shortened to 6 in. in length, to form a compact helix. Leave a sufficient length of wire to connect to the common binding post. Each anode compartment is a porous Alundum shell (extraction thimble obtained from the Norton Company, RA 84 or 360, 19 X 90 mm). The solution is stirred by a glass corkscrew stirrer or with a magnetic stirrer and bar.

Transfer aliquots containing 10-30 mg of copper into 400-ml beakers. To each, add 3 ml of concentrated nitric acid and 3 ml of concentrated sulfuric acid. Dilute to 200 ml. Warm the solution to 70°C.

Fill the anode compartments with a solution composed of nitric acid (3% v/v) and lead nitrate (5% w/v), and insert the anode compartments and the platinum cathode into the sample. Add 0.3 g of urea and short-circuit the electrodes.

Electrolyze for 15 min and then flush out the anode compartments with lead nitrate-nitric acid solution. Continue the electrolysis until the copper is completely deposited, as indicated by failure to plate on a fresh surface when the solution level is raised. Lower the beaker, and rinse the electrode with a stream of distilled water. Remove the cathode, dry, and weigh.

Experiment 25-3   Electrographic Spot Testing

A schematic diagram of an electrograph is given in Fig. 25-10. The sample is made the anode, in contact with a sheet of hardened filter paper, such as Schleicher & Schull Nos. 575 or 576 or Whatman No. 50, and backed by a thick, soft, backing paper such as S & S No. 601 or blotting paper.

General directions   Moisten the pad of the printing medium and backing paper with electrolyte, blot lightly, and place on the aluminum base plate with the printing surface upward. Place the specimen on the paper and make contact by bringing the other electrode onto the specimen by hand pressure or by a clamp. Close the electrical circuit for the length of time calculated by Eq. 25-25. Generally an exposure of 10 sec with a current of 15 mA/cm^2 of surface area is sufficient.

Examination of a print of pure metal   Moisten the pad with an electrolyte consisting of 0.5M sodium carbonate solution plus 0.1M sodium chloride solution. After printing, remove the upper sheet and cut it into 4 parts. Hold one part over a beaker of warm, concentrated aqueous ammonia, hold the second part over a beaker of warm, concentrated hydrochloric acid, and hold the third part over a warm surface until dry. Compare the original color and the colors of the treated portions of the test sheet with those listed.

Colors of Transfer Products of Certain Metals

Metal	Electrolyte: 0.5M Na$_2$CO$_3$ + 0.1M NaCl	Fuming Over NH$_3$	Fuming Over HCl	Exposure to Heat and Light
Copper	Greenish blue	Deep blue	Green-yellow	Green-blue
Silver	Colorless	—	—	Black
Iron	Brown	Brown	Orange-yellow	Brown
Nickel	Light green	Light violet	Green	Light green
Cobalt	Dirty brown	Brown	Blue	Deep blue
Molybdenum	Deep blue-violet	Gray	Gray	Gray
Chromium	Yellow	Yellow	Yellow	Yellow

SOURCE: H.W. Hermance and H.V. Wadlow, "Electro Spot Testing and Electrography," in *Am. Soc. Testing Materials, Spec. Tech. Pub.*, **98**, 25 (1950).

Use of color-producing reagents

1. Solder lugs may be lead- or tin-dipped. A test sheet impregnated with a 0.5M ammonium molybdate solution yields a blue color with tin; a zinc sulfide test paper, subsequently treated with warm, yellow ammonium polysulfide, reveals lead as a black stain.
2. A specimen of steel may be examined for nickel and chromium. A test sheet moistened with a saturated solution of barium hydroxide plus a 1% alcoholic dimethylglyoxime solution yields yellow barium chromate and red nickel dimethylglyoxime when these metals are present.
3. A flat laboratory spatula, labeled stainless, may be tested for pinholes in the coating over the underlying iron. A test sheet impregnated with 0.5M potassium ferricyanide

solution yields blue dots where holes exist in the coating unless the chromium plate has an underlying layer of nickel.

General Instructions for Constant-Current Coulometry Experiments

Assemble the titration apparatus as shown in Fig. 25-11. Connect a 45-V B battery through a 5000-Ω potentiometer (or bank of fixed resistors of different values), a 3000-Ω limiting resistor, the current-measuring device, the generator electrodes, and an ON–OFF switch. Measure the current with a calibrated 0-10 milliammeter or determine the iR drop across a precision 100-Ω resistor (for currents not exceeding 10 mA) with a student potentiometer. Time measurements made with a stopwatch or stopclock will provide results of moderate accuracy.

Arrange the generator electrodes and indicator system as shown. Positioning the generator electrode from which the reactant is derived adjacent to the indicator electrode, in the direction of stirring, gives a more rapid warning of the approach of the end point. For photometric indication, the electrolysis cell is positioned in the photometer in place of the usual cuvette.

When using amperometric indication for the end point, plot the amperometric signal vs. time. The end point will be signaled by an abrupt change in the amperometric current, which may be taken as the end point, or the coulometric titration may be terminated momentarily, the amperometric current and generation time noted, and then the generation continued for perhaps an additional 5-10 sec. This series of steps is repeated until 4 or 5 readings are obtained beyond the end point, which is then established by extrapolation of the two branches of the plot.

Pretitration of the supporting electrolyte should be done to remove impurities and to familiarize oneself with the end-point signal. Then the sample is added and the titration continued until the end-point signal reappears.

Experiment 25-4 Electrically Generated Hydroxyl Ion

Place 100 ml of 0.05M KBr solution (6 g/liter) in a 200-ml tall-form beaker. Add several drops of an appropriate indicator, or insert glass–calomel electrodes for potentiometric indication. The generator electrodes may be a platinum-foil cathode (10 cm^2) and a helix of silver (No. 6 gauge) wire as anode (or a second platinum-foil electrode isolated by a frit barrier).

Turn on the stirrer and generator current. Adjust the current by means of the potentiometer (or bank of resistors) to 10 mA or less. Titrate to the theoretical pH at the end point (adding a trace of acid, if necessary), then discontinue the current. Add the sample, then turn on the current and timer simultaneously. Adjust the variable rheostat whenever necessary to maintain the current at the selected value throughout the titration. Select an aliquot of the sample that will require about 200 sec (for example, 10-20 ml of 0.001N acid, transferred with a pipet or microsyringe if more concentrated acids are employed).

After every titration clean the anode with emery cloth or by dipping it into concentrated aqueous ammonia or thiosulfate solution. Several consecutive samples may be titrated without renewing the supporting electrolyte.

Experiment 25-5   Electrically Generated Bromine

Place 100 ml of 0.2M KBr solution (24 g/liter) and 3 ml of 18M H_2SO_4 in a 200-ml tall-form beaker. The generator electrodes are two platinum-foil electrodes, 10 cm^2 or larger in area. If amperometric indication is employed, insert a small platinum electrode and a large-area calomel electrode and apply 0.2 V positive with respect to the SCE. Pretitrate to an end-point signal, then add 1.00 ml of 0.005M As_2O_3 solution (0.987 g/liter) and titrate until the same signal is repeated.

Experiment 25-6   Electrolytically Generated Cerium(IV) Ion

Place 40 ml of 0.1M cerous ammonium sulfate and 10 ml of 9M H_2SO_4 in the electrolysis cell. Insert a platinum-foil working electrode in the cell and isolate a second platinum-foil electrode (the cathode) inside a tube with a fritted glass end and filled with 1.5M H_2SO_4. Purge the supporting electrolyte with nitrogen gas for 10 min, and maintain a stream of gas through the electrolyte during the titration.

Add a sample of iron(II) ammonium sulfate and titrate at 50 mA. The end point is ascertained potentiometrically with a platinum–calomel electrode pair or amperometrically with a platinum indicator electrode. A pretitration is recommended.

Experiment 25-7   Electrically Generated Iodine

Place 50 ml of 0.1M KI solution (16.6 g/liter) and 20 ml of 0.25M Na_2HPO_4 solution (36 g/liter) in the electrolysis cell. Add a few drops of 0.005M As_2O_3 solution (0.987 g/liter) and pretitrate to the end point (starch–iodide color, amperometric indication, or potentiometric indication).

Add 1.00 ml of 0.005M As_2O_3 solution and titrate once more to the end-point signal.

Experiment 25-8   Electrically Generated Silver Ion

For the titration of bromide and iodide, the supporting electrolyte is 0.5M KNO_3 (51 g/liter). The working electrode (anode) can be a clean silver foil (10 cm^2) or a silver rod;* the cathode is a platinum-foil electrode. For the titration of chloride, the supporting electrolyte is a nitric–acetic acid system (38-ml concentrated HNO_3 and 200-ml glacial acetic acid per liter) plus 0.05% gelatin.

Use 5.00-15.00 ml of 0.025M KBr (2.975 g/liter), or 0.025M KCl (1.864 g/liter), solution when the generating current is 30 mA. For amperometric indication, a large-area SCE may be short-circuited to the platinum indicator electrode through a suitable galvanometer.

---

*Any previous coating of silver halide must be completely removed (see Experiment 25-4).

Experiment 25-9 External Titration

Assemble an external generator, double-arm cell as shown in Fig. 25-18.

Connect a source of direct current, approximately 200 mA with 0.1% or less ripple, through a precision resistor, the pair of generator electrodes, and, if necessary, a 1250-Ω (100-W) rheostat and a 125-Ω (20-W) rheostat.

Feed a solution of the supporting electrolyte continuously into the generator cell to provide a delivery rate of 6 ml/min from each delivery tip. Use a 600- or 800-ml beaker for the titration.

Suggested supporting electrolytes: $1.0M$ sodium sulfate solution (adjusted to pH 7) for generation of hydroxyl or hydrogen ions; $0.05M$ potassium iodide solution in $0.1M$ boric acid (to neutralize the hydroxyl ion produced at the cathode) for generation of iodine; and $0.05M$ potassium bromide in $0.1N$ sulfuric acid for generation of bromine.

Experiment 25-10 Determination of Silver and Copper by Controlled Potential Coulometry

Dissolve a weighed sample of silver or copper metal in nitric acid, and remove the nitric acid by fuming with perchloric acid. Dilute to volume in a volumetric flask and take an aliquot containing several milligrams of alloy for analysis. To the aliquot in the analysis cell, add enough distilled water to cover the electrodes, set the potentiostat at +0.13 V with respect to a saturated calomel electrode, and electrolyze until completion. Read the number of coulombs. Reset the potentiostat to −0.12 vs. the SCE and electrolyze again. Read the number of coulombs. Calculate the percentage of silver and of copper in the alloy. Should the alloy contain other, more easily reducible metals, try pre-electrolyzing the sample at a potential insufficient to plate out either copper or silver.

Experiment 26-1 Titration of a Completely Ionized Acid

1. Transfer 50 ml of $0.01M$ hydrochloric acid to a 250-ml beaker and dilute to about 100 ml. Measure the conductance. Titrate with $0.1N$ sodium hydroxide added in 0.5-ml increments from a 10-ml buret. Measure the conductance (or resistance) after the addition of each increment.
2. Repeat the titration, using $0.001M$ and $0.0001M$ solution of hydrochloric acid and titrating with $0.01N$ and $0.001N$ sodium hydroxide, respectively.

Experiment 26-2 Titration of an Incompletely Ionized Acid

1. Transfer 50 ml of $0.1N$ acetic acid to a 250-ml beaker and dilute to about 100 ml. Measure the conductance. Titrate with $1N$ sodium hydroxide added in 0.5-ml increments from a 10-ml buret.
2. Repeat the titration, using $0.01N$ and $0.001N$ solutions of acetic acid and titrating with $0.1N$ and $0.01N$ sodium hydroxide, respectively.

968 EXPERIMENTS

3. Repeat the entire series of titrations, substituting aqueous solutions of ammonia as the titrant.
4. Plot all results on one piece of graph paper. The curves should resemble Fig. 26-7.

Experiment 26-3 Titration of Incompletely Ionized Acid: Partial Neutralization with Ammonia Followed by Titration with Sodium Hydroxide

1. Transfer 50 ml of $0.1N$ acetic acid to a 250-ml beaker. Dilute to about 100 ml with distilled water. Measure the conductance.
2. Add, in 0.5-ml increments, a total of 8 ml of $0.5N$ aqueous ammonia. Complete the titration with $1N$ sodium hydroxide added from a 10-ml buret.

Note The aqueous ammonia need not be standardized. The titration curve should resemble Fig. 26-8.

Experiment 26-4 Titration of a Mixture of Acids

1. Transfer 40 ml of $0.01N$ acetic acid and 10 ml of $0.01N$ hydrochloric acid to a 250-ml beaker. Dilute to about 100 ml with distilled water. Measure the conductance.
2. Titrate with $0.1N$ sodium hydroxide added from a 10-ml buret. Use 0.25-ml increments of titrant.
3. Repeat the titration, using 10 ml of $0.01N$ acetic acid and 40 ml of $0.01N$ hydrochloric acid.

Note If desired, a $0.1N$ aqueous ammonia solution may be used as titrant as was done in Experiment 26-2.

Experiment 26-5 Precipitation Titrations

1. Transfer 50 ml of $0.01N$ sodium chloride to a 250-ml beaker and dilute to about 100 ml. Measure the conductance.
2. Titrate with $0.1N$ silver nitrate added from a 10-ml buret.
3. Repeat the titration with $0.1N$ silver acetate as titrant.
4. Plot the results on the same graph and note the slope of the two branches of the titration curve for the different titrants.

Experiment 26-6 Measurement of Dielectric Constant

1. Connect the test-tube cell holder to a high-frequency instrument, such as the Sargent oscillometer. Balance the instrument with the cell empty.
2. Prepare a calibration curve of instrument response as a function of dielectric constant by successively measuring pure solutions of benzene, chloroform, acetone, ethanol, methanol, and water.

3. Determine the instrument reading of an unknown solution and report the dielectric constant of the unknown solution by reference to your calibration curve.

Note The instructor may wish to have the student prepare a calibration curve for a binary mixture—for example, nitrobenzene and benzene or aniline and nitrobenzene—and then determine the composition of an unknown.

Answers to Problems

Chapter 2

1. $\sin \theta = 3(3.8 \times 10^{-5} \text{ cm})/b < \sin \theta = 2(7 \times 10^{-5} \text{ cm})/b$, whatever the grating spacing.
2. From $17°14'$ (violet) to $33°6'$ (red). 3. (a) 1778 grooves/mm; (b) 239 nm; (c) $10.2°$; (d) 1184 grooves/mm; (e) 501 nm; (f) $20.7°$; (g) 591 grooves/mm; (h) 1.0 µm; (i) 296 grooves/mm; (j) $14.8°$; (k) 147 grooves/mm; (l) $21.7°$. 4. (a) Time constant is 1 sec, which would seriously degrade any transient signal whose lifetime was less than about 10 sec. (b) Remedies include shorter leads, lower R_L, or a combination of these. Preamplifier built into phototube base minimizes stray capacitance. 5. (a) 0.153 µm; (b) 0.165 µm; (c) 0.214 µm; (d) 0.278 µm. 6. Third-order passband is at 437.3 nm; the first order at 1312 nm. 7. About 750 and 375 nm. A cutoff filter lying midway between 500 and 750 nm would eliminate 750 nm; a cut-on filter lying midway between 500 and 375 nm would eliminate 375 nm. 555 nm. 8. 1.3%. Reflectance to 33%; since there is no absorption, transmittance is 67%. 9. In both case A and B the slit distribution is trapezoidal with a bandpass equal to 1.5 times the image and a spectral region isolated equal to twice the image. In last cases the slit distribution is rectangular and the slitwidth and spectral region isolated are equal. 10. (a) 999 grooves first order; 500 grooves second order; 333 grooves third order. (b) 0.59 nm is spectral bandwidth; 0.30 nm is bandpass. (c) 0.20 mm. 11. No; at base line an overlap of 1.7 nm exists. 12. (a) 5.0 nm/mm. (b) Yes; $\Delta\lambda$ available is 0.10 nm and required is 0.18 nm. 13. Opposing currents through galvanometer: $(i_g)_{\text{meas}} = kTI[a/(R_1 + r_g)]$ and $(i_g)_{\text{ref}} = kI[x/(R_2 + r_g)]$, where a and x are contactor positions on R_1 and R_2, respectively. At balance, currents are equal and since $R_1 = R_2$, then $x = Ta$, where T is the transmittance.

14.

TRUE ABSORBANCE	OBSERVED ABSORBANCE FOR STRAY LIGHT AT		
	0.1%	1.0%	5.0%
0.500	0.499	0.491	0.455
1.000	0.996	0.963	0.839
1.500	1.487	1.384	1.095
2.000	1.959	1.701	1.225

15. Effects of dark current are additive, affecting both sample and reference-beam intensities: $A = \log(I_0 + I_D)/(I + I_D)$. Extraneous factor would have to be multiplicative to cancel in double-beam operation.

Chapter 3

1. Stray light originated at 758 cm^{-1}, probably from second-order grating, because CaF$_2$ is opaque at 758 cm^{-1} as well as 379 cm^{-1}. If the stray light had originated from the third order (1137 cm^{-1}), the CaF$_2$ plate would also have shown the 379 cm^{-1} stray. 2. $\epsilon = 3120$ liter mol^{-1} cm^{-1}. 3. For Ti: $x = 5.38 A_{400} - 3.37 A_{460}$; for V: $y = 15.9 A_{460} - 7.93 A_{400}$. Sample 1, 0.053% Ti, 0.048% V; sample 2, 0.052% Ti, 0.394% V; sample 3, 0.099% Ti, 0.180% V; sample 4, 0.197% Ti, 0.186% V; sample 5, 0.293% Ti, 0.191% V; sample 6, 0.100% Ti, 0.574% V; sample 7, 0.139% Ti, 0.030% V; sample 8, 0.067% Ti, 0.043% V; sample 9, 0.114% Ti, 0.025% V. 4. Tyrosine at 294 nm and tryptophan at 280 nm. 5. First end point, 6.75 ml; second end point, 9.80 ml. Sodium acetate, 0.0682 M and o-chloroaniline, 0.0308 M. 6. First rise, extrapolated back to zero absorbance gives system blank. Net titrant volume: 2.70 − 0.23 = 2.47 ml; 78.0 μg of magnesium present. 7. Sketch should resemble a "flat-bottom U." First end point will be disappearance of reddish color of methyl orange; second end point will be appearance of pink phenolphthalein color. 8. For p-nitrophenol at 407 nm where anion absorbs, pK_a = 6.98; at 317 nm where undissociated acid absorbs, pK_a = 6.96. For papaverine (protonated cation), pK_a = 6.40 (average). 9. Bromophenol blue: pK_a = 4.06. Methyl red: pK_a = 4.93. Bromocresol purple: pK_a = 6.29.

10.

CONCENTRATION, M	dC/C (THERMAL NOISE)	dC/C (SHOT NOISE)[a]
0.0050	0.0330	0.0310
0.0100	0.0190	0.0167
0.0150	0.0145	0.0119
0.0200	0.0126	0.0096
0.0250	0.0116	0.0082
0.0300	0.0110	0.0073
0.0400	0.0111	0.0064
0.0500	0.0124	0.0058
0.0600	0.0126	0.0055
0.0800	0.0164	0.0054
0.1000	0.0224	0.0057
0.1100	0.0268	0.0060

[a] Assuming k = 0.004, a reasonable estimate for high-quality spectrometers.

11. Concentration, M	dC/C	12. Concentration, M	dC/C	13.	dC/C
0.0600	0.00255	0.0100	0.0170		
0.0700	0.00237	0.0200	0.0113		
0.0800	0.00232	0.0300	0.0103		
0.0900	0.00230	0.0400	0.0104		
0.100	0.00230	0.0500	0.0117		
0.110	0.00234	0.0600	0.0140		0.0084
0.120	0.00240	0.0700	0.0189		0.0059
0.130	0.00246	0.0800	0.0307		0.0061
0.140	0.00250	0.0900	0.0705		0.0081
0.150	0.00255				

14. (a) $dC/C = 5.12 dT$ by ordinary method (thermal noise) or $1.41 dT$ (shot noise). Using 5.0-mg reference for 100%T, $dC/C = 0.389 dT$, a 13.2- or 174-fold increase. (b) $dC/C = 0.113 dT$ by maximum precision method, a 45-fold precision increase. **15.** (a) FeL_3^{2+}. (b) $K'_f = 3.2 \times 10^{17}$ (in a medium approximately pH 4). **16.** (a) FeL_3^{2+}. (b) $\epsilon = 12{,}600$. **17.** Ratio of slopes: $0.353/0.117 = 3/1$; FeL_3^{2+}. **18.** (a) 0.333. (b) $K_f = 8.6 \times 10^9$. **19.** First structure has calculated $\lambda_{max} = (215 + 12) = 227$ nm; second structure contains only isolated double bonds. **20.** Calculated λ_{max} for structure I: $215 + 30 + 18 + 36 = 299$; for structure II: $215 + 12 = 227$. Structure I is the β-isomer. **21.** Calculated λ_{max} values permit identification: (a) 254 nm, (b) 219 nm, (c) 268 nm, and (d) 298 nm. Compounds a and b are dienes; while c and d are homoannular dienes.

Chapter 4

1. 1.3 µg/g. **2.** Eq. 4-6 valid up to 8 µg/50 ml. **3.** (a) With mercury lamp, lines at 303 or 313.5 nm fall near excitation maximum; emission filter should reject wavelengths below about 330 nm but pass longer wavelengths. (b) Use peak maximum near 295 nm for excitation and fluorescence emission at 350 nm. **4.** Excitation at 340 nm produces only phenanthrene emissions. Excitation at 275 nm maximizes the naphthalene fluorescence at 315 nm which is almost free from phenanthrene emission. Phenanthrene fluorescence at 370 nm almost free from naphthalene fluorescence. Phosphorescence offers no advantage. **5.** (a) Fluorescence of window material; scatter from scratches, bubbles, or etches. (b) Concentration quenching and scatter interference. **6.** (a) When excitation wavelength is varied, Rayleigh scatter will vary exactly as the excitation, whereas Raman shift is a constant energy shift. Any fluorescence peaks vary only in intensity with the wavelength maxima remaining unchanged. (b) Limits sensitivity by producing a high blank reading which drowns out weak fluorescence signals. (c) Change excitation wavelength; increase resolution to separate Raman scatter from the sample; use a polarizer; change solvents. (d) Raman shifts of four solvents at excitation wavelengths are:

	Excitation Wavelengths			
Solvent	254 nm	313 nm	365 nm	436 nm
Cyclohexane	274	344	408	499
Water	278	350	416	511
Ethanol	274	345	409	500
Chloroform	275	346	410	502

The wavelengths listed for each solvent are the Raman band maxima (in nm). 7. $(S/N)_{abs} = A(I_0)^{1/2}/k$ and $(S/N)_{fluor} = (C\phi_F A I_0)^{1/2}/k$, where C is collection efficiency for fluorescence or $f(\theta)g(\lambda)$. For same S/N, $A_{fluor} = A_{abs}^2/\phi_F C$, thus

CONCENTRATION, M	A_{abs}	A_{fluor}	SENSITIVITY INCREASE
10^{-4}	10^{-1}	5×10^{-2}	2
10^{-6}	10^{-3}	5×10^{-6}	200
10^{-8}	10^{-5}	5×10^{-10}	20,000

Few, if any, spectrophotometers can measure absorbance changes of less than 10^{-3}. 8. For aluminum, fluorescent peaks are at 417.1, 440.0, and 467.6 nm, giving $^1S^* \rightarrow {}^1S_0$ transitions to $v = 0, 2$, and 4 ground-state vibrational levels (23,975, 22,727, and 21,386 cm^{-1}, respectively). Phosphorescence peaks occur for $v = 0, 1, 2, 3, 4$, and 5 and involve transitions from the triplet state of 20,790, 20,202, 19,531, 18,857, 18,182, and 17,482 cm^{-1}, respectively. See also *J. Chem. Phys.*, **17**, 1182 (1949).

Chapter 5

1. Cement A: 0.14% Na_2O, 0.62% K_2O. Cement B: 0.37% Na_2O, 0.43% K_2O. Cement C: 0.22% Na_2O, 0.55% K_2O. 2. Emission from molecular band systems of CaO and CaOH. Blank would be larger because reading would increase approximately with the square of the bandpass of the filter or monochromator. 3. (a) 55 µg/ml. (b) 170 µg/ml. (c) 130 µg/ml. 4. (b) Ionization of strontium atoms in flame. (c) Ionization in (b) is repressed by large excess of calcium atoms. Addition of calcium also contributes a significant background emission due to CaOH. 5. 0.715 µg/200 ml. 6. 52.8 µg/ml. 7. 29.7 µg/ml. 8. Water: 27.0 µm. 50% methanol–water: 18.7 µm. 40% glycerol–water: 29.4 µm.

9.

FLOWRATE, ml/min	DROPLET DIAMETER, µm		
	WATER	50% EtOH–H_2O	MIBK
1.0	16.1	12.0	10.9
2.0	18.3	16.3	13.0
3.0	21.0	21.8	15.7
5.0	28.0	35.9	22.5

See also *Applied Optics*, **7**, 1353 (1968). 10. (a) 0.0094. (b) 0.094. (c) 0.29. 11. (a) 0.35. (b) 0.75. (c) 0.97. 12. To suppress the ionization of 0.23 µg/ml of sodium to 0.01 (1%), add:

ELEMENT	AT 2500°K	AT 2800°K
Cs	0.44 µg/ml	0.74 µg/ml
Rb	1.05 µg/ml	1.6 µg/ml
K	1.04 µg/ml	1.4 µg/ml
Li	22.50 µg/ml	19.5 µg/ml

13. 13.4°K for sodium, and 15°K for potassium. 14. (a) $S/N = 13.3$ (or 6.7 peak-to-peak) for 2 µg/ml. (b) Sensitivity = 0.22 µg/ml. (c) Detection limit = 0.30 µg/ml.

Chapter 6

1. Cadmium, magnesium, and zinc.   2. On double logarithmic paper, graph the reciprocal of the ratio (because these are density readings): Pb reading/Mg reading, against the lead concentration. (a) 0.26 mg/ml. (b) 0.34 mg/ml. (c) 0.41 mg/ml.   3. Gamma factor = 1.49; emulsion inertia is 0.155.   4. Densities of the tin and lead lines are the log (I_0/I) ratios for the two lines. Intensity values I_{Sn} and I_{Pb} are read from the characteristic curve of the emulsion (problem 3). Logarithm of the ratio I_{Pb}/I_{Sn} is then plotted against the logarithm of %Pb. Unknown contains 0.46% Pb.

Chapter 7

1. 910 cm^{-1}.   2. First overtone at 1108 cm^{-1}; second overtone at 1662 cm^{-1}. NaCl.   3. (a) 3060 cm^{-1}. (b) 3293 cm^{-1}. (c) 1128 cm^{-1}. (d) 1465 cm^{-1}. (e) 2143 cm^{-1}. (f) 1744 cm^{-1}.   4. (a) 2.22 µg/ml. (b) 1.52 µg/ml. (c) 0.41 µg/ml. (d) 0.328 µg/ml. (e) 0.455 µg/ml. (f) 1.98 µg/ml.   5. 1.72%.   6. Bands shift about 1 µm to longer wavelengths due to slowing down of allyl group vibration by a heavy atom.   7. (a) 2.00 × 10^{-4} M. (b) 5.00 × 10^{-4} M. (c) 1.70 × 10^{-3} M. (d) 1.24 × 10^{-4} M. (e) 5.90 × 10^{-5} M.   8. (a) Cell 1, 0.085 cm; cell 2, 0.13 cm; cell 3, 0.044 cm; cell 4, 0.022 cm. (b) 0.031 mm.   9. (a) meta. (b) para. (c) ortho.   10. p-Bromotoluene.   11. Hexachlorobenzene.   12. OH str (intermolecular H-bond) at 3380 cm^{-1}; CH str at 2940 cm^{-1}; CH$_2$ in-plane bend at 1460 cm^{-1}; nonsymmetrical breathing mode of phenyl ring at 690 cm^{-1}; out-of-plane bending of 5 adjacent hydrogen atoms at 740 cm^{-1}; C–O str at 1060 cm^{-1} of primary alcohol. Absent: C=O str at 1735 cm^{-1}; CHO str of aldehydes at 2720 cm^{-1}; C–CH$_3$ bend at 1380 cm^{-1}; C–O–C asymmetric stretching of ethers at 1182 cm^{-1}. Compound is benzyl alcohol.   13. 3-Methyl pentane.   14. Structure II.   15. Allyl cyanide.   16. Unusually strong overtone at 1848 cm^{-1} plus C=C group at 1650 cm^{-1} and vinyl group at 990 and 910 cm^{-1}; no methyl bending at 1390 cm^{-1} and no methylene rocking at 720 cm^{-1}. Thus, the limiting formula would have two terminal vinyl groups with no more than 3 methylene groups in between, and no chain branching. Compound is C=C–C–C–C=C.   17. p-Chlorobenzaldehyde.   18. (CH$_3$)$_3$C–CO–CH$_3$.   19. (CH$_3$)$_3$C–COOCH$_3$.   20. (CH$_3$)$_2$CH–CH$_2$–NH$_2$.   21. p-Cyanobenzaldehyde.

22.

Ph–C(t)=C(H)(H)–... CH$_2$–O–C(=O)–CH$_3$

23.

Ph–C(=CH$_2$)–CH$_3$

Chapter 8

1. There is a failure to appreciate the magnitude of the decrease in the efficiency of the photomultiplier tube coupled with the decrease in grating efficiency over the Raman range of 2960 cm^{-1} when the spectrum is excited by a He–Ne laser.   2. (a) 20 cm^{-1} per sec or 1200 cm^{-1} per min. (b) Scanning speed can be 13.4 cm^{-1}/mm so the slit can be

ANSWERS TO PROBLEMS   975

opened to 0.60 mm.   **3.** It is reduced by a factor of approximately 22.7   **4.** $(15,798/20,487)^4 = 0.35$.

5.

BENZENE, Δcm^{-1}	WAVELENGTH OF RAMAN LINES (nm) FOR EXCITING LINES				
	632.8 nm	488.0 nm	514.5 nm	568.2 nm	647.1 nm
606	6580	5029	5311	5885	6735
850	6688	5091	5380	5970	6848
991	6751	5128	5421	6021	6914
1176	6837	5177	5476	6089	7004
1584	7033	5289	5602	6244	7210
1605	7043	5295	5608	6252	7221
3047	7839	5732	6102	6872	8060
3063	7849	5738	6107	6879	8071

6. The C–H stretching frequencies at 3047 and 3063 cm^{-1} would overlap.   **7.** N_2O has no center of symmetry while CS_2 does have one. The structures must be N–N–O and S–C–S (type of bonds not intended to be indicated).   **8.** The molecule has a center of symmetry. If planar, the molecule would have three unpolarized Raman lines; if tetrahedral it would have but one. The spatial configuration is tetrahedral.

9.

UNKNOWN	VOLUME PERCENT, %		
	1,2,3-	1,2,4-	1,3,5-
A	33.3	33.3	33.3
B	40.0	26.0	34.0
C	25.0	37.0	38.0

10. Benzoyl chloride.   **11.** Dimethyl acetylene.   **12.** Nitrobenzene.

Chapter 9

1. 0.1240 Å; $Z = 87$.   **2.** (a) L_{II} to K, (b) M_{III} to K, (c) M_V to L_{III}.

3. and 4.

ELEMENT	K EDGE, eV	L_{III} EDGE, eV	$K\alpha_1$, Å
Al	1.559	0.073	8.34
Cr	5.990	0.599	2.30
Zr	18.0	2.22	0.786
Nd	43.51	6.22	0.333
W	69.66	10.20	0.209
U	115.9	17.17	0.126

5. For L_{III} spectra, $Z < 92$; for K spectra, $Z < 39$.   **6.** Operating voltage: 1300–1700 V.
7. Operating voltage: 1470–1545 V.   **8.** K edges of iodine and argon, respectively.

9.

CRYSTAL	Al	S	Ca	Cr	Mn	Co	Br	Sr	Ag	Mo	W
LiF	–	–	112.6°	69.4°	63.0°	52.6°	29.8°	25.1°	16.0°	20.4°	6.0°
CaF$_2$	–	115.8°	64.2°	42.4°	38.9°	32.8°	18.9°	16.1°	10.2°	12.9°	3.8°
EDDT	142°	71.8°	44.8°	30.2°	27.6°	26.5°	13.6°	11.4°	7.3°	9.7°	3.0°

10. Use of pulse height analyzer, regulation of voltage applied to source, use of appropriate filter, and incorporation of monochromator into the system.

11.
U, WT %	SLOPE, COUNTS SEC^{-1} %$^{-1}$ U	TIME (1.96σ), MIN
2	165	2.57
5	110	1.77
10	67	1.80
15	45	2.08
20	31	2.87

12. NaCl crystal; $2\theta = 143°36'$. Tube voltage: 3.5 kV. Counting times: background, 28 sec; sample, 84 sec for 0.4% S. 13. Topaz, 69.25°. LiF, 44.99°. Al, 38.48°. NaCl, 31.70°. CaF$_2$, 28.23°. Quartz, 26.65°. EDDT, 20.15°. ADP, 16.64°. Graphite, 13.19°. Gypsum, 11.64°. Mica, 8.87° (the minimum angle). 14. (a) Set base line at 6.5 V and window for 6 V. (b) Set base line at 15.5 V and window for 7.5 V. (c) Set base line at 17.5 V and window for 4.0 V. (d) Set base line at 27 V and leave window open (or set at 37 V). (e) Set base line at 9.0 V and window for 10.5 V. 15. (a) Use analyzing crystal of higher dispersion and/or finer collimation. Topaz will resolve these lines with a 0.48-mm collimator, whereas LiF fails. Use of Fe$K\alpha_1$ radiation would excite Cr spectrum but not that of Mn. (b) Maintain the X-ray tube voltage less than 10.4 eV. Use K edge of gallium salts as a selective filter. (c) Use of a thin sheet of nickel as a selective filter.

16. (a)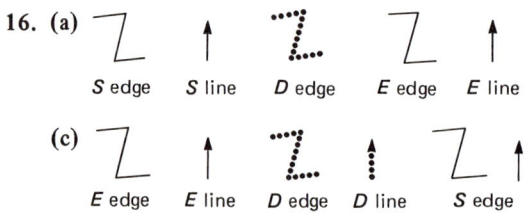

(b) Reverse positions of standard element and analysis element of part (a).

(c) (d) Reverse positions of standard element and analysis element of part (c).

17. (a) Sr: 16.10 kV. Y: 17.05 kV. (b) Sr: $K\alpha_1 = 25.2°$ and $K\beta = 22.4°$. Y: $K\alpha_1 = 23.8°$ and $K\beta = 21.2°$. (c) A: 0.250%. B: 0.177%. C: 0.060%. 18. $\Delta\lambda = 0.002$ Å. $\Delta\theta = 0.06°$. Use filter to reduce Zr$K\alpha_1$ intensity or, better, use a proportional counter with pulse height discrimination.

19.
	TOP LAYER			BOTTOM LAYER	
2θ	λ, Å	LINE	2θ	λ, Å	LINE
27.6°	2.53	Ti$K\beta$	34.0°	3.10	Ca$K\beta$
30.0°	2.75	Ti$K\alpha_1$	36.0°	3.28	Ca$K\alpha$
58.0°	5.15	Ti$K\beta$(2nd)	71.0°	6.17	Ca$K\alpha$(2nd)
63.0°	5.55	Ti$K\alpha_1$ (2nd)			

20. (a) 16 cm^2/g. (b) 52 cm^2/g. (c) 47 cm^2/g. 21. Decrease in intensity is 64% per 1-cm cell length vs. air, or 44% decrease vs. an octane blank. 22. Decrease in intensity is 97.5% per 1-cm cell length vs. air, or 59% decrease vs. an octane blank. 23. Sample 1, Ni$K\alpha_1$ on Mo$K\alpha_1$. Sample 2, Cu$K\alpha_1$ on Mo$K\alpha_1$. Sample 3, Au$L\beta_1$ on Ni$K\alpha_1$. Sample 4, Pd$K\alpha_1$ on Ni$K\alpha_1$.

24.

2θ:	111.0°	100.2°	57.8°	48.8°	45.1°	44.0°	40.4°
λ, Å:	3.320	3.090	1.947	1.664	1.545	1.509	1.392
Element:	Ni$K\alpha_1$(2nd)	Cu$K\alpha_1$(2nd)	Fe$K\alpha_1$	Ni$K\alpha_1$	Cu$K\alpha_1$	Ni$K\beta$	Cu$K\beta$

25. See G. L. Clark, Ed., *Encyclopedia of Spectroscopy*, Van Nostrand Reinhold, New York, 1960, pp. 704–711. 26. Elements with $Z < 25$, and especially good for Cl, P, and S in hydrocarbon matrices. See *Appl. Spectrosc.* 24. 557 (1970) and *Anal. Chem.*, 44 (14), 57A (1972). 27. 1.66 Å, Ni$K\alpha_1$; 2.10 Å, Mn$K\alpha_1$; 3.32 Å, Ni$K\alpha_1$(2nd); 4.20 Å, Mn$K\alpha_1$(2nd); 6.64 Å, Ni$K\alpha_1$(3rd). 28. 2.499 Å, 2.169 Å, and 1.539 Å. 29. A fully extended planar zigzag carbon chain is indicated since the C–C distance is essentially this value.

Chapter 10

1. 5.62×10^9 disintegrations/min.

2.

NUCLIDE	14 DAYS	A/A_0 FRACTION 30 DAYS	60 DAYS
^{32}P	0.506	0.233	0.054
^{131}I	0.299	0.0754	0.0057
^{198}Au	0.0272	4.45×10^{-4}	1.98×10^{-7}

3.

NUMBER OF COUNTS	P.E.(%)	1σ(%)	2σ(%)
3200	1.19	1.77	3.54
6400	0.84	1.25	2.50
8000	0.75	1.12	2.24
25,600	0.42	0.63	1.26
102,400	0.21	0.31	0.62

4. Dead time is 197 μsec; loss is 704 counts/min. 5. (a) 50%. (b) 80%. (c) 96%. (d) 99.9% efficiency. 6. For 99% counting efficiency, counting rate could not exceed 1% of reciprocal value of dead time. (a) 40,000 counts/sec. (b) 10,000 counts/sec. (c) 2000 counts/sec. (d) 37 counts/sec. 7. A_0 = 2250 counts/min; $t_{1/2}$ = 18 min; ^{80}Br. B_0 = 440 counts/min; $t_{1/2}$ = 264 min; ^{80m}Br. 8. E_{max} = 0.34 MeV; 11 mg/cm². 9. Range is approximately 574 mg/cm²; energy is 1.10 MeV. 10. 1.67 MeV. 11. μ = 0.0248 cm²/g; range (for 99.9% absorption) is 280 mg/cm²; energy is 0.714 MeV. 12. 200 mg. 13. 6.5 mg. 14. In air: 162 cm for ^{90}Sr and 910 cm for ^{90}Y. In iron: 0.25 mm for ^{90}Sr and 1.39 mm for ^{90}Y. 15. 0.0379N. 16. To absorb sodium activity, use aluminum, 634 mg/cm². 17. 71 min. 18. 0.26 μg. 19. 2.8 mg. 20. Cu: 62 sec; Mn: 96 sec; Ni: Impossible unless cooling period shortened or neutron flux increased tenfold; Co: 9.0 days. 21. In the 62 sec required to irradiate the sample for copper, only manganese concerns us. It will take 34.5 hr for manganese to decay to 1 count/min. An aluminum absorber (0.52 mg/cm²) will remove all manganese radiation which is the most energetic of the beta radiations. 22. Nuclides observed: ^{22}Na, ^{47}Ca, ^{59}Fe, ^{60}Co, ^{65}Zn, ^{85}Sr, ^{86}Rb, ^{95}Zr, ^{95}Nb, ^{103}Ru, ^{124}Sb, ^{134}Cs, ^{140}Ba, and ^{140}La. 23. Arsenic and antimony are present, typical of lead bullets.

Chapter 11

1. Bearing in mind the 1:2 relationship of the enolic hydrogen to the keto methylene group: % enol = [37.0/(37.0 + $\frac{1}{2}$ × 19.5)] × 100 = 79.1%.   2. As benzophenone contains 5.53% hydrogen, the sample contains: (0.8023/0.3055)(184/228)(5.53%) = 11.72%.
3. Since excess phenol was used, the average number of methylene bridges per chain is one less than the phenolic nuclei. The average numbers of aromatic and methylene protons per molecule are (3n + 2) and 2(n − 1), respectively, where n is the number of monomer units. One extra proton would be on each terminal phenolic group. Thus, (3n + 2)/2(n − 1) = 30/18 and n = 16. Average molecular weight is 1684 = (16 × 92) + (15 × 14) + 2.   4. The three signals are attributable to the methyl group in o-cyanotoluene, the methylene group in o-cyanobenzyl chloride, and the methine proton in o-cyanobenzal chloride. The relative molar proportions are 13/3, 20/2, and 10/1; and the proportions by weight are 1.0:3.0:3.7.   5. CH$_2$=C–CH$_2$ 6. In compound I the CH$_3$ resonance
$\quad\quad\quad\quad\quad\quad\quad\quad\quad\quad\quad\quad\quad\quad\quad\quad\quad$ | |
$\quad\quad\quad\quad\quad\quad\quad\quad\quad\quad\quad\quad\quad\quad\quad\quad\quad$ O–C=O
would be a doublet with J = 6 Hz, showing that there was a proton attached to the adjacent carbon atom. In compound II the CH$_3$ resonance, in addition to being further downfield, would be coupled very weakly with J ≃ 1 Hz.   7. HPO(OH)$_2$; H$_2$PO(OH).   8. First structure.   9. Second structure.   10. Structure III with intensity ratios of 4:3:1. Note that the methylene protons are not precisely equivalent, since two of them are *cis* to the methyl group and two are *trans*.   11. Peak at 7.3 δ assigned to olefinic proton in structure A and the peak at 6.7 δ to the olefinic proton in structure B. Isomer A: (50/72) × 100 = 69.4 mol %; and isomer B: (22/72) × 100 = 30.5 mol %.
12. O$_2$N–⟨⟩–CH(CH$_2$CH$_3$)–COOH   13. Ditolyl disulfide.   14. ⟨⟩–CH$_2$–C(CH$_3$)$_2$–Cl
15. CH$_3$O–⟨⟩–⟨⟩–S   16. CH$_3$–CH$_2$–CH(Br)–COO–CH$_2$–CH$_3$
17. Ethylchloroacetate.   18. CH$_3$–CH$_2$–O–C(=O)–CH$_2$–CH$_2$–C(=O)–O–CH$_2$–CH$_3$
19. HC≡C–CH$_3$ 20. On the delta scale, benzene is located at (418.4/60) = 6.97. The olefinic protons are located at 6.97 − (56.9/60) = 6.03 and at 6.97 − (87.5/60) = 5.52. Each is weakly coupled with the other (J = 1 Hz) and with the methyl protons. The methyl protons at 6.97 − (304.6/60) = 1.90 are coupled to two nonequivalent olefinic protons. Singlet at 3.69 is CH$_3$–O group.   21. CH$_3$–CH$_2$–O–⟨⟩–N(H)–C(=O)–CH$_3$

22. Use 4X multiplier yielding an output of 20 MHz plus a 0.7-MHz incremental oscillator; mix the two outputs in a single sideband modulator and select the lower sideband: 20 − 0.7 = 19.3 MHz.   23. If H_0 were varied, the effect would be the same as varying

the frequency of both the observing rf field, H_1, and the perturbing rf frequency, H_2, simultaneously.   24. (a) 1.988. (b) 0.126.

Chapter 12

1. H_0 = 12,500 G; ΔE = 2.3 × 10^{-16} erg/molecule.   2. $\Delta t < 10^{-6}$ sec.   3. Copper(II) and silver(II).   4. Spectrum 1 from dihydrofumaric acid; spectrum 2 from ascorbic acid; and spectrum 3 from reductic acid.   5. Spectrum A from semiquinone itself; B, trichloro-; C, monochloro-; D, 2,3-dichloro-; and E, tetrachloro-derivative.   6. (a) $\dot{C}H_3$, four peaks in 1:3:3:1 intensity ratio; (b) $C_6H_6^-$, seven lines; 1:6:15:20:15:6:1; (c) $(CH_3)_3\dot{C}$, ten lines, intensity ratio 1:9:36:81:126:126:81:36:9:1; (d) a quartet from CH_3 with coupling constant of 65 G which is further split into triplets with coupling constant of 50 G due to CH_2: the CH_3CH_2 · radical; (e) $\dot{C}H(OH)COOH$ radical with a_{CH}^H = 15.3 G and a_{OH}^H = 2.5 G; (f) X = CH_2Y.   7. Lines 2 and 4 result from proton splitting; each line in turn is split into three components due to interaction with nucleus X; line 3 results from overlap of triplet wings.   8. The predominant radical is $\dot{C}H(COOH)_2$ with a smaller amount of CH_2COOH.   9. (a) SO_2^- in equilibrium with $S_2O_4^{2-}$. (b) SO_2^- contains no nuclei with magnetic moments. (c) Enrichment with ^{33}S ($I = \frac{3}{2}$) would give a quartet (1:3:3:1) with SO_2^-.   10. Three lines arise from hyperfine splitting by ^{14}N; the additional lines arise from splitting by ^{15}N and their intensity corresponds to natural abundance of ^{15}N.   11. The principal intermediate is 2,6-di-t-butyl-4-methylphenoxy radical. The primary quartet arises from the three protons of the 4-methyl group (a_i = 11); the closely spaced triplets (a_i = 1.8) result from weaker coupling of electron with two *meta*-protons on the ring.   12. $-CH_2-\dot{C}H-CH_2-$; four equivalent protons are being split by the methine proton. The difference between the alpha proton and the four beta protons becomes observable. This interpretation assumes that the α- and β-hydrogens couple equally with the unpaired electron which means that all four β-hydrogens must be equally inclined to the carbon *p* orbital. When the stretched polyethylene samples were arranged with the direction of stretch perpendicular to the magnetic field, the polymer molecules were sufficiently oriented for the difference between the α-hydrogen and the four β-hydrogens to become observable and a spectrum of five doublets was obtained.   13. Larger triplet arises from electron coupling with protons 2 and 3 (a_i = 3); each triplet is split into five lines due to protons 5, 6, 7, and 8 (a_i = 0.3).   14. A quintet of quintets with the coupling of the alpha protons about three times larger than the coupling of the beta protons.   15. The radical $(CH_3)_2\dot{C}-OH$. The small hydroxyl–hydrogen splitting, likely to have considerable anisotropy, is not resolved but is responsible for the linewidths.   16. Spectrum A is from 2-methylcyclohexanone; spectrum B is from 4-methylcyclohexanone; spectrum C, a mixture of approximately three parts of B to one part of A, is from 3-methylcyclohexanone.   17. Structure I would give a quartet of triplets, whereas structure II would give a triplet of triplets. Coupling of methyl protons is 10.7 G, slightly larger than the 9.0 G observed for the methylene protons of the ethyl group.   18. Compound A is a triplet of triplets; compound B is a triplet; compound C is a septet of triplets; and compound D is a quartet of triplets.   19. In compound A, the *p*-methyl group was oxidized to COOH or CHO; the proton is in the plane of the ring and spin polarization will not be induced. In compound B the 4-ethyl group is oxidized to acetyl. In compounds C and D the 4-methyl group is oxidized to COOH or CHO.

Chapter 14

1. Allyl alcohol.   2. 26.28.   3. 1.552 (calculated).

4.

COMPOUND	SPECIFIC REFRACTION	MOLAR REFRACTION
Benzene	0.3334	26.05
Ethanol	0.2794	12.87
Ethyl acetate	0.2511	22.12
Toluene	0.3356	30.92
Nitrobenzene	0.2642	32.53
Water	0.2061	3.712

5. 66.7% D_2O and 33.3% H_2O.   6. 1.5415.   7. 8.41% by volume in the wine.   8. +66.50°.
9. 96.79%.   10. 55.77°.   11. 50%.   12. $[\theta] = \dfrac{3305}{bC} \log \dfrac{P_d}{P_l}$ 13. $N_d - N_l = 1.75 \times 10^{-7}$.   15. $[\theta] = -0.70$.

Chapter 15

1. and 2.

	VALUES OF k'			
	1	2	5	10
t_R, sec	40	60	120	220
V_R, ml	2	3	6	11
W_b, ml	0.44	0.66	1.32	2.41
Peak capacity	4	6	9	12
Zone velocity, cm/sec	1.25	0.83	0.42	0.23
Retardation factor	0.50	0.33	0.17	0.091

3. (a) k' values: toluene, 0.19; biphenyl, 0.61; anisole, 1.23; nitrobenzene, 4.08.

	COLUMN 1	COLUMN 2	COLUMN 3
Average linear velocity, cm/sec	0.48	0.40	0.56
Plate height, mm	0.16	0.030	0.018
Reduced velocity	32	13.3	12.2
Reduced plate height	7.8	3.0	2.8
Reduced column length	25,000	13,500	13,800
Free column volume, V_M, ml	1.18	1.99	1.33
Flowrate, F_c, ml/min	0.68	3.51	4.97

4. (a) $R = \text{const}\,[k'/(1 + k')]$. (b) $R = \text{const}\,[(\alpha - 1)/\alpha]$. Graph these equations.   5. In each instance resolution deteriorated from 1.3 to 0.9. In addition, with $N = 1600$ the peak width is increased. When k' is lowered, the peaks emerge earlier, the peak widths are less, but base line resolution is not achieved.   6. (a) Decrease carrier velocity; this means longer analysis time. Or, increase the mobile phase velocity provided that the column length is increased proportionately. (b) Decrease amount of stationary phase, but not to point where adsorption on the naked support begins to adversely affect plate

height. (c) Decrease size of support particle. Packing may become a problem to avoid wall effects and packing irregularities. Permeability decreases and higher inlet pressures are needed to drive the mobile phase. (d) In LC use a mobile phase having low viscosity; this also means lower inlet pressure. (e) Use a support where likelihood of large stagnant pockets of mobile phase are minimized, such as pellicular packings or microspheres with small pores.

Chapter 16

1. Volumetric flow at sample entrance and through sample loop is only 10 ml/min or 6 sec/ml. If loop volume is increased to 5 ml, it will take 30 sec to sweep sample into the column. No peak is narrower than 30 sec.   2. 324 ml.   3. (a) B-term is proportional to D_M, and C_{mobile}-term is function of D_M^{-1}. (b) See Fig. 16-15; $\bar{u}_{opt}$ (He) greater than $\bar{u}_{opt}$ (N$_2$). (c) Helium.
4. (a) and (c)

COLUMN LOADING	$\bar{u}$, cm/sec	H, cm	t_M, min	COLUMN LOADING	$\bar{u}$, cm/sec	H, cm	t_M, min
31%(H$_2$)	1.00	0.46	6.00	23%(H$_2$)	1.43	0.335	4.20
	1.65	0.34	3.63		2.81	0.243	2.14
	3.22	0.31	1.86		4.15	0.208	1.45
	6.16	0.41	0.97		5.43	0.218	1.10
	8.80	0.52	0.68		6.64	0.228	0.90
	13.28	0.73	0.45		12.26	0.303	0.49
13%(H$_2$)	2.49	0.247	2.41	23%(N$_2$)	0.71	0.176	8.45
	4.78	0.169	1.26		1.38	0.135	4.35
	6.86	0.161	0.88		2.72	0.127	2.21
	10.51	0.170	0.57		6.12	0.176	0.98

(b), (c), (e), (g)

COLUMN LOADING	B-TERM	C-TERM	$\bar{u}_{opt}$, cm/sec	H_{min}, cm	k'	$\bar{u}$ RANGE (cm/sec) FOR 90% EFFICIENCY
31%(H$_2$)	0.357	0.0479	2.73	0.307	5.61	1.6 to 4.6
23%(H$_2$)	0.376	0.0190	4.45	0.214	3.64	3.1 to 7.5
13%(H$_2$)	0.458	0.0072	7.99	0.160	2.44	4.2 to > 13
23%(N$_2$)	0.084	0.019	2.11	0.125	3.61	1.3 to 4.0

(f) $\bar{u}_{opt}$ increased as loading decreased; effective liquid film thickness decreased. (h) H = 0.240 cm and $\bar{u}$ = 7.5 cm/sec. 0.59 normal time (or 1.68 times faster).
5. (a) and (c)

$\bar{u}$, cm/sec	t_M, min	t_R, min	t'_R, min	N_{eff}	N_{theor}	N/t_R
6.5	3.85	31.7	27.9	12,500	14,100	445
12.2	2.05	16.9	14.9	16,500	18,600	1100
19	1.32	10.9	9.54	16,100	18,200	1680
27	0.93	7.64	6.71	13,900	15,700	2050
43	0.58	4.80	4.22	10,100	11,400	2380

(b) $\bar{u}_{opt} = 14.5$ cm/sec; $H_{min} = 0.90$ mm.

6.

COLUMN		1	2	3	4
k'	1-MeN	68	48	61	39
	2-MeN	61	43	55	35
N_{eff}	1-MeN	1970	2340	1810	2080
	2-MeN	2240	2380	1740	1460
N_{req}	$R = 1.0$	1440	1570	1340	1410
	$R = 1.5$	3230	3530	3010	3170
Resolution		1.3	1.3	1.2	1.2
Stationary phase, ml		6.5	1.6	0.44	0.065
Phase ratio		2.3	11.1	51	145
t'_R, min	1-MeN	4.8	4.5	3.2	2.4
	2-MeN	4.3	4.1	2.9	2.1
Relative retention		1.1	1.1	1.1	1.1

(d) Shortest analysis time obtained on column 4. Same would be true after increasing the column length 2.25 times to achieve resolution of 1.5.   **7. (a)** By definition nonane is 900, decane is 1000, undecane is 1100, and dodecane is 1200. Substitution in expression: $y(R.I.) = a + b \ln x(t_R)$ gives these R.I.: toluene, 791; 1-heptanol, 756; 4-octanol, 973; 1-octanol, 1056. Predicted R.I. are: 1-nonanol, 1156 and 1-decanol, 1256. **8. (a)** $t_M = 0.45$ min. **(b)** R.I. $= 1030 + 153.2 \ln t'_R$. For n-C$_{14}$, $t'_R = 11.15$ min.

(c), (d)

ALKANE	k'	N_{theor}	N_{eff}
n-C$_8$	0.49	44,300	4770
n-C$_9$	0.96	42,900	10,300
n-C$_{10}$	1.82	41,600	17,300
n-C$_{11}$	3.51	42,000	25,400
n-C$_{12}$	6.73	41,600	31,600
n-C$_{13}$	12.82	39,900	34,300

9. (a) R.I. $= 371 + 94.3 \ln t'_R$. R.I. values are: propane, 300; n-butane, 400; *trans*-butene-2, 466; *cis*-butene-2, 476; n-pentane, 500; 2-methylpentane, 580. **(b)** $t'_R = 11.33$ for n-hexane. **(c)** $N_{eff} = 2660$; $H = 0.56$ mm. **(d)** $R = 1.0$. **(e)** $k' = 7.8$ for *cis*-butene-2

ANSWERS TO PROBLEMS   983

and $k' = 23.5$ for 2-methylpentane. (f) $N_{req} = 3460$. (g) Increase column length 2.25-fold. Adjust carrier gas velocity to optimum value if not already using this value. Change to helium as carrier gas. (h) $\bar{u} = 6.4$ cm/sec. 10. (a) $N_{req} = 2800$, which is also theoretical plate number. Column lengths are 157 cm for case A and 303 cm for case B. (b) $H/\bar{u} = 3.1 \times 10^{-3}$ for case A and 1.8×10^{-2} for case B. Corresponding retention times are 21.0 sec and 12.2 sec, respectively. Thus almost twice the column length is required for case B to maintain the same plate number at the higher average linear velocity, but the analysis time is only 0.58 as long. (c) t_M is 8.72 sec for case A and 5.05 sec for case B. (a) Carbon No. $= -7.12 + 2.18 \ln V'_R$ for acetates on Carbowax. Carbon No. $= -2.20 + 1.09 \ln V'_R$ on Nujol. Corresponding equations for alcohols are: $C\# = -5.22 + 1.54 \ln V'_R$ (Carbowax) and $C\# = -1.23 + 1.21 \ln V'_R$ (Nujol).

12. (a) ΔH values, kcal/mol:

COLUMN	CH_2Cl_2	$CHCl_3$	CCl_4	$CCl_2=CCl_2$	C_6H_5Cl	C_6H_6
Paraffin	6.96	7.16	7.57	9.07	9.12	6.67
Tricresyl phosphate	7.50	8.23	7.50	8.82	9.42	7.18
Carbowax 4000	8.81	9.35	7.62	8.97	10.1	7.95

(b) $K_d = V_g \rho$. Specific retention volumes, given at column temperature per gram of liquid phase, need only be multiplied by density of liquid phase at given temperature to obtain K_d. (c) Values of V_g at 150°C for the solutes (left to right as listed): 7.0, 12.0, 7.5, 16.2, 39.9, and 11.1. (d) Values range from 19°–33°C with the median about 24°C. 13. For $t_R = 2.29$ min, $T_c \leqslant 514°K (241°C)$. 14. Yes; $t_R = 5.91$ min at $241°C (515°K)$. 15. (a) 120°C or less. (b) 952 cm. (c) 1.11 kcal.
16.

	PACKED COLUMN				WCOT COLUMN				SCOT COLUMN		
k'	N_{req}	L, cm	t_R, sec	k'	N_{req}	L, cm	t_R, sec	k'	N_{req}	L, cm	t_R, sec
1	17,400	1045	348	0.17	212,700	12,760	1241	0.33	69,800	4188	310
2	9800	588	294	0.33	69,800	4190	465	0.67	27,200	1630	151
3	7740	465	310	0.5	39,200	2350	294	1	17,400	1045	116
6	5930	356	415	1	17,400	1045	174	2	9800	588	98
12	5110	307	664	2	9800	588	147	4	6800	408	113
30	4650	279	1440	5	6270	376	188	10	5270	316	193

For minimum analysis time on each column, $k' = 2$.

17. (a)

		n-C_7	n-C_8	n-C_9	n-C_{10}	n-C_{11}
120°C	N_{theor}	822	866	964	958	1120
	N_{eff}	152	281	472	621	860
140°C	N_{theor}	1100	866	837	1040	1140
	N_{eff}	144	200	316	545	750
160°C	N_{theor}	1220	820	810	780	860
	N_{eff}	84	112	198	290	440
Retention temp., °C		84	102	124	143	163

Start program at room temperature; use a heating rate of 10°–12°/min. **(c)** Predicted retention temperatures: 183°C for n-C_{12}; 203°C for n-C_{13}; and 223°C for n-C_{14}. **18.** Estimated values: 58°C for methyl; 74°C for ethyl; 96°C for propyl; 107°C for butyl; and 127°C for pentyl acetate. Heating rate: 16°C/min. **19. (a)** 15.2 min and 8.81 min. **(b)** $R = 3.72$ and 2.75, more than adequate. Only a 7.45-m column (13,300 plate) required at lower velocity; at the higher velocity a 13.6-m column (14,130 plates) suffices. **(c)** $R = 14.5$ and 11.2. A column 0.49 and 0.82 m, respectively, will suffice. **20.** Peaks 2, 4, 6, and 9 are n-C_1 to C_4 formate esters; peaks 3, 5, 8, and 12 are the n-C_1 to C_4 acetates; peaks 7 and 11 are the iso-C_3 and iso-C_4 acetates; and peak 10 is sec-butyl acetate.

21.

SAMPLE	BENZENE	PERCENT p-XYLENE	TOLUENE
1	27.4	42.5	30.2
2	37.8	40.0	22.2
3	41.2	24.8	34.0
4	21.7	35.0	43.3
5	42.7	30.7	26.6

22. (a) C_5, 8.3%; C_6, 14.2%; C_7, 12.7%; C_8, 16.4%; C_9, 21.3%; C_{10}, 27.2%. **(b)** $R = 0.86$. For $R = 1.5$, the column length should be increased 3.0 times. **(c)** Relative retention should be increased from 1.31 to 2.00; column temperature should be decreased to 162°C.

Chapter 17

1. Resettling of packing with use, creating a void at top of column. Broad peaks with poor symmetry can result. **2.** $L/d_p = 20,000$ or $L = 20,000\, d_p$.

d_p, μm:	1	2	3	4	5	7	10	12	15	20
L, cm:	2.0	4.0	6.0	8.0	10.0	14.0	20	24	30	40

3. (a) The higher the pressure, the greater the number of plates per unit time, whatever the column. **(b)** Leads to too high an inlet pressure and too short an analysis time. **4.** $V_w = W_t V_{col} \epsilon_{total}/t_M$. $V_w = 4 V_{col} \epsilon_{total}(1 + k')/N^{1/2}$. **5.** 17% decrease. **6. (a)** 2.0 ml. **(b)** 5 mm. **7.** Column A: 10.4 μl; column B: 14.9 μl; column C: 9.3 μl.

8.

ELUENT	t_M, sec:	10	OPTIMUM PARTICLE DIAMETERS, μm 30	60	120	300
1-Propanol		0.7	1.2	1.6	2.3	3.7
Water		1.2	2.1	3.0	4.2	6.7
n-Hexane		2.1	3.7	5.2	7.4	11.7

9.

ELUENT	t_M, sec:	10	PRESSURE DROP (ΔP), IN ATM 30	60	120	300
1-Propanol		804*	268	134	67	26.8
Water		352	117	59	29.?	11.7
n-Hexane		123	41.1	20.6	10.3	4.1

For $\phi = 500$, divide all tabulated values by 2.

*Not achievable with usual instruments. $\Delta P \simeq 100$ atm is maximum pressure on "low pressure" chromatographs.

10.

	PARTICULAR DIAMETER, μm								
	2	3	4	5	7	9	10	11	12	15
L, cm	3.67	4.13	4.76	5.65	8.54	14.0	18.0	24.1	31.7	83.4
ΔP, atm	33.1	18.6	14.0	12.6	14.7	23.9	31.9	47.3	68.7	305

12.

ΔP	30 ATM				200 ATM			
d_p, μm	3	5	7	10	3	5	7	10
t_M, sec	22.5	34.5	51	81	8.7	17.4	28.7	50
L, cm	3.5	7.3	12.4	22.3	5.6	13.3	23.9	45

13. and 14.

	PARTICLE DIAMETER, μm							
	3	4	5	7	10	12	15	20
N	8280	8880	8510	7100	5280	4410	3440	2420
v	1.8	3.2	5.0	9.8	20.0	28.8	45	80
h	2.42	2.25	2.35	2.82	3.79	4.54	5.81	8.27
L, cm	6	8	10	14	20	24	30	40
$\bar{u}$, cm/sec	0.060	0.080	0.10	0.14	0.20	0.24	0.30	0.40
H, mm	0.0073	0.009	0.012	0.020	0.038	0.054	0.087	0.165

15. At high velocities the reduced plate height equation approximates to $h = Cv$, or $h/v = C$. Now $C = t_M D_M / N d_p^2$, thus $N = t_M D_M / C d_p^2$. Under conditions required to obtain N_{max}, pressure drop tends to infinity.

16.

	PARTICLE DIAMETER, μm					
20 ATM	3	4	5	7	10	15	20
L, cm	3.94	4.52	5.78	9.14	15.75	30.30	49.20
t_M, sec	85.2	62.9	65.7	83.9	122.3	201	298
100 ATM							
L, cm	3.84	5.90	8.40	14.56	26.6	54.0	90.1
t_M, sec	16.2	21.4	27.8	42.6	69.7	128	200

17. At low velocities, $hv = B$. Thus, $N_{max} = \Delta P d_p^2 / B \phi n D_M$. Under conditions required to obtain N_{max}, t_M becomes infinite. 18 (a) Reduced velocity is 67, from which reduced plate height is 7.44. Now $(R_2/R_1)^2 = h_1/h_2$ at constant column length, so $h_2 = 3.05$ and v_2 is 12.0 (also 1.0 on low-velocity side of h/v minimum). Finally, $(\Delta P_2/\Delta P_1) = v_2/v_1$) or 5.6-fold less (but $t_{R,2}$ is 5.6 times longer). (b) $(R_2/R_1)^2 = L_2/L_1 = 2.44$ times longer column; however, $t_{R,2}/t_{R,1} = (L_2/v_2)/(L_1/v_1) = 13.6$ and $t_{R,2}$ is 13.6 times longer. 19. Volume is 11.4 ml for maximum resolution and 9.4 ml for minimum analysis time at both flowrates; however, the time interval for each solvent is 8 min ($F_c = 1$ ml/min) and 4 min ($F_c = 2$ ml/min), respectively. 20. At higher temperatures, viscosity, and therefore pressure drop, would be reduced while efficiency would be increased.

Chapter 18

1. Due primarily to insufficient deactivation of the silica or alumina surface of the column.   2. Hexane and pentane are identical in eluent strength. Either 33:67 isopropyl chloride/hexane or 9:91 diethyl ether/hexane would provide the same $e°$ value among mixed solvents. Pure CS_2 also has the same value.   3. Slower.   4. Increase the "stronger" solvent content, in this case acetonitrile, in steps of 10% to a 40:60 ratio, and so on.   5. Bonded phase functioned in the normal mode since the more polar solvent, water, reduced the competitive interaction between the sugar and the polar bonded phase.   6. (a) Surface sulfonated styrene/DVB resin in separating column, $0.01F$ HCl as eluant, and Dowex 1X8 anion resin (OH⁻ form) in stripping column. Resin–OH⁻ + HCl → Resin–Cl⁻ + H_2O. (b) Surface sulfonated styrene/DVB in separating column, $0.002F$ $AgNO_3$ plus $0.0004F$ HNO_3 as eluant, and a dual stripper: Dowex 1X8 Cl⁻ in first column and Dowex 1X8 OH⁻ in second column. Resin-Cl⁻ + $AgNO_3$ → Resin–NO_3^- + AgCl(s); Resin–OH⁻ + HNO_3 → Resin–NO_3^- + H_2O. (c) Anion-exchange resin in carbonate form in separating column, $0.003F/0.0024F$ HCO_3^-/CO_3^{2-} as eluant, and cation-exchange resin (H⁺ form) in suppressor column. Resin–H⁺ + HCO_3^- (or CO_3^{2-}) → Resin–Na⁺ + CO_2 + H_2O.   7. $n = (V_B/V_A)(C_B/C_A)(1/K_y^x)$, where K_y^x is the selectivity coefficient of the ion x, which, in the series to be analyzed, has the greatest affinity for the separating resin relative to the eluting ion y.   8. With a methanol/water (actually 45/55 ratio) solvent containing heptyl sulfonic acid (pH about 3.5), one ensures complete ionization of the base and that the weak acid will be in its nonionic form. Thus, aspirin can be separated by ionic suppression while the ionized base is separated by RP–IPP.   9. A small counter ion would be most effective. Actually ammonium ion and TMA yield very similar resolution, but TBA displays a significant loss of resolution because its larger size dominated retention of the ion pair. That is, small chromatographic differences were masked by the more lipophilic TBA counter ion.   10. (a) By ion exchange one could use a cation-exchange column (low pH) and base the separation on the RNH_3^+ groups. Or at high pH and an anion-exchange column, the separation could be based on the RCOO⁻ groups. Cation exchange is preferred because the site of ionization is closer to the side chain of the amino acid, particularly if steric effects can play any role. (a) By RP–IPP, the amino acids can be chromatographed using either a quaternary amine (to pair with the RCOO⁻) at pH 7.5, or an alkyl sulfonic acid (to pair with the RNH_3^+) at a low pH.   11. The HCl forms $CuCl_4^{2-}$ and $CuCl_3^-$ species which will not bind to the cation-exchange sites.   12. Use a reducing agent (perhaps H_2O_2) in a dilute HCl/NaCl regenerant. Trivalent chromium will form and being positively charged will not be retained on the positive sites of the anion-exchange resin.   13. (a) $(D_c)_A = 7.0$. $(D_c)_B = 21.0$. (b) $\alpha = 3.0$.

14.

[H⁺], Aqueous Phase	$(D_c)_K$	$(D_c)_{Mg}$	Relative Retention
0.10	0.074	1.4×10^{-3}	0.1
0.50	0.015	5.6×10^{-5}	0.5
1.00	0.0074	1.4×10^{-5}	1.0
3.00	0.0025	1.6×10^{-6}	3.0

15. (a) In log M.W. units, range is 5–7. (b) $V_M = 3.55$ ml; $V_S = 2.38$ ml. (c) $V = 3.55$ ml ($K = 0$) and $V = 5.93$ ml ($K = 1$). (d) #1, 0.10; #2, 0.38; #3, 0.63; #4, 0.95. (e) R = 1.39 (peaks 1 and 2). (f) 2547 plates per 25 cm and 10,200 plates/meter. (g) Void porosity is 0.47; total porosity is 0.79.   16. Injection of an isotopically labeled solvent.

V_S corresponds to the subtraction of the void volume from the elution volume of the solvent. Labeled solvent can be detected by differential refractometry, mass spectrometry, or perhaps liquid scintillation counting.

Chapter 19

1. 2490 V for mass 18 and 224 V for mass 200.   2. 9.03 μsec for mass 44 and 8.9 μsec for mass 43.   3. 480.493.   4. For the pair, tridecyl benzene and phenyl undecyl ketone, resolution required is 7140; for 1,2-dimethyl-4-benzoyl naphthalene and 2,2-naphthyl benzothiophene, resolution required is 9320.   5. (a) 10,000. (b) 6500. (c) 10,000.   6. Mass = 237.1473 ± 0.0005; $C_{12}H_{19}N_3O_2$.   7. (a) $C_9H_8O_3$. (b) C_8H_8O. (c) $C_{14}H_{12}$. (d) $C_5H_6N_2$. (e) C_6H_7NO. (f) $C_6H_3Cl_2NO_2$. (g) $C_{13}H_{11}N$. (h) $C_{17}H_{18}O_5S$. (i) $C_9H_7NO_3$.   8. (a) C_3H_6NS; C_4H_8S. $C_3H_6NO_2$; $^{13}CC_3H_7S$. (b) The hydrogen 3 carbon atoms from the carbonyl group is transferred to the carbonyl oxygen (McLafferty rearrangement) with simultaneous cleavage of the C-2, C-3 bond. m/e 75.0267 is $CH_3-S-CH_2-CH_2^+$.   9. CH_3I; only one carbon is indicated plus a heavy monoisotopic element.   10. (a) 3 chlorine atoms. (b) 5 bromine atoms. (c) 4 bromine atoms. (d) 1 chlorine and 1 bromine atom. (e) 2 chlorine and 1 bromine atoms.   11. 90(P): 1 sulfur atom plus mass 58; therefore, $C_4H_{10}S$ because 5.61 − 0.78 = 4.58 and 4.58/1.08 = 4C's, or 58/14 = 4CH_2 groups plus 2H's. Compound is a dialkyl sulfide. 89(P): 1 chlorine, 1 nitrogen, plus residual mass 40. Since P + 1 indicates not over 2 carbon atoms, probable formula is C_2ClNO. 206(P): 2 sulfur atoms plus mass 142. P + 1 is 12.5% which indicates not over 10 carbon atoms [12.5 − 2(0.78) = 10.9]. Formula is $C_{10}H_{22}S_2$. 230(P): 2 bromine atoms plus mass 72; 72/14 = 5CH_2 groups plus 2H's. Formula is $C_5H_{12}Br_2$. 140(P): 1 sulfur atom. P + 1 is 9.54 − 0.78 = 8.76; therefore, not over 7 carbon atoms. Checking Table 19-3 uncovers possibility that 2 nitrogen atoms might be present. 151(P). 1 chlorine and 1 nitrogen atom; probably aromatic compound since base peak is parent peak. Residual mass is 102, and 102/12 = 8C's plus 6H's. Formula is C_8H_6ClN.   12. Appropriate masses: 32, 39, 46, and 59.

13.

UKNOWN	MeOH, %	EtOH, %	n-PrOH, %	i-PrOH, %
A	0.9	8.2	11.7	79.0
B	2.5	13.5	6.9	77.0
C	7.1	7.8	52.7	31.4

14. Metastable at 45.0 indicates the parent ion decomposes to mass 91 with loss of mass 93 neutral fragment(s). Metastable at 46.5 indicates that mass 91 decomposes further to mass 65 plus a neutral fragment of mass 26. These data, coupled with the peak intensities, indicate that we are dealing with a ring compound possessing the structure $C_6H_5-CH_2$ plus mass 93. Mass 93 is probably C_6H_5O. Molecule is tolylphenyl ether.   15. Metastable at 69.4 indicates the decomposition route is from mass 122 to mass 92 plus a neutral fragment of mass 30. Metastable at 46.5 indicates the decomposition of mass 91 to mass 65 plus a fragment of mass 26. Ion peak at mass 92, an even mass arising from an even mass molecular ion, suggests a rearrangement reaction. Coupled with the possible transition from mass 122 to 91, involving a loss of mass 31, this suggests the presence of a CH_2OH group which could participate in a McLafferty rearrangement to account for loss of CH_2O. Base peak at mass 91 suggests a tolyl group. Compound is 2-phenyl ethanol.   16. Metastable peaks indicate these transitions. 147.9: $200^+ \rightarrow 172^+ + 28$ (prob-

able loss of $CH_2=CH_2$). 121.7: $200^+ \to 156^+ + 44$ (probable loss of $CH_2=CH-OH$). 106.3: $108^+ \to 107^+ + 1$ (loss of H). 67.9: $173^+ \to 108^+ + 64$ (probable loss of SO_2). 53.5: $155^+ \to 91^+ + 64$ (probable loss of SO_2). 46.5: $91^+ \to 65^+ + 26$ (loss of $HC\equiv CH$ from tolyl). One sulfur atom is indicated by the $P + 2$ peak. The loss of $CH_2=CH_2$ and $CH_2=CHOH$, both as a result of rearrangement processes, plus the loss of mass 45 in the transition from mass 200 to mass 155, shows the presence of an ethoxy group which is linked to a sulfoxide group as shown by the subsequent loss of SO_2 from mass 155. Base peak at mass 91 suggests a tolyl group. These deductions account for the molecular weight of 200 and the presence of not more than 9 carbon atoms ($P + 1$ abundance). The compound is $C_2H_5-OSO_2-C_6H_5-CH_3$; ring substitution cannot be ascertained. **17.** 88(P): butyric acid. 86(P): methyl acrylate. 134(P): 3-phenyl propanaldehyde. 152(P): Formula is $C_8H_8O_3$; methyl salicylate. **18.** Metastable peaks indicate these transitions. 102.2: $153^+ \to 125^+ + 28$. 100.2: $151^+ \to 123^+ + 28$. 54.7: $103^+ \to 75^+ + 28$. All mass 28 losses involve ethylene fragment from an ethoxy group.

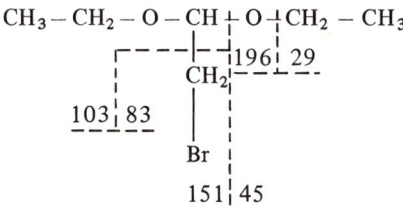

19. CH_3CCl_3. Apparent loss of 20 mass units (117 − 97) cannot be correlated with any known fragment except HF. However, mass 117 contains 3 chlorine atoms, and 117 − 3(35) leaves only 12 mass units, or 1 carbon atom. Thus, HF is ruled out and mass 117 is CCl_3. Now mass 97 contains 2 chlorine atoms, yet cannot arise from mass 117 by loss of 1 chlorine atom. Mass 97 is CH_3CCl_2. Thus, compound is CH_3CCl_3; no molecular ion peak is present. **20.** From empirical formula, two unsaturated bonds and/or rings are indicated. Mass 114, a rearrangement peak, arises from loss of water, indicating a COOH group. Mass 101 indicates loss of CH_3O or $HOCH_2$. Mass 59 suggests an ester, our second double bond. Pieces are: $CH_3O + CO$; COOH; and remainder of C_2H_4. Compound is methyl hydrogen succinate.

Chapter 20

1. $2NaHCO_3 \to Na_2CO_3 + CO_2 + H_2O$. **2.** $\Delta H = -(\text{slope})(2.303)(1.987) = 39.0$ kcal/mole, where slope is found from plotting $\log p_{CO_2}$ vs. $1/T$ (in °K). **3.** Plot specific heat vs. temperature; T_g, given by intersection of two linear segments, is 415°K. **4.** (a) 24.0%. (b) 80.0%. (c) 49.0%. (d) 40.0%. (e) 76.0%. **5.** Chain rotation, 13.7 cal/g; fusion, 38.0 cal/g. **6.** 89.1 cal/g of UO_2. **7.** (a) $w_{CaO} = (w_{600°} - w_{900°})(56/44)$; $w_{MgO} = 1.5(w_{300°} - w_{600°})(40.6/44)$. (b) $3MgCO_3 \to MgO \cdot 2MgCO_3 + CO_2$. $MgO \cdot 2MgCO_3 \to 3MgO + 2CO_2$. (c) 40.0% CaO and 13.5% MgO. **8.** In all cases, the first loss is the two moles of hydrated water. In an oxidizing atmosphere the final product is NiQ; in CO and N_2, it is nickel metal. **9.** $K_{sp}^{158°} = 3.0 \times 10^{-8}$. **10.** ΔH (neutralization) $= -13.5 + 3.3 = -10.5$ kcal/mole. ΔH (ionization) $= -RT \ln K_a + T\Delta S = 12.6 - 9.3 = 3.3$ kcal/mole. See *Chimia*, **17**, 102 (1963). **11.** Titration curve exhibits a rising portion (Ca^{2+} reacting) followed by a descending portion (Mg^{2+} reacting). For Ca: $\Delta H =$

−5.7 kcal/mole; for Mg: $\Delta H = 5.6$ kcal/mole.   **12.** (a) $MnCO_3 \xrightarrow{400°} MnO_2 \xrightarrow{550°} Mn_2O_3 \xrightarrow{900°} Mn_3O_4$. (b) In CO_2 atmosphere, decomposition is delayed; Mn_2O_3 is formed at 600°. Formation of MnO_2 requires presence of oxygen.   **13.** Glass transition at 40°C; melting of coating at 125°C.   **14.** The sharp endothermic peak at 70°C is superimposed upon a glass transition at 61°C, followed by the exothermic curing reaction which appears to be complete near 260°C. Cured sample manifests a glass transition at 116°C.   **15.** At a mole ratio of 1:1 there occurs an intermolecular substance; 110°C is transition point of hexamethylbenzene.

Chapter 22

1. (a) A: 14.08; B: 11.43; C: 12.15; D: 12.26. (b) pH < 9.6 at $[Na^+] = 0.2$; pH < 9.3 at $[Na^+] = 1.0$. (c) Below 0.02M.   **2.** Not a true buffer, only a stoichiometric mixture of K^+ and hydrogen phthalate ions.   **3.** A 10% error in the concentration would change the pH by only 0.01 unit.   **4.** Because of the increased dissociation of water at higher temperatures and because of the change in the liquid-junction potential caused by the presence of increasing numbers of highly mobile hydroxyl ions in the solution.   **5.** The several polymeric species of boric acid, and acids with varying degrees of hydration, each with its own temperature dependence, creates a complex set of equilibria.   **6.** Bromine is added to the sample to oxidize nitrite to nitrate; the amount of bromide ion produced is a measure of the nitrite ion originally present.   **7.** Electrode is 10 times more sensitive to fluoride ion than to hydroxide ion; therefore, a 0.001M OH^- concentration will produce a response equal to that from 0.01M F^-. To operate within a stipulated measurement the hydroxide concentration must be less than 0.001 × stipulated error.   **8.** pH > 5.2 to avoid complexation as HF; one could tolerate about pH 4.2 if standards and samples are adjusted to this pH beforehand.   **9.** Ratios of interferant to bromide ion: chloride, 330; iodide, 1.7×10^{-4}; hydroxide, 4000; cyanide, 4×10^{-4}.   **10.** Minimum calcium activity = $(1.0 \times 10^{-4})(0.7)^2/0.05 = 9.8 \times 10^{-4}M$.   **11.** Hydroxide, $10^{-4}M$; iodide, $5 \times 10^{-9}M$; nitrate, $10^{-6}M$; hydrogen carbonate, $2.5 \times 10^{-5}M$; and sulfate, $10^{-8}M$.   **12.** Total sulfide = $1.33 \times 10^{-3}M$.

Chapter 23

4.

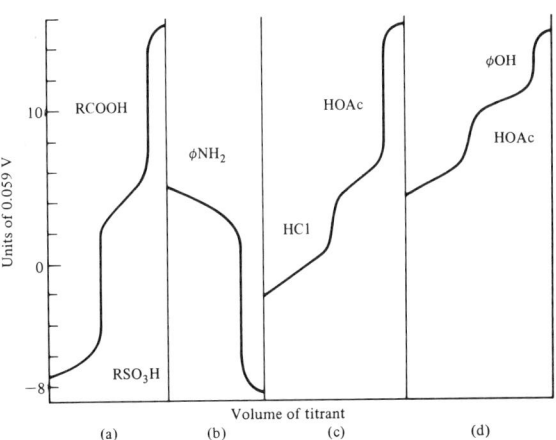

5. (a) 0.393 V. (b) 0.236 V. (c) 1.06 V. (d) 1.04 V.   6. (a) 0.017%. (b) 0.30%.
7. (b) 8.85. (c) 15.65 ml. (d) pK_a = 5.57.   8. (b) 10.0 ml.

Chapter 24

1. (a) Direct comparison method would be best. (b) and (c) Standard addition method would be best; sometimes for (b) the internal standard method might be applicable.   2. Factors not discussed in the text include kinetic currents and catalytic currents.   3. Polarography gives the entire course of the current–potential curve, of which a potentiometric measurement gives only one point. In quantitative polarographic analysis, the electric current is the reagent used to alter the composition of a layer of solution surrounding the microelectrode; the bulk composition is not altered nor are any equilibria affected, such as those involving complex ions. Potentiometry becomes difficult when solutions are more dilute than about 0.001 M initially, and when the inflection point is not large. Polarographic measurements can be used to carry out amperometric titrations in which data in the vicinity of the end point are not required; nor need one reactant be electroactive.   4. 1.2 × $10^{-3} M$.   5. 0.0012 M.   6. 0.0012 M.   7. 1.49 × $10^{-5} M$.   8. (a) Successive 2-electron steps for: $O_2 \rightarrow H_2O_2 \rightarrow 2H_2O$. (b) Uranium: VI to V, and finally to III. (c) Successive 1-electron and 5-electron reactions: $I^- \rightarrow \frac{1}{2} I_2 \rightarrow IO_3^-$.   9. To about 10^{-29} of its initial value.   10. Lead, 0.24 mM, and zinc, 1.10 mM.   11. Copper or bismuth; lead; indium; cadmium; nickel or zinc.   12. Copper (or bismuth); lead; thallium; indium; cadmium; zinc (or nickel).   13. Lead: 0.77 × $10^{-4} M$; zinc: 0.67 × $10^{-4} M$.   14. (a) Unnecessary to know the temperature or any of the variables in the Ilkovic equation. Accurate measurement of reagent volume is simpler than accurate measurement of a fluctuating electrical current. Regression analysis of linear portions of amperometric titration curve averages out random errors in individual current measurements. (b) Amperometric titrations can utilize all types of indicating electrodes, which extends the applicable range of potentials from the discharge of oxygen (anodic) to the discharge of hydrogen (cathodic) from the solvent. Rotating microelectrodes provide greater sensitivity; their nonreproducibility is no handicap in amperometric titrations. (c) Curvature of the linear segments in the vicinity of the end point can be ignored (due to slight solubility or dissociation of complexes) and the points sufficiently far before and after the end point (where the excess of one reactant forces the reaction virtually to completion) define the two straight lines which are extrapolated to their point of intersection. Nonelectroactive ions may be titrated.   15. (a) At −0.8 V, the titration curve is L-shaped; at −1.7 V, the curve is V-shaped. (b) −0.8 V. (c) −0.8 V to avoid the nickel ammine and zinc ammine waves.

Chapter 25

1. (a) Cathode potential: $\geqslant$ 0.26 V vs. SHE. (b) 10.2 min to remove 99.9% of the silver.   2. pH $\geqslant$ 2.2.   3. Residual copper concentration is $10^{-22} M$.   4. −0.46 V vs. SHE.   5. (a) 1.11 g. (b) 4.19 g. (c) 1.24 g. (d) 1.04 g.   6. (a) 0.50 V vs. SHE. (b) 0.38 V vs. SHE. (c) pAg = 5.93.   7. (a) 4.64 mmol. (b) ΔpH = 5.37.   8. (a) 0.109 M. (b) 0.128 V vs. SHE.   9. (a) 3.29 V. (b) Yes. (c) $10^{-12} M$.   10. Probably not completely since the cadmium-ion concentration is lowered only to $10^{-4} M$.   11. (a) 2.8 × $10^{-17} M$. (b) In a chloride solution containing hydrazine, the copper is involved in successive one-electrode reductions. Residual copper is 2.1 × $10^{-6} M$.   12. 7.9 × $10^{-17} M$

(assumes the anion is sulfate and not chloride). **13.** The overpotential term for hydrogen gas on copper shifts the point of incipient evolution of hydrogen gas more negative than the potential for initial deposition of zinc from $0.1 M$ solution, after which the still higher overpotential term for hydrogen gas on zinc takes over. **14.** One possibility is to use a Na_2CO_3–$NaNO_3$ electrolyte, wash with dilute acetic acid, and develop green color with 1% alcoholic solution of α-benzoinoxime. **15.** For sulfide, lead carbonate paper can be used with a sodium carbonate electrolyte. Specimen is cathodic. **16.** (a) 13.3 min. (b) 19.9 min. **17.** 0.028 mm. **18.** (a) Cd: -0.433 V and H: -0.118 V. (b) Cd: -0.61 V and H: -0.59 V. **19.** No, a cathode potential of -0.78 V vs. SCE would be required to lower the copper(I) concentration to $1 \times 10^{-5} M$. **20.** From a weakly acid solution, remove lead while controlling the cathode potential at -0.35 V vs. NHE, then cadmium at -0.70 V, and finally zinc at -0.91 V. The electrode is returned to the electrolysis cell after each plated metal is weighed without removing the accumulated deposits. **21.** A cyclic process is set up involving the oxidation of the copper(I) ammine complex at the anode and the reduction of the copper(II) ammine complex at the cathode. Also the copper(I) ammine reduces the silver ammine to metallic silver. **22.** (a) Ag | $Ag(CN)_2^-$, CN^-, $NaOH(1M)$ | $H_2(Pt)$. (b) Anode: $Ag + 2CN^- = Ag(CN)_2^- + e^-$; $E° = -0.31$ V. Cathode: $2H_2O + 2e^- = H_2 + 2OH^-$; $E° = -0.828$ V. (c) Although the spontaneous reaction is $Ag(CN)_2^- + H_2 + 2OH^- \rightarrow Ag° + 2CN^- + 2H_2O$, when the electrodes are first placed in the solution the potential of the silver electrode actually is more negative than that of the platinum electrode because the solution originally contains no $Ag(CN)_2^-$ and no hydrogen. Consequently, the above reaction proceeds from right to left although the spontaneous current decreases exponentially with time and approaches zero as equilibrium is established. **23.** (b) 0.46 min^{-1}. (c) Theoretically 15 additional min; total time 20 min. **24.** 18.8 mg. **25.** 1.98×10^{-5} microequivalents/count at 30 mA. **26.** $0.01087N$. **27.** $0.133N$. **28.** 1 mA sec corresponds to 0.0104 microequivalents. (a) 176 ng. (b) 632 ng. (c) 368 ng. (d) 331 ng. (e) 513 ng. **29.** 111.5 Q. **30.** $0.00145N$. **31.** (a) Cathode: $2H_2O + 2e = H_2 + 2OH^-$; anode: $Ag + Br^- = AgBr + e$. (b) Cathode: $2H_2O + 2e = H_2 + 2OH^-$; anode: $2Br^- = Br_2 + 2e$. Excess unused bromine reacted with $CuCl_3^{2-}$ generated at cathode: $Cu^{2+} + 3Cl^- + e = CuCl_3^{2-}$; anode compartment must be isolated. (c) Cathode: $Fe(CN)_6^{3-} + e = Fe(CN)_6^{4-}$; anode: $2H_2O = 4H^+ + O_2 + 4e$. **32.** $1.29 \times 10^{-4} M$. **33.** (a) 87.1 mg. (b) $E = 0.43$ V vs. SHE (a slightly more positive potential is desirable to allow for any iR drop). **34.** 0.319 mg. **35.** From a minimum of 0.00104 milliequivalents to a maximum of 0.0312. **36.** At a cathode potential of $+0.75$ V vs. SCE the manganese is reduced (VII to II); at 0.3 V vs. SCE vanadium is reduced (V to IV); and at -0.3 V vs. SCE the iron is reduced (III to II). **37.** Use the reversible vanadium(V/IV) system as a potential buffer. Iron(III) will be formed either by direct oxidation at the anode or indirectly by oxidation with the vanadium(V) formed at the anode. End point is signaled by the reaction between the first persistence of vanadium(V) and diphenylamine sulfonic acid which turns blue. **38.** Control the anode potential at 0.75 V vs. SCE and measure the coulombs involved in the oxidation of vanadium(IV). The cathode can be platinum (with H_2 and OH^- liberated). Reverse the working electrode potential and control the cathode value at 0.30 V vs. SCE and measure the coulombs required for the reduction of original and generated vanadium(V); the difference in number of coulombs between the cathodic reduction and anodic oxidation gives the original vanadium(V) concentration. **39.** 1930 sec. **40.** 154.4 g.

Chapter 26

1. (a) 118 Ω. **(b)** 11.8 Ω. **(c)** 2.35 Ω. **(d)** 1.18 Ω. Only (a) and (b); actually only (a) falls in desirable range. **2.** 26.7 to 318 Ω. **3.** Resistance should range from 546Θ to 273,000Θ. If Θ = 0.2, resistance readings will range from 100 to 54,600 Ω with 2740 Ω at midscale. **4. (a)** $R = 2\Theta$. Θ should be 50 to provide a resistance of 100 Ω (for 5% HCl) and 85,000 Ω (for 0.001M, assumed low concentration). **(b)** Resistance varies from 512Θ to 1620Θ. Θ could be 0.2. **(c)** Concentration ranges from 8.82 × $10^{-4} M$ to perhaps 1/600 of this value (0.1 ppm). Resistance varies from 108 Ω to 64,800 Ω for Θ = 0.01 (Θ could also be 0.1). **(d)** Case (c) repeated essentially. **(e)** Θ = 0.01R; Θ = 50 is often used. **(f)** Resistance varies from 2.5 to 250Θ. Θ = 20 is desirable. **5.** $\alpha = 0.0824$. $K_a = 1.79 \times 10^{-5}$. **6.** Solubility is 1.07×10^{-5} mol/liter. $K_{sp} = 1.15 \times 10^{-10}$. **7.** Specific conductance is 1.80×10^{-5} Ω^{-1} cm^{-1}. Resistance of saturated solution would be 110,000 Ω. **8.** 0.0325N; 0.130-g NaOH. **9.** 0.0625N. **10.** 0.0176M; individual end points for each replaceable hydrogen. **11.** Aromatic sulfonic acid: 0.630 mequiv; aliphatic carboxylic acid: 0.230 mequiv.
12.

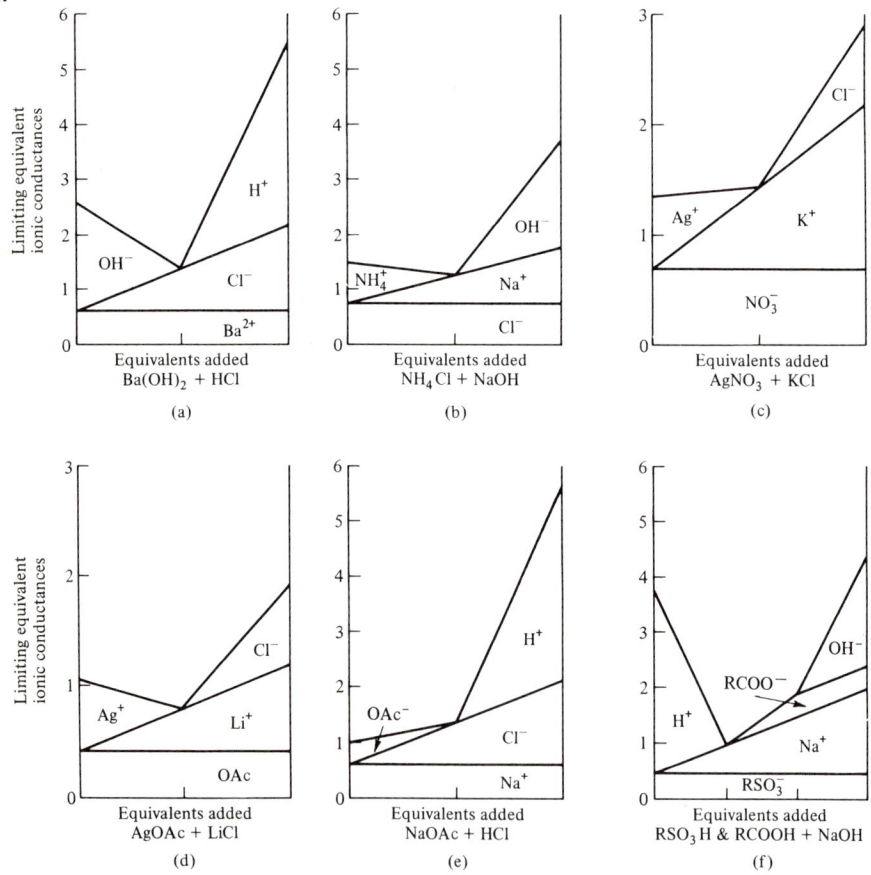

(g) Resembles graph for (f) except there is no sharp rise after the second end point since excess NH_3 contributes no conductance. First end point occurs after one equivalent of NH_3 has been added; second end point after three equivalents have been added.   **13.** Titrate with NaOH. Conductance rises up to first end point (neutralization of nicotinic acid), then falls more steeply as the ammonium ion is converted to ammonia. After the second end point the conductance rises abruptly with excess NaOH.   **14.** Conductance (in units of 10^{-4} Ω^{-1} cm^{-1}) after dilution by titrant: 0%, 21.26; 25%, 17.84; 50%, 14.40; 90%, 8.9; 100%, 7.5; 110%, 8.0; 150%, 12.4; 175%, 15.0.   **15.** Conductance (Ω^{-1}): 0%, 0.0288; 25%, 0.0282; 50%, 0.0272; 90%, 0.0266; 100%, 0.0264; 110%, 0.0310; 150%, 0.0489; 175%, 0.0594; 200%, 0.0695.   **16.** By linear regression: log $(1/R)$ = 1.139 + 0.974 log C. Use corrected conductance readings. A: 0.884 mequiv. B: 0.412 mequiv. C: 0.040 mequiv. D: 0.202 mequiv.   **17.** $\alpha = 1.18 \times 10^{-4}$. $K_a = 5.61 \times 10^{-10}$.

Chapter 31

1. A rugged continuous-flow-type pH electrode assembly is mounted in the line connecting the mixing tank and the initial filter. Phosphoric acid and soda ash are mixed at approximately 85° to 100°C and the hot solution filtered to remove iron and aluminum phosphates as well as silica. The signal from the electrode assembly is amplified and sent to a recorder. In an automated system, the amplified electrode signal would also be used to actuate a diaphragm motor valve controlling the addition of phosphoric acid (see D. M. Considine, Ed., *Process Instruments and Controls Handbook*, 2nd ed., McGraw-Hill, New York, 1974, p. 6–96.   **2.** Measurements to the fourth decimal place provide a precision of about 0.1%.   **3.** The design of nondispersive analyzers is less complicated, thus these instruments are easier to operate and maintain. The sensitivity of nondispersive analyzers is usually better. Nondispersive analyzers are not generally as selective or flexible as dispersive units and are thus dedicated to sampling light intensity at a limited number of predetermined wavelengths.   **4.** In process gas chromatography, the sample is always transferred from the process stream to the analyzer without human intervention. Process instruments are operated in a harsh plant environment by personnel with minimal analytical training. Process chromatography utilizes multicolumn techniques, such as stripping and backflushing, which are not typically used in laboratory chromatography. The data from process instruments are presented in formats, such as bar graphs, not generally used for laboratory data. The data are used immediately to control the monitored process. In process chromatography, the same components are usually in each sample (you know what to expect).   **5.** All can be handled by refractometric methods. See *J. Chem. Educ.*, **45**, A470 (1968).   **6.** Gas sample is drawn through the probe by means of a steam-operated aspirator. Steam–gas mixture is condensed in a cooling chamber and the resulting condensate, containing all the sodium-ion entrained in the original gas sample, is separated from the gas and passed through a rotameter ahead of the sodium analyzer (selective-ion electrode). Gas flow measured in second rotameter provides a multiplier, which may easily be corrected for absolute humidity, whereby the recorded sodium data can be reported on a dry gas basis.   **7.** At regular programmed intervals, a sample of treated water is drawn into the analyzer cell system. After the initial pH is measured and recorded, the sample automatically is titrated to pH 8.3, with the results fed to the analyzer computer and stored. Then the titration is continued to pH 4.3, the total alkalinity end point. This signal is fed to the computer and also the soda ash feed controller, and is combined with a raw water flow signal. The combined signal controls addition of soda ash to the unit. The residual hydroxide and/or bicarbonate-ion value is computed, recorded, and transmitted to the lime feed controller, where it also is combined with a raw water flow signal. This combined signal controls the addition of lime to the softener, or lime and soda ash are fed in a fixed ratio with residual hydroxide and/or bicarbonate-ion value providing the control signal.

8.

ANALYZER	MODE OF OPERATION	ADVANTAGES	DISADVANTAGES
Discrete	Batch, serial processing of sample	Runs selected test on limited number of samples, quick start-up	Expensive prepackaged reagents and cells, mechanically complex
Continuous flow	Continuous process of large number of samples for same series of tests	Fast, simple equipment, reagent cost minimized	Difficult to vary test doing a series of samples, tubing is critical
Centrifugal	Sample mixed with reagent by centrifugal force	Fast, all samples and controls measured under the same conditions	Loading samples is time-consuming. Can run only a limited number of tests.

9. All the parameters of the gas chromatography must be optimized and controlled. The method depends upon the use of small samples of about 0.6 mg, which allows for rapid combustion and the introduction of the products into the chromatographic column without excessive dilution by the carrier gas.   **10.** 78.99% C; 5.78% H; 15.29% N.

Appendixes

APPENDIX A Potentials of Selected Half-Reactions at 25°C

A summary of oxidation–reduction half-reactions arranged in order of decreasing oxidation strength and useful for selecting reagent systems.

Half-Reaction		$E°$, V
$F_2(g) + 2H^+ + 2e^-$	$= 2HF$	3.06
$O_3 + 2H^+ + 2e^-$	$= O_2 + H_2O$	2.07
$S_2O_8^{2-} + 2e^-$	$= 2SO_4^{2-}$	2.01
$Ag^{2+} + e^-$	$= Ag^+$	2.00
$H_2O_2 + 2H^+ + 2e^-$	$= 2H_2O$	1.77
$MnO_4^- + 4H^+ + 3e^-$	$= MnO_2(s) + 2H_2O$	1.70
$Ce(IV) + e^-$	$= Ce(III)$ (in $1M$ $HClO_4$)	1.61
$H_5IO_6 + H^+ + 2e^-$	$= IO_3^- + 3H_2O$	1.6
Bi_2O_4 (bismuthate) $+ 4H^+ + 2e^-$	$= 2BiO^+ + 2H_2O$	1.59
$BrO_3^- + 6H^+ + 5e^-$	$= \frac{1}{2}Br_2 + 3H_2O$	1.52
$MnO_4^- + 8H^+ + 5e^-$	$= Mn^{2+} + 4H_2O$	1.51
$PbO_2 + 4H^+ + 2e^-$	$= Pb^{2+} + 2H_2O$	1.455
$Cl_2 + 2e^-$	$= 2Cl^-$	1.36
$Cr_2O_7^{2-} + 14H^+ + 6e^-$	$= 2Cr^{3+} + 7H_2O$	1.33
$MnO_2(s) + 4H^+ + 2e^-$	$= Mn^{2+} + 2H_2O$	1.23
$O_2(g) + 4H^+ + 4e^-$	$= 2H_2O$	1.229
$IO_3^- + 6H^+ + 5e^-$	$= \frac{1}{2}I_2 + 3H_2O$	1.20
$Br_2(l) + 2e^-$	$= 2Br^-$	1.065
$ICl_2^- + e^-$	$= \frac{1}{2}I_2 + 2Cl^-$	1.06
$VO_2^+ + 2H^+ + e^-$	$= VO^{2+} + H_2O$	1.00
$HNO_2 + H^+ + e^-$	$= NO(g) + H_2O$	1.00
$NO_3^- + 3H^+ + 2e^-$	$= HNO_2 + H_2O$	0.94
$2Hg^{2+} + 2e^-$	$= Hg_2^{2+}$	0.92
$Cu^{2+} + I^- + e^-$	$= CuI(s)$	0.86
$Ag^+ + e^-$	$= Ag$	0.799
$Hg_2^{2+} + 2e^-$	$= 2Hg$	0.79
$Fe(III) + e^-$	$= Fe^{2+}$	0.771
$O_2(g) + 2H^+ + 2e^-$	$= H_2O_2$	0.682
$2HgCl_2 + 2e^-$	$= Hg_2Cl_2(s) + 2Cl^-$	0.63
$Hg_2SO_4(s) + 2e^-$	$= 2Hg + SO_4^{2-}$	0.615
$Sb_2O_5 + 6H^+ + 4e^-$	$= 2SbO^+ + 3H_2O$	0.581
$H_3AsO_4 + 2H^+ + 2e^-$	$= HAsO_2 + 2H_2O$	0.559
$I_3^- + 2e^-$	$= 3I^-$	0.545
$Cu^+ + e^-$	$= Cu$	0.52
$VO^{2+} + 2H^+ + e^-$	$= V^{3+} + H_2O$	0.337

APPENDIX A Continued

Half-Reaction		$E°$, V
$Fe(CN)_6^{3-} + e^-$	$= Fe(CN)_6^{4-}$	0.36
$Cu^{2+} + 2e^-$	$= Cu$	0.337
$UO_2^{2+} + 4H^+ + 2e^-$	$= U^{4+} + 2H_2O$	0.334
$Hg_2Cl_2(s) + 2e^-$	$= 2Hg + 2Cl^-$	0.2676
$BiO^+ + 2H^+ + 3e^-$	$= Bi + H_2O$	0.32
$AgCl(s) + e^-$	$= Ag + Cl^-$	0.2222
$SbO^+ + 2H^+ + 3e^-$	$= Sb + H_2O$	0.212
$CuCl_3^{2-} + e^-$	$= Cu + 3Cl^-$	0.178
$SO_4^{2-} + 4H^+ + 2e^-$	$= SO_2(aq) + 2H_2O$	0.17
$Sn^{4+} + 2e^-$	$= Sn^{2+}$	0.15
$S + 2H^+ + 2e^-$	$= H_2S(g)$	0.14
$TiO^{2+} + 2H^+ + e^-$	$= Ti^{3+} + H_2O$	0.10
$S_4O_6^{2-} + 2e^-$	$= 2S_2O_3^{2-}$	0.08
$AgBr(s) + e^-$	$= Ag + Br^-$	0.071
$2H^+ + 2e^-$	$= H_2$	0.0000
$Pb^{2+} + 2e^-$	$= Pb$	−0.126
$Sn^{2+} + 2e^-$	$= Sn$	−0.136
$AgI(s) + e^-$	$= Ag + I^-$	−0.152
$Mo^{3+} + 3e^-$	$= Mo$	approx. −0.2
$N_2 + 5H^+ + 4e^-$	$= H_2NNH_3^+$	−0.23
$Ni^{2+} + 2e^-$	$= Ni$	−0.246
$V^{3+} + e^-$	$= V^{2+}$	−0.255
$Co^{2+} + 2e^-$	$= Co$	−0.277
$Ag(CN)_2^- + e^-$	$= Ag + 2CN^-$	−0.31
$Cd^{2+} + 2e^-$	$= Cd$	−0.403
$Cr^{3+} + e^-$	$= Cr^{2+}$	−0.41
$Fe^{2+} + 2e^-$	$= Fe$	−0.440
$2CO_2 + 2H^+ + 2e^-$	$= H_2C_2O_4$	−0.49
$H_3PO_3 + 2H^+ + 2e^-$	$= HPH_2O_2 + H_2O$	−0.50
$U^{4+} + e^-$	$= U^{3+}$	−0.61
$Zn^{2+} + 2e^-$	$= Zn$	−0.763
$Cr^{2+} + 2e^-$	$= Cr$	−0.91
$Mn^{2+} + 2e^-$	$= Mn$	−1.18
$Zr^{4+} + 4e^-$	$= Zr$	−1.53
$Ti^{3+} + 3e^-$	$= Ti$	−1.63
$Al^{3+} + 3e^-$	$= Al$	−1.66
$Th^{4+} + 4e^-$	$= Th$	−1.90
$Mg^{2+} + 2e^-$	$= Mg$	−2.37
$La^{3+} + 3e^-$	$= La$	−2.52
$Na^+ + e^-$	$= Na$	−2.714
$Ca^{2+} + 2e^-$	$= Ca$	−2.87
$Sr^{2+} + 2e^-$	$= Sr$	−2.89
$K^+ + e^-$	$= K$	−2.925
$Li^+ + e^-$	$= Li$	−3.045

APPENDIX B Polarographic Half-Wave Potentials and Diffusion-Current Constants[d]

Generally, solutions contained 0.01% gelatin, and the data pertain to a temperature of 25°C. Half-wave potentials are referred to the saturated calomel electrode, and values of $i_d/Cm^{2/3}t^{1/6}$ are based on i_d in microamperes, C in millimoles per liter, m in mg sec^{-1}, and t in seconds.[a]

Ion	Supporting Electrolyte	$E_{1/2}$	I_d
Bi^{3+}	1M HCl	−0.09	
	0.5M tartrate (pH 4.5)	−0.29	
	0.5M tartrate + 0.1M NaOH	−1.0	
Cd^{2+}	0.1M KCl	−0.60	3.51
	1M NH$_3$ + 1M NH$_4^+$	−0.81	3.68
Co^{2+}	1M KSCN	−1.03	
	0.1M KCl	−1.20	
	0.1M pyridine + 0.1M pyridinium ion	−1.07	
CrO_4^{2-}	0.1M KCl (basic chromic chromate)	−0.3	
	($CrO_4^{2-} \to Cr^{3+}$)	−1.0	
	($Cr^{3+} \to Cr^{2+}$)	−1.5	
	($Cr^{2+} \to Cr°$)	−1.7	
Cu^{2+}	0.1M KCl (HCl)	+0.04	3.23
	0.5M tartrate, pH = 4.5	−0.09	2.37
	1M NH$_3$ + 1M NH$_4^+$ (1st wave)	−0.24	(Total
	(2d wave)	−0.50	3.75)
Fe^{3+}	0.5M citrate, pH = 5.8 (1st wave)	−0.17	0.90
	(2d wave)	−1.50	
	0.1M EDTA + 2M NaOAc (1st wave)	−0.13	
	(2d wave)	−1.3	
Fe^{2+}	0.05M BaCl$_2$	−1.3	
In^{3+}	0.1M KCl	−0.561	
Mn^{2+}	1M KCl	−1.51	
	1M KSCN	−1.55	
Ni^{2+}	0.01M KCl	−1.1	
	1M NH$_3$ + 1M NH$_4^+$	−1.09	3.56
	1M KSCN	−0.70	
	0.5M pyridine + 1M KCl	−0.78	
O_2	pH 1 to 10 ($O_2 \to H_2O_2$)	−0.05	(Total
	($H_2O_2 \to H_2O$)	−0.94	12.3)
Pb^{2+}	0.1M KCl	−0.40	3.80
	1M HNO$_3$	−0.40	3.67
	1M NaOH	−0.75	3.39
	0.5M tartrate + 0.1M NaOH	−0.75	2.39
Sb^{3+}	1M HCl	−0.15	
	0.5M tartrate + 0.1M NaOH	−1.32	
Sn^{2+}	1M HCl	−0.47	4.07
	0.5M tartrate + 0.1M NaOH (anodic)	−0.71	2.86
	(cathodic)	−1.16	2.86

APPENDIX B Continued

Ion	Supporting Electrolyte	$E_{1/2}$	I_d
Sn^{4+}	$1M$ HCl + $4M$ NH_4^+ (1st wave)	−0.25	2.84
	(2d wave)	−0.52	3.49
Zn^{2+}	$0.1M$ KCl	−1.00	3.42
	$1M$ NaOH	−1.50	3.14
	$1M$ NH_3 + $1M$ NH_4^+	−1.33	3.82
	$0.5M$ tartrate, pH = 9	−1.15	2.30

[a]Reproduced, by permission, from *Polarography*, by I. M. Kolthoff and J. J. Lingane, 2nd ed. Copyright 1952 by Interscience Publishers, Inc.

APPENDIX C Proton-Transfer Reactions of Materials in Water at 25°C

Substance	pK_1	pK_2	pK_3	pK_4
Acetic acid	4.76			
Ammonium ion	9.24			
Anilinium ion	4.60			
Arsenic acid	2.20	6.98	11.5	
Arsenious acid	9.22			
Ascorbic acid	4.30	11.82		
Benzoic acid	4.21			
Boric acid: meta-	9.24			
tetra-	4	9		
Bromocresol green	4.68			
Bromocresol purple	6.3			
p-Bromophenol	9.24			
Bromophenol blue	3.86			
Bromothymol blue	7.1			
Carbonic acid (CO_2 + H_2O)	6.38	10.25		
Chloroacetic acid	2.86			
Chlorophenol red	6.0			
Chromic acid		6.50		
Citric acid	3.13	4.76	6.40	
Cresol purple (acid range)	1.51			
(base range)	8.32			

APPENDIX C Continued

Substance	pK_1	pK_2	pK_3	pK_4
Cresol red	8.2			
Dichloroacetic acid	1.30			
Ethanolammonium ion	9.50			
Ethylammonium ion	10.63			
Ethylenediaminetetraacetic acid (EDTA)	2.0	2.67	6.27	10.95
Ethylenediammonium ion	6.85	9.93		
Ferrocyanic acid			2.22	4.17
Formic acid	3.75			
Glycine (protonated cation)	2.35	9.78		
Hydrazinium ion	−0.88	7.99		
Hydrocyanic acid	9.21			
Hydrofluoric acid	3.18			
Hydrogen peroxide	11.65			
Hydrogen sulfide	6.88	14.15		
Hydroquinone	10.0	12.0		
Hydroxylammonium ion	5.96			
N,N-bis (2-Hydroxyethyl) glycine (Bicine) (protonated cation)	8.35			
tris (Hydroxymethyl) aminomethane (TRIS) (protonated cation)	8.08			
N-2-Hydroxyethylpiperazine-N'-2-ethanesulfonic acid (HEPES)	7.55			
N-tris (Hydroxymethyl) methylglycine (TRIS) (protonated cation)	8.08			
Hypochlorous acid	7.50			
Methyl orange	3.40			
Methyl red	4.95			
2-(N-Morpholino) ethanesulfonic acid (MES)	6.15			
Nitrous acid	3.35			
Oxalic acid	1.27	4.27		
1,10-Phenanthrolinium ion	4.86			
Phenol	9.99			
Phenol red	7.9			
Phenolphthalein	9.4			
Phenylacetic acid	4.31			
Phosphoric acid: ortho	2.15	7.20	12.36	
pyro	1.52	2.36	6.60	9.25
o-Phthalic acid	2.95	5.41		
Pyridinium ion	5.21			
Salicylic acid	3.00	12.38		
Succinic acid	4.21	5.64		
Sulfamic acid	0.988			
Sulfuric acid		1.92		

APPENDIX C Continued

Substance	pK_1	pK_2	pK_3	pK_4
Sulfurous acid ($SO_2 + H_2O$)	1.90	7.20		
Tartaric acid: meso-	3.22	4.81		
Thymol blue	8.9			
Thymolphthalein	10.0			
Triethanolammonium ion	7.76			
Vanillin	7.40			
Veronal	7.43			

APPENDIX D Cumulative Formation Constants for Metal Complexes at 25°C

	$\log K_1$	$\log K_2$	$\log K_3$	$\log K_4$	$\log K_5$	$\log K_6$
AMMONIA						
Cadmium	2.65	4.75	6.19	7.12	6.80	5.14
Cobalt(II)	2.11	3.74	4.79	5.55	5.73	5.11
Cobalt(III)	6.7	14.0	20.1	25.7	30.8	35.2
Copper(I)	5.93	10.86				
Copper(II)	4.31	7.98	11.02	13.32	12.86	
Nickel	2.80	5.04	6.77	7.96	8.71	8.74
Silver(I)	3.24	7.05				
Zinc	2.37	4.81	7.31	9.46		
CHLORIDE						
Copper(I)		5.5	5.7			
Copper(II)	0.1	−0.6				
Tin(II)	1.51	2.24	2.03	1.48		
Tin(IV)						4
CITRATE (L^{3-} anion)						
Cadmium	11.3					
Cobalt(II)	12.5					
Copper(II)	14.2					
Iron(II)	15.5					
Iron(III)	25.0					
Nickel	14.3					
Zinc	11.4					

APPENDIX D Continued

	$\log K_1$	$\log K_2$	$\log K_3$	$\log K_4$	$\log K_5$	$\log K_6$
CYANIDE						
Cadmium	5.48	10.60	15.23	18.78		
Copper(I)		24.0	28.59	30.30		
Nickel				31.3		
Silver(I)		21.1	21.7	20.6		
Zinc				16.7		
ETHYLENEDIAMINE-N,N,N',N'-TETRAACETIC ACID						
Calcium	11.0					
Copper(II)	18.7					
Iron(II)	14.33					
Iron(III)	24.23					
Magnesium	8.64					
Mercury(II)	21.80					
Zinc	16.4					
1,10-PHENANTHROLINE						
Cadmium	5.93	10.53	14.31			
Cobalt(II)	7.25	13.95	19.90			
Copper(II)	9.08	15.76	20.94			
Iron(II)	5.85	11.45	21.3			
Iron(III)	6.5	11.4	23.5			
Nickel	8.80	17.10	24.80			
Zinc	6.55	12.35	17.55			

APPENDIX E Flame Emission and Atomic Absorption Spectra

Element	Wavelength, nm	Emission-Detection Limit, μg/ml/0.1 mV	Absorption Sensitivity, μg/ml/1% Abs
Aluminum	396.15	0.5 OAn	3.0 NAr
Antimony	217.58		0.6 AA
	259.81	0.6 OAn	
Arsenic	197.20		1.3 AA
	234.98	2.2 OAn	
Barium	553.55	0.03 OA	2.6 NA

APPENDIX E Continued

Element	Wavelength, nm	Emission-Detection Limit, µg/ml/0.1 mV	Absorption Sensitivity, µg/ml/1% Abs
Bismuth	223.06		0.7 AA
	227.66	6.4 OAn	
Boron	249.77	7.0 OAnr	10.0 NA
(as BO_2)	518.0	3.0 OAn	
Cadmium	228.80	4.0 OAn	0.03 AA
	326.11	2.0 OAn	20.0 AA
Calcium	422.67	0.07 OA	0.08 AA
Cesium	455.54	0.01 OH, OA	10.0 AP
	852.11	0.02 OH, OA	0.16 AP
Chromium	357.87	0.2 OAn	0.22 AAr, NA
	425.43	0.1 OAn	0.6 AAr, NA
Cobalt	240.72		0.02 AA
	242.49	1.7 OAr	0.2 AA
Copper	324.75	0.6 OA	0.1 AA
	327.40	0.01 NA	0.2 AA
Gallium	287.42		2.3 AA
	417.21	0.02 NA	3.7 AA
Gold	242.80	5.0 OH	0.3 AA, NA
Indium	325.61	2.2 OH	1.0 AA
	451.13	0.002 NA	2.8 AA
Iron	248.33		0.15 AAr
	371.99	0.12 OAn	1.0 AAr
Lanthanum (as LaO)	741.0	0.005 OAn	
Lead	217.00		0.23 AA
	283.31	6.0 OHn	
Lithium	670.78	0.0002 OH	0.02 AA
Magnesium	285.21	0.2 OAr	0.008 AA
Manganese	279.48	0.02 AA	0.06 AA
	403.08	0.005 NA	0.6 AA
Mercury	253.65	2.5 OAn	2.0 AA
Molybdenum	313.26	0.2 NA	0.8 NA
	379.83	0.5 OAn	2.0 NA
Nickel	232.00		0.15 AA
	352.45	0.2 OAn	0.6 AA
Niobium	405.89	1.0 NA	28.0 NA
Palladium	247.64		0.3 AA
	363.47	0.1 OAn	
Phosphorus (as HPO)	524.9	6.0 AH (reversed)	
Platinum	265.94	15.0 OAn	2.2 AA
Potassium	404.41	1.7 OH	3.7 AP
	766.49	0.0005 AA	0.01 AP
Rubidium	780.02	0.001 AA	0.04 AP
Silicon	251.61	4.0 OAnr	2.0 NA

APPENDIX E Continued

Element	Wavelength, nm	Emission-Detection Limit, µg/ml/0.1 mV	Absorption Sensitivity, µg/ml/1% Abs
Silver	328.07	0.2 OA	0.13 AA
	338.29	0.2 OA	0.22 AA
Sodium	589.00	0.0005 AA	0.004 AP
Strontium	460.73	0.06 OA	0.06 AA, NA
Sulfur (as S_2)	394.0	3.0 AH (reversed and shielded)	
Tellurium	214.28	7.0 OAn	0.5 AA
	238.58	2.0 OAn	43.0 AA
Thallium	276.79		0.1 AA
	377.57	0.02 NA	0.03 AA
	535.05	0.05 NA	
Tin	284.00	0.3 NA	
	286.33		10.0 AH
Vanadium	318.40	0.4 NA	0.4 NA
Zinc	213.86	50.0 OAn	0.025 AA

NOTE: The symbols used in this table and their meanings are as follows: AA, air–acetylene flame; AH, air–hydrogen flame; AP, air–propane flame; NA nitrous oxide–acetylene flame; OA, oxygen–acetylene flame; OH, oxygen–hydrogen flame; n, organic solvent containing solute aspirated directly into flame, solvent often is methyl isobutyl ketone; r, reaction zone of a fuel-rich flame.

APPENDIX F Values of Absorbance for Percent Absorption

%A	.0	.1	.2	.3	.4	.5	.6	.7	.8	.9
0.0	.0000	.0004	.0009	.0013	.0017	.0022	.0026	.0031	.0035	.0039
1.0	.0044	.0048	.0052	.0057	.0061	.0066	.0070	.0074	.0079	.0083
2.0	.0088	.0092	.0097	.0101	.0106	.0110	.0114	.0119	.0123	.0128
3.0	.0132	.0137	.0141	.0146	.0150	.0155	.0159	.0164	.0168	.0173
4.0	.0177	.0182	.0186	.0191	.0195	.0200	.0205	.0209	.0214	.0218
5.0	.0223	.0227	.0232	.0236	.0241	.0246	.0250	.0255	.0259	.0264
6.0	.0269	.0273	.0278	.0283	.0287	.0292	.0297	.0301	.0306	.0311
7.0	.0315	.0320	.0325	.0329	.0334	.0339	.0343	.0348	.0353	.0357

APPENDIX F *Continued*

%A	.0	.1	.2	.3	.4	.5	.6	.7	.8	.9
8.0	.0362	.0367	.0372	.0376	.0381	.0386	.0391	.0395	.0400	.0405
9.0	.0410	.0414	.0419	.0424	.0429	.0434	.0438	.0443	.0448	.0453
10.0	.0458	.0462	.0467	.0472	.0477	.0482	.0487	.0491	.0496	.0501
11.0	.0506	.0511	.0516	.0521	.0526	.0531	.0535	.0540	.0545	.0550
12.0	.0555	.0560	.0565	.0570	.0575	.0580	.0585	.0590	.0595	.0600
13.0	.0605	.0610	.0615	.0620	.0625	.0630	.0635	.0640	.0645	.0650
14.0	.0655	.0660	.0665	.0670	.0675	.0680	.0685	.0691	.0696	.0701
15.0	.0706	.0711	.0716	.0721	.0726	.0731	.0737	.0742	.0747	.0752
16.0	.0757	.0762	.0768	.0773	.0778	.0783	.0788	.0794	.0799	.0804
17.0	.0809	.0814	.0820	.0825	.0830	.0835	.0841	.0846	.0851	.0857
18.0	.0862	.0867	.0872	.0878	.0883	.0888	.0894	.0899	.0904	.0910
19.0	.0915	.0921	.0926	.0931	.0937	.0942	.0947	.0953	.0958	.0964
20.0	.0969	.0975	.0980	.0985	.0991	.0996	.1002	.1007	.1013	.1018
21.0	.1024	.1029	.1035	.1040	.1046	.1051	.1057	.1062	.1068	.1073
22.0	.1079	.1085	.1090	.1096	.1101	.1107	.1113	.1118	.1124	.1129
23.0	.1135	.1141	.1146	.1152	.1158	.1163	.1169	.1175	.1180	.1186
24.0	.1192	.1198	.1203	.1209	.1215	.1221	.1226	.1232	.1238	.1244
25.0	.1249	.1255	.1261	.1267	.1273	.1278	.1284	.1290	.1296	.1302
26.0	.1308	.1314	.1319	.1325	.1331	.1337	.1343	.1349	.1355	.1361
27.0	.1367	.1373	.1379	.1385	.1391	.1397	.1403	.1409	.1415	.1421
28.0	.1427	.1433	.1439	.1445	.1451	.1457	.1463	.1469	.1475	.1481
29.0	.1487	.1494	.1500	.1506	.1512	.1518	.1524	.1530	.1537	.1543
30.0	.1549	.1555	.1561	.1568	.1574	.1580	.1586	.1593	.1599	.1605
31.0	.1612	.1618	.1624	.1630	.1637	.1643	.1649	.1656	.1662	.1669
32.0	.1675	.1681	.1688	.1694	.1701	.1707	.1713	.1720	.1726	.1733
33.0	.1739	.1746	.1752	.1759	.1765	.1772	.1778	.1785	.1791	.1798
34.0	.1805	.1811	.1818	.1824	.1831	.1838	.1844	.1851	.1858	.1864
35.0	.1871	.1878	.1884	.1891	.1898	.1904	.1911	.1918	.1925	.1931
36.0	.1938	.1945	.1952	.1959	.1965	.1972	.1979	.1986	.1993	.2000
37.0	.2007	.2013	.2020	.2027	.2034	.2041	.2048	.2055	.2062	.2069
38.0	.2076	.2083	.2090	.2097	.2104	.2111	.2118	.2125	.2132	.2140
39.0	.2147	.2154	.2161	.2168	.2175	.2182	.2190	.2197	.2204	.2211
40.0	.2218	.2226	.2233	.2240	.2248	.2255	.2262	.2269	.2277	.2284
41.0	.2291	.2299	.2306	.2314	.2321	.2328	.2336	.2343	.2351	.2358
42.0	.2366	.2373	.2381	.2388	.2396	.2403	.2411	.2418	.2426	.2434
43.0	.2441	.2449	.2457	.2464	.2472	.2480	.2487	.2495	.2503	.2510
44.0	.2518	.2526	.2534	.2541	.2549	.2557	.2565	.2573	.2581	.2588
45.0	.2596	.2604	.2612	.2620	.2628	.2636	.2644	.2652	.2660	.2668
46.0	.2676	.2684	.2692	.2700	.2708	.2716	.2725	.2733	.2741	.2749
47.0	.2757	.2765	.2774	.2782	.2790	.2798	.2807	.2815	.2823	.2832
48.0	.2840	.2848	.2857	.2865	.2874	.2882	.2890	.2899	.2907	.2916
49.0	.2924	.2933	.2941	.2950	.2958	.2967	.2976	.2984	.2993	.3002

APPENDIX F Continued

%A	.0	.1	.2	.3	.4	.5	.6	.7	.8	.9
50.0	.3010	.3019	.3028	.3036	.3045	.3054	.3063	.3072	.3080	.3089
51.0	.3098	.3107	.3116	.3125	.3134	.3143	.3152	.3161	.3170	.3179
52.0	.3188	.3197	.3206	.3215	.3224	.3233	.3242	.3251	.3261	.3270
53.0	.3279	.3288	.3298	.3307	.3316	.3325	.3335	.3344	.3354	.3363
54.0	.3372	.3382	.3391	.3401	.3410	.3420	.3429	.3439	.3449	.3458
55.0	.3468	.3478	.3487	.3497	.3507	.3516	.3526	.3536	.3546	.3556
56.0	.3565	.3575	.3585	.3595	.3605	.3615	.3625	.3635	.3645	.3655
57.0	.3665	.3675	.3686	.3696	.3706	.3716	.3726	.3737	.3747	.3757
58.0	.3768	.3778	.3788	.3799	.3809	.3820	.3830	.3840	.3851	.3862
59.0	.3872	.3883	.3893	.3904	.3915	.3925	.3936	.3947	.3958	.3969
60.0	.3979	.3990	.4001	.4012	.4023	.4034	.4045	.4056	.4067	.4078
61.0	.4089	.4101	.4112	.4123	.4134	.4145	.4157	.4168	.4179	.4191
62.0	.4202	.4214	.4225	.4237	.4248	.4260	.4271	.4283	.4295	.4306
63.0	.4318	.4330	.4342	.4353	.4365	.4377	.4389	.4401	.4413	.4425
64.0	.4437	.4449	.4461	.4473	.4485	.4498	.4510	.4522	.4535	.4547
65.0	.4559	.4572	.4584	.4597	.4609	.4622	.4634	.4647	.4660	.4672
66.0	.4685	.4698	.4711	.4724	.4737	.4750	.4763	.4776	.4789	.4802
67.0	.4815	.4828	.4841	.4855	.4868	.4881	.4895	.4908	.4921	.4935
68.0	.4948	.4962	.4976	.4989	.5003	.5017	.5031	.5045	.5058	.5072
69.0	.5086	.5100	.5114	.5129	.5143	.5157	.5171	.5186	.5200	.5214
70.0	.5229	.5243	.5258	.5272	.5287	.5302	.5317	.5331	.5346	.5361
71.0	.5376	.5391	.5406	.5421	.5436	.5452	.5467	.5482	.5498	.5513
72.0	.5528	.5544	.5560	.5575	.5591	.5607	.5622	.5638	.5654	.5670
73.0	.5686	.5702	.5719	.5735	.5751	.5768	.5784	.5800	.5817	.5834
74.0	.5850	.5867	.5884	.5901	.5918	.5935	.5952	.5969	.5986	.6003
75.0	.6021	.6038	.6055	.6073	.6091	.6108	.6126	.6144	.6162	.6180
76.0	.6198	.6216	.6234	.6253	.6271	.6289	.6308	.6326	.6345	.6364
77.0	.6383	.6402	.6421	.6440	.6459	.6478	.6498	.6517	.6536	.6556
78.0	.6576	.6596	.6615	.6635	.6655	.6676	.6696	.6716	.6737	.6757
79.0	.6778	.6799	.6819	.6840	.6861	.6882	.6904	.6925	.6946	.6968
80.0	.6990	.7011	.7033	.7055	.7077	.7100	.7122	.7144	.7167	.7190
81.0	.7212	.7235	.7258	.7282	.7305	.7328	.7352	.7375	.7399	.7423
82.0	.7447	.7471	.7496	.7520	.7545	.7570	.7595	.7620	.7645	.7670
83.0	.7696	.7721	.7747	.7773	.7799	.7825	.7852	.7878	.7905	.7932
84.0	.7959	.7986	.8013	.8041	.8069	.8097	.8125	.8153	.8182	.8210
85.0	.8239	.8268	.8297	.8327	.8356	.8386	.8416	.8447	.8477	.8508
86.0	.8539	.8570	.8601	.8633	.8665	.8697	.8729	.8761	.8794	.8827
87.0	.8861	.8894	.8928	.8962	.8996	.9031	.9066	.9101	.9136	.9172
88.0	.9208	.9245	.9281	.9318	.9355	.9393	.9431	.9469	.9508	.9547
89.0	.9586	.9626	.9666	.9706	.9747	.9788	.9830	.9872	.9914	.9957

APPENDIX G Wave-Number/Wavelength Conversion Table

This table is based on the conversion: wave number (in cm^{-1}) = $\dfrac{10{,}000}{\text{wavelength (in } \mu\text{m)}}$.

For example, 15.4 μm is equal to 649 cm^{-1}.

Wavelength (μm)	0	0.1	0.2	0.3	0.4	0.5	0.6	0.7	0.8	0.9 cm^{-1}
1.0	10000	9091	8333	7692	7143	6667	6250	5882	5556	5263
2.0	5000	4762	4545	4348	4167	4000	3846	3704	3571	3448
3.0	3333	3226	3125	3030	2941	2857	2778	2703	2632	2564
4.0	2500	2439	2381	2326	2273	2222	2174	2128	2083	2041
5.0	2000	1961	1923	1887	1852	1818	1786	1754	1724	1695
6.0	1667	1639	1613	1587	1563	1538	1515	1493	1471	1449
7.0	1429	1408	1389	1370	1351	1333	1316	1299	1282	1266
8.0	1250	1235	1220	1205	1190	1176	1163	1149	1136	1124
9.0	1111	1099	1087	1075	1064	1053	1042	1031	1020	1010
10.0	1000	990	980	971	962	952	943	935	926	917
11.0	909	901	893	885	877	870	862	855	847	840
12.0	833	826	820	813	806	800	794	787	781	775
13.0	769	763	758	752	746	741	735	730	725	719
14.0	714	709	704	699	694	690	685	680	676	671
15.0	667	662	658	654	649	645	641	637	633	629
16.0	625	621	617	613	610	606	602	599	595	592
17.0	588	585	581	578	575	571	568	565	562	559
18.0	556	552	549	546	543	541	538	535	532	529
19.0	526	524	521	518	515	513	510	508	505	503
20.0	500	498	495	493	490	488	485	483	481	478
21.0	476	474	472	469	467	465	463	461	459	457
22.0	455	452	450	448	446	444	442	441	439	437
23.0	435	433	431	429	427	426	424	422	420	418
24.0	417	415	413	412	410	408	407	405	403	402
25.0	400	398	397	395	394	392	391	389	388	386
26.0	385	383	382	380	379	377	376	375	373	372
27.0	370	369	368	366	365	364	362	361	360	358
28.0	357	356	355	353	352	351	350	348	347	346
29.0	345	344	342	341	340	339	338	337	336	334
30.0	333	332	331	330	329	328	327	326	325	324
31.0	323	322	321	319	318	317	316	315	314	313
32.0	313	312	311	310	309	308	307	306	305	304
33.0	303	302	301	300	299	299	298	297	296	295
34.0	294	293	292	292	291	290	289	288	287	287
35.0	286	285	284	283	282	282	281	280	279	279
36.0	278	277	276	275	275	274	273	272	272	271
37.0	270	270	269	268	267	267	266	265	265	264
38.0	263	262	262	261	260	260	259	258	258	257
39.0	256	256	255	254	254	253	253	252	251	251
40.0	250									

APPENDIX H Four-Place Table of Common Logarithms

N	0	1	2	3	4	5	6	7	8	9
10	0000	0043	0086	0128	0170	0212	0253	0294	0334	0374
11	0414	0453	0492	0531	0569	0607	0645	0682	0719	0755
12	0792	0828	0864	0899	0934	0969	1004	1038	1072	1106
13	1139	1173	1206	1239	1271	1303	1335	1367	1399	1430
14	1461	1492	1523	1553	1584	1614	1644	1673	1703	1732
15	1761	1790	1818	1847	1875	1903	1931	1959	1987	2014
16	2041	2068	2095	2122	2148	2175	2201	2227	2253	2279
17	2304	2330	2355	2380	2405	2430	2455	2480	2504	2529
18	2553	2577	2601	2625	2648	2672	2695	2718	2742	2765
19	2788	2810	2833	2856	2878	2900	2923	2945	2967	2989
20	3010	3032	3054	3075	3096	3118	3139	3160	3181	3201
21	3222	3243	3263	3284	3304	3324	3345	3365	3385	3404
22	3424	3444	3463	3483	3502	3522	3541	3560	3579	3598
23	3617	3636	3655	3674	3692	3711	3729	3747	3766	3784
24	3802	3820	3838	3856	3874	3892	3909	3927	3945	3962
25	3979	3997	4014	4031	4048	4065	4082	4099	4116	4133
26	4150	4166	4183	4200	4216	4232	4249	4265	4281	4298
27	4314	4330	4346	4362	4378	4393	4409	4425	4440	4456
28	4472	4487	4502	4518	4533	4548	4564	4579	4594	4609
29	4624	4639	4654	4669	4683	4698	4713	4728	4742	4757
30	4771	4786	4800	4814	4829	4843	4857	4871	4886	4900
31	4914	4928	4942	4955	4969	4983	4997	5011	5024	5038
32	5051	5065	5079	5092	5105	5119	5132	5145	5159	5172
33	5185	5198	5211	5224	5237	5250	5263	5276	5289	5302
34	5315	5328	5340	5353	5366	5378	5391	5403	5416	5428
35	5441	5453	5465	5478	5490	5502	5514	5527	5539	5551
36	5563	5575	5587	5599	5611	5623	5635	5647	5658	5670
37	5682	5694	5705	5717	5729	5740	5752	5763	5775	5786
38	5798	5809	5821	5832	5843	5855	5866	5877	5888	5899
39	5911	5922	5933	5944	5955	5966	5977	5988	5999	6010
40	6021	6031	6042	6053	6064	6075	6085	6096	6107	6117
41	6128	6138	6149	6160	6170	6180	6191	6201	6212	6222
42	6232	6243	6253	6263	6274	6284	6294	6304	6314	6325
43	6335	6345	6355	6365	6375	6385	6395	6405	6415	6425
44	6435	6444	6454	6464	6474	6484	6493	6503	6513	6522
45	6532	6542	6551	6561	6571	6580	6590	6599	6609	6618
46	6628	6637	6646	6656	6665	6675	6684	6693	6702	6712
47	6721	6730	6739	6749	6758	6767	6776	6785	6794	6803
48	6812	6821	6830	6839	6848	6857	6866	6875	6884	6893
49	6902	6911	6920	6928	6937	6946	6955	6964	6972	6981
50	6990	6998	7007	7016	7024	7033	7042	7050	7059	7067
51	7076	7084	7093	7101	7110	7118	7126	7135	7143	7152
52	7160	7168	7177	7185	7193	7202	7210	7218	7226	7235
53	7243	7251	7259	7267	7275	7284	7292	7300	7308	7316
54	7324	7332	7340	7348	7356	7364	7372	7380	7388	7396

APPENDIX H Continued

N	0	1	2	3	4	5	6	7	8	9
55	7404	7412	7419	7427	7435	7443	7451	7459	7466	7474
56	7482	7490	7497	7505	7513	7520	7528	7536	7543	7551
57	7559	7566	7574	7582	7589	7597	7604	7612	7619	7627
58	7634	7642	7649	7657	7664	7672	7679	7686	7694	7701
59	7709	7716	7723	7731	7738	7745	7752	7760	7767	7774
60	7782	7789	7796	7803	7810	7818	7825	7832	7839	7846
61	7853	7860	7868	7875	7882	7889	7896	7903	7910	7917
62	7924	7931	7938	7945	7952	7959	7966	7973	7980	7987
63	7993	8000	8007	8014	8021	8028	8035	8041	8048	8055
64	8062	8069	8075	8082	8089	8096	8102	8109	8116	8122
65	8129	8136	8142	8149	8156	8162	8169	8176	8182	8189
66	8195	8202	8209	8215	8222	8228	8235	8241	8248	8254
67	8261	8267	8274	8280	8287	8293	8299	8306	8312	8319
68	8325	8331	8338	8344	8351	8357	8363	8370	8376	8382
69	8388	8395	8401	8407	8414	8420	8426	8432	8439	8445
70	8451	8457	8463	8470	8476	8482	8488	8494	8500	8506
71	8513	8519	8525	8531	8537	8543	8549	8555	8561	8567
72	8573	8579	8585	8591	8597	8603	8609	8615	8621	8627
73	8633	8639	8645	8651	8657	8663	8669	8675	8681	8686
74	8692	8698	8704	8710	8716	8722	8727	8733	8739	8745
75	8751	8756	8762	8768	8774	8779	8785	8791	8797	8802
76	8808	8814	8820	8825	8831	8837	8842	8848	8854	8859
77	8865	8871	8876	8882	8887	8893	8899	8904	8910	8915
78	8921	8927	8932	8938	8943	8949	8954	8960	8965	8971
79	8976	8982	8987	8993	8998	9004	9009	9015	9020	9025
80	9031	9036	9042	9047	9053	9058	9063	9069	9074	9079
81	9085	9090	9096	9101	9106	9112	9117	9122	9128	9133
82	9138	9143	9149	9154	9159	9165	9170	9175	9180	9186
83	9191	9196	9201	9206	9212	9217	9222	9227	9232	9238
84	9243	9248	9253	9258	9263	9269	9274	9279	9284	9289
85	9294	9299	9304	9309	9315	9320	9325	9330	9335	9340
86	9345	9350	9355	9360	9365	9370	9375	9380	9385	9390
87	9395	9400	9405	9410	9415	9420	9425	9430	9435	9440
88	9445	9450	9455	9460	9465	9469	9474	9479	9484	9489
89	9494	9499	9504	9509	9513	9518	9523	9528	9533	9538
90	9542	9547	9552	9557	9562	9566	9571	9576	9581	9586
91	9590	9595	9600	9605	9609	9614	9619	9624	9628	9633
92	9638	9643	9647	9652	9657	9661	9666	9671	9675	9680
93	9685	9689	9694	9699	9703	9708	9713	9717	9722	9727
94	9731	9736	9741	9745	9750	9754	9759	9763	9768	9773
95	9777	9782	9786	9791	9795	9800	9805	9809	9814	9818
96	9823	9827	9832	9836	9841	9845	9850	9854	9859	9863
97	9868	9872	9877	9881	9886	9890	9894	9899	9903	9908
98	9912	9917	9921	9926	9930	9934	9939	9943	9948	9952
99	9956	9961	9965	9969	9974	9978	9983	9987	9991	9996

Index

Abbé number, definition 405
Abbé refractometer 406, 407
Abbreviations (list) xiv
Absolute error 861
Absorbance, and bandwidth 71
  defined 66
Absorbance/percent absorption conversion (table) 1003–1005
Absorbance ratioing technique 509
Absorption/absorbance conversion (table) 1003–1005
Absorption, of alpha particles 289
  atomic 140–150
  of beta particles 290, 291
    experiment 947
  of gamma radiation 290, 292
    experiment 947
  of X rays 17, 262–264
Absorption band shapes and spectral bandwidth 40
Absorption bands (electronic) for chromophores (table) 95
Absorption coefficients, X-ray 262
Absorption and derivative signals in ESR 366
Absorption edge, of X rays 17, 242
  (table) 243
Absorption filters 30
Absorption flame photometry. See Atomic absorption
Absorption lines (atomic), of elements (table) 1001–1003
Absorption photometry, ultraviolet/visible, instruments for 51–63
Absorption sensitivity (atomic), of elements (table) 1001–1003
Absorption spectra, and molecular structure, in infrared 178–198
  in ultraviolet/visible 94–97

Absorption spectrophotometry, compared with fluorescence spectrometry 106
  radiation sources 20–22
Absorption spectrum from fluorescence measurements 107
Absorptivity, molar 68
  specific 67
Abundance, natural isotopic (table) 298
ac Arc, as spectroscopic source 159
Accelerating systems in mass spectrometry 568
Acceptance angle 38
Acceptance slit, of pulse-height analyzer 258
Accuracy, definition 862
  and instrumental calibration 864
  spectrophotometric 70–73
Acetaldehyde, molecular vibrations 178
Acetylene flames 131, 132, 137
Acid–base equilibria and ultraviolet/visible absorbance 70
Acid–base indicators, dissociation constant (table) 998–1000
  determination of 938
Acid–base titrations, in aqueous systems 682–684
  experiments 953
  in glacial acetic acid 953
  in nonaqueous solvents 684–687
  experiments 953
Acids, conductometric titration of 792–796
  coulometric titration of 771, 773
  dissociation constants (table) 998–1000
  measurement of 938
  infrared absorption bands 185

  potentiometric titration of, in aqueous systems 682–684
    in nonaqueous systems 684–688
  spectrophotometric titration of 940
  thermometric titration of 621
ac Polarography 711
Acrylonitrile, NMR spectrum 339
ac Spark, as spectroscopic source 159
Activation analysis, neutron 304–310
Active filters, electronic 832
Activity, of ions, measurement of 640
  optical 415
  of radionuclides 293
Addition evaluation method 379, 585
Address bus, computer 874
Adipate, 2,5-dimethyl, diethyl ester, infrared spectrum of 188
Adjusted retention time or volume 433
Adsorbents, for gas–solid chromatography 484
  for liquid adsorption chromatography 530–532
Adsorption chromatography 434, 530–536
Adsorption processes in liquid–solid chromatography 532–534
Aerosols, production of 127–129
Air-acetylene flame 131, 132, 137
Air-propane flame 131
Alcohols, infrared absorption bands 185
  (table) 186
Aldehydes, infrared absorption bands 186

1009

1010 INDEX

Alkali flame ionization detector 471
Alkali metals, flame photometric determination of 944
  ionization of, in flames 137
Alkaline earth metals, ionization of, in flames 137
Alkanes, infrared absorption bands 182
Alkenes, infrared absorption bands 183–185
Alkyl quaternary amines as counterions in ion-pair chromatography 542–545
Alkyl sulfonates as counterions in ion-pair chromatography 542–545
Alkynes, infrared absorption bands 185
Alphanumeric code 881
Alpha particles 289, 292
  counting 255, 290
Alternating-current arc, as spectroscopic source 159
Alternating-current spark, as spectroscopic source 159
Alumina, as solid adsorbent 531
  solvent strength parameter (table) 532
American Instrument Company thermogravimetric analyzer 610
American standard code for information interchange (ASCII) 881
Amides, infrared absorption bands 187
Aminco thermogravimetric analyzer 610
Amines, infrared absorption bands 179, 187, 188
Ammonia, formation constants for metal complexes (table) 1000
Ammonia gas-sensing electrode 647
Ammonia radical, ESR spectrum 368
Amperometric detectors for liquid chromatography 514
Amperometric titrations, apparatus for 723
  methodology 720–729
  with two-polarized electrodes 726–728
Amperostat 832
Amphiprotic solvents 685
Amplification factor of multiplier phototubes 25
Analog circuits, medium scale integration 829–832
  operational amplifier, basic 812
  for process control 922
Analog control of process streams 922
Analog-to-digital converters, counter converters 838
  data system components 844
  description of 838
  dual slope 840
  successive approximation 839
  voltage-to-frequency 839
Analysis time, optimization in liquid column chromatography 516

and resolution, in chromatography 444, 481
Analytical curve, as calibration technique 865
  and instrument response 847
Analytical wavelength, selection in ultraviolet/visible 76
Analyzer crystal, in X-ray spectrometry 247–249 (table) 249
Analyzer systems, magnetic-deflection type 578–580, 583–585
  in mass spectrometers 578–580, 583–585
Analyzers, automatic chemical 923–931
  optical, in polarimeters 421
AND gate 814
Angle, diffraction or reflection 43–46
  incidence 43
Angular dispersion 38
  of a grating 45
Angular momentum, spin 13
Angular momentum quantum number 14, 357
Anhydrides, infrared absorption bands 186
Anion interferences in atomic absorption and flame emission spectrometry 136
Anionic complexes of metals, ion exchange of 549
Anisotropic crystals 415
Anisotropy, in NMR spectroscopy 333
Annealing temperatures, from thermal analysis 616
Annihilation radiation 292
Anodic stripping analysis 715, 768
Answers to problems 971–994
Anthracene, excitation and fluorescence spectrum 106
Antimony, Auger MNN spectrum 391
Antimony(V), electrolytic reduction 766
Antireflection coatings 37
Anti-Stokes lines 218
Aperture, effective, of grating 46
  numerical 38
Aperture ratio 139
Aprotic solvents 685, 686
Arc, ac, as spectroscopic source 159
  dc, as spectroscopic source 155–159
Arc light sources 20
Area, surface, of adsorbent particles 531
Area normalization evaluation method 452
Area and quantity, correlation of 446–452
Argentometric titrations 621, 955
Argon (krypton) laser 226
Argon plasmas, as spectroscopic sources 161–164
Arithmetic-logic unit 871
ARL direct-reader emission spectrograph 165
ARL ion microprobe mass analyzer 388

ARL X-ray spectrometer 267
Aromatic compounds, infrared absorption bands 183, 184
  Raman lines 229
  ultraviolet/visible absorption bands 95
Arrays of photodiodes 29
Arsenic, amperometric titration with bromate 962
Ashing cycle in graphite furnace atomizers 149
Assembly language, computer 875
Astimatic image 41, 166
ASTM X-ray powder diffraction file 280
Asymmetrical stretching vibration 11
Asymmetrically substituted carbon atom 415
Asymmetry factor of chromatographic peak 433, 437
Asynchronous data transfer, application 845
  definition of 827
Atmosphere, effect in thermal analysis 613, 615
Atomic absorption flame spectrometer 142, 853, 889
Atomic absorption flame spectrometry 140–145
  comparison with flame emission 146
  experiments 945
  intensity relations 141
  light sources 143
  and observation height 132
  sensitivity of elements (table) 1001–1003
  spectra (table) 1001–1003
Atomic concentrations in flames, distribution patterns 132
Atomic emission 138–140
Atomic emission spectrometers 164–170
Atomic energy levels 4
Atomic fluorescence flame spectrometry 171
Atomic number/wavelength relationship in X ray 240
Atomic refractions (table) 405
Atomic term diagram 5
Atomization in flames 129, 131
Atomization step in graphite furnace atomizers 149
Atomizers 127–129
Attenuated total reflectance 206–209
Auger electrons 389, 390
Auger emission spectroscopy 380–382, 389–394
  instrumentation 391–393
  quantitative analysis by 393
Auger spectra 390, 391
AutoAnalyzer 927–930
Automated, definition 897
Automated process control system 921
Automatic, definition 897
Automatic chemical analyzers 923–931
Automatic clinical analyzer (DuPont) 926
Automatic diffractometers 279
Automatic gain control in spectrometers 54
Automatic laboratory analyzers 898

INDEX 1011

Automatic slit control in spectrometers 55
Automatic titrators 675–679
Automation strategy 898–899
Autoprotolysis ranges of solvents 685
Avalanche, of electrons in radiocounters 250
Axial diffusion in chromatography 441

Backflushing 918
 in gas–solid chromatography 485
Background, of flames 135
 in radiochemical measurements 258, 296
 in X-ray methods 247
Background absorption, of flames 135
Background correction methods, in flame spectrometry 135
 in radiochemical analysis 296
Back-reflection X-ray diffraction camera 278
Back-scattered energy, in gamma spectra 308
Back-scattering of beta particles 291
 experiment 948
Baird densitometer-comparator 172–173
Baird fluorescence spectrophotometer (double monochromation) 119
Baird microdensitometer comparator 172–173
Ball-and-disk integrator 447
Band broadening in chromatography 440–443
Bandpass, optical 37, 39
Bandpass filters, active 853
 passive 851
Bandwidth, in ESR spectrum 361
 and photometric accuracy 71
 spectral 36, 39, 71
Barn, definition 304
Barrier-layer photocell 23
Base line correction methods 135, 209, 451
Base line discriminator 260
Base line resolution 438
Base line width, in chromatography 433, 438
Base peak, in mass spectrometry 598
Base-ten logarithms, four-place table 1007–1009
Bases. See Acids
BASIC computer language 876
Basic solvents 685
Bates–Guggenheim convention 653
Bausch & Lomb Spectronic 20, 56
Beckman conductivity bridge 785–786
Beckman paramagnetic oxygen analyzer 911
Beckman pH meter, model G 659
Beckman spectrophotometer, model DU, 54
Beer's law 67–69
 application in dual path, dual frequency analyzer 902

chemical deviations from 69
 effect of refractive index upon 68
 effect of polychromatic radiation upon 69
 in X-ray absorption 262
Bending motions of molecules 11
Bendix time-of-flight mass spectrometer 581
Benzene ring substitution pattern of infrared combination bands 183
Berg–Barrett X-ray diffaction method 282
Beta particles 290, 291
 energy of (table) 298
 range-energy relationship 291
 experiment 947
Beta scintillation counter 304
Beta value, in gas chromatography 435, 478
Bimetallic electrode systems 671
Binary coded decimal counters 827, 881
Binary counter 826–827
Binary mixtures, and process analyzers 901
 simultaneous determination 77
Binary number system 880
Binding energies of core electrons 394
Bioanalytical Systems amperometric detector 514
Biot equations 418
Bipolar transistors 805–806
Birefringence, circular 414
Blackbody radiators 21, 189–191
Blaze angle 45
Blaze wavelength 45
Bleeding from gas chromatographic columns 461, 468
Blocking filter. See Cutoff filters
Bode plot 813
Bohr magneton 16
Boiler cap electrode 156
Bolometer 194
Boltzmann distribution, of nuclear energies 319
Boltzmann equation 139
Bond angle, and proton coupling 337, 338
Bonded stationary phases 461, 537–539
Boolean algebra, basic functions 813
 and computer operations 869
 identities 815
Borate pH reference buffer standard 655
Boric acid, thermometric titration curve 620
Boxcar averaging 855
Boxcar integration, hardware circuits 854
 software averaging 855
Bragg equation 247, 248
Brass analysis, by electrogravimetry 962
 by polarography 958
Bremsstrahlung 291
Brewster's angle 422
Bridge, conductance 785–786
 salt 638

Wheatstone 466–467
Bridge-potentiometer arrangement in photometers 55
Briggsian logarithms, four-place table 1007–1009
Broad band NMR spectrometer 329
Broad band NMR spectroscopy 323
Broadening of absorption lines 141, 144
Bromate, amperometric titration with arsenic 962
Bromine, coulometric generation of 771, 966
 energy-level diagram, X-ray 244
 X-ray spectrum 242
Brushes, type of bonded phase support 537
Buffers
 acid–base 652
 DAC circuit applications 837
 pH standard reference 652
  (table) 655
 potential 748
 radiation 137, 138, 158
 storage register 826
Buffer solutions 652, 655
Buffer value, Van Slyke 653, 655
Bulk properties of samples and continuous analysis 901–905
Buret electrode 671, 687
Burner-nebulizer systems for flame spectrometry 133
Burning velocity, fuel-oxidant mixtures (table) 131
Buses, computer 874
Butyl alcohols, mass spectral data (table) 593
Butyl ether, di-n-, NMR spectrum 345

Cadmium, Auger MNN spectrum 391
Calcium acetate monohydrate, DTA and TG curves 615
Calcium hydroxide, pH reference buffer 655
Calcium ion, standard reference values (table) 650
Calcium oxalate monohydrate, TG curve 611
Calcium-selective ion electrode 645
Calibrated sample loop and valves for chromatography 455–456, 501–502
Calibration curve 847
Calibration methods, analytical curve 864
 internal standard 867
 standard addition 865
Calomel reference electrodes 635, 637
 potential of (table) 636
Calorimetry, differential scanning 606–609, 611–616
 titration 619–622
Camera design, microradiographic 264
 X-ray diffraction 272–277
Capacitance, high-frequency 796
Capacitive current, in voltammetry 696, 708, 711

Capacitive effects, in conductance cells 784
Capacity factor 435
Capillary characteristics in polarography 697
Capillary columns, in gas chromatography 459, 460
Capillary injector-splitter 457, 458
Capture cross section of nuclei 304
  (table) 298
Carbonate pH reference buffer standard 655
Carbon electrodes, spectroscopic 156
Carbon monoxide, nondispersive infrared analyzer for 907
Carbon rod atomizer 148
Carbon/sulfur analyzer 886
Carbon tetrachloride, Raman spectrum 219
Carbon-13, chemical shifts (table) 335
  coupling with protons 331
  ensemble average with 855
Carbon-13 NMR spectroscopy 330–331
Carbonyl compounds, α, β, unsaturated, effect of substituents upon ultraviolet/visible absorption wavelength 96
Carbonyl group, infrared absorption bands 185
  (table) 187
Carboxyl group, infrared absorption bands 186
Carboxylic cation exchange resin 545–546
Carrier distillation excitation mode 157
Carrier gases, in chromatography 479
  for discharge lamps 143
Cary spectropolarimeter 426–427
Cathode surfaces, for phototubes 27
Cathodic stripping analysis 715
Cation-responsive glass electrodes 641
Cavity, sample, in ESR spectrometer 359
Cell, concentration 679
  conductivity 784, 787–789
  high-frequency conductance 797
  See also Cuvettes
Cell constant, in conductance 787
Cell electromotive force, defined 631, 738
Cell thickness, measurement optically 206
Central processing unit of computer 872
Centrifugal analyzers 930–932
Cerium(IV), coulometric generation of 770
  current-potential curve 770
Chain length, by NMR 347
Channel electron multiplier 385
Channel electron multiplier array 386, 575
Channel width of pulse-height analyzer 260

Characteristic curve of photographic emulsion 171
Charging current in voltammetry 696, 708, 711
Chelating resins 546, 549
Chelating site groups in ion-selective electrodes 646
Chelation, protective, in flame photometry 137
Chemical analyzers, automatic 923–931
Chemical coulometers 762
Chemical durability of glass electrodes 590
Chemical equilibria in flame gases 130, 138
Chemical exchange, effect on NMR spectra 336, 340
Chemical interferences, in flames 136
Chemical ionization, as mass spectrometer source 569
Chemical and physical changes in thermal analysis 148, 611–619
Chemical shifts, in Auger spectra 392
  of carbon-13 (table) 335
  in ESCA, 395–397
  in NMR, 332–336
  proton NMR (table) 324
Chemical vaporization for atomic absorption spectrometry 149
Chloride, amperometric titration 725
  formation constants for metal complexes (table) 1000
  potentiometric titration 681
  reference standard values (table) 650
Chlorine analyzer, continuous 916
Chloroform, Raman spectrum (partial) 233
CHN analyzers 931–935
Chopping, optical 30
Chopping frequency, in infrared 189
Chromatographic analysis, microcomputer network for 895
Chromatographic behavior of solutes 432–436
Chromatographic columns, construction 459, 502
Chromatographic methods, classification 430
Chromatography, adsorption 530–536
  exclusion 553–561
  gas–liquid 454–484
  gas–solid 484–486
  general principles 430–452
  ion-exchange 545–553
  liquid column 495–529
  liquid–liquid partition 536–545
Chromatography/mass spectrometry interfacing 585–591
Chromium(VI) amperometric titration with lead 721
Chromophore structure, resonance Raman probe of 220
Chromophores, electronic absorption bands (table) 95

Chronoamperometry 715–717
Chronopotentiometry 713–715
Circular birefringence 414
Circular-cage multiplier design 26
Circular dichroism 415–417
  applications 420
  instruments 427
Circularly polarized light 412, 413
Citrate, formation constants for metal complexes (table) 1000
  pH reference buffer standard 655
Clarity, measurement of 94
Clausius–Clapeyron equation 479
Clerget formula 419
Clinical analyzers, automatic 923–931
Clocked reset-set flip-flops 818
Closed loop control 921
Coatings, thin film 37, 768
Cobalt, potentiometric titration with ferricyanide 681, 954
  (III/II) hexammine system 634
Coincidence correction 297
Collimators for X-ray beams 246
Collimation of X-ray beams 246
Collisional impurity quenching in fluorescence 111
Color, specification of 89
Color comparators 87–89
Color difference 90
Color measurements 89
Colorimetry, laws of 66–70
  quantitative methodology 75–78
  sources of error 68–73
  terms and symbols 66–68
  See also Photometry, Spectrophotometry
Column, in gas chromatography 459–464
Column bleed 461, 468
Column efficiency, in chromatography 436–440
Column operation in liquid column chromatography, to achieve preselected retention time 517
  optimized in terms of time and cost 517
Column oven in gas chromatography 464
Column packings, for exclusion chromatography 553–555
  for liquid column chromatography 504
  for liquid–solid chromatography 531
Column permeability 516
Columns, in gas chromatography 459–464
  in liquid column chromatography 502–504
Column temperature, in gas chromatography 461–463, 479–484
Combination band, infrared 12
Combination pH/reference electrode 658
Common logarithms, four-place table 1007–1009

INDEX 1013

Common-mode-noise rejection 830
Comparative method of activation analysis 307
Complexation, effect on electrode potentials 633, 682, 747
  effect on overpotential 739
  in ion-exchange chromatography 549, 550
Complexes, evaluation of composition 941
  formation constants for metal (table) 1000
Complexometric titrations 681, 796
  experiment 956
Complier programs 877
Complimentary metal oxide semiconductors 808
  NAND gate 817
Compressibility factor in gas chromatography 434
Compton continuum 308-309
Compton edge, energy of 308
Computer/instrument interface, atomic absorption spectrometer 889-892
  carbon/sulfur analyzer 886-889
  centrifugal analyzer 931
  clinical analyzer 927
  hardware for 882
  software for 884
Computer network, for laboratory automation 894-895
  for process control 923
Computer organization, hardware 871-875
  software 875-879
Computer systems, for clinical analyzers 925-931
  for laboratory automation 892-895
  for process control 923
Computerized instrument systems 881-885
Computing integrator 448
Concave grating emission spectrometers 165-166
Concave holographic gratings 47, 165
Concentration cell measurements 679
Concentration distribution ratio 548
Concentration error in absorption spectrophotometry, relative 73-75
Concentration gradient 693, 700
Concentration overpotential 739
Condensed phase interferences in flame spectrometry 136
Conductance, definition 782
  electrolytic, instrumentation for 783-789
  equivalent 782
  of ions (table) 783
  ionic 782
  measurement of 783-789
  specific 782, 787
  of potassium chloride solutions (table) 787
  temperature effect on 783, 789

Conductance cells 784, 787-789
  high-frequency 797
Conductivity, electrolytic 781-783
Conductivity bridge 785-786
Conductivity detector, Coulson 476
Conductivity meters for process control 789
Conductometric titrations 790-796
Conformation change, effect on NMR spectra 336
Conformational interconversion and spin state 367
Conjugated dienes, effect of substituents upon absorption wavelength 96
Conjugation, effect on infrared spectrum 183
  effect on ultraviolet/visible spectrum 95-96
Constant-current coulometry 767-774
  instruments for 760
Constant-current electrolysis 742-745
Constant-current potentiometry 672-675
Constant-pressure pumps 498
Constant-voltage electrolysis 747
Continuous flow analysis 898
  clinical analyzers 927-931
  gas chromatographic analyzers 915-919
  industrial analyzers 899-901
  infrared analyzers 905-909
  oxygen analyzers 909-913
  potentiometric analyzers 913-915
  for process control 919-923
Continuous variation method 941
Continuous-wave NMR spectrometers 324-330
Continuum, X-ray 241
Control bus, computer 874
Controlled-potential coulometry 760-767
  applications 765-767
  instruments for 761-765
Controlled-potential electroseparations 745, 761-767
Control charts in data analysis 864
Control unit of digital computers 871
Convection in voltammetry 693
Convective mixing in chromatography 442
Copper, current-voltage curves 674
  electrodeposition of 748, 962, 963
  ESCA spectra 396
  separation by internal electrolysis 963
Copper-base alloy, polarographic analysis of 958
Core-electron binding energies 394
Corrected fluorescence spectra 118, 120
Corrected spectra computer 121

Correlation charts. See Molecular structure
Cotton effect 415, 420
Cottrell equation 711
Coulograms 766
Coulombs, evaluation of number 762-765
Coulometers, chemical 762
  electromechanical 764
  electronic 764
  gas 763
  Sargent-Welch 762
Coulometric methods 759-774
  at constant current 760, 767-774
  with controlled potential 751, 759, 761-767
  principles 750, 759
Coulometric titrations, primary 769
  secondary 769-773
Coulometric titrator, electronic 764
Coulson conductivity detector 476
Counter electrode, in spectrography 156-157
  in voltammetry 703
Counter geometry, experiment 947
Counterions, in ion-pair chromatography 542-545
Counters, digital, asynchronous 827
  construction from flip-flops 824
  modulus of 827
  synchronous 827
  for nuclear and X radiation 249-257, 304
Counting, background in 258, 296
Counting geometry 297
Counting statistics 294-297
Coupled column technique 485, 504
Coupling, proton-proton (table) 336
  spin-spin 363-365
Coupling constant, ESR spectroscopy 362
Critical-angle refractometer, for process streams 903
Critical ray 404
Cross assembler, computer 876
Crossed-coil NMR probes 326
Cross polarization for NMR 331
Cross-section, capture, of nuclei 304
  (table) 298
Cryogenic superconducting solenoids for NMR 325
Crystal, anisotropic 415
  for attenuated total reflectance 208
Crystal analyzers in X-ray spectrometry 247-249
  (table) 249
Crystal scintillators 255
Crystal topography 281
Cumulative formation constants for metal complexes (table) 1000
Curie, definition 294
Curie principle in pyrolysis 457
Current, capacitive or charging 696, 698
  diffusion 694, 695-700

factors governing in electroseparations 750
limiting, in voltammetry
   694
Current amplifier 829
Current density, effect on
   overpotentials 740
Current ion chambers 250
Current measurement 829
Current–time relationship in
   coulometry 750
Current–voltage relationships in
   voltammetry 692–695
Curved crystal X-ray spectrometer 267
Cutoff and cut-on filters, optical
   30, 115
Cuvettes, cleaning 72
   geometry, in liquid column
      chromatography 508
   for infrared spectroscopy
      204, 206
   infrared window material
      (table) 202
Cyanide ion formation con-
   stants for metal complexes
   (table) 1001
Cyanide-ion selective electrode
   649
Cyclic voltammetry 712
Cylindrical mirror (energy)
   analyzer 386, 392
Czerny–Turner mount 48

Damping, of pulsations from
   pumps 497
Dark current 28, 829
Data acquisition system 844
Data bus, computer 874
Data flip-flops, description
   818
   use in counters and registers
      825–829
Data representation 879–881
Data transfer, on computer bus
   874
   parallel 826
   serial 826
dc Arc, as spectroscopic source
   155–159
Deactivation, of gas chromatographic supports 460
   of silica and alumina surfaces
      533, 534
Dead-stop end point method
   726–728
Dead time, of counters 252,
   255
Debugging programs 877
Decay, radioactive 293
   resolution of mixtures 310
   experiment 948
Decay curves of phosphorescence 123
Decay schemes of radionuclides
   294
Decay time, of fluorescence 105
   of scintillators 255
Decomposition potential 737
Decoders 824
Decoupling in NMR spectroscopy 341–344
Decrement evaluation method
   651
Deflection-type refractometer
   511

Deformation vibrational mode
   11
Degenerate vibrational mode
   11
Deionizer, in flame spectrometry
   138
Delivery units for titrators
   676, 677
DeMorgan's theorems 815
Dempster design mass spectrom-
   eter 579
Demultiplexers, digital 824
Densitometers 172–173
Density of photographic emul-
   sions 172
Depletion region in semicon-
   ductor detectors 256
Depolarization ratio 231
Depolarized electrode in voltam-
   metry 692
Deposits, physical factors
   affecting 749
Depth profiling 382, 384
Derivative and absorption sig-
   nals in ESR 366
Derivative autotitrator 678
Derivative polarography 707
Derivative pulse polarography
   709–711
Derivative spectroscopy 82
Derivative titration curves 669
Derivatization methods in
   chromatography 458,
   464, 514
Deshielding zones, NMR spectroscopy 333
Detection limit 848
   flame emission of elements
      (table) 1001–1003
   photometric detectors 509
Detector windows for X-ray
   spectroscopy 253
Detectors, for ESR spectroscopy
   361
   in gas chromatography 464–
      478
   gas density 904, 906
   for infrared spectroscopy
      191–196
   in mass spectrometers 573–
      575
   in NMR spectrometers 326
   of nuclear radiation 249–258
   for Raman spectroscopy 226
   spectral response, determina-
      tion 937
   for ultraviolet/visible region
      22–29
   for X radiation 249–258
Deuterium discharge lamp 20
Deuterium hollow cathode
   lamp 135
Deuterium oxide, exchange with
   labile protons 340
Devising colorimetric procedures
   75–78
Dextrorotatory 412
Diacetyl, ESR spectrum of
   semiquinone 369
Diagnostic structural analysis.
   See Molecular structure
Dialyzer membrane 928
Diamagnetic shielding. See
   Shielding
Diatomaceous earth supports
   460
Diatomic molecule, energy-
   level diagram 8

Dichroic filter 34
Dielectric constant (table) 797
   effect on acid–base strength
      686–687
   measurement of 796–798
Diethyldisulfide, Raman spectrum 230
Diethyl fumarate, NMR spectrum 346
Diene, conjugated, effect of
   substituents upon absorp-
   tion wavelength 96
Difference spectroscopy 82
Differential detector mode, in
   gas chromatography 477
Differential electrolytic potentiometry 672–675
Differential refractometers
   511–513
Differential scanning calorimetry
   606–609, 611–616
Differential spectrophotometry
   78–82
   experiment 940
Differential thermal analysis
   606–609, 611–616
Differential thermogravimetry
   609
Diffraction gratings 43–47
Diffraction of X rays 247–
   249, 270–282
Diffuse reflection 87
Diffusion, in solution 693
Diffusion coefficient, in mobile
   phase 441
   in stationary phase 442
Diffusion current 693–695
   factors affecting 697–700
   measurement of 698
Diffusion-current constants
   (table) 997
Diffusion enrichment 587
Diffusion layer 693
Digital computers, buses 874
   components 871–879
   hardware vs. software 879
   interfacing to instruments
      868, 881–885
   laboratory automation systems
      892–895
   software 875–879
Digital filtering 855–861
   Fourier transforms 860
   weighted 856
Digital integrated circuits
   813–814
Digital logic 813
   DeMorgan's theorem 815
   gates 814
   identities 815
   truth tables 814
Digital-to-analog converter 835
   parameters 837
Dihedral angle, proton coupling
   337, 338
Dilution correction for linear
   titration methods 85
Dilution value, $pH_{1/2}$, 655
Dimethyl-2,5-diethyladipate,
   infrared spectrum 188
Diode-array rapid-scanning
   spectrometer 58
Diode-transistor logic 816
Diodes, semiconductor 804
   Zener 804
1,1-Diphenyl-2′-picrylhydrazyl
   365, 371
Dipole–dipole interaction 432

INDEX 1015

Direct comparison evaluation method in polarography 719
Direct conductance measurements 789
Direct-current arc, as spectroscopic source 155–159
Direct-current argon plasma 163
Direct digital control of processes 922
Direct memory access 885
Direct-reading pH and pI meters 660
Direct-reading spectrographs 165, 169
Direct-reading spectrophotometers 56–63, 174
Direct X-ray methods 260
Discharge lamps 20
Discrete analysis (batch) 898
  clinical analyzers 925–927
  industrial analyzers 899
Discrete automatic analyzers 925–927
Discriminator, in X-ray and nuclear methods 258–260
Disk, sector 171
Disk technique, in infrared 205
Dislodgement of mercury drops 704
Dispersion, optical 38, 43 403, 405
Dispersion mode signal in NMR 321, 327
Dispersion interaction 431
Dispersivity, in molecular weight determination 559
Dissociation, in flame gases 136
Dissociation constants, acids (table) 998–1000
  indicators (table) 998–1000
  potentiometric determination of 628
  spectrophotometric determination of 938
Dissolved oxygen electrode system 914
Distribution coefficient, in exclusion chromatography 555
Distribution patterns of atoms in flames 132
Distribution ratio, concentration 548
Doppler broadening 144
Double-beam filter photometers 55
Double-beam flame spectrometers 141, 142
Double-beam fluorescence spectrophotometer 119–121
Double-beam infrared spectrometer 198
Double beam-in-time optical spectrometer 57, 60
Double-beam ultraviolet/visible spectrometers 54–58
Double bond stretching vibrations 183
Double-focusing mass spectrometers 583–585
Double-junction salt bridge 639
Double monochromation 60, 118

Double-pass monochromator 61
Double resonance NMR spectroscopy 341
Double-wavelength spectrophotometer 61–63
Drench quenching 228
Drop detachment 704
Droplets (aerosol), production of 127–129
Dropping mercury electrode, assembly 703
  characteristics 695–700
Drop-size distribution of aerosols 128
Drude equation 416
Drying step with graphite furnace atomizers 149
DSC. See Differential scanning calorimetry
DTA. See Differential thermal analysis
Dual detectors, in gas chromatography 477
Dual-in-plane integrated circuit packages 810, 816
Dual slope analog-to-digital converter 840
Dual-wavelength photometer 507
Dual-wavelength spectrophotometer 61–63
Duane-Hunt equation 241
Duoplasmatron ion source 387–389
DuPont differential thermal analyzer 607
DuPont gradient elution system 500
Dyna-Cath mercury cathode unit 754
Dynamic mechanical analysis 619
Dynodes 25

$E°$, sign convention 630–632
  table of values 995
$E_{1/2}$, definition 700–702
  table of values 997
Eagle mount 166
Eberbach & Sons Dyna-Cath 754
Ebert mount 48, 167
Echelette grating 44
Echelle grating 46, 167
Echelle grating/prism spectrometer 167–170
Echelle spectral pattern 168
Eddy diffusion 441
Editor programs 877
EDTA, formation constants for metal complexes (table) 1001
Effective plate number 436
Efficiency and resolution in chromatography 436–440
Effluent splitter 586, 588
Effusion separator 586, 588
EG&G Princeton Applied Research Corporation Polarographic Analyzer 705–707
Elastic collision of noble gas ion and surface atom 383
ELDOR 371
Electric quadrupole moment 15

Electric sector in mass spectrometers 583, 584
Electrical conductivity and temperature 783, 789
Electroactive groups in voltammetry 717
  (table) 718
Electroanalytical methods in gas analysis 912
Electrocapillary curve of mercury 696, 698
Electrochemical cells 630, 634
Electrochemical detectors for liquid column chromatography 513
Electrode potentials 630–634, 745
  effect of complex formation upon 633
  effect of concentration upon 632
Electrode-solution interface 700
Electrodeless discharge lamps 144
Electrodeposition, basic principles 736–738
  equipment for 741
Electrodeposits, physical characteristics of 749
Electrodes, first kind 665
  glass membrane 641–643
  ion-selective 640–661, 682
  for nonaqueous solvent systems 687
  for pH measurements 641–643, 656–658
  for polarography 695, 702–704
  polarized 692
  redox 667
  reference 634–639
  second kind 666
  spectroscopic 156–158, 160
  third kind 666
  two indicating 671–675, 726–728
Electrography 757–759
Electrogravimetric analysis, electrolyte composition 747
  factors governing current 750
Electrogravimetry 742–757
Electrolysis, constant-current 742–744, 760, 767–774
  controlled potential 745, 759, 761–767
  internal 755–757
  mercury cathode 752–755
  separations by 736–759
Electrolysis cell 736–738
Electrolyte, supporting 693
Electrolytic cell 630
Electrolytic conductance cell 784, 787–789
Electrolytic conductivity 781–783
  instrumentation for 783–789
Electrolytic methods, classification 628–630
Electrolytic purification 759
Electrolytic separations, equipment for 741
Electrolytic solution pressure of an element 630
Electrolytic stripping 715–717, 768

Electromagnet, for ESR spectrometers 361
Electromagnetic radiation and energy changes involved 1, 4
Electromechanical coulometers 764
Electrometric methods, classification (table) 629
  introduction to 628–639
  steady state 628
  transient 628
Electromotive force series, basis for 631
  table of values 995
Electron, energy levels 4–6, 15, 357–358
  g-factor 16
  unpaired 357
Electron avalanche in Geiger and proportional counters 250
Electron beam probe 261
Electron behavior in magnetic field 4–9, 15, 357
Electron binding energies, core 394
Electron capture detector 473–475
Electron double resonance 371
Electron energy analyzers for ESCA 399
Electron impact ion source 568
Electron magnetic moment 357
Electron microprobe 261, 268, 380, 381
Electron multiplier phototube, for ion beams 386, 574
  for photons 25–28, 174
Electron nuclear double resonance 371
Electron paramagnetic resonance. See ESR spectroscopy
Electron probe microanalysis 261
Electron spectroscopy for chemical analysis. See under ESCA
Electron spin behavior 15
Electron spin resonance spectroscopy. See ESR spectroscopy
Electronic absorption bands for chromophores (table) 95
  correlation with molecular structure 94–97
Electronic coulometers 764
Electronic energy levels, of atoms 4
  of molecules 6
Electronic states, singlet 8
  triplet 8
Electroseparations 736–780
Electrostatically bonded ion-exchange resin 547
Electrothermal atomization 146–149
Elemental analyzers, automatic 931–935
Elemental imaging or mapping 328
Elements, atomic emission and absorption spectra (table) 1001–1003
  atomic weights, inside front cover
  ionization of, in flames 137
  mercury cathode separation of 755
  periodic chart of, inside back cover
Elliptically polarized light 415
Eluent, average linear velocity of 433
Eluent strength of solvents 539–541
  (table) 532
Eluotropic series. See Solvent strength parameter
Emf of half-cell 631
  table of values 995
Emission lines, of elements (table) 1001–1003
  mercury lamp 113
Emission spectra, origin 5
  photoluminescent. See Fluorescence, Phosphorescence
Emission spectrographs 164–170
Emission spectroscopy 154–174
  detection methods 170–174
  excitation methods 155–164
Empirical formula from isotopic contributions 595–597
Emulsions, photographic 170–171, 249
ENDOR 371
Endotherms in thermal analysis 612
End point, location of 667–671
Energy, of radiometric emission 290–293
  (table) 298
Energy analyzer 386, 392
Energy changes and electromagnetic spectrum 4
Energy dispersion methods in X-ray analysis 257, 268, 270
Energy filter in mass spectrometers 583
Energy interchange involved in surface analysis 380
Energy-level diagram, electronic 5
  spinning nucleus 14
  unpaired electron (ESR) 16
  X-ray 17
Energy levels, electronic 4–9
  of helium-neon laser 225
  nuclear 14
Energy/mass relationship 578, 580, 583
Energy/range relationship 291
Energy resolution with nuclear and X-ray detectors 255, 257, 259, 270
Energy transfer quenching in fluorescence 111
Enrichment devices for GC/MS interface systems 586–588
Ensemble averaging 855
Enthalpic titrations 619–622
Enzyme electrodes 647
EPR spectroscopy. See ESR spectroscopy
Equitransferent filling solution 638
Equivalence point, location of 667–671
Equivalent conductance 782
  of ions (table) 783
Erasable programmable memory 873
Error, relative concentration, in absorption spectrophotometry 73–75
Error analysis 861–864
ESCA 380–383, 394–401
  chemical shifts 395–397
  detector systems 399
  electron analyzers 399
  instrumentation 397–401
  quantitative analysis 401
  scanning modes 400
  spatially resolved 400
ESR spectra 362–372
  correlation with molecular structure 367–372
ESR spectrometer 358–362
ESR spectroscopy 357–378
  applications 357
  line widths 367
  sample handling 359
  sensitivity 361
ESR standards 365, 371
Esters, infrared absorption bands 186, 188
Ethers, infrared absorption bands 185
Ethyl radical, ESR spectrum 369
Ethyl trifluoroacetate, ESCA spectrum 397
Ethylene, control analysis during purification 923
Ethylene dibromide, X-ray absorption method 263
Ethylenediamine-N,N,N',N'-tetraacetic acid, formation constants for metal complexes (table) 1001
Europium shift reagents 344
Evaluation methods, direct comparison 719
  in gas chromatography 451–452
  internal standard 452, 720
  ion-activity 650–652
  standard addition 719
Evelyn photoelectric colorimeter 53
Excitation potential, critical for X rays 240–244
Excitation spectrum, photoluminescent 106
Exclusion chromatography 529, 553–561
  column calibration 559
  column packings 553–555
Exotherms in thermal analysis 612
Expanded scale spectroscopy 78–82
Expansion chambers 134
Expansion and extension measurement 616–618
External generation of titrant 772
External lock system (NMR) 327
Eye, human, as detector 23

Fabry-Perot filter 31
Fan-in digital integrated circuits 819
Fan-out digital integrated circuit 819
Faradaic diffusion current 696–699
Faraday, definition 774
Faraday cell, optical 426
Faraday cup collector 573
Faraday effect 418, 427

Periodic Chart

IA	IIA	IIIB	IVB	VB	VIB	VIIB	VIII		
1 **H** 1.0079									
3 **Li** 6.941	4 **Be** 9.01218								
11 **Na** 22.98977	12 **Mg** 24.305								
19 **K** 39.098	20 **Ca** 40.08	21 **Sc** 44.9559	22 **Ti** 47.90	23 **V** 50.9414	24 **Cr** 51.996	25 **Mn** 54.9380	26 **Fe** 55.847	27 **Co** 58.9332	28 **Ni** 58.71
37 **Rb** 85.4678	38 **Sr** 87.62	39 **Y** 88.9059	40 **Zr** 91.22	41 **Nb** 92.9064	42 **Mo** 95.94	43 **Tc** 98.9062	44 **Ru** 101.07	45 **Rh** 102.9055	46 **Pd** 106.4
55 **Cs** 132.9054	56 **Ba** 137.34	57 *****La** 138.9055	72 **Hf** 178.49	73 **Ta** 180.9479	74 **W** 183.85	75 **Re** 186.2	76 **Os** 190.2	77 **Ir** 192.22	78 **Pt** 195.09
87 **Fr** (223)	88 **Ra** 226.0254	89 †**Ac** (227)	104 **(Rf)** (260)	105 **(Ha)** (260)					

*Lanthanum Series

58 **Ce** 140.12	59 **Pr** 140.9077	60 **Nd** 144.2	61 **Pm** (147)	62 **Sm** 150.4	63 **Eu** 151.96

†Actinium Series

90 **Th** 232.0381	91 **Pa** 231.0359	92 **U** 238.029	93 **Np** 237.0482	94 **Pu** (244)	95 **Am** (243)

( ) Numbers in parentheses are mass numbers of most stable or most common isotope.

Atomic weights corrected to conform to the 1971 values of the Commission on Atomic Weights.

© by Fisher Scientific Company. Used by permission.

α,β-Unsaturated carbonyl compounds, effect of substituents upon absorption wavelength 96

Vacuum photoemissive tubes 24–28
Vacuum requirements for surface analysis methods 381
Vacuum system in mass spectrometers 576
Valinomycin membranes for potassium ion electrodes 645, 646
Valley definition of resolution 577
Valves, cutoff, in mass spectrometry 567
  multiport rotary 455–456, 501
Van Deemter equation 440, 442
  experiment involving 950
Van Slyke buffer value 653, 655
Vaporization of metal salts 129
Vaporization interferences in flame spectrometry 136
Vaporization methods for atomic absorption spectrometry 149
Variable path length infrared cuvette 204
Variable transmission interference filter 33, 35, 198
Varian carbon rod atomizer 148
Varian chemical ionization/electron impact ionization source 570
Varian double-beam spectrophotometer model 17D 60
  model 634 57
Varian double-pass monochromator model 219 61
Varian electron capture detector 474
Varian ESR spectrometer 360
Varian flame photometric detector 472
Varian NMR spectrometer 329
Varian thermionic emission detector 471
Velocity of eluent 433
Very large scale integration and single-chip computers 886
Vibrational energy levels 10
Vibrational modes in infrared region 11
Vibrationless transition 8
Vibronic excited state 220
Vignetting 47
Virtual excited state (Raman) 217
Viscosity, effect in liquid column chromatography 516, 520
Viscosity of solvents (table) 532
Viscous flow of gases 567, 587
Void volume in chromatography 433
Volatile memory 873
Volatilization, selective in emission spectroscopy 157
Voltage comparator circuits, in analog-to-digital converters 838
  level testor 833
  pulse counter 834
  Schmitt trigger 833

Voltage follower circuits 830
Voltage-to-frequency converter 839
  noise reducing property 852
Voltage measurement 829
Voltage/wavelength relationship for X rays 241
Voltammetry 691–735
Volume correction in linear titration methods 85
Volume-surface mean diameter of aerosol droplets 129
Volumetric phase ratio in chromatography 435

Wadsworth mount 166
Wagging vibrational mode 11
Wall coated open tubular (WCOT) column 459, 478, 480
Water, infrared determination of 178
  NMR determination 324, 348
  See also Moisture
Waters Associates radial compression module 503
Water treatment system, chlorine process analyzer 916
  process control 923
Wave guide 359
Wavelength, of atomic absorption lines (table) 1001–1003
  blaze 45
  defined 2
  of electronic absorption bands (table) 95
  of flame emission lines (table) 1001–1003
  nominal 31, 33, 36
  of X-ray lines and edges (table) 243
Wavelength/atomic number relationship for X rays 240
Wavelength shifter, for scintillation counting 303
Wavelength/voltage relationship for X rays 241
Wavelength/wave number conversion (table) 1006
Wave number 2
Wave number/wavelength conversion (table) 1006
WCOT columns 459, 478, 480
Wedge, interference filter 33, 35
Weight-average molecular weight 559
Weighted digital filtering 856
Weissenberg method 274
Well-type scintillation counter 254
Wheatstone bridge circuit 466–467
Whiteness 90
White noise 849, 850
Wide line NMR spectrometer 329
Wide line NMR spectroscopy 323
Window material, for infrared region (table) 202
  for X-ray region 252, 253
Window width of pulse-height analyzer 258–260
Working curve. See Analytical curve

X-Band frequencies 361
Xenon arc lamp, spectral radiance 21, 113
Xenon flash lamp 113
X-Ray absorptiometer, nondispersive 265
X-Ray absorption methods 262–264
X-Ray continuum 241
X-Ray diffraction 270–282, 946
  back-reflection camera 278
  cameras for 274–278
  powder data file 279
  radiation for 277
  sample handling 278
  of single crystals 271–274
  structural correlations 279–282
X-Ray emission lines (table) 243
X-Ray energy spectrometry 257
X-Ray energy levels 16–18, 243, 244
X-Ray fluorescence, applications 268–270
  methodology 265–270
X-Ray fluorescence spectrometer 266–268
X-Ray generating equipment 246
X-Ray methods 239–288
  instrumentation 245–270, 274–279
X-Ray microradiography 264
X-Ray photoelectron spectroscopy. See ESCA
X Rays, detectors for 249–258
  diffraction from crystal planes 247–249
X-Ray spectra, generation 240–245
  and oxidation state 245
  and wavelength relationship 241
X-Ray spectral range 4, 239
X-Ray spectroscopy, instrumental units 245–249
X-Ray tube 240, 246
XU (X unit), definition 245
o-Xylene, infrared spectrum 184

Zener diode 804
Zeroth order of grating 44, 46
Zirconia oxygen analyzer 912–913
Zone velocity in chromatography 444

INDEX 1029

Surface areas of adsorbent particles 531
Surface barrier detector 256
Surface scatter method 93
Surface scatter turbidimeter 93
Surface hydroxyl groups, in liquid–liquid chromatography 537
 in liquid–solid chromatography 531, 533
Surface ionization source for mass spectrometry 573
Surface species, distribution of 382
Surface spectroscopy 379–381
Symbols (list) xx
Symmetrical stretching vibrations 11
Synchronous digital counters 827
Syringe injection, in gas chromatography 455
 in liquid chromatography 500
Syringe-septum injection method 455, 500
Syringe-septumless injection system 501
Syringe-type pumps 498
Systematic errors 861

Tagging, with radionuclides 301
 with stable isotopes 593
Tailing in gas chromatography. See Asymmetry factor
Tartrate pH reference buffer standard 655
Tau scale, NMR chemical shifts 333
Temperature, effect on choice of liquid phase in gas chromatography 461–463
 effect on conductance 783, 789
 effect on electrolysis 748, 749
 effect in gas chromatography 479–484
 effect on liquid column chromatography 504, 524
 effect on overpotential 740, 749
 effect on pH measurements 655, 660
 effect on reference electrode potentials 636
 of flames 130
  (table) 131
 and relative retention 480
Temperature compensator on pH meters 660
Temperature programming in gas chromatography 481–484, 951
Temperature sensors in DTA and DSC, 608
Ternary solvent mixtures for liquid chromatography 541
Tesla, conversion from gauss 358
Tetramethylsilane 327, 332
Tetroxalate pH reference buffer standard 655
Theoretical plate number 436

Thermal analysis 606–627
 differential 606–609, 611–616
 methodology 611–616
Thermal conductivity detector 465–468
 in process chromatography 919
Thermal detectors in infrared 191–195
Thermal ionization source for mass spectrometry 573
Thermal neutrons 304
 capture cross section (table) 298
 sources 305
Thermal noise 849
 and relative error 73
Thermionic emission detector 471
Thermistors 192, 466
 detectors for gas density 905
 for thermometric titrimetry 620
Thermocouple 192
Thermogravimetry 609
 instrumentation 609
 methodology 611–616
Thermomechanical analysis 616–618
Thermometric titrimetry 619–622, 952
 instrumentation 619
 methodology 620–622
Thermopile 192
Thermostatted detector compartment 464
Thickness of films, by electrolytic stripping 768
Thin film coatings 37, 768
Thin film integrated circuits 809
Third class electrodes 666
Three-electrode potentiostat 704–707
Threshold wavelength 27, 241
Throwing power of electrolyte 749
Time–current relation in coulometry 750
Time-of-flight mass spectrometer 580
Time-resolution in phosphorescence 123
Tin, determination in brass 958, 963
Titration calorimetry 619–622
Titrations, amperometric 720–729
 conductometric 790–796
 at constant electrolysis current 672–675
 coulometric 769–773
 nonaqueous 621, 684–688
 photometric 83–86
 potentiometric 664–690
 thermometric 619–622
 to zero current 726
Titrator, automatic 675–679
 photometric 85
TMS, proton resonance standard 332
Topography, crystal 281
Tortuosity factor in chromatography 441
Total internal reflection 38
Total luminescence spectroscopy 121

Trace analysis method in spectrophotometry 81
Track-and-hold amplifiers 831
Transducers, input 801
 output 802
Transition times in chronopotentiometry 713–715
Transistor–resistor logic 816
Transistors, bipolar 805
 field effect 807
 metal oxide semiconductor field effect 807
Transistor–transistor logic 816–817
Transit time of nonretained solute 433
Translater programs 786
Transmission photocathode 27
Transmittance 66
 of filters (optical) 31, 33
 of infrared materials (table) 202
 of infrared solvents (table) 203
 photometric measurement error 72, 73–75
 of solvents for ultraviolet region (table) 77
Transmittance–absorbance conversion 66
Transverse relaxation processes 321
Trapping volatile samples 457
Triangulation evaluation methods 447
Trimethylpentanes, mass spectra 599
Triple bond stretching vibrations 185, 228, 230
Triple-headed reciprocal pump 497
Triplet electronic state 8, 9
Tristate integrated circuits 819
Tristimulus values 89
Truth table for digital logic circuits 814, 815
Tunable dye laser 225
Tuned amplifiers. See Active filters
Tungsten electrode 672
Tungsten–halogen incandescent lamp 21
Turbidimetric standards 92
Turbidimeters 92
Turbidimetry 90–94
Turner model 110 fluorometer 114
Turnkey computer system 893
Twisting vibrational mode 11
Two-channel double-beam atomic absorption spectrometer 142
Two indicating electrodes 671–675, 726–728

Ultraviolet spectral region, correlation with molecular structure 94–97
 detectors 22–29, 505–510
 dispersing devices 30–47
 light sources 20–22
 solvent cutoff values (table) 77
Universal asynchronous receiver/transmitter 845
Unpaired electron 357

for Raman spectroscopy 224–226
for surface analysis 380
for ultraviolet/visible region 20–22
for X-ray region 240–243
Spark, ac, as spectroscopic source 159
Spark microprobe 161
Spark source mass spectrometry 572
Specific absorption coefficient 67
Specific absorptivity 67
Specific activity of radionuclides 294
Specific conductance 782, 787
of potassium chloride solutions (table) 787
Specific ion electrode as process analyzer 914
Specific refraction 404
Specific retention volume 434, 479
Specific rotation 416
experiment 949
temperature effect on 418
Spectra, flame emission (table) 998–1000
Spectral bandwidth 36, 40
Spectral distribution curves of radiant energy sources 21
Spectral interferences in flame spectrometry 136
Spectral order 32, 44
Spectral response, of barrier-layer cell 23
of human eye 23
of infrared detectors 192
of photoemissive tubes 27
of photovoltaic cell 23
of p–n photodiode 29
Spectral transmittance characteristics, of absorption glass filter 31
of interference filter 33
Spectrametrics dc argon plasma source 164
Spectra-structure correlation, by ESCA 395–397
by ESR spectroscopy 362–371
by infrared spectroscopy 178–189
by mass spectrometry 595–600
by NMR spectroscopy 331–347
by photoluminescence 107
by Raman spectroscopy 221, 228–232
by ultraviolet/visible spectrophotometry 94–97
by X-ray diffraction 279
Spectrofluorometers 116–121
Spectrographs, direct-reading 165, 169
emission 164–170
Spectrometer, optical 19
Spectrophotometer 20
double-wavelength 61–63
operation 56
reversed optics type 59
Spectrophotometric accuracy 70–73

Spectrophotometric precision 73–75
Spectrophotometric titrations 83–86
experiments 940
Spectrophotometry, differential 78–82, 940
Spectropolarimeters 425–427
Spectroreflectometer 87
Spectroscopic buffer 137, 138, 158
Spectroscopic electrodes 156–158
Spectroscopic splitting factor 15, 361
Spectrum lines, emission intensity (table) 998–1000
Specular reflection 87
Speed of a spectrometer 40
Sphere of reflection 273
Spin, of electron 15, 357, 362
of nuclei 14, 317, 318
Spin angular momentum 317
Spin decoupling 328, 341
Spin label 370
Spin-lattice relaxation 15, 320, 322
Spinning sidebands 326
Spin quantum number, of electron 317–320
of nuclei 318
Spin–spin coupling 336–339
Spin–spin relaxation 321
Spin tickling 343
Splitting, of electron energy levels 362–366
of injected samples 457, 458
Spot testing, electrographic 757–759, 964
Sputter etching 382
Stability of light sources 22
Stagnant pools, effect in chromatography 442
Stallwood jet 158
Standard absorber method for cell thickness 206
Standard addition evaluation method 379, 651, 719, 866
Standard deviation 295
Standard hydrogen electrode 631, 635
Standards, for ESR spectroscopy 365, 371
for exclusion chromatography 559
for ion-activity measurements (table) 650, 655
for NMR spectroscopy 332, 333
for pH measurements (table) 655
Stationary liquid phases, for gas chromatography 461–464, 480
table of values 462
Stationary phases, for exclusion chromatography 553–555
in gas-solid chromatography 484
for liquid–liquid chromatography 536, 537–539
in liquid–solid adsorption chromatography 530–532

Statistical methods, in data processing 863
in noise reduction 846
in radiochemistry 294–297
Statistical weights of electronic states 138, 139
Stearic acid, infrared spectrum 185
Step-sector method 171
Stepwise elution in liquid column chromatography 521
Stereochemical effects, in electronic absorption 97
in NMR spectroscopy 337
in photoluminescence 107
Stigmatic arrangement 166
Stimulated emission (lasers) 224
Stockholm convention on electrode potentials 631
Stokes Raman lines 13, 218
Stopped-flow injection technique 501
Storage register 825
Straumanis method 277
Stray radiation, causes of 71
effects on photometric accuracy 72
Stretching vibrations 11
Stripper column 918
Stripping, electrolytic 715–717, 768
Structured computer programming 876
Structure determination. See Molecular structure or Spectral-structure correlation
Student's equations in statistics 863
Styrene monomer, Raman spectrum of 230
Successive approximation analog-to-digital converter 839
Succinic acid, ESR spectrum 367
Sucrose, determination by polarimetry 950
Sulfide ion-selective electrode 644
Sulfonate ion exchangers 545–547
Sulfur, X-ray spectra and oxidation state 245
Sulfur compounds, determination in gas streams, 933
infrared absorption bands 189
Raman spectroscopy of 228, 230
Supervisory computer control 922
Support-coated open tubular column 459, 478, 480
Supporting electrolyte 693 (table) 997
Supports in gas chromatography 460
Suppression of ionization in flames 138
Suppression of maxima, in polarography 699
Suppressor column, in ion chromatography 550–552
Surface analysis 379–401
contamination problems 381
with electron beam probe 261
energy interchange involved 380
sources for 380

INDEX   1027

Selectivity coefficient, in ion-exchange chromatography 548
  of ion-selective electrodes 952
Selenium photocell 23
Selective-ion electrodes. *See* Ion-selective electrodes
Self-absorption, in radiochemistry 290
Self-absorption broadening in hollow-cathode lamps 144
Self-absorption quenching in fluorescence 111
Self-generative photocell 23
Self-quenching counters 252
Selvidge 386
Semiconductor detectors, for infrared 195–196
  for nuclear radiation 255–258
  ultraviolet/visible 28
  for X rays 255–258
Semiconductors, complimentary 808
  metal oxide 807
  n-type 803, 807
  p-n junction 804
  p-type 804, 807
Semirigid cross-linked polymer gels 553
Sensitivity, definition 847
  of elements by atomic absorption (table) 998–1000
Sensors, temperature, in DTA and DSC 608
  weight, in TG 609, 610
Separation factor in ion exchange 549
Separation range and pore-size diameter in exclusion chromatography 554
Separators for GC/MS interface systems 586–588
Serial data transfer 826
  applications 845
  by RS-232 bus 874
Series-across-detector arrangement 485
Series/bypass detector technique 485
Set point 921
Settling time, of digital-to-analog converter 837
  of multiplexers 824
Seya-Namioka mount 49
Sharp-cut filters 30
Shave-recycle technique 524
Sheen 87
Shielded electrode 671
Shielded flames 131
Shielded hollow cathode lamps 143
Shielding constant in NMR 332
Shielding in radiochemistry 251, 254
Shielding zones in NMR 333
Shift reagents 344–345
Shift register 826
Shim coils 327
Shimadzu thermogravimetric analyzer 610
Shot noise 25, 73, 850
  and relative concentration error 74
Sieves, molecular 484

Signal-to-noise ratio 846
  improvement by digital filtering 855
Signal transformation module, operational amplifier as 802
Sign convention for electrode half-reactions 563–566
Significant figures 863
Silanization of supports 460
Silicon carbide, infrared source 191
Silica, as optical material 28
  as solid adsorbent 531
  solvent strength parameters on (table) 532
Silicate ester bonded phases 537
Silicon surface-barrier detector 256
Siloxane type of bonded phase support 537–539
Silver, Auger MNN spectrum 391
  current-voltage curves 725
Silver ion, coulometric generation of 966
Silver/silver chloride electrodes 637
  potentials of (table) 636
Simultaneous photometry of binary mixtures 77
Sine-bar linkage wavelength drive 49
Single-beam infrared spectrometer 198
Single-beam ultraviolet/visible spectrophotometer 52, 54
Single-channel analyzer for radionuclides and X rays 258
Single coil NMR probes 326
Singlet electronic state 7
Singlet (excited)-triplet transition 8
Single test analyzer (Technicon) 926–927
60° sector mass spectrometer 580
Skeletal vibrations, Raman spectroscopy of 231
Sliding plate valve 456, 501
Slit distribution function 36
Slit width, and resolution 35–37, 938
  spectral 31, 33
Slits, optical 35–37
Slope-ratio method 941
Slot burner 134
Small scale integration 817
"Smart" instruments 870
Soap film analyzer crystals 248 fn
Sodium, atomic term diagram 5
  decay scheme of ^{24}Na 294
  flame photometric determination 944
Sodium carbonate buffer solution 655
Sodium chloride X-ray powder diffraction pattern 277, 280
Sodium iodide scintillator 255
Sodium ion, correction nomograph for pH electrodes 662
  reference standard values (table) 650

Sodium ion-selective glass electrodes 641
Sodium tetraborate pH reference buffer 588
Solar-blind photomultipliers 28
Solids, infrared handling of 205–209
  mass spectrometry handling of 567, 572, 573
  Raman handling of 227–229
  ultraviolet/visible spectra of 86–89
Solid-state arrays 29
Solid-state ion sensors 643–645, 649
Solid-state mass spectroscopy 572
Solid-state reactions by thermal analysis 616
Solubility of precipitates, radiochemical experiment 949
Solutes, chromatographic behavior of 432–436
Solute–solvent interaction 431
  See also Gas liquid chromatography, Liquid column chromatography
Solutions, handling of, by emission spectroscopy 157, 160, 161–164
  by infrared spectroscopy 202–205
  by Raman spectroscopy 226
Solvent delivery systems for liquid column chromatography 495–500
Solvent programming 499, 520
Solvents, choice in liquid–solid chromatography 534–536
  effect on electronic absorption spectra 96
  effect on fluorescence 107
  for exclusion chromatography 555
  for infrared spectroscopy (table) 203
  for NMR spectroscopy 326
  for Raman spectroscopy 227, 228
  refractive index of (table) 532
  for ultraviolet/visible region 76
  table 77
  viscosity of (table) 532
Solvent strength parameter 534–536
  table of values 532
Solvent viscosity, effect in liquid column chromatography 516, 520
Source (lamp) operational modes 22
Source programs 876
Sources, for atomic absorption flame spectrometry 143
  for atomic emission spectroscopy 129–133, 155–164
  for ESCA 398
  for ESR spectroscopy 359
  for infrared region 189–191
  neutron 305, 306
  for photoluminescence 113, 122
  for polarimetry 421

Reflection losses 37
Reflection-type refractometer 511, 512
Reflectometers 87–89
Refraction, index of 2, 403
  specific 404
Refractive index 2
  effect on Beer's law 69
  of solvents (table) 532
Refractometers 406–408
  Abbé 406, 407
  differential 511–513
  immersion 406, 407
  for process streams 903
  recording 407–411
Refractometry, applications 406, 411
Refractory compounds, in flame spectrometry 137
Refractory metals, dissolution of 736
Registers, digital 825
Relative absorbance spectrophotometry 78–82, 940
Relative concentration error in absorption spectrophotometry 73–75
Relative error 862
Relative response factor 452, 951
Relative retention 435, 439, 480
  in ion exchange 549
Relative standard deviation 295
Relaxation processes in NMR spectroscopy 320, 322
Releasing agents 137
Repeat distance in polymers 281
Reset flip-flops 818
Residence time in flames 131
Residual current in voltammetry 697, 699
Resistance, electrical, of glass electrodes 657
Resistor-transistor logic 816
Resolution and analysis time in chromatography 444
Resolution, in chromatography 437–440
  in exclusion chromatography 558
  in mass spectrometry 577
  of monochromator 39
  of optical spectra 39
  and slit width 35–37
Resolving power of monochromators 39, 41
Resolving time of nuclear counters 297
Resonance lines in atomic absorption 140
Resonance Raman spectroscopy 219–221
Response, spectral, of human eye 23
  of infrared detectors 192
  of ultraviolet/visible detectors 23, 27, 29
Response time 24
  of infrared detectors 191–196
  of ion-selective electrodes 647
  of liquid chromatography detectors 505
Retardation factor in chromatography 435
Retention behavior, in chromatography 433

in exclusion chromatography 555–557
  and temperature 479–484, 504, 524, 950
Retention indices, Kovats 461
Retention time 433
  preselection in liquid column chromatography 518–520
  and resolution 444
Retention volume 433
Reversed-optics technique 59
Reverse phase ion-pair partition 543–545
Reverse phase liquid column chromatography 539–543
Reversible voltammetric reactions 702
Riboflavin, fluorometric determination of 943
Right-angle viewing mode in fluorescence 112
Rigidity, molecular, effect on fluorescence 108
Ring substitution, infrared absorption bands 183
  Raman spectroscopy 229
Rocking vibrational mode 11
Rohrschneider classification system 461–463
Rotary valve, multiport 455–456
Rotating can and shutter devices 121
Rotating crystal diffraction method 273
Rotating crystal method in X-ray diffraction 273
Rotating disk electrode as spectroscopic source 157, 160
Rotating microelectrode 695, 723
Rotating mirror-chopper 57
Rotating sector 57
Rotation, specific 416
Rotatory power and light absorption 415–417
Rowland circle mount 50, 165
RS-232 serial data bus 874
  use in atomic absorption spectrometer 892
Rudolph polarimeter 424–425
Rudolph spectropolarimeter 426

Saccharimetry 419
Saha equation 138
Salt bridge 638
  filling solutions for 570–572
Sample cavity in ESR spectrometers 359
Sample cell geometry in fluorescence 110, 112
Sample electrodes in spectroscopy 156–157
Sample and hold amplifier 830, 844
Sample illumination in fluorescence 110, 112
Sample injection, in gas chromatography 455–458
  in liquid column chromatography 500–502

Sample loops and valves 455–456, 501–502
Sampling problems with process streams 900
Sampling systems for mass spectrometers 566
Sand equation 714
Sargent-Malmstadt derivative autotitrator 678
Sargent-Welch coulometer 762
Sargent-Welch electrodeposition power supply 743
Saturated calomel electrode 637
  potential of (table) 636
Saturation factor in activation analysis 305
Sauter mean diameter of aerosol droplets 129
Scale expansion techniques in spectrophotometry 78–82
Scaling circuit 258
Scanning Auger microprobe 393
Scanning calorimetry, differential 606–609, 611–616
Scanning double-beam spectrophotometer 57
Scanning ESCA 400
Scanning spectrometer with diode array 58
Scan rate and photometric accuracy 71
Scattered radiation 71, 90, 135
  See also Raman
Scattered transmittance measurements 89
Scattering coefficients (Raman) 232
Scattering, in flames 135
  of X rays by atoms. See X-ray diffraction
Schmitt trigger 833
Scintillation counter for X-ray measurements 254
Scintillation liquid counting 302–304
Scissoring vibrational mode 11
SCOT columns 459, 478, 480
Scribing the surface 382
Second class electrode 666
Second derivative titration curve 669
Second order transition in thermal analysis 612
Secondary coulometric titrations 769–773
Secondary emitting coatings for dynodes 25
Secondary filters for fluorescence 114
Secondary ion mass spectrometry 380–382, 386–389
  instrumentation 387–389
Secondary scintillators 303
Sector mass spectrometers 579, 580, 583–585
Sector mirrors, rotating 57
Selection rules, in atomic energy transitions 6
  electron spin transitions 362
  in vibration transitions 10
Selective volatilization in emission spectroscopy 157
Selectivity in chromatography 438

INDEX 1025

Pressure broadening, in flames 141
Pressure-gradient correction factor 434
Pressure requirements in liquid chromatography 516, 518, 519
Primary coulometric titrations 769
Prints, electrographic 757–759
Priority interrupts 885
Prism, as dispersive device 40–43
 for infrared region 197
 polarizing and analyzing 421–425
Prism spectrophotometer 54
Probe unit in NMR spectrometers 325
Problems, answers 971–994
Process analyzers, industrial 899–901
 on-line process control 919–923
Process chromatographic systems 919
Process control analyzers 898
 gas chromatographic 915–919
 infrared 905–909
Process control methods 922–923
Process stream infrared analyzers 905–909
Process streams, handling and sampling 900
Programmed-temperature gas chromatography 481–484
Proportional counters 253
Proportioning pump 787, 788
Protective chelation in flame spectrometry 137
Proton bombardment for generation of X rays 268
Proton chemical shifts 333
 table of values 334–335
Proton magnetic resonance. See NMR
Proton–proton coupling constants (table) 336–337
Protons, coupling with other nuclei 336–338
Proton-transfer reactions, pK_a values (table) 998–1000
Pulsations, damping methods for in solvent delivery systems 497
Pulse amplitude/voltage relationship for ionization detectors 251
Pulse counter 834
Pulsed Fourier transform NMR spectrometer 330
Pulsed Fourier transform NMR spectroscopy 321–323
Pulsed neutron sources 306
Pulse formation in X-ray counters 250
Pulse height analyzer 258–260
Pulse polarography 708–711
Pulsing mode of lamp operation 22
Pumping a laser 226
Pumps for liquid chromatography 495–500

Purification, electrolytic 759
Pyroelectric detector 193
Pyrolysis of solids in chromatography 457

Quadrupole mass analyzer 387, 388, 581–583, 589
Quadrupole moment, electric 15, 320, 339
Quantitative analysis. See Evaluation methods
Quantitative methodology in ultraviolet/visible spectrophotometry 75
Quantity and area, correlation of 446–452
Quantum counter in fluorescence spectroscopy 120
Quantum efficiency, of photoemissive tubes 27
 of X-ray detectors 252
Quantum mechanical (chemical) zero of energy 394 fn
Quantum number, angular momentum 357
 spin of nuclei 317–320
Quartz, optical characteristics 28
 polarization by 421
Quasiexcited state (Raman) 12, 13
Quaternary amine ion exchangers 545–547
Quaternary amines (alkyl), as counterions in ion-pair chromatography 542–545
Quenching of photoluminescence 110, 228
Quenching agent in radiocounters 252
Quinine, absorption and fluorescence spectra 943
 fluorometric determination 943

Racemization 417
Radial compression module 503
Radiant energy, stray 71, 90, 135
Radiation buffer 137, 138, 158
Radiation sources, for ESCA 398
 for ESR 359
 for fluorescence 113
 for infrared 189–191
 light gathering from 35
 for phosphorescence 122
 for polarimetry 421
 for ultraviolet/visible 20–22
 for X-ray diffraction 277
 for X rays 240–245, 261, 263, 268
Radical, free 357, 362, 367–371
Radioactive decay 293
Radioactivity, measurement of 249–260, 294–297
 units of 294
Radiochemical measurements, counters for 249–260, 302–304
 precautions 300
 statistics in 294–297

Radio frequency spark source 572
Radionuclides, applications 297–310
 characterization of 289–294, 948
 decay schemes 294
 half-life 293
 table of values 298
 properties (table) 298
 sample handling 300
 in X-ray fluorescence 257, 268
Radioreagents 301
Radius of curvature of ion beam 578, 583
Raman effect 12, 217
Raman spectrometer 222–226
Raman spectroscopy, compared with infrared 221, 229–231
 diagnostic structural analysis 228–232
 instrumentation 222–226
 quantitative analysis 232
 sample handling and illumination 226–228
 solvents for 227, 228
 theory 217–221
Random access memory 873
 use in microcomputer 885, 890
Random errors 861
Range-energy relationship of beta particles 291
Ratio-type fluorometers 114–116
Ratio-type spectrofluorometer 118
Rayleigh scattering 12, 217–219
Read-only-memory 873
 in in-line computers 870
 in microcomputers 886, 890
Reagent gas for chemical ionization source 569, 571
Real time computer operation 870
Receptors. See Detectors
Reciprocal lattice concept 271–273
Reciprocal linear dispersion 38
Reciprocating pumps 496–498
Recording refractometers 407–411
Recording titrator 676
Recycle technique in liquid column chromatography 522–524
Redox buffers 748
Redox electrode 667
Redox nonaqueous titrations 688
Redox reactions, in potentiometry 680
Reducible organic functional groups (table) 718
Reduced mass for two atoms 10
Reduced variables in chromatography 443, 516
Reference electrodes 634–639
 for amperometry 723, 726
 potentials (table) 636
Reference materials. See Standards
Reflectance measurements 86–89
Reflection, attenuated total 206–209

in emission spectrography 171–173
in mass spectrometry 575
Photographic emulsion for nuclear radiation 249
Photographic speed 170
Photoionization detector 475
Photoluminescence intensity related to concentration 109
Photoluminescent methods 105–123
Photometer, definition 19
  double-beam 54–58
  filter 52, 53, 505–508, 510
  operation of 53, 55, 56
  single-beam 52–54
Photometric accuracy 70–73
Photometric analyzer, centrifugal 930–932
  for process streams 902
Photometric detectors for liquid chromatography 505–510
Photometric linearity 70
Photometric precision 73–75
Photometric titrations 83–86
  experiments 940
Photometry, laws of 66–70
  simultaneous 77
Photomultiplier tube 25–28, 174, 574
  in scintillation counting 254, 304
Photons 1
Photopeak on gamma ray spectra 307–310
Phototubes 24–28
Photovoltaic cell 23
Photovoltaic detector, for infrared 196
Phthalate pH reference buffer standard 655
Physical and chemical changes in thermal analysis 500–508
pI, electrometric measurement of 659–661
  reference standard values (table) 650
pI meters, direct-reading 660
Pilot ion evaluation method 720
pK_a values (table) 998–1000
Plane-polarized light 412, 413
Planimetry 446
Plasma emission sources 161–164
Plastic absorption filters 31
Plate height 436
Plate height/velocity relationship, in liquid chromatography 516
  in gas chromatography 440–443
Plate number 436
  in exclusion chromatography 558
Platinum microelectrode 695, 723
Platinum/10% rhodium electrode 672
PL1, computer language 876
Pneumatic infrared detector 195
Pneumatic nebulizer 127
p-n junction photodiodes 28, 256

Point-to-plane electrode configuration 157, 160
Point-to-point electrode configuration 157
Polarimeter 421–427
Polarimeter tubes 419, 423
Polarimetric measurements, standard conditions 417
Polarimetry, calculations in 419
  correlation with molecular structure 420
  light sources 421
  theory 412–415
Polarizability, molecular 12, 218
Polarization analyzer 222, 232
Polarization measurements by Raman spectroscopy 231
Polarization scrambler 222
Polarity, effect in liquid chromatography 530, 540
Polarized electrode 692
Polarized light 3
  rotation in magnetic field 418, 426
Polarizer, optical 421–425
Polarograph 703
Polarography 691–735
  alternating-current 711
  applications 717–719
  conventional (dc ramp) 702–704
  derivative 707
  evaluation methods 719
  half-wave potential 700–702
  table of values 997
  instrumentation for 702–707
  mass-transfer processes 693
  pulse 708–711
Polaroid filter 422
Polling I/O devices 884
Polyatomic molecule, vibrational modes 10–12
Polychromatic absorptiometry with X rays 265
Polychromatic radiation 2
  effect on Beer's law 69
Polyethylene, X-ray diffraction fiber diagram 281
Polymer gels, cross-linked, as packings for exclusion chromatography 553
Polymer, thermal studies 612, 616
  X-ray diffraction patterns 281
POPOP, secondary scintillator 303
Population inversion with lasers 224
Pore-size diameter and separation range in exclusion chromatography 554
Porous barrier separator 586, 588
Porous beads 484
Porous cup electrode configuration 157, 160
Porous glasses and silicas, for exclusion chromatography 554
Porous layer beads 531
Porous particles, stationary phase in adsorption chromatography 531
Porous polymer packings 484, 537–539, 553

Positive-filter nondispersive infrared analyzer 906
Positron 290, 292
Postcolumn derivatization in chromatography 514
Postirradiation radiochemical procedures 309
Potassium, atomic term diagram 5
  flame photometric determination 944
Potassium bromide pellet technique 205
Potassium chloride solutions, specific conductance of (table) 787
Potassium hydrogen citrate buffer solution 655
Potassium hydrogen phosphate buffer solution 655
Potassium hydrogen phthalate buffer solution 655
Potassium hydrogen tartrate buffer solution 655
Potassium ion-selective electrode 645, 646
Potassium tetraborate buffer solution 655
Potassium tetroxalate buffer solution 655
Potential, decomposition 737
  of half reactions (table) 995
  half-wave 700–702
  table of values 997
  liquid-junction 638
  of reference electrodes (table) 636
Potential buffers 748
Potential energy curves or surfaces 6, 7
Potentiometric on-line analyzer 913–915
Potentiometric titrations 664–690
  classes of 679–688
  experiments 953–955
Potentiometric titrators, automatic 609–612
Potentiometry, constant current 672–675
  null-point 679
Potentiostat 704–707, 746, 831
  three-electrode 704–707
Powder data file, X-ray diffraction 279
Powder diffraction method 274, 276, 278
Powders, by Raman spectroscopy 227, 229
PPO, primary scintillator 303
Praseodymium(III) shift reagents 344
Precession of nuclei 14, 317
Precipitation titrations, conductometric 755
Precision 862
  and significant figures 863
Precision in spectrophotometry 73–75
Precolumn derivatization in chromatography 464, 514
Pre-electrolysis concentration 715
Premix burner 133
Premixed flames, structure 130
  temperature (table) 131
Preresonance Raman spectra 220

Ohmic-pseudo-overpotential 739
Olefins, infrared absorption bands 183–185
Off-line computers 869
180° design mass spectrometer 579
On-line computers 869
On-line process control, continuous 919–923
Open tubular columns in GLC 459, 460, 478
Operational amplifiers, in active filters 852
 applications 812, 829
 basic circuit operation 810
 characteristics 810
 parameters 811
Operational definition of pH 653
Operating system software 878
OR gate 814
Optical activity 415
Optical-null system 58
Optical resonance (lasers) 225
Optical rotation, measurement of 417
Optical rotatory dispersion 415–417
 applications 420
Optical sensor for thermogravimetric analyzer 610
Optical speed 40
Optimization of gas chromatographic separations 478–484
Optimization of liquid column chromatographic methods 515–525
Orbital angular momentum 5
Order, spectral 32, 44
Order sorter, optical 60
Organic polarography 717–718
Organic scintillator liquids 311–314
Orifice in a diaphragm LC/MS interface 591
OSFET-type ion-sensitive electrode 645
Out-of-plane bending mode 11
Oven, column, in gas chromatography 464
Overpotentials 739–741
Overtone vibrations 12
Overvoltage 739–741
Oxidant-fuel mixtures for flame spectrometry 131
Oxidation potentials 632
Oxidation-reduction buffers 748
Oxidation-reduction halfreactions, potentials of (table) 995
Oxidation-reduction reactions, in potentiometry 680
Oxidation-reduction systems, three-dimensional representation 737
Oxidation-reduction titrations, experiments 954
Oxidation state, effect on X-ray spectra 245
Oxide films, determination by cathodic stripping 715
Oxygen, current-voltage curves 725

determination of dissolved 718
removal of dissolved 723
Oxygen-acetylene flame 131, 137
Oxygen analyzers 909–913, 933
Oxygen sensitive electrode 914
Overpotential 673–676

p-Type semiconductors 803, 807
Packed columns in gas chromatography 459–461, 478
Pair production 292
Paper web, moisture content determination 901
Parallel data transfer 825
 by IEEE-488 bus 874
Parallel photometric analyzers 930–932
Parallel wavelength acquisition technique 59
Paramagnetic metal ions, effect on photoluminescence 108
Paramagnetic oxygen analyzer 911
Paramagnetism 910
Parent-daughter metastable peak relation 597
Parent-daughter radioactive equilibrium 301
Particle diameter, of supports in chromatography 460, 516, 519, 520
Particle-induced X-ray emission 268
Partition coefficient 434, 480
Partition forces, nature of 431–432
Partition liquids, for gas chromatography 461–464
 table 462
 for liquid–liquid chromatography 536
Partition ratio 435, 439
PASCAL, computer language 876
Paschen-Runge mount 166
Passive filters 832
Path length, measurement optically 206
Pauli exclusion principle 7
Peak area integration 446–451
Peak asymmetry factor in chromatography 433, 437
Peak capacity 444–445
Peak switching in mass spectrometry 576
Pellet technique in infrared spectroscopy 205
Pellicular type of packing 530, 531, 547
Penetrating range of radiation 262, 290–293
Penetration depths of electrons and ions 380
Penetration mode of thermomechanical analysis 616–618
Percent absorption/absorbance conversion (table) 1003–1005
Periodic chart, inside back cover

Perkin-Elmer double-beam fluorescence spectrophotometer 120
Perkin-Elmer double-wavelength spectrophotometer 62
Perkin-Elmer filter-grating infrared spectrometer 198
Perkin-Elmer heated graphite atomizer 147
Perkin-Elmer thermomechanical analyzer 617
Permeability enrichment in GC/MS interface 586–588
Permeation range (selective) in exclusion chromatography 554, 556
pH, effect of, on electronic absorption spectra 70
 on fluorescence 108
 operational definition of 653
pH measurements 652–661
 interpretation of 654–656
pH meters, direct-reading 660
 null-balance 659
pH/reference electrode combination 658
pH scale, definition 652
pH standard reference buffers 654
 table of values 655
Phase difference, faradaic and capacitative currents 711
Phase ratio (volumetric) in chromatography 435, 478
Phase-sensitive amplifier. See Lock-in amplifiers
Phenanthrene, excitation and emission spectra 124
1,10-Phenanthroline, formation constants of metals (table) 1001
Phenol red, absorbancewavelength as function of pH 70
pH glass electrodes 640–643, 656–658
Philips Electronic Instruments X-ray powder diffraction cameras 275
Philips Electronic Instruments X-ray fluorescence spectrometer 266
Phosphate pH reference buffer standards 655
Phosphorescence 8, 9
 instrumentation for 122
 molecular structure correlations 107, 108
Phosphorescence lifetimes, effect of molecular structure 108
Phosphorus-32, decay scheme 294
Photocathode surfaces, spectral response 27
Photocell 23
Photoconductive detector 195
Photocurrent and light intensity, effect of lamp voltage 22
Photodetectors 22–29, 195
Photodiodes 28
Photoelectron effect 292
Photoemissive tubes 24–28
Photographic detection 170–173

by photoluminescence 107
by Raman 221, 228-232
by ultraviolet/visible 94-97
by X-ray diffraction 279
Molecular vibrations, in infrared region 10
intensity of 12
Molecular weight, by exclusion chromatography 559
from mass spectrum 481-483
Mole-ratio method 941
Monochromators 34-38
designs of optical 41, 42, 47-50, 164-170, 196-201
performance of 38-46
Monolithic integrated circuits 809, 830
Monostable, use in digital-to-analog converters 836
Moseley relationship between atomic number and wavelength 240
Moving belt LC/MS interface 589
Mulls, infrared handling of 205
Multichannel electron multiplier arrays 575
Multichannel spectrographic analyzers 165, 169
Multilayer interference filter 33
Multiple component analysis by mass spectrometry 592
Multiple internal reflection infrared analyzer 909
Multiple internal reflection technique 206-209
Multiplexers/demultiplexers, digital 823
with analog-to-digital converters 843, 844
Multiplicity, in atomic energy levels 6
Multiplier phototubes 25-28, 386, 574
Multiport rotary valve 455-456
Mutarotation 417

NAND gates 815, 824
Naphthalene, excitation and emission spectra 124
National Bureau of Standards pH reference buffers 587-589
table of values 588
Natural abundance of radionuclides (table) 298
Near-infrared region 178, 190
structural correlation in 178
Nebulization 127-129
Nebulizer, pneumatic 127
Nebulizer-burner system 133
Neon-helium laser 224-226
Nephelometric standards 92
Nephelometric turbidity unit 92
Nephelometry 90-94
Nerheim gas density detector 904, 906
Nernst diffusion layer 694
Nernst equation 632
Nernst glower, fabrication 190
spectral output 21
Net retention volume 434

Neutral extraction membranes for ion-selective electrodes 646
Neutron activation analysis 304-310
Neutron cross sections, thermal (table) 298
Neutrons, detection of 293
Neutron sources, fast 306
thermal 305
Nichrome coil as infrared source 190
Nicolet infrared Fourier transform interferometric spectrometer 200
Nicol prism 422
Nier-Johnson design mass spectrometer 584
Nier 60° sector mass spectrometer 580
Nitrogen (1s) electron binding energies for organic functional groups 396
Nitro group, infrared absorption bands 187
Nitrous oxide-acetylene flame 131, 137
NMR signals, absorption and dispersion 327
NMR spectra, correlation with molecular structure 331-340
elucidation of 340-347
NMR spectrometers, continuous-wave, 324-330
minimal type 328
multipurpose type 329
pulsed Fourier transform 330
wide line 329
NMR spectroscopy 316-349
basic principles 317-324
Fourier-transform 321-323
integration of spectra 339
molecular structure correlation 331-347
quantitative evaluation 339, 347-349
pulsed-wave 321-323
reference materials 332, 333
sample handling 326
shift reagents 344
solvent influence 334
solvents for 326
standards for 332, 333
wide line 323
Noble gas ions, in ion-scattering spectrometry 383-386
for sputter etching 382
Noise 848-850
shot 25, 74, 850
thermal 73, 849
Noise reduction 846
environmental 850
by hardware use 851
of shot noise 850
by software use 855
of thermal noise 849
Nominal wavelength 31, 33
Nonaqueous solvents, electrode systems for 687
in potentiometry 684-688
Nonaqueous thermometric titrations 621
Nonaqueous titrations, in alcohols 953
Nondispersive infrared analyzers 905-909

Nondispersive X-ray absorptiometer 265
Nondispersive X-ray fluorescence spectrometers 259
Nonflame atom cells 146-149
Nondestructive neutron activation analysis 307
Nonlinear integrated circuits 809, 813
Nonretained solute, transit time 433
Normal hydrogen electrode 631, 635
Normalization procedure 452
Normal phase liquid-liquid chromatography 537
Nonvolatile memory 873
NOR gate 815, 817
$n-p$ Semiconductor junction diode 28
n-Type semiconductor 804
Nuclear axis, precession of 14, 317
Nuclear induction NMR probes 326
Nuclear magnetic energy levels 14, 317-320
Nuclear magnetic moment 317-320
of nuclei (table) 320
Nuclear magnetic resonance spectroscopy. See NMR spectroscopy
Nuclear magneton, of proton 319
Nuclear properties of radionuclides (table) 298
Nuclear radiation, absorption of 290-293
interaction with matter 290-293
particles emitted 289
Nuclear orientation (NMR) 318
Nuclear Overhauser effect 342
Nuclear spin 13-15, 317, 318
Nuclei, interaction with radical electron 362-366
magnetic moments 12, 317-320
NMR properties 12
table of values 320
shielding of 332, 333
Nukiyama-Tanasawa equation 129
Null-balance pH meters 659
Null-balance principle in optical instruments 52, 55, 58
Null-point potentiometry 679
Number-average molecular weight 559
Numerical aperture 38
Nyquist sampling theorem 824

Object computer program 875
Obstructive factor in chromatography 441
Octadecyl alkyl bonded phase 538, 540, 543
Octal number notation 880
Octant rule 420
Octene-1, infrared spectrum 184
Off-axis ellipsoidal mirror 35

INDEX 1021

Magnetic-deflection analyzer systems 578–580
Magnetic electron multiplier 574
Magnetic field, in ESR spectroscopy 357
 in NMR spectroscopy 14, 319
 rotation of polarized light 418, 426
Magnetic moment, of electron 357
 nuclear 14, 318–320
Magnetic resonance condition for NMR 319
Magnetic susceptibility 910
Malonic acid, ESR spectrum of irradiated 364
Malus' law 422
Manganese-56, decay scheme 294
 gamma spectrum 308
Manganese phosphinate monohydrate, DTA and TG curves 613–614
Mannosan triacetate, NMR spectrum 341
Mark-Houwink equation 559
Masking agent. See Complexation, Complex ions
Mass absorption coefficient 262, 293
Mass difference, molecular formula calculation 597
Mass analyzer systems 578–585
Mass/energy relationships 578
Mass spectra, molecular structure correlation 595–599
Mass spectrometer, components 565–576
Mass spectrometer/chromatograph interface systems 585–591
Mass spectrometers 578–585
Mass spectrometry 565–605
 quantitative analysis 592
 sample handling 566
 use of stable isotopes 593
Mass spectrometry/gas chromatography interface systems 585–589
Mass spectrometry/liquid chromatography interface systems 589–591
Mass transfer processes, electrochemical 693
 in gas chromatography 441–443
 in polarography 693
Mattauch–Herzog geometry of mass spectrometer 583
Matte surface 87
Maxima, in polarography 699
Maximum precision method in spectrophotometry 81
Maximum suppressor 699
McReynolds classification system 461–463
 experiment 951
Medium scale integrated circuits, analog 829–832
 composite 833–843
 digital 823–829
 systems using 842
Melaven mercury cathode cell 752
Membrane type ion-selective electrodes 640–652

Membrane separator 587
Memory of computer 872
Mercury, hydrogen overpotential on 740, 752
Mercury arc, high-pressure 191
Mercury cathode 752–755
 separations involving (table) 755
Mercury drop detachment 704
Mercury electrode, dropping, assembly of 703
 characteristics of 695–700
Mercury lamp, characteristics 113
Mercury/mercury(I) chloride reference electrodes 637
 table of potentials 636
Mercury/mercury(II) iodide reference electrode 724
Mercury spectral emission lines 113, 507
Metal anionic complexes, ion exchange of 549
Metal complexes, formation constants (table) 1000
Metal compounds, dissociation in flames 136
Metal deposits, physical characteristics 749
Metal overpotentials 739
Metal oxide semiconductor field effect transistors, depletion mode 808
 enhancement mode 807
 operation of 807
Metal oxide semiconductors 807
Metals, ionization in flames 137
Metastable peaks 597
Methanol, ESR spectrum of irradiated 365
Methine protons, chemical shift values (table) 334–335
Method of additions for evaluation 379, 585
Methylene group, vibrational modes 181, 182
Methylene protons, chemical shift values (table) 334–335
Methyl protons, chemical shift values (table) 334–335
Microcomputer instrument system 886
 in atomic absorption spectrometer 889
 in carbon/sulfur analyzer 887
Microcomputers 886
 advantages 871
 in-line applications 870
 interfacing 885–892
 software vs. hardware 879
Microdensitometer comparator 172–173
Microelectrode, in voltammetry 694, 695–700
 rotating 694, 723
Microphotometer-comparator 172–173
Microprobe, laser 160
Microprobe mass analyzer, ion 388
Microprobe spectrometer, electron 261, 268, 380 381
Microprobes, as spectroscopic sources 160

Microprocessors 885
 added to existing instrument 888
 design of new instrument 888
Microradiography 264
Microwave bridge, for ESR spectrometer 360, 361
Microwave-excited discharge lamps 144
Microwave sources 359
Microwave spectrometer for ESR spectroscopy 358–362
Mid-infrared region 179, 190
 structural correlations 179–189
Midpoint potential of acids 685
Migration of ions 693
Minicomputers in laboratory automation systems 893
Mini-Press, wafers for infrared 206
Mnenomics 875
Mobile phase, average linear velocity of 433
Mobile phase selection, in liquid-liquid chromatography 538, 540, 541
 in liquid–solid chromatography 534–536
Modifiers for mobile phases in liquid–solid chromatography 536, 540
Modulated thermal conductivity detector 466, 469
Modulation, effect on photoresponse 22, 30, 191
Modulation coils for ESR spectrometer 359, 361
Modulation/demodulation in noise reduction 852
Modulation mode of lamp operation 22, 30
Modulators, in lock-in amplifier circuits 853
 in noise reduction 852
Modulus of counter 827
Moisture, determination in paper web 901
Molar absorption coefficient 68
Molar absorptivity, and circularly polarized light 415
 definition 68
 and single chromophores in ultraviolet/visible (table) 95
Molar refraction 404
Molecular absorption, in flames 136
Molecular ellipticity 415
Molecular electronic energy levels 6
Molecular flow, of gases 567, 587
Molecular formula from isotopic contributions 595–597
Molecular polarizability 217
Molecular rigidity, effect on fluorescence 108
Molecular rotation 419
Molecular separator 586–588
Molecular sieves 484
Molecular structure/spectra correlation, by ESCA 395–397
 by ESR 362–371
 by infrared 178–189
 with mass spectra 595–599
 by NMR 331–347

1020 INDEX

K-absorption edge 242-244
K-band ESR spectrometer 362
K-electron capture 242, 290
source of X rays 242
K emission lines 242-244
(table) 243
Ketones, infrared absorption bands 185
(table) 187
Kinetic energy spread of ion beam 567, 583
Klystron oscillator 359
Knox equation 516
Kovats retention indices 461, 951
Krypton laser 226

Labeled reagents tagged with isotopes 301
Laboratory analyzers, automatic 898, 923-935
Laboratory automation systems 892
Laboratory Data Control dual wavelength photometer 507
Laboratory Data Control filter fluorometer 510
Laboratory Data Control fixed wavelength photometric detector 506
Laboratory Data Control reflection-type (Fresnel) refractometer 512
Laboratory Data Control variable wavelength spectrophotometer 508
L absorption edges 242-244
Lambert's law 67
Laminar flow burner 133
Laminar flow flames, structure 130
Lamps, electrodeless discharge 144
hollow cathode 143
hydrogen discharge 20
mercury vapor 113
operational modes 22
xenon arc 21, 113
xenon flash 113
Langmuir-Saha equation 573
Lang X-ray diffraction method 282
Large scale integration 843-845
Larmor frequency of precession 317
Laser, as Raman source 224-226
as spectroscopic source 160
Laser Raman spectrophotometer 222-224
Lattice, reciprocal 271-273
Lattice spacing of crystals 247-249
table of values 249
Laurent half-wave plate 424
Lauryl alcohol, infrared spectrum 186
Laws of photometry 66-70
Layer lines 273, 281
Lead, amperometric titration 721
current-voltage curves 721, 722
Lead tetraethyl, by X-ray absorption 263
Lead-tin-tellurium infrared detectors 196

Leakage rate of salt bridge solutions 639
Leak detection by mass spectrometry 594
L emission lines 242
(table) 243
Leveling effect of solvents 684
Levorotatory 412
Lifetime, fluorescence 105
phosphorescence 123
Ligand exchange chromatography 549
Light, circularly polarized 412, 413
elliptically polarized 415
losses by reflection 37
plane-polarized 412, 413
stray 71, 72
velocity of 2
visible 4, 21
Light gathering from radiation sources 35, 40
Light interrupter 22, 30, 57, 122
Light scattering 71, 90, 135
in flame spectrometry 135
Light scattering photometer 92, 93
Light sources, for atomic absorption 143
for atomic fluorescence 144
fluorescence 113
infrared 189-191
for polarimetry 421
Raman 224-226
ultraviolet/visible 20-22
Limiting current in voltammetry 694
Limiting electrode potentials 745, 760-767
Limiting equivalent ionic conductances 782
table of values 783
Linear absorption coefficient 262
Linear capacity of chromatographic columns 502, 534
Linear diode arrays 29
Linear dynamic ranges of gas chromatographic detectors 465
Linear integrated circuits 809
Linear interpolation titration technique 679
Linear multiplier structure 26
Linear potential sweep stripping chronoamperometry 715-717, 768
Linear reciprocal dispersion 38
Linear titration method, amperometric 720-729
conductometric 790-796
photometric 83-86
thermometric 619-622
Linear variable differential transformer 609, 616
Line intensities, comparison in emission spectrography 172-173
Line intensity relations, spectral 174
Line widths, in ESR 361, 367
Lippich prism 423
Liquid column chromatography, detectors 504-514
instrumentation 495-515
mode selection guide 530

optimization of column performance 515-525
sample introduction 500-502
solvent delivery systems 495-500
Liquid column chromatography/mass spectrometry interface 589-591
Liquid exclusion chromatography 553-561
Liquid ion-exchanger membrane electrodes 645-647, 648
Liquid junction potential 638
Liquid-liquid partition chromatography 529, 536-545
Liquid membrane electrodes 645-647
Liquid partitioning liquids for gas chromatography 461-464
(table) 462
Liquids, infrared handling of 202-205
injection into gas chromatograph 455-458
injection into liquid chromatograph 500-502
mass spectrometry handling of 566
nebulization of 127, 133
Liquid scintillation counting 302-304
Liquid scintillators 303
Liquid-solid adsorption chromatography 530-536
Lithium atomic term diagram 5
Lithium-drifted germanium and silicon detectors 256
Littrow-echelle system 167-170
Littrow mirror 42
Littrow mount 41, 49
Lock-in amplifier 853
Logarithmic step-sector disk 171
Logarithms, common, four-place table 1007-1009
Logic families 816
London's dispersion interaction 431
Longitudinal diffusion 441
Longitudinal relaxation in NMR spectroscopy 320
Long-wave pass type of interference filter 34
Loop, calibrated sample 455-456, 501
Lorentz broadening 141
Lorentz and Lorenz equation 404
Low absorbance differential spectrophotometry 81
Luminescence, defined 105

Machine computer language 875
Macroporous type of ion-exchange resins 547
Magic angle spinning for NMR 331
Magic-T bridge 360
Magnesium atomic term diagram 5
Magnet, for ESR spectrometer 361
for NMR spectrometer 325

Infrared spectrometer, Fourier transform 199–201
Infrared spectrophotometers 196–201
 nondispersive 905–909
Infrared spectroscopy, compared with Raman 221, 229–231
 instrumentation 189–201
 principles 10, 177
 quantitative analysis 209
 sample handling 201–209
Infrared transmitting materials (table) 203
Infrared vibrational modes 10–12, 178–189
Inhomogeneity in surface analysis 382, 393
Injection of sample, in gas chromatography 455–458
 in liquid column chromatography 500–502
Injector-splitter for chromatography 457, 458
Inlet systems, mass spectrometer 566
In-line computers 870
Inner electron (K, L, M) shells 240
Inner filter effect in fluorescence 111
Inorganic scintillation crystal 255
In-plane bending 11
Input/output units of computers 873
 direct memory access 885
 interfacing 883
 interrupts 885
 polling 884
 relation to software 877
Instability constants. See formation constants
Instrumental amplifier circuit 830
 component of data acquisition system 844
Instrumentation Laboratories atomic absorption spectrometers 141, 142
Integer notation 880
Integrated circuits 802
 development of 822
 digital 813
 dual-in-plane 810
 large scale 843
 medium scale 843
 packages 809
 small scale 817
 symbols for common circuits 820
 very large scale integration 886
Integrated injection logic 817
Integrating sphere 88
Integration, current–time 750, 761–765
 of NMR spectra 339
 of peak areas 446–451
Integrators, boxcar circuit 854
 use in noise reduction 852
Interfacing gas chromatograph/mass spectrometer 585–589
Interference filter 31–34, 115
 multilayer 33
 variable transmission 33

Interference fringe method 206, 409
Interference wedge 33
Interferences, in flame spectrometry 134–138
 spectral 136
Interferometers 408–412
Interferometric infrared spectrophotometer 199–201
Intermolecular hydrogen bonding 203
Internal conversion electron 8, 290
Internal electrolysis 755–757
Internal lock system in NMR 327
Internal standard method 866
 in chromatography 452
 in emission spectroscopy 159
 in polarography 720
 in X-ray fluorescence 269
Interplanar crystal spacing (table) 249
Interrupts, computer 885
Intersystem crossing in photoluminescence 8, 9
Intramolecular hydrogen bonding 203
Iodide, potentiometric titration 955
Iodine, amperometric titration of 962
 coulometric generation of 966
Iodine-128, decay scheme 294
Ion accelerating system in mass spectrometry 568
Ion-activity evaluation methods 650–652
Ion-activity reference standards (table) 650
Ion chromatography 550–553
Ion-collection systems in mass spectrometers 573–575
Ion combination titrations, conductometric 796
 potentiometric 680
Ion-exchange chromatography 529, 545–553
Ion-exchange equilibrium 548
Ion-exchange packings 545–547
Ion-exchanger liquid membrane electrodes 645–647, 648
Ion exclusion in exclusion chromatography 557
Ionic conductances (table) 783
Ionic strength adjustment buffer 644
Ion inclusion in exclusion chromatography 557
Ionization, degree of in flames 138
 of metals in flame gases 137
 by nuclear radiation 290, 292
 suppression in flames 138
Ionization chamber 249, 250
Ionization constant, of acids (table) 998–1000
 of indicators (table) 998–1000
 of metals (gaseous) 138
Ionization detector, flame 468–471
Ionization detectors, in gas chromatography 468–473
Ionization effects, in flame photometry 137

Ionization limit of atom 6
Ionization potential (gaseous) of atoms (table) 137
Ionization sources for mass spectrometers 567–573
Ion microanalyzer 389
Ion microprobe mass analyzer 388
Ion-molecule chemical interactions in mass spectrometry 569
Ion-pair chromatography 541–545
Ion-pair formation in liquid-liquid chromatography 541–545
Ion-pair site groups in ion-selective electrodes 646
Ion scattering spectrometer 385–386
Ion scattering spectroscopy 380–382, 383–386
Ion selective electrodes 640–647, 682, 914
 interferences 648, 952
 use in complexometric titrations 682, 683
Ion source for ion scattering spectroscopy 385
Ion suppression in chromatography 541
Iron, current-potential curves 770
Iron(II)-1,10-phenanthroline complex, structure of 938, 942
Iron(III/II) system, electrode potential of 667, 770
Irradiation time in radiochemistry 306
Irreversible voltammetric reactions 702
Isosbestic point 70
Isotope dilution, in mass spectrometry 594
 in radiochemical analysis 302, 949
Isotope masses (table) 596
Isotope-ratio mass spectrometer 594
Isotopes, radioactive (table) 298
 stable, use in mass spectrometry 593
Isotopic abundance (table) 298, 596
Isotopic contribution to mass spectral peaks 595–597
Isotopic substitution 301, 593
ISS/SIMS systems 388
IUPAC convention for electrode potentials 631

Jarrell-Ash laser microprobe 161
Jeol Raman spectrophotometer 222
Jellett-Cornu prism 423
Jet/orifice separator 587
J-K flip-flops 819
 use in counters 827
Jobin–Yvon Raman spectrophotometer 223
Johnson noise 73
Junction potential 638

1018   INDEX

Gas-sensing electrodes   647, 718
Gas-solid chromatography   484–486
Gases, infrared handling of   201
  injection into gas chromatograph   455
  by mass spectrometry   566
  by Raman spectroscopy   227
Gates, digital logic   813–814
Gaussian distribution   864
Geiger counter   250–252
Gelatin, in polarography   699, 700
Gel permeation chromatography   553–561
General Electric X-ray absorptiometer   265
Generating electrodes, for coulometry   761, 772
Geometry of nuclear counters   297
Germanium (lithium) semiconductor detector   256
g-Factor   16, 365
Ghost lines   45
Glan-Thompson prism   421
Glass absorption filters   30
Glass electrodes, cation-sensitive   641
  pH responsive   641–643, 656–658
  use as reference electrode   672
Glass transition in thermal analysis   612
Globar   191
Gloss   87
Gloss meter   87
G-M counter   250–252
Golay detector   195
Goniometer of X-ray spectrometer   248, 266
Goniophotometer   87
Gradient elution, mobile-phase composition control   499–500, 520–521
Gran plots   670
Graphite electrodes, in electrochemistry   672
  in spectroscopy   156
Graphite furnaces   146–149
Grating formula   44
Grating intensity output   45
Grating monochromators, for infrared   197
  for ultraviolet/visible   47–50
Grating normal   43
Gratings, diffraction   43–47
Grating spectrographs   165–170
Gross errors   861
Grotrian atomic term diagram   4, 5
Group-selective adsorbents for HPLC   532
Guard column   495
Gyromagnetic ratio   15

Hach low range turbidimeter   92
Hach surface scatter turbidimeter   93
Half-cell, in electrochemistry   631
Half-life, of radionuclides   293
  (table)   298

Half-neutralization point of acids   685
Half-reactions, potentials of (table)   995
Half-shade optical devices   423
Half-thickness of nuclear absorber   292
Half-wave potential   700–702
  (table)   997
Halide ion electrodes   644
Halide pellets and wafers, in infrared   205
Halides, amperometric titration   724
  argentometric titration   681
  experiment   955
Harmonic of ac current in polarography   712
Harmonic vibrations   12
Hays Magno-Therm oxygen recorder   911
Haze, definition   94
Heated graphite furnaces   146–149
Heat of phase transition from thermal analysis   608
Heating rate in gas chromatography   483
Height of theoretical plate   436, 440–443
Height/velocity relationship, in gas chromatography   440–443
  in liquid column chromatography   516
Heisenberg uncertainty principle   1
Helium, doubly ionized. See Alpha particles
Helium gas chromatography detector   475
Helium leak detector   594
Helium–neon laser action   224–226
Herring oil, thermal analysis   618
Hersch galvanic cell   912
Hewlett-Packard ESCA spectrometer   398
Hewlett-Packard flame ionization detector   470
Hewlett-Packard GC/MS instrument   589, 590
Hewlett-Packard LC/MS interface   591
Hewlett-Packard modulated thermal conductivity detector   466, 469
Hewlett-Packard spectrometer with reversed optics   59
Hexidecimal number notation   880
n-Hexylamine, infrared spectrum   188
Hierarchial distributed control   921
High absorbance method   79
  experiment   940
Higher grating orders, removal of   30, 60
High-frequency methods   797, 798
High performance liquid chromatography methods   529–533
HNU systems photoionization detector   476
Holes   803

Hollow cathode lamps   143
Holographic gratings   47, 50
Host computer in laboratory automation system   892, 894
Hot-wire detector   465–468
Human eye, spectral response   23
Hydrides, generation for atomic absorption spectrometry   149
Hydrochloric acid, thermometric titration curve   620
Hydrodynamic radius of a molecule   559
Hydrogen bonding, effect on infrared spectra   203
Hydrogen bond interactions   432
Hydrogen discharge lamp   20
Hydrogen flame   131
Hydrogen gas electrode, standard   631, 635
Hydrogen hollow cathode lamp   135
Hydrogen overpotential (table)   740
Hydrogen stretching vibrations   182
Hydrophobic selectivity   540
Hydroxide pH reference buffer standard   655
Hydroxyl group, infrared absorption bands   185
  (table)   186
Hydroxyl ion, coulometric generation of   771, 772
Hyperconjugation   362
Hyperfine coupling constant   362
  (table)   363
Hyperfine splitting   362–365

IEEE-488 parallel data bus   874
Ilkovic equation   696–698
Immersion refractometer   406, 407
Incandescent filament lamps   20
Increment evaluation methods   651
  See also Standard-addition method
Index of refraction. See Refractive index
Indicator electrodes, classification   665–667
Indicators, pK_a values (table)   998
  spectrophotometric determination   938
Indium, Auger MNN spectrum   391
Indium-antimony infrared detector   196
Inductively coupled argon plasma   161–163
Inertia of photographic emulsion   171
Infrared absorption bands (table)   180–181, 186, 187
Infrared absorption and molecular structure   178–179
Infrared detectors   191–196, 511
Infrared process analyzer   905–909
Infrared radiation sources   189–191

INDEX 1017

Faraday second law 750, 758
Far-infrared region 179, 190
Farrand ratio fluorometer 115
Fast Fourier transform 857
Fastie mount 48
Fast linear sweep voltammetry 712
Fast neutrons, activation with 306
Fatigue, in photocells 23
　in photomultipliers 25
Feasibility, of acid–base titrations 683–687
　of ion-combination titrations 681
Fermi level 394 fn
Fiber optics 37
Fick's law 693
Field desorption as mass spectrometry source 571
Field-frequency lock 327
Field ionization, mass spectrometry source 571
Field-sweep NMR method 328
Filament lamps, incandescent 20
Filler gases, for Geiger and proportional counters 250, 252
　for hollow cathode lamps 143, 144
Films, infrared handling of 205
　thickness of 768
　thin coated in optics 37
Filter infrared analyzer 199
Filter fluorometers 114–116, 510
Filter-grating infrared spectrometer 197
Filter photometers 52, 55, 505–507
Filters, absorption glass 30, 31
　cutoff or blocking 30, 115
　for fluorescence, selection of 114
　hardware 851
　interference 31–34, 115
　Polaroid 422
　software 855–861
　for X rays 246
　　(table) 247
Filter wheel nondispersive infrared analyzer 908–909
Finnigan moving belt LC/MS interface 589, 590
First class electrodes 665
First derivative titration curve 668–670, 678
Fisher Titralyzer 85
Fission-spectrum flux 306
Flame absorption spectrometry. See Atomic absorption spectrometry
Flame atomization 129, 131
Flame emission spectra and detection limits of elements (table) 1001–1003
Flame emission spectrometer 139
Flame emission spectrometry 138–140
　comparison with atomic absorption 146
　experiments 944
　intensity relations 139
Flame gases, equilibrium attainment in 130

Flame ionization detector 468–471
Flame photometric detector 472
Flames, background absorption of 135
　in flame photometry 129–133
　ionization of elements in 137
　schematic structure of, 130
Flame spectrometry, interferences 134–138
　techniques 138–150
Flame temperatures 130 (table) 131
Flash lamps for laser pumping 226
Flash vaporizer injection port 456
Flicker noise 850
Flip-flops, digital 817–819
Floating point notation 880
Flow cells for photometric detectors 508
Flow programming in liquid column chromatography 522
Flow proportional counters 253
Flowrate, in chromatography 433
Flow resistance parameter 516
Fluorescence, atomic 8, 145
　contrasted with absorption spectrophotometry 106
　defined 8, 105
　effect on Raman spectrum 221
　effect on ultraviolet/visible absorption spectra 60
　experiments 943
　molecular structure correlation 107
　viewing modes 110, 112
　as wavelength shifter 255, 303
　X-ray 265–270
Fluorescence efficiency 109
Fluorescence indicators 108
Fluorescence intensity related to concentration 109
Fluorescence instrumentation 111–121
Fluorescence spectrophotometer 116–121
Fluorescence standards 116
Fluoride ion, reference standard values (table) 650
Fluorine coupling with protons 337, 338
Fluorometers, filter 114–116, 510
f/number 40
Focal length 36
Focal plane 36, 584
Focusing X-ray spectrometer 267
Force constant of chemical bonds 10
Fore-prism 197
Formation constants of metal complexes (table) 1000
FORTH, computer language 876
FORTRAN, computer language 876
Fourier transform infrared spectrometer 199–201

Fourier transform NMR spectrometer 330
Fourier transform NMR spectroscopy 321–323, 860
Fourier transformations, computer requirement 870
　description 857
　digital filtering with 860
　interferometers 860
　noise reduction by 860
Foxboro/Wilks infrared analyzer 199
Foxboro/Wilks multiple internal reflection attachment 207
Fractional volatilization in emission spectroscopy 157
Fragmentation patterns in mass spectrometry 599
Free-induction decay from nuclear spins 322
Free radicals 357, 362, 367–371
　g-factor 365
Frequency, defined 2
Frequency doubler for lasers 226
Frequency-sweep NMR method 328
Frequency synthesizer 328
Fresnel-type refractometer 511, 512
Fringes, interference 206, 409
Frontal (37°) viewing mode in fluorescence 112
Fuel-oxidant mixtures for flame spectrometry 131
Fuel-rich flames 131, 132, 137
Full-energy peak on gamma spectrum 308
Full-wave rectifier 742
Fundamental infrared absorption bands 10–12

GAB (green, amber, blue) system of color measurement 90
Gain of photomultiplier tube 25
Galvanic cell 630
Gamma radiation 290
　energy and intensity (table) 298
Gamma-ray spectrometry 307–310
Gas amplifier pump 498
Gas atmosphere, effect on thermograms 613, 615
Gas chromatograph 454–464
　process 915–919
Gas chromatograph/mass spectrometry interface 585–589
Gas chromatography, detectors 464–478
　gas–solid 484–486
　programmed temperature 481–484
　theory 454, 478–484
Gas coulometer 763
Gas density analyzers 904–905
Gas flow proportional counter 253
Gas–liquid chromatography. See Gas chromatography
Gas overpotential 740 (table) 740